International Conference on Innovative Research (Euroinvent ICIR 2018)

IOP Conference Series: Materials Science and Engineering
Volume 374

Iasi, Romania
17 - 18 May 2018

Editors:

Andrei Victor Sandu
Mohd Mustafa Al Bakri Abdullah
Petric Vizureanu
Che Mohd Ruzaidi Ghazali
Ion Sandu

ISBN: 978-1-5108-6487-0
ISSN: 1757-8981

Printed from e-media with permission by:

Curran Associates, Inc.
57 Morehouse Lane
Red Hook, NY 12571

Some format issues inherent in the e-media version may also appear in this print version.

Printed by Curran Associates, Inc. (2018)

For permission requests, please contact the Institute of Physics
at the address below.

Institute of Physics
Dirac House, Temple Back
Bristol BS1 6BE UK

Phone: 44 1 17 929 7481
Fax: 44 1 17 920 0979

techtracking@iop.org

Additional copies of this publication are available from:

Curran Associates, Inc.
57 Morehouse Lane
Red Hook, NY 12571 USA
Phone: 845-758-0400
Fax: 845-758-2633
Email: curran@proceedings.com
Web: www.proceedings.com

Table of contents

Volume 374

Euroinvent ICIR 2018

17–18 May 2018, Iasi, Romania

Accepted papers received: 30 May 2018
Published online: 12 June 2018

Preface

Papers

Synthesis and Characterization of Materials

Procedures and Technologies for Materials Engineering

Materials Application

Materials & Life Science

IOP Conf. Series: Materials Science and Engineering **374** (2018) 011001 doi:10.1088/1757-899X/374/1/011001

EUROINVENT
ICIR 2018

International Conference on Innovative Research

May 17th to 18th, 2018
Iasi – Romania
Palace of Culture

Organized by:

- Romanian Inventors Forum
- Faculty of Materials Science and Engineering, The "Gheorghe Asachi" Technical University of Iasi, Romania
- ARHEOINVEST Platform, Alexandru Ioan Cuza University of Iasi
- Centre of Excellence Geopolymer and Green Technology CEGeoGTech), Universiti Malaysia Perlis (UniMAP)

With support of:

- School of Fundamental Science, Universiti Malaysia Terengganu
- International Federation of Inventors' Associations - IFIA
- World Invention Intellectual Property Associations – WIIPA

Editors:

Andrei Victor SANDU, Mohd Mustafa Al Bakri ABDULLAH,

Petrică VIZUREANU, Che Mohd Ruzaidi GHAZALI, Ion SANDU

IOP Conf. Series: Materials Science and Engineering **374** (2018) 011001 doi:10.1088/1757-899X/374/1/011001

EUROINVENT ICIR 2018
International Conference on Innovative Research
Iasi, 17th–18th of May 2018

Foreword

This volume contains the information of the ICIR Euroinvent 2018 Conference and the abstracts of selected peer-reviewed papers from the 2018 International Conference on Innovative Research, which was held in Iaşi, România from 17 to 18th of May 2018.

The ICIR Conference is organized under the auspices of EUROINVENT. Euroinvent is a joint event promoting creativity in a European context, by displaying the contributions of consecrated schools from higher education and academic research and also of individual inventors and researchers.

The EUROINVENT International Conference on Innovative Research (ICIR) brings together leading researchers, engineers and scientists who will present actual research results in the field of Materials Science and Engineering.

The conference aims to provide a high level international forum for researchers, engineers and scientists to present their new advances and research results in the field of materials science and engineering.

The volume covers all the aspects of materials science and engineering, from synthesis and characterization of materials to procedures and technologies for materials engineering, as well as materials application and their involvement in the life sciences.

All the papers have been reviewed by two expert referees in their relevant topic disciplines. The papers selected for the volume depended on their quality and relevancy to the conference.

The editors hope that this volume will provide the reader a broad overview of the latest advances in the field of materials science and engineering, and that they will be a valuable references source for further research.

The editors would like to express their sincere appreciations and thanks to all the committee members of the ICIR 2018 for their tremendous efforts. Thanks also to IOP Conference Series for producing the volume with full articles.

Finally, the editors would like to thank all the authors for their contribution to this valuable volume.

IOP Conf. Series: Materials Science and Engineering **374** (2018) 011001 doi:10.1088/1757-899X/374/1/011001

EUROINVENT ICIR 2018 — International Conference on Innovative Research, Iasi, 17th–18th of May 2018

Scientific Advisory Board

Chairman

Event Coordinator

Technical and Organizing Committee

Lecturer Dr.Eng. Dragos Cristian ACHITEI
Gh. Asachi Technical University of Iasi, Romania

Assist.Dr.Eng. Madalina Simona BALTATU
Gh. Asachi Technical University of Iasi, Romania

Assist. Prof. Eng. Dumitru Doru BURDUHOS NERGIŞ
Gh. Asachi Technical University of Iasi, Romania

PhD student Eng. Diana Petronela BURDUHOS NERGIŞ
Gh. Asachi Technical University of Iasi, Romania

Assist.Prof. Dr. Eng. Bogdan ISTRATE
Gh. Asachi Technical University of Iasi, Romania

Lecturer Dr.Eng. Mirabela Georgiana MINCIUNA
Gh. Asachi Technical University of Iasi, Romania

PhD student Marius MUNTEANU
Al. I. Cuza University of Iasi, Romania

PhD student Vasile PELIN
Al. I. Cuza University of Iasi, Romania

Lecturer Dr.Eng. Manuela Cristina PERJU
Gh. Asachi Technical University of Iasi, Romania

PhD student Daniel POTOLINCA
Al. I. Cuza University of Iasi, Romania

Lecturer Dr.Eng. Ioan Gabriel SANDU
Romanian Inventors Forum & Gh. Asachi Technical University of Iasi, Romania

Assist. Dr. Eng. Catalin Andrei TUGUI
Gh. Asachi Technical University of Iasi, Romania

S.R. III. Dr. Viorica VASILACHE
Al. I. Cuza University of Iasi, Romania

IOP Conf. Series: Materials Science and Engineering **374** (2018) 011002 doi:10.1088/1757-899X/374/1/011002

Peer review statement

All papers published in this volume of *IOP Conference Series: Materials Science and Engineering* have been peer reviewed through processes administered by the proceedings Editors. Reviews were conducted by expert referees to the professional and scientific standards expected of a proceedings journal published by IOP Publishing.

Effects of Sm^{3+} on Luminescent Properties of $LiEu_{(0.55-x)}Y_{0.45}(WO_4)_2Sm_x$ Red Phosphor

N S Zailani[1], M F Ghazli[1,2], R Hussin[3], S Z Abd Rahim[1,2] and M N Mat Saad[1,2]
[1]School of Manufacturing Engineering, Universiti Malaysia Perlis (UniMAP), Perlis Malaysia
[2]Center of Excellence Geopolymer & Green Technology (CEGeoGTech), Universiti Malaysia Perlis (UniMAP), Perlis, Malaysia
[3]Department of Physics, Faculty of Science, Universiti Teknologi Malaysia, Skudai, Johor, Malaysia

E-mail: fathullah@unimap.edu.my

Abstract. Phosphor materials have been widely applied in lighting and displays. Rare earth elements are amongst important elements in giving luminescent properties to the phosphor. The needs to reduce rare earth element in phosphor is of interest due to cost and environmental issue. The red phosphor $LiEu_{(0.55-x)}Y_{0.45}(WO_4)_2$ were synthesized by substituting Sm^{3+} in the lattice using solid-state reaction. The effect on the substitution of Sm^{3+} which acts as an activator will be analysed using X-ray powder diffraction (XRPD), scanning electron microscope (SEM) and photoluminescence spectra (PL). The structure of the compound was found to be as a monoclinic which referred to the reference patterns. The optimum amount was recorded for the activator to be obtained in the host lattice for replacing the Eu^{3+} place as well as enhancing the intensity of the emission. The findings reveal that $LiEu_{(0.55-x)}Y_{0.45}(MoO_4)_2Sm_x$ can be a more efficient red phosphor candidate used white light emitting diodes (WLEDs) applications.

1. Introduction

White light emitting diodes (WLEDs) have attracted significant attention due to their advantages such as high energy efficiency, energy-saving, environmental friendly, and long life time [1, 5]. The first commercialized LED has been produced in year 1996 by exhibit a yellow emission (usually Ce^{3+} -doped $Y_3Al_5O_{12}$) under blue excitation and combining with an InGaN blue LED chip [1, 3, 4, 6-8]. However, the existing WLEDs have some limitations such as less colour rendering index (CRI) (<75), limited indoor usage and less luminous efficiency [1, 6]. It also found that the lack of red spectral component in the host compound will reducing the emission and CRI of the phosphor near-UV region [7, 9]. Hence, this is motivated for find a good red phosphor candidate which can provided the higher CRI, higher luminous efficiency as well as high chromatic stability [2, 5]. There is an increasing number of research on searching the scheelite-type structure material such as as $MR(BO_4)_2$ (M = Li, Na, K, Ag; R = lanthanides (Ln), Y, Bi; B = W, Mo) as a luminescent materials due to their thermal and chemical stabilities, self-activated nature, wide emission spectra in visible region, can transfer their absorbed energy to the dopant ions and high density [2, 5, 10]. Eu^{3+} has a good potential to a red phosphor candidate which offer a high quantum efficiency and photostability in the red region [10-12]. There are some of the previous study on the searching the luminescence properties based on the tungstate and introducing the Eu^{3+} to the lattice; $LiY(MoO_4)_{2-x}(WO_4)_x{:}Dy^{3+}$, Eu^{3+}, $LiEu(WO_4)_{2-}$

$_{x}(MoO_4)_x$, $LiLn(MO_4)_2:Eu^{3+}$ (Ln = La, Eu, Gd, Y; M = W, Mo) and $LiEu_{(1-x)}Y_x(WO_4)_2$ [1, 4, 13, 14]. The previous study by Fathullah et al. (2016) found the highest luminous efficacy of $LiEu_{(1-x)}Y_x(WO_4)_2$ found when the Eu^{3+} concentration at 55 mol% and Y^{3+} 45 mol% [14]. From this finding, it motivated to conduct this current work by introducing the Samarium (III) as a co-activator to the $LiEu_{(0.55-x)}Y_{0.45}(WO_4)_2Sm_x$ (x= 0 – 0.55) lattice. It expected that by introducing Sm^{3+} will help in reducing the usage of Eu^{3+} where the Eu^{3+} are more expensive. This work studied on the potential of the $LiEu_{(0.55-x)}Y_{0.45}(WO_4)_2Sm_x$ by analysing the effect on the crystal structure and the photoluminescence properties to be used as a red phosphor in WLEDs application.

2. Materials and methods

2.1. *Sample preparation*

The $LiEu_{(0.55-x)}Y_{0.45}(WO_4)_2Sm_x$ phosphors were prepared using solid-state reaction where changes the x concentration between 0 to 0.55. The starting materials were lithium carbonate 99.9%, yttrium, tungsten (VI) oxide 99+%, europium oxide (Eu_2O_3), and samarium oxide (Sm_2O_3). At first, the reactants were mixed with stoichiometric amount and then heated at 850 °C in open air for 3 hours. The compounds finally cooled at room temperature and gently ground for removing the sample powder from the crucible.

2.2. *Characterization*

The X-ray Powder Diffraction (XRPD) patterns were measured using a Bruker D2 Phaser with Cu Kα radiation (λ = 1.54184 Å). The morphology of the samples were observed using Hitachi TM3000 scanning electron microscope (SEM). Photoluminescence properties were measured from Perkin Elmer, LS 55 spectrometer. All the emission intensity were collected at room temperature with the changes of concentration activator (Eu^{3+}) and co-activator (Sm^{3+}).

3. Results and discussions

3.1. *XRPD analysis of $LiEu_{(0.55-x)}Y_{0.45}(WO_4)_2Sm_x$ samples*

The overlaid XRPD patterns of $LiEu_{(0.55-x)}Y_{0.45}(WO_4)_2Sm_x$ (x = 0, 0.05, 0.20, 0.40 and 0.55 mol) is presented in Figure 1.

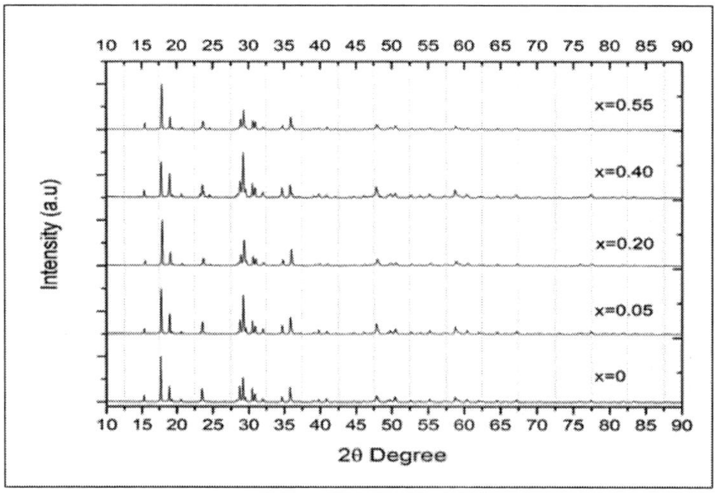

Figure 1. The XRPD stack patterns of $LiEu_{(0.55-x)}Y_{0.45}(WO_4)_2Sm_x$ (x=0, 0.05, 0.20, 0.40 and 0.55).

From this result, it clearly shows that there are no additional peaks or changes occurred as increases Sm^{3+} to the host lattice where indexed to the pure phases $LiEu(WO_4)_2$ (PDF 01-080-7695) and

$LiY(WO_4)_2$ (PDF cart 88-0854) by J. Postema et al., (2011) [15]. This confirms that the compound structure is a tetragonal scheelite structure with space group $I4_{1/a}$ and for sample $LiY_{0.45}(WO_4)_2Sm_{0.55}$ has found as a monoclinic with space group P2/n. This can be supported by referring to the refinement analysis to confirm their structures. Therefore, this findings also claimed that the present result clearly shown that there are two phases occurred since the concentration of Y^{3+} 0.45 mol which rejected the findings prepared by M. Fathullah et. al., (2016) [14] where found the compound $LiEu_{1-x}Y_x(WO_4)_2$ have two phases when the concentration of Y^{3+} at 55%. The sample $LiEu_{(0.55-x)}Y_{0.45}(WO_4)_2Sm_x$ were continued by refining the sample using the TOPAS software. From the refinement data, it seem like there have larger differences especially on the $LiY(WO_4)_2$ peaks. This confirms that the structure of $LiEu_{(0.55-x)}Y_{0.45}(WO_4)_2Sm_x$ samples have a tetragonal scheelite structure with space group $I4_{1/a}$ and the $LiY_{0.45}(WO_4)_2Sm_{0.55}$ was found as a monoclinic structure with space group P2/n with two phases. In Table 1 shows the results of the lattice parameters and cell volume of $LiEu_{(0.55-x)}Y_{0.45}(WO_4)_2Sm_x$. Based on this result, it shows the increasing of the lattice's volume about 0.33% from 300.336 Å^3 to 301.34 Å^3 as Sm^{3+} increases to the host lattice. This is due to the ionic radius of the Sm^{3+} is larger than Eu^{3+} and Y^{3+}. The highest R_{wp} calculated from the refinement of $LiEu_{(0.55-x)}Y_{0.45}(WO_4)_2Sm_x$ is 29.73. This result still can be accepted since the value of R_{wp} is better than other sample in this study and this phosphor has two phases that might have some problem to fitting with the reference models especially in the $LiY(WO_4)_2$ phase.

Table 1. Lattice parameters and cell volumes of $LiEu_{(0.55-x)}Y_{0.45}(WO_4)_2Sm_x$.

Eu^{3+}	Sm^{3+}	a(Å)	c(Å)	Vol(Å^3)	R_{wp}
0.55	0	5.19465(55)	11.1300(18)	300.336(81)	29.73
0.50	0.05	5.19807(61)	11.1300(18)	300.732(86)	36.93
0.35	0.20	5.20168(83)	11.1300(26)	301.15(12)	34.25
0.15	0.40	5.20184(81)	11.1300(24)	301.17(11)	33.41
0	0.55	5.20333(81)	11.1300(27)	301.34(12)	33.11

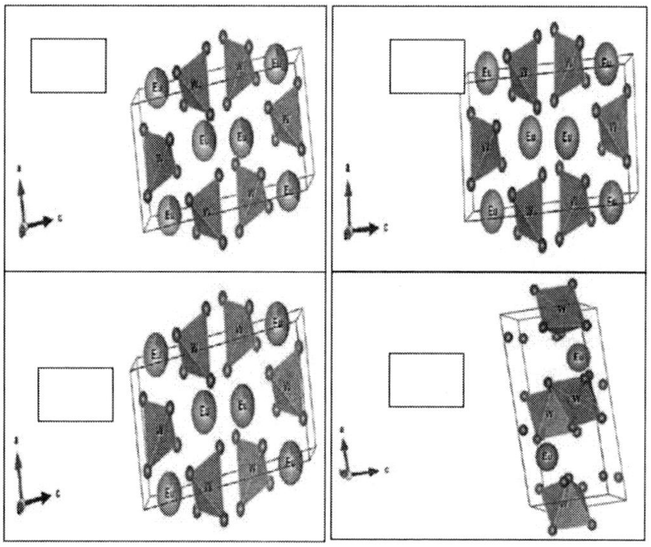

Figure 2. Modelled crystal structure of tetragonal (a) $LiEu_{0.55}Y_{0.45}(WO_4)_2$, (b) $LiEu_{0.40}Y_{0.45}(WO_4)_2Sm_{0.05}$, (c) $LiEu_{0.15}Y_{0.45}(WO_4)_2Sm_{0.40}$ and crystal structure of monoclinic (d) $LiY_{0.45}(WO_4)_2Sm_{0.55}$.

Figure 2 shows the modelled crystal structure of tetragonal (a) $LiEu_{0.55}Y_{0.45}(WO_4)_2$, (b) $LiEu_{0.40}Y_{0.45}(WO_4)_2Sm_{0.05}$, (c) $LiEu_{0.15}Y_{0.45}(WO_4)_2Sm_{0.40}$ and crystal structure of monoclinic (d) $LiY_{0.45}(WO_4)_2Sm_{0.55}$. From these figures, it clearly shows that the additional of Sm^{3+} are remain as the same patterns until Sm^{3+} had fully taking place of Eu^{3+} into the lattice. At 0.55 mol Sm^{3+}, the structure of the phosphor has changed as a monoclinic with space group P2/n. It can be seen that there are four oxygen coordination are surround each of the W^{6+} ions. Therefore, the ions in $LiY_{0.45}(WO_4)_2Sm_{0.55}$ are well-organized with octahedral coordination even though their lattice angles had slightly deformed.

3.2. Morphology of $LiEu_{(0.55-x)}Y_{0.45}(WO_4)_2Sm_x$ samples

Figure 3 shows the morphologies of $LiEu_{(0.55-x)}Y_{0.45}(WO_4)_2Sm_x$ with 1000% magnification of (a) 0 mol% (b) 0.05 mol% and (c) 0.55 mol% Sm^{3+} concentration. The structures show the angular particle shaped with irregular particle sizes between 1– 8μm.

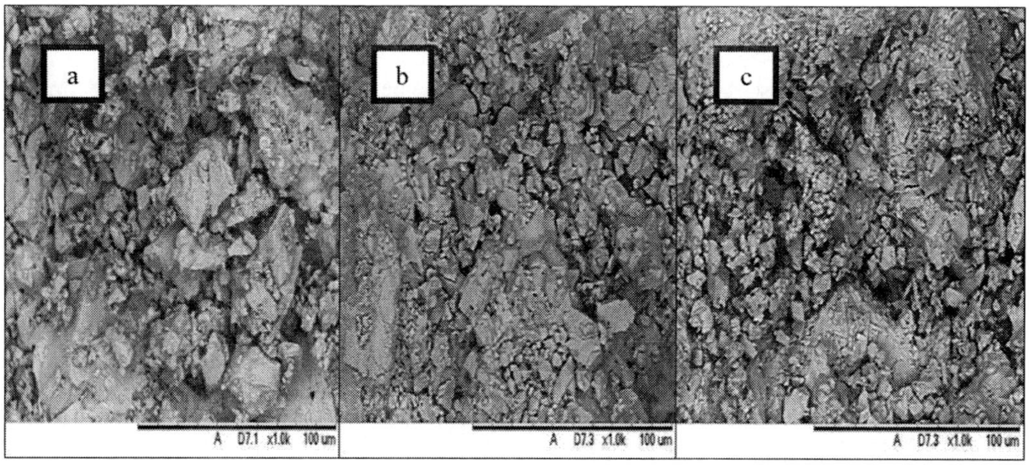

Figure 3. The morphologies of (a) $LiEu_{0.55}Y_{0.45}(WO_4)_2$, (b) $LiEu_{0.50}Y_{0.45}(WO_4)_2Sm_{0.05}$ and (c) $LiY_{0.45}(WO_4)_2Sm_{0.55}$ with 1000% magnification.

3.3. Photoluminescence properties

The emission spectra of $LiEu_{(0.55-x)}Y_{0.45}(WO_4)_2Sm_x$ in range 550 – 750 nm is shown in Figure 4.

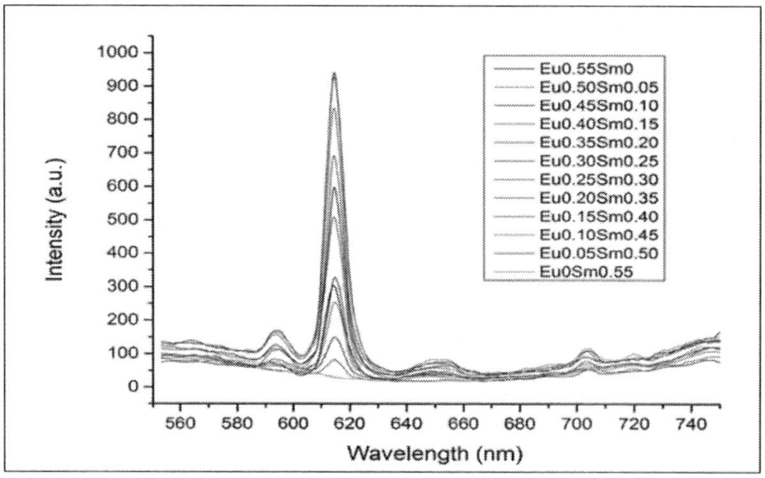

Figure 4. Emission spectra of $LiEu_{(0.55-x)}Y_{0.45}(WO_4)_2Sm_x$ (x= 0 until 0.55) from 550 nm to 750 nm.

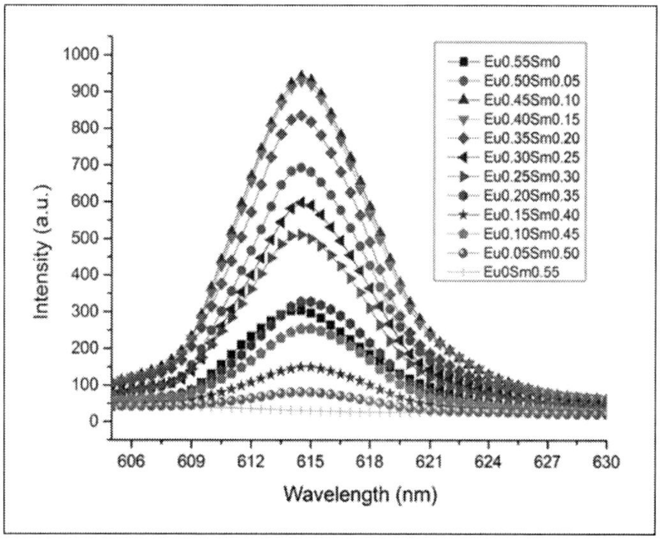

Figure 5. Emission spectra of $LiEu_{(0.55-x)}Y_{0.45}(WO_4)_2Sm_x$ (x= 0 until 0.55) from 605 nm – 630 nm.

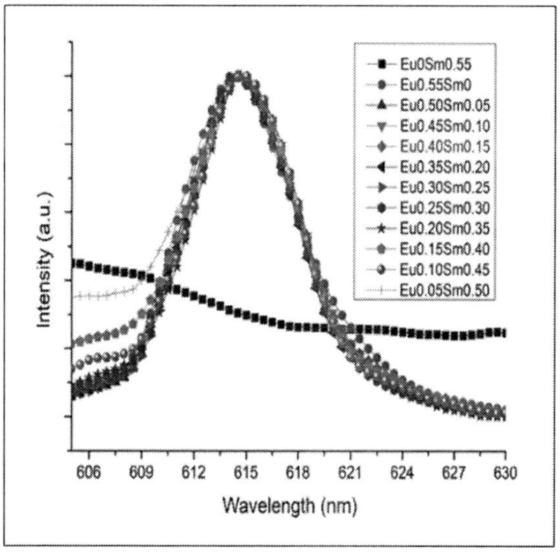

Figure 6. Normalized graph of emission spectra of the $LiEu_{(0.55-x)}Y_{0.45}(WO_4)_2Sm_x$ (x= 0 until 0.55) from 605 nm – 630 nm.

The emission bands were observed at 588 – 600 nm ($^5D_0 \rightarrow {}^7F_1$), 610 – 630 nm ($^5D_0 \rightarrow {}^7F_2$), 640 – 662 nm ($^5D_0 \rightarrow {}^7F_3$) and 681 – 710 nm ($^5D_0 \rightarrow {}^7F_4$) transitions which indicating that the excited state of the Eu^{3+} in the host phosphor. At 615 nm, the highest emission intensity occurred and clearly show the increases of Sm^{3+} in the host lattice did not altered the emission centre. It already known that at 615 nm, the domain electric dipole ($^5D_0 \rightarrow {}^7F_2$ transition) has taken place which make the emission intensity influenced strongly by ligand ions in the crystal [3].

By introducing the Sm^{3+} ions to the $LiEu_{(0.55-x)}Y_{0.45}(WO_4)_2$, it can be seen that an increasea on the intensity occurs. Figure 5 presents the emission spectra of $LiEu_{(0.55-x)}Y_{0.45}(WO_4)_2Sm_x$ from 605 – 630 nm. Based on this figure, it shows that the additional of Sm^{3+} with 0.10 mol% provides the highest emission intensity. Therefore, Figure 6 shows the normalized graph which proving the results have no broadening effects. This proved that this compound did not undergo any changes in the crystal symmetry. Interestingly, it clearly shows a huge trend on the intensity increment when introducing Sm^{3+} to the host lattice. It was about 210.2% increment was recorded on the intensity from 303.71 a.u. ($LiEu_{(0.55-x)}Y_{0.45}(WO_4)_2$) to 942.03 a.u. ($LiEu_{(0.55-x)}Y_{0.45}(WO_4)_2Sm_{0.10}$). Hence, it can be suggested that the current finding has a good potential as a red phosphor to be applied in phosphor converted LEDs applications.

4. Conclusions

In summary, the $LiEu_{(0.55-x)}Y_{0.45}(WO_4)_2Sm_x$ have successfully synthesized using a solid-state reaction with concentration changes (x = 0, 0.05, 0,10, 0.15, 0.20, 0.25, 0.30, 0.35, 0.40, 0.45, 0.50 and 0.55). The structure of the compound are tetragonal scheelite structure with space group $I4_{1/a}$ and the sample $LiY_{0.45}(WO_4)_2Sm_{0.55}$ has a monoclinic P2/n structure with two phase's lattices conducted by XRPD analysis. The morphologies were found in the angular shaped with particle sizes between 1–8μm. Substituting of Sm^{3+} to the $LiEu_{(0.55-x)}Y_{0.45}(WO_4)_2$ causes an enhancement on the emission intensity at 615 nm when Sm^{3+} 0.10 mol% without affecting the lattice symmetry. The findings of the study suggest that the phosphors have a good potential as a red phosphor candidate with better efficiency.

5. References

[1] Liang Y et al. 2017 *J. Alloys Compd.* **716** 56–64
[2] Zhou L et al. 2016 *Mater. Res. Bull.* **84** 429–436
[3] Shang M, Li G, Kang X, Yang D and Lin J 2011 *J. Electrochem. Soc.* **158**(5) H565
[4] Chiu C H, Wang M F, Lee C S and Chen T M 2007 *J. Solid State Chem.* **180**(2) 619–627
[5] Ramakrishna P V, Rao T L, Singh A, Benarji B and Dash S 2017 *J. Mol. Struct.* **1149** 426–431
[6] Zhao D, Shi J-C, Nie C-K and Zhang R-J 2017 *Opt. - Int. J. Light Electron Opt.* **138** 476–486
[7] Li L et al. 2016 *Mater. Res. Bull.* **78** 26–30
[8] Cao F 2012 *J. Lumin.* **132**(3) 641–644
[9] Wang Q F et al. 2015 *J. Alloys Compd.* **625** 355–361
[10] Kumari L S and Gopchandran K G 2015 *Mater. Today Proc.* **2**(3) 1007–1011
[11] Hwang K-S, Kang B-A, Hwangbo S, Kim Y-S and Kim J-T 2010 *Electron. Mater. Lett.* **6**(1) 27–30
[12] Ru J, Ying S, Zheng W and Chen J 2016 *Mater. Res. Bull.* **84** 468–473
[13] Liu Y et al. 2014 *RSC Adv.* **4**(9) 4754
[14] Postema J M, Fu W T and Ijdo D J W 2011 *J. Solid State Chem.* **184**(8) 2004–2008

Acknowledgments

We would like to acknowledge and thank you for funding and support from the Ministry of Higher Education Malaysia MOHE through the Fundamental Research Grant Scheme (FRGS) (Ref: FRGS/1/2016/STG07/UNIMAP/02/1), Application ID 218885-247922 and code 9003-00572, and also thank you to Universiti Malaysia Perlis.

IOP Conf. Series: Materials Science and Engineering **374** (2018) 012002 doi:10.1088/1757-899X/374/1/012002

SEM and EDX Study of Stainless Steels, Suggested as Human Body Implants

R N Simitchiiska[1], D Ivanova[1] and L Fachikov[1]
[1] Department of Inorganic and Electrochemical Production, University of Chemical Technology and Metallurgy, Bulgaria, Sofia 1756, 8, Kliment Ohridsky, Blvd.

E-mail: simitchiiska@uctm.edu

Abstract. The most important requirement for any material used as an implant is to be biocompatible and not to cause undesirable effects in the human body. Corrosion of implants is included in the topic of biocompatibility because it is a determining factor in their sustainability and seamless fulfillment of their functional purpose. The work presents the results obtained in the comparative study of two austenitic stainless steels (Cr18Ni9 and Cr18Mn12N) in two models of artificial saliva, accepted as environments for testing the corrosion behavior of materials for these aims. The nature of the corrosion attack, the composition of the corrosion products at 37°C and the pH 5.5 and pH 6.75 of the model media were determined using physical methods such as SEM and EDX of investigation and analysis. It was found that Cr18Mn12N steel (0.61 % N), exhibits higher corrosion resistance.

1. Introduction

Biocompatibility of dental material is a fundamental requirement of successful behavior in oral cavity. It draws on knowledge from biology, patient risk factors, clinical experience and engineering. There are two key factors that appear to be important in determining biocompatibility of any dental material – some involve various types of corrosion or material degradation and others include surface characteristics. Dental materials within the mouth interact continually with physiological fluids.

More often used orthodontic appliances are mainly made of stainless steels, nickel–titanium alloy, titanium and its alloys, or cobalt-chrome alloys, which have high corrosion resistance due to their stable passive films [1]. Later, stainless steel has proven to be a better material than noble metal alloys for orthodontic appliances. This superiority is primarily due to its low cost, greater strength, higher modulus of elasticity, good formability, and high corrosion resistance in the mouth. These advantages have been accepted by orthodontists, and as a result stainless steel is used extensively for bands, brackets, and arch wires [2].

Recently it has become apparent that there are needs for even higher corrosion resistance, high strength, improved formability, and lower cytotoxicity. Much effort has gone into satisfying these needs and into improving the properties of stainless steels. Super stainless steels that have higher corrosion resistances and greater mechanical strengths than the conventionally used stainless steels have been developed and these could overcome disadvantages [3-4].

Corrosion, the degradation of materials by electrochemical attack, is of concern particularly when orthodontic appliances are placed in the hostile electrolytic environment provided by the human mouth [5].

For all practical purposes, the metallic orthodontic wires cannot be isolated electrically from the tooth. Resistance to corrosion is critically important for orthodontic wires because corrosion can lead to roughening of the surface, weakening of the appliances, and liberation of elements from the metal or alloy. Release of elements can produce discoloration of adjacent soft tissues and allergic reactions in susceptible patients. Corrosion can severely affect the ultimate strength of the material leading to mechanical failure of the orthodontic materials.

Some alloys and metals are resistant to corrosion because of inherent nobility or the formation of a protective surface layer. However, many ions containing oral environment can cause corrosion during long time exposure [6-8].

In the present study the corrosion resistance and character of corrosion attack of two stainless steels Cr18Ni9 and HNS Cr18Mn12N in artificial saliva were investigated by SEM and EDX analyses.

2. Experimental

2.1. Materials and methods

The chemical content of the investigated steels are presented in Table 1.

Table 1. The chemical composition (wt.% and at.%) of the tested stainless steels.

Steel		Cr	Ni	Mn	C	N	Si	P	S
Cr18Mn12N	**wt.%**	16.50	0.05	12.00	0.04	0.61	0.36	0.011	0.023
	at.%	17.07	0.05	11.74	0.16	2.37	0.70	0.02	0.005
Cr18Ni9	**wt.%**	17.49	9.37	1.29	0.05	-	0.52	0.022	0.009
	at.%	18.74	8.85	1.23	0.23	-	1.03	0.04	0.016

The electrochemical treatments of the samples before the surface analyses were conducted in a conventional three electrodes cell in open air conditions at 37°C, with a platinum counter electrode and a saturated calomel electrode (SCE) as a reference one. The experiments were carried out with *EG&G Princeton Applied Research, Potentiostat/Galvanostat Model 263A*, provided with the specialized software package PowerCORR®. After the anodic polarization for 20 minutes at potential 100 mV more positive than the pitting potential determinate potentiodynamically for each steel in an artificial saliva.

Two physical methods for investigation of the surface were carried out.

Scanning electron microscopy (SEM). The morphology and structure of the surfaces were examined by scanning electron microscopy, using a *SEM/FIB LYRA I XMU, TESCAN* electron microscope, equipped with ultrahigh resolution scanning system secondary electron image (SEI).

Energy dispersive X-ray spectroscopy (EDX). The energy dispersive spectroscopy is a local X-ray spectral analysis which permits qualitative and quantitative determination of surface micro volume contents of the order of several μm^3. Apparatus *Quantax 200, BRUKER* with spectroscopic resolution at Mn-Ka and 1 kcps 126 eV is used.

2.2. Solutions

Two types of artificial saliva, presented in Table 2, are used. The second one is received by modification of the first with 1M HCl to acidify the solution. The parameters pH and σ, mS sm^{-1} of the solutions are pointed at Table 3.

Table 2. The chemical composition (g l-1) of the tested solutions.

Solution	Composition of the solution, g l^{-1}						
	NaCl	*KCl*	*$CaCl_2.2H_2O$*	*$Na_2HPO_4.2H_2O$*	*$Na_2S.9H_2O$*	*Urea,*	*Dist.H_2O*
Saliva	0.40	0.40	0.795	0.780	0.005	1.00	1000

Table 3. The parameters of the solutions.

Solution	pH	σ, mS sm^{-1}
Saliva -1	6.75	3.01
Saliva(HCl)-2	5.5	3.50

3. Results and discussions

3.1. SEM – analyses

The main form of corrosion on the steel surfaces after electrochemical treatment of the samples is basically pitting corrosion. In order to establish the type and size of the pits appeared on the surfaces of both investigated steels, the samples were observed by scanning electron microscopy. The microphotographs for both steels are presented on Figure 1 for saliva-1 and on Figure 2 for saliva-2.

The microscopy observations indicate more numerous pits on Cr18Ni9 steel surfaces which are larger in comparison to these on the surface of the Cr18Mn12N samples in both solutions.

The pits occurring at the nitrogen steel surface are spherical in shape and could be divided formally into two groups: *small pits* with diameters in the rage 20-60 µm and *large pits* with diameters of about 200 µm.

The pits on the nickel steel surface are also spherical in shape, but in contrast to the nitrogen steel samples they are larger (with 200-250 µm in diameter) and deeper.

Some of the pits are open and the others are under surfaces types.

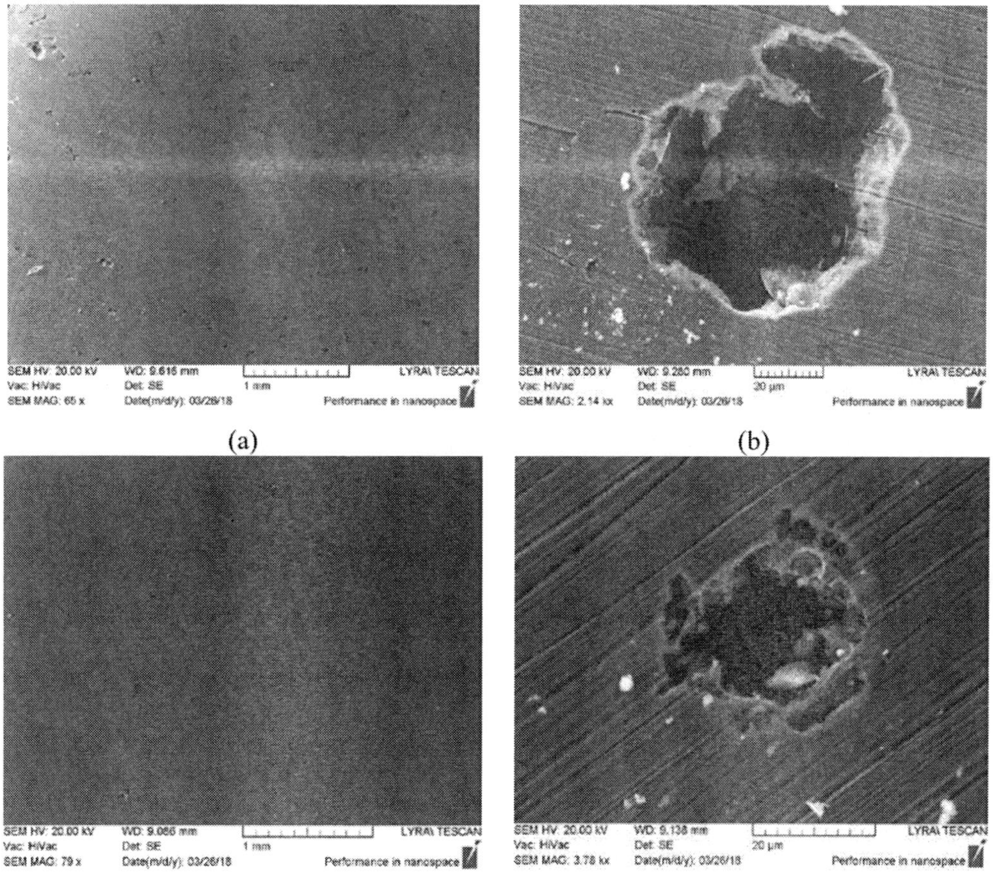

(a) (b) (c) (d)

Figure 1. SEM microphotographs of (a, b) Cr18Ni9 and (c, d) Cr18Mn12N steels after potentiostatic polarization in saliva-1.

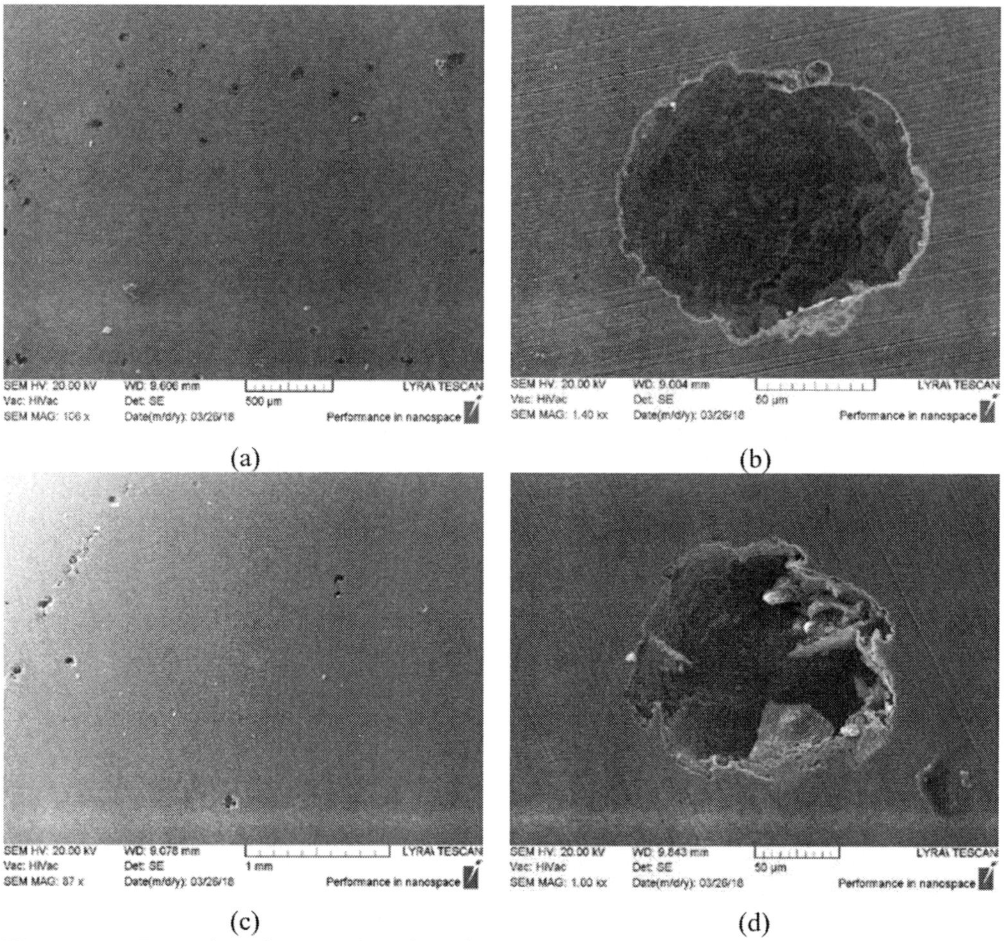

Figure 2. SEM microphotographs of (a, b) Cr18Ni9 and (c, d) Cr18Mn12N steels after potentiostatic polarization in saliva-2.

3.2. EDX analyses

EDX analyses of the surfaces on the bottom of the pits were performed after electrochemical treatment of the samples in two solutions: saliva-1 and saliva-2.

Table 4. EDX analyse of the elements saliva-1.

Elements	Concentration Cr18Ni9		Concentration Cr18Mn12N		Elements	Concentration Cr18Ni9		Concentration Cr18Mn12N	
	Wt. %	*At. %*	*Wt. %*	*At. %*		*Wt. %*	*At. %*	*Wt. %*	*At. %*
Fe	61.68	*56.68*	45.61	*36.32*	*Cl*	1.03	*1.49*	3.27	*4.1*
Cr	18.19	*17.95*	15.01	*12.84*	*P*	0.25	*0.42*	1.22	*1.75*
Ni	8.87	*7.76*	-	-	*Al*	0.2	*0.38*	1.54	*2.54*
Na	-	-	12.33	*23.86*	*S*	0.22	*0.34*	2.36	*3.27*
Mn	5.76	*5.38*	14.24	*11.53*	*Ca*	0.07	*0.09*	-	-
O	1.48	*4.75*	-	-	*Cu*	-	-	3.65	*2.55*
C	0.72	*3.08*	-	-	*Si*	0.04	*0.07*	0.78	*1.23*
Ti	1.48	*1.59*	-	-					

The analysis of the corrosion products (Table 4 and Table 5) indicates high content of O, P and Cl and other elements, which are components of the solutions, as well as Fe, Mn, Ni, Cr and so on. Therefore, apart from the oxides formation during surface dissolution there exist also metal chlorides and phosphates. The results presented in the tables indicate that the solution saliva-2 more aggressive to investigate steels.

Table 5. EDX analyse of the elements saliva-2.

Elements	Concentration Cr18Ni9		Concentration Cr18Mn12N	
	Wt. %	*At. %*	*Wt. %*	*At. %*
Fe	40.98	*22.71*	40.35	*31.67*
Cr	18.57	*11.05*	23.89	*20.14*
Ni	4.92	*2.60*	-	-
Mg	-	-	1.45	*2.61*
Mn	-	-	19.35	*15.44*
O	16.75	*32.40*	7.11	19.47
C	8.01	*20.64*	-	-
Ti	1.60	*1.03*	-	-
Cl	0.90	*0.78*	1.81	*2.24*
P	1.78	*1.77*	0.76	*1.08*
Al	3.70	*4.25*	2.33	*3.78*
S	1.39	*1.34*	0.71	*0.97*
Ca	0.43	*0.33*	1.42	*1.56*
Cu	-	-	0.26	*0.18*
Si	0.98	*1.08*	0.55	*0.86*

4. Conclusions

At the time of comparative investigation the corrosion behavior of two austenitic stainless steels (Cr18Ni9 and HNS Cr18Mn12N), suggested for the orthodontic appliances, in two model artificial saliva were defined the character (pitting corrosion, with different in size spherical pits) and the probable composition of the corrosion products. The results manifested that the nitrogen-containing steel possess higher résistance in model media and hence more suitable for orthodontic appliances.

5. References

[1] Kusy R P and Greenberg A R 1981 *Angle Orthod* **54** 325-341
[2] Wilkinson J V 1962 *Angle Orthod* **48** 192-206
[3] Halada G P, Kim D and Clayton CR 1996 *Corrosion* **52** 36-46
[4] Willenbruch R D , Clayton CR, Oversluizen M, Kim D and Lu YC 1990 *Corros Sci* **31** 179-190
[5] House K, Sernetz F, Dymock D, Sandy J R and Ireland A J 2008 *J. Orthop* **133** 584-592
[6] Eliades T, Athanasiou A E 2002 *Angle Orthod,* **72** 222-237
[7] Jia W, Beatty M W, Reinhardt R A, Petro T M, Cohen D M, Maze C R et al 1999 *J Biomed Mater Res* **48** 488-495
[8] Chaturvedi T P and Upadhayay SN 2010 *Indian J Dent Res* **21(2)** 275-284

Acknowledgments

This study has been financially supported by the Operational Programme "Science and education for smart growth" 2014-2020 of the European Union cofounded by the European Social Fund through the project BG05M2OP001-2.009-0015 "Support for the development of capacity of doctoral students and young researchers in the field of engineering, natural and mathematical sciences".

Synthesis of Self-Bonded Pellets of ETS-4 Phase by New Methodology of Preparation

P De Luca[1*], C Mastroianni[2] and J B Nagy[1]
[1]Dipartimento di Ingegneria Meccanica, Energetica e Gestionale
[2]Dipartimento Ingegneria per l'Ambiente ed il Territorio ed Ingegneria Chimica
Università della Calabria- Arcavacata di Rende -87036 Rende (Cs) Italy

E-mail: pierantonio.deluca@unical.it

Abstract. We hereby present the results of a research in order to prepare self-bonded pellets of ETS-4 phase by a new methodology of preparation. In particular, the pellets were prepared by kneading crystals of ETS-4 phase and a dry gel, the latter being the precursor for the synthesis of the ETS-4 phase crystals. One of the innovative aspects that is proposed and highlighted in this work is represented by the fact that the dry-gel acts as a "binder". It is characterized by the same chemical composition of crystals, thus avoiding contamination with other elements. In addition, dry-gel allows, the promotion of nucleation phenomena and thus the formation of new crystals of ETS-4 during the pellets baking phases. The pellets were characterized by X-ray diffraction (XRD), Electron microscopy (SEM) and mechanical strength by hardness tester.

1. Introduction

In this research the ETS (Engelhard Titanium silicate), object of study, is an important family of microporous materials. Two important molecular sieves belong to this family, ETS-4 and ETS-10, that were first synthesized from Kuznicki [1-3]. The ETS-10 phase is an extremely interesting titanosilicate microporous material, due to its high thermal stability and its wide open pores (pore apertures close to 0.8 nm). The structure of ETS-10 was proposed by Anderson et al. [4-6]. However, the ETS-4 phase is a small-pore microporous titanosilicate molecular sieve with an effective pore opening of about 3.7 A [7-8]. During the last decades, in addition to the study of their structure, many researches, have given particular attention to the synthesis of these materials [9-15]. Other studies regard the possibility of inserting hetero-atoms different from silicon and titanium in the structure of these materials to improve or modify their characteristics [16-19]. These materials are very versatile and can be applied in a variety of applications such as water purification and heavy metal removal [20-27], gas adsorption [28-31] and also for their photocatalytic action [32-35]. Generally these materials are obtained in the form of powders after synthesis. For many years, a great deal of research has been directed towards the synthesis of microporous materials in the form of pellets in as much as this guarantees better utilization and recovery after their use. The preparation of the pellets, however, requires the use of a binder, which can obstruct the porous structure resulting in decreased pores' accessibility and hence reducing property resulting from the porous structure such as ion exchange, molecular sieving, adsorption and catalysis [36, 37]. For this reason, some research has been focused on the preparation of pellets of microporous and zeolite materials without the use of a binder [38]. Even for ETS phases, the conditions, to obtain self-bonded pellets directly for synthesis without the use of a binder, have been studied. In 2008 Vuono et al. proposed the synthesis of self-bonded pellets ETS-4 and ETS-10

[39]. Further researches include eteroatoms inside pellets with the aim of modifying their characteristics [40, 41]. Recent studies report self-bonded ETS-10 and ETS-4 pellets with carbon nanotube insertion [42]. The aim of this research was to synthesize ETS-4 phase microporous materials in the form of pellets by an innovative process. In particular, the pellets were prepared by kneading crystals of ETS-4 phase with dry-gel, the latter is a precursor for the synthesis of the same ETS-4 phase crystals. One of the many innovative aspects that is proposed and highlighted by this work is represented by the fact that to acting as "binder", is dry-gel, that is characterized by the same chemical composition of crystals. In addition, dry-gel can lead to the promotion of nucleation phenomenon during the pellet baking phases and, thus, the formation of new crystals of ETS-4.

2. Materials and methods

This research was developed through different stages:

- preparation and pretreatment of the gel-dry precursor of the crystalline phase ETS-4;
- synthesis of crystals of the ETS-4 phases;
- preparation of pellets.

The characterization of the obtained materials was carried out through chemical-physical techniques. The RX diffractograms were obtained using the diffractometer model Philips PW 1730/10, equipped with a vertical goniometer PW 1830, with CuKα radiation type, scan speed of 1°/min and scanning angle 2θ between 5-45°. The samples were observed by Scanning Electron Microscope (SEM) FEI "Quanta" 200 FEG (E = 10kV, SE Image). Vanderkamp VK200 hardening tester was used for the test of mechanical resistance.

2.1. Preparation and pretreatment of the precursor gel of the selected crystalline phase

The preparation of precursor gels for the synthesis of the ETS-4 phase was carried out according to the procedures reported by Pavel et al. [43]. In particular, gels were prepared by the following molar systems: $2Na_2O$-$0.6KF$-$0.2TiO_2$-$2.56HCl$-$1.49SiO_2$-$39.5H_2O$. The gels were obtained preparing an acidic solution (A) by dissolving HCl (37 wt.%, Carlo Erba), $TiCl_4$ (50 wt.%, Merck) and KF (40 wt.%, Merck) in distilled water, and a basic solution (B) by dissolving sodium silicate (8 wt.% Na_2O, 27 wt.%SiO_2,Merck) in NaOH (50 wt.%, Carlo Erba). Solution (A) was poured into solution (B), and the gel obtained was stirred for 5 min. under stirring for 1 min. This preparation of the gel was followed by a drying procedure at 100 °C in a stove for 24 hours. Subsequently, it was finely pulverized in a mortar obtaining the so-called dry-gel. The X-ray diffraction spectra of the gel has already been reported in a previous work [39].

2.2. Synthesis and characterization of ETS-4 phase crystals

The ETS-4 phase crystals were obtained using gels prepared as described in the previous section but without performing the drying step. Once prepared, the gel was placed in Morey autoclaves (constituted externally by a steel jacket and internally by a teflon container) and subjected to hydrothermal synthesis in a stove at 190 °C for five days. After the baking time, the autoclaves were cooled down and then opened. The synthesis product was filtered and washed subsequently characterized by X-ray diffraction in order to verify its quality and to confirm the obtained phase. The RX diffraction spectra of the ETS-4 crystalline phase have already been reported in a previous work [39].

2.3. Preparation of pellets

Once the ETS-4 dry-gel and crystals were obtained, the pellets were prepared by a mixing phase crystals ETS-4 and dry-gel (precursor of the same crystalline phase) to different ratios crystals / dry-gel (%wt/wt) = 1, 3, 9 and by adding some mL of distilled water in order to make the system functional. This mixture was then preformed into pellets using a hydraulic press capable of applying 400 bars. The pellets obtained had a diameter of 1.3 cm and a thickness of 0.7 cm. For each system studied, three pellets were prepared. Subsequently the pellets, once prepared, underwent three different treatments: (a) ageing at 25 °C for 3 days; (b) drying at 100 °C in a thermo-ventilated stove for 3 days; (c) hydrothermal synthesis. In particular, in the latter, the pellets were placed in Morey type autoclaves where a support made of teflon was placed with the purpose of maintaining the pellet raised from 5 mm of distilled water in order to create a hydrothermal system. The autoclaves were then hermetically closed with screws and put in the oven at 190 °C for three days. The pellets extracted from the autoclave were filtered, washed with distilled water and dried in oven at 100 °C for 24 hours.

3. Results and discussion

As shown below, the results and discussion of: -preparation of phase pellets ETS-4 with new procedure, -activating of the ETS-4 phase pellets with addition of indicators, are reported.

3.1. Phase Pellets ETS-4 as made

The following Figure 1 shows the X-ray diffraction spectra of the pellets obtained by varying the crystals/ dry-gel ratio and the treatment modes. Figure 1 shows that the pellets are well formed, retain their shape even after hydrothermal synthesis at 190 °C and are white in color for all systems.

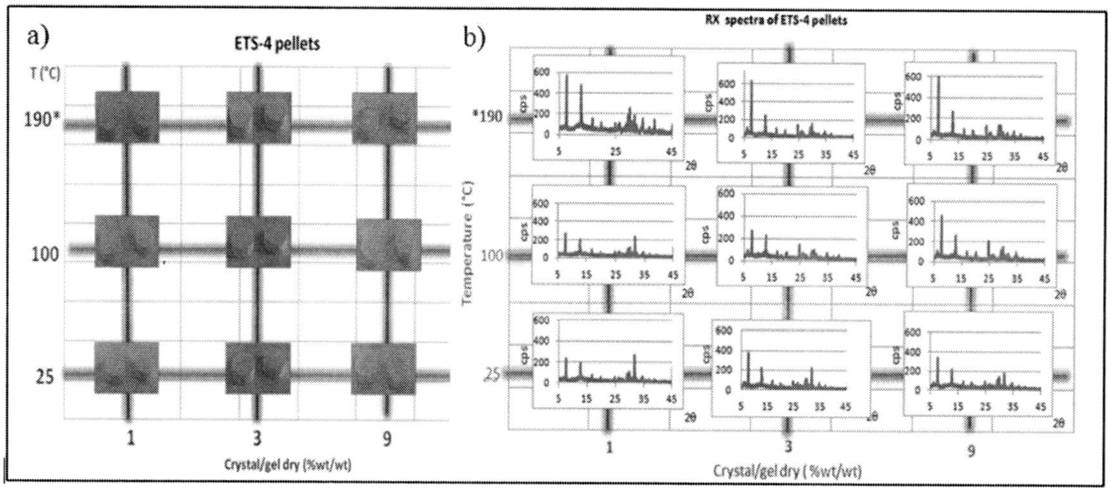

Figure 1. Imagines (a) and X-ray diffraction spectra (b) of the pellets obtained varying the crystals/ dry-gel ratio and the treatment modes.

X-ray diffraction spectra relating to ETS-4 phase pellets (Fig. 1b) show a marked increase in crystallinity, depending on the crystals/gel-dry ratio and the treatment mode. The increase in crystallinity by the augmentation in the crystals/dry-gel ratio is obviously due to higher amounts of crystals in the initial pellet preparation system. This occurs when the temperature of the treatment is increased due to crystallization of the dry gel. The characteristic peaks of the structure of the ETS-4 are easily observable in all the samples examined and, moreover, we must point out that the characteristic peak attributable to the dry-gel (32 2*theta*) tends to decrease both with the increase in the crystals / dry-gel ratio, and by increasing the temperature of post-pelletization treatment, confirming dry-gel crystallization phenomenon. On the whole this is more apparent for the system with crystals /dry-gel ratio equal to 50, in fact the characteristic peak of the gel is evident at 25 °C but disappears at 190 °C.

3.2. Mechanical strength of pellets

Figure 2 (see below) shows the mechanical resistance of the ETS-4 pellets as a function of the crystals / dry-gel ratio and the treatment temperature.

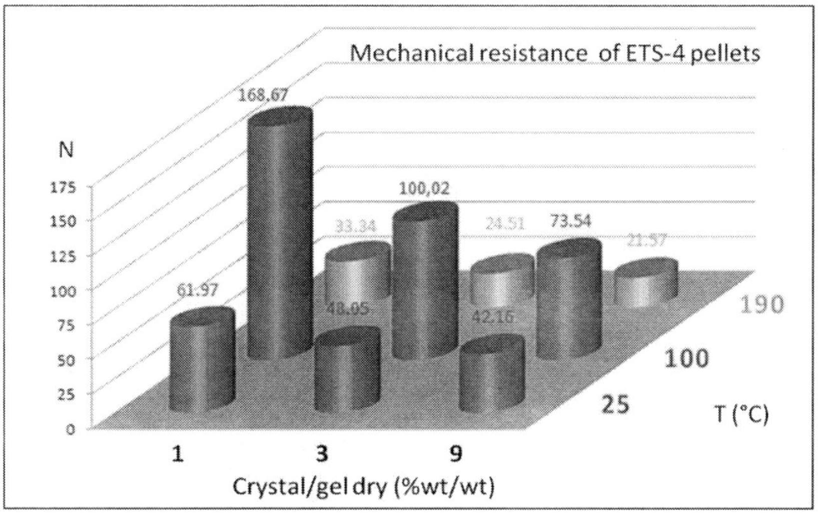

Figure 2. Mechanical resistance of ETS-4 pellets as a function of crystals / dry-gel ratio and temperature of treatment. (*) for T = 190°C - hydrothermal reaction.

We may observe in Figure 2 the mechanical resistance of ETS-4 pellets. All the pellets gave satisfactory results. Only in two cases, the resistance values drop below 30 N with a minimum corresponding to 21.57 N (crystal/gel-dry gel= 9 and T = 190 °C) which, however, is an acceptable result. The maximum resistance value, equal to 168.67 N, is recorded for T = 100 °C and for crystal / gel-dry= 1. If we analyze the graph carefully, it is clear that the change in the resistance is relative to the used variables (Crystal/ dry-gel ratio and mode treatment).

We may note by observing the graph from left to right, that while the crystal/gel-dry ratio increases, the resistance decreases and this is, of course, due to the decrease of the amount of dry-gel (binder).

Another aspect that can be highlighted is the increase in resistance from pellets subjected to baking for 24h at 100 °C as regard to pellets left at room temperature (T= 25 °C). However, by switching from 100 °C to the hydrothermal treatment at 190°C, the resistance decreases once more whilst because the crystallinity is increased due to the favorable conditions which allow the dry-gel to crystallize, though good values of resistance are always confirmed.

3.3 Mechanical resistance of the pellets after immersion in water

An important parameter to be evaluated in the ability of pellets is withstanding in water, since they may find applications in aqueous solutions for the particular peculiarities of the ETS-4 phase, such as heavy water purification, molecular sieving, etc.

For this reason, pellets obtained, according to the procedures reported in paragraph 2.3, were subsequently treated by immersion into distilled water for 10 minutes at a temperature of 25 °C. After the established time, they were dried and placed in a silica gel dryer for 24 hours. Subsequently, the mechanical compressive strengths were measured on pellets. Figure 3a shows the images of pellets obtained after immersion in water.

In particular, it is evident that despite the immersion into water, all samples retained their form with the exception of very few cases where slight splits around the border appeared. Overall, all the pellets retained their compactness and integrity as before being submerged in the water immersion cycle.

Figure 3 (b) reports data of the mechanical resistances of the ETS-4 phase pellets after immersion in water. Mechanical resistance data showed, in all cases, a lowering of compression resistances, although in many cases the resistance continues to be significant. In particular, it can be noted that the resistance value is reduced to a minimum of 7.85 N (for a crystal/dry-gel ratio equal to 9 and at T=

IOP Conf. Series: Materials Science and Engineering **374** (2018) 012003 doi:10.1088/1757-899X/374/1/012003

190 °C (hydrothermal synthesis) which, however remains an good result. The maximum resistance value of 133.37 N is, however, recorded for a T= 100 °C and a crystal / dry-gel ratio equal to 1. It is furthermore apparent that there is a variation in the resistance relative to the variables used (crystal / gel-dry ratio and treatment mode).

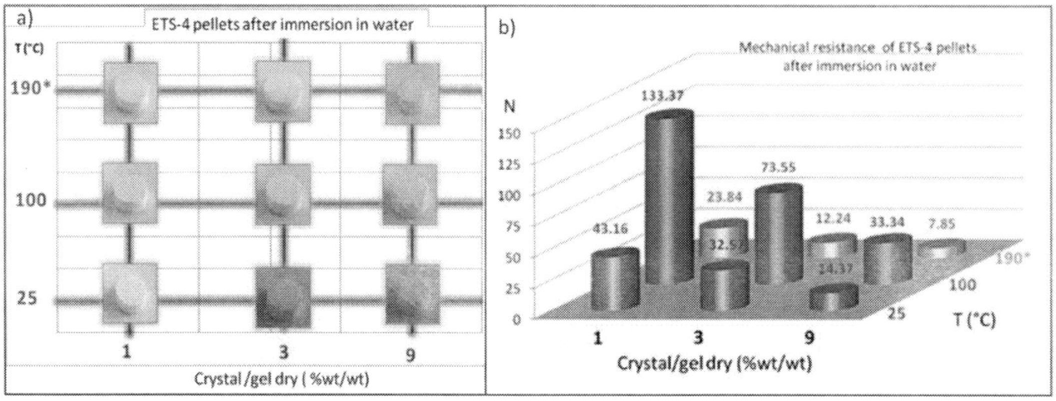

Figure 3. Image of ETS-4 pellets (a) and mechanical resistance (b) as a function of crystal/dry-gel ratio and treatment at temperature after immersion in water for 10 minutes at a temperature of 25°C followed by drying.

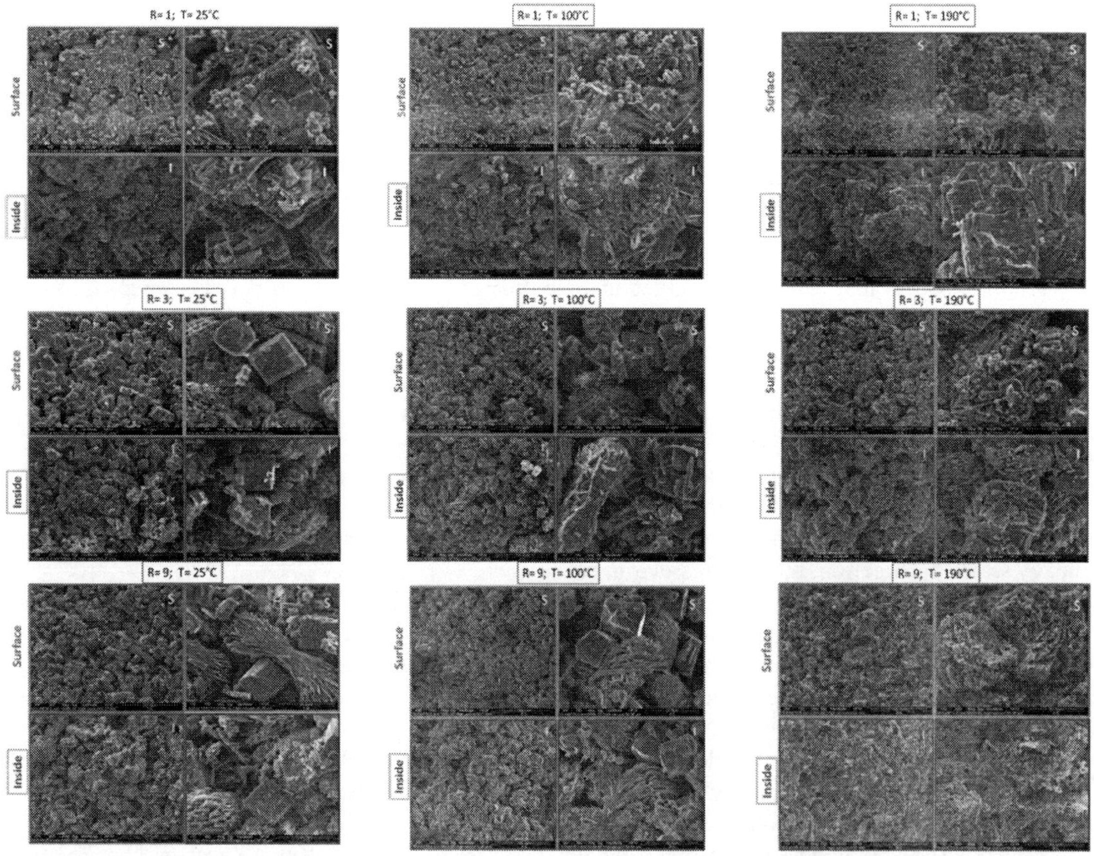

Figure 4. SEM images of the surface (s) and internal (i) of the pellets obtained with the variation of the ratio crystal/gel-dry (R) and the temperature at two different magnifications.

We may notice a decrease in resistance if we observe the graph from left to right, as the crystal / gel-dry ratio increases and the dry-gel (binder) decreases. Conversely, an increase in resistance can be observed for pellets subjected to baking for 24 hours at 100 °C respect to pellets left at room temperature (T = 25 °C). However, from treatment at 100 °C to treatment at T = 190 °C, the resistance decreases once more, due to the fact that when dry-gel is subjected to hydrothermal synthesis its crystallization is favored towards the ETS-4 phase decreasing its binding action.

3.4. Characterization by Scanning Electronic Microscope (SEM)

The interior and surface of the prepared pellets were observed by scanning electron microscopy at different magnification. The following Figure 4 shows the images of the pellets prepared in different temperature conditions and with different crystal / gel ratios. The SEM images show that, in all cases, the pellets have the same morphological appearance both inside and on the surface. As the treatment temperature increases, the hydrothermal treatment at 190 °C leads to the nucleation of the dry gel, in fact, an intercrescence of new crystals is visible on the surface of the crystals initially added. It is confirmed that the hydrothermal treatment leads to the crystallization of the dry gel in which it loses the binder function under these conditions. During treatment at 25 °C and 100 ° C this phenomenon is practically absent and the amorphous phase of the gel is visible. The surface of the crystals is clean without the intercrescence of new crystallization nuclei.

4. Conclusions

The aim of this work was the preparation of self-bonded ETS-4 pellets by a new procedure.

The results obtained allow us to draw the following conclusions:

The procedure of using crystals of ETS-4 mixed with a dry gel, precursor of the same crystalline phase, was found to be effective for the preparation of self-bonded pellets. The two variable parameters were particularly influential, i.e. the crystal/dry-gel ratio and the treatment modes after the pelletization phase.

- The obtained pellets were expected to be more crystalline by increasing the crystal/ dry-gel ratio but at the expense of resistance, owing to a lower percentual of dry-gel acting as a binder.
- The pretreatment mode after the pelletization phase induces the system to behave differently: at 25 °C gel-dry does not crystallize, in fact, X-ray diffraction spectra present the most intense peak of the dry gel (32 *2theta*), which is more apparent for low crystal / dry-gel ratios. Through this latter treatment, the dry gel performs to the binder function, in fact, significant compression resistances are obtained.
- Treatment at 100 °C is more advantageous than in the previous one because the pellets, obtained with the same variability, have greater resistance since the dry-gel cannot crystallize, even in this case, pellets undergo a degree of consolidation.
- Treatment of pellets subjected to hydrothermal synthesis at 190 °C indicates that under these conditions the dry gel crystallizes. This results in an increase in the crystallinity of the pellets but as a result of a lowering in the resistance.
- Water immersion resistance tests have shown that all ETS-4 pellets have the ability to resist immersion in water for ten minutes. The pellets were subjected once more to compression strength testing after water immersion, and after drying. Once again, resistance values were decreased, though equally elevated.

In summary, we may state that treatment at 100 °C has proven to be the best of those used as compared to the resistance of the pellets obtained.

In particular, the best ETS-4 pellets, which represent the best compromise between the characteristics considered are those obtained from crystal / dry-gel ratio =1 at 100 °C. Indeed, in these experimental conditions the pellets recorded the highest compression resistance value at around 168.67 N and good crystallinity (around 270 cps). Resistance dropped by only 20% after a 10 minute water immersion, and high values continued –i.e 133.37N.

5. References

[1] Kuznicki S M 1989 Large-pored crystalline titanium molecular sieve zeolites US Patent **4** 853 202

[2] Kuznicki S M 1990 Preparation of small-pored crystalline titanium molecular sieve zeolites US Patent **4** 938 939

[3] Kuznicki S M, Thrush K A, Allen F M, Levine S M, Hamil M M, Hayhurst D T, Mansour M 1992 in: M.L. Occeli and H.E. Robson (Eds.) Synthesis of Microporous Materials, Molecular Sieves **1** vanNostrand Reinhold, New York 427

[4] Anderson M W, Terasaki O, Ohsuna T, Philippou A, MacKay S P, Ferreira A, Rocha J and Lidin S 1994 *Nature* **367** 347-351

[5] Anderson M W, Terasaki O, Ohsuna T, Malley P J O, Philippou A, MacKay S P, Ferreira A, Rocha J and Lidin S 1995 *Philos. Mag. B* **71** 813.

[6] Rocha J and Anderson M W 2000 *Eur. J. Inorg. Chem.* 801-818.

[7] Chapman D M and Roe L 1990 *Zeolites* **10** 730

[8] Vuono D, Pavel C C, De Luca P, Nagy B J and Nastro A 2005 *Studies in Surface Science and Catalysis* **156** 417- 422

[9] Turta NA, De Luca P, Bilba N, B.Nagy J 2008 *Microporous and Mesoporous Materials.* 112 **1-3** 425-431

[10] Turta NA, Veltri M, Vuono D, De Luca P, Bilba N, Nastro A. 2009 *Journal of Porous Materials* **16** 527-536

[11] Pavel C C, De Luca P, Bilba N, Nagy B J and Nastro A 2005 *Thermochimica Acta* **435** 213-221

[12] Yilmaz B, Shattuck K G, Miraglia P Q, Warzywoda J and Sacco A 2004 *Studies in Surface Science and Catalysis* **154** Part A 763-769

[13] Yilmaz B, Miraglia P Q, Warzywoda J and Sacco A 2004 *Microporous and Mesoporous Materials.* **71** 167-175

[14] Yang X, Paillaud J L, van Breukelen H F WJ, Kessler H and Duprey E 2001 *Microporous and Mesoporous Materials* **46** 1 11

[15] Ismail N, Hamdy Abd El-Maksod I and Ezzat H 2010 *International Journal of Hydrogen Energy* **35** 10359-10365

[16] Pavel, CC, Vuono D, De Luca P, Bilba N, B.Nagy J and Nastro A 2005 *Microporous and Mesoporous Materials* **80** 263-268

[17] Catanzaro L, De Luca P, B.Nagy J and Nastro A 2004 *Studies in Surface Science and Catalysis* **154** 746-752

[18] Craveiro R and Lin Z 2012 *Journal of Solid State Chemistry* **190** 162-168

[19] Ferdov S, Lin Z, Ferreira R A S and Correia M R 2008 *Microporous and Mesoporous Materials.* **110** 436 441

[20] De Raffele G, Aloise A, De Luca P, Vuono D, Tagarelli A and Nagy B J 2016 *Journal of Porous Materials* **23** 389 400

[21] Sudheesh N and Shukla R S 2014 *Applied Catalysis A: General* **473** 116 124

[22] Figueiredo B R, Portugal I, Rocha J and Silva C M 2016 *Chemical Engineering Journal* **301** 276-284

[23] Vilardi D, De Luca P, Vuono D, B.Nagy J and Nastro A 2004 *Studies in Surface Science and Catalysis* **154** Part B 1929 1934

[24] Nikolova R P, Shivachev B L and Ferdov S 2013 *Microporous and Mesoporous Materials* **165** 121-126

[25] Ferreira T R, Lopes C B, Lito P F, Otero M, Lin Z, Rocha J, Pereira E, Silva C M and Duarte A 2009 *Chemical Engineering Journal* **147** 173 179

[26] Lopes C B, Lito P F, Otero M, Lin Z, Rocha J, Silva C M, Pereira E and Duarte A C 2008 *Microporous and Mesoporous Materials* **115** 98 105

[27] Lopes C B, Otero M, Lin Z, Silva C M, Rocha J, Pereira E and Duarte A C 2009 *Chemical Engineering Journal* **151** 247 254

[28] Liu Y, Xiao S, Bai P, Hu H and Jin L 2014 *Journal of Energy Chemistry* **23** 213 220
[29] De Luca P, Poulsen T G, Salituro A, Tedeschi A, Vuono D, Kònya Z, Madaràsz D and B.Nagy J 2015 *Journal of Thermal Analysis and Calorimetry* **122** 1257 1267
[30] Marathe R P 2005 *Journal of Colloid and Interface Science* **290** 69 75
[31] Bagnasco G, Turco M, Busca G, Armaroli T, Nastro A and De Luca P 2003 *Adsorption Science and Technology*. **21** 683 696
[32] De Luca P, Chiodo A and B.Nagy J 2011 *Transactions on Ecology and the Environment* **154** 155 165
[33] Guan G, Kida T, Kusakabe K, Kimura K, Abe E and Yoshida A 2005 *Applied Catalysis A: General*. **295** 71-78
[34] Guan G, Kida T, Kusakabe K, Kimura K, Abe E and Yoshida A 2004 *Inorganic Chemistry Communications* **7** 618 620.
[35] Guan G, Kida T, Kusakabe K, Kimura K, Fang X, Ma T, Abe E and Yoshida A 2004 *Chemical Physics Letters*. **385** 319 322
[36] Aiello R, Nastro A, Crea F and Colella C 1985 *Zeolites*. **2** 290 294
[37] Gilson J P 1985 US Patent **4** 537 866
[38] De Luca P, Crea F, Fonseca A and B.Nagy J 2001 *Microporous and Mesoporous Materials*. **42** 37 48
[39] Vuono D, De Luca P, B.Nagy J and Nastro A 2008 *Microporous and Mesoporous Materials* **109** 118 13
[40] Vuono D, Guzzo M, De Luca P and Nagy B J. 2014 *Journal of Thermal Analysis and Calorimetry* **116** 169 182
[41] De Luca P, Yuono D and Filice M 2009 *Environmental Engineering and Management Journal*. **8** 1009 1015.
[42] De Luca P, Nappo G, Siciliano C and Nagy B J 2018 *Journal of Porous Materials* **25** 283296
[43] Pavel C C, Vuono D, Catanzaro L, De Luca P, Nasto A, Bilba N and B.Nagy J 2002 *Microporous and Mesoporous Materials* **56** 227 239

Acknowledgments
We thank Dr. Mariano Davoli of the department of Biologia, Ecologia e Scienze della Terra (DiB.E.ST.). UNICAL. Rende (ITALY) for having offered his contribution to the characterization by Scanning Electron Microscope.

Corrosion Behaviour in Human Stimulation Media of a High Entropy Titan-Based Alloy

B Ghiban[1], G Popescu[1], C Lazar[1], L Rosu[1], I Constantin[2], M Olaru[2] and B Carlan[2]
[1]University Politehnica of Bucharest, Splaiul Independentei 313, Bucharest, Romania, 060042
[2]National R&D Institute for Nonferrous and Rare Metals – IMNR, 102 Biruinței Blvd., Pantelimon, Ilfov County, Romania, 077145

E-mail: gabriela.popescu@upb.ro

Abstract. The paper presents results on the corrosion behavior of high entropy alloys, commonly called BIOHEA in human physiological simulating media, respectively in the NaCl infusion solution and Ringer's lactate infusion solution. Corrosion tests were performed by potendiodinamic test using AUTOLAB type potentiostat equipped with specialized corrosion software including the PGSTAT302N, BA and SCAN250 modules. Three entropy alloy systems were investigated: FeTa0.5Nb0.5Ti1.5Zr0.5 (BIOHEA 1), FeMnNb0.5TiZr0.5 (BIOHEA 3), FeTa0.5Nb0.5TiZr0.5 (BIOHEA 4), and BIOHEA alloy 2 was obtained by remelting BIOHEA 1. A comparison of the results obtained in the present tests and the data from the literature shows, on the one hand, that the global results can be compared with the different results from the literature, and, on the other hand, the results are new, in the sense that in any work there are no combinations of alloys studied here or human simulating medians used for testing. The conclusion of the experimental investigations in the present paper is the fact that regardless of the simulation test environment, all the alloys experimental alloys have similar behaviors, there is a difference between the chemical composition of the experimental alloy and the displacement of the corrosion potential values at electropositive values, decreasing of corrosion current, and corrosion rates. The experimental results allow the corrosion resistance of the investigated alloys, alloy BIOHEA 2 having the best corrosion behavior in both test media, with very low corrosion rates (respectivelly 0.067 μm/year in NaCl infusion solution, and 0.021 μm / year in Ringer's lactate infusion solution).

1. Introduction

Mechanical biocompatibility of biomaterials is an important factor for development of new β-titanium alloys, which are advantageous for manufacturing orthopaedic implants.

The β stability can be described by a lot of alloying elements, such as niobium (Nb), tantalum (Ta), zirconium (Zr), molybdenum (Mo), vanadium (V), iron (Fe), etc [1].

Thus, in the last years Nb, Ta and Zr are considered the most favorable non-toxic and non-allergic alloying elements for β-titanium alloys for biomedical applications.

The advantages of the beta titanium alloys consist in the highest strength to weight ratios, lower modulus and very attractive combinations of strength, toughness and fatigue resistance compared to α-type and α+β-type Ti-alloys.

From β-type Ti-based alloys the most reported biomaterials are: Ti-Nb-Ta-Mo, Ti-Nb-Ta-Sn, Ti-Nb-Ta-Zr, Ti-Nb-Sn, Ti-Zr-Cr-Mo, Ti-Mo-Nb, Ti-Mo-Ta, Ti-Ta, Ti-Zr-Mo, Ti-Nb-Zr-Ta, [2÷10]. Many of these alloys contain more or less amounts of Nb, Ta, Zr, Mo, Sn and/or Cr.

These alloys are generally obtained by melting followed by heat treatment or sintering followed by hot-forging and cold rolling.

These alloys have elastic modulus between 42-93 GPa, tensile strength between 597 ÷ 1037 MPa, yield strength 547 -908 MPa and elongation 13-19% [11, 12].

Recently, a new Ti-based alloy (Ti-Nb-Zr-Ta-Fe) was reported with 2425 MPa yield strength and 52 Young modulus obtained after heat treatment at 960^{0}C, with dual β structure and grain size with average of 200-300nm [13].

Another method for obtaining Ti-based alloys more and more investigated in the last years is powder metallurgy, that use elemental or pre-alloyed powders, and leads to porous titanium parts at much lower temperature, with more precise control of process variables and pore size, as well as physical and chemical characteristics.

The most investigated Ti-based alloy obtained by this method are: Ti-Nb, Ti-Nb-Zr, Ti-Nb-Ta-Zr, Ti-Mo, Ti-Nb-Mo, Ti-Nb-Zr-Sn, Ti-Nb-Ta-Mn, Ti-Nb-Zr-Ta-Si etc [14÷23].

The paper presents study the corrosion behavior of high entropy alloys, commonly called BIOHEA in human physiological simulating media, respectively in the NaCl infusion solution and Ringer's lactate infusion solution.

2. Materials and experimental procedures

Chemical composition of the experimental BIOHEA alloys is given in Table 1.

Table 1. Chemical composition of the experimental alloys.

Alloy	Chemical composition of the alloy, %wt					
	Fe	Mn	Nb	Ta	Zr	Ti
BIOBEA 1,2	49.51	-	41.17	80.22	40.42	63.65
BIOHEA 3	66.83	65.74	55.58	-	54.56	57.28
BIOHEA 4	58.52	-	48.67	94.84	47.78	50.16

The corrosion behavior of the BIOHEA experimental alloys in human physiological simulating media, respectively, in the NaCl infusion solution and Ringer's lactate infusion solution, was performed by potentiodynamic testing using a potentiostat - galvanostat of type AUTOLAB type equipped with specialized corrosion software including the PGSTAT302N, BA and SCAN250, endowed with the Department of Materials Science and Physical Metallurgy, Polytechnic University of Bucharest.

3. Results and discussions

Results concerning corrosion resistance of the experimental BIOHEA alloys are given in Figure 1, Figure 2 and Figure 3.

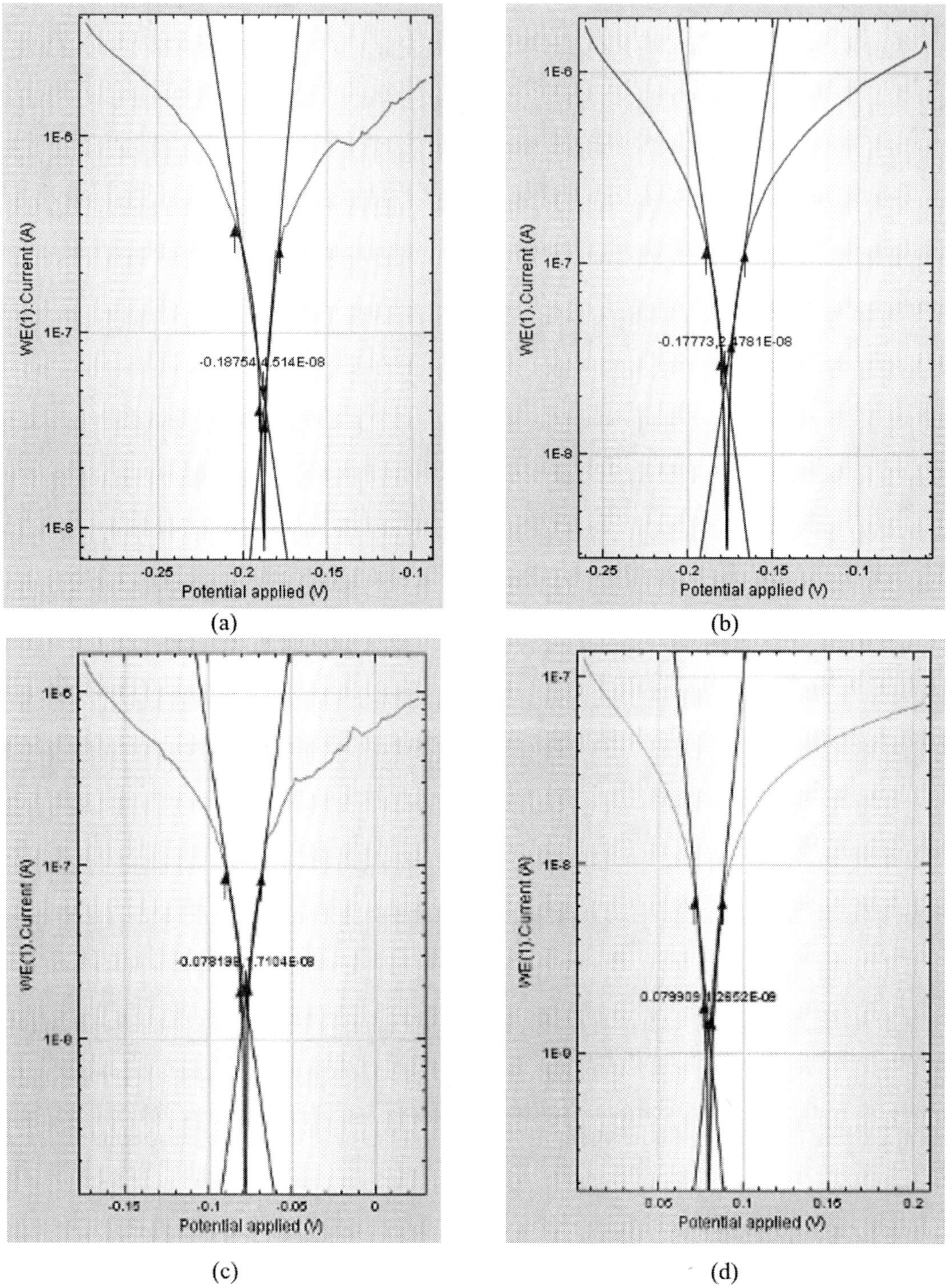

Figure 1. Tafel slope construction of the BIOHEA experimental alloys after testing in the NaCl infusion solution: a-alloy BIOHEA 3, b-alloy BIOHEA 4, alloy BIOHEA 1, alloy BIOHEA 2.

IOP Conf. Series: Materials Science and Engineering **374** (2018) 012004 doi:10.1088/1757-899X/374/1/012004

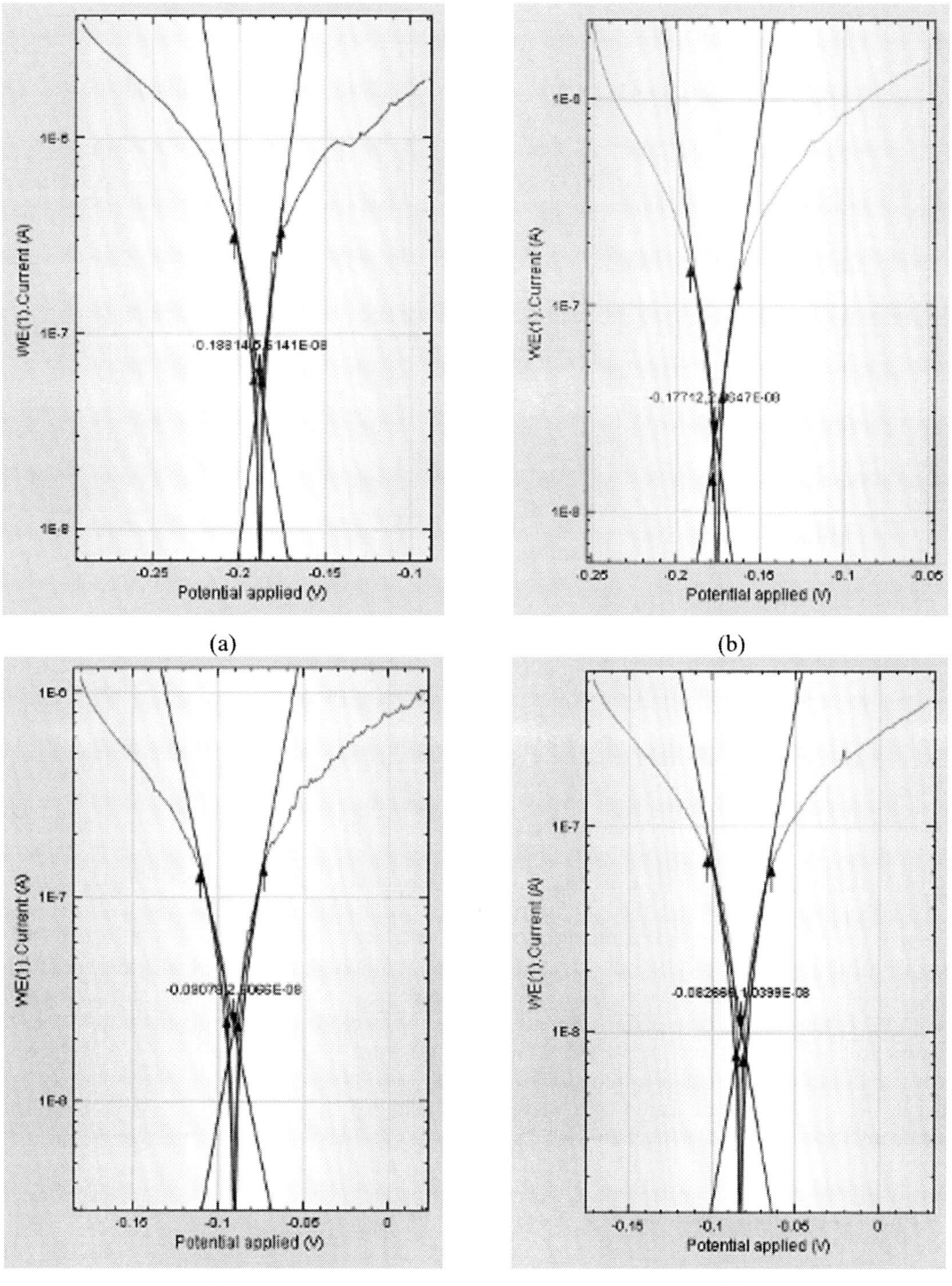

Figure 2. Tafel slope construction of BIOHEA experimental alloys after testing in Ringer's solution: a-alloy BIOHEA 3, b-alloy BIOHEA 4, c-alloy BIOHEA 1, alloy BIOHEA 2.

IOP Conf. Series: Materials Science and Engineering **374** (2018) 012004 doi:10.1088/1757-899X/374/1/012004

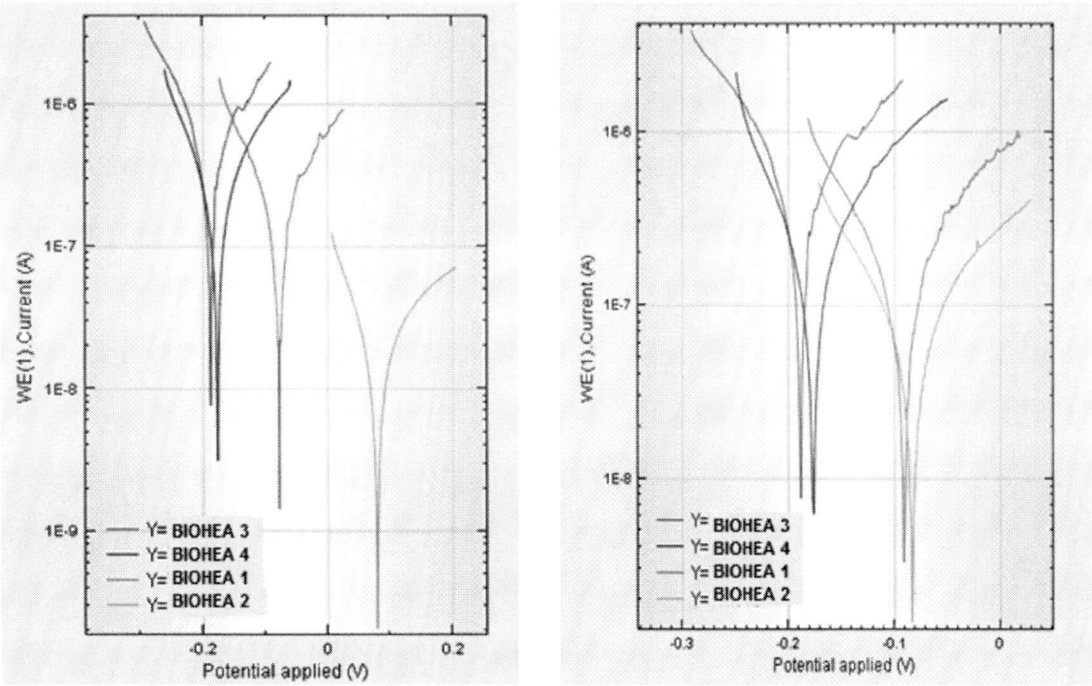

(a) (b)

Figure 3. Potentio-dynamic polarization curves of thermally untreated TiMo experimental alloys in NaCl (a) infusion solution and Ringer's solution (b) [114].

Analysis of comparative results by type of test environment led to the following observations:

3.1. NaCl solution environment for infusion (with 9 g NaCl per 1000 ml solution for infusion)
The comparative analysis of the experimental alloys in NaCl infusion solution is given in Table 2. For the experimental alloys, it is observed that the corrosion potential moves to more electropositive values depending on the type of experimental alloy, respectively from -187mV BIOHEA 3 alloy, at -177 mV (at BIOHEA 4 alloy), then at -70 mV (at BIOHEA 1 alloy), reaching to a positive value of +79 mV (for BIOHEA 2 alloy). Similarly, the values of corrosion currents, namely $1.99.10^{-3}$ μA / cm^2 (for BIOHEA 3 alloy), $1.13.10^{-3}$ μA / cm^2 (for BIOHEA 4 alloy), $5.89\text{-}10^{-5}$ μA / cm^2 (for BIOHEA alloy 1), having the lowest value on the reheated alloy, respectively 3.16. 10^{-6} μA / cm^2 (at BIOHEA 2 alloy). Also, the corrosion rate, which is related to the corrosion current density, decreases in the same manner from 40.36 μm / year (at BIOHEA 3 alloy) to 22.9 μm / year (at BIOHEA 4 alloy), 11.93 μm / year (at BIOHEA 1 alloy) and reaching 0.067 μm / year (at BIOHEA 2 alloy).
It can thus be concluded that in this test environment, all the experimental alloys have similar behaviors, there being a differentiation given by the chemical composition of the experimental alloy, respectively the displacement of the corrosion potential values at electropositive values, decreases the density of the corrosion current, and the decrease of the velocity of corrosion to a very low value, 0.067μm / year, as shown in Figure 4.

3.2. Lactate medium, Ringer's solution
Analysis of the comparative results in the Ringer Solution Test Medium of the experimental alloys is shown in Table 1. For the experimental alloys, it is observed the displacement of the corrosion potential from -188 mV (at BIOHEA 3 alloy) to -177 mV (at BIOHEA 4 alloy), -90 mV (at BIOHEA 1 alloy) reaching -80 mV (on BIOHEA alloy 2). Also, corrosion density values drop from 1.16. 10-3 μA / cm^2 (at BIOHEA alloy 3) at 2.01-10-4 μA/cm^2 (at BIOHEA 4 alloy), 8.24.10-3 μA/cm^2 (at

BIOHEA 1 alloy) and 3.71.105 μA / cm_2 (to BIOHEA alloy 2). In the same way, corrosion rate values also drop from 23.45μm / year (at BIOHEA 3 alloy) to 4.063 μm / year (at BIOHEA 4 alloy), at 1.66 μm / year (on BIOHEA 1 alloy) up to at 0.75 μm / year (at BIOHEA 2 alloy). Regarding the behavior of alloys in NaCl solution, the Ringer Solution Corrosion Parameters can be estimated to be approximately similar except for BIOHEA 2 alloy with a positive corrosion potential and a low corrosion rate relative to other alloys in the same state.

Table 2. Corrosion parameters and Tafel slope construction results for the experimental alloys after corrosion testing in human simulant solutions.

Alloy	Corrosion parameters					Tafel slope		
	Ecorr [mV]	i corr, [μA]	Rp [Ohm]	I corr [μA/cm^2]	Corrosion rate [μm/an]	bc [V/dec]	ba [V/dec]	charecter
Corrosion test in NaCl solution								
BIOHEA 3	-187	5,58.10^{-4}	40034	1,99.10^{-3}	40,36	0,010	0,016	Chatodic
BIOHEA 4	-177	3,18.10^{-4}	93189	1,13.10^{-3}	22,90	0,016	0,014	anodic
BIOHEA 1	-70	1,65.10^{-4}	119440	5,89.10^{-4}	11,93	0,013	0,014	Chatodic
BIOHEA 2	+79	1,27.10^{-6}	1780900	3,16.10^{-6}	0,067	0,010	0,010	mixt
Corrosion test in Ringer's solution								
BIOHEA 3	-188	5,61.10^{-5}	62118	1,16.10^{-3}	23,45	0,017	0,015	anodic
BIOHEA 4	-177	3,24.10^{-4}	98339	2,01.10^{-4}	4,063	0,014	0,017	Chatodic
BIOHEA 1	-90	2,34.10^{-5}	212540	8,24.10^{-5}	1,660	0,021	0,024	Chatodic
BIOHEA 2	-80	1,04.10^{-5}	440390	3,71.10^{-5}	0,75	0,020	0,021	Chatodic

It is also found in testing the Ringer's solution that the corrosion potential moves to more electropositive values, the corrosion current and the corrosion rates decrease in the same way depending on the chemical composition of the alloy. The same observation can be made for both the current density values, from 3.43.10-^{4} μA / cm^2 (for BIOHEA 3 alloy), 9.89.10^{-5} μA / cm^2 (for BIOHEA 4 alloy), 2.36.10-5μA / cm2 (for BIOHEA 1 alloy) and 1.23.10-6μA / cm2 (for BIOHEA 2 alloy) and for corrosion rate values, which are in the range of 6.9μm / year ÷ 0.026μm / year. Corrosion potential decomposed from -188mV (at BIOHEA 3 alloy), -70mV (at BIOHEA 4 alloy), -40mV (at BIOHEA 1 alloy), reaching -20mV (at BIOHEA 2 alloy). There is also a decrease in the value of the current ranges, from 4,67.10-3 μA / cm^2, to 1.26.10^{-3} μA / cm^2, to 1.13.10^{-3} μA / cm^2 and 9.83.10^{-7} μA / cm^2, and corrosion rates from 99.7μm / year to 22.5μm / year, 22.9μm / year and reaching 0.021μm / year for BIOHEA 2 alloy.

A comparison of the results obtained in the present tests and the data from the literature allow us to make some interesting observations. On the one hand, the overall results can be compared to the different results from the literature, on the other hand the results are new, meaning that in any paper there are no combinations of alloys studied here or the simulating human environments used for testing. Thus, Ho [24÷26] indicates corrosion potential values between -400 ÷ -200mV, but on Ti-7,5Mo alloys compared to the classical Ti6Al4V alloy, the tests here are in physiological simulant liquid, different from the two solutions treatment in this paper. Oliviera [27-30] studying the electrochemical behavior of a set of Ti-Mo alloys, obtain curves on another, that of the open stream of Na_2SO_4 and Ringer's solution.

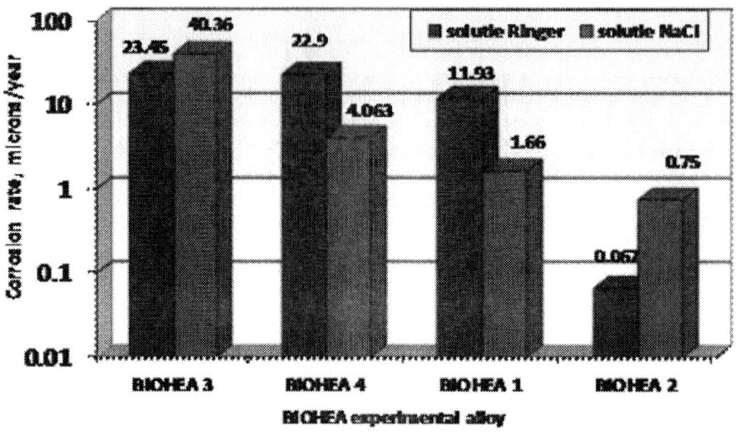

Figure 4. Comparative results concerning corrosion rates of the BIOHEA experimental alloys after testing in human simulating environments.

Comparable results also offer More [31], but on Ti-29Nb-13Ta-4,6Zr alloys, which compares with the Ti-12,5Mo alloy, resulting in 900mV (more electronegative potential than in our). Kumar [32÷35] studies the behavior of pure titanium in Ringer's solution to create maps of the occurrence of simultaneous corrosion-friction-wear phenomena. Gonzales [36] studying the corrosion behavior of titanium and titanium alloys as a biomaterial for dental applications concludes that molybdenum increases the stability of the passive phase, widening the passive alloy domain, by forming chlorides in molybdenum-soluble. All Kumar [32÷35] in other works obtains for Ti-15Mo alloys, with different concentrations of NaCl, the corrosion potential in the range -275 ÷ -457 mV, with corrosion currents of 0,31 ÷ 2.3 μA / cm^2 which increase with the concentration of floral ions in the solution. Mayouf [37] compares the Ti_6Al_4V to pure titanium and $Ti_3O_{1010}Ag$ in different flora-containing media, yielding low potency values ~ -500 mv. These authors mark the presence of corrosion located on the surface of the materials. Chapel [38] studies the repeatability of the corrosion parameters of Ti-Mo alloys in 0.9% NaCl solution, demonstrating the beneficial effect of molybdenum by lowering the corrosion susceptibility of Ti-Mo alloys in human simulating aggressive media containing 0.9% NaCl. Karthega [39] studies the electrochemical behavior of β-Ti alloys, but in the Hanck's solution. They show the high corrosion resistance of Ti-Mo alloys by forming compact layers on the surface of the metal.

4. Conclusions

The corrosion tests carried out in this paper confirm the possibility of obtaining HEA biocompatible alloys with superior characteristics and confer the judicious selection of selected chemical elements.

Alloys containing Nb and Ta have superior corrosion resistance characteristics, which confirm the beneficial effect of these elements on corrosion resistance;

The reheated alloy (BIOHEA 2) exhibits the best corrosion resistance due to the improvement of the alloy microstructure and the increase of the homogeneity of the chemical composition in its volume.

5. References

[1] Bania P 1993 Beta Titanium Alloys and Their Role in the Titanium Industry in: Beta Titanium Alloys in the 1990's D Eylon et al. (eds) TMS Warrendale PA USA
[2] Kuroda D, Niinomi M, Morinaga M et al. 1998 *Sci Eng A* **243** 244–249

[3] Nobuhito S, Niinomi M, Toshikazu A, Takashi S and Tadahiko F 2004 *Mater Sci Forum* **449–452** 1269–72
[4] Matsumoto H, Watanabe S and Hanada S J 2005 *Mater Trans* **46** 1070–1078
[5] Zhao X L, Niinomi M, Nakai M, Miyamoto G and Furuhara T 2011 *Acta Biomater* **7** 3230–36
[6] Gabriel S B, Panaino J V P, Santos I D, Araujo L S, Mei P R et al. 2012 *J Alloy Compd* **536** S208–S210
[7] Zhou Y L and Niinomi M 2008 *J Alloy Compd* **466** 535–542
[8] Zhao X L, Niinomi M, Nakai M, Ishimoto T and Nakano T 2011 *Mater Sci Eng C* **31** 1436–44
[9] Hynowska A, Pellicer E, Fornell J, González S et al. 2012 *Mater Sci Eng C* **32** 2418–2425
[10] Jung T K, Semboshi S, Masahashi N and Hanada S 2013 *Mater Sci Eng C* **33** 1629–1635
[11] Niinomi M 1998 *Mater Sci Eng A* **243** 231-236
[12] Zhang L C, Klemm D, Eckert J, Hao Y L and Sercombe T B 2011 *Scr Mater* **65** 21–24
[13] Li Y Y, Zou L M, Yang C, Li Y H and Li LJ 2013 *Mater Sci Eng A* **560** 857–861
[14] Bhupendra S S and Kumar K A 2016 *J Alloy Compd* **656** 978–986
[15] Mendez M W D, Agreda C G, Bressiani A H A and Bressiani J C 2016 *Mater Sci Eng C* **63** 671-677
[16] Hussein M A C and Suryanarayana N Al-Aqeeli 2015 *Mater Design* **87** 693-700
[17] Rao X, Chu C L and Zheng Y Y 2014 *J Mech Behavior of Biomed Mater* **34** 27-36
[18] Sakaguchi N, Niinomi M, Akahori T et al. 2005 *Mater Sci Eng C* **25** 363-369
[19] Xie F, He X et al. 2015 *Corros Sci* **95** 117-124
[20] Nazari K A, Nouri A and Hilditch T 2015 *Mater Design* **88** 1164-74
[21] Guo S, Chu A, Wu H et al. 2014 *J Alloy Compd* **597** 211-216
[22] Aguilar C, Guzman P, Lascano S, Parra C, Bejar L et al. 2016 *J Alloy Compd* **670** 346-355
[23] Li Y H, Yang C, Zhao H D et al. 2015 *Mater Design* **85** 7-13
[24] Ho W F et al. 2010 *Materials Science and Engineering: C* **30(6)** 904-909
[25] Ho W F et al. 2009 *Surface and Coatings Technology* **203(20–21)** 3142-50
[26] Ho W F 2008 *Journal of Alloys and Compounds* **464(1–2)** 580-583
[27] Oliveira N T C et al. 2007 *Materials Science and Engineering: A* **452–453(0)** 727-731
[28] Oliveira N T C and Guastaldi A C 2008 *Corrosion Science* **50(4)** 938-945
[29] Oliveira N T C and A C Guastaldi 2009 *Acta Biomaterialia* **5(1)** 399-405
[30] Oliveira N T C et al. 2009 *Electrochimica Acta* **54(5)** 1395-1402
[31] More N S et al. 2011 *Materials Science and Engineering: C* **31(2)** 400-408
[32] Kumar S T S N Sankara Narayanan and S Saravana Kumar 2010 *Corrosion Science* **52(5)** 1721-27
[33] Kumar S et al. 2010 *Tribology International* **43(7)** 1245-1252
[34] Kumar S and T S N 2009 *Journal of Alloys and Compounds* **479(1–2)** 699-703
[35] Kumar S et al. 2010 *Wear* **268(11–12)** 1537-1541
[36] González J E G and Mirza-Rosca J C 1999 *Journal of Electroanalytical Chemistry* **471(2)** 109-115
[37] Al-Mayouf A M, Al-Swayih A, Al-Mobarak A N and Al-Jabab A S 2004 *Materials Chemistry and Physics* **86(2–3)** 320-329
[38] Capela M V et al. 2008 *Journal of Alloys and Compounds* **465(1–2)** 479-483
[39] Karthega M, Raman V and Rajendran N 2007*Acta Biomaterialia* **3(6)** 1019-1023

Acknowledgment
The authors would like to acknowledge financial support from the MECS- UEFISCDI Romania, through PN-III-P2-2.1-PED-2016/project ctr.134PED/2017.

IOP Conf. Series: Materials Science and Engineering **374** (2018) 012005 doi:10.1088/1757-899X/374/1/012005

Thermal and Electrical Investigation of Conductive Polylactic Acid Based Filaments

R A Dobre[1*], A E Marcu[2], A Drumea[2] and M Vlădescu[2,3]

[1] Telecommunications Department, Politehnica University of Bucharest, Iuliu Maniu Blvd. no. 1-3, 061071, Bucharest, Romania

[2] Electronic Technology and Reliability Department, Politehnica University of Bucharest, Iuliu Maniu Blvd. no. 1-3, 061071, Bucharest, Romania

[3] Optoelectronics Research Center (UPB-CCO), Politehnica University of Bucharest, Iuliu Maniu Blvd. no. 1-3, Bucharest, Romania

E-mail: rdobre@elcom.pub.com

Abstract. Printed electronics gain momentum as the involved technologies become affordable. The ability to shape electrostatic dissipative materials in almost any form is useful. The idea to use a general-purpose 3D printer to manufacture the electrical interconnections for a circuit is very attractive. The advantage of using a 3D printed structure over other technologies are mainly the lower price, less requirements concerning storage and use conditions, and the capability to build thicker traces while maintaining flexibility. The main element allowing this to happen is a printing filament with conductive properties. The paper shows the experiments that were performed to determine the thermal and electrical properties of polylactic acid (PLA) based ESD dissipative filament. Quantitative results regarding the thermal behavior of the DC resistance and the variation of the equivalent parallel impedance model parameters (losses resistance, capacitance, impedance magnitude and phase angle) with frequency are shown.. Using these results, new applications like printed temperature sensors can be imagined.

1. Introduction

The 3D printing technologies are becoming available to the masses thanks to the decrease of the prices and to the multitude of technologies available. The oldest 3D printing technology is stereolithography (SLA), but currently the most popular 3D printers are using fused deposition modeling (FDM). Other ways to 3D print also exist like selective laser sintering (SLS), selective laser melting (SLM), electronic beam melting (EBM), and laminated object manufacturing (LOM) [1]. This increased availability and diversity open new areas of research for electrical engineers when conductive raw materials are used. 3D printing, in the early days, resumed to bringing the possibility to transform a raw material stored in a certain form (filament, liquid etc.) into a specific, previously designed, solid part. In the case of FDM, the raw materials have the form of filaments. The diameter of the filaments is typically 1.75 mm or 3 mm. Printing designs that implied hanging parts was a challenge because every new layer of raw material must be deposited on top of the previous layer or directly on the heated bed of the printer in the case of the first layer. The software used for generating the printer commands for extrusion and printer head movement is called "slicer". To allow the printing of hanging parts, auxiliary support pillars are also automatically considered by the software and added to the print [2]. In the first versions of 3D printers those were made of the same material as the main part and their detachment was done mechanically (breaking them) after printing and this operation could leave marks on the actual part or even break fragile hanging pieces. It is known that 3D printing is a slow process and the possibility of breaking the printed part in the process of removing auxiliary structures is not a desired feature. 3D printing has improved since then thanks to the new types of materials used as filaments [3]. For example, support structures can be built using soluble materials. The removal of those new auxiliary parts is easy and leaves no marks on the main part.

Conductive filaments also become available for 3D printing and electrical engineers manifest great interest in using them. There are two main categories of conductive filaments: electrostatic dissipative and properly conductive. The electrostatic dissipative materials typically have a resistance of 10^5 to 10^9 Ω/square and allow charges to flow to ground in a more controlled manner than conductive materials. They are used to avoid the accumulation of electrical charge on certain surfaces, charge that could lead to the failure of sensitive circuits. Conductive materials have the resistance in the range of 10^3 to 10^5 Ω/square and one application of this second category of filaments is building traces that can be used to directly interconnect electrical devices. Since FDM based 3D printing uses, in the case of PLA filaments [4], a temperature of the extruder between 180 °C and 220 °C, soldering parts could be a difficult operation. Conductive glues [5] could be a suitable candidate for this task. The studied material is part of the electrostatic dissipative category and the 3D printer that was used to build the samples to be measured was FDM based. Besides this introduction, the paper is organized as follows: in Section 2 are presented the measurement devices and the measurement procedures, in Section 3 the results are shown and discussed, and Section 4 concludes the paper.

2. Materials and methods

The studied material is an electrostatic dissipative PLA available as filament for 3D printing having a thickness of 1.75 mm. The 3D model for a material sample was designed using the TinkerCAD online 3D CAD design software. It consists of a straight line having the width equal to 4 mm and the length equal to 5 cm. To ease the electrical contact between the measurement devices and the filament structure, two annuli were designed in the continuity of the aforementioned straight line. The large circle in each annulus has the radius equal to 8 mm and the smaller one has the radius equal to 4 mm. The thickness of the sample was 1.2 mm assuring a quick print and being flexible in terms of printing resolution. For the available printing nozzle having a diameter of 0.4 mm, the recommended layer height is either 0.1 mm or 0.2 mm. The designed model was made freely available, for experiment repeatability reasons, at https://www.thingiverse.com/thing:2874794.

After design, six filament samples were printed using a HE3D Tricolor 3D printer. The nozzle diameter was 0.4 mm, the printing head temperature was 200 °C throughout the entire process, and the bed temperature was held at 60 °C. Those are typical values for 3D printing PLA.

To allow easier and more secure electrical connections, mechanical contact forms made of screws, bolts and metallic washers were used. Each annuli of the samples were tightened between two washers using an appropriate screw and bolt.

2.1. The procedure for measuring the material's DC resistance variation with temperature

The instruments that were used for measuring the variation of the DC resistance of the filament samples with temperature are a Espec SH-241 thermal chamber, a Keithley 2700 multimeter having attached a Keithley 7700 20 channel multiplexer allowing automatic measuring of multiple physical quantities (in the case presented in this paper 6 DC resistances were measured, the resistance for each of the 6 samples), and a personal computer running the provided Keithley acquisition software. The communication between the PC and the multimeter is done using a GPIB interface. The described measuring rig is illustrated in Figure 1.

Three kinds of experiments were done as follows: at first, the resistance of the filament samples was measured at fixed temperatures, starting from room temperature (in the moment of measurement it was 21 °C) to 100 °C with 5 °C increments between 25 °C to 100 °C. The measurement of the resistance was delayed by 10 minutes from the moment when the thermal chamber indicated the desired temperature value to allow the samples to reach thermal equilibrium.

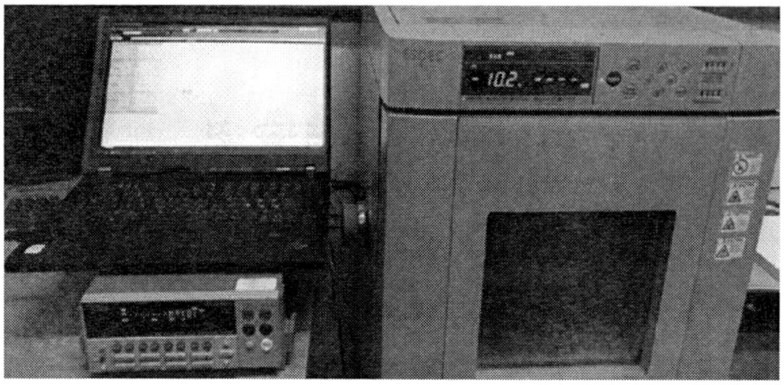

Figure 1. The technological rig used for measuring the DC resistance of the material samples at the temperatures of interest.

In the second type of experiment the samples were exposed to a cyclical temperature sweep between −10 °C to 100 °C to observe any permanent effects that high or low temperatures could have on the investigated material. Based on the results obtained after the first two experiments, a third experiment exposed the samples to 100 °C for 70 minutes to better observe any effects the high temperature could have on the material.

2.2. The procedure for measuring the material's impedance at various frequencies

The electrical characterization of the material samples was done using a Hioki 3532-50 LCR meter. The quadripolar probe was directly connected to the mechanical contacts fitted on the samples. An equivalent parallel model was considered for the impedance and the parameters that were measured are the capacitance (denoted with Cp), the parallel resistance (losses resistance denoted with Rp), the magnitude of the equivalent impedance (denoted with $|Z|$) and the phase angle (denoted with Θ) at frequencies between 100 Hz to 3 MHz [6]. Two measurements were made in each frequency decade (i.e., 100 Hz, 300 Hz, 1 kHz, 3 kHz etc.). All the electrical experiments were done at room temperature which in the moment of the experiment was 21 °C. The electrical measurements were done on the same samples that were used for thermal characterization of the material, after the thermal measurements. The samples used for electrical characterization are illustrated in Figure 2.

Figure 2. The material samples used for electrical characterization (after thermal investigation).

3. Results and discussions

3.1. Results regarding the thermal behaviour of the DC resistance of the investigated material

The results obtained after the first experiment described in the previous section are illustrated in Figure 3 and Figure 4. The graphs show the variation of the DC resistance of the studied filament samples with the temperature and it was observed a strong dependence, unlike in the case of thin metallic foils [7]. Because the span of the values is very large, vertical axes are logarithmic. The material manifested an increase of the DC resistance equal to three orders of magnitude in the investigated temperature range. Plotting all the results on the same graph would result in a very crowded and hard to read drawing, so the results were grouped and illustrated into two separate figures. The measured value at each temperature differed for each sample. This dispersion could be caused by the mechanical connection system and by the differences between the samples caused by the fabrication procedure (3D printing). It is important to observe that the allures of the graphs are very similar.

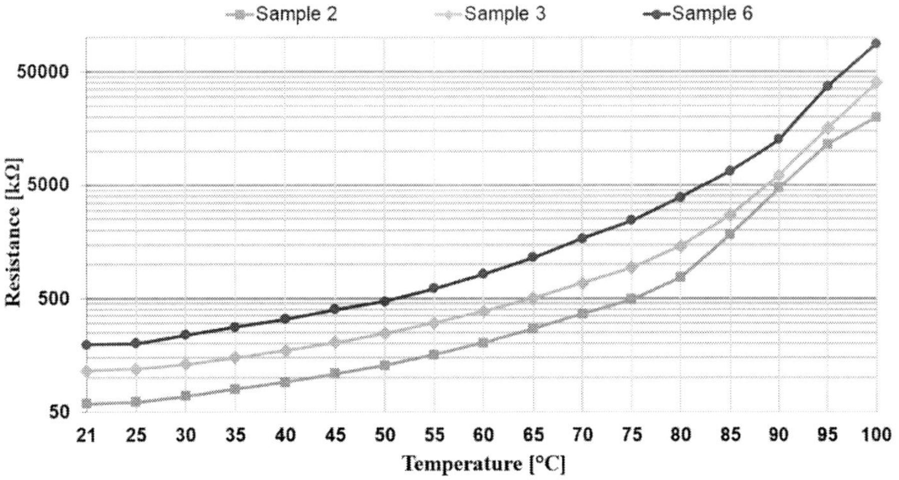

Figure 3. The variation of DC resistance of the material samples with the temperature (for Sample 2, Sample 3 and Sample 6).

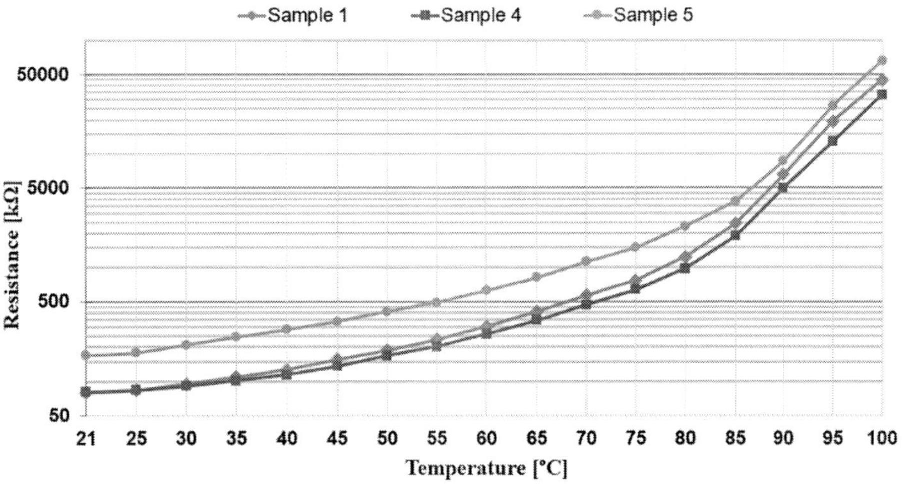

Figure 4. The variation of DC resistance of the material samples with the temperature (for Sample 1, Sample 4 and Sample 5).

The results obtained after the second type of experiment are shown in Figure 5 and Figure 6. The temperature started from 21 °C and was cyclically increased to 100 °C then decreased to −10 °C. The measured resistance was plotted against the time because the measuring system could not acquire information about the temperature. This parameter was displayed on the thermal chamber. Each extreme temperature conditions (−10 °C and 100 °C respectively) were held for at least 5 minutes. This can be observed on the graph in the areas where the measured resistance was maximum and

IOP Conf. Series: Materials Science and Engineering **374** (2018) 012005 doi:10.1088/1757-899X/374/1/012005

minimum. The six graphs were split into two plots, two which did not manifest noisy values shown, in Figure 5, and the ones who did, in Figure 6.

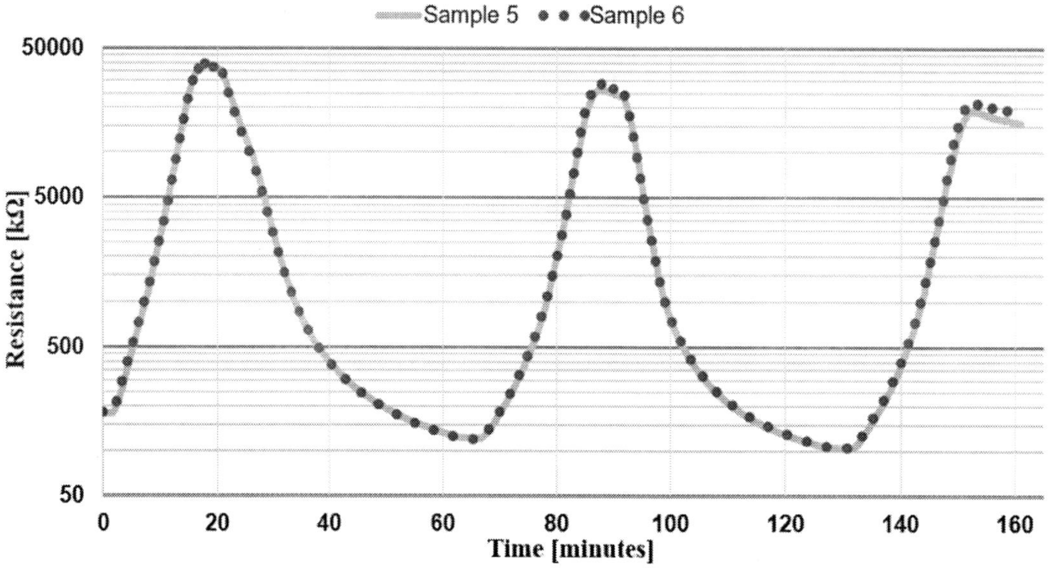

Figure 5. The variation of DC resistance of the material samples with the temperature in the case of temperature sweep (for Sample 5 and Sample 6).

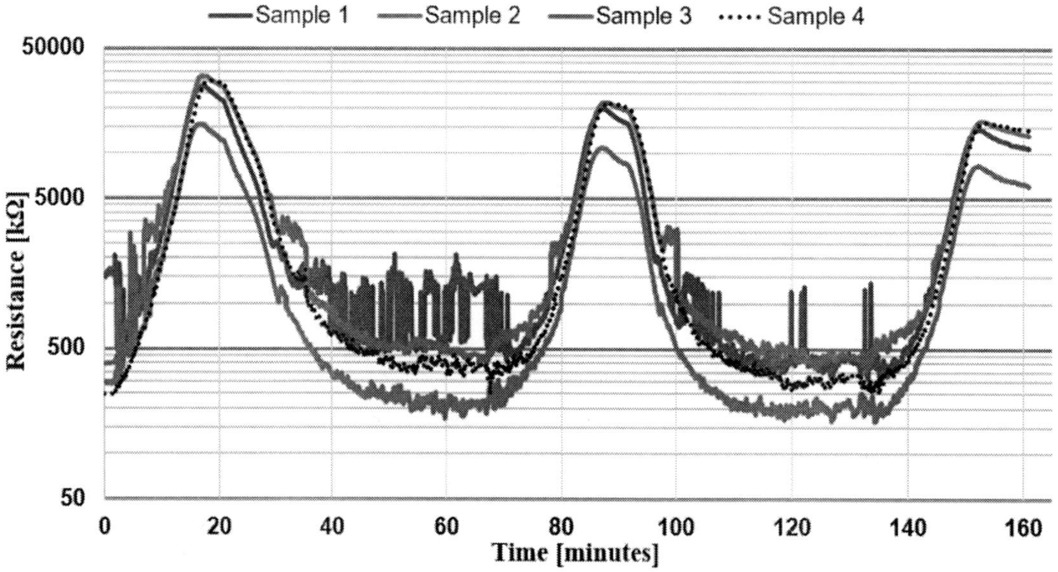

Figure 6. The variation of DC resistance of the material samples with the temperature in the case of temperature sweep (for Sample 1, Sample 2, Sample 3, and Sample 4).

One possible cause of the noisy behavior could be the mechanical contacting system that was used. It can be observed in Figure 4 and Figure 5 that the resistance slightly decreases when the filament is held at high temperatures and that the peak value for resistance becomes smaller after each temperature cycle. This behavior was further investigated by holding the samples at 100 °C for over an hour. The resistance decrease is shown in Figure 7.

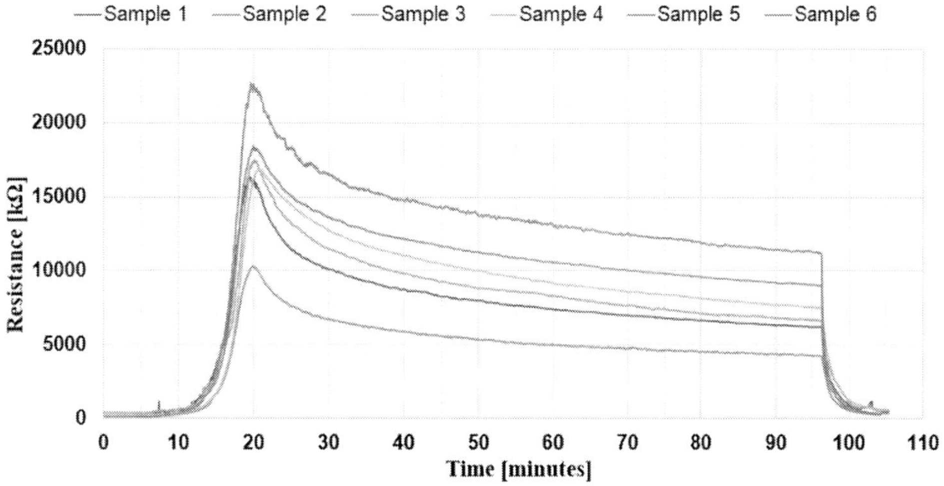

Figure 7. The variation of DC resistance of the material samples when held at a temperature equal to 100 °C.

3.2. Results regarding the electrical behaviour of the investigated material

The variation of the equivalent resistance (the parallel equivalent capacitor model was considered), denoted with Rp with the frequency is illustrated in Figure 8. In the equivalent capacitor model, this resistance represents the resistance of the dielectric and characterizes the losses of the capacitor. It can be observed that, for most of the samples, this parameter does not show a notable variation with the frequency. There is only one sample, Sample 6, presents a different behavior. This could be also caused by the mechanical contacts, especially because this experiment was done after the thermal characterization or because of degradation suffered during the heating.

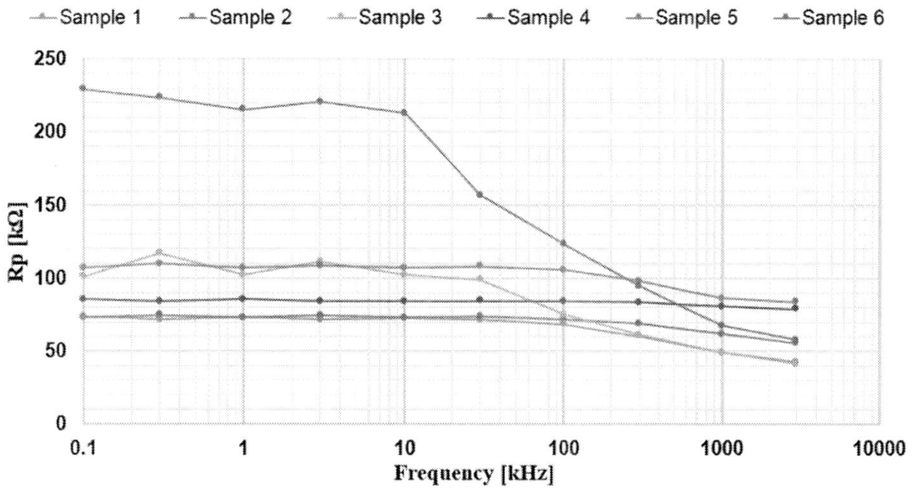

Figure 8. The variation of equivalent parallel resistance of the material samples with frequency.

The measured values for the capacitance of the material samples at the test frequencies are presented in Figure 9. It can be observed that the capacitance is almost constant over a large domain of frequencies (from 1 kHz to 100 kHz). Above 100 kHz, the capacitance slightly decreases.

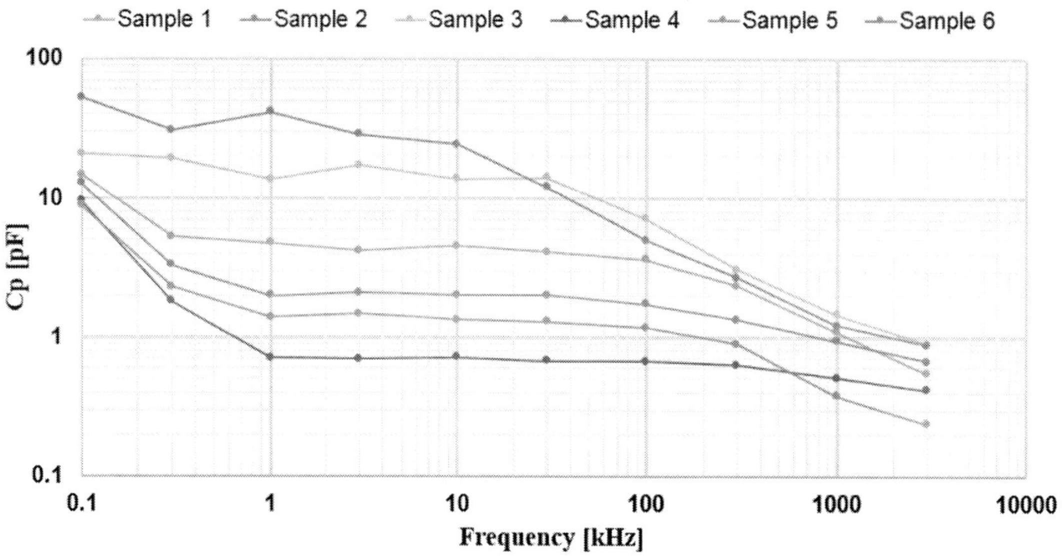

Figure 9. The variation of equivalent capacitance of the material samples with frequency.

The next measurements that were done investigated magnitude of the impedance. Because the value of the capacitance is very small (tens of pF), the impedance is dominated by the losses resistance (Rp) for frequencies lower than 1 MHz, as can be observed in Figure 10. The values are, on average, almost constant. At higher frequencies, the effects of the capacitance determine a decrease in the magnitude of the impedance. The same anomaly as in the case of measuring the equivalent parallel resistance can be observed for Sample 6.

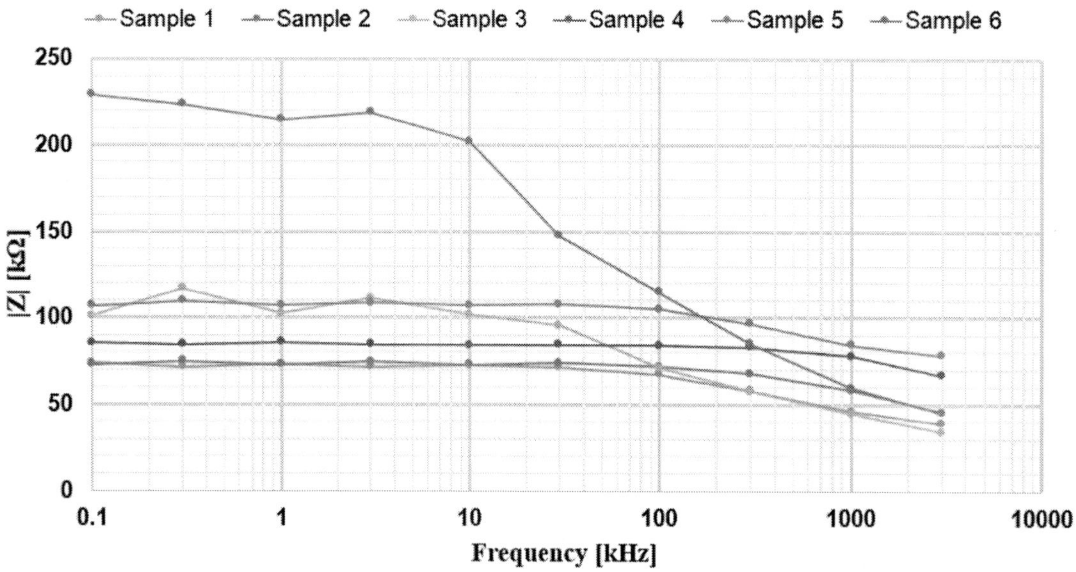

Figure 10. The variation of magnitude of the impedance of the material samples with frequency.

The last type of measurements investigated the phase angle between the voltage applied to the sample and the current flowing through it. As expected, the phase angle shows a capacitive behavior because it decreases with frequency. At low frequencies, the phase angle is very small because the equivalent impedance is dominated by the losses resistance Rp. Results are shown in Figure 11.

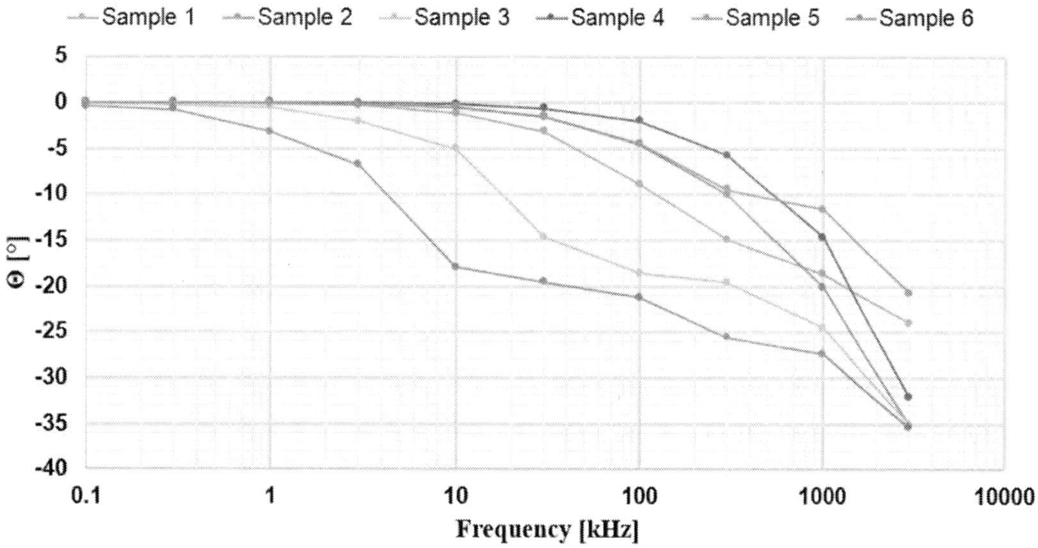

Figure 11. The variation of phase angle between the voltage applied to the sample and the current flowing through it with frequency.

4. Conclusions

This paper investigated the thermal and electrical properties of the polylactic acid based conductive filaments. The studied filaments are part of the electrostatic dissipative category. A FDM based 3D printer was used to realize six samples of material for further investigation. The model for the sample was designed using free CAD software and is publicly and free available for download, for easy repeatability of the conducted experiments.

The first experiments investigated the variation of the DC resistance of the studied material with the temperature. A strong dependence was observed, the resistance increased three orders of magnitude in the −10 °C to 100 °C interval. As the temperature was cycled, the resistance for each sample did not follow the exact dependency curve, instead the maximum value decreased with every cycle. It was further investigated and found that the material is affected by high temperatures and its resistance decreases. The frequency characterization unveiled a typical capacitive behavior.

The results in this paper can be used to develop new cost effective and easy to manufacture flexible thermal protection circuits, or new cases for electronic modules that could also act as a temperature fuse besides the default role of preventing ESD damaging. Also, care must be taken because if the temperature of the material increases, it could lose the electrostatic dissipation properties and fall into the anti-static material category.

5. References

[1] Jasveer S and Jianbin X 2018 International Journal of Scientific and Research Publications **8** 1-9
[2] Chua C K, Leong K F andLim C S 2003 *World Scientific Publishing* Singapore
[3] Jingjunjiao L, Hamideh G, Jun L, Craig B, Ali S, 2017 *Current Pharmaceutical Design* **23** 433-439
[4] Prashantha K and Roger F 2016 *Journal of Macromolecular Science Part A* 24-29
[5] Drumea A and Marghescu C 2015 Proc. 2015 *IEEE 21st International Symposium for Design and Technology in Electronic Packaging* (Braşov, RO, 22-25 October 2015), 297-300

[6] Pantazică M, Drumea A and Marghescu C 2017 Proc. 2017 *IEEE 23st International Symposium for Design and Technology in Electronic Packaging* (Constanţa, RO, 26-29 October 2015), 90-93
[7] Ionescu C, Bonfert D and Branzei M 2014 Proc. 2014 *37th International Spring Seminar on Electronics Technology* (Dresden, DE, 7-11 May 2014), 75-80

Acknowledgments
This work was supported by a grant of the Romanian National Authority for Scientific Research and Innovation, CNCS/CCCDI - UEFISCDI, project number PN-III-P2-2.1-PED-2016-1465, within PNCDI III.

IOP Conf. Series: Materials Science and Engineering **374** (2018) 012006 doi:10.1088/1757-899X/374/1/012006

Influence of Alloying on the Thermal Stability of Model Heat Resistant Compositions

A Tyagunov, E Baryshev, G Tyagunov and K Shmakova
Ural Federal University, 19 Mira st, 620002, Ekaterinburg, Russia

E-mail: e.e.baryshev@urfu.ru

Abstract. We studied the phase transformations at high temperatures in the carbonless heat-resistant nickel alloys by the differential thermal analysis and we calculated their thermal stability. For a relatively low alloyed compositions the greatest positive effect is achieved when it were alloyed by γ' -forming elements such as the niobium and titanium. The greatest thermal stability at temperatures below 1100°C is achieved for these alloys in the low alloyed composition. Multi-component carbonless heat resistant nickel alloys have the best characteristics when compositions contain the rhenium and a small amount of the tantalum. However, the most thermal stability at temperatures up to 1200°C has a composition that includes only 6% of tantalum.

1. Introduction

Foundry heat-resistant nickel alloys are intended for the manufacture of the most critical parts of turboshaft engines – the turbine blades that operating at high temperatures in an aggressive environment. Improving turboshaft engines, increasing their power, reliability and durability requires the development of new complex-alloyed compositions.

One of the important parameters is structure thermal stability information which is needed to create new highly effective compositions, assess the performance of alloys that are operated at extremely high temperatures [1, 2]. Thermostability of high-temperature nickel alloys is determined by the temperatures of the beginning and end of the dissolution of the secondary intermetallic γ'-phase – $t_{b.d.}$ and t_{solv}, by the kinetics of dissolution of the secondary γ'-phase in the temperature interval $t_{b.d.} - t_{solv}$, as well as the temperature of the solidus t_S. Temperature t_{solv} limits the region of the alloy existence in the form of a matrix that is strengthened by the secondary intermetallic γ'-phase embedded in it, and t_S is the temperature range for the alloy existence as a structural material.

Technological operations for turboshaft engines critical parts production from the majority of high-temperature nickel alloys are required thermal treatment. It includes homogenization operations, i.e. heating in the interval t_{solv} - t_S with prolonged exposure to 2-3 hours, hardening and subsequent aging. In this regard, the expansion temperature range t_{solv} - t_S allows to optimize the homogenization mode. For products obtained by foundry methods, temperatures of solidus t_S and liquidus t_L are important parameters for choosing optimal melting and casting temperatures, which significantly affects both the emerging structure and properties of parts and the output of the finished product [1].

2. Materials and methods

The most promising obtaining products method for modern turboshaft engines is single-crystal casting, which allows to get rid of grain boundaries and significantly increase operating temperatures

IOP Conf. Series: Materials Science and Engineering **374** (2018) 012006 doi:10.1088/1757-899X/374/1/012006

[2]. It is established [3] that carbon-free nickel refractory alloys possess the most favorable complex of structure and properties. In this case, the temperatures of phase transformations in some model carbon-free high-temperature alloys have been studied. Their chemical composition is given in Table 1.

Table 1. The studied chemical composition of the model high-temperature nickel alloys.

Sample №	The content of alloying elements, % mass.									
	Ni	Co	Cr	Al	W	Mo	Ti	Nb	Hf	Ta	Re
1	Base	5.00	4.50	6.56	8.95	-	-	-	-	-	-
2	Base	5.00	4.50	6.56	8.95	-	1.00	-	-	-	-
3	Base	5.00	4.50	6.56	8.95	-	-	1.00	-	-	-
4	Base	5.00	4.50	6.56	8.95	-	-	-	-	4.00	-
5	Base	5.00	4.50	6.56	8.95	-	-	-	-	2.00	4.00
6	Base	7.65	5.05	6.28	8.03	0.36	0.31	0.86	0.21	3.60	-
7	Base	7.65	5.05	6.28	10.00	0.36	0.31	0.86	0.21	3.60	-
8	Base	7.65	5.05	6.28	8.00	0.36	0.31	0.86	0.21	6.00	-
9	Base	7.65	5.05	6.28	8.00	0.36	0.31	0.86	0.21	4.00	2.00
10	Base	7.65	5.05	6.28	8.00	0.36	0.31	0.86	0.21	6.00	2.00
11	Base	7.65	7.05	6.28	8.00	0.36	0.31	0.86	0.21	6.00	-

The phase transformations temperatures are most often determined by the methods of thermal and differential-thermal analysis, the kinetics of dissolution of the secondary Z-phase is studied both by dilatometric methods and by high-temperature metallography methods.

The characteristic transformation temperatures in the studied carbon-free nickel alloys were detected on the VDTA-8M unit with monotonic heating and cooling of the thermographic block with speed of 20° C/min [4].

3. Results and discussions

Qualitative metallographic analysis makes it possible to reveal the structure of cast samples. After etching, the dendrites axes of the first and second order have a light color, the interdendritic space is colored dark, the eutectic of the globular morphology based on the intermetallic phase is white (figure 1). Both in the dendrites axes and in the inter-axial space, fine-grained inclusions of the secondary γ'-phase are visible. The secondary γ'-phase is distributed evenly, but in the interdendritic space its displacements are larger.

Analysis of the obtained DTA curves, published data [5-8], and the results of qualitative metallographic analysis makes it possible to identify the phase transformations founded during the research. The following phase transformations were revealed during heating and their temperatures were determined:

1. Particles dissolution of secondary γ'-phase in a metal matrix - doped γ-solid solution (t_{solv}).
2. The beginning of the alloy most fusible constituent melting - the eutectic on the basis of γ and γ'-phase (t_S).

3. The end of eutectic melting and the beginning of the metal matrix melting. Unfortunately, it is difficult to separate the heat effects associated with each of these processes on the DTA curves, so the temperature of these transformations was not taken into account in the results analyses.

4. Complete melting of the sample (t_L).

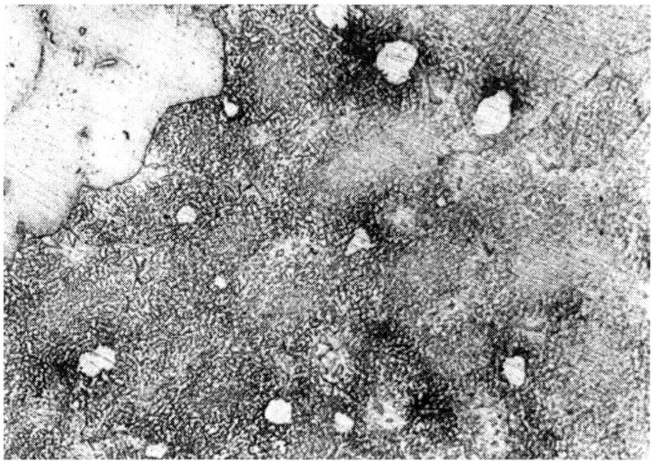

Figure 1. The structure of the cast carbonless superalloy.

The temperatures of complete dissolution of the secondary γ' - phase (t_{solv}), solidus (t_S), liquidus (t_L) determined from the DTA curves were used to calculate the melting temperature intervals ($t_L - t_S$), the existence of a doped solid solution ($t_S - t_{solv}$). Based on the liquidus temperatures during heating and cooling data, the supercooling of the melt was determined $\Delta T = t_{Lheat} - t_{Lcool}$ (Table 2).

Table 2. Temperatures and intervals of phase transformations.

№ Sample	$T_{Lheat.}$, ^0C	T_S, ^0C	T_{solv}, ^0C	$T_{b.d.}$, ^0C	T_L-T_S, ^0C	T_S-T_{solv}, ^0C	T_{Lcool}, ^0C	ΔT, ^0C
1	1480	1270	1260	780	210	10	1400	80
2	1480	1290	1275	725	190	15	1425	55
3	1475	1315	1265	685	160	50	1435	40
4	1465	1305	1250	675	160	55	1415	50
5	1470	1295	1245	740	175	50	1425	45
6	1455	1325	1270	775	120	55	1390	55
7	1455	1285	1250	770	175	35	1405	50
8	1440	1295	1280	820	145	15	1410	30
9	1440	1340	1275	715	100	65	1395	45
10	1445	1325	1265	730	120	60	1380	65
11	1455	1330	1260	670	125	70	1390	65

When a low-base base composition (composition 1) is added to 1 % mass. of titanium or niobium temperature t_{solv}, t_S is increased, the interval t_S- t_{solv} is widened, and the supercooling during the crystallization of the melt decreased.

The addition of tantalum and rhenium to the base composition leads to an increase in the solidus temperature, to a certain decrease in the temperature of complete dissolution of the secondary γ'-phase, to the expansion of the temperature interval t_S- t_{solv}, decrease of the melting interval and supercooling of the melt.

Thus, for the first series of samples characterized by a relatively low level of doping, the most favorable effect is exerted by elements that increase the thermal stability of the γ'-phase-titanium and niobium. Moreover, if high-temperature heat treatment (homogenization and subsequent aging) is not planned for the studied compositions, the best effect is obtained by alloying the base alloy with titanium. In this case, the dissolution temperature t_{solv} of the secondary γ'-phase is substantially increased, but the interval between the t_{solv} and the solidus temperature t_S is negligible.

If heat treatment is provided for the alloy, niobium alloying is most effective. For such an alloy, in comparison with other compositions, the thermostability of the alloy increases (increases t_{solv}) and the interval between t_{solv} and the temperature of the solidus is significantly increased.

The introduction of low-alloyed carbon-free nickel alloys of tantalum and rhenium does not increase their thermal stability, so it is ineffective.

Complex alloying of alloy 1, i.e. the introduction of niobium, titanium, hafnium and tantalum into it leads to a sharp increase in the temperatures of complete dissolution of the secondary γ'-phase and solidus. At the same time interval t_S- t_{solv} increased to 55°C, the melting interval as a result of growth of t_S and decrease of t_L decreased to 120°C.

Further, the effect of alloying components on the temperature of phase transformations in complex-alloyed alloys (compositions 6-11) was studied. It turned out that the introduction of tungsten or tantalum into the alloy leads to a decrease in the solidus temperature and, as a consequence, to a significant narrowing of the interval t_S- t_{solv} region of existence of a homogeneous solid nickel solution. The rhenium introduction into the alloy made it possible to substantially increase the temperatures of complete dissolution of the secondary γ'-phase and solidus, expand to 65 °C the existence interval of a single-phase alloy, and significantly narrow the melting interval.

Additions of tantalum and chromium in such a composition no longer lead to positive results (Table 2).

Table 3. The ratio V_T / V_0 at certain temperatures.

T,°C	Sample №										
	1	2	3	4	5	**6**	7	8	9	10	11
700	-	-	-	-	-	-	-	-	-	-	-
800	-	0.98	0.96	0.95	0.99	-	-	-	0.98	0.98	0.95
900	**0.94**	0.90	0.86	0.85	0.90	**0.94**	0.93	0.97	0.89	0.90	0.85
1000	**0.79**	0.75	0.71	0.68	0.74	**0.79**	0.77	0.85	0.74	0.75	0.69
1100	**0.56**	0.54	0.49	0.45	0.49	**0.57**	0.53	0.63	0.53	0.52	0.47
1200	**0.23**	0.25	0.21	0.17	0.17	**0.26**	0.20	0.32	0.25	0.23	0.19

To assess the thermal stability of a high-temperature nickel alloys Petrushin N.V., Logunov A.V. and Kovalev A.I. [9] proposed a formula relating the volumetric fraction of the γ'-phase at a given temperature V_T with its volume fraction at room temperature V_0 through the ratio of the temperatures of the beginning of dissolution $t_{b.d.}$ and end t_{solv} dissolution of the secondary 'phase:

$$V_T / V_0 = \alpha \cdot \left[t_{solv}^2 - t^2 \right] - \mu (t_{solv} - t_{b.d.}), \tag{1}$$

where $\alpha = (t_{solv} - t_{b.d.})^{-2}$,

$\mu = 2 \cdot t_{h.d.} / t_{solv} - t_{b.d.})^2$,

t – temperature.

A typical dependence of the amount of the secondary γ'-phase on temperature is shown in the Figure 2. The results of the calculations are summarized in Table 3. Calculation analysis shows that for relatively low-alloyed high-temperature alloys the greatest thermal stability at temperatures below 1100°C is achieved in the base sample (composition 1).

For heavily alloyed high-temperature alloys, the composition with 6 % mass. tantalum (sample 8) has the highest thermal stability at temperatures up to 1200°C. The addition of rhenium to 2% by weight slightly reduces the thermal stability.

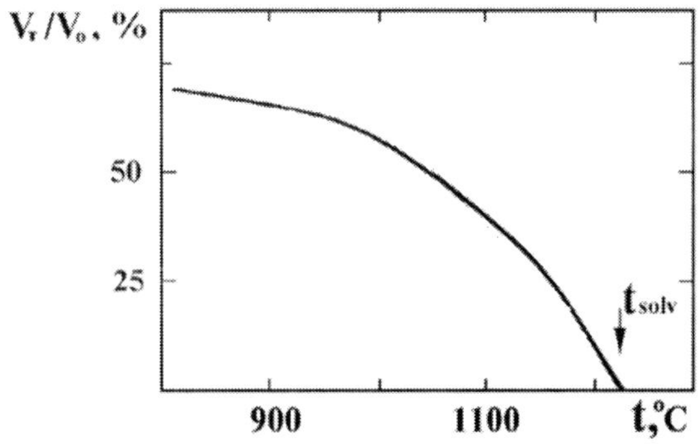

Figure 2. Changing secondary γ'-phase amount in the heat-resistant nickel alloy from the heating temperature.

4. Conclusions

1. New fundamental information has been obtained on the effect of doping on the carbon-free high-temperature nickel alloys phase transformations temperatures. Temperatures for liquidus, solidus, total dissolution of the secondary γ'-phase were determined for all compositions, the melting intervals and the existence of a single-phase nickel solid solution, supercooling of the melt upon cooling were calculated.
2. For relatively low-alloyed carbon-free high-temperature alloys, the greatest positive effect is achieved when they are doped with γ'-forming elements-niobium and titanium.
3. For multicomponent carbon-free high-temperature nickel alloys, compositions containing rhenium and a small amount of tantalum have the most favorable characteristics.
4. The increased content of tungsten, tantalum or chromium in complex-alloyed carbon-free high-temperature nickel alloys leads to a certain decrease in the temperatures of phase transformations.
5. For relatively low-alloyed high-temperature alloys, the maximum thermal stability at temperatures below 1100°C is achieved in the base sample (composition 1). For highly alloyed high-temperature alloys, the composition with 6 % mass. of tantalum has the highest thermal stability at temperatures up to 1200°C.

5. References

[1] Masoumi F, Shahriari D, Jahazi M, Cormier J, Devaux A 2016 *Sci. Rep.* **6** 28650
[2] Petrushin N V, Elyutin E S, Nazarkin R M, Pakhomkin S I, Kolodochkina V G, Fesenko T V, and Dzhioeva E S 2016 *Inorg. Mater: Appl. Research.* **7** 824–831
[3] Kuznetsov V P, Lesnikov V P, Moroz E V, Koryakovtsev A S 2008 *Met. Sci. Heat Treat.* **50** 171-174
[4] Lepihin S V, Baryshev E E, Tjagunov G V, Baum B A, Tsepelev V S 2005 *Zavodskaya laboratoriya: Diagnostika materialov* **71** 35-37
[5] Suwardiea John H, Artiagab R, Mierb Jose L 2002 *Thermochim. Acta* **392–393** 295–298
[6] D'Souza N, Dong H B 2007 *Scripta Mater.***56**41–44
[7] Kang S H, Deguchi Y, Yamamoto K, Ogi K, Shirai M 2004 *Mater. Trans.* **45** 2728-2733
[8] Liu G, Lin L, Han Z H, Zhang G J, Zhang J 2017 *Rare Metals* **36**(10) 792–798
[9] Petrushin N V, Logunov A V, Gorin V A 1984 *Met. Sci. Heat Treat.* **26**(5) 361-364

IOP Conf. Series: Materials Science and Engineering **374** (2018) 012007 doi:10.1088/1757-899X/374/1/012007

Research on Antiphonic Characteristic of AlMg10-SiC Ultralight Composite Materials

O Rusu[1] and I Rusu[1]
[1] Gheorghe Asachi Technical University of Iasi-Romania, Materials Science and Engineering Faculty, Blvd. Mangeron, No. 69, 700050, Iasi, Romania

E-mail: vrusu2003@gmail.com

Abstract. The paper presents the results on the absorption sound testing of an ultralight cellular composite material AlMg10-SiC, obtained by sputtering method. We have chosen this type of material because its microstructure generally comprises open cells (and relatively few semi-open cells), evenly distributed in the material, a structure that, at least theoretically, has a favorable behavior in relation to sound damping. The tests were performed on three types of samples, namely P11 – AlMg10 – 5%SiC, P12 – AlMg10 – 10%SiC şi P13 – AlMg10 – 15%SiC. The 15% SiC (P13) cellular material sample has the best sound-absorbing characteristics and the highest practical absorption degree.

1. Introduction

Sound is a form of mechanical energy which traverses the environment from the source of noise in all directions. The noise control process involves blocking its paths and eventually eliminating sound energy wherever possible.

The sum of all absorptions and reflections of sound waves, determined by the interactions with the environment through which the sound propagates creates a "picture", a soundtrack in the human ear. This phenomenon can be estimated by a parameter called NRC (Noise Reduction Coefficient). This is one of the most frequently evaluated and imposed features because it is closely related to the acoustic absorption curve. According to the American Standard ASTM C 423, this parameter is calculated as the average of the absorption values for 250 Hz, 500 Hz, 1000 Hz and 2000 Hz frequencies, with an approximation of 5%. The acoustic absorption degree (αW) is a parameter determined by a method indicated in EN ISO 11654 [1], the degrees of absorption measured in practice (according to EN ISO 354 [2]) being calculated in practical sound absorption degrees (αp) for each frequency band that lies on an octave.

According to the standard, the reference curve is plotted over the variation curve of the practical degree of absorption, indicating the existence of deviations zones between the theoretical and experimental values [3]. From graphical representations it can be deduced that if, in the range of the tested frequencies, the trace curve corresponding to the measured values is above the reference curve, then the tested material absorbs obviously more than is estimated by the αW value.

The fragility of many sound-absorbent materials has led to researches in order to eliminate this unwanted mechanical feature. As there is on the market a new class of sound-absorbent materials based on cellular composites, mainly used in the construction of trucks and heavy machinery, military industry and aeronautics industry [4-7], we chose to test an ultralight composite material type AlMg10-SiC.

2. Experiment

To perform the experiments we took samples (40x40x20 mm, Figure 1) from a AlMg10-SiC ultralight cellular composite material obtained by the bubbling method [8], which were then subjected to the sound absorption tests. We have chosen this type of material because it microstructure generally comprises open cells (and relatively few semi-open cells), evenly distributed in the material, structure that, at least theoretically, has a favorable behavior in relation to the sounds damping. The testing of a sound damping element is done on an installation (noise analysis cell), which simplified scheme is shown in Figure 2.

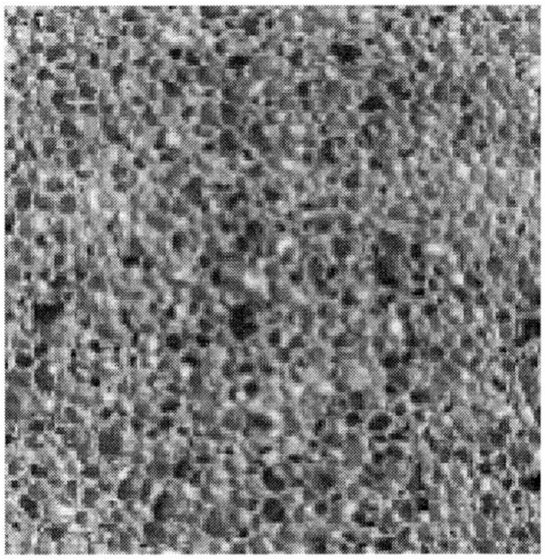

Figure 1. The overall appearance of the AlMg10-SiC composite test plates.

It includes an acoustic source, an enclosure with a window with an antisound element (the anti-noise element or component to be tested), a sound collector placed inside the enclosure and an automatic signal processing system. We used a NORSONIC Nor 850 [9] equipment, with a Klain & Hummel AK120 signal generator and amplifier [10] and a data acquisition system.

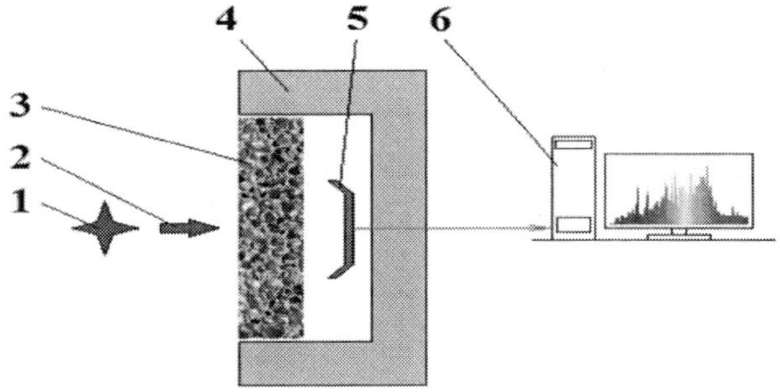

Figure 2. Schematic representation of sound damping measurement cell; 1 - primary sound emission source, 2 - emitted sound, 3 - sound damping element (made of composite material), 4 - test enclosure, 5 - sound receiving device, 6 - computerized recording and processing data system.

Because, in general, the most frequent noise and physical discomfort frequencies, for example, in the field of automotive and railway transport, are between 250 and 2000 Hz [11], we chose to perform the experiments in the frequency range 100 ÷ 2600 Hz.

3. Results analysis

The experimental obtained results regarding the level of sound absorption by AlMg10- SiC ultralight cellular composites are shown in Figure 3.

From the experimentally obtained data it is observed that the cellular materials best absorbs the sound, regardless of the percentage of SiC, in the frequency range of about 900 ÷ 2000 Hz, having, in principle, practical absorption degree values greater than 0.35. It is also clear that these materials are not recommended for damping low-frequency sounds (below 600 ÷ 700 Hz). The cellular material with best sound-absorbent characteristics is that with 15% SiC (P13) – the highest degree of absorption (0.52) at 1200Hz; in terms of this last parameter, the 10% SiC composite is quite close to the first (with a value of 0.47), but at a slightly higher frequency (1800Hz). The most unfavorable from the point of view of sound absorption behavior is the cellular composite with 5% SiC (whose practical degree of absorption does not exceed 0.37%), fact that is explained by its cellular structure more unfavorable to the absorption of sounds (smaller cells, lower relative porosity).

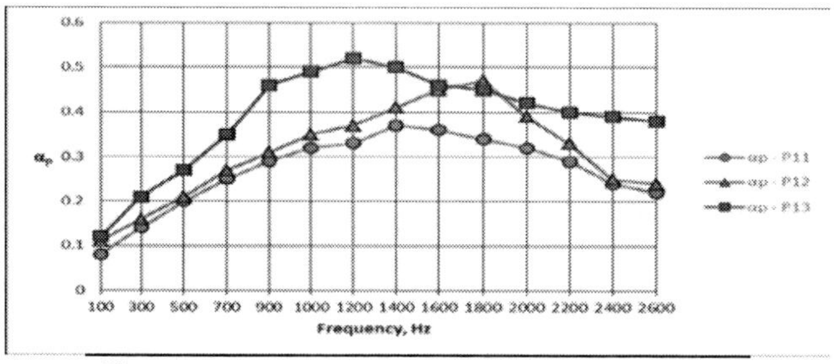

Frequency Hz	α_p - P11	α_p – P12	α_p – P13
100	0,08	0,11	0,12
300	0,14	0,16	0,21
500	0,20	0,21	0,27
700	0,25	0,27	0,35
900	0,29	0,31	0,46
1000	0,32	0,35	0,49
1200	0,33	0,37	0,52
1400	0,37	0,41	0,50
1600	0,36	0,45	0,46
1800	0,34	0,47	0,45
2000	0,32	0,39	0,42
2200	0,29	0,33	0,40
2400	0,24	0,25	0,39
2600	0,22	0,24	0,38

Figure 3. Experimental data concerning the variation of the practical absorption degree depending on frequency; P1i - cellular composite samples, i = 1, 2 or 3 corresponding to the percentage of SiC (5, 10 or 15%).

It is thus clear that, for these types of ultralight cellular composite materials, the technical field of interest in which the sound damping occurs is limited to frequency values in the 800-2200 Hz range (especially for 15% SiC material), the attenuation parameters depending on the degree of resonance damping that occur in the composite, especially for this frequency range [12-14]. In addition to these favorable sound absorbent characteristics, the experimented cellular composites combine these properties with mechanical strength ones, with a very good resistance to higher temperatures compared to other commonly used sound absorbent materials [15-17]. They are also completely recyclable, the absorption properties remaining constant over time. That is why I consider that these materials can be recommended for the damping of sounds and vibrations in the many fields like auto, railway, etc.

4. Conclusions

AlMg10-SiC ultralight cellular materials best absorb the sound, regardless of the percentage of SiC, in the frequency range of about 900 ÷ 2000 Hz, basically having practical absorbance values greater than 0.35.

Tested materials are not recommended for damping low-frequency sounds (less than 600 ÷ 700 Hz).

The best sound absorbent features were observed for the cellular material with 15% SiC, this one having the highest degree of absorption (0.52) at 1200Hz.

Experimental cellular composites combine sound absorbent properties with mechanical strength, with a high resistance to higher temperatures compared to other commonly used sound absorbent materials.

5. References

[1] * * *EN ISO 11654:1997 – Acoustics -- Sound absorbers for use in buildings -- Rating of sound absorption
[2] * * *EN ISO 354:2003 – Acoustics -- Measurement of sound absorption in a reverberation room
[3] http://www.scritub.com/biologie/biofizica/Unde-Acustice-I-Unde-Sonore 242182022.php
[4] Crocker MJ and Arenas JP 2007 *Use of Sound-Absorbing Materials,* in Handbook of Noise and Vibration Control editor Crocker M.J., John Wiley and Sons, New York
[5] Nick A, Becker U, Thoma W 2002 *Journal of Polymers and the Environment* **10**(3) 115–118
[6] Asdrubali F 2006 Survey on the Acoustical Properties of New Sustainable Materials for Noise Control *Proceedings of Euronoise*, Tampere, Finland
[7] Magrini U and Ricciardi P 2000 Surface Sound Acoustical Absorption and Application of Panels Composed of Granular Porous Materials *Proceedings of Inter-Noise* 27–30, Nice, France
[8] Bălţătescu O, Florea RM, Roman C, Rusu I, Carcea I 2013 *Journal of optoelectronics and advanced materials* **15**(7–8) 823–832
[9] http://www.norsonic.com/en/products
[10] https://www.neumann-kh-line.com/klein-hummel/globals.ns/resources
[11] Crocker M J, Li Z, Arenas J P 2005 *International Journal of Acoustics and Vibration* **10**(2) 52–60
[12] Champoux Y and Stinson M R 1992 *Journal of the Acoustical Society of America* **92** (2) 1120–1131
[13] Kidner M R F and Hansen C H 2008 *International Journal of Acoustics and Vibration* **13** (3) 112–119
[14] Allard J F 1993 *Propagation of Sound in Porous Media* Elsevier Applied Science, London
[15] Scheffler M and Colombo P 2005 *Cellular Ceramics: Structure, Manufacturing, Properties and Applications* Wiley-VCH, Weinheim
[16] * * *2007 *Determination of Weight Elasticity of Fuel Economy for Conventional ICE Vehicles, Hybrid Vehicles and Fuel Cell Vehicles* Report 55510, Forschungsgesellschaft Kraftfahrwesen mbH, Aachen
[17] Sandberg U and Ejsmont JA 2007 *Tyre/Road Noise – Generation, Measurement, and Abatement* in Handbook of Noise and Vibration Control, editor Crocker M.J., John Wiley and Sons, New York

The Influence of Mechanical Parameters on Dielectric Characteristics of Rigid Electrical Insulating Materials

G Buică[1], A E Antonov[1], C Beiu[1], R Dobra[2] and M Risteiu[2]
[1]National Research and Development Institute for Labor Protection "Al. Darabont", Bucharest, Blvd. Ghencea no. 35A, sect. 6, Romania
[2]1 Decembrie 1918", Gabriel Bethlen Str. No. 5, 510009, Alba Iulia, Romania

E-mail: gbuica@protectiamuncii.ro

Abstract. Rigid electrical insulating materials are used in the manufacture of work equipment with electric safety function, being mainly intended for use in the energy sector. The paper presents the results of the research on the identification of the technical and safety requirements for rigid electrical insulating materials that are part of the electrical insulating work equipment. The paper aims to show the behaviour of rigid electrical insulating materials under the influence of mechanical risk factors, in order to check the functionality and to ensure the safety function for the entire life time. There were tested rigid electrical insulating equipment designed to be used as safety means in electrical power stations and overhead power lines.

1. Introduction

Insulated work equipment used in the working process, according to the provisions of the Technical Instructions, which contain technical and safety instructions, must ensure the highest safety level for workers against the risk factors of accident and professional disease, without generating themselves a risk both for the user and for other people in the work environment [1].

This means, first of all, that protective equipment (equipment that is part of the work equipment category as defined by Executive Order no. 1146/2006) must form a barrier between the user and the environment, thus the risk on user to be as small as possible, under the permissible limit which might lead to injury or professional disease.

Also, protective equipment must not cause any danger to the user, such as: injuries, toxic emissions, emissions of corrosive substances, etc.).

It is known that most hazardous factors in the work environment act insidiously or abruptly, so it is impossible for the exposed person to seize the risk in a timely manner in order to avoid injury or professional illness [4, 8].

Studies on the behaviour over time of the protective equipment in use have made it clear that they lose their dielectric characteristics due to mechanical damage, the aging of materials, inappropriate use, etc.

Analysing the statistic weight of the factors that influence the safety characteristics over time it is found that the mechanical factors during use, followed by the aging of the protective materials, have a large share.

From the monitored work equipment categories it was found that 37% of them have scratches, 23% material bruises, 10% cracks.

The mechanical demands during the use of the protective equipment, require the choice, even from the design, of materials that are both electrically and mechanically resistant to technical characteristics and safety requirements to be tested.

2. Identification of the specific mechanical safety requirements

The identification of the specific technical and mechanical safety requirements that the rigid electro-electrical insulating materials, which are part of the electro-electrical insulating work equipment, have to be based on:

- field of use (type and technical characteristics of electrical installations to be used);
- the operating principle and the way of use;
- usage limits;
- unappropriated usage practices;
- environmental work requirements;
- technical criteria;
- the results of the safety tests for compliance certification, in order to certify the quality of the safety and the periodic tests (visual tests and verifications of electrical characteristics);
- the level of technical progress [7].
- user / consumer expectations regarding the safety of the work equipment in question [6, 7].

In Figure 1 the mechanical and safety requirements applicable to rigid electrical insulating materials are identified.

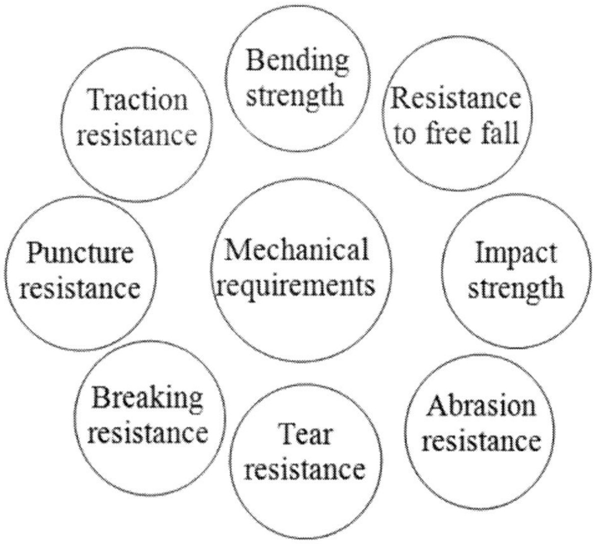

Figure 1. Mechanical and safety requirements applicable to rigid electrical insulating materials.

The influence of the mechanical factors on the dielectric characteristics of the rigid electrical insulating materials was studied on the following types of protective equipment: electrical insulating shutter. Electrical insulating shutters are protective equipment that applies between the disconnector's mobile (the disconnector is in open position) and fixed contacts [2].

A total of 9 types of electrical insulating shutters designs have been tested and monitored. Three are made of a material made by an extrusion process from non-plasticized polypropylene (PP), three other electrical insulating shutters designs are made from extruded material of non-plasticized PVC type S and three electrical insulating shutters designs are made from glass fibber reinforced polyester resin (GFRPR).

The three protection equipment models have been checked for compliance and security certification and have been monitored since commissioning for 5 years through examinations, visual inspections, and dielectric features. The three protection equipment models have been tested for compliance and safety certification and have been monitored since commissioning for 5 years, through examinations, visual inspections, and dielectric characteristics.

These characteristics are the maximum values of forces and stress that could be transmitted by the protective equipment or to which they might be subjected. Some of the mechanical tests, such as proofing of breakage and traction resistance, were performed on specimens obtained from the basic material, and some tests performed on materials cut from the body of the electrical insulating equipment presented as sample.

Part of the mechanical tests such as mechanical shock test with pendulum, bending, shock at free fall, were performed on the functional equipment, in order to assess the behaviour of the material after having undergone hot bending, clipping and bonding .

3. Experimental procedures

Mechanical tests were performed on specimens of protective materials and / or equipment that were conditioned by storing in a relaxed position for 24 hours at room temperature (24 ± 2) ° C and (50 ± 5) % RH, standard atmosphere B [10]. In table 1 shows the list of mechanical stresses in which samples of the material used to make the electrical insulating shutters were tested.

Table 1. List of mechanical stresses.

No. crt.	Mechanical stress	Maximum allowed limis
1	Traction	min. 12 N/mm^2
2	Bending	10 daN
3	Puncture	min. 45 N
4	Impact	20 J
5	Free falling	1,5 m
6	Endurance	100 mounting / dismounting
7	Abrasion	2000 cycles

For determination of mechanical stresses to which the material samples and the protective equipment were subjected, the provisions of the national and European standards were also taken into account [3].

4. Results and discussions

Protective equipment or samples of the collected material were tested for dielectric properties at the beginning of the study, and after each mechanical stress. For dielectric test, 50 kV in three points was applied, and held for 3 minutes. The electrodes used in this test are two cylinders of 25 mm in diameter and 25 mm in height and 75 mm in diameter and 25 mm in height, respectively.

Figure 2 shows the leakage current for all nine electrical insulating shutters made by the same manufacturer, before the mechanical stresses are performed.

The order in which the mechanical stresses were performed at the start of the study was random, during the study the order of the test was established.

Verification of mechanical resistance to puncture was performed on circular specimens with a diameter of 50mm. The longitudinal speed applied was 500 ± 50mm / min. It was measured the force at which the material was punctured. The results of the verification are shown in Figures 3 and 4.

The traction strength test was performed on dumbbells. The applied test speed was 50mm / min. The results for non-plasticized PVC type S are shown in Table 2, and for Fiberglass reinforced polyester resin (FGRPR) in Table 3.

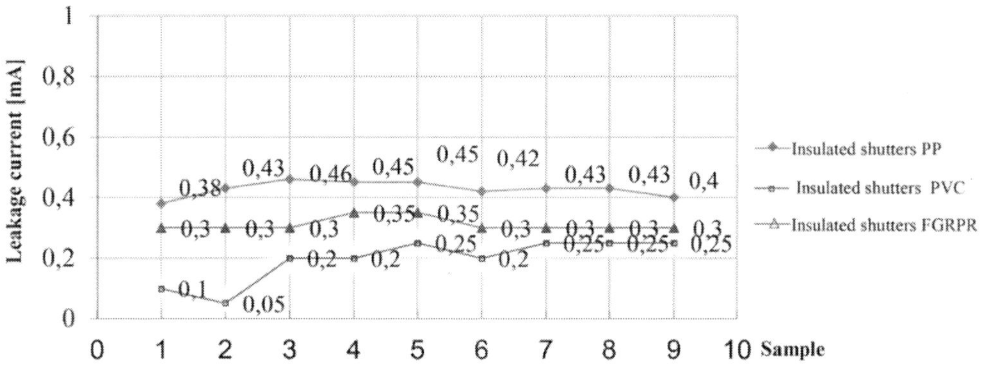

Figure 2. Leakage current values - before mechanical tests.

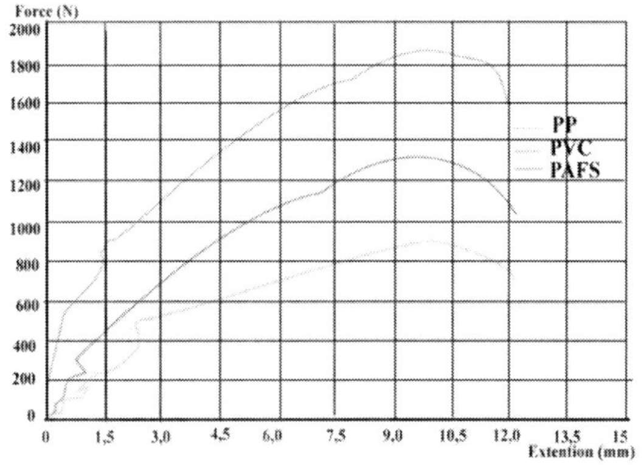

Figure 3. The behaviour of samples to puncture.

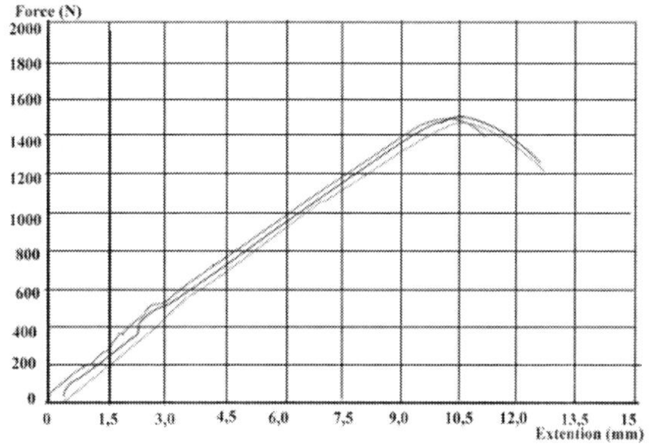

Figure 4. The behaviour of polypropylene (PP) samples, to puncture. Test on 3 specimens.

Table 2. Traction resistance test results for non-plasticized PVC type S.

Sample No.	Dimensions of dumbbell specimens (cm)		F_{tear} (N)	σ_{tear} (N/mm^2)	
i	b	h		$\sigma_{tear\ i}$	$\overline{\sigma}_{tear}$
1	4.1	6.1	1530	61.2	61
2	4.0	6.4	1520	59.4	
3	4.0	6.4	1490	58.2	
4	4.3	6.2	1710	64.1	

Table 3. Traction resistance test results for non-plasticized PVC type S.

Sample No.	Dimensions of dumbbell specimens (cm)		σ_{tear} (N/mm^2)	
i	b	h	$\sigma_{tear\ i}$	$\overline{\sigma}_{tear}$
1	4.0	6.2	65.3	62
2	4.2	6.2	59.3	
3	4.1	6.1	56.6	56
4	4.2	6.3	55.4	

The mechanical shock test was carried out on complete equipment, including accessories, with a shock test pendulum consisting of an oscillating arm around the horizontal axis and a hammer fixed to its lower part. The test consisted in the application of hits including on each corner (made by hot bending) and in areas as close as possible to soldering and fixing by gluing, with a mechanical shock effect of 2 Nm. After the mechanical shock test, dielectric rigidity was tested. Figure 5 gives the leakage current values. The dielectric characteristics are maintained.

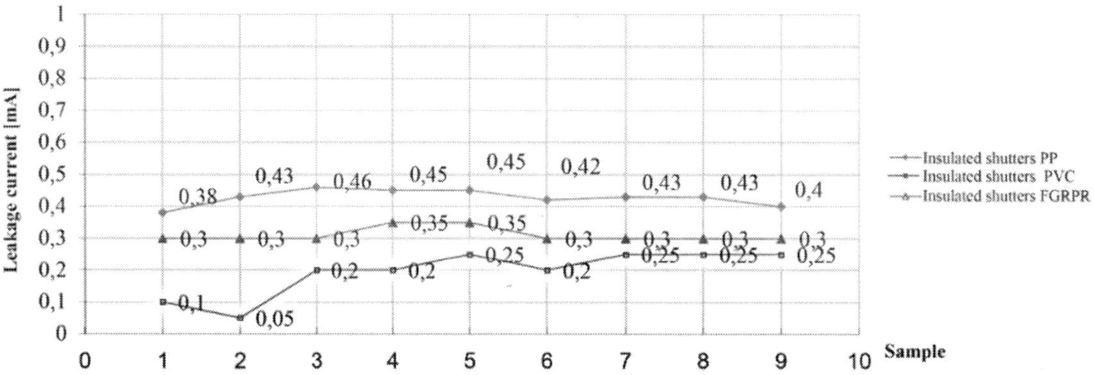

Figure 5. Leakage current values- after testing mechanical shock.

Verification of free fall resistance was performed on complete equipment, including accessories, the purpose of the verification being to check the material and the bending and binding of the material.

The equipment was dropped on a hard floor (concrete) from a height of 1.5 m, from different positions: vertical, horizontal, oblique. After the free falling test, dielectric rigidity was tested. Figure 6 shows the leakage current values. The Leakage current on the PVC specimen is above the 1 mA limit, due to bad processing of the corners. The PVC specimen was replaced.

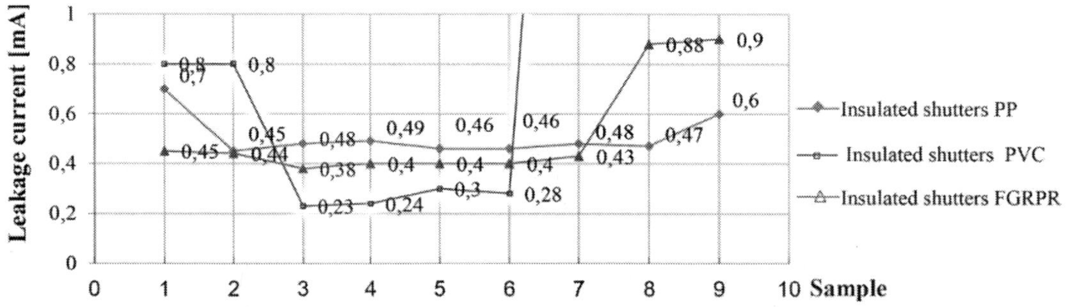

Figure 6. Leakage current values - after testing on free falling resistance.

Verification of endurance consisted in performing 100 mounting / dismantling of the electrical insulating shutters on the contacts of a 20 kV disconnector. At the end of the test, the 9 specimens had scratches on 30% of the surface. Immediately after this test, the dielectric rigidity was tested. Figure 7 shows the leakage current values. Decrease of dielectric characteristics is noted. As a result of the deterioration in the test, the materials were also tested for abrasion.

The study continued by performing mechanical tests after which the dielectric properties were tested at the end.

Test methods have been developed in the study to verify the mechanical safety characteristics of the electrical insulating shutters, which must not show on both sides harmful physical irregularities such as: protrusion that prevents the uniformity or smoothness of the surface, for example: cracks, cuts, foreign conductive particles embedded in the material, creases, bumps, air bubbles, protuberances or protruding protrusions of the mould. These harmful physical irregularities can't guaranty the dielectric characteristics of the protective equipment, with serious consequences on the user's health and safety and the risk of damage to the electrical panels in which they are used.

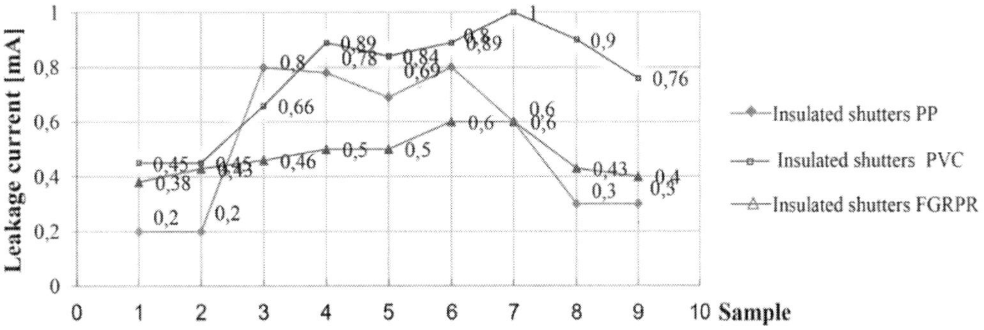

Figure 7. Leakage current values - after checking for resistance to endurance.

The samples of rigid electrical insulating material tested to mechanical puncture resistance were taken within 5 years from the manufacturers, in order to evaluate and verify the behaviour over time of various electrical insulating materials that are the basis of the manufacture of protective equipment for use in electrical installations.

The behaviour of rigid electrical insulating materials designed to make electrical insulating shutters was demonstrated by tests performed on specimens made from polypropylene (PP), PVC type S, and fibber glass reinforced polyester resin (FGRPR) respectively.

It is noted that the electrical insulating shutters made of polypropylene (PP) are more mechanically resistant than those made of PVC type S and FGRPR, maintaining the technical characteristic for a longer period of time, and guaranties the dielectric characteristics, both under normal and exceptional or unexpected working conditions [3, 4, 9].

The electrification plates made of non-plasticized polypropylene (PP) blends allow this to be used for a longer time than those made from PAFS, and PVC, which has led the manufacturers of rigid protective equipment to replace the composition and structure of the electrical insulating material during the period of the 5 years that were the subject of the study.

It is found that, depending on the content of the non-plasticized mixture from which the electrical insulating plate is made, the mechanical strength of the electrical insulating material is different, with effect on the preservation of the dielectric characteristics and the duration of use.

The inappropriate use of the materials for the protection equipment has a negative effect on the efficiency of the protection characteristic of the equipment in use and the duration of their use in the electrical installations, contributing to the increase of the non-security costs and of the investment expenses, having an effect on the competitiveness of the company and users' safety at work.

The electrical insulating shutters made of polypropylene (PP) allow to be used for a longer time than those made from FGRPR and PVC, which has led the manufacturers of rigid protective equipment to replace the composition and structure of the electrical insulating material within these 5 years, in which the study was performed.

It is found that, depending on the content of the non-plasticized mixture of electrical insulating shutters, the mechanical strength of the electrical insulating material is different, with effect on the preservation of the dielectric characteristics and the duration of use.

The inappropriate use of the materials has a negative effect on the efficiency of the safety characteristics of the equipment in use and the duration of their use in the electrical installations, contributing to the increase of the non-safety costs and of the investment expenses, having an effect on the competitiveness of the company and users' safety at work [1].

5. Conclusions

The choice of materials from which rigid protective equipment is made is a determining factor in ensuring occupational health and safety [2, 5]. Due to inappropriate use of materials from which protective equipment is made, they may suffer damage during use in electrical installations, losing their dielectric characteristics due to the mechanical parameters and the efficiency of the structure of electrical insulating material. Over time, it has been found that there is a need to apply procedures for testing different work equipment with protective functions, both at the level of the users and the installation in which they are used.

By eliminating these testing practices, in the last 15 years, it has been noted the need to improve the technical characteristics of rigid protective equipment, as a result of their behaviour over time, both in terms of reliability and safety features, in order to protect workers in electrical installations.

To guarantee the safety characteristics of work equipment with electrical insulating function, intended mainly for use in the energy sector, it must ensure a high degree of dielectric parameters for rigid electrical insulating materials and to reduce the mechanical risk factors resulting from the nature and structure of chosen electrical insulating material.

The behaviour of rigid electrical insulating materials is determined by the influence of mechanical risk factors in order to guarantee the functionality of the protective equipment and to ensure the safety function throughout their use as protective means in electrical power stations and overhead power lines.

6. References

[1] Antonov A, Buica G, Darabont D and Beiu C 2017 *Environmental engineering and management journal* **16**(6) 1401-08

[2] Buică G, *Contributions of diagnostic methods of safty and health at work in electrical installations*, PhD Thesis, Universitas Petroşani 70 – 149 2010

[3] Buică G, Dobra R, Păsculescu D and Tătar A IOP *Conference Series: Materials Science and Engineering*, **133**, conference 1

[4] Darabont D, Antonov A and Bejinariu C 2017 *International Conference on Manufacturing Science and Education* – MSE 2017, MATEC Web of Conferences 121:11007

[5] Dobra R, Pasculescu, G. Marc, M Risteiu and A Antonov IOP *Conference Series: Materials Science and Engineering*, **209**, conference International Conference on Innovative Research - ICIR EUROINVENT

[6] INCDPM, 2013 *Guide for the application of GD No.1146/2006 regarding the minimum safety and health requirements for the use of work equipment by workers (in Romanian)*, National Institute of Research and Development for Occupational Safety Alexandru Darabont, Bucharest, Romania, Online at: http://www.inpm.ro/files/publicatii/2013-05.01-ghid-t.pdf

[7] INCDPM, 2014 *Risk management within professional equipment maintenance activities to prevent work accidents and occupational diseases - a prerequisite for increasing the competitiveness of employers on the market, research study*

[8] Pasculescu V M, Vlasin N I, et. al. 2016 *Proceedings of the16th International Multidisciplinary Scientific Geoconference* SGEM 2016, Section Science and Technologies in Geology, Exploration and Mining, Volume II, Albena, Bulgaria, 181-188

[9] Dašić P, Hutanu, C Jevremović V, Dobra R, Risteiu M and Ileana I IOP *Conference Series: Materials Science and Engineering*, **209**, conference International Conference on Innovative Research - ICIR EUROINVENT

[10] SR EN 60212:2011 Standard conditions for use prior to and during the testing of solid electrical eletrical insulating materials

IOP Conf. Series: Materials Science and Engineering **374** (2018) 012009 doi:10.1088/1757-899X/374/1/012009

Investigation of Woven Characteristics on Electromagnetic Shielding Behaviour

M Javadi Toghchi[1,2,3,*], C Loghin[1], I Cristian[1], C Campagne[2], P Bruniaux[2] and A Cayla[2]

[1]Faculty of Textile, Leather and Industrial Management, Gheorghe Asachi Technical University of Iasi, Iasi, Romania

[2]Ecole Nationale Supérieure des Arts et Industries Textiles, Roubaix, France

[3]College of Textile and Clothing Engineering, Soochow University, Suzhou, China

E-mail: marzieh.javadi-togh@ensait.fr

Abstract. Textiles have been highly applied for electromagnetic shielding purposes due to the increasing concern about health issues caused by human exposure to radiation. Properties of conductive yarn, fabric structure, and garment design have extreme effects on the electromagnetic behaviour and comfort of the final product. Lots of electromagnetic shielding textiles are made of metallic yarns regarding their high electrical conductivity. Therefore, some researchers have worked on electromagnetic shielding textiles made of metals. For example; the shielding effectiveness of woven fabrics made of hybrid yarns containing stainless steel wire was investigated. As discussed earlier, the fabric structure has significant effects on electromagnetic protection. Consequently, woven samples were produced using two different commercial electroconductive yarns (PA12 coated with Ag and Inox) to investigate the effects of the fabric structure. The main purpose was to define the best pattern among three basic woven patterns leads to the highest electromagnetic shielding. Moreover, the different weft yarn densities were applied to examine the effects of yarn density on the level of electromagnetic shielding. The electromagnetic shielding effectiveness of all the 2-layer samples was evaluated in the frequency range from 0.8 to10 GHz in an anechoic chamber. The woven sample with higher yarn density of PA12 coated with Ag yarns shows higher protection against radiation. To conclude, the results show that the yarn properties play the main role in shielding as well as yarn density and fabric pattern.

1. Introduction

These days, electrical devices such as mobile cell phones, microwave ovens, televisions and etc. are used all around the world. These electrical devices emit non-ionizing electromagnetic waves into our living environment. The range of electromagnetic waves shows in Figure 1. There are two main categories for electromagnetic waves based on frequencies (ionizing and non-ionizing). Although the detrimental effects of non-ionizing radiation are less than ionizing radiation, it is a source of some health problems on a human body for example: raised body temperature, vertigo, headache, and fatigue and weight loss for both mother and fetus during pregnancy period. This work has focused on waves are emitted by microwave ovens and cell phones which belong to non-ionizing radiation category (1-10 GHz) as it can be seen in Figure 1 [1].

Textiles have been extremely considered in applications of electromagnetic interface shielding [2]. The electrically conductive materials which can generate and transport free charges lead to electromagnetic shielding effectiveness and higher conductive products generally have better electromagnetic shielding effectiveness behavior. This is the reason that metals have been applied to improve the electromagnetic shielding effectiveness of textiles. For instance, the shielding effectiveness of woven fabrics made of hybrid yarns was investigated [3]. Hybrid yarns containing stainless steel wire were produced and then eight different fabric patterns were manufactured using them. The shielding effectiveness measurements of these fabrics have been done in the frequency range of 30 MHz-9.93 GHz and it showed that the shielding effectiveness was affected by fabric structure, yarn directions, density and settlement type of conductive hybrid yarn in the fabric.

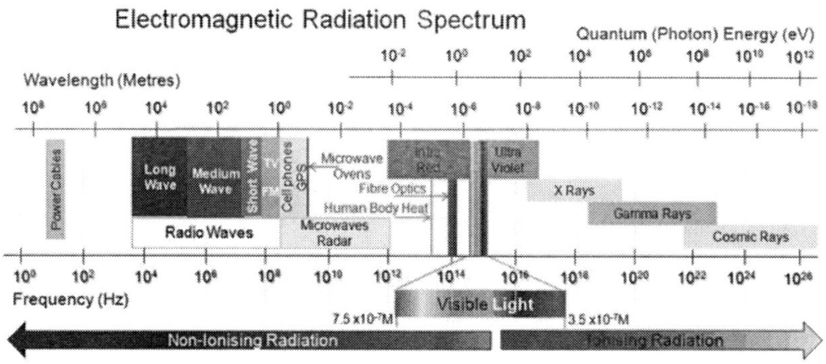

Figure 1. Electromagnetic radiation spectrum [1].

2. Experimental procedures

2.1. Electromagnetic shielding effectiveness tests

The method of determining the level of electromagnetic waves reduction for a medium often varies according to the particular shielding application. Some of the common techniques for testing shielding effectiveness for textiles are listed as below [4-7].

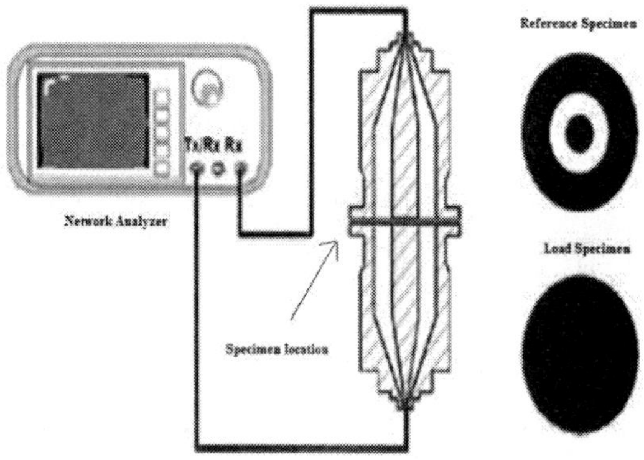

Figure 2. Measurement test setup for a method act in accordance with ASTM D4935-99 [6].

Coaxial transmission line test: This test method is used to define the shielding effectiveness of planar materials which measures plane-wave field electromagnetic wave radiation. A reference testing device is located in a holding unit and the voltage receives at a specific frequency is recorded. Then, the subject is replaced by a load device, which experiences the same test under the same conditions. A comparison between the reference and the load devices launches the ratio between power received with and without a shielding material. This test is based on ASTM D4935-10, 99 and the setup is shown in Figure 2 [6].

The results obtained in different laboratories with this method are comparable and it is considered the main benefit of this method. Also, this method can be used to resolve the records into reflection, absorption and transmission components. The main drawback of this method is that the procedure should be repeated at different frequencies. Obviously, this approach is time-consuming and takes several hours to make a spectrum [6].

Shielded box test: an anechoic chamber with a cutout portion is employed for shielded box test as it can be seen in Figure 3. A conductive coated shielding unit is positioned on the box's opening and transmitted and received radiations are measured. The electromagnetic signals from both inside and outside the box are recorded and the shielding effectiveness of specimen is defined using the ratio of signals in equation (1). This method has been applied to evaluate shielding effectiveness of textile products [6].

This method is used for comparative tests in order to compare the shielding effectiveness of different specimens. The main problems of this method are that a sufficient electrical contact between test specimen and the shielded box is challenging to achieve and the frequency range is limited. Also, it is possible to put all the outside facilities in another room while the specimen is located in the opening between two rooms to achieve more accurate results. This approach is known as shielded room method which has been developed to overcome the limitations of the shielded box method considering the frequency range and accuracy [7].

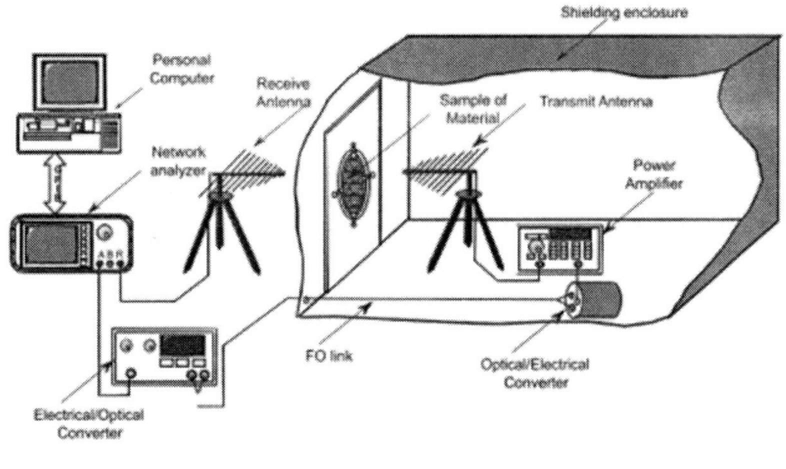

Figure 3. Adaptation of shielding effectiveness measurement method using the shielded box [6].

2.2. Experimental

The fabric structure and properties have significant effects on electromagnetic protection value. To investigate these effects, seven different woven samples have been made using two different commercial electroconductive yarns. The main purpose was to define the best pattern among three basic woven patterns to achieve the highest electromagnetic shielding. Also, the different weft yarn densities were applied to investigate the effects of yarn density on the level of electromagnetic shielding. Table 1 represents the characteristics of the samples. The shielding effectiveness of all the samples was evaluated using anechoic chamber in the frequency range (0.8-10 GHz).

The shielding effectiveness of 2-layer samples were measured since it is required to have conductive yarn in both horizontal and vertical direction to gain the shielding effectiveness.

Table 1. Different fabric samples properties.

Weft yarn**	Sample number	Pattern	Weft density (yarn/cm)	Areal density (g/m2)	Thickness (mm)	Volume Electrical conductivity (S/m)
PA12 coated with Ag	S1	Plain	16	102	0.58	18 ±2
	S2	Plain	8	121	0.46	250 ±24
	S3	Twill 2:2	16	105	0.68	310 ±30
	S4	Twill 2:2	8	118	0.54	428 ±13
Inox	S5	Plain	8	503	0.75	12 ±1
	S6	Twill 2:2	8	520	0.95	146 ±35
	S7	Satin 5	8	516	0.89	121 ±20

** Warp yarn is wool for all the samples

3. Results and discussions

As results display samples with Twill and Plain patterns and weft density (16/cm) shows the highest shielding effectiveness through all the samples. It is noted that the weft yarn for these samples was PA coated with Silver. It means that the higher yarn density leads to higher protection against radiation since sample S1 and S3 showed higher protection compared to S2 and S4.

To compare the pattern effects on shielding effectiveness, S5, S6 and S7 were compared and the results confirm that sample with Plain pattern shows higher protection followed by Twill pattern and the minimum shielding belongs to satin pattern.

Samples S4 and S6 were compared to see the effect of yarn properties on shielding. Although the results were really closed to each other, the shielding effectiveness of sample with PA coated with Ag as weft yarn showed higher shielding effectiveness. It is noted that the shielding effectiveness was between 40-50 dB for all the samples.

4. Conclusions

Textiles have been highly applied for electromagnetic shielding purposes due to the increasing concern about health issues caused by human exposure to radiation. The fabric structure has significant effects on shielding effectiveness. As a result, woven samples were produced using two different electroconductive yarns (PA12 coated with Ag and Inox). The main purpose was to define the best pattern among three basic woven patterns leads to the highest electromagnetic shielding. Also, the different weft yarn densities were applied to examine the effects of yarn density on the level of electromagnetic shielding. The electromagnetic shielding effectiveness of all the 2-layer samples was evaluated using anechoic chamber. The woven sample with higher yarn density shows higher protection against radiation. In addition to, the woven pattern and electrical conductivity of the yarn have influences on shielding effectiveness of the woven fabric. To conclude, the results show that the yarn properties, as well as yarn density, play the main role in shielding effectiveness of the final products.

5. References

[1] http://www.mpoweruk.com
[2] Maity S, Singha K, Debnath P, Singha P 2013 *Journal of Safety Engineering* **2**(12) 11-19
[3] Ortlek H G, Saracoglu O, Saritas O, Bilgin S 2012 *Fibers and Polymers* **13**(11) 63-67
[4] Lu Y, Jiang S, Huang Y 2010 *Surface and Coatings Technology* **204** 2829-2833
[5] Morari C, Bălan I 2015 *Electrotehnică, electronică, automatică* **63**(2) 127-137

[6] Šafářová V, Militký J 2011 *Comparison of methods for evaluating the shielding effectiveness of textiles* Chez strutex: structure and structural mechanics of textiles fabrics, Liberec
[7] Geetha S, Satheesh Kumar K, Rao R K, Vijayan M, Trivedi D C 2009 *Journal of applied science* **112**(4) 2073-2086
[8] Niembro-Martin A, Pistono E, Lemaître-Auger P, Vuong T P, Tourtollet G 2014 *Efficient Setup for the Measurement of Transmission- Coefficients with a Single-Room Anechoic Chamber* Antenna Measurements & Applications (CAMA), IEEE Conference 3-5

Acknowledgments
This work has been carried out as part of the Sustainable Management and Design for Textiles (SMDTex) Project, funded by the European Erasmus Mundus Program.

IOP Conf. Series: Materials Science and Engineering **374** (2018) 012010 doi:10.1088/1757-899X/374/1/012010

Influence of Selective Laser Melting Processing Parameters of Co-Cr-W Powders on the Roughness of Exterior Surfaces

M A Baciu[1], E R Baciu[2], C Bejinariu[3], S L Toma[3], A Danila[4] and C Baciu[*3]
[1]S.C. Color control S.R.L., Str. Orastiei 25A, Cluj-Napoca, Romania
[2]Faculty of Dental Medicine, Grigore T. Popa University of Medicine and Pharmacy Iasi, 16 Universitatii St., Iasi 700115, Romania
[3]"Gheorghe Asachi" Technical University from Iasi- Romania, Faculty of Materials Science and Engineering, Blvd. Dimitrie Mangeron, No. 71A, 700050, Iasi- Romania
[4]S.C. LAMAS MICROTECH S.R.L. Bucuresti, str. Aron Cotrus, no.51-59, Sector 1, Bucuresti, Romania

E-mail: constantin_baciu@yahoo.com

Abstract. Selective Laser Melting (SLM) represents an Additive Manufacturing method widely used in medical practice, mainly in dental medicine. The powder of 59% Co, 25% Cr, 2.5% W alloy (Starbond Co S Powder 55, S&S Scheftner C, Germany) was processed (SLM) on a Realizer SLM 50 device (SLM Solution, Germany). After laser processing and simple sanding with Al_2O_3 or two-phase sanding (Al_2O_3 and glass balls), measurements of surface roughness were conducted. This paper presents the influences exercised by laser power (P = 60 W, 80 W and 100 W), the scanning speed (vscan = 333 mm/s, 500 mm/s and 1000 mm/s) and exposure time (te = 20 μs, 40 μs and 60 μs) on the roughness of surfaces obtained by SLM processing. Based on the experimental results obtained for roughness (Ra), some recommendations regarding the choice of favorable combinations among the values of technological parameters under study in order to obtain the surface quality necessary for subsequent applications of the processed parts (SLM) have been made.

1. Introduction

The construction of parts through "layer by layer" technique while using Selective Laser Melting (SLM) method represents a modern processing technology CAD/CAM of Co-Cr alloys for diverse components specific to dental prosthetics [1].

The roughness of exterior surfaces obtained through SLM processing depends on numerous factors keeping both to metal powder and the technical features of the equipment. To improve the surface quality, two research directions have been approached: the study is influences by the technological processing parameters; the identification of some new technological processing possibilities that may turn the technical characteristics of the equipment used to good account.

Yasa, E. et al [2, 3] analysed the influences of hatch strategy on roughness by carrying out processing techniques "with two areas" also including a laser surface for the remodeling of the border area between the two successive passes of the laser beam. In their papers, Delagrdo et al [4] studied the influence of construction direction on the resistance to bending and elongation. Krauss, H. [5] investigated workability through SLM method following the characterization of the unique lines corresponding to the melting/solidification layout followed by the laser. Smurnov, I et al. [6] analysed the influence of thermal conductivity of the alloy used on the layer construction of parts. The exposure strategy of powder particles to the action of laser beam and the manner of distribution of heat in the molten metal bath represented lines of research for Fateri et al. [7]. The impact of alloy elements on the "bulling" effect has been analysed by Kruth et al. [8, 9]. They also studied the role of phosphorus additions on the superficial tensions existing at the contact between liquid metal and solidified metal. Gebhardt, A., [10] and afterwards Strano et al [11] investigated the influence of "stair stepping effect" in relation to the construction direction on the roughness of exterior surfaces.

2. Experimental method

To obtain the specimens we used Starbon CoS Powder 55 (S&S Scheftner C, Germany) metal powder having the chemical composition given in Table 1 and the particle size of 10…55 μm.

Table1. Chemical composition of Starbon CoS Powder 55.

Chemical composition [% mass]					
Co	Cr	W	Mo	Si	Other elements: C, Fe, Mn
59.0%	25.2%	9.5%	3.5%	max. 1%	max. 1.5

We may notice the absence of beryllium. The sizes of the 54 specimens were 40x30x0.5 mm, as they are going to be used in other experimental researches as well. Specimens were obtained on a Realizer SLM 50 device (SLM Solution Germany).The technological parameters for laser beam processing were as follows: laser power: P_1 = 60 W; P_2 = 80 W; P_3 = 100 W; scanning speed: v_1 = 333 mm/s; v_2 = 500 mm/s; v_3 = 1000 mm/s; exposure time: t_1 = 20 μs; t_2 = 40 μs; t_3 = 60 μs; powder layer thickness: g = 25 μm; hatch spacing: h = 60 μm; spot diameter: d = 50 μm. By combining these values 9 sets of technological parameters have been established:

Table 2. Parameter sets.

01	03	05	07	09	11	13	15	17
P_3	P_3	P_3	P_2	P_2	P_2	P_1	P_1	P_1
v_2	v_3	v_1	v_2	v_3	v_1	v_2	v_3	v_1
t_2	t_1	t_3	t_2	t_1	t_3	t_2	t_1	t_3

After SLM processing, all specimens were subjected to a thermal treatment by stress relief (T_{max} = 700°C, t = 45 min) as recommended by the power manufacturer.3 specimens were made for each set of parameters used: the first specimen remained non sandblasted - marking (NS), the second specimen was sandblasted only with aluminum oxide powder (d = 125 μm) – marking (1S) while the third specimen was sandblasted in two phases: once with aluminum oxide and then with glass balls (d = 25…50 μm) – marking (2S).

3. Experimental results

Roughness measurements were conducted on three distinct lines of the surface of each specimen.

Average values of experimental researches are given in Table 3.

Table 3. Roughness of SLM surfaces processed with different laser beam powers.

	Surface state	Technological parameters				
		P, [W]			$v_{scan,}$ [mm/s]	t_e,[μs]
		60	80	100		
Roughness Ra, [μm]	NS	8.80	6.43	8.99	333	60
	1S	4.04	4.17	4.81		
	2S	3.56	3.67	4.95		
Roughness Ra, [μm]	NS	8.28	6.14	5.55	500	40
	1S	5.23	3.82	3.79		
	2S	4.62	3.80	3.60		
Roughness Ra, [μm]	NS	11.34	8.44	7.85	1000	20
	1S	5.62	5.23	4.93		
	2S	5.36	4.06	4.85		

According to these results, we drew the variation curves Ra = f (P) for non sandblasted specimens (NS) – Figure 1, sandblasted once (1S) – Figure 2 and sandblasted twice (2S) – Figure 3.

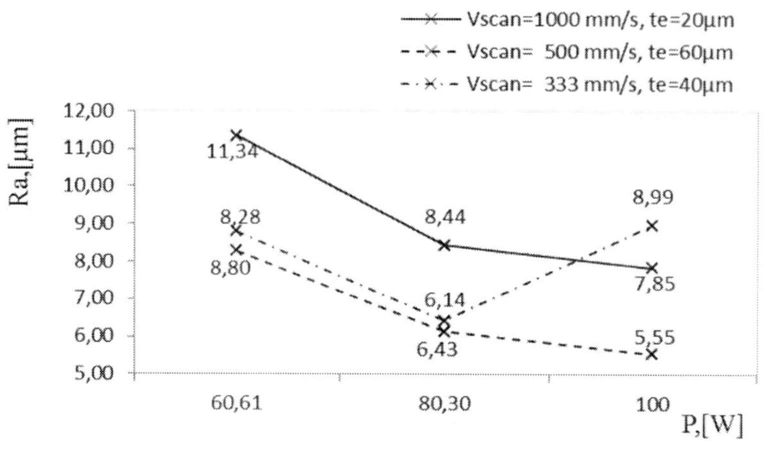

Figure 1. Variation curves Ra=f(P) for non sandblasted specimens (NS).

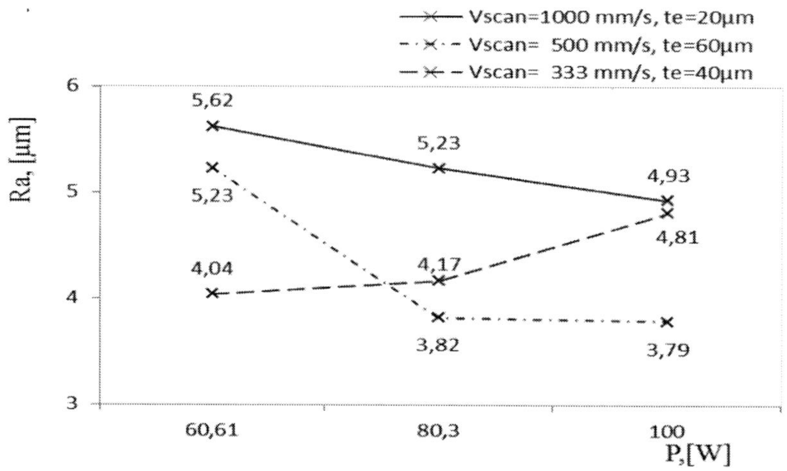

Figure 2. Variation curves Ra=f(P) for specimens sandblasted once (1S).

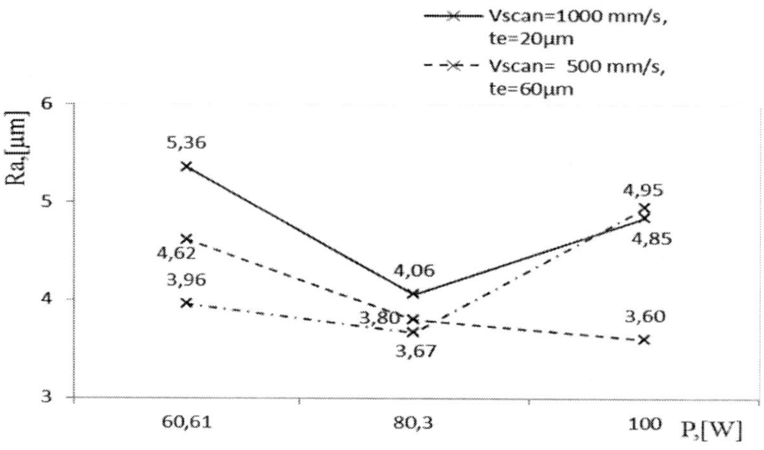

Figure 3. Variation curves Ra=f(P) for specimens sandblasted twice (2S).

4. Discussions

Based on the data in Table 3 and Figure 1 – 3 we may notice as follows:

- the maximum extension intervals for roughness (Ra) are:
  - Δ Ra_{max} = 11.34 – 5.55 = 5.79 μm – for (NS) specimens;
  - Δ Ra_{max} = 5.60 – 3.79 = 1.83 μm – for (1S) specimens;
  - Δ Ra_{max} = 4.62 – 3.15 = 1.76 μm – for (2S) specimens;
- the efficiency of sandblasting after SLM processing is highlighted through the reduction of interval for roughness values (Δ Ra);
  - by 31.60% after the first sandblasting;
  - by 48.72% after the two sandblasting phases.

A first variant for analysis of the experimental data may take into account laser power (P) and scanning speed (v_{scan}), a case where we may notice as follows:

- the lowest values of roughness for the non sandblasted specimens (NS) were obtained by SLM processing with cu P = 80 W or P = 100 W and v_{scan} = 333 mm/s and P = 80 W or P = 100 W and v_{scan} = 500 mm/s;
- the most recommended values of technological parameters for sandblasted specimens (1S) and (2S) are: P = 60 W or P = 80 W and v_{scan} = 333 mm/s, and P = 80 W or P = 100 W and v_{scan} = 500 mm/s. There is also the variant P = 80 W and v_{scan} = 1000 mm/s.

The synthetic analysis of this finding indicates that from roughness (Ra) viewpoint, the following value combinations are recommended for the two technological parameters (P) and (v_{scan}):

Table 4. Values combination.

Specimen surfacer	P, [W]			
	60	80	100	v_{scan} [mm/s]
NS		•		333
		• ·	• ·	500
				1000
1 S + 2 S	• ·	• ·		333
		• ·	• ·	500
		•		1000

The second variant for analysis of the experimental data may take into consideration laser power (P) and exposure time (t_e), a case where we may notice as follows:

- the lowest values of roughness for the non sandblasted specimens (NS) were obtained by SLM processing with cu P = 80 W or P = 100 W and t_e = 40 μs, and P = 80 W and t_e = 60 μs;
- the most recommended values of technological parameters for sandblasted specimens (1S) and (2S) are: P = 80 W or P = 100 W and t_e = 40 μs. There are also the variants: P = 60 W or P= 80 W and t_e = 60 μs and there is also the possibility P = 80 W or P = 100 W and t_e = 20 μs, P = 60 W respectively.

The synthetic analysis of this finding indicates that from roughness (Ra) viewpoint, the following value combinations are recommended for the two technological parameters (P) and (t_e):

Table 5. Values combination.

Specimen surface	P, [W]			
	60	80	100	t_e [μs]
NS				20
		• ·	• ·	40
		•		60
1 S + 2 S		•	•	20
		• ·	• ·	40
	•	•		60

The experiments conducted show that for SLM processing of the Co-Cr-W powder used, values of laser power P < 60 W, scanning speeds v_{scan}<300 mm/s or $v_{scan} \geq$1000 mm/s and exposure times t_e<40 μs are not recommended because the value of energy density (E_V) will be diminished, a fact that will negatively affect the selective melting/solidification processes of metal powder. A direct consequence of inadequate SLM processing is represented by the obtaining of some exterior surfaces with high roughness.

5. Conclusions

This scientific paper refers to the Selective Laser Melting processing method with laser beam of Co-Cr-W (Starbon CoS Powder 55, S&S Scheftner C, Germany) metal powders.

The variable technological parameters were laser power (P), scanning speed (v_{scab}) and exposure time (t_e) with whose values we made 9 different sets of specimens. Roughness (Ra) measurements were conducted on the surfaces sandblasted one – 1S or twice – 2S, or non sandblasted – NS.

On the basis of exterior surface roughness, we have reached the following conclusions:

(i) the most indicated combinations of parameters for non sandblasted (NS) specimens are: P = 80 W or P = 100 W and v_{scan} = 500 mm/s;

(ii) the most indicated combinations of parameters for (1S) and (2S) sandblasted specimens are: P = 80 W or P = 100 W and v_{scan} = 333 mm/s or v_{scan} = 500 mm/s; the combinations: P = 100 W and v_{scan} = 333 mm/s, and P = 60 W and v_{scan} = 1000 mm/s are less recommended;

(iii) from the viewpoint of volumetric energy density (E_v) necessary for SLM processing of Co-Cr-W powders, values of v_{scan}< 300 mm/s or v_{scan} = 1000 mm/s and values of exposure time t_e<40 μs are not recommended as they will result in high roughness values.

6. References

[1] Hong M–H, Bong K M and K won T–Y 2016 *Applied Sciences* **6** 1-10
[2] Yasa E and Kruth J P 2011 *Advances in Production Engineering & Management* **6** 259 – 270
[3] Yasa E and Kruth J P 2011*Procedia Engineering* **19** 389-395
[4] Delgado J, Ciurana J, Rodriguez C A 2012 *The International Journal of Advanced Manufacturing Technology* **60** 601 – 610
[5] Krauss H and Zaeh M F 2013 *Physics Procedia* **41** 815 – 822
[6] Yadroitsev I, Bertrand I and Smurov P 2007 *Applied Surface Science* **253** 8064 – 8069
[7] Fateri M, Hotter J S and Gebhardt A 2012 *Physics Procedia* **39** 464 – 470
[8] Kruth J P, Froyen L, Van Vaerenbergh J, Mercelis P, Rombonts M and Lawers B 2004 *Journal of Materials Processing Technology* **148** 616 – 622
[9] Gebhardt A 2013 Ed. 4, *Carl Hanser Verlag GmbH & Co*
[10] Strano G, Hao L, Everson R M and Evans K E 2013 *Journal of Materials Processing Technology* **213** 589 – 597

IOP Conf. Series: Materials Science and Engineering **374** (2018) 012011 doi:10.1088/1757-899X/374/1/012011

Prevention, Treatment and Tiagnosis of Pathogenic Infections by Using Pulsed Light Radiation Propagating Through Metamaterials

N Enaki [1]*, T Paslari [1], M Turcan[1], S Bazgan [1], C Ristoscu[2] and I N Mihailescu[2]
[1] Institute of Applied Physics, Chisinau, Republic of Moldova
[2] National Institute for Lasers, Plasma and Radiation Physics (INFLPR), Romania

E-mail: enakinicolae@yahoo.com

Abstract. We propose novel optical methods for prevention, treatment and diagnosis of infections by pathogens using metamaterials with various geometries consisting of microspheres (i.e. photonic crystals, photonic molecules) and optical fibers structures. Around the adjacent elements of metamaterials appear the evanescent zones of propagated pulsed light radiation overlapping each other. This effect gives us the possibility to significantly increase the decontamination volume especially in non-transparent media. The parking geometries of microspheres and optical fibers ensure the efficient contact zone between the pulsed light radiation with contaminated materials (gases, liquids, tissues, implant surfaces). The penetration depth of evanescent field in contaminated materials can achieve values comparable with pathogens dimensions. We propose an attractive antimicrobial strategy using combined action of ultrashort pulses with different frequencies and pulse duration to achieve the selective decontamination of microorganisms with minimal effects on the components of human cells and tissues. We take into consideration the intrinsic symmetries of microorganisms protein structures (inclusive virus capsids) and their possible resonant excitation in double frequencies induced Raman scattering. The development of nonlinear models of the excitation of vibration modes of biomolecules of viruses and bacteria are revised taking into consideration the multi-mode aspects of interaction of pulsed light with excited biomolecules of pathogens. This method opens new possibilities in decontamination and diagnosis of the new collective processes, which can take place in viruses, bacteria, or other cellular structures under the action of external light pulses. Exponential distribution of radiation in evanescent zone gives us the possibility to capture and trap the viruses and bacteria along the optical fibers or/and microsphere surfaces.

1. Introduction

In this paper we take into consideration the symmetry of protein structures of pathogens (inclusively virus capsids), which plays an important role in the proposed vibrational method of pathogen-destroying mechanism [1-3]. In this case, for manipulation of viruses and bacteria, we considered the evanescent zone of metamaterials consisting of microspheres and optical fiber structures with various topologies [4-6]. It is demonstrated that using metamaterials such as photonic crystals it is possible to significantly increase the decontamination volume especially in non-transparent media. This volume can fill all free space between the fibers or microspheres, when the evanescent fields of adjacent fibers (or microspheres) are overlapping each other. The action of the field on viruses and bacteria must be estimated theoretically and experimentally demonstrated [7-8]. For the moment, there exists the conceptual estimate model only. By developing such a technique one may be able to manipulate the microorganisms with the dimensions commensurable with the penetration depth of evanescent zone [9-12]. Taking into consideration that the depths of evanescent zone are not so large for short wavelength of radiation, one can be used long-wavelength radiation, for trapping viruses and bacteria near the surfaces of metamaterials elements so that the simultaneous actions of UVC and long-wavelength radiations effectively inactivate pathogens.

Most of all decontamination methods works on the UV lamps that emits nearly monochromatic light at 253.7 nm. Therefore, there is an increasing interest in pathogens decontamination not only for application in the industry but also for medical needs [14-15]. The bacterial contamination of any liquid or surface typically begins with the initial adhesion of only a few viruses that can then develop into a more structurally cohesive thin surface in a short time. Various methods with implication of UV radiation were elaborated for decontamination. For example, in the hospital, for decontamination of rooms and spaces are used germicidal UV lamps that work at 253.7 nm. This seems to be a good method for decontamination of viruses and micro-organism from air, however is not efficient. For large surfaces there are necessary high power sources, that are dangerous for patients. We show the possibility to use the evanescent zone created at the propagation of UV light through silica metamaterials, for disinfection of biological liquids.

The paper is organized as follows. A short description and characterization of the methods, experimental and theoretical approach. The setup and theoretical mechanism of the action of UV-radiation with viruses and bacteria are presented in section 2.

2. The Raman spectroscopy in the decontamination procedure

2.1. Objectives and methods of decontamination

Proposed efficient decontamination procedures using evanescent field of metamaterials opens up new perspectives and possibilities in inactivation of pathogens. The objectives of this paper are:

- To develop a method of decontamination using microspheres and optical fiber structures with various geometries.
- To demonstrate experimentally the efficient decontamination mechanisms which appear on the increasing of the contact surface of the evanescent zone in metamaterials.
- Different topological structures of meta-materials in order to increase the surface contact of pulsed light radiation with contaminated non-transparent media (implant surfaces, liquids (inclusively blood and plasma), gases)
- To find the new possibilities to trap and capture viruses, bacteria and other microorganisms from infected biological liquids along the fibers/microsphere surfases, using the effective pulsed light radiation
- The possibilities to selective decontamination of pathogens without destroying the human cell components.

2.2. Characterization methods

As we known coherent anti-Stokes Raman spectroscopy (CARS) [1] reported an attractive tool for rapid vibrational spectroscopy. For dappling the Ultra short pulsed (USP) lasers in Coherent Raman Scattering (CRS) or CARS in the decontamination procedure of viruses and bacteria in many cases it is necessary to know not only in the relative dimensions of viruses and bacterias, but their symmetry too [2, 3] (see Figure 1). So far, general evolutionary pressures shaping capsid design have remained elusive, even though an understanding of such properties may help in rationally impeding the virus life cycle and designing efficient nano-assemblies.

The authors of Ref. [4] uncover an unprecedented and species-independent evolutionary pressure on virus capsids, based on the notion that the simplest capsid designs are the best, which was shown to be true for all available virus capsids. The theories result in a physically meaningful periodic table of virus capsids that uncovers strong and overarching evolutionary pressures, while also offering geometric explanations to other capsid properties (rigidity, pleomorphic, auxiliary requirements, etc.) that were previously considered to be unrealatable properties of the individual virus.

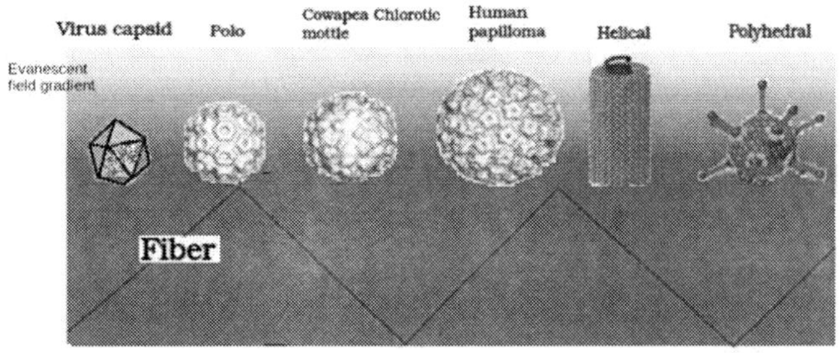

Figure 1. Relative dimension of viruses (bacteria) and their commensurability with the evanescent zone of a fiber optics.

The UV radiation is largely absorbed by matter, and the purification of water or other biological liquid needs small volumes. In order to exclude this question, i.e., decontamination of small volumes, an efficient method was proposed, based on evanescent field technique created around silica metamaterials. When UV light is propagated through an optical fibre, significant quantity of radiation is spread outside of the fibre as an evanescent field that interacts with surrounding medium. Optical techniques play an important role for localization of nanometric and submicronic particles. Such methods, based on optical metamaterials (e.g., photonic crystals, optical fibres or cavities) where the penetration of evanescent field in the contaminated zone is primordial, can be used as an efficient decontamination procedure for viruses and microorganisms [13]. In the following, we provide a theoretical description of the interaction of UV radiation with viruses, explaining the physical mechanism. We present the experimental results that confirm the efficiency of decontamination method published in our previous paper [15].

3. Methods of decontamination

3.1. The connection between the methods of decontamination

Let's start from the traditional method of decontamination of liquids using UV-C pulsed light (see Figure 2 (Lc_1)). If we have a cylinder with contaminated liquid and this cylinder is irradiated from all directions with UV radiation, the total decontamination surface is $S = 2\pi R(L+R)$, where the first term indicates the lateral surface and last term represents the surface of the bases, R, represents the radius of the base, L the length of the cylinder. So, if we use a classical method of decontamination, a big volume, of infected liquid remains contaminated $V_{cl} = \pi 2\pi R(L+R)d_p;\ V_{con} = \pi R^2 L - V_{cl} \gg V_{cl}$ where V_{cl} represent the efficient contamination volume, $d_p \sim \lambda$ is the penetration depth in liquid of UV radiation. Below we offer a method of decontamination using metamaterials for increasing of the decontamination volume. Sensing properties are expected to be related to nano-scale system dimensions. Let us firstly estimate the contact surface of flowing gas or liquid. In order to increase the contact surface of the contaminated liquid, below we proposed to examine the propagation of UV radiation through two types of meta-materials: type A corresponds to the packing of photonic-crystal fibers (PCF) and type B - photonic crystals (PC) (see Figues 2 (Lc_2) and (Lc_3) respectively), which both metamaterials are transparent for *UV* spectrum. If the PCF system is placed in a cylinder with contaminated liquid (see Figure 2 (Lc_1, Lc_2, Lc_3)), the liquid will fill all the space between the fibers.

The decontamination surface increases substantially $S_d = \pi r(r + 2rLN)$, where N is the number of fibers from PCF, r is the radius of one fiber. Last term represents the lateral surface of the fibers.

IOP Conf. Series: Materials Science and Engineering **374** (2018) 012011 doi:10.1088/1757-899X/374/1/012011

Here, it is considered that the UV radiation is inserted by fibers in the cylinder, see Figues 2 (Lc_2). The penetration depth of UV radiation (evanescent field) depends on the relative refractive indexes of fibers and contaminated liquid. The intensity of evanescent field is expressed by $I = I_0 \exp[-z/d]$, where I is the intensity of UV light of evanescent zone at distance z from fiber, d is characteristic exponential decay depth $d \sim \lambda / n_2$ where n_2 is refractive index of the liquid medium, λ is the wavelength of light radiation. Let's find the connection with classical method of decontamination. For this we need to express the decontamination area through the number of fibers. The estimations show that small radius of fiber r is proportional to $r \sim R/\sqrt{N}$. It is easy to obtain the following expression for decontamination area of N cylindrical fibers $S_d \sim 2\pi RL\sqrt{N}$. Here, we observe that the decontamination surface is proportional to the square root from the number of fibers. The decontamination volume of liquid in this case is proportional to the expression $V_d \sim 2\pi RLd\sqrt{N}$. It is not difficult to observe, that the increasing of decontamination volume is proportional to $\sqrt{N}$ where $(V_d / V_{cl} \sim \sqrt{N})$.

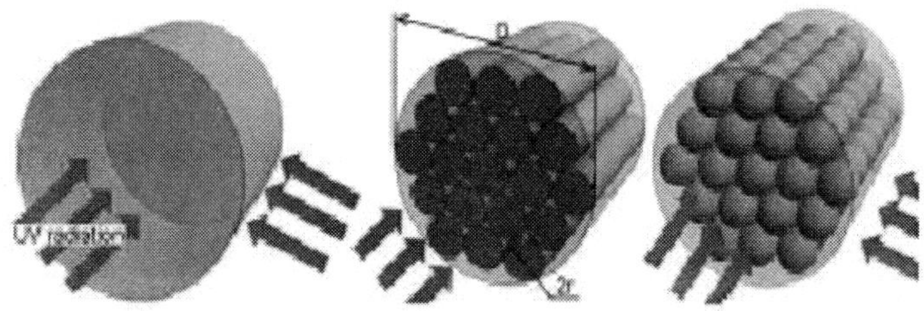

Figure 2. Methods of decontamination: (Lc_1) traditional decontamination; (Lc_2) decontamination using PCF in the hexagonal packed bundle; (Lc_3) metamaterial like photon crystal.

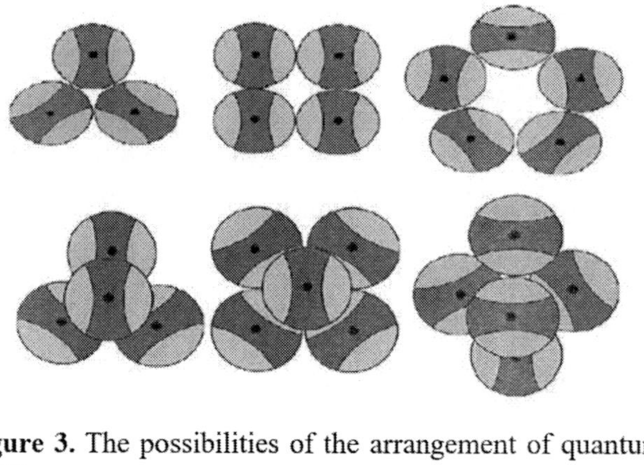

Figure 3. The possibilities of the arrangement of quantum cavities.

But it is not clear what happens with other free volume between the fiber, which is situated at larger distance in comparison with $\lambda / 2$. This volume may be involved in the contamination zone, if we continue to decrease the thickness of fiber. We may estimate the free volume between the fiber and the possibility to use all this volume with the decreasing of the fiber diameter. In the standard hexagonal packed bundle we may estimate the free volume between three fibers $v_f = r^2(\sqrt{3} - \pi/2)L \approx 0.18r^2L$. In this case the free volume in the big bundle doesn't depend on the fiber diameter and is equal to $V_f = \pi R^2 L(\sqrt{3} - \pi/2)$. The unused volume may be estimated $V_a = \pi RL(0.18R - \sqrt{N}\lambda)$. When this expression achieved zero value all volume between the fiber can be used for decontamination of fluids. This corresponds to the following expression of fiber radius $r \sim R/\sqrt{N} \sim \lambda/0.18$. The similar expression can be obtained for other type of packing of the fibers. Using the same method, we have estimated the decontamination surfaces of metamaterials $S_d = 4\pi r^2 N \sim \pi L^2 N^{1/3}$, (see Figure 2($Lc_1$, Lc_2, Lc_3)) like PC where L is the edge length of the cube, r is the radius of one micro-sphere, N is the number of micro-spheres of the metamaterial. The liquid will fill all space between micro-spheres. The decontamination volume can be expressed in following form $V_d \sim dS_d = 4\pi dR^2 N^{1/3}$. In this estimation we can mention that increasing of decontamination volume depends on the number of microspheres, described by the dependence $N^{1/3}$. Although, at first glance it appears that the decontamination volume is lower than in PCF case, this is only an illusion. Due to the fact that number of micro-spheres in metamaterial like PC is much larger than the number of fibers in PCF, the decontamination volume in the second case is much higher. Another priority of last metamaterial consists in the fact that this works in all directions symmetrically, in comparison with PCF. The free volume between balls in a photonic crystal can be presented in the form $v_{fr} = 8r^3(1 - \pi/6) \sim 0.48v_c$. The free volume between the large cube with dimension L have same proportion $V_{fr} = 0.48V$. In this case the difference between the $V_{fr} - V_d$ is proportional to $V[0.1/3/(2L)]$. When the $L < \pi N^{1/3}/0.98$, the increasing of surface becomes impossible and the classical aspects of evanescent zone is not acceptable. In this case, the volume between the balls may be regarded as a decontamination free volume. The wave is regarded as a quantum cavity for the standing wave (see Figure 3). The metamaterials, such as optical fibers or periodic photon structures open the novel possibilities to manipulate and kill viruses and bacteria in contaminated zones of liquids or organic tissue. For example, the good contact area between the implant and cells can be accomplished using such metamaterials on the surfaces. The guided UV radiation along the implant surface maintains the best medical assistance of contact surface against possible viruses or bacteria.

3.2. Experimental procedures and theory

The UV action against bacteria and viruses depends on the depth and volume of the evanescent zone of the periodical waveguide structures. In Figure 4, we represented such a periodical structure (fibers and spherical metamaterials), introduced into a cylinder through which the contaminated fluid flows. Taking into consideration the traditional conception of decontamination and possible decontamination volume in the evanescent zones of fibers or periodical bubble structures (see Figure 4), we may introduce the relative decontamination coefficient $\rho = V_d / V_c$. Considering that the lateral surface of the cylinder is larger than the surfaces of its two bases, we obtain the following expression for the relative decontamination volume $\rho \sim d\sqrt{N}/d_p$. The similar expression can be introduced and for metamaterials like PC if we filled the cylinder with periodical bubbles of SiO_2. In this case the relative decontamination coefficient is $\rho \sim dN^{1/3}/d_p$. Here the classical decontamination volume is considered the penetration of the radiation into the spherical elementary volume $4\pi R^2 d_p$ with the width $\Delta R \sim d_p$.

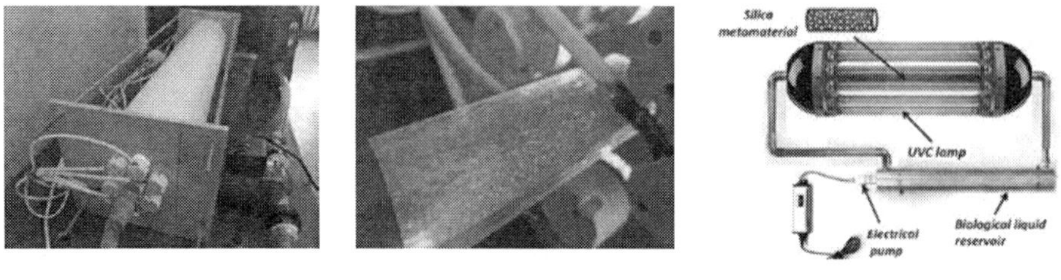

Lc_1 Lc_2 Lc_3

Figure 4. (Lc_1) Reactor for UV decontamination. The UV radiation is generated by 6 UVC lamps and it is reflected in the center of decontamination region, where it is placed a cylinder, filled up by quartz microspheres. (Lc_2) The cylinder filled up by quartz microspheres. (Lc_3) Schematic representation of reactor for UV decontamination.

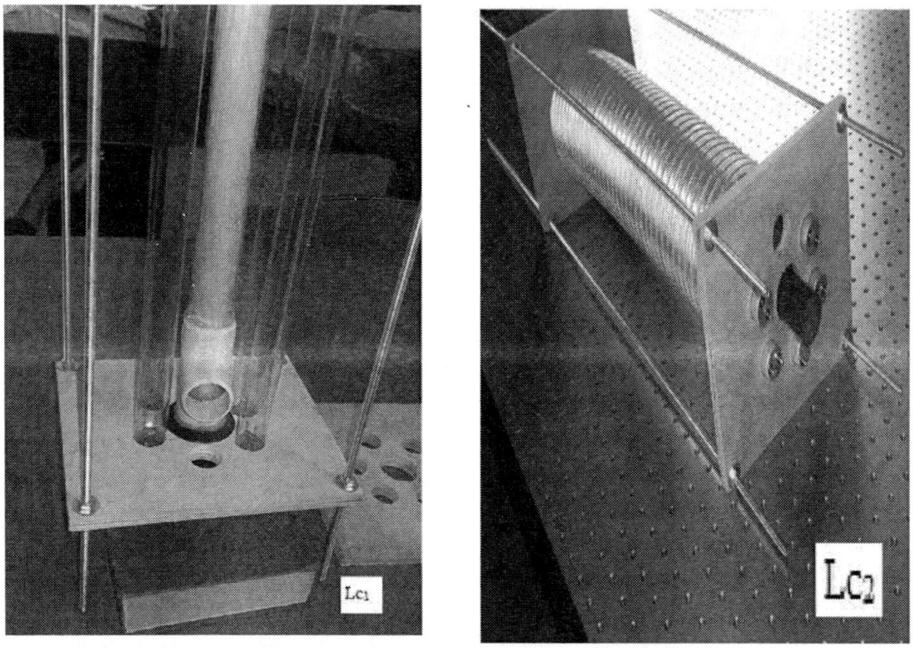

Figure 5. The preparation of mixed decontamination equipment, which use the UVC lamps and laser pulses. Lc_1. On the bottom part of the figure is applied the canal for laser beam represented by black circle. Lc_2. The final version of decontamination equipment in which laser pulses are sent through the black part of central core.

The combination of SiO_2 the PC with PCF improved the decontamination contact surface according to our preliminary investigations. The research of the efficient action of *UV* pulses on the chemical reactions, which take place in the microorganisms, is in the initial stage of our studies. Another effect which can appear in the process of decontamination is connected with the trapping of pathogen particles (viruses and bacteria) near the surface of fibers (or spheres). This effect is well known in the literature [5]. It consists in the attractive force acting on the particles with higher refractive indexes relative the refraction index of liquid index and appears due to the large gradient of EMF in the evanescent zone near the fiber Figure 5 (sphere Figure 4(Lc_2)).

The system of quartz fibers with periodical structures irradiated simultaneously with 5 UV-C lamp and nanosecond laser pulses effectively decontaminates the bacteria and fungi (yeast) present in translucent fluids. The process of development of this equipment which use both method of decontamination is presented in Figure 5. The last studies demonstrate that the UVC radiation becomes the effective method for inactivating bacteria, viruses, fungi and other micro-organisms. According to the literature [1-4] the special UVC radiation is an effective, environment-friendly and chemical-free method to dystrophy dangerous pathogens in any condition. On the other end, UVC doesn't pass the atmosphere, so it normally doesn't contribute to DNA damage, but it deserves to be mentioned that UVC lamps are used to kill bacteria and bedbugs. As a consequence, a local vibration energy of the modes may be coupled by an-harmonic non-linear term. For example, two vibration defects in cell replication and lead to cell death afterwards. The increasing of popularity of fiber system interaction with fabrics structures of clothing and the diversity of the optical schemes and methods of decontamination opens the opportunities for innovative research performing in this area. The scientific studies are stimulated also by a lot of incidence of short and long-term complications in the hazard situations, which took place decontamination of infected the fabrics (for example fabric clothing) used by people. One of them is when the fabric for cloth clothing adhesion with UVC is not in a good compatibility with decontamination sources (is not transparent or scatter the radiation so that the decontamination in the volume of material becomes impossible). Taking into consideration the metamaterials popularity, like PCF, it was decided to use these optical systems in their adhesion and penetration into a volume of fabrics killing the pathogens from fabrics and rooms inventory.

In our opinion, intrinsic topology of capsids may help to take into consideration more exactly the symmetry vibration modes of this virus structure, in order to estimate the possible anharmonic excitation of virus components by selectively distrusting them during the coherent Raman excitation.

Let us follow the excitation method of local vibration modes of biomolecules, like α or β tubulins for microtubules of dangerous bacteria [6] (see Figure 6) or capsids for viruses.

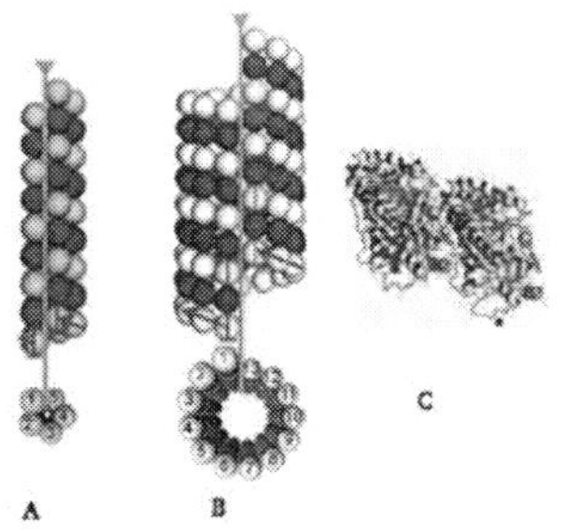

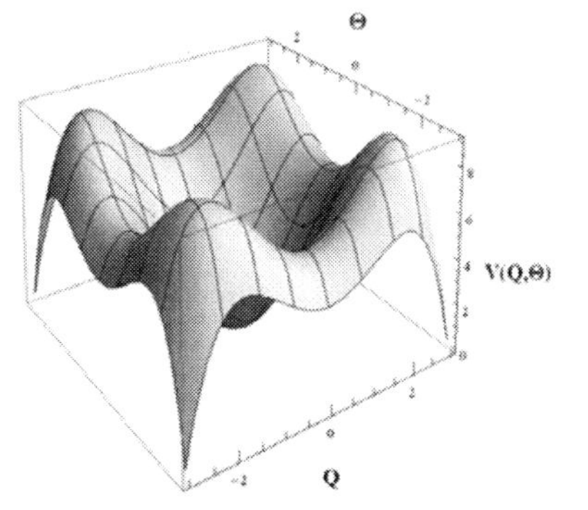

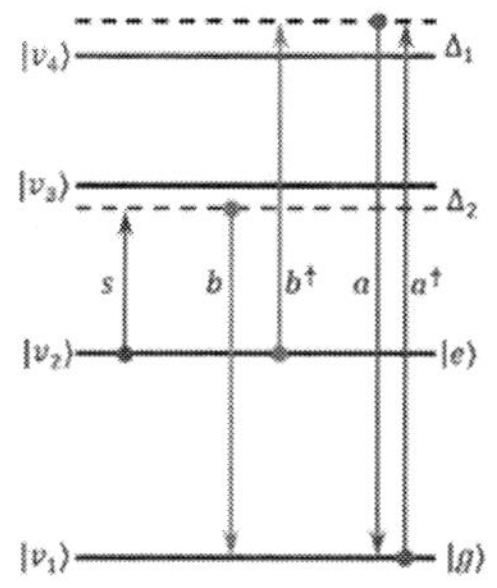

Lc_1 Lc_2 Lc_3

Figure 6. (Lc_1) Comparison of the architectures of a 5-proto-filament bacterial micro-tubule A represents the B and A tubs in dark - blue; B tub is in light-blue and B: a 13-proto-filament eukaryotic micro-tubule: α-tubulin in white β-tubulin in black. Seams and start-helices are indicated in green and red, respectively C form indicates the biomolecular structure of α and β tubulines. (Lc_2) The dependence of the potential energy of the non-linear oscillator on the two normalized modes $x = \sqrt{M/2\Omega Q}$, $y = \sqrt{M/2\Omega\Theta}$. ($Lc_3$) Energy scheme for nonlinear potential with Raman excitation.

Taking into consideration that a local vibration energy of the prin structures of pathogens Figure 6 may be described by anharmonic nonlinear term. For example two vibration modes Q and Θ can be represented as a symmetric function relative to the square value of these normal coordinates.

$$H_0 = \frac{M_q}{2}\left(\frac{dQ}{dt}\right)^2 + \frac{M_\theta}{2}\left(\frac{d\Theta}{dt}\right)^2 + \frac{M_q\Omega_q^2 Q^2}{2} + \frac{M_\theta\Omega_\theta^2\Theta^2}{2} - \kappa_q Q^4 - \kappa_\theta\Theta^4 - \kappa_{\theta,q}\Theta^2 Q^2. \quad (1)$$

Here M_i, Ω_j and κ_i are the effective mass, frequency and anharmonic parameter for the vibration mode i $(i \equiv q, \theta)$; $\kappa_{\theta,q}$ is nonlinear coupling of the normal modes of molecule oscillations. Anharmonic terms introduced in the Hamiltonian describe the possible destroying of this system by higher excitation. Let us consider two situations of the vibration of this molecular system represented by tubulins in protein packing in the microtubule of bacteria. We introduced collective nonlinear coupled modes like phonons in the condensed mater [7], described by above vibration model, for which two modes, Q and Θ become aperiodic according to the theory of catastrophe [8], for higher excitation of the system by short laser pulses [2, 9] with the pulse duration, $\tau_L < 1/\Omega_i$.

$$\begin{aligned}
&\frac{d^2}{dt^2}O(t) + \{\Omega_q^2 - 4\chi_q Q^2 - 2\chi_{q\theta}^2\Theta^2\}Q + 2\Gamma_q\frac{d}{dt}O(t) = F_q(t);\\
&\frac{d^2}{dt^2}\Theta(t) + \{\Omega_\theta^2 - 4\chi_\theta\Theta^2(t) - 2\chi_{\theta,q}O^2\}\Theta(t) + 2\Gamma_\theta\frac{d}{dt}\Theta(t) = F_\theta(t).
\end{aligned} \quad (2)$$

Here we have introduced the attenuation constant, Γ_i, of each coupling modes through nonlinear interaction, $\chi_{ij} = \kappa_{ij} / M_i$. The anharmonic term of each modes is described by the nonlinear constant $\chi_i = k_i / M_i$. The an-harmonic potential is described by the two dimensional localization potential. The external, short laser pulses interact with the molecular dipole of virus components, so that in the simple representation, this has the traditional form $H_I = -(\mathbf{P}(E), \mathbf{E}(t,z))$. Following the traditional representation [10, 11] the laser field induce the polarization of biomolecule, the components of which in the similar representation has tensor character, which depends on the symmetry of excited molecules (virus's or bacteria's components) $P_j(E) = \alpha_{jl}E_l$, where due to the tensor character of proposed two modes of oscillations, we decompose the recognizability $\alpha_{jl}(Q,\Theta)$ in Taylor serial decomposition relative the normal components Q and Θ.

$$\alpha_{jl} = \alpha_{jl}^0 + \frac{\partial\alpha_{jl}}{\partial Q}Q + \frac{\partial\alpha_{jl}}{\partial\Theta}\Theta + \varepsilon(Q^2;\Theta^2;Q\Theta) \quad (3)$$

Introducing the expression (3) in interaction Hamiltonian we observe that the interaction with the local vibration modes can be described by the function $U_I = \alpha'_{jlk}E_jE_lQ_k$, where $Q_1 = Q$; and $Q_2 = \Theta$ and the tensor $\alpha'_{jlk} = \frac{\partial\alpha_{jl}}{\partial Q_k}$ mast be maximally symmetrical according to the symmetry of the virus or bacteria biomolecules. Considering that laser pulse has same polarization, we represent it through the time dependent Gaussian function $E_i = E_0\exp[-(t/\sqrt{2}\tau_L)^2]\cos(\omega t)$. We substitute the generalized driving forces $F_j = -\partial U_I / \partial Q_j$ in the system of equation (2) and represent the solution in the linear approximation $Q_j = Q_{0j}\exp[-\Gamma_j t]\sin(\Omega_j t)$, where the amplitude has the following dependence on the intensity, E_0^2, pulse duration, τ_L, and local osculation frequency Ω_j of the normal mode j iprested below.

$$Q_{0j} = \frac{\sqrt{\pi}\tau_L E_0^2 \alpha_j^{'}}{2\Omega_j} \exp[-(\Omega_j \tau_L)^2 / 4]. \quad (4)$$

Observing the same exponential dependence on the $\Omega_j \tau_L$, as in the Refs. [2, 9], we address the problem so as to excite the system of coupling oscillators , so that the "nonlinear frequency" Formula 5.

$$\sqrt{\Omega_q^2 - 4\chi_q \langle Q^2 \rangle - 2\chi_{q\theta} \langle \Theta^2 \rangle} \quad (5)$$

Achieved zero value, which corresponds to destroying of the local vibration mode after the finite number of short laser pulses. For doing this, let us consider the set of consecutive pulses " n " which are generated in the time interval $T < \Gamma_j^{-1}$, so that the cumulated energy by the local oscillator after the predecessor pulses may be used in the next pulse excitations. For example, after first pulse the amplitudes of Q and Θ models are described by the expression above expression. We obtain the following mean values of Q^2 and Θ^2, Formula 6.

$$\langle Q_j^2 \rangle = \left(\pi\tau_L^2 E_0^4 \alpha_j^{'2} \Big/ 22^2 \Omega_j^2 \right) \exp[-(\tilde{\Omega}_q \tau_L)^2 / 2] \quad (6)$$

Where $\tilde{\Omega}_q = \sqrt{\Omega_q^2 - 4\chi_q \langle Q^2 \rangle - 2\chi_{q\theta} \langle \Theta^2 \rangle}$. This procedure of excitation of non-linear oscillator may continue till one of the amplitude of oscillation achieved the maximal separation line of nonlinear potential function $U_0(Q, \Theta)$ as represented in Figure 5 (Lc_2). It corresponds to the situation for which the one of frequency achieves the zero value after "n" short pulses $\Omega_q^2 - 4\chi_q \langle Q_n^2 \rangle - 2\chi_{q\theta} \langle \Theta_n^2 \rangle = 0$; or $\Omega_\theta^2 - 4\chi_q \langle \Theta_n^2 \rangle - 2\chi_{\theta q} \langle Q_n^2 \rangle = 0$.

4. Conclusions

We propose new methods of decontamination with applications in medicine and public health. These methods contribute to an important increasing of the rate and volume of decontamination by increasing the contact zone between the light radiation and the contaminated environment (gases, liquids and solid surfaces). Regarding implants, a fine metamaterial coating on the implant surfaces not only contributes to the decontamination of the implant-tissue contact, but can also be beneficial for improving both of mechanical contact between the optic network of the metamaterial - patient's organic tissue and the processes taking place in immediate proximity to the irradiation area. A special attention will be paid to the propagation of short light pulses through such metamaterials. They will provide a selective interaction with the pathogen biomolecules having a negligible destructive action on human tissue components. The paper contains both an application and a fundamental aspect. Many of the effects of decontamination and capture of viruses and bacteria along metamaterial surfaces are seen, for example, in the transport of nutrients to the surface of tubules. So the development of metamaterial contact techniques with organic tissue opens up new possibilities for the prevention, monitoring and treatment of pathogenic diseases in different environments, including implant ology. Here we can refer to different types of implants, such as implants for the heart, wrists implants, dental implants. Another field of application of the proposed decontamination methods directly addresses the issues of selective decontamination of the blood components. An important aspect of the fundamental part of the project is that we propose the development of the vibrational model of microorganism inactivation for a deeper understanding of the non-thermal selective action mechanisms on microorganisms taking into account both the duration, frequency and intensity of radiation pulses, as well as the basic characteristics of pathogens (dimensions, form, including capsid viruses, encapsulated or non-encapsulated viruses, internal symmetry of protein structures, types of bonds in biomolecules, single / double stranded nucleic acids, etc.).

5. References

[1] Begley R F, Harvey A B and Byer R L 1974 *Appl. Phys. Lett.* **25** 387-390
[2] Yan J Y-X, Gamble Jr E B and Nelson K A 1985 Chem. Phys. **83**(11) 5391-99
[3] Blombergen N 1998 *Nonlinear Optics (4th Edition)* Wold Scientific
[4] Mannige R V PLoS ONE **5(3)** e9423
[5] Kress H 2008 *Cuvillier Verlag* **140**
[6] Pilhofer M, Ladinsky M S, McDowall A W, Petroni G and Jensen G J 2011 *PLoS Biol* **9** e1001213
[7] Zhu T and Ertekin E 2014 Phys. Rev. B **90** 195209
[8] Zhu T and Ertekin E 2015 Phys. Rev. B **91** 205429
[9] Gilmore R 1981 *Wiley-Inter-science Publication* NY **1-2**
[10] Tsen S W D et all. 2012 *Journal of Biomedical Science* **19** 62
[11] Loudon R 2000 *The Quantum Theory of Light* Oxford University Press 448
[12] Bazgan S, Ristoscu C, Negut I, Hapenciuc C, Turcan M, Ciobanu N, Mihailescu I N and Enaki N 2015 *Rom. Rep. Phys.* **67**(4) 1602–07
[13] Enaki N, Bazgan S, Ciobanu N, Turcan M, Paslari T, Ristoscu C, Vaseashta A and Mihailescu I N 2017 *Journal of Applied Surface Science* **417** 40-47
[14] Enaki N, Profir A, Bizgan S, Paslari T, Ristoscu C, Mihailescu C, Badiceanu M and Mihailescu I N 2018 *Metamaterials for Antimicrobial Biofilm Applications: Photonic Crystals of Microspheres and Optical Fibers for Decontamination of Liquids and Gases*, ch. 13. 27, Handbook of Antimicrobial Coatings, 1st Edition, Authors: Atul Tiwari, eBook ISBN: 9780128119839, Elsevier 596
[15] Nahar Q, Fleibner F, Shuster J, Morawitz M, Halfpap C, Stefan M, Langbein U, Southam G and Mittler S 2014 *J. Biophotonics* **7(7)** 542–551
[16] Dai T, Vrahas M S, Murray C K and Hamblin M R 2012 *Expert Rev Anti Infect Ther* **10** 185–195

Acknowledgements
This work was supported by the projects: NATO G 4890, No. 15.817.02.07F, No. 18.80012.50.33A STCU 6140 and UEFISCDI 43 NATO/2017.

IOP Conf. Series: Materials Science and Engineering **374** (2018) 012012 doi:10.1088/1757-899X/374/1/012012

Microstructure, Friction and Wear of Aluminum Matrix Composites

R M Florea
"Gheorghe Asachi" Technical University of Iasi, Department of Materials Science and Engineering, Prof. Dr. D-tru Mangeron Street, 700050, Iasi, Romania

E-mail: raluca.m.florea@gmail.com

Abstract. MMCs are made by dispersing a reinforcing material into a metal matrix. They are prepared by casting, although several technical challenges exist with casting technology. Achieving a homogeneous distribution of reinforcement within the matrix is one such challenge, and this affects directly on the properties and quality of composite. The aluminum alloy composite materials consist of high strength, high stiffness, more thermal stability, more corrosion and wear resistance, and more fatigue life. Aluminum alloy materials found to be the best alternative with its unique capacity of designing the materials to give required properties. In this work a composite is developed by adding silicon carbide in Aluminum metal matrix by mass ratio 5%, 10% and 15%. Mechanical tests such as hardness test and microstructure test are conducted.

1. Introduction

The "composite" term refers to a material system which is composed of a discontinuous constituent (reinforcement) distributed in a continuous phase (matrix), and which has distinguishing characteristics like the properties of its constituents, the geometry and architecture of the constituents, and the properties of the boundaries (interfaces) between different constituents [1-3]. This paper has the subject of metal matrix composites and more specifically on the aluminium matrix composites (AMCs). In AMCs one of the constituent is aluminium alloy, the matrix phase. The other constituent is embedded in this aluminium alloy matrix and serves as reinforcement, which is ceramic, SiC_p. Properties of AMCs can be modified by varying the nature of constituents and their volume fraction.

The major advantages of AMCs compared to unreinforced materials are as follows [1, 4, 5]:

- Improved damping capabilities;
- Reduced density (weight);
- Greater strength;
- Thermal/heat management;
- Improved stiffness;
- Improved high temperature properties;
- Controlled thermal expansion coefficient;
- Control of mass (especially in reciprocating applications);
- Enhanced and tailored electrical performance;
- Improved abrasion and wear resistance.

The aim of this paper was the obtaining and the investigation of a metal matrix composite with three different percent of reinforcement.

2. Processing of AMCs

2.1. Liquid state processing

This involves incorporation of ceramic particle into liquid aluminium alloy. The crucial thing is to create good wetting between the particle reinforcement and the liquid aluminium alloy matrix. The simplest and most commercially used technique is known as Vortex technique or stir-casting technique.

The Vortex technique involves the incorporation of pre-heated ceramic particles into the vortex of molten alloy created by the rotating impeller. Lloyd [6] reports that vortex-mixing technique for the preparation of an aluminium matrix composites with ceramic particle dispersed was originally developed by Surappa & Rohatgi [7] at the Indian Institute of Science.

2.2. Preparation of Aluminium-Silicon Carbide Composite

Casting is one of the most ancient processes of manufacturing metallic materials. The metal matrix composite used in the present paper was prepared by the Vortex method. For the preparation of the Al/SiC_p, particles mass of 5%, 10% and 15% are taken. Figure 1 illustrates the schematic set up for Vortex technique. A three-phase electrical resistance furnace with temperature controlling device is used for melting. For each melting 500 g of alloy is used. The pre-heated molten metal is degassed at a temperature of 780 °C. SiC particles, preheated at 500 °C, are then added to the molten metal and mixed continuously by a mechanical stirrer at 720 °C.

The mixing time is between 5 and 8 minutes. During the mechanical agitation, Borax powder in small quantities was added to increase the wettability of SiC particles.

The metallic melt, with the reinforced particles, was poured into the cylindrical metallic mould. The pouring temperature was maintained at 680 °C. The melt was allowed to solidify in the mould in normal conditions.

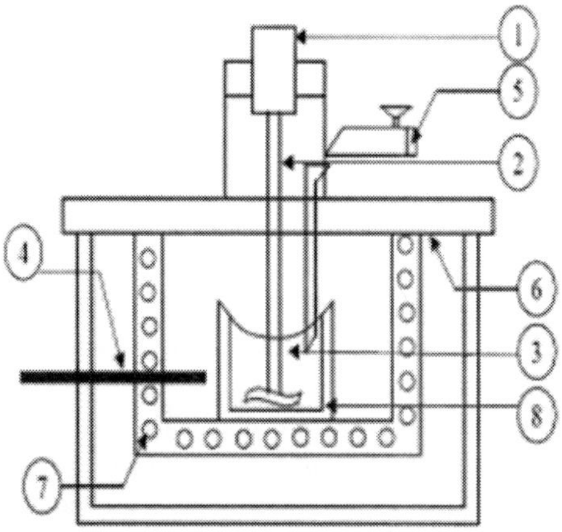

Figure 1. Schematic set up for Vortex technique [8]: 1. Motor; 2. Shaft; 3. Molten aluminium; 4. Thermocouple; 5. Particle injection chamber; 6. Insulation hard board; 7. Furnace; 8. Graphite crucible.

3. Investigation of Al/SiC_p

3.1. Microstructure Test

Metallographic samples were sectioned from the obtained composites. A 0.5 % HF solution was used to etch the samples. To observe the difference of SiC_p distribution in the aluminium matrix, microstructure of samples were taken on Optical type Metallurgical Microscope (Make: Nikon, Range-X50 to X1500).

All three samples were obtained by using Vortex technique by taking varying weight fractions of SiC_p. The various weight fractions were (5%, 10% and 15%) of SiC particles.

Figures 2a, 2b and 2c shows micrograph's of samples containing 5%, 10% and 15% SiC by weight. It is clearly showed the resulting homogeneous distribution of particles in the samples.

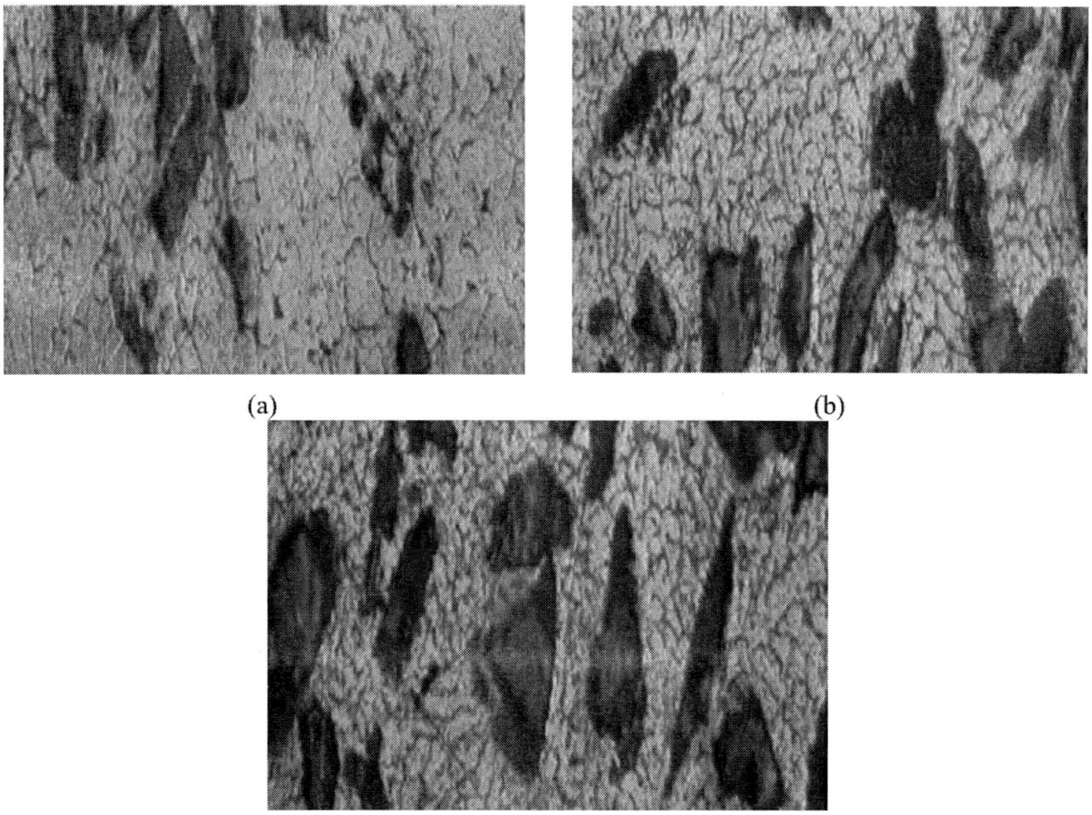

Figure 2. Microstructure of Al/SiC_p samples for different % of SiC: (a) 5%; (b) 10% and (c) 15%.

3.2. Hardness Test

The hardness test were conducted on Automatic Optical Brinell Hardness Tester Model OPFB.

Bulk hardness measurements were carried out on the aluminium alloy and composite samples using the standard Brinel hardness test. The Brinel hardness measurements were carried out in order to investigate the influence of SiC_p weight fractions on the composite hardness. The applied load was 250N, and the indenter was a steel ball of 5 mm in diameter. Hardness values obtained are given in Table 1.

Table 1. Hardness values of obtained sample with different SiC contents.

Sample No.	*% SiC_p by weight*	*Hardness in BH*
1.	0	32
2.	5	41
3.	10	47
4.	15	60

3.3. Friction and wear

The investigations were performed with a composite block sliding on cylinder surface made of hardened ground steel. The parameters of the tests are presented in Figure 3.

In some cases, the composites proved applicable, as sliding bearings infiltrated by oil (porous sliding bearing) or as a high-speed sliding bearing lubricated by oil. Are also useful as a material for pistons and cylinders of small engines running at very high rotation speed [9].

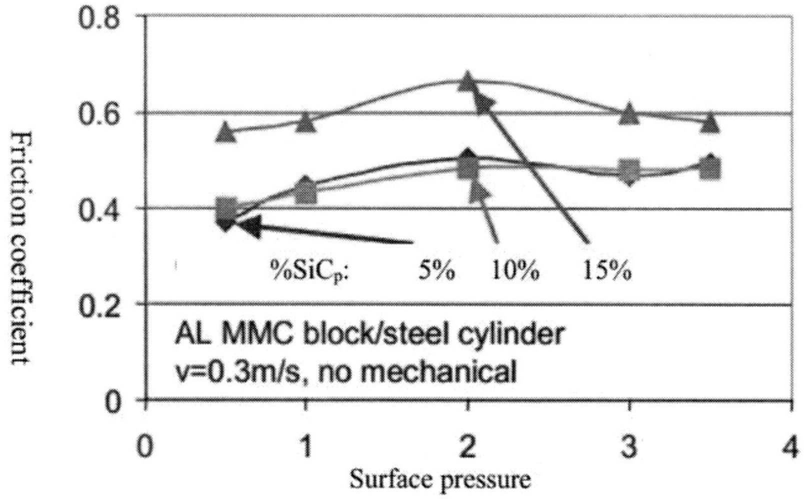

Figure 3. Coefficient of friction of different aluminium composites in function of SiC_p content.

Also, it was investigated the wear behavior of the pure Al alloy and of the metal matrix composites reinforced with SiC_p using a ball-on-disc test at room temperature under low loads and dry conditions. It was found that SiC particles increased the macro hardness (Figure 4).

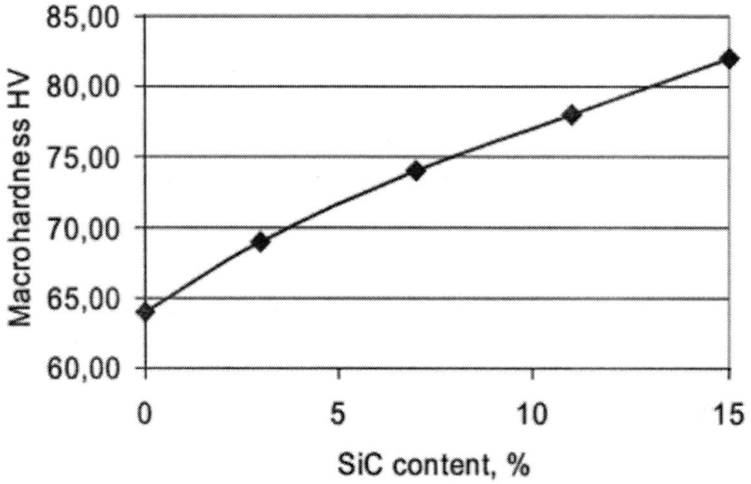

Figure 4. Variation of macro hardness of Al/SiC_p function of SiC content.

Metallographic investigations have revealed that hardened layers were developed on the sliding surfaces, in which fragmented Si phase was redistributed and aligned parallel to the sliding direction (Figure 5 and 6). It can be seen that the thickness of the hardened layer formed on the obtained composites was reduced by the incorporated SiC_p. The fragmentation of Si phase was found to be lesser magnitude in the Al/SiC_p leading to lower wear rate.

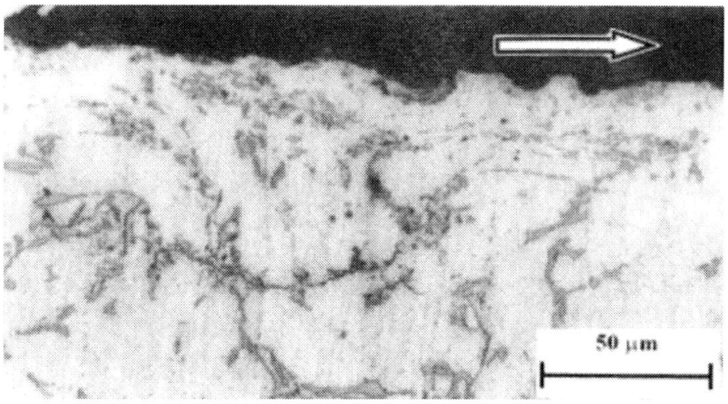

Figure 5. Subsurface layer of Al matrix after wear test with fragmented Si.

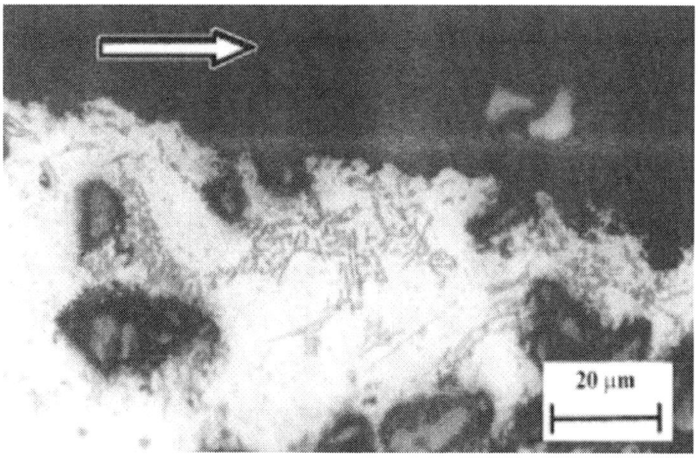

Figure 6. Subsurface layer of Al/SiC_p containing 15% SiC particles after wear test.

4. Conclusions

Silicon carbide particle reinforced aluminium matrix composite (Al/SiC_p) were prepared by Vortex technique casting with different particle weight (5%, 10%, and 15%) and the following conclusions can be drawn:

Metallographic samples show a homogeneous distribution of particles in the metal matrix;

Hardness of $Al-SiC_p$ is much better than the aluminum metal. In case of increased silicon carbide content, the hardness, and material toughness are enhanced and highest value is obtained at 15% SiC content;

SiC_p increase the macro hardness;

Obtained composites proved applicable, as sliding bearings infiltrated by oil (porous sliding bearing) or as a high-speed sliding bearing lubricated by oil. Are also useful as a material for pistons and cylinders of small engines running at very high rotation speed.

5. References

[1] Surappa M K 2003 *Sadhana* **28** 319
[2] Shulin L, Shusen W, Wei D, Chong L, Ping A 2012 *Journal of Materials Processing Technology* **212** 1281
[3] Behera R, Kayal S, Mohanta N R, Sutradhar G 2013 *Transactions of 61st Indian Foundry Congress*
[4] Jeevan V, Rao C S P, Sonraj N 2013 Development of New Aluminium Matrix Composite Reinforced with Silicon Carbide, ISSN 2321- 5747 1 34
[5] Bălţătescu O, Florea R M, Roman C, Rusu I, Carcea I 2013 *JOAM* **15** 7- 8 823
[6] Lloyd D J 1999 *Int. Mater. Rev.* **39** 23
[7] Surappa M K, Rohatgi P K 1981 *J. Mater. Sci.* **16** 983
[8] Pawar P B, Abhay A 2014 *Procedia Materials Science* **6** 1150
[9] Kozma M 2003 *The Annals of University "Dunărea de Jos" of Galaţi*, Fascicle VIII, Tribology ISSN 1221-4590

Evaluation of Sealing Process on the Surface Properties of Nanoporous Aluminum Oxide Layers Electrochemically Growth on 1050 Aluminum Alloy Surface

V M Dumitrascu[1]*, L Benea[1] and N L Simionescu[1]
[1] Competences Center: Interfaces-Tribocorrosion-Electrochemical Systems (CC-ITES), Faculty of Engineering, Dunarea de Jos University of Galati, 47 Domnească Street, RO-800008, Galați Romania.

E-mail: Lidia.Benea@ugal.ro

Abstract. Thanks to its low density, aluminum and its alloys are extensively used in the transport and construction industries. The improving of aluminum and its alloys corrosion resistance is an extensively studied topic. Different methods are applied to improve the corrosion resistance of aluminum and its alloy: painting, waxing, zinc-chromate priming, conversion coatings and also anodic oxidation. The anodic oxidation is the most used method to fabricate nanoporous aluminum oxide layers, with controlled thickness and pores diameters, on aluminum and its alloys surfaces. The aim of this work is to evaluate the influence of the sealing process of the nanoporous aluminum oxide layers obtained by anodic oxidation on 1050 aluminum alloy surface. Scanning electron microscopy (SEM) was used to observe the morphological structure of untreated and modified surfaces. Corrosion behavior of oxidized 1050 aluminum alloy, sealed oxide layer and untreated 1050 aluminum alloy were investigated using electrochemical methods such as open circuit potential (OCP), electrochemical impedance spectroscopy (EIS) and potentiodynamic polarization (PD). The experimental results show that the anodic oxidation process improve the corrosion resistance of 1050 aluminum substrate and the sealing process improve even more the anticorrosive properties.

1. Introduction

From economical and technical point of view, light weight materials, especially aluminum and its alloys are becoming increasingly important for automotive, aerospace and construction industries [1, 2].

The corrosion performances of aluminum and its alloys are an important industrial problem. Aluminum naturally develop a thin aluminum oxide layer at material-environment interface, after the exposure to air, but the formed native aluminum oxide layer is not uniform and do not offer enough corrosion protection against aggressive factors [1, 2].

In order to fabricate corrosion resistant ceramic coatings on aluminum and its alloys, a wide range of techniques have been employed: micro-arc oxidation, physical vapor deposition, chemical vapor deposition, heat treatments, anodic oxidation, etc. [3-8]. Anodic oxidation process is used to form on aluminum and its alloys surfaces aluminum oxide layers with complex structures and improved anticorrosive and wear resistance performances [9-11].

The anodic oxide layers structures depend by the applied parameters during the anodizing processes (current, potential, electrolyte stirring rate), the electrolyte type and also by the finishing methods used for the pretreatment of aluminum substrate [9, 11].

Depending by the electrolyte type, the oxide layer could present a barrier-type structure or a duplex structure composed by an outer porous-type layer and an inner barrier-type layer. A sealing process is required to improve the corrosion resistance of aluminum oxide layers. The anodic porous aluminum oxide layers can be sealed by boiling in hot water or using different cold sealing solutions [12-15].

S. Wang et al [13] have evaluated different sealing methods on corrosion resistance of anodized Al2024. F. Snogan et al [14] evaluate the corrosion properties of anodized aluminum sealed with 5.5 g/L Ni and Co acetate solution.

A comparison between different sealing methods was done by Y. Zuo et al [15] for different aluminum alloys anodized in 180 g/L H_2SO_4 solution. V. M. Dumitrascu and his co-workers [16] evaluate the influence of hot sealing process in 6% sodium molibdate for anodized Al1050 on corrosion resistance in 3.5% NaCl solution.

Electrochemical assays and especially electrochemical impedance spectroscopy are powerful methods to evaluate the corrosion resistance of nanoporous aluminum oxide layers obtained by anodic oxidation and exposed to aggressive environments [11, 15-17].

The electrochemical impedance spectroscopy data can be fitted to an appropriate equivalent circuit and allows the polarization resistance determination of porous aluminum oxide layers.

J.-P. Dasquet et al [17] evaluate the corrosion resistance of sealed aluminum oxide layers by EIS during the immersion in 0.2 M K_2SO_4 solution.

V.R. Capellosi and his co-workers [18] establish the corrosion resistance of sealed anodized Al2024 by fitting the EIS specters and determine the values of polarization resistances.

Therefore, the aims of this work are to evaluate and compare the corrosion resistance of untreated Al1050, anodized Al1050 and sealed in $K_2Cr_2O_7$ anodized Al1050 immersed in 3.5% NaCl solution using electrochemical assays. SEM and EDX analysis were used to investigate the structure of unsealed or sealed anodic oxide layers.

2. Experimental procedures

As received 2 mm thick Al1050 alloy sheet was cut into 35 x 30 mm size coupons. Prior to the anodizing processes, the Al1050 samples were mechanical wet-polished with 1000 and 1500 grit SiC emery paper, rinsed with distilled water, ultrasonically cleaned in alcohol and distilled water and finally dried under hot air in oven. The anodic oxidation processes were carried out in a standard two electrodes electrochemical cell in a solution of 1 M H_2SO_4 in which was added 1 g/L $Al_2(SO_4)_3$ x 18 H_2O at 300 rpm. The anodizing procedures were conducted for 15 minutes at an applied potential of 17 V at ambient conditions.

The anodized Al1050 samples were sealed by immersing in 6 g/L $K_2Cr_2O_7$ at 95°C for 60 minutes, cleaned with distilled water and dried under hot air in oven for 60 minutes.

The structure of anodic oxide layers before and after sealing process was evaluated using the scanning electron microscope coupled to X-ray energy dispersive analyzer FEI QUANTA 200.

The corrosion tests were done in 3.5% NaCl solution that simulates the aggressive marine environment. The standard electrochemical cell used in the corrosion assays was composed by a Pt-Rh grid as counter electrode, an Ag/AgCl electrode (saturated KCl solution, E=+199 mV vs. NHE) as reference electrode and as prepared Al1050 samples as working electrode. The electrochemical cell was connected to a potentiostat/galvanostat PGZ100. The experimental data were recorded with VoltaMaster 4 software.

The open circuit potential (OCP) were measured for 1 hour after 17 hours of immersion, in order to attain a steady state. The electrochemical impedance spectroscopy (EIS) tests were conducted over a frequency range of 10^5 down to 10^0 Hz with 10 points per decade, using 10 mV sin wave. The EIS specters were fitted using Zview 3.4f software and the quality of obtained results were evaluated using the chi-square (chi^2) values, with values bellow 10^{-3}. The potentiodynamic polarization (PD) curves were recorded from -1.45 vs. Ag/AgCl to -0.4 V vs. Ag/AgCl with a scan rate of 1 mV/s. Before the recording the potentiodynamic polarization curves, the analyzed Al1050 samples were cathodically treated at -2.1 V vs. Ag/AgCl for 3 minutes in 3.5% NaCl solution.

3. Results and discussions

3.1. Surface morphology and chemical composition

The SEM micrographs were recorded on Al1050 surfaces before and after the anodizing process and also, after the sealing process carried out in 6% $K_2Cr_2O_7$ at 95°C for 60 minutes.

The untreated Al1050 (Figure 1 (a)) show a rough surface, with numerous heights and valleys that present different sizes and forms. On untreated Al1050 surface can be identified nanopits with various shapes and diameters that will act further as location for nanopores developments.

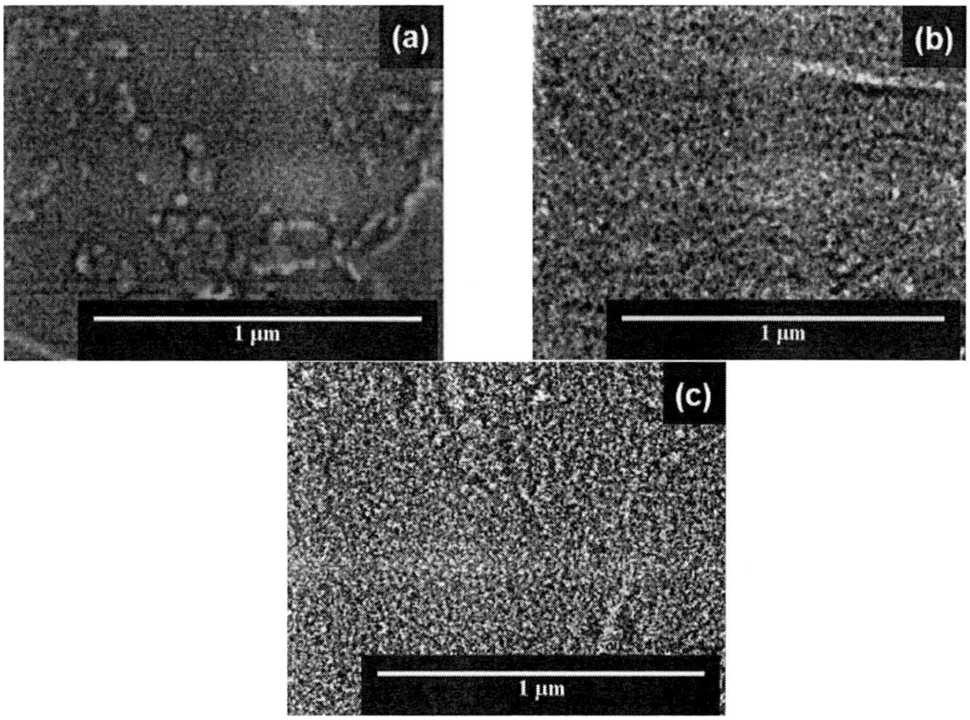

Figure 1. Top-view SEM image of (a) Untreated Al1050, (b) Anodized Al1050 and (c) Sealed anodized Al1050.

The recorded SEM image exposed in Figure 1 (b) reveal that the Al1050 surface morphology was changed after the anodic oxidation process. The heights were dissolved during the anodic oxidation process and an increased numbers of nanopores were formed. Due to the short anodizing process duration, the nanopores present various shapes and diameters and their distribution is not ordered.

After the sealing process, the SEM image (Figure 1 (c)) reveal that the morphology of anodized nanoporous aluminum oxide layer was changed and the nanopores present decreased diameters.

The chemical compositions of all 3 analyzed surfaces presented in figure 1 are exposed in Table 1. For the untreated Al1050, the chemical elements identified were: aluminum (92.19 wt%), oxygen (1.94 wt%) and carbon (5.97 wt%). The presence of carbon is caused by the alcohol washing of Al1050 before the morphological and compositional analyzes.

The anodic nanoporous aluminum oxide layer present a chemical structure composed by aluminum (43.98 wt%), oxygen (48.75 wt%), carbon (2.71 wt%) and sulphur (4.56 wt%). During the anodic oxidation process, the SO_4^{2-} anions were incorporated in the formed aluminum oxide layer.

After the sealing of anodized Al1050, the EDX analyzes revel that the chemical composition of aluminum oxide layers was changed. In addition to aluminum (42.57 wt%), oxygen (47.14 wt%), carbon (5.49 wt%) and sulphur (4.17 wt%) can be identified chromium (0.93 wt%). During the sealing process in 6% $K_2Cr_2O_7$ at 95°C, the anodic aluminum oxide was hydrated and the Cr^- ions were absorbed into the nanopores.

Table 1. Chemical composition of untreated Al1050, anodized Al1050 and sealed anodized Al1050.

Type of surface	Element				
	Al [wt%]	O [wt%]	C [wt%]	S [wt%]	Cr [wt%]
Untreated Al1050	92.19	1.94	5.87	-	-
Anodized Al1050	43.98	48.75	2.71	4.56	-
Anodized and sealed Al1050	42.57	47.14	5.49	4.17	0.93

3.2. Electrochemical assays

3.2.1. Open circuit potential (OCP). The potential-time measurements for untreated Al1050, anodized Al1050 and sealed anodized Al1050 in 3.5% NaCl solution are shown in Figure 2. The evolution of OCP in time was monitored in order to observe the changes that occur at the electrode surface (the electrode oxidation tendency) after the immersion in the corrosive electrolyte.

From figure 2 ca be observed that after 17 hours of exposure all the samples reach a steady state. The untreated Al1050 present slightly fluctuation around -700 mV vs. Ag/AgCl due to the dissolution of native aluminum oxide layer under the aggressive action of chloride ions from the electrolyte.

For the anodized Al1050 and sealed anodized Al1050, after 17 hours of immersion, the open circuit potential values remain almost unchanged being attained equilibrium states around -770 mV vs. Ag/AgCl and -795 mV vs. Ag/AgCl, respectively.

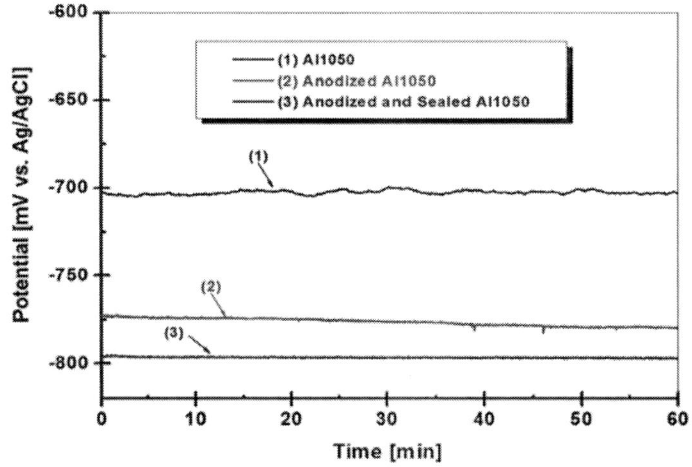

Figure 2. The evolution of open circuit potential recorded for (1) untreated Al1050, (2) anodized Al1050 and (3) sealed anodized Al1050 immersed in 3.5% NaCl.

The decreased values of steady state for anodized and sealed anodized Al1050 with immersion time could be related to the hydration of unsealed or sealed anodic aluminum oxide layers and the increasing of cathodically activity related to the partial dissolution of anodic oxide layers [19]. The stable values attained at the long exposure time might be attributed to the equilibrium between anodic and cathodic reactions that take place at electrode interface [18].

3.2.2. Potentiodynamic polarization (PD). The potentiodynamic polarization resistances curves recorded for untreated Al1050, anodized Al1050 and sealed anodized Al1050 in a linear representation are exposed in Figure 3 (a) and in a semi-logarithmic representation are exposed in Figure 3 (b). The recorded diagrams for all three analyzed surfaces were compared in order to evaluate the passive domains and the corrosion current of analyzed samples.

Analyzing the PD diagrams presented in Figure 3 (a) can be observed three potential domains: cathodic domain, passive domain and transpassive domain. From all three potential domains, the passive domain is related with the corrosion resistance of tested samples: the higher passive domain indicate the improved anticorrosive properties of the sample.

In Figure 3 (a) are displayed the PD diagrams recorded for the all 3 analyzed surfaces. The untreated Al1050 present the lowest passive domain, ranged between -1.25 V vs. Ag/AgCl and -0.68 V vs. Ag/AgCl. In the transpassive domain, the current density values increase exponentially, indicating the dissolution of aluminum substrate under the aggressive action of chloride ions from the electrolyte.

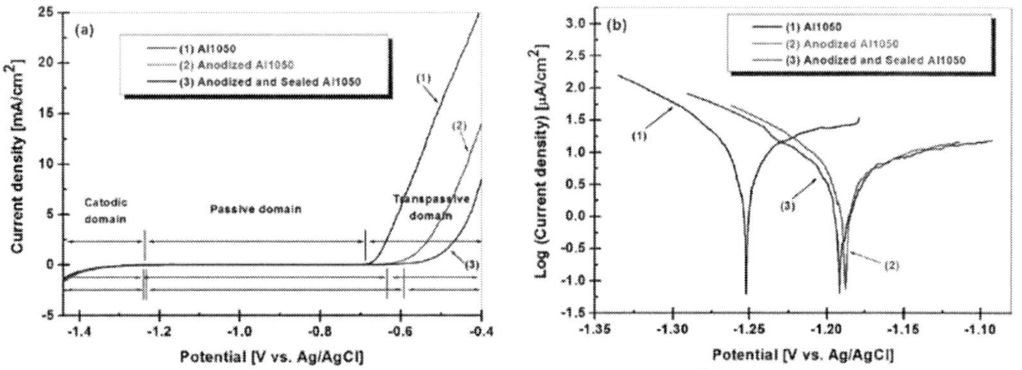

(a) (b)

Figure 3. (a) Linear representation and (b) Semi-logarithmic representation of potentiodynamic polarization diagrams recorded for (1) untreated Al1050, (2) anodized Al1050 and (3) sealed anodized Al1050 immersed in 3.5% NaCl.

For the unsealed or sealed anodized Al1050 samples the recorded passive domains are larger, being recorded between -1.18 V vs. Ag/AgCl and -0.63 V vs. Ag/AgCl and between -1.19 V vs. Ag/AgCl and -0.58 V vs. Ag/AgCl. Also, the current density recorded in the transpassive domains are lower in comparison with the values of current density recorded for untreated Al1050 sample.

The increasing of passive domains reveal that the anodizing and sealing processes improve the corrosion resistance of Al1050 samples exposed in 3.5% NaCl solution.From Figure 3 (b) quantitative information about the corrosion potential ($E_{(i=0)}$) and corrosion current density (i_{corr}) can be determinated using the Stern-Geary equation [20] and are summarized in Table 2.

$$i_{corr} = \frac{B}{R_p} \quad (1)$$

where i_{corr} is the corrosion current density, R_p is the polarization resistance and B is a constant, calculated using the equation:

$$B = \frac{b_a \, |b_c|}{2.303(b_a + b_c)} \quad (2)$$

where b_a and b_c are the Tafel slopes for anodic and cathodic reactions, respectively.

From Table 2 can be observed that the corrosion potential show an increasing tendency from untreated Al1050 to anodized Al1050 followed by a slightly decrease from anodized Al1050 to sealed anodized Al1050. Also, the anodizing and sealing processes decrease the corrosion current densities in comparison with the untreated Al1050.

Table 2. The corrosion potential and the corrosion current determinated for the analyzed samples immersed in 3.5% NaCl solution.

Type of surface	$E_{(i=0)}$ [mV vs. Ag/AgCl]	i_{corr} [$\mu A/cm^2$]
Untreated Al1050	-1251.9	16.1862
Anodized Al1050	-1186.2	6.2711
Anodized and Sealed Al1050	-1191.5	3.9103

The results confirm that the anodic oxidation process improves the anticorrosive performances of Al1050 and the sealing process improves even more the corrosion resistance of aluminum substrate.
3.2.3. Electrochemical impedance spectroscopy (EIS). Figure 4 shows the Nyquist representation of electrochemical impedance spectroscopy specters and their fitted results for untreated Al1050, anodized Al1050 and sealed anodized Al1050 immersed in 3.5% NaCl solution for 18 hours.

From Figure 4 (a) it is clear that the diameter of capacitive arc is higher for anodized Al1050 and sealed anodized Al1050, so, in Figure 4 (b) it is presented a zoom of Figure 4 (a) to observe the EIS specter of untreated Al1050.

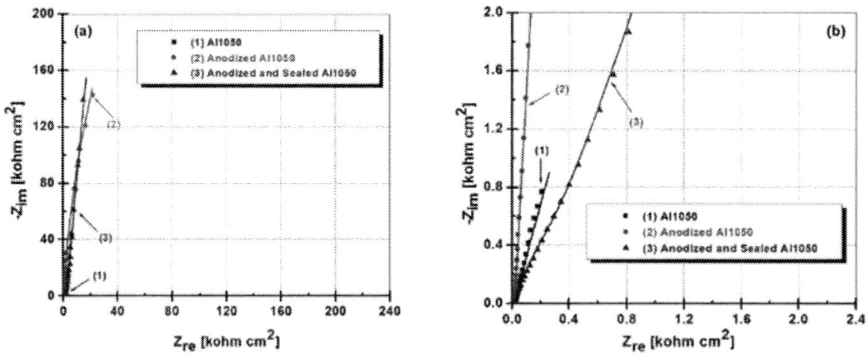

Figure 4. (a) Nyquist representation of EIS specters recorded for (1) untreated Al1050, (2) anodized Al1050 and (3) sealed anodized Al1050 immersed in 3.5% NaCl; (b) Zoom in a high frequency region to obtain a better view of EIS specter of untreated Al1050. The plain symbols represent the experimental data and the continuous lines represent the fitted results.

For the untreated Al1050 the recorded EIS specter (Figure 4 (b)) present a single capacitive loop that corresponds to the native aluminum oxide layer formed on Al1050 substrate. For anodized Al1050 and sealed anodized Al1050, from Figure 4 (a), 2 time constants can be identified, that corresponds to the complex structure of anodic aluminum oxide. The first time constant, recorded at high frequency is attributed to the porous aluminum oxide layer from the anodic oxide–electrolyte interface and the second time constant is attributed to the inner barrier-type aluminum oxide layer from the aluminum substrate–anodic oxide interface (barrier aluminum oxide layer from the bottom of nanopores).

Figure 5 presents the equivalent circuits used in the fitting of EIS experimental results recorded for untreated Al1050, anodized Al1050 and sealed anodized Al1050.

For the untreated Al1050, a simple equivalent circuit (Figure 5 (a)) was used to fit the EIS data and was composed by: R_e – electrolyte resistance, R_n – the native aluminum oxide layer resistance and CPE_n – the constant phase element.

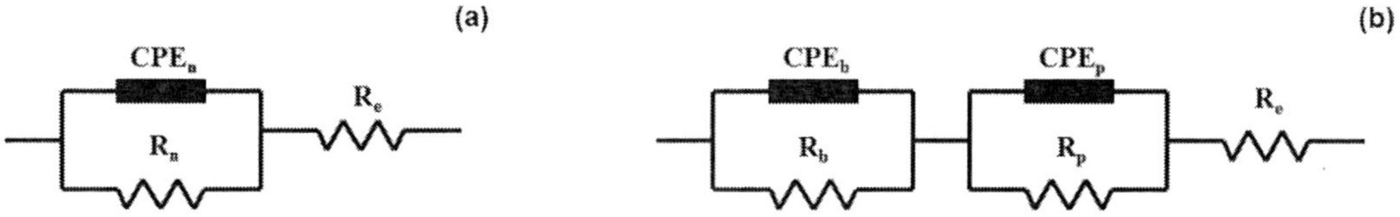

Figure 5. (a) The equivalent circuit used to fit the EIS specter recorded for (1) untreated Al1050 and (b) the equivalent circuit used to fit the EIS specters recorded for (2) anodized Al1050 and (3) sealed anodized Al1050 immersed in 3.5% NaCl.

Based on the complex structure of anodic aluminum oxide layer, the EIS specters of anodized Al1050 and sealed anodized Al1050 presented in Figure 4 (a) were fitted with the equivalent circuit shown in Figure 5 (b) made from 2 simple circuits connected in parallel, one for each component of anodic aluminum oxide layer. The R_e is the electrolyte resistance, the R_p and CPE_p are the polarization resistance and the constant phase element of porous aluminum oxide layer and the R_b and CPE_b are the polarization resistance and the constant phase element of the barrier aluminum oxide layer from the bottom of pores.

The CPE element was used to account for a non-ideal frequency response in equivalent circuits instead of capacitance elements, in order to explain the depression of capacitance semicircles caused by the surface heterogeneity, roughness, porosity, impurities, etc. [20].

The impedance of the constant phase element (CPE) was expressed accordingly with A. Lasia equation [21]:

$$Z_{CPE} = [Q(j\omega)^{\alpha}]^{-1} \tag{3}$$

where: Q is a frequency independent parameter, j is the imaginary number ($j = \sqrt{-1}$), ω is the angular frequency and α is an exponential factor ranged between -1 and 1. The CPE response correspond to those of a capacitor, a resistor or an inductor when the α=1, α=0 or α=-1, respectively.

Fitting the EIS experimental data using the proposed equivalent circuits, the specific polarization resistances of analyzed samples were determined. For the untreated Al1050, the specific polarization resistance has the lowest value, around 0.163 Mohm cm^2 and for the anodized and sealed anodized Al1050, it were determinated values of polarization resistance of 1.697 Mohm cm^2 and 9.356 Mohm cm^2, respectively.

Analyzing the specific polarization resistances values can be observed that the anodic oxidation process increase the corrosion resistance with more 10 grade of magnitude and the sealing procedure if anodic aluminum oxide layer increase even more the anticorrosive performances of Al1050 substrate.

The specific polarization resistances determinated by fitting the electrochemical impedance spectroscopy specters are in good agreement with the results of potentiodynamic polarization diagrams and the evolution of open circuit potential.

4. Conclusions

The untreated Al1050, anodized Al1050 and sealed anodized Al1050 were characterized morphologically by SEM-EDX and electrochemically by open circuit potential, potentiodynamic polarization and electrochemical impedance spectroscopy.

The anodic oxidation process changes the morphological structure of the untreated Al1050, the anodic oxide layer being strongly porous and disorganized. After the sealing procedure, the anodic aluminum oxide layer is still heterogeneous.

The corrosion behavior of analyzed samples was evaluated using electrochemical assays, after the immersion in 3.5% NaCl solution.

The anodized and sealed anodized aluminum oxide layers show higher reactivity after the immersion into electrolyte, the anodic layers become hydrated and the anticorrosive performances increase.

The potentiodynamic polarization measurements reveal that the passive domains increase for the anodized Al1050 and sealed anodized Al1050 in comparison with the passive domain of untreated Al1050. Also the recorded corrosion currents present a decreasing tendency concomitant with the increasing of treatment steps.

The electrochemical impedance spectroscopy diagrams indicate that the specific polarization resistance increase with 10 grade of magnitude for anodized Al1050 in comparison with the untreated Al1050 and the specific polarization resistance is increased with almost 8 grade of magnitude after the hot sealing process in $K_2Cr_2O_7$ solution.

The anodic oxidation process followed by the sealing procedure increase the corrosion resistance of Al1050 exposed to corrosive environments.

5. References

[1] Whelan M, Barton K, Cassidy J, Colreavy J, Duffy B 2013 *Surf. Coat. Technol.* **227** 75
[2] Li X, Nie X, Wang L, Northwood D O 2005 *Surf. Coat. Technol.* **200** 1994
[3] Ding H, Dai Z, Skuiry S C, Hui D 2010 *Tribol. Int.* **43** 868
[4] Ruppi S 2005 *Int. J. Refract. Hard. Met.* **23** 306
[5] Johnson R S, Lucovsky G, Baumvol I 2001 *J. Vac. Sci. Technol.* **19** 1353
[6] Sun T S, Chen J M, Venables J D, Hopping R 1978 *Appl. Surf. Sci.* **1** 202
[7] Ritala M, Leskela M, Dekker J-P, Mutsaers C, Soininen P J, Skarp J 1999 *Chem. Vap. Deposition* **5** 7
[8] Hu N, Dong X, He X, Browning J F, Schaefer D W 2015 *Corros. Sci.* **97** 17
[9] Donatus U, Thompson G E, Elabar D, Hashimoto T, Morsch S 2015 *Surf. Coat. Technol.* **277** 91
[10] Yoganandan G, Balaraju J N, Low C H C, Qi Guojun, Chen Z 2016 *Surf. Coat. Technol.* **288** 115
[11] Huang Y, Shih H, Huang H, Daugherty J, Wu S, Ramanathan S, Chang C, Mansfeld F 2008 *Corros. Sci.* **50** 3569
[12] Priet B, Odemer G, Blanc C, Giffard K, Arurault L 2016 *Surf. Coat. Technol.* **307** 206
[13] Wang S, Peng H, Shao Z, Zhao Q, Du N 2016 *Surf. Coat. Technol.* **286** 155
[14] Snogan F, Blanc C, Mankowski G, Pebere N 2002 *Surf. Coat. Technol.* **154** 94
[15] Zuo Y, Zhao P-H, Zhao J-M 2003 *Surf. Coat. Technol.* **166** 237
[16] Dumitrascu V M, Benea L, Danaila E 2017 *Pro. Int. Conf. SGEM2017 (Albena)* **17** 43
[17] Dasquet J-P, Caillard D, Conforto E, Bonino J-P, Bes R 2000 *Thin Solid Films* **371** 183
[18] Capellosi V R, Poelman M, Recloux I, Hernandez R P B, de Melo H G, Olivier M G 2014 *Electrochim. Acta* **124** 69
[19] Lundvall O, Gulppi M, Paez M A, Gonzalez E, Zagal J H, Pavez J, Thompson G E 2007 *Surf. Coat. Technol.* **201** 5973
[20] Ciubotariu A C, Benea L, Ponthiaux P 2016 *Arabian J. Chem.* DOI: 10.1016/j.arabjc.2016.07.002
[21] Lasia A. 2002 *Modern Aspects of Electrochemistry* vol 32, ed B E Conway, J O'M Bockris, R E White (New York, Kluwer Academic Publishers) chapter 2 p 143

Surface Modifications of Materials by Electrochemical Methods to Improve the Properties for Industrial and Medical Applications

Lidia Benea
Research (Competences) Centre: Interfaces-Tribocorrosion-Electrochemical Systems (CC-ITES), Dunarea de Jos University of Galati, 47 Domneasca Street, 800008 Galati, Romania.

E-mail: Lidia.Benea@ugal.ro

Abstract. There are two applied electrochemical methods in our group in order to obtain advanced functional surfaces on materials: (i) direct electrochemical synthesis by electro-codeposition process and (ii) anodization of materials to form nanoporous oxide layers followed by electrodeposition of hydroxyapatite or other bioactive molecules and compounds into porous film. Electrodeposition is a process of low energy consumption, and therefore very convenient for the surface modification of various types of materials. Electrodeposition is a powerful method compared with other methods, which led her to be adopted and spread rapidly in nanotechnology to obtain nanostructured layers and films. Nanoporous thin oxide layers on titanum alloys as support for hydroxyapatite or other biomolecules electrodeposition in view of biomedical applications could be obtained by electrochemical methods. For surface modification of titanium or titanium alloys to improve the biocompatibility or osseointegration, the two steps must be fulfilled; the first is controlled growth of oxide layer followed by second being biomolecule electrodeposition into nanoporous formed titanium oxide layer.

1. Introduction

The recent developments in technological processes and the increasing interest in the trivalent interrelationship of processing parameters, the structural and functional properties of materials and protective coatings have contributed to a substantial increase in the use of these coatings materials in the industry as well as in biomedical applications [1-20].

This increasing growth has also been stimulated as a response to the needs of advanced technology for better materials. Particularly in the field of surface technologies, the evolution of so-called "high tech" coatings is evident due to the development of more and more performing surface modification technologies such as ionic implants, plasma deposition, vacuum, composite electrodeposition, etc. An important reason to introduce materials with corrosion protection (protective coatings) in different areas is first of all the economic one [3].

Materials with coatings and or protective layers or films can have a much longer lifetime compared to the substrate and so it is possible to capitalize materials as a substrate with low prices. The economic impulse to use the coatings in corrosion control is obvious, if corrosion damage can be slowed or even stopped, the cost of the material is reduced [3].

The definition of surface functionalization has several variants, but can simply be considered as a method used to introduce new properties to a particular material in order to meet certain specific requirements [3].

A functional surface comprises new and / or chemically modified groups and / or morphological features that determine environmental interaction and subsequent specific behavior.

At present, advances in surface, nanotechnology, and nanomaterial engineering allow materials to be manipulated "up to the molecular scale", paving the way for new Hi-Tech functional materials.

Functional coatings (organic, inorganic or hybrid) are a class of materials that can be adapted for many applications where they should be able to perform well-defined functions. There are many functional coatings in many applications that are part of our everyday life.

In recent years, the development of nanotechnology and nanomaterials has promoted an explosion of new functional coatings and this trend will continue to grow in the coming years. Functional and smart coatings are undoubtedly a class of materials of enormous technological interest [3].

This paper aims to discuss and present the results of two applied electrochemical methods in the CC-ITES research center in order to obtain advanced functional surfaces on materials: (i) direct electrochemical synthesis by electro-codeposition process and (ii) anodization of materials to form nanoporous oxide layers followed by electrodeposition of hydroxyapatite or other bioactive molecules and compounds into porous film.

The two electrochemical methods are applied and perfected for surface functionalization, each with specific applications, but both with the potential to improve corrosion resistance and degradation of surfaces in specific working environments.

Electrochemical depositions of metals, alloys and hybrid layers, nanostructures and nanocomposites.

Controlled growth of oxide films on the surface of metals and alloys by electrochemical techniques, respectively the top-down nanotechnology method.

Both electrochemical methods for surface functionalization are associated with one or more of the material characterization methods in order to appreciate the obtained surface properties correlated with application field.

2. Electrodeposition

The deposition of layers by the electrochemical method, or in short electrodeposition, electro-crystallization, electroreduction, has been and is a long-time method used to fight corrosion of materials or for many other applications.

Electrochemical deposition (ELD) or electrodeposition is the process of depositing a coating (layer, film) of a metal or a composite on the surface of another metal to modify its surface morphology, structure and properties. ELD is a method of obtaining adherent layers to the surface of the material that is the essential aspect of a protective coating.

How, by definition, "corrosion" is the physico-chemical interaction between a metal and its environment of use, which leads to changes in the properties of the metal and can often lead to the deterioration of the metal, the environment, or the technical system to which they belong" (ISO 8044-1986), then if the integrity of the material is affected by the external action of the environment, it is necessary to protect it against this harmful action.

By making it possible to obtain the matter in the arbitrary geometry of the three dimensions, electrodeposition is a powerful method compared to other methods, which has made it adopted and spread rapidly in nanotechnology. Nanocomposites manufactured using electrodeposition include the magnetic and non-magnetic alternating layers, each of only a few nanometers thick to micrometric thick layer for industrial or biomedical applications [3].

A challenge for the process of electro-crystallization of metals and alloys is the electrodeposition of hybrid and composite layers [3].

In the last two decades, composite layers obtained by electrodeposition have also been developed in the field of advanced materials. The subject is particularly recent, given that lately the demands of composite materials have become larger and their electrochemical deposition can bring many advantages compared to other processes.

Composite coatings (AC) are part of the composite materials category, which are defined as systems composed of two or more materials in order to improve their properties compared to the separate attributes of components [3].

Composites are bi- or polyphase compacted materials formed by natural processes or artificially created by man, where there are boundaries of separation between the isolated components. The MATRIX is a continuous bonding element, which may be in the solid crystalline or amorphous state. The DISPERSED phase consists of one or more dispersed phases in the matrix, in any aggregation state. It may be present as dispersed particles, being discontinuous or fiber in this case, thus being continuous.

The manufacture of these composite coatings has begun to be investigated since 1972 by the electrolysis of a suspension [10]. This method generally employs a standard electrolyte in which the dispersed particles are uniformly held in solution by suitable stirring. In most cases, the codeposition of dispersed particles with metals requires the development of new electrolytes with different compositions relative to pure metal electrodeposition.

The new method of obtaining composite coatings realizes a principle known in nature, which essentially consists in the fact that the cumulative action of various materials gives an equivalent effect to the creation of a new material, the properties of which differ from the properties of each of the component materials.

The amount of incorporated particles is the essential parameter for composite production and property improvements, such as wear resistance, corrosion and oxidation. In addition, uniform distribution of the codeposited particles in the metal matrix has proven to be another essential parameter. However, it should be mentioned that both the morphological and structural characteristics of the metal matrix are strongly affected by the presence of the particles. Codeposition with a sufficient amount of particles should result in harder and more resistant coatings [3, 8, 13].

Schematically, the electro-codeposition process is shown in Figure 1 for examples of two microsized (micro ZrO_2 - 10 mm mean diameter, UHMWPE - 10 mm mean diameter) and two nanosized particles (nano-CeO_2 - 10 nm mean diameter, nano TiO_2 - 10 nm mean diameter) and a variety of electrolyte corresponding to the metal matrix [3, 11-20]. UHMWPE is a biocompatible polymer called ultrahigh molecular weight polyethylene.

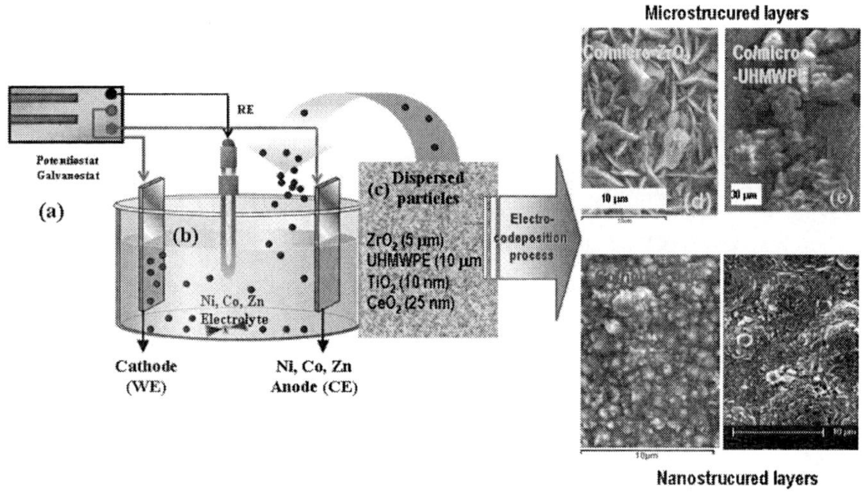

Figure 1. Schematic presentation of the electro-codeposition process of dispersed particles into metal matrix to obtain micro and nanocomposite and hybrid layers.

IOP Conf. Series: Materials Science and Engineering **374** (2018) 012014 doi:10.1088/1757-899X/374/1/012014

The schematic drown of electro-codeposition process from Figure 1 includes: (a) - electrochemical equipment; (b) - electrochemical cell with the working electrodes (WE), the counter electrode (CE), the reference electrode (RE), the electrolyte and the dispersed particles held in a homogeneous suspension by means of a stirring system; (c) - nanometer and micrometric dispersed particle types; (d) SEM micrograph of Co / micro-ZrO_2 composite layer; (e) SEM micrograph of Co / micro-UHMWPE microparticles; (f) - SEM micrograph of Co / nano-CeO_2 nanocomposite layer; (g) - SEM micrograph of Ni / nano-TiO_2 nanocomposite layer.

The variety of micro and nano-composite systems that can be obtained by electro-codeposition also provides various applications for these films and layers, from industrial applications to biomedical applications.

Although there are studies and proposals in the literature, a generally valid mechanism of electro-codepostion of dispersed particles with metals has not been achieved. The electro-codeposition mechanism can generally be considered to be characterized by the following steps [3, 9, 12]:

- At the time of introduction into the electrolyte, a layer of adsorbed ionic species is formed at the surface of the dispersed particles forming the double diffuse layer around each particle.
- The particles are transported by stirring and under the influence of electric field, they will flow to the surface of the cathode.
- When the adsorbed ionic species are reduced, the particles are embedded in the composite layer (in the metal matrix).

These ionic species can not directly pass through a charge transfer stage, so the first step in depositing a metal must be its chemical transformation into an electroactive species. This stage takes place during the diffusion transport of the ionic species to the surface of the electrode that takes place inside the hydrodynamic boundary layer or in some cases in the double electric layer [3, 12].

These layers, films and composite coatings can improve the corrosion resistance of various materials in specific environments [16-20]. For example, as the content of TiO_2 nanoparticles increases in the nickel layer, the corrosion current decreases and the corrosion potential passes to more positive values [20], a passive layer is formed, but the HNO_3 solution accelerates the degradation of the passive layer. A better corrosion resistance of Ni/nano-Al_2O_3 composite layers compared to pure nickel layer was obtained in experimental data by potentiodynamic polarization and electrochemical impedance spectroscopy [21].

Hybrid composite layers obtained by the codeposition of polymeric dispersed particles with metals can confer corrosion resistance in different specific environments, from industrial to biomedical (biological fluids) [15-22].

Thus, in Figure 2, are shown the micrograph of the Co / UHMWPE (b) hybrid layer (a) and cobalt (b) produced by the electro-codeposition of high molecular weight polyethylene micrometric particles (UHMWPE), special for biomedical applications [3].

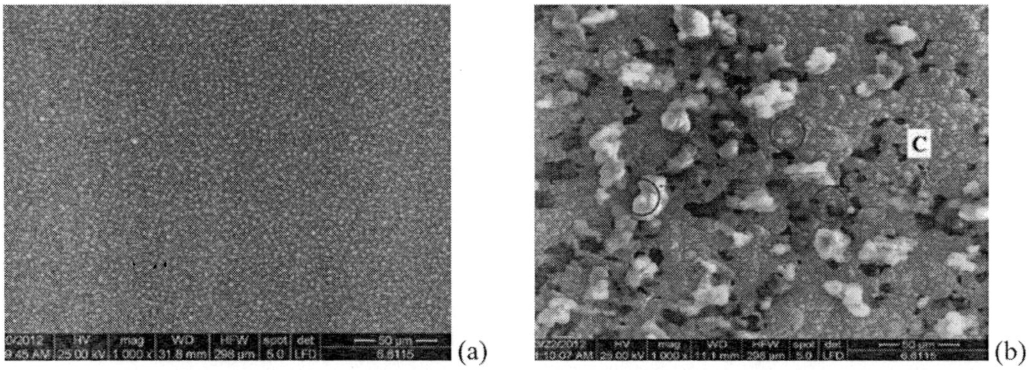

Figure 2. SEM micrographs of electrodeposited layers: (a) - pure cobalt; (b) - Co / UHMWPE hybrid coating obtained by electro-codeposition of UHMWPE (50 g / L) with cobalt.

Such hybrid composite layers have demonstrated better corrosion resistance and biocompatibility at their immersion in biological fluids, Figure 3 (a, b).

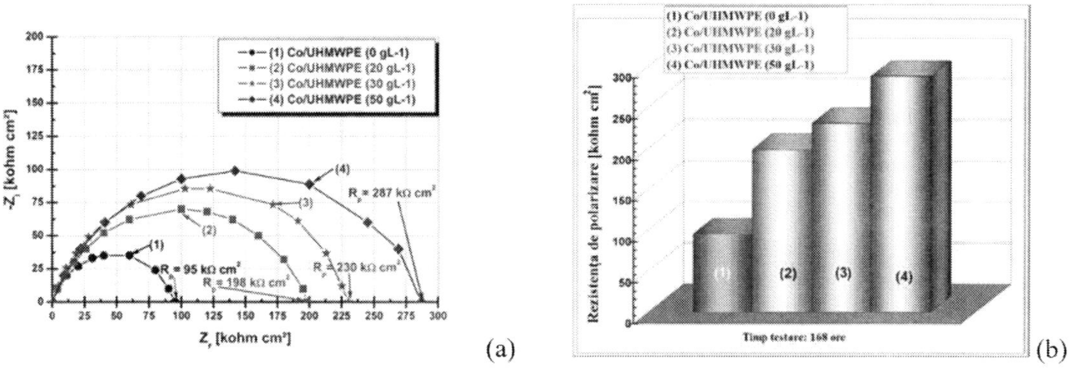

Figure 3. Corrosion resistance of Co / UHMWPE composite layers and pure cobalt layer immersed in Hank's biological solution after 168 hours: (a) Nyquist representation of electrochemical impedance spectroscopy data; (b) increasing the corrosion resistance of cobalt layers by increasing codeposited dispersed UHMWPE particles.

Figure 3 show the increase in polarization resistance, thus increasing the corrosion resistance, with the increase of the concentration of polymeric UHMWPE dispersed in the electrolyte. Hank's solution simulates the most aggressive fluids in the human body. An increase in polarization resistance is observed, thus decreasing the corrosion rate, with the increase of the UHMWPE polymeric particles in the cobalt electrolyte [15].

Ceramic oxide (CeO_2) nanoparticles, which are in turn biocompatible, can give also to cobalt matrix a higher corrosion resistance by codeposition with cobalt.

In Figure 4 the surface morphology of Co / nano-CeO_2 nanocomposite layer (a) and its uniform distribution in the nanocomposite layer (b) [16] is observed. The inclusion of cerium oxide (CeO_2) nanoparticles can be determined by chemical analysis or X-ray dispersion analysis on SEM provided with an X-ray analyzer, Figure 4 (c).

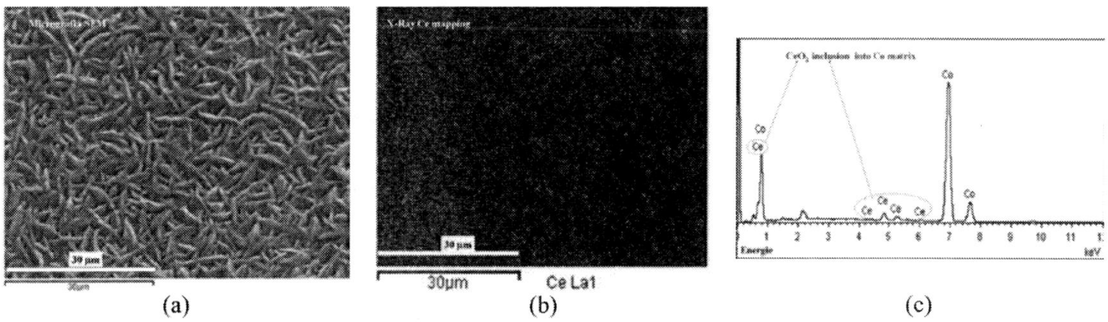

Figure 4. SEM-EDX analysis of Co/ nano-CeO_2 nanocomposite layers: (a) SEM surface micrograph; (b) SEM-EDX mapping for cerium (Ce); (c) SEM-EDX spectrum of the Co / nano-CeO_2 nanocomposite layer, with the peaks characteristic of the cerium element.

Figure 4 gives an EDX analysis spectrum for the Co / nano-CeO_2 nanocomposite layer, which confirms the inclusion of cerium oxide nanoparticles through the peaks characteristic of the cerium element (Ce).

The increase in the corrosion resistance in biological fluids (Hank's solution) of Co / nano-CeO_2 nanocomposite layers compared to pure cobalt layers is shown in Figure 5 [16]. Figure 5 shows both the morphological changes induced by the cerium oxide dispersed nanoparticles and their effects on corrosion resistance in specific biological environments. Corrosion resistance increases with increasing cerium oxide concentration.

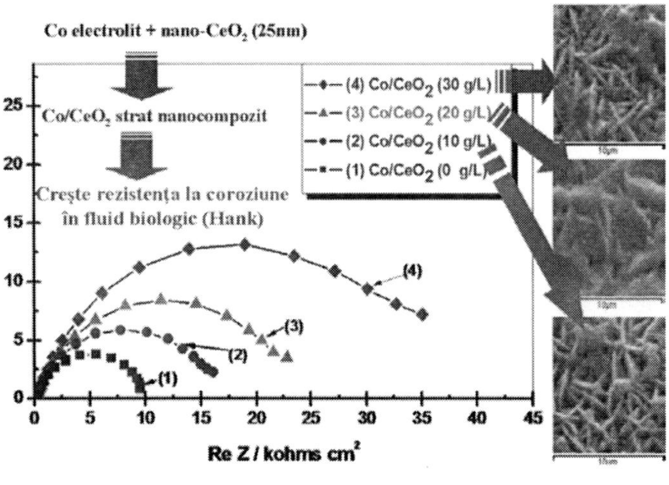

Figure 5. Increased corrosion resistance based on the percentage of dispersed phase embedded in the Co / nano-CeO_2 nanocomposite layer.

The effect of including dispersed nanoparticles into metallic matrices on their crystalline structure can be evidenced by X-ray diffraction. For example, inclusion of 10 nm titanium oxide (TiO_2) nanoparticles in the nickel matrix causes changes in the nickel crystallinity as seen in Figure 6 (a) [7, 8]. These structural changes confer also to Ni / nano-TiO_2 nanocomposite layers an improved corrosion resistance, Figure 6 (b) [3]. The inclusion of TiO_2 nanoparticles in the nickel matrix determines not only structural changes, but also determines an improvement in the corrosion resistance of Ni / nano-TiO_2 nanocomposite layers in NaCl solution, as can be seen in Figure 6 (b) [3, 8].

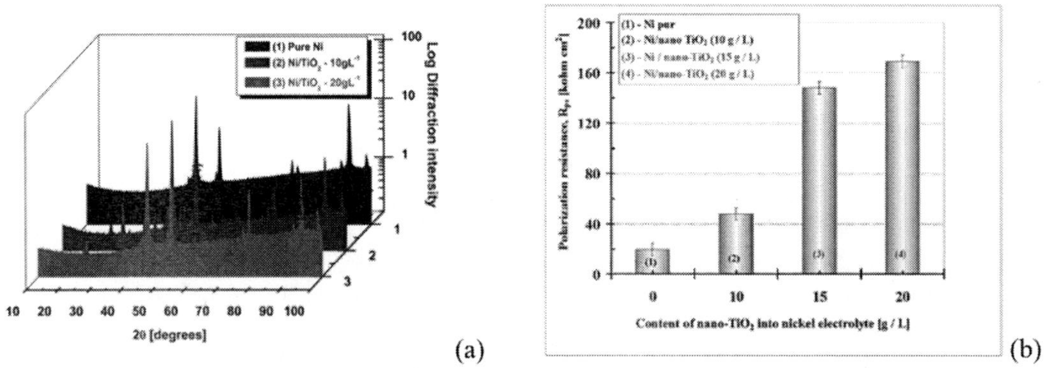

Figure 6. Ni/ nano-TiO_2 naocomposite layers; (a) X-Ray diffraction patterns and (b) increasing the corrosion resistance with increasing the content of dispersed particles in the electrolyte.

Ceramic carbides or oxide nanoparticles included in nanocomposite layers offer both corrosion resistance, but also improve wear resistance, resulting in better tribocorrosion resistance [3, 5, 15].

Electrodeposition can be combined with self assembled templates to prepare nanomaterials with exciting properties. Traditional applications of electrodeposition will continue, while new applications will emerge rapidly, especially in the areas of nanoelectronics, biotechnology, nanobiocorrosion, biomedicine, nanostrats, biosciences, and energy engineering. Electrodeposition of non-metallic materials will become more important, and the combination of electrodeposition with other processes will lead to a whole class of nanostructured materials.

3. Controlled growth of oxide films

Controlled growth of oxide layers or anodic oxidation is an electrochemical method for producing an oxide film on a metal substrate. It removes the electrons from a substance and oxidizes the anode.

Anodic oxidation increase the corrosion resistance and wear resistance, and provides better adhesion to primers for further paints or primer than does base metal.

Anodic films can also be used for a range of cosmetic effects, either with thick, porous layers that can absorb dyes or thin, transparent layers, which add interference to reflected light [3].

Anodic oxidation is an accelerated electrochemical oxidation process, which is intensified by the natural metal oxide film. This involves applying an electrical voltage to relatively low currents while the substrate is immersed in an acidic solution. The oxide films obtained can be very dense and stable with a variety of microstructure features. Normally, anodic oxidation is carried out in an electrochemical cell containing various aqueous solutions such as sulfuric acid, hydrogen peroxide and phosphoric acid. For example, anodic oxidation is used to modify the surfaces and properties of titanium and its alloys or aluminum and its alloys due to: the resulting mechanical properties, non-biotoxicity, biocompatibility properties and corrosion resistance.

3.1. Modification of titanium and titanium alloys by anodic oxidation

Implantology has made significant progress since the osteointegration of titanium implants and its alloys has been demonstrated to achieve remarkable therapeutic results. For the subjects who have suffered various traumas, those with malformations or other diseases of the skeleton, prosthetic restorations with implants proved to be particularly effective, their superior qualities being both functional and psychological [3].

In the area of implants, surface modification has become essential. At present, more emphasis is placed on alternative solutions, such as coatings with functional layers, to solve problems from the implant-tissue interface. Changes to surface properties can be made in two ways:

a) Mechanical, physical and chemical methods change the degree of smoothing of the material, the structure and the chemical composition (polishing, thermal treatments, thermochemical treatments, electrodeposition etc.).

b) By coatings. The implant is totally or locally coated with different compositional materials (plasma deposition, sol gel, electrodeposition, etc.).

The roughness of the surface of a biomaterial increases with an increase in the oxide film on its surface. Rough surfaces can adsorb a greater amount of fibronectin compared to smooth surfaces. Also, rough surfaces can provide better bone fixation than smooth surfaces and promote macro-phase attraction. Improving surface roughness and decreasing ion release can maximize the corrosion resistance of materials for medical implants [23]. The structural and chemical properties of anodic oxide may vary widely in changing process parameters such as the potential difference between the cathode and the anode and the current density required to achieve this potential difference value, pH of the electrolyte and temperature [23-25]. In Figure 7 is schematically presented the anodic oxidation process used for the modification of titanium and titanium alloy surfaces. Prior to anodic oxidation, titanium alloy samples, e.g. titanium alloy Ti-6Al-4V, are cut into 25 x 25 x 2 mm (or other dimensions depending on the electrochemical cell), an electrical contact is provided with copper wire. The non active surfaces are insulated with resin, thus the sample remains with a well-defined active surface of 625 mm^2 and subsequently polished successively with impermeable abrasive paper with a grain size of 320-4000 μm with a diamond paste size of 3 and 1 μm and a suspension of SiO_2 (particle size of 0.04 μm) resulting in a mirror surface.

The natural oxide layer possesses a low level of electronic conductivity and its thickness has been estimated to be in the order of several nanometers. This natural oxide layer for the Ti-6Al-4V alloy is composed of titanium oxide in various oxidation states (TiO_2, Ti_2O_3 and TiO) and oxides of aluminum (Al_2O_3) and vanadium, which are thermodynamically stable at the physiologically pH, thus leading to excellent localized biocompatibility [26]. However, in their native form, TiO_2 films that cover titanium and its alloys have poor mechanical properties and can easily deteriorate under wear or slip conditions.

Therefore, oxide films must be intentionally formed with the help of different surface modification techniques.

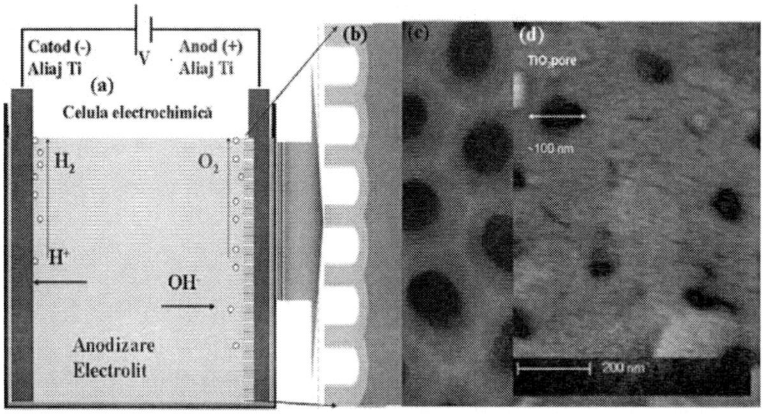

Figure 7. Schematic of the anodic oxidation process of titanium and titanium alloys: (a) the electrochemical cell with the appropriate electrodes and the anodic oxidation electrolyte connected to an adjustable voltage current source; (b) schematic representation of a porous oxide film in cross-section; (c) schematic representation of a porous oxide film on the surface; (d) the SEM micrograph of a nanoporous titanium oxide film obtained on the Ti-6Al-4V alloy by anodic oxidation.

By the anodic oxidation process, the metal to be treated is covered with a porous film beyond which the oxygen ions diffuse and oxidize the substrate thereby increasing the thickness of the film. This treatment aims to form a film which is hard (resistant to abrasion), which is chemically stable (to withstand corrosion), which reduces the release of ions and increases the adhesion and fixation properties on the surfaces of biomaterials.

Measuring the surface roughness parameters of an implant is very important because its value influences the adhesion, adsorption and differentiation of the cells. The most common roughness parameter used for implanted materials is the mean roughness R_a, which represents the mean arithmetic of the deviations of a rough 2D profile relative to the median line of the profile over the entire measured length [26], or S_a corresponding to a surface 3D.

In the literature, there are numerous reports that have demonstrated that the surface roughness of a titanium implant affects the rate of bone integration and biomechanical fixation [3, 26]. These reports have shown that titanium implants with rough surfaces have bone attachment larger than titanium implants with smooth surfaces.

Therefore, starting from a mirror surface with a roughness value corresponding to the 3D profile of 23.06 nm, by anodic oxidation at 100 V, higher values of the roughness profile corresponding to the 3D profile, e.g. equal to 123.35 nm was obtained [3, 26, 27].

Using Fusayama Meyer's artificial saliva solution as a biological medium, corrosion resistance tests of the untreated Ti-6Al-4V alloy and the nanoporous titanium oxide film obtained by the electrochemical treatment of this alloy were performed. Electrochemical methods for corrosion tests included open circuit potential (OCP), potentiodynamic polarization (PD) and electrochemical impedance spectroscopy (EIS).

The data from the electrochemical impedance records at the corrosion potential (open circuit potential), applying a sinusoidal signal of 10 mV, with the frequency sweep between 10^5 and 10^{-2} Hz, corresponding to the two types of Ti-6Al-4V alloy surfaces have been processed in the form of diagrams in Nyquist representations (the imaginary component of impedance versus on the real component). These Nyquist representations are used to deduce the polarization resistance that is equivalent to corrosion resistance, from the diameters of the semicircles obtained by simulation [3, 26-28]. By fitting and simulating the results, the polarization resistance can be determined.

Since most implant materials are subjected not only to an aggressive environment but are in motion, tribocorrosion tests (corrosion and wear) are more appropriate in their characterization. Thus, the influence of the nanoporous oxide film formed on the Ti-6Al-4V alloy is well emphasized on all the steps of measuring the electrochemical impedance spectroscopy diagrams, respectively before applying a force and applying the friction, during friction as well as after friction stopping, so as can be seen in Figure 8 (a, b).

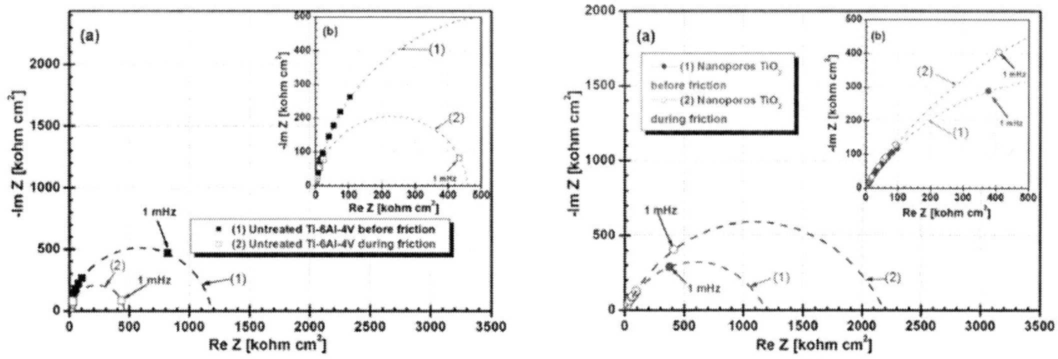

(a) (b)

Figure 8. Electrochemical impedance spectroscopy measured in saliva solution before friction (1) and during friction (2) for: (A) untreated Ti-6Al-4V alloy and (B) nanoporous TiO_2 layer formed anodically on Ti-6Al-4V alloy; (2) Ti-6Al-4V alloy. Layer a is drawn on entire EIS spectra, layer (b) represent the zoom of high frequency EIS spectra.

Figure 8 shows the beneficial effect of the nanoporous TiO_2 film on the alloy on the polarization resistance, especially during friction, which indicates a much better resistance of this oxide film during the friction or movement in an environment corrosive of the human body. For other titanium alloys, the electrochemical method of controlled oxide film formation can lead to improved corrosion resistance in specific biological use environments [3, 29-31]. As can be seen from Figures 8 (A, B), the polarization resistance of the untreated surface of the Ti-6Al-4V alloy decreases dramatically during the friction test (bidirectional fretting) compared with that obtained before friction. For the nanoporous titanium oxide layer, the polarization resistance value increases during friction compared to the measured value before friction, proving a better stability under tribocorrosion conditions.

Tribo-electrochemical characterization in simulated physiological fluid (Fusayama Meyer's saliva) showed that in their native form, TiO_2 films have poor tribological properties and are easily fractured under tribocorrosion conditions. The tribocorrosion performances of nanoporous TiO_2 films obtained by anodizing are better than untreated alloy surfaces under similar experimental conditions.

Under friction conditions the polarization resistance R_p of the untreated alloy dropped to 460 $\Omega\cdot cm^2$ from the saliva immersion value without application of the friction force when R_p had the value of 1180 $\Omega\cdot cm^2$. This is due to the removal of the oxide film in the friction area.

For the surface of nanoporous titanium oxide film, the polarization resistance measured during friction was greater having the value of 2200 $\Omega\cdot cm^2$ compared to the value obtained by saliva immersion prior to application of the friction force and which was of R_p = 1200 $\Omega\cdot cm^2$. Higher polarization resistance of nanoporous TiO_2 layer formed by anodization under friction and corrosion conditions in physiological solution is due to the superior titanium oxide insulation properties as well as the variation in roughness, thickness and morphological characteristics of its surface.

4. Conclusions

Electrochemical techniques are powerful tools for surface modifications and nanostructuration of materials as well as for their characterization in specific environments.

Electrodeposition and the combination of electrodeposition with other electrochemical processes (controlled oxide growth by anodization) can lead to the entire class of hybrid, composite coatings and nanostructured layers (films) on different support materials and structures necessary for a future based on nanotechnology and nanomaterials.

By electrochemical modifications the results show improved surface properties for corrosion and tribocorrosion resistance in specific environments for industrial as well as for biomedical applications.

Functionalizing the surface of materials and biomaterials is the way to reduces or by-pass degradation by corrosion and tribocorrosion and increases the life time.

5. References

[1] Plieth W 2011 *J. Solid State Electrochem.* **15** 1417
[2] Galus Z 2011 *J. Solid State Electrochem.* **15** 1777
[3] Benea L 2017 *Corrosion and Corrosion protection – from theory to practice*, Editura Academica Galati p 1
[4] Gurrappa I, Binder L 2008 *Sci. Technol. Adv. Mater.* **9** 1
[5] Danaila E, Benea L, Caron N, Raquet O 2016 *Met. Mater. Int.* **22** 924
[6] Benea L, Celis J P 2016 *Materials* **9** 269
[7] Pavlov A I, Benea L, Celis J-P, Vasquez L 2013 *Dig. J. Nanomater. Biostruct.* **8** 1043
[8] Benea L, Danaila E, Celis J-P 2014 *Mater. Sci. Eng.* A **61** 106
[9] Benea L, Bonora P L, Borello A, Martelli S, Wenger F, Ponthiaux P, Galland J 2001 *J. Electrochem. Soc.* **148** C461
[10] Guglielmi N 1972 *J. Electrochem. Soc.* **119** 1009
[11] Benea L 2015 *Solid State Phenom.* **227** 243
[12] Benea L, Danaila E 2016 *J. Electrochem. Soc.* **163** D655
[13] Benea L, Basa S B, Danaila E, Caron E, Raquet O, Ponthiaux P, Celis J-P 2015 *Mater. Design.* **65** 550
[14] Benea L 2009 *J. Appl. Electrochem.* **39** 1671
[15] Danaila E, Benea L, Bounegru I 2015 *Solid State Phenom.* **227** 507
[16] Benea L 2013 *Metall. Mater. Trans. A* **44** 1114
[17] Benea L, Ciubotariu A, Sand W 2013 *Int. J. Mater. Res.* **104** 489
[18] Benea L, Ponthiaux P, Wenger F 2011 *Surf. Coat. Technol.* **205** 5379
[19] Ciubotariu A, Benea L, Mitoseriu O, Ponthiaux P, Wenger F 2009 *J. Optoelectron. Adv. M.* **11** 892
[20] Baghery P, Farzam M, Mousavi A B, Hosseini M 2010 *Surf. Coat. Technol.* **204** 3804
[21] Ciubotariu A C, Benea L, Lakatos-Varsanyi M, Dragan V 2008 *Electrochim. Acta* **53** 4557
[22] Ciubotariu A, Benea L, Sand W 2011 *An. Univ. "Dunarea Jos" Galati, Fasc. IX* **1** 51
[23] Hossain M M, Gao W 2008 *Trends Biomater. Artif. Organs* **22** 144
[24] Liu X, Chu P K, Ding C 2004 *Mater. Sci. Eng. R Rep.* **47** 49
[25] Diamanti M V, Pedeferri M P 2007 *Corros. Sci.* **49** 939
[26] Benea L, Danaila E, Ponthiaux P 2015 *Corros. Sci.* **91** 262
[27] Benea L, Mardare-Danaila E, Celis J P 2014 *Tribol. Int.* **78** 168
[28] Benea L, Mardare E, Mardare M, Celis J P 2014 *Corros. Sci.* **80** 331
[29] Mardare E, Benea L, Celis J P 2012 *Dig. J. Nanomater. Biostruct.* **3** 933
[30] Benea L, Danaila E, Ponthiaux P 2017 *IOP Conf. Ser.: Mater. Sci. Eng.* **174** 012044
[31] Benea L, Danaila E, Dumitrascu V M, Ponthiaux P 2015 *E-Health and Bioengineering Conference (EHB)* **2015** 1

IOP Conf. Series: Materials Science and Engineering **374** (2018) 012015 doi:10.1088/1757-899X/374/1/012015

Optical and Morphological Properties of P3HT and P3HT: PCBM Thin Films Used in Photovoltaic Applications

L Hrostea[1,2,3], M Girtan[3], R Mallet[4] and L Leontie[1,2]
[1] Faculty of Physics, AlexandruIoanCuza University of Iasi, Bulevardul Carol I, Nr. 11, 700506, Iasi, Romania
[2] Integrated Center for Studies in Environmental Science for North-East Region (CERNESIM), AlexandruIoanCuza University of Iasi, Bulevardul Carol I, Nr. 11, 700506 Iasi, Romania
[3] Photonics Laboratory, (LPhiA) E.A. 4464, SFR Matrix, Université d'Angers, Faculté des Sciences, 2 Bd Lavoisier, 49000 Angers, France
[4] SCIAM, Angers University, Angers, France

E-mail: laura.hrostea@yahoo.ro

Abstract. This work is focused on the study of some physical properties of poly(3-hexylthiophene-2,5-diyl) (P3HT) and poly(e-hexylthiophene-2,5-diyl): Methanolfullerene Phenyl-C61-Butyric-Accid-Methyl-Ester (PCBM) blend thin films. Knowing the polymer advantages, such as ease of processing, high thermal stability, strong interaction with light, its properties have captured the attention regarding the changes that can occur in a polymer:fullerene blend in term of them. Polymer and polymer:fullerene blend (1:0.1, 1:0.2, 1:0.4 and 1:0.8 ratios) were deposited by spin coating on glass and SnO_2:F (FTO) coated glass. The optical properties were emphasized using spectrophotometry (300 – 2200 nm wavelength range) and spectroscopic ellipsometry models, to obtain the refractive index, extinction coefficient and the transmission (found higher than 80%). According to X-ray diffraction analysis, as-obtained films are amorphous. Investigation of the surface morphology of thin-film samples using Atomic Force Microscopy revealed a crystallite-like surface morphology with crystallite size in the nanometer range.

1. Introduction

Organic solar cells exhibit a huge potential in solar energy harvesting, in particular due to their great versatility in the manufacturing methods and low cost of raw materials. Polymers have become materials of great interest for photovoltaic applications due to the peculiarities and advantages they offer. These are cheap and inexhaustible materials and they can be easily processed even at low temperatures [1].

Found in a wide range of colours, the peculiarities of their interaction with light make them promise for use in solar cells (polymer-based photovoltaic cells). Being flexible, polymers can be deposited on different substrates. They show optical and electric properties that can be improved by doping with different materials (in the present study, fullerene). Disadvantages of polymer materials also include high sensitivity to oxygen, water etc., but these can be limited by device encapsulation. The cost of polymer solar cells remains high due to the incipient state of the study in this field [2].

This paper considers some characteristics of polymer thin films of poly(3-hexylthiophene) (P3HT) and Methanofullerene Phenyl-C61-Butyric-Acid-Methyl-Ester (PCBM). The interest in using these two organic materials is motivated by the fact that P3HT is a conductive π-conjugated polymer used in a bulk-heterojunction solar cell structure as p-type material active layer and PCBM is a fullerene used in solar cell active layer as a n-type material [3, 4].

2. Materials and methods

Highly regioregular (>98%) polymer, poly(3-hexylthiophene) (P3HT), purchased from Rieke Metals, and Methanofullerene Phenyl-C61-Butyric-Acid-Methyl-Ester (PCBM), purchased from Ses Research, were used in different weight ratio blends (1:0, 1:0.1, 1:0.2, 1:0.4 and 1:0.8), to obtain thin films deposited on glass and SnO_2:F (FTO) substrates by spin coating in normal atmosphere at room temperature.

The thicknesses of these samples (Table 1) were measured by profilometry, using a VEECO Dektak 6M Stylus Profilometer, and compared with the thickness values measured by spectroscopic ellipsometry technique, using a UVVISELTM ellipsometer from Horiba Jobin Yvon.

Table 1. Thickness of samples, deposited on glass, by profilometry and by ellipsometry.

Sample	Weight ratio (P3HT:PCBM)	Thickness by profilometry (nm)	Thickness by ellipsometry (nm)
Ps0	1:0	≈ 120	93 ± 3
Ps0.1	1:0.1	≈ 180	211 ± 1
Ps0.2	1:0.2	≈ 150	186 ± 3
Ps0.4	1:0.4	≈ 160	270 ± 5
Ps0.8	1:0.8	-	299 ± 1

The optical properties were emphasized by means of two methods of characterization, spectrophotometry and spectroscopic ellipsometry. The transmittance spectra of polymer and polymer: fullerene blend thin films were recorded in the spectral range from 300 to 2200 nm, by a double beam LAMBDA 950 UV/Vis/NIR spectrophotometer. Complementary, in the same wavelength range and at room temperature, a UVVISELTM ellipsometer with a 75 W high discharge Xenon lamp was used [5-7]. The structural and morphological properties were investigated using a Brucker D8 Advance X-ray diffractometer (nickel-filtered $CuK_{\alpha 1}$ radiation, λ=1.5406 Å) and a contact mode atomic force microscope (AFM), respectively [8–10].

3. Results and discussions

Graph 1 (left) represents the transmittance values for P3HT and P3HT:PCBM blend thin films. From this graph, it appears that the transmittance is influenced by the amount of fullerene, as follows: in the infrared spectral range, increasing the quantity of fullerene, the transmittance decreases visibly; in the visible spectral range, instead, there is a reversal so that the amount of fullerene improves the transmittance, which is between 20 and 60%.

Another specific characteristic determined by the processing of the experimental data recorded by spectrophotometry is the calculation of the optical band gap using Tauc relation. From the dependence of the absorption coefficient on the incident photon energy (Graph 1 - right), the optical band gap for P3HT was found equal to 1.90 eV.

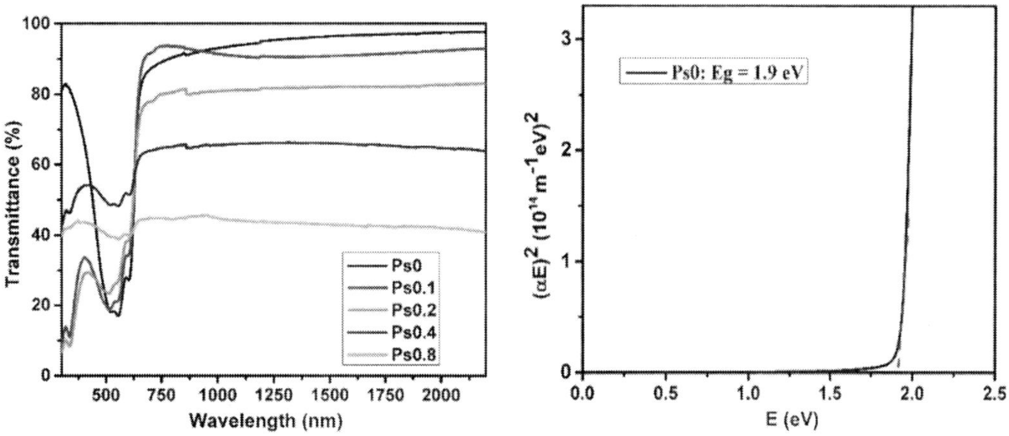

Figure 1. Transmittance spectra of P3HT and P3HT:PCBM thin films (left) and the optical band gap determination of P3HT polymer (right).

Ellipsometry is an optical characterization technique, indirect and nondestructive. The experimental data obtained are compared with those generated by the ellipsometric model based on new amorphous dispersion formula (1):

$$n(\omega) = n_{\infty} + \frac{B\cdot(\omega-\omega_j)+C}{(\omega-\omega_j)^2+\Gamma_j^2}, \quad k(\omega) = \begin{cases} \frac{f_j\cdot(\omega-\omega_j)}{(\omega-\omega_j)^2+\Gamma_j^2}; for\ \omega > \omega_g \\ 0; \quad for\ \omega < \omega_g \end{cases}; \qquad (1)$$

where:

$$\begin{cases} B_j = \frac{f_j}{\Gamma_j}\cdot(\Gamma_j^2 - (\omega_j - \omega_g)^2) \\ C_j = 2\cdot f_j \cdot \Gamma_j \cdot (\omega_j - \omega_g) \end{cases}. \qquad (2)$$

In expression (1) and (2), n denotes the refractive index, k represents the extinction coefficient, f_j is the oscillator strength (amplitude of the extinction coefficient) ($0< f_j<1$), Γ_j is broadening factor of absorption peak ($0.2<\Gamma_j<8$), ω_j represents the energy for which the extinction coefficient is maximum ($1.5<\omega_j<10$), while ω_g is the energy at which the extinction coefficient is minimum ($\omega_g<\omega_j$). All these parameters are expressed in electron-Volts (eV).

Relation (3) describes the fundamental relation of ellipsometry which governs Graph 2, containing the reflection ellipsometric ratio (ρ), amplitude ratio, Ψ, and phase difference, Δ, by which the refractive indices, extinction coefficients and dielectric constants corresponding to each sample will be extracted, depending on the fullerene amount:

$$\rho = \frac{r_p}{r_s} = \tan\Psi\, e^{i\Delta}\ . \qquad (3)$$

Here, r_p and r_s represent the reflection coefficients of light polarized parallel, respective perpendicularly to the plane of incidence.

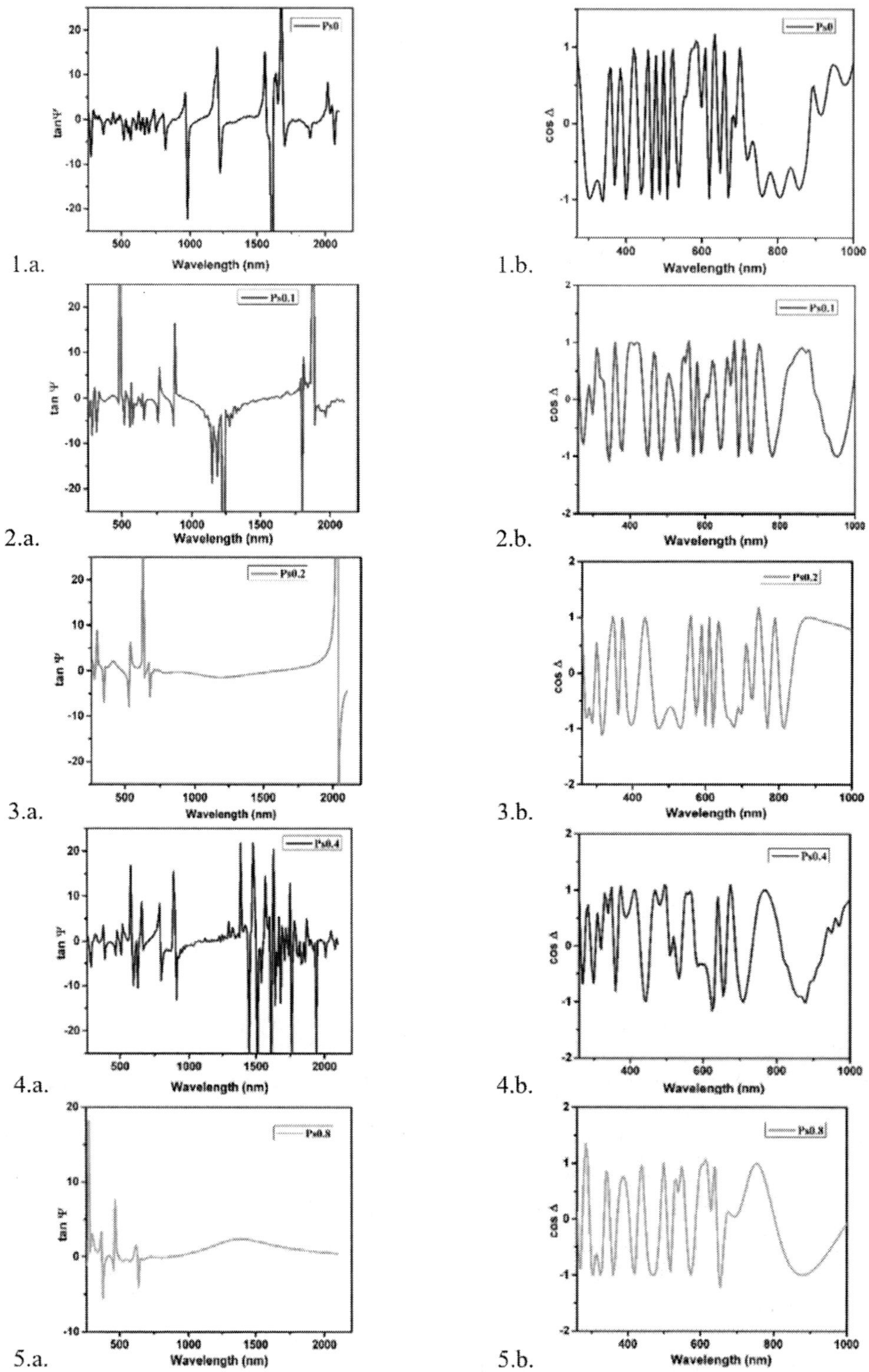

Figure 2. Spectral dependences of the amplitude ratio (a) and phase differences (b).

Applying definitions of refractive index and coefficient of extinction, from the dispersion formula (1), we obtain Graph 3, representing the dependence of the refractive index, n, on the wavelength

IOP Conf. Series: Materials Science and Engineering **374** (2018) 012015 doi:10.1088/1757-899X/374/1/012015

(Graph 3 - left) and spectral dependence of the extinction coefficient, k, (Graph 3 - right), considering the dopant amount. Thus, in the visible range, both the refractive index and the extinction coefficient decrease with the increase in the fullerenes amount. In the graph of the extinction coefficient, the fingerprints of the two constituents (polymer and fullerene) is clearly visible, given the relationship between this coefficient and the absorption coefficient [7, 11].

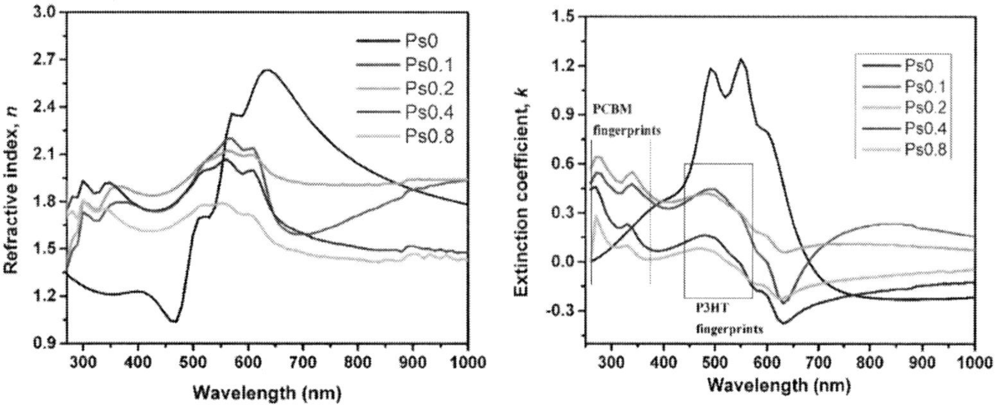

Figure 3. Spectral dependences of the refractive index, n, (left) and the extinction coefficient, k, (right).

Starting from the definition (formula 4) of the complex refractive index, the dielectric constant, ε, can be determined (formula 5), in which $\varepsilon_r$is the real part and $\varepsilon_i$is the imaginary part of the it.

$$\tilde{n} = n - ik; \tag{4}$$

$$\tilde{\varepsilon} = (\tilde{n})^2 = \varepsilon_r - i\varepsilon_i. \tag{5}$$

Like the refractive index, the dielectric constant (Graph 4) depends on the amount of fullerene, decreasing together with the increasing fullerene concentration.

From the X-ray diffractograms, a predominantly amorphous nature was found, irrespective of the amount of fullerene, with a small intensity peak corresponding to the polymer, at $2\theta \approx$ 5 degrees, and another peak located at $2\theta \approx$ 10 degrees, corresponding to the fullerene.

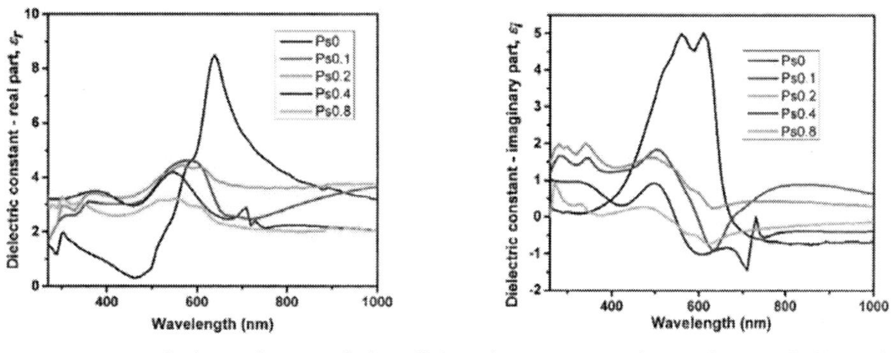

Figure 4. Spectral dependence of the dielectric constant: the real part (left) and the imaginary part (right).

According to AFM analysis, there are not significant changes in polymer and polymer:fullerene blend morphologies. But regarding thin-film deposition on different substrates, glass and FTO, the influence of the FTO substrate on the thin film morphology was clearly manifested: measured roughness values range between 3 and 10 nm for deposits on the glass and between 25 and 30 nm for those on FTO (with a FTO roughness of about 34 nm).

4. Conclusions

In summary, polymer thin films deposited by spin coating at room temperature, with a thickness between 150 and 200 nm, were optically, morphologically and structurally characterized using different techniques. Depending on the P3HT: PCBM weight ratio, they exhibit peculiarities in the optical properties variation (decreasing the refractive index, extinction coefficient and dielectric constant, depending on the amount of fullerene). Considering the influence of both weight ratio and substrate nature, from the point of view of samples morphology, the roughness increases with the increase in the fullerene amount and it is influenced by the substrate roughness.

5. References

[1] Berger P R and Kim M 2018 *Journal of Renewable and Sustainable Energy* **10**
[2] Majumder C, Rai A and Bose C 2018 *Optik* **157** 924
[3] Antohe S, Iftimie S, Hrostea L, Antohe V A and Girtan M 2017 *Thin Solid Films* **642** 219
[4] Sun C, Pan F, Bin H, Zhang J, Xue L, Qiu B, Wei Z, Zhang Z G and Li Y 2018 *Nature Communications* **9**
[5] Girtan M, Hrostea L, Boclinca M and Negulescu B 2017 *AIMS Materials Science* **4** 594
[6] Hajduk B, Bednarski H, Jarzabek B, Janeczek H and Nitschke P 2018 *Beilstein Journal of Nanotechnology* **9** 1108
[7] Kasikov A, Kahro T, Matisen L, Kodu M, Tarre A, Seemen H and Alles H 2018 *Applied Surface Science* **437** 410
[8] Collins B A, Tumbleston J R and Ade H 2011 *Journal of Physical Chemistry Letters* **2** 3135
[9] Campoy-Quiles M, Ferenczi T, Agostinelli T, Etchegoin P G, Kim Y, Anthopoulos T D, Stavrinou P N, Bradley D D C and Nelson J 2008 *Nature Materials* **7** 158
[10] Yin W and Dadmun M 2011 *ACS Nano* **5** 4756
[11] Fujiwara H 2007 *Wiley*, Chichester

Aspects Regarding Instantaneous Corrosion of Nodular Iron in Household Wastewater

M C Perju[1,2], C Savin[1], C Nejneru[1], M Axinte[1], D C Achiţei[1,2] and C Bejinariu[1]
[1]Technical University "Gheorghe Asachi" of Iasi-Romania, Department of Technologies and Equipments for Materials Processing, Blvd. Mangeron, No. 51, 700050, Iasi, Romania
[2]Centre of Excellence Geopolymer & Green Technology (CEGeoTech), School of Materials Engineering, Universiti Malaysia Perlis, Kompleks Pengajian Jejawi 2, 02600 Arau, Perlis[3] Department, University, Address

E-mail: nejnerucarmen@yahoo.com

Abstract. Waste water pumps operates under high corrosion and wear conditions due to the working environment containing sand particles, suspension particles, sulfuric acid and various dissolved salts, etc. Because of the operating conditions, the rotor and sometimes the pump stator (especially immersion ones) suffer various aggressions, such as material avulsion, corrosion, etc. This paper represents a study on the nodular cast iron samples corrosion from the rotor of a single-chamber pump, which were tested in synthetic corrosion environments that mimic the characteristics of natural environments.

1. Introduction
Nodular cast iron is a cast iron with spheroidal graphite. Due to the nodular aspect of the graphite, the rejection effect (crack propagation) is minimized and high tensile and tearing strength is obtained. Thanks to these properties, these materials have multiple uses, such as cylinder heads, engine shafts, gears, grills, parts for pumps, presses, or electric machines. In this study we have chosen a nodular cast iron used for the construction of pumps used in pumping stations for domestic wastewater transfer into the sewerage network [1-4].

The study aims to quickly assess instantaneous corrosion, corrosion that occurs on direct contact between metal and wastewater, and the pH influence on this process.

2. Materials and methods
The composition of the alloy was determined by optical emission spectrometry with a Foundry Master spectrometer and is shown in Table 1. In addition to these alloy elements, Mo, V, Co, Sn are present in less than 0.01%. The microstructure was determined using the Zeiss metallographic inverse optical microscope and is shown in Figure 1.

Considering the use of nodular cast iron in the construction of pumps used in wastewater transfer, in this study we decided to study the corrosion of the material in a series of domestic waste water simulating solutions.

To this end, we started from Boeije's studies [5] on the average composition of decanted water from municipal wastewater from municipal sewage. Choosing synthetic solutions components is based on the expected composition of domestic water discharges. Compared to the composition proposed by Boeije, minor modifications have been made to the solution used here, especially on the ratio of the mineral components.

Table 1. Chemical composition for the nodular cast iron, [%].

Element	Fe	C	Si	Ni	Mg	Mn	P	S	Cr	Ti	Al	Cu	Nb
Percent	balance	4.50	2.28	0.12	0.10	0.09	0.05	0.04	0.02	0.02	0.01	0.01	0.01

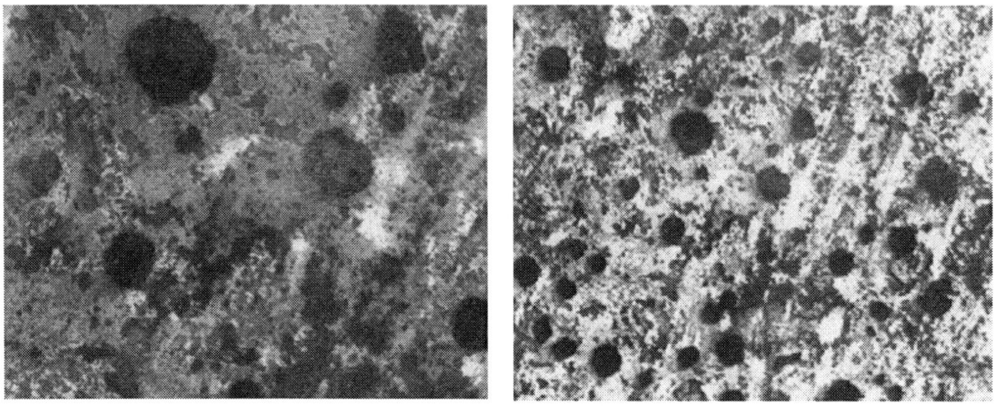

(a) (b)

Figure 1. Nodular cast iron microstructure with and ferrite-pearlite metal base: (a) 200X; (b) 100X.

The composition of the basic synthetic residual water, abbreviated as DWW-1, is shown in Table 2. Inorganic components are responsible for the corrosive properties of the solution, while food components can function as protectors or inhibitors. The DWW-1 solution presents a pH value close to neutral: pH = 6.5.

To study the influence of pH on corrosion behavior, base solution (DWW-1) was used. The pH was modified by the addition of sodium hydroxide. Thus the following solutions were obtained:

DWW-2: pH = 3.0 - by adding 0.1M HCl solution to the basic solution,

DWW-3: pH = 11.0 - by the addition of 0.1M NaOH to the basic solution.

Table 2. The basic residual wastewater composition (DWW-1).

Chemical components	mg/l	Food components	mg/l	Metallic residues	mg metal/l
NH_4Cl	15	Powdered milk	118	$Cr(NO_3)_3.9H_2O$	0.10
$CH_3COONa.3H_2O$	142	Beer yeast	54	$CuCl_2.2H_2O$	0.20
$MgSO_4.7H_2O$	32	Starch	122	$MnCl_2$	0.05
$CaHPO_4$	20	Soybean oil	15	$NiSO_4.7H_2O$	0.08
$K_2HPO_4.3H_2O$	56			$PbCl_2$	0.07
$FeSO_4$.7H2O	14				
Urea	98				
Peptone*	15				

Electrochemical study methods are a quick and effective means of obtaining information about the following: the thermodynamic corrosion probability of a metal immersed in a liquid, the instantaneous corrosion rate (corrosion rate at simple metal immersion in the corrosion environment), the corrosion type (generalized or localized) and its intensity, as well as factors that can influence the corrosion process (temperature, pH level, accelerators or corrosion inhibitors), [6-8]. Electrochemical corrosion is an oxidation-reduction process in which metal anodic oxidation occurs simultaneously with a cathodic reduction process in which the electrons generated in the anodic reaction are consumed. The two processes are coupled and carried out at the same speed, at a common potential called mixed potential, or ***corrosion potential*** (E_{cor}). The speed of the two processes is directly proportional to the current. The thermodynamic tendency of a metal or alloy oxidation, submerged in an electrolytic environment can be expressed by the ***corrosion potential***, E_{cor}, while the corrosion rate is expressed by the anode → cathode electrons transfer. At the corrosion potential, the anode current (I_a) is equal to the cathodic current (I_c) and the current in the outer circuit is zero. Under these conditions, the metal/solution system is at thermodynamic equilibrium, $(I_a\text{-}I_c)_{E=Ecor}=0$, and no apparent reaction takes place on the surface of the metal. However, metal degradation takes place only in the anodic process by oxidation, so that corrosion occurs at a rate proportional to the anode current (generally equal to the cathode current): $I_{cor}= (I_a)_{Ecor}= (I_c)_{Ecor}$ proportional with the ***corrosion current***, I_{cor}, respectively the ***current density***, $j_{cor} = I_{cor}/S$. The current density measured at the corrosion potential is a measure of the equilibrium oxidation process speed at the simple immersion of the metal or alloy in the solution and is called instantaneous corrosion rate.

The corrosion behavior of the alloy has been studied on the basis of the linear polarization curves obtained with a Dynamic Electrochemical System, VoltaLab 40 (PGZ 301) (Radiometer Analytical SAS - France).

The measurements were carried out in a three-electrode cell. Samples used for studies (work electrodes) are circular discs of 10 mm diameter, which are mounted in the special cell support and provide a flat surface of 6 mm in diameter. The metal surface exposed to the electrochemical treatment is circular without edges or corners and has in all cases the value of 0.283 cm^2. Prior to mounting in the electrochemical cell, each sample was polished onto SiC paper to 2500 grit, degreased in acetone and washed with water. The auxiliary electrode, which polarizes the work electrode, was platinum and is supplied by the appliance manufacturer.

The reference electrode, the one used to measure the the potential of the working electrode, was a saturated calomel electrode (ECS) that exhibits good stability over time. Usually, when the potential value of the electrode is indicated, the reference electrode used to measure it is also indicated: E mV (ECS) or E mV/ ECS. In the present paper, in order not to complicate the writing, the potentials were expressed in mV or V without indicating the type of electrode to which they were measured. The determinations were performed at 25°C.

Linear polarization measurements were performed in the potential range (+ 200 ... -200 mV) against the open circuit potential with a scanning potential rate of dE/dt = 0.5 mV/s. The acquisition and processing of the electrochemical data was done with the VoltaMaster 4 software.

3. Results and discussions

The density of the instantaneous corrosion current can also be evaluated on the basis of the linear polarization curve using the ***polarization resistance method***, this one representing the slope of the tangent to the potential-current density curve [$E = f(j)$] in the equilibrium point ($E = E_0=E_{cor}$ or $\eta = 0$, where $\eta=E-E_0$ is the overpotential), that means at the corrosion potential: $R_P = (dE/dj)_{E_0}$.

The polarization resistance, Rp, is evaluated on the basis of the Butler-Volmer equation. If the electrode is corroding the metal through a single electrode controlled charge reaction can be written as [9, 10]:

$$j = j_{cor}\left[\exp\left(\frac{2{,}303(E - E_{cor})}{b_a}\right) - \exp\left(-\frac{2{,}303(E - E_{cor})}{b_c}\right)\right] \quad (1)$$

where: b_a şi b_c represent the slopes of the linear portions of the anode and cathode sections in the diagram E= f(log J) close to the corrosion potential, known as the Tafel slopes:

$$b\,b_a = \left(\frac{\partial E}{\partial \log j}\right)_a = \frac{RT}{\alpha nF} \quad \text{respectively} \quad b_c = \left(\frac{\partial E}{\partial \log j}\right)_c = \frac{RT}{(1-\alpha)nF} \tag{2}$$

It has been experimentally observed that *j* variation is approximately linear according to the applied potential (E), starting at about 50...60 mV with respect to the corrosion potential and only on a range of about 10...20 mV. Stern şi Geary [11, 12] have simplified the Butler-Volmer equation for the case of low overvoltages relative to E_{cor} , linearizing this equation by developing logarithmic terms in a series ($e^x = 1 + x + x^2/2! + x^3/3! + \ldots$), and neglecting the terms of superior order.

The simplified ecuation it may be given:

$$R_p = \left(\frac{dE}{dj}\right)_{(E_{cor})} = \frac{b_a \cdot b_c}{2{,}303 . J_{cor} . (b_a + b_c)} \qquad (\text{ohm}\cdot\text{cm}^2) \tag{3}$$

After re-arranging the equation we can obtain for instantaneous corrosion intensity:

$$J_{corr} = \frac{b_a \cdot b_c}{2.303(b_a + b_c)\cdot R_p'} \qquad (\text{mA}\cdot\text{cm}^2) \tag{4}$$

The corrosion rate, expressed as penetration velocity, can be evaluated with the [12-19] relation:

$$v_p = 3{,}27 . \overline{\left(\frac{A}{z}\right)} . \frac{J_{cor}}{\rho} \qquad (\text{mm/an}) \tag{5}$$

where $\overline{\left(\frac{A}{z}\right)} = \sum g_i \frac{A_i}{z_i}$ represents the average electrochemical equivalent of the alloy; g_i represents the gravimetric fractions (mass fractions) of the alloy components and (A_i/z_i) the corresponding chemical equivalents.

The corrosion potential was evaluated from the linear polarization curves, using the Evans diagram, which represents the logarithm of the current density according to the potential of the electrode over a overpotential range of ±50....60 mV. In these coordinates at the intersection of the linear portions of the anodic and cathodic sections of the polarization curve, on the potential axis is obtained the corrosion potential value, $E_{cor} = E_{(I=0)}$.

The Evans diagram for the nodular casting behavior in the three corrosion environments and the Tafel slope evaluation module are shown in Figure 2. All measures necessary to calculate the corrosion rate, including corrosion rate, were evaluated using the features of the VoltaMaster 4 software. The results are shown in Table 3. The calculated area taken into consideration was 120 mV (± 60 mV around the corrosion potential) and the linearity range on the Evans curves was 25 mV.

The analysis of these data allows to highlight the following observations:

IOP Conf. Series: Materials Science and Engineering **374** (2018) 012016 doi:10.1088/1757-899X/374/1/012016

(a) The corrosion potential values indicates that from the thermodynamic point of view the most aggressive medium is the solution with almost neutral pH and the least aggressive solution is the alkaline solution.

(b) The polarization resistance has the highest value for the neutral solution, this being reflected in a low value of current density and corrosion rate. In acidic environment the very low value of the polarization resistance leads to the highest corrosion rate. In basic medium, the rate of corrosion has an average value of the solution between the one corresponding to the neutral and the acidic solution. Apparently, there is a contradiction between the conclusions that can be highlighted on the basis of the corrosion potential and those resulting from polarization resistance or current density analysis. This can be explained by the fact that the corrosion potential is a purely thermodynamic balance of the redox reactions in the system, which does not take into account the kinetic factors that influence differently the two coupled reactions, the oxidation of the metal and the reduction of hydrogen or oxygen.

(c) The fact that in all cases, b_a is lower than b_c can explain the contradiction above. If the anodic and cathodic reaction rate is equal, the two Tafel slopes are equal. When the rate of reaction is controlled by the transfer rate of the ions from the solution to the metal surface, polarization of the concentrations may occur. In addition, the pH of the solution can have a significant influence on electrode reactions;

- in acidic environment:
  - metal oxidation occurs at the anode: $M \rightarrow M^{+z} + ze^-$
  - hydrogen reduction occurs at the cathode: $zH^+ + ze^- \rightarrow z/2\ H_2$
  - the global reaction is: $M + zH^+ \rightarrow M^{+z} + z/2\ H_2$
- in neutral or bazic environments:
  - at the anode: $M \rightarrow M^{+z} + ze^-$
  - at the cathode: $zH_2O + ze^- \rightarrow zOH^- + z/2H_2$
  - the global reaction is: $M + zH_2O \rightarrow M^{+z} + zOH^- + z/2\ H_2$

Hydrogen is released in both the acidic and the basic environment at the cathode. In the case of the studied alloy, the oxidation component is iron, and the global reactions are:

- in acidic environment: $Fe + 2H^+ \rightarrow Fe^{+2} + H_2$
- in neutral or bazic environments: $Fe + 2H_2O \rightarrow Fe^{+2} + 2OH^- + H_2$

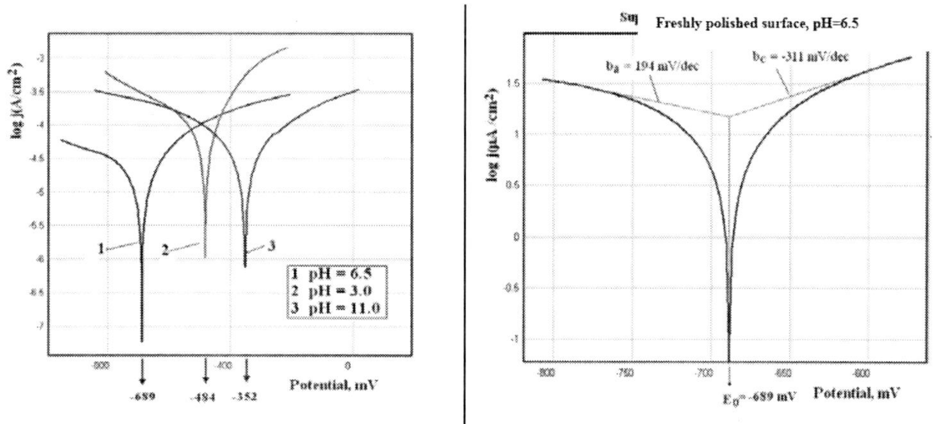

(a) (b)

Figure 2. (a) Evans diagram for nodular cast iron behavior for those three corrosion environments; (b) Evaluation of Tafel slopes and corrosion potential for the standard sample.

When the diffusion of oxygen or hydrogen through the corrosion medium is not fast enough, it can be argued that the reaction is kinetically controlled and $b_a < | b_c |$, with the reaction rate being different from the thermodynamic predicted one. The formation of oxide on the surface of the metal, which may or may not lead to passivation, alters the surface of the studied sample, and this changes the data of the Butler-Volmer equation and thus the values of the Tafel constants. Mixed control of the corrosion process can occur when hydrogen and oxygen are simultaneously reduced in the cell, which further complicates the global corrosion reaction [13-24].

Table 3. Electrochemical parameters of instantaneous corrosion in simulated wastewater.

Corrosion Agent	pH	E(I=0), mV	R_P, ohm.cm^2	j_{cor}, mA/cm^2	v_{cor}, mm/an	b_a mV/dec.	b_c mV/dec.
DWW-1	6.5	-689	2410	0.0150	0.176	194	-311
DWW-2	3.0	-484	467	0.0660	0.773	135	-252
DWW-3	11.0	-352	1310	0.0301	0.351	238	-250

4. Conclusions

Using the method of polarization resistance, was analyzed the instantaneous corrosion of a nodular cast iron used in the construction of pumps used in the circulation of waste water in simulated domestic waters, and the influence of pH on the corrosion rate was also studied.

The studies showed that the corrosion rate is about 4.5 times higher in acidic environment and only twice as high in the alkaline environment than in neutral wastewaters.

5. References

[1] Agape I, Gaiginschi L, Gaiginschi R, Baltagiu D 2016 *Automobile Engineering* **38** 22-25
[2] Aelenei N, Lungu M, Mareci D and Cimpoesu N 2011 *Environmental Engineering and Management Journal* **10**(12) 1951-1958
[3] Burlacu-Arsene M C, Leonte C, Antonescu M C, Sandu A V and Minciuna M G 2015 *Rev. Chim.* **66**(4) 487-491
[4] Tugui C A and Vizureanu P 2017 *IOP Conference Series-Materials Science and Engineering* **209** 012020
[5] Boeije 1999 Chemical fate prediction for use in geo-referenced environmental exposure assessment, *PhD. Thesis, Faculty of Agricultural and Applied Biological Sciences. Ghent University* 225
[6] Psyllaki P P, Pantazopoulos G and Pistoli A 2013 *Engineering Failure Analysis* **35** 418–426
[7] Vanaei H R, Eslami A and Egbewande A 2017 *International Journal of Pressure Vessels and Piping* **149** 43-54
[8] Hernandez-Sandoval J, Gonzalez-Lopez R, Hernandez-Rodriguez M A L and Guzmán A M 2015 *Engineering Failure Analysis* **53** 124–131
[9]. Bockris J O'M and Reddz A K N 1970 *Modern Electrochimistry* **2** New York, Plenum Press
[10] Gileadi E 1993 *Electrode Kinetics for Chemists, Chimical Engineers and Material Scientists* VCH
[11] Stern M and Geary A L 1957 *J. Electrochem. Soc.* **104** 56
[12] Stern M and Geary A L 1958 *J. Electrochem. Soc.* **105** 638
[13] Aelenei D M 2009 Corrosion study of some dental materials *PhD Thesis, Faculty of Materials Science and Engineering, Gh. Asachi" Technical University of Iasi*
[14] Lungu (Costea) Maria 2012 Metals corrosion in the natural marine environment *PhD Thesis, Faculty of Chemical Engineering and Environmental Protection , Gh. Asachi" Technical University of Iasi*

[15] Mareci D 2004 Electrochemical behavior of some biomaterials *PhD Thesis, Faculty of Chemical Engineering and Environmental Protection, Gh. Asachi" Technical University of Iasi*
[16] Florea C D, Munteanu C, Cimpoesu N, Sandu I G, Baciu C and Bejinariu C 2017 *Rev. Chim.*, **68** 2582–2587
[17] Lazar P, Bejinariu C, Sandu A V, Cazac A M, Corbu O, Sandu I G and Perju M C 2017, *Iop Publishing Ltd: Bristol* **209** 12025
[18] Florea C D, Carcea I, Cimpoesu R, Toma S L, Sandu I G and Bejinariu C 2017 *Rev. Chim.* **68** 2397–2401
[19] Florea C, Bejinariu C, Munteanu C and Cimpoesu N 2017 *Advanced Materials Engineering and Technology* **1835** 020053
[20] Sandu A V, Bejinariu C, Nemtoi G, Sandu I G, Vizureanu P, Ionita I, Baciu C 2013 *Revista de chimie* **64** 825-827
[21] Toma S L 2013 *Surf & Coat. Technol.* **220** 261–265
[22] Toma S L, Badescu M, Ionita I, Ciocoiu M and Eva L 2014 *Applied Mechanics and Materials* **657** 296-300
[23] Toma S L, Bejinariu C, Baciu R, Radu S 2013 *Surf & Coat. Technol.* **220** 266–270
[24] Voiculescu I, Geantă V, Stern A, Ashkenazi D, Cohen M and Cvikel D 2017 *Metallography, Microstructures And Analysis* **6(1)** 1-20

Researches on the Improvement of the Bioactivity of TiO_2 Deposits, Obtained by Magnetron Sputtering - DC

B F Toma[1], R E Baciu[2], C Bejinariu[3], N Cimpoieşu[3], B M Ciuntu[4*], S L Toma[3*], D P Burduhos-Nergis and D Timofte[4]

[1] Grigore T.Popa University of Medicine and Pharmacy, Faculty of Medicine, Department of Obstetric&Ginecology, 16 Universitatii Str., 700115,Iasi, Romania.

[2] Grigore T.Popa University of Medicine and Pharmacy, Faculty of Dental Medicine, Department of Dental Materials, 16 Universitatii Str., 700115,Iasi, Romania.

[3] "Gheorghe Asachi" Technical University from Iasi, Faculty of Science and Materials, Bd. D. Mangeron, No. 64,700050, Romania.

[4] Grigore T.Popa University of Medicine and Pharmacy, Faculty of Medicine, Department of Surgery, 16 Universitatii Str., 700115,Iasi, Romania

E-mail: stl_toma@yahoo.com, bogdanmciuntu@yahoo.com

Abstract. In this paper, layers of TiO_2 were deposited, by magnetron sputtering, on a glass support. The parameters of the deposition process were kept constant except for the O_2/(Ar + O2) ratio that varied on three levels. The physical and mechanical properties of the layers obtained were investigated by SEM optical microscopy, electronics, AFM and X-ray diffraction. The bioactivity of TiO_2 surfaces was investigated by growing M3C3-E1 osteoblast cells produced by RIKEN Cell Bank (Japan) for a period of 5 days. The modification of the working environment in the enclosure determines both the phasic modification in the TiO_2 film, respectively the amount of the anatase or rutile phase and the decrease of the average roughness of the film from 112.3nm to 56.7nm. The research has demonstrated that the finer layers with a high content of anatase promote the growth of M3C3-E1 cells.

1. Introduction

Due to its physical, chemical and mechanical properties titanium has applications in various fields such as aeronautics, machine building, chemical, and nuclear industries, etc., [1-4]. It is known that in the air, the titanium alloys form on the outer surface a thin layer of TiO_2 with a rutile or anatase structure, which offers a very good corrosion resistance, high thermodynamic stability and reduced sensitivity to ion formation - in aqueous solutions [5-8]. All of these properties contribute to increasing the biocompatibility of Ti and its alloys leading to better bone integration compared to other metallic implants [9-11]. It is known that, any titanium alloy introduced into the human body reacts at interface with the osteoblast cells through an osteointegration process, [12]. The osteointegration capacity of the implant is dependent on surface characteristics, such as: structure, chemical composition, rugosity, topography, and morphology of the implant surface, [13].

In some medical applications, titanium is used as titanium oxide (TiO_2), deposited as a thin layer obtained by various techniques: resistive evaporation, thermal spraying, physical vapor depositions (PVD), magnetron sputtering-reactive DC, reactive magnetron sputtering -RF, vapor electron beam, ion beam, plasma electrolytic deposition, etc., [14-16] obtaining, each time, films which dependent on the preparation techniques and deposition conditions. Of all these techniques, the method reactive magnetron sputtering - DC reactive is the most used due to its high productivity and due to fact, that allow obtaining of dense and uniform films [17, 18].

In this paper, TiO_2 layers were deposited on a glass substrate - by magnetron sputtering -DC method and the influence of the $O_2/(Ar + O_2)$mixture on the layer structure and bioactivity of TiO_2 films was studied.

2. Experimental procedure

TiO_2 layers were deposited in vacuum by magnetron sputtering, using an experimental installation - similar to the installation used by Toma et al. [7], whose sketch is represented in Figure 1. Thin films of TiO_2 have been deposited on glass microscope lamellas, having the dimensions of 25x75x1.5mm^3.

Before performing the deposition, the glass lamellas were ultrasonically cleaned in an acetone bath and ethanol. At the end, they were rinsed with distilled water and dried by blowing of dry argon (99,999% purity). The argon and oxygen flow rate introduced into the enclosure have varied on three levels and the substrate temperature was 150°C.

The experimental parameters used are presented in Table 1. The target on which deposition was performed was Ti, of 99.9% purity, with a diameter of 50.8mm and a thickness of 4mm [7]. The glass lamellas were fixed on the anode made of the steel plate, cut in the central area, having a thickness of 2 mm.

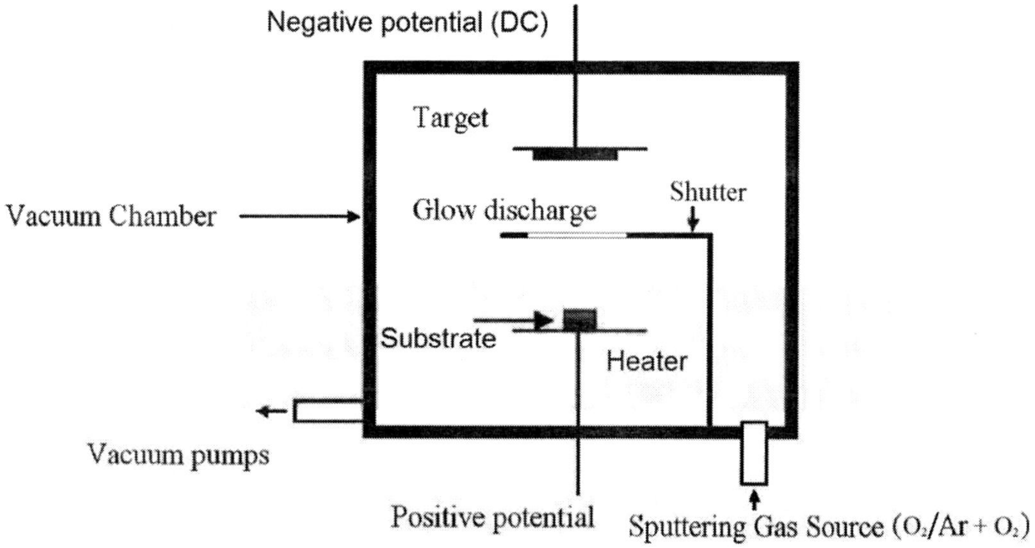

Figure 1. Sketch of the installation for deposit of TiO_2 films by DC magnetron sputtering system [7].

Table 1. Technical parameters of magnetron sputtering.

Parameters	Value
Sputter target	Titanium target
Target to substrate distance (mm)	80
Sputter power (W)	250
Substrate temperature (°C)	150
$O_2/Ar + O_2$ ratio	0.1/0.2/0.3
Sputter pressure (mbar)	$3.6 \cdot 10^{-3}$
Sputter time (sec)	26

2.1. Characterization of films

The crystalline structure of TiO_2 depositions was highlighted by X-ray diffraction using X-ray diffractometer - type X'PERT PRO MRD product by PANalytical Holland, using Cu Ka radiation (λ = 0.15406 nm) which works with acceleration voltages of 40kV and an emission current of 40mA. The SEM investigations carried out using the electronic scanning microscope: FEI - Quanta 200 SEM highlighted the deposition morphology. The surfaces of TiO_2 films were obtained by atomic sheet microscopy using the microscope: AFM Easyscan2, Nanosurf Switzerland - who operated in tapping mode. The osteoblast cells of MC3T3-E1 type, produced by RIKEN Cell Bank firm (Japan) are placed in the Dulbencco modified Eagle's medium containing 10% fetal Bovine Serum, 500 units/ml penicillin and 500 units/ml streptomycin. All incubations of cells were made in the Co atmosphere at 37°C, with relative humidity of 100%. Medium was changed every 2 days of experiment. The cells investigation was performed by carrying out the proliferation test after a day, 3 and 5 days since of cell cultivation. Proliferated cells were quantified by counting them with a haemocytometer.

3. Results and discussions

In figure 2 are presented diffractograms of TiO_2 layers deposited by magnetron sputtering at different working reports. The diffractograms present two diffraction peaks, well defined at 2θ values, corresponding to the planes (101) and (200), of the crystalline phase of the anatase. The films have a crystal structure - in accord to other works using magnetron sputtering deposition techniques, [19, 20]. It can be seen in Figure 1 that the intensity of the anatase peaks increases with the increase in O_2 content. It can also be observed that the excess presence of O_2 quantity in the enclosed workplace impedes the crystallization of the anatase phase.

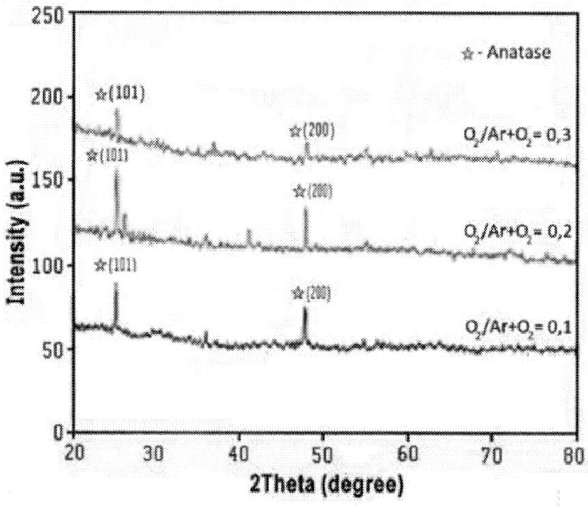

Figure 2. XRD patterns of TiO_2 thin films deposited at p=3.6·10^{-3}mbar and T=150°C.

In Figure 3 are presented AFM images of the three samples prepared at different O_2/(Ar+O_2) ratios, in the Table 2 there are presented the average grain size (D), average roughness (R_{ave}), and root mean square roughness (R_{rms}) obtained by five determinations - by AFM analysis. It is observed an is an obvious increase of the physical dimension of the grains with the addition of O_2. In addition, by analyzing the roughness variation, it can be observed that when increasing the O_2/(Ar + O_2) ratio from 0.1 to 0.3, the average roughness decreases from 112.3nm to 56.7nm.

To investigate the interaction of cells with the surface of TiO_2 film, the culture layers were scanned using the electron microscope. Figure 4 are presented the morphologies of the culture deposited on the surface of the TiO_2 film on after 3 day.

Table 2. Experimental result.

No	$O_2/Ar+O_2$ ratio	D[nm]	R_{ave}[nm]	R_{rms}[nm]
1	0.1	46.5	112.3	128.2
2	0.2	122.4	86.4	104.5
3	0.3	182.2	56.7	69.4

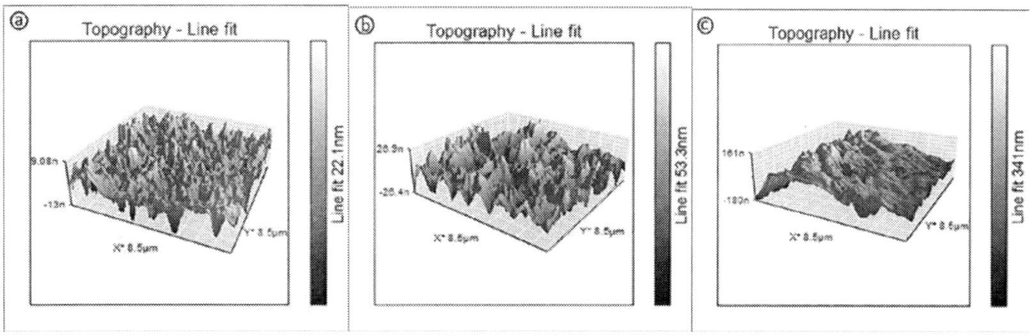

Figure 3. AFM images of TiO_2 thin films deposited at different ratio $O_2/(Ar+O_2)$: (a) 0.1; (b) 0.2; (c) 0.3.

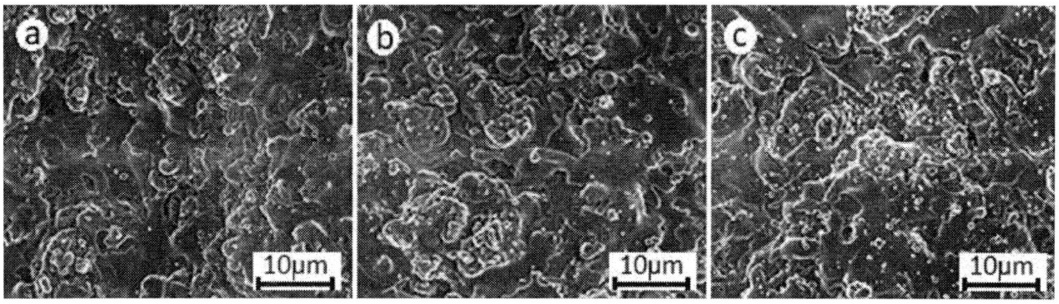

Figure 4. SEM images of osteoblasts cultured directly on TiO_2 layers obtained at different ratio $O_2/(Ar+O_2)$: (a) 0.1; (b) 0.2; (c) 0.3.

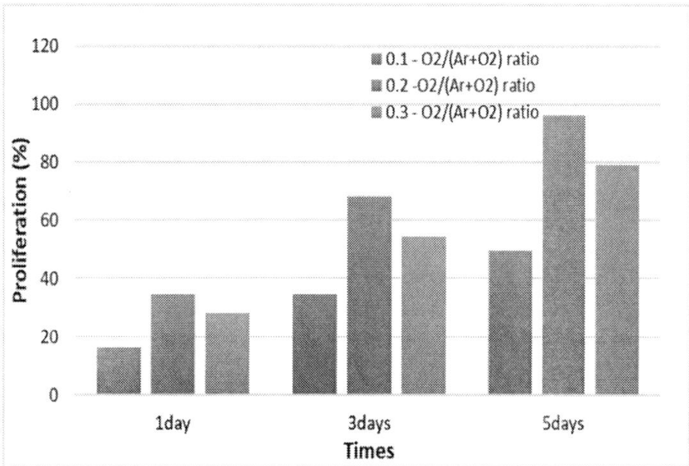

Figure 5. Proliferation cells MC3T3 after a day, 3 and 5 days since the cell cultivation.

As it can have observed, the film obtained is porous and the cells have the morphology rounded. Increasing of larger coatings are recorded on the TiO_2 surface, obtained at a ratio of $O_2/(Ar+O_2) = 0.2$. It can be suggested that the anatase phase present in the TiO_2 film favours cell growth.

The result of MC3T3 cell proliferation on the surface of TiO_2 film after a day, 3 and 5 days since the cell cultivation it is presented in Figure 5. As it is observed, the surface of TiO_2, obtained at 0.1 value of the $O_2/(Ar + O_2)$ ratio has the lowest cell proliferation.

A higher cell proliferation has been recorded on TiO_2 surfaces, obtained at an $O_2/(Ar + O_2)$ratio of 0.2. It can be suggested that the anatase phase - present in the TiO_2 film and the existence of a smaller roughness - Table 2 and Figure 5, favours cell growth.

4. Conclusions

Following experimental research on the bioactivity of thin films of TiO_2 obtained in vacuum by magnetron spraying we can state:

- the increase in oxygen content in the working chamber used to obtain the TIO_2 layers - favours the increase of the phase of the anatase – see figure 1. However, the increase in excess of the $O_2/(Ar + O_2)$ ratio, prevent the crystallization of anatase phase;
- the grain size and the roughness of the TiO_2 deposition decrease with the increase of the O_2 content in the enclosure – see figure 2;
- the presence of the anatase phase in the TiO_2 layer and of a smaller roughness - see Table 2, favour the growth of osteoblast cells of MC3T3-E1 type.

5. References

[1] Akl A A, Kamal H and Abdel-Hady K 2006 *Applied Surface Science* **252** 8651-8656
[2] Yang W and Wolden C A 2006, *Thin Solid Films* **515** 1708 -1713
[3] Tugui C A, Nejneru C, Galusca D G, Perju M C, Axinte M, Cimpoesu N and Vizureanu P 2015 *Journal of Optoelectronics and Advanced Materials* **17**(11-12) 1855-1861
[4] Perju M C and Vizureanu P 2014 *Revista de chimie* **65**(6) 694-696
[5] Zitter H, Plenk H J 1987 *J. Biomed. Mater. Res.* **21** 881
[6] Wieser E, Tsyganov I, Matz W, Reuther H, Oswald S, Pham T, Richter E 1999 *Surf. Coat. Technol.* **111** 103
[7] Toma S L, Bejinariu C, Eva L, Sandu I G, Toma B F 2015 *Key Engineering Materials* **660** 86-92
[8] Toma S L, Badescu M, Ionita I, Ciocoiu M and Eva L 2014 *Applied Mechanics and Materials* **657** 296-300
[9] Lee K, Ko Y M, Choe H C, Kim B H 2012 *J. Nanosci. Nanotechnol.* **12** 822
[10] Liua Z, Wang W, Liu H, Wang T, Qi M 2013 *Appl. Surf. Sci.* **266** 250
[11] Toma S L 2013 *Surf & Coat. Technol.* **220** 261–265
[12] Crawford G A, Chawla N, Das K, Bose S, Bandyopadhyay A 2007 *Acta Biomater.* **3** 359
[13] Le Gué hennec L, Soueidan A, Layrolle P, Amouriq Y 2007 *Dent. Mater.* **23** 844
[14] Tillmann W, Walther F, Luo W F, Haack M, Nellesen J, Knyazeva M 2018 *Journal of Thermal Spray Technology* **27** 50-58
[15] Li S J, Wang Y, Xiang D, Meng W J 2016 *Rare Metal Materials and Engineering* **45** 2555-2560
[16] Nechachea R, Nicklaus M, Diffalah N, Ruediger A, Rosei F 2014 *Appl Surf Sci* **313** 48–52
[17] Suhail M H, Mohan G, Mohan S 1992 *J. Appl. Phys.* **7**(3) 1421-1427
[18] Irimiciuc Ş A, Agop M, Nica P, Gurlui S, Mihaileanu D, Toma S, Focsa C 2014 *Jpn. J. Appl. Phys.* **53** 11
[19] Hou Y Q, Zhuang D M, Zhang G, Zhao M and Wu M S 2003 *Applied Surface Science* **218** 98-106
[20] Nedeff V, Mosnegutu E, Panainte M, Ristea M, Toma S, Agop M 2012 *Powder Technology* **221** 312–317

IOP Conf. Series: Materials Science and Engineering **374** (2018) 012018 doi:10.1088/1757-899X/374/1/012018

Mechanical and Microstructural Characterization of a New Corrosion Resistant Stainless Steel

I Voiculescu[1*], V Geantă[1], R Stefănoiu[1], C Cotruț[1], R Ciocoiu[1] and M Ionescu[2]
[1] POLITEHNICA University of Bucharest, 313 Splaiul Independenței, 060042, Bucharest, Romania
[2] SC IMA METAV SA, 67-77 Biharia street, 013981, Bucharest, Romania

E-mail: ioneliav@yahoo.co.uk

Abstract. The paper investigates the manner in which the chemical composition and delivery status of a new type of stainless steel, highly alloyed with Ni and Cr, affect mechanical properties, microstructure and corrosion resistance. The results obtained during the mechanical test (tensile, compression, Charpy test and micro-hardness) have revealed promising values. During the corrosion test, the preferential attack of the reagent (Aqua regia) located on the grain boundaries, inclusions or polyhedral precipitates have been observed. On the corroded surfaces, some localized pitting effects on grain boundaries have been revealed. The analyses of the parameter values recorded during the corrosion test revealed that the corrosion current density had a low value, comparable to that of other specific types of stainless steels. The actual I_{corr} (1.089 $\mu A/cm^2$) value measured for the experimental alloy proves good resistance to corrosion in 3% NaCl saline solution. The estimated rate of corrosion presented acceptable values (0.011 mm/year).

1. Introduction

Austenitic microstructure refractory alloys, class X10NiCrAlTi 32-20, containing 32% Cr and 20% Ni, are currently used for the manufacture of components integrated in industrial equipment and plants operating in corrosive environments at high temperatures.

The main requirements imposed on these alloys are related to their microstructural stability at high temperatures (creep resistance) and corrosion resistance [1].

There are currently known grades of similar refractory stainless steel (X 10 NiCrAlTi 32 20*/X 5 NiCrAlTi 31 20**/X 10 NiCrAlTi 32 20 H***) - manufactured by specialized companies - serviceable in the best conditions up to temperatures of 500°C in humid corrosive environments. Similar alloys with high resistance to high stress, high temperatures and corrosive environments are manufactured on the specialized market under different trade names (Dicaloy, A286) used for turbine rotors, engines or chemical reactors components [2].

EN 14532-2 provides X10NiCrAlTi 32-21 and X0NiCrAlTi 32-21 (RK) as serviceable in the 600 - 1000°C temperature range for up to 300000 hours of operation. Unfortunately, super-austenitic steels do not behave well at temperatures above 680°C, and nickel based super alloys (Inco 617, INCOLOY 800, 800T, 8000HT, Waspaloy) [3-7] are required for such applications.

The Asian market can be considered the main competitor in the special steels market at global level. Some of the most famous manufacturers of special steels and super alloys are: SPECIAL Metals Corporation - USA; Bibus Metal - Switzerland; Santok Steel - India; Piyush Steel - India; Tools&Die

Steels - China; Aubert Duval-France. The first paragraph after a heading is not indented (Bodytext style).

According to the Romanian National Institute of Statistics, the total annual steel consumption in Romania (an accurate indicator of the GDP evolution) in 2017 amounted to approx. 3.4 million tons, of which, the special alloys required amounted to approx. 40,000 t/year, quantity fully imported.

The X10NiCrAlTi 32-20 refractory alloy manufactured in Romania is classified as stainless steel due to the very high Ni content (32.0 - 38.0%), although its chemical composition is very close to the one of a super alloy (Ni-based), taking from it not only the advantages (corrosion resistance, high working temperature etc.), but also the disadvantages, of which the most important are the low technological properties, thus creating great problems for obtaining desired quality parts.

The high content of Cr (19 - 23%) also creates great problems during production and casting, requiring smart solutions for placing coolers in zones with high contractions and sections to ensure a low level of impurities (by refining and micro-alloying with the Ti, Al, Ce stabilizing elements).

Since the main problem in the processing of this alloy is the loss of microstructural stability and the hot cracking [9, 10], elements such as Ti, Al and lanthanides (Ce) are added to the chemical composition in order to finish the granulation and limit the effects of precipitation of chromium carbides, causing the alloy to become brittle at high temperatures.

The aggregate content of the micro alloying elements should be less than 0.8% (Al + Ti + Ce) in order not to generate additional hardening effects and to maintain the tenacity of the metal matrix.

The aim of the paper is to present the results for the execution and testing of a new grade of refractory austenitic steel, highly alloyed with 32% Cr and 20% Ni and stabilized with titanium, in order to obtain parts which can be used in various niche fields in the industry. The development of the production technology combined with the casting technology for this steel grade will bring economic benefits and prestige to the manufacturer, IMA METAV, one of the few SMEs in Romania whose activity is the manufacture of castings made of special alloys.

2. Materials

The experimental X10NiCrAlTi 32-20 refractory stainless steel were obtained in the RAV plant of IMA METAV, Bucharest.

The chemical composition of the new austenitic refractory stainless steel was determined by optical emission spectrometry using the SpectromaXxM spectrometer and the Fe-30-M (Cr-Cr/Ni-steel – M) program.

For inter-comparison, the chemical composition analysis was conducted by IMA-METAV as well, using the Fe-30 program, and the results are summarized in Table 1.

Table 1. Chemical composition of X10NiCrAlTi 32-20, wt. %.

C	Si	Mn	Cr	Ni	Mo	Al	Cu	Ti	P	S
UPB										
0.104	1.09	0.42	19.98	32.28	0.453	0.432	0.160	0.175	0.014	0.0011
IMA METAV										
0.094	1.22	0.689	19.0	31.69	0.435	0.46	0.191	0.151	0.018	0.012

3. Testing methods

3.1. Mechanical tests

The tensile tests were carried out using the ”W+b” Walter+Bai AG traction/compression testing machine, using specimens taken from the cast melts, processed by milling (Figure 1).

The results of the tensile test performed on a batch of 3 identical specimens are shown in Figure 2 and Table 2.

The elasticity modulus was determined by linear regression E = 104608.98±3946.2 Mpa.

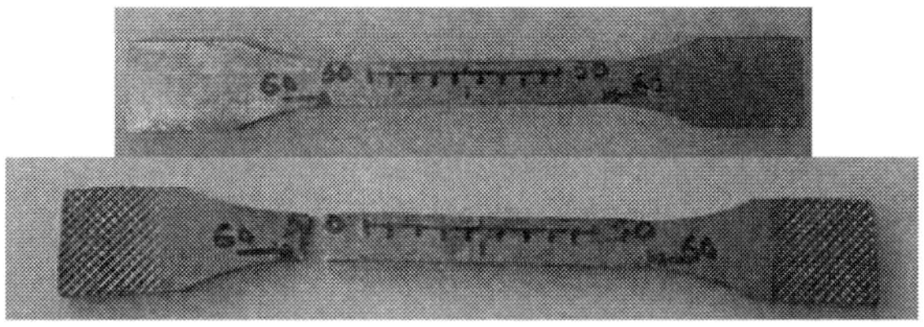

Figure 1. Continuous tensile test specimen, dimensions: 9.54x3.99 mm and Lc=105 mm, before and after fracture.

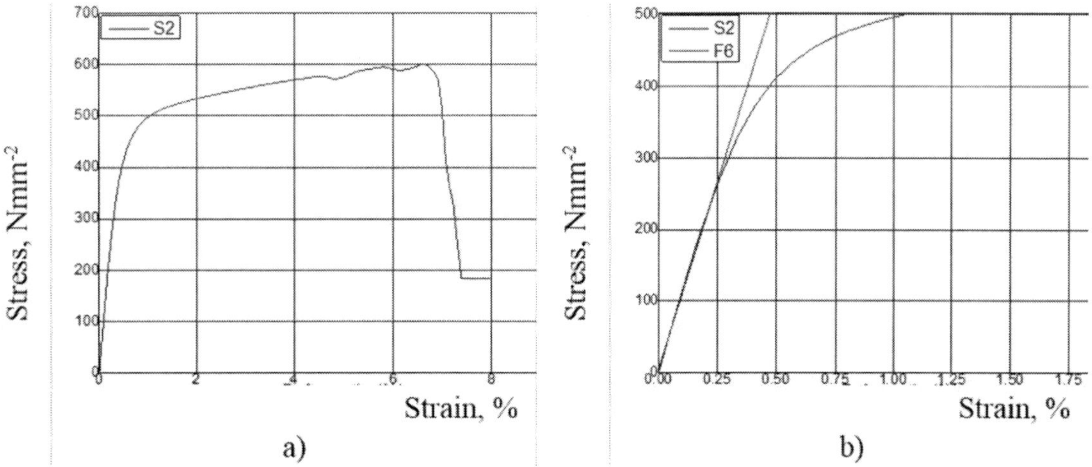

Figure 2. The tensile stress-strain diagrams: (a) global diagram; (b) proportionality zone for 0.25% proportional deformation.

Table 2. The average values of mechanical characteristics for experimental refractory stainless steel.

Material	σ_e [MPa]	$\sigma_{0.2}$ [MPa]	σ_{max} [MPa]
X10NiCrAlTi 32-20	260	450	601

The compression test results are presented in Figure 3.

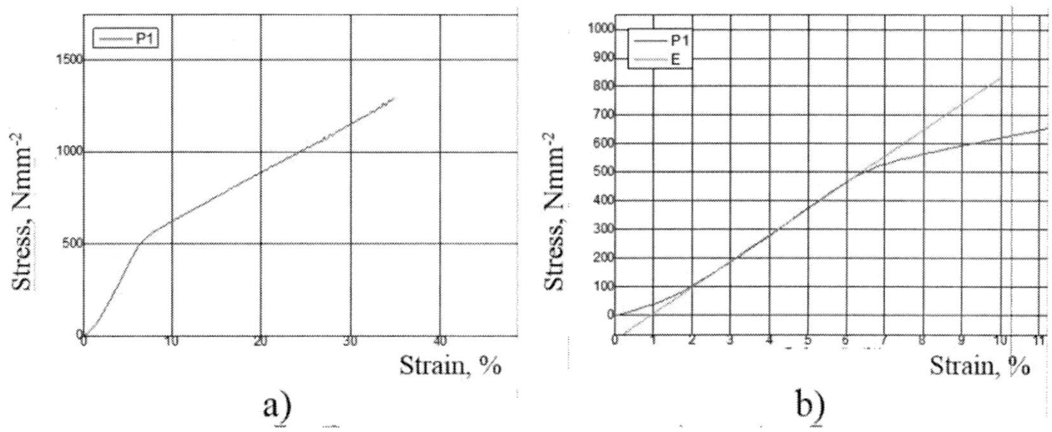

Figure 3. Stress-strain diagrams for the compression test: (a) Stress-strain diagram; (b) Determination of the elastic limits value: 498MPa with 6.34% deformation.

The impact tensile test was conducted at ambient temperature (23 °C. 45% humidity), using the W+b (Walter+Bai AG) testing machine. Four identical specimens, with small cross-section, were tested, and the results are shown in Figure 4 and Table 3, indicating the mixed-mode fracture (ductile/brittle) and the existence of sliding planes.

Test 1

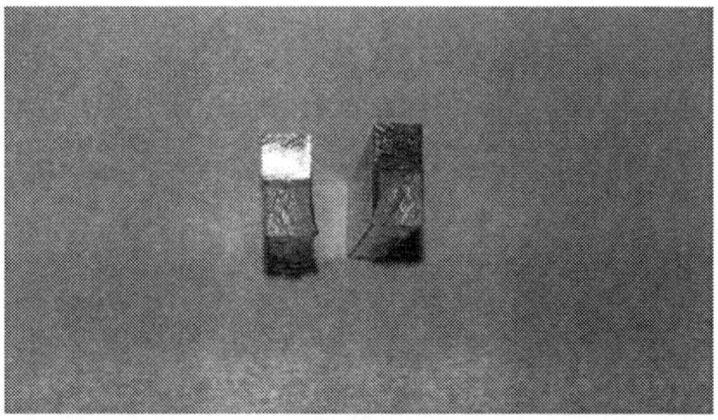
Test 2

Test 3

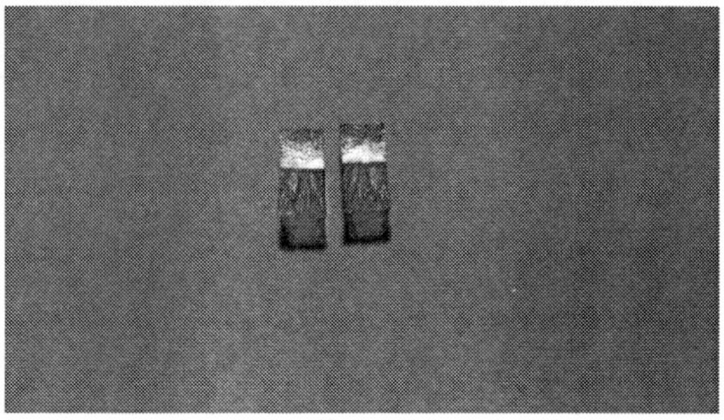
Test 4

Figure 4. Macroscopic appearance of impact fracture surfaces.

Table 3. Absorbed energy values during the impact test.

Sample	**1**	**2**	**3**	**4**	**Average value**
Absorbed energy, J	14.4	11.1	8.9	9.9	11

3.2. Corrosion resistance assessment of stainless steel

The corrosion resistance was determined using *the linear polarization technique,* consisting of the following steps: measurement of open circuit potential (E_{OC}) over 6 hours; plotting the potential-dynamic polarization curves from -1 V (vs OCP) to +1 V (vs SCE), at a 1 mV/s scan rate. The corrosion resistance evaluation tests were conducted using a potentiostat / galvanostat (PARSTAT 4000 model, manufactured by Princeton Applied Research, USA), and the potential dynamic curves were acquired using the Versa Studio v.2.50.3 software.

The tests were conducted using an electrochemical cell (Figure 5) consisting of a calomel saturated electrode (CSE) - a reference electrode, a platinum electrode - a recording electrode and a working electrode, represented by the stainless steel samples investigated.

The surface of the samples was prepared before the tests were carried out by: grinding using 400- to 1000-grit metallographic papers, polishing with a particle suspension of Al_2O_3 with an average diameter of 1 μm. In order to ensure the electrical contact, the surface of the samples was coated with an epoxy resin and the surface exposed to corrosion was of approximately 1 cm^2.

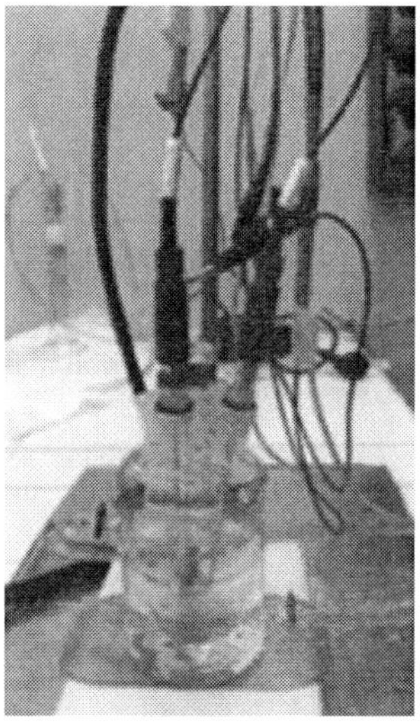

Figure 5. Electrochemical cell used in corrosion tests.

The tests were conducted in 3.5% NaCl saline solution at 25 °C. Before the tests, the samples were immersed in electrolyte until the open circuit potential (OCP) stabilized. The open circuit potential variation (E_{oc}) of the stainless steel is shown in Figure 6, and the potential dynamic curve is shown in Figure 7. The polarisation resistance, used to characterise the corrosion resistance, was calculated using the parameters determined by the Tafel technique. The main parameters of the electrochemical corrosion process are presented in Table 4.

Table 4. The main parameters of the corrosion process.

Material	**E_{oc} (mV)**	**E_{cor} (mV)**	**i_{cor} (μA/cm^2)**	**β_c (mV)**	**β_a (mV)**	**Rp (kΩxcm^2)**	**CR (mm/year)**
Stainless steel	-179	-369	1.089	103.13	1333	38.21	0,011

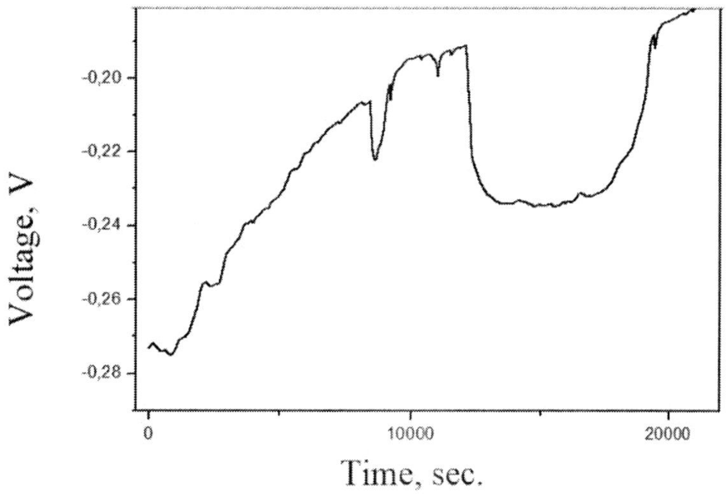

Figure 6. The evolution of the open circuit potential (E_{oc}).

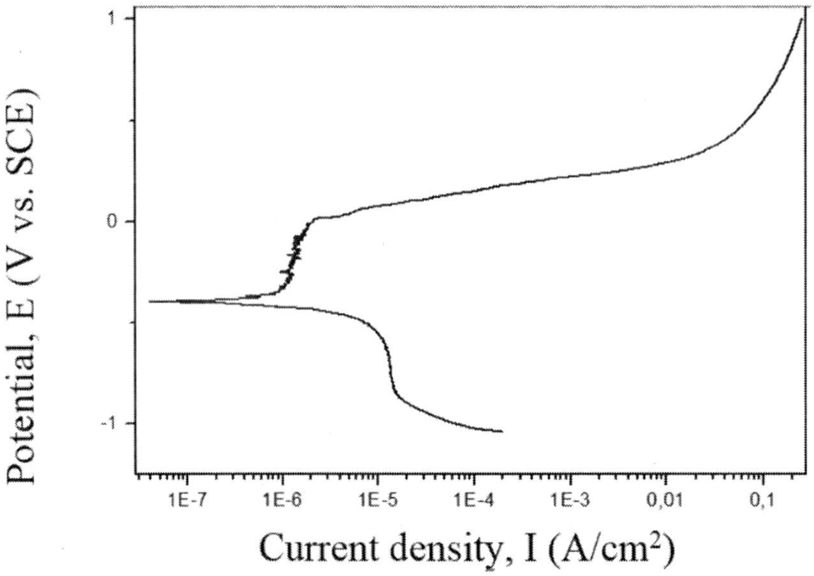

Figure 7. The potential dynamic curve of experimental stainless steel.

The analysis of recorded and calculated parameters reveals that the corrosion current density is low and close to other types of stainless steel. This i_{cor} (1.089 $\mu A/cm^2$) value of the material being investigated gives it good resistance to corrosion in 3% NaCl saline solution.

Furthermore, in terms of corrosion rate, there can be seen that it has acceptable values (0.011 mm/year), but account must be taken of the operating mode and the application.

The variation of the potential in open circuit reveals its evolution to more electro-positive potentials (passivation), but with periods when the potential decreases strongly (the protective oxide layer on the surface deteriorates), followed by re-passivation.

The shape of the potential dynamic curve reveals in the cathodic curve a passivation trend between - 300 mV and 0 V with a small variation in the density of the corrosion current followed by a decrease in the current density, which indicates sensitivity to pitting.

3.3. Metallographic analysis

The samples selected for the metallographic analysis were sanded using abrasive paper (600-1000 grain size) and then polished with fine Al_2O_3 abrasive powder, 1.0 to 0.25μm grain size.

Sample surfaces were ultrasonically degreased for 30 minutes in ethanol then chemically attacked with metallographic reagent (Aqua regia) and subjected to metallographic analysis (Figure 8).

The microstructural aspects highlighted by metallographic analysis indicate the dendritic microstructure, with pitting effects, as a result of the preferential attack of the chemical reagent on the grain boundaries.

There can also be noted the intra-granular separation of polyhedral inclusions and precipitations.

The fracture surfaces reveal the presence of polygonal inclusions and sliding or flattening planes, alternating with cup-type ductile fracture zones.

The metal matrix micro hardness was measured using a Shimadzu HMV 2T micro hardness tester, in longitudinal section (198 HV_1) and in cross section (194 HV_1).

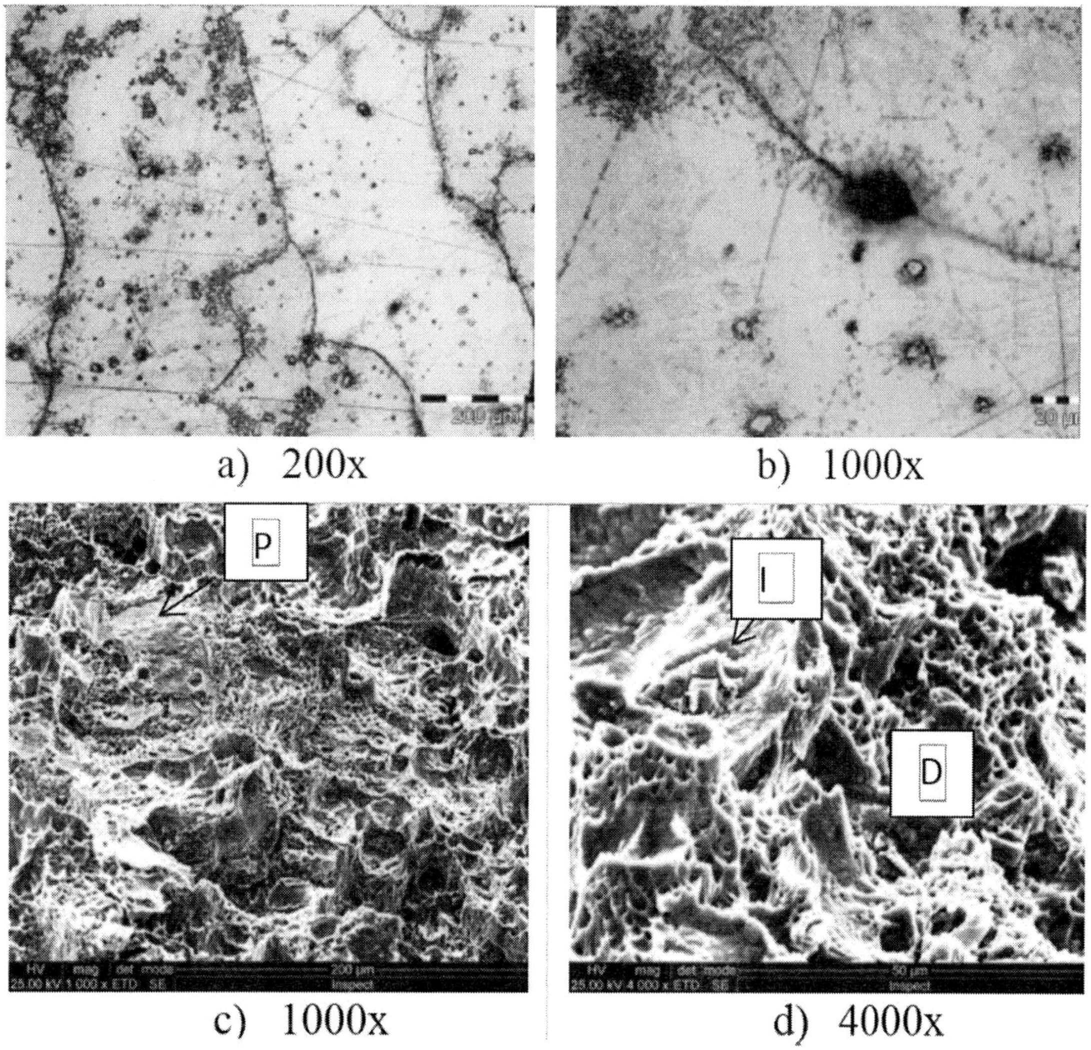

Figure 8. Microstructural aspects of experimental refractory stainless steel; (a), (b) dendritic microstructure and precipitation of inter-metallic compounds (optical micrograph); (c), (d) Fracture surfaces: flattening and sliding planes (P) alternating with cup-type ductile fracture zones (D) and polygonal inclusions (I) (SEM micrograph).

4. Conclusions

The microstructural results are in good agreement with the impact fracture behaviour, the values obtained being inferior to a steel of the same alloying class, but in laminated state.

The hardness values determined by the Vickers method fall within the usual limits for the steel class of the analysed experimental material (200HV), with no significant differences between the values determined in cross section or longitudinally, attesting the homogeneity of the metal matrix.

The compression test behaviour of the analysed alloy is satisfactory, the value of the longitudinal elasticity coefficient being of maximum 9855.99 ± 37.17MPa and the value of the elastic limit being 498MPa. The fracture energy values fall within the margin of medium resistance steels.

The analysis of the results obtained in the corrosion tests reveals that the density of the corrosion current has a low value, close to the one of other types of stainless steel (i_{corr} = 1.089 $\mu A/cm^2$), giving it good corrosion resistance in 3% NaCl saline solution. In terms of corrosion rate, the experimental steel features acceptable values (0.011 mm/year).

5. References

[1] Buechsenschuetz R, Poweleit B and Thiessen W 1981 *Mat. Sci. B* **12** 14
[2] Abe F, Kern T U, Viswanathan R 2008 *Creep-resistant els* (CRC Press) p 135
[3] Fabritius H, Wessling W 1980 *Materialwissenschaft und Werkstofftechnik* **11** 172
[4] Geantă V, Ştefănoiu R 2008 *Ingineria producerii oţelului (Engineering of steelmaking)* (BREN, Bucharest)
[5] Geantă V, Tripşa I, Stefănoiu R 2007 *Tratat de Ştiinţa şi Ingineria Materialelor, Vol. 2., Bazele teoretice şi ingineria obţinerii materialelor metalice, Cap. 7 – Elaborarea oţelurilor (Materials Science and Engineering, Vol. 2., Theorethical basis and engineering of obtainment of metallic materials, Cap. 7 – Steelmaking)* (Academia de Ştiinţe Tehnice din România, AGIR Bucharest)
[6] Van der Voort G F 1999 *Metallography: Principles and Practice* (ASM International, Materials Park, OH)
[7] Folkhard E 1988 *Welding Metallurgy of Stainless Steels* (Springer-Verlag Wien)
[8] Henderieckx I G 2006 *Schaeffler Diagram. Calculation of structure for stainless steels* (GIETECH BV)
[9] Adolfs R, Detert K 1992 *Crack initiation and low-cycle fatigue behavior of materials* (Springer Netherlands) p 582
[10] Voiculescu I, Geantă V, Vasile I M 2016 *Aliaje feroase pentru structuri sudate (Ferrous alloys for welded structures)* (BREN Bucharest) p 212
[11] Parrot R, Pitts H 2011 *Chloride stress corrosion cracking in austenitic stainless steel, Health and Safety Executive* (RR 902, London)
[12] Hamada A S, Karjalainnen L P, Somani M C 2006 *Mat. Sci. & Engineering A* **431** 211

Acknowledgments
This work was supported by the Romanian National Authority for Scientific Research, CNDI–UEFISCDI, through the project number **PTE 48/2016** (NONCORO).

IOP Conf. Series: Materials Science and Engineering **374** (2018) 012019 doi:10.1088/1757-899X/374/1/012019

Geopolymers and Their Uses: Review

D D Burduhos Nergis[1], M M A B Abdullah[2], P Vizureanu[1*] and M F M Tahir[2]
[1]Technical University "Gheorghe Asachi" of Iasi-Romania, Department of Technologies and Equipments for Materials Processing, Blvd. Mangeron, No. 51, 700050, Iasi, Romania
[2]Center of Excellent Geopolymer and Green Technology, School of Materials Engineering, University Malaysia Perlis, P. O Box 77, D/A Pejabat Pos Besar Kangar, Perlis 01000, Malaysia

E-mail: peviz2002@yahoo.com

Abstract. Outlining the past-present history of the study of alumino-silicate materials, it is well known that geopolymers are inorganic polymers obtained from chemical reaction, also known as geopolymerisation, between an alkaline solution and a solid reach in aluminium and silicone. There is still some controversy surrounding the alkaline activators used to create geopolymer concrete, because homogeneous mixture composed of two (NaOH and Na_2SO_3) or more chemical in varying proportions are usually highly corrosive and hard to handle. In order to overcome Portland cement many wastes have been used in recent studies to create "friendly" cements by geopolymerisation. In this short review we present basic information's about how to create and use geopolymers, alkaline activators and raw materials that can be used and conclusions. One question that needs to be asked: Can those materials replace on large scale Portland cement?

1. Introduction

The geopolymer concept was introduced by Joseph Davidovits in 1976 for alkali aluminosilicate binders. [1] The transformation of kaolinite into tridimensional tecto-aluminosilicates used for the polycondensation of organic resins at low temperature is very similar with the thermosetting method of these materials. The result of the process is a nanocomposite that looks like an artificial rock. The geosynthesis take place itself in great abundance in nature. The crust of Earth is composed, in 55% of volume, from siloxo-sialates and sialates, yet, just only 12% are pure silica or quartz. The geosynthesis process is based on changes induced in crystallography of silica backbone by the aluminium ion (6-fold or 4-fold coordination) and on the chemical changes produced by the same aluminium ion [2].

These materials possess chemical composition similar to zeolites consisting of a polymeric Si–O–Al framework, different properties and amorphous structure. Their properties are highly depending of: the Al-Si source, the activator, the aggregate source and its grading, the water source, the mix quantities of each material, the hardening/curing time, the temperature, the dimensions of particles, the calcium concentration, the heat treatment, if applied [3-5].

Depending of cost and the final desired properties, there are a wide range of raw materials that can be used to create geopolymers, such as: metakaolin [6], red mud [7], fly ash [8, 9], different wastes [10], etc. During manufacturing, the emission of CO_2 and energy consumption is very low comparative with Portland cement, this new binder offer a strong reduction of the global warming by releasing just 169 kg CO_2/m^3 when ordinary Portland cements release 306 kg CO_2/m^3 for the same mechanical properties, which represents a decrease of emission by 45% [11].

2. Obtaining of geopolymers

Geopolymers are inorganic polymeric materials obtained by mixing of a dry solid (aluminosilicate) with an alkaline solution and other constituents, if necessary. The major constituent is the source material, this one must be rich in silicone (Si) and aluminium (Al), they could be natural minerals such as clays, kaolinite, etc. or red mud, fly ashes, slags, etc. known as "waste" materials. The liquid is usually based on sodium (hydroxide or silicate) or potassium (hydroxide or silicate) with are soluble alkali metals. [12] The time of hardening is very short, in first 4 hours of setting they obtain 70% of the final compressive strength [13].

The structures of geopolymers consist in three dimensional silico-aluminate links (sialate) of SiO_4 and AlO_4 tetrahedra with share all the oxygen atoms, designed as in Table 1. Geopolymer matrix presents an empirical Formule 1, as following:

$$M^{+}_{v}\{-(SiO_2)_x\text{-}AlO_2\text{-}\}_v*aH_2O \qquad (1)$$

Where: M^+ - Alkali cation of K^+, Na^+; v – degree of polymerization; x – Si/Al ratio, a – the amount of water, if necessary (when the alkali activator is solid, a specific amount of water is necessary to create the reaction environment, if the activator is liquid, there is no need of supplementary water). The properties of the geopolymer are highly depending of the value of z (z = 1-15, up to 300), for high value of z (z>3) geopolymers exhibit adhesive and rubbery properties, their structure being a linear 2-D network, when the value of z is low ($z \leq 3$) the material presents stiff and brittle properties and a cross-linked 3-D network.

Table 1. Types of poly(sialates) [14].

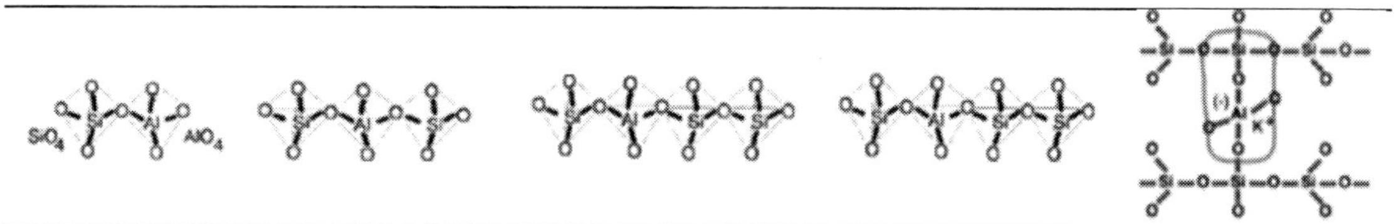

Poly(sialate) -Si-O-Al-O-	Poly(sialate-siloxo) -Si-O-Al-O-SI-O-	Poly(sialate-disiloxo) -Si-O-Al-O-Si-O-Si-O-	Poly(sialate-disiloxo) -Si-O-Al-O-Si-O-Si-O-	Sialate link
Si:Al = 1	Si:Al = 2	Si:Al = 3	Si:Al = 3	Si:Al > 3

The geopolymer result after the exothermic process carried out through oligomers, process known as geopolymerization, is a polymer with very long reticular network, where the specific tetragonal network of aluminates (AlO_4) and groups of silicates (SiO_4) are, also, present. The bond between these tetrahedrons is equilibrated by alkali ions of: K^+, Na^+ or Li^+.

In general, geopolymerization can be divided into three stages, see Figure 1. First of them is dissolution, when the solid alumino-silicates material is dissolved because of the water and alkali activator presence. After eliminating a small amount of water, the reorientation starts, now the group atoms take their place in the structure. During the solidification (at 20 Celsius degree or higher approximate 1000 ^{0}C) the water is almost totally eliminated and the material shows his final form. [15-17].

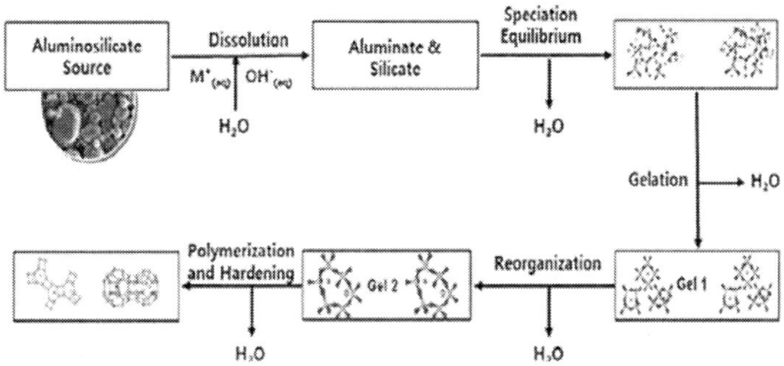

Figure 1. Stages of geopolymerisation process [18, 19].

2.1. Materials for geopolymers

Because of the fact that any material rich in aluminium and silicone see Figure 2, can be used as raw materials, there are many materials that can be used to create geopolymers. Kaolinite was the first material widely used in geopolymer synthesis [20] after successfully uses of this new material, the scientists start developing new raw materials, such as calcinated clays [21] or industrial waste (e.g. slag [22], ashes [23], waste glass [24], aluminium mine tailings [25], etc.) and natural silico-aluminates (e.g. pure Al2O3–2SiO2 powder) [26], zeolites [27], etc.).

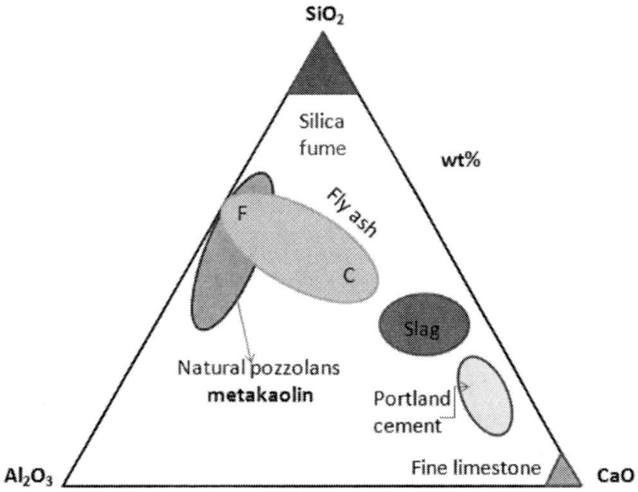

Figure 2. CaO–Al2O3–SiO2 ternary diagram of cementitious materials [28].

2.2. Geopolymers obtained using metakaolin

The first use of geopolymers was as building products [29], this products include fire-resistant chip-board panels, made from a wooden core faced with two coatings and ceramic tiles water-resistant at temperatures lower than 450°C, without firing. At 150°C kaolinite that is one of the clay components reacted with caustic soda. The industrial applications of this kaolinite reaction with alkali began in the ceramic industry with Olsen in 1934.

The alumino-silicate kaolinite reacts with NaOH at 100°C - 150°C and polycondenses into hydrated sodalite (a tectoaluminosilicate), or hydrosodalite, Formule 2.

$$\underset{\text{Kaolinite}}{Si_2O_5,Al_2(OH)_4 + NaOH} \Rightarrow \underset{\text{Hydrosodalite}}{Na(\text{-}Si\text{-}O\text{-}Al\text{-}O)_n} \quad (2)$$

IOP Conf. Series: Materials Science and Engineering **374** (2018) 012019 doi:10.1088/1757-899X/374/1/012019

Kaolin, after being calcinated at temperatures between 650°C and 750°C, is the main source of metakaolinite. Kaolin is a phyllosilicate, consisting of alternate layers of silica and alumina in tetrahedral and octahedral coordination, respectively. This electrically neutral crystalline layer structure, which is a common characteristic of clay minerals, leads to a fine particle size and plate like morphology and allows the particles to move readily over one another, giving rise to physical properties such as softness, soapy feel and easy cleavage. Kaolinite is the mineralogical term for hydrated aluminium disilicate, $Al_2SiO_5(OH)_4$, the primary constituent of kaolin (40–70%). Other minerals comprising kaolin include quartz, muscovite-like micas and rutile. When kaolin heated to temperature of 650–900 C, it loses 14% of its mass in bound hydroxylions. This heat treatment, or calcination, breaks down the structure of kaolin such that the alumina and silica layers become puckered and lose their long-range order. Resulting from this dehydroxylation and disorder is metakaolinite, a highly reactive transition phase, amorphous material with pozzolanic and latent hydraulic reactivity, suitable for use in cementing applications [30].

Metakaolin-based geopolymers can be manufactured consistently with a high level of predictability in hardened characteristics, see Figure 3, displays compressive strengths and porosities for both metakaolin geopolymers and metakaolin/fly ash mixtures.

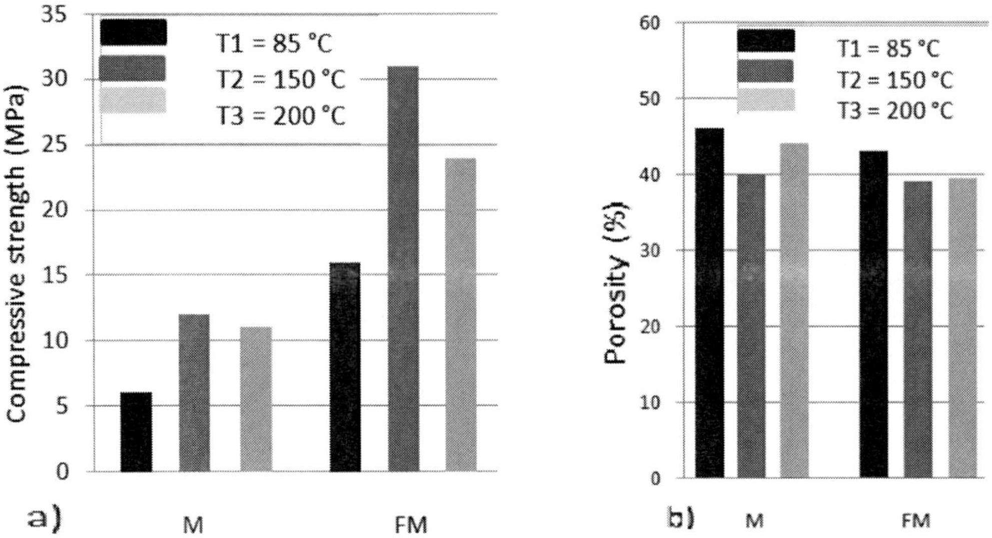

Figure 3. Mechanical Strength and Porosity of Metakaolin Geopolymers [31].

2.3. Geopolymers obtained using ashes

The need of new construction materials with high properties is in a continuous growth and so is the need of reducing of using the natural minerals for this purpose. In this condition the use of ashes in cements or concretes [32] is a great "green" idea. Beside the high chemical, physical and mechanical properties, the positive effect on environment and low emission of CO_2, the economical aspect is also important, playing an important role when this material is chosen.

Fly ashes are spherical glassy particles obtained by solidification of coal by-product volatile material. Due to the fusion-in-suspension these fly ash particles are mostly minute solid spheres and hollow ecospheres with some particles even being plerospheres, which are spheres containing smaller spheres.

Fly ash can be divided into two classes based on its source of origin and composition. Class F is a fly ash is obtained from burning of bituminous or anthracite coal and meets this chemical composition $SiO_2 + Al_2O_3 + Fe_2O_3 \geq 70\%$. Class C is normally produced from burning of sub-bituminous or lignite coal and has the following chemical composition $SiO_2 + Al_2O_3 + Fe_2O_3 \geq 50\%$ [33].

2.4. Geopolymers obtained using red mud

There are multiple uses of this highly alkaline (pH of the order of 11 or higher) waste generated by aluminium production plants (for 1 tone of aluminium, 2 tones of red mud are deposited) however the reuse rate is very small and this material become a huge problem for the environment [34].

Red mud is a mixing of minerals with complex chemical composition, see Figure 4, where iron has the highest concentration followed by aluminium, silica, calcium, sodium and low concentration of rare earth chemical elements, like: Dy,Er, Yd, Nd, Sm, Eu, Gd, Sc, La, Ce, Y, Pr. Bayer Process consist in dissolving of bauxite with (NaOH) at 175°C resulting aluminium hydroxide $Al(OH)_3$ and 50% -70% red mud. Red mud is usually used to create geopolymer cements with low CO_2 emission, small price and excellent quality [35].

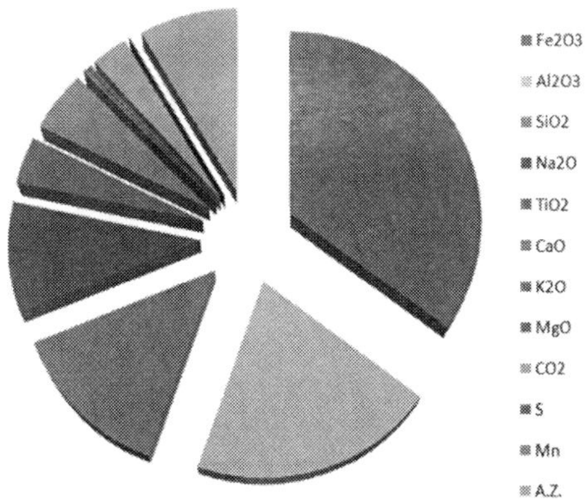

Figure 4. Typical red mud chemical composition.

2.5. Geopolymers obtained using slags

Blast furnace slag is a foundry residue result from production of cast iron. In this process, the iron is melted and mixed with different additions in order to eliminate the impurities from metal bath, when the slag is expelled on the bath surface it is fast removed and cooled down in water. The result material, slag, has amorphous structure and chemical composition reached in calcium aluminates and silicates. [36] The slags can be used in manufacture of different type of materials [37] and by geopolymerization is used to create binders [38], mortar [39] and different types of concretes with good mechanical properties, see Figure 5 [40].

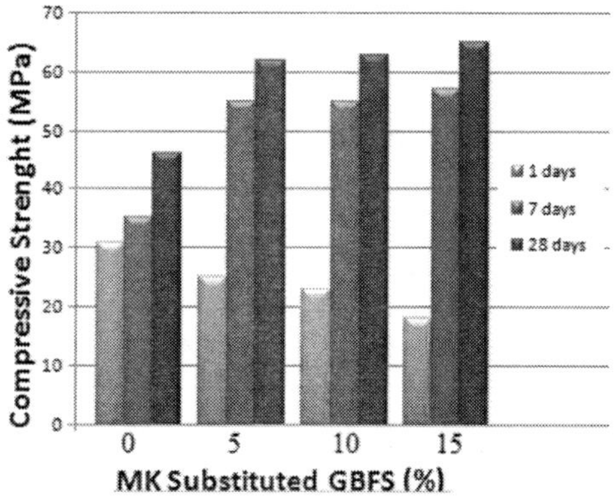

Figure 5. Compressive strength value for different types of slags (GBFS)-metakaolin (MK) based geopolymers [41].

2.6. Alkaline Activators

Activation of the selected pozzolanic material is the most significant factor in producing a mechanically-sound cementitious material via the geopolymerization process. The activators prompt the precipitation and crystallization of the siliceous and aluminous species present in the solution. OH- acts as a catalyst for reactivity, and the metal cation serves to form a structural element and balance the negative framework carried by the tetrahedral aluminum. The initial mechanism of reaction is driven by the ability of the alkaline solution to dissolve the pozzolanic material and release reactive silicon and aluminum into solution. There are multiple activating agents that can be used for geopolymerisation (M-alkali ion): Alkalis, MOH; Strong Salt Acids, M_2SO_4, Aluminosilicates, M_2O * nAl_2SO_3 * (2-6)SiO_2; Aluminates, M_2O * $nAlO_3$; Silicates, M_2O * $nSiO_3$; Weak acid salts, M_3PO_4, M_2SO_3, M_2CO_3, MF; [42]

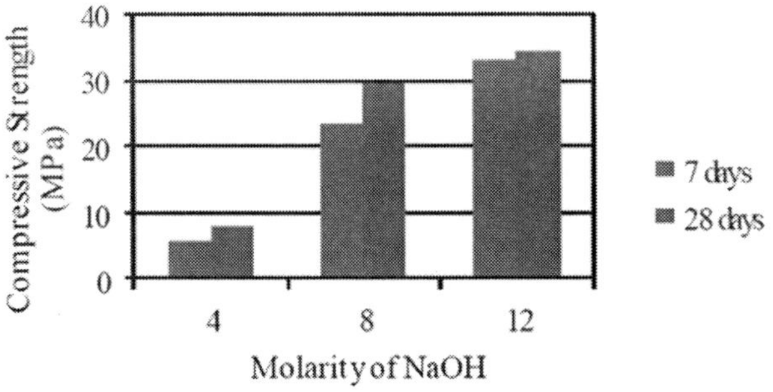

Figure 6. Compressive strength values depending of NaOH concentration.

Sodium Hydroxide (NaOH) is a geopolymer activator commonly used as alkaline solution. The sodium cations promote better geopolymerisation, compared with potassium cations, due to the fact that they can migrated easy thru the network in the gelatin stage being smaller. The strength value of geopolymers increase with the increasing of NaOH concentration, see Figure 6 at early stages. Yet, the aged material will be compromised due to excessive OH that modifies morphology in a non-uniform one. These types of activated geopolymers are more stable in sulfates and acids environments by having greater crystallinety. Potassium hydroxide (KOH) activator improves the compressive strength of the geopolymers cements and their porosity. K^+ will offer a denser final structure compared with other activators because of its high reactivity with the alumino-silicate material [44].

Sodium silicates (Na_2SO_3) are rare single used activator for geopolymers being usually mixed with NaOH to achieve enough activation potential. It is used mostly for its increasing strength and alkalinity properties. The most used geopolymerisation activator is a combination between Na_2SO_3 and KOH with different value ratio [45].

2.7. The processing parameters effect

Because of the fact that, the final properties of the geopolymer material are highly depending of the processing parameters, the effect of different processing factors was studied by many scientists [46-49] studied the influencing of four of the main factors: curing temperature and molar ratios of SiO_2/Al_2O_3, Na_2O/SiO_2 and H_2O/Na_2O as can be observed in the Figure 6. They conclude that, by increasing this all of these major factors, the mean value of compressive strength decreases. Also, they observed that, the curing temperature have no significant effect on this properties for short period of time. After passing of a limit value of SiO_2/ Al_2O_3 (2.9 %) and Na_2O/SiO_2 (0.2 %) molar ratio, the system will be saturated with Na^+ ions and the formation of oligomers is prevented [50]. The molar ratio of H_2O/ Na_2O presents the highest negative effect by prevention from dissolving of alumino silicate and by high porosity introduced by water withdrawal, presenting high decrees of properties between concentration of 13.75% and 21.75 % [51].

3. Geopolymers application

Thanks to this low-temperature transformation from kaolinite into hydrosodalite, since 1979 (J. Davidovits) this „mineral" showed great interest for manufacturing of: thermal insulation, [52] fire resistant materials as can be seen in Figure 7, [53] low-tech building materials, [54] foundry industry, [55] decorative stone artifacts, [56] thermal shock refractories, [57] bio-technologies (medicinal applications), [58] infrastructures repair and strengthening composites [59] low energy ceramic tiles, [60] aircraft interior and automobile composites, [61] refractory items, [62] cements and concretes [63] radioactive and toxic waste containment [64].

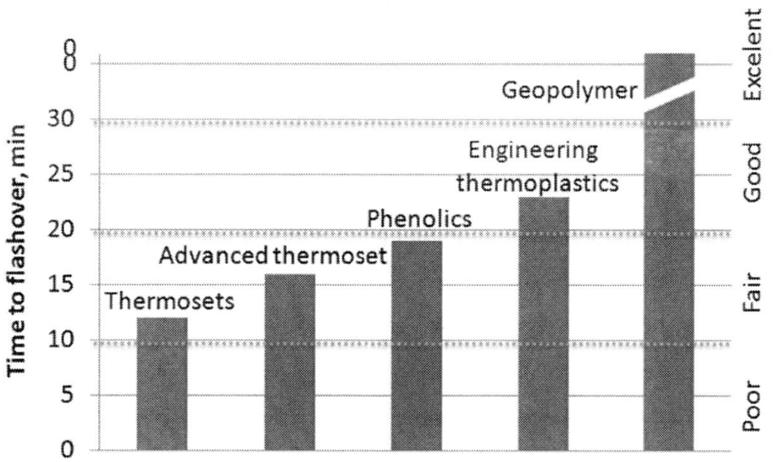

Figure 7. Time to flashover (minutes) for various organic resins compared to geopolymer resin [65].

Related to properties, they possess very good volume stability, 4/5 lower shrinkage than Portland cement, as high strength gain in short time, in first 4 hours of setting they obtain 70% of the final compressive strength. Geopolymers, also, present excellent durability, their properties are almost the same after decades, low thermal conductivity and high fire resistance (fig. 6), geopolymer keep their functions at 1000 °C - 1200 °C. The heat conductivity of geopolymer varies from 0.24 w/m·k to 0.3 w/m·k, compared well with lightweight refractory bricks (0.3 w/m·k to 0.438 w/m·k) [66-68].

Depending on the Si/Al ratio and structural chemical network, geopolymers used for low CO_2 cements and concretes and radioactive and toxic waste encapsulation are thus with Si/Al ratio 2:1 and 3D network, for bricks and ceramics are thus with Si/Al ratio 1:1 and 3D network, for fire resistant and heat resistant fiber composites are used thus with Si/Al ratio 3:1 and a 2D structure, for sealants used in industry in the range of 200 °C and 600 °C are thus with Si/Al ratio 2:1 and 2D structure, for fire protection, fiber glass composites, foundry equipments, heat resisters composites in the range of 200°C - 1000°C and tooling for aeronautics titanium proc. applications are thus with Si/Al ratio 1:1 and 2D structure.

4. Conclusions

The sources of raw materials are very wide, any source of silicates, pozzolanic compound or alumino-silicates that can be dissolved in alkaline solution can be used as a source of production of geopolymers.

Geopolymers can be considered green materials, because to create this new material can be used different industry wastes and they have low emission of CO_2 during manufacturing related to the emission of CO_2 to manufacture the Portland cement. Geopolymers do not require large energy consumption, for the same quantity of geopolymer cement and Portland cement, the ratio of energy used is 3/5.

Some of them are easy to prepare cooperated with the conventional material, they can be obtained by mixing alumino-silicates materials with strongly alkaline solutions, then curing at room temperature. A reasonable value of strength is gained is short time.

The properties of the final material are highly depending of different factors, like: Si/Al ratio, particle dimension, alkaline activator concentration, calcium quantity, time of hardening/drying and synthesis temperature.

Because of possibility to use local wastes without multiple previous operations to create geopolymers, they present a strong economic advantage.

Geopolymers can be considered as difficult to create, because manufacturing process include corrosive chemical substance that can harm humans, such as sodium hydroxide.

5. References

[1] Davidovits J1991 *J Therm Anal* **37 (8)** 1633-1638
[2] Davidovits J 1994 *Journal of Materials Education* **16** 91-137
[3] Jaishree D and Rekha G 2018 *International Journal of Pure and Alied Mathematics ume* **118 (20)** 2139-2146
[4] Burduhos Nergis D D, Abdullah M M A B and P Vizureanu 2017 *European Journal od Materials Science* **2 (4)** 111-118
[5] Swanepoel J C and Strydom C A 2002 *Alied Geochemistry* **17 (8)** 1143-1148
[6] Lahoti M, Wong K K, Yang E H and Tan K H 2018 *Ceramics International* **44 (5)** 5726-5734
[7] Krivenk P, Kovalchuk O, Pasko A, Croymans T, Hutt M, Lutter G, Vandevenne N, Schreurs S and Schroeyers W 2017 *Construction and Building Materials* **151** 819-826
[8] Pan Z, Tao Z, Cao Y F, Wuhrer R and Murphy T 2018 *Cement & Concrete Composites* **86** 9-18
[9] Fan F H, Liu Z, Xu G J, Peng H and Cai C S 2018 *Construction and Building Materials* **160** 66-81
[10] Novais R M, Ascensao G, Ferreira N, Seabra M P and Labrincha J A 2018 *Ceramics International* **44 (6)** 6242-6249
[11] Habert G, d'Espinose de Lacaillerie J B and Roussel N 2011 *Journal of Cleaner Production* **19** 1229-1238
[12] Lloyd N A and Rangan B V 2010 Curtin University of Technology "Geopolymer Concrete with Fly Ash"
[13] Hardjito D, Cheak C C and Ing C H L 2008 *Modern Alied Science* **2 (4)** 3-11
[14] Duxson P, Fernández-Jiménez A, Provis J I, Lukey G C, Palomo A and Van Deventer J S J 2007 *J Mat Sci* **42 (9)** 2917–2933
[15] Provis J L, Kilcullen A, Duxson P, Brice D G and van Deventer J S J 2012 *Industrial & Engineering Chemistry Research* **51**(5) 2483-2486
[16] Buchwald A, Zellmann H D and Kaps C 2011 *Journal of Non-crystalline Solids* **357 (5)** 1376-1382
[17] Davidovits J 1989 *Journal of thermal analysis* **35**(2) 429-441
[18] Jun N H, Minciuna M G, Abdullah M M A, Jin T S, Sandu A V, Ming L Y 2017 *Revista de chimie* **68**(10) 2367-72
[19] Ogundiran M B and Kumar S 2015 *Alied Clay Science* **108** 173-181
[20] Romisuhani A, AlBakri M M, Kamarudin H and Andrei S V 2017 International Conference of Alied Science and Technology for Infrastructure Engineering 267 Article Number: UNSP 012013
[21] Tuyan M, Andic-Cakir O and Ramyar K 2018 *Composites Part b-engineering* 135 242-252
[22] Alcamand H A, Borges P H R and Silva F A 2012 *Ceramics International* **44 (5)** 5037-5044
[23] El-Wafa M A and Fukuzawa K 2018 *Journal of materials in civil engineering* **30 (4)** Article Number: 04018040
[24] Tho-In T, Sata V and Boonserm K 2018 *Journal of Cleaner Production* 172 2892-2898
[25] Lemougna P N, Kai-tuo W and Qing T 2017 *Construction and Building Materials* **156** 486-495

[26] Ducman V and Korat L 2016 *Materials Characterization* **113** 207-213
[27] Papa E, Medri V and Amari S 2018 *Journal of Cleaner Production* **171** 76-84
[28] Khatib J M 2016 *Sustainability of Construction Materials (Second Edition) a ume in Woodhead Publishing Series in Civil and Structural Engineering* 415–457
[29] Davidovits J 1989 *Chemical & Engineering News* **67**(27) 4-5
[30] Rashad M A 2013 *Construction and Building Materials* **41** 303-318
[31] Jeffrey C, Petermann C and Saeed A 2010 *Alied Research Associates Inc* AFRL-RX-TY-TR-2010-0097
[32] Liew K M, Sojobi A O and Zhan L W g 2017 *Construction and Building Materials* **156** 1063-1095
[33] AASHTO M 295 2007 Standard Specification for Coal Fly Ash and Raw or Calcined Natural Pozzolan for Use in Concrete American Association of State Highway and Transportation Officials Washington DC
[34] Suchita R, Wasewar K L, Mukhopadhyay J, Yoo C K and Uslu H 2012 *Arch Environ Sci* 6 13-33
[35] Liu Y, Lina C and Wu Y 2007 *Journal of Hazardous Materials* **146**(1-2) 255-261
[36] Qin Y, Liu H and Yang Y 2018 *Metallurgical Research & Technology* 115 (1)
[37] Bobirica C, Shim J H and Park J Y 2018 *Ceramics International* **44**(6) 5886-5893
[38] Jianwei S, Zhe W and Zhonghui C 2018 *Journal of Thermal Analysis and Calorimetry* **131**(3) 2291-2301
[39] Zhao J, Wang D and Yan P 2017 *Construction and Building Materials* **140** 248-256
[40] Cho B S, Lee H H and Choi Y C 2017 *Construction and Building Materials* **140** 293-300
[41] Gao X and Yu Q L 2015 *Brouwers, Cem. Concr. Compos.* **59** 119–128.doi:10.1016/j.cemconcomp.2015.01.007.
[42] Aboshia A M A, Rahmat R A M and Zain F M 2017 *International Journal of Building Pathology and Adaptation* **35** **(5)** 438-455
[43] Reddy J H and Brahmini A 2016 Performance of 12 Molarity Based Geopolymer Mortar with Fly Ash and GGBS
[44] Fasihnikoutalab M H, Pourakbar S, Ball R J and Richard J 2017 *International Journal of Geosynthetics and Ground Engineering* **3** **(2)** Article Number: UNSP 12
[45] El Hafid K and Hajjaji M 2018 *Construction and Building Materials* **159** 598-609
[46] Aziz I H, Abdullah M M A B and Yong H C 2017 Conference: International Conference on Innovative Research (ICIR Euroinvent) Location: Iasi, Romania
[47] Antunes Boca Santa R A, Soares C and Riella H G 2017 *Construction and Building Materials* 157 459-466
[48] Catauro M, Bollino F and Cattaneo A S 2017 *Journal of the American Ceramic Society* **100**(5) 1919-1927
[49] Onutai S, Jiemsirilers S and Kobayashi T 2017 Alied Environmental Materials Science for Sustainability Book Series: Advances in Environmental Engineering and Green Technologies 165-185
[50] Roviello G, Menna C and Tarallo O 2017 *Composites Part b-engineering* **128** 225-237
[51] Chen X, Sutrisno A and Zhu L 2017 *Journal of the American Ceramic Society* **100**(5) 2285-2295
[52] Zaidi S F A, Ul Haq E and Nu K r 2017 *Construction and Building Materials* **157** 994-1000
[53] Chiu Y P, Lu Y M and Shiau Y C 2015 *Materials Research Innovations* **19**(5) 642-649
[54] Komnitsas K A 2011 *Procedia Engineering* **21** 1023-1032
[55] Thaarrini J and Ramasamy V 2016 *Periodica Polytechnica-civil Engineering* **60**(2) 159-168
[56] Won J and Kang S 2017 *Journal of Ceramic Processing Research* **18**(2) 166-171
[57] Yahya Z, Abdullah M M A, Hussin K, Ismail K N, Sandu A V, Vizureanu P and Abd Razak R 2013 *Revista de chimie* **64** 1408-1412
[58] Naghsh M and Shams K 2017 *Alied Clay Science* **146** 238-245

[59] Laskar S M and Talukdar S 2017 *Construction and Building Materials* **154** 176-190
[60] Namkane K, Naksata W and Thiansem S 2016 *Environmental Earth Sciences* **75 (5)** Article Number: 386
[61] Nair Q, Zhao B G and Coope R F r 2007 *Journal of Materials Science* **42 (9)** 3083-3091
[62] Bernal S A, Bejarano J and Garzon Cristhian C 2012 *Composites Part b-engineering* **43 (4)** 1919-1928
[63] Law D W, Adam A A and Molyneaux T K 2016 *Materials and Structures* **48 (3)** 721-731
[64] Cozzi A D, Bannochie C J, Burchet P R, Crawford C L and Jantzen C M 2011 Immobilization of Radioactive Waste in Fly Ash Based Geopolymers World of Coal Ash (WOCA) Conference Denver USA
[65] Davidovits J Geopolymer 2002 Conference October 28-29 2002 Melbourne Australia -1- 30 Years of Successes and Failures in Geopolymer Alications Market Trends and Potential Breakthroughs
[66] Ismail N and El-Hassan H 2018 *Journal of Materials in Civil Engineering* **30 (4)** Article Number: 04018029
[67] Pilehvar S, Cao V D and Szczotok A M 2017 *Cement and Concrete Research* **100** 341-349
[68] Jez B, Nabialek M, Gruszka K, Pietrusiewicz P, Jez K, Rzacki J, Szota M, Bloch K, Gondro J 2017 *European Journal of Materials Science and Engineering* **2**(2-3) 57-62

Aluminum Coating Influence on Nitride Layer Performance Deposited by MO-CVD in Fluidized Bed on Austenitic Stainless Steel Substrate

D G Găluşcă[1], M C Perju[1]*, C Nejneru[1], D D Burduhos Nergiş[1] and I E Lăzărescu
[1]Technical University "Gheorghe Asachi" of Iasi-Romania, Department of Technologies and Equipments for Materials Processing, Blvd. Mangeron, No. 51, 700050, Iasi, Romania

E-mail: cryss_ela@yahoo.com

Abstract. The modification of surface properties by duplex treatments, involving the overlapping of two surface treatment techniques, has been established as an intelligent solution to create new applications for the substrate metallic material. There are driveline components operating under very tough wear and corrosion conditions, with high temperature and humidity variations. Such components are usually made of high Cr and Ni stainless steel and for the hardening of surfaces it is recommended a thermo chemical treatment. Since stainless steels, especially austenitic stainless steels, are difficult to nitride, experimental studies focus on increasing the depth of the nitride layer and surface hardness. Achieving the goal involves changing active layer chemical composition by introducing aluminum in the surface layer. In order to find a solution, a new surface treatment technique is produced by combining aluminum thin films by MO-CVD in a fluidized bed using a triisobutylaluminum precursor with a thermo chemical nitriding treatment.

1. Introduction

The surfaces of components used in some equipments construction must withstand stresses where the combination of multiple properties, such as over-wearing resistance, over bearing capacity, fatigue and corrosion, is absolutely necessary. Unfortunately, such properties cannot always be possessed by the base metal used.

The surface properties modification by thermo-chemical/deposition treatments has been established as an intelligent solution to create new applications for the substrate metallic material. Surface properties that directly meet the requirements of the mode and conditions of use for treated component can be obtained.

Experimental studies focus on increasing the depth of the nitride layer and surface hardness for nitride austenitic stainless steels by duplex heat treatments [1-4]. Duplex treatments involve overlapping of two surface treatment processes on the same base material to obtain superior mechanical properties compared to the mechanical properties of the substrate.

Achieving the objective involves changing surface chemical composition by introducing in active layer aluminum (Al), an residual element in the stainless steel, knowing the fact that aluminium favor the formation of a hardly nitride surface layer. To find solutions, we approach a new surface treatment technique by combining thin aluminum layer deposition by MO-CVD in a fluidized bed using a triisobutylaluminum (TIBA) precursor with thermo-chemical nitriding treatment [5-7].

IOP Conf. Series: Materials Science and Engineering **374** (2018) 012020 doi:10.1088/1757-899X/374/1/012020

2. Experimental

2.1. Materials

Choosing material for experiment (raw material, deposition element – aluminum, gas precursor – triisobutylaluminum) and procedures.

2.1.1. Raw material. For this experiment stainless steel X5CrNi18-10, W1.4301, was used, which chemical composition is shown in Table 1. The composition was determined by means of a Foundry Master spectrometer. The samples were cut with laser beam at 12 mm x 10 mm with a thickness of 2 mm.

Table 1. Chemical composition of X5CrNi18-10, W 1.4301 steel.

Element	C	Si	Mn	Cr	Ni	S	P	Fe
Percent	0.12	0.71	0.42	16	9.41	0.026	0.042	Balance

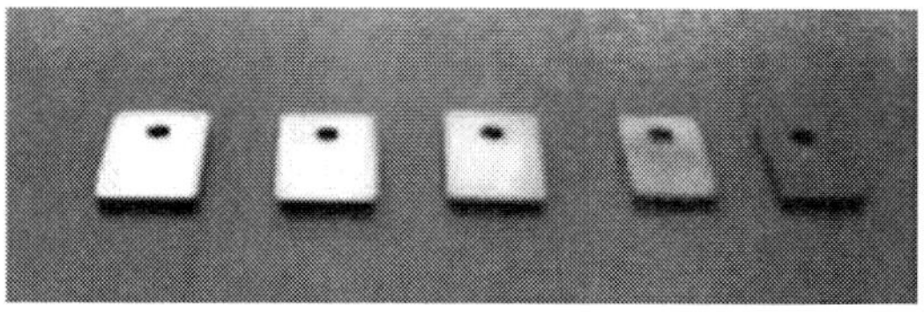

Figure 1. Samples used for experiment.

2.1.2. Aluminum deposition by MO-CVD- precursor. In order to obtain aluminum deposits on austenitic stainless steel supports, a triisobutylaluminum precursor has been used, TIBA is an metalorganic chemical agent with the molecular formula $[(CH_3)CHCH_2]_3Al$ having molecular weight 198.32, boiling point 86°C, melting point at 4 - 6°C and a density of 0.786 g/l at 25°C. The precursor will be heated to the vaporization temperature (86 °C) and the vapor-carrier gas (nitrogen) mixture is introduced into the fluidized bed heating system where the steel samples are positioned [8]. During contact with the fluidized bed particles at temperatures between 250-500°C, the triisobutylaluminum molecules dissociate, resulting in active Al +, which will be deposited on the surface of the metallic samples immersed in the fluidized bed. The thermal decomposition of TIBA can be represented by the reaction from Figure 2 [9]:

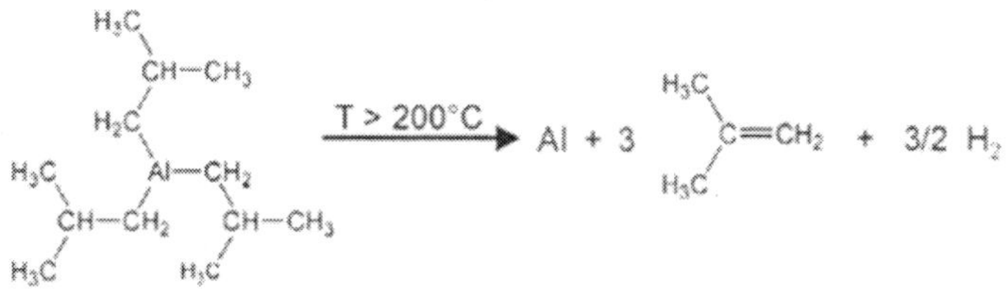

Figure 2. The decomposition reaction for Tributylaluminum.

2.2. Aluminum deposition by MO-CVD with fluidized bed heating

2.2.1. Experimental procedure. The technological solution we are focus is CVD process based on metalorganic compounds, MO-CVD (Metalorganic CVD), in this way the reaction temperature can be greatly reduced. Thereby aluminum (Al) layers can be obtained at temperatures just below 400°C [10].

The use of the fluidized bed deposition method has a many advantages especially temperature uniformity and very efficient heat transfer. However, the feed rate of the precursor is limited by the saturation capacity of carrier gas with precursor vapor and in the same time by the relative volatility of precursor which, until reaching the reaction zone, must be maintained below reaction or decomposition temperature [11-15].

2.2.2. Deposition installation. An experimental MO-CVD deposition installation with fluidized bed heating was used to perform the experiment. The fluidization chamber is made of a quartz tube with a diameter of 30 mm and a length of 300 mm and exterior heating resistance. The external protection is made by another quartz tube with a diameter of 75 mm and a length of 400 mm (see Figure 3).

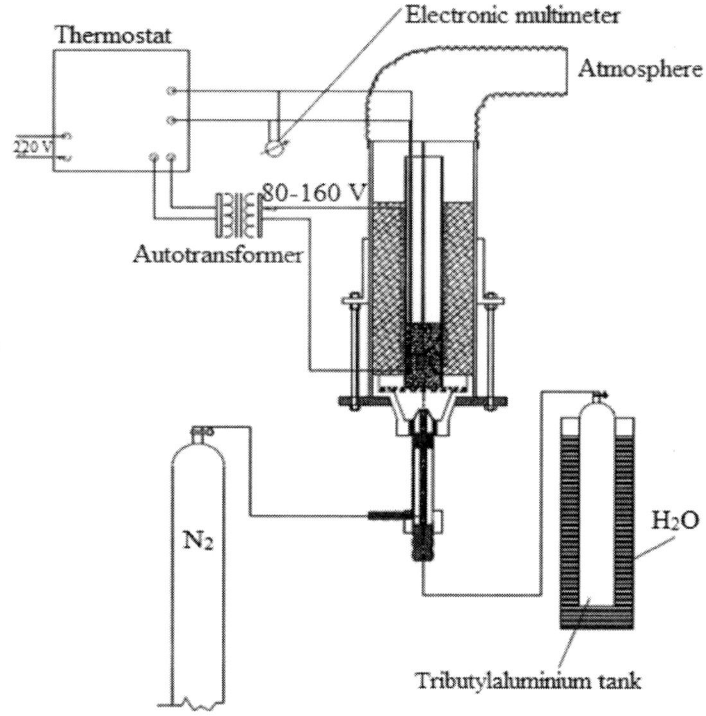

Figure 3. Schematic presentation of MO-CVD in fluidized bed deposition installation.

The feeding of TIBA was made directly from a special cylinder which was introduced into a temperature control tank heated up to 86°C. The carrier gas (nitrogen) pressure was 1.2 bar, sufficient to ensure the fluidized bed operation, the flow was measured by means of a rotameter. Vapor-gas mixture is introduced after the reactor reaches the constant temperature regime. The reactor temperature was maintained at 320°C for 35 minutes during the experiment.

2.3. Sample analyzes after aluminum deposition

The samples were analyzed by microstructure point of view by means of a scanning electron microscope (SEM and EDX) after deposition (2D and 3D). Rugosity influence on MO-CVD deposition was also presented in this study.

2.4. Analyses of nitride stainless steel with aluminum deposition by MO-CVD

- SEM Microstructural analysis of nitride samples; XPS analysis of nitride samples; Microhardness analysis.

3. Experimental results

3.1. Aluminium deposition by MO-CVD on stainless steel support

Stainless steel samples were heated in fluidized bed installation from boundary temperature to 320°C. The maintaining time was 35 minutes. Since the surface absorption of aluminum on the samples during MO-CVD operation is very important, the roughness parameter influence to deposition (absorption) degree has also been observed. In case of a ragged sample, the surface irregularities will favor the deposition of aluminum (increasing the active surface). For this reason the samples were prepared in terms of different average surface roughness.

3.1.1. Control sample analysis. The control sample was analyzed by EDX by means of a scanning electron microscope. The presence of aluminum is not highlighted.

Table 2. Chemical composition of control sample - percentage by weight.

Element	% C	% Cr	%Ni	% O_2	% Fe
Percent	0.14	16.93	8.14	0.84	balance

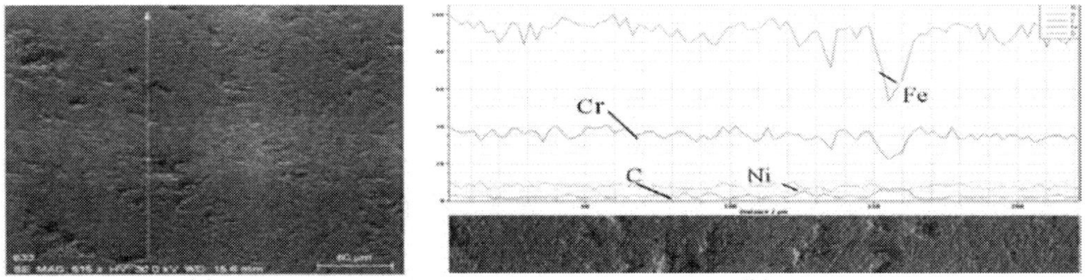

(a) (b)

Figure 4. Line analysis for the control sample to highlight the distribution of surface chemical elements (a) Marking of the EDX analyzed area; (b) Line analysis with percentages of Fe, C, Cr and Ni.

3.1.2. Sample Rz = 0.6 μm. For sample Rz = 0.6 μm Al was deposited in 0.478 % on the surface, small percentage due to low surface roughness which does not favor MO-CVD deposition.

Table 3. Chemical composition of sample 1 superficial layer.

Element	C	Cr	Ni	Si	Al	Fe
percent	0.12	18.44	7.33	0.84	0.47	balance

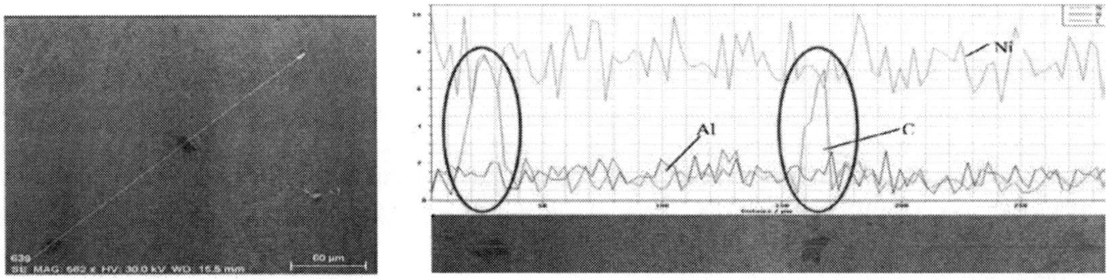

(a) (b)

Figure 5. Line analysis for sample Rz = 0.6 μm to highlight the surface chemical elements distribution: (a) Marking of the EDX analyzed area; (b) Line analysis with percentages of C, Al and Ni.

3.1.3. Sample Rz = 3.2μm. For sample Rz = 3.2 μm Al was deposited in 1.363 % on the surface, high percentage due to high surface roughness which does not favor MO-CVD deposition.

Table 4. Chemical composition of sample R_z = 3.2 μm surface layer.

Element	C	Cr	Ni	Al	Fe
Percent	0.13	15.65	8.03	1.36	balance

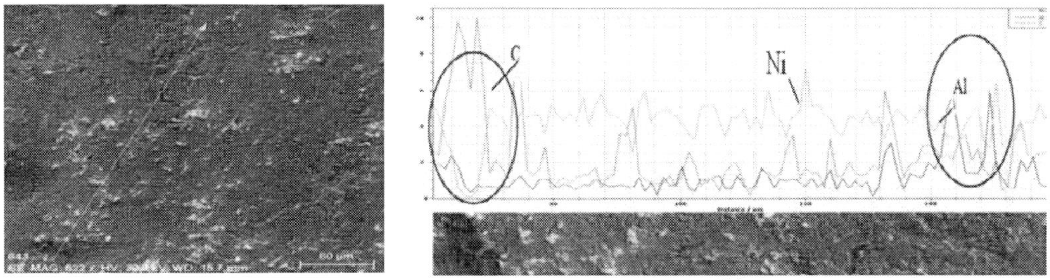

(a) (b)

Figure 6. Line analysis for sample R_z=3,2 μm to highlight the surface chemical elements distribution: (a) Marking of the EDX analyzed area; (b) Line analysis with percentages of C, Al and Ni, possible Al_4C_3 formation highlighted in rounded zone.

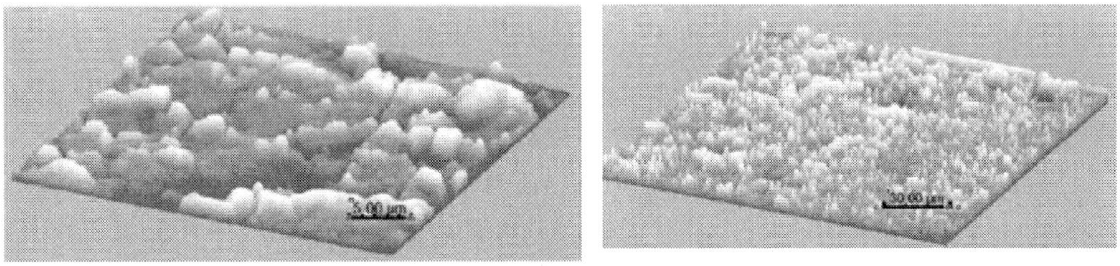

(a) (c)

Figure 7. 3D surfaces for sample Rz = 3,2 μm after deposition of Al in fluidised bed by MO-CVD technology: (a) sample 5 μm scale (b) 30 μm.

3.2. XPS analysis of samples after duplex heat treatement (MO-CVD and nitride)

For this analysis an X-ray photoelectron spectrometer, PHI 5000 Versa Probe-XPS, ULVAC-PHI-INC, is used.

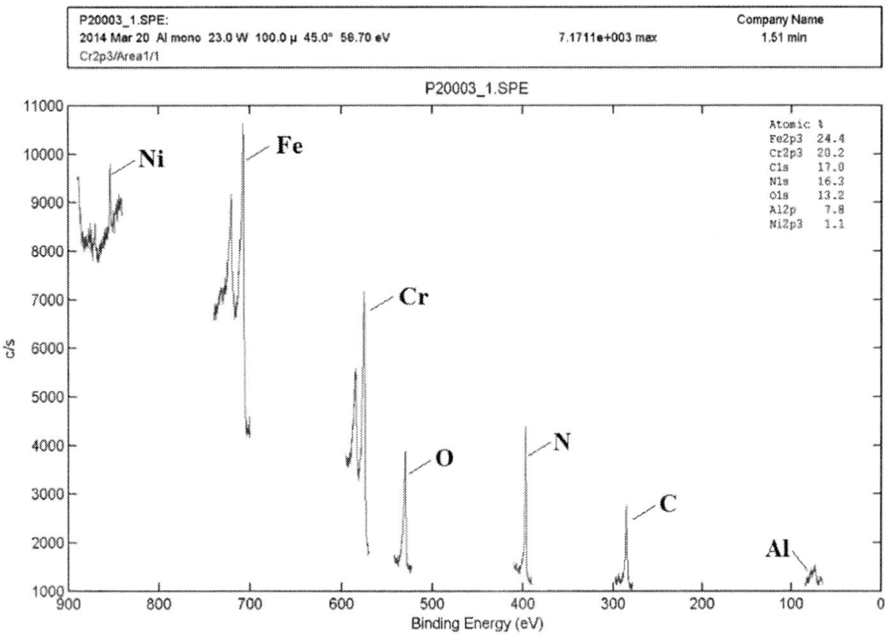

Figure 8. Characteristic XPS spectrum and orbitals.

The exposure time has been increased to remove the effects of contamination because elements such as Si, C and O occur in large quantities, due to surface absorption, either during deposition or subsequent contamination.

For each chemical element the energetic bonds were studied, i.e. the chemical compounds appeared in the layer, and for this purpose enlarged areas from the energy scale were made. Example:

Analyze for Al2p [15], with the band energy between 65 ÷ 85 eV, the peaks in this energy range are observed corresponding to the following chemical compounds: $Al_{25}Ni_{75}$, $AlN_{0,097}$, AlN , Fe / Al_2O_3, Al_2FeO_4, Al_2O_3, Al_2NiO_4.

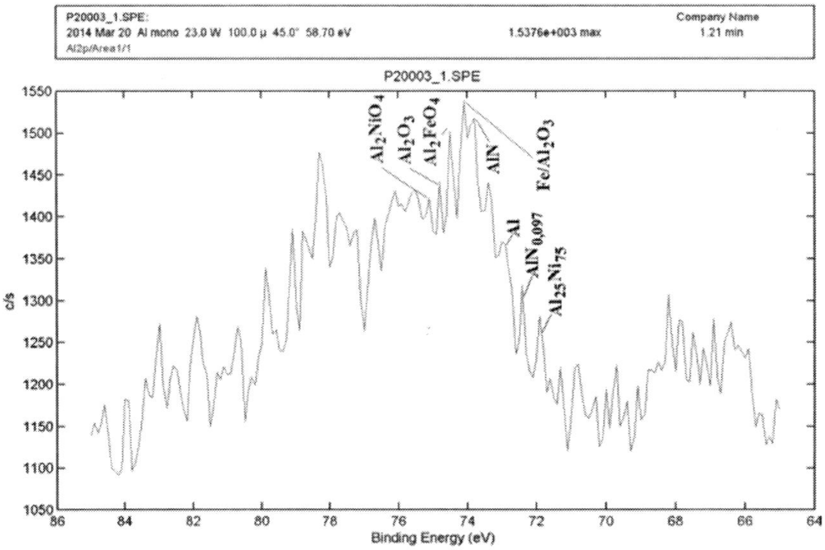

Figure 9. XPS analysis for Al2p for stainless steel sample R_z = 3,2 µm, highlighted peaks for following chemical compounds: ($Al_{25}Ni_{75}$, $AlN_{0,097}$, AlN, Fe/Al_2O_3, Al_2FeO_4, Al_2O_3, Al_2NiO_4).

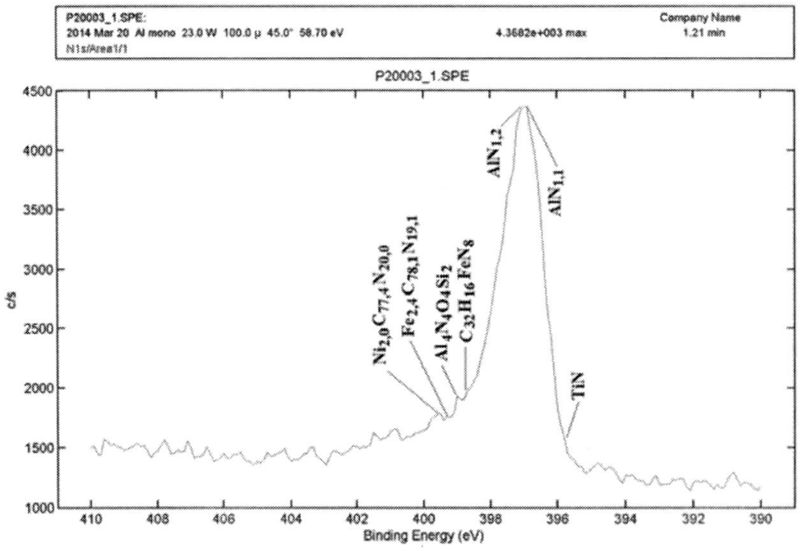

Figure 10. XPS analysis for N1s for stainless steel sample highlighted peaks for following chemical compounds: TiN, $AlN_{1,1}$, $AlN_{1,2}$, $C_{32}H_{16}FeN_8$, $Al_4N_4O_4Si_2$, $Fe_{2,4}C_{78,1}N_{19,1}$, $Ni_{2,0}C_{77,4}N_{20,0}$.

Analyze for N1s [12], with the band energy between 390 ÷ 410 eV, the peaks in this energy range are observed corresponding to the following chemical compounds: TiN, $AlN_{1,1}$, $AlN_{1,2}$, $C_{32}H_{16}FeN_8$, $Al_4N_4O_4Si_2$, $Fe_{2,4}C_{78,1}N_{19,1}$, $Ni_{2,0}C_{77,4}N_{20,0}$.

Aluminum participates in the formation of simple and complex nitrides, $AlN_{0,097}$, AlN, $AlN_{1,1}$, $AlN_{1,2}$, $Al_4N_4O_4Si_2$, in surface structure. The nitrated surface layer is formed of nitrogen and alloying elements present in steel compounds, namely Al, Cr, Ni, Fe ($Al_4N_4O_4Si_2$, Cr_2N, $Fe_{2,4}C_{78,1}N_{19,1}$, $Ni_{2,0}C_{77,4}N_{20,0}$), not of α (nitroferrite), γ′ (Fe_4N), ε ($Fe_{2-3}N$) and ξ (Fe_2N) as in classical ionic nitration, which give a excellent tribological behavior.

3.3. *Microhardness testing*

Experimental microhardness determinations were performed by means of NAMICON CV Instruments 400 DM. Measurements was performed with 50 gf for control and MO-CVD samples and 200 gf for duplex treated samples (MO-CVD + nitride).

Because the thickness of the deposited layer was very small, indentations were made on the sample surface. This type of measurement gives information about deposited aluminum layer on stainless steel substrate. For each sample, three indentations were performed. The results are shown in Table 5.

Table 5. Vickers microhardness measurements.

Sample	Microhardness after Heat treatment/MO-CVD HV50	Average HV50	Microhardness after nitride HV200	Average HV200
Heat treatment control sample + NI	239.00	241.00	595.00	594.66
	259.00		602.00	
	225.00		587.00	
MO-CVD control sample	218.10	216.67		
	208.70			
	223.20			
Sample R_z = 0.6 μm	218.10	216.67	964.30	981.77
	208.70		957.90	
	223.20		1023.10	
Sample R_z = 3.2 μm	216.90	213.37	988.30	950.63
	222.20		914.40	
	201.00		949.20	

4. Conclusions

The steel used is a high Cr and Ni alloy stainless steel with austenitic structure and about 0.1 % C. Alloying elements create a good austeniticity slowing the penetration of nitrogen.

In order to improve the stainless steel behavior at nitrating, an MO-CVD deposition with a triisobutylaluminum (TIBA) precursor was performed to create aluminum / compact microarea wich favor the absorption of nitrogen in the surface layer.

It has been found that the surface roughness influence the deposition quality, the deposited layer depth increase with the increase of surface roughness.

XPS analysis of duplex-treated austenitic stainless steel components (MO-CVD aluminum deposition and nitride) indicates the presence of complex nitrides and carbonitrides, TiN, $AlN_{1,1}$, $AlN_{1,2}$, $C_{32}H_{16}FeN_8$, $Al_4N_4O_4Si_2$, $Fe_{2,4}C_{78,1}N_{19,1}$, $Ni_{2,0}C_{77,4}N_{20,0}$.

According to Table 5 thermally treated and nitride steel presents low hardness values. After Al deposition by MO-CVD, microhardness decreases up to 10%, but after MO-CVD + nitride duplex treatment, surface hardness has a clear increase.

The thermal regimes applied for the duplex treatment used has a great advantage, because it dosen't affect the physico-mechanical properties of heat treatead substrate.

5. References

[1] Podgornik B P, Vizintin J, Wanstrand O, Larsson M, Hogmark S 1999 *Surface and Coatings Tehnology* **120** 502-508
[2] Ueda M, Gomez G F, Kostov K G, Reuther H, Lepienski C M, Soares Jr. PC, Takai O, Silva M M 2004 *Brazilian Journal of Physics* **34**
[3] Alsaran A, Altun H, Karakan M, Celik A 2004 *Surf. Coat. Technol.* **176** 344
[4] Braga D, Dias J P, Cavaleiro A 2006 *Surface & Coatings Technology* **200** 4861–4869
[5] Ţugui C A, Nejneru C, Găluşcă D G, Perju M C, Axinte M, Cimpoeşu N, Vizureanu P 2015 *Journal of Optoelectronics and Advanced Materials* **17**(11-12) 1855–1861
[6] Tugui, C A, Axinte M, Nejneru C, Vizureanu P, Perju M C, Chicet D 2014 *Applied Mechanics and Materials* **657** 369
[7] Axinte M, Nejneru C, Agop M 2012 *Metalurgia International* **17**(12) 87-92
[8] Liu Y, Compson C, Liu M 2004 *Journal of Power Sources* **138**(1–2) 194–198
[9] Czok G S, Werther J 2006 *Power Tehnology* **162** 100-110
[10] Surdu I E; Nejneru C, Galusca D G, Agop M 2013 *Metalurgia International* **18**(4) 5-11
[11] Surdu I E; Nejneru C, Galusca D G, Agop M 2013 *Applied mathematics and physics* **75**(1) 181-192
[12] Achitei D C, Vizureanu P, Minciuna M G, Sandu A V, Buzuianu A, Dana D I 2015 *Materiale Plastice* **52**(2) 165-167
[13] Tugui C A, Vizureanu P, Iftimie N, Steigmann R 2016 *7th International Conference on Advanced Concepts in Mechanical Engineering* **147** UNSP 012040
[14] Perju M C, Tugui C A, Nejneru C, Axinte M, Vizureanu P 2016 *International Conference on Innovative Research - ICIR EUROINVENT*, Book Series: IOP Conference Series-Materials Science and Engineering **133** UNSP 012025
[15] Barca E S, Plaiasu A G, Abrudeanu M, Istrate B, Luca D and Munteanu C 2015 *Journal of Optoelectronics and Advanced Materials* **17** 1522-27
[16] www.lasurface.com/database/elementxps.php

Magnetic Properties of Rapid Cooled FeCoB Based Alloys Produced by Injection Molding

M Nabialek[1*], B Jeż[1], K Jeż[1], P Pietrusiewicz[1], K Gruszka[1], K Błoch[1], J Gondro[1], J Rządki[1], M M A B Abdullah[2], A V Sandu[3] and M Szota[4]
1 Institute of Physics, Faculty of Production Engineering and Materials Technology, Czestochowa University of Technology, 19 Armii Krajowej Str., 42-200 Czestochowa, Poland
2 Gheorghe Asachi Technical University of Iasi, Faculty of Materials Science and Engineering, 41 D. Mangeron Blvd., 70050, Iasi, Romania
3 Geopolymer & Green Technology (CeGeoGTech), School of Material Engineering, Universiti Malaysia Perlis (UniMAP), 01000 Kangar, Perlis Malaysia
4 Institute of Materials Engineering, Faculty of Production Engineering and Materials Technology, Czestochowa University of Technology, 19 Armii Krajowej Str., 42-200 Czestochowa, Poland

E-mail: nmarcell@wp.pl

Abstract. The paper presents the results of investigations of the structure and magnetic properties of massive rapid cooled $Fe_{50-x}Co_{20+x}B_{20}Cu_1Nb_9$ alloys (where x = 0, 5). Massive alloys were made using the method of injecting a liquid alloy into a copper mold. Samples were obtained in the form of 0.5 mm thick plates. The structure of the obtained samples was examined using an X-ray diffractometer equipped with a CuKα lamp. The phase composition of the alloys formed was determined using the Match program. By using Sherrer's dependence the grain sizes of the identified crystalline phases were estimated. Using the Faraday magnetic balance, the magnetization of samples as a function of temperature in the range from room temperature to 850K was measured. Magnetization of saturation and value of the coercive field for the prepared alloys were determined on the basis of magnetic hysteresis loop measurement using the LakeShore vibration magnetometer.

1. Introduction

Amorphous materials are a relatively new group of functional materials, known for over 60 years. Originally, materials with amorphous structure were produced in the form of thin strips and layers [1]. Quick cooling of the liquid alloy allowed to block the crystallization process. In this way, it is possible to maintain a chaotic arrangement of atoms in the volume of material. This configuration of atoms is suitable for liquids. Amorphous materials combine the properties of solids and liquid phases, which implies their extraordinary properties. Taking the form of a solid with a liquid structure, amorphous alloys show better properties compared to their crystalline counterparts [2-6].

However, amorphous alloys produced in the form of thin strips have considerably limited applicability. Few μm for many years was the maximum thickness of this type of materials. Thickness of amorphous strips reaching several dozen μm. Numerous studies of scientists from around the world and, above all, the works of A. Inoue have contributed to the emergence of a new group of materials, massive amorphous materials. Three empirical principles formulated by a Japanese scientist are some kind of guidelines for the production of amorphous materials [7, 8]. The multi-componentity of the alloy, the negative heat of mixing of the components and the correspondingly large differences in the length; of the atomic diameters of the main components allow the inhibition of the crystallization process during the cooling of the liquid alloy. Atoms of individual components can be treated as spheres with different atomic radii. Appropriate differences in their radii cause their mutual blocking during diffusion processes occurring during solidification of the liquid melt. The negative heat of mixing< influences the increase in the viscosity of the alloy, which additionally hinders the migration of atoms. Preventing the movement of atoms significantly slows down the crystallization process. Cooling of the alloy at a suitable rate allows solidification of the alloy in the configuration of atoms suitable for liquids. Obtaining volumetric amorphous alloys is possible during cooling at a speed of 10^0 - 10^3 K / s. Well-known methods for producing massive amorphous materials are suction and injection methods. Using these methods, one can create a rapid-cooled alloy in the form of cores, rods or plates.

Amorphous alloys are often heat treated to obtain a nanocrystalline structure. Materials of this type are often characterized by better properties than their amorphous precursor. Depending on the chemical composition used, it is possible to manufacture partially crystallized alloys in a one-stage process as soon as it solidifies [9, 10]. The selection of the appropriate chemical composition and cooling rate allows to slow the crystallization process in a way that allows only a certain amount of crystal phase grains to be distributed in the amorphous matrix.

Amorphous and nanocrystalline materials are particularly interesting group of materials due to their good magnetic properties. Alloys with high content of iron and cobalt are characterized by the so-called magnetically soft properties, characterized by among others: high saturation magnetization and low coercive field [11-18]. In addition, alloys with a high cobalt content of ten have a high Curie temperature.

The purpose of the work was to produce volumetric two-phase rapid cooled alloys with chemical composition $Fe_{50-x}Co_{20+x}B_{20}Cu_1Nb_9$ (gdzie x = 0, 5) and to study their structure and magnetic properties.

2. Methods and material

The base alloy with the chemical composition $Fe_{50-x}Co_{20+x}B_{20}Cu_1Nb_9$ (where x = 0, 5) was made using a plasma furnace. For this purpose, the alloy components were measured to within 0.001 grams. Five gram weight was allocated to the polycrystalline ingot. Chemical elements with high purity, over 99.99%, were used. The ingot was made in an arc furnace under a protective atmosphere of argon. The temperature of the plasma arc was regulated by the intensity of the current flowing through the electrode. Before the actual melting of the batch, pure titanium was melted which has a significant impact on the purity of the atmosphere in the working chamber. The batch was melted several times, each time turning the ingot over to the other side using a manipulator, which ensures better mixing of the alloy components. Purified from external contamination of ingots, crushed into smaller pieces.

The ingot served as a batch material to produce a rapid cooled alloy by the method of injecting a liquid alloy into a copper mold. The alloys were produced in the form of 5mm x 5mm x 0.5mm tiles. Pieces of a polycrystalline ingot were placed in a quartz capillary. The batch was inductively melted at a constant current. The liquid alloy was forced into the copper mold through a hole in a 1mm diameter quartz capillary. During the process, constant argon pressure was maintained. The entire casting process of the high-temperature melt was carried out under a protective gas (argon) after reaching a vacuum in the working chamber (10-5 mbar.) To obtain a clean atmosphere, the working chamber and the injection system were subjected to additional argon washing. The liquid melt was then injected into copper water cooled molds with a hollow core in the shape of a cylinder The scheme for the production of rapid cooled alloys is shown in Figure 1.

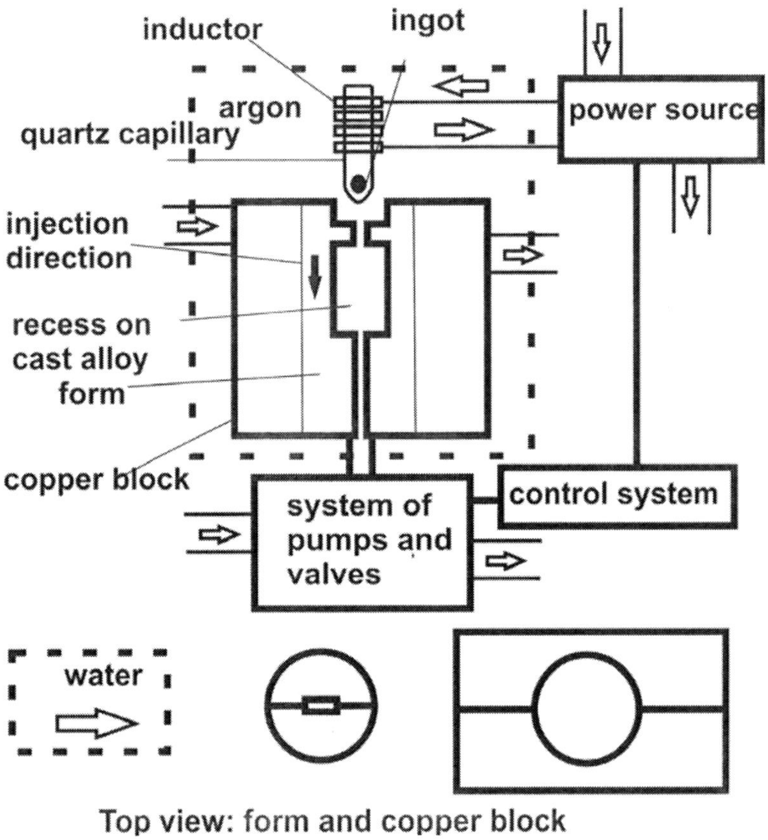

Figure 1. Scheme of the system for the production of rapid cooled alloys by the injection method.

The presented method of producing quick-cooled alloys allows the production of bar materials. The structure of the produced $Fe_{50-x}Co_{20+x}B_{20}Cu_1Nb_9$ alloy samples (where x = 0.5) was investigated using X-ray diffraction (Brucker apparatus, Cu-Kα lamp). Advance 8 model. The thermomagnetic properties of the produced samples were examined by Faraday's magnetic weight. The size of crystallites formed during the alloying process was estimated using the Sherrer formula (1):

$$D=(\lambda*K)/2\beta_0 cos\Theta \qquad (1)$$

K –Scherrer shape coefficient (K = 0,91),
λ – characteristic radiation wavelength,
B_0 – half-width intensity (background was included In analysis),
Θ – Bragg angle.

3. Results

Figure 2 shows diffraction patterns for $Fe_{50-x}Co_{20+x}B_{20}Cu_1Nb_9$ in as cast state.

The investigations were carried out in the 2 theta angle range from 30° to 100°. On recorded diffractograms, narrow peaks of considerable intensity can be distinguished from crystalline phases: Fe_2B, BCo and αFe. In the samples, there is one more crystalline phase that could not be identified. In addition, the wide maximum in the angle of 2 theta from 40° to 50° is visible on the diffractograms. The maximum called the amorphous halo is derived from X-rays scattered on chaotically distributed atoms in the volume of the alloy samples.

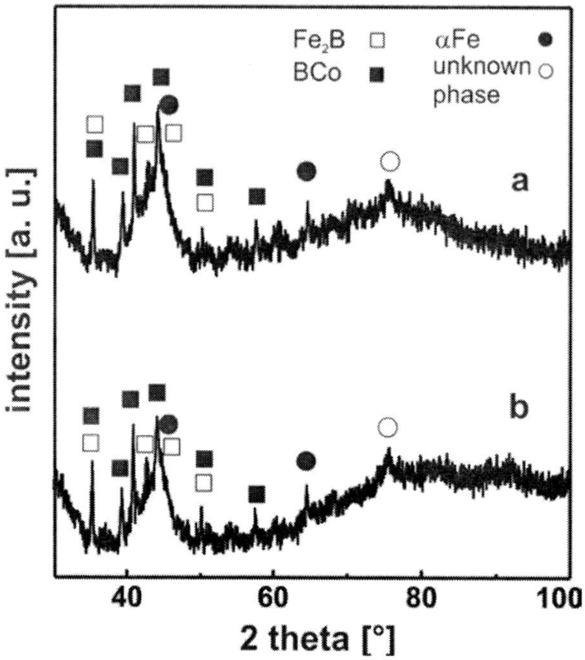

Figure 2. X-ray diffraction patterns for the alloy after solidification in powder form: (a) $Fe_{50}Co_{20}B_{20}Cu_1Nb_9$, (b) $Fe_{45}Co_{25}B_{20}Cu_1Nb_9$.

The diffractograms are typical for two-phase alloys consisting of an amorphous matrix and crystal grains distributed in its volume. Crystalline phases were identified using the Match program. Based on the diffractograms using the Sherrer formula, the average sizes of the resulting crystallites for particular phases were determined (Table 1).

Table 1. Estimated average crystallite sizes produced during rapid cooling of the liquid alloy $Fe_{50-x}Co_{20+x}B_{20}Cu_1Nb_9$.

Crystalline phase / Alloy	Fe_2B [nm]	BCo [nm]	αFe [nm]
$Fe_{50}Co_{20}B_{20}Cu_1Nb_9$	30,2	23,2	15,6
$Fe_{45}Co_{25}B_{20}Cu_1Nb_9$	29,1	30,6	16,1

In the volume of the alloys, the share of the crystalline phases was identified, and the estimated average grain sizes for the individual phases are similar. It can be stated that the dynamics of the crystallization process in this case nucleation of crystal grains is similar for both alloys. This means that the process energy for the transformation of the second type is similar. Estimated mean grain sizes for all identified crystalline phases do not exceed the 100nm dimension, which allows to qualify the obtained samples as a nanocrystalline material [19]. During the solidification process, crystalline phases rich in iron, cobalt and boron were formed. As it is known, the presence of copper and niobium atoms in appropriate proportions affects the formation of fine-grained structure, which in turn has a significant impact on improving the magnetic properties of the material produced [20-23]. A small addition of copper made it possible to create a material with a nanocrystalline structure. Most likely the copper or niobium placed on the grain boundaries were small enough that their identification with XRD was impossible.

Samples of $Fe_{50-x}Co_{20+x}B_{20}Cu_1Nb_9$ in the form of powders were measured using magnetic weight In the fucntion of temperature using Faraday balance.

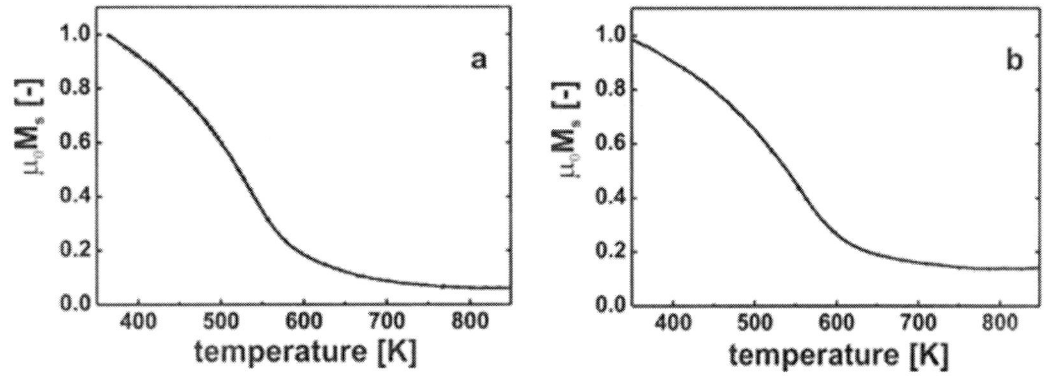

Figure 3. Magnetic polarization curves of saturation as a function of temperature: (a) $Fe_{50x}Co_{20}B_{20}Cu_1Nb_9$, (b) $Fe_{45}Co_{25}B_{20}Cu_1Nb_9$.

Figure 3 shows reduced magnetic saturation polarization curves. The measurement was carried out in the range from room temperature to 850K. In the examined range, one of the inflections from the transition of the amorphous phase from the ferro to the paramagnetic state is observed for the magnetization curves. As the temperature increases, the magnetization of the test sample decreases but does not drop to 0, which may be a confirmation of the presence of other magnetic phases occurring in the volume of the alloy (αFe, BCo, Fe_2B). The Curie temperature of these phases is outside the measurement range. The measured magnetic saturation polarization curves were subjected to numerical analysis. To determine the Curie temperature of the amorphous phase, a critical coefficient of 0.36 was used for ferromagnetics that met the Heisenberg assumptions. Curie temperature curves are shown in Figure 4.

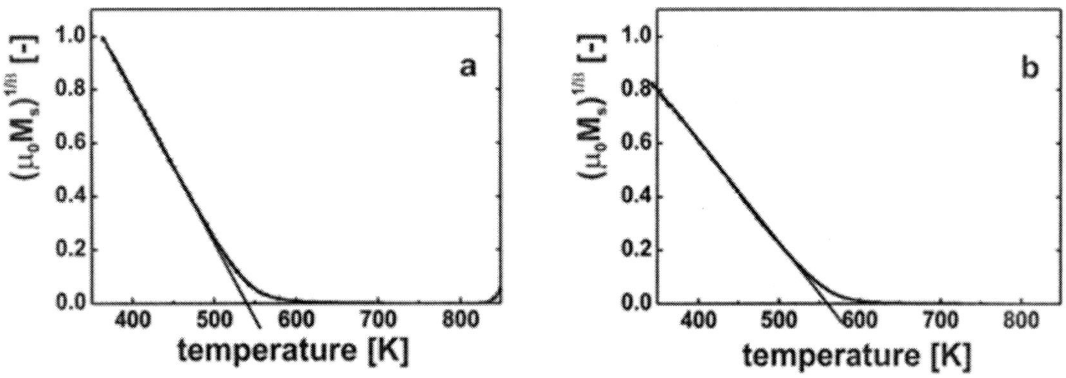

Figure 4. Dependence $\mu_0M)^{1/\beta}(T)$ for alloys: a) $Fe_{50}Co_{20}B_{20}Cu_1Nb_9$, b) $Fe_{45}Co_{25}B_{20}Cu_1Nb_9$.

The Curie temperature for the $Fe_{50}Co_{20}B_{20}Cu_1Nb_9$ alloy is 540K while the $Fe_{45}Co_{25}B_{20}Cu_1Nb_9$ alloy is 557K. These results are in line with expectations, it is known that increasing the cobalt content in the alloy increases the Curie temperature value. However, the Curie temperature increase is small compared to increasing the cobalt content of the alloy at the expense of iron. In work [24], a 2% change in the cobalt content at the expense of iron raises the Curie temperature by 20K. Smaller than expected, the increase may be related to the larger average grain size of the cobalt rich phase (BCo) for the $Fe_{45}Co_{25}B_{20}Cu_1Nb_9$ alloy. It is possible to absorb more cobalt to the crystalline phase BCo in the case of $Fe_{45}Co_{25}B_{20}Cu_1Nb_9$ alloy, which may be a reason for reducing the content of these elements in the amorphous matrix. Therefore, a slight increase in Curie temperature may be observed depending on the cobalt content in the alloy. It should be remembered that for amorphous materials, the Curie temperature is not a discrete value but rather a temperature range in which the amorphous matrix passes from the ferro-to-paramagnetic phase. This fact is closely related to the structure of the amorphous structure. The volume of amorphous material contains areas of similar chemical composition, which is not the case with the crystalline state. These small differences affect local differences in the Curie temperature value, therefore the value determined is not a discrete value.

Figure 5 contains static magnetic hysteresis loops. The measurement was made with the LakeShore vibration magnetometer in the magnetic field up to 2T.

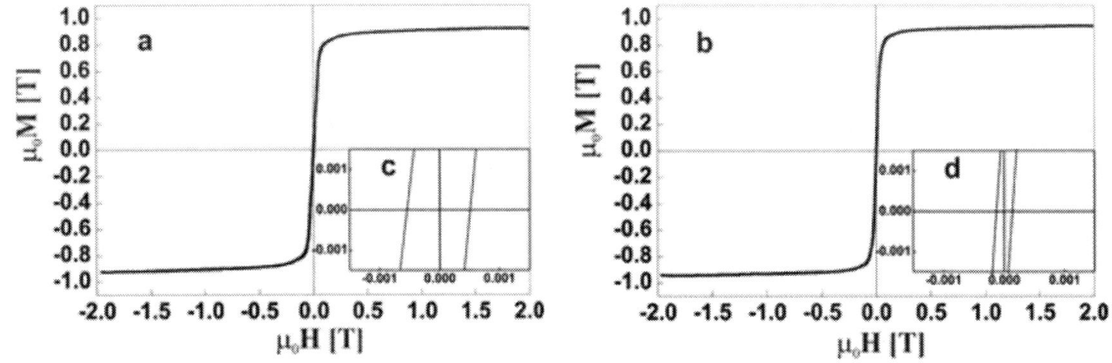

Figure 5. Static hysteresis loops with zoom in the center of M-H for: (a) and c) $Fe_{50}Co_{20}B_{20}Cu_1Nb_9$, (b) and d) $Fe_{45}Co_{25}B_{20}Cu_1Nb_9$.

The shape of static magnetic hysteresis loops indicates the magnetically soft properties. The shape of the wasp-like is not observed in the recorded loops, which excludes the presence of magnetic hard and semi-hard phases in the samples. The saturation magnetitization for the $Fe_{50}Co_{20}B_{20}Cu_1Nb_9$ alloy is 0.92T, while the value of the coercive field read from the enlarged coordinate system is 98A/m. For $Fe_{45}Co_{25}B_{20}Cu_1Nb_9$ alloy saturation magnetization equal to 0.95T and coercive field value 418A/m were measured. The measured saturation magnetization values can be considered relatively high. The coercive field values hold the created alloys in the group of so-called soft magnetic materials, assuming that these materials have a coercive field value below 1000A / m [25-26].

4. Conclusions

The purpose of the work was to create a material with soft magnetic properties by the method of rapid cooling of a liquid alloy. A volumetric rapid-cooled alloy was produced. The pressing method was used for this. The studies carried out allow to draw the following conclusions:

It is possible to produce a nanocrystalline material in a one-step manufacturing process, i.e. without heat treatment of the material. The impact on the formation of grains of crystalline phases not exceeding the size of 100nm is possible by introducing appropriate proportions of elements forming nucleation centers and inhibiting the crystallization process, copper and niobium.

The cooling rate of 10^2 K / s is sufficient to produce a nanocrystalline structure in the volume of the alloy with the chemical composition $Fe_{50-x}Co_{20+x}B_{20}Cu_1Nb_9$ (where x = 0, 5). FeCoB based rapid cooled alloys are characterized by a relatively high Curie temperature value, and the effect of cobalt content on the increase in Curie temperature is confirmed.

The increase in the cobalt content at the expense of iron in the tested alloy affects a slight decrease in the saturation magnetization value with a significant decrease in the coercive field value.

IOP Conf. Series: Materials Science and Engineering **374** (2018) 012021 doi:10.1088/1757-899X/374/1/012021

5. References

[1] Klement W, Willens R H, Duwez P 1960 *Nature* **187** 869-870
[2] Nabiałek M 2016 *Arch. Metall. Mater.* **61** 439–444
[3] Inoue A, Yano N, Masumoto T 1984 *J. Mater. Science* **19** 3786-3795
[4] Chen H S, Miller C E 1970 *Review of Scientific Instruments* **41** 1237
[5] Mchenry M E, Willard M A, Laughlin D E 1999 *Prog. Mater. Sci.* **44** 291
[6] Gruszka K *Materiali in Tehnologije* **50(5)** 707-718
[7] Błoch K, Nabiałek M 2015 *Acta. Phys. Pol. A* **127** 442–444
[8] Inoue A, Kato A, Zhang T, Kim S G, Masumoto T 1991 *Materials Transaction JIM* **32** 609-616
[9] Yanai T, Yamasaki M, Nakano M, Fukunaga H, Yoshizawa Y 2003 *Soft Magnetic Materials* **10** 737-741
[10] Herzer G 2013 *Acta Mater* **61** 718
[11] Błoch K 2015 *J. Magn. Magn. Mater.* **390** 118–122
[12] Jeż B, Nabialek M, Pietrusiewicz P, Gruszka K, Błoch K, Gondro J, Rzącki J, Abdullah M M A B, Sandu A V, Szota M, Jeż K, Sałagacki A 2017 *IOP Conf. Series: Materials Science and Engineering* **209** 012023
[13] Roy R K, Panda A K, Mitra A 2016 *Journal of Magnetism and Magnetic Materials* **418** 236–241
[14] Zhang Y,Wang Y, Makino A 2018 *Aip Advances* **8** 047703
[15] Herze G 1995 *Scripta Metallurgica et Materialia* **33** 1741-1756
[16] Han Y, Wang Z, Xu Y, Xie Z, Li L 2016 *Journal of Non-Crystalline Solids* **442** 29–33
[17] Wang W H, Dong C H 2004 *Materials Science and Engineering R* **44** 45-89
[18] Li W, Yang Y Z, Xu J 2017 *Journal of Non-Crystalline Solids* **461** 93–97
[19] Leonowicz M 1998 *Nanokrystaliczne materiały magnetyczne*, Warszawa
[20] Yoshizawa Y, Fujii S, Ping D H, Ohnuma M, Hono K 2003 *Scripta Mater* **48** 863
[21] Hono K, Ping D H, Ohnuma M, Onodera H 1999 *Acta Mater* **47** 997
[22] Pradeep K G, Herzer G, Choi P, Raab D 2014 *Acta Materialia* **68** 295-309
[23] Hono K, Inoue A, Sakurai T 1991 *Apel. Phys. Lett.* **58(19)** 2180-2182
[24] Jeż B 2017 *Revista de Chimie* **68(8)** 1903-1907
[25] Liebermann H 1993 *Rapidly Solidified Alloys* New Jersey

Thermal Analysis of a New Glass Fiber-Reinforced Bismaleimide Composite Material Used for Firefighter Helmets

P Ciubotariu–Ana, C A Micu, N M Lohan, B Pricop, L G Bujoreanu* and C Bejinariu
"Gheorghe Asachi" Technical University from Iasi- Romania, Faculty of Materials Science and Engineering, Blvd. Dimitrie Mangeron, No. 41, 700050, Iasi- Romania

E-mail: lgbujor@tuiasi.ro

Abstract. Safety helmets represent essential Personal Protection Equipment (PPE) used in firefighter protection and emergency situations. They protect firefighter's face and eyes against flames, heat and flying debris. When temperature levels are high, user's thermal comfort is affected. A glass fibre-reinforced bismaleimide composite material has a number of improved properties in terms of mechanical and thermal characteristics, as compared to the materials that are currently used. The present paper aims to comparatively analyse the thermal behaviour of an injection moulded polypropylene helmet and the newly developed hot modelling material, under the form of a glass fibre-reinforced bismaleimide composite material. Thermal analysis was performed using Differential Scanning Calorimetry (DSC) and Dynamic-Mechanical Analysis (DMA). DSC and DMA thermograms were corroborated and discussed, from the point of view of the consecutive solid state transitions occurring during heating, mostly in the second heating-cooling cycle. The isothermal behaviour of glass fibre-reinforced bismaleimide composite material, during strain sweeps performed by DMA, enabled the determination of internal friction and storage modulus, under vibratory loads, at different temperatures. The advantages of the newly developed glass fibre-reinforced bismaleimide composite material were highlighted.

1. Introduction

Personal protective equipment (PPE) play an essential role in the work system of firefighters and are the tools that ensure the basic health protection and safety of user's, from fire, chemicals or physical impact [1]. Fire protection helmets provide head protection against high temperatures [2] and are equipped with glass protection screens against smoke that can damage the eyes. Protective helmets must also provide head protection against impact and must be able to withstand penetration and absorb the impact force of a blow [3]. In some cases, the helmets should protect against electrical shocks as well.Protective helmets differ as a function of purpose and design but share the same component: the shell [4].

Currently, the most used materials for the production of shells are polyethylene, synthetic resin, ABS [4] and hybrid glass/jute reinforced epoxy composite [5]. The bismaleimide resins are recognized for high-temperature performance [6], outstanding thermomechanical and flammability behavior on the finally cured state [7] nonvolatility and low cost [8]. Firefighter helmet materials are essential for an effective intervention during rescue / accident operations. In order to have superior mechanical properties, the deflective helmet shell surface can be produced from continuous fiber composites processed by long fiber thermoplastic processing techniques [9]. A long fiber thermoplastic structure can comprise a thermoplastic or thermoset polymer matrix (such as polyaryl amide, polyethylene terephthalate, bismaleimide) and a continuous reinforcing fiber (such as glass, aramid, carbon fiber) wherein the continuous reinforcing fiber is hot melt impregnated with the thermoplastic resin.

A glass fiber-reinforced bismaleimide composite material has a number of improved properties in terms of mechanical and thermal characteristics as compared to currently available materials [10]. The purpose of the present paper is to study the thermal behavior of a new composite material based on bismaleimide reinforced with fiberglass, using differential scanning calorimetry (DSC) and dynamic mechanical analysis (DMA) investigations and to compare the results with those obtained oncommercial materials that are currently used for firefighter helmets.

2. Materials and methods

The new material was obtained, by hot molding process, as a composite of glass fiber reinforced bismaleimide. The bismaleimide resin precursor (PABMI) were prepared by reaction of 4,4-bismaleimidodiphenyl methane with 4,4-diaminodiphenyl methane according to the reported method [11]. Silica cloth was impregnated with a 50 wt. % solution of PABMI pre-polymer in N-methylpyrolidone (NMP) and allowed to dry at room temperature for 24 h and in a circulating air oven for 10 min at 140°C and 5 min at 170°C.

The pre-pregs dimension 10x10 cm were stacked over the female part of the mild steel mold coated with silicone release agent. The male counterpart of the mold was placed between the preheated plates (140°C). The first type of samples, (laminate samples) were obtained in sheets consisting of a single layer. Subsequently, by pressing several layers followed by heating according to the following scheme1h at 150°C, 2h at 180°C for, 4h at 200°C for and 4h at 230°C, after curing for 4h at 200°C, the samples were allowed to cool down to room temperature gradually under a pressure of 10 Kgf/cm^2 and the second type, of multilayer samples, were obtained.

2.1. DSC analysis

Up to 5-mm long sample fragments were cut from both sample types, weighing less than 50 mg, for DSC experiments. For this purpose, a NETZSCH differential scanning calorimeter type DSC 200 F3 Maya was used, with sensitivity: < 1 W, temperature accuracy of 0.1 K and enthalpy accuracy—generally <1%. The device was calibrated with Bi, In, Sn, and Zn standards.

Temperature scans were performed between 20 and 300 °C, in two cycles, with a heating rate 10 K/min, under Ar protective atmosphere. In order to compare the obtained DSC results with those characteristic to currently available material, a 12.4 mg sample was cut from acommercial firefighter protection helmet. The sample was heated from 20 to 130 °C, in single cycle, to avoid melting. The DSC thermograms recorded during heating were evaluated with Proteus software, provided by NETZSCH.

2.2. DMA analysis

For the DMA tests, one 20 mm-long fragment was cut from type 2 multilayer sample and another from a commercial firefighter protection helmet. The experiments were performed by means of an analyzer type DMA 242 Artemis NETZSCH, with force resolution of 0.0005 N, amplitude range: ± 0.1 up to 240 μm and amplitude resolution: 0.0005 μm. A three-point bending specimen holder was used and a maximum static pre-loading force of 12 N. The specimens were tested at a constant frequency of 1 Hz, and amplitude of 20 μm, under Ar protective atmosphere using a 5 K/min heating rate.

2.3. STA analysis

The thermal experiments dealing with degradation of polymers were carried out by means of an integrated TGA-FTIR NETZSCH STA 449F1 system in nitrogen atmosphere within 30-600°C range at 10°C/min heating rate.

3. Results and discussions

The pre-polymer PABMI, synthesized by powder mixture heating of the 4,4-bismaleimidodiphenyl methane (0.1 mol.) and 4,4-diaminodiphenyl methane (0.05 mol.) at 150°C for 20 min, having M_n=8400 and polydispersity index of 1.4, was dissolved in NMP leading to a solution of 50 wt. % with which the glass fabric was impregnated.

The thermal behavior of pre-polymer was monitored by DSC measurements, in isothermal regime at 130°C (Figure 1) and in dynamic regime with 10°C/min (Figure 2). As it can be seen from the two figures, the exothermic curves are due to breaking of the unreacted double bonds.

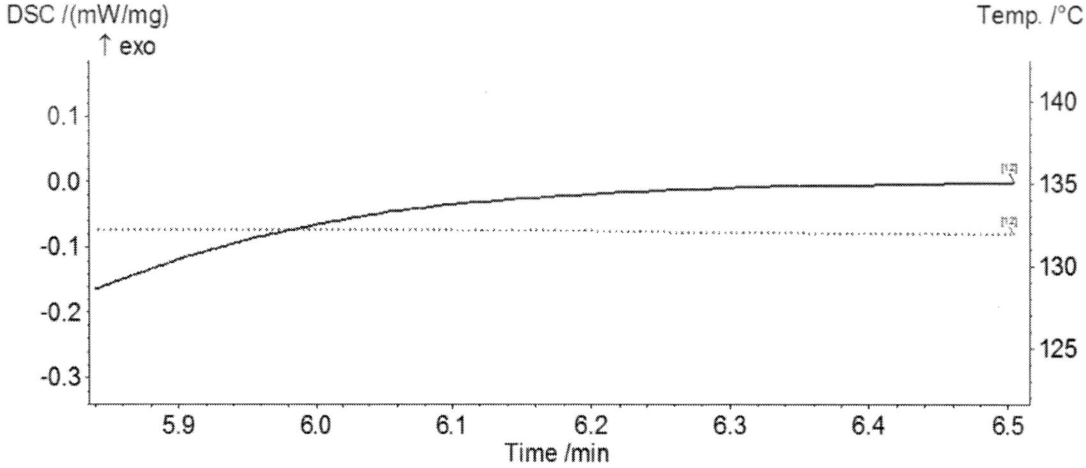

Figure 1.DSC thermogram of the pre-polymerin isothermal regime.

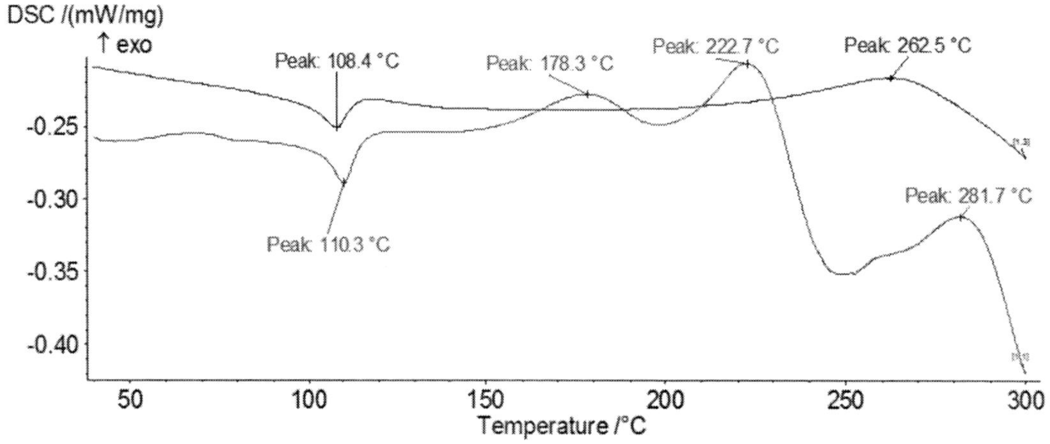

Figure 2.DSC thermogram of the first type sample in dynamic regime with 10°C/min, two heating cycles.

During the first heating cycle, an endotherm minimum at the 110 °C was observed, a scribed to the melting point of the PABMI pre-polymer, and three exothermic peaks specific to the secondary amide group polyaddition reactions to the BMI group (at 178.3 °C), to the homo-polymerization of the molecular groups (at temperature of 222,7 °C) and to other crosslinking reactions (at 281,7 °C), respectively. During the second heating cycle, a retention of a minimum of endotherm is observed, and the crosslinking reaction is higher due to the increase of the viscosity of the material. During the second heating cycle, a preservation of the endothermic minimum is observed and the crosslinking reaction tends to move to higher temperatures due to the increase in viscosity of the material.

The thermal stability of composite based on PABMI, compared with a sample collected from a helmet based on polyamide, are presented in the DGA thermograms from Figure 3. The two materials begin to decompose around 380°C but the material resulting from the helmet loses all the mass in a 50°C temperature range, while the PABMI-based composite has a residue of 84% even at 650°C. The decomposition products of PA_{BMI}-based material consist mainly of: benzene, aniline, toluidine, diethylamine, in varying proportions, and they can be seen at temperatures above 350°C.

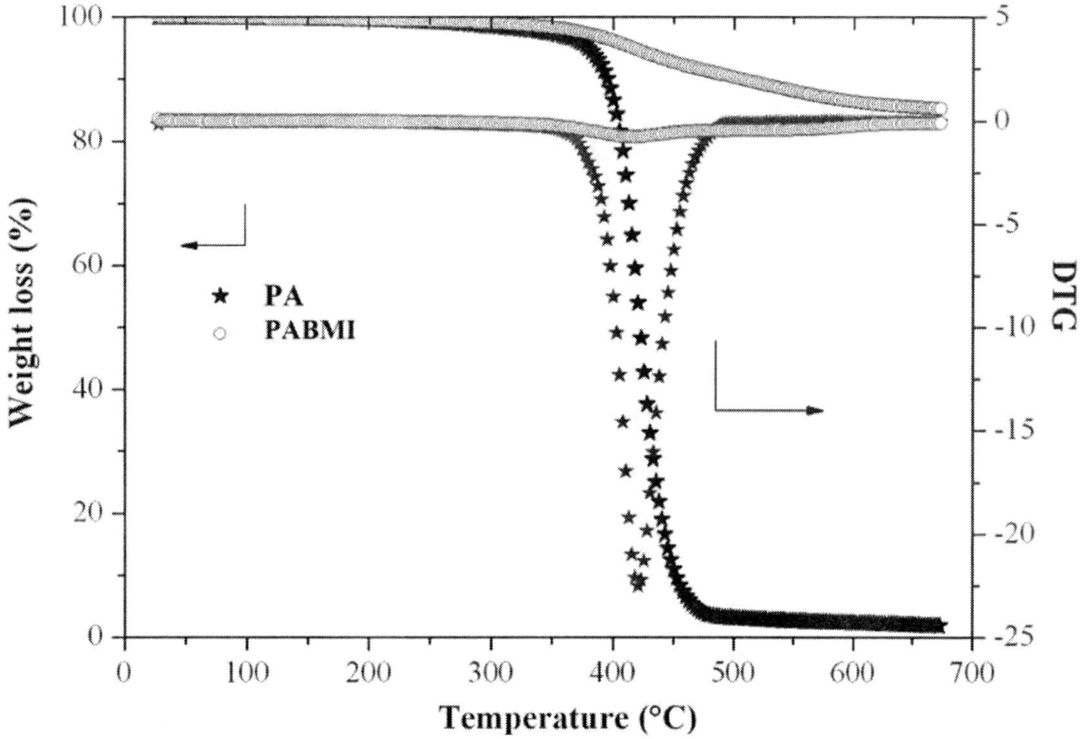

Figure 3. DTG thermogram of PABMI and PA.

The DSC curves of the two materials are presented in Figure 4. The sample from helmet based on polyamide shows an exothermic peak at 110.8°C and is probably due to the softening or melting of the material. The composite based on PABMI presents an increase in heat flow, probably due to the continuation of crosslinking reactions and the breakdown of chemical bonds at higher temperatures.

The two materials are solid at room temperature and exhibit high elastic moduli, above 2 GPa for sample from helmet and around 10.7 GPa for PABMI-based composite. The sample from helmet begins to melt at 98.3°C and ends melting at 104.1°C, while the PABMI-based composite has a modulus loss of about 1.5 GPa in the temperature range of 167-190°C. A small drop in E' of PAMBI-based composite is noticed between 50 and 120°C that may be associated with a β transition.

During the glass transition process the elastic modulus E' abates till 1.5 GPa, that means that it decreases one order of magnitude. For an amorphous linear polymer the decline of E' in the glass transition range amounts three orders of magnitude in a narrow temperature span. Consequently, we can infer that the compound contains crosslinking that obstruct the large movement of ample chain segments and it acts as a rigid material [12].

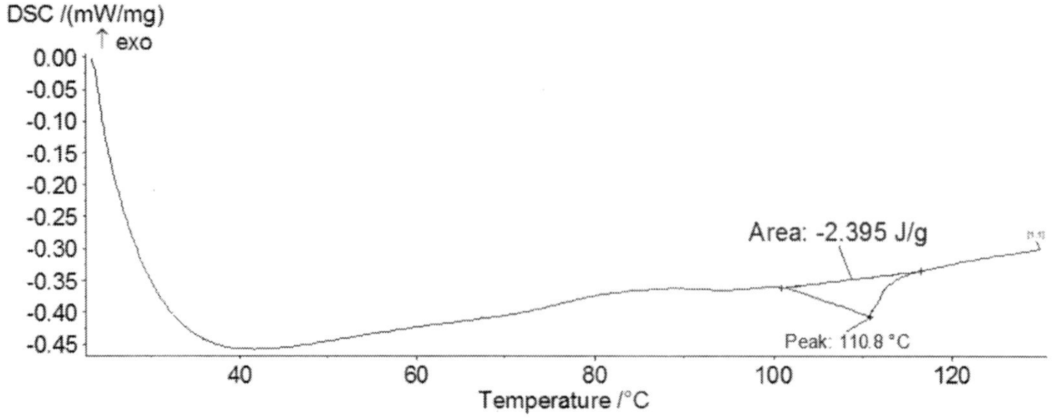

(a)

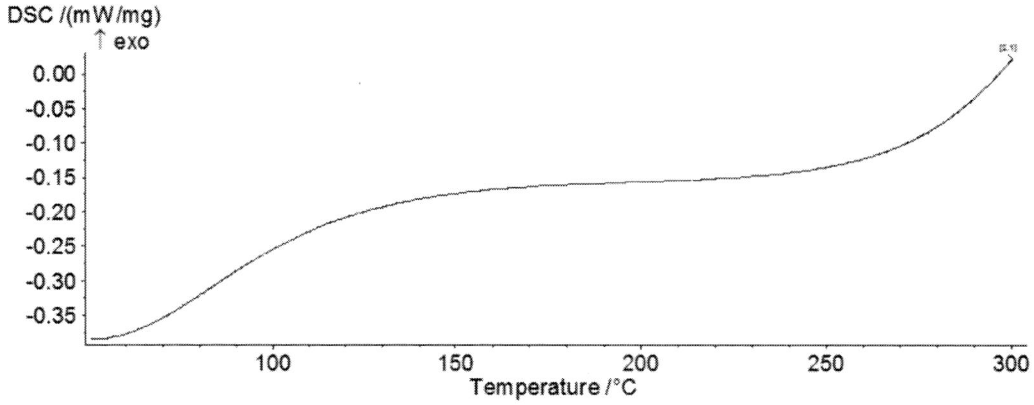

(b)

Figure 4.DSC thermogram of polyamide (a) and second type samples (b).

The viscoelastic behavior of the two materials was recorded by DMA (Figure 5).

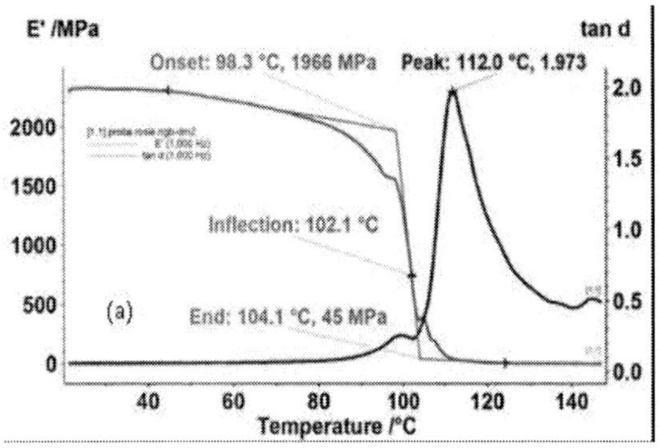

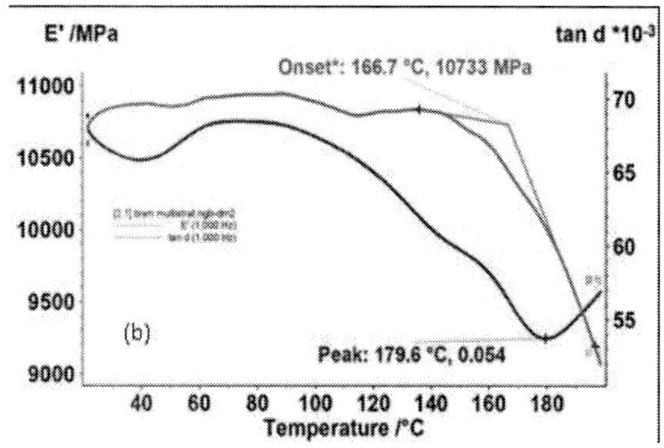

(a) (b)

Figure 5.DMA thermogram of polyamide (a) and second typesamples (b).

4. Conclusions

The newly developed glass fiber-reinforced bismaleimide (PAMBI) composite material experienced a superior behavior as compared to commercially available fireman helmet based on polyamide (PA) with regards to the following aspects:

- after a first heating cycle, the material reticulated and became thermally inert and no solid state transitions are noticeable during subsequent heating;
- mass losses were only 16 % at PAMBI, when heated to 650^{0}C, while the PA base helmet totally decomposed, when heated to 380^{0}C;
- PA based helmet totally softened above 110^{0}C (storage modulus decreased to 0), while PAMBI maintained a value of 10.7 GPa, at the same temperature.

All these preliminary results recommend the newly developed glass fiber-reinforced bismaleimide composite material a promising candidate for a new generation of fireman helmets.

5. References

[1] Sutton I 2017 *Plant Design and Operations*, Chapter 14 – Personal Protective Equipment 401–415
[2] Ghani S, Ali E M, El Bialy A, Bakochristou F, Gamaledin S M Ah and Rashwan M Mo 2017 *Applied Thermal Engineering* **111** 624–637
[3] Mills N J and Gilchrist A 1993 *Safety Science* **16**(3–4) 221-238
[4] Jachowicz M 2013 *Journal of Sustainable Mining* **12**(1) 27–33
[5] Bajpai P K, Ram K, Gahlot L K and Kumar Jha V 2018 *Materials Today: Proceedings* **5**(2) Part **3** 8699-8706
[6] Fink J K 2018 *Reactive Polymers: Fundamentals and Applications (Third edition)*, A Concise Guide to Industrial Polymers, *A volume in Plastics Design Library* 367–402
[7] Airinei A, Fifere Ni, Homocianu M, Gaina C, Gaina V and Simionescu B C 2011 *International Journal of Molecular Sciences* **12** 6176-93
[8] Ursache O, Gaina C, Gaina V and Musteata V E 2012 *Journal of Polymer Research* **19** 9968
[9] Vaidya U2012 Patent US 2012/0047634 A1
[10] Mandric Gh, Chiriac C, Gherasim G, Sava M and Pecincu S 1980 *Materiale Plastice* **18** 133-138
[11] Gaina V, Gaina C, Stoleriu A, Timpu D, Sava M and Rusu M 1999 *Polymer Plastic Technologies and Engineering* **38**(5) 927-38
[12] Cristea M, Gaina C, Ionita D and Gaina V 2008 *Journal of Thermal Analysis and Calorimetry* **93**(1) 69-76

Preliminary Tests for Ti-Mo-Zr-Ta Alloys as Potential Biomaterials

M S Bălțatu[1,*], P Vizureanu[1,2,*], T Bălan[3], M Lohan[1] and C A Țugui[1]
[1]"Gheorghe Asachi" Technical University of Iasi, Faculty of Materials Science and Engineering, 41 "D. Mangeron" Street, 700050, Iasi, Romania
[2] Faculty of Engineering Technology, Universiti Malaysia Perlis, (UniMAP), Perlis, Malaysia
[3] School of Environmental Engineering, Universiti Malaysia Perlis, (UniMAP), Perlis, Malaysia
[4]Faculty of Materials Science and Engineering, Gheorghe Asachi Technical University of Iasi, Romania

E-mail address: peviz2002@yahoo.com

Abstract. Nowadays, there is a continuing concern for the research and development of alloys for medical and biomedical applications. In order to check the biocompatible character of a new Ti-Mo-Zr-Ta alloys, it is necessary to carry out preliminary laboratory tests to follow how a biomaterial surface would interact with the host. The paper presents tests for Ti-Mo-Zr-Ta alloys like contact angle and DSC test to identify biocompatible character. Contact angle measurement is an experimental technique used to assess the hydrophilic or hydrophobic character of surfaces by reference to the 90° contact angle value and to characterize the thermal behavior, for temperature range between 36.5-37.2°C, interval which a biomaterial works inside the healthy human body, was used DSC test.

1. Introduction

Biomaterials play an important role in many aspects of the contemporary medical field through considerable progress over time. These are products of organic or inorganic nature that find various uses: prostheses or implants in biological tissues in humans or animals. In order to obtain these materials used for biological purposes, are used metals, ceramics, organic polymers etc. [1-5].

The common element of all biomaterials is their biocompatibility with human tissue, so that the response to the interaction between the implant and human tissue is positive without causing adverse reactions [1, 6, 7].

Pure titanium and Ti6Al4V alloy were the first titanium-based materials for commercial use, used as biomaterials. Titanium and its alloys have lately become an important alternative in medical applications being considered the safest material for implantation. There has been a growing interest in their application in the medical field taking into account the properties they present (mechanical properties, low density, good biocompatibility, increased corrosion resistance, etc.) [6-9].

As the alloying elements are added to the titanium, there is a modification of temperature at which occurs the phase transformation. Alloying the pure titanium with stabilizing elements α, increases the temperature range in which there is phase α, respectively the alloy with elements β, leads to an increase of the β phase domain. Other elements such as Zr or Sn have a neutral contribution to the temperature domains in which the two phases exist [1, 6-11].

Studies have shown that β-type alloys, due to molybdenum stabilizing, tantalum and niobium stabilizers, have the advantage of increasing mechanical strength and an elastic modulus close to that of the human bone, important aspects of the use long time of biomaterials in the medical field [6-9].

Ti-Mo alloys with various biocompatible elements such as zirconium and tantalum show superior mechanical properties such as high tensile strength and a much lower modulus of elasticity close to that of human bone compared to other classical biomaterials [1].

This research work is focus on tests for two alloys, Ti15Mo7Zr15Ta and Ti20Mo7Zr15Ta (TMZT), tests like contact angle and DSC test to identify biocompatible character.

Biocompatible materials, such as titanium alloys, have been used and are used in the form of implants that come in contact with tissues. The wettability of biomaterial surfaces is a physico-chemical property of the important surface in the realization / optimization of cell adhesion and proliferation. Thus, the surface properties of the materials are essentially to be analyzed to determine interactions with host tissues.

Differential Calorimetric Analysis (DSC) refers to the quantitative measurement of heat exchange during a thermodynamic process. To characterize the thermal behavior, the temperature range to be watched is between 36.5-37.2°C, the interval within which a biomaterial works inside the healthy human body.

2. Methodology

The alloys analyzed were obtained with a vacuum arc vacuuming equipment MRF ABJ 900 using as raw materials high purity elements such as Ti-99.8%, Mo-99.7%, Zr-99.2% and 99.5% Ta.

One of the biomaterial requirements is cellular adhesion on the surface of the material, depending on surface energy. Specialty studies have indicated that contact angle measurement is important for the study of cell adhesion to the surface of the studied material [12].

Measurement of the contact angle (Figure 1) is an experimental technique used to assess the hydrophilic or hydrophobic character of the surfaces. If the angle of contact is between 0-90°, the material is hydrophilic and hydrophobic, if the angle of contact is between 90-180° [13].

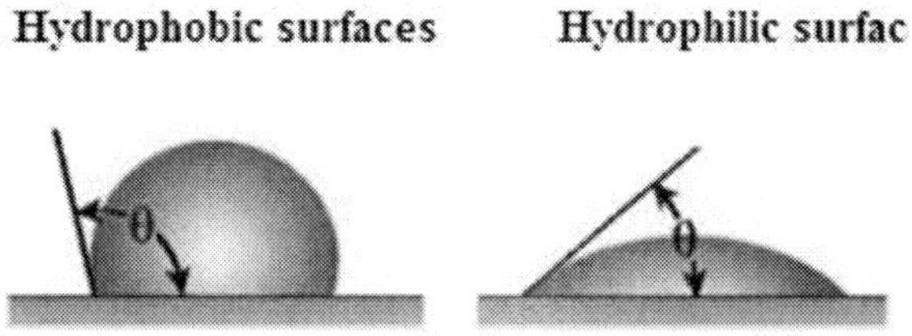

Figure 1. Measurement of contact angle for hydrophilic and hydrophobic surfaces [12].

The choice to use distilled water as a control sample in assessing contact angle for experimental TMZT alloys is based on the fact that water represents 70% of the human body.

The value of contact angle of water in contact with TMZT alloys was determined by the Sessile Drop [13]. Method (Sessile Drop Method), determining the hydrophilic or hydrophobic character of investigated experimental alloys . The principle of measuring the angle of contact is by placing a drop of water with a microsurge, with the drop volume of 4 microlitres. Drop lighting is made from behind and recorded from the opposite side with a digital camera. The image obtained is further analyzed by the Famas software, an integrated software for KYOWA goniometers [13, 14].

In order to measure the contact angle of TMZT alloys, 10 mm x 10 mm x 5 mm samples were used.

The DSC allows to obtain information on the processing temperatures associated with the alloying processes. The results of the calorimetric measurements are represented by a curve called a thermogram, conventionally denominated in the DSC curve [15, 18].

Both the sample crucible and the standard (which is an empty crucible) are placed inside the DSC system cell. The temperature difference between the sample and the reference is measured and recorded as a heat flux.

The temperature is measured with three thermocouples, one indicating the sample temperature, the second reference temperature, and the final temperature of the furnace. The output signal from DSC is taken over by a computer that performs data analysis with a suitable software (NETZSCH PROTEUS Thermal Analysis). DSC analysis equipment also includes a cylinder with inert gas (argon) and one with coolant (liquid nitrogen).

In view of the thermal analysis by differential calorimetric analysis of TMZT alloys, samples were prepared at sizes corresponding to the mass up to 50 mg, the dimensions being imposed by the dimensions of the crucibles where the samples were tested. From the measurements made after the sampling, it was found that the actual mass is in a range of 42-49 mg.

3. Results and discussion

The water drop on the sample surface was recorded using a digital camera. The images obtained were then analyzed through the Famas program. Figure 2 shows the water droplet image on the surface of the TMZT alloys. These differ depending on the contact angle the water forms on the surface of the sample.

Research in the field, affirm that materials with moderately hydrophilic surfaces improve adhesion, cell growth and ultimately biocompatibility, and a hydrophobic surface can lead to decreased adhesion and loss of biocompatibility [16-19].

10 measurements of the contact angle (θ) for each experimental alloy were performed and the value presented is the average of the measured measurements with the maximum error of ± 1°. The mean value of contact angle for each alloy is shown in Table 1.

Table 1. Values of contact angle with water at the surface of TMZT alloys.

Alloy	Ti15Mo7Zr15Ta	Ti20Mo7Zr15Ta
Liquid used	water	
Contact angle (degrees)	57,93	56,11

(a) (b)

Figure 2. Images of water droplet on the surface of TMZT alloys: (a) Ti15Mo7Zr15Ta, (b) Ti20Mo7Zr15Ta.

The two investigated alloys have a contact angle of less than 90° (Figure 2), thus exhibiting a hydrophilic character, which means high cell adhesion to the surface of the alloys. From the data obtained for the surfaces of the two TMZT alloys analyzed, the highest arithmetic mean value is recorded at the water contact angle on the Ti15Mo7Zr15Ta alloy surface and the lowest at the Ti20Mo7Zr15Ta alloy, this alloy having a more pronounced hydrophilic character.

From the results obtained it was observed that the alloy with a percentage of 20% Mo, shows the best value. The values of TMZT alloys obtained from the surface moisture analysis are certified to be that titanium alloys have a good interaction with tissues host. Thus, we have the certainty of good cellular adhesion, which could allow for the "in vitro" analysis of the determination of good biocompatibility for the use of alloys developed in the human body.

Thermograms recorded following the DSC analysis performed with PROTEUS are shown in Figure 3. The thermal regime to which the analyzed samples were subjected was:

a. heating from room temperature to 40° C at a rate of 10° / min;

b. cooling at 10° / min to room temperature.

The thermograms of the investigated TMZT alloys did not record phase transformations in the 30-40° C temperature range. As a result of the thermal analysis, the investigated alloys show stable values for the analyzed interval, interval which a biomaterial works in a human body.

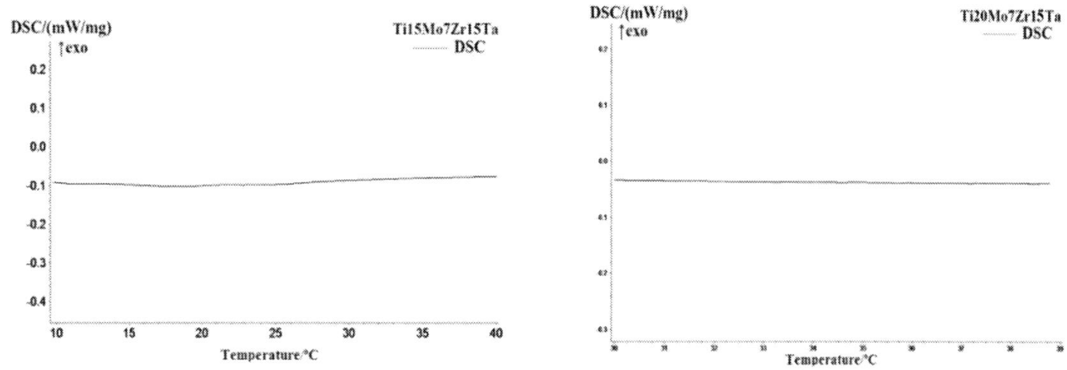

(a) (b)

Figure 3. Thermograms obtained for experimental alloys: (a) Ti15Mo7Zr15Ta, (b) Ti20Mo7Zr15Ta.

Following thermal analysis by DSC, TMZT alloys showed thermal stability at the temperature of the human body, showing good interaction with the cellular host for use as medical implants.

4. Conclusions

Biocompatible materials, such as titanium alloys, have been used and are used in the form of implants that come into contact with living tissues, thus the surface properties of these materials are essentially to be analyzed to determine interactions with host tissues.

This research work is focus on tests for two alloys, Ti15Mo7Zr15Ta and Ti20Mo7Zr15Ta, tests like contact angle and DSC test to identify biocompatible character.

The values determined for the contact angle of the TMZT alloys are hydrophilic, which means a high adhesion of the cells to the surface of the alloys, the Ti15Mo7Zr15Ta alloy have a pronounced hydrophilic character, the alloy of a less pronounced hydrophilic character is Ti20Mo7Zr15Ta .

Thermal analysis by differential scanning calorimetry revealed that TMZT alloys did not show phase transformations in the range of 30-40°C, being stable at the temperature of the human body.

5. References

[1] Bălţatu M S, Vizureanu P, Ţierean M H, Minciună M G and Achiţei D C 2015 *Advanced Materials Research* **1128** 105-111
[2] Geetha M, Singh A K, Asokamani R and Gogia A K 2009 *Mater. Sci.* **54** 397-425
[3] Minciuna M G, Vizureanu P, Achitei D C, Ghiban N, Sandu A V and Forna N C 2014 *REV. CHIM. (Bucharest)* **3** 335-338
[4] Istrate B, Munteanu C, Matei M, Oprisan B, Chicet D and Earar K 2016 *IOP Conference Series: Materials Science and Engineering* **133** 12010
[5] Istrate B, Mareci D, Munteanu C, Stanciu S, Luca D, Crimu C I and Kamel E 2015 *Journal of Optoelectronics and Advanced Materials* **17**(7-8) 1186-1192
[6] Niinomi M 1998 *Mater. Sci. Eng.* **243** 231-236
[7] Elias C N, Lima J H C, Valiev R and Meyers M A 2008 *Biological Materials Science* 46- 49
[8] Bombac D M, Brojan M, Fajfar P, Kosel F and Turk R 2007 *Materials and Geoenvironment* 54 (**4**) 471-499
[9] Elias C N, Lima J H C, Valiev R and Meyers M A *2008 Biological Materials Science* 46- 49
[10] Minciuna M G, Vizureanu P, Achitei D C, Sandu A V, Berbecaru A and Sandu I G 2016 *Journal of optoelectronics and advanced materials* **18**(1-2) 174-178
[11] Minciuna M G, Vizureanu P, Geanta V, Voiculescu I, Sandu A V, Achitei D C and Vitalariu A M 2015 *Revista de chimie* **66**(6) 891-894
[12] Vogler E A 1998 *Adv. Colloid Interface Sci.* **74** 69-117
[13] Baier R E *2006 Journal of material science: Materials in medicine* **17**(11) 1057-1062
[14] ***http://www.ramehart.com/contactangle.htm
[15] Rusu I 2011 *Tehnici de analiza in ingineria materialelor* Ed. PIM, Iaşi
[16] Kwork D Y, Lam C N C, Li A, Leung A, Wu R, Mok E and Neumann A W *1998 Colloids and surfaces A.* **142** 219-235
[17] Minciuna M G, Vizureanu P, Achitei D C, Sandu A V, Berbecaru A and Sandu I G 2016 *Journal of Optoelectronics and Advanced Materials* **18**(1-2)
[18] Ştefănoiu R, Geantă V, Voiculescu I, Csakin I and Ghiban N 2014 Revista de Chimie 65(7 819-821
[19] Geantă V, Voiculescu I and Stanciu E M 2016 International Conference on Innovative Research, Iop Conf Series: Materials Science And Engineering 133 1-8

Acknowledgements
This work was supported by a grant of the Romanian Ministery of Research and Innovation, CCCDI – UEFISCDI, project number PN-III-P1-1.2-PCCDI-2017-0239 / 60PCCDI 2018, within PNCDI III.

Preliminary Results on Thermal Shock Behavior of CuZnAl Shape Memory Alloy Using a Solar Concentrator as Heating Source

C Tudora[1,3], M Abrudeanu[1,4], S Stanciu[2], D Anghel[1], G A Plaiaşu[1], V Rizea[1], I Ştirbu[2] and N Cimpoeşu[2,*]
[1]University Pitesti, 1 Targu din Vale Str, Pitesti, Arges, Romania
[2]Technical University "Gheorghe Asachi" of Iasi, Mangeron 44 blv., 700050, Iasi, Romania
[3]Institute for Nuclear Research Pitesti, 1Campului str., 115400, Mioveni, Arges, Romania
[4]Technical Science Academy of Romania, 26 Bd. Dacia, sector 1, Bucuresti, Romania

E-mail: nicanornick@yahoo.com

Abstract. It is highly accepted that martensitic transformation can be induced by temperature variation and by stress solicitation. Using a solar concentrator, we manage to increase the material surface temperature (till 573 respectively 873 K) in very short periods of time in order to analyze the material behavior under thermal shocks. The heating/cooling process was registered and analyzed during the experiments. Material surface was analyzed before and after thermal shocks by microstructure point of view using scanning electron microscopy (SEM) and atomic force microscopy (AFM). The experiments follow the material behavior during fast heating and propose the possibility of activating smart materials using the sun heat for aerospace applications.

1. Introduction

Shape memory alloys (SMA abreviated) have a number of great properties compared to common metallic materials. Of these, the feature is the ability to change geometric shape when switching from low to high. Under certain conditions, the shape change can be reversible so that the material can store two geometric shapes, namely both high temperature (warm form) and low temperature (cold form) - these transformations are performed as a result of a memory effect form (through this effect the material can also perform a mechanical work during the transition from cold to hot form) [1-5].

Alloys with Cu-Zn-Al shape memory are alloys derived from ordinary brass. The chemical composition of Cu-Zn-Al AMPs is established taking into account the equilibrium diagram and critical point variation with the alloying element concentration. Thus, for a binary Cu-Zn alloy, a temperature Ms = 0 °C is obtained at a concentration of 38.5% (at) [6]. It is necessary to apply a very high-speed cooling so that phase b is retained to ambient temperature and martensite transformation without which the form memory effect cannot exist [7]. Aluminum alloy reduces the braze quenching speed, making martensitic transformation easier, besides this aluminum also increases corrosion resistance, mechanical strength and plasticity. Alloying with 4-8% Al ensures the formation of martensite 9R, which gives the transformation a good reversibility with a 5-15% hysteresis. Conventional copper Cu-Zn-Al alloys have an MS ranging from -200 ° C to +100 ° C and the chemical composition of shape memory brasses is selected according to the desired MS value within the limits: 62-72% Cu, 14-30 % Zn, 4-8% Al [8].

In this article we analyze the effects of 10 thermal shocks till 300 respectively 600 °C on the microstructure of a CuZnAl shape memory alloy. The thermal shocks were realized using a solar concentrator furnace reaching heating rates of 70 °C/ms.

2. Materials and methods

The solar concentrator used is part of the Promes France facility. PROMES-CNRS at Font-Romeu Odeillo host a whole range of high and very high flux solar furnaces and 10 min nearby in Targassonne the Themis solar tower [13]. Shape memory alloys samples were heated using solar light reflection from a metallic window (5x12 m) situated at the bottom of the building, through a shuters system (used to control the intensity of the light beam) and a concentrator, as presented schematically in figure 1 a) [9, 10]. The light is forward transmitted at 1-1.5 m down to a center and spread near that. In order to locate the sample in the center of light concentration and to benefit of the biggest intensity and temperature we use an aluminum trolley, figure 1 b) that can be moved on X-Y axis.

The support system is always chilled with water (the temperature can increase really fast till 3000°C at a full opening of the shuters). The samples, 10x10 mm, were fixed using a mechanical system and under the sample, in the middle part, a K thermocouple was used to retrieve data information connected to a Graphtec Corporation equipment type GL220 [11]. Both heating and cooling stages were registered and analyzed. The shocks were performed moving the trolley from and under the sun light. The heating experimental temperatures were 300 and 600°C ±(10-25).

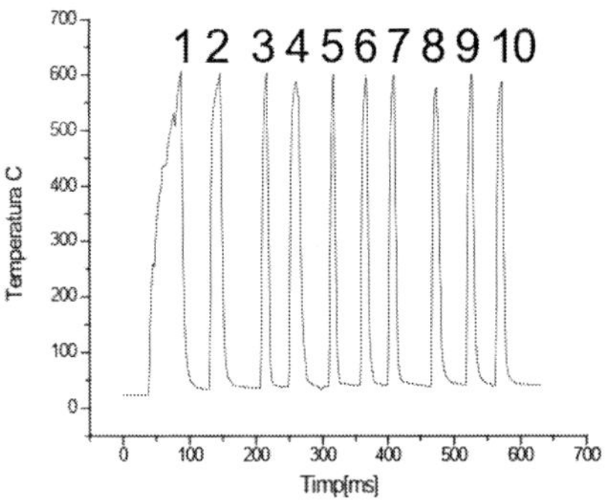

(a)

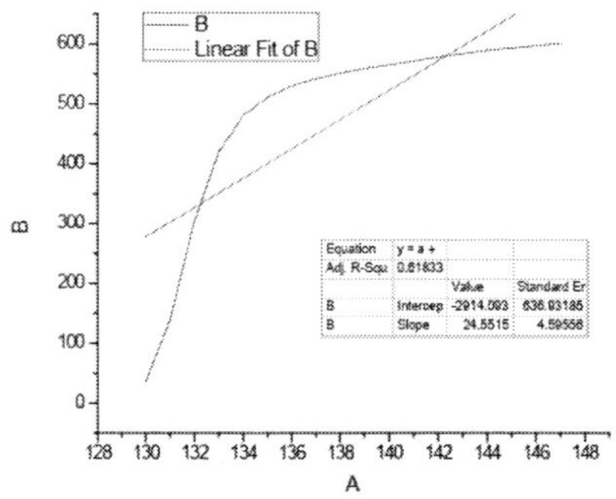

(b)

Figure 1. (a) Temperature variations on CuZnAl sample in time (b) during the solar heating and room cooling and linear fit curve of cooling.

Copper based shape memory alloy (Cu17.5Zn7.2Al) as 10x10 mm square were used for exposure to heat (10 cycles) at 300 and 600°C. The flux was applied in the middle part of the sample by situating the sample in the middle of the solar flux. The temperature increasing and decreasing was registered using a K type thermocouple and the heat was analyzed from the middle of the sample. In Figure 1 a) and b) is represent the temperature variation in time for 10 heating/cooling cycles.

The heating process consists of a mirror that collect the sun light, a concentrator of light through a shuters system and a trolley for positioning of the sample in the middle of the concentrated light. The heat intensity is given from the shuters system opening and can be control based on the clearness of the sky. Small variations near 300 and 600°C temperatures were observed. The appearance of these variations is based on the human factor that move and sustain the trolley and can be eliminated using a motorized system. In tables 1 and 2 are presented the heating/cooling rates registered during the thermal shock experiments till 300 respectively 600°C.

Table 1. Heating and cooling rates of shape memory alloy for thermal shocks tests till 600°C.

Thermal shocking parameters	Cycle 1		Cycle 2		Cycle 3		Cycle 4		Cycle 5	
	H	C	H	C	H	C	H	C	H	C
Fitting rate (°C/ms)	9.4	16.9	24.6	22.9	59.4	19.4	45.5	33.9	73.6	46.8
Calculated rate (°C/ms)	12.1	24.7	33.2	29.5	51.6	28.3	45.8	28.6	62.3	50.5
Thermal shocking parameters	Cycle 6		Cycle 7		Cycle 8		Cycle 9		Cycle 10	
	H	C	H	C	H	C	H	C	H	C
Fitting rate (°C/ms)	56.2	37.7	53.7	30.4	67.1	27	65.7	20.1	55	21.7
Calculated rate (°C/ms)	50.5	39.3	46.6	36.8	67.2	36.8	62.1	25.2	54.7	28.5

Before and after the thermal shocks the metallic material surface was analyzed through scanning electron microscopy (SEM VegaTescan LMH II) and atomic force microscopy (AFM EasyScan II).

3. Results and discussions

The experimental samples, before and after the thermal shocks, were investigated without any surface polishing in the area affected by solar light through scanning electron microscopy, Figure 2, and atomic force microscopy, figure 3. By microstructural point of view, we follow the material behavior after a fast heating to 300 respectively 600°C and a cooling at room temperature, 25°C. As in the case of Cu-Al alloys, the solid β solution (based on the electron-intermetallic compound equiatomic, CuZn, crystalline structure A_2 is ordered, it results in ordered β2 (B2) austenitic which is formed between 454 and 468°C.

At slow cooling, the α (cfc) and γ solid solutions (based on the inter-metallic compound Cu_5Al_9, cubic complex with 52 atoms per elemental cell) are precipitated from β_2. On sudden cooling or by applying an external stress, the pink colorant martensite α_2 (3R) is obtained [13, 14]. SEM images, Figure 2, present an affected surface in both experimental cases b) and c) comparing with the initial alloy surface state.

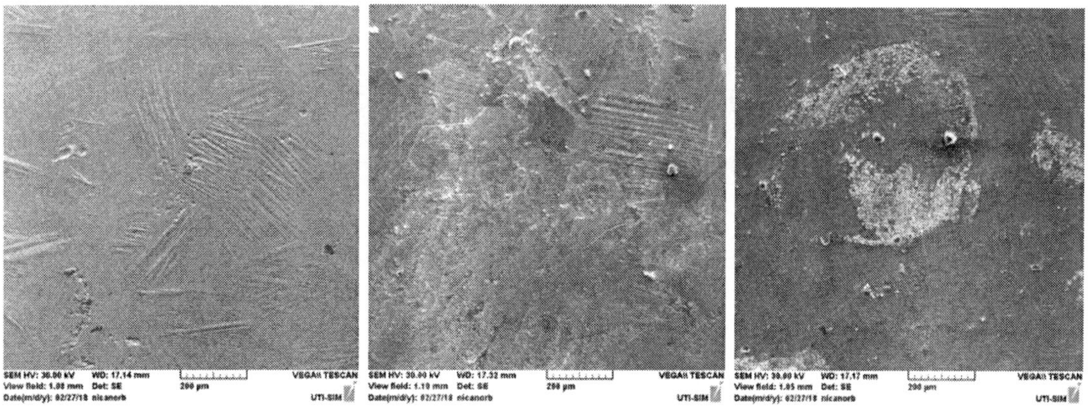

(a) (b) (c)

Figure 2. SEM micrographs of CuZnAl surface: (a) initial state, (b) after 10 thermal shocks at 300°C and (c) after 10 thermal shocks at 600°C.

IOP Conf. Series: Materials Science and Engineering **374** (2018) 012024 doi:10.1088/1757-899X/374/1/012024

After heating till 300 respectively 600°C the material surface is affected structurally and chemically by formation of oxides on the surface and local thermal modification. The martensite plates are presented in similar orientation after the thermal shocks as in the initial state. The effect of the thermal shocks on the shape memory properties will be analyzed in the near future. Excepting the local surface affected areas the structure keeps a similar aspect with grains of the same dimension and primary and secondary plates. The thermal shocks effect must be analyzed also in depth in order to establish the influence of a higher temperature on the shape memory properties.

In Figure 3 are presented atomic force microscope images of CuZnAl surface 2D and 3D a) and d) initial state, b) and e) after 10 thermal shocks at 300°C and c) and f) after 10 thermal shocks at 600°C. Although Cu-Zn-Al-AMF martensitic plates have an internal defect substructure created by inverse planar shear, the acoustics of the plates in the austenitic matrix are made by twining to maintain the coherence of the crystal lattice on the austenite-martensite interface.

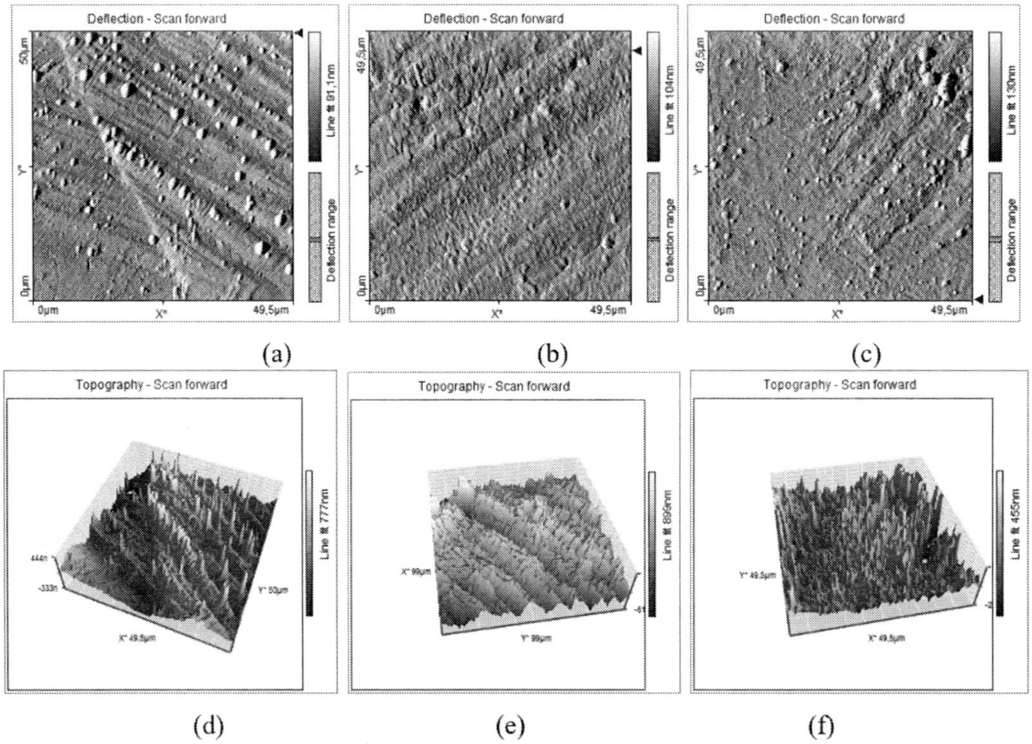

Figure 3. AFM images of CuZnAl surface 2D and 3D: (a) and d) initial state, (b) and (e) after 10 thermal shocks at 300°C and (c) and (f) after 10 thermal shocks at 600°C.

Macles (twins) observed in the microstructure of these alloys are accommodating rather than transforming. In addition, it should be noted that approx. 50% of these macules are type II, as well as AM-based Cu-Al-Ni [15]. Because of the twinning process, martensite from AMF Cu-Zn-Al present a shallow relief. The influence of the thermal shocks can be considered as a decrease of the relief of the martensite plates, Figure 3 b) c) and e) and f). The growth and reversal of martensite plaques is one of the most intensely studied phenomena, characteristic of the thermo-elastic martensite, for which a wide variety of characterization methods were used [16, 17]. The relief is visible also in the secondary plates, in the initial phase, the relief characteristic of the primary plates - which passes all the micrograph and plate fields secondary martensite. The latter are shorter and finer because they did not have any time or space to increase to the size of the main plates. In the case of samples subjected to thermal shock, the formation of the secondary plates is no longer observed.

4. Conclusions

In conclusion after the experimental results analyze:

- Activation of shape memory elements can be realized using sun light,
- A large domain of temperatures and heating rates can be reached including the high temperature shape memory alloys,
- Solar concentrators represent a cheap and viable solution for shape memory alloy functionalizing,
- Thermal shocks till 300 and 600 °C modify the chemical composition of the surface by forming oxides,
- SEM and AFM results present no modification at the primary plates dimensions and orientation also after the thermal shocks no secondary plates formation was observed.

5. References

[1] Lohan M N, Pricop B, Bujoreanu L-G, Cimpoesu N 2011 *International Journal of Materials Research* **102** 1345-1351
[2] Bujoreanu L G, Lohan N M, Pricop B, Cimpoeşu N 2012 *Materials Science Technology* **28** 658-667
[3] Cimpoeşu N, Stanciu S, Doroftei I, Ioniţă I, Radu V, Paraschiv P 2010 *Optoelectronics and Advanced Materials-Rapid Communications* **4** 2028 – 2031
[4] Paun M-A, Cimpoesu Hanu R, Cimpoesu C, Agop M, Baciu C, Stratulat S, Nejneru C 2010 *Materiale Plastice* **47** 209-214
[5] Bujoreanu L G 2015 *Journal of Optoelectronics and Advanced Materials* **17** 1437-1443
[6] Stošić Z, Manasijević D, Balanović L, Holjevac-Grgurić T, Stamenković U, Premović M, Minić D, Gorgievski M, Todorović R 2017 *Materials Research* **20** 1425-1431
[7] Dasgupta R 2014 *Journal of Materials Research* **29** 1681-1698
[8] Coseri S, Spatareanu A, Sacarescu L, Socoliuc V, Stratulat I S, Harabagiu – Pullulan V 2016 *Journal of Applied Polymer Science* **133** 42926
[9] Crane N B 2009 *Journal of Solar Energy Engineering* **132** 11007
[10] Alxneit I 2011 *Solar Energy* **85** 516-522
[11] Wieckert C, Frommherz U, Kra¨upl S, Guillot E, Olalde G, Epstein M, Sante´n S, Osinga T, Steinfeld A 2007 *Journal of Solar Energy Engineering ASME* **129** 190 – 196
[12] Cimpoesu I, Stanciu S, Cimpoesu N, Munteanu C, Istrate B, Dragos A U, Dana D, Alexandru A, Nejneru C 2013 *Journal Of Optoelectronics And Advanced Materials* **15** 1392-1398
[13] Jani J M, Leary M, Subic A, Gibson M A 2014 *Materials and Design* **56** 1078-1113
[14] Liang S M, Schmid-Fetze R 2016 *Calphad* **52** 21-37
[15] Adachi K, Perkins J, Wayman C M 1986 *Acta metallurgica* **34** 2471-2485
[16] Velten B 1992 *Methods for characterization of transformation–and shape memory properties*, Progr.Shape Mem.All., (Eucken, S. ed.), DGM Informations gesellschaft Verlag, Bochum, pp 23-46
[17] Achitei D C, Vizureanu P, Dana D and Cimpoesu N 2013 *Metalurgia International* **18** 104-109

Acknowledgement

This article was made with support of project "Thermal shock and fatigue on shape memory alloy from CuZnAl and CuAlNi systems – HTSMAs" FP7-INFRA-312643, SFERA user research proposal.

Characteristics of Thermoplastic Potato Starch/Bentonite Nanocomposite Film

N H Zakaria[1, 2, *], N Muhammad[1, 2], M M A B Abdullah[2, 3], I G Sandu[2,4] and C L Mei Wan[1]

[1]School of Manufacturing Engineering, Universiti Malaysia Perlis, Perlis, Malaysia
[2]Center of Excellence Geopolymer & Green Technology (CEGeoGTech), School of Materials Engineering, Universiti Malaysia Perlis, Perlis, Malaysia
[3]Faculty of Engineering Technology, Universiti Malaysia Perlis, Perlis, Malaysia
[4]Gheorghe Asachi Technical University of Iasi, Faculty of Materials Science and Engineering, Blvd. D. Mangeron 41, 700050, Iasi, Romania

E-mail: *nurulhusna.zakaria@gmail.com, noorhafiza@unimap.edu.my, mustafa_albakri@unimap.edu.my, wan94@hotmail.com

Abstract. The aim of this study is to investigate the effect of bentonite towards thermoplastic potato starch nanocomposite films on the mechanical, microstructure and physical properties. The nanocomposite films were prepared using bentonite nano filler (0, 1, 5, 10, 15 and 20%) through solution casting technique. Obtained result indicate that, tensile strength increased significantly with increasing bentonite content and the highest tensile strength was recorded for nanocomposite film with 20% bentonite content. Meanwhile, elongation at break increased as the bentonite content increased from 0 to 15%, however significantly decreased at 20% bentonite content due to ductile structure and anti-plasticizing effect. Besides, good dispersion between bentonite nano filler and starch matrix with slightly remaining anglomerates was evident in scanning electron microscopy (SEM) image. Overall result shows that the addition of bentonite nano filler in potato starch film significantly influenced the properties of the films.

1. Introduction

Nowadays, there has been vast applications in biomaterials for biomedical engineering, and one of the growing interest is biodegradable starch based biopolymer. Starch is a natural source which can easily extracted from agricultural raw materials and known as the most abundant, renewable, non-toxicity, biocompatibility and readily biodegradable with thermoplastic behaviour at low cost [1–3]. Besides, starch possess similar properties to synthetic polymer and thus they are one of the potential candidates for replacing current synthetic materials.

Starch possible to process into thermoplastic materials through disruption and plasticization of starch macromolecules by heating in the presence of water and plasticizer, such as glycerol. Nonetheless, thermoplastic starch (TPS) products are water sensitive and would lead to poorer mechanical and physical properties, due to their hydrophilic nature. In order to improve the mechanical properties and water resistance, starch can be altered by incorporation of an inorganic reinforcing material to form composite products [4, 5]. Recently, special attention has been focused on the use of nano sized fillers into starch matrix to form nanocomposites. The application of nanocomposites could yield significant enhanced mechanical properties of starch based materials as compared to conventional micro composites. This is because the mutual effect between filler at nanometer size permits them to form molecular bridges in polymer matrix.

Nanoclays are the most frequently employed nanofiller for polymers because of its natural source and high modulus of clay platelets which mains a significant role to improve the mechanical properties in nanocomposites film. Furthermore, nanoclay is a naturally abundant toxin-free mineral and relatively low cost which can be used as one of the components in biomedical application [6]. Several studies have investigated the development and characterization of starch nanocomposites film suingbentonite clay. The results showsignificant improvement in mechanical propertiesof potato starch nanocomposite films [7]. Hence, resolving the barrier of starch-based film, usage of clays has proved to strengthen the mechanical properties of TPS.

Therefore, in this study natural bentonitewas introduced as nanofiller to prepare potato starch nanocomposites film through solution casting method. Nanocomposites film with different bentonite content were prepared to investigate the effect on the tensile and microstructure properties.

2. Materials and method

2.1. Materials

Commercially available potato starch, $(C_6H_{10}O_5)_n$ (HmbG Chemicals) was used as matrix material andbentonite powder from (Multifilla Bhd) was used as nanofiller. Glycerol, $(C_3H_8O_3)$ obtained from (HmbG Chemicals) was used as plasticizer. The selection of the materials were determined according to the requirement of biomedical application, as the investigations reported that:

i. Potato starch have ability to improve its durability, good tissue integration and did not induce a pathological foreign body reaction.
ii. Bentonite is from natural source and have high modulus clay platelets to attain mechanical properties. The bentonite also have barrier properties in which able to inhibit the diffusion of water and oxygen.
iii. Glycerol have better performance in producing biodegradable film compared to sorbitol.

2.2. Preparation of the thermoplastic starch films

Starch nanocomposites films were prepared via solution casting technique. About 5g of potato starch was first dispersed in 100 ml deionized water with (30% relative to dry starch) of glycerol content according to the corresponding run of experiments. The mixture was then heated to 85°C in a water bath for 30 min under constant stirring. Specified amount of dried Bentonite and HA were dispersed at room temperature in 40 ml deionized water with the following mixing modes:

i. Mechanical mixing at 1000 rpm for 30 minutes, using a RW 20 mixer (IKA,USA) in a 100ml beaker.
ii. Sonication mixing for 30 minutes, perform on a HD 2200 ultrasonic homogenizer (Bandelin, Germany) with the high frequency power at a frequency of 20 kHz and 6 mm probe diameter.

Table 1. Formulation of thermoplastic potato starch nanocomposites film.

Material	B1	B5	B10	B15	B20
Potato starch (g)	5	5	5	5	5
Distilled water (mL)	100	100	100	100	100
Glycerol (%)	30	30	30	30	30
Mixing temperature (°C)	85	85	85	85	85
Bentonite (%)	1	5	10	15	20

The nanofiller suspension was added to the aqueous dispersion of starch and the mixing was continued for 30 minutes at 85°C. Subsequently, the mixture was poured homogeneously onto petri dish of 14 cm diameter. The mixture were then dried at 45°C in an oven with air force circulation. The dry films were removed from the petri dish and stored in a dessicator (40% of relative humidity) in a plastic bag for a week prior to testing. Thermoplastic starch films with varies contentof bentonitewas

IOP Conf. Series: Materials Science and Engineering **374** (2018) 012025 doi:10.1088/1757-899X/374/1/012025

denoted according to bentonite content as (B1, B5, B10, B15 and B20). Table 1 shows the formulation of thermopalstic potato starch nanocomposites film..

2.3. Tensile Test
The tensile test were performed as suggested by Sanyang et al. [8], and the tensile strength and elongation at break were determined according to the ASTM standard (D882-02) by using a 5kN INSTRON tensile machine. Film strips were cut into 70mm x 10mm sections are subjected to crosshead speed of 2mm/min and initial grip separation of 40mm. The test was conducted with three replication for each sample and the average value was calculated.

2.4. Microstructure Analysis
The microstructure of the film was observed by using scanning electron microscopy (SEM) model (Hitachi TM-3000) operated at accelaration voltage of 15kV.

3. Results and discussion
3.1. Tensile Test
The mechanical properties of thermoplastic potato starch nanocomposites films were studied through a tensile test. This test was conducted in order to investigate the influence of bentonite content onto the mechanical behaviour of the starch film. The results of tensile strength and elongation at break are illustrated in Figure 1. As shown in Figure 1(a), it is clearly indicated that the tensile strength of the starch films increased significantly with increasing bentonite content. The highest tensile strength value of 5.74MPa was recorded for films prepared with 20% bentonite content, meanwhile neat starch films prepared without bentonite filler showed the lowest tensile strength value of2.5MPa. The improvement in tensile strength as the bentonite content increased were reported by several studies [9-11]. This behaviour proved that bentonite filler have reinforcing capability, which can optimize the host polymer properties when added at high content.

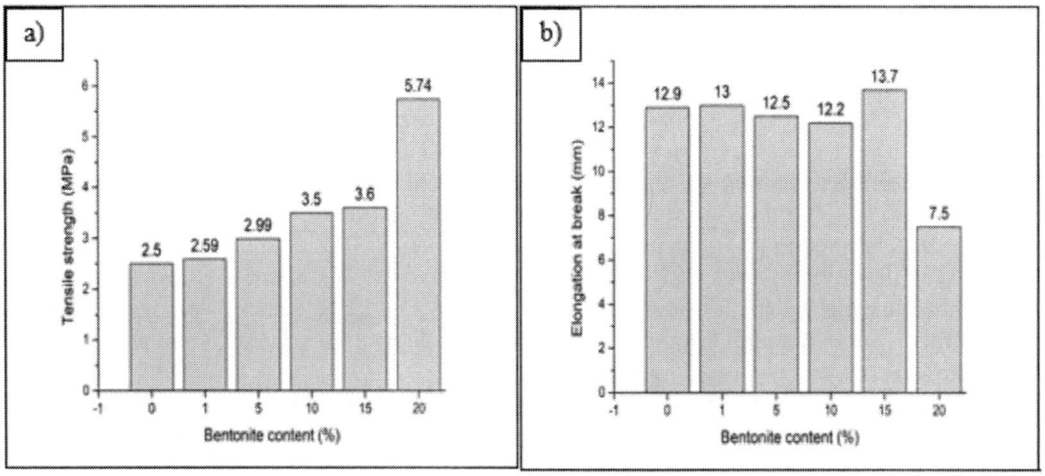

Figure 1. Tensile properties of potato starch film with different bentonite content, (a) tensile strength and (b) elongation at break.

Figure 1(b) shows the effect of mixing temperature on the elongation at break of the potato starch film. The significant drop in the in elongation at break (13 mm to12.2mm) was observed as the increased in bentonite content (0% to 10%). This fact is probably due to poor dispersion of bentonite in the starch matrix that create high localized stresses in the matrix during tensile loading. However, a sudden increased in elongation (13.7 mm) was observed at 15% of bentonite content. This finding might be associated with better dispersion of bentonite into starch matrix and thus improve strength

IOP Conf. Series: Materials Science and Engineering **374** (2018) 012025 doi:10.1088/1757-899X/374/1/012025

and flexibility of the films. On the other hand, the starch nanocomposites film with higher bentonite content shows reduction in elongation at break. This could probably due to reduce quality of bentonite dispersion which lead to the formation of bentonite tactoids.

3.2. Microstructure Analysis

Gelatinization process and dispersion of nanofillers into thermoplastic starch nanocomposites produces measurable changes by microstructures characterization. The changes between shape and size, as well as contacts between the phases are commonly observed during gelation process. Figure 2 shows the microstructure of potato starch nanocomposites films containing bentonite filler with different content of (0%, 1%, 5%, 10%, 15% and 20%).

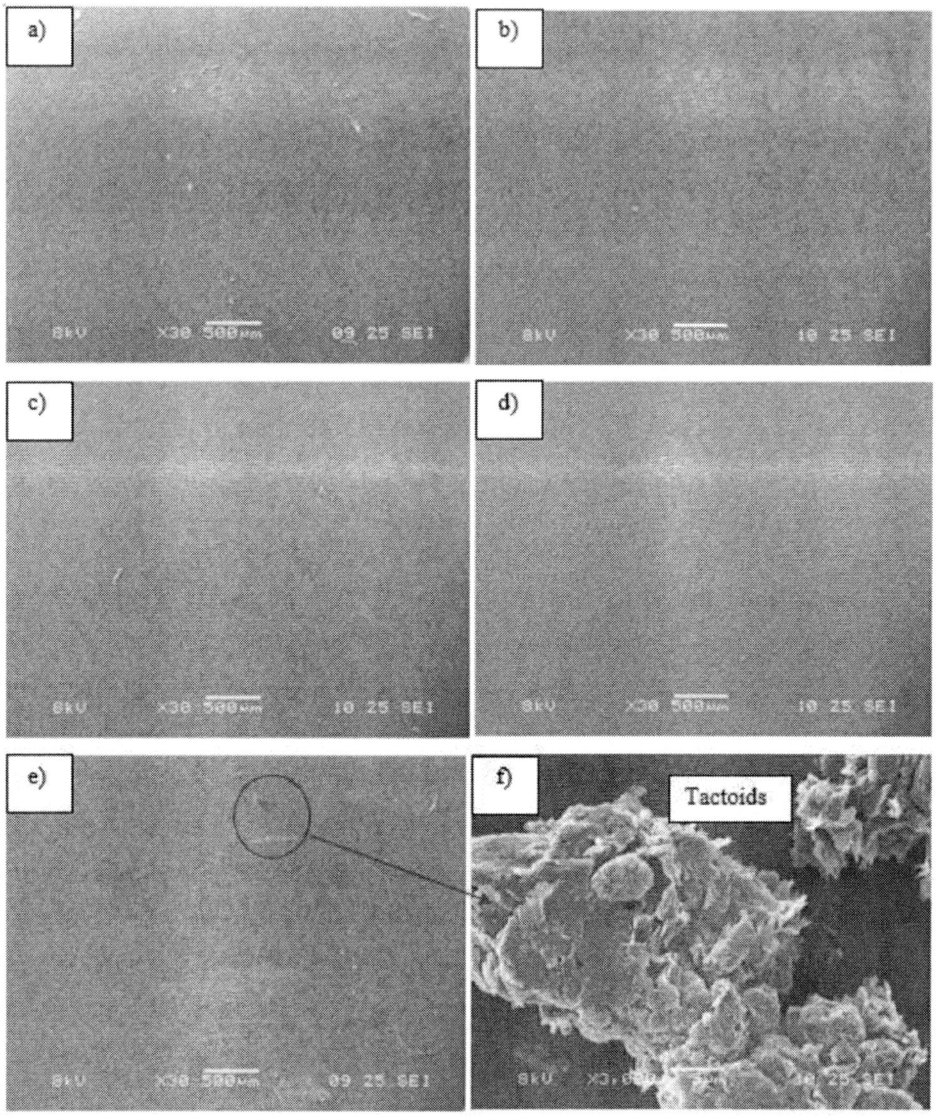

Figure 2. SEM image of potato starch nanocomposites film with different bentonite content: (a) B1, (b) B5, (c) B10, (d) B15, (e) B20 and (f) B20 tactoids.

Observation from SEM image reveal the changes in microstructure development with increasing bentonite content. The surface morphology of potato starch nanocomposites film for all nanofillers content illustrated homogenous distributionstructure on the surface area with no remaining starch granules. Figure 2(a) showed rougher surface with slightly agglomeration of nanofiller on the certain surface area which could be due to poor dispersion of bentonite filleronto starch matrix. Meanwhile, the nanocomposites film with 15% of bentonite content showed good dispersion and gradual smoothing with less visible agglomerates which can be observed in Figure 2(d). This is in good agreement with tensile result, which identified the nanocomposites film with 15% of bentonite content showed higher flexibility and strength. However, several agglomerates and tactoids in the size range of 55 μm was observed in the nanocomposites film with 20% of bentonite content due to poor dispersion as showed in Figure 2 (e) and 2 (f). This is in line with the elongation result that exhibited the lowest value.

4. Conclusions

Potato starch nanocomposites films, reinforced with different amounts of natural bentonite was successfully fabricated through a solution casting technique. The varying bentonite content significantly affected the properties of potato starch film by increasing the mechanical properties. The results displays that thermoplastic potato starch nanocomposites prepared with 15% of bentonite nanofiller shows the better strength and flexibility properties, and thus identified as desired nanofiller content in this study. This result was supported by microstructure analysis of the nanocomposites, which revealed more smooth and homogeneous surface of the film. These findings presents starch based nanocomposites film as a potential biopolymer for the development of biodegradable materials.

5. References

[1] Maitz M F 2015 *Biosurface and Biotribology* **1**(3) 161–176
[2] Tian H, Tang Z, Zhuang X, Chen X, and Jing X 2012 *Prog. Polym. Sci.* **37**(2) 237–280
[3] Maria Guadalupe Lomeli Ramirez T S F-S, Satyanarayana K G, Iwakiri S, Bolzon de Muniz G, Tanobe V 2011 *Carbohydr. Polym.* **86**(4) 1712–1722
[4] Velasquez D, Pavon-djavid G, Chaunier L, Meddahi-pellé A, and Lourdin D 2015 *Carbohydr. Polym.* **124** 180–187
[5] Olad A 2011 *Adv. Divers. Ind. Appl. Nanocomposites* 113–138
[6] Bakraji E H and Karajou J 2003 *J. Trace Microprobe Tech.* **21**(2) 397–405
[7] Avella M, De Vlieger J J, Errico M E, Fischer S, Vacca P, and Volpe M G 2005 *Food Chem.* **93(**3) 467–474
[8] Sanyang M L, Sapuan S M, Jawaid M, Ishak M R, and Sahari J 2015 *Polymers (Basel)* **7** 1106–1124
[9] Xie F, Pollet E, Halley P J, and Avérous L 2013 *Prog. Polym. Sci.* **38**(10–11) 1590–1628
[10] Lendvai L, Kmetty Á, and Karger-Kocsis J 2017 *Mater. Sci. Forum* **885** 129–134
[11] Borsos Z, Paun V P, Botez I C, Stoica C M, Vizurean P and Agop M 2008 S *Revista de chimie* **59** 1169-71

Synthesis and Luminescent Properties of $LiEu_{(0.50-x)}Gd_{0.50}(WO_4)_2Sm_x$ Red Phosphor

N S Zailani[1], M F Ghazli[1,2], R Hussin[3], S Z Abd Rahim[1,2] and M N Mat Saad[1,2]
[1]School of Manufacturing Engineering, Universiti Malaysia Perlis (UniMAP), Perlis Malaysia.
[2]Center of Excellence Geopolymer & Green Technology (CEGeoGTech), Universiti Malaysia Perlis (UniMAP), Perlis, Malaysia
[3]Department of Physics, Faculty of Science, Universiti Teknologi Malaysia, Skudai, Johor, Malaysia

E-mail: fathullah@unimap.edu.my

Abstract. Red phosphor is widely used in enhancing white light emitting diodes. Recent trends show there are several efforts in improving the efficiency of lighting particularly using Europium (III) as activators in Gadolinium tungstate phosphors. In this work, $LiEu_{(0.50-x)}Gd_{0.50}(WO_4)_2Sm_x$ red phosphor was synthesized using solid-state reaction. The crystal structure, morphology and photoluminescence properties of the sample were analysed. The structure of the phosphor was found to be tetragonal with space group I41/a by characterizing the phosphor using the X-ray powder diffraction (XRPD) application. The morphologies present the angular-shaped particles. The highest emission peak occurred at 615 nm corresponding to the $^5D_0 \rightarrow {}^7F_2$ transition under near-UV region. The optimum amount of Sm^{3+} that enhanced the intensity of the host lattice at x = 0.15 mol. The improvement on the photoluminescence properties of these phosphors without any changes on the structure and crystal symmetry shows a good potential of the red emitting phosphor to be used in the light emitting diodes (WLEDs) applications.

1. Introduction

In solid state lighting application, phosphor converted white light emitting diodes (pc-WLEDs) are very important to be used due to their many advantages such as energy saving, long lifetime, environmental protection, high brightness and high luminescent efficiency [1-6]. Currently, the commercial WLEDs are made by combining a blue-emitting InGaN-based LED chip (450-470 nm) with a yellow-emitting phosphor $(YGd)_3(AlGa)_5O_{12}:Ce^{3+}$ ($YAG:Ce^{3+}$) [1, 3, 6-8]. However, there are some weaknesses occurred in these WLEDs which are low colour rendering index (CRI), less luminous efficiency, low colour reproducibility and limited indoor use [1, 6, 7, 9]. Another approach that can generate white light is by combining the red, green and blue emitting phosphors near-UV LED chip [5, 9]. Most commercially red emitting phosphor used nowadays is sulfide-based materials such as $CaS:Eu^{2+}$, $Y_2O_2S:Eu^{3+}$, $SrY_2S_4:Eu^{2+}$, and ZnCdS:Cu,Al. Unfortunately, these phosphors cannot be efficiently excited in the near-UV region (350 – 420 nm), chemical instability and high FWHM [4, 5, 9, 10]. Therefore, there is a need to find a red emitting phosphor which offers a higher efficiency that can be excited greatly in the near-UV region. In recent years, there has been an increasing interested in searching for high efficiency of the red-emitting phosphor. Among the rare earth compound, Eu^{3+} has found as an excellent candidate to be used as a red element phosphor since Eu^{3+}

ions can occupy the lattice very sharp around 615 – 617 nm ($^5D_0 \rightarrow {}^7F_2$ transition) and can be excited at 395 nm and 465 nm [9, 11, 12]. The tungstate compound have been extensively studied for WLEDs application due to their excellent thermal and chemical stabilities, strong absorption bands near-UV regions, high luminescence efficiency and as well as high density [2, 4, 6, 10, 13-15]. As highlighted by Fathullah et al. in year 2016, the author found the optimum amount of Gd^{3+} can be employed in the $LiEu_{(1-x)}Gd_x(WO_4)_2$ is 0.50 mol to produce the highest intensity and highest luminous efficacy so far. Based on this finding, it motivated to introducing the Sm^{3+} to the $LiEu_{(0.50-x)}Gd_{0.50}(WO_4)_2Sm_x$ as x increase from 0 to 0.50 mol. Since the ionic radius of Sm^{3+} (1.079 Å) have almost similar with the ionic radius of Eu^{3+} (1.066 Å) ion with eight coordination numbers, it expected that the Sm^{3+} can easily substituted in the Eu^{3+} site without changing the crystal symmetry and enhancing the efficiency of the phosphor. In the present work, the $LiEu_{(0.50-x)}Gd_{0.50}(WO_4)_2Sm_x$ were prepared by using solid-state reaction and their crystallography, morphology and photoluminescence properties were recorded.

2. Materials and methods

A series of $LiEu_{(0.50-x)}Gd_{0.50}(WO_4)_2Sm_x$ (x = 0, 0.05, 0,10, 0,15, 0.20, 0.25, 0.30, 0.35, 0.40, 0.45 and 0.50 mol) were synthesized using solid-state reaction. Lithium carbonate 99.9%, gadolinium oxide 99.99%, tungsten (VI) oxide 99+%, europium oxide (Eu_2O_3), and samarium oxide (Sm_2O_3) were mixed homogeneously and calcined at 850 °C in open air for 3 hours. Then, the compound were cooled at room temperature and gently ground the sample for removing from the crucible. The X-ray powder diffraction (XRPD) patterns of the sample phosphors were characterized by a Bruker D2 Phaser with Cu $K\alpha$ radiation ($\lambda = 1.54184$ Å) in range of $2\theta = 10 - 90°$. A scanning electron microscope (SEM), Hitachi TM3000 were used for characterized the morphologies of the phosphors. The emission spectra were recorded by using the Perkin Elmer, LS 55 spectrometer. All the sample were analysing at room temperature.

3. Results and discussions

3.1. XRPD analysis of $LiEu_{(0.50-x)}Gd_{0.50}(WO_4)_2Sm_x$ samples

The XRPD pattern of $LiEu_{(0.50-x)}Gd_{0.50}(WO_4)_2Sm_x$ where x=0, 0.05, 0.25 and 0.50 is presented in Figure 1.

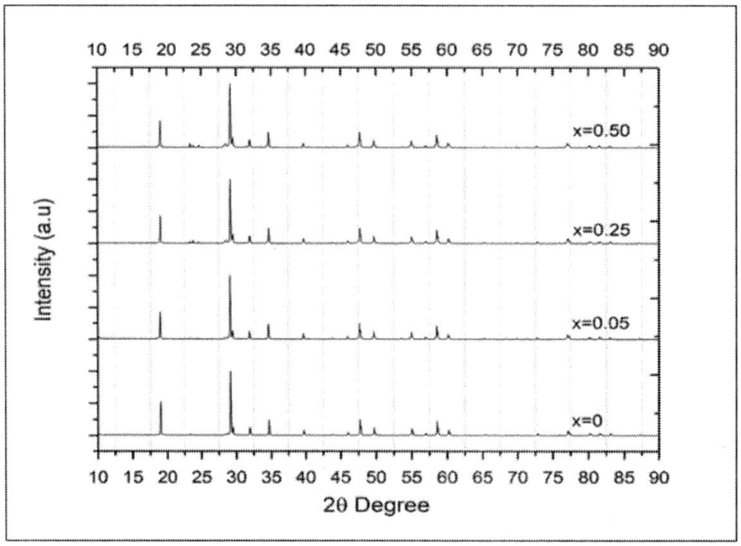

Figure 1. The XRPD stack patterns of $LiEu_{(0.50-x)}Gd_{0.50}(WO_4)_2Sm_x$ (x=0, 0.05, 0.25 and 0.50).

From this result, it shows the crystalline behaviour and the pattern are consistent with the pure standard tetragonal structure pattern $LiEu(WO_4)_2$ (PDF 01-080-7695) [16]. It also shows that there are no additional peaks and do not undergo any phase transformation occurred as Sm^{3+} increases in the host lattice. Therefore, it confirms that the structure of current compound are remain as a tetragonal scheelite structure with space group $I4_{1/a}$.

The sample $LiEu_{(0.50-x)}Gd_{0.50}(WO_4)_2Sm_x$ phosphors then were brought for refinement analysis using the TOPAS software and the $LiEu(WO_4)_2$ as the reference sample. All the peaks in the samples were closely matched to the reference sample which agreed that Sm^{3+} and Gd^{3+} were substituted well for the Eu^{3+} in the $LiEu(WO_4)_2$ lattice. Table 1 provides the lattice parameters and cell volumes for $LiEu_{(0.50-x)}Gd_{0.50}(WO_4)_2Sm_x$ samples. It clearly shows that there a trend on the increasing of the cell volume as Sm^{3+} increases in the lattice where it is about 0.39% from 305.1626 Å^3 to 306.355 Å^3. This phenomenon occurred as the ionic radius of the Sm^{3+} (1.079) with eight coordination number is larger than ionic radius of Eu^{3+} (1.066). Therefore, for the sample $LiEu_{(0.50-x)}Gd_{0.50}(WO_4)_2Sm_x$, the best R_{wp} can be calculated is 12.57 with the addition of Sm^{3+} 0.05 mol. The refinement data for all sample were also looking good as the collected data of the R_{wp} have found with small total errors.

Table 1. Lattice parameters and cell volumes of $LiEu_{(0.45-x)}Gd_{0.55}(WO_4)_2Sm_x$.

Eu^{3+}	Sm^{3+}	a(Å)	c(Å)	Vol(Å^3)	R_{wp}
0.50	0	5.203776(64)	11.26886(22)	305.1526(95)	14.28
0.45	0.05	5.204444(59)	11.27072(20)	305.2814(88)	12.57
0.25	0.25	5.206657(95)	11.27730(32)	305.720(14)	15.93
0	0.50	5.21014(11)	11.28566(38)	306.355(17)	17.26

Figure 2 shows the modelled crystal structure of tetragonal (a) $LiEu_{0.50}Gd_{0.50}(WO_4)_2$, (b) $LiEu_{0.45}Gd_{0.50}(WO_4)_2Sm_{0.05}$ and (c) $LiGd_{0.50}(WO_4)_2Sm_{0.50}$. From these figures, it clearly shows that there is no changes occurred on the structure as Sm^{3+} taking place in the lattice. Each of the W^{6+} is coordinated with four oxygen ions with in line with the finding prepared by Postema et al., (2011) where found the sample $LiEuW_2O_8$ had eight coordination number for Li^+ and four oxygen coordination for W^{6+} [16]. Due to this findings, it confirms that the structure of $LiEu_{(0.45-x)}Gd_{0.55}(WO_4)_2Sm_x$ are remain as a tetragonal scheelite structure since they are located at the same place with the previous finding of tetragonal structure.

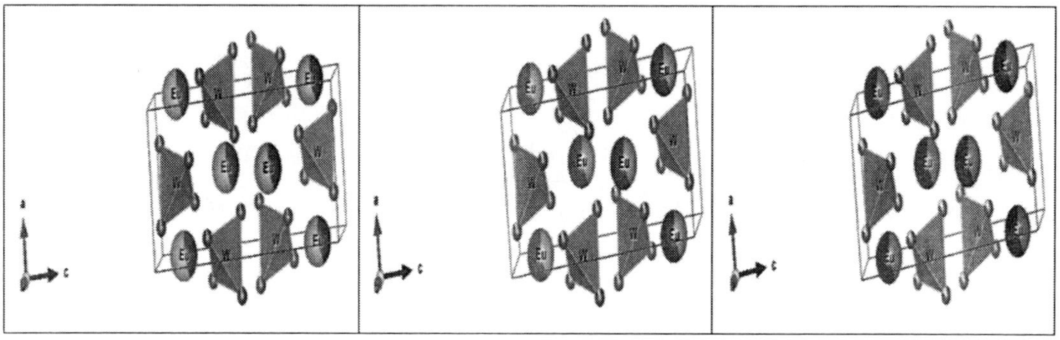

Figure 2. Modelled crystal structure of tetragonal (a) $LiEu_{0.50}Gd_{0.50}(WO_4)_2$, (b) $LiEu_{0.45}Gd_{0.50}(WO_4)_2Sm_{0.05}$ and (c) $LiGd_{0.50}(WO_4)_2Sm_{0.50}$.

3.2. Morphology of $LiEu_{(0.50-x)}Gd_{0.50}(WO_4)_2Sm_x$ samples

The morphologies of $LiEu_{(0.50-x)}Gd_{0.50}(WO_4)_2Sm_x$ where x = (a) 0, (b) 0.05 and (c) 0.50 with 2000% magnification are presented in Figure 3. From the figure, it shows that the compound have the angular shaped surface and the particle sizes ranging between 1– 6 μm. However, compare to the molybdate compound, it found that the tungstate compound have less adherence on their structure.

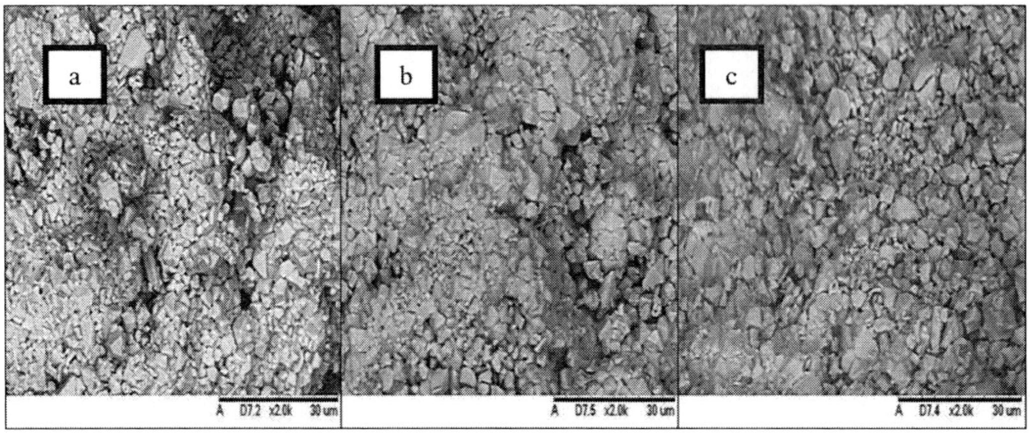

Figure 3. The morphologies of (a) $LiEu_{0.50}Gd_{0.50}(WO_4)_2$ (b) $LiEu_{0.45}Gd_{0.50}(WO_4)_2Sm_{0.05}$ and (c) $LiGd_{0.50}(MoO_4)_2Sm_{050}$ with 2000% magnification.

3.3. Photoluminescence properties

The emission spectra of sample $LiEu_{(0.50-x)}Gd_{0.50}(WO_4)_2Sm_x$ as function of increasing Sm^{3+} from 0 to 0.50 mol are resulted in Figure 4. The highest emission intensity was recorded at 615 nm where refer to the 610 – 630 nm ($^5D_0 \rightarrow {}^7F_2$ transition). Another emission peaks were appeared at 588 – 600 nm ($^5D_0 \rightarrow {}^7F_1$), 640 – 662 nm ($^5D_0 \rightarrow {}^7F_3$) and 681 – 710 nm ($^5D_0 \rightarrow {}^7F_4$). This is consistent with the findings reported by Fathullah et al., (2016) and Rasu et al., (2017) where found the highest intensity of the $LiEu_{(1-x)}Gd_x(WO_4)_2$ and $LiGd(WO_4)_2$:Eu^{3+} in the same transitions [10, 15]. Therefore, the spectra peak shapes are in the same pattern and it believed that the substitution of Sm^{3+} to the lattice did not affected the symmetry of the emission centre.

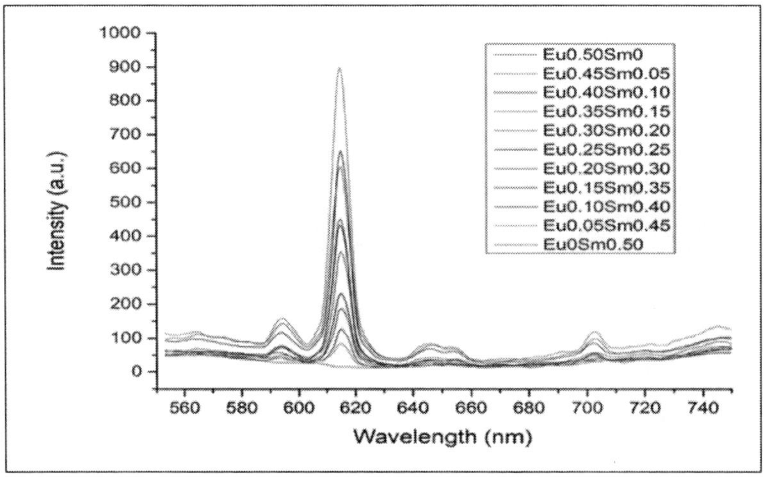

Figure 4. Emission spectra of $LiEu_{(0.50-x)}Gd_{0.50}(WO_4)_2Sm_x$ (x= 0 until 0.50) from 550 nm to 750 nm.

The emission in range 605 – 630 nm of the $LiEu_{(0.50-x)}Gd_{0.50}(WO_4)_2Sm_x$ is shown in Figure 5. From this figure, it clearly shows that the intensity was enhanced until the concentration of x = 0.15 mol.

This proves that the emission of the Eu^{3+} has been taking place by the Sm^{3+} until 0.15 mol and started to have a quenching effect when more Sm^{3+} concentration added to the lattice.

Figure 6 shows the normalized graph of the $LiEu_{(0.50-x)}Gd_{0.50}(WO_4)_2Sm_x$ phosphor. Since there are no broadening effects occurred on the phosphor, it confirms that the current phosphor did not undergo any phase transformation which is consistent with the XRPD result in Section 3.1.

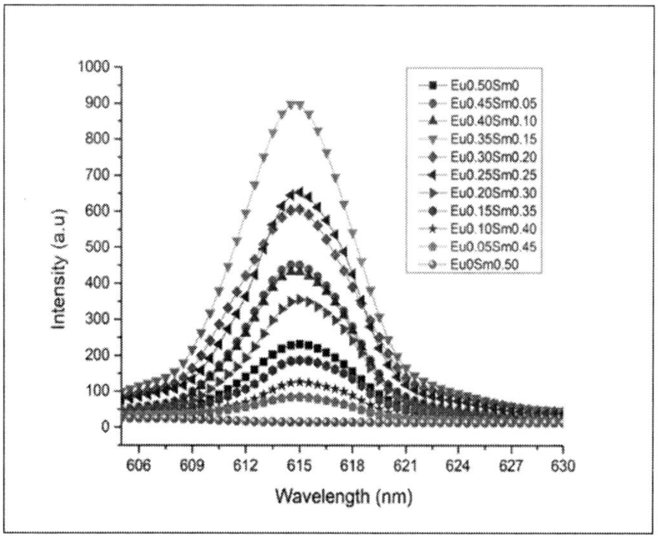

Figure 5. Emission spectra of $LiEu_{(0.50-x)}Gd_{0.50}(WO_4)_2Sm_x$ (x= 0 until 0.50) from 605 nm – 630 nm.

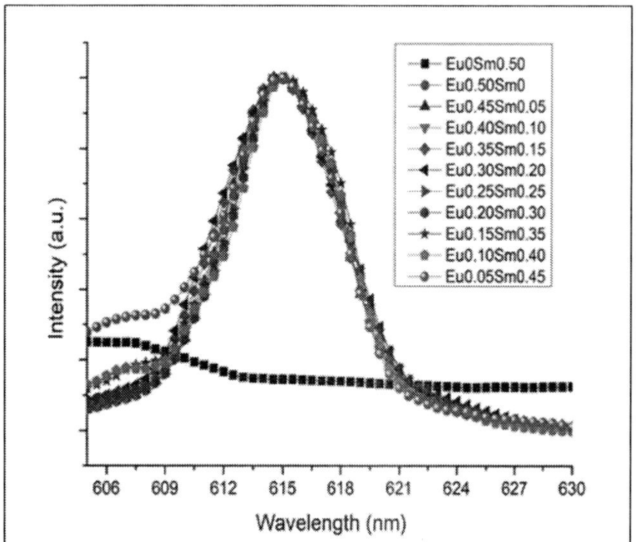

Figure 6. Normalized graph of emission spectra of the $LiEu_{(0.50-x)}Gd_{0.50}(WO_4)_2Sm_x$ (x= 0 until 0.50) from 605 nm – 630 nm.

This findings also suggest that the additional of Sm^{3+} in the host lattice do not block the energy transfer between Eu^{3+} neighbour ions which might affect the host symmetry.

4. Conclusions

$LiEu_{(0.50-x)}Gd_{0.50}(WO_4)_2Sm_x$ were successfully synthesized. The XRPD patterns confirm the structure of the phosphors to have the tetragonal scheelite structure with space group $I4_{1/a}$. The morphologies indicate the angular-shaped structure with the particle sizes ranging between 1– 6 μm. The intensity of the phosphor enhanced until 0.15 mol% when adding the Sm^{3+} concentration and produced the highest emission at 615 nm corresponding to $^5D_0 \rightarrow {}^7F_2$ transition. There are no broadening effect occurred on the emission peaks proving that the additional of Sm^{3+} to the host lattice did not affect the emission centre. All the results collected indicate that the $LiEu_{(0.50-x)}Gd_{0.50}(WO_4)_2Sm_x$ phosphor hasbetter efficiency which make it to be a potential red phosphor for fabricating in the WLEDs.

5. References

[1] Liang Y et al. 2017 *J. Alloys Compd.* **716** 56–64
[2] Zhou L et al. 2016 *Mater. Res. Bull.* **84** 429–436
[3] Shang M, Li G, Kang X, Yang D and Lin J 2011 *J. Electrochem. Soc.* **158**(5) H565
[4] Chiu C H, Wang M F, Lee C S and Chen T M 2007 *J. Solid State Chem.* **180**(2) 619–627
[5] Xie A, Yuan X and Wang F 2011 *Sci. China Technol. Sci.* **54**(1) 70–75
[6] Zhao D, Shi J-C, Nie C-K and Zhang R-J 2017 *Opt. - Int. J. Light Electron Opt.* **138** 476–486
[7] Li L et al. 2016 *Mater. Res. Bull.* **78** 26–30
[8] Benoît G, Véronique J, Arnaud A and Alain G 2011 *Solid State Sci.* **13**(2) 460–467
[9] Yi L, He X, Zhou L, Gong F, Wang R and Sun J 2010 *J. Lumin.* **130**(6) 1113–1117
[10] Rasu K K, Balaji D and Babu S M 2017 *J. Cryst. Growth* **468** 159–161
[11] Hwang K-S, Kang B-A, Hwangbo S, Kim Y-S and Kim J-T 2010 *Electron. Mater. Lett.* **6**(1) 27–30
[12] Ru J, Ying S, Zheng W and Chen J 2016 *Mater. Res. Bull.* **84** 468–473
[13] Wang Q F et al. 2015 *J. Alloys Compd.* **625** 355–361
[14] Liu Y et al. 2014 *RSC Adv.* **4**(9) 4754
[15] Fathullah J S M 2016 *An Investigation into Red Emitting Phosphors for Display and Lightings* Brunel University London
[16] Postema J M, Fu W T and Ijdo D J W 2011 *Solid State Chem.* **184**(8) 2004–2008

Acknowledgments

We would like to acknowledge and thank you for funding and support from the Ministry of Higher Education Malaysia MOHE viathe Fundamental Research Grant Scheme (FRGS) (Ref: FRGS/1/2016/STG07/UNIMAP/02/1), Application ID218885-247922 andcode 9003-00572,and also thank you to Universiti MalaysiaPerlis.

Preliminary Results on the Corrosion Behaviour of a New Biodegradable Metallic Material Based on Zinc

S Dobriţa, B Istrate, N Cimpoeşu, S Stanciu, V Apostol, R Cimpoeşu*, I Ioniţă and P Paraschiv
[1] Gheorghe Asachi" Technical University of Iasi, 44 Mangeron blv., 700050, Iasi

E-mail: ramonahanu@yahoo.com

Abstract. The class of biomaterials (metallic, polymer or ceramics) has applications as implants with a temporary function. Among magnezium (high corrosion rate) and iron (slow corrosion rate) based biodegradable alloys a new class based on zinc try to fullfill the gap between the first two alloy systems concerning the degradation rate. The role of these alloys is to provide some specific characteristics for a limited period of time. In this article preliminary results based on microstructural and chemical characteristics of a Zn-Mg alloy are presented. The results were obtained using optical and electronic microscopy (MO and SEM) and dispersive energy spectroscpy (EDS) and X-ray difraction (XRD).

1. Introduction

Ideally, biodegradable coronary stents should provide a compromise between mechanical integrity and degradation [1, 2]. Degradation should begin at a very low speed to maintain the optimal mechanical integrity of the stent until the process of complete remodeling takes place completely, a process lasting from 6 to 12 months, and then with the progression of degradation, the mechanical properties [3, 4]. Once the recovery process has been completed, the degradation rate must be sufficiently high so that no intolerable accumulations or degradation material occur around the deployment area or the main organs. An example for the temporary use of an implant that was introduced to open a narrowed artery. These temporary implants are based on the initial concept of stents, but introduced into different physiological environments to have different specific functions. Temporary cardiovascular implants should open the thickened artery and keep it open until the blood vessel reshapes and then degenerate gradually, being replaced by new blood vessel tissues [5, 6].

In the case of temporary orthopedic implants, such as fixing a bone with bolts or rivets, the material must support the bone fractured sufficiently until it reoccurs and then degrade and be replaced by the new bone tissues. A reasonable degradation period for these stents may be between 12 and 24 months after implantation [2, 7]. These periods are also supported by sufficient data or evidence for all clinical cases and may vary from case to case.

In this sense, the concept of biodegradable material is: "temporarily creating support for the healing process of diseased tissue and progressive degradation after this process." From this definition appear two major directions: 1) temporary support; 2) degradation. Biodegradable metallic materials must also be bioactive, to interact positively during the healing process. Both the metals and the degradation process and their products must not cause adverse effects in the healing process [8].

The mechanical properties of metals include the use of alloying elements, thermomechanical treatments and new processing methods. The use of alloying elements that can form an anticorrosive coating on the surface of metals, for example Cr for stainless steels, should be avoided in order to make the alloys degradable or less resistant to corrosion [9].

Biodegradable metals must provide adequate mechanical support to support the healing process during implantation. However, it is difficult to define a support period as close as possible to medical reality depending on the various interventions that take place.

Mechanical integrity of biodegradable metals can theoretically be predicted by performing "in vitro" degradation tests. Kannan et al. performed low-tension tension tests on Mg-Al alloys by correlating mechanical properties and material degradation [3, 10]. The results showed that the elongation at alloy AZ 91% Ca in SBF (simulated human fluid) decreases by approximately 15-20% compared to initial air properties. Biodegradable metals have to decompose into a complex physiological environment that belongs to the human body by combining the kinetics of degradation with the healing period. Degradation products should be transported and removed from the body in order not to cause local or systematic accumulation.

In this paper we analyzed through linear and cyclic potentiometry the corrosion behavior in a SBF solution of two zinc-based alloys with applicative potential in the medical field.

2. Materials and methods

An experimental alloy Zn-Mg was obtained through classical melting in induction furnace. The behavior of two experimental samples: zinc and zinc-magnesium at electrocorosion in a simulated body fluid [11-14] with the chemical composition shown in Table 1 was analyzed.

Table 1. The chemical composition of the electrolyte solution.

The chemical composition (ions)[mmol/dm^3]	Na^+	K^+	Mg^{2+}	Ca^{2+}	Cl^-	HCO_3	HPO_4^2	SO_4^2
SBF	142	5	1.5	2.5	147.8	4.2	1	0.5
Human blood plasma	142	5	1.5	2.5	103	27	1	0.5

The samples were analyzed for electro-corrosion resistance using a potentiostat with a three-electrode cell. Teflon embedded samples were investigated at linear and cyclic potentiometry. After potentiometry sample surfaces were analyzed by electronic microscopy and EDS chemical analysis.

3. Results and discussions

For the experiments, a laboratory potentiostat with a 3-cell electrode was used: calomel, platinum and the third electrod was the experimental alloy mounted in a teflon support. The results of linear (Tafel) and cyclic potentiometry are shown in Figure 1 (a) and (b) respectively.

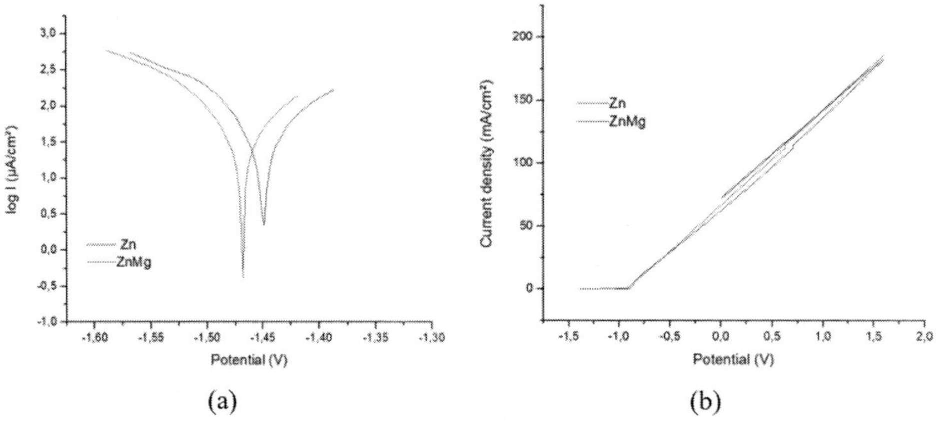

Figure 1. Electro-chemical resistance analysis (a) Tafel diagram; (b) cyclic diagram.

The behavior at electro-corrosion in SBF of the two metallic materials is similar to the generalized corrosion type, resulting from the form of cyclic potentiometry curves. Both alloys fall within a corrosion potential range of 1.425 to 1.475 V with an increased corrosion (the potential is of the order of volts).

From the characteristic Tafel diagrams of the two alloys, a higher corrosion rate of the Zn element and a decrease (as a conclusion of the degradation rate) with the addition of Mg element is observed. Although magnesium has a higher corrosion rate than pure zinc, the alloy exhibits a moderate decrease in corrosion rate over the pure zinc element. Parameters obtained from the electrocorosion resistance tests of the experimental samples are shown in Table 2.

Table 2. Parameters obtained from the electro-corrosion resistance tests.

Sample/ Parameters	E(i=0) [mV]	Current intensity [μA/cm²]	Polarization resistance [ohm*cm²]	Beta a [mV]	Beta c [mV]	Corrosion rate [cm/Y]
Zn	1469	107.6	377.98	25.4	160.1	1.352
ZnMg	1450	98.8038	394.34	41.7	159.7	1.241

Parameters obtained from the corrosion resistance test are close to the two SBF samples with corrosion rates ranging from 1.352 cm/year for Zn and 1.241 cm/year for ZnMg respectively. These speeds are actually a corrosion of 1.12 to 0.75 mm per month, parameters that cover many metallic requirements for medical applications.

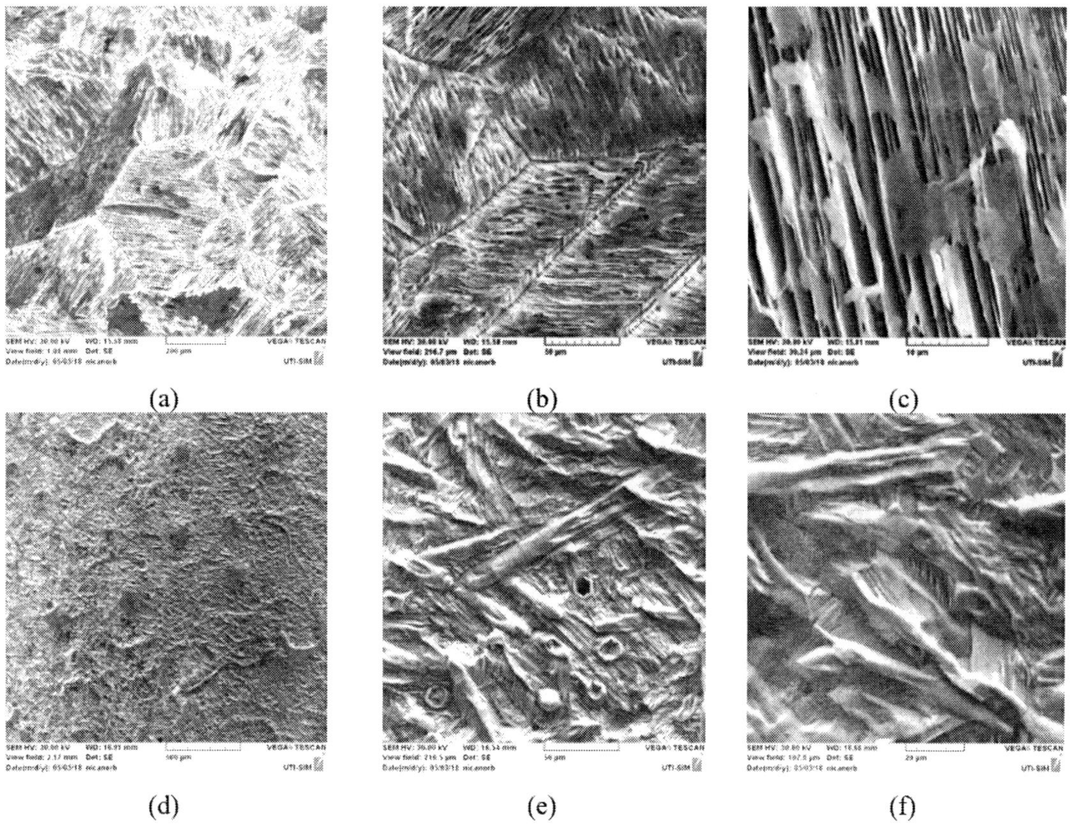

Figure 2. State of the surface of the experimental alloys after the electro-corrosion resistance tests (a) Zn 100x; (b) Zn 500x; (c) Zn 2000x; (d) ZnMg 100x; (e) ZnMg 500x; (f) ZnMg 2000x.

The values of the parameters obtained for the two materials are very close and the order of the corrosion rates can be easily influenced by the experimental conditions such as the surface condition of the 2 materials, favoring the formation of gas bubbles during the experiment, etc. For these reasons, the experiments on the two alloys must be repeated and at this time the corrosion rates obtained or their order can not be confirmed using a magnetic stirrer to eliminate gas during the experiment.

After the corrosion resistance tests, the surface state of the three experimental materials by SEM (VegaTescan LMH II) electron microscopy was analyzed at different amplification powers (100x, 500x, and 2000x).

Figure 2 shows the state of the surface of the experimental alloys after electro-corrosion resistance tests a) Zn 100x, b) Zn 500x, c) Zn 2000x, d) ZnMg 100x, e) ZnMg 500x, f) ZnMg 2000x.

Both samples corroded all over the surface confirming the generalized corrosion character of the corrosion cyclic curves. The electro-corrosion test carried out in the case of pure zinc underlines its multigranular structure, Figure 2 (a) and (b). At the micrometric level it is observed the formation of some compounds on the surface of the material above the zinc sheets, Figure 2 (c). The compounds resulted from the interaction of the material with the electrolyte solution and showed no mechanical stability on the surface in appearance. Also, both samples also show stronger pitting areas, areas that have favored the corrosion process and where impurities, diffused material phases or microstructural defects have been found.

The surfaces of the metallic samples were investigated by X-ray energy spectroscopy using a Bruker EDS detector (EDAX). The analyzes were performed on a surface area of 4 mm^2 per sample. Qualitatively identified elements are passed, depending on the energies they participate in the compounds on the surface, in the spectra of Figure 3 for (a) Zn and (b) ZnMg.

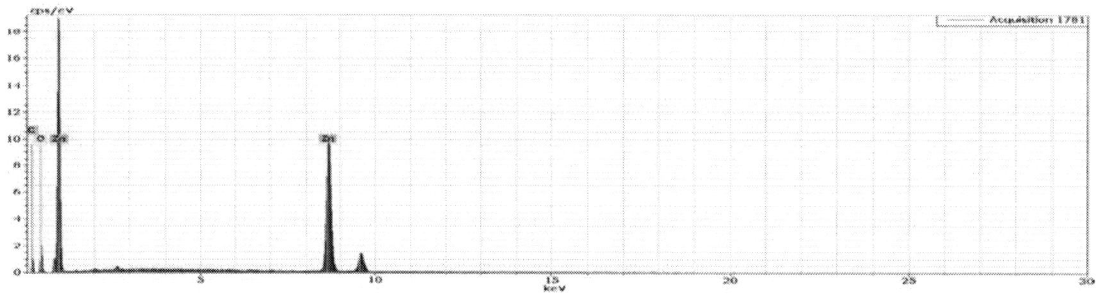

(a)

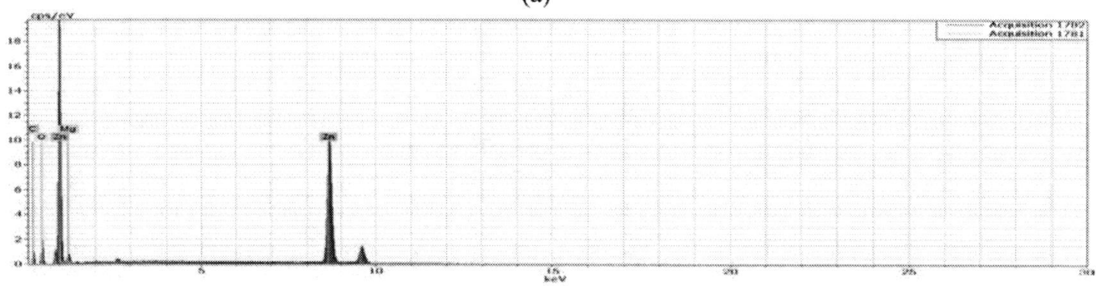

(b)

Figure 3. X-ray energy specter (a) Zn; (b) ZnMg.

Table 3 shows the chemical composition of the experimental alloys after electrocorosion tests in mass and atomic percentages. The error of EDAX equipment for each element is also given. The presence of oxygen and carbon on the surface is due to the passage of these elements from the electrolyte solution to the surface of the metallic material.

Table 3. Chemical composition of experimental alloys after electro-corrosion tests.

Alloy/element	Zn		O		Mg		C	
	wt%	at%	wt%	at%	wt%	at%	wt%	at%
Zn	74.77	40.19	19.21	42.22	-	-	6.01	17.59
ZnMg	68.12	34.55	19.04	39.46	6.78	9.24	6.06	16.74
Error EDAX %	2		1.26		0.56		1.39	

All the samples show the oxidation of the surface of the alloys and formation of oxides on the surface of the metallic material to be passed into the electrolyte solution. It is noted that chemical analyzes will be resumed after cleaning the samples with technical alcohol and ultrasonography. At this time, it was desired to capture the surface of the experimental materials immediately after the electrocorosion tests to analyze the behavior of the materials during the corrosion process.

4. Conclusions

Zinc alloys can be a suitable solution for medical or industrial applications of degradable materials that provide degradation periods ranging from those given by magnesium alloys and those derived from iron-based alloys.

Zinc alloying with different chemical elements, also degradable and in various mass or atomic percentages, leads to the modification of its behavior in electrolyte solutions.

According to the experimental results obtained from the electrocorosity tests, zinc alloying with Mg leads to a decrease of the corrosion rate of pure zinc in SBF solution (human fluid similar to that of human). Aligning with various degradable chemical elements can be considered a method of controlling the rate of degradation of these alloys in various biological or electrolyte media.

5. References

[1] Alhamed M and Abdullah W 2013 *Journal of Electron Devices* **18** 1563-1567
[2] Hermawan H, Dubé D and Mantovani D 2009 *Acta Materialia* **6** 1693–1697
[3] Stratulat S I, Dombici C, Forna D 2013 *Rev. Med. Chir. Soc. Med. Nat.* **117** 875-879
[4] Kannan M B and Raman R K S 2008 *Biomaterials* **29** 2306–14
[5] Waksman R, Pakala R, Baffour R, Seabron R and Hellinga D 2008 *Journal of Interventional Cardiology* **21** 15–20
[6] Cimpoeşu N, Trincă L C, Dascălu G, Stanciu S, Gurlui S O and Mareci D 2016 *Journal of Chemistry* Article ID 9520972
[7] Schinhammer M, Hänzi A C, Löffler J F and Uggowitzer P J *Acta Biomaterialia* 2009 **6** 1705–13
[8] Izquierdo J, Bolat G, Cimpoesu N, Trinca L C, Mareci D and Souto R M 2016 *Applied Surface Science* **385** 368-378
[9] Lupu V V, Ignat A, Paduraru G, Ciubara A, Ioniuc I, Ciubara A B, Gheonea C and Burlea M 2016 *Revista de Chimie* **67** 2501-03
[10] Moravej M, Purnama A, Fiset M, Couet J and Mantovani D 2010 *Acta Biomaterialia* **6** 1843–51
[11] Stanciu S, Ursanu A, Trincă L C, Trofin A E, Solcan C, Munteanu C, Cimpoesu N, Acatrinei D, Sindilar E V, Stanciu T, Fântanariu M and Topliceanu L 2016 *Environmental Engineering and Management Journal* **15** 973-980
[12] Achitei D C, Vizureanu P, Dana D and Cimpoesu N 2013 *Metalurgia International* **18** 104-109
[13] Rambu A P, Sirbu D, Sandu A V, Prodan G and Nica V 2013 *Bulletin of materials science* **36** 231-237
[14] Kokubo T, Kushitani H, Sakka S, Kitsugi T and Yamamuro T 1990 *Journal of Biomedical Materials Research* **24** 721–734

Corrosion Resistance of a Cast-Iron Material Coated With a Ceramic Layer Using Thermal Spray Method

C D Florea, C Bejinariu, C Munteanu, B Istrate, S L Toma, A Alexandru and R Cimpoesu*
"Gheorghe Asachi" Technical University of Iasi, Mangeron 44 blv., 700050, Iasi, Romania

E-mail: ramonahanu@yahoo.com

Abstract. Cast-iron 250 used for breake systems present many corrosion signs after a mean usage time based on the environment conditions they work. In order to improve them corrosion resistance we propose to cover the active part of the material using a ceramic material. The deposition process is an industrial deposition system based on thermal spraying that can cover high surfaces in low time. In this articol we analyze the influence of a ceramic layer (40-50 μm) on the corrosion resistance of FC250 cast iron. The results were analyzed using scanning electron microscopy (SEM), X-ray energy dispersive (EDS) and linear and cyclic potentiometry.

1. Introduction

Automotive industry represents a higher metallic materials consumer. Millions of cars still present a high usage of metallic materials. The necessity of quality materials is observed in all the metallic materials applications [1, 2]. Safety represents an important issue of cars manufacturers. The brake system is designed to slow down and halt the motion of the vehicle. Various components are used in braking system to convert the momentum (kinetic energy) in to heat energy using friction. Two forms of friction play important role in automotive braking:

– kinetic friction,
– static friction.

Braking action creates kinetic friction kinetic friction in the brakes and static friction between the tire and road to slow the vehicle. When brakes are applied, the vehicle's weight is weight is transferred to the front wheels and is unloaded on the rear unloaded on the rear wheels [3, 4]. There are four basic factors determine the braking power of the system. Pressure: through pressure – amount of friction generated between moving surfaces contacting one another depends in part on the pressure exerted on the surfaces. The hydraulic force is used to move brake pads or brake shoes brake shoes against spinning spinning rotors or drums mounted on the wheels. Coefficient of friction in automotive brakes, the COF expresses the frictional relationships between pads and rotors or shoes and drums and is carefully engineered to ensure maximum performance (COF 0.25 to 0.55) [5]. Frictional contact surface it is the amount of surface or area frictional contact surface – it is the amount of surface, or area, that is in contact. Bigger brakes stop the car more quickly [6]. Heat dissipation is a large amount of heat is produced in brakes. The weight and the speed of the vehicle determine the braking mechanism [7, 8].

In this article we analyze the corrosion resistance of FC250 cast iron material with and without a ceramic layer Al2O3. The ceramic layer was deposited using a plasma deposition automatic arm from 2-5 μm Al2O3 powders.

2. Materials and methods

A usual cast iron material (FC250) used for automotive breaking system was covered with a ceramic material (105NS-1 aluminum oxide) using an industrial deposition system (Sulzer Metco). The reason was mainly to improve the corrosion resistance properties. Plasma jet was covered by Ar (pressure 5.2 bar and gas flow 39 NLPM) and H (pressure 3.4 bar and gas flow 6.6 NLPM). We apply 4 passes (around 12-15 μm by layer) [8, 9]. Samples for corrosion resistance analyze were prepared by mechanical grinding, sand blasting and mechanical fixation for deposition [8-10]. The experimental set up is presented in Figure 1 and is form by a rotational support, automatic arm for pulverization and samples holder. The experimental equipment presents a high covering area with a good potential of applications in the industrial field.

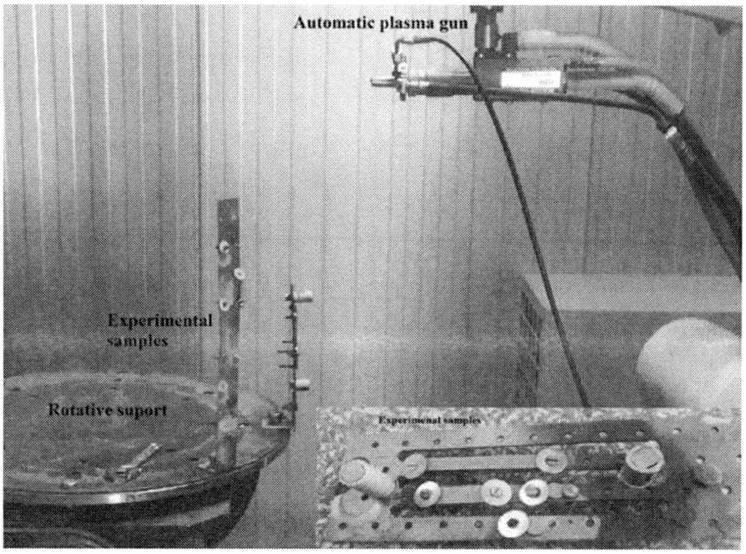

Figure 1. Experimental set up for deposition of Al_2O_3 ceramic layer.

After the samples preparation and deposition the materials were cooled down to room temperature. The material surface after deposition of ceramic material Al_2O_3 via plasma jet and after electro-chemical corrosion test was analyzed using VegaTescan LMH II scanning electron microscope and Bruker detector for X-ray energy spectroscopy. The electro-chemical corrosion resistance test was realized using potentiostat equipment with three electrodes, one being the experimental sample.

3. Results and discussions

In Figure 2 we present the electro-chemical corrosion test results a) Tafel linear potentiometry and b) cyclic potentiometry curves. A considerable improve of corrosion resistance is observed for the metallic sample, 0.8 V potential, covered with the ceramic layer that have a protective role.

Both materials present a generalized corrosion type, Figure 2 b), with a smaller slope for the metallic material covered with the ceramic layer. The corrosion resistance behaviour present an improve for the metallic material with a ceramic layer but further investigations on the interface between the metallic and the ceramic layer are necessary. The experiments were realized in rain water meet on the streets of city Iasi. In Table 1 the experimental electro-corrosion resistance tests parameters are presented.

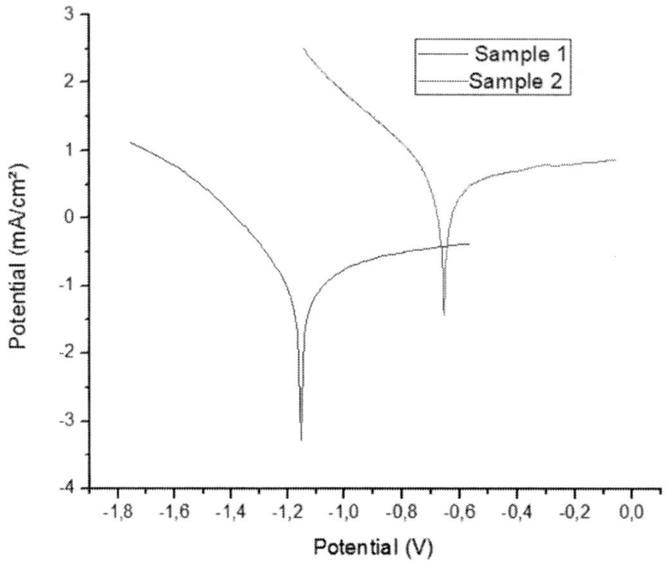

(a)

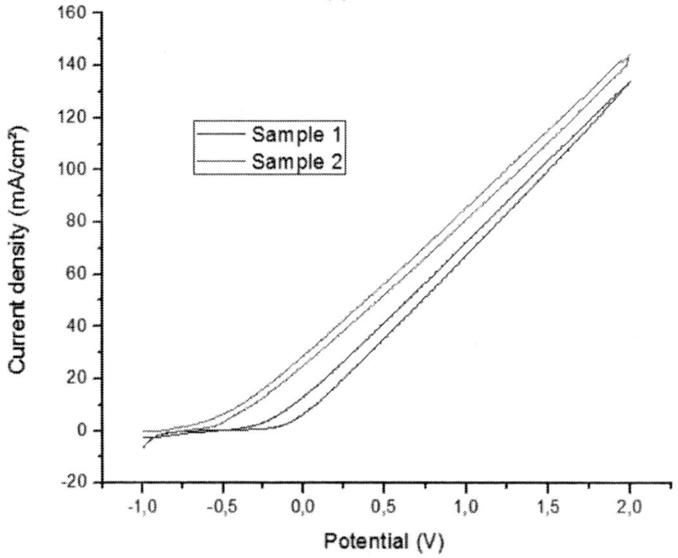

(b)

Figure 2. Electro-chemical corrosion test results (a) Tafel linear potentiometry and (b) cyclic potentiometry.

Table 1. Experimental electro-corrosion resistance test parameters.

Sample	E_0 mV	b_a mV	b_c mV	R_p ohm.cm²	J_{cor} mA/cm²	V_{cor} mm/year
Sample 1	1150.2	730	765	694.57	2.16	1.61
Sample 2	654.1	350	389	23.84	1.87	0.38

The ceramic layer has a compact structure without micro-pores or micro-cracks. After the electro-corrosion test no compounds were observed on the surface also no pitting holes were observed in the ceramic layer.

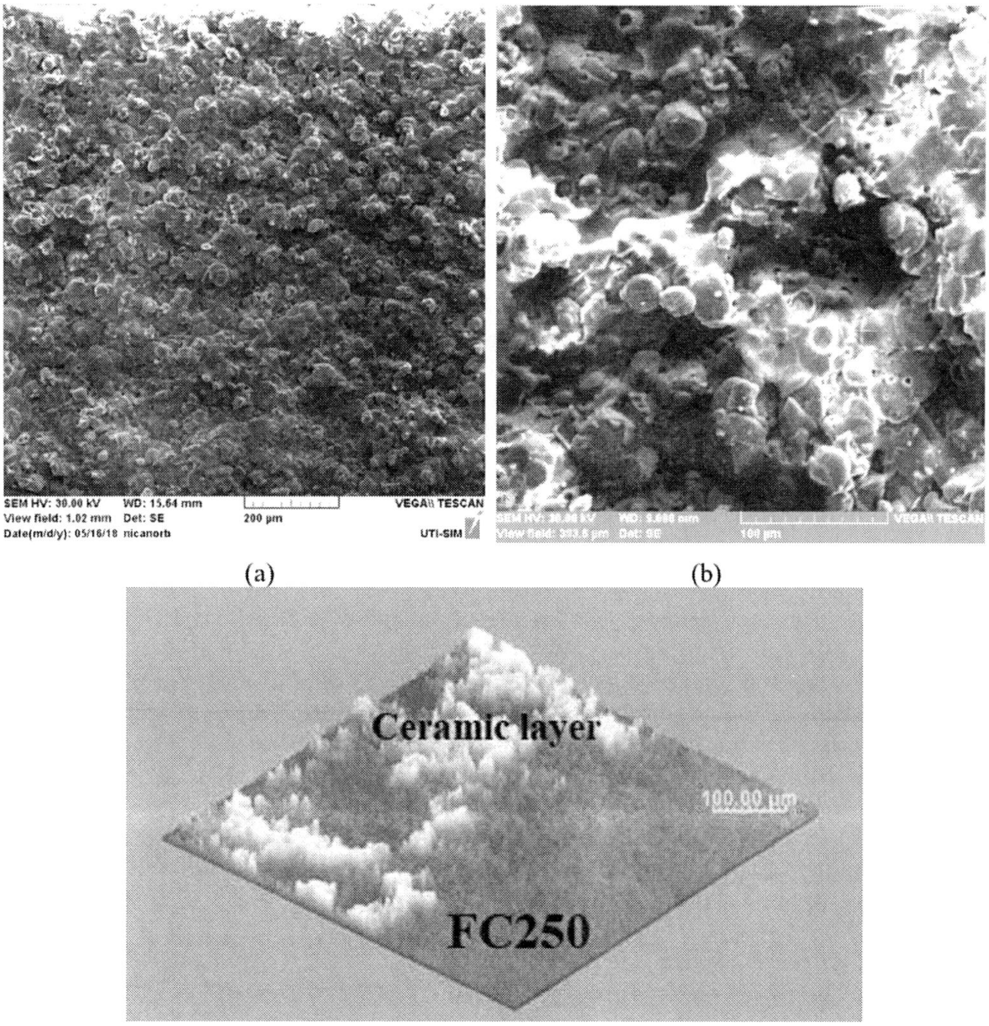

(c)

Figure 3. SEM micrographs (a) ceramic layer, (b) ceramic layer after electro-corrosion test and (c) 3D image of ceramic layer at the interface with the metallic material.

The difference between covered and un-covered samples is more than 1.2 mm/year corrosion rate. The corrosion resistance is obviously improved using the ceramic layer (Figure 3 c). Chemical composition of the surface was not affected by the electro-chemical tests so no oxides or other compounds were observed on the surface [11-18]. Ceramic layers present a very good behaviour at electro-corrosion tests.

4. Conclusions

Corrosion resistance of a metallic material, usual use as breaking discs, FC250 cast iron was improved through a plasma jet deposition of a ceramic layer. In order to improve the ceramic layer adhesion to the metallic material a sandblasting process was applied to all elements. A thick Al_2O_3 layer was obtained after four passes with plasma jet. The corrosion resistance was highly improved after the ceramic layer deposition. No oxides or other compounds were observed on the ceramic surface after de electro-corrosion test. The electro-corrosion tests were performed in city rain water. The electrolyte solution contains dust and contaminants from the city environment.

5. References

[1] Yang K, Rong J, Feng J, Zhuang Y, Tao S, Ding C 2016 *Journal of the European Ceramic Society* **36** 4261-4267
[2] Song B, Dong S, Liao H, Coddet C 2015 *Surface and Coatings Technology* **268** 24-29
[3] Górny Z, Sobczak J 2005 *Modern casting materials based on nonferrous metals*. Kraków: ZA-PIS
[4] Krupiński M, Dobrzański L A, Sokołowski J H 2008 *Archives of Foundry Engineering* **8** 71-74
[5] Smola B, Stulikowa I, Pelcora J, Mordike B 2003 *Magnesium alloys and their applications*. Proceeding of the 6-th International Conference., Wiley-VCH Verlag, pp. 43–46
[6] Carmo S N, Damásio F Q, Alves V N, Marques T L, Coelho M M 2013 *Microchemical Journal* **110** 320–325
[7] Neto J A G, Filho V R A 2008 *Analytical Letters* **41** 1555–1570
[8] Florea C, Bejinariu C, Munteanu C, Cimpoesu N 2017 *Advanced Materials Engineering and Technology* V, Book Series: AIP Conference Proceedings **1835**, Article Number: UNSP 020053, DOI: 10.1063/1.4983793
[9] Florea C D, Munteanu C, Cimpoesu N, Sandu I G, Baciu C, Bejinariu C 2017 *Revista de Chimie* **68** 2582-2587
[10] Cimpoesu N, Stanciu S, Tesloianu D, Cimpoesu R, Popa R F, Moraru E 2017 *Metal Science and Heat Treatment* **58** 11 – 12
[11] Lupu V V, Ignat A, Paduraru G, Ciubara A, Ioniuc I, Ciubara A B, Gheonea C, Burlea M 2016 *Revista de Chimie* **67**(12) 2501-2503
[12] Florea C D, Carcea I, Cimpoesu R, Toma S L, Sandu I G, Bejinariu C 2017 *Revista de Chimie* **68** 2397–2401
[13] Lazar P, Bejinariu C, Sandu A V, Cazac A M, Corbu O, Sandu I G, Perju M C 2017 Corrosion Evaluation of Some Phosphated Thin Layers on Reinforcing Steel. In *International Conference on Innovative Research - Icir Euroinvent*; Iop Publishing Ltd: Bristol **209** 12025
[14] Nejneru C, Perju M C, Sandu A V, Axinte M, Quaranta M, Sandu I, Costea M and Abdullah M M A 2016 *Revista de chimie* **67**(6) 1191-94
[15] Perju M C, Nejneru C, Galusca D G and Stefanica R G 2010 *Modtech: New face of tmcr proceedings Conference: 14th International Conference on Modern Technologies Quality and Innovation* 455-458
[16] Bălțatu M S, Vizureanu P, Cimpoeșu R, Abdullah M M A B and Sandu A V 2016 *Revista de chimie* **67**(10) 2100-2002
[17] Perju M C and Vizureanu P 2014 *Revista de chimie* **65**(6) 694-696
[18] Irimiciuc Ș A, Agop M, Nica P, Gurlui S, Mihaileanu D, Toma S, Focsa C 2014 *Jpn. J. Appl. Phys.* **53** 11

Assessment of Hard Thin Layers Deposited by Plasma Spray on Hydroboration

C A Tugui[1,2,*], P Vizureanu, M C Perju[1], C Savin[1], C Nejneru[1], S M Baltatu[1], C Bejinariu[1] and M Benchea[3]
[1]Technical University "Gheorghe Asachi" of Iasi-Romania, Department of Technologies and Equipments for Materials Processing, Blvd. Mangeron, No. 51, 700050, Iasi, Romania
[2]National Institute of Research and Development for Technical Physics, Nondestructive Testing Department, 47 D. Mangeron Blvd., Iaşi, 700050, Romania
[3]Technical University "Gheorghe Asachi" of Iasi-Romania, Faculty of Mechanical Engineering, Blvd. Mangeron, No. 51, 700050, Iasi, Romania

E-mail: tzugui.andrei@yahoo.com

Abstract. Chemical mixers used in pharmaceutic industry are mostly constructed from a series of iron, chrome and nickel alloys. These mixers are used for mixing corrosive chemicals (acid or base) that often contain solid particles acting as the abrasive particles on the mixer blades. In this paper, martensitic stainless steel AISI 415 is used as base material and powder based on tungsten carbide is used for deposition of hard thin film resistant at common action of abrasive particles. The properties of metallic materials can be improved by deposition of layers with special material by plasma spraying method. Microstructure of thin films made of tungsten carbide and base material was performed with Scanning Electron Microscopy (SEM). In order to determine the micro hardness of the base material and the deposited layer, Vickers micro hardness testing have carried out. For determination of roughness of the deposited layer, tests were carried out using the equipment Mitutoyo SJ 301.

1. Introduction

Along classical technologies for obtaining special thin films [1-2], modern techniques to realize thin films by chemical and physical procedures [3-6] were perfected and developed, assuring high hardness and adhesion.

The base material is martensitic stainless steel AISI 415 which has low resistance to hydro abrasive wear. In industry, corrosion and wear are responsible for the maintenance and loss of productivity resulting from the shortened life of components, and the catastrophic failure of structures leading to massive costs of replacement and litigation [7-9].

This type of steel is used for realization of blades from mixers for fluids used in chemical, petrochemical and pharmaceutical industries, of blades from mixers from biogas installation in order to maintain the homogeneity of the substrate, so that the gas shall be released equally from the processed mass.

Erosive wear is caused by hard particles entrained in a fluid or a gas stream striking the surface [10]. Therefore, use of wear resistant materials in the form of coating/cladding is a pragmatic solution to combat wear [11-15].

IOP Conf. Series: Materials Science and Engineering **374** (2018) 012029 doi:10.1088/1757-899X/374/1/012029

2. Materials and methods

The base material has been cropped from a chemical mixer's blade from pharmaceutical industry using Metacut M250 equipment in order to obtain specimens for deposition, roughness and micro hardness tests. The specimens have 25 mm width, 25 mm length and 5mm thickness. The chemical composition of martensitic stainless steel AISI 415 was determined with Foundry Master Spectrometry in order to achieve the best precision and is Fe – 80.6%; C – 0.10%; Cr - 13%; Ni – 4.38%; Mn – 0.63%; others – 1.8%.

The analysis by SEM were carried out in order to analyses the surface and the microstructure of the base material. The SEM image is presented in Figure 1 showing elongated grains of Fα with acicular martensitic structure.

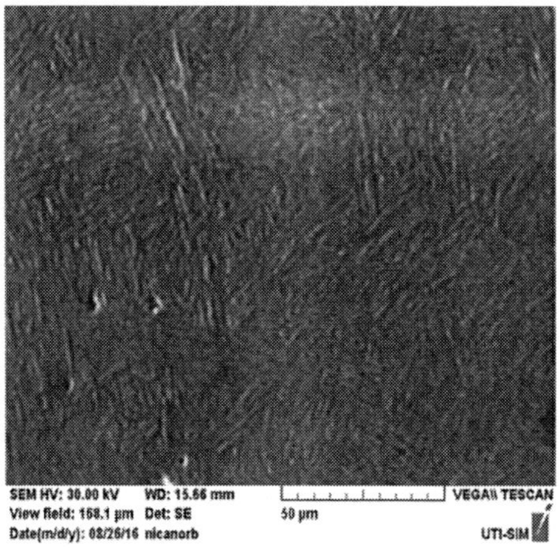

Figure 1. SEM analysis of AISI 415 stainless steel specimen.

Tungsten carbide powder is used for deposition because are compact, with average roughness, high hardness and wear resistant. These depositions have high resistance to hydro abrasion and friction. Chemical composition of the powder is: Co ≈13.0%; C ≈ 4%; W ≈ 83%. Due to cobalt percent over 12%, the depositions are more ductile but with higher hardness and good resistance to abrasion and friction.

Figure 2. SEM analysis of WC powders and their dimensions.

The plasma spraying methods is carried out using Sulzer Metco 9 MCE equipment from the Faculty of Mechanical Engineering, Technical University Gh. Asachi Iasi. Due to high temperature of plasma jet, between 10000-16000°C, the tungsten is melted well realizing compact surface and due to high speed of plasma jet over 3000 m/s, compact layers can be obtained [16].

3. Results and discussions

In this paper, thin films based on tungsten carbide powder were deposited by plasma spraying on a stainless steel AISI 415.

In order to increase the lifetime of chemical mixer's blades [18-20] made from AISI 415 stainless steel used in pharmaceutical industry, hard thin films were deposited by plasma spraying to resist to hydroabrasion. Roughness tests, adherence and micro hardness tests were carried out to determine how these parameters influence the hydro abrasive wear of chemical mixer's blades.

SEM analysis of the deposited layer has been carried out using VEGA II LMH microscope produced by TESCAN having an EDX module – TAX QX2 type with which the microstructure, the thickness of deposited film as well as the area of droplets formed on the deposited surface were determined.

(a)

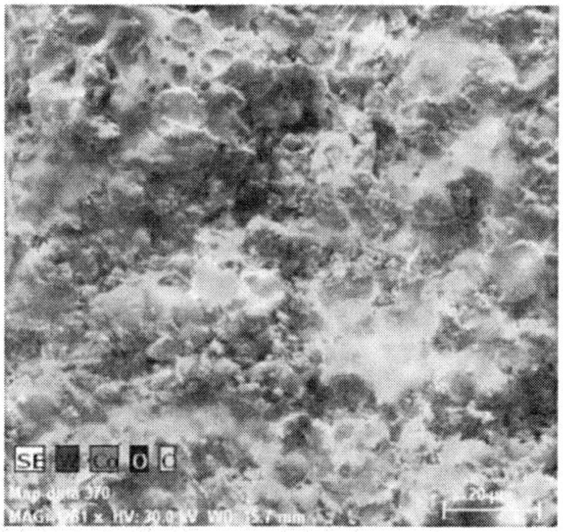

(b)

Figure 3. EDX analysis of the surface area on which the distribution of chemical elements in the sample deposited with tungsten carbide (WC) powder by plasma spray method has been determined: (a) the sample area; (b) distribution of elements.

From SEM image, it can be observed that the aspects are not very rough surface profile, having few detachments, micro cracks and un-melted powders.

Table 1. Chemical composition of the sample deposited by plasma spray methods using WC based powder determined by EDX.

Elements	%
Wolfram	76.37
Carbon	9.84
Cobalt	6.97
Oxygen	4.40
other	2.42

The analysis by EDX presented in Figure 3 emphasizes a compact deposition of tungsten carbide. On the surface, large areas with oxides (dents) and flattened droplets of cobalt appears, offering information about the fact that the powder contains separately particles of tungsten carbide and cobalt [21-25].

The plasma spraying method create no transitional structures of melted material layer-substrate but only structures formed by flattened droplets due to the dynamic of spraying which are soldered to the surface of the specimen. In Figure 4 is presented the cross section of deposited film.

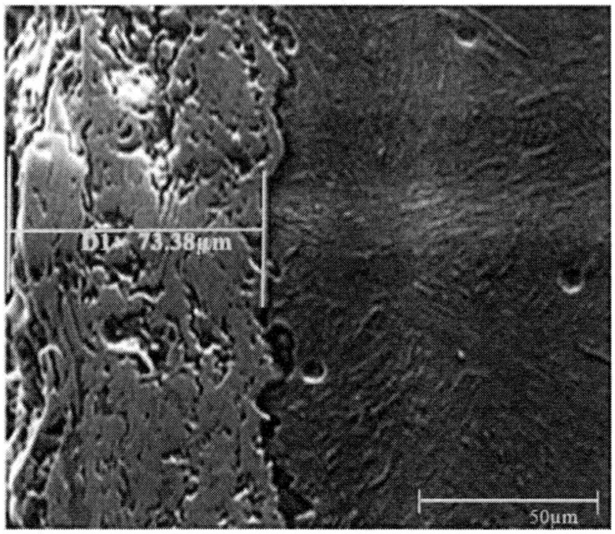

Figure 4. SEM analysis of thin film thicknesses deposited with tungsten carbide by plasma spraying method: 1 - layer, 2 - a substrate.

From the profile analysis made with Scan Explorer Nanofocus Profilometer, on the sample deposited with tungsten carbide powder by plasma spraying, presented in Figure 5, it can be observed that the deposition has not very rough surface, having few peaks of maximum roughness, closed as magnitude.

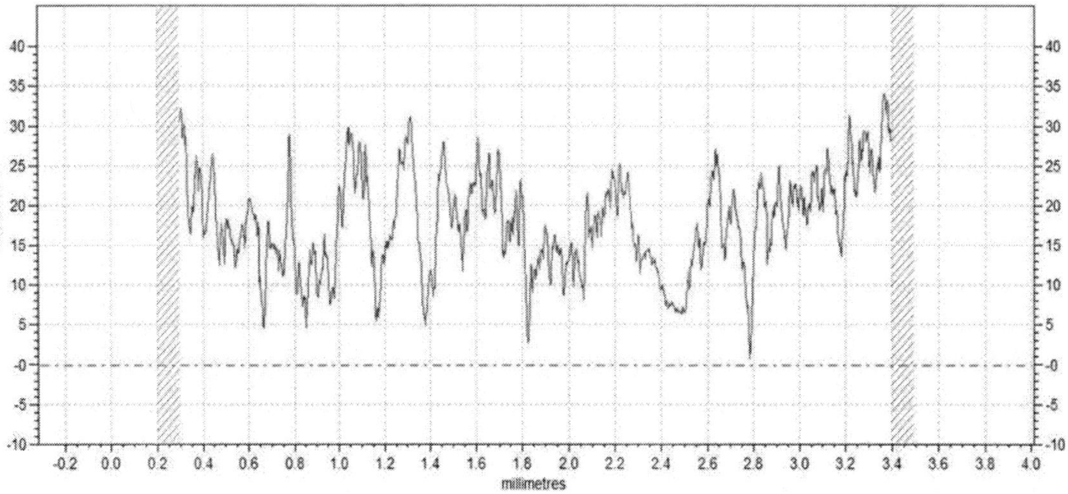

Figure 5. Roughness analysis of film deposited with tungsten carbide (WC) by plasma jet spraying.

The small value of roughness Ra of 3.86μm is due to the powders of tungsten carbides that have small grains dimensions and which can be deposited as very fine surface, reducing or eliminating the finishing operations.

4. Conclusions

The SEM analysis of layers deposited with powder WC on the basis material AISI 415 present a surface with fine granular aspects with micro particles non-melted embedded into the surface without micro exfoliating and evident micro adherence.

The roughness at deposition with plasma jet is influenced most often by the powder granulation. At deposition with powders based on WC, small roughness have been obtained (Ra= 3.86μm), because the grains of WC powders are very small (i.e. 8μm).

From the SEM and optical microscopy analysis of cross section, it can be observed that the specimens with deposition by plasma jet method presents intermetallic compounds, almost compact between the layer and substrate, leading to a good adherence of the coating.

By plasma jet spraying, hard layers can be deposited, at different dimensions, depending on the number of plasma jet passing over the surface to be deposited. The deposition process is automatic, obtaining large deposited surface in short time.

5. References

[1] Adamska A M, Lawrence E, Bright, Sutcliffe J, Liu W, Payton O D, Picco L and Scott T B 2015 *Thin Solid Films* **597** 57-64
[2] Mohamed M and Abdel-Rahim M A 2015 *Vacuum* **120** 75-80
[3] Ţugui C A, Axinte M, Nejneru C, Vizureanu P, Perju M C and Chicet D 2015 *Applied Mechanics and Materials* **657** 369-373
[4] Ţugui C A, Nejneru C, Găluşcă D G, Perju M C , Axinte M, Cimpoeşu N and Vizureanu P 2015 *Journal of Optoelectronics and Advanced Materials* **17** 1855-1861
[5] Hung C and Wu J Z 2015 *Journal of Alloys and Compounds* **651** 357-362
[6] Laatar F, Hassen M, Maaloul N K, Khirouni E and Ezzaouia H 2016 *Journal of Alloys and Compounds* **658** 337-347
[7] Levy A V 1988 *Surface & Coatings Technology* **36** 387-389
[8] Ninkam A J and Levy A V 1988 **Wear 347** 121
[9] Saengera R, Martina D and Gabriellib C 2005 *Surface & Coatings Technology* **194** 335-343
[10] Finnie I 1960 *Wear* **3** 87-103
[11] Ma N, Guo L, Cheng Z, Wu H, Ye F and Zhang K 2014 *Appl. Surf. Sci.* **320** 364-371
[12] Balla V K, Bose S and Bandyopadhyay A 2010 *Mater. Sci. Eng. A* **527** 6677-82
[13] Gupta D and Sharma A K 2011 *Wear* **27** 1642-1650
[14] Oladijo O P, Sacks N, Cornish L A and Venter A M 2012 *Int. J. Refract. Met. Hard Mater.* **35** 288-294
[15] Zafar S and Sharma A K 2016 *Wear* **29-45** 346-347
[16] Axinte C, Mareci D, Iacob S, Strugaru I and Munteanu C 2012 *Academia Romana* 1049-1050
[17] http://www.lineagealloys.com/tungsten-carbides
[18] Le Peshev A A, Karpov I V, Uschakov A V and Nagibin G E 2016 *Journal of Alloys and Compounds* **663** 631-635
[19] Toma S L, Bejinariu C, Gheorghiu D A and Baciu C 2013 *Advanced Materials Research Proceedings of ISCS13* **814** 173-179
[20] Fasano J, Janz E E and Myers K 2012 *International Journal of Chemical Engineering* **171** 838
[21] Verdian M M 2016 *Vacuum* **132** 5-9
[22] Toma S L, Bejinariu C, Baciu R, Radu S 2013 *Surf & Coat. Technol.* **220** 266–270
[23] Achitei D C, Vizureanu P, Minciuna M G, Sandu A V, Buzuianu A, Dana D I 2015 *Materiale Plastice* **52**(2) 165-167
[24] Achitei D C, Vizureanu P, Dana D and Cimpoesu N 2013 *Metalurgia International* **18** 104-109

IOP Conf. Series: Materials Science and Engineering **374** (2018) 012030 doi:10.1088/1757-899X/374/1/012030

Noninvasive Evaluation of Special Alloys for Prostheses Using Complementary Methods

A Savin[1,*], P Vizureanu[2], Z Prevorovsky[3*], M Chlada[3], J Krofta[3], M S Baltatu[2], B Istrate[4] and R Steigmann[1]
[1]National Institute of R&D for Technical Physics, NDT Department, Iasi, Romania
[2]Gh.Asachi Technical University, Faculty of Science and Materials Engineering, Iasi, Romania
[3]Institute of Thermomechanics, NDT Laboratory, Prague, Czech Republic
[4]Gh.Asachi Technical University, Faculty of Mechanical Engineering, Iasi, Romania

E-mail: asavin@phys-iasi.ro

Abstract. Ti-Mo-Si alloys have gained the attention of biomedical industry due to specific strength and corrosion resistance and the best biocompatibility among metallic materials used in medical prostheses. In order to characterize the material, the experimental determination of elastic matrix, mechanical wear and the probability of appearance and propagation of thin cracks are imposed. Thus, resonant ultrasound spectroscopy and acoustic emission as non-invasive methods and complementary methods as SEM, EDX are involved, to choose the best concentration of elements with the aim of mechanical properties improvement.

1. Introduction

Modern medical prosthesis as dental implant, knee prosthesis and hip stem are made from high strength brittle materials such ceramics, intermetallic and their composites for structural applications where cyclical loading is critical. Over the last decades, total hip arthroplasty (THA) has become one of the most successful surgical operations. Symptoms such as pain and loss of functions indicate THA as the solution. These symptoms can appear in different forms, as rigidity, deformity, limb shortening or movement reduction. Candidates for joint arthroplasty include patients who either exhibit excessive pain due to dysfunction in the hip or knee due to fracture of the femoral head, arthritis, or other conditions. THA improves quality of life in patients with end-stage osteoarthritis of the hip [1]. More than 500,000 joint replacements are performed each year in the United States alone, with many more hundreds of thousands being performed worldwide. Due to the change in demographics, the total demand will increase about 30% during the next 20 years. A total hip prosthesis is surgically implanted to replace damaged bone within the hip joint. Most of the implant is nowadays modular[2] (Figure 1a).

That assures the possibility to adapting the geometry of the prosthesis to the joint morphology of the patient. The total hip prosthesis consists of three parts:

- a cup that replaces the hip socket. The cup is usually plastic or ceramic metal
- a metal or ceramic ball that will replace the fractured head of the femur
- a metal stem that is attached to the shaft of the bone to add stability to the prosthesis.

Modularity assures the possibility of adapting the geometry of the prosthesis to the joint morphology of the patient. The femoral head has a spherical shape, slightly flattened both anteriorly and posteriorly. The femoral head is covered of articular cartilage, with the exception of the fovea. On the acetabular face, fovea covers a horseshoe shaped zone and it is set in the central region. The fovea is thick, not articulated and it is filled with some fat, the sinusoidal membrane and the ligamentum rotondum [3]. The articular capsule is inserted on the acetabular edge proximally and on the femur distally. This solution provides more flexibility during primary surgery and simplified revision procedures[4]. Modular components (metal augments) of various size and shape are used to restore bone defect to contained one capable of supporting a revision cup. The size and placement of augments is highly dependable on the bone loss pattern. Theses augments are secured with multiple screws for rigid bone fixation. In the last decade, revision hip arthroplasty has been the center of orthopedic surgeons' attention because these difficult and hazardous surgeries are growing in numbers. Nowadays, the research is focused on improving surgical techniques, implants design as well as development of biomaterials and bearing surfaces with a significant contribution for obtaining favorable outcome after revision hip arthroplasty, without having complex solution.

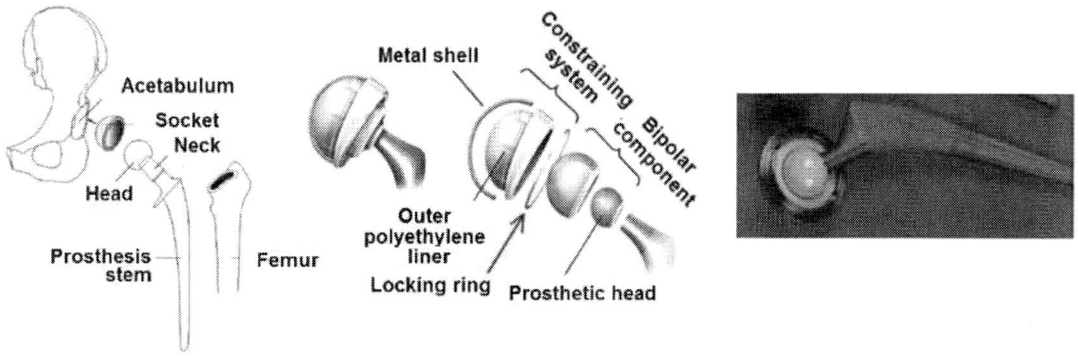

(a) (b)

Figure 1. Hip implants: a) modular principle; b) photo.

Since the 1960s, titanium and its alloys are used as medical devices due their unique properties. Titanium properties (i.e. tenacity, low density, resistance to corrosion, stiffness, etc.) made it together with its alloys, in combination with alloying elements (Fe, V, Mo, through other elements) a good replacement of steel pieces (in automotive industry in engine components –valves, connecting rods, drive shafts, crankshafts) [5-7]. Titanium alloys have the advantage of being lightweight and can be used in the aerospace and naval industry, military use, chemical and petrochemical machinery and other high performance applications [8, 9]. Due the biocompatibility, titanium alloys have broad applications in medical implants and prosthetic devices. The elastic modulus of α and $\alpha+\beta$ Ti alloys is much higher [10] than of natural bone (10-30 GPa). Intense researches are known in the domain of development of new bio-functional alloys based on Titanium [11], considered more appropriate of human bones as mechanical bio-functioning compared with polymers and ceramics biomaterials [12-14]. Development of a microstructure with optimal mechanical properties represents a difficult problem in this type of alloys domain. Methods for improving the properties of titanium alloys have been developed, in order to increase the capacity of joining bone tissue-metal implant and to give longevity to implant.

This paper proposes to examine mechanical properties of biomaterials based on Ti used in medical prosthetic with different ratios between alloying elements, i.e Mo 10-20%, Zr 6-8%, Si 1-5% to improve the bio-functional characteristics of the materials. For this investigation, we used alternative methods such as scanning electron microscopy and microstructure characterization, acoustic emission (AE) [15] and resonant ultrasound spectroscopy (RUS) [16] as noninvasive methods. EDX are involved, to choose the best concentration of elements with aim of improvement of mechanical properties.

2. Theoretical principle of resonant ultrasound spectroscopy

The elasticity of a material is the predisposition of a material to return to a minimum energy configuration. The change in energy, ΔU, under the influence of deformation strain can be described by an expansion in the size of the deformation strains e_{ij} around the local equilibrium (stress free interval state)[17].

$$\Delta U = \frac{1}{2} C_{ijkl} e_{ij} e_{kl} + \frac{1}{6} C_{ijklmn} e_{ij} e_{kl} e_{mn} + \ldots \quad (1)$$

where the C_{ijkl} are constants form the elastic modulus tensor; this is the sum of the effects of the local elastic restoring forces and conforming to the overall symmetry of the system.

The C_{ijklmn} are the third order elastic constants resulting from the third strain derivatives of the second order elastic constants C_{ijkl}. The restoration force implied in (1) can be written in a linearized form as Hooke's law with stress tensor τ_{ij}

$$\tau_{ij} = C_{ijkl} e_{kl} \quad (2)$$

RUS [18] is based on the fact that the resonance frequencies of an homogeneous object are determined by the geometry, density, and restoring forces of the object, described in (1). The resulting small strain elastic vibrations can be described by the wave equation solution using the linear elastic constants tensor. For a homogeneous and isotropic body, the matrix C can be written as

$$C = \begin{bmatrix} K+\frac{4G}{3} & K-\frac{2G}{3} & K-\frac{2G}{3} & 0 & 0 & 0 \\ K-\frac{2G}{3} & K+\frac{4G}{3} & K-\frac{2G}{3} & 0 & 0 & 0 \\ K-\frac{2G}{3} & K-\frac{2G}{3} & K+\frac{4G}{3} & 0 & 0 & 0 \\ 0 & 0 & 0 & G & 0 & 0 \\ 0 & 0 & 0 & 0 & G & 0 \\ 0 & 0 & 0 & 0 & 0 & G \end{bmatrix} \quad (3)$$

where K is the bulk modulus, $K = \frac{E}{3(1-2\nu)}$ and G is the shear modulus, $G = \frac{E}{2(1+\nu)}$, E represents the Young's modulus and ν is Poisson's ratio[19].

The calculation of vibration modes of elastic objects with free boundaries is a classic problem in mechanics. Unfortunately, exact analytical solutions exist only for few cases such as isotropic spheres, cylinders and certain models of a parallelepiped [20]. In most cases, approximation methods as finite element method (FEM) or Rayleigh-Ritz method must be used for estimating the eigen-frequencies of normal modes. The Young modulus, shear modulus and Poisson's ratio were determined based on the propagation speed of longitudinal and transversal ultrasonic waves using impulse - echo method. More on RUS technique can be find in papers of Miglori [21], Zadler [22] and Damarest [23].

3. Materials and methods

Hip prostheses are presently manufactured from Ti and its alloys due to their superior biocompatibility compared with other representative metallic biomaterials (SUS 316L stainless steel, Co Cr Mo alloys, etc), while the Young modulus is smaller comparatively with those, other characteristics such as corrosion resistance, strength and ductility are superior of other metallic materials. The evaluation of biocompatible materials using nondestructive evaluation techniques (NDE) is often resumed in literature [24-26]. Fatigue cracks in materials used for implants are characterized by their small opening in incident phase, they begin by the appearance of a small irregular crack located on the border of a crystalline grain and propagate then when the mechanical tension exceeds the threshold value [19]. This behavior of biocompatible materials used in realization of prostheses requires the application of complementary NDE techniques, leading to safety in exploitation.

3.1. Samples

When the prostheses are implanted to reconstruct the bone, to prevent bone resorption and to obtain adequate remodeling, their Young's modulus must be close to that of the bone. An efficient way to obtain a low elastic modulus in metallic materials is to realize a porous structure. The porosity has double effect, beside the one reminded, it allows the thickening of the bone in the pores and transferring the load from implant to the bone, helping the biological fixation. Ti-Mo alloys accomplish this task, solid titanium alloys being a viable alternative for prostheses, compatible biomechanical and biochemical, having mechanical features compatible to the human bones, the best being the βTi alloys with central cubic crystalline structure (bcc), because the titanium atoms are not dense packaged as in the case of α with dense closed packed (hcp) structure. Molybdenum has been selected as alloying element, secondary as ratio, because it not producing toxic reaction and is non allergic, being in balance with the human body. The mechanical properties are improved by alloying Ti with β isomorph stabilizers (i.e Mo), alloying that improve the resistance to corrosion.

Using high purity elements Ti-99.8%, Mo-99.7% and Si-99.2% as starting materials, the Ti 84 Mo 15 Si 0.75 alloy was prepared with vacuum arc melting furnace, in argon atmosphere. The advantages of using this equipment are very high melting temperatures can be achieved, the possibility of vacuum melting the metallic samples under a protective atmosphere and can create alloys with uniform composition. For the acoustic emission and RUS analysis, from the studied probe, a cube sample with 10mm side has been cropped using MAXIEM 1530 device which provides high-pressure water jetting with a precision of 0.1 mm/ml. Cutting quality has not been thermal affected and the material has not been mechanically deformed, nor included internal stresses, micro fissures or structural changes.The sample for metallographic observations was polished with SiC paper and Al_2O_3 particles with water. The polished sample was etched in an erodent with composition of 10 ml HF, 5 ml HNO_3, 85 ml H_2O, requiring 30 s immersion time.

Figure 2. SEM analysis of TiMoSi alloy.

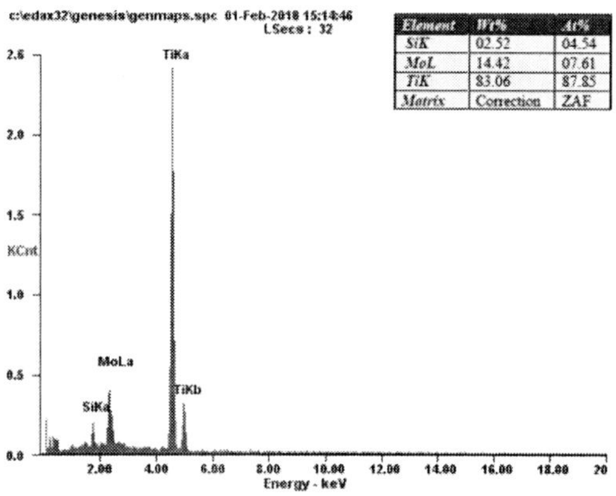

Figure 3. EDAX of specimen.

3.2. SEM evaluation

In order to better characterize of the materials, SEM and EDX analysis have been performed. Taking into account the structure of Ti 84 Mo 15 Si 0.75 (TiMoSi), to obtain relevant information about the

IOP Conf. Series: Materials Science and Engineering **374** (2018) 012030 doi:10.1088/1757-899X/374/1/012030

influence of Mo concentration over structure, Secondary Electrons (SE) images, as well as BackScattered Electrons (BSE) images have been taken. A Scanning Electron Microscope (QUANTA200 3D) operating at an acceleration voltage of 20 kV has been used of the topographical characterization of the specimens. SEM images emphasize that the Ti alloys is compact (Figure 2). A further confirmation of the local atomic percentage of the elements in sample composition is given in Figure 3.

3.3. Microhardness measurement

The measurements of microhardness were performed by Vickers method, with a load of 100 gF (HV100) in 25 seconds, on faces of the sample after cutting with method mentioned above. The Vickers microhardness remain approximately constant at all faces of the sample (Figure 4).

Table 1. Results of microhardness measurements.

Zone	HV (GPa)
1	353.9
2	359.8
3	356.6

The average values of hardness obtained from three different areas of this, are presented in Table 1. According to the literature [27, 28] and the results obtained from the experimental researches on TiMoSi alloys, a comparative study was carried out also with the classical biomaterials of the Ti6Al4V alloy, respectively the class of CoCrMo alloys (Figure 5).

Figure 4. The fingerprint left by the Vickers microhardness.

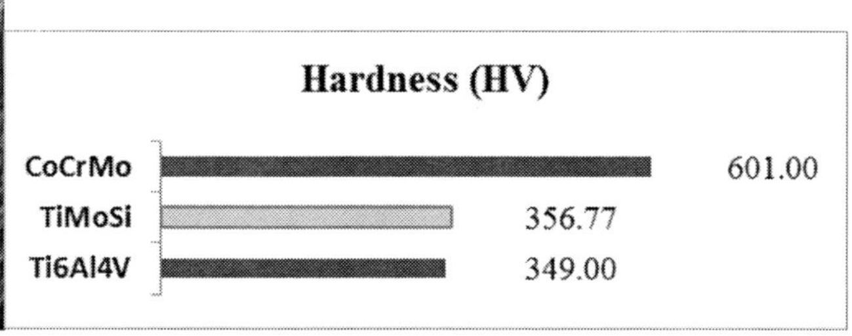

Figure 5. Graphical comparison of the hardness values of TiMoSi alloy with other biomaterials.

Compared to other biomaterials, the TiMoSi alloy had a similar hardness value as the Ti6Al4V alloy, which is the most used in implantology [27,29,30].

3.4. Ultrasound characterization of samples

For sample Ti84 Mo15 Si0.75 cube, having 10 mm length, in order to determine mechanical parameters such as Young modulus E, shear modulus G and Poisson ratio, the propagation speed of ultrasound waves were determined using impulse - echo method. Longitudinal velocity wave was measured using a sensor G5KB GE with central frequency of 5 MHz, the coupling being assured by coupling gel. The transversal wave velocity was determined with a sensor MB4Y GE with central frequency of 4 MHz. The PanametricsPR 5073 Pulser Receiver is used for the emission impulses and the reception of the signals. The digitizing of the signals and the measurements of the time of flight was made with the digital oscilloscope Le CroyWave Runner 64Xi.

The mechanical characteristics are presented in Table 2.

Young's modulus was estimated using the method described above but according with presence of porosity. Future research in this area should concentrate on the porous structures implants with different levels of porosity (15-70%) to reduce the Young's modulus.

Table 2. Mechanical characteristics of sample taken into study.

Metallic sample	Length [mm]	Density [kg/m^3]	Young modulus [GPa]	Shear modulus [GPa]	Poisson ratio	C_l [m/s]	C_t [m/s]	Observations
TiMoSi Ti84.2 Mo 15 Si0.75	10	5.063	88.81 -	33.93 -	0.3 -	5514	2589	With cracks detected by cutting and microscopic analyzed

4. Experimental Set-up

4.1. Acoustic emission

Acoustic emission (AE) is a passive ultrasonic nondestructive testing and evaluation (NDT / NDE) method for monitoring mechanically, thermally, etc., stimulated materials and structures. Transient elastic waves are generated in material due to the rapid release of energy from localized sources within material. The sources can originate from different mechanisms. The primary sources of AE are mechanical deformation and fracture, phase transformations, and other processes. Elastic waves propagate through a structure from the source to detectors (mostly piezoelectric transducers), which convert wave energy to electric signals. Weak electrical signals are amplified and characterized by parameters of their wave forms like energy, intensity, frequency spectrum and others [31, 32]. Due to the high sensitivity of piezo-transducers is AE technique widely used to identify failure initiation, the initial stages of damage, damage propagation, and catastrophic failure of the material. The recording of values when microcracks are initiated is a more accurate way of predicting the real strength of materials used in critical biomedical applications [33-35].

The sample was subjected to compression loading on INSTRON 1195 electromechanical testing machine with Bluehill 2 control software at the constant displacement velocity rate 0.2 mm/min. The maximal machine loading capacity is 100 kN, thus, the compression test was pre-programmed to stop at 80 kN load. AE was continuously monitored throughout the whole test using two piezoelectric AE transducers DAKEL IDK09 (diameter 9 mm) with sintered alumina wear plate. The experimental setup is presented in Figure 6.

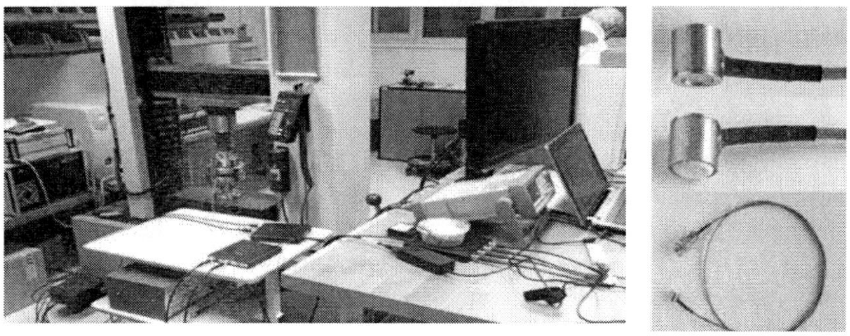

(a) (b)

Figure 6. Experimental setup for compression test with AE monitoring: (a) equipment; (b) AE transducers.

The frequency band of transducers was relatively flat (10 dB) in the range of 30 – 500 kHz. Figure 7 presents the detailed layout of the sample and AE transducers.

The small dimensions of the sample didn't allow direct attaching of transducers. The transducers were glued with cyanoacrylate onto a hard metal plate cover playing the role of wave guide from transducers, toward the sample. A support plate was also used on the bottom, to avoid damage the surface of loading clamps by the small and very hard specimen.

Instead of standard AE analyzer for AE signals treatment, two AE transducers with different signal amplification and USB HS5 oscilloscope connection were used for AE signals treatment in order to entire processing of their large dynamic range (more than 80 dB). Signals from both transducers S1 and S2 placed on opposite sides of the waveguide were amplified in PAC 20-40-60 preamplifiers (PA1 by 20 dB, and PA2 by 40 dB, respectively) and then were input to two channels of the 14 bit ADC's of the USB digital oscilloscope Tie-Pie HS5 for continuous signal sampling by 2.5 MHz frequency without thresholding. This assembly allows to record AE signals with full dynamic range including noise. The full ranges of both oscilloscope channels were adjusted differently: channel with S1 at 2 V and S2 at 400 mV respectively, enabling variable threshold settings of AE hits and events definition [32] in post-processing of AE signals, which is not possible with standard analyzers.

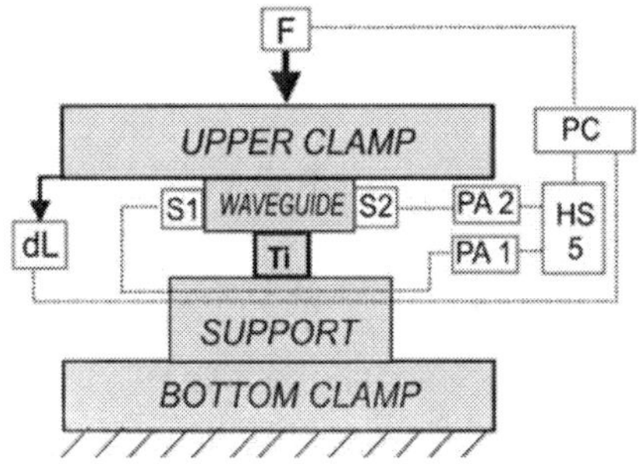

(a)

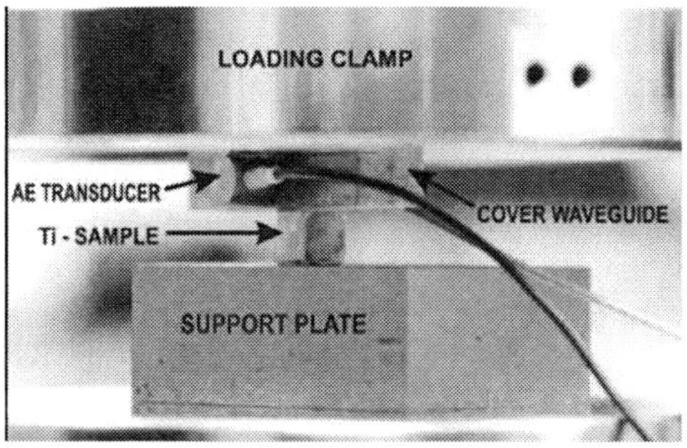

(b)

Figure7. Detailed layout of the sample and AE transducers: (a) scheme; (b) photo.

4.2. Resonant Ultrasound Spectroscopy - RUS

According with [18] sample geometry affects data acquisition. In this case the used sample was a cube with 10 mm length. For this we have few excitation modes and the spectrum requires more analysis than at a long bar. The increasing of spectral complexity supposes adjacent analyses of the sample before concluding about this sample.

In order to determine parameters, more resonance frequencies must be searched. This requires large time of computational calculation and repeated tests. The same number of normal modes for a short sample comparing with another long requires a narrow frequency range. The measurements are carried out with transducers in contact. The coupling of the transducer with the specimen influences which modes are measured. When the sample is pinned on its edge, more modes are excited and the modes are better defined than when the transducers are placed at the cube surface (Figure 8).

The probe is fixed between the emission transducers (excites an elastic waves with constant amplitude and variable frequency for specimen) and reception transducers (that detects response in US frequency of sample) in order to accomplish the condition of stress free surface. The equipment allows the setting so that for the established position of the cylindrical sample, the contact on the edge assures the excitation of a maximum number of possible resonances for the fixed geometry.

A Network/Spectrum/Analyzer Agilent 4395A generates a sweep frequency between 120 kHz and 200 kHz in 1 kHz step. The signal is amplified by Power amplifier AC 1012 AG&TC Power Inc. USA and applied to an US transmitting transducer P111.O.06P3.1 type, selected for the large bandwidth. The signal from receiving transducer, identical to the transmitting one, is applied to the B port of the 4395A Agilent, the spectrum being acquired by a PC that is used also to program the functioning of the equipment with a numerical code developed in Matlab, via PCIB interface. The command of the power amplifier is made with the same PC via RS232 interface.

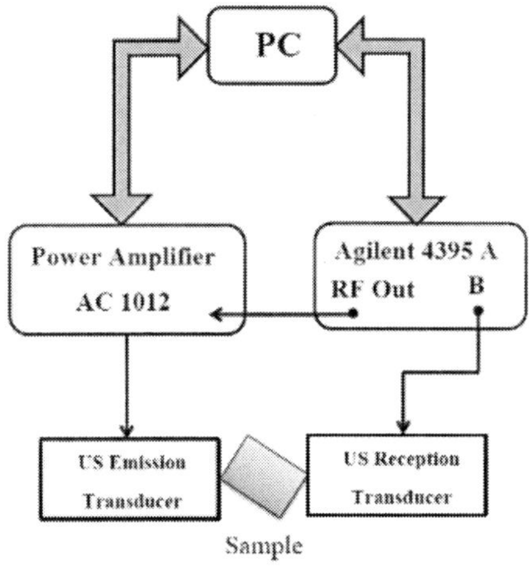

(a)

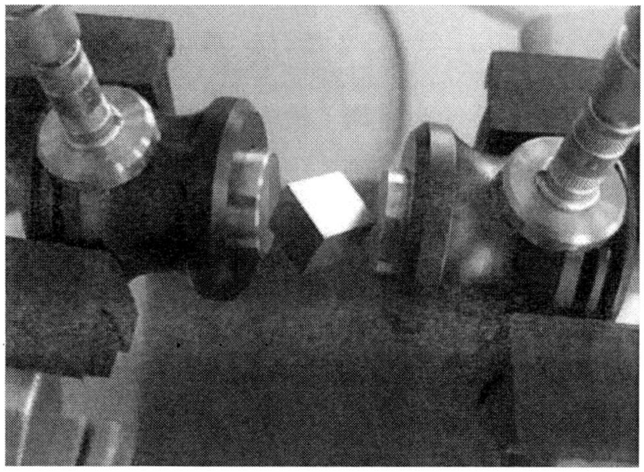

(b)

Figure 8. Experimental set-up for RUS: a) scheme; b) sample placement.

5. Experimental results

The presence of microcracks inside the materials used in the construction of medical prostheses has long been controversial due to the disturbing factors and the fact that it is difficult to observe and quantify their dimensions. The novel non-destructive techniques as RUS and AE have improved detection capability by increasing the confidence coefficient in the test analysis. The micro cracks accumulate with cyclic loads, the preexistence of a pore can be transformed into the propagation crack. Transient reduction of bone mass involves the use of a prosthetic material that can bear the loads by slowing the accelerated remodeling in response to loads. The weakening of a prosthetic implant is also generated by the ratio between porosity and damage. It is very important to know, in order to obtain and characterize the materials that the perturbations of material properties (density and especially the elastic constants tensor) affect the resonance frequencies and especially how they affect. In the manufacturing process, it is necessary to investigate the pore geometry, which has a strong effect on cell attachment and matrix formation, especially in the conditions of their realization in 3D technology. To evaluate what resonances will be observed in an experiment of the sample is very important to anticipate the theoretical response. The real case shows that it is useful to use resonance eigenvalues and eigenvectors in prediction.

5.1. AE test results and their interpretation

Figure 9 presents the compression loading diagram with AE activity of Ti alloy sample. Maximal attained compression stress wasσ= 800 MPa as loading machine was automatically stopped at 80 kN, yet before the final fracture exceeding loading force 100 kN. Crosshead displacement controlled by the loading machine (0.2 mm/min) was strictly linear up to its maximum 2.1 mm. Direct measurement of the strain was not realized due to small sample dimensions, so that maximal strain was only estimated after correction on overall machine stiffness as 1.44 %, which implies compression modulus in linear part approx. 89 GPa. AE signals from both channels were continuously sampled with 2.5 MSPS, which resulted in two about 1.6 GB long records. The post-processing of these records was carried out after correction on amplification set at 10 dB_{AE} counting threshold (i.e. approx. 3μV, about 2dB above a noise level) and count rate dNc/dt proportional to the energy of AE activity was evaluated from higher sensitive second AE channel. A total number of counts Nc was computed by summation of count rates.

AE activity is connected with various processes taking place in the compressed samples. The first AE hits recorded at about the test time t = 160 s are caused by the first contact of upper machine clamp with the sample.

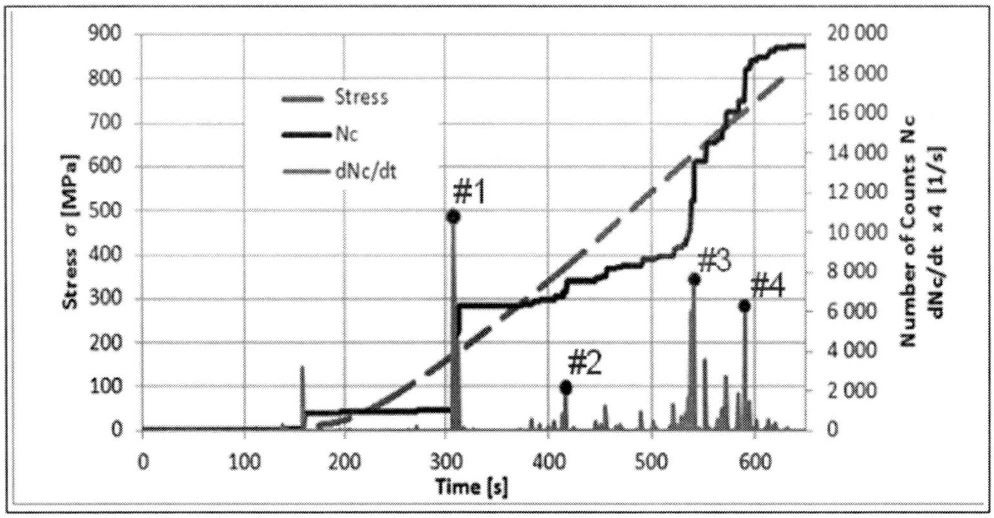

Figure 9. Compression loading diagram of Ti sample with recorded AE activity.

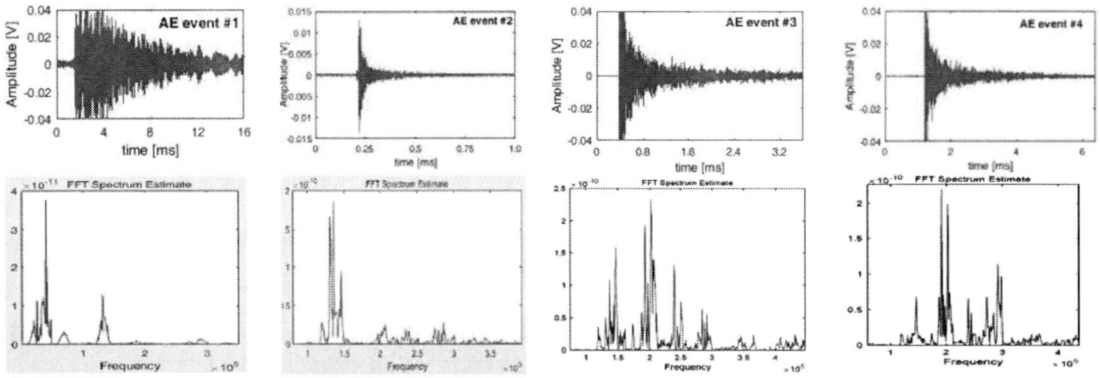

Figure 10. AE signal waveforms of events no. #1 to #4 (up) and their corresponding spectra (down).

The upper and lower surfaces of the sample were not polished and also not perfectly plan-parallel, which resulted in the fracture of one salient corner manifested by the high AE event at t = 310 s marked as #1 in Figure 9. After that begins linear stress growth in the sample. At about σ = 100 MPa starts formation of small cracking in the sample accompanied by AE activity resulting in bigger event #2 representing probably coalescence of smaller cracks. Internal cracking continues under approx. constant count rate 20 counts/s up to higher energy events ended by the event #3, representing longer crack jumps. Due to the high elastic energy stored in the sample, the following AE activity accompanying further cracking is characterized by the average count rate of 78 counts/s. It was ended with internal crack jump manifested by event #4, which represents release of a large amount of stored energy. The test was stopped at σ = 800 MPa under continuing smaller AE activity. Additional compression loading under lower displacement rate was carried up to $\sigma = 900$ MPa (not recorded in Figure 9) when it stops before the overall sample failure signalized by again growing AE activity. Different mechanisms of fracture processes taking place during various periods of compression loading are identified by comparing frequency spectra of individual large events marked in Figure 9 as #1 to #4. The signal waveforms of those events are plotted along with their corresponding spectra in Figure 10. Large event #1 representing the break off of the salient sample corner that exhibits long duration at very low maximal frequency only 40 kHz. It can be interpreted as a vibration resonance of the whole sample clamped in the testing machine due to large energy release and falling down of the broken off part. This resonance is perfectly hearable when the signal of that event is played down-sampled. The event #2 occurred with started internal cracking has the frequency maximum around 140 kHz. A higher crack velocity at higher stored elastic energy in the sample is accompanied with higher maximal frequency 193 kHz of the event #3. A similar feature is observed at event #4 where the frequency maximum is again around 200 kHz but another high peaks are at 300 kHz. The frequency spectra were calculated from signals recorded by the first channel, as amplitudes of signals from the more sensitive second channel were limited.

5.2. RUS

The resonance spectrum has been measured on the cube of alloy taken in this study. In the range 10 kHz-100 kHz, the obtained results are in conclusive and therefore, only the resonance spectrum in the range 120 kHz-200 kHz will be described. For a better understanding of the alloy used in manufacturing of femoral heads vibration modes, the eigen vibrations have been calculated using Solid Works 2014, Simulations Toolbox. For titanium alloy sample, axisymmetric, isotropic and homogeneous, a mesh of 62280 nodes and 43215 elements has been used. Therefore it can be concluded that *every mode must fall into one of the three classes* according to [22]: *extensional* axisymmetric mixtures of compression and shear motions, *flexural* modes through along pass that are tilted with respect to the axis, (the flexural modes occur in pairs named doublets, both members of which have the same resonance frequency) and *torsional* (the frequencies of these modes depends entirely upon the samples shear velocity) mode. The oscillation eigenfrequencies were calculated using/based on the parameters set presented in table 1. Figure 11 shows the vibration modes obtained by simulation, for the frequencies between 120kHz and 200kHz.

These figures represent the results of simulation of deformations for two extensional modes (a, b), and two flexural modes (c, d), for the cube obtained using the finite element method using SolidWorks 2014. The amplitude of signal in the swept frequency range is in tight connection with properties (especially to the density) of the sample. Red color represents maximal displacements and the minimal ones are presented in blue. The simulated information is very important to determine which of the resonances are observables for an investigated spectrum.

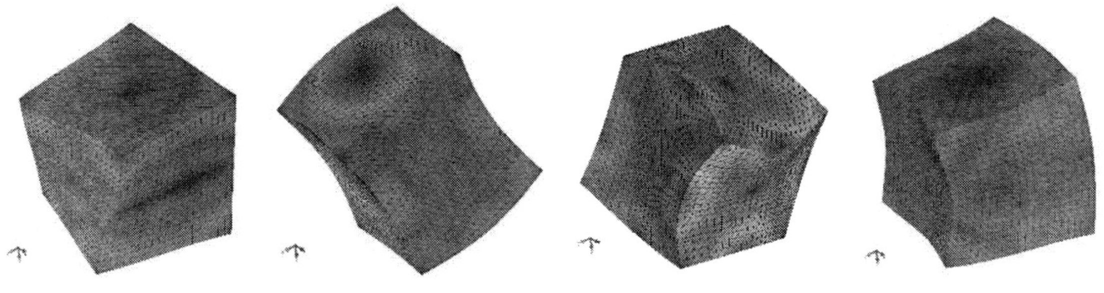

(a) (b) (c) (d)

Figure 11. RUS spectrum for the main vibrational modes:(a) extensional mode – 122 kHz; (b) extensional mode – 190 kHz; (c) flexural mode – 197 kHz; (d) flexural mode – 200kHz.

The resonance frequencies obtained by simulations correspond to those experimentally obtained.

The in homogeneities inside the material can be identified from the resonance spectrum, by changes of the resonance frequency, splitting of peaks/increasing of their width, and modifications in amplitude. In order to determinate, in the basis of experimentally measured resonance spectra, of the main elastic properties of the sample, the inversion of data was used, implying conjugate gradient method [22], minimizing the objective function.

$$F = \sum_i w_i \left(f_i^{(p)} - f_i^{(m)} \right)^2 \tag{4}$$

Where $f^{(p)}$ are the computed frequencies, $f^{(m)}$ are the measured frequencies, w_i are the weights, which characterize the confidence we have in the measurements. Using Matlab 2014 and because the number of peaks corresponding to the frequencies is small, the data inversion could be achieved for determination of elastic modulus and shear modulus. The variation of element concentrations from sample modifies the position of the peaks from spectrum as well as their amplitude. The presence of other discontinuities leads to modification of the signal phases. Figure 12 presents the resonance spectrum between the 120 kHz and 200 kHz, where d1 to d4 are cube's diagonals.

The resonance spectrum is specific for each object, comparing the resonance spectrum for two elastic objects, a difference between the elastic parameters of each object can be observed. For the studied sample, the cube having 10 mm sides, 4.467 g mass, the shape of the spectrum, depends on the position of discontinuities with respect to diagonal contact direction of transducers (polarization orthogonal or parallel to discontinuity).

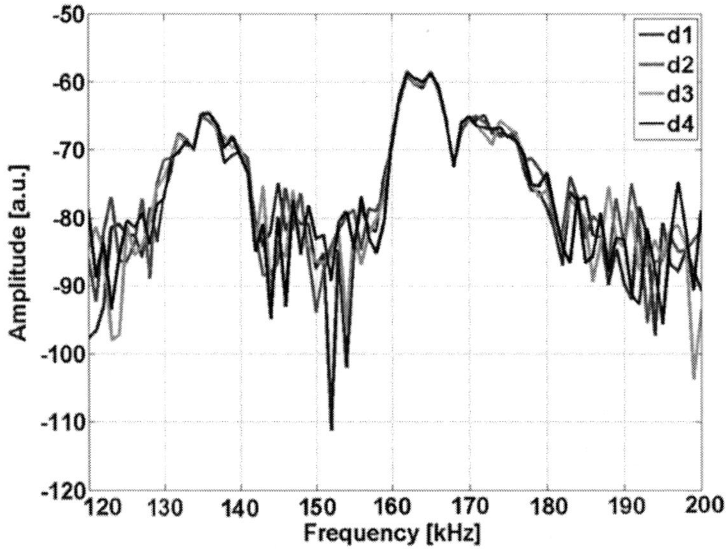

Figure12. Resonance ultrasound spectra of the sample.

6. Conclusions

The compression test results may be concluded by an ascertainment that compressive strength of Ti alloy sample is higher than 900 MPa. Never the less, its internal cracking manifested by accompanying AE activity begins at the stress about 300 MPa and more rapid cracking starts at stress about 600 MPa, which may be considered as an application limit. AE at higher compressive loading means warning before the catastrophic failure of a component. The frequency information and other parameters from AE data are useful for identification of failure mechanism.

RUS has been employed to detect the presence of discontinuities at surface/inside the sample. In both situations, the results are different beside the material without discontinuity. For the studied sample with dimensional ratio around 1, the interpretation is favorable because the torsion modulus is smallest, well separated by the rest for ν>0, allowing the determination of shear modulus. In the case in which the elements are incorrect alloyed, with a smaller density than the prescribed one, and the elastic and shear modulus are smaller, important modification in the spectrum shape and resonance frequencies appear. The changing in resonance frequencies towards elements concentrations and the dependency of vibration modes by the alloy's base element will be following.

7. References

[1] Ethgen O, Bruyere O, Richy F, Dardennes C and Reginster J Y 2004 *The Journal of Bone and Joint Surgery* **86**(5) 963-974

[2] Dzaja I, Lyons M C, McCalden R W, Naudie D D, Howard J L 2014 *The Journal of Arthroplasty* **29**(8) 1594-97

[3] Standring S ed 2015 Gray's anatomy e-book: the anatomical basis of clinical practice (London: Elsevier Health Sciences)

[4] Kinov P and Tivchev P 2013 Revision hip arthroplasty: management of bone loss. *Arthroplasty-Update.* Chapter 15 301 – 338 (London: InTech)

[5] Uhlmann E, Kersting R, Klein T B, Cruz M F and Borille A V 2015 *Procedia CIRP* **35** 55-60

[6] Cuddihy M A, Stapleton A, Williams S and Dunne F P E 2017 *International Journal of Fatigue* **97** 177-189

[7] Kumar J, Srivathsa B and Kumar V 2009 *Materials & Design* **30**(4) 1118-1123

[8] John A A, Jaganathan S K, Supriyanto E and Manikandan A 2016 *Current science* **111**(6) 1003

[9] Nnamchi P S, Obayi C S, Todd I and Rainforth M W 2016 *Journal of the mechanical behavior of biomedical materials* **60** 68-77

[10] Xie F X, He X B, Cao S L, Lu X and Qu X H 2013 *Corrosion Science* **67** 217-224

[11] Niinomi M 1998 *Materials Science Engineering A* **243** 231-236

[12] Li Y, Yang C, Zhao H, Qu S, Li X and Li Y 2014 *Materials* **7**(3) 1709-1800

[13] Majumdar P, Singh S B and Chakraborty M 2008 *Materials Science and Engineering: A* **489**(1-2) 419-425

[14] Coakley J, Isheim D, Radecka A, Dye D, Stone H J and Seidman D N 2017 *Scripta Materialia* **128** 87-90

[15] Kober J, Prevorovsky Z and Chlada M 2016 *Sensors and Actuators A: Physical* **240** 50-56

[16] Savin A, Craus M L, Turchenko V, Bruma A, Dubos P A, Malo S, Konstantinova T E and Burkhovetsky V V 2015 *Applied Sciences* **5**(4) 1665-82

[17] Landau L D and Lifshitz E M 1982 *Theory of Elasticity* 2nd ed (New York, N Y: Pergamon)

[18] Migliori A and Sarrao J 1997 *Ultrasound Spectroscopy* (New York, N Y: Willey)

[19] Lemaitre J and Chaboche J L 1994 *Mechanics of solid materials* (Cambridge: Cambridge University Press)

[20] Mindlin R D 1956 *Journal of Applied Physics* **27**(12) 1462-66

[21] Migliori A, Sarrao J L, Visscher W M, Bell T M, Lei M, Fisk Z and Leisure R G 1993 *Physica B: Condensed Matter* **183**(1-2) 1-24

[22] Zadler B J, Le Rousseau J H, Scales J A and Smith M L 2004 *Geophysical Journal International* **156**(1) 154-169

[23] Demarest Jr H H 1971 *The Journal of the Acoustical Society of America* **49**(3B) 768-775

[24] Nelson T S and Cripton P A 2008 Inducing head motion with a novel helmet during head-first impact can mitigate neck injury metrics: an experimental proof-of-concept investigation using mechanical surrogates *Proceedings of the International Research Conference on the Biomechanics of Impact* Bern Switzerland.

[25] Cormier J, Manoogian S, Bisplinghoff J, Rowson S, Santago A, Mcnally C, Duma S and Bolte J T 2011 *Journal of Biomechanical Engineering* **133**(2) 021004

[26] Ereifej N, Silikas N and Watts D C 2008 *Dental materials* **24**(9) 1289-95
[27] Geetha M, Singh A K, Asokamani R, Gogia A K 2009 *Mater. Sci.* **54** 397-425
[28] Chen Q, Thouas G A 2015 *Materials Science and Engineering R* **87** 1–57
[29] Niinomi M 2003 *Science and Technology of Advanced Materials* **4(5)** 445-454
[30] Zaman H A, Sharif S, Idris MH and Kamarudin A 2014 *Applied Mechanics & Materials* **735** 19-25
[31] AE Testing – General principles, European Standard No. EN 13554:2011
[32] Terms used in AE Testing, European Standard No. EN 1330-9:2000
[33] Ereifej N, Silikas N and Watts DC 2008 *Dental materials* **24(9)** 1289-95
[34] Mavrogordato M, Taylor M, Taylor A and Browne M 2011 *Medical Engineering and Physics* **33(4)** 395-406
[35] Cormier J, Manoogian S, Bisplinghoff J, Mcnally C and Duma S 2008 *Biomedical Sciences Instrumentation* **44** 147-152

Acknowledgments
This work was supported by a grant of the Romanian Ministry of Research and Innovation, CCCDI-UEFISCDI, project number PNCDI III-P1-1.2-PCCDI-207-03239/60PCCDI 2018, within PNCDI III and Nucleus Program PN 18 06 01 02/2018".

Institutional support RVO: 61388998 of the Institute of Thermomechanics of the Czech Academy of Sciences, v.v.i., and by the Czech Science Foundation grant no. GACR 17-22615S.

IOP Conf. Series: Materials Science and Engineering **374** (2018) 012031 doi:10.1088/1757-899X/374/1/012031

Oxidative Degradation of Methyl Orange Solution by Fe-MKSF Catalyst: Identification of Radical Species

N H Abdullah[1], M K A Selamat[1], N Nasuha[1,2], H Hassan[1] and N A Zubir[1,2*]
[1]Faculty of Chemical Engineering, Universiti Teknologi MARA (UiTM), Pulau Pinang, Malaysia
[2]Hybrid Nanomaterials, Interfaces & Simulation (HYMFAST), Faculty of Chemical Engineering, Universiti Teknologi MARA (UiTM), Pulau Pinang, Malaysia

E-mail: noraida709@ppinang.uitm.edu.my

Abstract. Iron–immobilized montmorillonite KSF (Fe-MKSF) has been recognized as promising catalyst in degrading persistence organic contaminants. However, detailed mechanistic insight during the catalysis which involving the formation and identification of radical species were remained indeterminate due to complex reaction. Inspiring by this gap, iron-immobilized clay (Fe-MKSF) was synthesized and used as heterogeneous catalyst in the oxidative degradation of methyl orange (MO) solution. Identification of radical species were determined through the inclusion of different types of radical scavenging agent during the Fenton-like reaction at optimum condition. Interestingly, dominant radical species were found to be hydroperoxyl radicals (•OOH) which subsequently followed by hydroxyl radicals (•OH) during the catalysis. Based on the percentage of MO removal, it was suggested that approximately 88% of the •OOH radicals existed at the interface of catalyst while 39% presence in bulk solution. Meanwhile, the interface •OH radicals promoted 38% of MO removal, whilst 4% by the bulk •OH radicals. Hence, these findings have conveyed novel insight on detailed radicals' identification as well as its' interaction during the catalysis.

1. Introduction

The mechanism behind heterogeneous Fenton process is intriguing due to unsolved, complex interaction of radical species, organic compound and the catalyst itself. In heterogeneous Fenton-like, there are two main radical species that typically discussed in literature which are hydroxyl (•OH) radicals and hydroperoxyl (•OOH) radicals. The ·OH radicals are the stronger radical than the ·OOH as the standard reduction potential of the former one is 2.59 V and the latter is 1.5 V [1].

With the OH and OOH high reduction potential, many organic contaminant were able to degrade at near- diffusion controlled rate [2]. The radicals weaken the structure of organic contaminant by abstracting H-atom from C-H, N-H, or O-H bonds and adding OH to unsaturated bonds like C=C or aromatics rings [3]. In general, it was reported that the dominant radical's presence during the heterogeneous Fenton-like catalysis is •OH radical. However, the coexistence of •OOH radical was also contributing towards the enhancement of organic's degradation during the catalysis whilst synergic effect of both radicals were confirmed [4].

Since the lifespan for these radicals are short (microseconds), the radical only can be detected indirectly at the present. The electron spin resonance spectroscopy (ESR) [5], chemiluminescence (CL) [6] or photoluminanescence (PL) fluorescent [7] are among the common methods which were used to identify the types of radical presence during catalysis. However, the utilization of these methods are either time consuming or expensive. Therefore, a direct in-situ measurement method which offers reasonably cost effective is desired in determining the types of radical via the utilization of specific types of radical scavenging agent during catalysis.

The reaction of •OH radical with alcohols [8] like methanol, butanol and inorganic anions [9] like HCO_3^- are used in •OH dominated degradation reaction. The alcohols and inorganic anions consume the •OH very fast, thus it served well as OH scavengers in radical identification reaction. The •OOH may be deduced by the quenching and scavenging of benzoquinone [10].

The additions of the radical scavengers were reported to reduce [8, 10] the catalytic efficiency indicating minimal radical radicals remains after the scavenging effect. This scavenging radical method may not be the most accurate one but is fairly less complicated, cost effective and tailored towards the specific types of radical which presence either on the surface of the active sites or in bulk solution during the catalysis [8].

Hence, the aims of this work is to investigate on plausible mechanistic insight of heterogeneous Fenton-like reaction by identifying the types of radical species presence through the inclusion of specific radical scavenging agent during the catalysis at optimum reaction condition.

2. Materials and Methods

2.1. Materials

All materials that were used in this study were purchased from Merck and used as received.

2.2. Preparation of iron-immobilize MKSF (Fe-MKSF) catalyst

The iron-immobilized MKSF (Fe-MKSF) composites catalyst were synthesized from a reaction solution consisting of iron precursor and dispersion of MKSF clay as describe elsewhere [11]. Na_2CO_3 was added slowly to vigorously stirred 0.2 M of $FeCl_2.4H_2O$ solution at a molar ratio of 1:1 for $[Na^+]/[Fe^{2+}]$. The solution was continuously stirred for 3 h. Then, the MKSF was added to the mixture and stirred for another 3 h. The ratio of iron precursor to MKSF was set at 20 mmol of Fe/g of clay. After that, the mixture was aged at 100°C for 48 h in an autoclave reactor. The resulting precipitants were centrifugally recovered and washed several times with deionized water and dried at 60°C overnight. The dried samples were calcined in air at 350°C for 20 h at ramping rate of 10°C/min.

2.3. Identification of radical species

Four types of chemicals were used as radical scavenging agents during the catalysis were shown in Table 1. The scavengers were added 30 min prior to the addition of H_2O_2. The concentrations of scavengers were fixed at 6 mM. The catalysis procedure was performed at optimized conditions of pH 2.8, catalyst dosage of 0.8 g/L, concentration of oxidant of 15 mM and initial methyl orange (MO) concentration of 25 mg/L. The changes in MO's colour were observed during 180 mins of reaction. The samples were periodically withdrawn, filtered and analyzed using UV-Visible Spectrophotometer (Lamda 25, Perkin Elmer) at λ_{max} = 464 nm.

Table 1. List of radical scavengers.

Chemicals	Type of quenching radicals	Ref
Potassium iodide	$^{\bullet}OH$ radicals on surface of catalyst	[8]
T-butanol	$^{\bullet}OH$ radicals on bulk solution	[8]
Sodium azide	$^{\bullet}OOH$ radicals on surface of catalyst	[5,12]
P-benzoquinone	$^{\bullet}OOH$ radicals on bulk solution	[5,12]

2.4. Characterization of catalyst

X-ray diffraction analysis (XRD Bruker D8 Advance) for determination of catalyst's phase. CuKα was used as source of radiation (λ=1.54Å). The diffractometer operated at 40 kV and 40 mA and scan range of 2θ from 10° to 80°. The concentration of iron leached was analyzed using Inductively Coupled Plasma Mass Spectrometry (ICP-MS NexION ® 300).

3. Results and discussions

Figure 1 shows the MO removal after quenching with scavenging agents. The sodium azide, p-benzoquinone, potassium iodide and t-butanol were added into the reaction mixture to selectively quenched surface-hydroperoxyl radical ($^{\bullet}OOH_S$), bulk-hydroperoxyl radical ($^{\bullet}OOH_B$), surface-hydroxyl radical ($^{\bullet}OH_S$) and bulk-hydroxyl radical ($^{\bullet}OH_B$), respectively.

Interestingly, the dominant radical species were found to be hydroperoxyl ($^{\bullet}OOH$) radicals and subsequently followed by hydroxyl radicals ($^{\bullet}OH$). Based on the MO percentage removal, it was suggested that approximately 88% of the $^{\bullet}OOH$ radicals existed at the interface of the catalyst while 39% presence in the bulk solution. On the other hand, only 38% of MO removal was promoted by $^{\bullet}OH$ at the interface and 4% removal at the bulk solution. It is contradict from the previous work [13] that reported the dominant radical detected Fenton reaction was $^{\bullet}OH$ radicals.

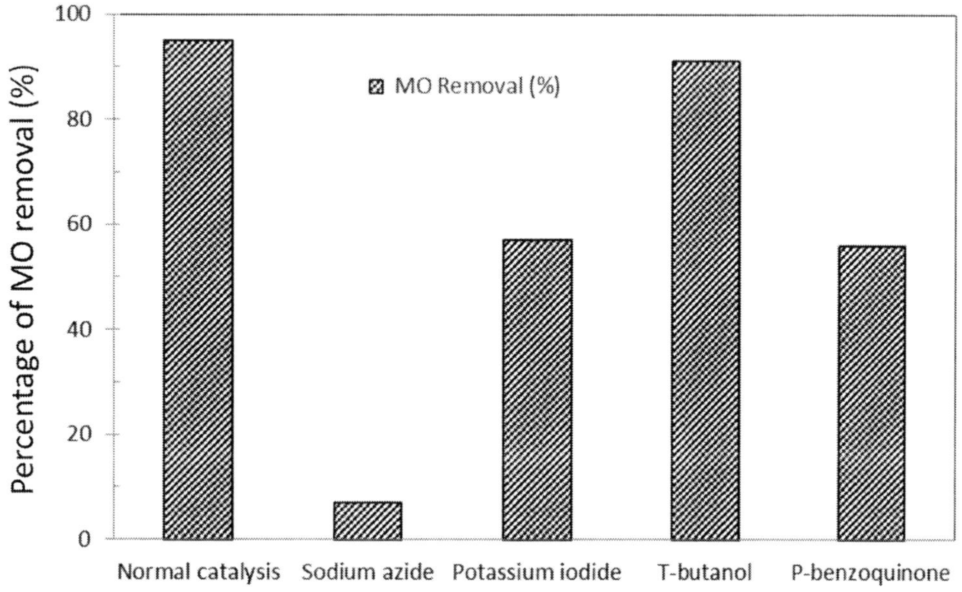

Figure 1. The influence of radical's scavengers on the removel of MO in heterogeneous Fenton –like reaction.

On possible explanation is that the iron species that bind to the MKSF is mainly governed by Fe^{3+} ions instead of iron Fe^{2+} ions. The different oxidation number of iron species may lead to different reaction initiation of Fenton reaction. For heterogeneous Fenton-like reaction using Fe-MKSF catalyst, the reaction started with the activation of H_2O_2 by Fe^{3+} and producing $^{\bullet}OOH_S$ radicals and Fe^{2+} as shown in the Eqn 1. The slower rate of reaction of Eqn 2 results in slower production of Fe^{2+}, hence only capable to generate reserved quantity $^{\bullet}OH_S$ radicals.

$$\equiv Fe^{3+} + H_2O_2 \rightarrow \equiv Fe^{2+} + {}^{\bullet}OOH_s + OH^- \quad (1)$$

$$\equiv Fe^{2+} + H_2O_2 \rightarrow \equiv Fe^{3+} + {}^{\bullet}OH_s + H^- \quad (2)$$

This postulation was strongly supported by the XRD results analysis. The Fe-MKSF was ascribed to the hematite phase, α-Fe_2O_3, a rhombohedral iron (iii) oxides as shown as Figure 2. The crystal structure of hematite,α-Fe_2O_3 (indexed to PDF-01-087-1165) was presented in the iron oxide structure with peaks at 2θ = 25.12°, 33.08° and 35.63° assigned by (012), (104), and (110) crystal planes respectively. The α-Fe_2O_3 spinel structures detected in the catalysts were in evident with the calcination temperature of 350°C. The phase formation as function of temperature were reported by

Farbod (2012) [14] and the iron oxide thermodynamically transform to hematite phase at temperature above 300°C [15]. Hence, the immobilized Fe^{3+} ions ($\equiv Fe^{3+}$) in the system can be ascribed to this hematite phase.

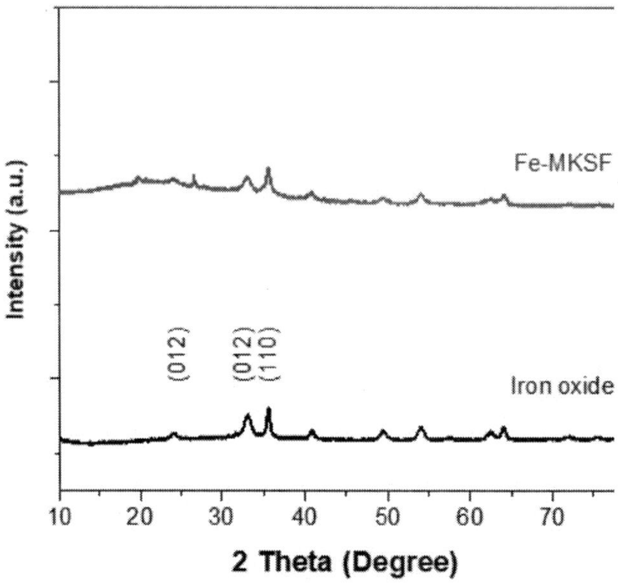

Figure 2. XRD patterns for iron oxide and Fe-MKSF catalysts.

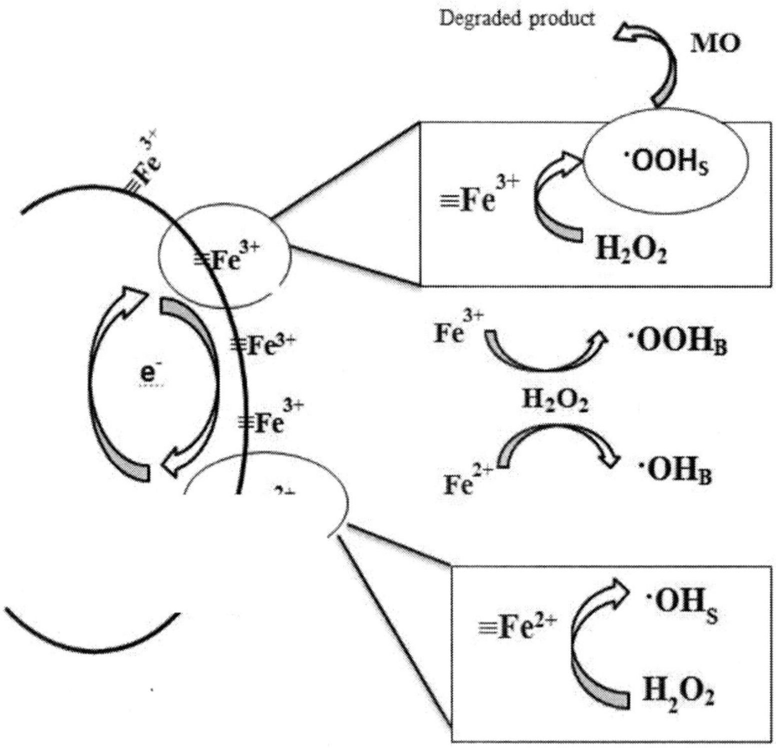

Figure 3. Proposed mechanism during heterogeneous Fenton-like reaction using Fe-MKSF catalyst.

The reaction initiate with the Fe^{3+} ions activated by the presence of H_2O_2 to produced $^{\bullet}OOH$ radicals. Figure 3 demonstrates the proposed mechanism based on the identified radicals special during the catalysis. The $\equiv Fe^{3+}$ ions was mainly presence at the interface of the catalyst rather than the $\equiv Fe^{2+}$ ions that resulted to the detection of $^{\bullet}OOH_S$ and $\bullet OH_S$ radicals, respectively. However, a minimal amount of immobilised ions ($\equiv Fe^{3+}$ and/ or $\equiv Fe^{2+}$) may also leached to bulk solution homogeneously in the form of Fe^{3+} and Fe^{2+} ions, which subsequently resulted to the detection of $^{\bullet}OOH_B$ and $\bullet OH_B$ radicals. Leaching of immobilised ions ($\equiv Fe^{3+}$ and/ or $\equiv Fe^{2+}$) was due to the solution acidity during catalysis.

The amount of leached iron in the solution after undergoing 180 mins of reaction was found to be 0.3 mg/L. The concentration of iron is below limit of standard effluent discharge of 2 mg/L as stated in European legislation [16]. The relatively minimal concentration of leached-out iron indicates that negligible contribution of homogeneous Fenton process in the reaction. It strongly suggests a solid deposition of iron ions ($\equiv Fe^{3+}$ and/ or $\equiv Fe^{2+}$) onto the MKSF matrix and the decolourization may dominantly attributed to the heterogeneous Fenton-like reaction that take place on the surface of Fe-MKSF catalyst.

4. Conclusions

The degradation of MO was mainly contributed by the surface-hydroperoxyl radical ($\bullet OOH_S$) which subsequently facilitated by surface-hydroxyl radical ($\bullet OH_S$) as secondary radicals during the heterogeneous Fenton-like reaction using Fe-MKSF catalyst. The contribution of these radical on the degradation of MO can be represented in the sequence of $\bullet OOH_S > \bullet OH_S > \bullet OOH_B > \bullet OH_B$. Such findings can be attribute to hematite phase found in the Fe-MKSF catalyst. This new findings provide a novel insight in understanding the underlying MO's degradation mechanism which directly relates towards identification as well as interaction of radical species presence during the heterogeneous Fenton-like reaction.

5. References

[1] Remucal C K and Sedlak D L 2011 *Aquatic Redox Chem.* 1071 177–197
[2] Keenan C R and Sedlak D L 2008 *Environ. Sci. Technol.* 42 1262–1267
[3] Xu L and Wang J 2016 *Appl. Catal. B Environ.* 123–124 117–126
[4] Pignatello J J, Oliveros E and MacKay A 2006 *Crit. Rev. Environ. Sci. Technol.* 36 1–84
[5] He J, Yang X, Men B and Wang D 2016 *J. Environ. Sci.* 39 97–109
[6] Soh N 2006 *Anal. Bioanal.Chem.* 386 532–543
[7] Yu F, Xu D, Lei R and Li K 2008 *J. Agric. Food Chem.* 56 730–735
[8] Arshadi M, Abdolmaleki M K, Mousavinia F, Khalafi-Nezhad A, Firouzabadi H and Gil A 2016 *Chem. Eng. Res. Des.* 112 113–121
[9] Fukuchi S, Nishimoto R, Fukushima M and Zhu Q 2014 *Appl. Catal. B Environ.* 147 411–419
[10] Zhang L, Nie Y, Hu C and Qu J 2012 *Appl. Catal. B Environ.* 125 418–424
[11] Feng J, Hu X, Yue P L, Zhu H Y and Lu G Q 2003 *Ind. Eng. Chem. Res.* 42 2058–2066
[12] Gan P P and Li S F Y 2013 *Chem. Eng. J.* 229 351–363
[13] Wang Y, Sun Y, Li W, Tian W and Irini A 2015 *Chem. Eng. J.* 267 1–8
[14] Farbod M, Movahed A and Kazeminezhad I 2012 *Mater. Lett.* 89 140–142
[15] Bhavani P, Reddy N R, Reddy I V S and Sakar M 2017 *IEEE Trans. Magn.* 53 1–5
[16] Ersöz G 2014 *Applied Catal. B, Environ.* 147 353–358

Acknowledgements

The authors gratefully acknowledges the Ministry of Higher Education Malaysia (MOHE) and Universiti Teknologi MARA (UiTM) for financial support under Fundamental Research Grant Scheme (600-IRMI/FRGS 5/3 (110/2016)).

Effects of Partial Replacement of Eggshell Powder by Halloysite Nanotubes on the Properties of Polyvinyl Alcohol Composites

N F Alias[1] and H Ismail[1,*]
[1]School of Materials and Mineral Resources Engineering, Universiti Sains Malaysia, 14300, Nibong Tebal, Penang, Malaysia.

E-mail: ihanafi@usm.my

Abstract. Polyvinyl alcohol (PVA)/eggshell powder (ESP) were prepared via solution casting method. The effects of gradual replacement of ESP by halloysite nanotubes (HNTs) were investigated based on tensile properties, physical properties and biodegradability. The main objective is to study the effect of hybrid fillers and also to compare the properties of PVA/ESP composite with conventional filler, HNT. The tensile properties decreased with increasing HNT loading. Scanning electron microscopy (SEM) studies showed that agglomeration of filler were present throughout the composites. Due to the presence of hydroxyl group on the outer and inner surface of HNT, the water absorption and water vapor transmisibility were found to increase with increasing HNTs loading. The biodegradability of film filled with HNT is lower compared to the film filled with ESP.

1. Introduction

The tremendous production and use of plastics in every segment of our life have increased the plastic waste in huge scales. Since polymeric materials do not decompose easily, disposal of waste plastics is a serious environmental problem. Plastic waste is now regarded as a worldwide environmental problem [1]. Over the years, public concern over waste management has increased due to the decreasing availability of landfills, environmental concern and public perception. Due to the waste management problems, degradable polymeric materials have become one of the major focuses for research.

Polyvinyl alcohol (PVA) is a useful polymer, which have been synthesised with a backbone consist primarily of –OH bond. The hydroxyl group makes it biodegradable and highly soluble and therefore widely used for packaging application. The advantages of PVA are good film formation, strong conglutination and high thermal stability. Generally, PVA films have been used for packaging application, where the contents need to be dissolved or dispersed in water because it is water soluble.

Recently, most researchers are interested to utilize agricultural wastes such as corn cob, palm kennel shell and eggshell. Eggshell is considered to be used as bio-filler because it is abundant, inexpensive and environmental friendly [2]. Basically, eggshell is an aviculture byproduct that has been listed as one of the worst environmental problems especially when egg product industry is well developed [3]. Eggshell is a natural bio-ceramic composite as it consists of both inorganic and organic components. Calcium carbonate is the main constituent of eggshell about 94% by weight. The unique chemical composition and substantial availability makes eggshell a potential candidate for bio-based calcium carbonate [4].

Halloysite nanotubes (HNTs) are eco-friendly nanotubes with low cost compare to carbon nanotubes. HNTs have a lot of commercial applications such as, additives in polymer and plastic, electronic components, drug delivery vehicles, cosmetics and in home and personal care products. They are naturally formed in the earth as the result of strain caused by lattice mismatch between adjacent silicon dioxide and aluminium oxide layers. HNTs have been reported to be used in several researches as filler to enhance final properties of composites [5-6].

Development of ternary composite either by blending the matrices or using hybrid fillers is currently gaining interest among researchers [7-9]. HYBRID fillers provide the advantages to modify the overall properties by balancing the properties that lack in one filler by incorporating another filler. From previous study, PVA/ palm kernel shell powder composites showed to have low tensile strength after addition of the natural filler. However, when HNTs was incorporated, the tensile properties were enhanced [10]. There is no work has been found in literature on PVA/ESP/HNTs ternary composites. The purpose of this series is to study the effect of partial replacement of ESP with HNTs on mechanical, physical and biodegradability properties of composites.

2. Experimental procedures

2.1. Materials

PVA with average molecular weight of 72000 and density of 1.3 g/cm3 was supplied by Sigma-Aldrich (M) Sdn. Bhd (Kuala Lumpur, Malaysia). Eggshell waste was collected from Kafe Staff in Universiti Sains Malaysia (Penang, Malaysia). Glycerol (1,2,3-propanetriol) used as a plasticizer and was supplied by Merck (M) Sdn. Bhd (Selangor, Malaysia). Hexamethylenetetramine (HTMA) was supplied by Sigma-Aldrich (M) Sdn. Bhd (Kuala Lumpur, Malaysia). It is a reagent with molecular weight of 140.19 g/mol. Tween 80 with density of 1.070 g/cm3 was supplied by Merck (M) Sdn. Bhd. (Selangor, Malaysia). De-ionized water is used as a solvent. Halloysite nanotube was supplied by Imerys Tableware New Zealand Ltd. (Auckland, New Zealand). The HNT has typical dimension in range of 150 nm-2 μm long, 20-100 nm outer diameter and 5-3 nm inner diameter.

2.2. Preparation of eggshell powder (ESP)

The chicken eggshell was thoroughly cleaned with tap water to remove the chicken dung and the residual of albumen (egg white) and yolk adhering to the eggshell. After cleaning, the eggshell was subsequently dried in an open air for 24 hours. The ESP was prepared by grinding the eggshell by using mini crusher from Chyun Tseh Industrial Co. LTD, Taiwan. Sieving process was carried out by using 3 sieving plates with size of 1000, 600, and 75 micron. After that, the ESP was dried for 24 hours at 70°C. The ESP was kept in dessicator for subsequent composites preparations.

2.3. Preparation of composites

The composites were prepared by solution casting method. Deionized water (200 ml) was used to dissolve PVA. The mixture was stirred for 10 minutes at 95 °C. The filler was added and continued to stir for 10 minutes. The additives were added and let to stir for another 10 minutes to ensure homogenous mixture. The solution was casted over glass plates and dried at room temperature for 24 hours. The film was then further dried in oven for 2 hours at 70°C. The film was peeled off and reserved for further testing. PVA/ESP/HNTs composite was prepared at five different ratios, varying the ratio of ESP and HNT, while keeping the PVA/filler composition constant at 80/20 wt%.

2.4. Characterization methods

2.4.1. Tensile properties. The tensile properties of the composite films were evaluated by using Instron 3366 universal testing machine (Instron, Petaling Jaya, Malaysia). Five dumbbell shaped samples (Type IV) (ASTM D638) (2008). The crosshead speed used for tensile test was 50 mm/min and the load cell was 10 KN. The gauge length was set at 50mm. The test was carried out at room temperature and an average value five repeated tests were taken for each composition.

2.4.2. Morphological study. The surface morphology of composite was observed by using scanning electron microscope (SEM) Supra 35VP (ZEISS, Oberkochen, Germany). The composite film was mounted on aluminium stubs with carbon tape and then sputter coated with thin gold layer to avoid charging and poor resolution during examination.
2.4.3. Water absorption. The water absorption (W_a) of films was performed as described by Yun et al., (2008). Dried plastic films were immersed in distilled water at room temperature (25 °C). After the 24 hours, moisture on the surface of the films was removed, and the weight of the films was measured. The water absorption was calculated as:

$$W_a\,(\%) = [(W_e - W_o) / W_o]\ \text{x}\ 100 \tag{1}$$

W_e is final weight (g), and W_o is the initial weight (g).
2.4.4 Water vapour transmissibility. Water vapour transmissibility, WVT was measured according to ASTM E96 (1998), modified by Gontard et al. A container of about 30ml distilled water was sealed on top by film sample. The initial weight of container sealed with film sample was weighed and recorded. Then, containers were all placed in the desiccator at room temperature. The containers were weighted daily using analytical balance until constant weight. The WVT was then calculated using:

$$WVT = G/t / A= (G / t) / A \tag{2}$$

Where G is the weight gain, t is the time and A is the area of test area.
2.4.5. Natural weathering. The samples were cut into dumbbell shape (Type IV) (ASTM D638) (2008) from the film of each composition. The samples were subjected to weathering test for fourteen days. The test was carried out according to the ASTM D1435. The weathering resistance and the degradation of the samples were determined by observation on the changes in dimension, physical appearance and the weight changes.
2.4.6. Soil burial. The samples were cut into dumbbell shaped (Type IV) (ASTM D638) (2008) and buried in the natural soil at School of Materials and Mineral Resources Engineering, Universiti Sains Malaysia, Nibong Tebal, Seberang Perai Selatan, Penang for a period of fourteen days. The degradation of the samples was determined by observation on the changes in dimension, physical appearance and the weight changes.

3. Results and discussion

3.1. Tensile properties

Figure 1 shows the effects of HNT loading on the tensile strength of PVA films. The film filled with only ESP shows the highest tensile strength and apparently, incorporation of HNT into the film reduced the tensile strength. HNT is usually added into polymer to enhance the mechanical properties [11]. However the result obtained shows that addition of HNT reduces the chain mobility, resulting lower tensile strength of film as compares to PVA film with ESP. It is believed to be due to the poor wettability and interfacial interaction between filler and matrix. The continuous of filler addition can cause the phase separation problem arising from particles agglomeration [12].

When two fillers is used in the PVA film, it is vital to ensure that both fillers are compatible to each other, so that, high filler-filler interaction can be achieved. The decrement in tensile strength can be explained by the imperfect distribution of the filler through the polymer matrix and poor adhesion between matrix and filler [13]. Tensile strength of film filled with hybrid filler, which is ESP and HNT is showed to be low due to the incompatibility of the fillers and the matrix.

On the other hand, the elongation at break is related to the elasticity of the sample as it is measured through the extension under traction. Figure 2 displays the elongation at break of PVA/ESP/HNT films. As HNT loading increased, the elongation at break of the film decreases. This is due to increasing HNT loading in PVA matrix resulted in the stiffening and hardening of the film. This reduces the resilience and toughness, and causes the decrement in elongation at break. This decrement is due to the fact that addition of filler promotes brittleness and increases rigidity of the composite.

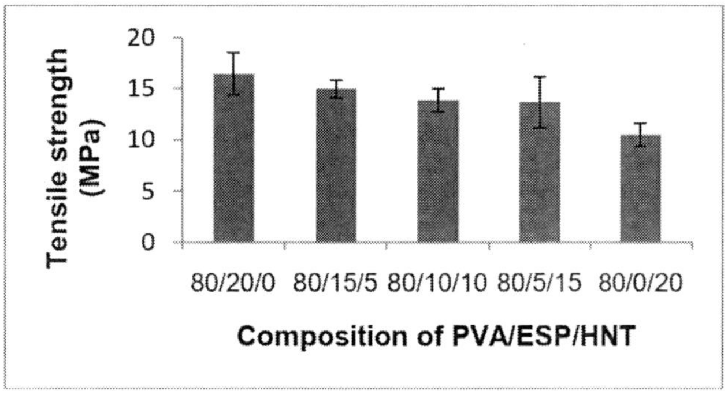

Figure 1. Tensile strength of PVA/ESP/HNT composite at different compositions.

Another factor that may explain the decrement of tensile properties as HNT is incorporated to the matrix is the high specific area to volume ratio of the filler itself. According to Bhushan et al. [14], due to the high surface area to volume ratio, the nano-sized filler has the tendency to agglomerate as the loading increase, which could prevent homogenous dispersion. As agglomeration occurs in the matrix, it will act as the stress concentration area that may lead to premature failure.

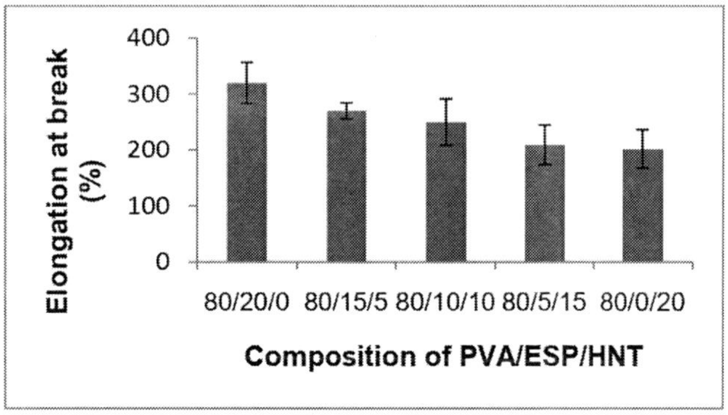

Figure 2. Elongation at break of PVA/ESP/HNT composite at different compositions.

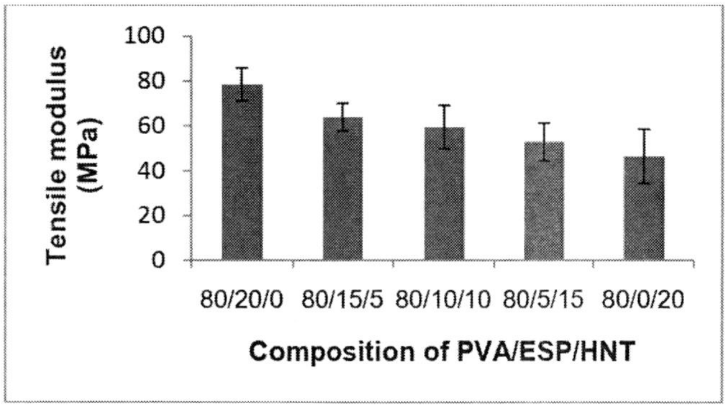

Figure 3. Tensile modulus of PVA/ESP/HNT composite at different compositions.

Tensile modulus is defined as the measure of force required to deform the sample at specific amount and it actually a measure of the stiffness of the sample [15]. Figure 3 shows the tensile modulus of PVA/ESP/HNT film at different compositions. It is observed that the tensile modulus decreased as the loading of HNT increased. Tensile modulus exhibited a similar trend in tensile strength. The highest tensile modulus is exhibited by the film filled with ESP, while the lowest tensile modulus is obtained from film with HNT. This might be due to the weak interfacial adhesion between the PVA matrix and HNT. Besides that, the decrement also may arise due to the agglomeration of the filler as the homogenize filler distribution is not achieved.

3.2. Surface Morphology

Figure 4 shows the surface morphology of composites at different compositions. From the micrograph, agglomeration occurs throughout the surface. Some of the fillers are not properly wetted by PVA matrix, therefore indicates the poor interfacial adhesion between filler and matrix. Agglomeration of HNT was observed throughout the surface. The keys to fabricate composites with enhance mechanical properties are by achieving a good dispersion of the filler in polymer matrix and strong filler-matrix interfacial bonding.

This may explained the decrement of tensile properties as HNT is added. Nanotubes have the tendency to gather and form bundles or micron-sized agglomeration due to their high specific surface area, which adversely affect the mechanical properties of nanocomposites [16]. Agglomeration of HNT will cause accumulation or concentration of stress that eventually promote the rupture of the composite.

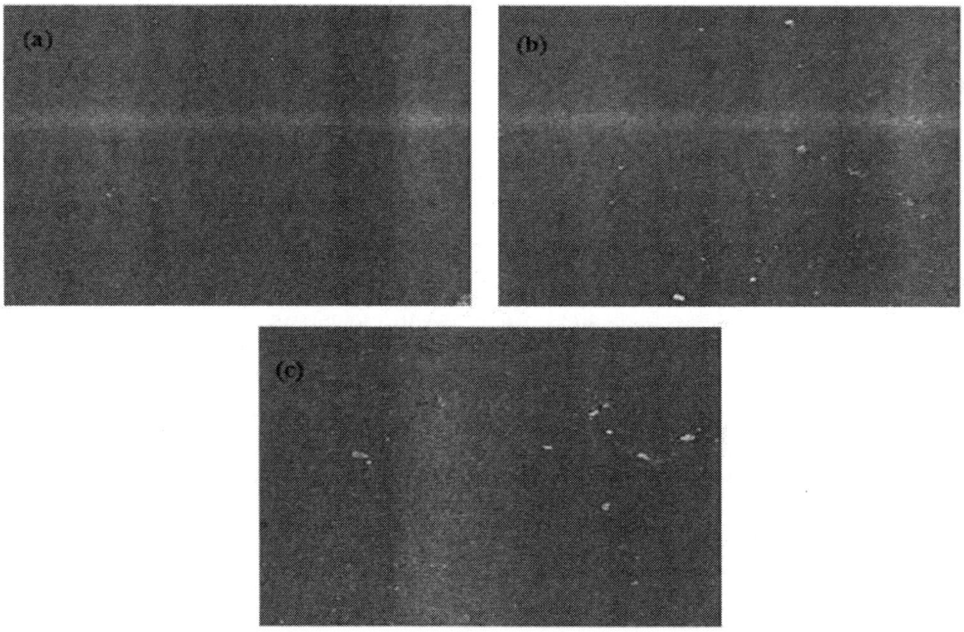

Figure 4. Surface morphology of PVA/ESP/HNT (wt%) composite (a) 80/20/0; (b) 80/10/10; (c) 80/0/20 at 500X magnification.

3.3. Water absorption

The water absorption test are essential for empathetic the performance of PVA composites, since the moisture pickup intimately relates to the composite properties as mechanical strength, dimensional stability and its appearance [17]. Figure 5 shows the percentage of water absorbance for different composition of PVA/ESP/HNT films. The films with higher HNT loading show more water absorption because higher content of filler that can absorb more water.

Water absorption percentage increased with higher content of HNT due to the properties of HNT itself whereby HNT is naturally hygroscopic. Another factor that may contribute to the increment of water absorption is the agglomeration of filler. As higher HNT loading is used, the formation of agglomerations increases due to the difficulties of achieving homogenous dispersion of filler at high filler loading. The agglomeration of filler basically increases the water absorption of the film because there will be poor interfacial adhesion, allowing water easily penetrate into the film.

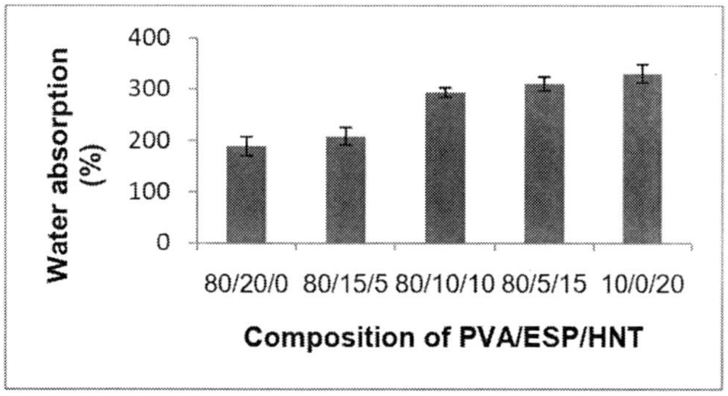

Figure 5. Water absorption percentage of PVA/ESP/HNT composite at different compositions.

3.4. Water Vapour Transmissibility (WVT)

Water vapour transmission of PVA/ESP/HNT film at different composition is shown in Figure 6. As illustrated, the WVT increases as the HNT loading into the film increases. The sensitivity of film matrix towards water can affect the WVT rate. HNT is capable to absorb moisture due to the structure, in which inner and outer surface of HNT is covered by OH groups and oxygen [18].

Another factor that may contribute to the increment of WVT as HNT loading increase is the agglomeration. These agglomerations will cause poor interaction between matrix and filler. Besides that, the poor compatibility also will increase the WVT because the interchain spacing will increases, allowing more water molecules to penetrate or diffuse through the film. This is similar to observation made by Ooi et al. [19].

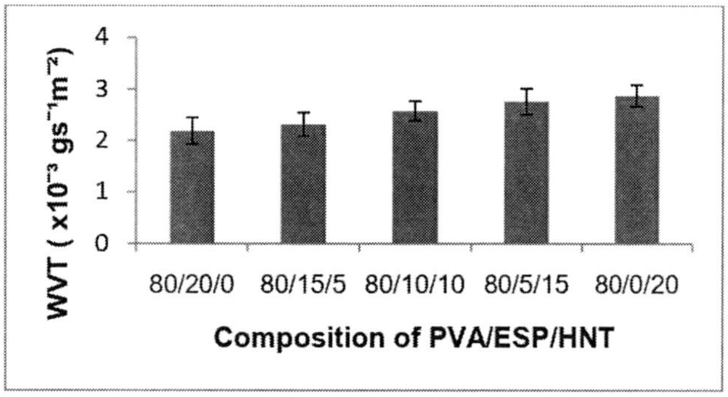

Figure 6. Water vapour transmission of PVA/ESP/HNT composite at different compositions.

3.5. Natural Weathering

Degradation by UV may occur as films were exposed to natural weathering and it may be indicated as the weight loss [20]. Figure 7 depicts the weight loss after weathering of PVA/ESP/HNT film at different compositions. The weight loss decreases as the loading of HNT increases. Highest weight loss was exhibited by the film filled with ESP, while the lowest weight loss was shown by film filled with HNT. Therefore, it implied that the addition of ESP causes the film to degrade easier due to weathering as compared to HNT.

The weight loss can be related to molecular weight changes. The weight loss indicated that PVA underwent chain scission after natural weathering. The same observation was reported by Sam et al. [21] in their study in effect of cobalt stearate on natural weathering of LLDPE/soya powder blends. In general, the weight losses are found to be small as the film may have absorbed water and humidity upon rainfall and weathering.

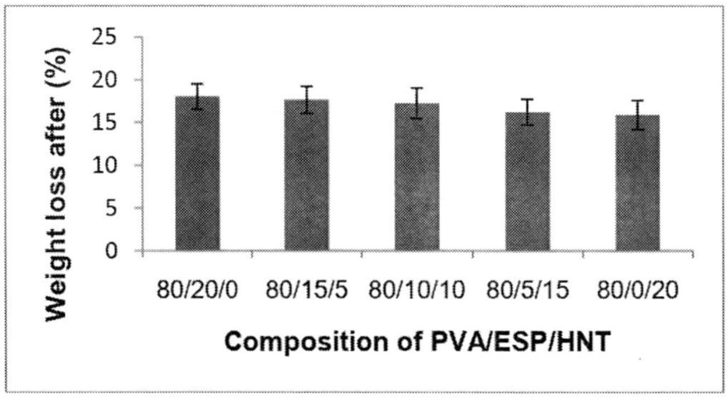

Figure 7. Weight loss after weathering of PVA/ESP/HNT film at different compositions.

3.6. Soil Burial

In order to evaluate the degradation of the films, the weight loss of films was evaluated. Figure 8 shows the dependence of weight loss after soil burial test on the HNT loading on PVA films. The weight loss of the film showed a decrement as higher HNT loading is incorporated. In soil burial test, the possibility of microbial degradation is high. The lowest weight loss was depicted by film with 20% HNT. This may be due to the fact that HNT has higher resistance toward microbial, causing it to take longer time to degrade.

Microbial activities enhance the degradation of plastic films and decompose them into smaller fragments. Highest weight loss was exhibited by the film with 20 wt% ESP. This implies that ESP may have better hydrolysis capability to convert hydroxyl for esterification [22, 23]. These results suggest that the addition of ESP to the films increased their degradation more than the addition of HNT. ESP has better attraction for microbe to penetrate and initiate enzymatic attack and continue the degradation process. It is shown that PVA filled with ESP can increase the degradability of film compared to PVA filled with HNT.

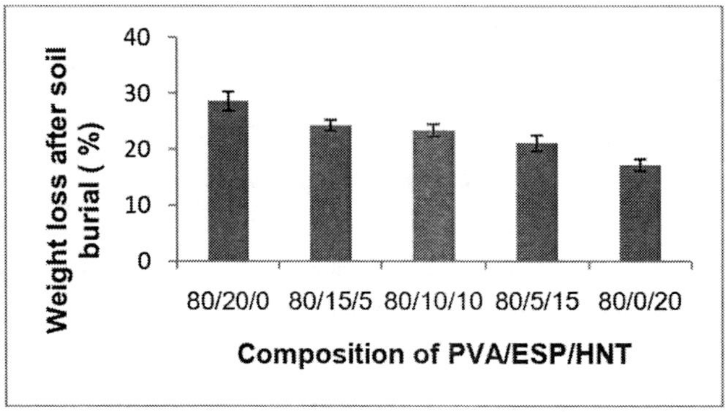

Figure 8. Weight loss after soil burial of PVA/ESP/HNT composite at different compositions.

4. Conclusions

In summary, the PVA/ESP/HNT composite show lower tensile strength, modulus and elongation at break as HNT loading increases. The SEM study reveals that poor dispersion and present of HNT agglomeration on film surface. This is due to the high specific surface area of HNT, which prone to agglomeration. Based from micrograph obtained from SEM, it is proven that agglomeration of HNT is present throughout the surface. Water absorption and water vapour transmission of film increase with increasing HNT loading. The biodegradability properties tested in soil burial and natural weathering is showed to decrease with higher HNT loading. In comparison to the film filled with ESP, the weight loss of HNT filled film is lower, indicating that the microorganism and UV attack is not favourable toward HNT.

5. References

[1] Nawang R, Danjajaji I D, Ishiaku U S, Ismail H and Mohd Ishak A M 2001 *Polymer Testing* **20** 167-172
[2] Kang D J, Pal K, Park S J, Bang D S and Kim J K 2010 *Mater. Des.***31** 2216–2219
[3] Siti Shuhadah and A G.Supri 2009 *Journal of Physical Science* **20(1)** 87–98
[4] Hassan S B, Aigbodion V S and Patrick S N 2012 *Tribology in Industry* **34(4)** 217–225
[5] Prashantha K, Lacrampe M F and Krawczak P 2013 *Advanced Materials Manufacturing & Characterization* **3(1)** 11–14
[6] Fernández A, Muniesa M and González J 2013 *Journal of Mechanical Engineering* **59** 183–192
[7] Samotu I A, Dauda M, Anafi F O and Obada D O 2015 *Tribology in Industry* **37(2)** 142–153
[8] Majdzadeh-Ardakani K and Nazari B 2010 *Composites Science and Technology* **70(10)** 1557–1563
[9] Taghizadeh M and Sabouri N 2013 *International Nano Letters* **3** 51
[10] Alias N F, Ismail H and Ab Wahab M K 2017 *BioRes* ***12*(4)** 9103-9117
[11] Zhou X Y, Cu Y F, Jia D M and Xie D 2009 *Polymer-Plastic Technology and Enginnering* **48(5)** 489-495
[12] Ismail H and Zaaba N F 2012 *Journal of Elastomers and Plastics* **46(1)** 96-109
[13] Buzarovska A, Bogoeva-Gaceva G, Grozdanov A, Avella M, Gentile G and Errico M 2008 *Australian Journal of Crop Science* **1(2)** 37-42
[14] Bhushan B, Luo D, Schricker S, Sigmund W and Zauscher S 2014 *Handbook of Nanomaterials Properties*. Berlin, Heidelberg: Springer.
[15] Toh W Y, Lai J C, Wan A and Wan A R 2011 *Journal of Composite Materials* **45** 1201-07
[16] Bregg R *New Frontiers In Polymer Research* 2006 Nova Science Publishers: New York
[17] Messiry M El, Deeb R El and Tarfawy S. El 2015 *Recent Manufacturing Technologies Conference* 234–244.
[18] Liu F, Wu J, Chen K and Xue D 2010 *Microscopy: Science, Technology, Applications and Education* 1781–1792.
[19] Zhong O X, Ismail H, Abdul Aziz N A and Abu Bakar A 2011 *Polymer-Plastics Technology and Engineering* **50** 705–711
[20] Borsos Z, Paun V P, Botez I C, Stoica C M, Vizurean P and Agop M 2008 *Revista de chimie* **59** 1169-71
[21] Sam S T, Ismail H, Ahmad Z 2011 *Polymer-Plastics Technology and Engineering* **50** 957–968.
[22] Gawdzinska K, Nabialek M, Bryll K, Szymanski P and Sandu A V 2017 *Materiale Plastice* **54(3)** 539-542
[23] Borsos Z, Paun V P, Botez I C, Stoica C M, Vizurean P and Agop M 2008 *Revista de chimie* **59** 1169-71

Acknowledgements
The authors acknowledge the support from Jabatan Perkhidmatan Awam, Malaysia and School of Materials and Mineral Resources Engineering, Universiti Sains Malaysia.

IOP Conf. Series: Materials Science and Engineering **374** (2018) 012033 doi:10.1088/1757-899X/374/1/012033

Phosphating of Carbon Steels in Solutions Containing Zinc and Zinc-Manganese Phosphates

D Ivanova
Department of Inorganic and Electrochemical Production, University of Chemical Technology and Metallurgy, Bulgaria, Sofia 1756, 8, Kliment Ohridsky, Blvd.

E-mail: dimkaivanova@uctm.edu

Abstract. The formation of phosphating coatings on carbon steels by novel baths containing zinc and zinc-manganese phosphates has been investigated. By gravimetric, chemical, electrochemical and physical methods are determined the characteristics of preparations (density, pH, conductivity, total and free acidity) and those of the produced coatings (thickness, phase and chemical compositions, structure, protection ability). The concentration range has been varied from 5 to 20% vol., while the temperature effects have been studied in the range from 20 to 80°C. The thicknesses of the coatings and the amount of the substrate dissolved have been determined. It is proved that Mn-phosphate decreases the coating thickness for all working solution concentrations and temperatures, at the same time increases the mass of dissolved metal of the substrate. The results obtained indicate that the coatings developed in zinc and manganese baths contain the following phases: hopeite, phosphophyllite, quasihopeite, strunzite and mixtures of them.

1. Introduction

Phosphating of ferrous and non-ferrous materials is the most widely used surface treatments in practice, due to its easy production, good corrosion resistance, low-cost and excellent adhesion with the next paint or polymer coatings.

The process can be defined as the treatment of a metal surface so as to give a reasonably hard, electrically non-conducting surface coating of insoluble phosphate which is contiguous and highly adherent to the underlying metal and is considerably more absorptive than the metal. The coating is formed as a result of a topochemical reaction, which causes the surface of the base metal to integrate itself as a part of the corrosion resistant film.

The formation of a phosphate coating is a complex process involving consequently dissolution, crystal nucleation and growth with rates and mutual influences affecting the crystallization kinetics as well as the coating structure and properties [1-3].

The most commonly used phosphate coatings are iron phosphate, zinc phosphate and manganese phosphate. In zinc phosphating, the corrosion protection is key, which is increased by the zinc content included. But zinc phosphate is also suitable as a sliding surface. Resilient sliding properties are the main feature of manganese phosphate. The layer reduces friction and is very absorptive for oil, so that it is used for plain bearings and bushings for example. For highly loaded sliding surfaces it is used in preference to the other phosphating methods [4, 5].

The present work reports data on phosphating process of mild steel surfaces with zinc and manganese phosphating solutions (ZnPh and MnPh). The main characteristic such as density, pH, and conductivity as well as the total and free acidity of the phosphating concentrates were determined. The effects of solution concentration and temperature on the coating thicknesses, phases, structures and elemental compositions were studied. The corrosion resistance and the protective properties of the coatings were determined, too.

2. Experimental

2.1. Materials and methods

2.1.1. Materials and samples. Experimental probes had the form of a square (0.005 m^2) and were made of sheet mild steel (0.17% C) 1.0 mm thick. The electrochemical experiments were carried out with plate-shaped specimens prepared in accordance with ISO/FDIS 17475 (ISO 17475:2005) and a fixed working surface area of 1 cm^2. In addition, samples with a form of a disk (0.001 m^2) were used as specimens in case of all physical methods applied in the study.

2.1.2. Solutions. The phosphating solutions are liquid concentrates of zinc and manganese hydrogen orthophosphates and phosphoric acid, containing stabilizing agents and accelerators.

The working media for phosphating were aqueous solutions of the liquid concentrates with concentrations 5.0; 10.0; 15.0 and 20 % vol. The experiments were carried out at temperature 20.0; 40.0; 60.0 and 80°C for 5.0; 10.0; 15.0 and 20 min. The working/model media used for the corrosion experiments is 0.6 M NaCl.

2.1.3. Methods. Gravimetric method. The Gravimetric method was used for studying kinetics of forming and determining conditions of the processes for increasing of the coating's thicknesses depending on the influence of different factors. The method allows determining mass alteration of experimental samples after forming and removing coatings:

$$M = \frac{m_1 - m_2}{S},\ g\ m^2$$

where: M, $g\ m^2$ is mass or as accepted to call a thickness of the obtained coating,; m_1 and m_2 are the sample mass after forming and removal of coating, g; S is the sample surface area, m^2.

Electrochemical methods. Open circuit potential determination. Metals immersed in liquid electrolytes media resulted in establishment of non-equilibrium, i.e. the so-called corrosion potentials. The corrosion potentials are not indicative with respect to the resistance attained but their values and time variations provide enough information about the character of the corrosion process, the behavior of both the metal and metal coating in different media under various conditions.

Potentiodynamic polarization method. The polarization curves recorded potentiostatically or potentiodynamically allow determination of various corrosion characteristics such as corrosion rate, corrosion potential, etc. The experiments were carried out with EG&G Princeton Applied Research, *Potentiostat/Galvanostat, Model 263A*, provided with the specialized software package PowerCORR®.

Physical methods. Optical microscopy (OM). The microstructure of the coatings were determined by optical microscopy, using a microscope OPTIKA® *Model:* B-500Bi.

Scanning electron microscopy (SEM). The morphology and structure of the coatings were examined by scanning electron microscopy, using a *SEM/FIB LYRA I XMU, TESCAN* electron microscope, equipped with ultrahigh resolution scanning system secondary electron image (SEI).

Energy dispersive X-ray spectroscopy (EDX). The energy dispersive spectroscopy is a local X-ray spectral analysis which permits qualitative and quantitative determination of surface micro volume contents of the order of several μm^3. Apparatus Quantax 200, BRUKER with spectroscopic resolution at Mn-Ka and 1 kcps 126 eV is used.

X-ray structural method (XRD). It was used for determining the phase composition of phosphate coatings. The investigations were carried out by TUR-M-62 diffractometer with focusing system Bragg-Brentano, using Co-Kα (1.79021 Å), with interval $5° < \Theta < 55°$.

3. Results and discussions

Table 1 shows the values of the main indicators concerning the concentrates for phosphating: ρ – density; pH; σ – conductivity; Ko – total and Kc – free acidity.

Table 1. Main characteristics of the phosphating concentrates.

Solutions	ρ, g/cm^3	pH	σ, mS/cm	K_o	K_c
ZnPh	1.290	0.79	98.9	300	32
MnPh	1.305	0.51	118.9	334	70

3.1. Gravimetric investigations

The gravimetric method was used to elucidate the effect of the operating conditions (concentration and temperature of the phosphating solutions as well the duration of the process) on the coating thickness M_1. Figure 1 shows typical relationships thickness/mass of phosphate coatings, M_1 – time, min obtained at optimal conditions (the surfaces of the treated steel samples are coated by homogenous dense films): 80°C temperature, different working concentrationc and duration of the process.

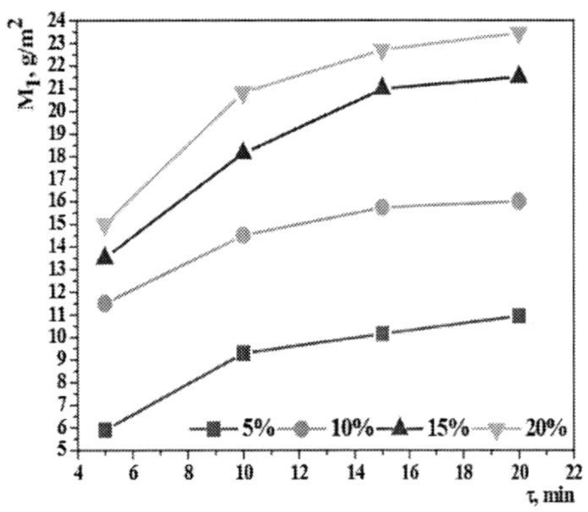

(a)

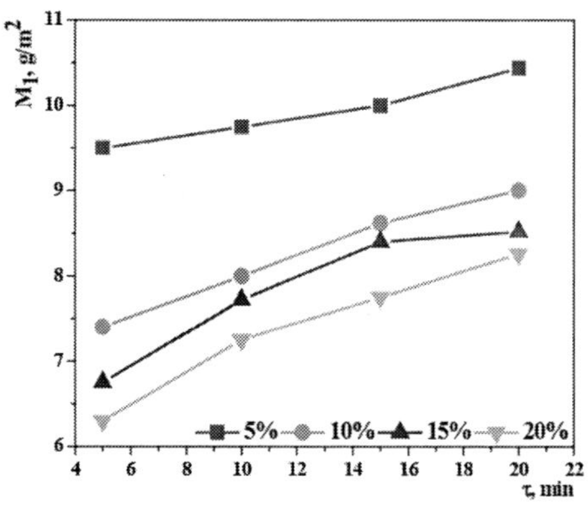

(b)

Figure 1. Effect of the phosphating duration, τ, min on the thickness/mass of the phosphate coatings obtained in the phosphating solutions, M_1, g m^{-2}: (a) ZnPh; (b) MnPh.

It is obvious that with increase of the solution concentrations in the range of 5 ÷ 20 %vol for both phosphating concentrates the mass/thickness of the phosphate films increases. The results show also, that the thicknesses of the coatings obtained in ZnPh solutions are significantly higher.

3.2. Electrode potentials measurement

The corrosion resistance of the phosphate coatings was determined in model aqueous solutions of 0.6NaCl. The tests itself includes an immerse procedure for about 1 h of two samples in the NaCl solution. The coating was determined as resistant if for 1 h (an empirically chosen time interval) no changes in the surface as colour spots or colorization of the solutions occur. Meanwhile, in parallel to these tests the corrosion potential of the samples were measured that allowed to compare the behaviour with that of uncoated steel samples.

The potential behaviour shown in Figure 2 is relevant to the corrosion resistance of phosphate coatings in 0.6 M aqueous solutions of NaCl. The plots reveal that just after the sample immerse in the corrosion medium the potentials shift sharply in the negative directions and after a certain accommodation time interval (about 500 seconds) a stable state is attained. The corrosion potential of an uncoated steel sample is more negative than the phosphate coated ones.

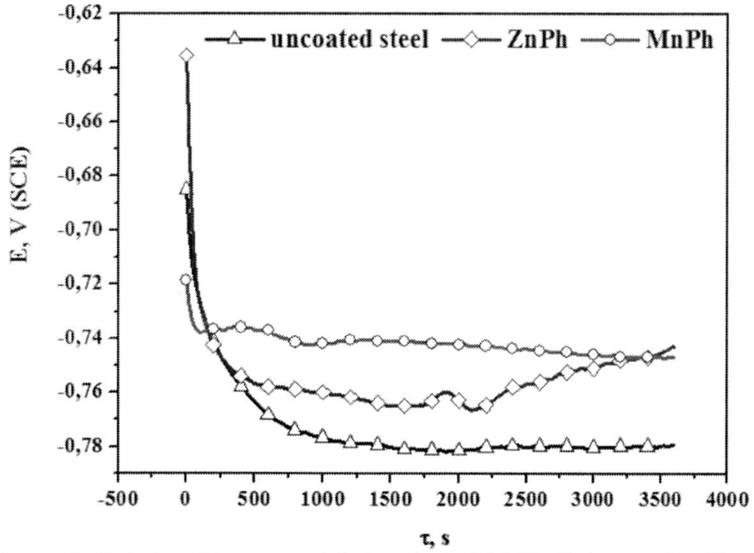

Figure 2. Relationships potential-time in 0.6 M NaCl (open circuit): 20 % wt.; 80°C; 15 min.

3.3. Potentiodynamic polarization investigations

The potentiodynamic polarization studies of the phosphate coatings were carried out at room temperature (20°C) in model aqueous solutions of 0.6 M NaCl. The scanning rate of 10 mV s^{-1} was chosen based on preliminary test runs. The target of these experiments was to obtain information about the corrosion resistance of the phosphate coatings formed in the corresponding model media in course of their cathodic and anodic polarization and consequently to compare them.

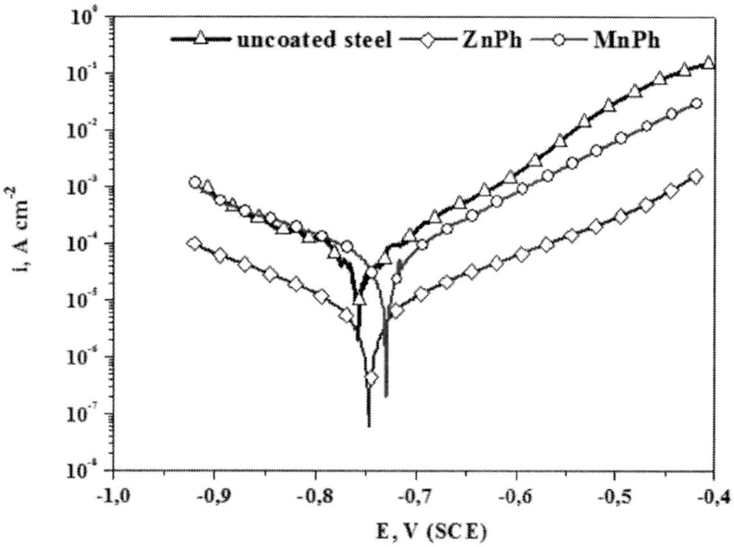

Figure 3. Potentiodynamic polarization relationships of the phosphate coatings in 0.6 M NaCl: 20 % wt.; 80°C; 15 min.

The potentiodynamic polarization relationships of the phosphate coatings (see Figure 3) obtained in the model solutions of 0.6 NaCl allow estimating both the corrosion potentials and the corresponding currents, summarized in Table 2. The potential of the coating, obtain in the MnPh solution is 30 mV more positive and the potential of the coating, obtain in the ZnPh solution is 15 mV more positive then the potential of the uncoated steel. In spite of these little differences, the more positive potentials and the low values of the coated samples show better corrosion resistance vs. uncoated steel surface.

Table 2. Corrosion parameters of the coatings in 0.6 M NaCl.

Model solution	ZnPh solution		MnPh solution		Uncoated steel	
	E_{corr}, *V (SCE)*	i_{corr}, $A\ cm^{-2}$	E_{corr}, *V (SCE)*	i_{corr}, $A\ cm^{-2}$	E_{corr}, *V (SCE)*	i_{corr}, $A\ cm^{-2}$
0.6 M NaCl	-0.745	$1.1x10^{-6}$	-0.730	$8.0x10^{-6}$	-0.760	$2.0x10^{-5}$

3.4. Optical microscopy

The micrographic pictures on Figure 4 demonstrates surface of the phosphate coatings obtain into ZnPh and MnPh solutions at 20 % wt.; 80°C and 15 min duration time. Surfaces of the coatings are homogeneous with dense fine film crystals of dark grey and black colour. The crystals in ZnPh coating are bigger, with comparison to these in MnPh coating.

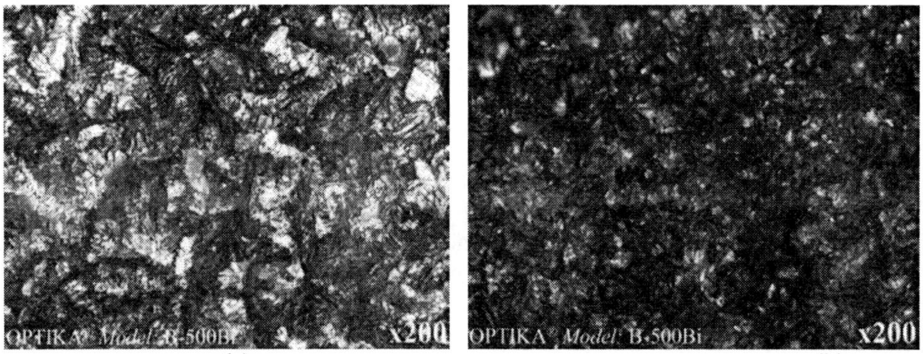

(a) (b)

Figure 4. Optical microscopy of the phosphate coatings: (a) ZnPh solution; (b) MnPh solution.

3.5. SEM and EDX analyses

The microphotographs in Figure 5 and Figure 6 demonstrate the morphology of the phosphate coatings obtained on steel surfaces into phosphating solutions. It follows from the figures that the habit of coatings is constant – i.e. the crystals initiate from one centre and increase radially (like spherulite). Crystals have length between 5 and 50 μm.

Table 3. EDX analyses of the elements in the coatings.

Elements	**ZnPh solution**		**MnPh solution**	
	Weight %	***Atomic %***	***Weight %***	***Atomic %***
O	26.26	53.97	19.11	43.26
Zn	44.00	22.13	29.05	16.09
P	13.53	14.36	13.45	15.73
Fe	16.21	9.55	36.43	23.63
Mn	–	–	1.97	1.30

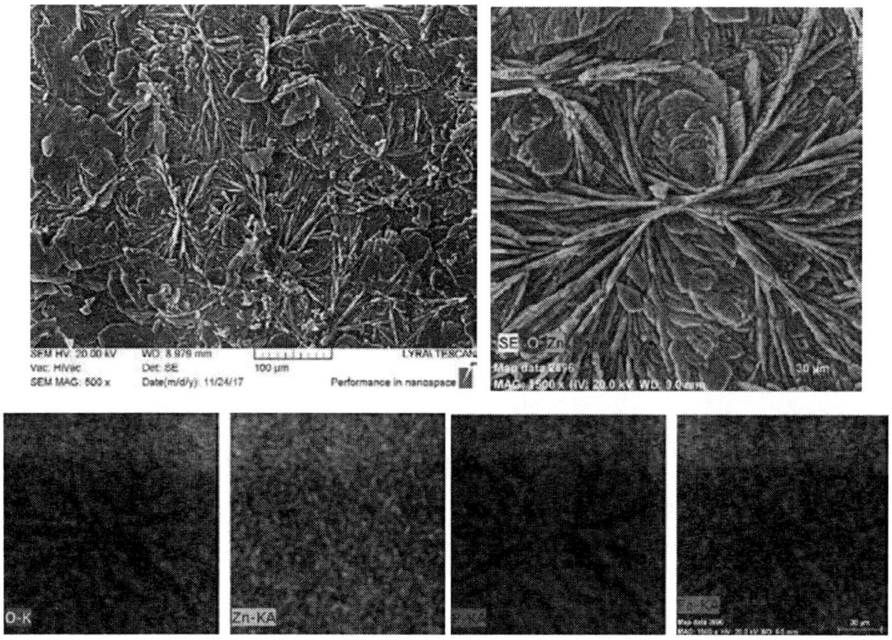

Figure 5. SEM – microphotograph and distribution of the elements in the coating obtained in ZnPh solution.

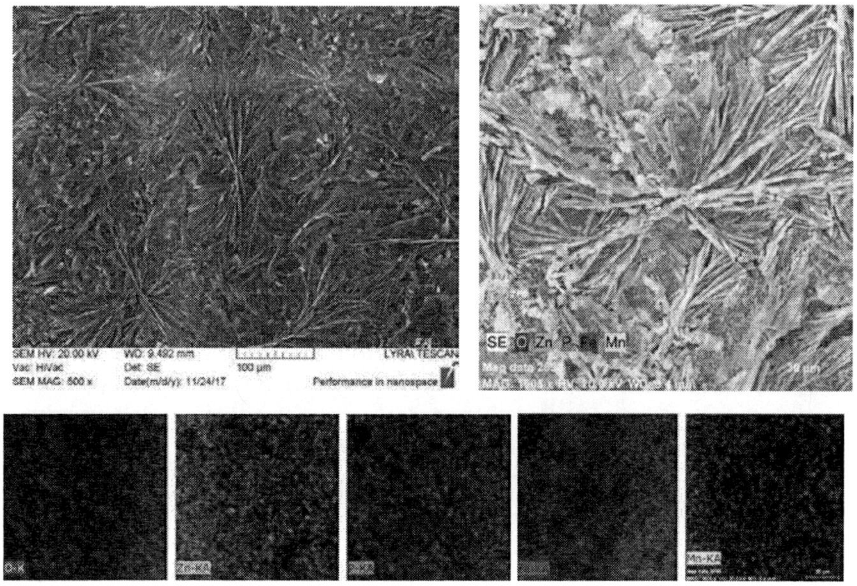

Figure 6. SEM – microphotograph and distribution of the elements in the coating obtained in MnPh solution.

The surfaces in different colours illustrate the distribution of the elements containing in the coating.

The EDX-analysis is performed on specimens treated in phosphate solutions: 20 % wt.; 80°C and 15 min. The basic elements containing in the coatings determined by means of this analysis are O, Zn, P, Fe and a quantity of Mn in the coating produced in MnPh solution. The results of EDX-analysis are presented in Table 3.

3.6. X-ray structural method

The XRD-spectra of the crystalline coatings grown in the ZnPh and MnPh phosphating solutions are shown in Figure 7.

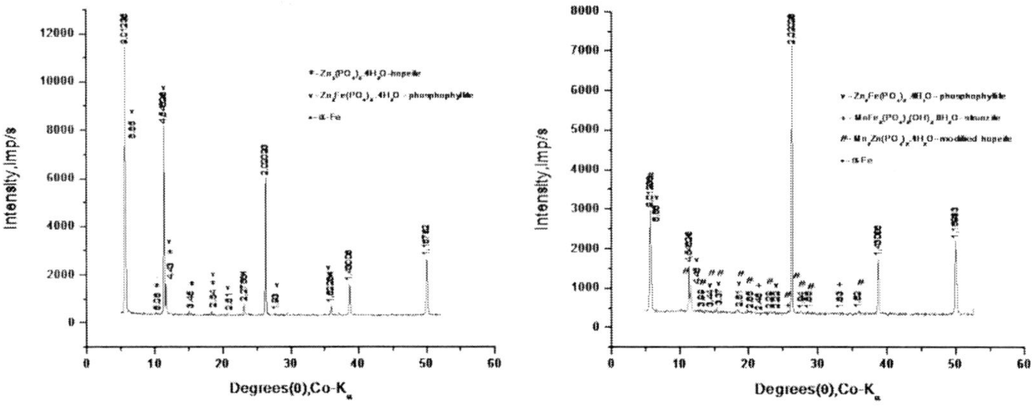

(a) (b)

Figure 7. XRD-spectra of the crystalline phosphate coatings grown in: (a) ZnPh; (b) MnPh.

The XRD analysis principally indicates presence of hopeite; phosphophyllite, quasihopeite, strunzite and mixtures of them in the coatings obtain in two phosphating solutions.

4. Conclusions

With the help of chemical, electrochemical and physical-analytical methods the processes of heavy crystal phosphating of carbon steels have been investigated. The properties of the coatings produced in ZnPh and MnPh solutions were determinate.

The addition of manganese phosphates to the ZnPh solution leads to increase the conductivity, decrease the pH value and to increase the value of it's free acidity. With increase of the solution concentrations in the range of 5 ÷ 20 % vol. for both phosphating concentrates the mass/thickness of the phosphate films increases. The results show also, that the thickness of the coatings obtained in ZnPh solutions is significantly higher.

The electrode potential during the formation of the coatings was more positive when in the phosphating solution there're manganese phosphates. In the media of 0.6 M NaCl electrode potential of the coating produced in MnPh solution is again more positive than the other. The potentiodynamic polarization relationships of the phosphate coatings, the more positive potentials and the low values of the coated samples show better corrosion resistance vs. uncoated steel surface.

Surfaces of the coatings are homogeneous with dense fine film crystals of dark grey and black colour. The crystals in ZnPh coating are bigger, with comparison to these in MnPh coating.

The XRD analysis principally indicates presence of hopeite [$Zn_3(PO_4)_2.4H_2O$] and phosphophyllite [$Zn_2Fe(PO_4)_2.4H_2O$] in the coating obtain in the ZnPh solution. And in the coating produced in the other MnPh solution there is a presence of quasihopeite [$Mn_2Zn(PO_4)_2.4H_2O$], strunzite [$MnFe_2(PO_4)_2(OH)_2.6H_2O$] and phosphophyllite [$Zn_2Fe(PO_4)_2.4H_2O$].

5. References

[1] Rausch D 1990 *The Phosphating of Metals*, Finishing Publications Ltd, Teddington, Middlesex, England

[2] Freeman D B 1986 *Phosphating and Metal Pretreatment*, Woodhead-Foulkner, Cambridge

[3] Lorin G 1974 *Phosphating of metals*, Finishing Publications Ltd, Middlesex

IOP Conf. Series: Materials Science and Engineering **374** (2018) 012033 doi:10.1088/1757-899X/374/1/012033

[4] Sankara T, Narayanan S N 2005 *Rev.Adv.Mater.Sci.* **9** 130-177
[5] Tamilselvi M, Kamaraj P, Arthanareeswari M, Devikala S and Arockia Selvi J 2015 *Inter. J. of Adv. Chem. Sci. and Appl. (IJACSA)* 26-41

Acknowledgments
This study has been financially supported by the Operational Programme "Science and education for smart growth" 2014-2020 of the European Union cofounded by the European Social Fund through the project BG05M2OP001-2.009-0015 "Support for the development of capacity of doctoral students and young researchers in the field of engineering, natural and mathematical sciences.

IOP Conf. Series: Materials Science and Engineering **374** (2018) 012034 doi:10.1088/1757-899X/374/1/012034

Technical Parameters Modeling of a Gas Probe Foaming Using an Active Experimental Type Research

A M Tîtu[1*], A V Sandu[2], A B Pop[3], C Ceocea[4] and S Tîtu[5]
[1] Lucian Blaga University of Sibiu, 10, Victoriei Street, Sibiu, România
[2] Gheorghe Asachi Technical University, 67, Dimitrie Manjeron, Iasi, România
[3] SC TEHNOCAD SA, 72, Vasile Alecsandri Street, Baia Mare, România
[4] SIF MOLDOVA SA, 94C, Pictor Aman Street, Bacău, România
[5] The Oncology Institute "Prof. Dr. Ion Chiricuță", 34-36, Republicii Street, Cluj Napoca, România

E-mail: mihail.titu@ulbsibiu.ro

Abstract. The present paper deals with a current and complex topic, namely - a technical problem solving regarding the modeling and then optimization of some technical parameters related to the natural gas extraction process. The study subject is to optimize the gas probe sputtering using experimental research methods and data processing by regular probe intervention with different sputtering agents. This procedure makes that the hydrostatic pressure to be reduced by the foam formation from the water deposit and the scrubbing agent which can be removed from the surface by the produced gas flow. The probe production data was analyzed and the so-called candidate for the research itself emerged. This is an extremely complex study and it was carried out on the field works, finding that due to the severe gas field depletion the wells flow decreases and the start of their loading with deposit water, was registered. It was required the regular wells foaming, to optimize the daily production flow and the disposal of the wellbore accumulated water. In order to analyze the process of natural gas production, the factorial experiment and other methods were used. The reason of this choice is that the method can offer very good research results with a small number of experimental data. Finally, through this study the extraction process problems were identified by analyzing and optimizing the technical parameters, which led to a quality improvement of the extraction process.

1. Introduction

Natural gas has been known since antiquity, both in the form of unspoiled fires and in the form of near-surface sprays. These were used by the Chinese with 2000 years of heating and were transported by bamboo strains. Natural gas entered the energy balance very late in Romania. The same has happened in the case of their use as raw material in the chemical industry. These have a great importance also for the caloric power they have up to 13,500 kcal/m^3. The methane gas discovery in our country was made in 1909 with a helping probe in the Sarmasel town. This was prospecting for potassium salts. Indications of the presence of natural gas existed before their discovery through the manifestations of "muddy volcanoes" or "live fires" located in different areas of Romania.

The first chapter involves the theoretical considerations regarding experimental research and data processing. Statistical processing of experimental data is the object of a specific branch of mathematics and it is called mathematical statistics. This is related to the probability theory. The second chapter refers to the active experiment - a beneficial solution for experimental testing and data processing. The factorial experiment, was choose to carry out this research because factorial experimentation can be achieved most easily by addressing and solving problems that appear relatively simple, but they are intuitive [1, 2]. In the third chapter introducing, the composition and importance of natural gas, was presented. I also have been highlighted their importance and made a short history of natural gas in Romania [3]. Natural gas is a complex mixture of saturated hydrocarbons (mainly methane and different amounts of ethane, propane, butane or other higher molecular weight hydrocarbons), small quantities of hydrocarbon-free gases (hydrogen, sulphide, carbon dioxide) and even gases neutral ones such as nitrogen and helium. For the exploitation of natural gas fields, a complex surface infrastructure is needed and it develops with the change of the number of productive wells, the dynamics of the pressure of the wells and the pressure from the collection system to which the deposit is granted. The natural gas extraction process brings to the surface a mixture of products in gaseous, liquid and solid form, which requires producers to condition their natural gas before handing them over to the distributor, consumer or carrier.

Generally formed from gaseous hydrocarbons in which carbon has an important part, with 80-99% of its composition, natural gas, as it is known, especially under the name of "methane gas", is found to accumulate in large quantities in the bark of the globe, or in most cases in the interplay with oil deposits [4]. Methane is a first-rate fuel, a renewable energy source, a material already considered in the third decade of the century - ideal for hygienic lighting and heating, easy to transport through pipelines, uninfluenced by external weather, exploited by its own its tension in the reservoir, and as a raw material of overwhelming importance for the chemical industry, that is methane gas, or the natural gas encountered in considerable quantities in the basement of Transylvania [5].

2. The factorial experiment method

Factorial experiments show their effectiveness where they want to study the influence of two or more factors on an objective function. Identifying the main concepts and notions with which the factorial experiment is currently practiced can be easily achieved by addressing and solving relatively simple but intuitive problems [6]. The optimal strategy provided by the EFC process involves the use of a minimal number of attempts, which is done after an optimal strategy, and the regression polynomial corresponding to this type of experiment provides information on the direction to be optimized [7]. In the first step, modeling with the help of factorial experiments refers to the correct and complete definition of the problem to be solved, the specification of the interesting objective function for the researcher regarding the description of the object, the phenomenon or the process to be analyzed. In the second important step in the application of factorial experimentation is the identification of the influence factors of each objective function of the ones to be analyzed [8].

3. The factorial experiment method-performing the experiment, processing and data optimizing

Identifying the main concepts and notions with which the factorial experiment is currently practiced can be easily achieved by addressing and solving relatively simple but intuitive problems [7]. In 1918 a pipeline was built between the Bazna and Medias localities that transformed the locality of Târnava Mare into a powerful industrial center. Methane gas is one of the most important factors in the economic development of Mediaşul in the period between the two world wars. The need to modernize and streamline the oil and gas industry in Romania has led to the conclusion of partnerships with world-recognized companies in the field. Romgaz has formed an association with the largest service company in the field, Schlumberger, for the optimization of the Laslau Mare gas field in Mureş County. Schlumberger Limited is the world leader in industrial and gas industry services, providing technology, solutions and integrated project management that optimize the performance and exploitation of customer reserves in the field. Laslau Mare gas field has been explored since 1960. 88 exploration and exploitation drills have been excavated. The field includes 54 probes, out of which 48 assets. The probes exploit productive horizons grouped into 4 operating packages. These are referred to as the deadlines of each productive layer. These depths range from 1,800 m to 3,000 m. Each

IOP Conf. Series: Materials Science and Engineering **374** (2018) 012034 doi:10.1088/1757-899X/374/1/012034

mining package has characteristics of nearby reservoir formations. Between packages, these features are differentiated but not radically. Active probes can also be divided into exploration wells and exploration wells that have been converted to exploitation [9].

3.1. Object definition and measurement mode

After more than 40 years of exploitation, the Laslău Mare reservoir has reached an advanced degree of depletion, current flows are maintained only with regular or exceptional interventions. Regular activities include:

- Spumogenic intervention;
- Measurement of probe parameters - Tube pressure, column pressure, daily flow and deposit water production;
- Gas drying and compression in order to optimize for transportation in the gas-gating system and to consumers;
- Separating, measuring and removing deposit water from the gas stream.

The object of this study is to make regular interference with wells with various sputtering agents. This procedure makes hydrostatic pressure reduced by forming a foaming water and the scrubbing agent that can be removed from the surface of the gas flow produced. On the Laslau Mare gas field are used foaming agents of several types:

- Solid density solids that sink into water and form foam on the bottom of the well;
- Solid lightweight agents floating on the water foam on the surface of the liquid column;
- A sparkling liquid, injected into column wells and used in the case of mechanical restraints that do not allow the solid foaming agent to reach the liquid level in the well.

Next it is wanted to analyze the results of probe spraying using experimental and data processing methods. Depending on the results, the investigation to other sample specimens will be extended, so that in the end this process can be optimized according to the exploitation package and the type of foam agent for the entire Laslau Mare gas field. The result should be reflected in the maximization of gas production and discharge of deposit water wells.

3.2. The factors selection and their values to be tested

Consider an active deep-water well with a medium gas flow and medium water storage for a period of one year. Given the limitations of the program of analysis, the data will be taken every 5 calendar days. Tubing pressure, column, difference between column pressure and tube (indicates the degree of water load of the probe), the amount of daily sparkling will be considered as inputs or factors, and daily flow and daily water production will be considered as objective, resulting. The data submitted were previously verified and validated by a pre-selection process.

Table 1. General data.

Pt (bar)	Pc (bar)	Pc-Pt (bar)	Spum. (m^3)	Dt (m^3/h)	Apa (m^3)
9.9	12.8	2.9	0.01	9.03	20
8.6	12.3	3.7	0.01	9.64	40
7.4	11.5	4.1	0.01	9.67	0
6.5	11.5	5	0.01	9.99	0
6.9	11.3	4.4	4	10.67	250
6.8	10.9	4.1	1	10.16	50
6.4	10.9	4.5	4	9.4	240
6.6	10.9	4.3	1	10.51	100
7	11.6	4.6	0.01	10.23	0

IOP Conf. Series: Materials Science and Engineering **374** (2018) 012034 doi:10.1088/1757-899X/374/1/012034

Table 1. General data.

Pt (bar)	Pc (bar)	Pc-Pt (bar)	Spum. (m^3)	Dt (m^3/h)	Apa (m^3)
8.1	11.5	3.4	1	10.35	70
7.5	11.4	3.9	1	10.26	40
7.2	11.2	4	1	10.32	120
7	11.6	4.6	1	10.52	20
6.6	11.4	4.8	1	10.74	0
6.5	11	4.5	1	10.77	160
6.8	11.7	4.9	1	10.71	50
7.8	10.7	2.9	1	10.04	50
10	13.5	3.5	1	9.51	40
6.5	10.7	4.2	1	10.58	60
7.2	11.2	4	1	9.82	0
7.9	11.3	3.4	1	10	20
7.2	11.6	4.4	1	9.86	80
7.4	11	3.6	2	9.74	100
6.3	10.4	4.1	2	9.95	100
6.4	11.1	4.7	2	10.46	150
6.5	11.2	4.7	0.01	10.31	0
7.2	11.4	4.2	0.01	9.87	20
6.3	11.3	5	0.01	9.79	0
6.2	11.6	5.4	0.01	9.49	40
8.2	11.6	3.4	4	9.69	190
6.6	11	4.4	4	10.2	200
6.9	11.4	4.5	4	10.21	320
7	11.7	4.7	0.01	10.8	20
6.7	11.1	4.4	0.01	10.86	60
7.2	12.2	5	0.01	10.55	0
7.2	11.6	4.4	3	10.53	110
7.7	11.9	4.2	3	10.11	160
6.8	12.2	5.4	0.01	9.6	0
7.2	13.4	6.2	0.01	9.98	0
7.1	11.6	4.5	0.01	9.34	0
7	12.8	5.8	0.01	8.48	0
7.3	12.3	5	0.01	8.35	0
7.2	13	5.8	0.01	8.08	0
7.2	13.2	6	0.01	7.88	0
7.6	10.4	2.8	0.01	8.84	60

Table 1. General data.

Pt (bar)	Pc (bar)	Pc-Pt (bar)	Spum. (m^3)	Dt (m^3/h)	Apa (m^3)
7.8	10.6	2.8	1	8.57	70
7.6	10.5	2.9	1	8.55	30
7.8	10.4	2.6	1	8.26	60
7.5	10.1	2.6	1	8.15	70
8.8	11.2	2.4	1	7.87	50
8.4	11.1	2.7	1	8.02	50
8.2	10.8	2.6	1	7.96	90
7.4	12.7	5.3	1	7.67	100
7.6	12	4.4	1	7.78	40
6.7	10	3.3	1	8.53	20
7.2	9.9	2.7	1	8.51	60
8.1	11	2.9	3	7.94	170
7	9.9	2.9	3	8.28	170
7.9	11.2	3.3	3	7.65	170
8.5	11.8	3.3	3	11	170
8	11.2	3.2	3	11.57	180
8.4	11.3	2.9	3	11.44	170
7.7	14.3	6.6	3	11.64	100
8.2	12	3.8	3	10.88	150
8.8	11.9	3.1	3	11.16	100
8	11.9	3.9	3	10.92	140
8.2	12	3.8	3	10.61	90
8.1	11.1	3	3	10.3	120
8.5	11.6	3.1	3	10.02	120
8.3	11.3	3	3	9.69	130
8.8	12	3.2	3	9.87	80
8.4	11.3	2.9	3	9.31	80
7.5	9.7	2.2	3	9.79	80
7.4	10.2	2.8	3	9.41	100
10.4	12.1	1.7	3	8.65	70
7.9	10.7	2.8	3	8.02	100
8.6	10.9	2.3	3	8.71	90
7.8	10.9	3.1	3	8.73	90
8.2	11.2	3	3	8.58	70
8.5	11	2.5	3	8.42	80
8.5	11.3	2.8	3	8.52	100

Table 1. General data.

Pt (bar)	Pc (bar)	Pc-Pt (bar)	Spum. (m^3)	Dt (m^3/h)	Apa (m^3)
7.9	10.9	3	3	8.55	80
8.7	11.3	2.6	3	8.55	70
8.4	11.2	2.8	3	8.58	110

3.3. The experience plan selection

From the analysis of the wells production data, the candidate for this research was obtained. Due to the intensified depletion of the gas field, the flow of the wells and the start of their loading with deposit water were recorded. At that time, it was necessary to smooth the wells of the wells, to optimize the daily production flow and to eliminate the accumulated water in the wellbore. A research plan with 4 factors and 2 objective functions, with 84 levels, representing a set of data every 5 days, over a calendar year, is resulting. Given the existence of measured data, try to adapt the experimental plan to analyze it as they are, without imposing limits or thresholds of attention. The descriptive statistical indicators of the process are set in the table. It can be noted that all 84 cases are valid. For each parameter, the mean, median, or minimum and maximum values were generated, the variance and standard deviation were calculated, and skewness (karyotype) and kurtosis (skewness) were also calculated. The skewness value denotes an inclination to the left, except for the flow function, which is tilted to the right. The small, near-zero values have a distribution close to the median. The exception is the deposit water function, where it can be observed a very large standard deviation, due to the large range of values of this objective function. The kurtosis value has a more plastic distribution than the normal distribution, with greater disproportion around the mean.

Table 2. Descriptive statistical indicators of the process.

	Valid N	Mean	Median	Minimum	Maximum	Variance	Std. Dev.	Skewness	Kurtosis
PT	84	7.610	7.500	6.200	10.400	0.730	0.854	0.732	0.838
PC	84	11.409	11.300	9.700	14.300	0.678	0.823	0.813	1.620
PC-PT	84	3.798	3.750	1.700	6.600	1.079	1.038	0.477	-0.370
Spum	84	1.692	1.000	0.010	4.000	1.774	1.332	0.128	-1.484
Flow	84	9.554	9.765	7.650	11.640	1.105	1.05133	-0.138399	-1.054
Water	84	80.47619	70.00000	0.000000	320.0000	4380.493	66.18530	0.970108	1.22320

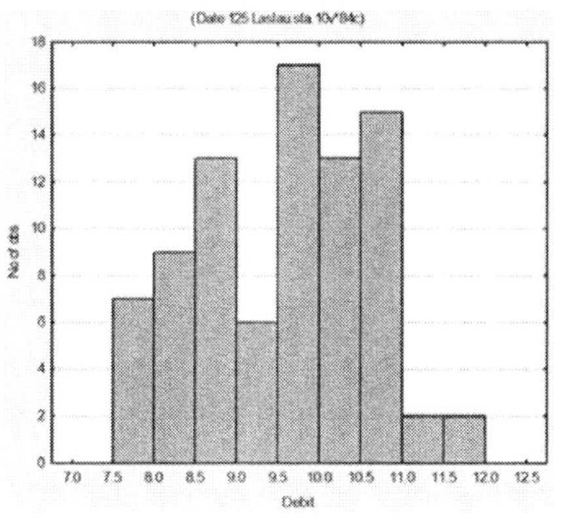

Figure 1. Gas Flow Histogram.

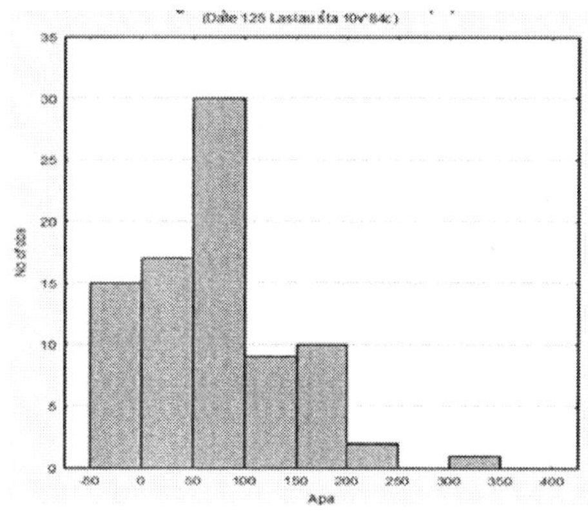

Figure 2. Water Production Histogram.

IOP Conf. Series: Materials Science and Engineering **374** (2018) 012034 doi:10.1088/1757-899X/374/1/012034

The first histogram, as far as the gas flow is concerned, can be seen in Figure 1 that it was measured in meters per hour, and after the graph the values are in the range of 7.5 to 12.0 but focus more on the intervals 9.5-10.0 and 10.5 -11.0. In the second histogram (Figure 2) the production of deposit water can be seen that was measured in m^3 and it is evident that the recorded values of the deposit of water production are concentrated in the range of -50 to 350, but most relevant samples contained values of water production breakdown in two ranges, namely 50-100 and 0-50.

The above histograms show the statistical variation of the parameters analyzed over the data unfolding interval. From the histogram analysis it can be seen the distribution of values, the minimum and maximum values, and the number of observations of each value.

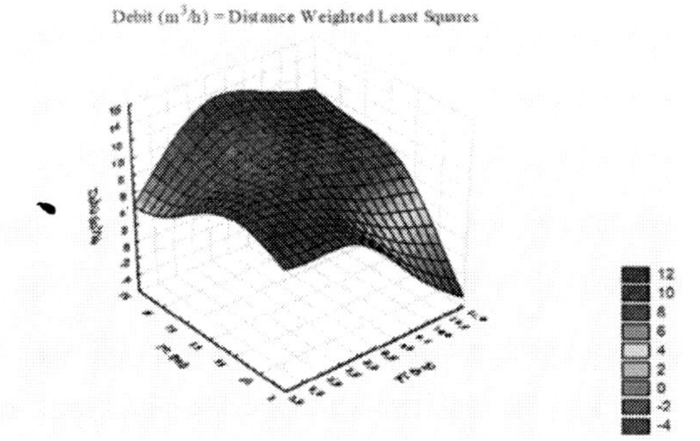

Figure 3. 3D Graph Flow Function vs. Factors Pressure Tubing and Pressure Column.

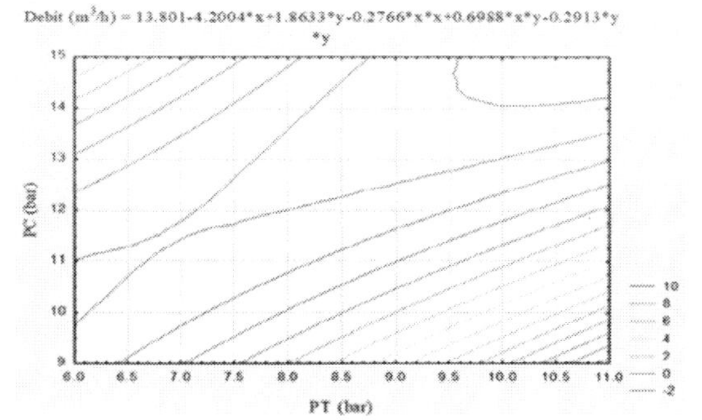

Figure 4. Section Flow Function vs. Factors Pressure Tubing and Pressure Column.

In Figure 3 is graphically represented the function of the flow target being influenced by the pressure and the column pressure. The maximum flow rate in this case is 12 and the minimum value is -4 depending on the tube pressure and the column pressure. The highest recorded flow rate corresponds to 15 bar and 8.5 bar, respectively. It can be seen in Figure 4 maximizing the gas flow at column pressures above 10 bar and tube pressures up to 7 bar. The section presented brings added value and a relaxation of the 3D graphics rendering.

In figure 5 is graphically represented the function of the flow objective being influenced by the pressure and the pressure difference. The maximum flow rate is 10 and the minimum value of 6 is based on the pressure and refinement difference. The highest flow record is 4.5 m^3 and 5 bar differential pressure. We notice increasing gas flow to a differential pressure of between 4 and 5 bar. It is also noted that a plateau area is formed at foams of between 1 and 4.5 units. Under these conditions, it is recommended to use up to 2 units of foam / day to avoid creating other mechanical problems due to over-spraying.

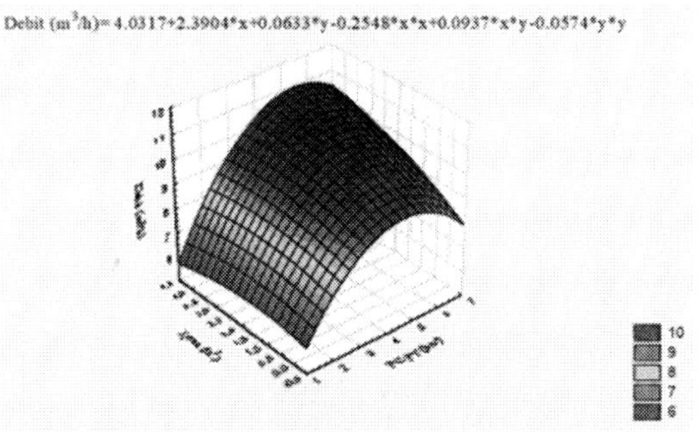

Figure 5. 3D Graph Flow Function vs. Differential Pressure and Sparking Factors.

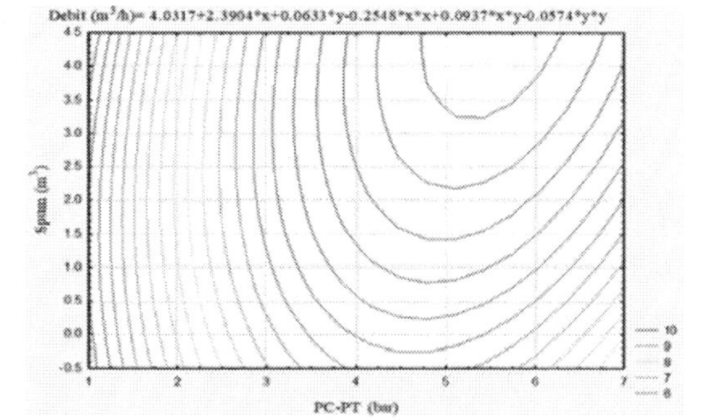

Figure 6. Section Function Flow Rate vs. Difference of Pressure and Sparkling.

3.4. The results analysis and the results configuration defining

Considering all the data presented, continuous spraying and large amounts of foam agent would give the best results. However, the need not to immerse the probe in spumogenic substances and the variation in the pressure of the tube caused by the whole field pressure leads to a cautious approach to this conclusion. Continuous monitoring of parameters and prompt response to accumulation of water in the wellbore, as well as repeating the analysis will lead to continuous improvement of the process. The analysis should also be extended to several probes with the same characteristics to confirm or invalidate the conclusion of the particular analysis of the probe presented.

4. Conclusions

Natural gas fuels come from free natural gas fields, they do not contain other heavy hydrocarbons in liquid state. Condensed gas deposits contain, in addition to natural gas and light liquid hydrocarbons, separate in the extraction process.

The exploitation of the deposits is done by means of probes. The gas probes are very similar to the oil wells, the difference being that at the gas probes, the columns inserted are cemented to the surface.

In order to exploit and monitor under optimal conditions the components of the surface installations related to the production wells within the gas-generating structures, they are grouped spatially in number, varying in particular the distance between them, forming so-called groups of probes.

The analysis of the studied process was based on data from Romgaz, which formed an association with the largest service company in the field, Schlumberger, for optimizing and improving the Laslau Mare gas field in Mureş County.

The subject of this study is regular interference with probes with different sputtering agents. This procedure causes the hydrostatic pressure to be reduced by the formation of foam from the deposit water and the scrubbing agent which can be eliminated at the surface by the gas flow produced.

The probe sputtering results using experimental research methods and data processing, were analyzed. Depending on the results the investigation to other representative wells will be extended so that in the end this process can be optimized according to the exploitation package and the spumogen type of the entire Laslau Mare gas field. The result should be reflected in the maximization of gas production and the discharge of deposit water wells.

Tubing pressure, column, difference between column pressure and tubing (indicates the degree of water load of the probe), the amount of daily frothing will be considered input parameters or factors, and the daily flow and daily water production will be considered objective, resulting functions. The data submitted were previously verified and validated by a pre-selection process.

Following the realization of their graphs and their analysis, it was found that continuous spraying with large amounts of foam agent would yield the best results. However, the need not to immerse the probe in spumogenic substances and the variation in the pressure of the tube caused by the whole field pressure leads to a cautious approach to this conclusion. Continuous monitoring of parameters and prompt response to accumulation of water in the wellbore, as well as repeating the analysis will lead to continuous improvement of the process. The analysis should also be extended to several probes with the same characteristics to confirm or invalidate the conclusion of the particular analysis of the probe presented.

5. References

[1] Montgomery D C 2013 *Design and Analysis of Experiments* (8th ed.), Wiley
[2] Box G, E. Hunter W G and Hunter J S 2005 *Statistics for Experimenters: Design, Innovation, and Discovery* (2nd ed.). Wiley. ISBN 0-471-71813-0
[3] Foidaş I 2014 *Testarea şi producerea zăcămintelor de gaze naturale* Editura Universităţii Lucian Blaga din Sibiu 220
[4] Basgan I 1940 *Petrolul şi gazele naturale în România* Bucureşti 81
[5] Laţiu V 1927 Gazul metan din bazinul Transilvaniei în: *Industria şi bogăţiile naturale din Ardeal si Banat* Cluj 17

[6] Freedman A, Pisani R and Purves R 2007 *Statistics* 4th edition. W.W. Norton & Company ISBN 978-0-393-92972-0
[7] Țîțu A M, Oprean C and Boroiu A 2011 *Cercetare experimentală aplicată în creșterea calității produselor și serviciilor* Editura AGIR București 410
[8] Țîțu A M and Oprean C 2006 *Cercetare experimentală și prelucrarea datelor-Partea I* Editura Universității Lucian Blaga din Sibiu 212
[9] Giura L 1998 *Contribuții la istoria gazului metan din România* Editura Universității Lucian Blaga din Sibiu 107

IOP Conf. Series: Materials Science and Engineering **374** (2018) 012035 doi:10.1088/1757-899X/374/1/012035

Parameters Affecting the Mechanical Properties of Fly Ash-Based Geopolymer Binders – Experimental Results

A Lăzărescu[1,2], H Szilagyi[1], A Ioani[2] and C Baeră[3,*]
[1] N.I.R.D. URBAN-INCERC, Cluj-Napoca Branch, Romania
[2] Technical University of Cluj-Napoca, Faculty of Civil Engineering, Romania
[3] N.I.R.D. URBAN-INCERC, Timişoara Branch, Romania

E-mail: adrian.lazarescu@incerc-cluj.ro

Abstract. As the demand for concrete and the needs to satisfy development of infrastructure facilities increase, it is essential to find alternatives to create environment-friendly concrete. The particular procedure of alkaline activation of fly ash - in which ash resulting from a power plant is combined with a specific alkaline activator in order to create a solid material, then dried at a certain temperature - opened new opportunities for this new material to get attention worldwide. In order to obtain a material with similar properties of ordinary Portland cement concrete and to obtain desirable compressive strengths, the parameters that affect this type of binders should be fully understood. The aim of this paper is to study the main parameters affecting the mechanical strength of the fly ash-based geopolymer paste and their interactions. Parameters such as molarity of sodium hydroxide (from 8M to 12M) and alkaline activators ratio (from 0,5 to 2,5) were analysed to observe how they affect the mechanical properties of the geopolymer paste. Experimental results show that the compressive strength of the fly ash-based geopolymer paste produced using Romanian local raw materials increases with the increase of the concentration of sodium hydroxide and higher ratios of $Na_2SiO_3/NaOH$ solution.

1. Introduction

Large amounts of fly ash, resulting from burning coal power-plants, in Romania, can create new opportunities to use this type of waste as a total substitute for Portland cement in the production of new types of materials. When partially used in the production of concrete, mixed with Portland cement, fly ash helps obtaining concrete with improved performances. The concept of alkali-activated materials (AAM) has been known since the early 20th century, but more recently, since the 1990s, research in the field of AAM and geopolymer materials (which are a subclass of the AAM) have experienced an impressive progress. The vast majority of the research is focused on the microstructure of these materials and much less on establishing or verifying their implementation process, service life, sustainability or advanced engineering properties. This type of materials are seen more as niche applications and not necessarily as ordinary Portland concrete (OPC) substitutes [1].

The substantial progress made in the field of alkali-activated materials and geopolymer materials is particularly highlighted by the growing interest in the environmental problems, particularly regarding the carbon dioxide emissions into the atmosphere. More and more countries are aligning themselves to the international standards on environmental issues, this creating an important niche for civil engineering research and beyond.

A geopolymer binder is generally formed by the interaction of an alkaline solution (activating solution or activator) with a highly reactive aluminosilicate "powder" [2]. In the usual practice, the production of geopolymer materials is based on a mixture between fly ash and the alkaline-activator. When in contact with the alkaline liquid, the fly ash partially dissolves and its components are rearranged, forming an alumino-silicate gel, which hardens in the presence of heat forming the geopolymer matrix (or binder). By adding sand and aggregates it is possible to obtain a concrete with similar or even better properties than conventional Portland cement concrete.

The microstructure of the geopolymer binder depends mainly on the chemical composition of the constituent materials, their specific ratios in the proposed mixtures and the thermal curing of the obtained material [3]. A suitable heat curing is essential for geopolymeric materials, as without it the obtained material develops a porous microstructure and, as consequence, low mechanical performances [4]. Addition of rich silicon-containing materials into the alkaline liquid can lead to the formation of a homogeneous and uniform microstructure of the geopolymer [5], which can also determine improved mechanical performances.

Mechanical behavior represents basic property of a material used in the construction industry, making it suitable for specific applications. Compressive strength is crucial, especially for a concrete type material. The early 1950 geopolymeric mixes proved good compressive strength results, reliable workability and durability even superior to those of OPC concrete. However, the compression behavior of the geopolymer materials strongly varies depending on the raw materials used, namely their chemical properties, and also on the used production methods [7]. In order to obtain good compressive strength properties, one must take into account the type and molar ratios of the oxides in the source material rich in Al-Si, the type, the pH and the molar properties of the alkaline solution and also the solubility of the source material in the activating alkaline-solution [8, 9].

Regarding the importance of the dissolution of the Al-Si-rich materials in the alkaline solution and the actual geopolymerisation process, it was proved that the mechanical properties of the material are directly affected by the reorganization of the source material at the time of the chemical reaction. Flexural strength, compressive strength and apparent density of the material increase with the increase in the NaOH solution concentration [10], and the amorphous content of the product has also increased [11].

The production of geopolymer paste for the current study this research is carried out using specific, state-of-the-art methods and it is based on the reaction between the silicon and aluminum in the fly ash (FA) and the alkaline liquid (sodium silicate solution and sodium hydroxide solution). In this paper, various factors which influence the compressive strength of the alkali-activated fly ash-based geopolymer paste have been studied and the results are discussed by means of graphical representation corresponding to the compressive strength of the batches. Therefore, the actual goal was developing various geopolymer paste mixes, with 100% OPC replacement by processed FA resulted from Romanian local sources and to evaluate the influence of different parameters affecting the mechanical properties of the material. Hence the effect of various parameters affecting the compressive strength of the alkali-activated fly ash-based geopolymer paste (i.e. ratio of sodium silicate solution to sodium hydroxide solution, concentration of sodium hydroxide solution, curing temperature) should be carefully controlled for optimizing and improving the overall performances of this materials developed by using Romanian local raw materials.

2. Materials and methods

Similar to the OPC traditional concrete mixes, the basic constituents of the fly ash-based geopolymer concrete are the cement paste (in this case geopolymer paste) and the aggregates. In case of the geopolymer concrete, the geopolymer paste (or binder) results from combining the fly ash and the alkaline liquid (a mixture of sodium hydroxide solution and sodium silicate solution).

2.1. Fly ash (FA)

Ashes from power-plants (fly ash, FA) can be defined as fine powders of spherical vitreous particles, derived from the combustion of pulverized coal [12, 13]. According to the classification of the

American Society for Testing and Materials (ASTM) [14], there are identified various FA types that can be used in the production of fly ash-based geopolymer binders. The present research used a typical Class F FA powder with good pozzolanic properties and characterized by $SiO_2+Al_2O_3+Fe_2O_3 > 70\%$, $SO_3 < 5\%$ and lost on ignition (L.O.I.) < 6%.

The mineral composition of the FA sample used in the production of the alkali-activated fly ash-based geopolymer paste with Romanian local raw materials was determined by X-Ray fluorescence analysis (XRF) and it is presented bellow (Table 1).

Table 1. XRF analysis data for the fly ash batch.

Oxides	%
SiO_2	53.61
Al_2O_3	26.16
Fe_2O_3	7.58
CaO	2.42
MgO	1.49
SO_3	0.26
Na_2O	0.59
K_2O	2.60
P_2O_5	0.12
TiO_2	1.04
Mn_2O_3	0.08
L.O.I.	3.57

2.2. Alkaline liquid

The used alkaline solution is a mix between sodium silicate solution (Na_2SiO_3 solution) and sodium hydroxide solution (NaOH solution). The sodium silicate solution was purchased from the market and its chemical oxide composition is SiO_2=30%, Na_2O=14% and H_2O=56%. The sodium hydroxide solution is prepared dissolving NaOH, 98% purity, flakes in water to obtain the desired concentration of the final solution. The NaOH solution concentration for the mixes in this study is set to 8M, 10M and 12M. To obtain 8M NaOH solution, 320g of NaOH flakes were dissolved in one liter of solution (40g x 8M = 320g NaOH/ liter of solution, where 40g is the molecular weight of NaOH). It was measured that for one kilogram of NaOH solution, the mass of the NaOH was 262g (1 kg of NaOH solution, 8M = 26,2% NaOH and 73,8% water). Similarly, the mass of NaOH flakes per kg of solution for the other concentrations were: for 10M: 314 grams (31,4% NaOH and 68,6% water) and for 12M: 361 grams (36,1% NaOH and 63,9% water). It is very important to note that the mass of the NaOH solids is only a fraction of the mass of the NaOH solution, water being the major component that affects the water to geopolymer solids influence.

2.3. Mix design, moulding and curing treatment

In order to study the effect of different parameters affect the mechanical properties of fly ash-based geopolymer paste using Romanian local raw materials, the alkaline activator/fly ash ratio was set to 0,5 because, based on the preliminary results obtained on geopolymer paste, it proved good workability, [15]. The sodium silicate solution to sodium hydroxide solution ratio (Na_2SiO_3/NaOH) was varied between 0.5 and 2.5, according to the literature specifications, in order to study the effect of the alkaline activator ratio on the compressive strength of the alkali-activated fly ash-based geopolymer paste.

The effect of the concentration of the sodium hydroxide solution (NaOH solution) was studied by increasing the molar concentration of the solution, from 8M to 12M, for all the proposed Na_2SiO_3/NaOH solution ratio. For all the studied mixtures, their molar constituents ratios were calculated to obtain the water to geopolymer solids ratio and to evaluate this effect on the paste fresh properties. Details of the alkali-activated fly ash-based geopolymer paste are presented in Table 2.

The preparation of the alkaline activator by combining the sodium silicate solution with the sodium hydroxide solution took place 24 hours prior mixing. The fly ash and the activator were mixed together for 5 minutes, until a homogeneous paste was obtained. The alkali-activated geopolymer paste was then placed in 40mm x 40mm x 160mm moulds and exposed for 24 hours to 70°C heat curing. The effect of heat curing temperature over geopolymer pastes. Namely compressive strength variation, was studied by varying from 60°C to 70°C, the heat regime curing. During the heat curing a glass film was applied on top of the molds to prevent the water quick release from the pastes. After samples removing from the molds, 24 h after casting, the prismatic specimens were kept in the climatic chamber, at (20±1)°C and (60±5)% humidity, until the 7 days compressive strength tests.

Table 2. Mixture details of the geopolymer paste.

Alkaline liquid / Fly ash	Na_2SiO_3 / NaOH ratio	NaOH solution concentration
0.5	0.5	8M 10M 12M
	1.0	
	1.5	
	2.0	
	2.5	

2.4. Testing methods
The compressive strength of the studied mixtures, for all the alkali-activated fly ash-based geopolymer paste prepared using Romanian local raw materials, was determined using three specimens for each mixing batch and the corresponding average value was considered as relevant. Tests were conducted according to SR EN 196-1:2006 standard method for evaluation of mechanical performances of OPC paste and standard type mortar. The testing was performed at the age of 7 days.

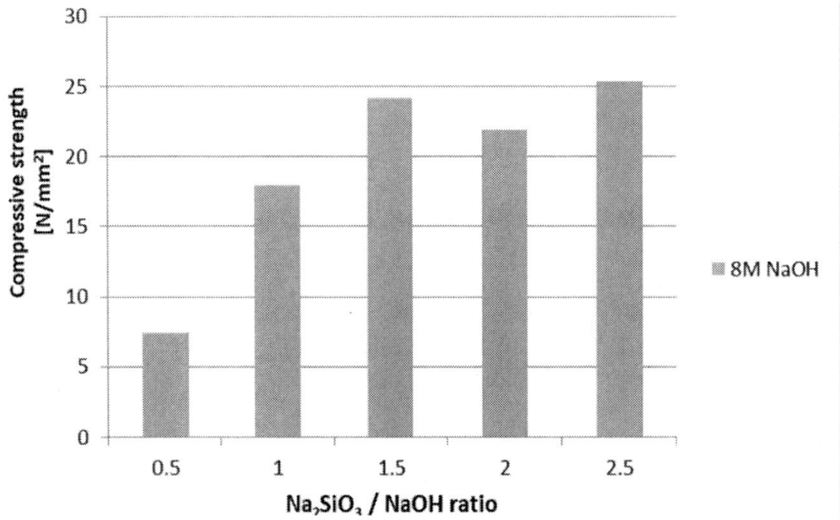

Figure 1. Influence of the Na_2SiO_3/NaOH ratio on the compressive strength (8M concentration NaOH solution).

3. Results and discussions

3.1. Influence of sodium silicate solution to sodium silicate solution ratio on the compressive strength
The effect of the sodium silicate solution to sodium hydroxide solution ratio effect was evaluated considering all mixes developed in the current experimental study. The results are graphically presented in Figure 1, Figure 2 and Figure 3. It has been noticed that, as the ratio value increases, the compressive strength of the alkali-activated fly ash-based geopolymer paste increases as well. The

increasing values of the compressive strength of the samples is considered to mainly depend on the chemical process governing the geopolymerization. Increasing this ratio leads to more SiO_2 species, therefore more Si-O-Si bonds are formed, creating a stronger material [16]. Some mixes presenting a decreased compressive strength may happen because the geopolymerisation rate of them was lower, therefore, the compressive strength of the batches was lower.

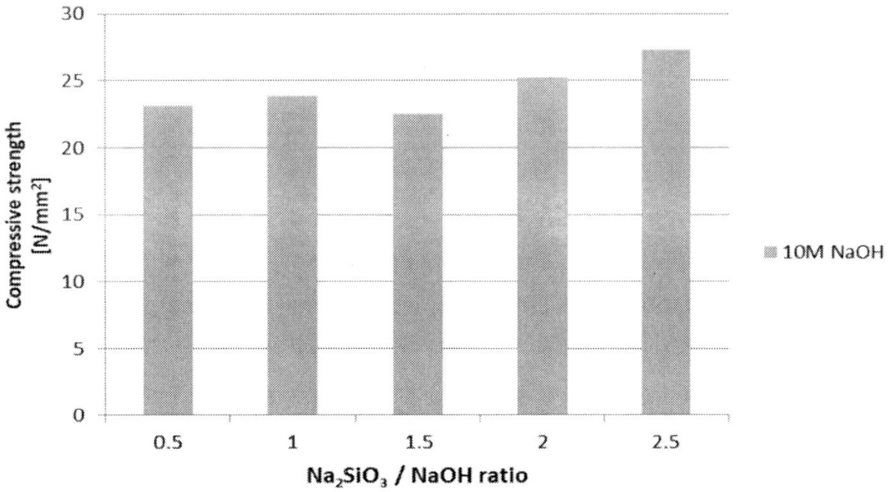

Figure 2. Influence of the Na_2SiO_3/NaOH ratio on the compressive strength (10M concentration NaOH solution).

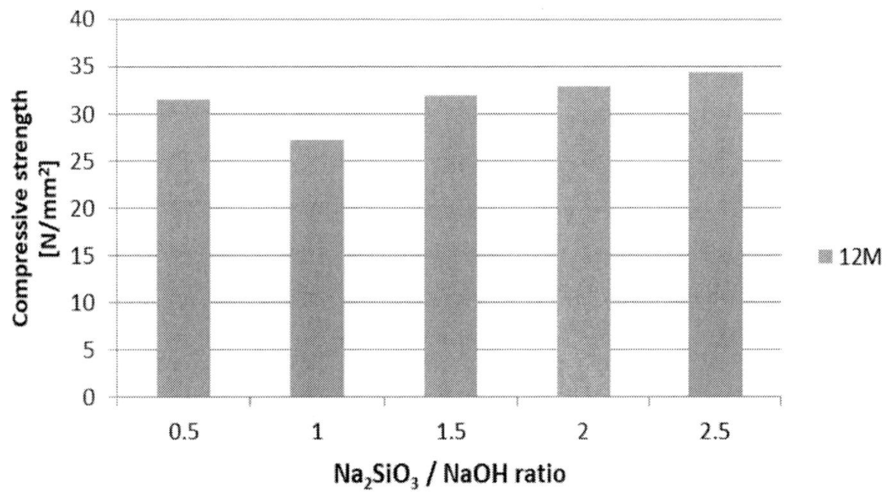

Figure 3. Influence of the Na_2SiO_3/NaOH ratio on the compressive strength (12M concentration NaOH solution).

3.2. Effect of sodium hydroxide solution on the compressive strength

The effect of the sodium hydroxide concentration on the alkali-activated fly ash-based geopolymer paste samples is analysed on all the developed mixtures and the testing results are presented in Figure 4. It has been noticed that higher concentration of the sodium hydroxide solution induces better compressive strengths, when the alkaline activator to fly ash and alkaline activator ratio is kept constant.

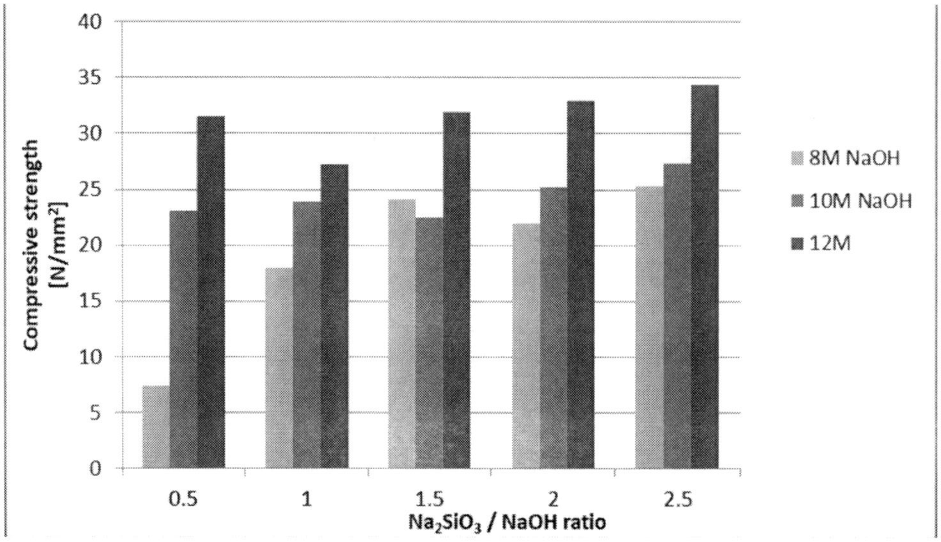

Figure 4. Influence of the NaOH solution concentration on the compressive strength for all the studied mixtures of fly ash-based geopolymer paste.

4. Conclusions

All the desirable properties of geopolymer paste developed by using fly ash combined with alkaline solution, mainly mechanical performances, can be enhanced only by carefully monitoring and controlling the relative quantities of used reactants and also by controlling the overall conditions where the geopolymerisation process takes place. The chemical composition of the used raw materials is one of the most important, influencing parameters of the the geopolymerization process, therefore, it is considered that before conducting any experimental study regarding geopolymer materials, these properties should be well evaluated and known.

When keeping the alkaline liquid to fly ash mixing ratio constant and varying the Na_2SiO_3 to NaOH solution ratio and also the NaOH solution concentration, it was possible to study the effect of the two main parameters on the compressive strength of the alkali-activated fly ash-based geopolimer binder. Results show that by increasing the sodium silicate solution content in the mixtures, a significant increase in the compressive strength of the materials was induced. This effect was considered to happen because the solution provides extra silicon (Si) species, which play an important role in the geopolymerization process, thus creating a much performed, well packed matrix of the material. Also, by increasing the concentration of the NaOH solution in the current study, increased compressive strength values were obtained, therefore, increasing the concentration of the solution from 8M to 12M increased the mechanical properties of the material.

As the "green" building materials started to become more desirable in comparison to traditional OPC, and they also became more and more present, both technologically and commercially, in the construction industry, a need for standard methods for material testing and general evaluation becomes more than necessary. The initial and long-term properties of these materials must be well understood, so they could be properly designed and controlled from the initial step at their best. Large scale acceptance of the geopolymer cement and concrete could reduce Portland cement demands and represent a green solution for worldwide maintaining concrete the most used building material.

The above results obtained on alkali-activated fly ash-based geopolymer paste using Romanian local source materials have shown not only the possibility of producing this type of material, but also the opportunities for further research in this topic. The aim of this research was both to produce alkali-activated fly ash-based geopolymer paste using Romanian local source materials and to study the influence of two main parameters on the compressive strength of the material.

IOP Conf. Series: Materials Science and Engineering **374** (2018) 012035 doi:10.1088/1757-899X/374/1/012035

5. References

[1] Provis J L, van Deventer Alkali J S J *Activated Material*, State-of-the-Art Report RILEM TC 224-AAM

[2] Duxon P, Fernandez-Jimenez A, Provis J L, Lukeym G C, Palomo A, van Deventer J S J 2007 *Journal of Materials Science* **42**(9) 2917-2933

[3] Hardjito D, Rangan B V 2005 Development and properties of low calcium fly ash-based geopolymer concrete *Curtin University of Technology, Research Report GC1,* Curtin University, Australia

[4] Davidovits J 1988 Geopolymers of the first generation: SILFACE-Process *Geopolymer '88, First European Conference on Soft Mineralogy* Compiegne, France, p. 49-67

[5] Duxon P, Provis J L, Lukey G C, Mallicoar S W, Kriven W M, van Deventer J S J 2005 *Colloids and Surfaces: A – Physicochemical and Engineering Aspects* **269**(1-3) 47-58

[6] Davidovits J 1999 Chemistry of geopolymeric systems, terminology *Geopolymer '99 International Conference, France 9-44*

[7] Mhidi B 2009 *Materials Technology* **24**(2) 79-87

[8] Xu H *Geopolymerisation of aluminosilicate minerals*, PhD. Thesis Department of Chemical Engineering, University of Melbourne, Australia

[9] Mustafa Al Bakri Abdullah A M, Kamarudin H, Binhussain M, Nizar K, Mastura W I W 2011 Mechanism and Chemical Reaction of Fly Ash Geopolymer Cement – A Review *Article in Asian Journal of Scientific Research*

[10] Wang H, Li H, Yan F 2005 *Colloids Surf. A: Physicochem. Eng. Aspects* **268** 1-6

[11] Luz Granizo M, Blanco-Varela M T, Martinez-Ramirez S 2007 *J. Mater. Sci* **42** 2934-2943

[12] EN 450-1 Fly ash for concrete. Part 1: Definition, specifications and conformity criteria

[13] EN 197-1 Cement. Part 1: Composition, specifications and conformity criteria for common cements

[14] Davidovits J 1994 Properties of geopolymer cements, *Alkaline Cements and Concretes,* Kiew, Ukraine, p. 131-149

[15] Lăzărescu A V, Szilagyi H, Baeră C, Ioani A 2017 The Effect of Alkaline Activator Ratio on the Compressive Strength of Fly Ash-Based Geopolymer Paste *International Conference on Innovative Research – ICIR EUROINVENT 2017, IOP Conf. Series: Materials Science and Engineering,* 209 012064

[16] Mustafa al Bakri Abdullah A M, Omar A K A, Kareem A, Myint S 2009 Study On The Effect Of Alkaline Activators Ratio In Preparation Of Fly Ash-Based Geopolymer *RAAM & ASMP 2009* Conference Paper

Experimental Characterization of Aluminum-Based Hybrid Composites Obtained Through Powder Metallurgy

D F Marcu[1], M Buzatu[1], V G Ghica[1], M I Petrescu[1], G Popescu[1], F Niculescu[1] and G Iacob[1,*]
[1] Faculty of Materials Science and Engineering, University POLITEHNICA of Bucharest, 313 Splaiul Independentei Blvd, 060042, Bucharest, Romania

E-mail: iacob_gh@yahoo.com

Abstract. The paper presents some experimental results concerning fabrication through powder metallurgy (P/M) of aluminum-based hybrid composites - Al/Al_2O_3/Gr. In order to understand the mechanisms that occur during the P/M processes of obtaining Al/Al_2O_3/Gr composite, we correlated the physical characteristics with their micro-structural characteristics. The characterization was performed using analysis techniques specific for P/M process, SEM-EDS and XRD analyses. Micro-structural characterization of the composites has revealed fairly uniform distribution this resulting in good properties of the final composite material.

1. Introduction

There are numerous researches [1-8] to obtain metal matrix composites and one type of hardening material (ceramic particles of Al_2O_3, SiC, Gr, TiC, B_4C, etc.). Regarding hybrid composite materials, the international literature contains a series of data to demonstrate the scientific applicability of composites with aluminum metallic matrix hardened with SiC, Al_2O_3 and nickel-coated graphite Al/SiC + Al_2O_3, Al/SiC + Gr, Al/Al_2O_3 + Gr or Al/SiC + Gr-Ni. The major advantage of hybrid composites lies in the possibility of modulating the properties and thus obtaining a wide range of materials, the use of which can be extended to almost all areas of technical activity [5, 7, 9, 10].

Aluminum matrix hybrid composites hardened with ceramic particles (Al_2O_3, SiC, Gr) have much better mechanical properties than unalloyed aluminum alloys and have for some time been used as tribological parts due to their high strength/density ratio, showing a significant improvement in tribological properties, including wear resistance, abrasion resistance and galling resistance [7, 9-14].

By varying the nature of the components and their volume fraction, the properties of the hybrid composites can be adjusted. Increasing the content of the hardening element generally leads to an appreciable increase in the durability of semi-finished products. Any of the matrix composite production techniques involves, in a first step, mixing the components (the metal matrix and the hardening particles) so as to obtain a uniform distribution of the material, a high resistance interface and a controlled microstructure, and in a second stage the forming of the mixed material by various technologies (casting, powder metallurgy, plastic deformation etc) [15-18].

Solid state techniques have several advantages, such as a low processing temperature, which results in a low interaction of matrix-hardener and therefore good mechanical properties can be obtained. Solidification defects (shrinkage, porosity, segregation) are avoided and, in general, a more uniform distribution of the hardener can be obtained. The most common technique in solid state is powder metallurgy. Powder metallurgy is an important branch of modern technology, the emergence of this technique being determined by the necessity of obtaining materials with special characteristics that can not be achieved by classical processes. It should be noted that at present, the automotive industry, the household appliance industry, electrical measuring and control instruments, sports equipment, television sets and even rockets can not be designed without the use of powder metallurgy [19-21].

One of the successful methods applied to obtain uniform and homogeneous powders is the method of mechanical alloying. This is a solid state processing technique that allows the synthesis of new alloys that cannot be obtained by any other technique. Mechanical alloying is a process of grinding with high velocities and kinetic energies that can be carried out in a dry or wet environment, and is used to manufacture a variety of advanced materials and alloys (over-saturated solid solutions, amorphous materials, intermetallic compounds and MMCs). Mechanical alloying is a simple and useful technique for achieving a homogeneous distribution of inert fine particles in a fine grain matrix [20-22]. The main purpose of this paper was to obtain hybrid composites with aluminum metal matrix using mechanical alloying followed by compaction and sintering processes. Alumina was chosen as hardening material and graphite was chosen to improve the tribological properties. The characterization of samples was performed using analysis techniques specific for P/M process, SEM-EDS and XRD analyses.

2. Materials and methods

Powders of aluminum, alumina, graphite and zinc stearate were homogenized and mechanically alloyed in a RETSCH PM 400 high energy planetary mill. The experiments were carried out in a dry environment using zinc stearate as a process control agent to avoid powder agglomeration as well as depositing it on the walls of the vials and on the balls used for milling. The speed of the mill was kept constant, with a rotational speed of 300 rpm. The grinding time was 4 hours. The size of the elemental powders from which the mechanical alloying process is started is of the order of the microns, the powders having a fairly high purity, as can be seen from Tables 1-4, following the grinding process resulted a very small granulations, as will be demonstrated (even the nanometric order), which ensures a uniform and homogeneous structure of the hybrid composites.

The compositions selected to investigate the effects of the amount of hardening elements in the $Al/Al_2O_3/Gr$ hybrid composites are shown in Table 1.

Table 1. Compositions selected for experiments.

Notation	Al (wt. %)	Al_2O_3 (wt. %)	Graphite (wt. %)
HyCo 1	89	10	1
HyCo 2	84	15	1
HyCo 3	79	20	1
HyCo 4	87	10	3
HyCo 5	82	15	3
HyCo 6	77	20	3

The mechanical alloying process started from raw materials in the form of powders of micron size and with high purity (Al - 44 μm and 99.5%, Al_2O_3 - 44-150 μm and 98.5%, Gr < 32 μm and 95%), their average size decreasing by almost half. The zinc stearate powder was 1% of the total mass of the powder. For the production of $Al/Al_2O_3/Gr$ hybrid composites the Al powder was used as a ductile matrix, Al_2O_3 powder was used as a hardener and the graphite powder as lubricant. The chemical compositions and specifications of powders according to the supplier's technical sheet are shown in Table 2 to 5.

Table 2. Chemical composition of aluminum powder (Al).

Element	wt.%
Fe	0.11
Si	0.04
Cu	0.01
Mn	< 0.01
Cr	< 0.01
Ni	< 0.01
Zn	< 0.01
Ti	< 0.02
Ga	< 0.01
Fe	0.11

Table 3. Chemical composition of alumina powder (Al_2O_3).

Element	wt.%
SiO_2	max. 0.015
Fe_2O_3	max. 0.020
Na_2O	max. 0.450
CaO	max. 0.050
P_2O_5	max. 0.0012
K_2O	max. 0.018
Li_2O	max. 0.017
ZnO	max. 0.0015
V_2O_5	max. 0.0025
TiO_2	max. 0.0025

Table 4. Chemical composition of graphite powder (Al).

Component	wt.%
C	95
Fe	max. 0.5
S	max. 0.1
Ash	max. 2

Table 5. The specifications of the raw material for the zinc stearate.

Component	wt.%
Zn content	10.3 – 11.3
Fat acids	max. 0.5
Humidity	max. 0.8
Density	1.095
Melting temperature	120 – 130

Secondary electron image analysis (SEI) was performed at a resolution of up to 2.98 μm, in a magnification range up to 100,000X, at a 30 kV acceleration voltage in high vacuum, with the use of the scanning electron microscope Quanta Inspect F50, with a field emission gun (FEG) with 1.2 nm resolution and an Energy Dispersive X-ray Spectrometer (EDXS). XRD data were obtained using a Panalytical X'PERT MPD X-ray diffractometer with a Cu K-a radiation source (λ = 1.5418 Å) in the range of 2θ = 10–90°.

The hardness of compressed samples from composite powders obtained by mechanical alloying and subsequently sintered was determined by the Leeb Rebound Hardness Test (LRHT), using the NAMIKON hardness tester.

3. Results and discussions

3.1. Technological properties of the Al/Al_2O_3/Gr hybrid composites

For the samples obtained after pressing (Atlas^{TM} 15T Manual Hydraulic Press from SPECAC) and sintering processes (was realized in a heating induction furnace, Balzers type, sintering temperature: 620°C; 60 minute exposure time) the technological characteristics (density, compactness, porosity, etc.) were determined. The physic-technological properties determined on the hybrid composites samples with different compositions, determined according to the specific powder metallurgy procedures, and are shown in Figures 1 to 3, the notation of the samples being presented in Table 1.

The densities of the hybrid composites after sintering show a significant increase compared to the densities after pressing. As can be seen from Figure 1, densities after sintering show a decreasing trend with increasing the Al_2O_3 content from 10 wt. % to 20 wt.% and maintaining a constant content of 1 wt. % Gr, and an oscillating tendency with increasing the Al_2O_3 content from 10 wt.% to 20 wt.% and maintaining a constant content of 3 wt.% Gr.

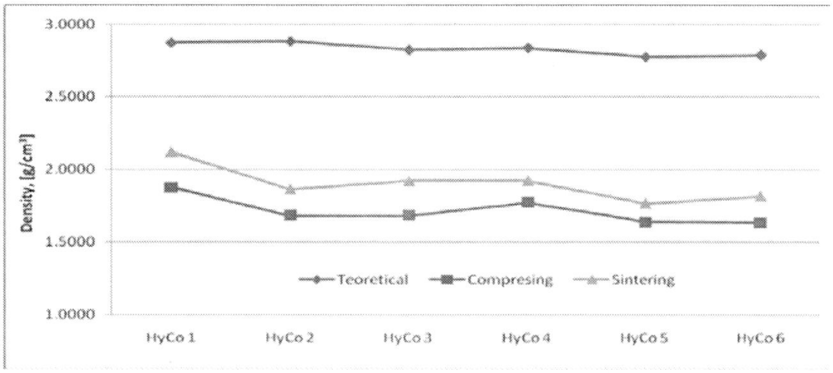

Figure 1. Density of the $Al/Al_2O_3/Gr$ hybrid composites.

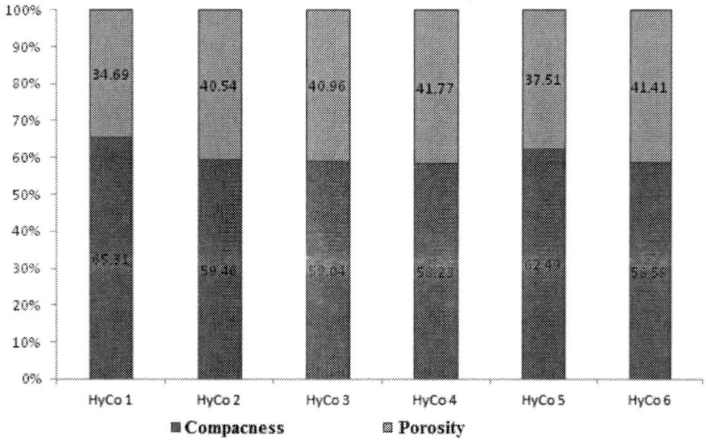

Figure 2. Compactness and porosity after pressing the $Al/Al_2O_3/Gr$ hybrid composites.

The compactness of the $Al/Al_2O_3/Gr$ after pressing process shows a decreasing trend with the increase of Al_2O_3 content. Regarding the variation of the values according to the graphite content we can note a slight decrease, except for HyCo 5. The maximum value corresponds to HyCo 1 and the minimum value corresponds to HyCo 4, both having a 10% Al_2O_3 content but differing graphite content. The porosity values of the tablets are complementary to the compactness values.

After the sintering process of the $Al/Al_2O_3/Gr$ tablets we can observe an improvement in values. Similar to the data presented above they shows a decreasing trend with the increase of Al_2O_3 content. Regarding the variation of the values according to the graphite content we can note a slight decrease, except for HyCo 6. The maximum value corresponds to HyCo 1 and the minimum value corresponds to HyCo 3. The porosity values of the tablets are complementary to the compactness values.

The degree of porosity of the samples is calculated according to the initial density of the tablets and increases with the increase of the reinforcement content of the hybrid composites. From the data presented above, it is observed that the degree of porosity after sintering of the samples is much lower than that of the pressed tablets. This reduction is due to the elimination of voids between the particles and the contraction after sintering.

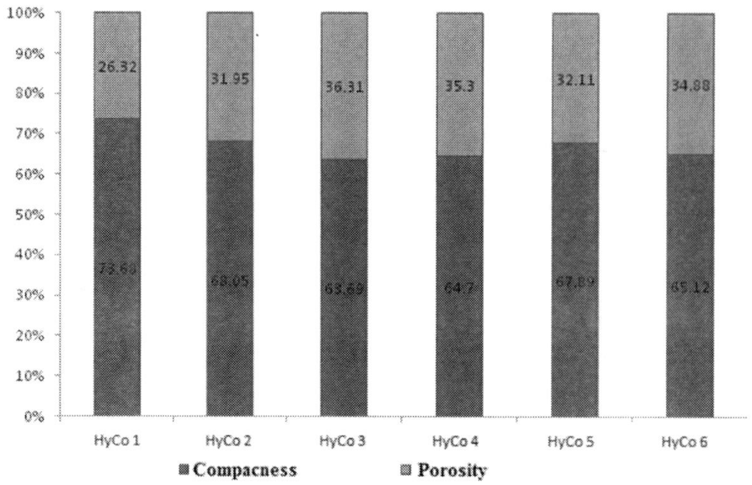

Figure 3. Compactness and porosity after sintering of the $Al/Al_2O_3/Gr$ hybrid composites.

3.2. Micro-structural investigations
These analyzes were performed in order to observe the formation, appearance and diameter of the grains, as well as particle identification at very high magnifications. At the same time, it was possible to determine precisely the distribution of the constituent elements of the examined samples. For this purpose we chose sample 6 with maximum content of Al_2O_3 (20 wt. %) and graphite (3 wt. %). The micro-structural aspect of the analyzed sample is shown in Figures 4 and 5.

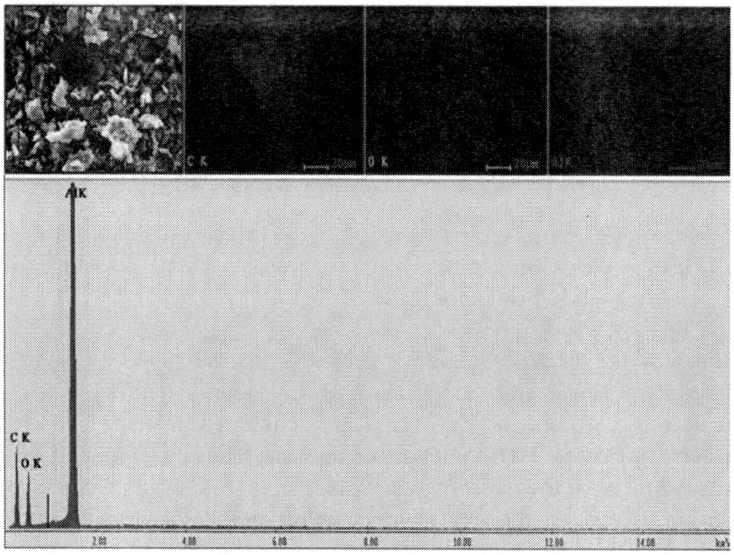

Figure 4. SEM-BSED images and EDAX analysis of the HyCo 6.

The distribution of the main elements in the composition of the analyzed sample is presented in the images associated with the chemical analysis. The images highlight the C (red), O (yellow) and Al (blue) content of HyCo 6 powder composition after 4 hours of mechanical alloying, obtained by analyzing the hue differences in the images. It can also be seen in the micro-analyzed area the presence of a particle of Gr on which grains of Al_2O_3 are trapped.

IOP Conf. Series: Materials Science and Engineering **374** (2018) 012036 doi:10.1088/1757-899X/374/1/012036

The presence of only the Al, C and O elements indicated by the spectrum of analysis, and the fact that no other elements are indicated, confirms that the powders are not contaminated with other elements that may come from other sources.

Also by means of electronic microscopy, the size of microcrystalline and even nanocrystalline compounds formed between the Al and Al_2O_3 and Gr powders shown in Figure 5 was determined. Determination of the grain size in the analyzed samples was very difficult to determine due to the very fine particle agglomeration. It is recommended that particle-dispersing equipment be used before analyzes are performed. The grain size was approximately between 295.0 nm and 1.63 μm.

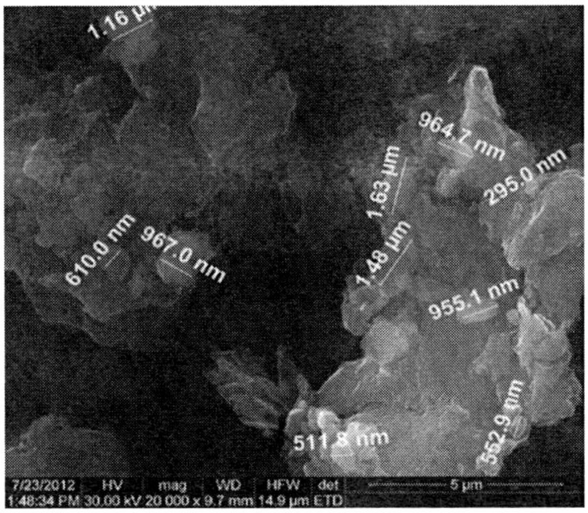

Figure 5. SEM-ETD image of the HyCo 6.

The diffraction patterns of the hybrid composites after 4 hours of mechanical alloying exhibit various peaks corresponding to the face centered cubic phase of Al. The α-Al_2O_3, β-Al_2O_3 and δ-Al_2O_3 phases were observed in the diffraction patterns (Figure 6), these being phases with great stability.

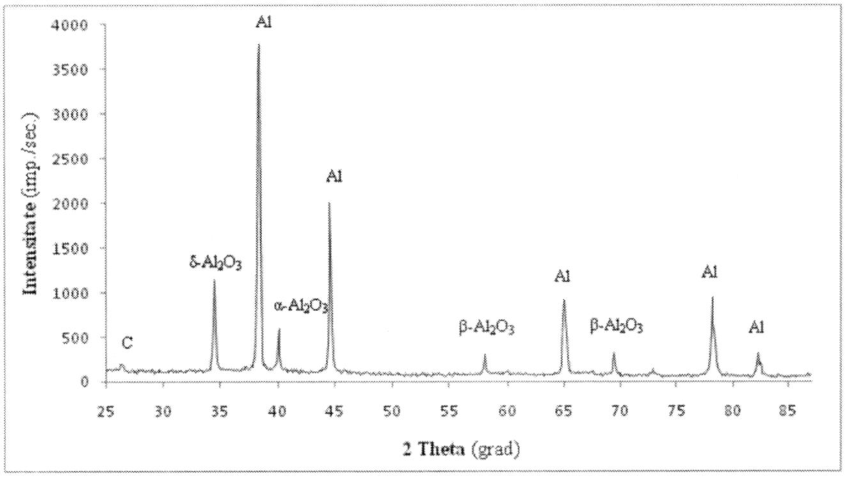

Figure 6. XRD analysis of the HyCo 6.

Through micro-structural analyzes it was observed at very large scale, provided by SEM microscopy, forming a very fine sub-grain structure with diameters of hundreds of nanometers, beneficial for mechanical stress behavior

3.3. *Hardness testing*

Determination of the hardness of hybrid composites after pressing and sintering was done by the Leeb method to illustrate the effect of alumina (10, 15 and 20 wt. %) and graphite (1 and 3 wt. %). Figure 7 presents the results from the determination of the samples belonging to the HyCo 1-6 group, which have been studied in this paper.

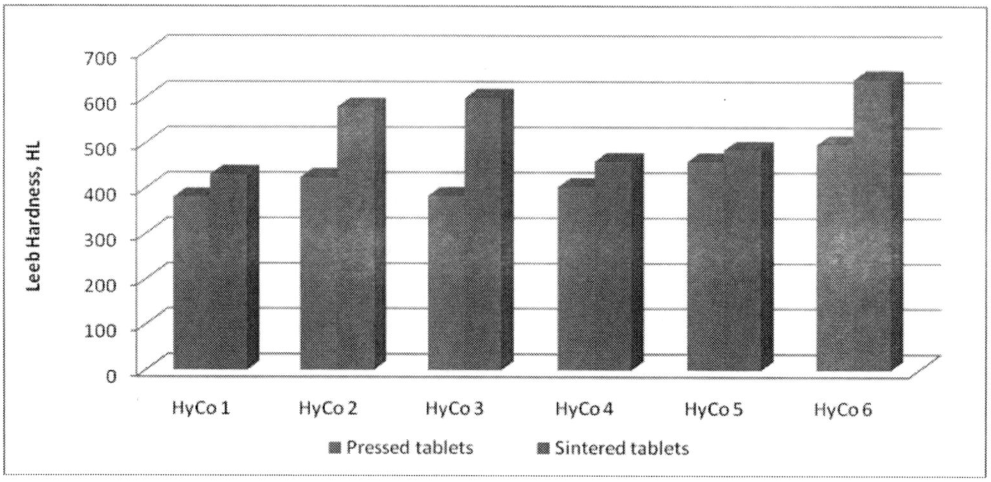

Figure 7. Comparison between hardness results on tablets and after sintering.

Leeb hardness of hybrid composite tablets pressed or sintered shows an increasing trend by rising the content of Al_2O_3 and Gr, there is also a significant increase in values of sintered tablets compared to pressed tablets, as demonstrated by the analysis of the test results by other methods of determining the hardness [13].

4. Conclusions

Following the experimental process of mechanical alloying, various compositions of $Al/Al_2O_3/Gr$ composite powders were produced, which were pressed and subsequently sintered, thus completing a complete powder metallurgy cycle.

The physico-technological, microstructural and mechanical properties were investigated demonstrating promising results. The result obtained from present work can be summarized in the following points:

- Densities of the HyCo 1-6 samples after sintering show a significant increase compared to the densities after pressing; these can be greatly improved if higher compaction pressures are used.
- The results of the compactness and porosity of the HyCo 1-6 samples shows an improvement after sintering process of the $Al/Al_2O_3/Gr$ tablets we can observe an improvement in values.
- The porosity values of the tablets are complementary to the compactness values. They are calculated according to the initial density of the tablets.
- The composition of the analyzed sample indicates only the presence of Al, C and O elements, which means there is no contamination of the powders.
- According to the SEM analysis the grain size in the investigated micro-zone was approximately between 295.0 nm and 1.63 μm, meaning that was formed a fine and homogeneous structure.
- The hardness results of of HyCo 1-6 samples tablets and shows an increasing trend with the content of Al_2O_3 from the 10 wt. % to 20 wt. % and Gr from 1 wt. % to 3 wt. %, there is also a significant increase in values of sintered tablets compared to pressed tablets.

5. References

[1] Mendoza-Duarte J M, Estrada-Guel I, Carreño-Gallardo C and Martínez-Sánchez R 2015 *Journal of Alloys and Compounds* **643** S172-S177
[2] Tousi S S R, Rad R Y, Salahi E, Rahimipour M R, Kazemzade A and Razavi M 2009 *International Journal of Engineering, Transactions B: Applications* **22** 169-178
[3] Ibrahim I A, Mohamed F A, Lavernia E J 1991 *Journal of Materials Science* **26** 1137-1156
[4] Singla M, Dwivedi D D, Singh L, Chawla V 2009 *Journal of Minerals & Materials Characterization & Engineering* **8** 455-467
[5] Włodarczyk-Fligier A, Dobrzański L A, Kremzer M, Adamiak M 2008 *Journal of Achievements in Materials and Manufacturing Engineering* **27** 99-102
[6] Kok M 2005 *Journal of Materials Processing Technology* **161** 381-387
[7] Güler O, Çuvalci H, Gökdağ M, Çanakçi A, Çelebi M 2017 *Particulate Science and Technology*, http://dx.doi.org/10.1080/02726351.2017.1326994
[8] Yin S, Xie Y, Cizek J, Ekoi E, Hussain T, Dowling D, Lupoi R 2017 *Composites Part B* **113** 44-54
[9] Ekka K K, Chauhan S R, Goel V 2015 *Proceedings of the Institution of Mechanical Engineers, Part L: Journal of Materials Design and Applications* **230** 537-549
[10] Moghadam A D, Omrani E, Menezes P L, Rohatgi P K 2015 *Composites Part B: Engineering* **77** 402–420
[11] Han Q, Geng Y, Setchi R, Lacan F, Gu D, Evans S L 2017 *Composites Part B* **127** 26-35
[12] Ahmad Z, Khan S 2014 *International Journal of Current Engineering and Scientific Research (IJCESR)* **1** 1-8
[13] Iacob G, Ghica V G, Buzatu M, Buzatu T, Petrescu I M 2015 *Composites: Part B* **69** 603–611
[14] Zabihi M, Toroghinejad M R, Shafyei A 2016 *Materials Science and Engineering: A* **667** 383-390
[15] Gürbüz M, Şenel M C, Koç E 2018 *Journal of Composite Materials* **52** 553-563
[16] Chen S H, Wang T C 2002 *Acta Mechanica* **157** 113-127
[17] Ramesh B, Senthilvelan T 2010 *IACSIT International Journal of Engineering and Technology* **2** 1-6
[18] Kittali P, Satheesh J, Anil Kumar G, Madhusudhan 2016 *International Research Journal of Engineering and Technology (IRJET)* **03** 2412-2416
[19] Whittaker D 2008 *European Powder Metallurgy Association* 1-31
[20] Suryanarayana C 2001 *Progres in Materials Science* **46** 1-184
[21] El Eskandarany M S 2001 *William Anderson Publishing* New York USA
[22] Iacob G., Popescu G., Miculescu F., Buzatu M 2012 *U.P.B. Sci. Bull., Series B* **74** 221-230

Acknowledgments

This work has been supported by a grant of the University Politehnica of Bucharest, Program for research - Excellence Research Grants – GEX, project number 27/2017.

We hereby acknowledge the research founds project UPB-GEX 2017 (internal no. SM 35-17-04, ctr. No 70/2017) for providing the infrastructure used in this work and the project.

Spinning Disc Technology – Residence Time Distribution and Efficiency in Textile Wastewater Treatment Application

E T Iacob Tudose[1] **and C Zaharia**[1,*]
[1] "Gheorghe Asachi" Technical University of Iasi, Faculty of Chemical Engineering and Environmental Protection, 73 Prof.Dr.Docent D.Mangeron Blvd, 700050, Iasi, Romania

E-mail: czah@ch.tuiasi.ro

Abstract. The spinning disc (SD) technology has received increased attention in the last years due to its enhanced fluid flow features resulting in improved property transfers. The actual study focuses on characterization of the flow within a spinning disc system based on experimental data used to establish the residence time distribution (RTD) and its dependence on the feeding liquid flowrate and the disc rotational speed. To obtain these data, an inert tracer (sodium chloride) was injected as a pulse input in the liquid stream entering the disc and the salt concentration of the liquid leaving the disc was continuously recorded. The obtained data indicate that an increase in the liquid flowrate from 10 L/h to 30 L/h determines a narrower RTD function. Also, at rotational speed of 200 rpm, the residence time distribution is broader than that for 500 rpm and 800 rpm. The RTD data suggest that depending on the needed flow characteristics, one can choose a certain flowrate and rotational speed domain for its application. Also, the SD technology was used to process textile wastewater treated with bentonite (as both coagulation and discoloration agent) in order to investigate whether the quality indicators such as the total suspended solid content, turbidity and discoloration, can be improved. The experimental results are promising since the discoloration and the removals of suspended solids attained values of over 40%, and respectively, 50 %, depending on the effluent flowrate (10 l/h and 30 L/h), and the disc rotational speed (200 rpm, 550 rpm and 850 rpm) without any other addition of chemicals, or initiation of other simultaneous treatment processes (e.g., advanced oxidative, or reductive, or biochemical processes). This recommends spinning disc technology as a suitable and promising tool to improve different wastewater characteristics.

1. Introduction

Spinning disc (SD) technology involves the use of centrifugal force created by the rotation of the surface of a disc on which shaft a fluid is fed, and the formation on its surface of a thin film that travels predominantly radially to the edge of the disc.

Spinning disc (SD) or Spinning Disc Reactor (SDR) technology has received a lot of attention over the past few years due to the net superiority compared to conventional technologies [1-4].

Although SD technology has been studied for a long time, there are a number of issues that have not been sufficiently investigated and some new areas in which SD can be extremely efficient [5-6]. For example, the hydrodynamics of the film on the disc and, in particular, the achievement of a more uniform flow, close to the plug flow, but also some parameters that cause or, on the contrary, disrupt such an ideal flow, were not enough explored. Knowing such parameters can cause the reactor to operate at optimal parameters so as to obtain a higher final product quality.

In this study, it was sought to gather information on the distribution of residence times in such a rotating disc (RD) system and the influence of some factors such as the liquid supply rate on the disc as well as the rotational speed. These data were used as a starting basis to investigate the SD technology efficiency on treatment of a textile effluent without initiation of any other process (mechanical-physical, chemical, or biochemical), or operation.

2. Experimental installation and measurement procedure. Data analysis and preliminary performance estimation methodology

2.1. Experimental installation laboratory setup

The set-up used to perform all experiments is presented in Figure 1. A rotating disc 1 of 20 cm diameter, connected to a variable speed, adjustable electric motor 2, encapsulated in an acrylic housing 3, was fed with water/wastewater from a tank 4, using a pump 5, passing a flow meter 5, with a preset flow rate from tap 6.The rotational disc speed was measured using a laser tachometer with an accuracy of ± 0.1 rpm. For the residence time distribution, a pre-calibrated WTW 315i Conductometer 8 with a probe 9 has been used to measure the water conductivity. The outlet liquid was collected for spectral measurements by means of tap 10.

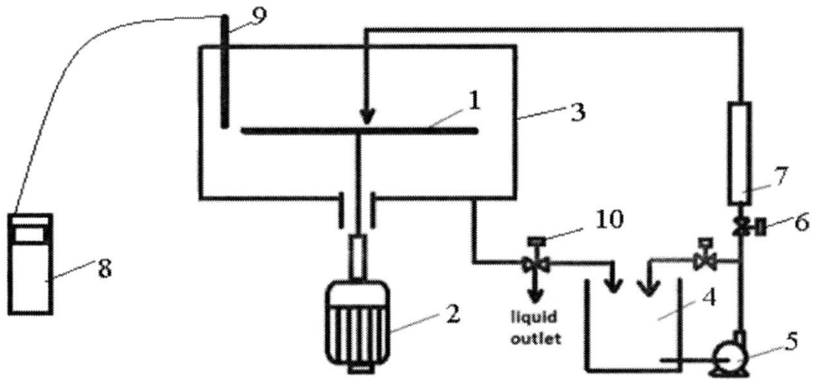

Figure 1. Experimental set-up for RTD and wastewater quality measurements.

The working liquid was tap water for the residence time distribution and textile wastewater for the effluent quality indicators investigation, and also performance estimation of the SD experimental installation in textile effluent treatment.

The temperature was measured using a calibrated thermometer and was found to be practically constant, with a value of 20 °C±0.1°C during each test run.

The measured conductivity was recorded and rendered the solution salt concentration at the disc outlet, while for the wastewater data, samples from the liquid outlet were collected and analyzed using a DR/2000 UV-VIS spectrophotometer (HACH Company).

2.2. Studied wastewater characterization

It has been worked with real industrial effluents, i.e. textile wastewaters, produced in a private NE Romanian company due to the chemical finishing stage of textile fabrics (cotton) after dyeing, and also 1st and 2nd rinsing steps [7, 8].

2.2.1. Wastewater analytical analysis. It depends firstly on the initial and after a few specific times (t_i) supervision and sampling of untreated and/or treated textile wastewater from SDR outlet (exit of the experimental installation) and secondly on the analysis methods applied for characterization of wastewater samples, especially consisting in some specific quality indications such as pH, suspended solids content, turbidity, colour, organic matter expressed as chemical oxygen demand (COD) and biochemical oxygen demand (BOD), etc.

All analyses were performed according with recognized standard methods, internationally approved, as well as specific reference materials and kits of analytical purity. In this work, it was used as coagulation agent only local indigen bentonite powder, that can be considered also a potential discoloration agent for the studied textile wastewater [9].

2.2.2. PH determination. It was measured directly at HACH One Laboratory pH meter (HACH Co.).

Colour determination. It was measured the absorbance related to a blank with distilled water, measured at three different standard characteristic wavelenghts (SR ISO 7887-97) meaning 436, 525, and 620 nm, especially at 436 nm for industrial wastewaters (*apparent colour* in supernatant, or *real colour* in filtrate), or Hazen colour index (Hazen units – HU, i.e. an absorbance of 0.069 at 456 nm corresponds to 50 HU).

2.2.3. Suspended solids determination. It was measured directly at 630 nm (in mg/L) using a DR/2000 spectrophotometer-based methodology (test program #630) under a blank with distilled water.

2.2.4. Turbidity determination. It was measured directly at 450 nm (in FTU) using the DR/2000 spectrophotometer-based methodology (test program #750) under a blank with distilled water.

2.2.5. COD_{Cr} determination. It was used the COD reactor' methodology, based on a 2h-oxidation at 150°C in the presence of a strong oxidation agent (H_2SO_4 conc. and $K_2Cr_2O_7$ 0.25 N) and catalyst ($HgSO_4$) and chloride inhibitory agent (Ag_2SO_4). The calibration of test program was performed with potasium acid phtalate as COD-Cr calibration reagent solution.

The performed experimental results will permit the estimation of SDR technology performance in terms of suspended solids and turbidity removals, and also of colour removal, or textile wastewater discoloration.

The higher the value of studied species removal the better the performance of SD technology is. It can be determined the minimal and maximal removal efficiency of wastewater treatment based on SDR technology without any other chemical adding, or can be initiated supplementary operations or processes for SD technology performance improvement. That is why, these basic results are significant for the further treatment optimization possibilities.

2.3. Data analysis methodology

The data has been processed in Excel files using the following equations for calculating the RTD values:

- the residence time distribution indicating the variation of tracer concentration at the disc exit:

$$E(t) = \frac{C(t)}{\int_0^\infty C(t)dt} \tag{1}$$

- the dimensionless residence time distribution:

$$E(\theta) = \tau E(t) \tag{2}$$

with τ - the mean residence time:

$$\tau = \frac{\sum t_i C_i \Delta t_i}{\sum C_i \Delta t_i} \tag{3}$$

where C_i is the measured salt concentration at the time t_i.

The variance σ^2 and the normalized variance σ_θ^2 are determined as

$$\sigma^2 = \frac{\int_0^\infty t^2 c dt}{\int_0^\infty c dt} - \tau^2 \tag{4}$$

$$\sigma_\theta^2 = \frac{\sigma^2}{\tau^2} \tag{5}$$

and the dispersion number, defined by $D/u_m R$, is calculated from the normalized variance [10], for open-open boundary conditions with the equation:

$$\sigma_\theta^2 = 2\frac{D}{u_m R} + 8\left[\frac{D}{u_m R}\right]^2 \quad (6)$$

where D is the dispersion coefficient (m^2s^{-1}), u_m is the radial mean velocity across the disc (ms^{-1}), R is the disc radius (m).

The dispersion number is a measure of the ratio of the rate of transport by diffusion and the rate of transport by convection, thus if the dispersion number is very small, the dispersion is negligible, hence the flow attains plug hydrodynamics.

The efficiency of SD technology applied in the case of the studied textile wastewaters was established based on the calculation of the removal efficiency (%), R, using the well known relation:

$$R\ (\%) = \frac{C_i - C_f}{C_i} * 100 \quad (7)$$

where, C_i is the initial value of studied quality indicator (mg/L, or FTU, or HU) and C_f is the final (after t time) value of studied quality indicator (mg/L, or FTU, or HU).

3. Results and discussions

3.1. Liquid flowrate and disc rotational speed influence on the residence time distribution

3.1.1. Feeding flow rate influence on the residence time distribution. Figure 2 a, b, c shows the normalized distribution functions of the non-dimensional residence times E(θ) calculated with the formulas presented above, at two different flow rates of 10 and 30 L / h, at a constant rotational speed, with values of 200, 500 and 800 rpm, respectively.

Absolutely all distributions have a characteristic, relatively symmetrical shape, however the presented experimental data show that an increase in the liquid flow rate for a constant disc speed generally results in a narrower distribution of the residence times and a higher maximum value E(θ) for the higher liquid flowrate, indicating closer hydrodynamics to plug flow.

One can observe that, at the lowest liquid flowrate of 10 L/h, the E(θ) function presents some two other smaller peaks indicating some less active areas on the disc.

Comparison to some other literature data [11], even though performed for some different flowrates and rotational speeds, indicate similar trends in RTD curves.

3.1.2 Rotational speed influence on the residence time distribution. Figure 3 a and b indicates the influence of the disc rotational speed on the residence time distribution at 10 l/h and respectively, 30 l/h. One can observe that an increase in the rotational speed renders a narrower distribution of E(θ) function, suggesting a disc flow hydrodynamics closer to the ideal plug flow.

However, the experimental data obtained at a rotational speed of 800 rpm were not included, since, especially at the larger flowrate of 30 L/h, a slight liquid atomization occurred, so the measured liquid salt content at 800 rpm was smaller and the rendered conductivity signal was slightly less than the 500 rpm conductivity signal.

Estimations of the dispersion number, using the equation (6), are presented in Table 1 and indicate furthermore that at low flowrates such as 10 Lph, the dispersion number decreases with the increase in the disc rotational speed. The trend is followed also at 30 Lph, for a disc rotational speed increase from 200 rpm to 500 rpm, however an increase at 800 rpm renders a lower dispersion, possibly due to the already mentioned slight liquid atomization.

IOP Conf. Series: Materials Science and Engineering 374 (2018) 012037 doi:10.1088/1757-899X/374/1/012037

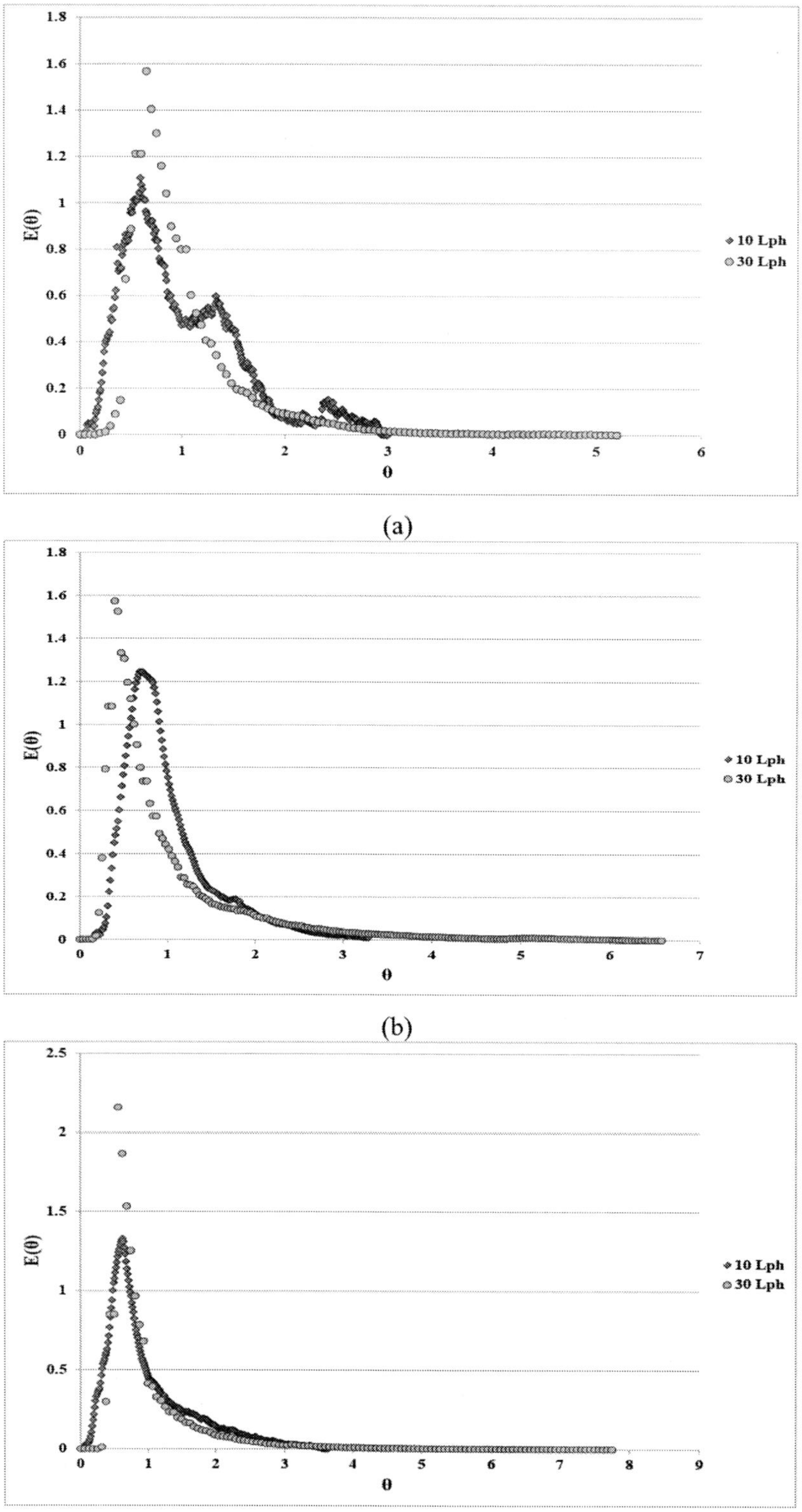

(c)

Figure 2 (a-c). Water flow rate influence on residence time distribution for different disc rotational speeds (a) 200 rpm, (b) 500 rpm, (c) 800 rpm.

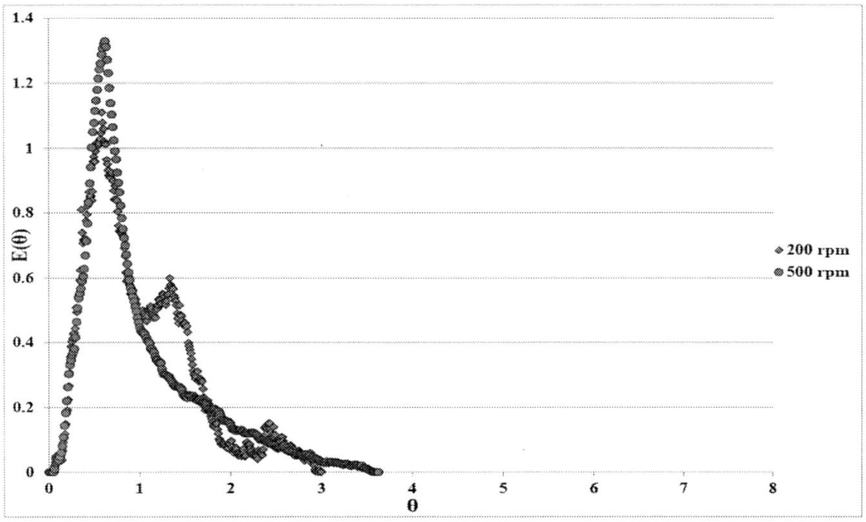

(a)

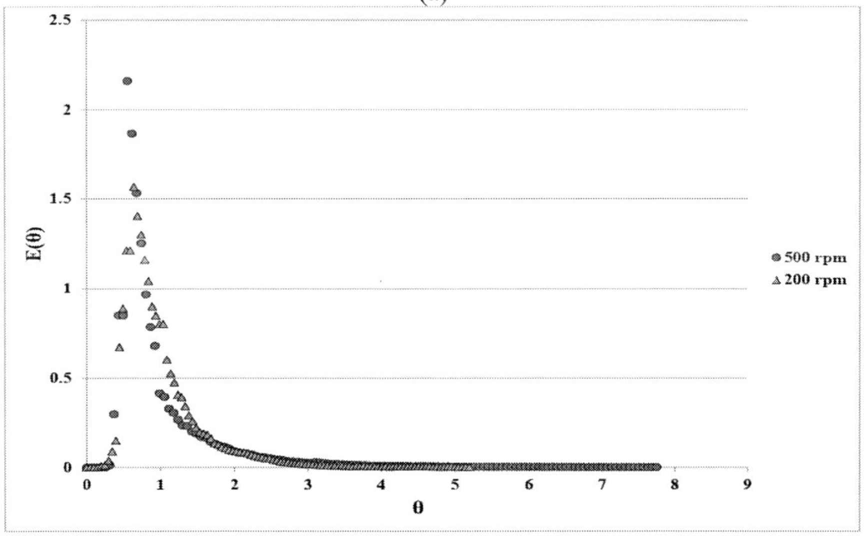

(b)

Figure 3 (a-b). Disc rotational speed influence on residence time distribution for different water flowrate values (a) 10 L/h, (b) 30 L/h.

Table 1. Dispersion number for different water flow rates and disc rotational speeds.

Liquid flow rate [L/h]	Disc rotational speed [rpm]		
	200	*500*	*800*
10	0.430811	0.417594	0.261875
30	0.323889	0.569083	0.168079

3.2. Textile wastewater treatment performance using SD technology laboratory setup

The main characteristics of the textile wastewater analyzed in the SD system and the maximum admissible concentrations are presented in Table 2 [7-8]. The NE Romanian textile company is manufacturing cotton fabrics and the studied wastewaters are being generated from fabrics coloured from yellow to red.

Table 2. The main characteristics of the studied textile wastewater and the maximum admissible concentrations.

Quality indicators	Measured value [mg/L]	M.A.C.* [mg/L]	Quality indicators	Measured value [mg/L]	M.A.C.* [mg/L]
pH	7.49-7.75	6.50-8.50	Total P	5.72-6.39	1
Colour, HU (A_{436})	871-4447 (1.203-2.57)	50	Extractible substances	25.44-31.67	20
Suspended solids	386-900	35 (60)	Total N	8.34-9.80	10
Turbidity, FTU	185-808	-	Ammonia	2.10-2.45	2
Fixed residues	3600-4000	1000	Sulfates	782.34-810	600
COD, mg O_2/L	564.12-650.24	125	Chlorides	98.56-138.23	70
CBO_5, mg O_2/L	322.10	25	Phenol index	2.63-3.26	0.30
Synthetic detergents	1.78-2.10	0.50	Total heavy metal ions	< 4	< 2 (max 5)

* M.A.C.- maximum admissible concentration, according with Romanian Government Decision No. 352/2005-Technical Norms for Treated Wastewater Discharged in Natural Water Resources (NTPA 001)

3.2.1. Flowrate influence on the effluent treatment efficiency using SD technology setup. Considering the obtained values of all studied quality indicators measured at a rotation speed of 550 rpm, when SD technology was applied to textile effluents, the highest removals of colour, suspended solids and turbidity were, respectively, of 21.449 % (10 L/h), 13.004 % (30 L/h) and 11.30 % (10 L/h), as it is illustrated in Figure 4.

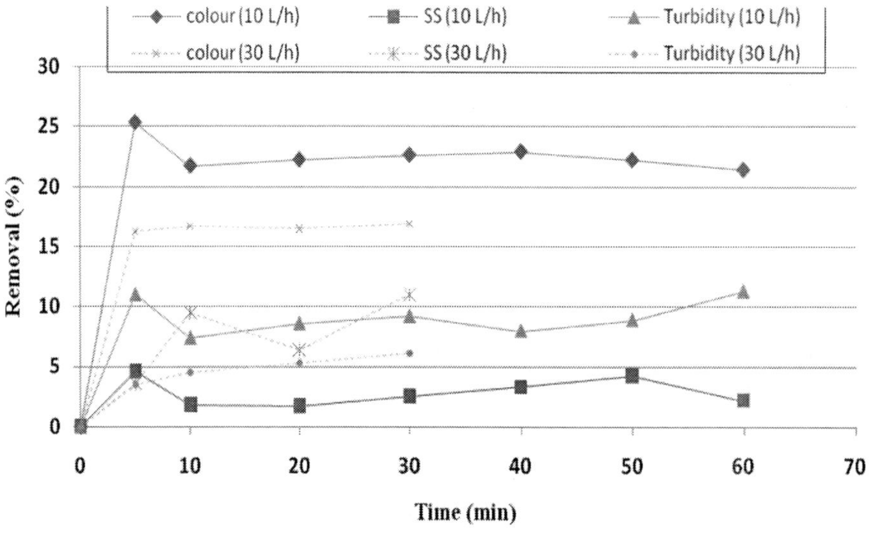

Figure 4. Variation in time of studied removal efficiencies at 550 rpm for two flowrates (10, 30 L/h).

The 10 l/h flowrate value was the most indicated to perform the highest removal in terms of colour, however, for the suspended solids and turbidity the highest removals were obtained for the higher flowrate of 30 L/h. Thus, at a rotational speed of 550 rpm a clear influence of flowrate cannot be stated at this point. At a rotation speed of 850 rpm, the experimental results presented in figure 6 indicated that the highest removals were performed at the high flowrate (30 l/h) for all studied quality indicators, i.e. colour (48.6%, after 5 min), suspended solids (22.1%, after 5 min) and turbidity (22.02%, after 5 min).

3.2.2. Rotation speed influence on the treatment efficiency using SD laboratory setup. At a low flow rate of 10 l/h and 500 rpm, a relatively constant efficiency in colour removal (between 20.229 - 22.988 %) was found during the first tested 40 minutes, afterwards, a slight decrease was registered, as presented in figure 6a.

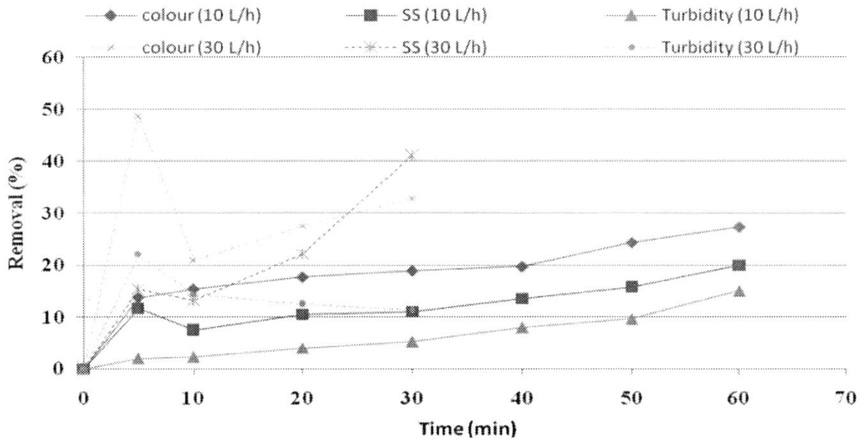

Figure 5. Variation in time of studied removal efficiencies at 850 rpm for two flowrates (10, 30 L/h).

At the same flow rate of 10 l/h and a rotational disc speed of 250 rpm and respectively, of 850 rpm, lower values for colour removal, namely 2.725 - 19.967 % and, respectively, 13.84 - 18.919 %, were obtained, during the first 30 minutes measuring time interval. However, after 50 min, the discoloration degree increased up to 27.260 % for a flow rate of 10 l/h (Figure 6a) and up to 32.93 % for a flow rate of 30 l/h (Figure 6b). All experimental results concerning the maximal efficiency in suspended solids and turbidity removal are summarized in Table 3. One can observe that the highest removal for suspended solids (54.29 %) and turbidity (49.07 %) are obtained at a flow rate of 10 L/h and a rotational speed of 250 rpm, after 60 min.

The effluent treatment efficiency by SD technology in the laboratory setup is relatively good considering that no other chemical, mechanical-physical, or biological treatment processes or operations were applied, and, thus, this can be used to minimize the suspended solids and turbidity loads within a relatively short period of time (no more than 30 min for a flow rate of 30 L/h, or 60 min for a flow rate of 10 l/h, as seen in Table 3).

Table 3. Maximal values of suspended solids (SS) and turbidity removal for SD technology setup.

Rotation speed [rpm]	Flow rate [L/h]	SS removal [%]	Operating time [min]	Turbidity removal [FTU]	Operating time [min]
250	10	54.29	60	49.07	60
	30	1.52	30	2.36	30
550	10	4.23	50	11.32	60
	30	13.03	30	6.12	30
850	10	19.93	60	14.95	60
	30	41.15	30	22.02	5

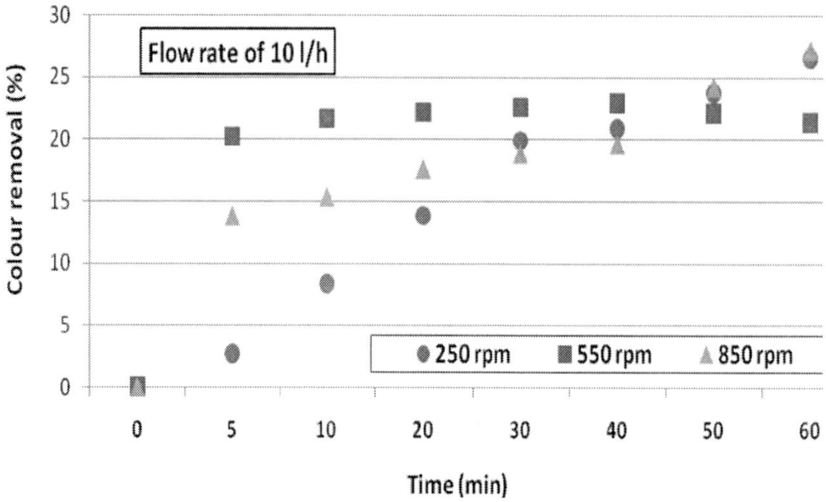

(a)

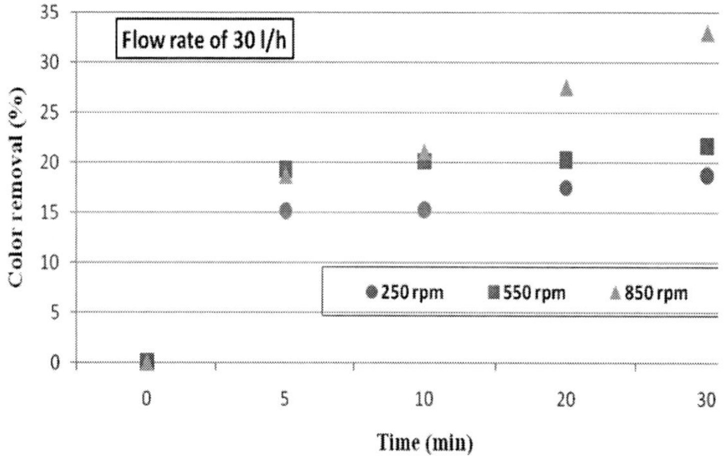

(b)

Figure 6 (a-b). Variation in time of colour removal efficiencies at three different rotational speeds.

These results provide significant reference data to initiate a few new technological improvements of textile wastewater treatment based on SD technology and also, process optimization.

4. Conclusions

The experiments performed to obtain the residence time distribution in order to characterize the flow on a spinning disc indicated that an increase in the liquid flowrate from 10 L/h to 30 L/h determines a narrower RTD function and, also, at a rotational speed of 200 rpm, the residence time distribution is broader than that for 500 rpm. The RTD data suggest that depending on the needed flow characteristics, either plug or stirred flow, one can choose a certain flowrate and rotational speed domain for its application.

The maximal obtained efficiency of SD technology applied for textile wastewater treatment without any supplementary process or operation is very good for the flowrate of 10 L/h and rotational speed of 250 rpm, after 60 min of application (22.99 % for colour, 54.29 % for SS, and 49.07 % for turbidity), and also for the flowrate of 30 L/h working, at a rotational speed of 850 rpm, for 30 min of application (48.06 % for colour after only 5 min, 41.15 % for SS, and 22.02 % for turbidity after 5 min).

These experimental results determined on textile wastewaters demonstrate the feasibility of SD technology application for removal of some polluting species from different systems.

5. References

[1] Chin S F, Iyer K S, Raston C L, Saunders M 2008 *Adv.Funct.Mater.* **18** 922–927
[2] Chang C Y, Wu N L 2010 *Ind.Eng.Chem.Res.* **49** 12173–12179
[3] Feng X, Patterson D A, Balaban M, Fauconnier G, Emanuelsson E A C 2013 *Chem.Eng.J.* **221** 407-417
[4] Feng X, Patterson D A, Balaban M, Emanuelsson E A C 2013 *Chem.Eng.Res.&Des.* **91**(9) 1684-1692
[5] Zhang A Y, Zhou M H, Han L, Zhou Q X 2011 *J.Hazard.Mater.* **186** 1374–138
[6] Lazar M A, Varghese S, Nair S S 2012 *Catalysts* **2** 572-601
[7] Zaharia C, Suteu D, Muresan A 2012 *Environ.Eng.Managem. J.* **11**(2) 493-509
[8] Zaharia C, Amarandei V, Muresan A 2014 *Adv.Mater.Res.* **1036** 58-64
[9] Bertea A, Bertea A-P 2008 *Discoloration and recycling of textile wastewaters* (in Romanian: *Decolorarea şi reecircularea apelor uzate textile*), Performantica Ed., Iasi
[10] Levenspiel O 1999 *Chemical reaction engineering* third edition, JohnWiley & Sons, New York
[11] Mohammadi S, Boodhoo K V K 2012 *Chem.Eng.J.* **207–208** 885–894

Influence of Nanostructuration on the Sound Velocity in Aluminum Al_99.50

A M Cazac[1], A Alexandru[1], C Baciu[1] A V Sandu[1] and C Bejinariu[1*]
[1]"Gheorghe Asachi" Technical University from Iasi- Romania, Faculty of Materials Science and Engineering, Blvd. Dimitrie Mangeron, No. 41, 700050, Iasi- Romania

E-mail: costica.bejinariu@tuiasi.ro

Abstract. The paper proposes is a multidisciplinary study on the influence of nanostructured material obtained by cyclic closed die forging process, in this case the aluminum with a purity of 99.50% (Al_99.50), on the sound velocity. The study of nanomaterials is a branch of material science on the basis of which nanotechnology can be approached. Severe plastic deformation (SPD) is a generic term describing a group of metal and alloy processing techniques involving very high stresses without including significant changes in the overall dimensions of the model or workpiece. The sample is of a regular quadrangular prism shape with the side square of a = 10 mm and the height of h = 16 mm, so with a dimensional factor h / a = 1.6. For each sample, a number of 7 determinations were performed to establish a mean value for the sound velocity. As a result of the microstructure analysis, it is observed that at the deformation cycle 4 the grains have an average size between 250 and 500 nm.

1. Introduction

The paper proposes a multidisciplinary study on the influence of the nanostructuring of a material obtained by the cyclic closed die forging process, in this case the aluminum with a purity of 99.50% (Al_99.50), on the sound velocity. The process was patented by A.K. Ghosh in 1988 [1] and is schematically represented in the Figure 1.

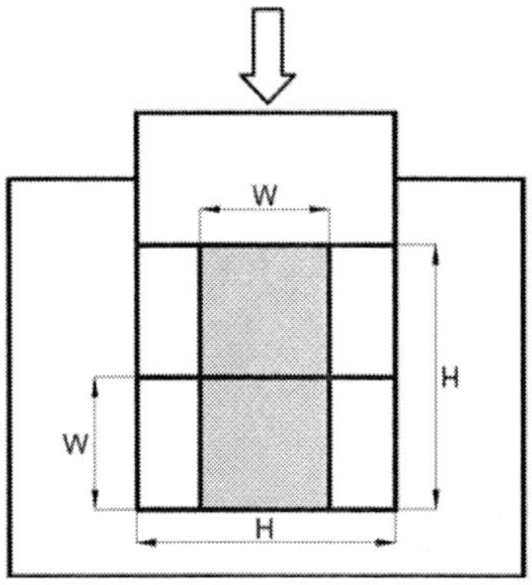

Figure 1. Gosh scheme for multiaxial forging [2].

where: W is the side of the initial square section of the blank;

H is the initial height of the blank.

Plastic deformation is the method of machining where permanent deformation of solid materials without macroscopic cracking is achieved in order to obtain semi-finished products or finished parts [3]. Multiaxial forging is a discontinuous deformation process consisting of a series of deformation processes defining the severe plastic deformation cycle, with the consequence of the nanostructure of the deformed material at a certain number of passes [4-11].

The shape and dimensions of the blank are kept after each pass, as shown in the Figure 2 at the pass "i" (SPD_i), where the deformation scheme is presented.

2. Materials and methods

The deformation process is discontinuous and comprises deformation processes defining a severe plastic deformation cycle [12].

Figure 2 shows the phases of multiaxial forging at a passage of the parallelepiped blank.

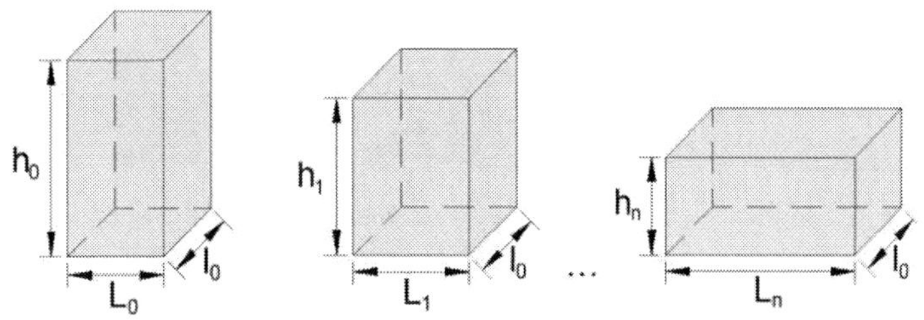

Figure 2. Stages of multiaxial forging for a single passage: (a) blank; (b) blank deformation; (c) deformed part.

in whitch: h_0 este initial height of the blank, $h_0=20\ mm$;

L_0 – initial length of the blank, $L_0=10\ mm$;

l_0 – initial width of the blank is kept constant throughout the deformation, $l_0=10\ mm$;

$h_1,h_2,...,h_{n-1}$ – current height of the workpiece deformed values $h_i[mm]\in(10,20),\ i=\overline{1,n-1}$;

$L_1,L_2,...,L_{n-1}$ – current length of the workpiece deformed values $L_i[mm]\in(10,20),\ i=\overline{1,n-1}$;

h_n – height of the piece at the end of severe plastic deformation process, $h_n=10\ mm$;

L_n – length of the piece at the end of severe plastic deformation process, $L_n=20\ mm$.

The samples were subjected to 12 deformation cycles by cyclic closed die forging process. We used the device designed on the principle of A.K. Ghosh, (Figure 1). It was mounted on a hydraulic press of 750 kN. The press is equipped with a force transducer of 1000 kN and a displacement one having a stroke of 0 ÷ 100 mm.

The nanostructure device for metallic materials, Figure 3, was designed and made on the Ghosh principle, the blank and the piece having the same shapes and dimensions, Figure 2a.

The device according to figure 3 consists of an active plate (1), placed on a pressure plate (2), the assembly being guided on the inner surface of the support (3).

The blank (4) is centred in the recess of active plate over a counterpunch (5) placed in turn on a tablet pressure (6) positioned in the holder (3).

The deformation of workpiece is achieved by applying a force over the punch (7) moving with the given value of stroke limiter (8).

The deformation force applied is dynamically registered with an acquisition system, not shown, by means of a load cell (9) mounted between the holder (3) and a pressure plate (10).

After deformation, the pressure plate (6), counterpunch (5) and limiter (7) are eliminated and the punch is pressed (7) using another stroke limiter until the extraction of the deformed workpiece. The workpiece is rotated by 90° vertical and 90° horizontal, and introduced into the initially described device in order for the next pass and so on until nanostructuring the blank material.

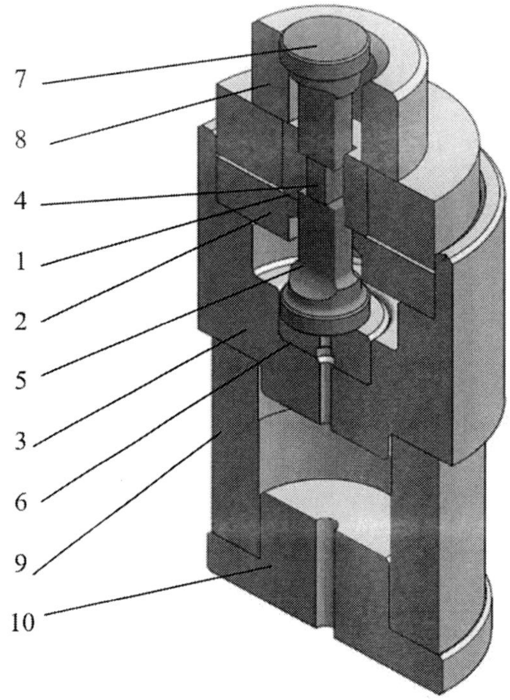

Figure 3. The constructive functional scheme of multiaxial forging device [2].

2.1. Material

To determine the chemical composition of the sample material used in the investigations, we performed the chemical analysis test with the GNR Metal Lab 75/80 V spectrometer, as shown in Table 1.

Table 1. The chemical composition of the aluminum 99.50.

Element	Si	Fe	Cu	Mn	Mg	Cr	Zn	Ni	Ti
Percent	0.143	0.213	0.021	0.004	0.050	0.003	0.021	0.008	0.006

Element	Pb	Sn	B	Ca	Co	V	Na	P	Al
Percent	0.005	0.011	0.000	0.003	0.002	0.011	0.004	0.000	99.495

The sample is of a regular quadrangular prism shape with a square base of a = 10 mm and a height of L = 16 mm, so with a dimensional factor L / a = 1.6, Figure 4.

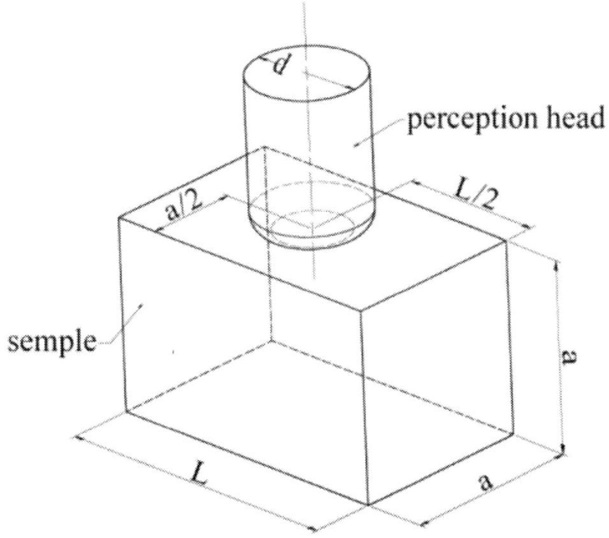

Figure 4. Sound rate measurement scheme.

3. Results and discussions

On each sample we performed 7 determinations and in Table 2 we present the average values of the sound velocity, where:

N is the number of passes (deformation), *[-]*;
L - length of the blank, *[mm]*;
l - width of the blank, *[mm]*;
$V_{Al\_99.50}$ - the average of the sound velocity for Al_99.50 samples, *[m/s]*.

Table 2. Valorile medii ale vitezei sunetului pentru cele 12 treceri a semifabricatelor din Al_99.50.

N [-]	*L* [mm]	*l* [mm]	$V_{Al\_99.50}$ [m/s]
0			6400
1			6230
2			6080
3			5930
4			5820
5			6040
6	16	10	6180
7			6350
8			6370
9			6450
10			6530
11			6540
12			6550

Based on the values in Table 2, we have plotted the variation of the sound velocity according to the number of passes, Figure 5.

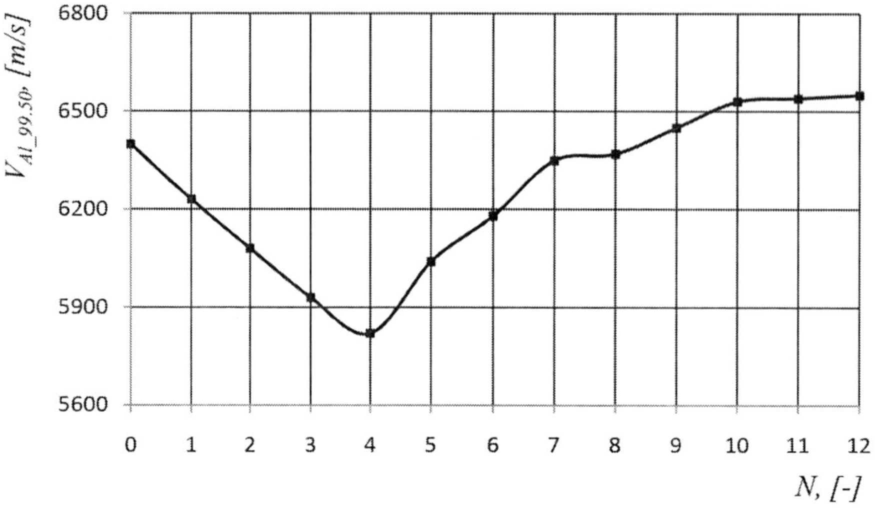

Figure 5. Variation of the sound velocity according to the number of passes for aluminum Al_99.50.

4. Conclusions

The variation of the sound velocity was plotted against the number of deformation cycles, a variation describing a convex line having a minimum in the area of the deformation cycle number 4.

As a result of the microstructural analysis, it is observed that at the deformation cycle 4 the grains have an average size between 250 and 500 nm.

On the basis of the above, it can be deduced that the area of passes 3, 4 and 5 is, in fact, precisely the transition zone between micrometric granulation and mesoscopic (ultrafine) granulation, which is nothing but an intermediate area between micrometric granulation and nanometric granulation.

5. References

[1] Ghosh A K 1988 patent US4721537
[2] Cazac A M, Bejinariu C, Ionita I, Toma S L and Rodu C 2014 *Applied Mechanics and Materials* **657** 193-197
[3] Valiev R Z, Islamgaliev R K and Alexandrov I V 2000 *Progress in Materials Science* **45** 103–189
[4] Bejinariu C, Cazac A M, Darabont D C, Birlescu V, Craus M, Girtu M, Ghizdovat V, Predescu C, Agop M and Constantinescu S 2017 *Journal of Computational and Theoretical Nanoscience* 1744-1750
[5] Florea C D, Carcea I, Cimpoesu R, Toma S L, Sandu I G and Bejinariu C 2017 *Rev. Chim.* **68** 2397–2401
[6] Cazac A M, Bakri Abdullah M M, Predescu C, Sandu A V and Bejinariu C 2015 *Materials Science Forum* **803** 216-221
[7] Bejinariu C, Cazac A M, Baciu C, Bernevig-Sava M A, Coman G, Nica P E, Forna N C, Ghizdovat V, Agop M, Constantinescu S, Postolache P, Ianosi S E and Jimborean G 2018 *Journal of Computational and Theoretical Nanoscience* **15** 895-907
[8] Florea C D, Munteanu C, Cimpoesu N, Sandu I G, Baciu C and Bejinariu C 2017 *Rev. Chim.* **68** 2582–2587
[9] Perju M C, Nejneru C, Galusca D G, Stefanica R G 2010 Modtech 2010: new face of tmcr proceedings Conference: *14th International Conference on Modern Technologies Quality and Innovation (ModTech)* 455-458

[10] Avram P, Imbrea M S, Istrate B, Strugaru S J, Benchea M, Munteanu C 2014 *Indian journal of engineering and materials sciences* **21** 315-321
[11] Perju M C and Vizureanu P 2014 *Revista de chimie* **65**(6) 694-696
[12] Cazac A M, Bejinariu C, Baciu B, Toma S L and Florea C D 2014 *Applied Mechanics and Materials* **657** 137-141

IOP Conf. Series: Materials Science and Engineering **374** (2018) 012039 doi:10.1088/1757-899X/374/1/012039

The Effect of Heat Treatment and Corrosion Behavior of AISI420

M G Minciuna[1,2,3], D C Achitei[1,2,3*], P Vizureanu[1,2,3], M Benchea[4] and A V Sandu[1,2,3]

[1] Gheorghe Asachi Technical University of Iasi, Faculty of Materials Science and Engineering, Blvd. D. Mangeron 41, 700050, Iasi, Romania
[2] Romanian Inventors Forum, Str. Sf. P. Movila 3, Iasi, Romania
[3] Univ Malaysia Perlis, Sch Mat Engn, Ctr Excellence Geopolymer& Green Technol CEGeoTec, Kompleks Pengajian Jejawi 2, Arau 02600, Perlis, Malaysia
[4]Gheorghe Asachi Technical University of Iasi, Faculty of Mechanical Engineering, Blvd. D. Mangeron 41, 700050, Iasi, Romania.

E-mail: dragos_adc@yahoo.com

Abstract. The present paper shows the effect of heat treatments applied to an AISI 420 steel. The characterization of corrosion behavior of AISI 420 steel, in 0.5M NaCl aqueous solution, was made in an electrochemical cell with 3 electrodes, connected at PG STAT 302N Autolab Potentiostat, at laboratory temperature. The medium used was a NaCl 0.5M solution with a pH of 6.26 and an electro-conductivity by 49.9 mS/cm.

1. Introduction

The foundation of modern industry is stainless steel due to it's mixt application in aqueous corrosive environments in a significant number of industries [1, 2]. High volume production and advance manufacturing techniques of stainless steels has resulted in the availability of cost effective and corrosion resistant steels [3]. Austenitic stainless steels grades generally have strong corrosion resistance in mildly corrosive acid, industrial and marine environments [4, 5]. Martensitic stainless steels have good mechanical properties and strong resistance to surface deterioration.

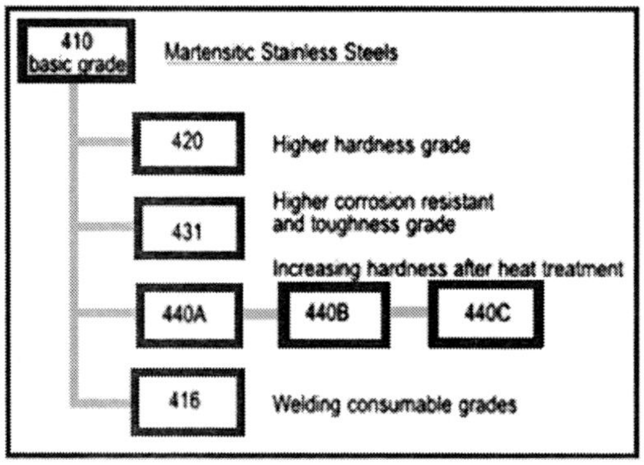

Figure 1. Martensitic stainless steels [6].

The need for superior corrosion resistant properties in specific industrial applications necessitates performance improvement on the surface property and metallurgy of these steels [7-9]. Altering the microstructure of stainless steel significantly improves their corrosion resistance. Microstructural constituents such as grain size, phases, precipitates, flaws and inclusions are strongly modified by heat treatment to effect changes in their mechanical, chemical and surface properties based on the martensite/austenite formation, changes in grain size and defects [10, 14-16]. Choi et al (2007) studied the relationship between different austenitizing temperature of 0.3C-14Cr-3Mo type stainless steel and their electrochemical performance in Cl containing neutral aqueous solutions. He observed that increase in austenitizing temperature caused a proportionate increase in corrosion and pitting potential of martensitic steel due to precipitation of fewer types of carbide among the steel matrix after high temperature austenitization [11].

2. Materials and methods

Previous researches on the effect of heat treatment processes on the corrosion resistance of stainless steel have given mixed results. Nasery et al (2011) studied the effects of astenitizing temperature, tempering temperature and time, and on the microstructure, mechanical and corrosion properties of AISI420 martensitic steel, see Figure 2.

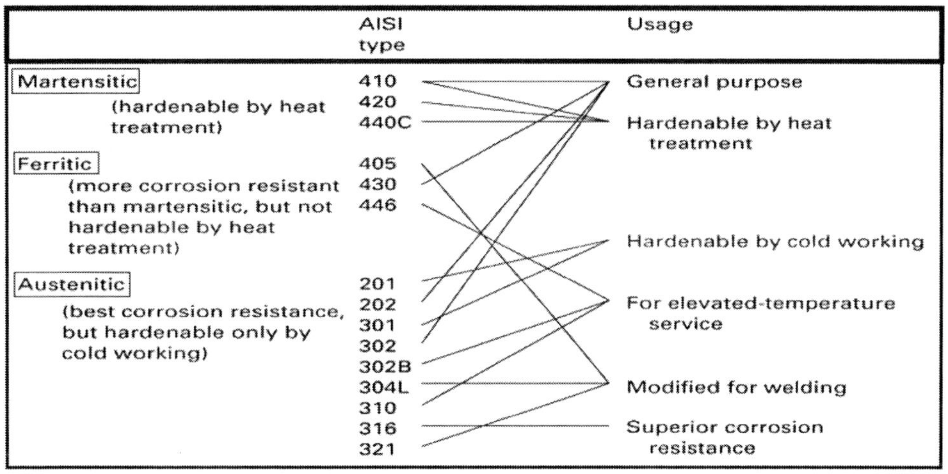

Figure 2. Usage stainless steels.

Results showed that the temperature significantly influenced the mechanical properties of the steel. The effects of heat treatment on microstructure, hardness uniformity, surface roughness and corrosion resistance of a 13Cr-type martensitic stainless steel was studied by means of optical microscopy, scanning and transmission electron microscopy and polarization resistance. Observation showed that the corrosion resistance of the steels reduced after tempering in comparison to quenching [12].

The cycle control of heat treatments, on the quenching and tempering operation of AISI 420 stainless steel, is essential for improved material performance. The adequate choice of heat treatment parameters, can lead an optimization on its mechanical properties and corrosion resistance. Thus, this paper aims to investigate the effects of quenchants medium, and austenitizing and tempering temperatures, on the microstructure and mechanical properties of AISI 420 steel.

The composition was determined with an optical spectrometer, Foundry Masters, using iron analysis base, with ± 3% accuracy of results. The sample subjected to analysis was processed from laminated bar and prepared by cutting on lathe. The determination of chemical composition, by quantitative spectral analysis, was made in 3 various points from prepared surface. Based on measured percents, the specific software made an average of registered values, see Table 1, with are present in analysis bulletin.

Table 1. Chemical composition determined by spectral analysis.

Fe	*Cr*	*Mn*	*Si*	*C*	*Ni*	*Mo*	W
86.50	**12.10**	**0.29**	0.24	0.23	0.18	0.10	0.07
Cu	V	Co	P	S	Al	Nb	Ti
0.07	0.05	0.02	0.01	0.01	< 0.01	< 0.01	< 0.01

Based on analysis bulletin obtained, the studied alloy is part of martensitic stainless steels classes, type AISI 420 (20Cr120) or 1.4021.

Based on percents of alloying elements and it is established the following heat treatments parameters [9-11], which can be applied to AISI 420 steel according to Table 2.

Table 2. Parameters of heat treatments applied to AISI 420 steel.

Symbol	*Heat treatment*	*Heating temperature* [°C]	*Heating speed* [°C/min]	*Heating medium*	*Maintaining time* [min]	*Maintaining temperature* [°C]	*Cooling speed*	**Cooling medium**
AISI 420 (I)	Initial state	Laminated bar, obtained by continuous casting and calibrated at standard dimensions imposed by beneficiaries.						
AISI 420 (A)	Annealing	770	10	Air	20	770	Slow	Furnace
AISI 420 (Q)	Martensitic quenching	1000	10		30	1000	Rapid	Water
AISI 420 (T)	High tempering	700	10		60	700	Free	Air

The heat treatments was made in laboratory conditions, using standard samples, with dimensions which permit their utilization both in preparation operations by polishing and chemical attack, also by SEM, X-ray diffractometry and corrosion behavior in 0.5M NaCl aqueous solution.

The heating and maintaining operations of effectuated heat treatments, was realized in an electric furnace, with fixed hearth, CE12 UTTIS model, controlled according the heat treatment diagram, with a specific software Protherm.

3. Results and discussions

The electrochemical behavior of AISI 420 steel was evaluated using a electrochemical cell with 3 electrodes, connected at Autolab PG STAT 302N potentiostat. The medium used was a NaCl 0.5M solution with a pH of 6.26 and an electro-conductivity by 49.9 mS/cm.

In the Table 3 it is shown the extension of each potential domain for each sample and the mentioned characteristics on voltametric behavior.

Table 3. The characteristics of voltametric behavior on extended potential domain of samples by AISI 420 steel, in 0.5M NaCl solution.

Sample	*Domain I* [V]	E_{BD} [V]	*Domain II* [V]	p_1 [mA/cm^2·V]	p_2 [mA/cm^2·V]
A	- 0.3876 … 0.0792	0.1256	0.1549 … 3.000	2.4214	75.02
C	- 0.3360 … 0.1012	0.1280	0.1695 … 3.000	1.0399	76.00
D	- 0.3226 … 0.1878	0.2586	0.1695 … 3.000	1.2356	59.63
B	- 0.2576 … 0.0906	0.1077	0.1297 … 3.000	0.4545	44.52

Based on the presented data from Table 1 we can draw the following conclusions:

- The value of potential in open circuit (inferior limits of domain I) show that the order of corrosion resistance in 0.5M NaCl for the 4 samples is the following: sample B > sample D > sample C > sample A, while the breakdown potential for AISI 420 steel is small, for A and C is similarly, and the higher is for D sample.
- The value of p1 which considers the behavior on first potential domain concerning the variation of current density with potential modifications applied to sample, the speed of anodic process increase in the following order: B < C < D < A;
- The value of p2 highlight the corrosion resistance of the 4 samples after passing the breakdown potential, when the density of anodic current increase in the order: B < D < A < C.

On the OCP to + 1.2 V … -1.2 V to OCP interval the cyclic voltammograms were drawn, starting from the potential in open circuit in anodic direction until 1.2V, then in return scanning is modified the potential at -1.2V, then return at the potential from which are going.

Based on the obtained data from potentiodynamic courves [13], using the corrosion program from Nova 1.10 software, we have obtained the Tafel coefficients (ba and bc), the corrosion potential (Ecor,calc) defined at the intersection of Tafel courves (anodic and cathodic branch), which is moving at negative values (with 11 to 26 mV) between OCP, this is due to the medium aggressivity, density of corrosion current (jcor), corrosion speed (vcor) and polarization resistance (Rp), all being presented in Table 2. In the evaluation of the corrosion speed, it is considered the current density (j), defined by Butler-Volmer relation that being:

$$j = j_{cor} \cdot \left[\exp\left(\frac{2.303 \cdot (E - E_{cor})}{b_a} \right) - \exp\left(-\frac{2.303 \cdot (E - E_{cor})}{b_c} \right) \right] \quad (1)$$

In which ba and bc are Tafel slopes:

$$b_a = \frac{R \cdot T}{\alpha \cdot n \cdot F} \quad (2)$$

Respectively:

$$b_c = \frac{R \cdot T}{(1-\alpha) \cdot n \cdot F} \quad (3)$$

Table 4. Corrosion parameters of analyzed samples on 0.5M NaCl aqueous solution.

Parameters	*Measurement units*	Samples			
		AISI 420 (A)	AISI 420 (B)	AISI 420 (C)	AISI 420 (D)
S	[cm^2]	2.1372	2.7264	2.5491	4.6434
- OCP	[mV]	384	323	321	240
b_a	[mV/dec]	296.0	141.8	197.3	115.5
b_c	[mV/dec]	207.8	200.6	284.4	185.8
- $E_{cor,calc}$	[mV]	395.9	334.5	336.2	265.6
j_{cor}	[$\mu A/cm^2$]	32.156	6.445	10.872	3.262
v_{cor}	[mm/an]	0.3736	0.0749	0.1263	0.0379
R_p	[Ω]	772	2053	1336	2042

The data presented in Table 4, both by OCP value and the corrosion speed, confirm that the AISI 420 (B) sample is much resistant at corrosion, sample A is much attackable; the order of the 4 samples according with the increase of corrosion resistance being: B > C > D > A.

The value of polarization resistance from Table 2 confirm that A sample is much attackable, also the higher value of corrosion resistance is for C, but similarly with B sample.

For characterize the behavior of the 4 steel sample is used and the electrochemical impedance spectroscopy by drawing the spectrums, after their immersion in aggressive medium and achievement of a stationary state for system (at potential in open circuit).

The obtained spectrums at potential in open circuit, at amplitude of alternative current by 10 mV and in a frequency domain by 10-1 to 104 Hz, are present in Figure 3.

For interpretation of data is used an equivalent electric circuit giving by Boukamp model by data fitting with "Electrochemical circle fit", being the circuit by which the experimental data are similar with the fitted data and in which the calculated sizes are: Rs ohmic fall from solution or uncompensated resistance; Rp polarization resistance; CPE element of constant phase, defined by:

$$Z_{CPE} = \frac{1}{Y_0 \cdot (j \cdot \omega)^N} \tag{4}$$

Where: is admission, ω angular frequency of alternative current, j=(-1)1/2 and N is the exponent of the element by constant phase.

The CPE presence in equivalent circuit is the proof of surface heterogeneity and the diffusion processes which is taking place.

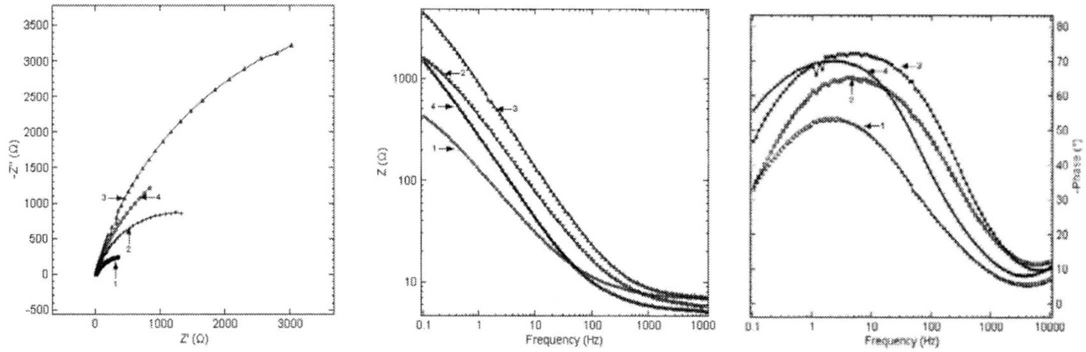

(a) (b) (c)

Figure 3. Spectrums of electrochemical impedance for the 4 steel samples: (a) Nyquist representations; (b) and (c) Bode representations.

The polarization resistances calculated by electrochemical impedance spectroscopy are different of the potentiodynamic polarization curves, due to the different time of immersion both to realize the stationary state for OCP achievement and the time necessary to drawing the spectrums of electrochemical impedance, this explains the high values obtained by electrochemical impedance spectroscopy.

Table 5. Experimental data obtained by electrochemical impedance spectroscopy.

Sample	CPE	N	R_p	R_s	φ
	[μF]	-	[Ω]	[Ω]	[°]
AISI 420 (A)	1960.20	0.9946	812	7.48	53.7
AISI 420 (C)	570.12	0.9962	2453	6.42	65.2
AISI 420 (D)	167.73	0.9968	9489	6.90	72.2
AISI 420 (B)	311.73	0.9965	5106	5.18	70.1

The order of corrosion resistance increase from electrochemical impedance spectroscopy is the following: A < C < B < D.

The presence in given medium of Cl- ions amplify the corrosion process due to their absorption both in active surface and passive in electrode surface, the proof being and the 2 control mechanisms of corrosion highlighted by electrochemical impedance spectroscopy.

A quantitative appraisal of corrosion process for steel in considered medium can be realized by drawing of potentiodynamic curves on ± 0.15 V domains towards by OCP at a small speed of potential.

4. Conclusions

The polarization resistances calculated by electrochemical impedance spectroscopy are different of the potentiodynamic polarization curves, due to the different time of immersion both to realize the stationary state for OCP achievement and the time necessary to drawing the spectrums of electrochemical impedance, this explain the high values obtained by electrochemical impedance spectroscopy.

The comparative study realized on AISI420 stainless steel, after heat treatments, highlight that by modern methods can be establish the link between microstructure – phases – hardness – corrosion resistance.

5. References

[1] Nasery Isfahany A, Saghafian H and Borhani G 2011 *Journal of Alloys and Compounds* **509** 3931-3936
[2] Xi Y, Liu D and Han D 2008 *Applied Surface Science* **254** 5953-5958
[3] Lin C H and Duh J G 2009 *Surface and Coatings Technology* **204** 784-787
[4] Angelini V, Boromei I, Martini C, Scheuer C J and Ceschini L 2016 *Tribology International* **103** 555-565
[5] Çetin A, Tek Z, Öztarhan A and Artunç N 2007 *Surface and Coatings Technology* **201** 8127-8130
[6] Krakhmalev P, Yadroitsava I, Fredriksson G and Yadroitsev I 2015 *Materials & Design* **87** 380-385
[7] Achitei D C, Vizureanu P, Minciuna M G, Sandu A V, Buzaianu A and Dana D I 2015 *Materiale Plastice* **52** 165-167
[8] Minciuna M G, Vizureanu P, Achitei D C, Sandu A V, Berbecaru A and Sandu I G 2016 *Journal Of Optoelectronics And Advanced Materials* **18** 174-178
[9] Istrate B, Mareci D, Munteanu C, Stanciu S, Crimu C I, Trinca L C and Kamel E 2016 *Environmental Engineering And Management Journal* **15** 955-963
[10] Nejneru C, Perju M C, Sandu A V, Axinte M, Quaranta M, Sandu I, Costea M and Abdullah M M A 2016 *Revista De Chimie* **67** 1191-1194
[11] Bakhsheshi-Rad H R, Haerian B, Najafizadeh A, Idris M H, Kadir M R, Hamzah E and Daroonparvar M 2013 *Canadian Metallurgical Quarterly* **52** 449-457
[12] ASM International – Editor MF Rothman High Temperature property data: Ferrous Alloys Metal Park OH 44073 ISBN 0-87170-243-6
[13] Barlow L D and Du Toit M 2012 *Journal of Materials Engineering and Performance* **21** 1327-1336
[14] Nejneru C, Perju M C, Sandu A V, Axinte M, Quaranta M, Sandu I, Costea M and Abdullah M M A 2016 *Revista de chimie* **67**(6) 1191-94
[15] Nejneru C, Cimpoesu N, Stanciu S, Vizureanu P and Sandu A V 2009 *Metalurgia International* **14**(7) 95-98
[16] Perju M C and Vizureanu P 2014 *Revista de chimie* **65**(6) 694-696

IOP Conf. Series: Materials Science and Engineering **374** (2018) 012040 doi:10.1088/1757-899X/374/1/012040

Structural Analysis of Carabiners Materials Used at Personal Protective Equipments

D P Burduhos Nergiş[1], C Nejneru[1], D C Achiţei[1], N Cimpoieşu[1] and C Bejinariu[1*]
[1]"Gheorghe Asachi" Technical University from Iasi- Romania, Faculty of Materials Science and Engineering, Blvd. Dimitrie Mangeron, No. 69, 700050, Iasi- Romania

E-mail: costica.bejinariu@tuiasi.ro

Abstract. Thus specialized shackles are used in severe environment condition, such as: temperature variation (firefighters, mountain climbers), highly corrosive (construction, caving, sailing), etc. so it must be manufactured from materials with good corrosion resistance and refractoriness. Because the carabiners are used at height, besides the properties already exposed, they must possess properties of resilience. This paper presents a metallographic analysis using Scanning Electron Microscope and Energy Dispersive Analysis X-ray to highlight the structural characteristics in order to increase resilience and improve the wear and corrosion resistance properties.

1. Introduction

A carabiner is a special type of shackle with a spring-loaded closure used to connect components quickly and reversibly, mainly at safety systems and to attach elements to a belt or band [1].

Using range of carabiners includes: mountaineering, caving, climbing utility, construction, rescue, arboriculture [2], canoeing and yachting.

Due to their low specific weigh, almost all shackles currently used for climbing are made of aluminum alloys. [3] For situations where weight is not an important factor, such as fixed anchors, steel carabiners are used due to their high wear resistance and tensile strength.

When used for fire rescue or mountaineering courses, steel carabiners are excellent. Refractory steel keeps its integrity at temperatures above 500 °C, making it ideal for evacuation and rescue operations including high temperatures [4]. However, steel carabiners as fire safety connectors are not indicated, because steel loses its resistance properties long before it melts. The steel is hard, high tensile strength and relatively elastic, making it resistant to cracking or deformation [5]. Carbon steel carabiners will rust quickly if exposed to fresh or salt water, while stainless steel is not as wear-resistant, but has much better corrosion properties [6].

The main aluminum carabiners advantage is the low specific weight. Aluminum alloys based on Al-Cu, Zn, Mg, Cr systems are hardened by solution annealing and quench-ageing [7]. Typically, the 7075-T6 is used for the carabiners body and locking element [8].

2. Linking elements analysis

2.1. Carabiners classification

According to SR EN 12275 *Mountaineering equipment – Connectors – Safety requirements and test* [9] standard, carabiners are classified as can be seen in Table 1.

IOP Conf. Series: Materials Science and Engineering **374** (2018) 012040 doi:10.1088/1757-899X/374/1/012040

Table 1. Types of carabiners.

Type	Name	Scheme	Rl_1[1]	Rl_2[2]	Rt[3]
B	Basic carabiner	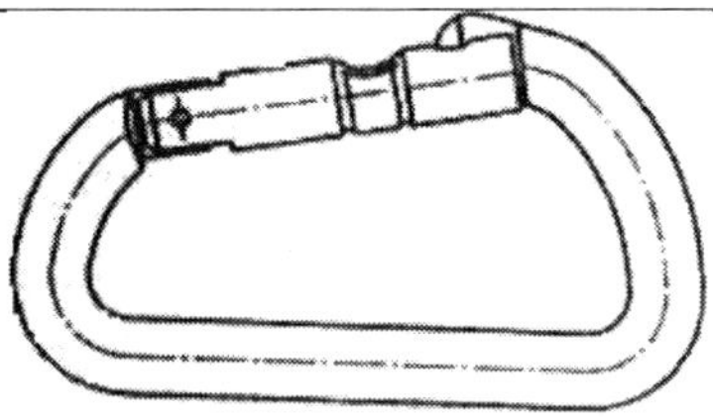	20KN	7KN	7KN
H	HMS carabiner	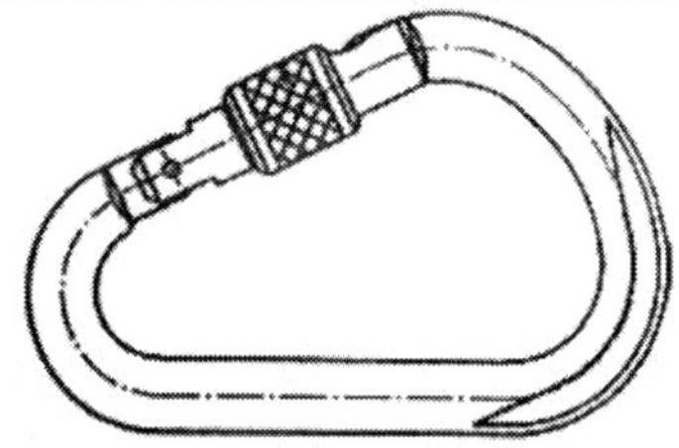	20KN	6KN	7KN
K	Klettersteig carabiner	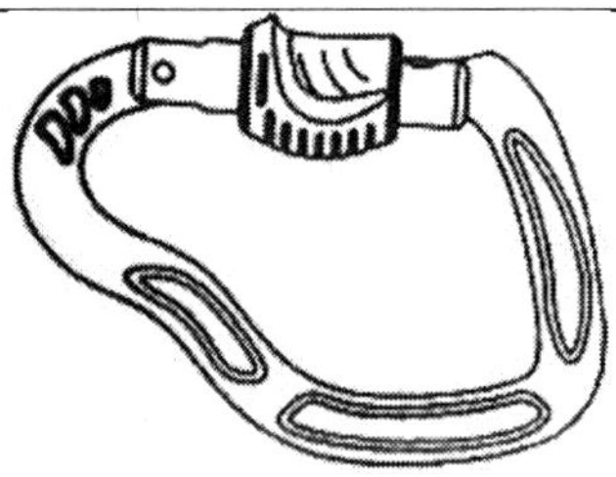	25 KN	8KN	7KN
A	Special piton carabiner	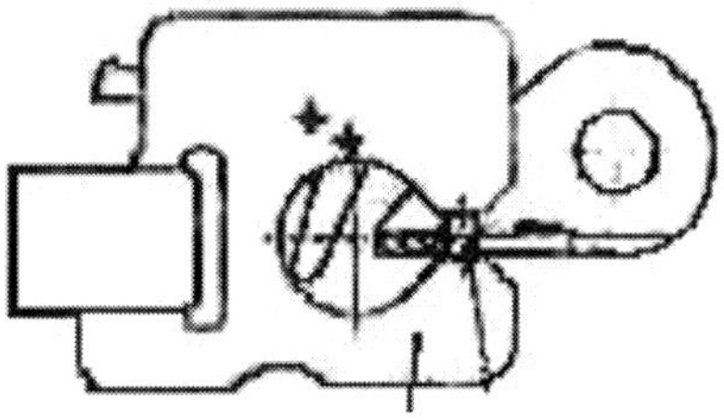	20KN	7KN	
T	With secured position	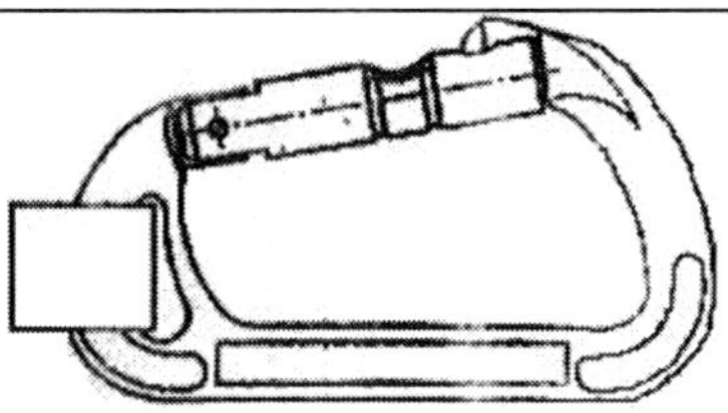	20KN	7KN	
Q	With threaded gate	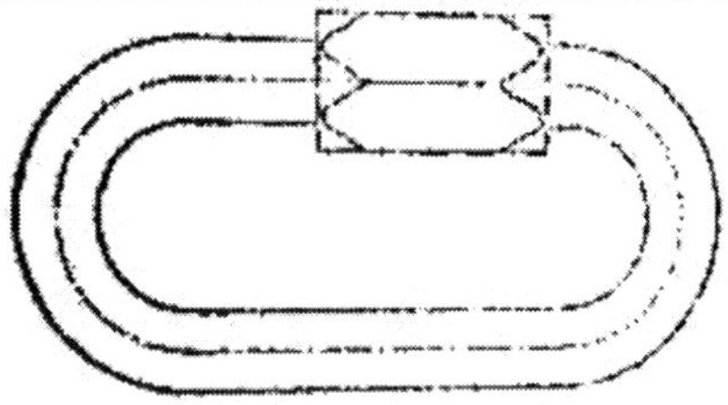	25KN		10KN

IOP Conf. Series: Materials Science and Engineering **374** (2018) 012040 doi:10.1088/1757-899X/374/1/012040

Table 1. Types of carabiners.

Type	Name	Scheme	Rl_1 [1]	Rl_2 [2]	Rt [3]
X	Oval carabiner	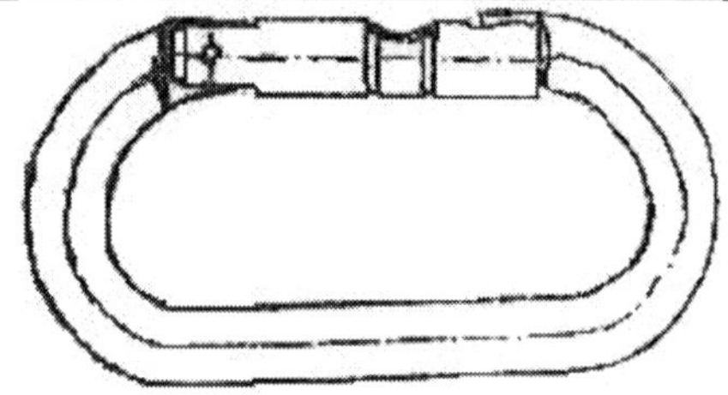	18KN	5KN	7KN

[1]Static resistance on longitudinal axis with gate closed
[2]Static resistance on longitudinal axis with gate open
[3]Static resistance on transversal axis

2.2. Standards and requests

In mountain climbing, thus are used to connect safety rope to anchors in order to stop the climber fall, if necessary. Carabiners are often sold in pairs, attached together through a nylon strap. While one is attached to the anchor, the other one is rope clamped. The requirements of a carabiner can be grouped after loading, geometry or environmental conditions [10].

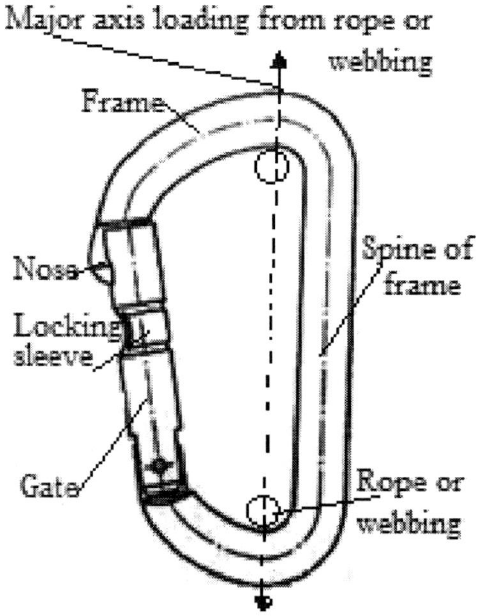

Figure 1. Schematic presentation of a carabiner.

Minimum load requirements for climbing connectors are presented by international and European standards. For current aluminium alloys carabiners, EN and UIAA standards described the safety tests. [9, 11, 12].Carabiners should not be thrown or dropped from heights because scratches can occur on their surfaces due to impacts and may affect their physical properties. Thus must always work correctly, the gate should automatically and completely shut off after releasing, locking may be automatically or by screwing, any geometrical modification are not allowed. Environmental characteristics are very important because humidity and temperature variation can have significant effects on material properties. Carabiners can be exposed at very high temperature ranges, from -40 °

C (Everest peak) to + 80 ° C (due to friction during descent) [8]. Besides these, it is necessary to know the material chemical resistance properties.

3. Structural analysis

Due to the carabiners operating conditions, the manufacturing materials must have good corrosion and shock elasticity properties.

This paper focuses on structural analysis of x type carabiner (Figure 2) used for low loads to attaché the rope to the piton. It is made of steel, has a minimum tensile strength of 22 kN and a mass of 175 g. Due to its design, the load is always in the middle of the carabiner body.

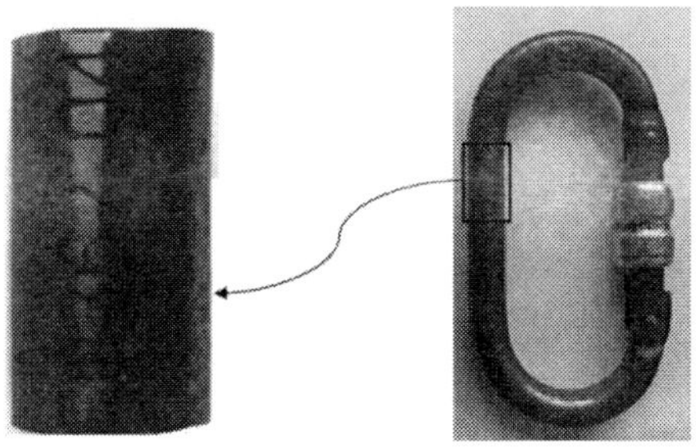

(a) (b)

Figure 2. X type carabiner, (a) studied sample, (b) full carabiner.

The chemical composition (Table 2) of the sample has been determined by optical emission spectrometry using a Foundry Master spectrometer. According to values obtained, the steel grade is C45 (SREN 10083-2 (2006)).

Table 2. Chemical composition.

Element	Fe	C	Si	Mn	P	S	Ni	Cr
Percent	balance	0.47	0.13	0.75	0.03	0.03	0.20	0.15

Metallographic investigations have been carried out on the carabiner material by cutting multiple segments for samples. The microstructure presented in Figures 3, 4a and b consists of equiaxed pearlitic-ferrite grains with well-defined grain boundary and relatively uniform distribution of the pearlitic grains. The EDX analysis performed shows the structure uniformity also from chemical point of view (Figure 4 c).

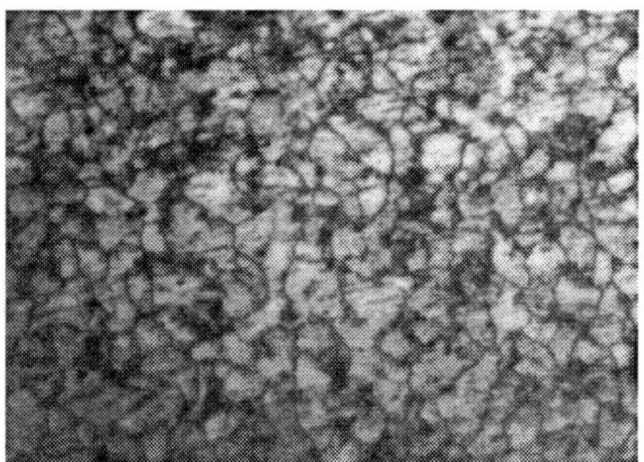

Figure 3. Optical microstructure, X50. Equiaxed pearlitic-ferrite grains.

IOP Conf. Series: Materials Science and Engineering **374** (2018) 012040 doi:10.1088/1757-899X/374/1/012040

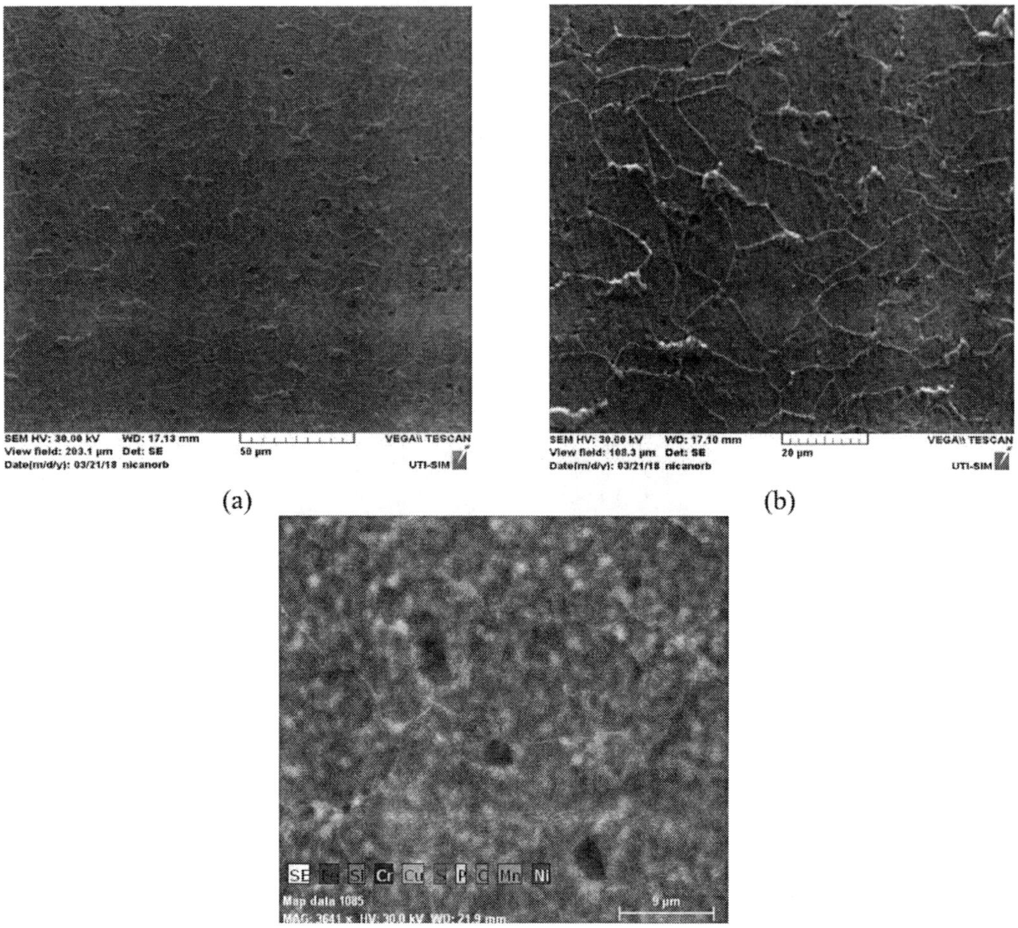

(c)

Figure 4. Carabiner material microstructure obtained using the electron scanning microscope, (a) SEM microstructure, 50 μm; (b) SEM mictrostructure, 20 μm; (c) EDX, 9 μm.

In order to obtain the standard required properties, the material high hardness and wear resistance necessary have been obtained by a final thermal treatment of quenching. The 3400 MPa surface hardness value has been obtained by Vickers test method. Quenching was followed by annealing to increase tenacity.

4. Conclusions

The material has good resistance properties (resilience) and medium corrosion resistance. For carabiner that operates under tough corrosion, wear and temperature variations, a material with better properties (or better quality material) must be used.

In order to keep the final product at a low price (not to use refractory or stainless materials), the surface properties can be improved by thermochemical treatments or coatings with metallic or polymeric substances.

5. References

[1] "Climbing Dictionary & Glossary" *Mountain Daysnet Archived from* the original *on 2007-01-03* Retrieved 2006-12-05
[2] Statham J M and Roebuck B Sc 2003 *Health and Safety Laboratory Buxton*

[3] Blackford J R 2003 Jenkins M (ed) *Materials in sports equipment* Woodhead, Cambridge 279-325
[4] Florea C D, Munteanu C, Cimpoesu N, Sandu I G, Baciu C and Bejinariu C 2017 *Rev Chim* **68** 2582–87
[5] Lazar P, Bejinariu C, Sandu A V, Cazac A M, Corbu O, Sandu I G and Perju M C 2017 *International Conference on Innovative Research - Icir Euroinvent 2017* Sandu A V, Abdullah M M A, Vizureanu P, Ghazali C M R, Sandu I Eds Iop Publishing Ltd: Bristol **209** 120-125
[6] Florea C, Bejinariu C, Munteanu C and Cimpoesu N *Advanced Materials Engineering and Technology* Abdullah *V,* AbdRahim M, BinGhazli S Z, Tahir M F, Yong M F M, Ahmad H and Eds Amer R 2017 *Inst Physics:* Melville **1835** UNSP 020053 ISBN 978-0-7354-1505-8
[7] Achitei D C, Vizureanu P, Dana D and Cimpoesu N 2013 *Metalurgia International* **18** 104-109
[8] Bright C M 2014 *A Hystory of Rock Climbing Gear Technology And Standards* Mechanical Engineering Udergraduate Honor Theses
[9] British Standards Institution 1998 *EN 12275 Mountaineering equipment – Connectors – Safety requirements and test,* British Standards Institution London
[10] Scott V 2008 *Design of a Composite Carabiner for Rock Climbing* Final Year Project Mechanical Engineering Imperial College London
[11] UIAA 2004 *UIAA 121 - Mountaineering and Climbing Equipment – Connectors,* Union Internationale des Associations d'Alpinisme
[12] Pintilei G L, Crismaru V I, Abrudeanu M, Munteanu C, Luca D and Istrate B 2015 *A lied surface science* **352** 169-177
[13] Nedeff V, Mosnegutu E, Panainte M, Ristea M, Toma S, Agop M 2012 *Powder Technology* **221** 312–317

IOP Conf. Series: Materials Science and Engineering **374** (2018) 012041 doi:10.1088/1757-899X/374/1/012041

Dynamic Impact Behaviour of High Entropy Alloys Used in the Military Domain

V Geantă[1], I Voiculescu[1], R Stefănoiu[1*], T Chereches[2], T Zecheru[3], L Matache[3] and A Rotariu[4]
[1] Politehnica University of Bucharest, 313 Splaiul Independenţei, 060042, Bucharest, Romania
[2] UPS PILOT ARM, 2 Laminorului Street, 137210, Târgoviste, Romania
[3]Scientific Research Center For CRBN Defence And Ecology Bucharest, 225 Oltenitei Road 041309, Bucharest, Romania
[4]Military Technical Academy, 39-49 George Coşbuc Boulevard, 05014, Bucharest, Romania

E-mail: radustefanoiu@yahoo.com

Abstract. AlFeCrCoNi high entropy alloys (HEA) feature significant compressive strength characteristics, being usable for severe impact applications in the military domain. The research paper presents the results obtained by testing the impact resistance of four HEA samples of different chemical compositions at perforation with 7.62 mm calibre incendiary armour-piercing bullets. The dynamical behaviour was modelled by numerical simulation based on the results of the dynamic tests conducted in the firing range, thus allowing the development of more efficient high entropy alloys, to be used for collective/personal protection.

1. Introduction

The materials used for ballistic protection have the role of collective or personal defence in the military tactical field. The main characteristics of the ballistic protection materials are the strong resistance to impact with ammunition, as well as the high capacity to absorb the kinetic and explosive energy developed. One of the classes of alloys that can ensure these performances is the high entropy alloys class (HEA) [1-4].

High entropy alloys are composed of n major alloying elements, $n \geq 5$, introduced in equimolar or nearly equimolar ratios, easily leading to the formation of simple solid phase solutions with BCC or FCC, nano-structures or even amorphous states as cast [5-7].

The decision to use a certain high entropy alloy in military defense applications is adopted after a scientific analysis of the chemical composition, of the influence of the alloying elements and of the heat treatments which determine the impact resistance capacity. The design of ballistic protection systems is initially validated by simulations followed by experiments conducted in firing ranges. The application of numerical simulation methods results in the streamlining of the technical solutions, in reducing the time required for the completion of the products and in increasing the economic efficiency [8-11].

Numerical simulations are applied from the first steps of the investigation - the design of physical models, followed by the modelling of technological processes and, in the end, by the modelling of the ballistic impact behaviour of the composite structures analysed.

The accuracy of the solutions provided by numerical simulations is guaranteed by the correctness of the physical models used. At present, the mathematical support of numerical simulation methods - the mathematical model - is sufficiently developed. In this respect, the use in numerical simulations of current software packages: ABAQUS, ANSYS, AUTODYN, COSMOS M, LS-DYNA, NASTRAN etc. guarantees the correctness of the solutions, provided that the user correctly designs the physical models using discrete elements [12-15].

Given the diversity of high entropy alloys and their properties, a first step in putting in place is the design of physical models, which are defined by physical consistent patterns expressed by mathematical expressions in the form of constitutive equations. As rapid deformations are accompanied by local thermal effects as well, the material models also include the influence of the thermal regime on the behaviour of the material [4, 16, 17].

The elasticity of the material is generally expressed by the generalized Hooke's law as a linear function between the elastic strain tensor and the stress tensor. The constitutive equations expressing the plastic behaviour of the materials have different forms, according to the plasticity models used. Currently, the plasticity model most commonly used is the plastic yield model [18]. In the case of high-speed deformation, the most commonly used model is the visco-elasto-plastic model. The constitutive equations of this model contain constants specific to each material and which can be determined experimentally [4].

The paper presents the results obtained by testing some high entropy alloys, used for collective/personal protection, subjected to dynamic stresses with high deformation speeds (impact resistance behaviour by perforation with incendiary armour-piercing bullet). The impact behaviour was numerically simulated based on the results of the dynamic tests conducted in the firing range, in order to choose the best alloy to be used for the above-mentioned applications.

2. Materials and methods

The high entropy alloys were obtained using a vacuum arc remelting equipment (MRF ABJ 900), located in the Laboratory ERAMET - Materials Science and Engineering Faculty from Politehnica University of Bucharest [1, 4]. For the experiments, there were considered four different compositions of high entropy alloys, varying the quantity of some elements (Al, Ni) considered as of significant influence on the properties. The chemical composition of the high entropy alloys obtained and selected for the experimental program is presented in Table 1 [4].

Table 2 presents the values determined for the compressive strength and the hardness of the high entropy alloys selected for the tests [4].

Table 1. The chemical composition of high entropy alloys used in the experimental program.

Element	Chemical composition, %			
	HEA-1 **AlCrFeCoNi**	HEA-5 **$Al_{0.8}$CrFeCoNi**	HEA-12 **AlCrFeCo$Ni_{1.4}$**	HEA-14 **AlCrFeCo$Ni_{1.8}$**
Al	10.67	8.72	9.76	8.99
Cr	20.55	21	18.8	17.32
Fe	22.13	22.61	20.24	18.65
Co	23.32	23.82	21.33	19.65
Ni	23.33	23.85	29.86	35.38

The full characterization of high entropy alloys subjected to dynamic stresses with high deformation speeds involves performing a number of tests, including the simplest tests, conducted in mechanical testing laboratories, as well as special tests for analysing the material response to rapid stresses.

Table 2. The mechanical properties of the high entropy alloys used in the experimental program.

Property	HEA-1 **AlCrFeCoNi**	HEA-5 **$Al_{0.8}CrFeCoNi$**	HEA-12 **$AlCrFeCoNi_{1.4}$**	HEA-14 **$AlCrFeCoNi_{1.8}$**
Compressive strength, MPa	2200	1350	1600	1200
Hardness $HV_{0.1}$	563	427	456	316
Hardness HB	535	406	432	300

The purpose of the test experiments is also to determine the dynamic behaviour of HEA when impacted by incendiary armour-piercing bullets. The need for experimental approaches is due to the following main reasons:

- HEA are produced with the goal of building multilayered ballistic packages, consisting also of other materials such as ceramic, steel, aluminium, Dyneema;
- The phenomena associated with the impact of bullets/ballistic fragments/shock waves and various ballistic packages are extremely complex and occur at very high speeds;
- Numerical simulation is a mandatory stage for the design and delivery of optimal ballistic packages;

The execution of a simulated model, as close as possible to the real one, requires the dynamic characterization of the materials involved in the model.

Experimental tests were performed in the Ballistic Personal Protection Equipment Laboratory (Figure 1) of the CBRN and Ecology Defence Scientific Research Centre in accordance with the approved military procedures. There was tested the behaviour of the HEA structures under investigation to 7.62 mm incendiary armour-piercing bullet impacts. The HEA samples were placed on a polystyrene parallelepiped support placed in a metal frame, 5 m from the barrel's muzzle.

Figure 1. Firing room picture.

A chronograph with photocell frames was inserted in the firing direction to determine the initial speed of the bullets. The test method assumed the variation in the bullets' initial speed so as to obtain the speed at which all the kinetic energy of the bullet is consumed for its plastic deformation and total or partial penetration of HEA-type samples.

The test configuration is shown in Figure 2. The amount of powder in the cartridge was varied in order to obtain the desired initial speed. The limit speed was calculated as an average penetration speed for 10 firings in the steel discs and its value was 437 m/s.

The effect the impact with the target plate had on the bullets, required for the numerical simulation, was also monitored.

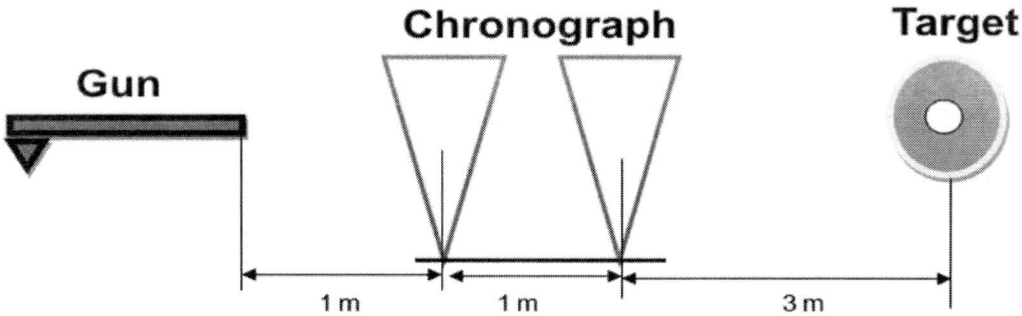

Figure 2. Test configuration.

The impact with the incendiary armour-piercing bullet resulted in the partial or total penetration of the samples, with or without total fragmentation. In the case of partial penetration without total fragmentation, it is possible to measure the characteristic dimensions of the sample, i.e. the depth of penetration H_1 and the maximum thickness of the sample following deformation H_2, shown schematically in Figure 3.

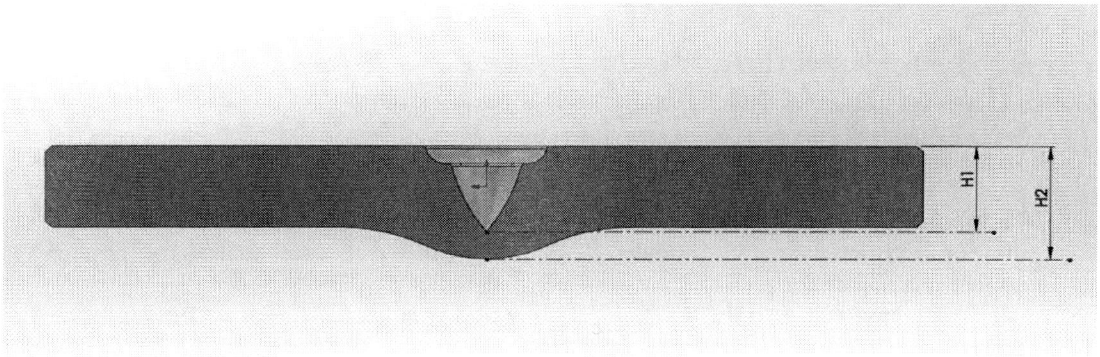

Figure 3. Characteristic appearance and dimensions of a sample after impact with the 7.62 mm calibre incendiary armour-piercing bullet (partially penetrated).

3. Results and discussions

The conditions under which the tests were performed and the configurations of the HEA-type samples obtained after firing are shown in Figures 4 – 7. The results obtained after firing 7.62x39 mm steel core incendiary armour-piercing bullets in the firing range are shown in Table 3.

IOP Conf. Series: Materials Science and Engineering **374** (2018) 012041 doi:10.1088/1757-899X/374/1/012041

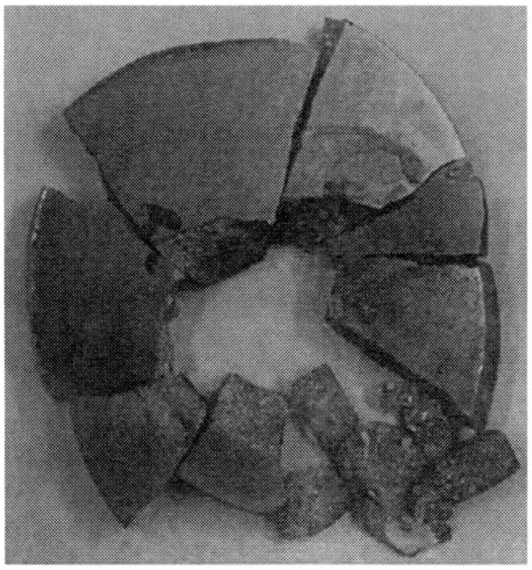

Figure 4. HEA 1 plate after the firing of a 7.62x39 mm steel core incendiary armour-piercing bullet: **Alloy plate made of AlCrFeCoNi - HEA 1:**
Outer diameter Ǿ=65.80 mm
Thickness g=6.70 mm
7.62x39 mm steel core incendiary armour-piercing bullet:
Bullet 8: Bullet speed V_0 = 412 m/s.

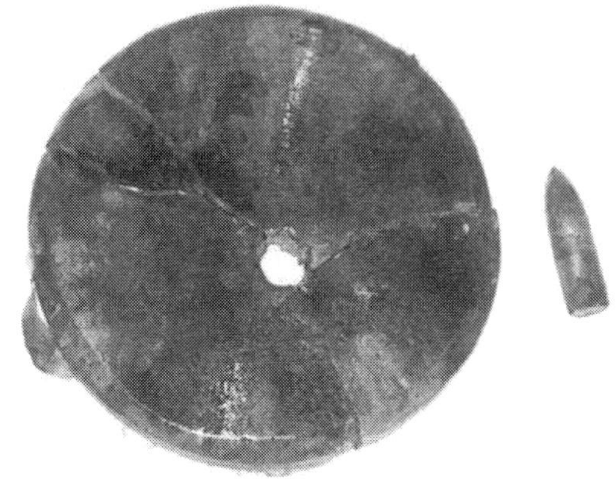

Figure 5. HEA 5 plate after the firing of a 7.62x39 mm steel core incendiary armour-piercing bullet: **Alloy plate made of $Al_{0.8}$CrFeCoNi - HEA 5:**
Outer diameter Ǿ=65.80 mm
Thickness g=6.60 mm
7.62x39 mm steel core incendiary armour-piercing bullet:
Bullet 9: Bullet speed V_0 = 451 m/s.

Figure 6. HEA 12 plate after the firing of a 7.62x39 mm steel core incendiary armour-piercing bullet:
Alloy plate made of AlCrFeCoNi1.4- HEA 12:
Outer diameter Ǿ=65.80 mm
Thickness g=6.80 mm
7.62x39 mm steel core incendiary armour-piercing bullet:
Bullet 11: Bullet speed V0 = 443 m/s.

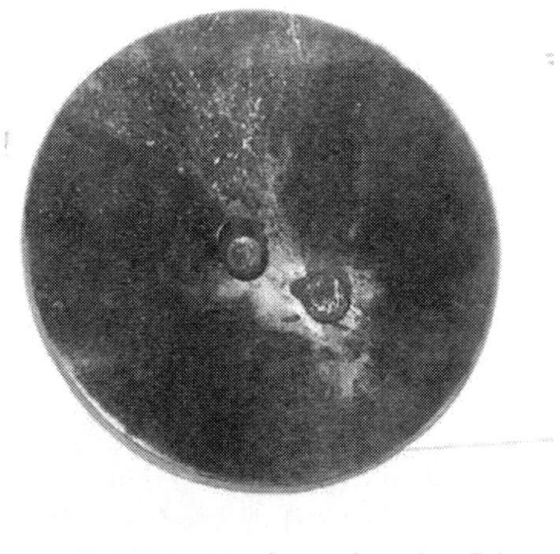

Figure 7. HEA 14 plate after the firing of 2 7.62x39 mm steel core incendiary armour-piercing bullets:
Alloy plate made of AlCrFeCoNi1.8- HEA 14:
Outer diameter Ǿ=65.80 mm
Thickness g=6.60 mm
7.62x39 mm steel core incendiary armour-piercing bullet:
Bullet 12: Bullet speed V0 = 457 m/s
Bullet 13: Bullet speed V0 = 419 m/s.

Table 3. The results of the tests performed in the firing range.

	Armoured steel plate 1	
Bullet 1 impact effect, 553 m/s speed	Pierced	**YES**
	H1	-
	H2	-
Bullet 2 impact effect, 360 m/s speed	Pierced	**NO**
	H1	5.6 mm
	H2	8.2 mm
Bullet 3 impact effect, 370 m/s speed	Pierced	**NO**
	H1	6.5 mm
	H2	8.8 mm
	Armoured steel plate 2	
Bullet 1 impact effect, 388 m/s speed	Pierced	**NO**
	H1	5.5 mm
	H2	7.8 mm
Bullet 2 impact effect, 426 m/s speed	Pierced	**NO**
	H1	6.8 mm
	H2	8.8 mm
Bullet 3 impact effect, 427 m/s speed	Pierced	**NO**
	H1	7.2 mm
	H2	9.1 mm
	Armoured steel plate 3	
Bullet 1 impact effect, 491 m/s speed	Pierced	**YES**
	H1	-
	H2	-
	HEA1 plate	
Bullet 1 impact effect, 412 m/s speed	Pierced	**YES** - the sample broke into 12 parts as detected
	H1	-
	H2	-
	HEA 5 Plate	
Bullet 1 impact effect, 451 m/s speed	Pierced	**YES** - the sample broke into 4 parts
	H1	-
	H2	-
	HEA 12 Plate	
Bullet 1 impact effect, 443 m/s speed	Pierced	**YES** – the sample broke into 4 parts
	H1	-
	H2	-
	HEA 14 Plate	
Bullet 1 impact effect, 457 m/s speed	Pierced	**YES**
	H1	-
	H2	-
Bullet 2 impact effect, 419 m/s speed	Pierced	**YES**
	H1	-
	H2	-

Following the ballistic tests, there was found as follows:

- The steel samples were not pierced at low speeds during adjustment firings;
- The steel samples were perforated without fragmentation following impact at the set limit speed;
- The HEA 1, HEA 5 and HEA 12 samples were pierced with total fragmentation;
- The HEA 14 samples were pierced without total fragmentation.

4. Numerical simulation of the impact behaviour of high entropy alloys

The results of the ballistic firing into the samples made of high entropy alloys were interpreted using the numerical simulation method. For the four HEAs with known mechanical properties, simple dynamic material models were formulated. In order to formulate the dynamic plasticity model, the following coefficients were sufficient:

- σ_c hypothetical yield point, initial;
- E_T average tangent plasticity modulus;
- C deformation speed scaling coefficient;
- P deformation speed exponent.

The data obtained using statistic methods in previous research [4] was used for the measures σ_c and E_T. The p exponent is given the value 1, considering a linear viscosity law. The C coefficient values are determined using successive numerical simulation methods. The C coefficient is adjusted until the result of the simulation overlaps the result obtained by experimental firing in the firing range.

The axial-symmetric model shown in Figure 8 was used both for calibrating the simulation method on the control interval and for the simulation of thresholds on the samples made of high entropy alloys. The firings were simulated in accordance with the experimental firing conditions.

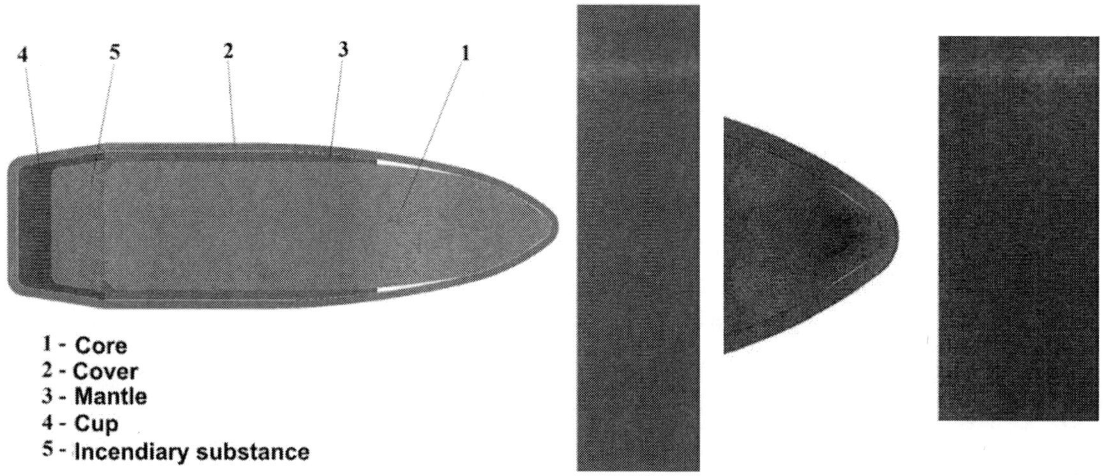

(a) (b)

Figure 8. Axial-symmetric model: (a) Model; (b) Discretization detail.

Figure 9 shows an example of the calibration simulation results.

The figure shows the bullet steel core speed variation graph when crossing the target and the section by model in the last recorded state.

After comparing the results of the simulations with the firing ones, it was found that the simulation model and method used needed no changes.

The dynamic characterization of high entropy alloys consisted in regulating the effects of the viscosity (parameter C) and of the actual plastic deformation following yielding when achieving the similar final states.

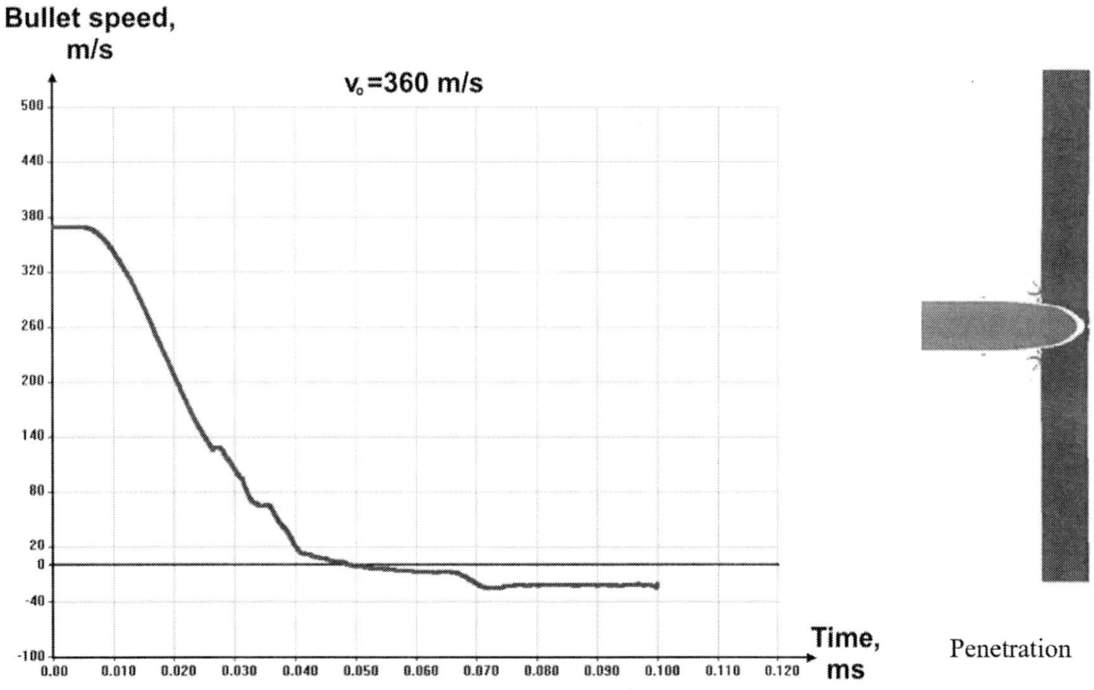

Figure 9. Simulation of calibration firing onto the steel sample.

Figures 9 - 13 show the results of the numerical simulations by bullet steel core speed variation graphs and the final shapes of the assembly.

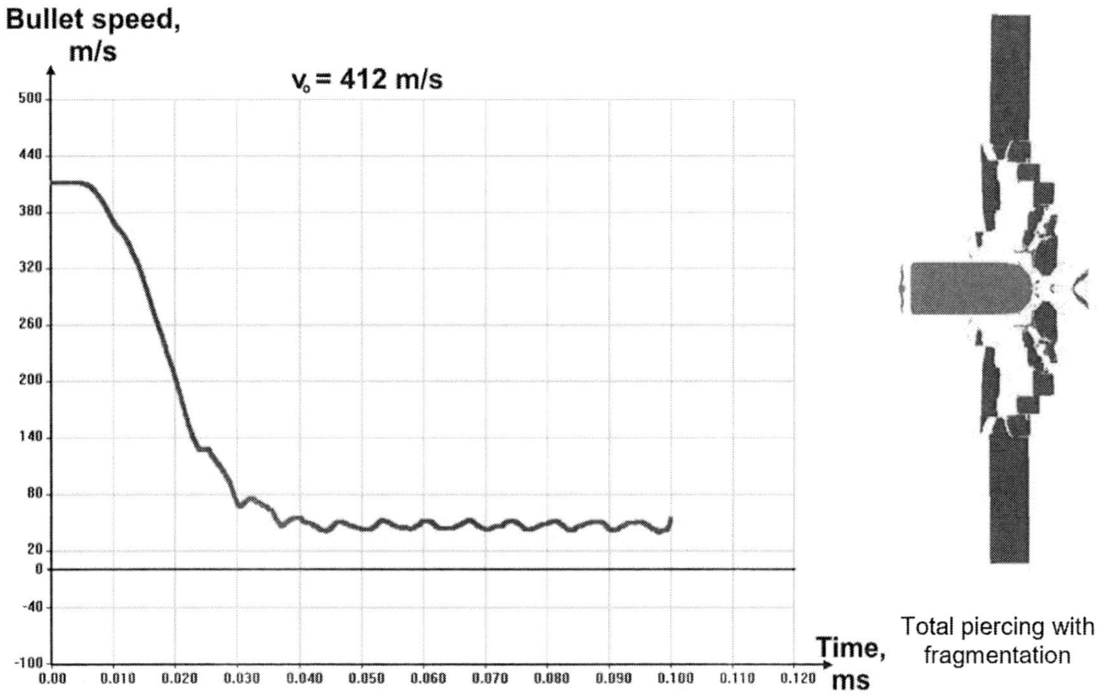

Figure 10. Firing no. 1 into the sample made of HEA 1.

IOP Conf. Series: Materials Science and Engineering **374** (2018) 012041 doi:10.1088/1757-899X/374/1/012041

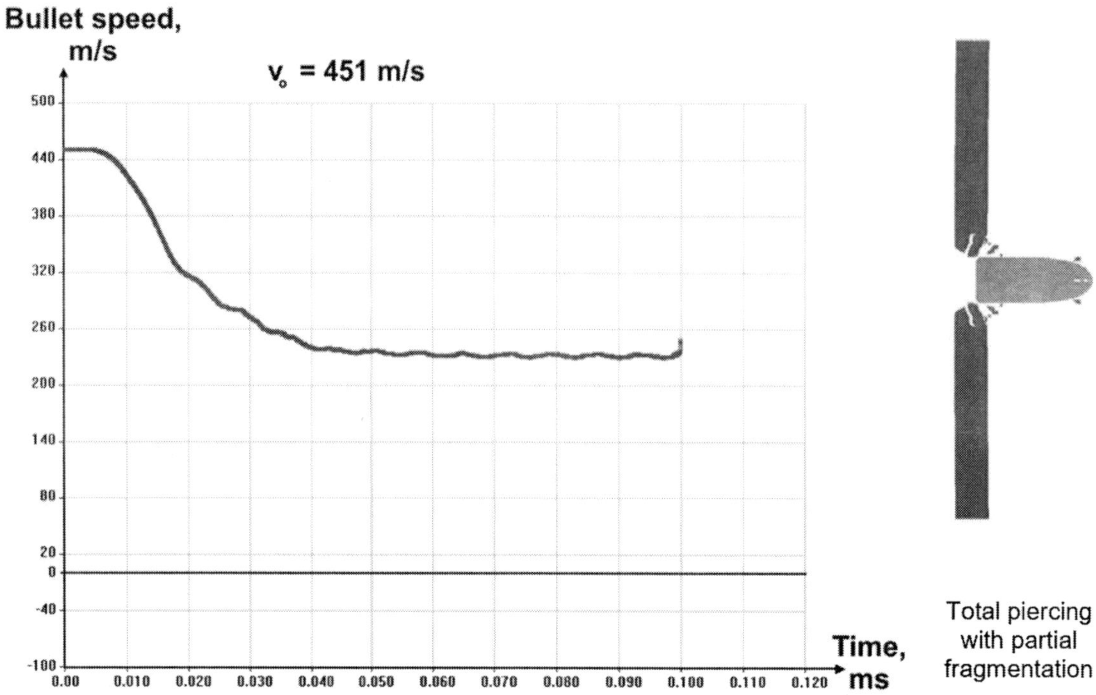

Figure 11. Firing no. 2 into the sample made of HEA 5.

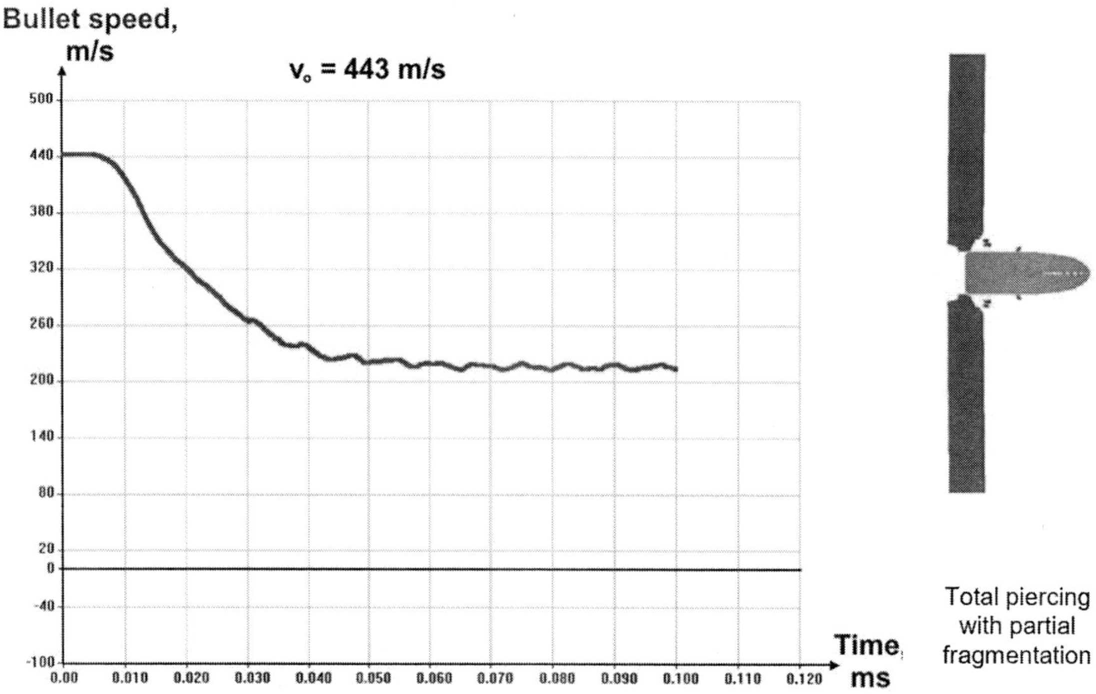

Figure 12. Firing no. 4 into the sample made of HEA 12.

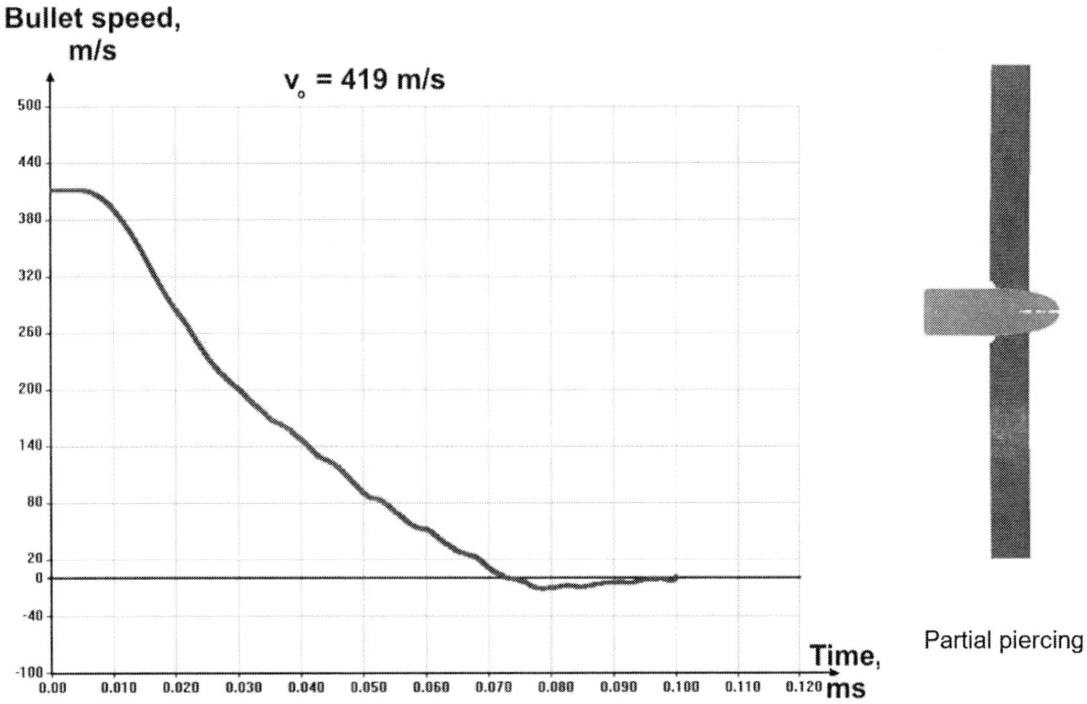

Figure 13. Firing no. 5 into the sample made of HEA 14.

5. Conclusions

The parallel analysis of the results of the experimental firings in the firing range and of the simulated ones leads to some conclusions regarding the behaviour of high entropy alloys under dynamic stresses at high deformation speeds:

- Very high variations in behaviour are observed for the four types of alloys tested;
- The materials of the fragmented samples exhibit good resistance characteristics, although their tenacity is very high (HEA-1, HEA-5, HEA-12);
- The material of the pierced but not fragmented samples has lower resistance characteristics and better tenacity (HEA-12);
- The resistance and tenacity parameters of the HEA-5 and HEA-12 materials can be improved by appropriate heat treatments;
- The high entropy materials, HEA-5 and HEA-12, with the necessary improvements, especially for increased tenacity, can be used to build composite structures in different combinations, designed for ballistic protection.

6. References

[1] Geantă V, Voiculescu I, Ştefănoiu R, Savastru D, Csaki I, Patroi D and Leonat L 2013 Optoelectron. Adv. Mat. **7** 874
[2] Voiculescu I, Geantă V, Ştefănoiu R, Pătroi D and Binchiciu H 2013 *Rev. Chim* **64** 1441
[3] Geantă V, Chereches T, Lixandru P, Voiculescu I, Ştefănoiu R & all. 2017 *Metals* **7** 496
[4] *Composite structures resistant at dynamic loadings applied with high deformation speeds used in the field of collective protection – HEAMIL.* Research Project – PCCA 209/2012
[5] Zhou Y J, Zhang Y, Wang Y L and Chen G L 2007 *Mat. Sci. Eng. A-Struct.* **260** 454
[6] Hsua U S, Hunga U D, Yeh J W, Chen S K, Huanga Y S and Yang C C **2007** *Mat. Sci. Eng. A-Struct.* **77** 457
[7] Lee C P, Chang C C, Chen Y Y, Yeh J W and Shih H C 2008 *Corros. Sci.* **50** 2053

[8] Bhatnagar A *Lightweight ballistic composites, Military and law-enforcement applications* 2006 Woodhead Publishing and Maney Publishing, Cambridge, England
[9] Ben – Dor G., Dubinsky A. and Elperin T 2002 *Compos. Struct.* **56** 243
[10] Cepus E 2003 An experimental investigation of the early dynamic impact behaviour of textile armour systems: Decoupling material from system response, PhD Thesis The University of British Columbia
[11] Duan Y, Keefe M, Bogetti T A and Cheeseman B A 2005 *Compos. Struct.* **68** 331
[12] Hansbo P, Hermansson J and Svedberg T *Comput. Methods Appl. Mech. Engng.* 2004 **193** 4195
[13] Hiermaier S, Riedel W, Clegg R and Hayhurst C 1999 *Advanced material models for hypervelocity impact simulations*, Tech. Rep., ESA/ESTEC Contract no. 12400/97/NL/PA(SC).
[14] Lim C T, Shim V P W and Ng Y H 2003 *International J. of Impact Engineering* **28** 13
[15] Moaveni S 2003 *Finite Element Analysis: Theory and Application with ANSYS*, 2nd Edition (Prentice – Hall)
[16] Jovicic J, Zavaliangos A and Ko F 2000 *Composites: Part A* **31** 773
[17] Leigh S P and Porwal P K 2003 *Int. J. of Solids and Structures* **40** 6723
[18] Jacobs M J N and Van Dingenen J L J 2001 *J. Mate. Sci.* **36** 3137

Acknowledgements
This work was supported by a grants of the Romanian Ministery of Research and Innovation, CCCI - UEFISCDI, PCCA 209/2012 "Composite structures resistant to dynamic loadings applied at high deformation speeds used in the field of collective protection – HEAMIL" and PN-III-P1-1.2-PCCDI-2017-0875 / 20PCCDI 2018 - "Individual and collective protection systems for the military domain based on high entropy alloys - HEAPROTECT", within PNCDI III.

Research Regarding Membrane Filtration Capacity of Water Collected from Siret River

I Mihalache, Ş C Pintilie, I G Bîrsan, E Dănăila and Ş Baltă*
"Dunărea de Jos" University of Galați, str. Domnească nr. 111, 800201, Galați, Romania

E-mail: stefan.balta@ugal.ro

Abstract. In the past decade, the high demand and strict legislations regarding pure and potable water production and quality require finding new treatment technologies with higher effectiveness. When compared with conventional treatment technologies, membrane technology is a viable option in water and wastewater treatment due to high performance, ease in implementation, cost-efficiency among other advantages, also, leading to a rapid expansion in use in almost all areas of industry. Polymeric ultrafiltration membranes have been successfully used in various industries since 1969, and in later years they were studied in the water purification sector, mainly as a pre-treatment step to reduce severe fouling that could occur in reverse osmosis filtration stage. Polysulfone (PSf) was the polymer of choice in this study with two concentrations, 25 wt.% and 30 wt.%. Surface SEM morphology, roughness and water affinity were analyzed for the studied membranes. Water from Siret river was used in the permeation tests in order to analyze the retention capacity and anti-fouling ability. The present study revealed higher retention for the 30 wt.% PSf membranes, from the physico-chemical and microbiological point-of-view and lower fouling, also.

1. Introduction

Membrane technology arisen from the need for cleaner water to provide areas with freshwater scarcity or shortage, areas that do not have a well-established treatment plant or areas with low access to clean water. A viable option is to use river water in order to withstand high water demand, is the use of membrane filtration technology of watercourse sources [1]. Conventional treatment plants for potable water production consists in multiple stages with complex processes (coagulation, flocculation, sedimentation, sand filtration, pH correction, chlorination, etc.) for producing the desired water quality, which in almost all stages various chemicals are used (aluminium sulphate, sulphuric acid, lime, chlorine gas, etc.) [2], each requiring strict monitorization and safety precautions. Also, due to increase of pollution in natural waters and rigorous regulation for drinking water quality demand continuous modifications and improvements in conventional treatment plants [3]. Membrane technology offers superior advantages when compared with conventional treatments, such as high water quality, modular construction, majority of chemicals needed are for membrane cleaning meaning very low by-product and residue production [4]. Polymeric membranes fabricated though the phase inversion method are the most promising in water and wastewater treatment. Due to ease in controlling the fabrication parameters, membranes are used in different sectors of industry [5]. Depending on the area of use of membranes, the water permeate quality (boiler feed water, cooling water, ultra-pure water, potable water, water for chemical production, etc.) can be easily controlled to meet customer requirements through various types of membranes, most notably, in the order of retention performance, are microfiltration (MF), ultrafiltration (UF), nanofiltration (NF) and reverse osmosis (RO) [5]. Usually, when treating surface water though membrane technology, three stages are needed: prefiltration, ultrafiltration and reverse osmosis [6] in order to prevent severe fouling in the latter stage. The major drawbacks in membrane filtration are caused by the fouling phenomenon and concentration polarization. Fouling occurs when colloidal materials are adsorpted or deposited on the surface and/or in the membrane structure, whereas concentration polarization appear as an accumulation of retained particles on the membrane surface [7]. These issues cause a decrease in water permeation and change in retention capacity, resulting in a lower water production with undesired quality [8] and higher costs due to pressure increase stabilize the flux of permeate. According to Benavente et al. (2016), fouling is related with hydrophilicity and roughness parameters [9]. A membrane with low contact angle and low roughness possess superior anti-fouling properties.

This study will observe membrane performance, depending on two polymer concentrations of 25 wt.% and 30 wt.%, in the retention of natural water collected from Siret river near Galați city. Increasing the polymer concentration will lead to higher retention properties such as permeate quality and fouling resistance, but lower flux.

2. Methods and analysis

2.1. Materials and membrane preparation

Polysulfone (PSf, average Mw. 35 000) and N-methyl-2-pyrrolidone (NMP, C5H9NO, 99%) were purchased from Sigma Aldrich and used without further modifications. Support layer of non-woven polyester fabric (Novatexx 2471) was kindly supplied from Freudenberg (Winheim, Germany). The polymer solutions were obtained through mixing PSf with NMP solvent, Table 1, at constant stirring at room temperature for at least 24 hours.

Table 1. Membrane composition.

Membrane	Polymer type	Polymer concentration [wt.%]
25_PSf	PSf	25
30_PSf		30

After complete mixing, the homogeneous polymeric solution was casted on the support layer, that was fixed on a glass plate, with a casting knife with an air gap of 250 μm. Subsequently, the casted solution was immersed in a coagulation bath (distilled water) until phase inversion was complete. The formed thin film membranes were stored in distilled water until testing.

2.2. Contact angle measurements

Surface hydrophilicity is inversely proportional with contact angle. The prepared PSf membranes were analyzed using contact angle goniometer (OCA 15EC, DataPhysics). The dropping was repeated minimum 5 times and the average was reported. The software error was below 0.9 μm and the standard deviation was maximum +/- 5.

2.3. Morphological analysis

Membrane surface morphology was analyzed with a scanning electron microscope SEM Quanta 200 (Fei, Philips) after the samples were dried.

Roughness data and 3D images were determined using SEM images of the membrane top-view and ImageJ software (Imagej software, https://imagej.nih.gov/ij/) with two plug-ins. The *SurfCharJ* plug-in used for roughness values transforms the variation of grey tones in peaks or valleys, lower roughness values corresponding to lower grey tones and vice-versa, resulting in average roughness (Ra). *Interactive 3D surface plot* plug-in uses a similar concept of data acquisition that transforms the greyness values into three-dimensional peak and valleys [10]. Pore density was studied using SEM images and Imagej software by pore counting on a defined surface. A total membrane surface of approx. 450 μm^2 per membrane was carried for total pore counts.

2.4. Permeation test
All permeation tests were carried using a dead-end stirred cell, Sterlitech HP4750 and an operation pressure of 10 bar. The water transport through the studied membranes was calculated with the flux [L m^{-2} h^{-1}] equation:

$$J_{0,1} = V / A \times t \quad (1)$$

where J_0 is the pure water flux [L m^{-2} h^{-1}], J_1 is the river water flux [L m^{-2} h^{-1}], V is the permeate volume [L], A is the membrane active surface [m^2] and t represents the filtration time [h].

The fouling tendency is calculated with the relative flux equation, expressed as the ratio between river water flux and pure water flux, as follows:

$$Rf = J_1 / J_0 \quad (2)$$

2.5. Physical and chemical analysis of studied water
The values of chemical oxygen demand (COD-Cr), suspended matter, extractable substances and chlorides indicators were analyzed in a specialized laboratory. pH was analyzed with HQ40d portable multiparameter (Hach Lange GmbH, Germany), turbidity was analyzed with 2100P ISOTurbidimeter (Hach Lange GmbH, Germany) and total nitrogen, total ionic iron, lead, zinc, chromium and indicators were analyzed with UV-Vis Spectrophotometer HACH DR5000 (Hach Lange GmbH, Germany).

2.6. Total bacteria count
Total bacteria count was made with a Most Probable Number Method (MPN) with Hach test kit. The procedure consists in transferring a predefined amount of water sample into a tube (A) containing a reaction liquid, followed by hand agitation. Afterwards, 1 mL of the resulted liquid is transferred into another tube (B) of the same reaction liquid.

Table 2. Total bacteria count test chart [www.hach.com].

Tube A	Tube B	Tube C	Tube D	Water quality
-	-	-	-	< 1 bacterium/mL
+	-	-	-	~1-10 bacteria/mL
+	+	-	-	~10-100 bacteria/mL
+	+	+	-	~100-1000 bacteria/mL
+	+	+	+	>1000 bacteria/mL

The procedure is repeated a total of 4 times (A, B, C and D) for repeated dilution and the tubes are incubated at 35°C for 24 - 48 h. After incubation, the cloudy tubes indicate bacterial growth whilst the clear tubes indicate no bacterial growth, as shown in Table 2. The "+" symbol represents cloudy tubes that showed bacteria growth, while the "– "symbol represents clear tubes which showed no bacteria growth.

3. Results and discussion
3.1. Contact angle measurements
Water contact angle is an important parameter in order to determine the membranes water affinity. It is seen in table 3 that higher concentration of polysulfone results in higher contact angle. Polysulfone is

hydrophobic by nature and thus, increasing its concentration will result in a higher hydrophobic membrane.

Table 3. Membrane properties.

Membrane	Contact angle [°]	Pure water flux [L m^{-2} h^{-1}]	Pore density [pores/μm^2]
25_PSf	61.33	147.418	1.0788
30_PSf	68.27	76.146	0.6413

3.2. Morphological analysis

From top-view SEM images, as show in Figure 1, it can be seen that the membrane with lower PSf concentration (25_PSf) contain a higher pore number, compared with 30 wt.% PSf membrane, as depicted from table 3, also. However, higher polymer concentration lead to smaller pores, meaning that, although flux could have a lower value, retention will be superior. According to Holda et al. (2013), pore size and surface porosity is directly proportional with polymer concentration due to slower solvent (NMP)/non-solvent (water) diffusion during phase inversion which lead to delayed demixing resulting a more closed membrane surface [11]. According to Zinadini et al. (2017), membrane roughness and surface water affinity determines the fouling tendency of membranes [12].

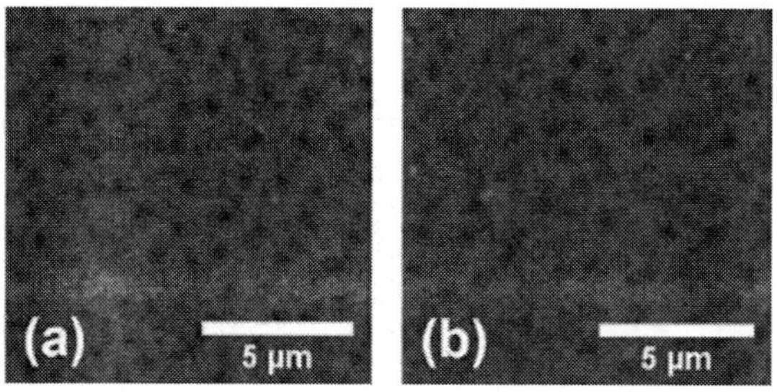

Figure 1. Top-view SEM images for (a) 25_PSf and (b) 30_PSf membranes.

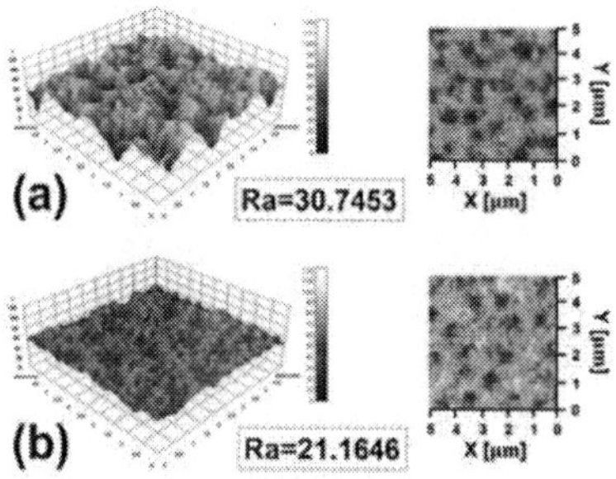

Figure 2. 3D surface (left) and 2D surface (right) images for "Ra" roughness calculation for (a) 25_PSf and (b) 30_PSf membranes using SEM imagery and ImageJ software.

From Figure 2, it is observed that higher polymer concentration lead to lower surface roughness, from 30.7453 for 25 wt.% PSf to 21.1646 for 30 wt.%. One explanation could be the decrease in pore density of the higher polymer concentration membrane, as show in Table 3, meaning lesser valleys.

3.3. Permeation tests

Figure 3 shows the behavior of membranes during the filtration of the Siret River water, from the flux point of view. At the beginning of the retention test, the initial retention flux should ideally start at the last value of pure water flux (Table 3) or even higher due to membrane loosening after compaction with a certain pressure. The pure water flux of 30 wt.% PSf membrane is 76.146 L m^{-2} h^{-1} and an initial retention flux of 74.3 L m^{-2} h^{-1}, while for the 25 wt.% PSf membrane the initial retention flux starts at approx. 122.7 L m^{-2} h^{-1}, a difference of approx. 24.7 L m^{-2} h^{-1} when compared with pure water flux. The reason of immediate decrease in flux is due to instantaneous deposition of particles on the membrane surface, blocking approx. 16.7 % of the total pores (ratio between J_0 and initial J_1 value).

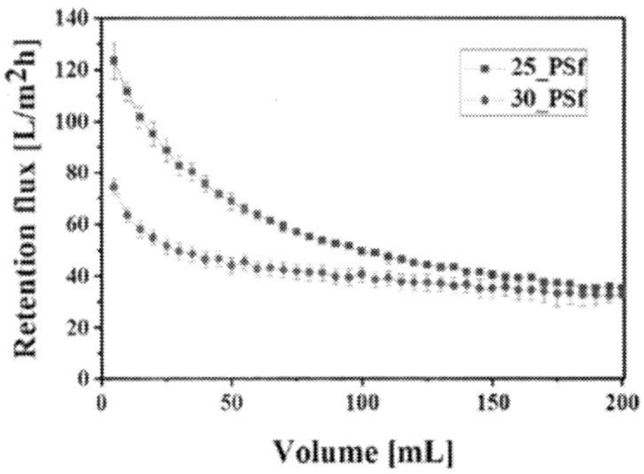

Figure 3. Retention flux of Siret water for 25 and 30 wt.% PSf membrane.

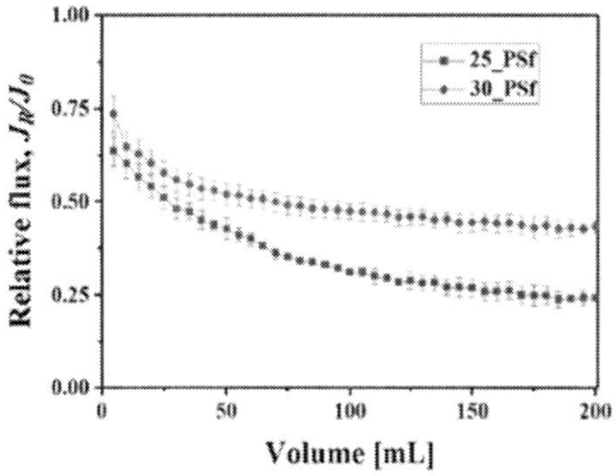

Figure 4. Relative flux for 25_PSf and 30_PSf membranes.

As presented in Figure 3, the retention fluxes of the two membrane types show a highly different initial flux (124 L m^{-2} h^{-1} for 25_PSf and 75 L m^{-2} h^{-1} for 30_PSf) and low difference at final flux (approx. 40 L m^{-2} h^{-1} for both membranes). The 25 wt.% PSf membrane show a 66 % decrease in Siret River water flux, whilst 30 wt.% PSf membrane shows a decrease of 46 %.

Correlating Figure 3 and 4 with the roughness values, it is observable that the lower roughness of 30 wt.% PSf membrane presents a higher anti-fouling ability. The fouling of 25 wt.% membrane is more obvious, reaching to a final value of approx. 0.25.

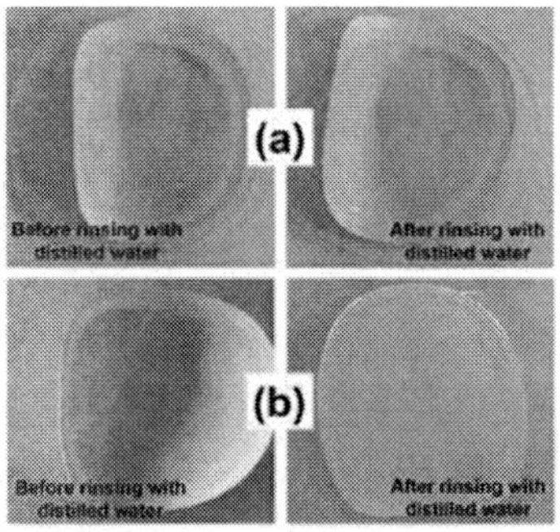

Figure 5. First observations of membrane fouling after channel water filtration for (a) 25_PSf and (b) 30_PSf membranes.

The low fouling tendency is also observed after a simple rinse with distilled water, as show in Figure 5. After rinsing the samples with distilled water, the higher observable fouling for 25 wt.% membrane could signify that particles from Siret water have adhered on membrane surface and structure more prominent.

Physical and chemical indicators for Siret River water and permeate water quality

The permeates were analyzed though physico-chemical methods. The most notable differences were for COD-Cr, turbidity, suspended matter, iron and zinc indicators.

The pH value has not changed sufficient, due to the fact that both raw and permeate water pH values are in acceptable range of water quality [13].

Turbidity is an optical property of water, in which suspended or dissolved materials (organic and inorganic material, plankton, microorganisms, etc.) absorb or scatter light. Turbidity decreases from 245 NTU to 1.39 and 1.19 NTU for 25 wt.% PSf and 30 wt.% PSf membrane, respectively. The turbidity of the membrane with higher concentration of PSf show a lower turbidity, an explanation being that it has smaller pores, thus retaining more matter that increases turbidity value.

Table 4. Physico-chemical analysis of raw water and filtrated raw water with 25_PSf and 30_PSf membranes.

No.	Indicator	Unit	Siret raw water	25_PSf permeate	30_PSf permeate
1	pH	pH units	6.9	7.3	7.3
2	Chemical Oxygen Demand (COD-Cr)	mgO_2/L	259.6	0	0
3	Turbidity	NTU	245	1.39	1.19
4	Suspended matter	mg/L	283	0	0
5	Extractable substances	mg/L	4	0	0
6	Chlorides	mg/L	143.230	141.812	141.103
7	Total nitrogen	mg/L	4	4	4
8	Total ionic iron (Fe^{2+} and Fe^{3+})	mg/L	5.23	3.20	2.28
9	Lead	µg/L	0.11	0.1	0.1
10	Zinc	µg/L	0.32	0.19	0.1
11	Chromium	µg/L	0.06	0.05	0.05

The COD-Cr parameter is understood as the indirect measurement of organic and inorganic compounds oxidability in reaction of a strong acid, such as potassium dichromate (KCr_2O_7) [13]. The membranes from this study have reduced to zero the amount of organic matter that can be chemically oxidised. Although COD test is widely used in water and wastewater analysis, its main limitation is the inability in differentiating biodegradable from biologically inert organic matter [14], approved by the results from total bacteria count shown in Table 5.

Suspended matter is composed of fine particles originating from natural sources (plankton, fine plant debris, minerals, etc.) and from human activity (organic and inorganic matter), which could have a negative influence on the ecosystem [15]. Both types of membranes have eliminated the suspended matter. The extractable substances value after membrane filtration were 0, also, for both membrane types, meaning that their sizes were large enough to be rejected by the membranes.

Notable differences were reported for iron and zinc indicators, but were not eliminated from permeate, the membrane with 30 wt.% PSf concentration showing higher permeate quality.

No difference in filtration were reported for chloride, lead, chromium and total nitrogen tests.

3.4. Total bacteria count

Bacteria growth in membrane treatment plants are of high concern, due to the appearance of biofouling on membrane surface and inside its pores, especially for reverse osmosis membranes, which requires additional treatment upstream [16]. The raw water and permeate waters from 25 wt.% and 30 wt.% PSf membranes were subjected to a simple bacteria test. Each sample were prepared in 4 tubes, and noted as follows: for the Siret river water tubes 1, 2, 3 and 4 corresponding with the A, B, C and D tubes explained at the methods section, see *2.6. Total bacteria count test*. The permeate samples were prepared in the same manner, tubes 5, 6, 7 and 8 for 25 wt.% PSf membrane and tubes 9, 10, 11 and 12 for 30 wt.% membrane, respectively, as shown in Figure 6.

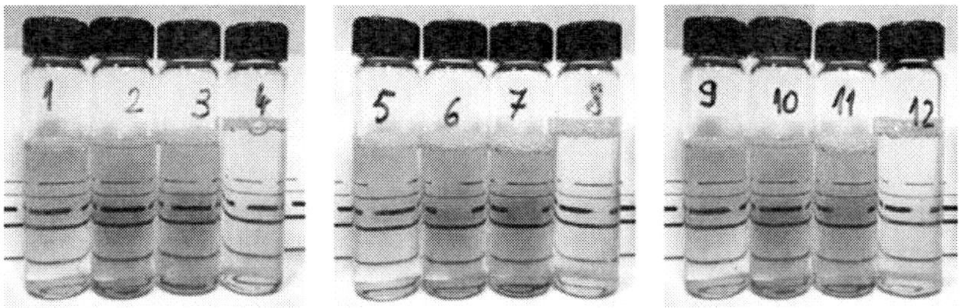

Figure 6. Water sample preparation for total bacteria count test for (1, 2, 3, 4) Siret water, (5, 6, 7, 8) 25_PSf membrane permeate and (9, 10, 11, 12) 30_PSf membrane permeate.

After 24-48 hours of incubation time of the test tube, it was observed (Figure 6) that the raw water has developed in all tubes bacterial growth, due to the fact that all tubes are cloudy, corresponding to high bacterial presence of >1000 bacteria/mL of sample, as shown in Table 5.

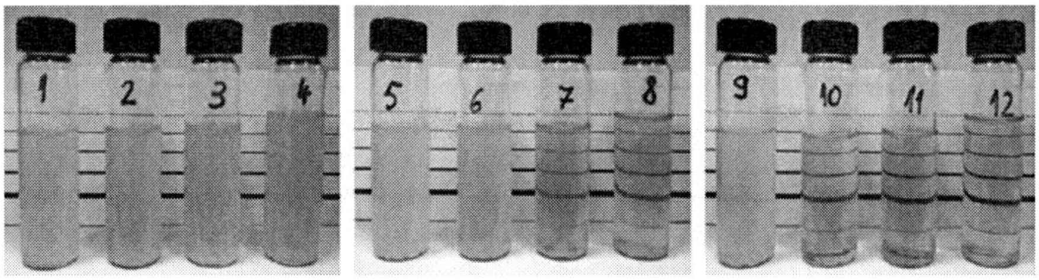

Figure 7. Water samples after 48 hours of incubation for bacteria growth.

For the 25_PSf membrane permeate samples, it is observed that tubes 5, 6 are cloudy and tubes 7, 8 are clear, which means that the permeate show a total of ~10-100 bacteria/mL of water. The membrane with the lower polymer concentration have managed to reduce bacteria approximately 50 %. In the case of the membrane with the highest polymer concentration and the lowest pore sizes, 30 wt.% PSf, the permeate bacteria count was noticeably reduced to ~1-10, approximately 99%.

Table 5. Water quality for studied water samples.

Water sample	Water quality [bacteria/mL]
Siret water	>1000
25_PSf permeate	~10-100
30_PSf permeate	~1-10

Despite the fact that the permeate quality is 99% free of bacteria, it still poses the risk of illness due to the presence of bacteria, regardless of quantity. According to Romanian law no. 458/2002 regarding the quality of drinking water [17], the maximum value for bacteria presence in potable water is 0.

Although the above stated, the 30 wt.% PSf membrane could be a viable option for the pretreatment of river water upstream in order to reduce high fouling in the reverse osmosis filtration modules, regardless of the destination of water use.

Correlating all the methods of membrane characterization, it is observed that roughness, pore size and density (Figure 2, Figure 1 and Table 3, respectively) determines the membrane fouling behavior (Figure 3, Figure 4 and Figure 5) and retention capacity (Table 4 and 5).

For 25 wt.% PSf membrane, the higher roughness causes higher deposition of particles on the surface and possibly inside the membrane structure, resulting in higher fouling during filtration. Also, the total bacteria from the permeate suggests that possibly a considerable amount of bacteria have attached on the surface and inner pores, which also approve higher fouling. In the case of 30 wt.% PSf membrane, due to lower roughness and smaller surface pores, all retention performance and anti-fouling property are superior to lower polymer concentration membranes.

4. Conclusions

Retention and fouling capacity of 25 wt.% and 30 wt.% PSf membranes were tested successfully. In terms of pure water flux and water affinity, the 25 wt.% PSf membrane exhibited the highest values. Nevertheless, the membrane hydrophilicity does not influence the fouling tendency during retention of Siret River water, and pure water flux is merely a parameter of standardization, which is an orientative value in membrane science. On the other hand, 30 wt.% PSf membrane show superior stability in flux during retention, higher retention values for removal of suspended matter, extractable substances, total ionic iron and zinc. The chemical oxygen demand values show that the permeate eliminated all oxidizable pollutants from the river water. Turbidity was reduced 99.4% for 25_PSf membrane and 99.5% for 30_PSf membrane. The key result on the bacteria removal capacity demonstrated that the use of 30 wt.% PSf membrane in the filtration of river water is a viable option in the prefiltration stage of reverse osmosis water treatment.

5. References

[1] Ang W L, Mohammad A W, Hilal N, Leo C P 2015 *Desalination* **363** 2-18
[2] Chiru E 2006 Water supply of Bucharest - past, present, future: A study case (*S.C. Apa Nova Bucuresti S.A: Bucharest, Romania) 119-130*
[3] Xia S, Li X, Zhang Q, Xu B, Li G 2007 *Desalination* **204** 351– 8
[4] Biron D S, Santos V, Zeni M 2018 Ceramic membranes applied in separation processes *(Springer International Publishing AG: Gewerbestrasse, Switzerland)*
[5] Ursino C, Castro-Muñoz R, Drioli E, Gzara L, Albeirutty M H, Figoli A 2018 *Membranes* **18** 2-40
[6] Clever M, Jordt F, Knauf R, Rabiger N, Rtidebusch M, Hilker-Scheibel R 2000 *Desalination* **131** 325-36
[7] Kenneth S Y Ng, Malavika H, Dalton J E H, Dave E D, Gregory J O M 2017 *J. Membr. Sci.***523** 144-162
[8] Ahmad A L, Abdulkarim A A, Mohd Shafie Z.M H, Ooi B S 2017 *Desalination* **403** 53–63
[9] Benavente L, Coetsier C, Venault A, Chang Y, Causserand C, Bacchin P, Aimar P 2016 *J. Membr. Sci.* **520** 477–489
[10] Pintilie S C, Tiron L G, Lazar A L, Vlad M, Birsan I G, Balta S 2018 *Materiale Plastice* **55** 54-62
[11] Hołda A K, Aernouts B, Saeys W, Vankelecom I F J 2013 *J. Membr. Sci.* **442** 196–205
[12] Zinadini S, Zinatizadeh A A L, Rahimi M, Vatanpour V 2017 *Journal of Industrial and Engineering Chemistry* **46** 9–18
[13] Popa P, Timofti M, Voiculescu M, Dragan S, Trif C, Georgescu L P 2012 *The Scientific World Journal* **2012** 1 – 10
[14] Bourgeois W, Burgess J E, Stuetz R M 2001 *J Chem Technol Biotechnol* **76** 337–348
[15] Chapman D 1996 Water Quality Assessments – A Guide to Use of Biota, Sediments and Water in Environmental Monitoring – Second Edition *(UNESCO/WHO/UNEP, UNITED NATIONS ENVIRONMENT PROGRAMME)*
[16] Li N N, Fane A G, Ho W S W, Matsuura T 2008 Advanced Membrane Technology and Applications *(John Wiley & Sons, Inc., New Jersey, USA)*
[17] Romanian Law 458, 2002, in Romania, on the quality of drinking water, republished in Romanian Official Monitor, part I, no. 875, 2011

Protection of Tempered Aluminum Alloy in Contact with the Environment

A O Araoyinbo[1]*, M A A Mohd Salleh[2], A Rahmat[2], A I Azmi[1], W M F Wan Abd Rahim[1], M C Perju[2,3] and T S Jin[1]
[1]Faculty of Engineering Technology, Universiti Malaysia Perlis, Perlis Malaysia
[2]Center of Excellence Geopolymer and Green Materials (CEGeoGTech), School of Materials Engineering, Universiti Malaysia Perlis (UniMAP), Perlis, Malaysia
[3]Gheorghe Asachi Technical University of Iasi, Faculty of Materials Science and Engineering, Blvd. D. Mangeron 41, 700050, Iasi, Romania

Email: alaba@unimap.edu.my

Abstract. In many service applications an increasing temperature or inadequate protections often give rise to localized forms of corrosion in an initially free and unprotected system. This research understudy the corrosion chemistry, the effect of chromium as the inhibitor, Vickers hardness test, and weight loss on tempered aluminium alloy 7075 in corrosive mediums. The tempers of the aluminium alloy used are T6 and T73 where obtained by solution heat treatment at 470°C and quenched before immersion test in acidic (pH3), and slightly alkaline (pH7.5) solutions. The results obtained were characterized by conventional weight loss process and morphology observation with a microscope. The surface morphology shows exfoliation form of corrosion and the weight loss analysis shows the as received sample experience more weight loss when compared with the other heat treated samples.

1. Introduction

Corrosion is known to be an electrochemical process in which a metal comes in contact with its environment to form an oxide or other compounds. As a result of this interaction, the resulting destructive attack of the material due to this corrosion process has become a problem of world-wide significance [1, 2]. In addition to the regular encounters with this form of degradation, which often leads to plants shut down, waste of valuable resources, loss or contamination of product, reduction in efficiency, costly maintenance, and expensive over design, it also jeopardizes safety and inhibits technological progress [3-5]. The electrochemical process has three essential constituents: an anode, a cathode and an electrically conducting solution. The anode is the site at which the metal corrodes; the electrolyte solution is the corrosion medium; and the cathode forms the other electrode of the cell which is not consumed during the corrosion process [6].

At the anode the corroding metal passes into solution as positively-charged ions, releasing electrons which participate in the cathodic process. Thus the rate of corrosion is equivalent to the electron current flowing between anode and cathode of the corrosion cell. The distribution of anodic and cathodic areas is one of the most important factors determining the type of corrosion which occurs, but in the simplest case the corrosion cells are very small and numerous, and distributed in a random manner over the surface of the metal, and the effect is more or less uniform attack on the surface [7,

8]. The overall rate of the process under these circumstances may depend on a number of factors but the important criterion is the balance between rates of anodic and cathodic reactions required to maintain electro neutrality, and the overall rate of corrosion is controlled by the rate of whichever reaction occurs least readily. Although many metals display active corrosion behaviour when immersed in a corrosive solution, others including many engineering alloys carry a stable film of solid corrosion product, frequently an oxide. If this film has good cohesive strength and adhesion to the metal, and low ion conductivity, the rate of metal dissolution is limited to the rate at which metal ions can pass through the film [9-11]. This is the phenomenon of passivity, exhibited by stainless steels, nickel alloys, titanium and many similar materials and responsible for their utility as engineering materials in corrosive environments. This study aims to determine how effective the chromate inhibitor can be when the as received specimen, peak age and over age tempers are immersed in the solution with/without chromate and to also compare the hardness of the heat treated specimens with the as received sample.

2. Materials and methods

An aluminum alloy 7075 was purchased commercially for the sole purpose of this research. Specimens with dimensions of 25 mm x 30 mm x 15 mm were cut from the as received aluminum alloy 7075. The surface was ground and polished to mirror finish. This was accompanied by using 1200 grit abrasive papers and 0.5 µm alumina powders. Heat treatment process was initiated by solution heat treatment at 470 °C for 60 minutes and then quenched in water (at room temperature) for 60 seconds. The process was then followed by precipitation heat treatment for different durations of tempers for the aluminum alloy. The specimen in the T6 temper was subjected to solution heat treatment at 470 °C for 60 minutes, quenched for 60 seconds and followed by precipitation heat treatment or artificial ageing at 138 °C for 960 minutes. The specimen in the T73 was subjected to solution heat treatment at 470 °C for 60 minutes quenched for 60 seconds and followed by two precipitation heat treatments processes at 113 °C for 480 minutes and 182 °C for 720 minutes respectively.For the hardness test two heat treated and one as received specimens having dimensions of 20 mm x 20 mm x 10 mm were cut with diamond cutter from the aluminum alloy 7075. They were polished using 1200 grit abrasive papers and diamond paste to produce a mirror finish surface prior to the hardness test. The tests were conducted using a Vicker's hardness tester (model: FV-7) and measurements were carried out with a 10 kgF load.

2.1. Immersion Test

Two heat treated specimens and one as received specimen were immersed in different solutions for a period of time. Two different pH environments were used which is an acidic environment with pH3, and a slightly alkaline medium of pH7.5.

2.2. Acidic Medium (pH3)

The peak aged T6, and over aged T73 with as received AR (Al-alloy 7075) were used in the acidic medium containing 4 M sodium chloride, 0.5 M potassium nitrate and 0.1M nitric acid diluted to make one litre with deionised water for 6 days according to exfoliation test of aluminum and aluminum alloy ASTM G112-92 (2003) and ASTM G34-01 for visual and optical observations. The dimension of these specimens are 51 mm X 17 mm X 6 mm, they were polished with 1200 grit abrasive papers and 0.5 µm diamond paste to produce a mirror finish surface. Keller's reagent was used to clean the furnace of the specimens prior to the immersion test.

2.3. Alkaline Medium (pH7.5)

The dimensions of specimens used are 51 mm X 17 mm X 6 mm respectively they were polished with 1200 grit abrasive papers and 0.5 µm diamond paste to produce a mirror finish surface. Keller's reagent was used to clean the furnace of the specimens prior to the immersion test. Two separate solutions were prepared for the alkaline medium the first solution contains 3.56 wt% of sodium chloride diluted with deionised water to make one litre without any inhibitor according to ASTM standards G31 72 (2004) while the second solution contains 3.56 wt% of sodium chloride, and 0.2 wt% of sodium chromate as an inhibitor and diluted with deionized water to make one litre.

3. Results and discussion

Solution heat treatment involves heating to a high temperature very close to the solidus temperature and above the solvus line. It is vital and important to stick to the specific temperatures because if the temperature is too low, the resulting mechanical properties will be low, whereas too high temperature will over heat the product (formation of hot spots), leading to irreversible damage on the aluminum alloy. This is due to partial fusion of complex eutectics and elements precipitated at the grain boundaries [12]. Peak aged T6 specimen produced by solution heat treatment has the highest mechanical strength shown in Table 1 while the Over aged specimen produced also have high strength but lower in value to the peak age strength. The solution heat treatment strongly increased the volta potential difference between the intermetallics and the surrounding matrix [13].

Table 1. Heat treatment and hardness value.

Temper Designation	Solution Heat treatment	Precipitation Heat treatment	Hardness Value
T6	470°C/60 min	138°C/960 min	193 HV
T73	470°C/60 min	113°C/480 min 182°C/720 min	188 HV
AR	-	-	183HV

3.1. Hardness Test

The Hardness test measures the resistance of the aluminum alloy 7075 to an applied load or indentation. The Vickers hardness testing machine that was used for the test on the as received sample with an applied load of 10 KgF indentation on the polished surface of the aluminum alloy gives a value of 183 HV based on the index number which is determined by the depth or penetration level of the indentation. Solution heat treatment and subsequent quenching of AA 7075 causes dissolution of $MgZn_2$ strengthening particles, which are formed during aging to the T6 temper and provide high-strength to the alloy [14].

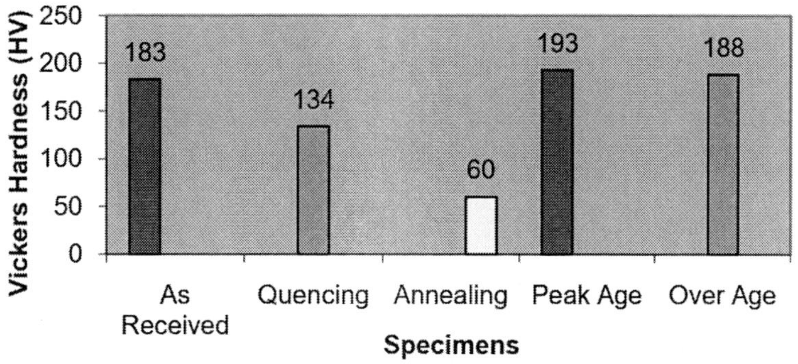

Figure 1. Vickers hardness test.

The dissolution of the $MgZn_2$ strengthening particles causes enrichment of the matrix in solute atoms like Zn, Mg and Cu resulting in the presence of a supersaturated solid solution when the alloy is fast quenched from the solution heat treatment temperature to room temperature this microstructure changes occurring in the alloys causes a reduction in the alloy strength. Characterization of the specimen strength by Vickers hardness testing machine gives the value 134 HV showing a reduction in strength which can be attributed to the dissolution of the strengthening intermetallics $MgZn_2$ in the microstructure [14]. Peak age (T6) in Figure 1 shows the increased strength which is attributed to a high density metastable precipitation distributed homogeneously in the Al matrix while the T73 (Figure 1) strength is attributed to lower density of thicker precipitates more stable in nature which accounts for the inferior strength to the peak aged specimen but it should be noted that there is no change in the microstructure [15].

Both specimens produced by the heat treatment have mechanical strength higher than the as received sample. The hardness test result shown in Figure 1 as a tendency for it to increase this can be attributed to the fact that being a high strength alloys it continually undergo natural ageing.

3.2. Heat treated specimen at pH3

Figures 2 (a), (b), & (c) shows exfoliation corrosion which is one of the localized corrosion on the specimens after immersion into the solution by optical observation. Exfoliation is a form of intergranular corrosion that can occur in wrought high strength aluminum alloys, the attack takes several forms ranging from pitting, powdering or flaking of the surface to the development of blisters several millimetres in diameter. It was observed that overageing has beneficial effects on the susceptibility to exfoliation corrosion by reducing the rate of intergranular corrosion one difference between these two tempers' microstructures is that the η-$MgZn_2$ hardening phase is lined up in beads along the grain boundary in the T6 temper with small spacing, whereas in the T73 temper the η-$MgZn_2$-particles spacing are larger [14]. Large sheets of the surface had loosened from the substrate in all immersed specimen, the flakes had loosened, which floated in the testing solution or lay near the sample at the bottom.

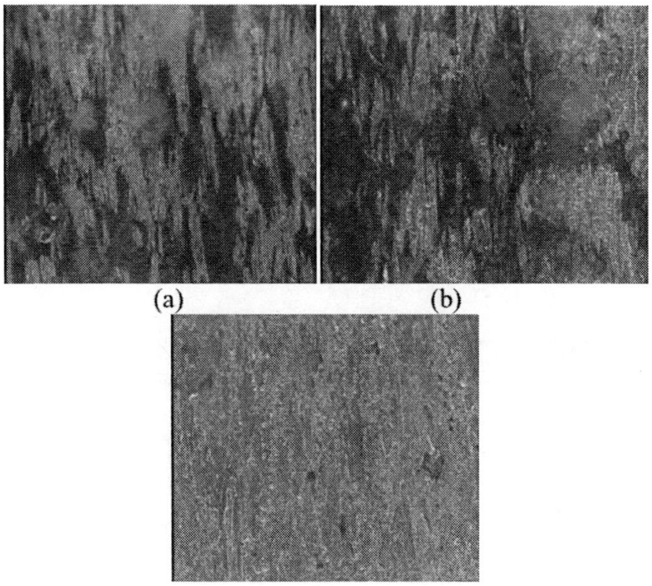

(a) (b) (c)

Figure 2. (a) AR Exfoliation test observed from an opticalmicroscope at 10X, (b) T6 Exfoliation test observed from an optical microscope at 10X, and (c) T73 Exfoliation test from an optical microscope at 10X.

The galvanic interaction between grain boundary precipitates and adjacent solute depleted zones cause the intergranular corrosion [16]. The corrosion products have a larger volume than the original aluminum alloy and the expansion causes stresses that lift the surface grains, exposing the underlying material as shown in the visual observation of the specimens Figures 3 (a), (b), & (c). The grain shape, the degree of cold deformation and the heat treatment condition are key factors influencing the exfoliation corrosion [17].

(a) (b)

(c)

Figure 3. (a) T73 exfoliation corrosion by visual observation, (b) AR exfoliation corrosion by visual observation, and (c) T6 Exfoliation Corrosion by Visual Observation.

Table 2 shows the weight losses and corrosion rate observed for AR, T6, and T73 immersed in a test solution at pH3 for 6 days. The weight loss obtained for the three different specimens in the acidic solution is 0.8242 g, 0.7968 g and 0.7734 g which are all high. These show that regardless of the heat treatment solution if the material is unprotected, the deterioration and weight loss becomes rapid that can lead to material failure when in an acidic environment.

Table 2. Weight losses and corrosion rate test result at pH3.

Specimen	Initial Weight (g)	Duration (days)	Final Weight (g)	Weight loss (g)	Solution	Corrosion Rate (mm/yr)
AR	17.2143	6	16.3901	0.8242	NaCl KNO_3 HNO_3	20.65
T73	18.3232	6	17.5498	0.7734	NaCl KNO_3 HNO_3	19.38
T6	15.7532	6	14.9564	0.7968	NaCl KNO_3 HNO_3	19.97

3.3. Heat treated specimen at pH7.5

Table 3 and 4 shows two and four weeks immersion test of the different specimens showing its weight losses and corrosion rate in a slightly alkaline environment. AR has the highest susceptibility to corrosion This susceptibility is partially decreased by the precipitation of the $MgZn_2$ particles at the grain boundaries when the alloy is aged to the T6 temper. However, the susceptibility to intergranular corrosion remains rather high for the T6 temper since the grain boundaries present an almost continuous precipitation of particles. Aging to the T73 temper strongly improves the resistance to intergranular corrosion because the size of the $MgZn_2$ particles at the grain boundaries and their interparticle distance increases during overaging [16].

The Table 4 shows the corrosion rate result after 4 weeks of immersion, it is observed that the corrosion rate decreases with time these is most likely due to the ability of the alloy to undergo a form of passive film repair which further decreases or slows down the rate of electrochemical reaction taking place in the solution.

Table 3. Weight losses and corrosion rate for 2 weeks test result at pH7.5.

Specimen	Initial weight (g)	Duration (days)	Final weight (g)	Weight loss (g)	Solution	Corrosion rate (mm/yr)
AR	9.3804	14	9.2842	0.0962	NaCl	1.03
T6	9.1950	14	9.1079	0.0871	NaCl	0.94
T73	10.0513	14	9.9796	0.0717	NaCl	0.77

Table 4. Weight losses and corrosion rate for 4 weeks test result at pH7.5.

Specimen	Initial Weight (g)	Duration (days)	Final Weight (g)	Weight loss (g)	Solution	Corrosion Rate (mm/yr)
AR	11.7512	30	11.5803	0.1709	NaCl	0.87
T6	9.7110	30	9.5504	0.1606	NaCl	0.80
T73	9.2586	30	9.1266	0.1320	NaCl	0.66

3.4. Heat treated specimen at pH7.5 with chromate inhibitor

The unique chemical and electronic properties of the oxo-compounds of chromium, Cr (valence electronic structure $3d^5\ 4s^1$), gives them a seemingly unique ability to inhibit corrosion in ferrous and nonferrous materials. The electronic properties of Cr allow for very different behaviours for the oxo- and hydroxo- compounds of Cr^{6+} and Cr^{3+}. The tetrahedral, d^0, hexavalent oxoanion compounds of chromium (chromate, dichromate, bichromate, and chromic acid) dissolve as stable complexes in water, transport easily, and adsorb on oxide surfaces.

Chromate used as inhibitor in aluminum alloy 7075 effectively protects it against localized corrosion attack because chromate is a powerful cathodic inhibitor even at high chloride, chromate ratios are excellent inhibitors of oxygen reduction in near neutral and alkaline solutions [13, 16]. In these environments, they can stifle corrosion by suppressing this cathodic partial reaction. The inhibition appears to involve reduction of Cr^{6+} to Cr^{3+} at a metal surface and formation of an irreversible adsorbed Cr^{3+} hydroxide surface layer of near monolayer thickness.

The Table 5 shows the low rate of weight loss and corrosion rate of the AA 7075 protected by chromate inhibitor because the treatment of AA 7075 with chromate in the sodium chloride solution causes a significant reduction of cathodic activity. This was attributed to rapid formation of a thin chromium oxide film on the α-Al (Cu, Fe,) particles resulting in significant cathodic passivation of the phase, inhibition of the oxygen reduction reaction at cathodic intermetallic particles.

Table 5 Weight losses and corrosion rate for 4 weeks in chromate inhibited solution.

Spe.cimen	Initial weight (g)	Duration (days)	Final weight (g)	Weight loss (g)	Solution	Corrosion rate (mm/yr)
AR	7.6583	30	7.6539	0.0044	NaCl Na_2CrO_4	0.022
T73	6.4742	30	6.4709	0.0033	NaCl Na_2CrO_4	0.017
T6	6.7052	30	6.7021	0.0031	NaCl Na_2CrO_4	0.016

The performance of the chromate inhibitor has been attributed to several factors providing barrier protection of the aluminum matrix and providing stability over a wide pH range [13].

4. Conclusions

The corrosion behaviour of tempered aluminium alloy in two corrosive mediums were studied. The hardness test shows increase in strength during the solution heat treatment which is higher for both T6 and T73 when compared with the as-received material. For T6 and T73 the increased strength is due to high density metastable precipitation distributed homogenously and in the case of T73 it is due to low density of thicker precipitate more stable in nature. The corrosion test shows T6 and T73 experience marginal weight loss and the introduction of the chromate inhibitor further reduces the weight losses and the corrosion rate per year. The AR specimen the surface is with only a thin oxide layer on it, thus corrosive attack initiates rather fast because this oxide layer can be dissolved easily by the aggressive test solution and the resulting hydrogen evolution which leads to severe embrittlement of the material. While the oxide layer of the T6 and T73 are much thicker because of the heat treatment at elevated temperatures – including solution heat treatment and different artificial ageing processes which might significantly delay the onset of corrosion.

5. References

[1] Chinwko E C, Odio B O, Chukwuneke J L and Sinebe J E 2014 *International Journal of Scientific and Technology Research* **3**(7) 306-310
[2] Boudalia M, Sebbar N K, Bourazmi H, Lahmidi S, Ouzidan Y, Essassi E M, Tayebi H, Bellaouchou A, Guenbour A and Zarrouk A J 2016 *Journal of Materials and Environmental Science* **7**(3) 878-888
[3] Malarvizhi M, Dheenadhayalan S, Brindha T and Mallika J 2016 *Advances in Physical Chemistry* 1-12
[4] Obot I B and MadhankumarA J 2015 *Journal of Industrial and Engineering Chemistry* **25** 105-111
[5] Odewunmi N A, Umoren S A and Gasem Z M 2015 *Journal of Environmental Chemical Engineering* **3** 286-296

[6] Vasyliev G S 2015 *Corrosion Science* 33-39
[7] Coelho L B, Mouanga M, Druart M E, Recloux I, Cossement D and Olivier M G 2016 *Corrosion Science* **110** 143-156
[8] Abdel-Gaber A M, Abd-El-Nabey B A, Khamis E and Abd-El-Khalek D E 2011 *Desalination*, **278**(1-3) 337-342
[9] Kazem S B, Changiz D and Saeed Y 2017 *Corrosion Science* **126** 272-285
[10] Abosrra L, Ashour A F, Mitchell S C and Youseffi M 2014 Bradford School Of Engineering, University Of Bradford, Uk.
[11] Zakowski K 2011 *Anti-Corrosion Methods and Materials* **58**(4) 167-172
[12] Andreatta F, Lohrengel M M, Tenyn H and De Wit J H W 2002 *Corrosion Science* **45**(8) 1733-1746
[13] Andreatta F, Lohrengel M M, Tenyn H and De Wit J H W 2003 *ElectrochimicaActa* **48**(20-22) 3239-3247
[14] Christian V 2004 "Corrosion of Aluminum," Elsevier Ltd, the Boulevard Langford Lane Kidlington Oxford OX 5 IGB, UK
[15] Buzea C Gh, Bejinariu C, Boris C, Vizureanu P and Agop M 2009 *International journal of nonlinear sciences and numerical simulation* **10** 1399-1414
[16] Lyberatos G and Kobotiatis L 1991 *Corrosion Science* **47** 820-824
[17] Robert K G, John S R, David S W and Rudolph B G 2003 1st Edition, Mercel Dekker Inc, New York

Acknowledgements
I would like to express my gratitude to Universiti Malaysia Perlis (UniMAP) and Center of Excellence Geopolymer and Green Materials (CEGeoGTech) for giving me the opportunity to be involved in this project.

The Effect of Aggressive Corrosion Mediums on the Microstructure and Properties of Mild Steel

A O Araoyinbo[1,*], M A A Mohd Salleh[2], A Rahmat[2], A I Azmi[1], W M F Wan Abd Rahim[1], D C Achitei[2,3] and T S Jin[1]
[1]Faculty of Engineering Technology, Universiti Malaysia Perlis, UniCity Alam, Perlis Malaysia
[2]Center of Excellence Geopolymer and Green Materials (CEGeoGTech), School of Materials Engineering, Universiti Malaysia Perlis (UniMAP), Perlis, Malaysia
[3]Gheorghe Asachi Technical University of Iasi, Faculty of Materials Science and Engineering, Blvd. D. Mangeron 41, 700050, Iasi, Romania

Email: alaba@unimap.edu.my

Abstract. Mild steel is known to be one of the major construction materials and have been extensively used in most chemical and material industries due to its interesting properties which can be easily altered to suit various application areas. In this research, mild steel is exposed to different aggressive mediums in order to observe the effect of these interactions on its surface morphology and properties. The mild steel used was cut into dimensions of 7 cm length and width of 3 cm. The aggressive mediums used are 100 mls of aqueous solution of hydrochloric acid, sodium hydroxide (40 g/L), and sodium chloride (35 g/L) at room temperature. The characterizations performed are the hardness test with the Rockwell hardness tester, the surface morphology by optical microscope, surface roughness and the weight loss from the immersion test. It was observed that the hardness value and the weight loss for the different cut samples of mild steel immersed in the different aggressive mediums reduces with prolong exposure and severe pitting form of corrosion was present on its surface.

1. Introduction

Corrosion is a worldwide phenomenon and have been acknowledged by many researchers to be a problem that strongly influences our industrial environments.

Corrosion process is sometimes misunderstood and if not properly identified and characterized at certain times this can result to in service material degradation leading to negative impact on the safety of the environments. In many cases, corrosion has acted as life span limiting factor forboth structural and fabricated components [1]. Corrosion and some forms ofpollutions can be said to be interrelated as it is known that there is the possibility of harmful processes acelerating a corrosion process [2, 3].

Since rust from metal iron can act as a source of pollutants when it comes in contact with water bodies, both processes have the capability to influence the quality of the environment, and the efficiency of the industries. Therefore, it is essential to be able to apply corrosion engineering knowledge, methods and techniques to reduce the corrosion activity [4].

Corrosion is the deterioration of materials by the chemical interactions with the different environments.

Most metals have the tendency to interact with its environment when it is exposed under certain conditions, this interractions often leads to the formation of metal oxides with visible corrosion products when the passive films layer breaks down due to prolonged exposure and lack of protection [5-8]. Steel materials can be susceptible to attack in aggressive media when exposed to form rust. Since steel find use in a wide area of applications it becomes even more important to avoid any catastrophic failure and huge financial losses when in use [9, 10].

In this study, the corrosion behaviour of mild steel was investigate when exposed to different aggressive mediums.

2. Materials and methods

The raw materials and chemicals used in this project are mild steel which was locally purchased, and three aggressive mediums of aqueous sodium chloride (neutral), sodium hydroxide (alkaline) and hydrochloric acid (acidic).

2.1. Sample preparation

The mild steel specimens were cut into dimensions of 7 cm x 3 cm and polished with abrassive papers to remove dust, stains and any oxidation films on its surface. The 27 cut samples of mild steel were weighed before and after the immersion test and the data were recorded.

2.2. Aggressive medium preparation

The aggressive corrosion mediums were divided into three different aqueous solutions of near neutral, alkaline and acidic solutions. The near neutral solution was prepared by adding 35 g of sodium chloride in 1 litre of water, the alkaline solution was prepared by adding 40 g of sodium hydroxide in 1 litre of water, and the acidic solution by adding 36 mls of hydrochloric acid in 1 litre and using 100 mls of the prepared aqueous solution.

The 27 test samples were separated into 9 samples each and carefully immersed into the beakers containing the different corrosive mediums for 1, 3, and 5 weeks in order to observe the weight loss, hardeness and surface roughness of the metal in this three environments.

The characterization of the mild steel test samples were by optical microscope to observe the surface morphology, the weight loss analysis technique for the immersed samples and surface roughness of the specimens.

3. Results and discussion

3.1. Weight Loss of Mild Steel

Figure 1, shows the weight loss graph of mild steel immersed in three different aggressive corrosion mediums of NaCl, NaOH, and HCl for 1 to 5 weeks duration.

The results obtained shows that the mild steel samples experienced weight loss in all the aggressive mediums it was exposed to.

For one week immersion test the mild steel specimens experience weight loss of 0.141 g for HCl test solution, 0.026 g for the NaCl solution, and 0.121 g for the NaOH solution. The acid solution produce the highest weight loss for the first week due to the presence of H^+ ions and the Cl^- ions which leads to severe attack on the mild steel surface.

As the immersion test of the mild steel samples are increased for three and five weeks more weight loss is further experienced by all the three aggressive mediums.

The data obtained for week three and five weight losses shows 0.082 and 0.163 g from NaCl solution, 0.211 and 0.273 g from NaOH solution, and 0.213 and 0.275 g from HCl solution.

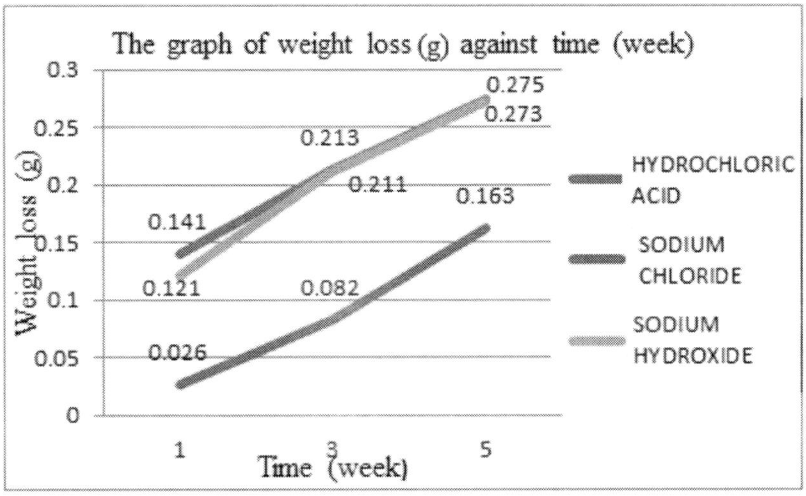

Figure 1. The weight loss graph of three types of corrosion mediums.

3.2. Mild steel hardness

Figure 2 shows the hardness test result of mild steel samples immersed in three aggressive mediums.

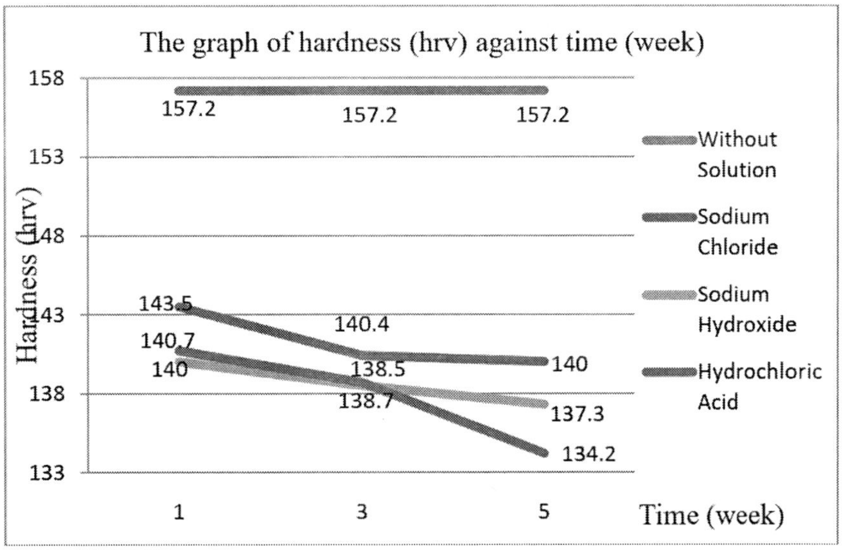

Figure 2. The graph of hardness test result of mild steel immersed in three aggressive mediums.

The graph shows the decrease in hardness of mild steel against time (week) in different corrosion mediums. Mild steel without solution immersion test have hardness value of 157.2 hrv. The other samples immersed in the aggressive solutions shows the decrease of hardness which starts at 143.5 hrv for the NaCl, 140.0 hrv for the NaOH and 140.7 hrv for the HCl for the first week. The third week of hardness testing from the immersed samples indicates slight decrease in the hardness value. The NaOH solution records 138.5 hrv for the hardness value and 138.7 hrv for HCl. The NaCl solution showed a decline in the first to third week at 143.5 hrv to140.4 hrv, but in the fifth week it decreased slightly to 140.0 hrv. The HCl solution shows a very drastic drop from 140.7 hrv to 134.2 hrv when compared to others. It is clear that the acid and the alkaline solutions affect the hardness strength of the mild steel more when compared to the near neutral solution.

3.3. Surface roughness of mild steel

Figure 3 shows the surface roughness comparison of the mild steel immersed in the aggressive mediums for 1, 3, and 5 weeks. All the data obtained for the three solutions showed an increase in surface roughness for all the specimens. Mild steel without immersion in a solution indicates the level of surface roughness value to be 0.311 μm. This is due to its surface not being exposed to elements that can speed up oxidation process on the surface of the mild steel. The graph shows the mild steel surface roughness from the NaCl solution for the first week increased, and the surface roughness rate of mild steel starts at 5.006 μm, and NaOH records 1.365 μm for the surface roughness, and 4.070 μm for the HCl solution. Surface roughness of the mild steel samples from NaCl solution shows the highest rate when compared with NaOH and HCl solutions. The surface roughness rate of mild steel for NaOH solution increased from 1.365 μm in the first week to 2.022 μm in the third week and 2.500 μm in the fifth week. While the surface roughness recorded for mild steel in the HCl solution shows an increase from 4.07 μm for the first week to 6.022 μm in the third week, and 10.525 μm in the fifth week.

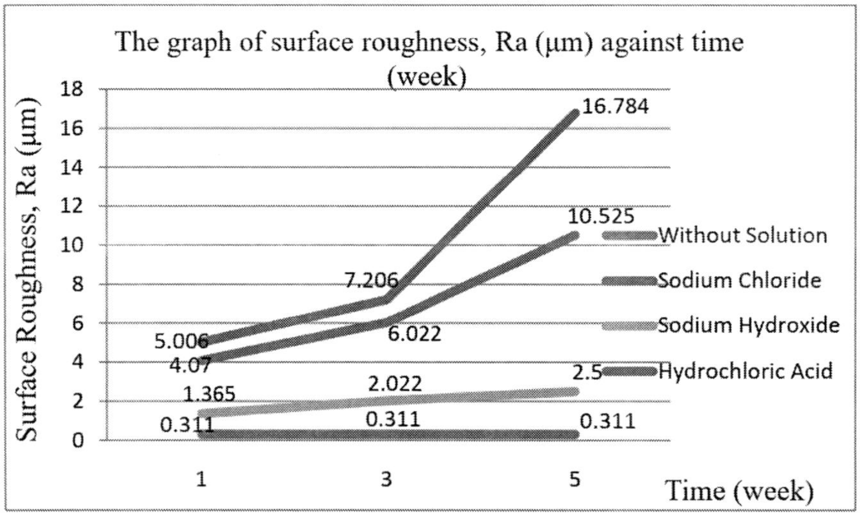

Figure 3. The graph of surface roughness of mild steel.

It was observed that the NaCl solution gave the highest value for the surface roughness probably due to the rapid chloride ion attack on the mild steel surface. The rate of surface roughness of mild steel in NaCl increases from 5.006 μm for the first week to 7.206 μm in the third week, and 16.784 μm in the fifth week.

3.4. Surface morphology

Figure 4 shows the surface morphology of mild steel before and after immersion test in HCl solution for 1, 3 and 5 weeks.

The other surface morphologies from NaCl and NaOH as observed by the optical microscope produce similar micrographs to the HCl solution and only HCl solution is shown here. Figure 4 (a) shows surface morphology of the mild steel without immersion in the HCl solution. The surface is observed to be clear and free from any corrosion product. After immersion for 1 week Figure 4 (b) shows different surface coloration which starts to appear on the surface of the metal at a rapid rate. The color indicates the presence of rust from the electrochemical oxidation process. As the immersion is prolonged Figure 4 (c) and (d) indicates continuous presence and spreading of rust which grows with prolong exposure to the corrosion medium.

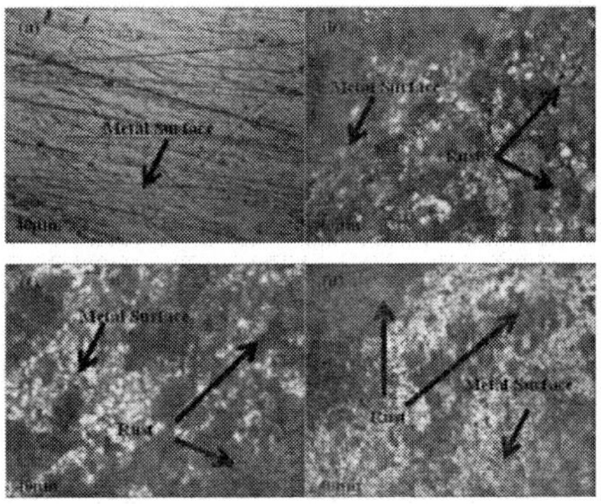

Figure 4. Morphologies of mild steel after immersed in HCl solution for various periods. (a) No immersion (b) 1 week (c) 3 weeks (d) 5 weeks.

4. Conclusions

Steels find use in many application areas and this has made it a material of choice in many structural applications. The mild steel test samples exposed to three aggressive mediums corrodes as the attack on its surface becomes severe with time. The weight loss analysis as observed shows reduction in the weight for all the different mediums and with prolonged exposure the weight continues to reduce. The hardness test result also confirms reduction in the hardness as the exposure is prolonged. This investigation shows that mild steel cannot withstand aggressive mediums because the attacks are severe with loss of weight and hardness strength. The surface morphology also indicates high surface roughness from the corrosion products.

5. References

[1] Saranya J, Sounthari P, Kiruthuka A, Parameswari K and Chitra S 2015 *Journal of Materialsand Environmental Science* **6**(2) 425-444
[2] Okafor P C, Ikpi M E, Uwaha I E, Ebenso E E, Ekpe U J and Umoren S A 2008 *Corrosion Science* **50** 2310-2317
[3] Xiao K, Zhang X, Dong C, Wei D, Wang C and Li X 2012 *Journal of Wuhan University of Technology Material. Science Edition* **27**(1) 27–32
[4] Chinwko E C, Odio B O, Chukwuneke J L and Sinebe J E 2014 *International Journal of Science and Technology Research* **3**(7) 306–310
[5] Chauhan L R and Gunasekaran G 2007 *Corrosion Science* **49** 1143-1161
[6] Durowaye S I, Durowaye V O and Begusa B M 2014 *International Journal of Engineering and Technology* **4**(8) 469-475
[7] Montemor M F 2014 *Surface and Coatings Technology* **258** 17–37
[8] Bentiss F, Traisnel M and Lagrenee M 2000 *Corrosion science* **42**(1) 127-146
[9] Loewenthal R E, Morrison I and Wentzel M C 2004 *Water Scienceand Technology* **49**(2) 9-18
[10] Pillay C and Lin J 2014 *Corrosion Science* **80** 416-426

Acknowledgements

I would like to express my gratitude to Universiti Malaysia Perlis (UniMAP) and Center of Excellence Geopolymer and Green Materials (CEGeoGTech) for giving me the opportunity to be involved in this project.

IOP Conf. Series: Materials Science and Engineering **374** (2018) 012045 doi:10.1088/1757-899X/374/1/012045

Effect of Solid to Liquid Ratio on Heavy Metal Removal by Geopolymer-Based Adsorbent

N Ariffin[1*], M M A B Abdullah[1,2], M R R Mohd Arif Zainol[3], M S Baltatu[4] and L Jamaludin[1,2]

[1]Center of Excellence Geopolymer and Green Technology (CEGeoGTech), School of Materials Engineering, Universiti Malaysia Perlis (UniMAP), Perlis, Malaysia

[2]Faculty of Engineering Technology, Universiti Malaysia Perlis (UniMAP), Perlis, Malaysia

[3]School of Civil Engineering, Engineering Campus, Universiti Sains Malaysia (USM), Nibong Tebal, Pulau Pinang, Malaysia

[4]Gheorghe Asachi Technical University of Iasi, Faculty of Materials Science and Engineering, Blvd. D. Mangeron 41, 700050, Iasi, Romania

E-mail: nurliyana.arifin@gmail.com& mustafa_albakri@unimap.edu.my

Abstract. Microstructure of three-dimensional aluminosilicate which similar to zeolite cause geopolymer based adsorbent accepted in the treatment of wastewater. This paper presents an investigation on the copper removal from the wastewater by varying the solid to liquid ratio in the fly ash, kaolin and sludge-based geopolymer adsorbent. The adsorption test was conducted to study the efficiency of the adsorbent and the copper concentration was examined by using Atomic Adsorption Spectrometry (AAS). The optimum solid to liquid ratio with the highest percentage removal were 1.0, 0.5 and 0.8 for fly ash-based geopolymer, kaolin-based geopolymer and sludge-based geopolymer adsorbent.

1. Introduction

Great concern raises to water pollution since water constitutes a basic necessity for daily life. Rapid growth of industrial society leads to a significant increase in the demand of water. Discharge of wastewater without detail and adequate treatment give negative impact via contamination to the land and water. Among other issues, water contaminations by heavy metals are more pronounced especially when heavy metals are exposed to the natural ecosystem. Therefore, it is very necessary to treat heavy metal in wastewater from industrial. Adsorption become excellent alternative and central research focus due to effectiveness, simplicity and low cost [1].

Geopolymers, inorganic polymers or zeolites precursors was introduced to replace Ordinary Portland Cement (OPC) due to their unique properties in mechanical, chemical or fire resistance, high temperature resistance and superior durability toward acid attack [2-5]. In addition, compared to OPC, geopolymer being low emission of greenhouse gases and low energy consumption manufacturing process thus proposed as alternative building material to OPC. Reaction between aluminosilicate with an aqueous alkaline activator (mixing of sodium silicate and sodium hydroxide) produce a geopolymer structure consist of an amorphous network of SiO_4 and AlO_4 tetrahedral units similar to zeolite connected by oxygens and charge-balanced by hydrated alkali cations [6-8].

The advantage of geopolymerization method are environmental protecting, energy saving, no more high temperature techniques, compare to zeolite is the process of geopolymerization is simpler than that of zeolite and the temperature used is below 100°C [9, 10]. However, the fundamental different between zeolites and geopolymer, no sufficient time and space for paste to grow into a well crystallized structure during geopolymerization since the paste quickly transform into a hard geopolymer [11]. Nevertheless, decision choice of the source materials for making geopolymers should be take into account the cost, type of application, availability and demand in the market [12].

In the geopolymer, the process of alkali activation of alumino-silicate materials has not been fully described yet. The understanding by now, reaction of aluminosilicate in a strong alkaline environment create disintegration of siliceous bonds. Then, the phase start containing aluminum (Al) parts are formed result in producing zeolite precursor [13]. Geopolymer framework consist of Si-O-Al, negative charge of Al^{3+} balanced by the positive charge of alkali (Na^{+}) ions [14]. Al Bakri reported that the amount of Al in the aluminosilicate gel indicates the reactivity level of fly ash [15]. Thus, the ratio of raw material to alkaline activator affected the adsorbent efficiency.

Million tons of fly ash generated from power station creating the disposal problem. Utilization of fly ash in geopolymer is an excellent and beneficial alternative to overcome an abundant of fly ash byproduct [16, 17].There are two types of fly ash, class F and class C. Fly ash with class F considered as pozzolanic materials act as a binder which activated by alkaline solution [18]. Kaolin is a naturally occurring material composed primarily of fine-grained minerals; it is the most abundant mineral found in the earth. Kaolin is one source of alumino-silicate oxide source to synthesize geopolymer. Kaolinite possesses a slow dissolution of Al and require sufficient time for interactions among the source materials [19, 20]. Sludge contain settleable solid, organic chemical compound, biological flocs, inorganic and others. A few types of sludge such as primary sludge, secondary sludge, tertiary sludge, chemical sludge, humus sludge, surplus activated sludge and attached growth system sludge. Due to the its characteristic which is quite similar to the properties of kaolin, sludge having a potential to be raw material for geopolymer [21].

The mix design of geopolymer are divided into solid to liquid (S:L) ratio and sodium silicate to sodium hydroxide (Na_2SiO_3:NaOH) ratio. Geopolymer paste with low S:L ratio tend to have high viscosity and increase in the dissolution rate. The high content of alkaline activator lead to excessive OH- in the system, contribute to weakening the geopolymer structure [22]. Increase in the alkaline activator (containing sodium waterglass) will increase the Si content thus result in stronger Si-O-Si [23]. This is supported by Al Bakri, increasing the waterglass in the activator will increase the geopolymerisation reaction rate. In certain limit, the rate is negatively affected [24]. Alkaline activator used is to activate the binder in the geopolymerisation process which the combination of alkaline solution with waterglass. Alkali hydroxide is a strong base consist of metal cation and hydroxide ions. Alkali silicate comprise with metal cation used as activator in alkali activated binder usually sodium and potassium [25]. This research presents a performance of fly ash, kaolin and sludge-based geopolymer adsorbent. It proposes new technology of adsorbent in Malaysia which contribute to environmental friendly and economic.

2. Materials and methods

The materials used in this research are fly ash, kaolin, sludge. Geopolymer is a result of mixing raw material and alkaline activator solution. The adsorption test was conducted to study the efficiency of adsorbent.

2.1. Material preparation

For the raw material, fly ash was obtained from ManjungCoal-Fired Power Station, Lumut, Perak, Malaysia. Kaolin was purchased from Associated Kaolin Industries Sdn.Bhd while sludge was collected in the Arau Water Treatment Plant, Perlis, Malaysia. Alkaline activator consists of sodium silicate and sodium hydroxide. Sodium silicate supplied from South Pacific Chemical Industries SdnBhdwith chemical composition of 30.1% SiO_2, 9.4% Na_2O and 60.5% H_2O with the modulus

SiO_2/Na_2O of 3.2, density at 20°C of 1.4g/cm^3 and viscosity at 20°C of 0.4 Pa-s. Sodium hydroxide pallet with 98% purity supplied from Sigma-Aldrich Pty.Ltd.

2.2. Experimental procedure
According to the Figure 1, alkaline activator was produced by the result mixing of sodium silicate and 10M sodium hydroxide together for 5 minutes. Before mixing, NaOH pallet was diluted with distilled water then let it cooled down at room temperature for 1 day. The alkaline activator then was added and mixed homogenously with raw material to developaluminosilicategeopolymer adsorbent paste according to formulation in Table 1. The pastes were cured at temperature 100°C for 24 hours. The process was proceeded to crushing and sieving to obtain powder in size.

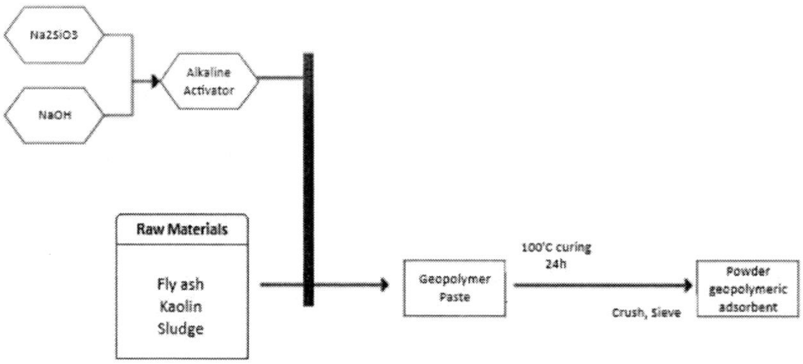

Figure 1. Preparation process of geopolymer-based adsorbent.

Table 1. Formulation use to produce geopolymeric adsorbent.

Raw material	**NaOH:Na_2SiO_3**	**S:L**
Fly ash	1.5	1.0
		1.5
		2.0
		2.5
		3.0
Kaolin	1.5	0.5
		0.6
		0.7
		0.8
		0.9
		1.0
Sludge	1.5	0.5
		0.6
		0.7
		0.8
		0.9
		1.0

2.3. Absorption test

This study performed an adsorption test to investigate the effectiveness of adsorbents to adsorb heavy metal. The initial pH of heavy metal solution was controlled to be lower than 5 for every adsorption test. Adsorption of 100mL of metal ions solutions was conducted by using orbital shaker. After adsorption, the supernatant was filtered by using filter paper. The changes in the concentration of heavy metal was analysed by using Atomic Adsorption Spectrometry (AAS).

Percentage removal efficiency was calculated using the following equation:

$$E(\%) = \frac{Co - Ce}{Co} \text{x } 100 \quad (1)$$

E = Removal efficiency
c_o = Initial concentration
c_e = Final concentration

3. Results and discussions

Reaction in the adsorption occurs due to the ion exchange between the surface and the inside positions of the adsorbent. Figure 2 shows the percentage removal of heavy metal by different materials used and various solid to liquid ratio.

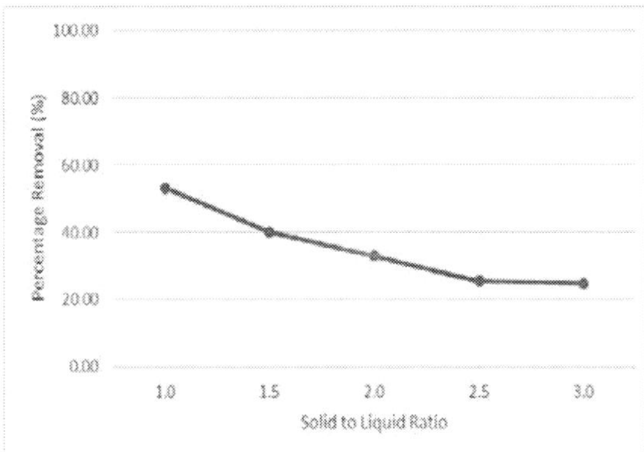

(a) Fly ash-based geopolymer.

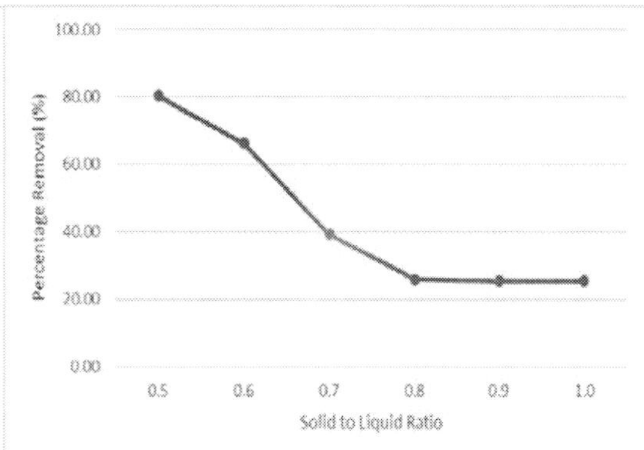

(b) Kaolin-based geopolymer.

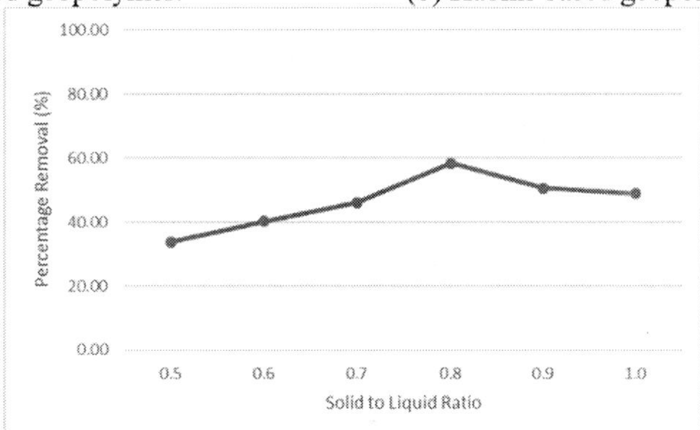

(c) Sludge-based geopolymer.

Figure 2. Percentage removal of copper removal by different geopolymeric adsorbents.

The optimum of solid to liquid ratio of fly ash-based geopolymer are 1.0 which 53% are taken out from the wastewater solution and 46% left behind. The graph 2(a) show the adsorbent efficiency decreased with the increase of solid to liquid ratio. Study by Liew reported that NaOH-fly ash geopolymer observed crystalline and granular structure produced [26]. This show that, increase in the alkaline activator will increase in the adsorbent efficiency. The geopolymer phase thought to be reactive, dissolve by alkali metal. Some cations must be present to keep the structure neutrality or actively bonded into the matrix. Furthermore, the cation bonded via Al-O or Si-O bond by the substitution [27]. This result supported by Cheng and Luukkonen which use S:L ratio 1.3 and removed about 40.9 mg/g and 21.07mg/g [9, 28].

Kaolin-based geopolymer shows highest metal ions uptake among the adsorbents used. This may due to the properties of low reactivity of kaolin. Balancing of Graph 2(b) shows removal decrease with the increase of solid to liquid ratio then slightly different from 0.8 to 1.0. This show that the minimum solid to liquid ratio can be used is 0.8. About 80 percent of copper ions were removed from the solution at 0.5 ratio of solid to liquid. Yang reported that addition of alkali activated concentration is an important factor on other Si gel generation [29,30]. Furthermore, the percentage removal of copper is high. Although decrease the ratio will lead to increase in the removal, the ratio of kaolin/alkaline activator cannot be reduce because of its high workability [31].

Sludge contain 54% SiO_4 and 29% Al_2O_3, high percentage of Si and Al make this material suitable to be treated by alkaline activator to develop geopolymer structure. Solid to liquid ratio of sludge-based geopolymer adsorbent was studied from 0.5 until 1.0. Figure 2(c)shows0.8 (58 percent) is the best solid to liquid ratio copper removal from wastewater solution by adsorbent from sludge-based geopolymer. This indicated that at ratio 0.8, high chemical binding occurs in geopolymer or silicate phases. The minimum percentage heavy metal removal by adsorbent is 33%.This may due to lack alkaline activator as reported by El-Eswed that metal contaminations either by physical or covalent bonds (metal is bonded to the silicate chain or hydroxide links) [32].

4. Conclusions

The present work focused on the removal of copper heavy metal by using fly ash, kaolin and sludge-based geopolymer adsorbent. Different materials having different solid to alkaline activator ratio. The best solid to liquid ratio for every fly ash, kaolin and sludge-based geopolymer adsorbent is 1.0, 0.5 and 0.8 respectively.

5. References

[1] Andrejkovicova S, Sudagar A, Rocha J, Patinha C, Hajjaji W, Da Silva E F and Rocha F 2016 *Applied Clay Science* **126** 141–152

[2] Kolezynski A, Krol M and Zychowicz M 2018 *Journal of Molecular Structure* **1163** 465–471

[3] Shahedan N F, Abdullah M M A, Hussin K, Sandu I, Ghazali C M R, Binhussain M, Yahya Z, and Sandu A V 2014 *MaterialePlastice* **51**(3) 258-262

[4] Abdullah M M A, Jamaludin L, Kamarudin H, Binhussain M, RuzaidiGhazali C M and Ahmad M I 2013 *Advanced Materials Research* **686** 227-233

[5] Izzat A M, Al Bakri A M M, Kamarudin H, Moga L M, Ruzaidi C M, Faheem M T M, Sandu A V 2013 *Materiale Plastice* **50**(3) 171-174

[6] Papa E, Medri V, Amari S, Manaud J, Benito P, Vaccari A and Landi E 2018 *Journal of Cleaner Production* **171** 76–84

[7] Heah C-Y, Liew Y-M, Abdullah M M A and Hussin K 2017 *Scientific Reports* **7**

[8] Yahya Z, Abdullah M M A B, Hussin K, Ismail K N, Sandu A V, Vizureanu P and Razak R A 2013 *Revista de Chimie* **64**(12) 1408-1412

[9] Cheng T W, Lee M L, Ko M S, Ueng T H and Yang S F 2012 *Applied Clay Science* **56** 90–96

[10] Wang H, Li H, Wang Y and Yan F 2015 *Ceramics International* **41**(9) 11177–11183

[11] Al-Zboon K, Al-Harahsheh M S and Hani F B 2011 *Journal of Hazardous Materials* **188**(1–3), 414–421

[12] Razak R A, Abdullah M M A, Kamarudin H, Ismail K N, Sandu I, Hardjito D and Yahya Z 2013 *Revista de Chimie* **64**(6) 593-598

[13] Błaszczynski T Z and Krol M R 2017 *IOP Conference Series: Materials Science and Engineering* **245**(2)
[14] Liew Y M, Kamarudin H, Al Bakri A M M, Luqman M, Nizar I K and Heah C Y 2011 *Australian Journal of Basic and Applied Sciences* **5**(9) 441-449
[15] Al Bakri A M M, Kamarudin H, Nizar I K, Bnhussain M, Zarina Y and Rafiza A R 2012 *Advanced Materials Research* **341-342** 189-193
[16] Nordin N, Abdullah M M A, Tahir M F M, Sandu A V and Hussin K 2016 *International Journal of Conservation Science* **7**(1) 161-166
[17] Abdullah M M A, Tahir M F M, Hussin K, Binhussain M, Sandu I G, Yahya Z, Sandu A V 2015 *Revista de Chimie* **66**(7) 1001-1003
[18] Ibrahim W M W, Abdullah M M A, Sandu A V, Hussin K, Sandu I G, Ismail K N, Kadir A A and Binhussain M 2014 *Revista de Chimie* **65**(11) 1340-1345
[19] Heah C Y, Kamarudin H, Al Bakri A M M, Luqman M, Nizar I K and Liew Y M 2011 *Australian Journal of Basic and Applied Sciences* **5**(7) 1026-1035
[20] Heah C Y, Kamarudin H, Al Bakri A M M, Bnhussain M, Luqman M, Nizar I K, Ruzaidi C M, and Liew Y M 2013 *International Journal of Minerals, Metallurgy and Materials* **20**(3) 313-322
[21] Abdullah M M A B, Nordin N, Tahir M F M, Kadir A A and Sandu A V 2016 *International Journal of Conservation Science* **7**(3) 753-758
[22] Yahya Z, Abdullah M M A B, Hussin K, Ismail K N, Razak R A and Sandu A V 2015 *Materials* **8**(5) 2227-2242
[23] Al Bakri A M M, Kamarudin K, Karem O A K A, Ruzaidi C M, Rafiza A R and Norazian M N 2012 *Applied Mechanics and Materials* **110-116** 734-739
[24] Al Bakri A M M, Kamarudin H, Omar A K, Norazian M N, Ruzaidi C M and Rafiza A R 2011 *Australian Journal of Basic and Applied Sciences* **5**(9) 1916-1922
[25] Faris M A, Abdullah M M A B, Sandu A V, Ismail K N, Moga L M, Neculai O and Muniandy R 2017 *MaterialePlastice* **54**(1) 145-154
[26] Liew Y-M, Heah C-Y, Al Bakri M M and Kamarudin H 2016 *Progress in Materials Science* **83** 595-629
[27] Khale D and Chaudhary R 2007 *Journal of Materials Science* **42**(3) 729–746
[28] Luukkonen T, Sarkkinen M, Kemppainen K, Rämö J and Lassi U 2016 *Applied Clay Science* **119** 266–276
[29] Burduhos Nergis D D, Abdullah M M A and Vizureanu P 2017 *European Journal of Materials Science and Engineering* **2** 111-118
[30] Yang W and Li Q 2012 *Applied Mechanics and Materials* **204–208** 4101–4104
[31] Faheem M T M, Al Bakri A M M, Kamarudin H, Ruzaidi C M, Binhussain M and Izzat A M 2013 *International Review of Mechanical Engineering* **7**(1) 161-166
[32] El-Eswed B L, Yousef R I, Alshaaer M, Hamadneh I, Al-Gharabli S I and Khalili F 2015 *International Journal of Mineral Processing* **137** 34-42

The Influence of Pre-Heated Treatment to Improve Adhesion Bond Coating Strength of Fly Ash Based Geopolymer Ceramic

L Jamaludin[1], M M A B Abdullah[1], K Hussin[1] and A Abdul Kadir[3,1]
[1]Centre of Excellence Geopolymer & Green Technology (CEGeoGTech), School of Materials Engineering, Universiti Malaysia Perlis (UniMAP), Perlis, Malaysia
[2]Faculty of Engineering Technology, Universiti Malaysia Perlis (UniMAP), Perlis, Malaysia
[3]Faculty of Civil and Environmental Engineering, Universiti Tun Hussein Onn Malaysia, Johor, Malaysia

E-mail: liyanajamaludin@unimap.edu.my

Abstract. The study focus on effect of pre-heated ceramic surface on the adhesion bond strength between geopolymer coating coating and ceramic substrates. Ceramic substrates was pre-heated at different temperature (400 °C, 600 °C, 800 °C and 1000 °C). Fly ash geopolymer coating material potential used to protect surface used in exposure conditions after sintering at high temperature. Fly ash and alkali activator (Al_2O_3/Na_2SiO_3) were mixed with 2.0 solids-to-liquid ratios to prepare geopolymer coating material at constant NaOH concentration of 12M. Adhesion test was conducted to determine the adhesion bond between ceramic substrates and fly ash coating material. The results showed the pre-heated ceramic substrates effect the adhesion bond of coating compared with untreated substrates with increasing of strength up to 20 % for temperature 600 °C.

1. Introduction

Alkali activated materials from the waste such as fly ash, kaolin, slag, rice husk and volcano ash has been considered by many researchers as an alternative to cementitious materials for protective coating materials especially in ceramic industry [1]. Alkali activated materials or geopolymer are formed when alumino-silicates dissolve in a strong alkaline solution, reorganize, and precipitate in a hardened state [2, 3]. Good corrosion resistance, imperviousity to water and air, good chemical and physical stabilities at moderately high temperature and high strength keeping in view of using this method to protect ceramic surface. This is due to surface deteriorations and corrosion become major problem for wear of ceramic products. Fly ash coating material was deposited as forming layer on ceramic substrates. Fly ash is a finely solid waste generated from coal-fired thermal power stations during the combustion of coal. At present, about 80% of total power generation in the world based on thermal power plants which consume CO_2 to atmosphere which effect the greenhouse emissions. Due to environmental concern, new ways of utilizing fly ash being explored in order to protect the environment and provide useful ways for its utilization and disposal.

In the ceramic industry, several problems posed by slips and engobes are prone to pinholes and crawling [4]. Pinholes refers as very small opening that appear on the surface of ceramic glaze and will result on reveal of clay body [4]. Ceramics are finding an increasing range of applications, however, the limitations for their use often relate to surface phenomena such as fracture initiation effects, environmental degradation and complex tribological interactions [5]. Commonly in ceramic industry, clay are pre-heated at temperature more than 900 °C to increase strength of ceramic [6]. Fly ash with geopolymer formulations prepared with mixing alumino-silicate with the alkaline activator solution has been applied as protective coating material that suitable for high temperature applications such as fire resistant panel [7]. Geopolymer has good bonding structure of amorphous inorganic polymer which has excellent bonding agent properties and shows high bond strength in an early stage [8]. The geopolymerisation process occurs due to the mixing of fly ash, sodium silicate and sodium hydroxide as the alkaline activator, which produces aluminosilicate gel that acts as a binder [9].In order to curb this problem, geopolymeric application is used to replace traditional method of ceramic glaze processing by possessing several better characteristics especially increasing of adhesion strength. This paper summarizes the influence of pre-heated ceramic substrates on the adhesion strength of geopolymer coating substrates.

2. Materials and methods

2.1. General procedures

Ceramic substrates with dimensions of 9 mm in thickness, 55 mm in width and 110 mm in length were used. The pre-heated treatment process was performed by heating ceramic substrates at temperature range from 400 °C to 1000 °C for 1 hour. The fly ash was supplied by YTL Cement Corporation, Malaysia. Class F fly ash used as aluminosilicate source to produce geopolymer coating material by using sodium hydroxide added to sodium silicate solutions. The best ratio of fly ash/alkaline activator choose is 2.0 at specific constant ratio of Na_2SiO_3/NaOH solution of 2.5. The NaOH solution was prepared by dissolving sodium hydroxide pellets in deionised water, the concentration of the NaOH was kept constant at 10 M.

2.2. Coating sample preparation

Geopolymer coating material was prepared by mixing alkaline activator with fly ash powder. The mixture was applied at ceramic substrates before and after pre-heated treatment by brushing method. After coating, samples were placed at room temperature for drying process for 1 hours.

2.3. Testing and analyzing

The chemical compositions of material was analyzed by using X-ray Flourescence (XRF). Adhesion strength was evaluated for characterizes the performances of ceramic substrates after applying with fly ash geopolymer coating materials. The adhesive strength of the coated samples was measured with an Elcometer F106/3, adhesion tester according to ASTM D4541. The average of three measurements was evaluated as adhesion strength. A minimum of three specimens were tested to evaluate the strength gain for the specimens after pre-heated ceramic substrates.

3. Results and discussions

3.1. Chemical compositions analysis

Table 1 shows the main composition of fly ash and ceramic substrate. Fly ash contains silica dioxide (SiO_2) with 35.7 wt % and calcium oxide (CaO) contributes 20.8 wt % in the fly ash. The composition analysis indicated that ferum (III) oxide (Fe_2O_3) has 19.91 wt % followed by aluminium oxide (Al_2O_3) has 15.4 wt %. Ceramic substrates analysis contains high silica dioxide (SiO_2) with 66.5 wt % and 19.1 wt% of alumina (Al_2O_3). The high SiO_2 content proved to silica availability lead to faster conversion of solid aluminosilicate sources to geopolymer gel [10] in order to determine geopolymer gel structure and reaction mechanisms. As geopolymer source materials, the rich of chemical composition of SiO_2 and Al_2O_3 is important in synthesised in geopolymerization process which link to the strength performance such as adhesive strength [11].

3.2. Effect of pre-heated ceramic substrate on adhesion strength bond of coating samples

Figure 1 shows adhesion strength analysis of coating samples on the pre-heated ceramic substrates. The geopolymer coating with unheated ceramic substrates shows lowest adhesion strength which is 1 MPa while highest adhesion strength contributes at pre-heated ceramic substrates at temperature 600 °C with 4.7 MPa strength. At temperature 700 °C to 1000 °C, the adhesion strength starts decreased from 4.3 MPa to 3.2 MPa. The decreasing strength of coating due to the ceramic sample tend to bend due to excessive of high temperature. Pre-heated treatment of ceramic substrates with different temperature will adequate different strength of interfacial interactions with geopolymer material coating due to the good penetration of coating material with ceramic substrates after pre-heated process compared with unheated substrates [11-13]. The ceramic substrates pores minimizes the surface deteriorations and will results successful adhesion bond strength between substrates and coating material. From the results obtained, the temperature affecting the adhesion strength due to bonding between matrix and geopolymer which indicates high strength of coating surface substrates. This phenomenon occurs due to coating material cannot diffuse very well on the overly high sintered substrate surface [14-16].

Table 1. Chemical compositions of fly ash and ceramic substrate.

	Chemical Compositions (wt%)							
	Al_2O_3	SiO2	CaO	Fe_2O_3	SO_3	K_2O	TiO_2	MnO
Fly ash	15.4	35.7	20.8	19.91	3.55	1.85	1.13	0.11
Ceramic substrate	19.1	66.5	8.91	1.12	-	1.39	0.67	1.42

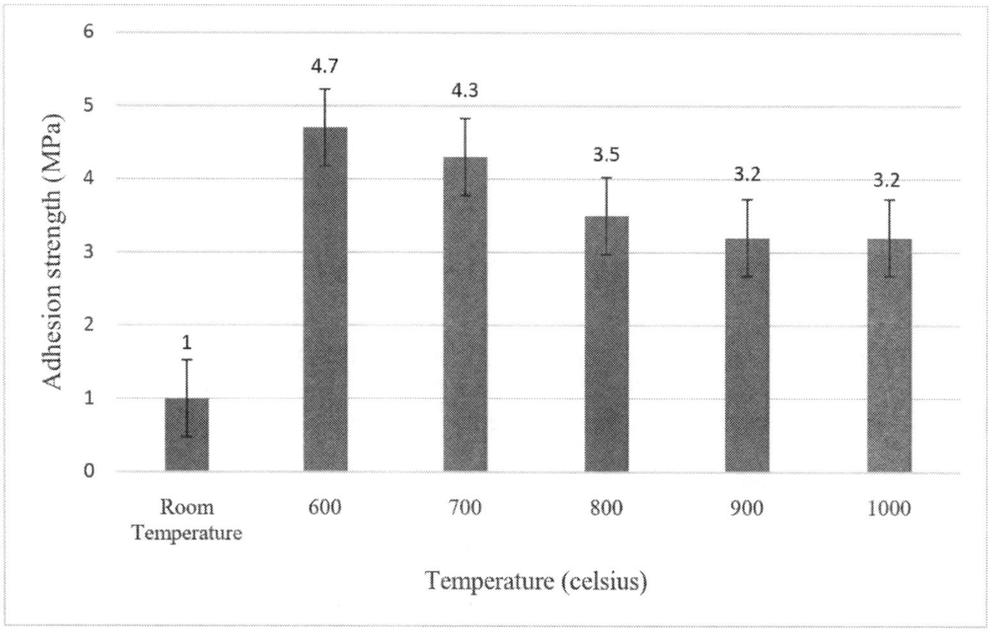

Figure 1. Adhesion strength of coating samples on the pre-heated ceramic substrates.

4. Conclusions

The optimum pre-heated temperature of ceramic substrates for fly ash geopolymer coating in this study is 600 °C which lead to high adhesion bond strength is 4.7 MPa. Pre-heated treatment of ceramic substrates with different temperature will adequate different strength of interfacial interactions with geopolymer material coating due to the good penetration of coating material with ceramic substrates after pre-heated process compared with unheated substrates.

5. References

[1] Jing Z, Matsuoka N, Jin F, Yamasaki N, Suzuki K and Hashida T 2006 *J Mater Sci* **41** 1579–84
[2] Duxson P, Fernandez-Jimenez A, Provis J L, Lukey G C, Palomo A and van Deventer J S J, 2007 *Journal of Materials Science* **42** 2917-33
[3] Davidovits J 1991 **37** 1633-56
[4] Elimbi A, Tchakoute HK and Njopwouo D 2011 *Constr Build Mater* **25** 2805–12
[5] Al Bakri Abdullah M M, Jamaludin L, Kamarudin H, Binhussain M, Ruzaidi Ghazali C M and Ahmad M I 2013 *Advanced Materials Research* **686** 227-233
[6] Lee W K W and van Deventer J S J 2002 *Colloid and Surfaces A* **211** 49–66
[7] Mustafa Al Bakri A M, Kamarudin H, Khairul Nizar I, Bnhussain M, Zarina Y and Rafiza A R 2012 *Advanced Materials Research* **341-342** 189-193
[8] Shahedan N F, Abdullah M M A B, Hussin K, Sandu I, Ghazali C M R, Binhussain M, Yahya Z and Sandu AV 2014 *Materiale Plastice* **51**(3) 258-262
[9] Rowles M and O'Connor B 2003 *Journal of Materials Chemistry* **13** 1161–1165
[10] Ruzaidi C M, Mustafa Al Bakri A M, Binhussain M, Siti Salwa M S and Alida A, Muhammad Faheem M T and Azlin S S 2014 *Key Engineering Materials* 540-545
[11] M A Bakri A M, Kamarudin H, Khairul Nizar I, Bnhussain M, Zarina Y and Rafiza A R 2012 *Advanced Materials Research*, **341-342** 189-193
[12] Burduhos Nergis D D, Abdullah M M A and Vizureanu P 2017 *European Journal of Materials Science and Engineering* **2** 111-118
[13] Temuujin J, Minjigmaa A and Rickard W 2011 **107** 287-292
[14] Jun N H, Minciuna M G, Abdullah M M A, Jin T S, Sandu A V and Ming L Y 2017 *Revista de chimie* **68**(10) 2367-72
[15] Popovici A, Corbu O, Popita GE, Rosu C, Proorocu M, Sandu AV and Abdullah MMAB 2015 *Materiale Plastice* **52**(4) 588-592
[16] Puskas A, Corbu O and Kollo SA 2017 *Modern technologies for the 3rd millennium* p. 239-244

Acknowledgements
The financial support from Center of Excellence Geopolymer & Green Technology (CEGeoGTech), Universiti Malaysia Perlis is gratefully acknowledged.

Degradation of Organic Matter from Stabilized Leachate by Using Zinc Sulphate as Coagulant Agent

M A Kamaruddin[1,3,*], MS Yusoff[2], N H Adam[2], M R R Maz[2,3], M M A B Abdullah[3], R Alrozi[4] and M H Zawawi[5]
[1]Environmental Technology Division, School of Industrial Technology, Universiti Sains Malaysia, Penang, Malaysia
[2]School of Civil Engineering, Universiti Sains Malaysia, Penang, Malaysia
[3]Center of Excellence Geopolymer and Green Technology (CEGeoGTech), School of Materials Engineering, Universiti Malaysia Perlis (UniMAP), Perlis, Malaysia
[4]Faculty of Chemical Engineering, Universiti Teknologi Mara, Permatang Pauh, Penang
[5]Department of Civil Engineering, Universiti Tenaga Nasional, Kajang, Selangor, Malaysia

E-mail: anuarkamaruddin@usm.my

Abstract. Stabilized landfill leachate often contains higher organic fractions than the young one. The organics require several sequential treatments to render the leachate parameters concentrations to permissible discharge limits before being discharged to receiving water. This study focused on the application of Zinc Sulphate ($ZnSO_4$) as coagulant agent followed with microfiltration of 0.45 μm pore size under different condition of landfill leachates. The results indicated that the sludge volume index (SVI), soluble COD and turbidity concentrations were inter-related to each other when compared under different $ZnSO_4$ dosages. However, that was not the case when correlation between stabilized and young leachate were compared side by side. To conform the finding, one-way analysis of variance (ANOVA) was conducted and the results were further explained by the adequacy and significant of confidence interval. Finally, it was proven that, soluble and particulate COD had significant CI of 95% applicable for stabilized leachate alone.

1. Introduction

Leachate is generated when water or moisture inherent within the waste percolates through the landfill [1]. Leachate is considered highly complex and contaminated wastewater by high concentrations of literally wide ranges of compounds present in waste including organic and inorganic compounds both in dissolved and suspended forms [2]. Among many discharge limit criteria's set, chemical oxygen demand (COD) is one of the major parameters to ensure safe discharge of wastewater [3]. COD is a measure of the oxygen equivalent of the organic matter content of a sample that is susceptible to oxidation by a strong chemical oxidant. COD test uses a strong chemical oxidant (potassium dichromate), acid, and heat to oxidize organic carbon to carbon dioxide and water [4]. COD is the main parameter analyzed to determine the quality of a wastewater. The dichromate reflux method is preferred over procedures using other oxidants (potassium permanganate) because of its superior oxidizing ability, applicability to a wide variety of samples and ease of manipulation. Oxidation of most organic compounds is 95-100% of the theoretical value [5]. However, there is no documented nor published COD fractionation of Pulau Burung Sanitary Landfill (PBSL) leachate has been done. Thus, in order to establish appropriate behavioral mechanisms, COD fractionation was carried out in this study to compare the stabilized and young landfill leachate collected from the same landfill. Coagulation and flocculation method using $ZnSO_4$ as coagulant was applied and coded as Method 1 followed by filtration using 0.45 μm filter paper. Another method, which was a modified from Method 1 in determining the effect of coagulant amount, coded as Method 2. A simple mathematical formula was used in finding particulate COD value.

2. Materials and methods

Leachate was collected through from an oxidation pond which retains 'stabilized leachate' whereas the leachate produced from the active phase of the landfill which was defined as 'young leachate'. The leachate from PBSL is classified as standard B that refers to discharges outside catchment area Environmental Quality Control of Pollution from Solid Waste Transfer Station and Landfill Regulations (2009). Samples were collected by using polyethylene bottles. The bottles were washed and rinsed using distilled water before sample collection. Samples were kept in a cold room (4°C) to avoid any changes. The following parameters were systematically monitored; COD, BOD_5, ratio of BOD_5/COD, ammonia (NH_3), iron (Fe^{3+}), conductivity, suspended solid (SS), pH, colour and turbidity.

2.1. Analytical

Soluble COD (SCOD) is one of the main fractions in total COD. Soluble COD contains readily biodegradable COD and unbiodegradable soluble COD. Determination of soluble COD is important in determining the other fractions. In this research, determination for SCOD is carried out using $ZnSO_4$ coagulation and flocculation process. Coagulation is the destabilization of colloid impurities and transfer small particles into larger aggregates by neutralizing the forces that keep them apart. Cationic coagulant provides positive electric charges to reduce the negative charge of the colloid to form larger particles [6]. Coagulation process is using as a main method in this analysis. Different volume of ZnSO4 was used in determine the optimum dosage for this process. The filtration process using 0.45 μm filter paper was applied in all methods. Fig. 1 shows the references for the methods.

2.2. Description of method

Method 1. 1 mL of a 0.6 M $ZnSO_4$ solution was added to each of the three 200 mL beaker containing 100 mL of sample. Sample was mixed for 1 min at 200 rpm followed by slow mixing (30 rpm) for 5 min. Samples were then allowed to settle for an hour and supernatant was withdrawn at the end of settling period. The supernatant were filtered using a pre-rinsed 0.45 μm mixed cellulose ester membrane filter (Milipore, MA). The aliquots of the filtrate was taken and measured for COD, turbidity and SVI value in triplicate samples. pH was adjusted to 10.5±0.3 using 6M NaOH.

Method 2. All the procedures were repeated from Method 1 in different volume of $ZnSO_4$ until the graph of $ZnSO_4$ volume verses COD value is constant. The constant graph shows the optimum dosage of ZnSO4 in both leachate samples. The maximum removal of COD, turbidity and SVI also related with optimum dosage of coagulant. SVI (sludge volume index) can be calculated using Equation. 1.

$$SVI = \frac{1000\, x\, sludge\ volume\ (SV)}{Suspended\ solids\ (SS)} \tag{1}$$

3. Results and discussions

3.1. Leachate characteristics

COD. Based on Table 1, young leachate (3100-3310mg/l) shows higher value in terms of COD as compared to stabilized leachate (2610-2830 mg/L). The COD value was due to degradation organic matters that contain in solid wastes and then contribute to leachate formation [7]. In the methanogenic phase in stabilized leachate, the production of methane achieved maximum concentrations and slowly decreased. Thus, the degradation of organic matters in the leachate also reduced. At this point, the oxygen demand was also decreased. However, young landfill leachate undergo acid phase contributed to the higher concentrations in organic matters [8]. COD oxidized all organic matters using strong oxidation agents in absence of oxygen. However, stabilized leachate exerts high molecular weight and became difficult to breakdown. As comparison, young leachate is a fresh leachate which is has a high level of O_2. The O_2 is used in oxidized organic matters in this leachate and increase the oxygen demand in COD value.

BOD_5/COD ratio. The BOD_5/COD ratio represents the proportion of biodegradable organics in landfill leachate. Some information on the wastewater biodegradability can be gained where a high ratio of BOD_5/COD showed that the wastewater is relatively biodegradable whereas a low ratio indicates that the wastewater slower to be biodegraded. In this research, the ratio of BOD_5/COD in young leachate (0.102-0.105) was higher than stabilized leachate (0.086-0.090) indicated that young leachate was more biodegradable than stabilized leachate. Thus, the suitable treatment was suggested to be flocculation and coagulation. Stabilized leachate shows the lower ratio in BOD_5/COD. This value indicated that this landfill had reached a stable status and probably were not suitable for biological treatment processes [9]. Results from BOD5/COD ratio in this experiment evaluated that the leachate from PBSL in different ages contain low percentage in biodegradable fraction of COD.

Table 1. Leachate characteristics.

	Stabilized leachate		**Young leachate**	
Parameter	**Range**	**Mean**	**Range**	**Mean**
COD(mg/l)	2610-2830	2744	3100-3310	3195
BOD_5(mg/l)	239-250	243	324-336	329.333
BOD_5/COD(mg/l)	0.086-0.090	0.089	0.102-0.105	0.103
Fe(mg/l)	1.4-1.6	1.5	2.13-2.33	2.259
Conductivity(mS/cm)	22.2-23.2	22.7	13.3-13.9	13.7
Ammonia(mg/l)	2190-2270	2227.5	1220-1280	1240
Suspended solid(mg/l)	87.4-95.9	93	231-249	240
Colour(PtCo)	2450-2740	2526	1480-1630	1559
Turbidity(NTU)	55.6-59.2	57.567	95.3-99.3	97.625
pH	8.16-8.27	8.22	7.43-7.65	7.57

3.2. Soluble COD and particulate COD.

Particulate organic commonly feature as a large particle size which could not pass through cell wall and needed to undergo extracellular hydrolysis prior to adsorption [10]. Both of soluble and particulate COD contain biodegradable and unbiodegradable matter. In this research, SCOD values determined by after the determination of soluble COD, the particulate COD can be calculated as shown in Equation 2:

$$Particulate\ COD\ (PCOD) = COD_{Total} - COD_{soluble} \quad (2)$$

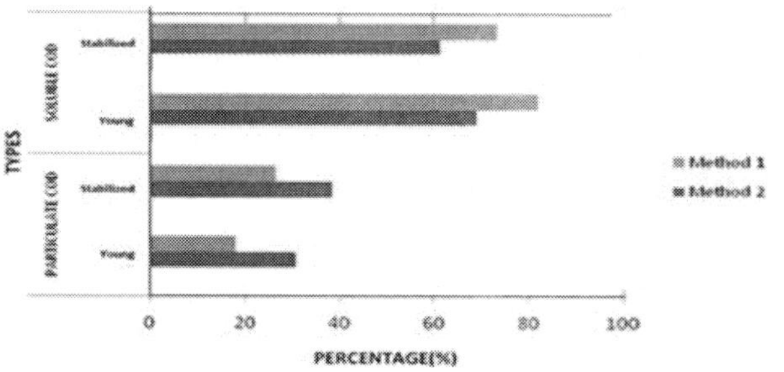

Figure 1. The correlation between PCOD and SCOD for two methods.

Method 1 utilized 5ml/l of zinc sulphate to coagulate and flocculate all the particulate matter in both leachate before filtered the sample was filtered with 0.45µm filter paper. The experimental results in Figure 1 indicated that Method 1 exhibited the highest percentage of soluble COD (stabilized leachate = 72.28%; young leachate = 82.19%) compared with Method 2. However, analysis on COD removal based on Figure 2 shows that Method 1 able to removed 27.50% of COD from stabilized leachate sample whereas young leachate showed smaller percentage of COD removal which is 19.70%. The analysis from Method 1 identified that the COD value in particles size 0.45µm contains more in young leachate compare in stabilized leachate.

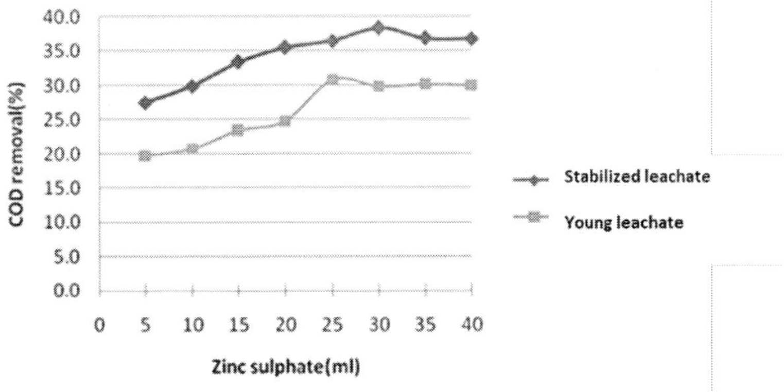

Figure 2. Correlation between COD removal and amount of Zinc sulphate.

As seen in Figure 2, the graph shows the correlation between COD removal percentage and dosage of zinc sulphate in flocculation and coagulation process. The graph also implied that the COD removal increased with increasing of zinc sulphate dosage until the point of maximum removal (stabilized leachate = 38.4%; young leachate = 30.8%), at about 30ml/l of coagulant in stabilized leachate and 25 ml/l in young leachate. The graph started to be constant after the optimum dosage was achieved. At the optimum dosage, the coagulant had no significant effect on the increasing dosage of coagulant [11]. The COD removal percentage in both leachate in Figure 2 was reduced after optimum dosage because of the destabilization of particulates. Zinc sulphate shows the effectiveness in remove more COD in stabilized leachate compare in young leachate. This result might have been caused by different molecular weight of leachate. Based on previous study, the stabilized leachate contains high molecular weight (humics like substances) whereas young leachate contains low molecular weight [12]. Therefore, the coagulation and flocculation process in this method could remove the organic materials including humics like substances that contains more in stabilized leachate. Consequently, the removal of COD in stabilized leachate is higher than young leachate.

Figures 3 (a) and Figure 4 (a) show the effect of turbidity in stabilized and young leachate by increasing the dosage of zinc sulphate. The turbidity data for stabilized leachate was decreased when the volume of zinc sulphate was increased. It implied that the value of turbidity reduced from 38.336 NTU to 3.085 NTU when the dosage of zinc suphate was increased from 25 ml/l to 30 ml/l. Leachate from young leachate showed similar pattern; when the dosage of zinc sulphate was increased to 25 ml/l, the turbidity of leachate was decreased. At this stage, the volume of zinc sulphate was already optimum for the sample.

The sludge volume index (SVI) was calculated by using simple formulation. From Figures 3(b) and 4 (b), the results revealed that the highest SVI value was achieved by using 30 ml/l of zinc sulphate in stabilized leachate and 25 ml/l in young leachate. Determination of optimum dosage of zinc sulphate as coagulant was defined by analyzing the turbidity and SVI data. The research found that the optimum dosage for coagulant (zinc sulphate) in coagulation and flocculation process is 30ml/l in stabilized leachate and 25ml/l in young leachate.

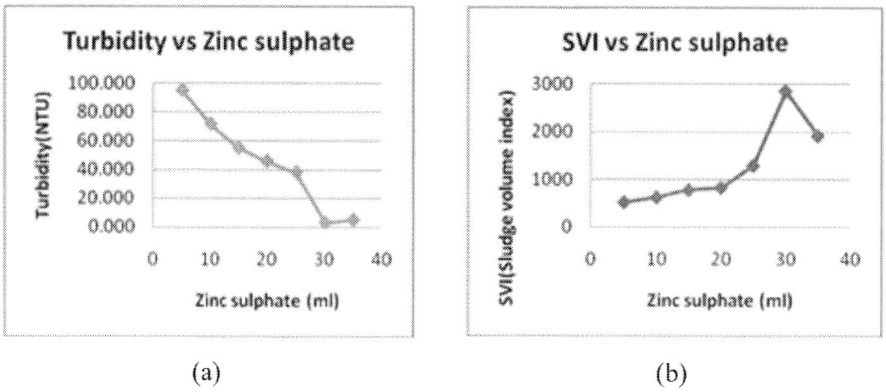

Figure 3. (a) Correlation between turbidity and amount of Zinc Sulphate; (b) Correlation between SVI and amount of Zinc Sulphate.

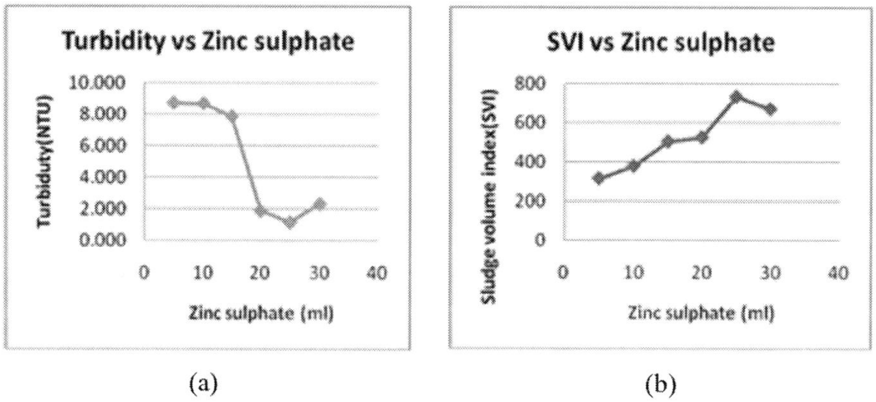

Figure 4. (a) Correlation between turbidity and amount of Zinc Sulphate; (b) Correlation between SVI and amount of Zinc Sulphate.

The optimum dosage of coagulant in stabilized leachate indicated the higher value compare to young leachate due to the characteristics of leachate. The results may affected by colour content. Colour usually related to COD and turbidity of sample. Stabilized leachate contains high value of colour (2526.667 Pt-Co) while young leachate contains lower value (1559.667 Pt-Co) of colour. Therefore, stabilized leachate need more coagulant in coagulated and flocculated all the organic materials (humic like substance) that cause the contamination of colour [13].

3.3. Biodegradable and Unbiodegradable COD

The biodegradable COD in both leachate sample from PBSL show a small value as compared to unbiodegradable COD. The results for stabilized and young leachate samples came out with same pattern in both leachates. Figure 5 (a) implied that stabilized leachate contains only 27% of biodegradable COD compare to 73% of unbiodegradable COD. The young leachate sample configured 32% of biodegradable COD higher than biodegradable COD in stabilized leachate in Figure 5 (b). Generally, biodegradable COD contains more two fractions including readily biodegradable COD (Sbsi) and slowly biodegradable COD (Sbpi). It was separated by physical properties which is soluble and particulate.

Biodegradable COD can be removed from leachate by aeration process. However, stabilized and young leachate were recognized as moderate stable and stable landfill from the characteristics analysis. The data from characteristics for both leachates prove that the leachate from PBSL contains more unbiodegradable COD and not classified as biodegradable landfill based on ratio of BOD_5/COD.

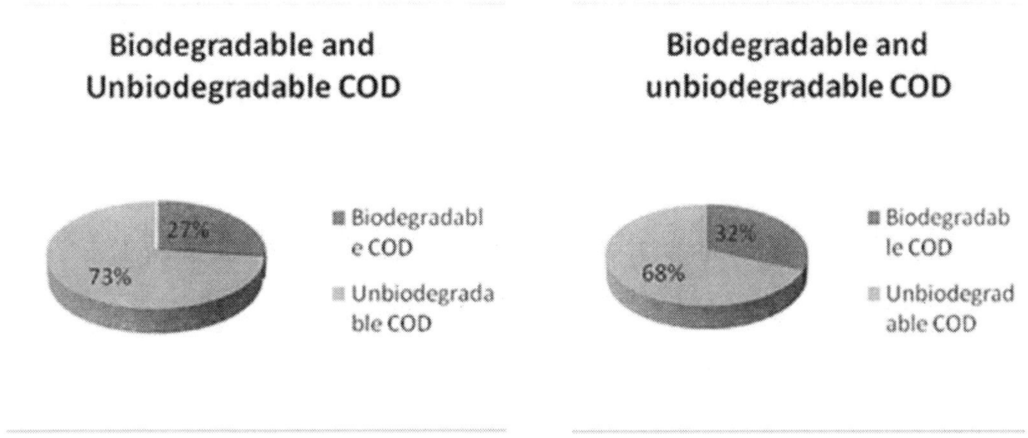

(a) (b)

Figure 5. (a) The percentage of biodegradable and unbiodegradable in stabilized leachate and (b) The percentage of biodegradable and unbiodegradable in young leachate.

4. Conclusions

In this study, COD fractions in leachate were investigated by using different methods based on physical and chemical separation process. Leachate with different ages was categorized as stabilized and young leachate from PBSL. Characterization of leachate shows that the COD value in stabilized leachate in range 2610-2830 mg/L while young leachate value was 3100-3310mg/L. The BOD_5/COD ratio for stabilized leachate is 0.086-0.09 and young leachate was 0.102-0.105. Young leachate has high COD and BOD_5/COD ratio compared to stabilized leachate because of the age of landfill. Although the COD and BOD_5 value of PBSL was lower than typical landfill, the analysis on characteristics of this landfill is continuous analysed in treat the leachate.

5. References

[1] Alrozi R, et al 2017 *Removal of organic fractions from landfill leachate by casuarina equisetifolia activated carbon: Characteristics and adsorption mechanisms* in *AIP Conference Proceedings* AIP Publishing
[2] Kamaruddin M A, et al 2017 *Environmental Science and Pollution Research* 1-33
[3] Kamaruddin M, et al 2017 *Coagulation-Flocculation Process in Landfill Leachate Treatment: Focus on Coagulants and Coagulants Aid* in *IOP Conference Series: Materials Science and Engineering* IOP Publishing
[4] Apha, A, *WPCF 2005* Standard Methods For Examination of Water And Wastewater, 21th Ed New York: APHA, AWWA, WPCR **1**
[5] Lu J, et al 2006 *Chemosphere* **62(2)** 322-331
[6] Jiang J-Q and Lloyd B 2002 *Water research,* **36**(6) 1397-1408
[7] Kamaruddin M A, et al 2016 *Current status of Pulau Burung Sanitary Landfill leachate treatment, Penang Malaysia* in *AIP Conference Proceedings* AIP Publishing
[8] Kamaruddin M A, et al 2015 *Applied Water Science* **5**(2) 113-126
[9] Kamaruddin M A, Yusoff M S and Ahmad M A 2011 *Optimization of durian peel based activated carbon preparation conditions for ammoniacal nitrogen removal from semi-aerobic landfill leachate*

[10] Orhon D and Çokgör E U 1997 *Journal of Chemical Technology & Biotechnology: International Research in Process, Environmental AND Clean Technology* **68(3)** 283-293
[11] Amuda O and A Alade 2006 *Desalination* **196**(1-3) 22-31
[12] Calace N, et al 2001 *Environmental pollution* **113**(3) 331-339
[13] Tatsi A, et al 2003 *Chemosphere* **53**(7) 737-744

IOP Conf. Series: Materials Science and Engineering **374** (2018) 012048 doi:10.1088/1757-899X/374/1/012048

Enhancement of Energy Conversion Efficiency for Dye Sensitized Solar Cell Using Zinc Oxide Photoanode

N Jamalullail[1], I Smohamad[1,2,3*], M Nnorizan[1,2,3] and N Mahmed[2,4]
[1]School of Microelectronic Engineering, Universiti Malaysia Perlis, Pauh Putra Campus, 02600 Arau, Perlis, Malaysia
[2]Center of Excellence Geopolymer and Green Technology (CEGeoGTech), Universiti Malaysia Perlis, Perlis, Malaysia
[3]Advanced Multi-Disciplinary MEMS-Based Integrated NCER Centre of Excellence (AMBIENCE), Universiti Malaysia Perlis, Perlis, Malaysia
[4]School of Material Engineering, Universiti Malaysia Perlis, Perlis, Malaysia

E-mail: ilisalwani@unimap.edu.my

Abstract. Dye sensitized solar cell (DSSC) is a third generation solar cell that is well known for its low cost, simple fabrication process and promised reasonable energy conversion efficiency. Basic structure of DSSC is composed of photoanode, dye sensitizer, electrolyte that is sandwiched together in between two transparent conductive oxide (TCO) glasses. Each of the components in the DSSC contributes important role that affect the energy conversion efficiency. In this research, the commonly used titanium dioxide (TiO_2) photoanode has previously reported to have high recombination rate and low electron mobility which caused efficiency loss had been compared with the zinc oxide (ZnO) photoanode with high electron mobility (155 $cm^2V^{-1}s^{-1}$). Both of these photoanodes had been deposited through doctor blade technique. The electrical performance of the laboratory based DSSCs were tested using solar cell simulator and demonstrated that ZnO is a better photoanode compared to TiO_2 with the energy conversion efficiency of 0.34% and 0.29% respectively. Nanorods shape morphology was observed in ZnO photoanode with average particle size of 41.60 nm and average crystallite size of 19.13 nm. This research proved that the energy conversion efficiency of conventional TiO_2 based photoanode can be improved using ZnO material.

1. Introduction

Photovoltaic technology had shown rapid growth in the past two decades [1]. It is known that this type of technology supports the use of renewable energy source from sun to produce electricity. There are many types of photovoltaic cell such as gallium arsenide (GaAs) based solar cell [2], silicon (Si) based solar cell [3] and dye sensitized solar cell (DSSC) [4]. Each type of solar cell hastheir own speciality with difference energy conversion efficiency. Based on research that had been done, GaAs based solar cell has remarked the energy conversion efficiency of 30% [5], Si solar cell with 25% [6] and 12.3% for DSSC [7]. Even though GaAs and Si based solar cell have recorded higher energy conversion efficiency compared to DSSC, these type of solar cells required complicated fabrication process and high manufacturing cost [8,9].Thus, DSSC is suggested as an alternative renewable energy supply as it promised low cost, simple fabrication process and reasonable energy conversion efficiency for low power appliances [10, 11].

DSSC is a third generation solar cell that is first introduced by O'Regan and Gratzel in 1991 which inspired by the photosynthesis process to absorb solar energy or also known as photon and convert it into electrical energy [12, 13]. Basic structure of DSSC is composed of semiconductor photoanode layer, dye sensitizer, electrolyte which is sandwiched together in between two transparent conductive oxide (TCO) glassesthat will work as working electrode and counter electrode as shown in Figure 1 [14, 15]. Each of the components in the DSSC contributes an important role that affect the energy conversion efficiency.When DSSC is exposed to the photon, dye molecules that are mounted on the surface of semiconductor photoanode layer will absorb the incoming photon. The photoexcitation of the dye will cause an injection of electrons into semiconductor photoanode layer. The injected electrons will disperse through the semiconductor photoanodestructure and emits through the working electrode, external load and proceed to the counter electrode. Next, electrolyte that contains iodide–triiodide (I^- / I_3^-) will act as the mediator to fetch the electrons which arrived at the counter electrode and regenerate the oxidized electrons in the dye molecules.The cycle will continue repeated with the presence of photon in the environment.

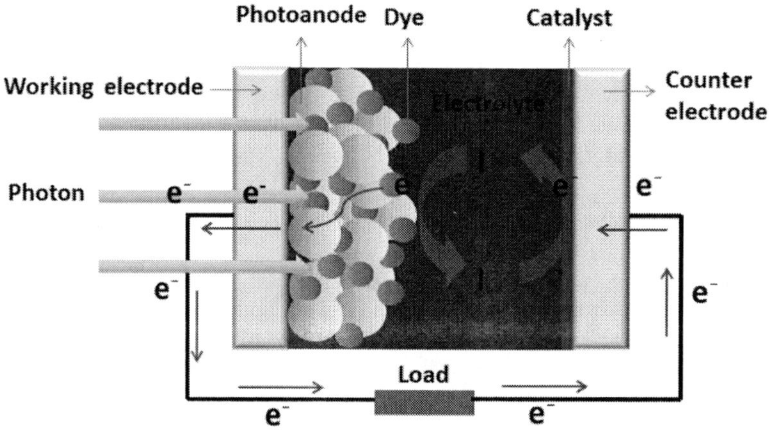

Figure 1. Basic structure and operating principle of DSSC.

The process of photon converted into electrical energy was mainly happen in semiconductor photoanode layer. Semiconductor photoanode layer is the agent that will provide the pathways for electrons to move through and also provide the surface for the dye to be attached. Thus, the selection of the semiconductor photoanode material for the DSSC needs to be taken seriously so that the DSSC will work efficiently. The most commonly used semiconductor photoanode material in DSSC is titanium dioxide (TiO_2) [16, 17]. However, TiO_2 have the drawback of low electron mobility (0.1 – 4 $cm^2V^{-1}s^{-1}$) [18] that causing it to have high recombination rate [19, 20] which leads to energy conversion efficiency loss [21]. Thus, in this research, the performance of the DSSC with commonly used TiO_2photoanode had been compared with the zinc oxide (ZnO) photoanode that had been reported to have high electron mobility (155 $cm^2V^{-1}s^{-1}$) [22].

2. Experimental methods

2.1. Fabrication of DSSC

The fabrication process started with cleaning the transparent conductive oxide glass (TCO, 25 mm x 25 mm, ±15 Ω/cm^2) in ultrasonic bath using ethanol, acetone and deionized water for 10 minutes sequentially. Next, the glasses were dried using air blower. There are two types of photoanode paste prepared in this research. First, TiO_2 photoanode paste was prepared by mixing 1 g of TiO_2 powder (99% anatase) with 1.5 ml of 0.1M nitric acid solution in a mortar. The solution was stirred using pestle until a soupy solution was formed. Next, ZnO photoanode paste was prepared by mixing 1 g of ZnO powder (98% wurzite) with 6 ml of ethanol mixed with deionized water (7:3) and 0.5 ml of 0.1M nitric acid solution. Same as the preparation of TiO_2 photoanode paste, the ZnO solution also need to be stirred using pestle until a fine and soupy solution was formed. Before applying the photoanode paste on the TCO glass, the sheet resistance of the TCO was tested using multimeter to determine which side of the glass is conductive. Then, the conductive surface of the TCO glass was placed faced up and coveredthe edgesof the glass using scotch tape leaving 1.0 x 1.5 cm^2 as the active area as shown in Figure 2. After that, the photoanode paste was deposited on the conductive surface of the TCO glass using doctor blade technique. Then, the samples were annealed at 450 °C in the furnace for 1 hour.

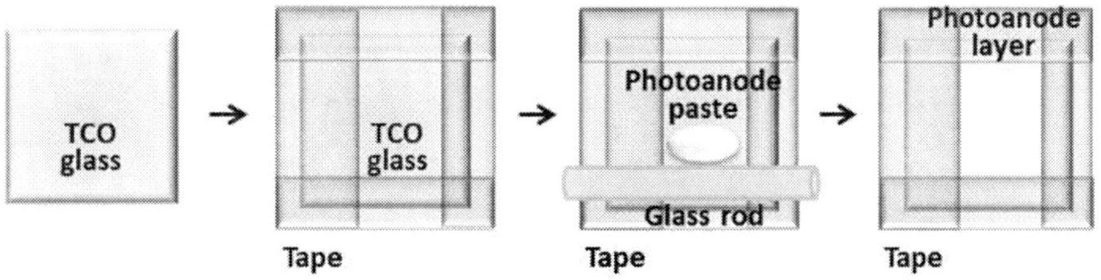

Figure 2. Deposition photoanode paste using doctor blade technique.

After the annealing process of the photoanodes, the substrates weresoaked in 0.3 mM ruthenium (Ru N719) dye that had been synthesized using 25 ml ethanol solution for 24 hours. In order to control the environment effect, the samples were stored in a closed space without the presence of light.Figure 3 shows the image of photoanode substrate after the dye soaking.Meanwhile, the counter electrode was prepared by applying the soot of the candle on the conductive surface of the ITO glass as shown in Figure 4. Then, the complete DSSC structure was done by sandwiching the working electrode and counter electrode together using a binder spacer. Next, a drop of iodine solution was dropped in between the two glasses as a mediator for the redox process.

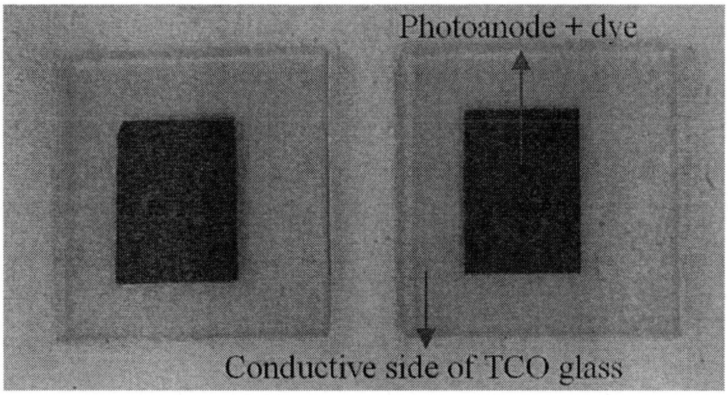

Figure 3. Phoatoanode substrate coated with Ru dye.

2.2. Characterization
In order to analyze the performance and efficiency of the fabricated DSSC, characterization process was done. Scanning Electron Microscope (SEM-EDX, JOEL JSM-6010LV, Oxford Instrument) was used to observe the surface morphology of each type of synthesized photoanode and the average particle size was measured using ImageJ software. The crystallite size and structure of the photoanode was observed through x-ray crystallography technique using X-Ray Diffraction instrument (D2 Phaser, Bruker). Meanwhile, the absorption ability of the Ru N719 dye was analyzed using UV-Visible Spectroscopy (Lambda 950, Perkin Elmer). Lastly, for the electrical characteristics and energy

conversion efficiency of the fabricated DSSC was investigated using Solar Simulator (SMU 2450, Keithley). During the electrical testing process, the DSSC was exposed under 1000 W/m^2 light supplied by Xenon Lamp Power Supply (XPS-1600, Solar Light) imitating the real solar energy. From the current-voltage (IV) characteristics, the energy conversion efficiency can be calculated.

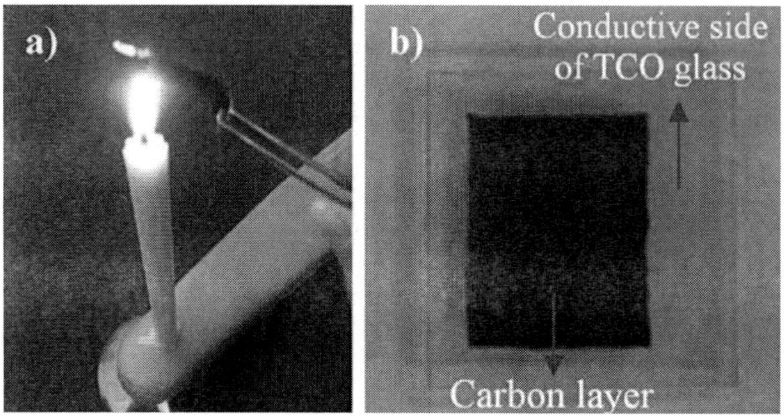

Figure 4. (a) Preparation of counter electrode (b) Carbon coated layer.

3. Results and discussions

3.1. Surface morphology

The surface morphology observed for TiO_2 and ZnO photoanode were shown in Figure 5. Observed in surface morphology of TiO_2 photoanode is the spherical agglomeration of the nanostructure with the average particle size of 19.0 nm. As for the ZnO photoanode, the surface morphology observed is in nanorods shape with the average particle size of 41.6 nm that is bigger than TiO_2 photoanode. The bigger particle size of the photoanode nanostructure will offer bigger surface for the dye attached and thus more light can be absorbed and harvested [23]. In addition, the nanorods structure of ZnO photoanode can provide direct pathways for the electrons to move through the cell and thus increase the energy conversion rate that is in good agreement with the past research [24, 25].

(a) (b)

Figure 5. Surface morphology of (a) TiO_2 (b) ZnO photoanode X-ray diffraction analysis.

Figure 6 presents the XRD pattern of the TiO_2 and ZnO deposited on the glass substrate. For the TiO_2 photoanode, all of the peaks were indexed with the anatase phase (JCPDS card no 00-021-1272). Meanwhile, the ZnO photoanode were indexed in hexagonal ZnO (JCPDS card no 00-036-1451) indicating that the nanorods structure were crystallize in wurzite structure.The crystallite size of the TiO_2 and ZnO photoanode were calculated by XRD line broadening measurement using the Scherrer's equation:

$$D = 0.9\lambda / \beta cos\theta \quad (1)$$

The XRD data were recorded by using Cu Kα radiation (λ = 1.5406 Å) over the 2θ range of 20° – 80°.The calculated average crystallite size for the TiO_2 and ZnO photanode is 18.76 nm and 19.13 nm respectively.

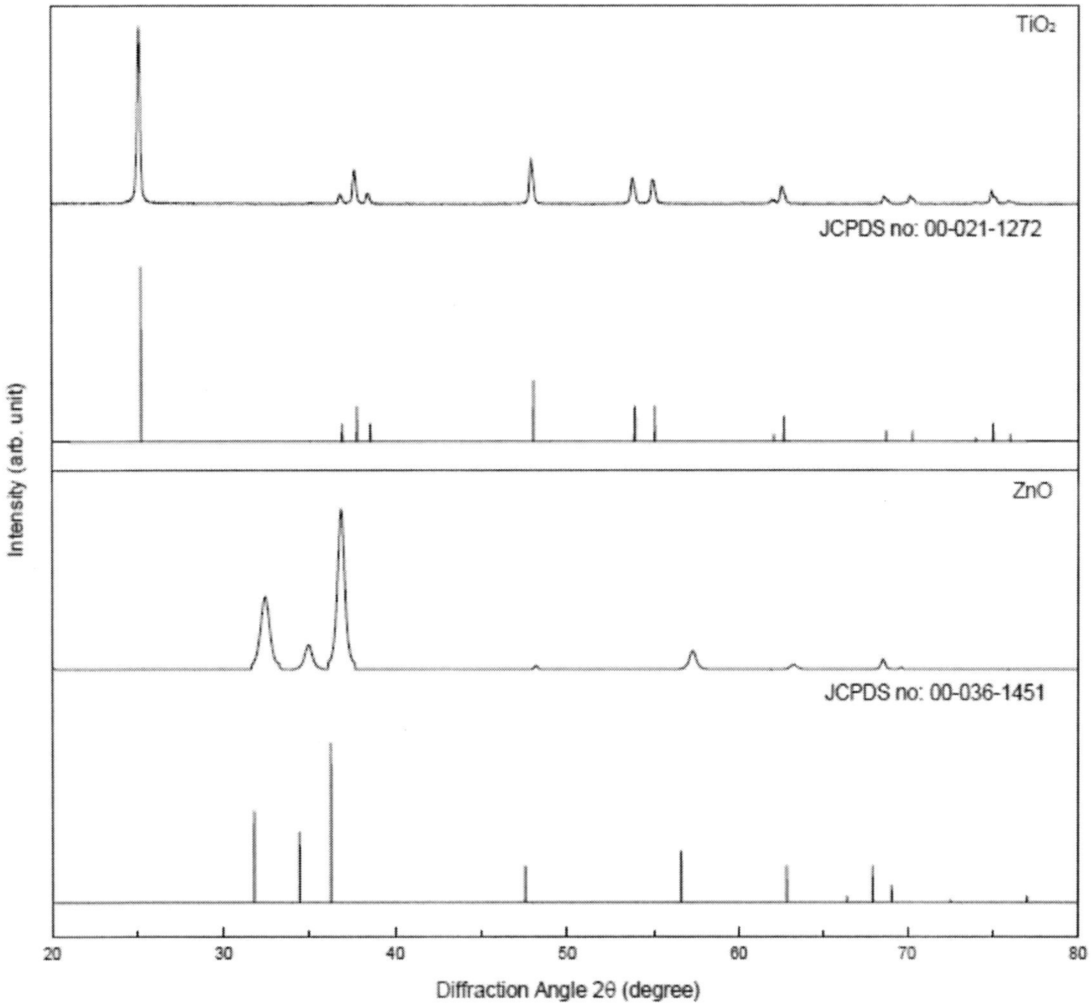

Figure 6. XRD pattern for TiO_2 and ZnO photoanode.

3.2. Optical performance

The ability of a DSSC to absorb sunlight is depending on the dye sensitizer. Thus, it is important to measure the absorption capability of the dye viaoptical characterization. In this research, the 0.3 mM Ru N719 dye had been tested under 200 – 800 nm light spectrum. Figure 7 shows the absorption spectrum of 0.3 mM Ru N719 dye. The absorption peak (λ),λ_1 was observed at 211 nm followed by$\lambda_2$at 312 nm, λ_3 at 393 nm and λ_4 at 536 nm.Thus, it can be conclude that the absorption spectrum of the 0.3 mM Ru N719is broad to absorb the visible part of the solar spectrum.

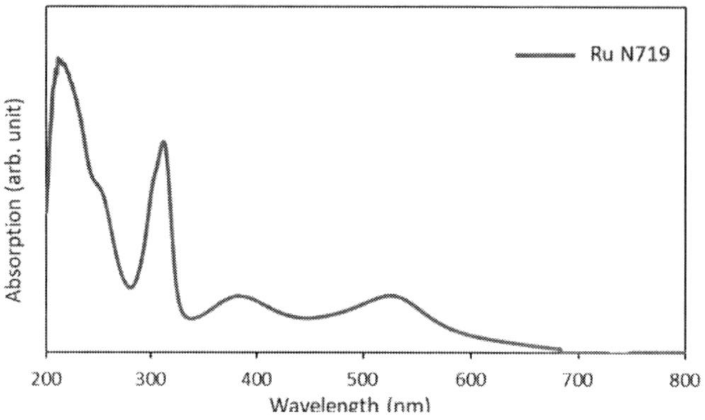

Figure 7. Absorption spectrum of 0.3 mM Ru N719 dye.

3.3. Electrical performance

Table 1 listed the comparative performance between TiO_2 and ZnO photoanode based DSSC in terms of short circuit current density (J_{sc}), open circuit voltage (V_{oc}), fill factor (FF) and energy conversion efficiency (η). The I-V curve of the TiO_2 and ZnO photoanode based DSSC were also shown in Figure 8. According to the electrical characterization, the highest energy conversion efficiency recorded is 0.34% is obtained from ZnO photoanode based DSSC. By default, the ZnO photoanode had shown better performance as the photoanode compared to TiO_2 (0.29%).

Table 1. Electrical performance of the DSSC with different types of photoanode.

Types of photoanode	J_{sc} (mA/cm2)	V_{oc} (V)	FF	η (%)
TiO_2	2.42	0.46	0.26	0.29
ZnO	2.85	0.43	0.28	0.34

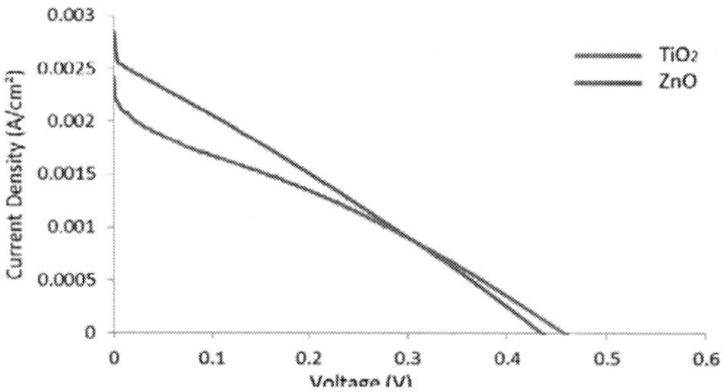

Figure 8. I-V curve for the TiO_2 and ZnO based DSSC.

4. Conclusions

The enhancement of energy conversion efficiency for DSSC using ZnO had been done in this research. The highest energy conversion efficiency was achieved by ZnO photoanode with the energy conversion efficiency of 0.39% compared to 0.29% by TiO_2 photoanode. The nanorods structure obtained in the surface morphology shows that bigger particle of photoanode nanostructure can help to provide bigger surface area for the dye attached and thus more light can be absorbed and harvested. The ZnO nanorods structure also helps to provide the direct pathways for the electrons to move through the cell that can increase the energy conversion rate. The high energy conversion efficiency achieved by the ZnO photoanode is also supported by the chemical properties of ZnO itself where it has the high electron mobility (155 $cm^2V^{-1}s^{-1}$) and large free excitation binding energy (60 meV).This research proved that the energy conversion efficiency of the ZnO based DSSC can be further boosted in future and it is a suitable candidate for the photoanode in DSSC.

5. References

[1] Sharma S, Siwach B, Ghoshal S K and Mohan D2017 *Renew. Sustain. Energy Rev.***70** 529
[2] Lee H J, Lee J W, Kim H J, Jung D-H, Lee K-S, Kim S H, Geum D, Kim C Z, Choi W J and Baik J M 2016*Phys. Chem. Chem. Phys.***18**2906
[3] Shanmugam V, Cunnusamy J, Khanna A, Boreland M B and Mueller T2013 *Energy Procedia***33** 64
[4] Hegazy A, Kinadjian N, Sadeghimakki B, Sivoththaman S, Allam N K and Prouzet E 2016 *Sol. Energy Mater. Sol. Cells***153** 108
[5] Norizan M N, Zahari S M, Mohamad I S, Osman R A M, Shahimin M M and Murad S A Z 2017*IOP Conf. Ser. Mater. Sci. Eng.***209**12029
[6] Blakers A, Zin N, McIntosh K R and Fong K2013 *Energy Procedia***33**1
[7] Akbar Z A, Oh J H, Hadmojo W T, Yang S J, Do Y R and Jang S-Y 2015 *Opt. Express***23** A1280
[8] Shakeel Ahmad M, Pandey A K and Abd Rahim N2017 *Renew. Sustain. Energy Rev.***77** 89.
[9] Abdin Z, Alim M A, Saidur R, Islam M R, Rashmi W, Mekhilef S and Wadi A2013 *Renew. Sustain. Energy Rev.***26** 837
[10] Prabavathy N, Shalini S, Balasundaraprabhu R, Velauthapillai D, Prasanna S and Muthukumarasamy N2017*Int. J. Energy Res.***41**1372
[11] Xie K, Guo M and Huang H 2015*J. Mater. Chem. C***3** 10665
[12] Brian Oand Gratzel M1991 *Nature***353** 737
[13] Cavallo C, Di Pascasio F, Latini A, Bonomo M and Dini D 2017*J. Nanomater.***1**
[14] Ye M, Wen X, Wang M, Iocozzia J, Zhang N, Lin C and Lin Z2015 *Mater. Today***18** 155
[15] Klein M, Pankiewicz R, Zalas M and Stampor W2016*Sci. Rep.***6** 30077
[16] Nebi M and Peker D 2016*J. Phys. Conf. Ser.***766**120-26
[17] Liu J, Huo J, Zhang M and Dong X 2017*Thin Solid Films***623** 25
[18] Chandiran A K, Abdi-Jalebi M, NazeeruddinM K and Grätzel M2014 *ACS Nano***8** 2261
[19] Zhang J, Zhou P, Liu J and Yu J2014 *Phys. Chem. Chem. Phys.***16** 20382
[20] Basu K, Benetti D, Zhao H, Jin L, Vetrone F, Vomiero A and Rosei F 2016*Sci. Rep.***6**23312
[21] M V, K P and M R G2018*Appl. Surf. Sci.***436** 708
[22] Lai J, Jian D, Lin Y, Ku M and Jian W 2018*Phys. B Condens. Matter***532** 135
[23] Mohamad I S, Ismail S S, Norizan M N, Murad S A Z and Abdullah M M A 2017*IOP Conf. Ser. Mater. Sci. Eng.***209** 12028
[24] Nabialek M, Bloch K, Szota M and Sandu A V 2017 *Materiale Plastice* **54(3)** 491-494
[25] Rouhi J, Mamat M H, Ooi C H R, Mahmud S and Mahmood M R2015 *PLoS One***10** e0123433

Acknowledgments

The authors would like to thank School of Microelectronic Engineering, Centre of Excellence Geopolymer and Green Technology (CeGeoTech), Centre of Renewable Energy (CERE) and University Malaysia Perlis for providing the research facilities. Besides, the authors also would like to acknowledge the support from the Research Acculturation Grant Scheme (RAGS) under a grant number of RAGS/1/2015/SG0/UNIMAP/03/3 from the Ministry of Higher Education Malaysia.

IOP Conf. Series: Materials Science and Engineering **374** (2018) 012049 doi:10.1088/1757-899X/374/1/012049

Sulphated Electric Arc Furnace Slag Asfenton-Like Catalyst for Degradation of Reactive Black 5

N A Zubir[1,2], N Nasuha[1,2*] and R Alrozi[1,2]
[1]Faculty of Chemical Engineering, Universiti Teknologi MARA (UiTM), Pulau Pinang, Malaysia
[2]Hybrid Nanomaterials, Interfaces & Simulation (HYMFAST), Faculty of Chemical Engineering, Universiti Teknologi MARA (UiTM), Pulau Pinang, Malaysia

E-mail: norhaslinda.nasuha@ppinang.uitm.edu.my

Abstract. Sulphated electric arc furnace slag (S-EAFS) was obtained through a facile chemical and thermal treatment method. The S-EAFS was evaluated as a Fenton-like catalyst for the oxidative degradation of reactive black 5 (RB5). The S-EAFS was characterized by XRD, SEM-EDX and nitrogen adsorption analysis. The highest RB5 degradation efficiency obtained in this study was above 90% which was maintained across seven successive cycles with minimum iron leaching. This was achieved at a RB5 concentration of 0.15 gL^{-1} (50 ppm) with 8 mM of H_2O_2 and a pH of 4.5. Characterization revealed that the presence of sulphated groups (SO_4^{2-}) within the EAFS improved the surface acidity of the material and corresponded to an increase in the catalytic activity for the degradation of RB5 at mild pH.

1. Introduction

The presence of low concentrations of certain toxic, carcinogenic and mutagenic pollutantsenable to change the colour of receiving water. Thus, colour of water has been included in water quality standards due to the potential for these pollutants to impart plausible negative impact towards public health and the hydrosphere [1]. Among the typical types of coloured pollutant; reactive and acidic cationic dyes are responsible for the majority of physical, chemical and microbial damage to the environment. Hence, it is imperative to put in place for effective treatment technologies in removing these pollutants prior to discharge into natural water bodies. Among the available technologies, advanced oxidation processes (AOPs) are known to be highly effective in removing these pollutants from waterways. Thus, an investigation into these processes, particularlyto Fenton process, is warranted to improve environmental outcomes in this context.

The development of AOP catalysts have been plagued with deactivation issues under the acidic environment of the Fenton reaction. Of the catalysts tested, the iron oxide-silica oxide [2,3] is reported as promising catalyst to overcome this key limitation. In fact, a recent study [4] found that an iron oxide-silica oxide heterogeneous catalyst derived from steel waste electric arc furnace slag (EAFS) is highly stable under Fenton reaction conditions. The generation of this catalyst from waste product, coupled with its high stability make it a high performing and environmentally friendly alternative to the other catalysts in this area. Interestingly, the synergistic effect of multiple iron phases, mainly maghemite and magnetite, played a dominant role in accelerating the redox cycles of active sites in the generation of Fe^{2+}. This allows a continuous oxidative degradation of RB5 to be achieved. However, the performance of activated electric arc furnace slag (A-EAFS) in the degradation of RB5, AB29 and MB were only excellent under acidic pH reaction (pH 3), even when assisted by visible light [5].

Due to the limited range of high performance of the EAFS catalyst, there is a gap in improving its performance under different reaction conditions. This is particular towards the treatment of dye contaminated wastewater which is often less acidic than the previously reported conditions [4,5]. In this study, it is hypothesized that by improving the surface acidity of the EAFS will subsequently increases its catalytic activity under mild pH reaction conditions. The attachment of a sulphated group (SO_4^{-2}) on the surface of catalysts has attracted much attention in academic literature due to their strong acidity, nontoxicity and high catalytic activity [6]. Li et. al., 2016[7], reported that the acidic sites on the catalyst surface formed by the sulphated functional groups were highly active in the photo reduction of pollutants. Therefore, this study will focus on the performance of sulphated–EAFS for degradation of RB5 under mild pH conditions.

2. Materials and methods

2.1 Materials

All materials that were used in this study were purchased from Merck and were used without further alteration. The three purchased materials are: high-quality (98%) RB5, sulfuric acid (95%) and hydrogen peroxide (30%).The Electric Arc Furnace Slag (EAFS) was collected from a local steel factory in Penang, Malaysia. Acidic (0.1M HCl) and basic (0.1M NaOH) solutions were used to regulate pH.

2.2 Catalyst preparation and characterization

Preparation of sulphated electric arc furnace slag (S-EAFS) catalyst. Raw EAFS (R-EAFS) collected from site was ground and sieved into the 40– 50 μm size range. The R-EAFS particles were then soaked in 20 wt% H_2SO_4 for 4 h and washed with distilled water until the pH of the supernatant was neutral. This washing process was performed to remove adhering dust and fine materials. The resulting samples were centrifuged at 500 rpm for 10 minutes, filtered, and dried at 100°C for 24 h. The last stage of preparation involved the calcination of the dried samples at 400 °C for 2 h. Samples that have been prepared up to this step are termed as S-EAFS.

Characterization of catalysts. The structure and crystal phase of S-EAFS catalyst were recorded by X-ray diffraction (XRD) patterns in a D/max-2500/PC X-ray diffractometer (Rigaku, Japan) using Cu Kα radiation, operated at 40 KV and 30mA from 10 to 90°. The morphologies of samples were observed by scanning electron microscope (SEM) equipped with energy dispersive X-ray (EDX). The Brunauer–Emmet–Teller (BET) specific surface area of S-EAFS was determined at liquid-nitrogen temperature (77 K) using Micromeritics ASAP 2020 system. The total pore volumes were calculated based on N_2 adsorption at a relative pressure of 0.99. Pore-size distributions were calculated from the adsorption branch of the isotherm, according to the Barrett–Joyner–Halenda (BJH) model.

2.3 Catalytic activity

The oxidative degradation of RB5 was performed in a 250 mL conical flask with 200 mL of solution. The solution pH (2-7) and initial RB5 concentration (25–150 mg L^{-1}) were varied in accordance to the study performed by Nasuha et al., 2016 [4]. To initiate each test a certain amount of H_2O_2 (2–12 mM) was added to the aqueous solution. The adsorption experiments were performed in dark conditions for half an hour to obtain the equilibrium adsorption of RB5. Samples were taken at regular intervals to be analyzed by UV–vis spectrometry (UV-1700 PharmaspecShidmadzu). The RB5 concentrations was determined by measuring the absorbance intensity of the solution at λ_{max} 599 nm.

2.4 Reusability and leaching test

Oxidative degradation of each solution was carried out for 180 minutes. At the conclusion of the run, spent S-EAFS catalyst was separated, washed with distilled water, and dried at 60 °C before being used in the next cycle. Each of the seven cycles were operated at the same optimal conditions using a pH of 5. After the completion of all cycles the EAFS sample was designated as used SEAFS (US-EAFS). S-EAFS stability in degrading RB5 over each cycle was monitored using UV–vis spectroscopy. Meanwhile, the filtrate solution was tested for the presence of homogeneous Fe ions. The filtrate solution was submitted for leaching tests performed by an inductively coupled plasma (ICP-OES), Optima 7000 DV, Perkin Elmer.

3. Results and discussions

3.1 S-EAFS characterization

3.1.1. XRD. Figure 1 exhibits the X-Ray Diffraction (XRD) pattern of the optimized S-EAFS sample. It can be seen from the intensity of the peaks that the sample is highly crystalline. The XRD pattern of the S-EAFS sample contains a variety of crystalline phases such as magnetite, maghemite, hematite and calcium sulphate. Crystalline larnite, anorthite and calcium ferrate phases are also present in the sample but at much reduced intensities. It is hypothesized that the CaO compound (highest element) reacted with H_2SO_4during the synthesis resulting in the formation of new crystalline phases i.e. larnite anorthite.

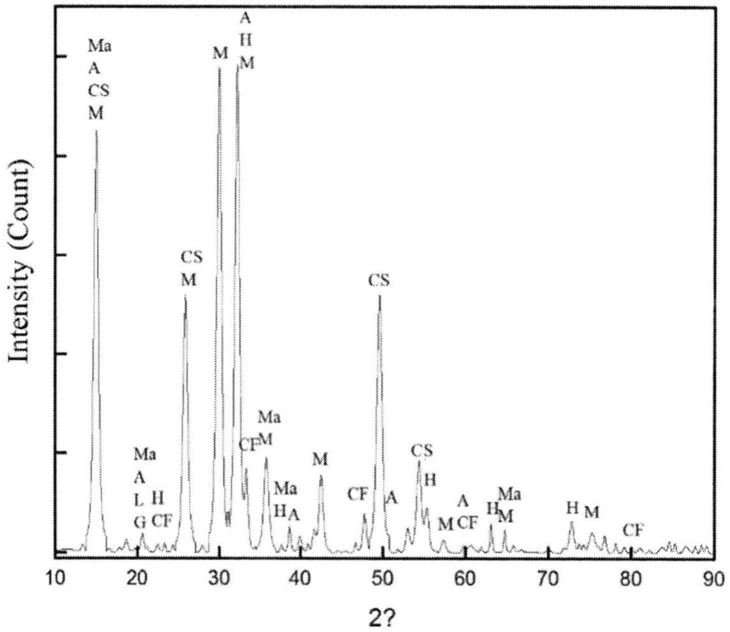

Figure 1. XRD pattern for S-EAFS. L=Larnite ($2CaO_4SiO_2$), A=Anorthite ($Al_2Ca_1O_8Si_2$), M=Magnetite (Fe_3O_4), Ma=Maghemite (γ-Fe_2O_3), H=Hematite (α-Fe_2O_3), CF=Calcium Ferrate (IV) ($Ca_1Fe_1O_3$), CS=Calcium sulphate (IV) (CaO_4S_1).

3.1.2. SEM-EDX. The physical structure of the S-EAFS sample is displayed in Figure 2. It has been shown that the fresh EAFS (R-EAFS) sample has a combination of cluster shapes which include fractured and amorphous particles [4]. Meanwhile, upon exposure to chemical and thermal treatment, the structure significantly changes into bulk breaking particles with some pores presence on the surface of catalyst. It can clearly be seen that the structure of S-EAFS changed markedly after the chemical and thermaltreatment. The S-EAFS sample exhibits crushed crystal and rod like shaped particles with a width in the range of 10-20 μm. The S-EAFS sample is also seen to consist of fractured bulk particle agglomerates. Furthermore, EDX analysis confirmed that the S-EAFS sample contains the elements Iron (Fe), Oxygen (O), Calcium (Ca), Silica (Si) and Sulfur (S).

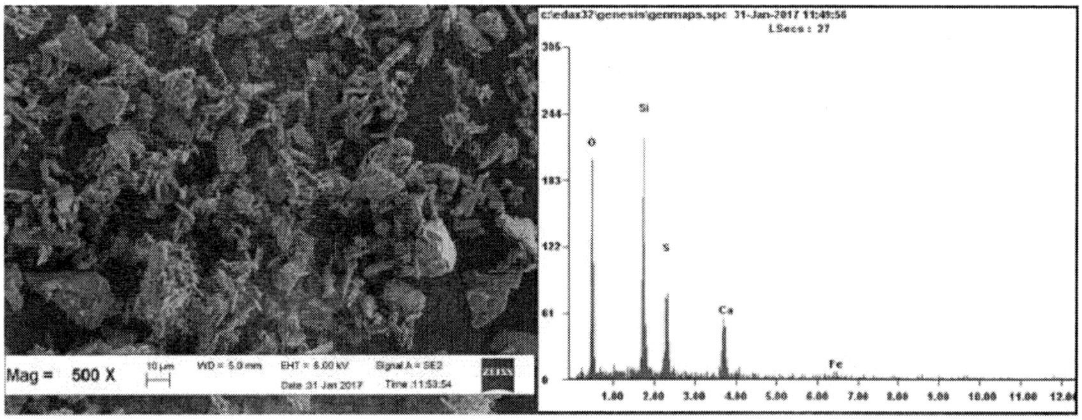

Figure 2. SEM images of S-EAFS at 500x magnification and EDX spectrum of S-EAFS.

3.1.3. Nitrogen sorption analysis. Nitrogen adsorption and desorption isotherms of S-EAFS sample before and after calcination are shown in Figure 3a. The isotherms in Figure 3a conform to a type IV classification under IUPAC, which represents the characteristics of mesopores materials (2-50 nm). Interestingly, the S-EAFS isotherm that has been calcined adsorbs more nitrogen than the sample that hasn't been calcined. This result agrees with Figure 3b which shows that the calcined sample has a higher pore volume. Moreover, the pore size distribution, illustrated in Figure 3b, revealed that acid treatment resulted in a narrower pore size distribution between 6.5-7 nm and between 18 – 120nm compared to the previously reported pore size distribution [4, 5, 15]. It can be seen that the surface area of the S-EAFS increases during the calcination process (100 vs 81 m^2/g). The increase in surface area is further evidence that acid modification increases sample porosity. The samephenomenon was reported by Ayodele et al., 2012 [8]. Moreover, the pore width of S-EAFS was slightly reduced after undergoing calcination. Mantri et al., 2013 [9] postulated that decreasing pore width was likely due to formation of micro channels after calcination.

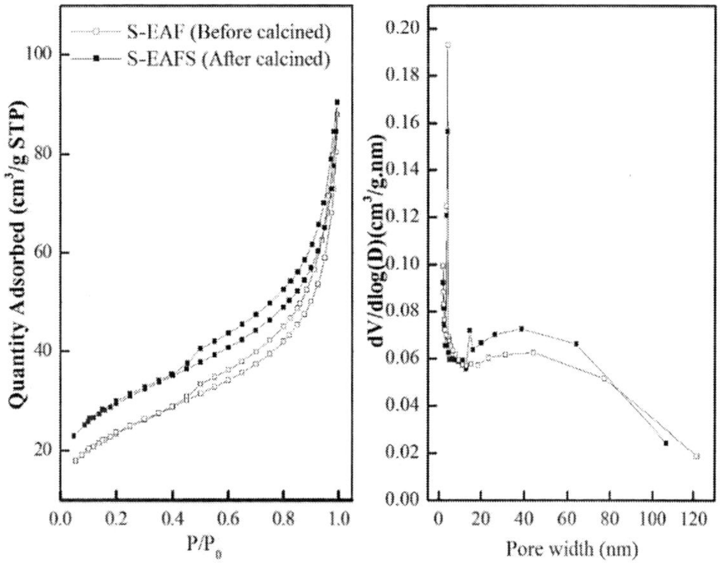

Figure 3. (a) Nitrogen adsorption-desorption isotherm and (b) pore distribution of S-EAFS before calcined and after calcined.

3.2. Catalytic activity

3.2.1 Effect of catalyst dosage. Figure 4 ashows the effect of catalyst dosage on RB5 degradation. It can be clearly seen in this figure that the dosage of S-EAFS positively influences the performance efficiency of RB5 degradation under mild pH conditions (pH 4.5). Degradation of RB5 was increased from 80 % to 94 % with an increase in catalyst dosage from 0.01 g to 0.03 g under dark Fenton-like conditions. The results suggest that an increase in catalyst dosage results in more adsorption due to the presence of more active sites. This behavior leads towards accelerated decomposition of H2O2 into •OH radicals as report by Wan et al., [10], even though this only occurs up to a certain range of catalyst dosage [16]. Once the optimized dosage is reached, further addition results in possible scavenging of excess generation of •OH radicals. Therefore, the performance of degradation of dyes remained relatively constant at higher catalyst dosages.

3.2.2 Effect of H_2O_2 dosage. The dosage of H2O2 was varied to understand the impact of oxidant on RB5 degradation in the presence of S-EAFS catalyst. The effect of H2O2 was studied by varying the dosage (2 - 12 mM) while keeping all other parameters constant. The degradation of RB5 gradually increases with H2O2 dosage from 2 mM to 8 mM. Further increases in H2O2 dosage above 8mM resulted in a negligible change of RB5 degradation. It is thought that the hydroxyl radicals generated at H2O2 concentrations in excess of 8mM producing hydroperoxyl radicals, HOO•. Hydroperoxyl radicals are less reactive compared to the hydroxyl radicals and only slightly contributedtowards oxidative degradation of organic pollutants [11, 12].

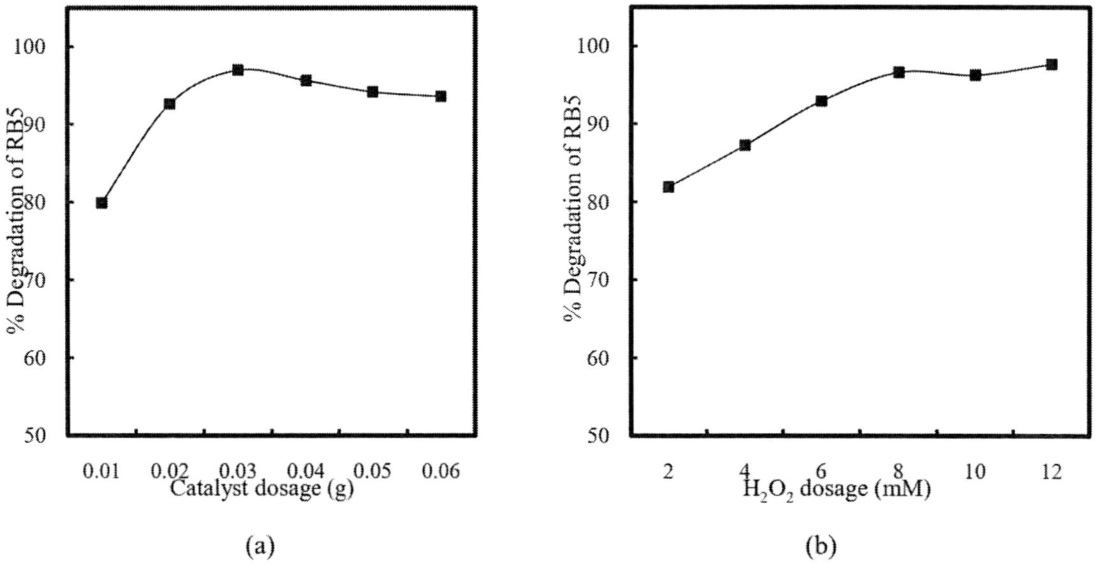

Figure 4. Effect of (a) catalyst dosage and (b) H_2O_2 dosage on the oxidative degradation of RB5.

3.2.3 Effect of pH solution. The effect of pH solution on the degradation of RB5 under dark conditions was tested between pH 2 and pH 7 (Figure 5a). It can be seen that a pH of 2 – 5 gives this highest degradation of RB5. Interestingly, the degradation performance was maintained from acidic untilmild pH solution. Such findings is suggested due to the nature of acidic surface sites of the sulphated catalyst which confer micro acidic environment within the actives sites during the catalysiseven at neutral pH solution [6]. The colour removal of RB5 at pH 6 decreased significantly afterwards. This phenomenon can be explained by decomposition of more H_2O_2 into molecular oxygen and water on the active sites without the formation of appreciable amounts of HO• radicals [13].

3.2.4. Effect of initial concentration. The effect of RB5 degradation on RB5 initial concentration was studied and the results are shown in Figure 5b. It can be seen that degradation performance reduces with an increase in the initial concentration of RB5. At a low RB5 initial concentration (25 mg L^{-1}), complete degradation was achieved while at high initial concentration (150 mg L^{-1}) approximately half of the RB5 was degraded. The reduction in degradation performance can be attributed to an insufficient amount of •OH radicals which is required for the degradation of dyes. Further, as RB5 initial concentration is increased, the ratio of catalyst dosage to RB5 concentration decreases. This is mainly due to the fact that competitive adsorption may occur in the presence of adsorbed dyes molecules and oxidant molecules (H_2O_2) [14].

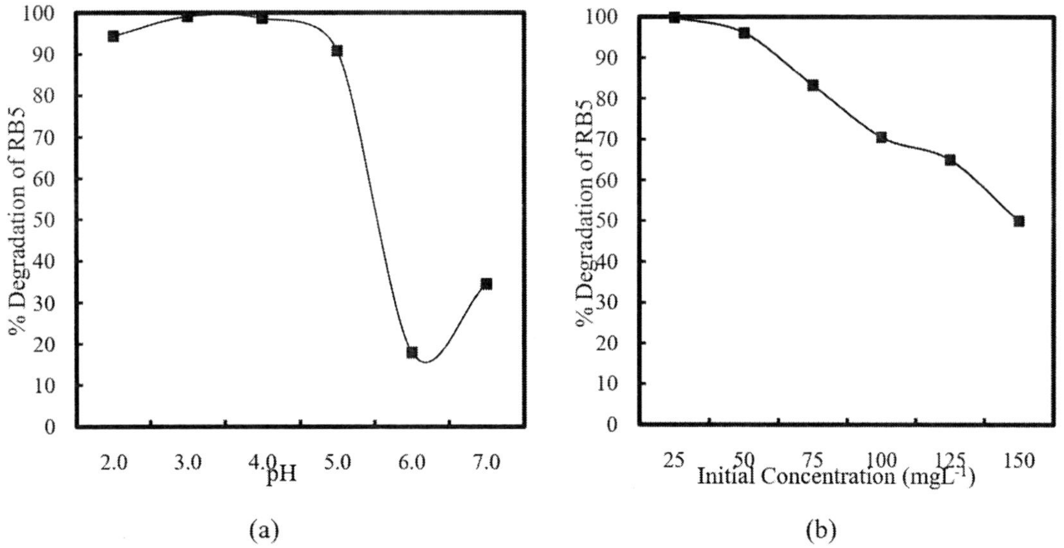

Figure 5. Effect of (a) pH solution and (b) initial dye concentration on the oxidative degradation of RB5.

3.2.5 Catalyst stability

The S-EAFS catalyst has shown high degradation efficiencies, reusability and stability as presented in Figure 6. The catalyst was able to sustain its' catalytic activity with more than 90% of RB5 degradation over seven successive cycles of reactions. It is anticipated that this high performance can be attributed to the strong interaction and combination between the oxidative radical species and the acid treated surface of the S-EAFS catalyst. The maximum iron leaching concentration after seven cycles was approximately 0.15 mgL^{-1}, which can be neglected because it was below the European Union directives value (<2 ppm) [6].

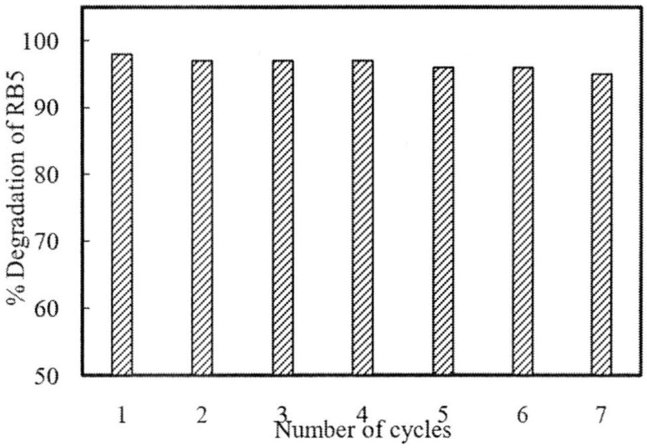

Figure 6. Reusability of S-EAFS for degradation of RB5.

4. Conclusions

This study has shown that S-EAFS as an effective Fenton-like catalyst for degradation of RB5. The collective findings of characterization analysis and catalytic studies have proven that the presence of sulphated sulphated groupson the EAFS catalyst surface able to enhance the resultant catalytic performance. The modified S-EAFS catalyst provided with higher degradation of RB5 at mildly acidic pH than any other studies. Such finding is plausible due to the nature of acidic surface sites of the sulphated catalyst which confer micro acidic environment within the actives sites during the catalysis. Interestingly, the resultant catalyst have shown modest catalytic stability of more than 90% RB5 degradation in seven successive cycles at optimised reaction conditions (50ppm, 8 mM of H_2O_2 and pH 4.5).

5. References

[1] Yagub M T, Sen T K, Afroze S and Ang H M 2014 *Advances in Colloid and Interface Science* **209** 172–184
[2] Szegedi Á, Popova M, Lázár K, Klébert S and Drotár E 2013*Microporous and Mesoporous Materials* **177** 97–104
[3] Ursachi I, Stancu A and Vasile A 2012 *Journal of Colloid and Interface Science* **377(1)** 184–190
[4] Nasuha N, Ismail S and Hameed B H 2016 *Journal of the Taiwan Institute of Chemical Engineers* **67** 235-243
[5] Nasuha N, Ismail S and Hameed B H 2017 *Journal of Environmental Management* **196** 323–329
[6] Ahmed I, Khan N A, Mishra D K, Lee J S, Hwang J S and Jhung S H 2013 *Chemical Engineering Science* **93** 91–95
[7] Li Y, Liu L, Guo M and Zhang M 2016 *Journal of Environmental Sciences* **47** 14–22
[8] Ayodele O B, Lim J K and Hameed B H 2012 *Chemical Engineering Journal* **197** 181–192
[9] Mantri K, Selvakannan P R, Tardioand J and Bhargava S K 2013 *Colloids and Surfaces A: Physicochemical and Engineering Aspects* **429** 149–158
[10] Wan Z, Hu J and Wang J 2016 *Journal of Environmental Management* **182** 284–291
[11] Wang H L, Liang W Z, Zhang Q and Jiang W F 2010 *Chemical Engineering Journal* **164(1)** 115–120
[12] Daud Y M, Hussin K, Osman A F, Ghazali C M R, Abdullah M M A B and Sandu A V 2017 *Materiale Plastice* **54(3)** 543-545
[13] Ramirez J H, Costa C A, Madeira L M, Mata G, Vicente M A, Rojas-Cervantes M L, López-Peinado A J and Martín-Aranda R M 2007 *Applied Catalysis. B: Environmental* **71** 44–56
[14] Vinita M, Praveena J D R and Palanivelu K 2010 *Solar Energy* **84(9)** 1613–18
[15] Faris M A, Abdullah M M A, Sandu A V, Ismail K N, Moga L M, Neculai O and Muniady R 2017 *Materiale Plastice* **4(1)** 145-154
[16] Burduhos Nergis D D, Abdullah M M A and Vizureanu P 2017 European Journal of Materials Science and Engineering **2** 111-118

Acknowledgements

The authors gratefully acknowledges the Ministry of Higher Education Malaysia (MOHE) and UniversitiTeknologi MARA (UiTM) for financial support under Fundamental Research Grant Scheme (600-IRMI/FRGS 5/3 (110/2016)).

Method of Analysis of the Topic of Doctoral Thesis in the Field of Castings Production. Case Study on the Situation in Romania

V F Soporan[1], V Samoilă, T R Lehene, S Pădurețu, M D Crișan and M M Vescan
[1]Technical University of Cluj-Napoca, Department of Environmental Engineering Entrepreneurship and Sustainable Development, Romania

E-mail: vfsoporan@gmail.com

Abstract. The paper presents a method of analysis of doctoral theses in castings production, elaborated in Romania, the analysis period ranging from 1918 to 2016. The procedure, based on the evolution of the analyzed problem, consists of the following steps: establishment of a coding system for the domains and subdomains established in the thematic characterization of doctoral theses; the establishment of the doctoral organizing institutions, the doctoral specialties, the doctoral supervisors and the time frame for the analysis; selecting the doctoral thesis that will be included in the analysis; establishing the key words for characterization of doctoral theses, based on their title; the assignment of theses to the domains and subdomains according to the meaning of the keywords, to the existing groups of the coding system; statistical processing of results and determination of shares for each domain and subdomain; conclusions on the results obtained and their interpretation in the context of economic and social developments. The proposed method being considered as general, the case study is carried out at the level of the specific field of castings production, the territory of the analysis refers to the institutions organizing doctoral studies.

1. Introduction

From the analysis of the specialized literature, we found that there is no particular presentation of a method of analysis of doctoral theses at the level of the castings, especially in Romanian literature. There are general studies showing the evolution of the Romanian technical education [1] or the development of science and technology in the interwar period [3]. Among the monographic works there are some from the most representative unit in the Romanian technical education, the Polytechnic University of Bucharest [6] or the Technical University of Cluj-Napoca [5]. At an international level, we note the synthesis paper on the evolution of casting processes in the French industry [4] or a complete history of castings present at the global level [9].

2. Considerations on the evolution of engineering training at the manufacturing stage of castings

The spatial location of the study at the level of Romania is not coincidental, because on December 1st 2018, one hundred years will have passed since the Great Union, when the provinces with a major Romanian population (Transylvania, Banat, Crișana, Bucovina, Basarabia) decided to unite with Romania. In the analyzed period, the country went through a period of construction of the market economy according to the specifics of the interwar period (1918-1940); the war period (1940-1945);

IOP Conf. Series: Materials Science and Engineering **374** (2018) 012050 doi:10.1088/1757-899X/374/1/012050

the period of centralized economy (1945-1989); the transition period of the construction of the market economy (1989-present).

In a primary assessment, from the point of view of general engineering training, it is estimated that there are four periods: the period of preliminary accumulation structured in the beginning of the engineering training, without the recognition of the engineering diplomas issued in Romania - up to the level of 1890; the period of accumulation and structuring of engineering training according to Western models - positioned between 1890 and 1948; the period of engineering formation that coincides with the period of Romania's high-speed industrialization in a centralized economy - positioned between 1948 and 1990; the period of engineering training during the transition from the "centralized economy" to the "market economy" - positioned between 1990 and the present.

From a historical perspective, the evidence of the development of casting production and the structuring of the training tools in the manufacturing of molded parts is represented by the "foundry manual" of engineer Petre Dumitraşcu [2], of which the cover and the first page are presented in Figure 1. Speaking about historical events, in 1936, the first "engineer doctor" diploma was awarded by "Politehnica of Bucharest" to the American engineer W.J. Crook, the leadership being provided by Prof. Traian Negrescu, the presentation being carried out in French (Figure 2). The theme of the thesis is from the metallurgical field, which proves the level achieved by this specialty in the Romanian higher technical education.

Figure 1. Foundry manual, Eng. Petre Dumitraşcu (1939).

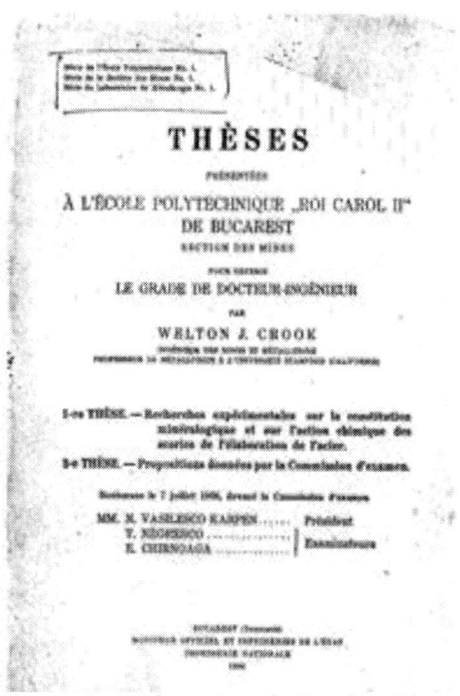
THÈSES

À L'ÉCOLE POLYTECHNIQUE „ROI CAROL II"
DE BUCAREST

LE GRADE DE DOCTEUR-INGÉNIEUR

WELTON J. CROOK

Figure 2. Welton Joseph Crook - First PhD thesis in engineering in Romania (1936).

3. Consideration on the casting industry

In the characterization of the production of the castings we tried to present the evolutions existing at the global level in relation to the situation existing in Romania. The evolution of casting production is presented using the "World Castig Production" annual reports, produced by WFO and published in each December in Modern Casting.

So far, there have been presented 50 global analyzes "World Castig Production", the last one being presented in December 2016, about the situation in 2015. For the year 1996, the data included in the analysis of Professor Iulian Riposan were used [7]. The study also mentions the situation in 1989, when Romania was producing 1.5 million tons of castings, ranked as 9th in the world. In order to have a correct picture of the evolution of the production of castings in Romania over the last years (2000-2015) the following table shows the evolution of the production of castings in Romania compared to the production achieved globally (Table 1).

Table 1. Evolution of castings production between 2000-2015, tons.

Year	2000	2005	2012	2015	Production ratio-PR 2015:2000
Global situation	64.750.239	85.741.076	91.673.839	104.129.257	1,608
Romania	360.120	154.319	101.427	129.053	0,358

Therefore, from the correlation of the data, it is estimated that the molded casting industry has reached a peak of production in 1990, amounting to one million tons of molded parts, and a minimum of 80,000 tons was generated in 2009. At present, it is stabilized at between 100,000 and 150,000 tons. The domains served over time include to means of transport, metallurgical equipment, machinery for the chemical and petroleum industry, machinery and equipment for the extractive and energy industry, textile machinery and others.

4. Methodology of analysis

The knowledge of the topics of doctoral thesis is important from two perspectives, a historical one determined by the establishment of a natural correlation between the researches made and the existing technological problems and the second one related to the suggestion of the expertise accumulated at the level of the main institutional acts validating the activities carried out, the title of Doctor of Engineering. Therefore, the establishment of this methodology is important in order to be able to assess the evolution of doctoral thesis in Romania throughout an entire century (1918-2016).

The method, based on the evolution of the analyzed problem, consists of the following stages: establishment of a coding system for the domains and subdomains established in the thematic characterization of PhD thesis; the establishment of the doctoral organizing institutions, the doctoral specialties, the doctoral supervisors and the time frame for the analysis; selecting the doctoral thesis that will be included in the analysis; establishing the key words for characterization of doctoral theses, established on the basis of the title they have; the assignment of theses to the domains and subdomains, according to the meaning of the keywords, to the existing groups of the coding system; statistical processing of results and determination of weights for each domain and subdomain; conclusions on the results obtained and their interpretation in the context of economic and social developments. The detailed presentation of the method is carried out at the level of the doctoral thesis elaborated within the PhD thesis [8].

5. Identifying the spatial and temporal framework and its characterization

The spatial location of the study is made at the level of Romania, and the analyzed period is related to the events that took place from 1918 to the present. In the analyzed period, the following organizing centers of doctoral studies have functioned (using the current name) in the field of casting manufacturing: "Politehnica" University of Bucharest started in 1936 - the year of the first PhD thesis, "Politehnica" University of Timisoara, Technical University from Cluj Napoca since 1960, "Transilvania" University of Braşov, "Gheorghe Asachi" Technical University of Iaşi, "Valahia" University of Târgovişte, "Dunărea de Jos" University of Galaţi. The fields of doctoral studies in which doctoral theses have been elaborated, refer to materials technology, metallurgy, materials engineering etc.

6. Establishing domains and subdomains for identifying the subject

The keyword method was used to assign PhD theses in the subject area of a doctoral thesis. The topic, identified from the title of the thesis, falls within the established domains, and at their level, within the constituent subdomains. The domains will be similar to the first group of figures and, given the purpose of the present paper, each field will be divided into subdomains, which will be characterized by the second group, the subdomains being considered in this case as the first rank. The division process can continue, in order to refine it, so that for each subdomain the development continues, the resulting subdomains will be called the second rank. The procedure can continuethe same way. Within the framework of this study, 27 domains of the molding process were identified and each field is divided into subdomains. By way of the construction of the coding system, there are possibilities to enter 99 subdomains for each domain.

The keyword method was used to assign PhD theses in the subject of a doctoral thesis. The topic, identified from the title of the thesis, falls within the established domains, and at their level, through the constituent subdomains. The domains will be similar to the first group of figures and, given the purpose of the present paper, each field will be divided into subdomains, which will be characterized by the second group, the subdomains being considered in this case as the first rank. The division process can continue, in the desire to refine it, so that for each subdomain the development continues, the resulting subdomains will be called the second rank. The procedure can continue in the same way. Within the framework of this study, 27 domains of the molding process were identified and each field is divided into subdomains. By way of construction of the coding system, there are possibilities to enter 99 subdomains for each domain.

The domains identified by the keyword identification and characterization groups of the PhD thesis are as follows: Group of cast alloy family, coded with (01); Group of cast alloy characteristics, coded with (02); Group for the characterization of casting applications, coded with (03); Group ofcasting functionality, encoded with (04); Group of casting weight characterization, encoded with (05); Geometrical shape characterization group of castings, coded with (06); Group for characterizing the complexity of the castings construction, coded with (07); Group for the characterization of casting production type, coded with (08); Gropup of physical and chemical phenomena present in the development and casting of alloys, coded (09); Technological elements characterization group, coded with (10); Process elaboration characterization group, coded with (11); The alloy casting characterization group, coded with (12); Casting procedures characterization group, coded with (13); Group of characterization of the model gaskets and the SDV necessary for making shapes, coded with (14); Group for the characterization of auxiliary operations in casting manufacturing processes, coded with (15); Group for the characterization of the eco responsibility shown in casting production, encoded with (16); Grouping of equipment and installations used in foundries, coded with (17); Group for the characterization of the control, investigation and quality assurance activities of casting parts, coded with (18); Group with reference to the laboratory techniques used in the casting process, coded with (19); Group for the characterization of standardization and normalization activities of the casting industry, coded with (20); Group regarding the modeling of casting processes and the use of information tools at the manufacturing stage of the molded parts, coded with (21); Group on the intelligent use of information incasting manufacturing, coded with (22); Group with reference to the preparation of castings production, coded with (23); Group with reference to the manufacture of materials with special properties obtained by casting, coded with (24); Group with reference to technological processes competing with the manufacture of castings, coded with (25); Group relating to training in the manufacture of castings, coded with (26); Group with reference to the history of the manufacturing processes of the cast parts, coded with (27).

The subdomains that characterize the first domain have the following form: Ash Fonts - (01). (01); Fonts with nodular graphite - (01). (02); Malleable Fonts - (01). (03); Cast steels - (01). (04); Copper alloys - (01). (05); Alloys based on Aluminum - (01). (06); Zinc alloys - (01). (07); Magnesium alloys - (01). (08); Other types of alloys - (01). (99). Other domains have been divided in a similar manner.

7. Calculation of the appreciation indicators of the frequency of the structured topics in the PhD thesis

The calculation procedure follows the following steps: selection of doctoral thesis with significance for the manufacturing of cast pieces; allocation of domain and subdomain codes based on the keywords set at the level of each thesis selected in a primary and secondary characterization; frequency calculation for each characterization code; the calculation of the share indicators at the level of each characterization code.

In the analysis made, that of doctoral thesis with significance for the manufacturing of castings, out of the total theses analyzed, determined at the doctoral studies organizing centers, a number of 76 were selected, those which are relevant to the field of casting production. The indicators of appreciation for the doctoral thesis analyzed are related to the following aspects: the share of belonging to a field; the share of belonging to a subdomain within the domain; the dissemination of theses at the level of the doctoral studies organization centers and the annual distribution of the doctoral thesis defence over the analyzed period of time. Table 2 shows the number of frames for each domain coding structure at the level of doctoral thesis analyzed, as well as their calculated weights.

Table 2. Appreciation indicators fot the PhD thesis analyzed.

Nr. crt.	Domain coding	Number of frames	Framing Frequency %	Nr. crt.	Domain coding	Number of frames	Framing Frequency %
1.	01	34	44.73	**15.**	15	0	0
2.	02	0	0	**16.**	16	5	6.57
3.	03	3	3.94	**17.**	17	0	0
4.	04	0	0	**18.**	18	0	0
5.	05	0	0	**19.**	19	2	2.63
6.	06	0	0	**20.**	20	0	0
7.	07	0	0	**21.**	21	5	6.57
8.	08	0	0	**22.**	22	0	0
9.	09	9	11.84	**23.**	23	1	1.31
10.	10	3	3.94	**24.**	24	8	10.52
11.	11	1	1.31	**25.**	25	0	0
12.	12	1	1.31	**26.**	26	0	0
13.	13	2	2.63	**27.**	27	0	0
14.	14	2	2.63	**28.**	**Total**	**76**	**100 %**

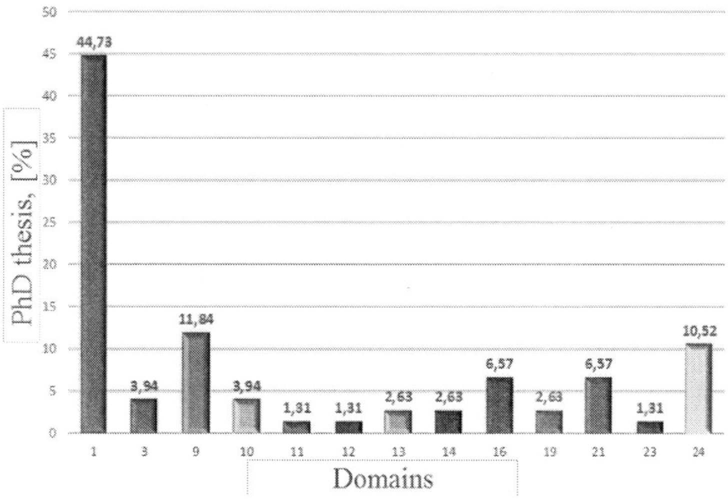

Figure 3. Frequency of assignment of doctoral theses in domains.

The topic distribution of the doctoral theses on the established domains is presented in Figure 3. The distribution analysis on the subdomains will be made on the characterization group with the highest share, that of the characterization of the family of the alloy. The results of sub-domain assignment are shown in Table 3.

Table 3. Distribution analysis on subdomains of the alloy family characterization (01).

Nr. crt.	Coding the subdomains of the group (01)	Number of frames	Subdomain share (%)
1	(01) (01) gray cast iron	5	14.70
2	(01) (02) cast iron with nodular graphite	5	14.70
3	(01) (03) malleable cast iron	1	2.94
4	(01) (04) cast steels	8	23.52
5	(01) (05) copper based alloys	1	2.94
6	(01) (06) alloys based on aluminum	7	20.58
7	(01) (07) zinc alloys	1	2.94
8	(01) (08) magnesium alloys		
9	(01) (99) other types of alloys	6	17.64
	Total	**34**	**100%**

From the point of view of the shares, the PhD thesis dealing with the issue of the cast iron in the first group (the family of alloys) is 32.34%, that of steels is 23.52% and the aluminium-based alloys is 20.58%. Within the cast iron, gray cast iron has the same share, that of 14.70%, with that of the nodular graphite cast iron, and the malleable cast iron has a weight of 2.94%. The graphical presentation of this distribution is shown in Figure 4.

Regarding the distribution for the periods mentioned at the beginning of the paper, this is how it looks like: II (1918-1940) 1 PhD thesis; III (1940-1989) - 5 PhD thesis; (1989-present) - 70 PhD thesis. The analysis shows that the share of theses sustained in the last analyzed period is 92.10%. This period coincides with the decrease in the production of castings to 100,000 tons. The production period reached 1 million tons, that of the years 1945-1989, the share of the PhD thesis was only 6.57% of the total of the PhD thesis defended during the 100 years analyzed, from the perspective of castings.

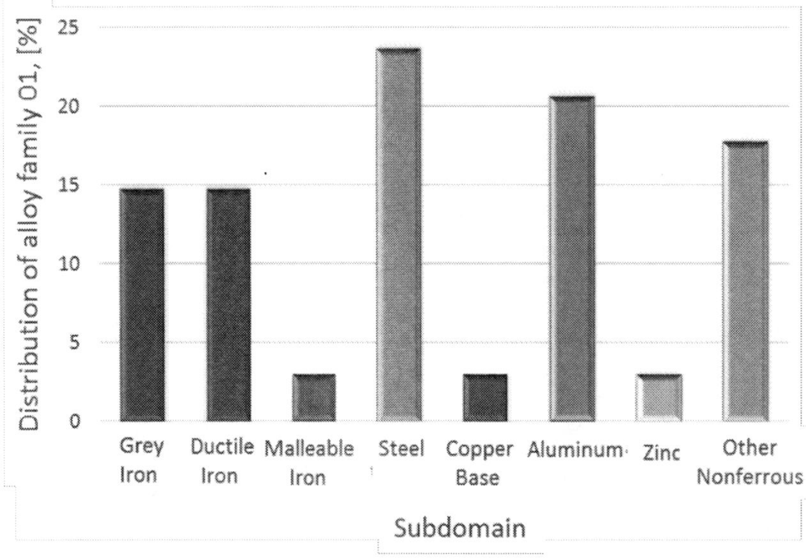

Figure 4. Subdomain distribution of alloyfamily01.

8. Conclusions

From the presentation we can see that the doctoral thesis, which are part of the "Alloy family characterization group" - coded with (01), have a share of 44.73% and occupy the first position within the established hierarchy. The following are the results obtained from the succession within the domains: "Group of characterization of the physicochemical phenomena present in the elaboration and casting of alloys" - the codified group (09) occupies the second position, with a share of 11.84%; the "Group referring to the production of materials with special properties obtained by casting", coded as (24) occupies the third position, with a share of 10.52%; the "Group of characterization of the ecoresponsibilities of casting production", coded as (16) and the "Group referring to modeling of casting processes and the use of information tools in the manufacturing of casting parts", coded as (21) occupies the 4th position with a share of 6.57%.

At the same time, we notice that a number of key areas in the casting industry have not seen any approach to the doctoral thesis over 80 years, for example the following can be noted: the alloy casting characterization group, the group characterizing the functions of the casting parts, characterization group of auxiliary operations, characterization group of the machines and installations in the foundries, the group with reference to the laboratory techniques and the investigation of the quality of the cast parts, the group with reference to the preparation of the production of castings, etc.

From the analysis we find that the subjects of doctoral theses in the field of castings manufacturing have dealt more with the problems of the alloys from the point of view of the elaboration and casting, their specific phenomena, the modeling and the use of the IT tools at the level of the technological elements.

Unfortunately, there aren't any doctoral theses that cover the casting in terms of the functionalities, the characteristics that it possesses, through the properties embedded in the cast alloy, auxiliary operations, means of investigation of quality, technological infrastructure and activities related to the preparation of the manufacturing.

By following the tendencies at the level of the manufacturing, which mostly coincide with the fields not covered by the PhD thesis, it is found that the problems approached were most often chosen randomly among the tendencies of the moment, it did not coincide with the solving of the major theme, which brings about development and modernization by using the new tools of the general technologies of the moment.

9. References

[1] Colan H 2003 *Dezvoltarea ştiinţelor tehnice după Marea Unire* (198 - 1940). Noema, ianuarie, **II(1)** 101 - 116
[2] Dumitraşcu P 1939 *Manual de turnătorie.* Bucureşti: Fundaţia pentru literatură şi artă "Regele Carol II".
[3] Iancu Ş 2003 Dezvoltarea ştiinţei şi tehnicii în perioada interbelică. Noema, Februarie **II(2)** 117 - 154.
[4] Marseille J 2002 *Fonderie, une industrie du XXI-e siecle.* Paris: Editions Monsieur-le-Prince.
[5] Nistor I S 2008 *Istoria învăţământului tehnic superior clujean.* Cluj-Napoca: Press.
[6] Popescu I M and Dumitrache I G 2011 *Istoria Universităţii "Politehnica" din Bucureşti.* Bucureşti: Editura Paideia.
[7] Riposan I 1997 Starea actuală şi tendinţele fabricaţiei de piese turnate. *Revista de turnătorie* **3(3)**
[8] Samuilă V 2017 *Contribuţii la analiza sistemelor de formare inginerească în domeniul fabricaţiei pieselor turnate*, Cluj-Napoca: s.n.
[9] Ştefănescu D 2009 Handbook **15** *Casting.* s.l.:ASM International.

IOP Conf. Series: Materials Science and Engineering **374** (2018) 012051 doi:10.1088/1757-899X/374/1/012051

Critical Thinking on the Introduction of Digitization Within Engineering Training Systems in the Manufacturing Stage of Cast Parts

T R Lehene[1], V Samoilă, V F Soporan[1*], S Pădurețu[2] and M M Vescan
[1]Technical University of Cluj-Napoca, Department of Environmental Engineering Entrepreneurship and Sustainable Development, Romania
[2]Technical University of Cluj-Napoca, Department of Modern Languages and Communication, Romania

E-mail: vfsoporan@gmail.com

Abstract. The paper aims to present a methodology for the analysis of the engineering training systems at the manufacturing stage of castings through critical engineering thinking. Its use [4, 5] requires the development of procedures capable of responding to the problems faced by engineering training in terms of acquiring the tools and procedures. The structure of the analysis took into consideration the following aspects: the motivation to use the proposed procedure, considerations on the engineering behavior, the design of the reasoning adapted to the analysis of the engineering training systems, the determination of the correlations in the processes of obtaining the cast products, the definition and calibration of the digital experiment, the definition and analysis of the factors influencing the last solidification area (the nature of the alloy, the shape of the mold and the casting geometry).

1. Introduction

1.1. Motivation and presentation of the method of critical engineering thinking

At present, at the manufacturing stage of the molded parts, the specific processes are carried out by using the computer products, both at the level of the constructive conception and at the level of the technological solutions. The justification for including the mentioned aspects at the level of engineering training programs and the modalities of implementation are proposed to be achieved through an analysis in critical engineering thinking.

In view of the importance of critical thinking in solving the specific problems specific of casting manufacturing, the methodology and the spirit of the analyzes carried out by the Foundation for Critical Thinking [4, 5] have been adapted to this paper. The motivation for using this method was determined by the fact that within the foundation there were conducted analyzes of disciplines in "aerospace engineering", "electrical engineering" and "mechanical engineering". From the analysis we have found that there is no "critical thinking" approach to the analysis of the educational programs at the manufacturing level of casting engineering in the "engineer's reasoning guide" [1].

2. Conceiving the reasoning adapted to the analysis of engineering training systems in the field of castings manufacturing

Taking into account the elements of the engineering behavior (the characteristics of the reasoning, the attributions of the engineering activities, the engineering behavior, the way of thinking and the structuring of the judgments), in this paper we carried out a synthesis of the phasing stage in the study of the use of the engineering reasoning in the modeling of the superior technical training structures in the production of castings [7].

The synthesis used mainly in the reasoning has the following questions structure: 1. What is the essential phenomenon in the process of obtaining the casting?, 2. What is the major problem facing the pieces obtained by casting?, 3. What are the correlations between the areas of last solidification and the compaction of the molded parts?, 4. What are the correlations between the last solidification area and their compactness from the perspective of the constructive and technological solutions?, 5. What are the determinations of the macro-solidification process in the expression of the influence factors? What is the conception of a digital experiment in highlighting the influence factors?, 7. What are the objectives of the digital experiment in validating the results of the analysis?, 8. What are the determinations of the technological influences and how to include them in the experiment? 9. Which are the constructions at the level of the digital experiment?, 10. What are the results obtained in complying with the established objectives?, 11. What are the conclusions of the digital experiment in adapting it to the manufacturing of molded parts?

From the answers formulated, the following aspects are taken into consideration: the appreciation that the solidification represents the essential phenomenon in the process of obtaining castings; compactness is one of the major issues, even if it is not unique; the last solidification area determines, in view of the phenomenology of the process, the occurrence of a lack of compactness in the mass of the alloy; the lack of compactness leads to the appearance of defects in the molded parts, which is why an assessment of the importance of the factors of influence on the macrosolidification process is necessary; the conclusions reached at the level of influence factors can be used at the level of constructive solutions and those of technological nature; the verification of the digital information tool used in the digital experiment is important in the veracity of the conclusions that will result in the analysis; the influences taken into account in the analysis are grouped into the following categories: the nature of the alloy, the nature of the form; casting geometry and casting conditions; the objectives of the experiment aim at visualizing the positioning of the last solidification area in various constructive and technological variants; the experiment aims at tracking - under certain conditions - the influences determined by the nature of the alloy, the nature of the shape and the geometry of the cast piece.

3. Determination of the correlation between the last solidification area and the compaction of cast parts

We considered the area in which the alloy passes from a liquid state into a solid one, within a cast piece or the last solidification area, as the potential for defects primarily linked to the compactness of the material in that area.

In order to determine the correlation mentioned, following the stages established within the critical engineering reasoning, the digital experiment and the practical one were connected to the same constructive-technological structure. In order to verify this hypothesis, some determinations were made on castings in the form of **T** intersections [2].

Initially three situations were computationally analyzed, where two walls of different or equal thicknesses intersect perpendicularly within a casting. The three situations, where S was noted as the thicknesses of the two intersecting walls, are shown in Figure 1.

From the simulation of the solidification of the parts in the three analyzed cases it can be clearly seen that the place of the last solidification is placed in different places depending on the thickness of the intersecting walls, respectively on the ratio between them.

In order to verify in practice how to place the last solidification site in the case of castings made of an Aluminum alloy in wet cast molds, polystyrene designs have been made, with the shape and dimensions of sections as shown in Figure 2 for the three analyzed cases and lengths equal to 100 mm.

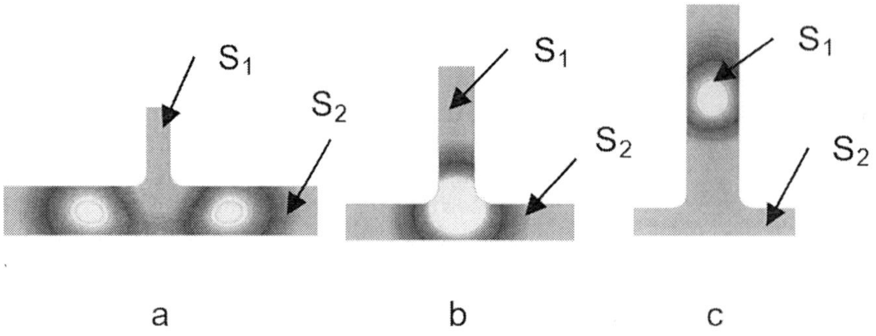

Figure 1. Tshaped wall intersections;a) $S_1<S_2$, b) $S_1=S_2$, c) $S_1>S_2$.

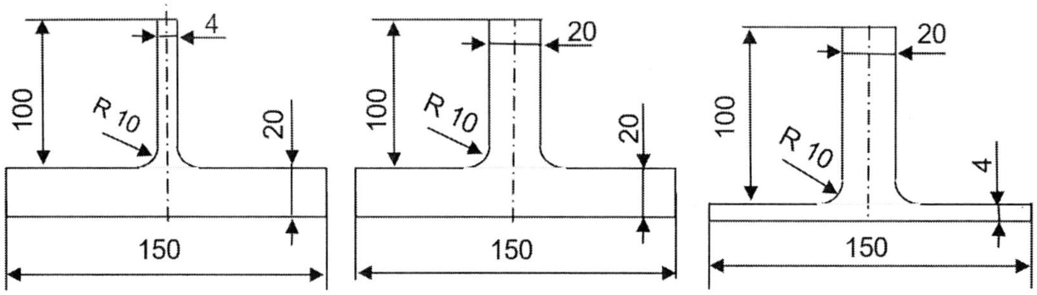

Figure 2. Form and dimensions of castings sections.

After casting, the pieces were sectioned longitudinally in several pieces on each side of the wall intersection area. One of the cast pieces is shown as an example in Figure 3, and the samples obtained by cutting such a piece in Figure 4. For these samples, the specific mass of each of them was determined, considering that for the same material, the differences are mostly determined by the degree of compactness of the piece in the respective area, i.e. the presence of voids in the volume of the analyzed sample. This determination was made by measuring the mass of samples in air and immersed in distilled water (resulting in sample volume) and dividing the two values resulting in specific mass or sample density.

Figure 3. Aluminum-based alloy **T** -type casting piece.

Figure 4. Samples obtained by longitudinal sectioning of the parts.

IOP Conf. Series: Materials Science and Engineering **374** (2018) 012051 doi:10.1088/1757-899X/374/1/012051

The diagram of the cast pieces sections from the three categories of ratios between the wall thicknesses and the determined densities for each sample in the intersection area are shown in Figure 5.

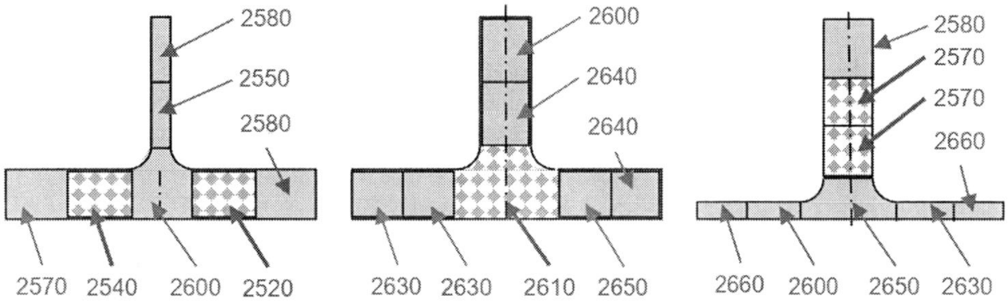

Figure 5. Sections of the pieces and specific mass values determined in kg/m^3.

By analyzing the results obtained, it can be easily observed that the smallest density zones, determined on castings, are placed identically with the last solidification areas on the computationally analyzed samples, by the theoretical simulation of the solidification of some parts of the same type of alloy and cast in wet forming molds, thus confirming the premise that the areas with the lowest density within the parts are also the last solidification areas, or that the last solidification areas are the areas with the lowest compactness in a casting.

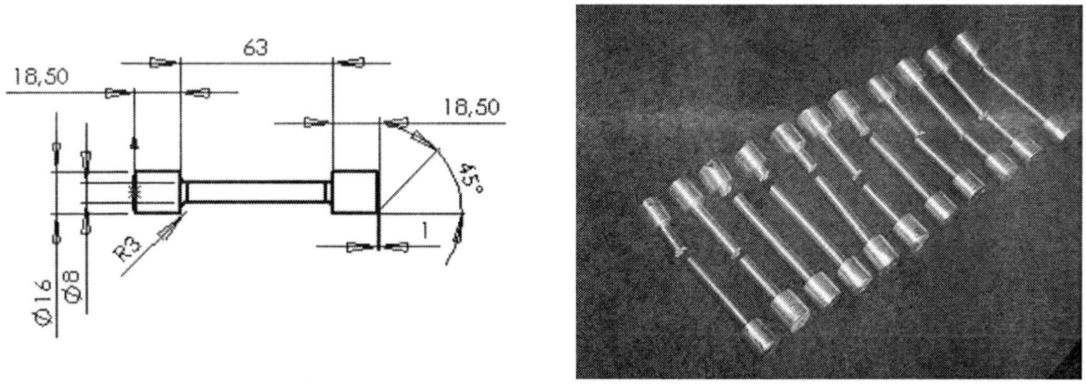

Figure 6. Shape and dimensions of the samples for the determination of tensile strength.

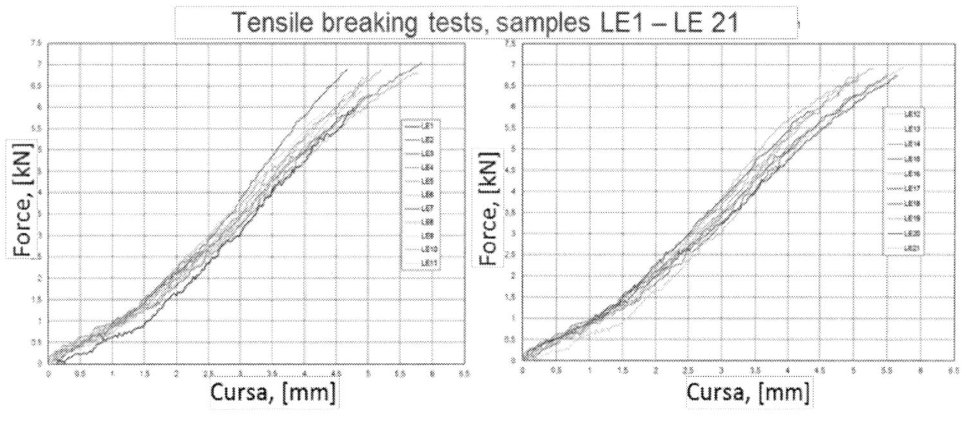

Figure 7. Tear deformation charts for samples LE1 – LE21.

In order to establish a correlation between the degree of compactness of the cast alloy and its mechanical characteristics, from the pieces resulting from the sectioning of the cast parts, 21 cylindrical samples having the dimensions shown in Figure 6 were made, to determine the tensile breaking strength.

The tear deformation diagrams of the 21 samples, numbered from LE1 to LE21, are shown in Figure 7, and the tensile breaking strength values, according to sample density, are shown in Table 1 [3].

The results of these determinations [3] clearly demonstrate that in the areas of the last solidification of a cast piece, the degree of compactness is diminished, and that this degree of compactness ultimately determines the breaking strength of the material, in the sense of decreasing the values the resistance at the decrease in density, respectively the degree of compactness of the material, this being reflected by the trend of the values of the diagram presented in Figure 8.

Table 1. The values of the breaking strength of samples LE1 - LE21, depending on their density.

No.	ρ, [kg/m^3]	σ_R [N/mm^2]	No.	ρ, [kg/m^3]	σ_R [N/mm^2]	No.	ρ, [kg/m^3]	σ_R [N/mm^2]
1.	2520	119	2.	2640	129	3.	2600	109
4.	2520	137	5.	2650	141	6.	2600	117
7.	2550	140	8.	2650	134	9.	2600	135
10.	2570	125	11.	2660	123	12.	2610	127
13.	2580	137	14.	2670	122	15.	2610	138
16.	2580	135	17.	2670	142	18.	2630	129
19.	2580	52	20.	2650	128	21.	2640	135

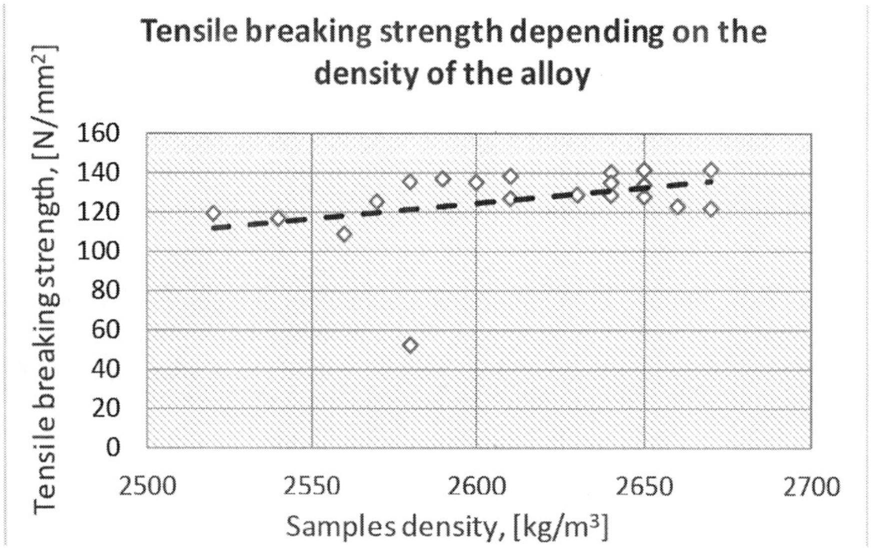

Figure 8. Tensile breaking strength depending on density.

4. Digital experiment for determining the last solidification area

The digital experiment is carried out for a constructive-technological structure for the constitution of a tubular piece, on which there is the possibility of tracking the solidification front and positioning of the last solidification areas for the different variants followed in the analysis. Figure 9 shows the location of the last solidification area for the variant if the inner diamond is kept constant, and in Figure 10 if the thickness of the piece is kept constant. The data present in the presentation of the results of the digital experiment will be used to analyze the factors that influence the last solidification area.

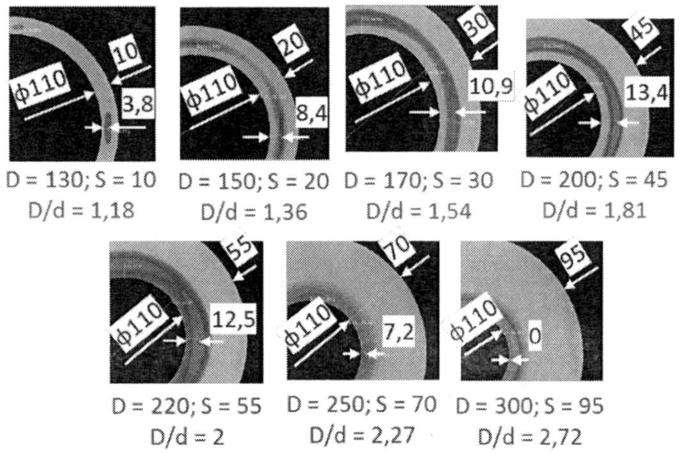

Figure 9. Moving of the last solidification area in tracking the influence of the geometry expressed by providing a constant inner diameter of the cast piece.

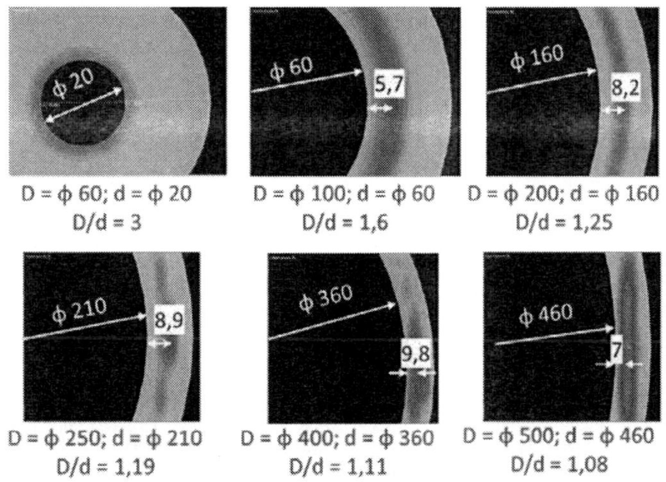

Figure 10. Moving the last solidification area in tracking the influence of the geometry expressed by providing a constant thickness of the cast piece.

5. Factors which influence the area of last solidification

The area in which the solidification of the last alloy quantities takes place, in a cast piece, known as the last solidification area, is determined by several factors, the most important of which are: the nature of the cast alloy, the nature of the mold, the geometry of the part, geometry of casting, position at casting, and so on.

5.1. Nature of the alloy

In order to determine the influence of the alloy on the last solidification area, simulations of the solidification process of three aluminum alloys with different thermophysical characteristics were made. These alloys with their characteristics are shown in Table 2, the meaning of the notations being: TL – liquidus temperature, [°C]; TS – solidus temperature, [°C]; ΔT – solidification interval, [°C]; λ – thermal conductivity of the alloy [W/m·°C], cp – specific heat, [J/kg·°C]; ρ – specific mass, [kg/m^3]; a - coefficient of thermal diffusivity, [m^2/s].

Table 2. The values of the thermophysical characteristics of the analyzed alloys at the temperatures corresponding to the solidification interval.

Aliajul	T_L [°C]	T_S [°C]	ΔT [°C]	λ [W/m·°C]	c_p [J/kg·°C]	ρ [kg/m^3]	a [m^2/s·10^{-5}]
AC-Al Si_7Mg	617	566	51	181,5	1157	2500	6,25
AC-Al Si_{12}	581	558	23	168	1225	2500	5.48
AC-Al Si_7Cu_4	610	523	87	108	1142	2570	3,67

In the digital experiment for the study of the influence of the nature of the alloy, the elements of characterization of the nature of the shape, the geometry of the part and the casting conditions were kept constant. The evolution of the solidification front and the positioning of the last solidification area for the three types of alloys analyzed is shown in Figure 11.

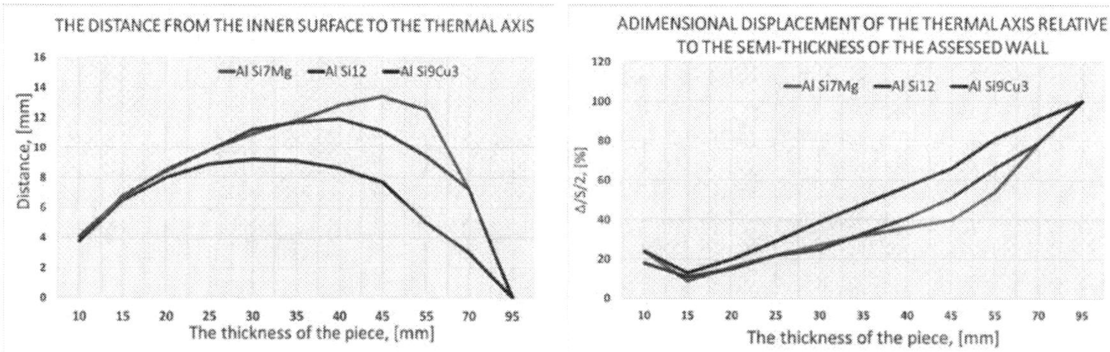

(a) (b)

Figure 11. The influence of the alloy on the position of the last solidification area, for the situation of the constant inner diameter of the piece: (a) the distance from the inner surface to the thermal axis; (b) adimensional displacement of the thermal axis relative to the semi-thickness of the assessed wall.

In the case where the internal diameter is constant and the thickness of the piece increases, the nature of the alloy, expressed by diffusivity (Table 2), influences the positioning of the last solidification area (Figure 11.a) by the fact that the lower diffusion alloy (Al9SiCu3) has a greater constant of the distance of the last solidification area relative to the inner wall of the piece, in relation to the thickness. It is observed that the other alloys (AlSi7Mg and AlSi12) have a higher increasing slope, which determines a variation in the distance of the last solidification area proportional to the thickness of the wall. From a technological point of view, in alloys with a higher diffusivity coefficient, the location of the last solidification area is established with great difficulty, which can create additional technological problems for the piece obtained by the casting. Adimensional reporting of the difference between the thermal axis and the geometric axis (Figure 11.b) of the alloy, using the data of the digital experiment, shows that the alloy with the highest thermal diffusivity (AlSi7Mg) has a lower approximation between the two axes than the alloys with lower diffusivity (AlSi12 and AlSi9Cu3). At the same time, it is observed that for a piece thickness close to 70 mm, the diffusivity does not influence the position of the last solidification zone, the thermal axis coinciding with the geometric axis.

5.2. Nature of the shape

One of the factors which significantly influences the conditions of cooling and solidification of the cast alloys is obviously the nature of the mold, which, by the thermophysical characteristics, firstly determines the cooling and solidification rate of the cast alloys. An important coefficient in the cooling processes of cast alloys is the coefficient of heat accumulation by the shape, b_f, which is a coefficient which takes into account the values of thermal conductivity, the specific heat and the specific mass of the material from which the casting is made, $b_f = \sqrt{\rho_f \cdot c_{pf} \cdot \lambda_f}$ [W·s$^{1/2}$/m^2·°C].

In order to show that these parameters, respectively the nature of the mold, have an influence on the position of the last solidification area, simulations have been made, of the solidification of an Aluminum-based alloy in three types of shapes: classical - wet molding, metallic (gray iron) and graphite. The characteristics of these materials taken into consideration for the analysis are shown in Table 3.

Table 3. The values of the thermophysical characteristics of the materials for the analyzed molds.

Material of the mold	**λ** [W/m·°C]	**c_p** [J/kg·°C]	**ρ** [kg/m^3]	**b_f** W·s$^{1/2}$/m^2·°C
Green Sand	0,9	549	1550	875
Grey iron	50	500	7200	13 416
Graphite	135	600	1922	12 477

While maintaining the alloy characterization elements and casting geometry, the evolution of the solidification front and the location of the final area were observed for each situation. Their presentation, for the analyzed cases, is shown in Figure 12.

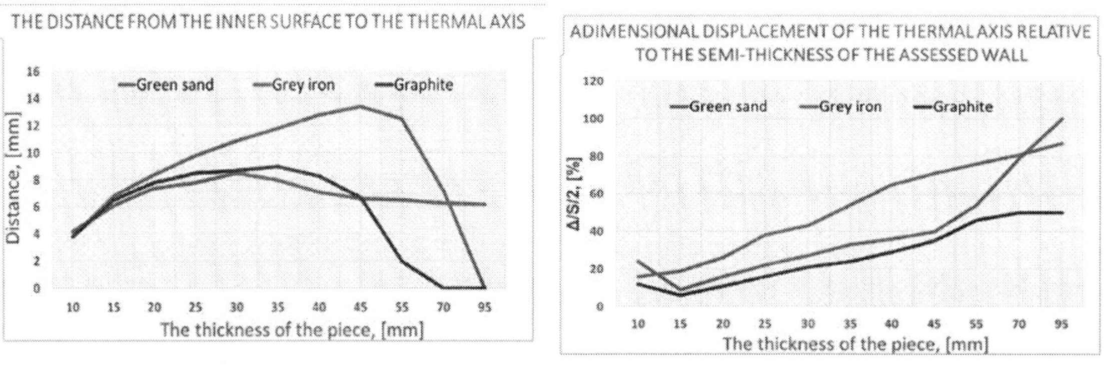

(a) (b)

Figure 12. The influence of the nature of the shape on the position of the last solidification area, for the situation of the constant inner diameter of the piece: (a) the distance from the inner surface to the thermal axis; (b) adimensional displacement of the thermal axis relative to the semi-thickness of the assessed wall.

From the analysis of the graphs shown, we notice that for the shapes in the forming mixture the positioning of the last solidification zone –"LSZ" is approximately constant at the dimensional level with the increase of the wall thickness, instead of the gray cast iron and graphite forms there is a maximum of the "LSZ" positioning, after which the location approaches the inner edge of the wall, the one that is in contact with the core. The technological conclusion shows that for the molds in the mixturethe final positioning area is constantly located at about one third of the thickness of the cast piece. In contrast, for cast iron (metallic) shapes, the variation in the position of the last solidification area is much higher, basically from the geometric axis to the edge of the wall. From the practical point of view, it is found that the pieces obtained in forms of the mixture of formation are less sensitive to the effects of the discontinuities. Instead, castings in metallic or graphite forms are very sensitive to the presence of discontinuities in the active areas to be processed in the surface areas. Therefore, in the technological design of castings in the molds, particular care must be taken to position the discontinuities in areas that do not affect their safety or machinability that provide the final dimensions, or to provide higher processing additions.

Under the same conditions, the analysis of the case where the wall thickness of the piece is constant, the two diameters (inside and outside) are modified accordingly. The results of these solidification simulations are shown in Figure 13 for the influence of the alloy and in Figure 14 for the influence of the mold material on the position of the last solidification area.

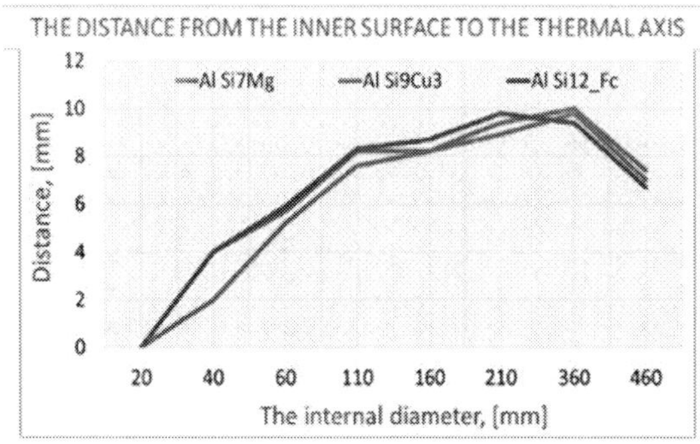

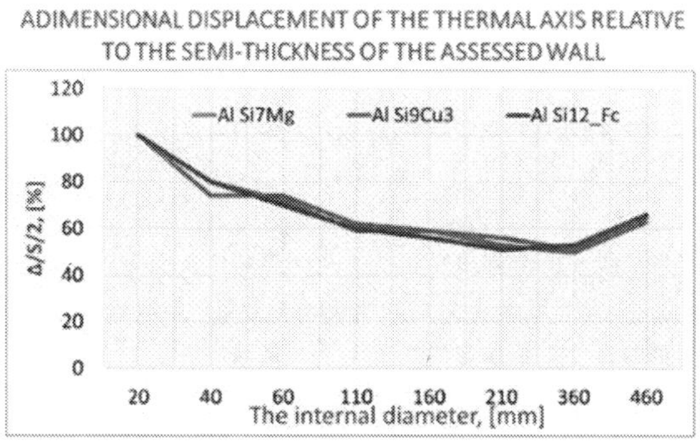

(a) (b)

Figure 13. The influence of the alloy on positioning of the last solidification area, for the situation the wall thickness of the constant piece: (a) the distance from the inner surface to the thermal axis; (b) adimensional displacement of the thermal axis relative to the semi-thickness of the assessed wall.

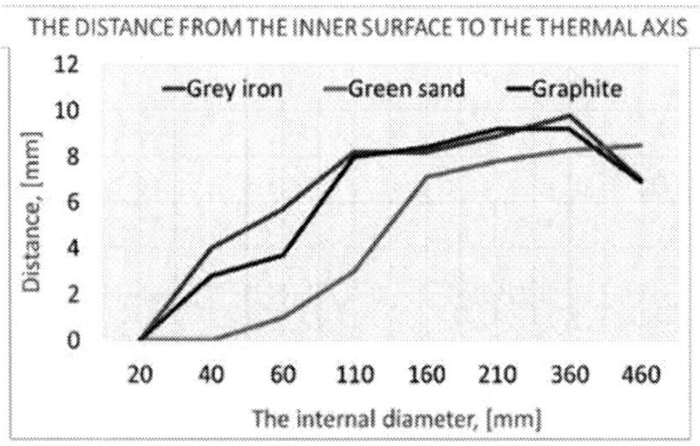

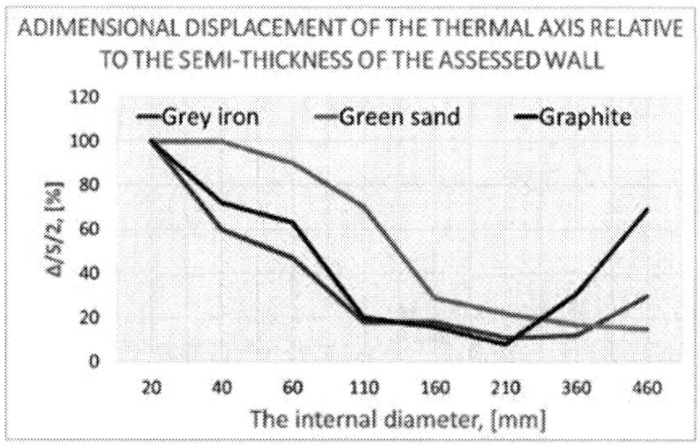

(a) (b)

Figure 14. Influence of the nature of the mold on the position of the last solidification zone, for a constant thickness of the piece wall: (a) the distance from the inner surface to the thermal axis; (b) adimensional displacement of the thermal axis relative to the semi-thickness of the assessed wall.

In the case of a constant wall thickness (Figure 13), the influence of the alloy is not a major one on the positioning of the last solidification area. Therefore replacing the alloy and keeping the thickness constant should not cause problems in positioning the area of last solidification, with the potential for defects to occur. Regarding the influence of the mold material, we notice that it is much more pronounced than that of the alloy, with being significant differences in the location of the last solidification zone, especially between the shapes in the mixture of formation as compared to the other two.

5.3. Geometry of the piece

The influence of casting geometry is followed by the digitization of the following situations: in the first case by maintaining the inner diameter of the piece constant and the variation of the outer diameter so that its thickness varies, and in the second case by modifying the two diameters so as to keep constant the wall thickness of the piece. Analyzing the results of the two variants presented above for different alloys as well as materials of different shape, it can be seen that in parts with the same wall thickness the differences between the last solidification areas for the alloys analyzed are almost insignificant but as the thickness of the walls increases, the difference between the positioning areas becomes larger. Regarding the influence of the formation mixture in the two situations, it can be seen that by increasing the thickness of the wall of the cast piece, the area of last solidification distances more and more from the geometric axis of the wall, instead when increasing the inner diameter, the area of last solidification

Comes closer to the geometrical axis, respectively the radius being increasingly bigger, the situation tends towards the behavior of a planar wall.

6. Conclusions

From the point of view of the proposed objective, the paper highlights the fact that the introduction of digitization at the level of the engineering training programs is important now, because the manufacturing of the castings cannot be achieved without a good knowledge of the way in which IT applications influence the development of processes.

Highlighting this aspect does not favor the replacement of the phenomenological knowledge with the tools of informatics, the optimal solution is expressed in this paper by the need to optimize and harmonize the educational offer with the two tendencies that are manifested, the traditional emphasis on the phenomenological part and the one determined by the strict use of the simulation software for the physical phenomena present in the given technologies. The engineering critical thinking, currently used, represents a tool in optimizing the two tendencies. We believe that the model proposed in the paper can lead to a real knowledge of the casting process, the phenomenological interpretation made by the use of critical engineering thinking in the design of the digital experiment gives us the opportunity to acquire the skills for accurate constructive and technological solutions.

7. References

[1] FCT2006 *Le guide de raisonnement de l'îngenieur* s.l.: s.n.
[2] Lehene T 2014 *Studii asupra solidificăriiI intersecţiilor în formă de T a pereţilor pieselor turnate.* Sebes AGIR 567-574
[3] Lehene T R. 2015. *Cercetări privind îmbunătăţirea caracteristicilor mecanice ale pieselor turnate din aliaje de aluminiu* Teza de doctorat Cluj-Napoca Universitatea Tehnică
[4] Paul R and Elder L 2008 *Mini-guide de la pensee critique* s.l.:Fondation for Critical Thinking.
[5] Paul R and Elder L 2008 *Mini-guide de la pensée critique Concepts et Instruments* s.l.:Fondation for Critical Thinking
[6] Samuilă V 2017 *Analiza critică asupra evoluţiei sistemelor de formare inginerească în domeniul fabricaţiei pieselor turnate* (manuscr. În: Teza de doctorat „Cercetări asupra sistemelor de formare inginerească la nivelul fabricaţiei pieselor turnate. Cluj-Napoca Universitatea Tehnică
[7] Samuilă V 2017b *Contribuţii asupra utilizării raţionamentului ingineresc în modelarea structurilor de formare tehnică superioară la nivelul fabricaţiei pieselor turnate* Cluj-Napoca s.n.
[8] Soporan V, Vamoş C and Pavai C 2003 *Modelarea numerică a solidificării,* Cluj-Napoca: Editura Dacia

Development of a New Optical Measuring Set-Up

I P Miroshnichenko[1,*] and I A Parinov[2]
[1] Don State Technical University, 1 Gagarin Square, Rostov-on-Don, Russia
[2] Southern Federal University, 105/42 Bolshaya Sadovaya street, Rostov-on-Don, Russia

E-mail: ipmir2011@yandex.ru

Abstract. The paper proposes a description of the developed optical measuring set-up for the contactless recording and processing of measurement results for small spatial (linear and angular) displacements of control surfaces based on the use of laser technologies and optical interference methods. The proposed set-up is designed to solve all the arising measurement tasks in the study of the physical and mechanical properties of new materials and in the process of diagnosing the state of structural materials by acoustic active methods of nondestructive testing. The structure of the set-up, its constituent parts are described, and the features of construction and functioning during measurements are discussed. New technical solutions for the implementation of the components of the set-up under consideration are obtained. The purpose and description of the original specialized software, used to perform *a priori* analysis of measurement results, are present, while performing measurements, for *a posteriori* analysis of measurement results. Moreover, the influences of internal and external disturbance effects on the measurement results and correcting measurement results directly in their implementation are determined. The technical solutions, used in the set-up, are protected by the patents of the Russian Federation for inventions, and software is protected by the certificates of state registration of computer programs. The proposed set-up is intended for use in instrumentation, mechanical engineering, shipbuilding, aviation, energy sector, etc.

1. Introduction

At the present time, the appearance of a significant number of new materials and their wide use in various fields of technology leads to the need to develop and use new high-precision measuring instruments that allow solving scientific and production problems. In particular, it is related to the study of the physico-mechanical properties of these materials and the process of diagnosing the state of construction materials under their production and operation by using acoustic active methods of nondestructive testing.

One of the actual and promising directions for the creation of new high-precision measuring instruments is the development of contactless optical measuring devices based on modern laser technologies and new methods of optical interferometry, as well as systems, devices and software that support their operation in solving various measurement problems, accuracy, quality and informative of measurement results [1-3].

The aim of the research is the development of an optical measuring set-up for recording and processing the results of measurements of small spatial displacements of the surfaces of control objects. It allows one to solve all the arising measurement problems by studying the physical and mechanical properties of new materials and during the diagnostics of the state of construction materials by acoustic active methods of nondestructive testing.

2. Structure and components of the proposed optical measuring set-up

The structure and components of the proposed optical measuring set-up are schematically shown in Figure 1.

The optical measuring set-up includes the following components:

(1) control system for the optical measuring set-up;

(2) optical interference device for contactless measurement of small spatial displacements of control object surfaces;

(3) device for protection against the influence of internal disturbing (destabilizing) effects and correction of measurement results;

(4) device for protection against the influence of external disturbing (destabilizing) effects and correction of measurement results;

(5) system of *a priori* analysis and preparation for conducting measurements;

(6) system for recording measurement results;

(7) system for processing and *a posteriori* analysis of measurement results.

3. Brief description of the set-up components

The control system of the optical measuring set-up is designed to control and provide preparation for the measurement, registration and correction of measurement results during measurements and processing of the obtained measurement results after their completion.

The pointed system is implemented on the base of modern personal computer technology using the original specialized software, developed for the proposed measuring set-up by solving various problems arising in the measurement process.

Optical interference device for contactless measurement of small spatial displacements of control object surfaces is intended for contactless measurement of small spatial (linear and angular) displacements of these surfaces by measuring the intensity of the optical field uniquely associated with the pointed small displacements in a given region of the interference pattern, produced by the laser interferometer. This device is developed on the base of a laser two-way interferometer with combined branches, modified to solve measurement problems.

The appearance of typical interference patterns, generated by the laser interferometer in the measurement of small linear displacements for their different values, is shown in Figure 2. Figure 3 shows the appearance of a typical interference pattern by measuring small spatial (linear and angular) displacements of the control object surfaces.

For the proposed set-up, a number of devices have been developed that make it possible to measure small linear displacement [4], small linear and angular displacements [5], small linear and all angular (spatial) displacements [6] of the control object surface. It can be used as part of both the stationary and mobile diagnostic stations [7]. By this, they use acoustically active methods of monitoring and diagnosing the state of materials, applied in instrument engineering, mechanical engineering, shipbuilding, aircraft building, fuel and energy industry, etc. Devices [4-7] are grounded scientifically in the process of carrying out theoretical calculations and experimental studies of their functional characteristics, as well as in experimental operation (the reasoning is given in [8-11]).

The device for protection against the influence of internal disturbing (destabilizing) influences is intended for recording the influence of internal disturbing influences (arising in the measuring device itself, for example, power supply voltage jumps during measurements, etc.) on the measurement results for their correction directly in the process of measurements.

A new technical solution was proposed in [12] providing protection from the influence of internal disturbing influences. It is based on recording the total intensity of the optical field of the interference pattern, which must be unchanged during the measurements, regardless of the values of the recorded small linear and angular displacements. The technical solution [12] is grounded scientifically, and the technical result achieved with its help is confirmed by carrying out a number of test studies of the functional characteristics [13] and by its experimental operation.

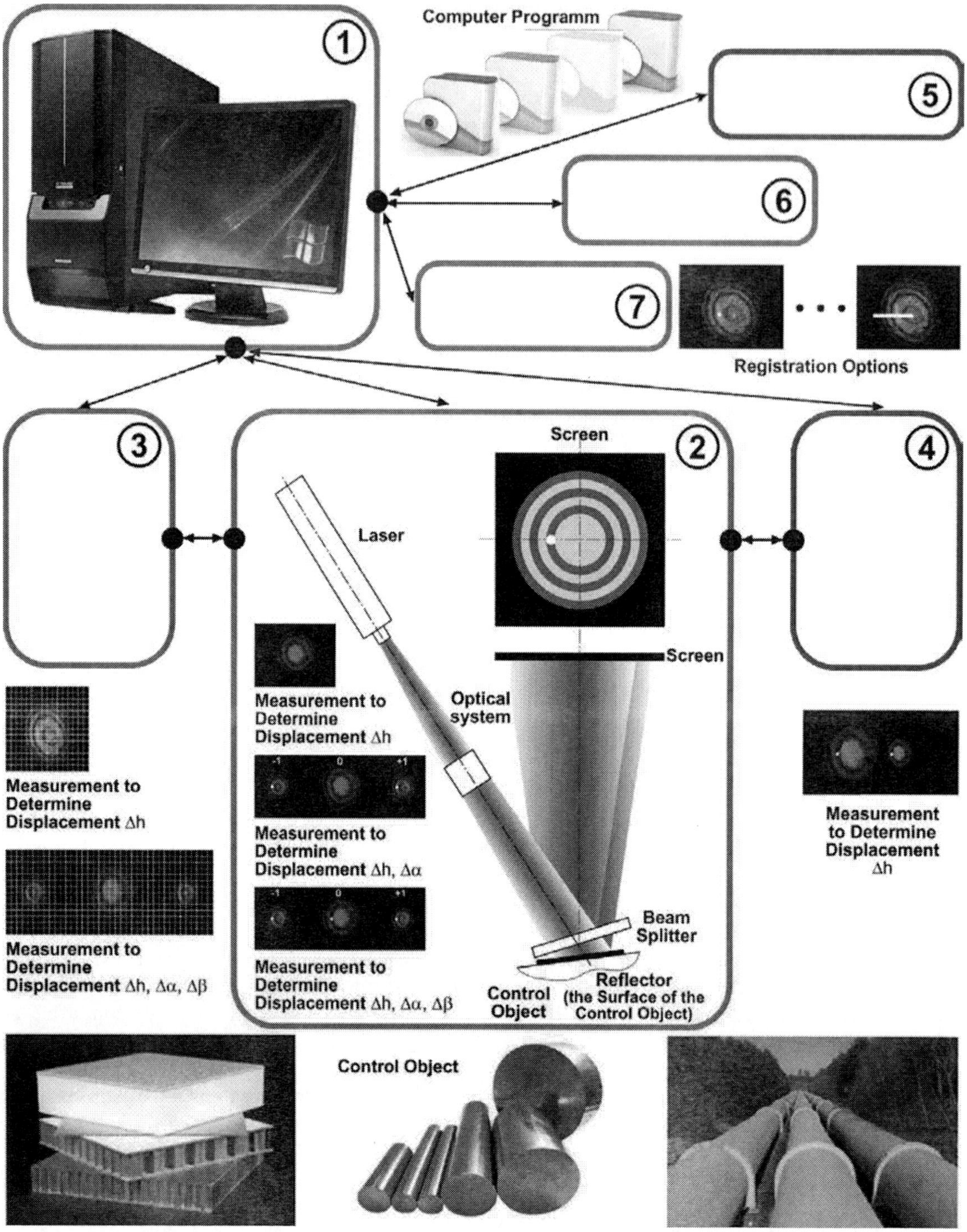

Figure 1. Structure and components of the optical measuring set-up.

The device for protection against the influence of external disturbing (destabilizing) effects is intended for recording the external disturbing effects (arising from outside the measuring device, for example, impacts, vibrations, oscillations of a different nature, etc.) on the results of measurements. It allows one to correct the measurement results directly in the process of the fulfillment of experiments.

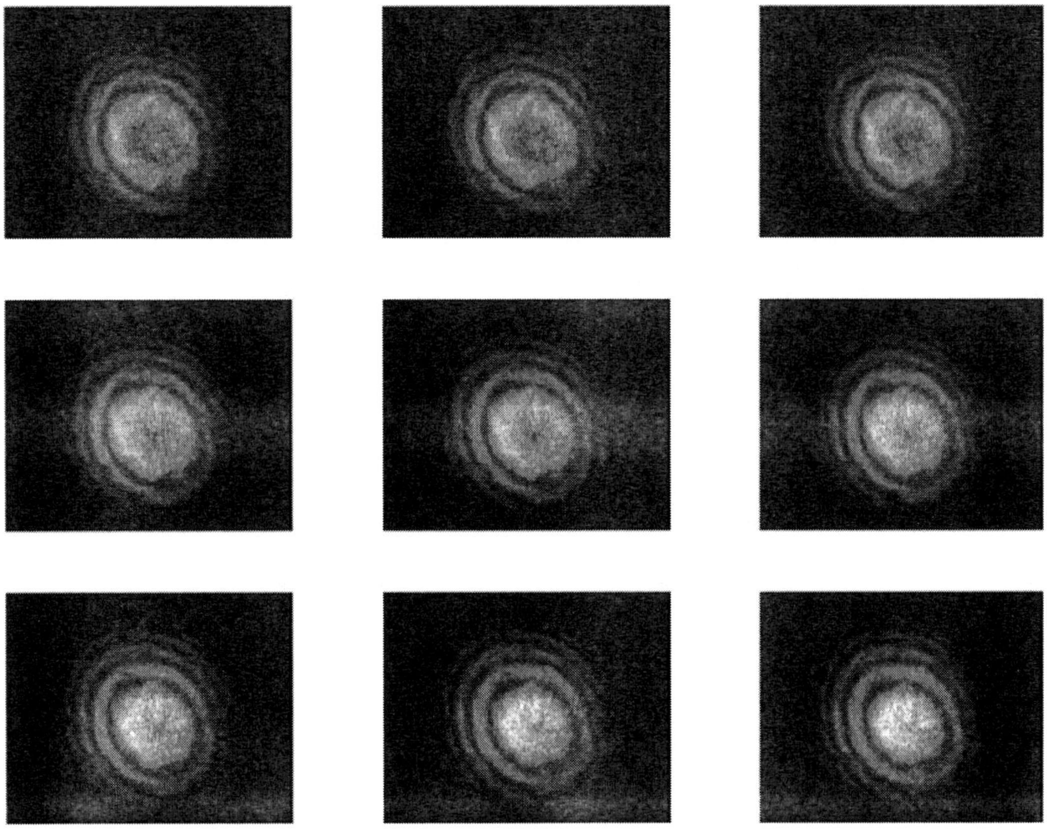

Figure 2. Appearance of typical interference pictures (measurement of small linear displacements).

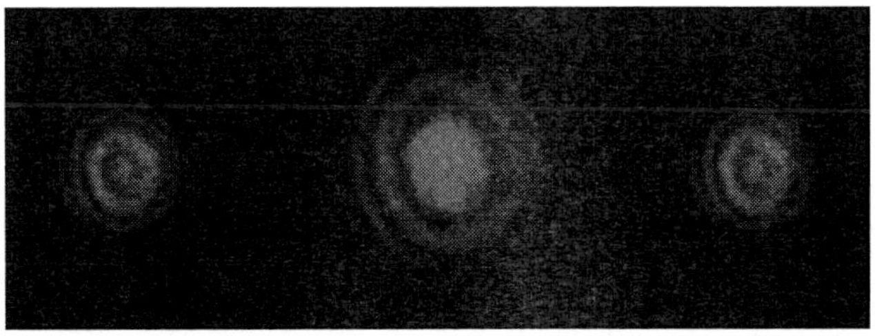

Figure 3. Appearance of typical interference pattern (measurement of small linear and angular displacements).

A new technical solution was proposed in [14] providing protection from the influence of external disturbing influences, based on the use of the measuring capabilities of the optical interference device itself for the contactless measurement of small spatial displacements of the control object surfaces without the use of additional measuring devices.

The technical solution [14] constructively represents two combined single-type laser interferometers. It provides simultaneous recording the intensity of the optical field in a given region of the interference pattern, as well as the registration of the increment of the intensity arising under external disturbing effects. The intensity of the optical field includes the total intensity, corresponding to the resulting small displacement of the surface of the test object, and the intensity increment arising from the influence of external disturbing effects. The result of the measurements is the difference between the pointed intensity values. The technical solution [14] is grounded scientifically, and the technical result achieved with its help is confirmed in the course of experimental studies of the device functional characteristics [15].

The system of *a priori* analysis and preparation of measurements is intended for preliminary predictive modeling of the small displacements of the control object surfaces by using the active acoustic method of nondestructive testing. Moreover, this system defines the distribution of the optical field intensity of interference patterns during measurements.

The system for recording the results of measurements is designed to record the optical field intensity of the interference pattern in certain areas during measurements.

The system for processing and *a posteriori* analysis of the measurement results is intended to processing the results of measurements after their completion and recognizing the obtained measurement results. It uses the results of preliminary predictive modeling of the dependencies of small displacements of the control object surfaces for the active acoustic method of nondestructive testing applied in *a priori* analysis.

4. Description of the software included in the set-up

The software of the proposed optical measuring set-up includes the following:

(i) Computer programs for modeling the intensity distributions of the optical fields of interference patterns, taking into account the geometric characteristics of the measuring system used, longitudinal or transverse polarization and the type of beam splitter (semitransparent mirror, amplitude sinusoidal array, amplitude zone plate, phase sinusoidal grid with uniform period, phase zone plate or amplitude holographic diffraction grating) (preliminary predictive modeling of the measurement process).

(ii) Computer programs for modeling the dependencies of changes in small displacements of control surfaces by using the active acoustic method of nondestructive testing, taking into account the spatial-temporal distribution of the sources of probing, the type of the material of control object (monolayer, layered), its physical-mechanical characteristics (isotropic, transverse-isotropic), types of elastic waves propagating in this material and the features of the arising wave processes (preliminary predictive modeling the control object condition).

(iii) Computer programs for registering and processing the intensity of optical fields, obtained for certain areas of the interference pattern, by recording the intensity by a single photoreceiver with specific geometric characteristics. This photoreceiver is disposed at the selected ring of the interference pattern (see, Figure 4a). By using two photoreceivers with prescribed geometric characteristics, they are installed in adjoining rings of interference pattern with different intensity (Figure 4b). Figure 4c corresponds to vertical and horizontal selected regions of the interference pattern and Figure 4d relates to the selected area of the interference pattern, characterized by the maximum contrast in accordance with the technical solution [4].

(iv) Computer program for determining the total intensity of the optical field of the interference pattern and correcting the results of intensity measurements directly by performing measurements of small linear displacements (Figure 5a) or small linear and angular (spatial) displacements (Figure 5b) of control object surfaces.

(v) Computer program for correcting the results of measurements, obtained from the influence of external disturbing effects. It determines the results of measurements as the difference between the total intensity, corresponding to the arising small displacement of the control object surface, and the intensity increment arising from the influence of external disturbing influences, as well as the intensity increment arising from external disturbing influences. The appearances of interference patterns for this case are shown in Figure 6 by measuring small linear displacements, respectively, for each of the components of the difference from left to right).

The software for modeling the intensity distributions of the optical fields of interference patterns are developed by using new mathematical models, describing the functioning of various versions of the proposed optical interference device for the contactless measurement of the small spatial displacements of control object surfaces with help of a two-way interferometer with aligned branches (see [16-19]).

The software for modeling the dependencies of the changes in the small displacements of the control object surfaces by using the active acoustic method of non-destructive testing are developed with help of new methods. These methods are based on the tensor relations of the generalized method for the scalarization of dynamic elastic fields in transverse-isotropic media. These constitutive relations allow us to determine displacements, stresses and deformations in certain regions of monolayer and layered, flat, cylindrical and elliptical constructions for isotropic and transverse-isotropic materials of the layers (see [8-9, 11]).

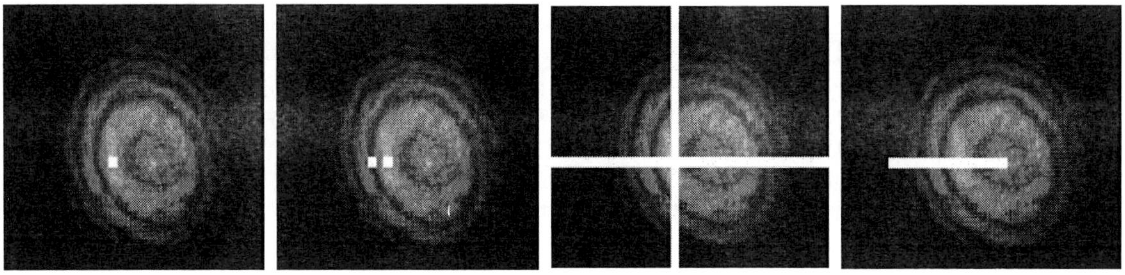

(a) (b) (c) (d)

Figure 4. Variants of recording the intensity of optical fields, obtained from certain areas of interference patterns.

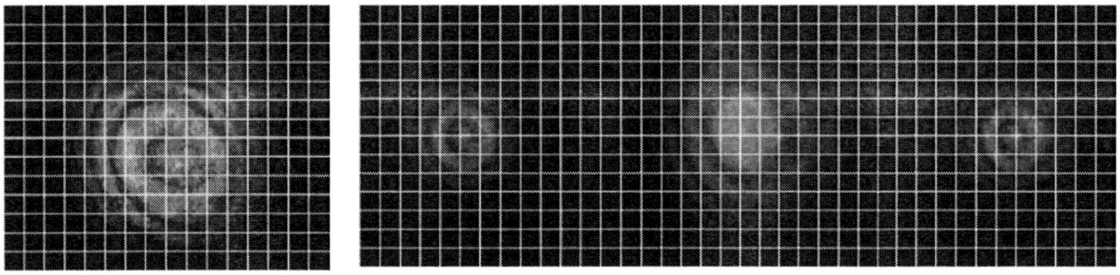

(a) (b)

Figure 5. Variants of recording the total intensity of the optical field of interference pattern.

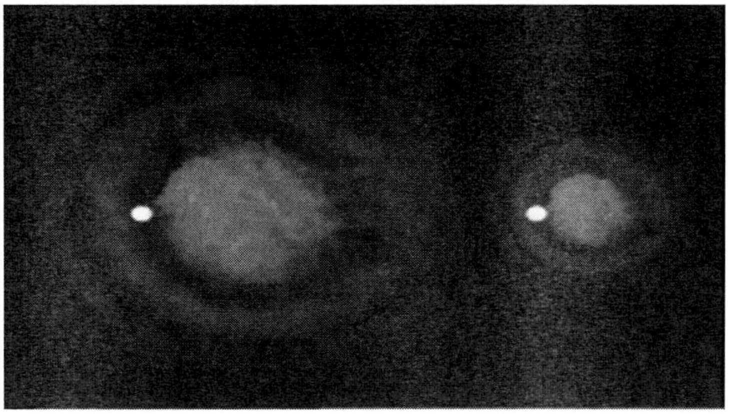

Figure 6. Appearance of interference patterns for correction the results of measurements, caused by the influence of external disturbances (measurement of small linear displacements).

5. Conclusions

Based on the above results, we can make the following conclusions:

(i) an optical measuring set-up has been developed that allows solving all arising measurement tasks by studying the physical and mechanical properties of new materials and during diagnosing the state of construction materials by acoustic active methods of nondestructive testing;

(ii) the technical solutions and the original specific software, included in the proposed measuring set-up, are grounded in the process of carrying out theoretical and experimental studies, and they also confirmed the achievement of the expected technical and functional characteristics in the solution of various actual measurement tasks in the test operation;

(iii) by using the developed measuring complex, it is possible to increase the information content and reliability of the measurement results to 20 – 30%, and also to correct the measurement results directly during measurements without using additional measuring instruments (by using their own measuring capabilities), which allows one to improve the quality of measurement results to 20 – 40%;

(iv) the technical solutions, implemented in the proposed measuring set-up, are protected by patents of the Russian Federation for inventions, and software is protected by the certificates of state registration of computer programs;

(v) the proposed optical measuring set-up is intended for using in the composition of both stationary and mobile diagnostic stations in instrument engineering, mechanical engineering, shipbuilding, aircraft building, fuel and energy industriy, etc.;

(vi) the results of the paper were presented and approved at the International Conferences on "Physics and Mechanics of New Materials and Their Applications" (PHENMA-2015, Rostov-on-Don, Russia; PHENMA-2016, Surabaya, Indonesia and PHENMA-2017, Jabalpur, India), International Conference on Innovative Research "EUROINVENT ICIR 2017" (Iasi, Romania), etc. Moreover, they were discussed at the 45 International Salon of Inventions "Inventions Geneva-2017" (Geneva, Swiss), the 9-th European Exhibition of Creativity and Innovation "EUROINVENT 2017" (Iasi, Romania), the 31-th World Genius Convention and Education «Expo 2017» (Tokyo, Japan) and the XI International Warsaw Invention Show «IWIS 2017» (Warsaw, Poland), where were awarded gold medals.

6. References

[1] Miroshnichenko I P, Parinov I A, Rozhkov E V, Serkin A G, Sizov V P 2012 *Piezoelectrics and Related Materials: Investigations and Applications* 239-282

[2] Miroshnichenko I P, Parinov I A, Rozhkov E V, Sizov V P, Shevtsov V A 2013 *Physics and Mechanics of New Materials and Their Applications* 145-154

[3] Miroshnichenko I P, Serkin A G 2010 *Metallurgist* **54(3-4)** 189-190

[4] Miroshnichenko I P, Parinov I A, Rozhkov E V, Serkin A G, *Optical device for measuring displacements* Patent RU No. 2373492/ 20.11.2009

[5] Miroshnichenko I P, Serkin A G, Sizov V P, *Measuring method of linear and angular displacements* Patent RU No. 2388994/ 10.05.2010

[6] Miroshnichenko I P, Parinov I A, Sizov V P, *Measuring method of linear and angular components of the small displacements of control object surfaces* Patent RU No. 2606245/ 10.01.2017

[7] Miroshnichenko I P, Parinov I A, Rozhkov E V, Serkin A G, *Optical device for measuring displacements* Patent RU No. 2407988/ 27.12.2010

[8] Miroshnichenko I P, Parinov I A, Rozhkov E V, Chang S-H 2016 *Advanced Materials Manufacturing, Physics, Mechanics and Applications. Springer Proceedings in Physics* **175** 341-356

[9] Miroshnichenko I P, Parinov I A, Wu J-K, Hong W-L, Yeh M-Y 2017 *Advanced Materials - Techniques, Physics, Mechanics and Applications. Springer Proceedings in Physics* **193** 591-605

[10] Miroshnichenko I P, Parinov I A 2017 Scientific Ground of a New Optical Device for Contactless Measurement of the Small Spatial Displacements of Control Object Surfaces *IOP Conference Series: Materials Science and Engineering* **209**
[11] Miroshnichenko I P, Parinov I A 2018 *Advances in Structural Integrity* 663-671
[12] Alekhin V E, Miroshnichenko I P, Serkin A G, Sizov V P, *Method of recording displacements by optical sensors* Patent RU No. 2343403/ 10.01.2009.
[13] Miroshnichenko I P, Serkin A G *Sensors and Systems* **3** 28-30
[14] Miroshnichenko I P, Nesterov V A, Serkin A G, Sizov V P, Shevtsov V A, *Interference meter of small displacements* Patent RU No. 2410642/ 27.01.2011.
[15] Miroshnichenko I P, Shevtsov V A *Polzunov Vestnik* **2** 66-71
[16] Miroshnichenko I P *Advanced Materials. Physics, Mechanics and Applications, Springer Proceedings in Physics* **152** 163-179
[17] Miroshnichenko I P, Sizov V P 2015 *Advanced Materials - Studies and Applications* 353-390
[18] Miroshnichenko I P 2016 Method and Computer Software for Definition of Stress-Strain State in Layered Anisotropic Constructions at Pulse Loading *Physics Research and Technology. Proceedings of the 2015 International Conference on "Physics, Mechanics of New Materials and Their Applications'* (Nova Science Publishers, New York) 509-514
[19] Miroshnichenko I P 2017 *Advanced Materials - Techniques, Physics, Mechanics and Applications* **193** 435-443

Acknowledgements
The study was performed under partial support of the Russian Foundation for Basic Research (No. 16-08-00740). I. P. Miroshnichenko and I. A. Parinov acknowledges financial support of the Russian Ministry of Education and Sciences in the framework of "Organization of Scientific Research" Government Assignment.

IOP Conf. Series: Materials Science and Engineering **374** (2018) 012053 doi:10.1088/1757-899X/374/1/012053

Validation of Ion Chromatographic Method for Determination of Standard Inorganic Anions in Treated and Untreated Drinking Water

V Ivanova[1,2], A Surleva[1] and B Koleva[2]
[1]University of Chemical Technology and Metallurgy, 8, Kliment Ohridski Blvd., Sofia 1756, Bulgaria
[2]Bulgarian Institute of Metrology, National Center of Metrology, Department of Chemical Measurements and Ionazing Radiation, 52B, G. M. Dimitrov Blvd., Sofia 1040, Bulgaria

E-mail: v.ivanova@bim.government.bg

Abstract. An ion chromatographic method for determination of fluoride, chloride, nitrate and sulphate in untreated and treated drinking waters was described. An automated 850 IC Professional, Metrohm system equipped with conductivity detector and Metrosep A Supp 7-250 (250 x 4 mm) column was used. The validation of the method was performed for simultaneous determination of all studied analytes and the results have showed that the validated method fits the requirements of the current water legislation. The main analytical characteristics were estimated for each of studied analytes: limits of detection, limits of quantification, working and linear ranges, repeatability and intermediate precision, recovery. The trueness of the method was estimated by analysis of certified reference material for soft drinking water. Recovery test was performed on spiked drinking water samples. An uncertainty was estimated. The method was applied for analysis of drinking waters before and after chlorination.

1. Introduction

During the last decades ion chromatography (IC) have gain much attention due its unique capability to detect anions in different matrices in very precise and accurate manner offering low detection limits, low sample and reagent consumption, as well as short analysis time [1, 2]. Nowadays the main challenges in environmental analysis by ion chromatography are related to new analytes separation and determination at trace levels. Due to the intensive technological development new molecules appeared in the environment, which imposes determination of new analytes in common matrices or standard anions in new matrices. Novel stationary phases with appropriate selectivity and high separation efficiency have been proposing din attempt to respond to this challenge [3, 4]. The environmental and health authorities, as well as modern industrial technologies impose stringer and stringer demands to the water quality and methods for determination at trace levels are highly demanded [5-9]. Ion chromatography appears to be widely studied method for anion determination which successfully addressed this issue [1, 2, 7-12].

It should be emphasised the need of validation of analytical methods in order to ensure quality of the results and their fit for purpose according to the demands of customers. A line of studies on optimisation and validation of ion chromatographic methods for anions determination have been recently published [7-13]. The validation procedure was profoundly described in the international legislation to ensure comparable laboratory results [14-17].The validation should prove that the results obtained by given laboratory were not significantly different from the reference results [16, 17].Validation of a method for chloride, nitrate and sulphate ions determination in drinking water was reported [12]; all contributions involved in the analysis were considered in estimation of the uncertainty. A validation of an ion chromatographic methods for determination of fluoride, chloride, nitrite, bromide, nitrate, phosphate and sulphate anions in wet depositions and other meteoric water sources with precision less than 2 % was described [11]. A sensitive and reliable ion chromatographic method with limit of detection at low ppb level was validated for determination of corrosive anions (fluoride, chloride, nitrate, phosphate and sulphate) in power plants water [7]. Single-run determination of several anions using different separation columns was proposed and the results from method validation in environmental samples were reported [8,13].A thorough review on ion chromatographic methods for ox halide ions determination has been recently published [1], the results obtained with different columns, eluents and suppression modes were compared. As can be seen from the literature survey the most intensively studied ion chromatographic determination is based on conductivity detection using different suppressors.

This paper presents the results from a validation study of ion chromatographic method with conductivity detection after chemical and carbon dioxide suppression for determination of standard anions in treated and untreated drinking water.

2. Experimental

An ion chromatographic, metal-free system (IC-850 Professional model, Metrohm AG, Switzerland) controlled by Metrodata MagIC Net™ software and equipped with 858 Professional Sample Processor, sample filtration system with a 0.2 μm regenerated cellulose membrane, six channel injection valve, low pulsation high-pressure pump, chemical suppression and CO2 suppression, eluent degasser and conductivity detector was used. The separation was performed on a Metrosep A Supp 7-250 column (250 x 4 mm, polyvinyl alcohol with quaternary ammonium groups, 5 μm particle size), Metrosep RP 2 Guard/3.5 (polymer with pore size 0.2 μm). In order to keep the baseline signal as low as 0.99 μS/cm, 0.1 M sulphuric acid solution was used for regenerated chemical suppression. The optimized chromatographic parameters are presented in Table 1.

Table 1. Optimized parameters for ion chromatographic measurement.

Parameter	Value
Column	Metrosep A Supp 7 – 250/4.0
Precolumn	Metrosep RP 2 Guard/3.5
Eluent	Na_2CO_3 (3.6mmol/L)
Eluent flow rate	0.7 mL/min
Elution mode	Isocratic
Sample volume	20 μL
Column temperature	45 °C
Detection	Suppressed conductivity
Full-scale range	0 – 15000 μS/cm single range
Suppressor	Metrohm Suppressor Module (MSM)
CO_2 Suppressor	Metrohm CO_2 Suppressor (MCS)
Determination mode	Peak area

Eluent solution was prepared by sodium carbonate superpose Na2CO3(CertiPUR®, Merck, Darmstadt, Germany). Multi ion standard solutions of fluoride, chloride, nitrate and sulphate were prepared by appropriate dilution of their anion standard stock solution (Fluka 1000 mg/L for IC, TraceCERT®, Sigma–Aldrich, Buchs, Switzerland). All solutions were prepared gravimetrically in

Milli-Q water, (Mill-Q Direct 8, Merck Millipore; resistivity >18.2 MΩcm, equipped with a Millipack 0.22μm filter). Certified Reference Material for soft drinking water UK-Anions №ERM® - CA016a, LGC, Teddington, Middlesex was used to confirm the trueness of the method. Sulphuric acid (96% Suprapur®, Merck KGaA, Darmstadt, Germany) was used in conductivity suppression regeneration system after appropriate dilution. Before measurements the samples were filtered by MF-Millipore™ membrane filter with pore size of 0.45 μm (Merck KGaA, Darmstadt, Germany).

3. Results and discussion

The studied anions F^-, Cl^-, NO_3^- and $SO_4^{2-}$are separated by 3.6 mmol/lNa_2CO_3 eluent under the optimized parameters (Table 1). Examples of the obtained chromatograms are presented on Figure 1. All anions were separated completely in a run with total analysis time of 30 min.

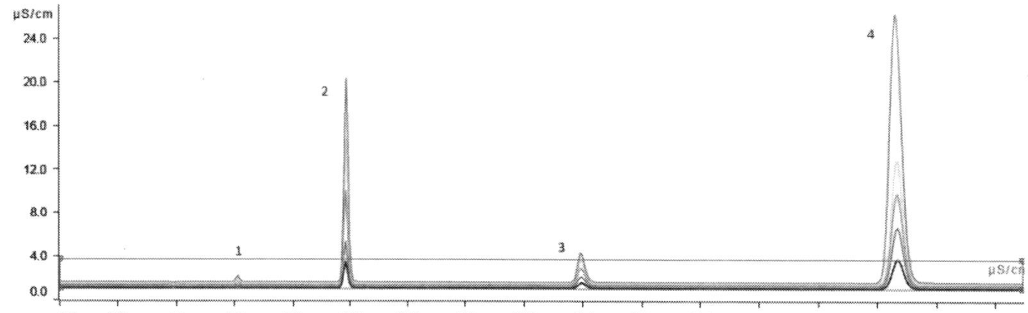

Figure 1. Chromatograms for standard solutions of F^-, Cl^-, NO_3^- and SO_4^{2-}at different concentration. Peak 1: F^- at 0.02, 0.06, 0.1, 0.14, 0.18 mg/L; Peak 2: Cl^- at 1.5, 2.5, 5.0, 7.5, 10.0 mg/L; Peak 3: NO_3^- at 1.0, 2.0, 3.0, 4.0, 5.0 mg/L; Peak 4: SO_4^{2-} at 4.0, 8.0, 12.0, 16.0, 20.0 mg/L.

The retention time of the studied anions and their relative standard deviation obtained by spiking untreated water sample are presented in Table 2. The standard deviation is estimated for 5 injections at 5 different concentration levels varying as follows: 0.02-0.18 mg/L F^-, 1-5 mg/L NO_3^-, 4-20mg/L SO_4^{2-} and 1.5-10 mg/L Cl^-. Therelative standard deviation was less than 1 %and it proved that the increased concentration do not affect the peak position. Additionally, the retention time of each anion in the standards and in the sample were compared. The relative difference was not exceeding 2 % and thus fulfilled the tolerance criterion established by the European Commission Decision 2002/657/EC [6]. According to the cited document the retention time of analyte in standard solution must correspond to that in the sample within tolerance of 2.5% [6].

Table 2. Retention time for studied anions atoptimized ion chrotomagraphic conditions in untreated water sample (n=25).

	Retention time (min)	s (min)	s_r (%)
F^-	6.11	0.0045	0.07
Cl^-	9.87	0.0100	0.10
NO_3^-	17.92	0.0114	0.06
SO_4^{2-}	28.64	0.0460	0.16

3.1. Linearity

Calibration curves were obtained in the concentration interval 0.4 -12.0 mg/L for F-, Cl-, NO3- and 0.4-10 mg/L for SO42- . The measurements were run in quadruplicate for 7 concentration level sof multi-ion standards and the means were used for calibration curves. The regression analysis and "lack-of-fit test" were used according to the IUPAC validation guidelines [15].The results are presented in Table 3. It should be pointed out that the method showed wider working range if we consider that good linearity is observed in the concentration range 0.1-1.0 mg/L of each studied anion (N=7, r=0,9997) and 10-25 mg/L (N=7, r=0.9996). However, in the concentration range 10-25mg/L a memory effect was observed which needs long washing periods for the signal to reach base line. Hence, a dilution of more concentrated samples was recommended. The quality requirement for the acceptance of the calibration function was: correlation coefficient $r^2 \geq 0.995$ [18].The linearity in all 3 studied concentration regions were additionally proved by lack-of-fit test. All standard curves fulfilled the criterion $F_{tabulated} > F_{calculated}$ and the obtained models were considered as "fit to data" [15].

Table 3. Calibration curves for F^-, Cl^-, NO_3^-at 0.4-12 mg/L (N=7, P=95%; F_{tab} = 2.9582) and SO_4^{2-} at 0.4-10 mg/L, (N=6, P=95%; F_{tab} = 3.2592).

	Slope	Slope stdev	Intercept	Intercept stdev	r^2	$F_{calc.}$
F^-	0.4579	0.0056	0.0729	0.0372	0.9992	0.1234
Cl^-	0.3071	0.0004	0.0303	0.0025	0.9999	1.8520
NO_3^-	0.1711	0.0004	0.0062	0.0022	0.9999	0.8158
SO_4^{2-}	0.2398	0.0003	0.0039	0.0013	0.9999	1.6806

3.2. Limits of detection (LOD) and limits of quantification (LOQ)

LOD and LOQ were estimated using two approaches: (1) calculation of LOD and LOQ based on 3.3 or 10 times, respectively, standard deviation of the lower concentration obtained from standard curve. The standard deviation (SD) was calculated in the repeatability conditions [13]. Three spiked reagent blanks at the low concentration levels were prepared for this study, each sample was analyzed in triplicate and the means are used to calculate standard deviation. The standard deviation (in mg/L) was divided by the square root of the number of measurements; (2) calculation based on 3.3 or 10 times the standard deviation of residuals divided by the slope of standard curve [16,17].

In both cases the calibration curves were prepared in the µg/L region: 0.038-0.057 mg/L F- (N=9, r=0.9962); 0.019-0.057 mg/L Cl-(N=5, r=0.997); 0.0087-0.077 mg/L NO3-(N=5, r=0.997); 0.0093-0.047 mg/L SO42-(N=5, r=0.991).The obtained values of LOD and LOQ differ according depending on the calculation strategy. The obtained results are presented in Table 4. For comparison some literature data are included in the Table. As can be seen the ion chromatography with chemical and carbon dioxide suppression offers low detection limits down to µg/L level. The achieved LOD and LOQ are lower than LOD for water samples reported in the literature [11-13, 18]. Zivojnovicet al. [7] reported remarkably low LOD for fluoride and chloride 0.077 and 0.082 µg/L, respectively, in ultra-pure water but using1000µL sample volume. Hence, the method LOD and LOQ could be additionally lowered if higher injection volume was used.

Table 4. Limits of detection and limits of quantification of IC method.

	Calculation based on the SD of the lowest concentration [14]		Calculation based on SD of the residuals [16,17]		LOD and LOQ in water samples mg/L						
	LOD (mg/L)	LOQ (mg/L)	LOD (mg/L)	LOQ (mg/L)	LOD [18]	LOQ [18]	LOD [11][a]	LOQ [11][a]	LOD [13]	LOD [12][a]	LOQ [12][a]
F^-	0.0023	0.007	0.0043	0.013	0.080	0.050	0.001	0.004	0.02	-	-
Cl^-	0.0042	0.013	0.0032	0.010	0.160	0.068	0.010	0.04	0.11	0.14	0.47
NO_3^-	0.0161	0.049	0.0062	0.019	0.180	0.015	0.006	0.02	0.04	0.22	0.72
SO_4^{2-}	0.0039	0.012	0.0054	0.016	0.200	0.023	0.008	0.03	0.09	0.13	0.45

[a]the results are calculated by multiplying the standard deviation of the lower concentration by 3 or 10.

3.3. Repeatability and intermediate precision

The repeatability and intermediate precision tests were made according to the recommendation of Eurachem Guide 2014 [14]. The repeatability test was made for a short period, by the same user with the same instrument. Measurements were made at 5 concentration levels in 5 replicates. The intermediate precision was estimated by analyzing 5replicates in the same concentration range for the period of 4 months, each measurement was made in 5 replicates. Relative standard deviations are presented in Table 5.

Table 5. Repeatability and intermediate reproducibility of the ion chromatographic method.

	Repeatability (RSD %)				Intermediate precision (RSD %)			
Concentration(mg/L)	0.2	0.8	1.0	8.0	0.2	0.8	1.0	8.0
F^-	0.83	0.17	0.48	0.21	1.81	0.35	1.58	0.15
Cl^-	0.79	0.50	0.43	0.11	1.68	0.58	0.51	0.12
NO_3^-	1.45	0.90	0.30	0.20	2.83	1.66	0.44	0.22
SO_4^{2-}	1.10	0.33	0.44	0.17	2.35	0.49	0.60	0.17

3.4. Trueness

The trueness of the method was estimated by analysis of certified reference material for soft drinking water (UK-Anions № ERM® - CA016a) in four replicates with external calibration. Zeta-score was used as an estimate of the method trueness and performance indicator. Zeta-score was calculated as:

$$\zeta = \frac{X_{lab} - X_{ref}}{\sqrt{u_{lab}^2 + u_{ref}^2}} \quad (1)$$

where X_{lab} is the mean value obtained in the laboratory; X_{ref} is the actual value of CRM; u_{lab} is the uncertainty of the result obtained by the laboratory; u_{ref} uncertainty of the reference value from CRM. The obtained zeta–score was less than 2.0 and the method performance was considered as "satisfactory" [18, 19].The recovery of analytes in CRM was below the acceptance limits recommended in the Council Directive 98/83/EC [5]. The results are presented in Table 6.

Table 6. Recovery test results from CRM soft drinking water.

	Certified Value (mg/L)	Uncertainty (mg/L) k=2	Measured Value (mg/L)	Uncertainty (mg/L) k=2	Recovery (%)	Trueness Zeta-score $\lvert\zeta\rvert \leq 2.0$
F^-	1.5	± 0.1	1.52	± 0.16	101.0	0.08
Cl^-	250	± 7	247.3	± 3.2	98.9	0.36
NO_3^-	48	± 3	49.0	± 4.4	102.1	0.20
SO_4^{2-}	254	± 10	248.6	± 2.5	97.9	0.52

Additionally, a recovery of spiked treated and untreated drinking water samples was estimated at 3 concentration levels of standard additions. The results are presented in Table 7. As an acceptance criterion, a tolerance of±10% of actual value was implemented [5, 18]. To study matrix influence multiple standard addition method was used. The slope of standard and multiple addition curves in treated and untreated water samples coincided well and the ratio of the slopes was in the range 99-102

IOP Conf. Series: Materials Science and Engineering **374** (2018) 012053 doi:10.1088/1757-899X/374/1/012053

%. As can be seen from the results the validated method was not influenced by the matrix composition.

Table 7. Recovery test results from spiked water samples.

	Concentration of spike (mg/L)	Recovery in untreated drinking water (%)	Recovery in drinking water after chlorination (%)
	0.02	105.5	98.7
F^-	0.04	98.3	99.8
	0.08	104.1	99.3
	1.25	111.5	114.9
Cl^-	2.5	105.8	108.0
	5.0	103.6	104.4
	0.75	112.1	113.7
NO_3^-	1.5	104.9	106.7
	3.0	103.3	104.4
	3.0	109.0	110.2
SO_4^{2-}	6.0	104.6	105.6
	12.0	102.8	103.7

3.5. Uncertainty

The uncertainty was estimated according to the procedure recommended by EURACHEM/CITAC Guide [20]. The expanded uncertainty calculated using a coverage factor of 2 (k=2) and 95% confidence level, was presented with: U = 2.u_c, where u_c was a combined uncertainty. Three sources of uncertainty were included in the uncertainty budget. Linear least squares fitting procedure was applied to estimate the uncertainty of concentration of each of studied anions from the standard curve [20]. Standard stock solution was used in the estimation. The results are presented in Table 8.

Table 8. Uncertainty determination of F^-, Cl^-, NO_3^- and SO_4^{2-} in CRM-№ ERM - CA016a.

Anion	Sources of uncertainty	Value x	Standard Uncertainty u(x)	Relative Standard Uncertainty u(x)/x	Measured Value (mg/L)	Combined Uncertainty u_c	Expanded Uncertainty U
F-	Linear least squares calibration	0.075 mg/L	0.004 mg/L	0.0533	1.52	0.0809 mg/L	0.16 mg/L
	Balance (Dilution factor)	2.0406 g	0.0004 g	0.0002			
	Concentration of the standard stock solution	1001 mg/L	4.0 mg/L	0.002			
Cl-	Linear least squares calibration	5.016 mg/L	0.031 mg/L	0.0062		1.618 mg/L	3.24 mg/L

Table 8. Uncertainty determination of F^-, Cl^-, NO_3^- and SO_4^{2-} in CRM-№ ERM - CA016a.

Anion	Sources of uncertainty	Value x	Standard Uncertainty u(*x*)	Relative Standard Uncertainty u(*x*)/*x*	Measured Value (mg/L)	Combined Uncertainty u_c	Expanded Uncertainty U
	Balance (Dilution factor)	0.5183 g	0.0004 g	0.0008	247.3		
	Concentration of the standard stock solution	1001 mg/L	4.0 mg/L	0.002			
	Linear least squares calibration	0.9941 mg/L	0.045 mg/L	0.0453			
NO_3	Balance (Dilution factor)	0.5201 g	0.0004 g	0.0008	49.0	2.222 mg/L	4.44 mg/L
	Concentration of the standard stock solution	1001 mg/L	4.0 mg/L	0.002			
	Linear least squares calibration	5.041 mg/L	0.023 mg/L	0.0046			
SO_4^{2-}	Balance (Dilution factor)	0.5183 g	0.0004 g	0.0008	248.6	1.253 mg/L	2.51 mg/L
	Concentration of the standard stock solution	1001 mg/L	4.0 mg/L	0.002			

3.6. Drinking water analysis

The validated method was applied for analysis of treated and untreated drinking water. The results are presented in Table 9and a chromatogram on Figure 2. The obtained concentrations are far below the recommended chemical and indicator values in drinking water [5].

Table 9. Results from analysis of drinking water by validated IC method (n=3, P=95%).

	Concentration (mg/L)		
	Recommended values [5]	Drinking untreated water	Drinking water after chlorination
F^-	1.5	0.037 ± 0.004	0.037 ± 0.004
Cl^-	250	2.33 ± 0.04	2.80 ± 0.04
NO_3^-	50	1.57 ± 0.08	1.53 ± 0.08
SO_4^{2-}	250	6.05 ± 0.03	6.06 ± 0.03

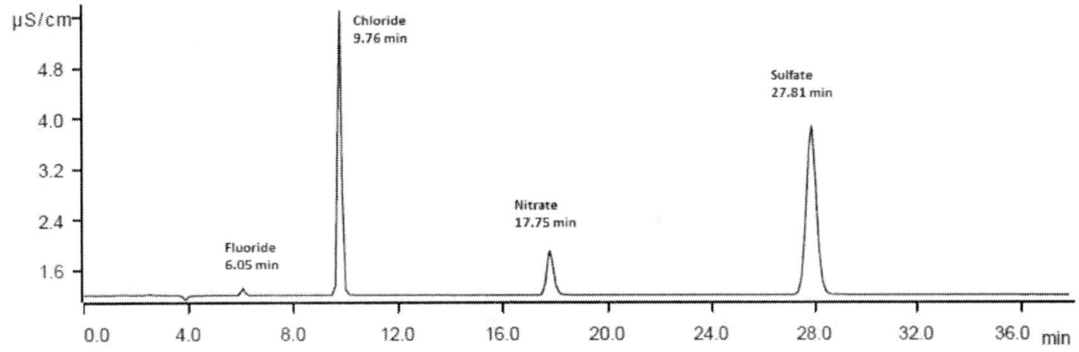

Figure 2. Chromatogram of untreated drinking water. The operating parameters are presented in Table 1.

4. Conclusions

An ion chromatographic method was validated for determination of fluoride, chloride, nitrate and sulphate ions in drinking water before and after chlorination. The method characteristics were evaluated, the trueness was confirmed by analyzing certified reference material and the uncertainty budget was built up. The obtained values for method characteristics satisfied the requirements of water authorities. The validated method was accepted in the routine work of the Department of Chemical Measurements and Ionazing Radiation ofthe Bulgarian Institute of Metrology, Bulgaria.

5. References

[1] Gilchrist E, Healy D, Morris V and Glennon J 2016 *Anal. Chim. Acta* **942** 12-22
[2] Lopez-Ruiz B 2000 *J. Chrom. A* 881 607–627
[3] Uzhel A, Zatirakha A, Shchukina O, Smolenkov A and Shpigun O 2016 *J. Chrom. A* **1470** 97–103
[4] Linda R, Lim L and Takeuchi T 2013 *J. Chrom. A* **1294** 117– 121
[5] European Commision 2014 *Off. J. Eur. Commun.* **2014** L221, 1-101.
[6] European Commission 2002 *Off. J. Eur. Commun.* **L221** 8-36
[7] Živojinović D and Rajaković L 2011 *Desalination* **275** 17–25
[8] Cengiz M and Bilgin A 2016 *J. Pharm. Biomed. Analysis* **120** 190–197
[9] Verrey D, Louyer M, Thomas O and Baurès E 2013 *Microchemical J.* **110** 608–613
[10] Abudiak H, Robinson C, Duggal M, Strafford S and Toumba K 2012 *J. Dentistry* **40** 632-638
[11] Bajenaru I, Minca I, Guran C, Josceanu A and Bacalum F 2013 *Rev. Chim.(Bucharest)* **64** 1229-13
[12] Lopez-Moreno C, Viera I and Urbano A 2010 *Desalination* **26** 111–116
[13] Krata A, Kontozova-Deutsch V, Bencs L, Deutsch F and Van Grieken R 2009 *Talanta* **79** 16–21
[14] Eurachem Guide: *The Fitness for Purpose of Analytical Methods* - A Laboratory Guide to Method Validation and Related Topics, Magnusson B and Örnemark U (eds.) 2014 2nd ed.
[15] Thompson M, Ellison S and Wood R 2002 *Pure Appl. Chem.* **74** 835–855
[16] Kruve A, Rebane R, Kipper K, Oldekop M, Evard H, Herodes K, Ravio P and Leito I 2015 *Anal. Chim. Acta* **870** 29-44
[17] Kruve A, Rebane R, Kipper K, Oldekop M, Evard H, Herodes K, Ravio P and Leito I 2015 *Anal. Chim. Acta* **870** 8-28
[18] Miskaki P, Lytras E, Kousouris L and Tzoumerkas P 2007 *Desalination* **213** 182–188
[19] Eurachem 2011, *Selection, Use and Interpretation of Proficiency testing (PT) Schemes*, 2nd Ed., 30

[20] EURACHEM/CITAC Guide CG4, 2012, *Quantifying Uncertainty in Analytical Measurement*, 3rd Ed., 74-80 and 115-116

Acknowledgments
This study has been financially supported by the Operational Programme "Science and education for smart growth" 2014-2020 of the European Union cofounded by the European Social Fund through the project BG05M2OP001-2.009-0015 “Support for the development of capacity of doctoral students and young researchers in the field of engineering, natural and mathematical sciences.

IOP Conf. Series: Materials Science and Engineering **374** (2018) 012054 doi:10.1088/1757-899X/374/1/012054

The Taguchi Method Application to Improve the Quality of a Sustainable Process

A M Titu[1*], A V Sandu[2], A B Pop[3], S Titu[4] and T C Ciungu[5]
[1] Lucian Blaga University of Sibiu, 10, Victoriei Street, Sibiu, România
[2] Gheorghe Asachi Technical University, 67, Dimitrie Manjeron, Iasi, România
[3] SC TEHNOCAD SA, 72, Vasile Alecsandri Street, Baia Mare, România
[4] The Oncology Institute "Prof. Dr. Ion Chiricuță", 34-36, Republicii Street, Cluj Napoca, România
[5] SC Construcții SA, 51, Morilor Street, Sibiu, România

E-mail: mihail.titu@ulbsibiu.ro

Abstract. Taguchi's method has always been a method used to improve the quality of the analyzed processes and products. This research shows an unusual situation, namely the modeling of some parameters, considered technical parameters, in a process that is wanted to be durable by improving the quality process and by ensuring quality using an experimental research method. Modern experimental techniques can be applied in any field and this study reflects the benefits of interacting between the agriculture sustainability principles and the Taguchi's Method application. The experimental method used in this practical study consists of combining engineering techniques with experimental statistical modeling to achieve rapid improvement of quality costs, in fact seeking optimization at the level of existing processes and the main technical parameters. The paper is actually a purely technical research that promotes a technical experiment using the Taguchi method, considered to be an effective method since it allows for rapid achievement of 70 to 90% of the desired optimization of the technical parameters. The missing 10 to 30 percent can be obtained with one or two complementary experiments, limited to 2 to 4 technical parameters that are considered to be the most influential. Applying the Taguchi's Method in the technique and not only, allowed the simultaneous study in the same experiment of the influence factors considered to be the most important in different combinations and, at the same time, determining each factor contribution.

1. Introduction

By analyzing the evolution of agricultural production during the last years, it has been observed that traditional agriculture, still used on a large scale by Romanian farmers, is currently unsatisfactory and insufficient. The production is met at minimum quotas, sometimes even lacks due to drought and low precipitations, the lack of crop irrigation systems or insufficient ones, the human effort made for sustaining the harvest is immense and the quality of the finite product is poor, the soil is degraded, the fertility is low and there are significant economic losses [1].

Taking into consideration the effects of traditional agriculture presented before, is considered the fact that it is the time of applying the sustainable agriculture concept, that represents the future and can be defined by long-term action aimed at overcoming the problems and constraints confronting conventional agriculture and even society in general in order to ensure economic viability, good environmental status and acceptance of the alternative farming system [1].

Agricultural experiences are all the more necessary as agriculture is more developed and the more numerous and more varied the factors are, because the harmonic intertwining of the factors action produces the maximum productions. Farming experiences are of permanent importance, as new plants varieties and hybrids are developed annually and their production capacity needs to be studied under different technological and environmental conditions. Modern experimental technique allows the simultaneous study, in the same experience of several factors in different combinations and the precise determination of each contribution.

Global corn production is ranked third and this is a basic food source for most people as well as for domestic animals [2]. Respecting clear agronomic principles and the proper crop irrigation, it has been noticed that good harvest can be achieved, but the varied and numerous factors influencing a maize crop are preventing it from reaching maximum gains. There is a close interdependence between the development of research and the progress of agricultural practice, as production reflects the level of research.

In this paper, an experimental agricultural research has been carried out, i.e. the optimization of the main parameters directly influencing a corn crop in an irrigated system in the Baragan Plain, in order to have the desired production and profit.

Modern experimental technique can be applied in any field [3] and this case study reflects the benefits of the interaction between the principles of sustainable agriculture and the application of experimental research methods, respectively the goals are achieved in a very short time, the working method is fast, simple and does not entail costs, and economic losses are non-existent, thus providing an efficient and lasting agricultural management system.

2. Research method

2.1. Experimental methods used in research

As the success of an experiment depends essentially on its preparation, the experimental method used in this case study is the Taguchi Method.

The Taguchi method is effective because it allows 70-90% of the desired optimization of parameters to be reached quickly, the missing 10-30% can be obtained with one or two complementary experiments, limited to 2-4 parameters considered to be those more influential [4, 5].

The experimental method used in this case study is the Taguchi Method, which consists of combining engineering and statistical technologies to achieve rapid improvement in quality costs, seeking optimization at existing processes and key parameters [6].

An experience plan is a series of attempts to determine with a minimal number of attempts and with maximum precision the influences of different parameters in order to optimize the performance of a process or system. Taguchi experience plans use the average and variability of measured values. They use the Signal / Noise ratios, which simultaneously account for: the desired value (signal) to be reached and the undesired variable of this counter-noise value (3) as performance indicators [7]. The use of this performance indicator makes it possible to find "the first shot" to combine the levels of controlled factors that prove to be the least sensitive to noise factors.

2.2. Research objectives defining

In the present case study, two major objectives were sought. The first was to upgrade a 50 ha agricultural land in the Baragan Plain to meet the requirements of agricultural good practice to ensure the need for corn crops and to build the concept of sustainable agriculture. Given that natural resources are found to be ever more limited, quantitative and qualitative, it is only natural to have perspectives that sustain and develop this sector, which is the insurer of the basic needs of human existence.

This first objective has provided with the culture from an agricultural point of view, ie it will result a medium production. In order to have a production close to the maximum value, the main factors influencing a maize harvest must be brought to an optimal value. Traditional methods commonly used to optimize factors that influence a maize harvest involve soil sampling and practical experiments on it, pedological, chemical, bacteriological, biological research, all of which add up to high costs, increased human effort but also a long time in reaching objectives, for years.

That is why this optimization was carried out by an experimental method, using the Taguchi method, which allowed the simultaneous study of the most important factors in different combinations and, at the same time, the determination of the contribution of each factor [8]. This process was the second objective of this paper, and more precisely, the purpose of applying the experimental method means a profit of 500 € / ha and a production of 14-15 t / ha of corn grain.

The most important factors directly affecting a maize harvest are the plant density due to the fact that it has been shown that greater density is inevitably linked to increased production [9, 10], the watering standard must be applied correctly to ensure the water demand and the distance between rows is also the key to big productions. The amount of fertilizer and the amount of herbicide applied are two factors that if left unchecked or applied correctly, can pollute the aquifer, the hydrological networks in the area and inevitably reach the body of animals.

3. Research method planning

In this scientific paper, it has been decided to optimize the main factors directly affecting a maize harvest on a 50ha plot of the Baragan Plain in order to achieve the desired goals, a profit of 500 € / ha and a production of 14-15 t / ha corn grain.

The most important 5 controllable factors were identified, according to Table 1, the numbering being from A to E as well as the decision to adopt 2 levels.

Table 1. Controllable factors.

No	Controlled factor	Level 1	Level 2
A	Density of plants	65.000 pl./ha	75.000 pl./ha
B	The watering standard	155 mm/mp	175 mm/mp
C	Distance between rows	65 cm	70 cm
D	The amount of fertilizer applied	500 kg/ha	700 kg/ha
E	The amount of applied herbicides	6 l/ha	8 l/ha

A complete factorial experience plan has been adopted, resulting in 25 attempts because there are 5 factors at 2 levels.

The superiority of the Taguchi method in relation to the classic method of experience plans results from the use of the performance indicator: Signal / Noise ratio. The signal noise coefficient is an indicator of variability, its use leads to obtaining of products that are robust enough to be able to function / develop accordingly to the conditions of the variations to which they are subjected.

One of the objectives of this study is to achieve a target profit of 500 € / ha, and in Table 2 the value of the Signal / Noise ratio for this measured feature was calculated.

Expression of the Signal / Noise (dB) ratio for a commonly used target criterion is: S / N = 10 log [(y2 / s2) - 1 / n] (dB) where:

y = arithmetic mean of measured values;

s = standard deviation of measured values;

n = the number of measurements made.

The higher the S / N algebraic value, the lesser the loss will be, resulting a better performance of the product or of the process to be optimized.

Regarding the corn production, the criteria that should be maximized at 14-15 t / ha value, in Table 3 the value of the Signal / Noise ratio was calculated using the following formula: S / N = - 10 log [(1 / y2) (1 + 3 * (s2 / y2))] (dB), where:

y = arithmetic average of measured values;

s = standard deviation of measured values;
n = the number of measurements made.

Table 2. The value of the Signal / Noise ratio.

No. tries	CONTROL FACTORS					Measured values			Average	Standard Deviation	S/N dB
	A	B	C	D	E	Nr. 1	Nr. 2	Nr. 3			
1	1	1	1	1	1	313	326	329	322.67	6.94	5.02
2	1	1	1	1	2	320	337	319	325.33	8.26	5.02
3	1	1	2	2	1	330	345	328	334.33	7.59	5.05
4	1	2	1	1	1	323	337	319	326.33	7.72	5.03
5	1	1	1	2	1	346	324	338	336.00	9.09	5.05
6	2	1	1	1	1	319	325	343	329.00	10.20	5.03
7	1	2	2	1	1	346	369	357	357.33	9.39	5.11
8	2	1	2	1	1	386	362	379	375.67	10.08	5.15
9	1	2	2	1	2	410	427	399	412.00	11.52	5.23
10	2	1	1	2	1	386	416	420	407.33	15.17	5.22
11	1	2	2	1	2	437	421	440	432.67	8.34	5.27
12	2	1	1	2	1	428	400	432	420.00	14.24	5.25
13	1	2	2	1	2	448	468	451	455.67	8.81	5.32
14	2	2	1	2	2	478	489	446	471.00	18.24	5.35
15	2	1	2	1	2	462	449	486	465.67	15.33	5.34
16	2	2	2	2	1	492	475	499	488.67	10.08	5.38
17	1	2	1	2	2	501	480	496	492.33	8.96	5.38
18	2	2	2	2	1	503	486	499	496.00	7.26	5.39
19	2	1	2	1	1	507	511	498	505.33	6.66	5.41
20	2	2	1	1	2	489	493	513	498.33	10.50	5.40
21	1	2	2	2	2	519	501	500	506.67	8.73	5.41
22	1	2	2	1	2	521	517	511	516.33	4.11	5.43
23	2	1	2	2	2	531	528	503	520.67	12.55	5.43
24	2	2	1	2	2	519	526	542	529.00	9.63	5.45
25	2	2	2	2	2	548	522	515	528.33	14.20	5.45
			Overall answer rate						434.11		5.26

Table 3. The value of the Signal / Noise ratio.

No. tries	CONTROLED FACTORS					Measured values			Average	Standard Deviation	S/N dB
	A	B	C	D	E	Nr. 1	Nr. 2	Nr. 3			
1	1	1	1	1	1	10	10.4	10.1	10.17	0.17	-0.04
2	1	1	1	1	2	10.4	10.8	10.2	10.47	0.25	-0.07
3	1	1	2	2	1	10.5	11	10.3	10.60	0.29	-0.10
4	1	2	1	1	1	10	10.7	11.2	10.63	0.49	-0.24
5	1	1	1	2	1	10.3	11.1	10.7	10.70	0.33	-0.12
6	2	1	1	1	1	10.9	11.3	11.7	11.30	0.33	-0.12
7	1	2	2	1	1	11.1	11.8	10.9	11.27	0.39	-0.16
8	2	1	2	1	1	11.7	12	11.2	11.63	0.33	-0.12
9	1	2	2	1	2	11.3	11.8	12	11.70	0.29	-0.10
10	2	1	1	2	1	11.7	12.4	12.1	12.07	0.29	-0.10

Table 3. The value of the Signal / Noise ratio.

No. tries	CONTROLED FACTORS					Measured values			Average	Standard Deviation	S/N dB
	A	B	C	D	E	Nr. 1	Nr. 2	Nr. 3			
11	1	2	2	1	2	12.5	12.1	11.9	12.17	0.25	-0.07
12	2	1	1	2	1	12.4	12.8	11.9	12.37	0.37	-0.15
13	1	2	2	1	2	12.7	12.5	13.2	12.80	0.29	-0.10
14	2	2	1	2	2	13.9	14	12.8	13.57	0.54	-0.28
15	2	1	2	1	2	13.7	12.9	13	13.20	0.36	-0.14
16	2	2	2	2	1	14.1	13.6	13.4	13.70	0.29	-0.10
17	1	2	1	2	2	13.5	13.3	14.3	13.70	0.43	-0.19
18	2	2	2	2	1	14.2	13.7	13.9	13.93	0.21	-0.05
19	2	1	2	1	1	14	13.4	13.9	13.77	0.26	-0.08
20	2	2	1	1	2	13.8	13.5	14.1	13.80	0.24	-0.07
21	1	2	2	2	2	14.2	14.7	13.9	14.27	0.33	-0.12
22	1	2	2	1	2	14.8	14.4	14.7	14.63	0.17	-0.04
23	2	1	2	2	2	14.4	14.8	14.9	14.70	0.22	-0.06
24	2	2	1	2	2	14.7	14.8	15	14.83	0.12	-0.02
25	2	2	2	2	2	15	14.8	15	14.93	0.09	-0.01
Overall answer rate									12.68		-0.11

To evaluate the average effect of each factor on the noise coefficient, calculate the arithmetic average of all experiments corresponding to the same level, values that are found in Table 4 for the measured profit characteristic and in Table 5 for corn harvest.

Table 4. Profit.

S / N ratio effect		No. factor	Effect on measured value	
Level 1	Level 2		Level 1	Level 2
-0.07	0.06	A	-32.63	30.12
-0.08	0.07	B	-39.38	30.94
-0.06	0.05	C	-28.89	22.70
-0.05	0.06	D	-24.70	26.75
-0.09	0.08	E	-42.55	39.28

Table 5. Corn harvest.

S / N ratio effect		No. factor	Effect on measured value	
Level 1	Level 2		Level 1	Level 2
-0.01	0.01	A	-0.75	0.69
0.01	-0.01	B	-0.77	0.60
-0.02	0.02	C	-0.53	0.42
0.00	0.00	D	-0.56	0.60
-0.01	0.01	E	-0.83	0.77

In order to choose the optimal variant of the studied factors, it was chosen for each one the level which, for each factor, optimizes this criterion and the corresponding S / N ratio.

Table 6. The configurations.

Controlled factor		PROFIT	CORN HARVEST
Plant Density	Optimal level	2	2
	Contribution	0.06	0.01
Watering standard	Optimal level	2	1
	Contribution	0.07	0.01
Distance between rows	Optimal level	2	2
	Contribution	0.05	0.02
The amount of applied fertilizer	Optimal level	2	2
	Contribution	0.06	0.00
The amount of applied herbicides	Optimal level	2	2
	Contribution	0.08	0.01

Analyzing the obtained results by Table 6, the configuration that ensures the achievement of goals is: **A2B2C2D2E2**.

For the presented case study, the following working model was recommended:
Plant Density - 75,000 plants / ha;
Watering standard - 175 mm / square meter;
Distance between rows - 75 cm;
Amount of fertilizer applied - 700 kg / ha;
The amount of applied herbicides - 8 l / ha.

4. Conclusions

The importance of agriculture to humans has been known since ancient times and continues to remain a vital field for humans in nowadays society, a source of food and raw material [11, 12]. The problems encountered in practicing agriculture can be overcome by respecting good agricultural practices, irrigating the right culture and implementing the principles of sustainable agriculture. It describes a system of production capable of maintaining productivity, used by society at infinity and requires the recycling of resources, protecting the environment and being commercially competitive.

However, this system is not fully economically satisfactory for farmers because production and profit do not reach the maximum threshold. The factors that influence a corn crop are numerous and varied, but optimizing the most important factors is the key to rich production and a profitable gain. Using a method of experimental research, the Taguchi Method, it was possible to simultaneously study the five factors in different combinations and it was possible to determine the contribution of each factor. The objective of this optimization process, 14-15 t / ha of corn and a profit of 500 € / ha, was achieved in accordance with the resulting configuration, ie all factors will be at level 2.

In this scientific paper we can see the beneficial effect of the experimental methods in the agricultural field. Traditional methods commonly used to optimize factors that influence a maize harvest involve soil sampling and practical experiments on it, pedological, chemical, bacteriological, biological research, all of which add up to high costs, increased human effort but also a long time in reaching objectives, expanding for years [1].

Noteworthy is that level 2 means a higher profit of 200 € / ha and about 4 t / ha of extra corn grain, and the process has had insignificant costs and has reached notable results in a short amount of time. Experimental techniques can be applied in any field, and agriculture, due to its development, requires as many experimental research as possible to provide the resources needed for future generations. What we are looking for today, we'll be picking tomorrow!!!

5. References

[1] Constantin E 2006 *Dezvoltare rurala* Editura Bren Bucureşti 973-648-591-9

[2] *http://www.agrifarming.in/maize-farming/ 22.01.201*

[3] Tîţu M, Oprean C and Tomuţă I 2007 *Cercetarea experimentală şi prelucrarea datelor" – studii de caz* Editura ULBS

[4] Wu C F and Hamada J 2002 *Experiments: Planning, Analysis, and Parameter Design Optimization* Wiley

[5] Ghosh S, Rao C R 1996 *Design and Analysis of Experiments. Handbook of Statistics* 13 North-Holland ISBN 0-444-82061-2

[6] Oprean C and Ţîţu M 2007 *Cercetarea experimentală şi prelucrarea datelor Partea a II-a*, Editura Universităţii „Lucian Blaga" din Sibiu

[7] Ţîţu M and Oprean C 2006 *Cercetarea experimentală şi prelucrarea datelor Partea I* Editura Universităţii „Lucian Blaga" din Sibiu

[8] Montgomery D 2013 *Design and Analysis of Experiments* Eighth Edition ed. Hoboken: John Wiley & Sons, Inc.

[9] Jugenheimer R W 1976 *Corn improvement seed production and uses* John Wiley & Sons Inc, 278-345

[10] Moen R D, Nolan T W and Provost L P 1991 *Improving Quality Through Planned Experimentation* ISBN 0-07-042673-2

[11] *http://www.environmentalscience.org/history-agriculture*

[12] *Improvements in modern corn and soybean production*, Demonstration report. Learning Center at Monmouth Ilinois 2012

Two Dimensional (2D) P-Aramid Dry Multi-Layered Woven Fabrics Deformational Behaviour for Technical Applications

M A Abtew[1,2,3,4,*], C Loghin[1], I Cristian[1], F Boussu[2,3], P Bruniaux[2,3] , Y Chen[4] and L Wang[4]
[1]Faculty of Textiles, Leather and Industrial Management, Gheorghe Asachi Technical University of Iasi, Iasi, Romania
[2]University of Lille 1 Nord de France, Lille, France
[3]Ecole Nationale Supèrieure des Arts et Industries Textiles, GEMTEX, Roubaix, France
[4]Soochow University, College of Textile and Clothing Engineering, Suzhou, China

E-mail: mulat-alubel.abtew@ensait.fr

Abstract. In today's scenario for the various technical applications, from composites to body armour, the material mouldability along with its mechanical property become very important. In the present study, two dimensional (2D) woven fabrics made of para-aramid high performance fibres in multi-layer dry structure were used for investigating different forming characteristics. The different layers were arranged with 0°/90° orientation for deep drawing formability test to analyse the effect of number of layers and blank-holder pressure (BHP) during the test. Specific preforming device with low speed forming process and predefined hemispherical shape of punch has been applied. Using fine photographic analysis, some important 2D multi-layer fabrics forming characteristics i.e., material drawing-in, surface shear angle etc. from the imposed deformation have been observed, measured and analysed for better understanding and co MPa rison. The result revealed that the mouldability behaviour of the multi-layered dry textile fabric preforms is directional, and closely dependent on blank-holding pressure and number of layers. This indicates both parameters should be carefully considered while material deformation to avoid the formation of wrinkling and maintain other mechanical properties on final application.

1. Introduction

Different stacked layers of two dimensional high performance fabrics are widely used for a considerable amount of applications [1]. Besides using flat 2D fabrics especially for conventional products like apparels, some technical applications such as protective clothing, i.e. military helmet, female body armour, and textile reinforcement and composite for aeronautical, naval and automotive applications need to mould the fabric into 3D shapes. In order to achieve this particular 3D shapes, the formability property of the specific textile material plays a great role. "Mouldability", is generally defined as the performance of a planar textile structure to be directly deformed at macro scale by changing the angle between two yarns systems at crossover points, to fit three-dimensional components without the formation of wrinkling's, kinks or tears [2-4]. In today's more complicated and advanced manufacturing process, using proper mouldability, cutting of planar textile fabrics to fit three-dimensional shapes could be eliminated. This will not only increase the productivity of the material fabrication, but also enhance the mechanical properties of the specific parts due to the maintained integrity of the material. Many researchers have tried to carry out various experimental, numerical and theoretical investigations to understand the mouldability behaviour of the material using different methods. For example, the mouldability property and its parameter analyses of different 2D [5-9] and 3D [8] textile reinforced laminated composite have been conducted and quantified by many researchers. Apart from textile reinforced laminated composite, dry multi-layer woven fabrics are also increasingly employed in the field of many technical applications. For example, in soft body armour design, different kinds of multi-layer high performance woven fabrics are stacked together with different methods to give a good ballistic protection. However, the material should have not only good mechanical performance but also flexibility. Meanwhile, only few research works have been published related to mouldability and its analysis related to dry multilayer textile structures [10]. Even the situation becomes more complex when it is required as to designed curved shapes i.e., female body armour for accommodating the bust area. In order to accommodate different curvy parts of females including the bust area [11] different methods such as traditional cut-and-sew with darts, stretch folding and fabric folding have been used so far. Developing and assembling of different parts of panels using those methods with seams and stitching could affect the ballistic performance, well-being and comfort performance of the final garment. To overcome those problems, different researchers also mention to use good mouldability material, new design process and latest technological solutions [12-15]. In general the mouldability of woven fabrics greatly depends on fabric shear rigidity, which mainly comes from abrasion between the warp and weft yarns at the cross-over points. Besides shear rigidity, significant number of study also urged different factors should be considered while mouldability processes. One of the research work [16] publicized that thick and dense multilayer fabric has a uniform surface as co MPa red to low density and thin fabric, whereas tow-aspect ratio of external layers was found greater than the tow-aspect ratio at internal layers for moulded angle interlock fabric. Another study has also designed and manufactured low shear rigid angle interlock fabrics by considering various structural parameters, i.e. warp density, weft density and weft layers in order to assess the parameters effect on their mouldability behaviour [17]. Another study [18] also tried to investigate the effects of the blank holder force (BHF) on the mouldability behaviour of non-crimp fabrics with chain stitches using varying blank holder force from 0.5 to 61.4 kgf.

The current research work is to investigate the effect of blank-holding pressure (BHP) and numbers of layers on the moulding behaviour of 2D aramid weave multi-layered panels. Using a slow process forming process, this investigation tried to observe and measure to understand various important forming parameters i.e. Force-deformation value, surface shear angles, material drawing-in etc. of the different developed sample specimens.

2. Material and Experimental Procedures

2.1. Materials

2.1.1. Sample fabrics. So far many technical products including soft body armour have been designed and developed by using high performance 2D woven fabric. For this particular research work, Twaron CT-709 2D plain weave fabric made with linear density of 930dtex and yarn density 105 yarns/10cm for both warp and weft yarns. The areal density of the fabric is 200 g/m^2. The properties of the fabrics are presented in Table 1.

Table 1. The mechanical properties of the material.

Fibre type	Linear density, Dtex	Weight, g/m2	Fabric density		Tenacity at break, mN/tex	Strength at Break, N	Elongation at fracture, %
			Warp/ 10cm	weft/ 10cm			
Aramid	930	200	105	105	2.35	225	3.45

The four different sample panels were composed of one (01), five (05), ten (10) and fifteen (15) fabric layers. Each sample panels were also prepared for testing three different blank-holder pressures.

Fabric layers were arranged one over another without using any kind of stacking methods, but griped at the edges using tapes to stabilize the layer and protect the edge from yarn frying while performing as showing in Figure 1.

2.1.2. Sample preparation. Dry multilayer panel samples made of Twaron CT-709 2D plain woven fabric were developed with 0°/90° orientation in hybrid forms as presented in Table 2.

Table 2. Sample preparation, designation and applied blank-holder pressure (BHP).

No.	Sample designation	Fabric type	No. of layer used	Applied blank-holder pressure (MPa)
1	2DWF-L-01	2D plain weave	01	0.1
				0.2
				0.3
2	2DWF-L-05	2D plain weave	05	0.1
				0.2
				0.3
3	2DWF-L-10	2D plain weave	10	0.1
				0.2
				0.3
4	2DWF-L-15	2D plain weave	15	0.1
				0.2
				0.3

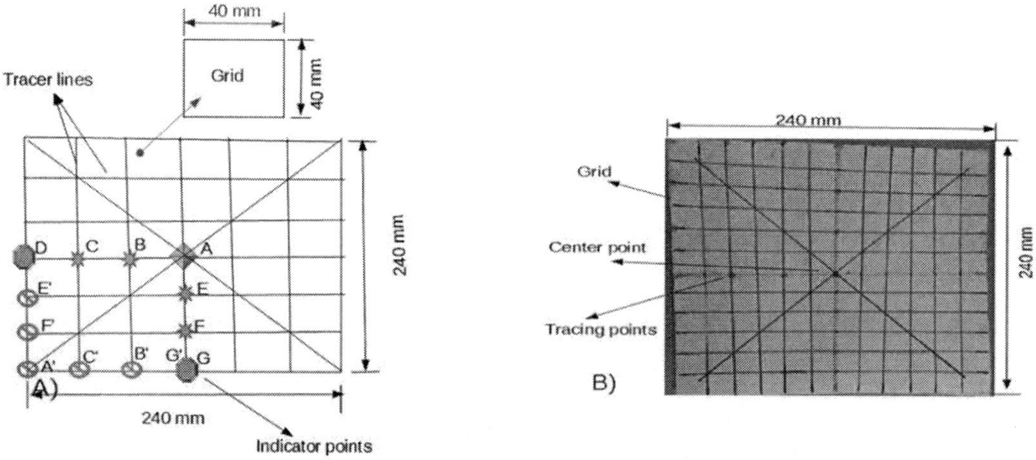

Figure 1. (a) Model and (b) Photographic views of sample preparations for performing.

The sample fabric were cut and prepared in 240×240 mm^2 surface dimensions. As it is indicated also in Figure 1, the different guide lines with coloured pens, known as "tracers", and different points "indicators" has been illustrated on warp and weft directions of the preform surface. Basically this grid structure will assist for easy follow up, measure, understand and analyse the different set of measurements and mechanical forming characteristics, i.e. material drawing-in, material deformation recovery, thickness of the fabrics in different positions, surface shear angles etc. of the perform while and after forming process.

2.1.3. Experimental equipment and procedures. Experimental equipment set up. The formability testing machine, based in GEMTEX laboratory, is used for conducting this research. The machine is a modified pneumatic based testing machine adapted to a fast, safe and ambient temperature stamping process. The forming using a pre-defined hemispherical punch is performed on a specific formability device.

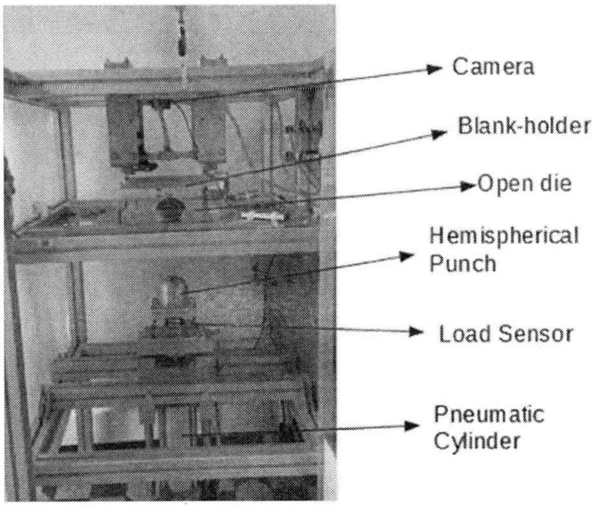

Figure 2. Modified pneumatic based formability testing machine [2].

A device specifically designed to perform the forming and used to observe the formability properties while forming process of different sample is presented on Figure 2. This particular forming bench is composed of both a static blank holder system and an open die which distribute pressure provided by four jacks to the edges of the preform. A non-heating punch gives the desired shape through a vertical and controlled motion driven by a pneumatic jack. The die can be opened to access direct observation of the different characteristics during the forming process. The two most important parameters which have to be considered during forming process are the blank holder pressure and the velocity of the punch. While punching process, the sample should appropriately mount on the proper sample holding position. Moreover, the pressure applied on the blank holder must be properly set in all direction of the sample and sufficient enough to maintain the good preform. Too high or too low blank holder pressure may result sample folds and yarn breakage respectively. Meanwhile the position of the punch which is found below the holder is controlled by a position sensor and equipped with a stress sensor to measure forces applied by the punch to the performing material. Besides its motion is controlled by using a piloted four jacks mounted below the punch. This device has been also developed to adapt and preform different shape and sizes of punch and die.

Beside direct observation, a camera, located on the top of the forming bench, also helps to observe and capture the forming behaviour of the sample during the forming process. Even though various shapes of punch are available for analysing the different forming characteristics of the material, we have preferred to use a 30 mm diameter semi hemispherical shaped punch which ensures a symmetric double curvature deformation during the forming process. This particular hemispherical shape punch has been mostly used to observe the local and global deformations of the different preformed samples by following the different path of yarns in the structure both in the warp and weft directions.

Beside direct observation, a camera, located on the top of the forming bench, also helps to observe and capture the forming behaviour of the sample during the forming process. Even though various shapes of punch are available for analysing the different forming characteristics of the material, we have preferred to use a 30 mm diameter semi hemispherical shaped punch which ensures a symmetric double curvature deformation during the forming process. This particular hemispherical shape punch has been mostly used to observe the local and global deformations of the different preformed samples by following the different path of yarns in the structure both in the warp and weft directions.

2.1.4. Experimental Procedures. While testing, the different 2D layers of sample woven fabrics turn by turn are placed between the blank-holder and an open die. While adjusting the sample on the bottom holder, on the two sides the light will apply on the surface for finding the proper arrangements and position of the perform material. Later the four jacks, governed by pneumatically and found at the top part of the machine which will be connecting to the blank holder, will apply a sufficient adjustable pressure on the fabric. In the forming process, different important forming parameters such as material draw-in, the in-plane shear, the inter-layer sliding, help to measure the material behaviour.

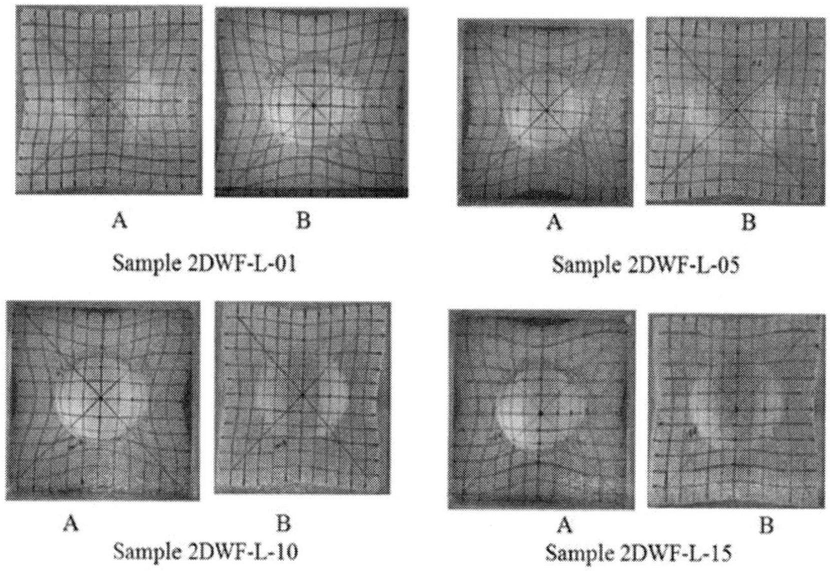

Figure 3. Some deformed shape of samples at 0.2 MPa (a) While forming, (b) After forming.

Another electric jack linked to the punch imposes a movement (punch displacement) and a load sensor machine (in different units) will acquire the punch force while the forming process. In punching of the different samples of the 2D woven sample fabrics as shown in Figure 3, the initial corresponding punch displacement for all sample were 45 mm and applied pressures on the blank-holder were 0.1, 0.2 and 0.3 MPa . In Figure 4, for a given pressure of 0.2 MPa , different samples of 2D fabrics with respectively 1, 5, 10 and 15 layers have been observed during and after the forming process. In order to improve the understanding of the given sample and predict the suitable material for double-curved shape forming, the investigation of stamping behaviour of each sample will be performed by analysing the different characterizations of in-plane, out of plane and through-the-thickness of the formed samples.

3. Results and discussion

3.1. Evolution of the punching force with the stamping time

While forming process the force applied by the installed punch was governed by a load sensor of 500 N (±0.3 %) which is directly connected with the punch. The punching force has been slowly applied in the bottom to the forming material with the help of pneumatic methods. The evolution of the punching force in relation to the stamping time for the different samples in specified holding pressure by the blank holder has been recorded. The corresponding average forming force versus forming duration relationships for the different samples is illustrated in Figure 4. As it can be seen from the figures all the curves of the preform samples at the different holding pressure show more or less a similar trend

IOP Conf. Series: Materials Science and Engineering **374** (2018) 012055 doi:10.1088/1757-899X/374/1/012055

for all samples. At the beginning, the force shows an increase while deformation until it reaches the maximum displacement. Later, a constant force of punch has applied to maintain the final shape of the deformed sample. However, it is clearly shown that as the numbers of layers in the sample become more, it needs high punching force in order to achieve the maximum displacement. Moreover, it is evidently revealed that for the given sample panels, as the capacity of blank-holder pressure increases, the force required to achieve the maximum deformation for the specific number of layers has been increased. This is due to the punching force applied to reach its maximum displacement has been shared with the corresponding number of layers. Generally we can conclude that, the punching force is directly related with both the number of layers and the blank-holder pressure applied in the panels during forming.

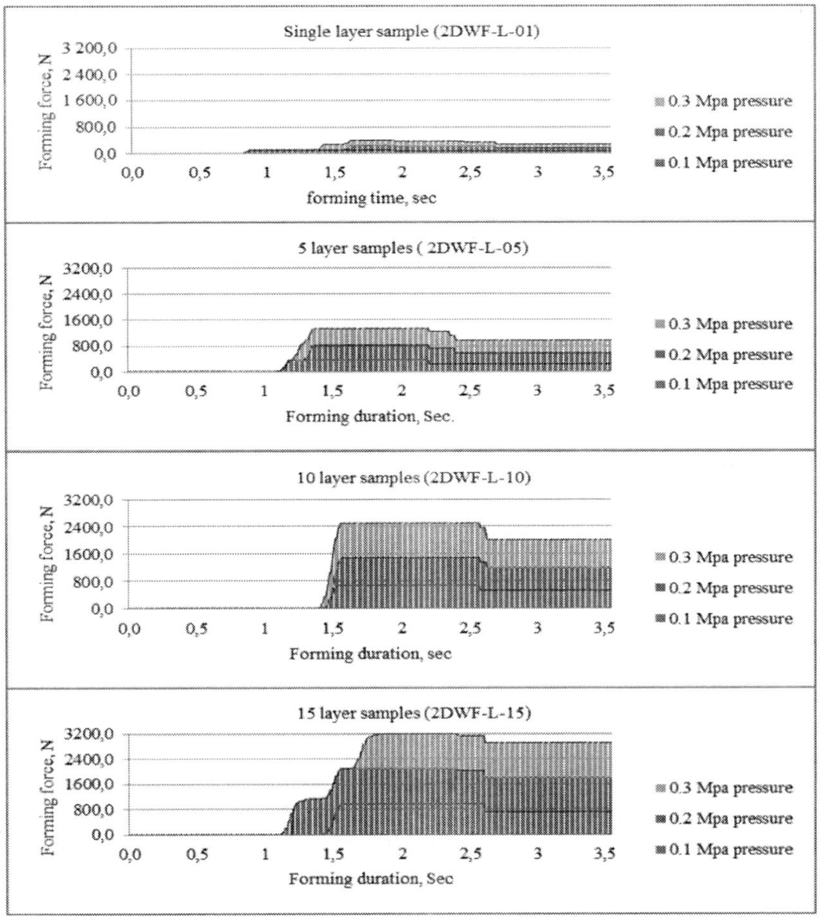

Figure 4. Punching force while forming in different blank-holder pressures (BHP) for different number of sample layers.

3.2. Different forming characteristics after forming process

3.2.1. Material drawing-in measurement in weft and warp direction. The material drawing-in values are mostly expressed in terms of the consumed length of the sample fabrics both in the warp and weft directions during the forming process. Our research has used the quarter portion of the sample region for analysing and determining the material drawing-in values of the sample at different points both in warp and weft directions. This is due to the force applied by the hemispherical punch and the formability properties are nearly uniform on the divided four quarter sample regions shown in Figure 5. The sample drawing-in values were measured using ImageJ software by taking an appropriate and

IOP Conf. Series: Materials Science and Engineering **374** (2018) 012055 doi:10.1088/1757-899X/374/1/012055

precise picture at the top view of central point and perpendicular to the surface of the preform. While taking pictures, sufficient distant from the sample has been considered in order to reduce errors due to shooting. Figure 5 indicates the schematic diagrams of the sample material drawing-in at the different points while deformation.

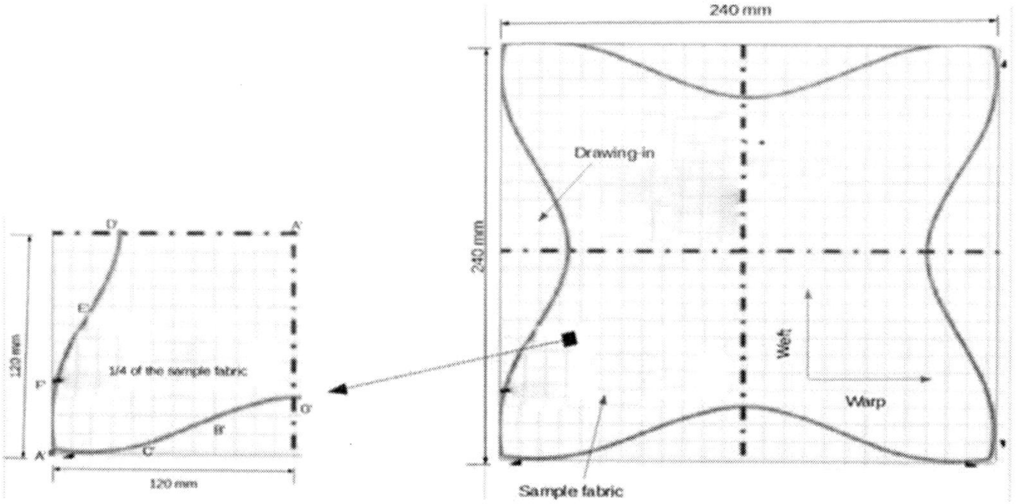

Figure 5. The schematic diagrams of material drawing-in while deformation.

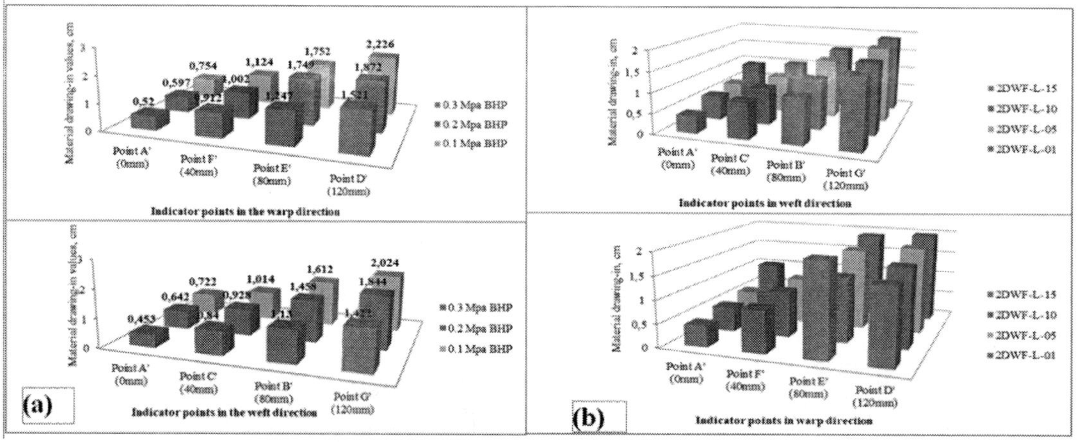

Figure 6. (a) Material drawing-in values of sample (2DWF-L-05) in warp and weft direction at different BHP (b) Material drawing-in values of different samples at various point in weft and warp direction at 0.2 MPa BHP.

In order to investigate the effects of different BHP on sample drawing-in values, sample 2DWF-L-05 were chosen. As it can be shown from Figure 6(a), the higher blank-holder pressure, the drawing-in value of the sample to be drawn has been reduced. This is mainly when the blank-holder pressure increases, the edge of the sample will be firmly held and prevented from drawing-in. However, in order to reach to the maximum deformation, the other part of the sample region which was not held by the blank-holder will be solely responsible for the deformation. Moreover, it is also noticed that the drawing-in value of the sample has been increased from edge to centre of the sample. In general it was also observed that the material drawing-in value in the warp direction is slightly higher than in weft direction, mainly due to lower crimp value of weft yarns co MPa red to warp yarns. Moreover, an

IOP Conf. Series: Materials Science and Engineering **374** (2018) 012055 doi:10.1088/1757-899X/374/1/012055

investigation has been also carried out in order to see the number of layer effects in the sample panel drawing-in values. We have selected and applied 0.2 MPa holding pressure for all the samples and measure the values at various indicator points in both weft and warp direction.

The result is illustrated in Figure 6(b), and depicts that for the specified value of blank holding pressure (BHP), the majority drawing-in value reduces as the number of layer increases. This is due to the increase of layers; the force which was applied on the sample panel fabrics by the blank holder will be distributed to the individual samples. From the measurement, the maximum and minimum drawing-in values achieved in weft direction were 2.11 & 0.476 mm, 1.89 & 0.427 mm and 1.602 & 0.382 mm at 0.1, 0.2 and 0.3 MPa BHP respectively. Whereas in warp direction 2.2 & 0.52 mm, 1.97 & 0.448mm and 1.803 & 0.39 mm at 0.1, 0.2 and 0.3 MPa BHP respectively. Most of the maximum and minimum drawing-in values were found at the centre (point D' and G') and edge corner (point A') respectively.

3.2.2. Surface shear angle measurements of the material. During forming processes of the different materials, wrinkling is one of the most common faults which often leading to unexpected failures. This common and important parameter has always a strong relationship with the shear angle of the material. In general shear angle is the angle between the warp and weft yarns. In the deformation research work it is usually captured and optically measured with the help of camera. In our particular study, due to the hemispherical punching system, one fourth of useful zones of the deformed plies as presented in Figure 7 were used to determine the shear angle measurement. Using the marked region tracking method and with the help of the dedicated ImageJ software at the measurement precision of ±0.25°, it is possible to measure and study the in-plane surface shear angle during the whole preforming stage.

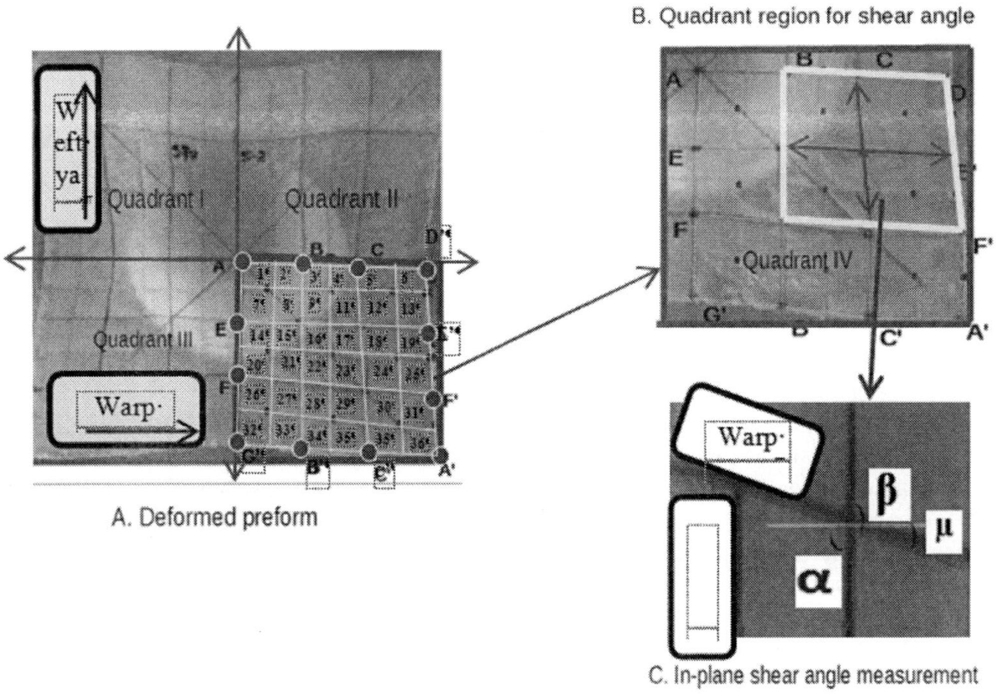

Figure 7. Surface shear angle measurement.

While forming, the shear angle is only calculated for areas where no wrinkling was found. In this particular research for analysing the effects of number of layers in the preform, all the prepared samples (01, 05, 10 and 15 layers) were investigated with the same blank-holder pressure (0.2 MPa). In general, from the measured shear angle values of each region of different layer samples, the minimum shear values has been observed on the top and corner edge of the regions (region 1, 2, 29,

30, 34, 35, 36), whereas the maximum shear values was found around the diagonal direction on the neck/edge of the deformed regions (region 8, 10, 11, 12, 15, 16, 17, 18, 20, 21). Besides, along all the measurement, the maximum shear angle was found 22° in region 20, 25° at region 20, 24° at region 16 & 21 and 34° in region 16 for 2DWF-L-01, 2DWF-L-05, 2DWF-L-10 and 2DWF-L-15 respectively. Whereas the minimum shear angle value was achieved 2° in region 34, 0° at region 1, 0° at region 1 and 34 and 1° in region 34 for 2DWF-L-01, 2DWF-L-05, 2DWF-L-10 and 2DWF-L-15 respectively.

Table 3. Measurements of surface shear angle for different samples at 0.2 MPa BHP.

No.	Surface shear angle range	Severity	Number of marked region for specified shear angle in different sample			
			2DWF-L-01	2DWF-L-05	2DWF-L-10	2DWF-L-15
1.	0° - 4°	Extremely less	14	14	13	13
2.	5° - 9°	Very Less	15	8	7	8
3.	10° - 14°	Less	3	7	6	9
4.	15° - 19°	Moderate	3	4	8	4
5.	20°- 24°	Severe	1	3	2	0
6	25 °- 29°	Very severe	0	0	0	0
7	30° - 34°	Extremely severe	0	0	0	2
Total			36	36	36	36

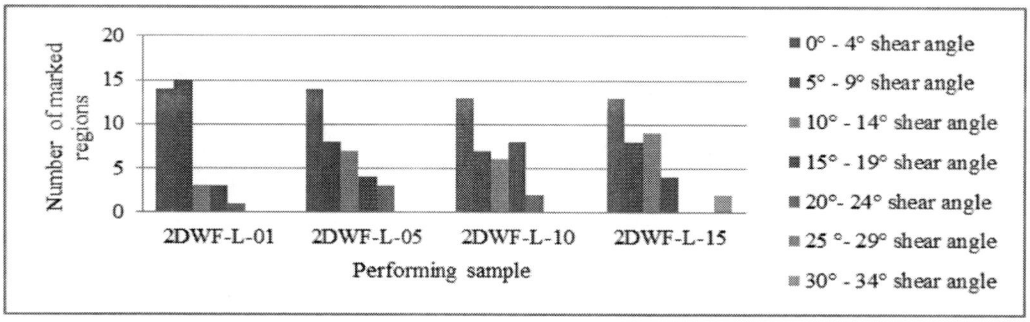

Figure 8. Surface shear angle values distributions for different number of layer samples of quarter regions at 0.2 MPa BHP.

However, in order to clearly understand the shear angle analysis of the different samples, thanks to the accuracy of the measurements, we tried to segregate the different shear angle values of the specified marked region (36 in number) into seven categories with the categories of 5° for the different samples, as recorded in Table 3.

Even though the values in the table did not reveal the shear angle of each and every region of the marked area, the analysis result shows that except some fluctuation and quite unpredicted values, in majority the relationship between the numbers of layers in the sample has a linear relation with the amount of the shear angle in the surface. As it is presented from Figure 8, the sample with small number of layer has faced less amount of surface shear angle (less severe), whereas large amount of shear angle (more severe) occurred in the high number of layer samples. Besides, in order to understand the blank-holder pressure (BHP) influences on the shear angle values, the sample with same layer number (2DWF-L-05) but different BHP (0.1 MPa, 0.2 MPa and 0.3 MPa) was considered. Figure 9 shows the effects of BHP on the given sample panels in the specified regions. The result shows clearly that the shear angle value has higher values as the blank holding pressure increase.

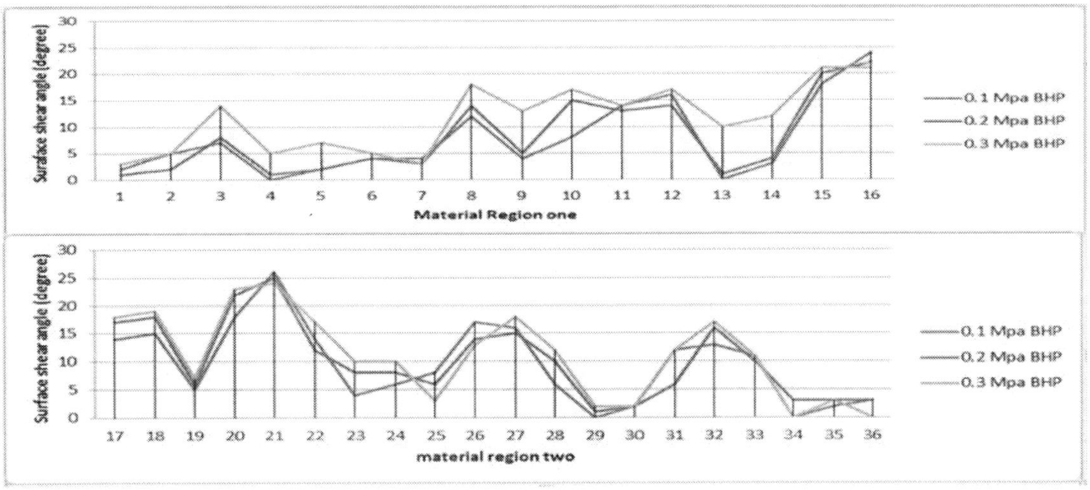

Figure 9. Surface shear angle of sample 2DWF-L-05 at different BHP.

4. Conclusions

In the present study, different layers (01, 05, 10 and 15) of stacked 2D woven aramid fabrics were investigated for their different moulding behaviour using different blank-holder pressures (0.1, 0.2 and 0.3 MPa). The various moulding behaviour of the sample, i.e. through-the-thickness variation, material drawing-in, surface shear angle etc. were considered for the study. In general, it is noticed that formability behaviour in deformation are very important to be considered in which it will have a significant role on ultimate mechanical properties. General result shows that in preforming, both the blank holder pressure applied and number of layers in the sample have significant role on the forming properties of the samples. For instance, for all applied blank holder pressures, the drawing-in values of the different samples in the warp direction are slightly higher than in the weft direction due to the slight bending of weft yarns co MPa red to the warp yarns. As the holding pressure of the blank holder increases, the drawing-in value has been reduced and the region which was not maintained by the holder is solely responsible for the deformation. Regarding the surface shear angle, it becomes more severe as the number of layers and pressure applied increased. The result will give a hint for different researcher, scientist, designers and manufacturer to consider blank holder pressure and number of layers used while forming in the application of technical product such as soft body armour.

5. References

[1] Bilisik K 2017 *Text. Res. J.* **87(18)** 2275–2304
[2] Zhong T and Hu H 2007 *Autex Res. J.* **7(4)** 245–251
[3] Bekampiene P and Domskiene J 2010 *Int. J. Mater. Form.* **3(1)** 211–214
[4] Mozafary V, Payvandy P, Bidoki S and Bagherzadeh R 2013 *Fibers Polym.* **14(9)** 1535–40
[5] Ye L and Daghyani H R 1997 *Compos. Part A* **28A** 869–874
[6] Haanappel S P, Ten Thije R H W, Sachs U, Rietman B and Akkerman R 2014 *Compos. Part A Appl. Sci. Manuf.* **56** 80–92
[7] Duhovic M and Bhattacharyya D 2006 *Compos. Part A Appl. Sci. Manuf.* **37(11)** 1897–1915
[8] Zhu B, Yu T X, Zhang H and Tao X M 2008 *Composites* **1(1)** 969–972
[9] Abtew M A *et al* 2018 *Compos. Part A Appl. Sci. Manuf.* **109** 517–537
[10] Hu J and Jiang Y, 2002 *Compos. Part A Appl. Sci. Manuf.* **33(5)** 725–734
[11] Abtew M A, Bruniaux P, Boussu F, Loghin C, Cristian I and Chen Y 2018 *Comput. Ind.* 100
[12] Chen X and Yang D 2010 *Text. Res. J.* **1** 1–8
[13] Mahbub R, Wang L and Arnold L, 2014 *Int. J. Fash. Des. Technol. Educ.* **7(3)** 198–207
[14] Hussein M and Parker G 2001 US6281149 B1

[15] Dufour C, Wang P, Boussu F and Soulat D 2014 *Appl Compos Mater* **21** 725–738
[16] Nawab Y, Legrand X and Koncar V 2012 *J. Text. Inst.* **103(12)** 1273–79
[17] Chen X, Lo W Y and Tayyar A E 2002 *Text. Res. J.* **72(3)** 195–200
[18] Lee J S, Hong S J, Yu W R and Kang T J 2007 *Compos. Sci. Technol.* **67(3–4)** 357–366

Acknowledgement
This work has been carried out as part of the Sustainable Management and Design for Textiles (SMDTex) Project, funded by the European Erasmus Mundus Program.

Tactile Comfort Evaluation of Conductive Knitted Fabric Using KES-FB

M G Tadesse[1, 2, 3], Y Chen[3], L Wang[3], V Nierstrasz[2] and C Loghin[1]
[1]Gheorghe Asachi Technical University of Iasi, Faculty of Textiles, Leather and Industrial Management, 53, D.Mangeron Blv. Iasi-700050, Romania
[2]University of Borås, Textile Materials Technology, Department of Textile Technology, The Swedish School of Textiles, Faculty of Textiles, Engineering and Business, 501 90, Borås, Sweden
[3]Soochow University, College of Textile and Clothing Engineering, Gan Jiang Dong Road, 215021 Suzhou, China

E-mail: tadesse.melkie@tuiasi.ro

Abstract: Tactile comfort has a strong relation with low-stress mechanical properties of textile fabrics having close contact with the human skin. In this work, we attempt to analyse the low-stress mechanical properties of the functional knitted fabric obtained using Kawabata's fabric evaluation system (KES-FB). The measured results were compared with those of the controlled polyester fabric. The bending ability of the product increased from 0.2448 to 0.8010gf.cm^2/cm and hence the rigidity influenced when copper yarn is introduced. However, the compressibility increased from 0.173 to 0.449gf.cm/cm^2 and hence the compressibility slightly boosted. The surface roughness (SMD) highly increased from 7.196 to 14.258 μm. It was observed that the incorporation of conductive copper yarn during knitting brought an effect on the tactile comfort of the fabrics and reduced by 69%. The overall comfort properties of the conductive textile fabric were reduced due to the introduction of copper yarn during knitting operations. Focus should be given when functional fabric developed which has close contact to the human skin.

1. Introduction

Now a days, textiles are serving the human being beyond the traditional use (shelters to the skin) such as wearable smart materials [1, 2]. Smart and functional textiles are fabric which replace the functions of electronics and simultaneously behave physically like traditional textiles which enable computing, digital components; electronics to embed into wearable into them. They have wide application areas such medical, sport clothing, protection from danger, and many other uses. The smart textile fabrics can be produced in many ways such as coating [3] and incorporation of smart fibre via knitting [4]. However, most researchers have been focused on the functionality aspect only. On the other hand no or little work has been done on the interaction of the wearer to the fabric. That means a lot of work should be required to assess and evaluate the tactile comfort of the smart fabrics as customers need always comfort in addition to quality of the product. This work intended to solve the quality dimensions in terms of tactile comfort.

After the birth of the objective evaluation of textile products by Pierce in 1930 [5], a lot of works have been done such as measuring the low-stress mechanical properties of the textile product using SiroFAST (Fabric Assurance by Simple Testing) and [6] FTT (Fabric touch tester [7] for the

IOP Conf. Series: Materials Science and Engineering **374** (2018) 012056 doi:10.1088/1757-899X/374/1/012056

realization of the objective evaluation and inspection of the tactile comfort of textile products. The major advance in the objective evaluation was the work done by Kawabata and his co-workers in 1991 [8-11]. The authors addressed the quality of the textile product by measuring the mechanical properties at low-load regions. The major findings in this work was the measurement of tensile properties, shearing properties, bending properties, compressional properties and surface and frictional properties of the textile product under low-load regions. This means the mechanical properties will be analysed without damaging the sample. Hence the orders of measurement doesnot have much influence on the obtained results. The KES-FB measurement system has been developed primarily for the objective evaluation of textile products [12]. They are ideally suited for the tactile hand evaluation of rigid (woven) fabrics. This initiates to apply the Kawabata's evaluation systems to rigid knitted fabric which maintains its rigidity on the applications of low-load tensile forces. A fully automatic fabric test system was employed for the measurement of low-load mechanical properties of the knitted fabric.

In this work, we attempted to measure the comfort of the functional textile product using Kawabata's evaluation system and tried to compare the results with conventional textile product. In addition to the objective evaluation, functional textile can also be assessed subjectively using human subject [13]. In this work, the authors addressed the subjective assessment of functional fabrics using bipolar attributes and found a permissible result.

2. Experimental

2.1. Materials

Plain woven polyester fabric from Almedahl-Kinna AB, with 159 g/m^2 was used as a control fabric. Composite of polyester (76/47/1 Dtex; 47 filaments; 10,000 m: weigh 76 gm and 1 thread only) and copper (0.1 mm ϕ; 2.5Ω/m) was used for KES-F measurements. The surface resistance of the conductive fabric was 0.158Ω/m.

2.2. Methods

Tensile, shear, bending, compression, surface, and friction properties of the functional textile fabrics were measured using tensile and shear tester (KES-FB1), bending tester (KES-FB2), compression tester (KES-FB3), surface and friction tester (KES-FB4) with KES FB-Auto (KATO TECH CO., LTD). All low stress-mechanical properties were measured in both warp and weft directions except the compression property; the average of the two were reported. Sixteen low-stress mechanical properties were measured and reported. The measurements were made under high sensitivity conditions as of reference [14].

3. Results and discussion

An important aspect of the current work is to observe the effect of incorporating conductive copper yarn on the tactile comfort of the functional fabric. All the low-stress mechanical properties of the knitted conductive fabric and controlled fabric were measured using Kawabata's evaluation systems under controlled environment and the results are discussed as follows:

3.1. Tensile properties

The load elongation graph obtained by Kawabata's evaluation system is shown in Figure 1. As shown in the figure, the results are displayed up to elongation of 51%. The conductive sample can extend beyond this limit. This could be attributed to the knitting structure introduced. We only compared the results up to this limit.

The load-elongation curves of the controlled fabric in this figure is a clear indication of the knitted fabrics show a maximum elongation even with at low load applications at low-load regions. This is a clear indication of how much sensitive is the Kawabata's evaluation system to detect the change in the tensile properties at low stress regions.

a)
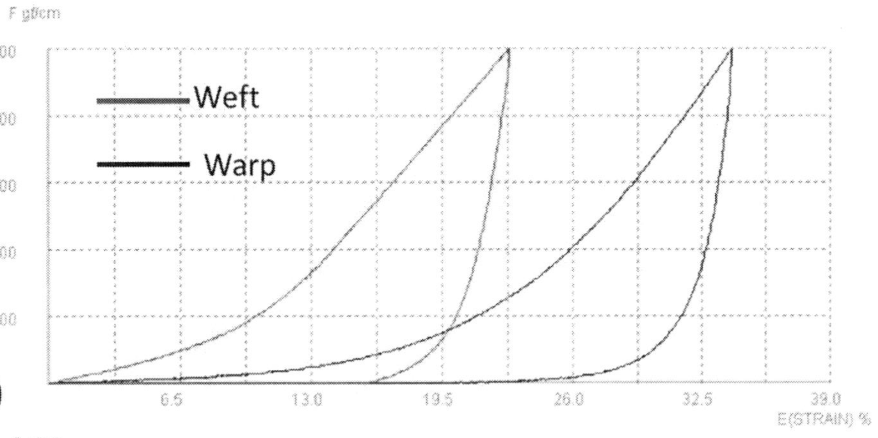

b)
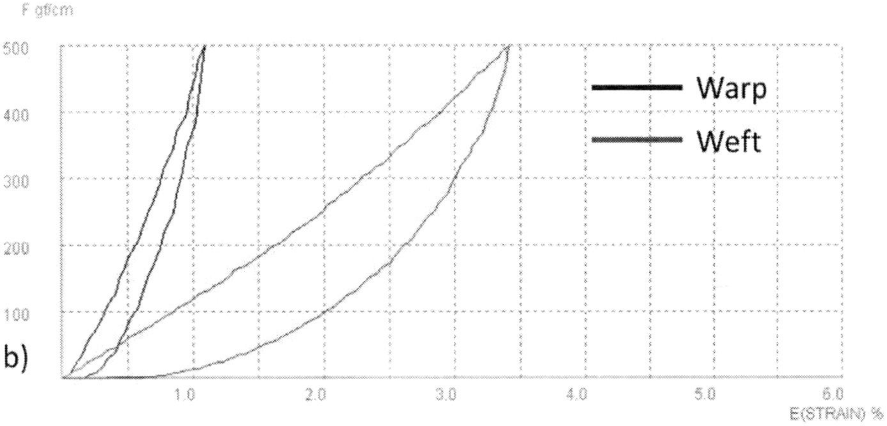

Figure 1. Tensile properties of (a) conductive and (b) reference samples.

Furthermore, there can be shown there is great difference between the warp and the weft yarn behaviours under load-extension curve for the tow fabric samples. This could be attributed to the structure difference between the fabrics.

Other low-stress tensile properties are displayed in Figure 2. Under this, extensibility, EMT [-], linearity, LT [-], tensile energy, WT [gf.cm/cm^2], and tensile resilience, RT [%] were measured using KES-FB1 and analysed. As shown in the figure, the linearity values (LT) of the conductive sample is a little bit lower than that of the reference sample. This indicates that the incorporation of copper fabric has no any effect on the linearity of the conductive sample. As the value approaches to 1, the sample becomes harder. As shown in the figure, the fabric extensibility in the initial strain range is high in the cases of conductive fabrics. Theoretically, the lower the linearity is, the higher the comfort of the fabric in wearing. However, in our cases the low elongation recorded at low load is due to the structure of the fabric. Therefore, practically, the incorporation of the copper fabric reduces the comfort. Hence, linearity (LT) did not show the variation in the comfort of the samples.

On the other hand, incorporation of conductive copper to polyester fabric during knitting brought a great influence on the tensile energy (WT) of the conductive sample. The value for the conductive sample is higher than that of the reference sample. This could be attributed to the conductive sample in produced using knitting operations. Therefore, the stretchability increases when compared to that of weaving.

Similarly, the tensile resilience (RT) of the conductive sample is highly influenced due to incorporation of conductive coper fabric. As the value approaches to 100%, it becomes highly resilience, hence comfortable.

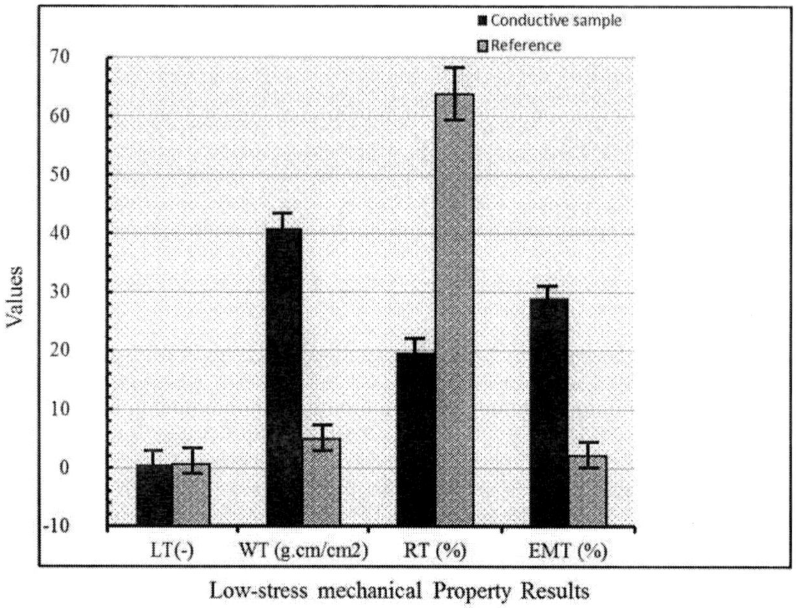

Figure 2. Low-stress mechanical properties of conductive and reference samples.

As per this conclusion, the tactile comfort of the conductive sample reduces by 69% when compared to the reference sample. The strain (EMT) value of the conductive sample is much higher than that of the reference sample. This could be attributed to the increase in tensile energy due to its structure (Knitting structure). Therefore, for these samples, results of the tensile properties doesnot support the idea theoretical conclusions about the tactile comfort of textile product due to structure difference.

3.2. Shearing, bending, compression, and surface friction properties

Other mechanical properties of the samples measured under low-load regions using Kawabata's evaluation are illustrated in Table 1 and Figure 3.

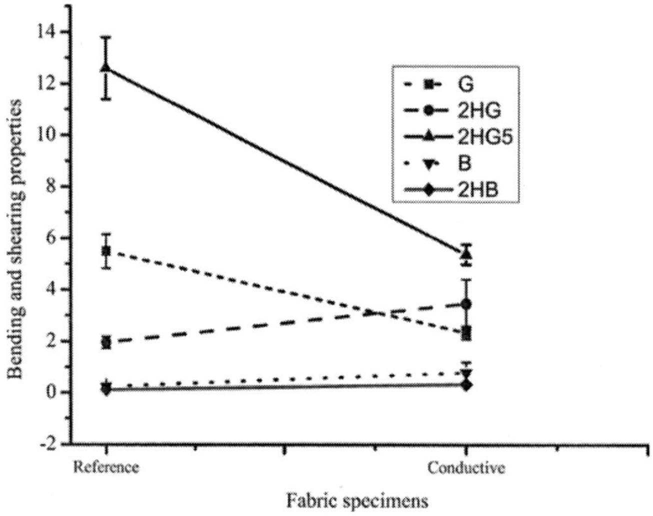

Figure 3. Bending and shearing property results.

Table 1. Low-stress mechanical properties of the conductive and reference sample.

Mechanical properties		Samples	
		Reference	Conductive
Compression	LC [-]	0.29	0.179
	WC [gf.cm/cm2]	0.173	0.449
	RC [%]	60.93	66.99
Surface friction	MIU [-]	0.181	0.232
	MMD [-]	0.070	0.067
	SMD [μm]	7.196	14.238

In the compression measurement, the compressional energy (WC) is related to the comfort of the textile product. The higher the compressional energy is, the easier the compressibility. The compressional energy of the sample increased from 0.173 to 0.449. This indicates that the textile material becomes softer when copper fabric is introduced. This may be due to the structural change from woven to knitted product.

In the measurement of the surface frictional properties of the samples, the mean deviation of surface friction (MMD) is highly related to the comfort of the product. It tells us about whether the fabric is rough or smooth. The higher the MMD is, the harder the sample. The MMD result confirmed that the reference sample is rougher than that of the conductive sample. This is an indication of incorporating the conductive yarn using knitting operation doesnot have an effect on tactile comfort.
Finally, the control sample and the conductive knitted sample were tested if they can fit for some purpose by calculating the total hand value (THV) and by drawing and calculating the hand values for respective application areas. The control sample was tested for men's summer suit, while the conductive knitted sample was tested for knit outwear (winter) and the results are shown in Figure 4.

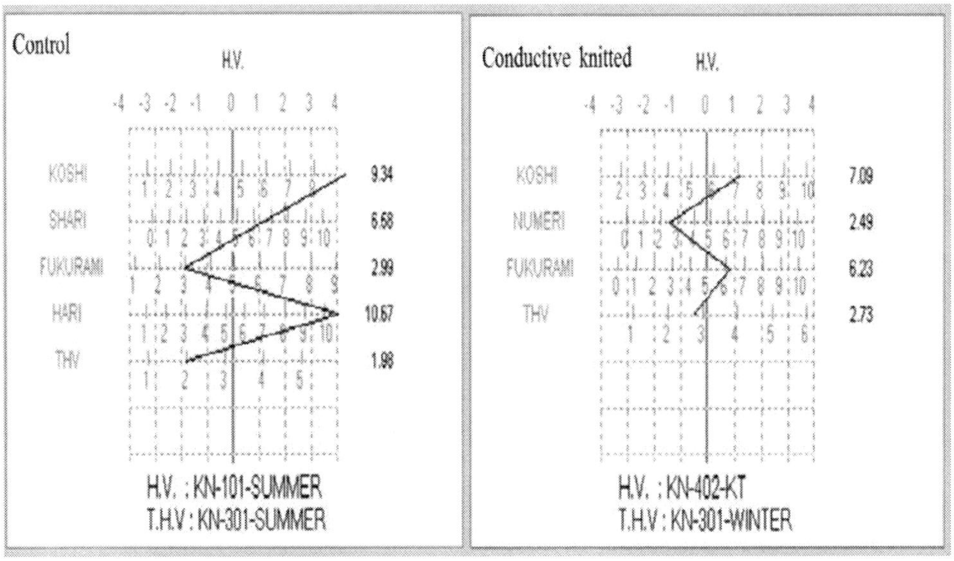

Figure 4. Primary and THV values.

As shown in the figure, the primary hand values (KOSHI, SHARI, FUKURAMI, and HARI) are within the range of values for men's winter suit in the case of control fabric. Similarly, the primary hand values for conductive sample for knit outwear are also within the range for its specific applications. The calculated total hand value (for which THV: 5; excellent and THV: 1; poor) is another indication for which the product can fit for purpose or not. In this case, the total hand value for conductive fabric (2.73) is higher than that of control (1.98) even though they have different application due to structure difference.

4. Conclusions

At low-load regions, polyester knitted fabrics with copper incorporated are rougher than that of the pure polyester fabric. This could be due to copper yarns property. In similar way, the bending ability of the knitted fabric was improved due to the characteristic features of the knitted fabrics. The results of the shearing test also indicate that the conductive sample with copper yarn incorporated was highly deformable due the change in structure from weaving to knitting. The tensile test result showed that the tactile comfort reduced by 69% due to the incorporation of the copper yarn during knitting. The overall result confirmed that it is possible to make an analysis on the comfort of the functional fabric based on the results of the low-stress mechanical properties obtained by Kawabata's evaluation system. To the end, we can conclude that tensile, shearing, and bending properties are very difficult to compare the comport properties of fabric which have different structure. However, other low-stress properties such as surface friction and compressional properties are used to distinguish the tactile comfort properties of textile products with different fabric structures.

5. References

[1] Stoppa M, Chiolerio A 2014 *Sens.* **14** 11957
[2] Edmison J, Jones M, Nakad Z, Martin T 2002 Sixth International Symposium 41-48
[3] Shim BS, Chen W, Doty C, Xu C, Kotov N A 2008 *Nano Lett.* **8** 4151
[4] Jost K, Stenger D, Perez C R, McDonough J K, et al. 2013 *Energy Environ. Sci.* **6** 2698
[5] Peirce F T 1930 *J. Text. Inst. Trans.* **21** T377
[6] Tokmak O, Berkalp O B, Gersak J 2010 *Fibres.Text.East.Eur.* **18** 79
[7] Hu J Y, Hes L, Li Y, Yeung K W, Yao B G 2006 *Polym.Test.* **25** 1081
[8] Kawabata S, Niwa M 1991 *Int. J. Cloth. Sci.Technol.* **3** 7
[9] Kawabata S, Niwa M 1989 *J.Text. Inst.* **80** 19
[10] Kawabata S, Niwa M, Wang F1994 *Text. Res. J.* **64** 597
[11] Kawabata S, Niwa M, Yamashita Y 1999 *Int.J. Cloth.Sci.Technol.* **11** 134
[12] Singh J P, Behera BK, Matsudaira M 2014 *J. Text. Inst.***105** 467
[13] Melkie Getnet T, Harpa R, Chen Y, Wang L, Nierstrasz V, Loghin C 2018 *J. Ind. Text.* DOI: 10.1177/1528083718764906
[14] Mahar T J, Wheelwright P, Dhingra R C, Postle R 1990 *Text. Res. J.* **60** 7

Acknowledgments

An author wishing to acknowledge the Erasmus Mundus Joint Doctorate Programme as this work was supported financially by SMDTex-Sustainable Management and Design for Textile [grant number n°2015-1594/001-001-EMJD]. The authors would like to thank Technical university of Liberec for conducting the research.

IOP Conf. Series: Materials Science and Engineering **374** (2018) 012057 doi:10.1088/1757-899X/374/1/012057

The Effect of Some Key Changes in the Chemistry of Water in Relation to Copper and Brass Corrosion Control

M C Gorovei[1] **and L Benea**[1,*]
[1] Competences Center: Interfaces-Tribocorrosion-Electrochemical Systems (CC-ITES) Faculty of Engineering, Dunarea de Jos University of Galati, 47 Domneasca street, Galati, Romania

E-mail: lidia.benea@ugal.ro

Abstract. Corrosion means the degradation of the metals or their alloys, under the action of chemical or electrochemical agents from the environment. The complex corrosion phenomenon has a destructive action, generating undesirable economic consequences: metals and labor losses, appreciable reduction in the lifetime of various metal constructions, insecurity in the operation of industrial machinery. Under the current conditions of accelerated growth in the production of material goods, one of the most important issues is the economy of raw materials and materials, energy and labor force. Copper, having a purity of over 99%, is used in the manufacture of gas and water pipes, roofing materials, utensils and ornamental objects. Brass is used in the manufacture of flexible tubes, pipes, coils, cartridges, various electrochemical parts, jewelry, etc. The aim of this research work was to evaluate the corrosion resistance of copper and brass in various solutions: with different chloride ions as 35 g/L NaCl, waste water and tap water. The corrosion behavior of copper and brass was analyzed by electrochemical methods, such as: open circuit potential (OCP), electrochemical impedance spectroscopy (EIS) and cyclic voltammetry (CV). Pure copper exhibits more noble potential values than its alloy (brass), according to the evolution of free potential in all tested solutions. After performing the electrochemical assays, ex-situ investigations, by optical microscopy, were made and the results confirm that the chloride ions affect the corrosion behavior of copper and brass. Corrosion of materials is a very important process to consider when choosing a material that has to operate in a specific environment.

1. Introduction

Copper and brass are materials especially used in cooling water systems, air conditioning systems and drinking water systems because they have good corrosion resistance and high thermal conductivity [1-6].

In addition to excellent conductivity, copper has ideal mechanical properties at low, ambient and high temperatures, it is easy to manufacture or mold and can be easily machined. It has excellent resistance to oxidation and corrosion [1, 2]. Copper is the fifth most widely used metal in the world to be useful in pure or alloy form. The most commonly used copper alloys are brass, copper-nickel and bronze. The presence of corrosion products on the copper surface leads to a decrease in the performances of a system built with this material [5].

Copper has a low corrosion rate. The good corrosion resistance of copper and its alloys is due to the development of protective layers of degradation products that diminish the attack rate [6].

Brass is copper alloyed with zinc. In industry are used alloys with up to 50% Zn. Brass with 30-32% Zn content possess good plasticity characteristics [2]. Brass is an alloy tough and easy to be processed used to make pipes, valves, fittings objects [3].

When chloride ions are present in aqueous solutions, copper and brass are susceptible to pitting corrosion [7].

The aim of this work is to evaluate the corrosion behavior of copper and brass in 35 g/L NaCl solution, tap water from Galati water plant and a low concentration waste water from water distribution system in Galati. The corrosion behavior of copper and brass was evaluated using electrochemical methods: open circuit potential (OCP) and electrochemical impedance spectroscopy (EIS) and the surfaces of analyzed samples, before and after the corrosion assays, were evaluated using the Optical Microscope OPTIKA XDS3 MET.

2. Experimental procedures

2.1. Materials and methods

The chemical composition of the tested samples was Cu-99.97% for copper and for brass: Cu-57.8%, Zn-39.02%, Pb-2.99%, Fe- 0.19%.

In order to assess the corrosion behavior, the analyzed samples were connected with copper wire for electrical contact and insulated with epoxy resin, leaving a working area of 2 cm^2. The working area was cleaned with HNO_3 and rinsed with distilled water before the corrosion assays.

For the electrochemical measurements it was used a conventional three-electrode cell, consisting of: a working electrode made of copper or brass, a Pt-Rh grid that serve as auxiliary electrode and an Ag/AgCl reference electrode (with saturated KCl solution, E= +199 mV vs. standard hydrogen electrode–SHE). The corrosion experiments were done using a potentiostat / galvanostat PGZ 100 and the experimental data were recorded with VoltaMaster4 software.

Corrosion assays were performed for copper and brass in three different electrolytes: tap water from Galati water plant, alow concentration waste water from water distribution system in Galati and 35 g/L NaCl aqueous solution. The chemical composition of these electrolytes, by point of view of chloride concentration, is shown in Table 1.

In Table 2 are shown the physico-chemical characteristics of these electrolytes which have been measured with a multi-parameter analyzer CONSORT C 533.

Before and after the corrosion assays, the surfaces of analyzed samples of copper and brass were analyzed using the optical microscope OPTIKA XDS-3MET in order to evaluate the corroded area.

Table 1. The chemical composition of the electrolytes used in corrosion assays.

Electrolytes	NaCl solution	Tap water	Waste Water
Chloride ions (Cl^-)	21.27 g /L	0.0360 g /L	0.801 g /L

In Table 2 are shown the physico-chemical characteristics of these electrolytes which have been measured with a multi-parameter analyzer CONSORT C 533.

Table 2. Electrolytes physico-chemical parameters.

Solution type	pH	Conductivity [mS]	Salinity [‰]
NaCl solution	5.71	41.2	26.2
Tap water	7.52	1.04	0.5
Waste water	6.07	6.57	3.6

The solutions used for corrosion tests are different concentration of chloride ions and differ slowly in pH value, which are the key components in an aggressive corrosive environment.

3. Results and discussion

3.1. Open circuit potential

The corrosion tests began with monitoring the evolution of open circuit potential after the immersion of copper and brass into the electrolytes. This free potential varies with time due to the oxidation trend of the samples surfaces.

From Figure 1 (a) it can be seen that copper immersed in 35 g/L NaCl aqueous solution showed a small shift tendency of the free potential to the more negative (active) direction having at the end of 60 minutes the value of -235.81 mV vs. Ag / AgCl, compared to the initial value of free potential at the immersion time of -233.81 mV vs. Ag / AgCl.

On the other hand, for the copper immersed in waste water from Figure 1 (a) it can be observe that, the open circuit potential show a slow shift in the more positive direction, from +0.44 mV vs. Ag/AgCl up to +18.07 mV vs. Ag/AgCl at the end of monitoring period.

For the copper sample immersed in tap water, from Figure 1 (a) it can be seen a slow shift of open circuit potential in the more positive direction, having at the end of the monitoring period a value of -36.06 mV vs. Ag/AgCl, compared with the initial immersion time value of -73.31 mV vs. Ag/AgCl.

Comparing the evolution of OCP recorded for copper in all three solutions (Figure 1 (a)) it can be observed that due to the high concentration of chloride ions in 35 g/L NaCl aqueous solution, the immersed sample react with the chloride ions and the passive layer formed at material-electrolyte interface is dissolved immediately.

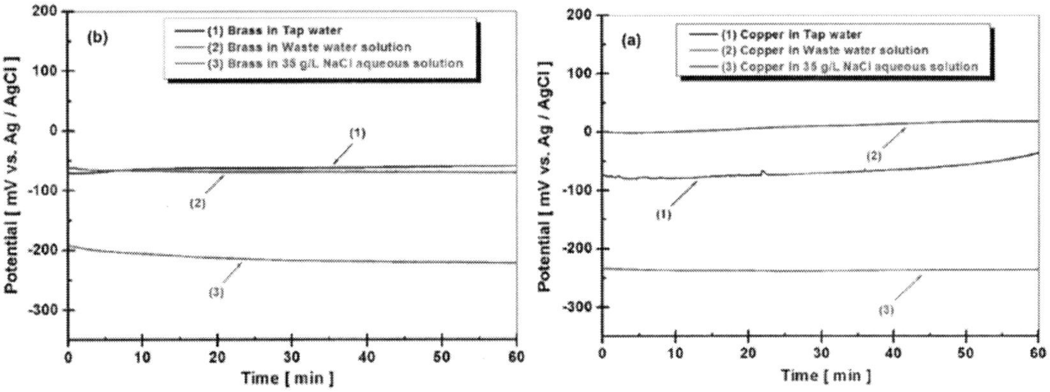

Figure 1. Open circuit potential (OCP) recorded for (a) copper and (b) brass immersed in 35 g/L NaCl solution waste water and tap water for 60 minutes.

For the copper samples immersed in artificial waste water or tap water, the free potential shows a shift tendency, to more noble values, that could be explained by the fact that at the material-electrolyte interface a passive layer is formed and protects the copper substrate against the aggressiveness of chloride ions from the electrolytes.

The open circuit potential recorded for the brass sample immersed in 35 g/L NaCl aqueous solution presented in Figure 1 (b) shows a shifted tendency to more negative value from -192.25 mV vs. Ag/AgCl at the initial immersion time to the value of -221.56 mV vs. Ag/AgCl at the end of the measurement.

For brass sample immersed in waste water, the open circuit potential exposed in Figure 1 (b) shows a very small shifted value to more negative value of free potential from -60.87 mV vs. Ag/AgCl at initial immersion time to -69.43 at the end of measurement. For brass immersed in tap water, the open circuit potential shows a shifted trend to more positive (noble values), from -70.43 mV vs. Ag/AgCl at initial immersion time to -58.62 mV vs. Ag/AgCl at the end of the monitoring period.

The shifted values of free potential during immersion time of brass confirms the formation on the brass surface of a more stable passive oxide layer in tap water solution while the shifted trend to more negative values of brass in waste water and 35 g/L NaCl aqueous solution indicates a higher instability and a degradation of the passive film formed on brass surface in these solutions due to higher concentration of chloride ions.

3.2. Electrochemical Impedance Spectroscopy

Electrochemical impedance spectroscopy (EIS) is a non-destructive method used to investigate the corrosion performance metals that are exposed to an aqueous environment. Experimental impedance is the result of the serial binding of resistances and capacitances.

The electrochemical impedance spectroscopy measurements were performed to characterize the corrosion behaviour of copper and brass immersed in NaCl solution, tap water and waste water solutions.

The EIS spectra were recorded in the frequency range from 100 kHz to 10 mHz, with amplitude of the sine-wave signal of 10 mV. The impedance data recorded for copper and brass immersed in all three solutions were displayed in the form of a Bode plots in (Figure 2 and Figure 3).

The EIS spectra were fitted to equivalent circuit (Figure 4) using ZView 3.4f software and the quality of the fitted results was evaluated with the chi-square value that was lower than 10^{-3}.

Figure 2 (a) and (b) shows the EIS results, in the form of the Bode plots for copper immersed in three electrolytes after 1 h of immersion.

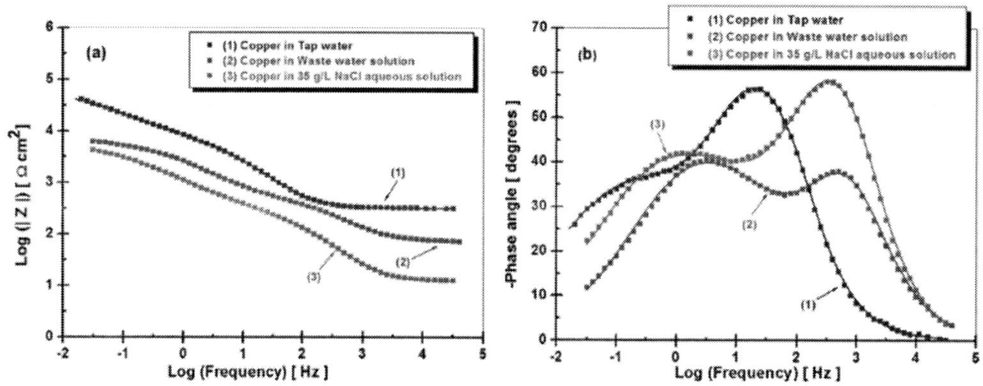

Figure 2. Bode impedance plots of copper immersed in: (1) tap water (2) waste water and (3) NaCl solution. (a) impedance modulus vs. frequency plots and (b) phase angle vs. frequency plots. The plain symbols represent the experimental data and the continuous lines represent the fitted results.

From the Bode plots between log |Z| and log (f) from Figure 2 (a), it could be observed that the curve corresponding to copper immersed in tap water displayed a higher impedance modulus over the entire frequency range. This remark indicates a nobler electrochemical characteristic for copper immersed in tap water as compared with copper immersed in waste water and NaCl solution. The corrosion attack is much more aggressive for copper immersed in waste water and NaCl solution as compared with copper immersed in tap water which could be easily detected by small phase angle in the lower-frequency ranges, as it is shown in Figure 2 (b)). This behaviour can be explained by to the lower chloride ions concentration in tap water as compared with the waste water and NaCl solution.

Figure 3 (a) and (b) shows the EIS results, in the form of the Bode plots for brass immersed in three electrolytes after 1 h of immersion time.

From the Bode plots showed in Figure 3 (a) and (b) it is observed a similar behaviour for brass immersed in tap water as it was observed for copper immersed in the same solution. The corrosion attack is much more aggressive for brass immersed in waste water and NaCl solution as compared with brass immersed in tap water. Also here, the higher concentration of chloride ions from waste water and NaCl solution is responsible for more pronounced corrosive attack.

For the copper and brass electrodes immersed in NaCl aqueous solution, tap water and the and waste water solution, an equivalent electrical circuit is proposed and shown in Figure 4.

The proposed electrical circuit was used to fit the EIS experimental spectra where R_1 was designated as the resistance of the electrolyte, CPE_1 and R_2 are the constant phase element and the polarization resistance respectively of the formed oxide layer on the copper and brass surfaces during oxidation of the base metal.

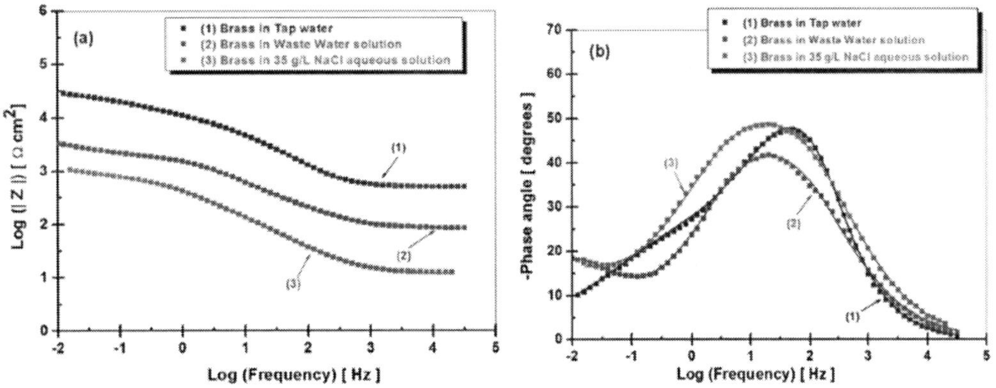

Figure 3. Bode impedance plots of brass immersed in (1) tap water, (2) waste water and (3) 35 g/L NaCl: (a) impedance modulus vs. frequency plots and (b) phase angle vs. frequency plots. The plain symbols represent the experimental data and the continuous lines represent the fitted results.

CPE_2 and R_3 designate the constant phase element and the resistance respectively of the base metal.

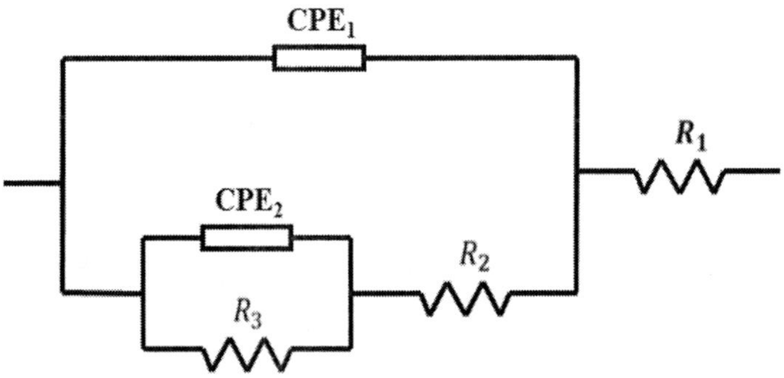

Figure 4. Schematic representation of the equivalent electrical circuit used to fit and simulate the experimental data of the electrochemical impedance spectra recorded for copper and brass surfaces in three aggressive environments.

Following the fitted electrochemical impedance spectroscopy diagrams the specific polarization resistance values for both surfaces in the three corrosive environments were obtained. For the copper electrode immersed in waste water solution the specific polarization resistance value is approximately 7231 kohm cm^2. On the other hand, for the brass electrode immersed in waste water, the specific polarization resistance value is about 5450 kohm cm^2.

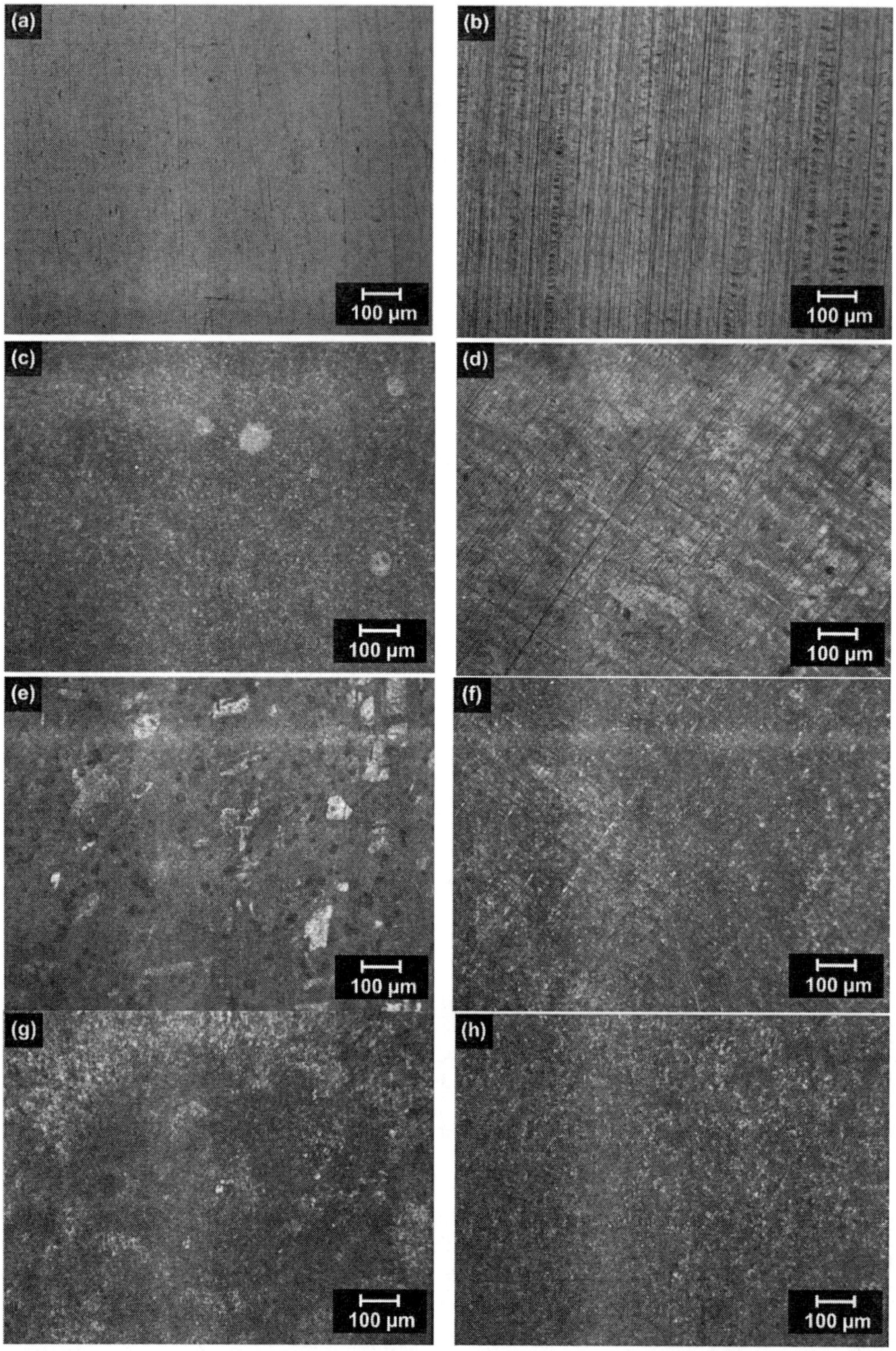

Figure 5. Optical microscopy images recorded before the corrosion assays for (a) brass and (b) copper. The micrographs after the corrosion assays in tap water for (c) brass and (d) copper. After the corrosion assays in waste water for (e) brass and (f) copper. After the corrosion assays in NaCl aqueous solution for (g) brass and (h) copper.

The specific polarization resistance determined using the proposed equivalent circuits shows that for the copper electrode immersed in tap water the specific polarization resistance value is about 69297 kohm cm^2. On the other hand, for the brass electrode immersed in the tap water the specific polarization resistance value is approximately 58837 kohm cm^2.

The specific polarization resistance was determined using the proposed equivalent circuit and shows that for the copper electrode immersed in NaCl aqueous solution, the specific polarization resistance value is 6348 kohm cm^2. On the other hand, for the immersed brass electrode of NaCl the specific polarization resistance value is about 3330 kohm cm^2.

3.3. Optical Microscopy analysis before and after corrosion

Copper and brass surfaces were investigated using the optical microscope before and after the corrosion test in NaCl aqueous solution, tap water and waste water in order to analyze the corrosive effects on their surfaces and the recorded images are shown in Figure 5.

Before the electrochemical assays, on the copper and brass surfaces can be observed lamination tracks, Figure 5 (a) and (b). The copper and brass surfaces, Figure 5 (c) and (d) immersed in tap water, on the analyzed surfaces a few corrosion spots (localized corrosion) can be identified due to the small quantity of chloride ions in the electrolyte.

For the copper and brass samples immersed in artificial water, Figure 5 (e) and (f) and for immersion in NaCl solution, Figure 5 (g) and (h) a generalized corrosion can be identified. Following the increasing concentration of chloride ions into the electrolytes, the corroded areas increase also on copper and brass surfaces.

The optical micrographs confirm the observed corrosion behaviours of copper and brass from electrochemical measurements as open circuit potential and electrochemical impedance spectroscopy.

4. Conclusions

The aim of this paper was to evaluate the corrosion resistance of copper and brass in various solutions, with different chloride ions, such as 35 g/L NaCl solution, waste water and tap water.

According to the evolution of open circuit potential of copper in all tested solutions, it can be observed that due to the high concentration of chloride ions in 35 g/L NaCl aqueous solution, the immersed sample react with the chloride ions and the passive layer formed at material-electrolyte interface is dissolved immediately. On the other hand, copper samples immersed in artificial waste water or tap water, the free potential shows a shift tendency, to more noble values, that could be explained by the fact that at the material-electrolyte interface a passive layer is formed and protects the copper substrate against the aggressiveness of chloride ions from the electrolytes.

The electrochemical impedance spectroscopy measurements reveals that the copper and brass samples immersed in tap water show a higher impedance modulus in comparison with the brass samples immersed in waste water and NaCl solution.

The specific polarization resistance shows that for the copper electrode immersed in NaCl aqueous solution, the specific polarization resistance value is 6348 kohm cm^2. On the other hand, for the copper electrode immersed in waste water solution the specific polarization resistance value is approximately 7231 kohm cm^2 and for the copper electrode immersed in tap water the specific polarization resistance value is about 69297 kohm cm^2.

The specific polarization resistance shows that for the brass electrode immersed in NaCl aqueous solution, the specific polarization resistance value is 3330 kohm cm^2. On the other hand, for the brass electrode immersed in waste water solution the specific polarization resistance value is approximately 5450 kohm cm^2 and for the brass electrode immersed in tap water the specific polarization resistance value is about 58837 kohm cm^2.

From the analysis of optical images, it can be observed that the increasing the chloride ions content, the analyzed surfaces present larger corroded area. Also, it can be observed that the copper and the brass samples present a generalized when were immersed in electrolytes with an increased quantity of chloride ions.

The electrochemical impedance spectroscopy measurements reveal that a nobler electrochemical characteristic is achieved for copper immersed in all tested solutions as compared with brass, which describes a more protective passive film. The specific polarization resistance is higher for copper in all tested solutions.

5. References

[1] Anaum N, Deen K M, Farook A, Ahmed R B 2015 *Appl. Mater. Today* **2** 5170
[2] Tuck C D S, Powell C A, Nuttall J 2010 DOI: 10.1016/B978-044452787-5.00094-9
[3] Hulya K, Suleyman A 2015 *Arab. J. Chem.* http://dx.doi.org/10.1016/j.arabjc.2015.02.007
[4] Andrew D, Igor T, Judson D, Jochen A, Adolf G, Uwe H 2013 *Corros. Sci.* **76** 453
[5] Fateh A, Aliofkhazraiei M, Rezvanian A R 2017 *Arab. J. Chem.* http://dx.doi.org/10.1016/j.arabjc.2017.05.021
[6] Procaccini R, Vázquez M, Ceré S 2009 *Electrochim. Acta* **54** 7324
[7] Procaccini R, Schreiner W H, Vázquez M, Ceré S 2013 *Appl. Surf. Sci* **268** 171

Acknowledgements

This work was supported as part of the research projects and expertise of Competences Center for Interfaces - Tribocorrosion and Electrochemical Systems (CC-ITES) from Dunarea de Jos University of Galati, Romania. www.cc-ites.ugal.ro.

IOP Conf. Series: Materials Science and Engineering **374** (2018) 012058 doi:10.1088/1757-899X/374/1/012058

The Synergistic Effect of Proteins and Reactive Oxygen Species on Electrochemical Behaviour of 316L Stainless Steel for Biomedical Applications

N Simionescu[1], L Benea[1] and V M Dumitrascu[1]
[1] Competences Center: Interfaces-Tribocorrosion-Electrochemical Systems (CC-ITES) Faculty of Engineering, Dunarea de Jos University of Galati, 47 Domneasca street, Galati, Romania

E-mail: lidia.benea@ugal.ro

Abstract. The stainless steels, especially 316L type is the most used metallic biomaterials for biomedical applications due to their good biocompatibility, low price, excellent corrosion resistance, availability, easy processing and high strength. Due to these favorable properties 316L stainless steel has become the most attractive biomaterial for dental implants, stents and orthopedic implants. However an implant material in the human body is exposed to an action effect of other molecules, including proteins (such as albumin) and reactive oxygen species (such as hydrogen peroxide - H_2O_2) produced by bacteria and immune cells. In the literature there are few studies to follow the effect of proteins and reactive oxygen species on 316L stainless steel used as implant material and are still unclear. The degree of corrosion resistance is the first criterion in the use of a metallic biomaterial in the oral or body environment. The aim of this research work is to investigate the influence of proteins (albumin) and reactive oxygen species (H_2O_2) in combination, taking into account the synergistic effect of these two factors on 316L stainless steel. Albumin is present in the body near implants and reactive oxygen species could appear in inflammatory processes as well. The study shows that the presence of albumin and reactive species influences the corrosion resistance of 316L stainless steel in biological solutions. In this research work the corrosion behavior of 316L stainless steel is analyzed by electrochemical methods such as: open circuit potential (OCP), Electrochemical Impedance Spectroscopy (EIS). It was found that, the electrochemical results are in a good agreement with micro photographs taken before and after corrosion assays. The albumin and reactive oxygen species have influence on 316L stainless steel behavior.

1. Introduction

Metallic biomaterials used in biomedical applications include stainless steel, especially 316L type, which is the most used metallic biomaterial for biomedical applications due to their good biocompatibility, low price, excellent corrosion resistance, availability, easy processing and high strength.

Due to these favorable properties 316L stainless steel has become the most attractive biomaterial for dental implants, stents and orthopedic implants. The primary limitation of this metallic biomaterial is the release of toxic metal ion, which can lead to various side tissue reactions [1].

Understanding the chemical and biological requirements for metallic biomaterials is an essential condition, so it is necessary to discuss possible reactions between corrosion products and the human body. In the event of corrosion products occurring on the metallic implant introduced into the human body, the flux of electrons in the metal leads to an ion flux around the tissue, resulting in a nerve cell disruption, which is undesirable because the metal ions released from the cause of inorganic corrosion reactions is transported to internal organs such as the liver or kidney.

In the continuous corrosion process, these ions accumulate in the internal organs, leading to other possible diseases, if the toxicity limit is exceeded. An organic reaction, for example a metal reaction with proteins, can lead to allergic or inflammatory reactions of the surrounding tissue. Albumin is the most common protein in the peri-implant environment and ROS (reactive oxygen species) such as H_2O_2 are produced in the presence of inflammation reaction, which are common after surgery, or in response to infection or corrosion products [2].

In the literature there are few studies to follow the effect of proteins and reactive oxygen species on 316L stainless steel used as implant material and are still unclear [3].

The purpose of this study was to evaluate the synergistic effect of proteins and reactive oxygen species on electrochemical behavior of 316L stainless steel exposed in Phosphate Buffered Saline solution (PBS) pH 7.32 with addition of albumin (1 mL/L) , H_2O_2 (1 mL/L) and proteins (albumin) and reactive oxygen species (H_2O_2) in combination in view of biomedical applications.

2. Experimental procedures

2.1. Materials and methods

The corrosion experiments were performed on 316L stainless steel medical grade. The chemical composition of 316L stainless steel is presented in Table 1.

Table 1. The chemical composition of 316L stainless steel (wt %).

Element / wt%	C	Cr	Ni	Mo	Si	Mn	Fe
316L	0.03	16.5-18.5	11-14	2.0-2.5	1.0	2.0	Balance

Table 2. The chemical composition of biological solutions.

Nr. Crt.	Compound	Phosphate Buffered Saline (PBS) g / L	Phosphate Buffered Saline (PBS) g / L	Phosphate Buffered Saline (PBS) g / L	Phosphate Buffered Saline (PBS) g / L
1.	NaCl	8.0	8.0	8.0	8.0
2.	KCl	0.2	0.2	0.2	0.2
3.	Na_2HPO_4	1.14	1.14	1.14	1.14
4.	KH_2PO_4	0.27	0.27	0.27	0.27
5.	30% H_2O_2	-	1 mL / L	-	1 mL
6.	Albumin	-	-	1 mL	1 mL
7.	Deionized water	Balance	Balance	Balance	Balance
8.	pH	7.32	7.32	7.32	7.32

In order to asses the corrosion behavior of 316L stainless steel a conventional three-electrode cell was used, consisting of: the working electrode (WE) made of 316L stainless steel, an Ag/AgCl reference electrode with KCl saturated solution (E = +199 mV vs. standard hydrogen electrode – SHE) and a Pt-Rh grid as auxiliary electrode. Before corrosion experiments, the squares specimens for electrochemical measurements were connected to a copper wire for electrical connection and insulated with epoxy resin leaving a working area of 2.25 cm^2. The working surface was cleaned with alcohol and acetone and rinsed with distilled water.

The corrosion experiments were done using a Potentiostat–Galvanostat PGZ 100 and the experimental data were recorded with VoltaMaster4 software. The electrochemical measurements such as Open Circuit Potential (OCP), and Electrochemical Impedance Spectroscopy (EIS) were carried out to access the anticorrosive characteristics of the 316L stainless steel medical grade.

Corrosion tests were performed for 316L stainless steel in Phosphate Buffered Saline (PBS) solution [4]. The chemical composition of biological solution it is shown in Table 2.

In Table 3 are shown the physico-chemical characteristics of biological solution which was measured with the multi-parameter analyzer CONSORT C 533.

Table 3. Biological solution characteristics.

Biological solution	**pH**	**Salinity [‰]**	**Conductivity [mS]**
Phosphate Buffered Saline	7.32	8.20	14.5

3. Results and discussions

3.1. Open circuit potential (OCP)

The open circuit potential measurements indicate the tendency of a material to electrochemical oxidation in a corrosive environment. The potential time measurements of 316L stainless steel medical grade immersed in Phosphate Buffered Saline solution (PBS) are shown in (Figure 1).

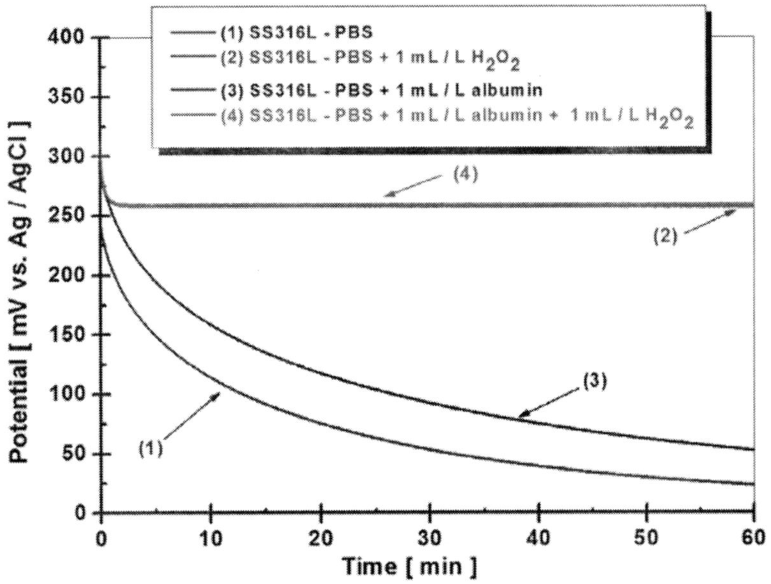

Figure 1. Open circuit potential plots of the 316L stainless steel immersed in: (1) Phosphate Buffer Saline, (2) PBS + 1 mL/L H_2O_2, (3) PBS + 1 mL/L albumin and (4) PBS + 1 mL/L H_2O_2 + 1 mL/L albumin during 60 minutes of immersion time.

From (Figure 1) it can be seen that 316L stainless steel immersed in PBS solution, showed a slow shifted value of open circuit potential in the more negative (active) direction, having at the end of 60 min. the value +23.19 mV vs. Ag/AgCl, compared with the initial immersion time value of +239.88 mV vs. Ag/AgCl. For stainless steel immersed in PBS solution with addition of 1 mL/L albumin, the open circuit potential shows a decreasing value from +288.25 mV vs. Ag / AgCl at initial immersion time to +53.07 mV vs. Ag / AgCl at the end of measurement.

The shift of the potential to a negative direction indicates that the material is not able to form a protective oxide layer on its surface after the exposure under the biological solution.
For stainless steel immersed in PBS solution with addition of 1 mL/L H_2O_2, the open circuit potential show the same shifted trend to more active values from +294.38 mV vs. Ag/AgCl measured at initial immersion time to +256.57 mV vs. Ag/AgCl at the end of monitoring time.

The OCP values of samples immersed in PBS solution with addition of albumin (1 mL/L) and H_2O_2 (1 mL/L) in combination, showed a very slow shifted value in the more negative (active) direction, having at the end of 60 min. the value +259 mV vs. Ag/AgCl, compared with the initial immersion time value of +298.13 mV vs. Ag/AgCl.

Also it can be seen from (Figure 1) that the open circuit potential steady state is attained for 316L stainless steel immersed in PBS solution with addition of mixed H_2O_2 (1 mL/L) and proteins (albumin) (1 mL/L) after 5 minutes of immersion into electrolyte.

3.2. Electrochemical impedance spectroscopy
The electrochemical impedance spectroscopy measurement (EIS) is a non-destructive technique and was performed to characterize the corrosion behavior of 316L stainless steel immersed in PBS solution, PBS solution with addition of albumin (1 mL/L), H_2O_2 (1 mL/L) and albumin and reactive oxygen species (H_2O_2) in combination.

The EIS spectra were recorded in the frequency range from 100 kHz to 0.2 Hz, with amplitude of the sine-wave signal of 10 mV. The impedance data recorded for 316L stainless steel immersed in all four solutions were displayed in the form of a Nyquist plot in (Figure 2).

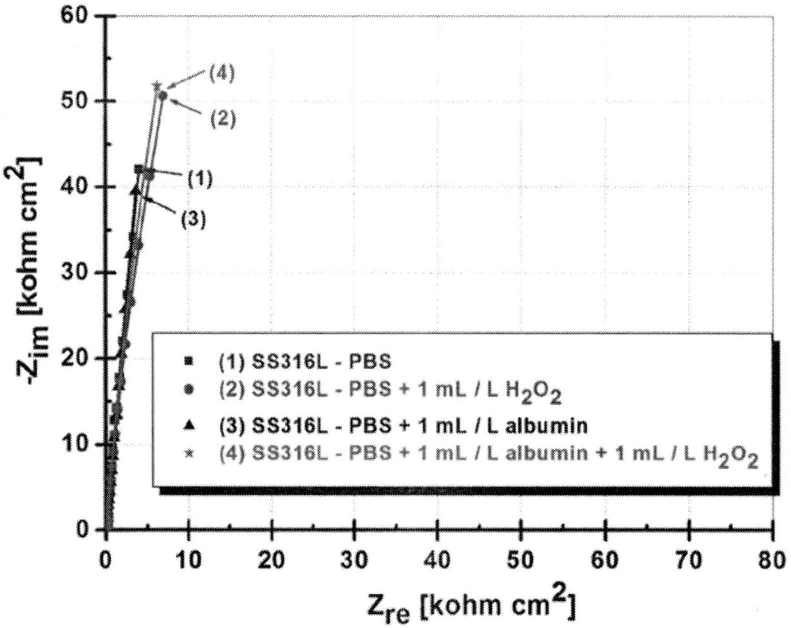

Figure 2. The Nyquist representation of EIS specters recorded for SS316L immersed in (1) Phosphate Buffer Saline, (2) PBS + 1 mL/L H_2O_2, (3) PBS + 1 mL/L albumin and (4) PBS + 1 mL/L H_2O_2 + 1 mL/L albumin. The plain symbols represent the experimental data and the continuous lines represent the fitted results.

The EIS spectra were fitted using ZView 3.4f software and the quality of the fitted results was evaluated with the chi-square value that was lower than 10^{-3}.

The EIS spectra recorded for 316L stainless steel immersed in PBS electrolyte can be characterized by a large semicircle capacitive loop. The EIS specters recorded for 316L stainless steel tested in PBS with 1 mL/L H_2O_2, PBS with 1 mL/L albumin and PBS with 1 mL/L H_2O_2 and 1 mL/L albumin present a small capacitive semicircle at high frequency and a large semicircle capacitive loop at low frequency.

For 316L stainless steel immersed in PBS solution, the EIS spectra was fitted using a simple Randles equivalent circuit (showed in Figure 3 (a)) composed by R_s is the electrolyte resistance, CPE_r and R_r are the constant phase element and the resistance of 316L stainless steel base metal.

In order to fit the EIS specters recorded for 316L stainless steel immersed in PBS with 1 mL/L H_2O_2, PBS with 1 mL/L albumin and PBS with 1 mL/L H_2O_2 and 1 mL/L albumin it was used a complex equivalent circuit presented in (Figure 3 (b)) where R_s was designated the electrolyte resistance, CPE_o and R_o are the constant phase element and polarization resistance of oxide layer formed in the presence of H_2O_2 or the albumin layer adsorbed on 316L stainless steel surface during the oxidation of [5], CPE_b and R_b are the constant phase element the resistance of the base metal.

The constant phase element was used instead of pure capacitor because the non-homogeneity of analyzed samples and non-uniform current distribution caused by the surfaces roughness [6].

Accordingly with A. Lasia [7], the impedance of constant phase element is defined as:

$$Z_{CPE} = \frac{1}{Q(j\omega)^{\alpha}} \tag{1}$$

where: $j = \sqrt{-1}$ is an imaginary number, ω represent the angular frequency, Q is a frequency independent parameter and α is an exponential factor ranged between -1 and 1.

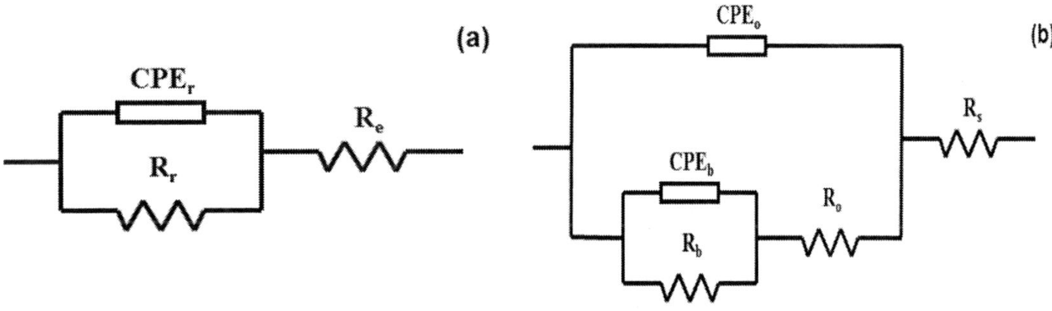

Figure 3. Schematic representation of equivalent electric circuits of 316L stainless steel surfaces for fitting and simulation of electrochemical impedance spectroscopy experimental data: (a) Randles equivalent circuit and (b) complex equivalent circuit.

The specific polarization resistance determinated using the proposed equivalent circuits reveal that for 316L stainless steel immersed in PBS solution, the specific polarization resistance value is around 19930 kohm cm^2. The addition of 1 mL/L H_2O_2 into PBS electrolyte increase the rate of anodic and cathodic reactions and the specific polarization resistance decrease up to 1222 kohm cm^2. On the other hand, the addition of 1 mL/L albumin in PBS electrolyte cause the inhibition of anodic and cathodic reactions and the specific polarization resistance increase until 24970 kohm cm^2. In the presence of both, H_2O_2 and albumin in PBS electrolyte, the specific polarization resistance of 316L stainless steel decrease up to 1431 kohm cm^2 in comparison with the specific polarization resistance of 316L stainless steel immersed in PBS electrolyte. The H_2O_2 and albumin added into PBS electrolyte increase the rate of anodic reaction simultaneously with the inhibition of cathodic reaction rate and increase the corrosion rate of 316L stainless steel.

IOP Conf. Series: Materials Science and Engineering **374** (2018) 012058 doi:10.1088/1757-899X/374/1/012058

Similar results were obtained by F. Yu and his co-workers [3] when evaluated the synergetic effect of bovine serum albumin and H_2O_2 added in physiological solution on corrosion behavior of Ti6Al4V.

3.3. Optical Microscopy analysis before and after corrosion

The surfaces of 316L stainless steel were investigated before and after corrosion tests in Phosphate Buffer Saline, PBS + 1 mL/L H_2O_2, PBS + 1 mL/L albumin and PBS + 1 mL/L H_2O_2 + 1 mL/L albumin in order to estimate the corrosive effects and are presented in (Figure 4).

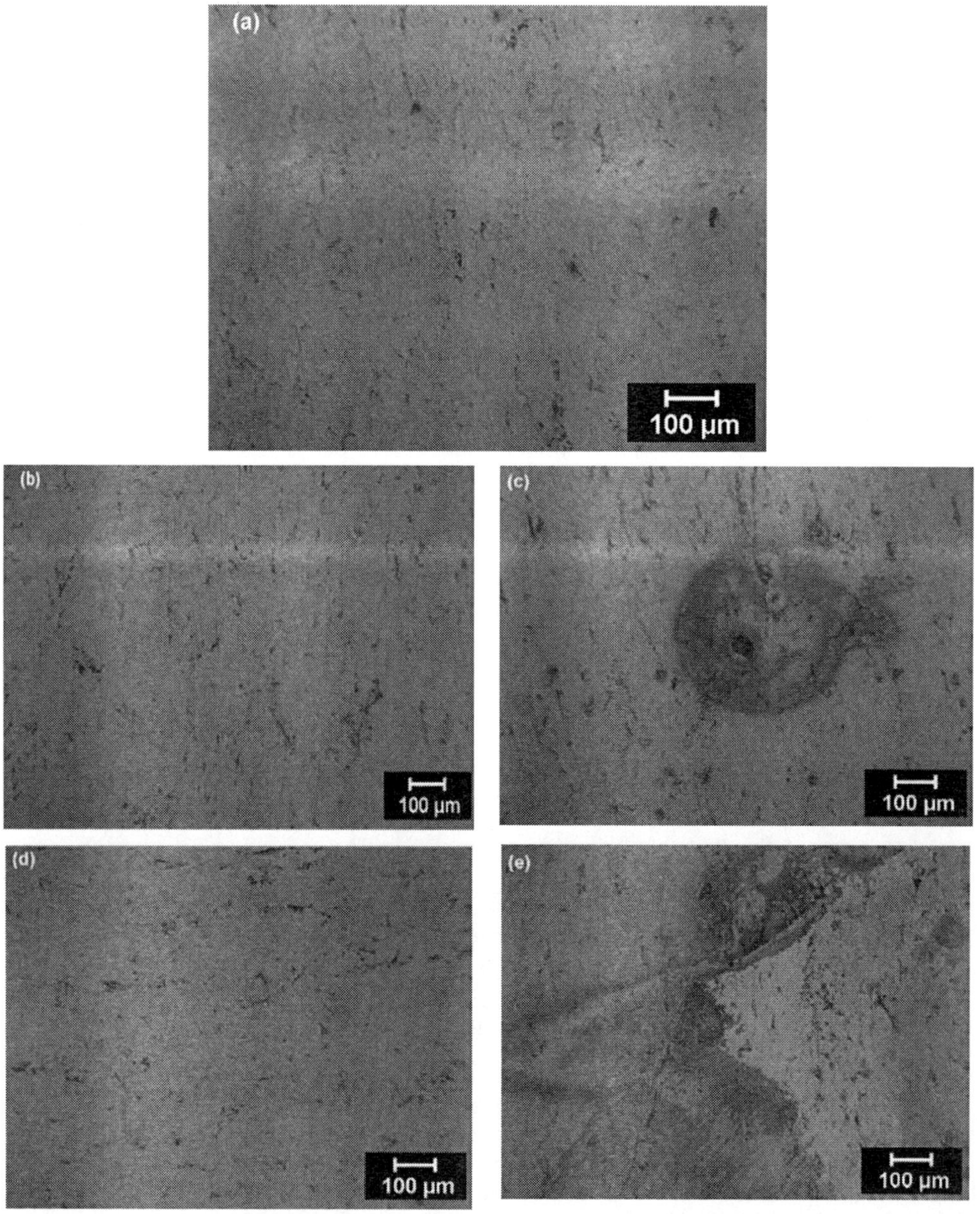

Figure 4. Optical microscopy of stainless steel 316L (a) before corrosion, (b) after corrosion in Phosphate Buffer Saline, (c) after corrosion in PBS + 1 mL/L H_2O_2, (d) after corrosion in PBS + 1 mL/L albumin and (e) after corrosion in PBS + 1 mL/L H2O2 + 1 mL/L albumin.

Figure 4 (a) shows that before corrosion tests the 316L stainless steel has a uniform surface with no defects. From Figure 4 (b) after corrosion in Phosphate Buffered Saline and 4 (d) after corrosion in PBS + 1 mL/L albumin 316L stainless steel surface are the least affected by pitting corrosion.

The samples of 316L stainless steel immersed in PBS + 1 mL/L H_2O_2 + 1 mL/L albumin (Figure 4 (e)) suffer a corrosion at the surface of the material while the 316L stainless steel surface immersed in PBS + 1 mL/L H_2O_2 present pitting damage (Figure 4 (c)), the pitting damage covers a high surface of the 316L stainless steel sample and the pits are deeper into the substrate. The optical microscopy images are in good agreements with electrochemical measurements.

4. Conclusions

The aim of this research paper is to investigate the influence of proteins (albumin) and reactive oxygen species (H_2O_2) and as well as their influence in combination, taking into account the synergistic effect of these two factors on 316L stainless steel in Phosphate buffered saline solution.

The experimental results reveal that the addition of 1 mL/L H_2O_2 into PBS electrolyte increase the rate of anodic and cathodic reactions and the specific polarization resistance decrease up to 1222 kohm cm^2. On the other hand, the addition of 1 mL/L albumin in PBS electrolyte cause the inhibition of anodic and cathodic reactions and the specific polarization resistance increase until 24970 kohm cm^2. In the presence of both, H_2O_2 and albumin in PBS electrolyte, the specific polarization resistance of 316L stainless steel decrease up to 1431 kohm cm^2 in comparison with the specific polarization resistance of 316L stainless steel immersed in PBS electrolyte.

The H_2O_2 and albumin added into PBS electrolyte increase the rate of anodic reaction simultaneously with the inhibition of cathodic reaction rate and increase the corrosion rate of 316L stainless steel.

From the optical microscopy analysis of 316L stainless steel surfaces after corrosion immersed in Phosphate Buffered Saline solution and PBS + 1 mL/L albumin are the least affected by pitting corrosion. The samples of 316L stainless steel immersed in PBS + 1 mL/L H_2O_2 + 1 mL/L albumin suffer a corrosion at the surface of the material while the 316L stainless steel surface immersed in PBS + 1 mL/L H_2O_2 present pitting damage and the pits are deeper into the substrate.

5. References

[1] Romeu C, Daniel M, Corneliu M, Georgiana B, Carmen C and Irina Z 2014 *Dig. J. Nanomater. Biostruct.* **4** 1349
[2] Biehl V and Breme J 2001 *Mat.-wiss. u. Werkstofftech* **32** 137
[3] Yu F, Owen A and Alison J D 2015 *Acta. Biomater.* **26** 355
[4] Handzlik Pand Fitzner K 2013 *Trans. Nonferrous Met. Soc.* **23** 866
[5] Lewis A C, Kilburn M R, Papagiorgiou I, Allen G C and Case C P 2005 *J. Biomed. Mater. Res. A*. https://doi.org/10.1002/jbm.a.30368
[6] Shima K, Tirdad N and Akram A 2011 *Corros. Sci.* **53** 3262
[7] Lasia A 1999 *Modern Aspects of Electrochemistry* vol 32, ed R E White, B E Conway, J O'M Bockris (Kluwer Academic/Plenum Publishers, New York) p 143

Research on Hydrophilic Nature of Polyvinylpyrrolidone on Polysulfone Membrane Filtration

L G Tiron, M Vlad and Ş Baltă*
"Dunarea de Jos" University of Galati, str. Domnească nr. 111, 800201, Galați, Romania

E-mail: stefan.balta@ugal.ro

Abstract. The membranes used in wastewater filtration are obtained from polymers, this technique is widely applied because of the small installations and low costs as against conventional systems. The polymeric membranes have high mechanical strength and flexibility, but is a challenge to improve in the same time the permeability and retention capacity of the membranes. A process that can improve the membrane properties is the addition of additives to the polymer solution, resulting in noticeable changes in the resulting membrane structure. Polyvinylpyrrolidone (PVP) is a highly hydrophilic polymer, used as a food additive that acts as stabilizer and thickening agent, which brings improvements in membrane properties. This study analyses the effect of polyvinylpyrrolidone (PVP) on the casting solution of the prepared membranes. The polymer solution was prepared from polysulfone (PSf) and N-methyl-2-pyrrolidone (NMP) at different concentrations. The membranes were obtained by phase inversion method. The PSf/PVP/NMP membranes with different concentrations were characterized by contact angle measurements, surface roughness, morphological structure and permeation tests. The results show that the hydrophilic nature of PVP improve the pure water flux, the contact angle and exhibit a higher anti-fouling property.

1. Introduction

Membrane technology is an efficient process in water treatment due to its high removal quality of toxic components and recovery of important elements from industrial effluents. This technique is widely used in food industry, chemical technology and pharmaceutical industry, also for a large area of applications [1].

This domain is of interest for researchers focused to improve the membranes durability, stability in time and high permeation properties. Most of the researches are based on high performance modification in wastewater treatment using hydrophilic membranes, polymers or hydrophilic additives incorporated in membranes [2].

Polysulfone is a thermoplastic polymer often used to prepare polymeric membranes, which contains in the main chain structure the sulfone group (SO_2) and the isopropylidene or ether groups [3-6], defined as aromatic constituents that provide high resistance to oxidation and heat [5, 6]. Polyethersulfone (PES) is another type of polymer used in membrane fabrication due to its thermal, chemical, mechanical resistances and hydraulic stability, applied in ultrafiltration process. Mahendran et al. improved the PES membrane properties by adding polyvinylpyrrolidone (PVP) in polymer casting solution and used as an organic additive [7]. Polyvinylpyrrolidone (PVP) is a hydrophilic polymer used in polymer solutions as additive to fabricate membranes by phase inversion technique [8]. PVP has the role of forming pores and water affinity increase of membranes, also [9-10].

The membranes obtained by phase inversion technique is produced by a change in the composition of constituents. A method of this technique is by immersing a polymer/solvent solution in a non-solvent coagulation bath in which a demixing of the solvent from the polymer solution occurs [11].

A disadvantage of membrane technology is the fouling phenomenon which affects retention performance of the membrane and permeation properties [12]. The fouling mechanism consist in blocking the membrane pores with pollutant particles [13]. This particles (foulants) can be organic or inorganic solutes, biological or colloids solids. There are many studies on fouling resistance improvements by changing the membrane fabrication technique or by applying different treatments on membranes already obtained [14]. Another way to avoid the membrane fouling is applied a regularly backwashing and chemical cleaning [15].

Han et al. have studied the effect of PVP addition on 15 wt.% PSf membranes and observed that the additive accelerates the phase separation of the solvent from the film of polymer solution [9]. Saljoughi et al. observed the influence of PVP concentration on pure water flux and on insulin rejection. The study shows that increasing the concentration of PVP to 1.5 wt.%, the water flux increases but the rejection capacity decreases because of macrovoid formation. Further increase in PVP concentration to 9 wt.% produces suppressed macrovoids and water flux decrease [1].

2. Methods and analysis

2.1. Flat-sheet membrane preparation

The membrane samples were obtained from Polysulfone (PSf) and N-methyl-2-pyrrolidone with 2 wt.% addition of Polyvinylpyrrolidone (PVP). Neat PSf membranes with four different concentrations of polymer (23, 25, 27 and 30 wt.%) and modified membranes with PVP additive with 2 wt.% addition in casting solution were of interest in this study, as shown in Table 1. The membranes were prepared by phase inversion method induced by immersing the casted solutions in a non-solvent coagulation bath (distilled water).

Table 1. Membrane composition.

Membrane name	PSf [wt.%]	PVP [wt.%]	NMP [wt.%]
23_PSf/NMP	23	0	77
25_PSf/NMP	25	0	75
27_PSf/NMP	27	0	73
30_PSf/NMP	30	0	70
23_PSf/PVP/NMP	23	2	75
25_PSf/ PVP/NMP	25	2	73
27_PSf/ PVP/NMP	27	2	71
30_PSf/ PVP/NMP	30	2	68

2.2. Permeation test

The pure water permeability (*PWP*, L m^{-2} h^{-1} bar^{-1}) was determined by using a dead-end cell. The permeation is obtained by measuring the pure water flux at four different pressures (P), from 6 to 12 bar, obtaining a linear regression of the pure water flux (*Jw*, L m^{-2} h^{-1}). The PWP was calculated by the following equation:

$$PWP = Jw/P \tag{1}$$

IOP Conf. Series: Materials Science and Engineering **374** (2018) 012059 doi:10.1088/1757-899X/374/1/012059

2.3. Morphological analysis
The surface membrane images were obtained by Philips FEI, QUANTA 200, a scanning electron microscope (SEM). The samples were dried for 24 hours at 50 inCdried maintained until the analysis was performed. The membranes were prepared by coating with gold to form a conductive surface.

When analyzing the membrane surface, it can be determined the roughness represented by the average of the peaks and valleys. The surface roughness is an important membrane property because it can be correlated with fouling membrane resistance [16]. The surface roughness average (Ra) was determined by using the top view SEM images in the ImageJ software [ImageJ, http://rsb.info.nih.gov/ij/]. The mechanism to obtain the roughness using ImageJ is based on using the images from SEM analysis that convert the grey contrasts in peaks or valleys, darker grey represents membrane valleys while lighter grey represents membrane peaks. The average of the valleys and peaks it is the roughness average (Ra).

3. Results and discussion
3.1. Permeation tests
Membrane permeability is inversely proportional to the increase of polymer concentration in the casted solution, as seen in Figure 1. Increasing polymer concentration reduces permeability due to decreased solvent ratio (N-methylpyrrolidone), which means that when the membranes are obtained by phase inversion (the transition from the liquid phase to the solid phase), the demixing of the solvent in the water of the coagulation bath is slower and fewer pores are formed in the membrane structure.

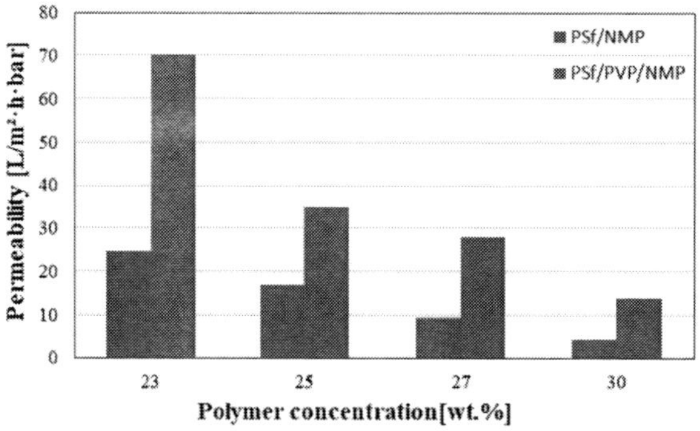

Figure 1. Permeability of PSf/NMP and PSF/PVP/NMP membranes.

The explanation for the permeability increasing when PVP additive is blended in the PSf/NMP cast solution is found in the properties of hydrophilic polyvinylpyrrolidone (increased affinity for water) and in the fact that the additive used is a hydrogen donor.

3.2. Fouling resistance
The relative flux is the ratio between the flux of distilled water and the flux of dye solution, if the value is close to 1 it approaches the ideal state of a membrane where the fouling of the membrane is insignificant.

In Figure 2 it shows that the addition of polyvinylpyrrolidone gives the membrane a low resistance to fouling. On the beginning of the test the flux is higher than at the end, therefore the relative flux is decreasing with the increased dye volume tested and the relative flux values are not close to 1. This phenomenon is explained by the density of the large pores, which means that their size is smaller and the fouling resistance decreases, which was also observed in the morphological analysis (SEM).

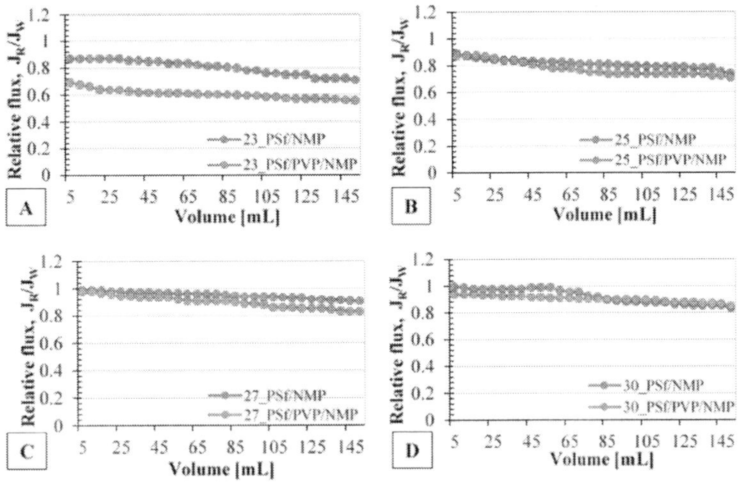

Figure 2. Influence of PVP additive in different PSf/NMP membrane concentration on the relative flux.

The highest fouling resistant membranes with PVP additive in casting solution are the membranes with 27 and 30 wt.% PSf due to their similar relative flux with the membranes without PVP addition. More than that, the membranes with 30 wt.% PSf have the same relative flux at the end of the test. The relative flux is lower for the PSf/PVP/NMP membranes because of their small pore size from the membrane surface and the fouling phenomenon appears easier because the particles from the dye solution are blocking easier in membrane pores.

3.3. Morphological analysis

Surface morphology studies were carried out by SEM analysis to evaluate surface membrane changes since the polymer casting solutions were different (Table 1).

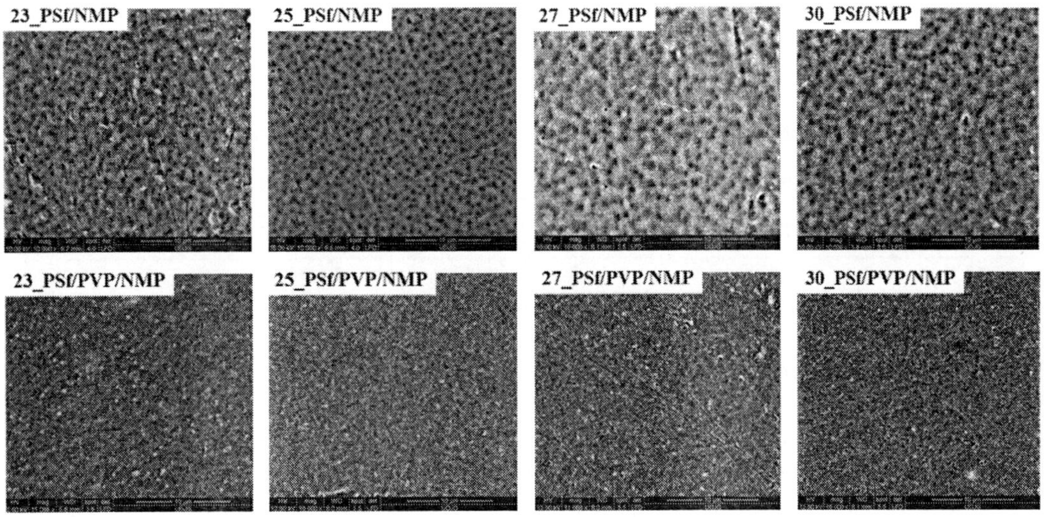

Figure 3. Top-view SEM images for neat and composite membranes.

Figure 3 shows the pure PSf membranes with the four PSf concentrations and the membranes with 2 wt.% of PVP used as additive. The PVP additive in the membrane composition has led to smaller pore sizes and is no longer as visible on the surface of the membrane, while the membranes without additive show larger pores. This tendency of pore formation in membranes containing polyvinylpyrrolidone has been well studied by Ochoa et al., 2001; Yoo et al., 2004; Susanto and Ulbricht 2009 [17-19].

This aspect can be explained by the double role of the PVP additive, the main effect being to increase the affinity of the membrane to water and the second is to act as a pore forming agent that can increase the surface of the pores or decrease in pore size depending on its concentration in the casting solution [20].

The surface roughness of the neat membranes is obviously lower than that with PVP additive membranes. It is well established that a membrane with smoother surfaces has a higher resistance to fouling.

Table 2. Roughness average of membrane surface.

PSf [wt %]	PVP [wt %]	Ra
23	0	35.2961
	2	39.7519
25	0	31.1571
	2	37.9759
27	0	28.1696
	2	30.7582
30	0	20.1573
	2	23.4298

The membrane surface show a higher roughness due to the addition of PVP, consequently it has a greater number of pores on the membrane surface, explaining the high flux of the PSf/PVP/NMP membranes compared to the flux of the pure membranes. The roughness is also influenced by the concentration of the polymer resulting in a higher roughness. The recorded surface roughness values of pure and composite membranes are shown in Table 3.

Sample morphology was investigated using the Quanta 200 electronic scanning microscope equipped with an EDX elemental composition analyzer at an acceleration voltage of 12.5 kV. The presence of the polyvinylpyrrolidone additive in membranes with different concentrations of PSf was confirmed by the EDX spectra, as shown in EDX mappings from Figure 4 and the concentration values of the elements detected on the surface of the membrane, detailed in Table 3, also.

Table 3. EDX analysis of PSf/PVP/NMP membranes, [wt.%].

Element	23_PSf/PVP/NMP	25_PSf/PVP/NMP	27_PSf/PVP/NMP	30_PSf/PVP/NMP
C	79.59	80.67	79.21	80.87
O	10.75	10.59	9.55	9.46
S	6.98	7.04	7.81	9.34
N	2.62	1.9	1.85	1.76

The EDX analisys was done on the surface of the membranes. For the studied samples the elements of interest are sulfur (S) derived from PSf polymer and nitrogen (N) from PVP additive. The sulfur element is observed at 2.3 keV and the peak for nitrogen (N) is observed at 0.392 keV, respectively.

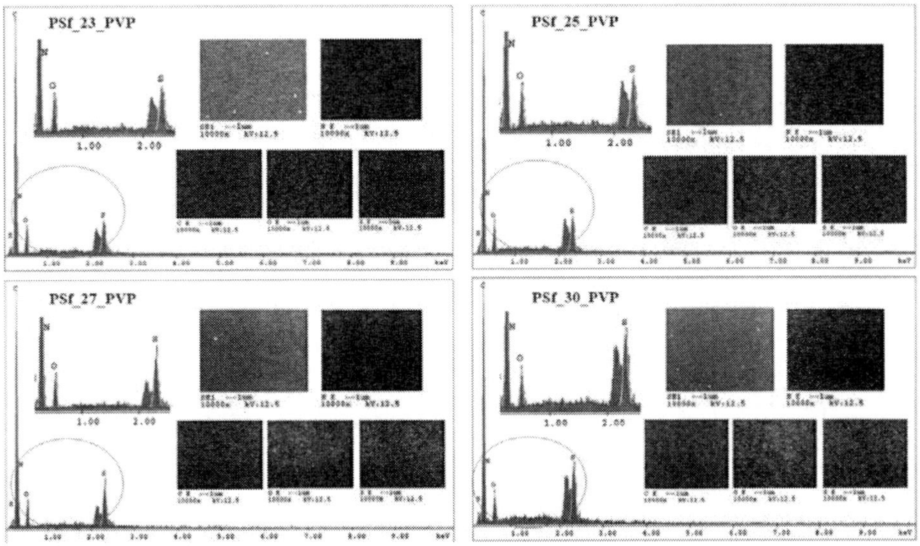

Figure 4. Spectrum and distribution of elements in EDX membrane analysis.

On the map of nitrogen (N) it can be observed a uniform distribution for all the membranes, that means that the PSf polymer concentration do not have a negative influence on the distribution of PVP additive on membrane structure.

4. Conclusions

The presence of the PVP additive in the membrane composition produces a significant effect on the performance of the polysulfone ultrafiltration membranes.

The main aspect that has been observed is that it improves the permeability of the membrane. Thus, PSf membranes with 2 wt (%) PVP addition show an increase in flux of approximately 180%, 118%, 195% and 275% for 23 wt.%, 25 wt.%, 27 wt.% and 30 wt.% PSf membranes, respectively.

Due to the ability to form the hydrogen donor complexes it has been observed that the hydrophilic property of the membranes has been improved.

5. References

[1] Saljoughi E, Amirilargani M, Mohammadi T 2009 *J. Appl. Polym. Sci.* **111** 2537–44
[2] Nigiz F U 2017 Complete desalination of seawater using a novel polyvinylidene fuoride/zeolite membrane (*Environmental Chemistry Letters, Springer*: Switzerland)
[3] Marino T, Blasi E, Tornaghi S, Di Nicolò E, Figoli A 2018 *J. Membr. Sci.* **549** 192-204
[4] Abu-Thabit N Y, Ali S A, Zaidi S M J, Mezghani K 2012 *J. Mater. Res.* **27** 1958-68
[5] Gong G, Wang J, Nagasawa H, Kanezashi M, Yoshioka T, Tsuru T 2013 *Mat. Lett.* **109** 130-33
[6] Tran A T, Patterson D A, James B J 2012 *J. Food Eng.* **2** 38-49
[7] Mahendran R, Malaisamy R, Mohan D R 2004 *Polym. Adv. Technol.* **15** 149–57
[8] Rafizah W A W, Ismail A F 2008 *J. Membr. Sci.* **307** 53–61
[9] Han M-J, Nam S-T 2002 *J. Membr. Sci.* **202** 55-61
[10] Boom J P, Punt I G M, Zwijnenberg H, De Boer R, Bargeman D, Smolders C A, Strathman H 1998 *J. Membr. Sci.* **138** 237-58
[11] van de Witte P, Dijkstra P J, van den Berg J W A, Feijen J 1996 *J. Membr. Sci.* **117** 1-31
[12] Abdel-Karim A, Gad-Allah T A, El-Kalliny A S, Ahmed S I A, Souaya E R, Badawy M I, Ulbricht M 2017 *Sep.Purif. Techno.* **175** 36-46
[13] Shultz S, Bass M, Semiat R, Freger V 2018 *J. Membr. Sci.* **546** 165-72
[14] Mohammad A W, Teow Y H, Ang W L, Chung Y T, Oatley-Radcliffe D L, Hilal N 2015 *Desalination* **356** 226-54

[15] Warsinger D M, Chakraborty S, Tow E W, Plumlee M H, Bellona C, Loutatidou S, Karimi L, Mikelonis A M, Achilli A, Ghassemi A, Padhye L P, Snyder S A, Curcio S, Vecitis C, Arafat H A, Lienhard J H, 2018 A review of polymeric membranes and processes for potable water reuse, *Prog. Polym. Sci.*, ***in press***
[16] Vrijenhoek E M, Hong S, Elimelech M 2001 *J. Membr. Sci.* **188** 115–28
[17] Ochoa N A, Prádanos P, Palacio L, Pagliero C, Marchese J, Hernández A 2001 *J. Membr. Sci.* **187**(1–2) 227-37
[18] Yoo S H, Kim J H, Jho J Y, Won J, Kang Y S 2004 *J. Membr. Sci.***236** 203-07
[19] Saljoughi E, Mohammadi T 2009 *Desalination* **249** 850–54
[20] Kourde-Hanafi Y, Loulergue P, Szymczyk A, Van der Bruggen B, Nachtnebel M, Rabiller-Baudry M, Audic J-L, Pölt P, Baddari K 2017 *J. Membr. Sci.* **533** 261–69

IOP Conf. Series: Materials Science and Engineering **374** (2018) 012060 doi:10.1088/1757-899X/374/1/012060

"In vitro" Implantation Technique Based on 3D Printed Prosthetic Prototypes

D Tarnita[1]*, C Boborelu[2], I Geonea[1], R Malciu[1], L Grigorie[1] and D N Tarnita[2]
[1]University of Craiova, 107, Al.I. Cuza Street, Craiova, Romania
[2]University of Medicine and Pharmacy Craiova, 2-4, Petru Rares Street, Craiova, Romania; Emergency Hospital of Dolj County

E-mail: tarnita.daniela@gmail.com

Abstract. In this paper, Rapid Prototyping ZCorp 310 system, based on high-performance composite powder and on resin-high strength infiltration system and three-dimensional printing as a manufacturing method are used to obtain physical prototypes of orthopaedic implants and prototypes of complex functional prosthetic systems directly from the 3D CAD data. These prototypes are useful for in vitro experimental tests and measurements to optimize and obtain final physical prototypes. Using a new elbow prosthesis model prototype obtained by 3D printing, the surgical technique of implantation is established. Surgical implantation was performed on male corpse elbow joint.

1. Introduction

Rapid Prototyping (RP), a relatively new class of manufacturing technology, is the name for several techniques for reading the data provided by computer-aided design (CAD), three-dimensional (3D) automatic design and for manufacturing of virtual objects layer by layer, depending on the design [1].

It started to develop at the beginning of 1980s and many papers were written in this new research field [1-6].

A prototyping process can be called rapid prototyping [3] if:

- The process is based on using 3D CAD data.
- The prototype part is created automatically.
- The model is ready in a few minutes, few hours or few days.

Current basic application of RP systems consists in reducing the time in which a new product is (almost) finished: right before making large expenditures on creating manufacturing equipment [1]. Testing a completely functional prototype part at this moment offers the opportunity of localizing the design errors and their correction while the costs of change are still small, errors that may have been unnoticed in testing only CAD 3D model [1-3]. The important aspect of RP process is the automatic translation of 3D CAD model to the physical model, given that the technology used does not have too much importance [5, 6].The key idea of this new rapid prototyping technology is based on decomposing 3D model in thin cross sectional layers, followed by a physical form of the layers and their bonding "layer after layer" [3].

The objective of this study is to obtain prototypes for human bones and for different types of elbow prosthesis using 3D Printing in order to apply them in experimental testing and surgical techniques establishing of in vitro implantation, developing methods that increase the process flexibility and reduce production costs.

IOP Conf. Series: Materials Science and Engineering **374** (2018) 012060 doi:10.1088/1757-899X/374/1/012060

2. 3D Printing

The three-dimensional printing (3DP) is a RP technology, used to create complex 3D parts directly from a computer model of the part, with no need for tooling. Advances in both process capabilities and RP materials have led to the possibility of rapid manufacture of finished components directly from 3D CAD data [7-10, 12-14]. In recent years 3DP has become a very competitive process in terms of cost and speed, creating high accuracy filler structures for the fabrication of complex 3D prototypes from powdered materials. A large number of models can be produced during the product development phase, taking into account the comparatively high speed and low operational cost of 3D printers [7-19].

The general scheme of the 3D printer fabrication principle is shown in Figure 1 [9]. A multichannel head deposition is moved in xy coordinates on the surface of the platform. The prototypes are highly accurate in the XY plane, but they have a scale effect, step by step in z direction. If the model is disposed with very thin layers having only some microns in size, the model looks like the original [2, 13-20].

Instead of printer ink, this head of deposit leaves drops of adhesive in the tank, which will strengthen in contact with particles of material deposited in a thin layer on the working platform [7, 9]. The cavity 1 with moving piston1 contains the initial quantity of powder Figure 1 (a). The process begins by spreading of a thin layer of powder taken from cavity 1 over a build platform 2 Figure 1 (b). Parts are created inside the cavity that contains a powder bed supported by the moving piston 2, by a layered printing process. Computer software splits the 3-D CAD data into a series of thin horizontal cross-sections. Each new layer is fabricated through lowering of the piston by a layer thickness and filling the resulting gap with a thin distribution of powder [7, 9]. An inkjet printing head prints a binder solution onto this layer of powder to form a slice of the 3-D CAD file Figure 1 (c). The layering process is repeated until the part is completed Figure 1 (d). Following a heat treatment, which consolidates the bonded material, the unbound powder is removed, leaving the fabricated part behind [7, 9].

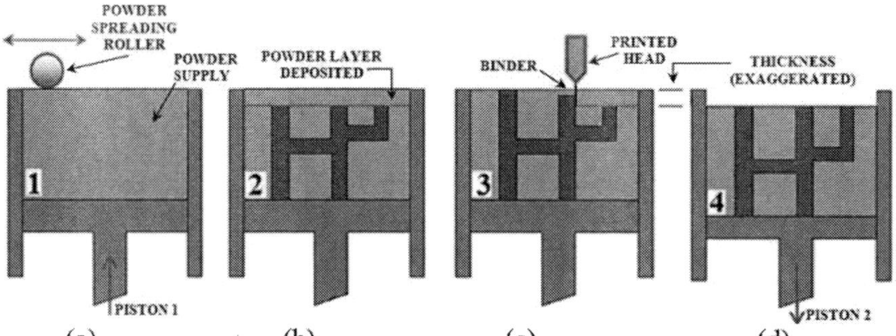

(a) (b) (c) (d)

Figure 1. A schematic representation of the 3D printer fabrication principle [9].

3. Material and method

Using the Rapid Prototyping 3D Zcorp 310 Printer system Figure 2, from the Faculty of Mechanics, Craiova, we manufactured prototypes of human bones, orthopaedic implants, endoprosthesis, all of this being necessary for research, for in vitro simulations, experimental tests, experimental measurements etc. Printing System Z 310 Plus creates physical models directly from digital data with a construction volume of 203 mm x 254 mm x 203 mm and a thickness of 0.076 mm to 0.254 mm [7, 9, 11].

3.1. Stages of 3D printing

In rapid prototyping process of 3D printing type, the following stages are passed through:

- 3D geometric modeling.
- Transfer of CAD model to the sectioning processor.
- Prototyping proper.

- Extraction of prototypes from the powder tank.
- Keeping the prototype in the tank for a period of time imposed by the machine in order to dry the bonding liquid and to harden the prototype.
- Cleaning and finishing.
- Infiltration with epoxy resin.

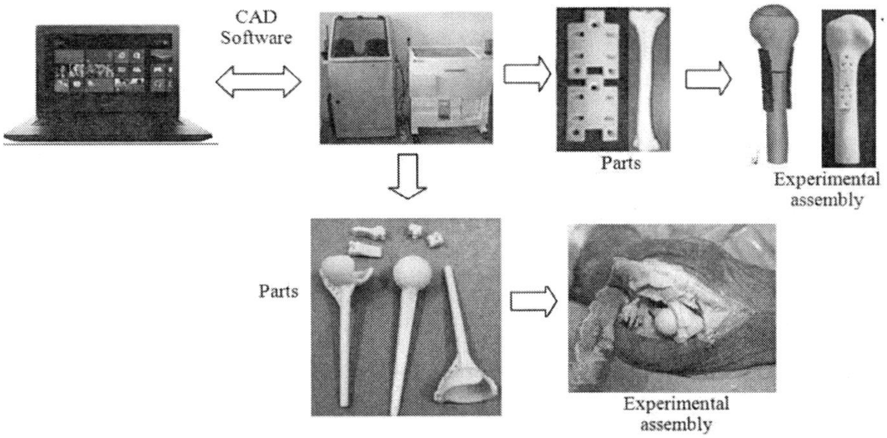

Figure 2. The Zcorp 3D printer and the methodology in order to obtain an experimental assembly using RP.

3.1.1. 3D geometric modeling. For bones prototypes Solid Works was used to geometrically reconstruct them based on a large number of electronic cross sections obtained using CT imaging device. Newly designed components, implants or prosthesis were designed as 3D virtual models using Solidworks or ANSYS software through original design or based on existing models on the medical market and continuing to optimize them [14-19]. In order to generate and design the new virtual geometric model of elbow prosthesis with ball coupling, the conditions imposed by elbow joint anatomy, by surgical prosthetic techniques and by the analysis of the cinematic parameters of unaffected elbow biomechanics were analyzed. It was also intended to obtain prosthesis with fewer components, with simpler technologic process, after athorough analysis of the constructive or functional advantages, disadvantages, shortcomingsand deficienciesof the elbow prostheses currently used elbow prostheses currently used for elbow joint reconstruction [20].Solid Works program was used for the parametric modeling of ball coupling prosthesis, the model being designed to allow the modification of dimensions [20] for its adjustment to groups of patients using the complete geometric parameterization property.

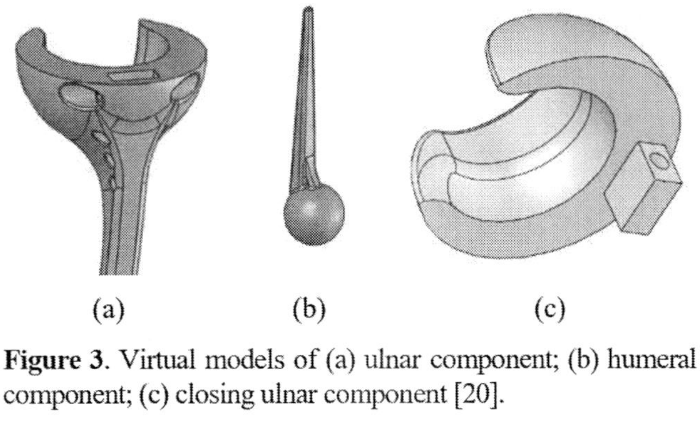

Figure 3. Virtual models of (a) ulnar component; (b) humeral component; (c) closing ulnar component [20].

IOP Conf. Series: Materials Science and Engineering **374** (2018) 012060 doi:10.1088/1757-899X/374/1/012060

The model of elbow prosthesis with ball coupling Figure 4 was designed using the assembly module of the program (Assembly) and specific commands that cancel certain degrees of freedom.

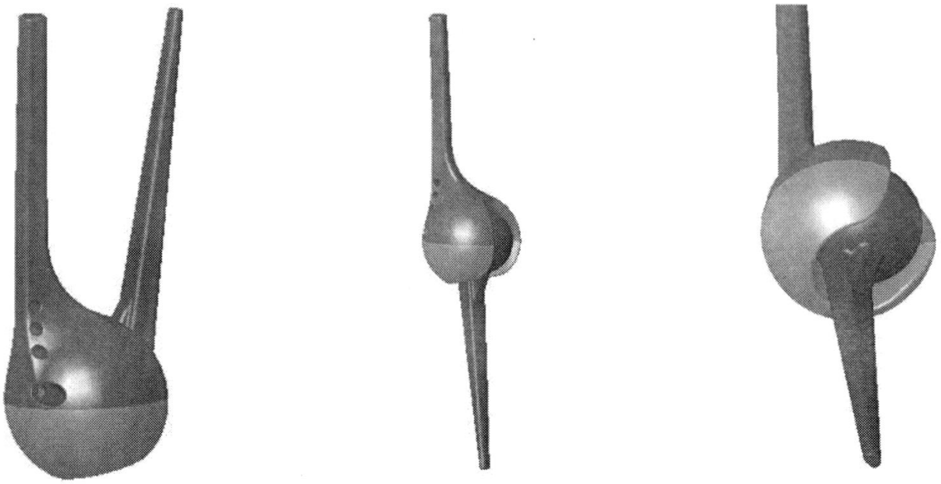

Figure 4. Model of elbow prosthesis with ball coupling [20].

3.1.2. Transfer of CAD model to the sectioning processor. In many cases, sectioning can be made by the CAD program used for modeling. The most common method for sectioning and construction is the approximation of the model with plane triangular elements Figure 5 and Figure 6.

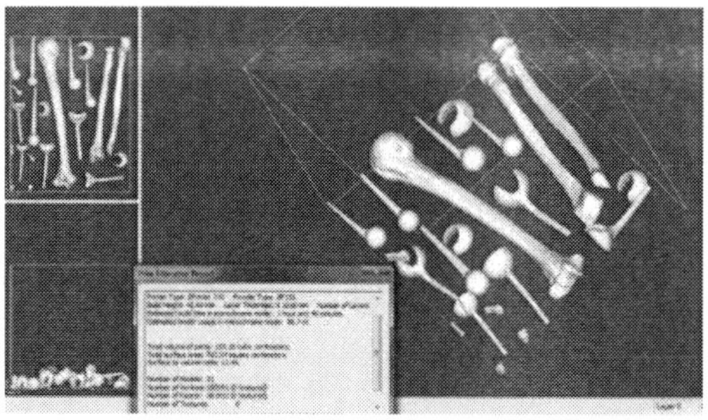

Figure 5. Transfer of CAD models in Zcorp processor.

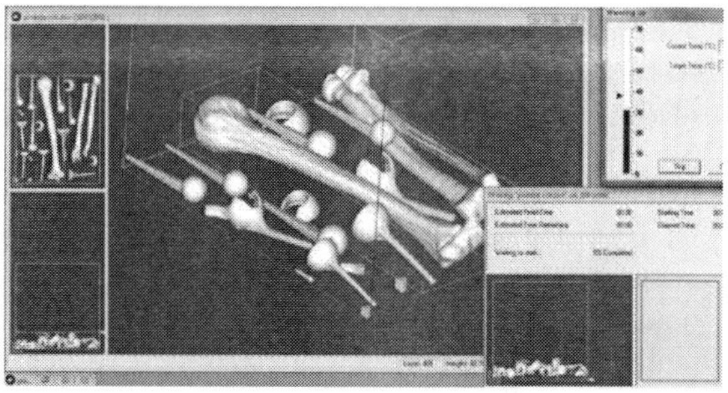

Figure 6. Verification of temperature increase stage to the value imposed for starting prototyping process.

3.1.3. Prototyping process

Figure 7. Printing process.

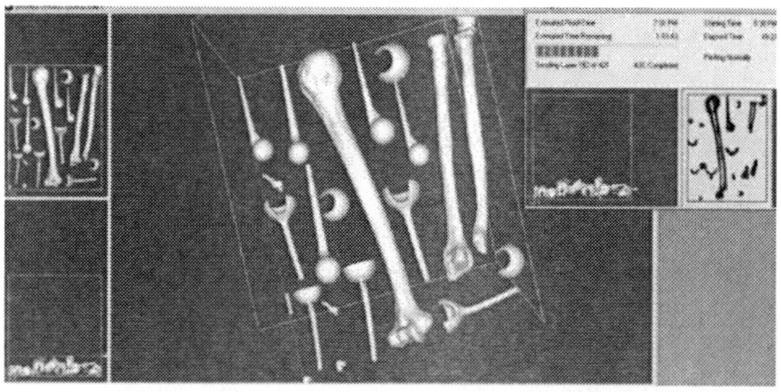

Figure 8. Verification of prototyping stage.

IOP Conf. Series: Materials Science and Engineering **374** (2018) 012060 doi:10.1088/1757-899X/374/1/012060

3.1.4. Extraction of prototypes from the powder tank

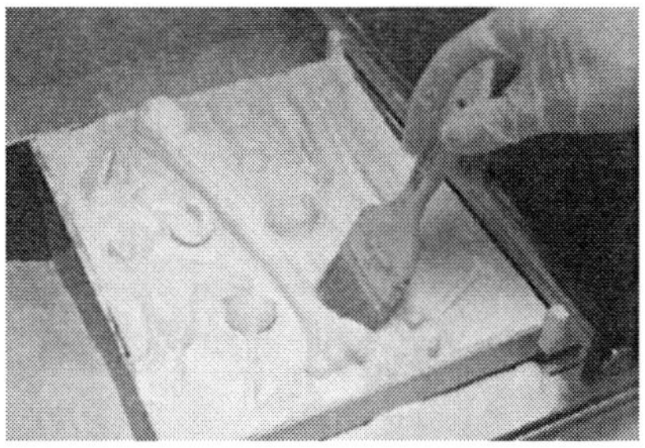

Figure 9. Extraction of prototypes from the powder tank.

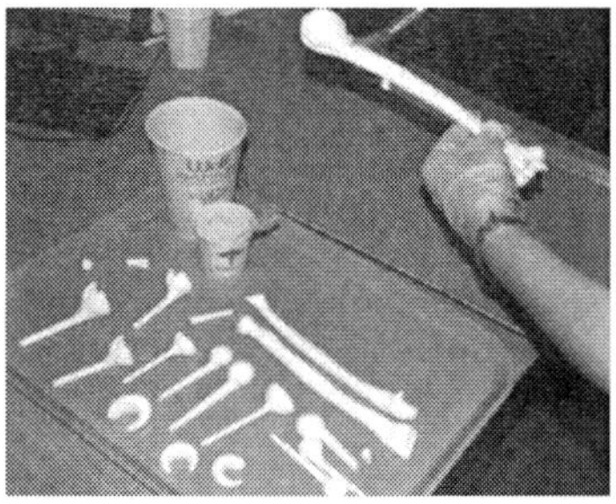

Figure 10. Infiltration with epoxy resin.

3.1.5. Keeping the prototype in the tank for a period of time imposed by the machine in order to dry the bonding liquid and to harden the prototype.

3.1.6. Cleaning and finishing. These are operations in which the supports used in construction and the excess material are eliminated or other processing operations (including mechanical) aimed to improve the dimensional precision and the quality of surfaces.

3.1.7. Infiltration with epoxy resin. 3 DP components have low mechanical properties, and they must be infiltrated with resin to improve their characteristics Figure 10. Infiltration is the process of applying a liquid resin to a printed part to provide strength and impart specific properties. After infiltration with Z-Max ™ which is a two-part system: Z-Max™ Infiltration Resin and Z-Max™ Infiltrant Hardener, the parts are hardened at room temperature, within 12-24 hours [7, 9]. Once infiltrated, the parts can be easily processed, exploited, burnished and processed. The material Zp131 is a high-performance powder with composition containing ingredients such as: Crystalline Silla Plaster containing <1% vinyl polymer, carbohydrate, salt and sulphate produced consumables Technologies Ltd, (www.technologysupplies.co.uk). 60 Zb clear binder mixture is a formula to be used to fuse the powder to achieve 3D models in rapid prototyping [7, 9].The process achieve in a few minutes to several hours, the three-dimensional physical models of the objects starting directly from the electronic format 3D CAD/CAM, the only equipment used being the 3D printer [7, 9].

4. Results

The result consists in very strong models, up to 43 MPa of flexural strength and up to 98 MPa of compression strength. Parts made with Z-Max™ are hard and rigid so they dont deform under load [7, 9]. As results, we present some examples of human bones like humerus and radius Figure 11 and two models of elbow prosthesis: the new spherical model Figure 12, the elaboration of the virtual model being presented in [20] and the Latitude prosthesis Figure 13, a well known actual elbow prosthesis, usually used in elbow orthopedic surgery [9].

The prosthetic system is of the ball head type consisting of three parts: a part implanted in the humerus, a part implanted in the ulna bone and a third part transforms the assembly from an unconstrained prosthesis in a semi-constrained prosthesis. The metallic implant is fixed through cementation in the humerus and ulna bones.

The newly created prosthesis fulfils all the criteria in order to be considered a prosthesis superior to those used so far [20], namely:

- criterion of simplicity – regarding the design and the implantation; the implantation kit is simple and it does not use special tools.
- criterion of conservation – meaning the ability to sacrifice a minimal amount of healthy tissue;
- criterion of manufacturing – the simplicity of the prosthetic components allows them to be made with classical tools with no special equipment;

IOP Conf. Series: Materials Science and Engineering **374** (2018) 012060 doi:10.1088/1757-899X/374/1/012060

- criterion of cost – due to the simplicity of the components, the costs are much smaller than in the case of current prostheses;
- criterion of safety –the implantation in a human corpse proved the safe operation of the prosthetic components.

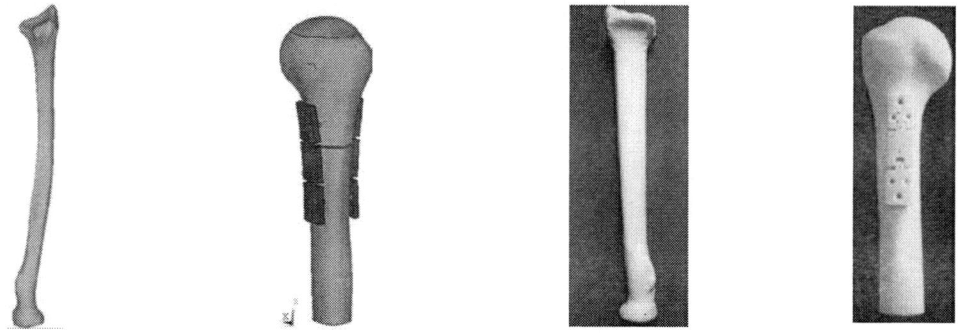

Figure 11. Virtual and prototiped bones: radius, humerus and phalangies [7, 9].

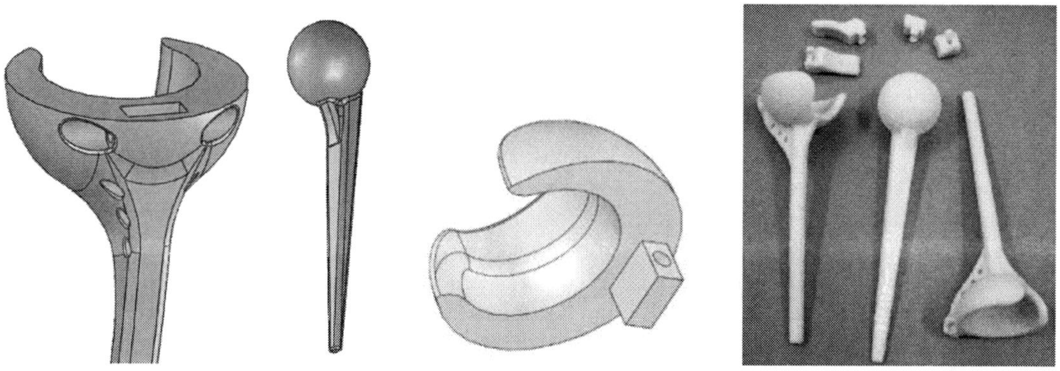

Figure 12. The virtual model and the prototype of the components of the spherical prosthesis.

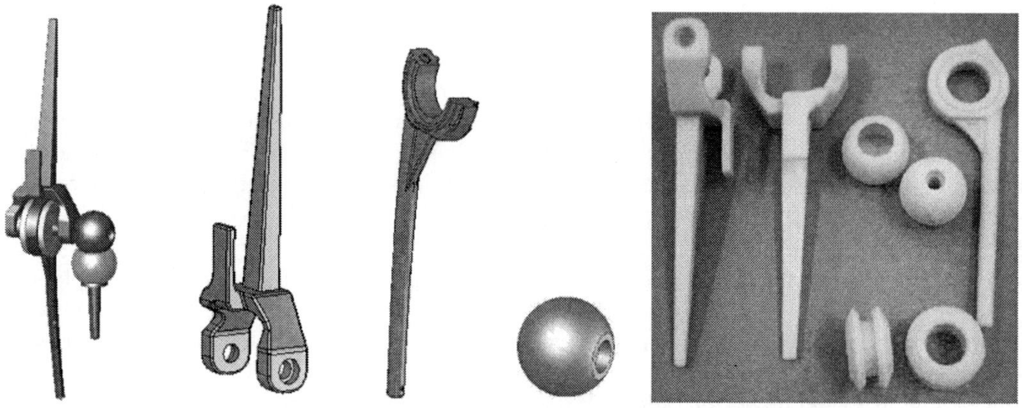

Figure 13. The virtual model of Latitude elbow prostheses, the virtual models of some components and some of the prototyped components [9].

The experimental part of the new model of elbow humero-ulnar prosthesis implantation on the elbow joint was carried out in one of the dissection laboratories belonging to the Department of Human Anatomy of the Faculty of Medicine of Craiova.

Surgical implantation was performed on the elbow joint of a male body preserved by embalming by injection with alcohol and glycerol solution, stored in a refrigerator dedicated to body preservation.

The surgical technique for implanting elbow prosthesis according to the invention involves the following steps:

- an incision is made on the dorsal face of the elbow, long about 10 cm, 5 cm on the cubital edge of the forearm and 5 cm on the dorsal face of the arm, Figure 14.

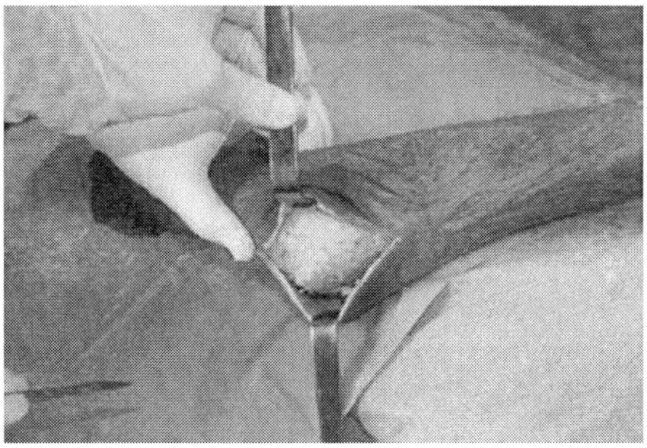

Figure 14. Incision on the dorsal face of the elbow.

- a longitudinal incision of the cubitus periosteum along the whole length of the tegument incision is made.
- the periosteum of the cubitus proximal epiphysis is removed using a cutting specific instrument.
- the olecranon is transversely cut with an oscillating saw, Figure 15.
- two longitudinal incisions are made using a scalpel at the level of the triceps brachii muscle tendon and the trochlea of the humerus pallet is pointed out, Figure 16.

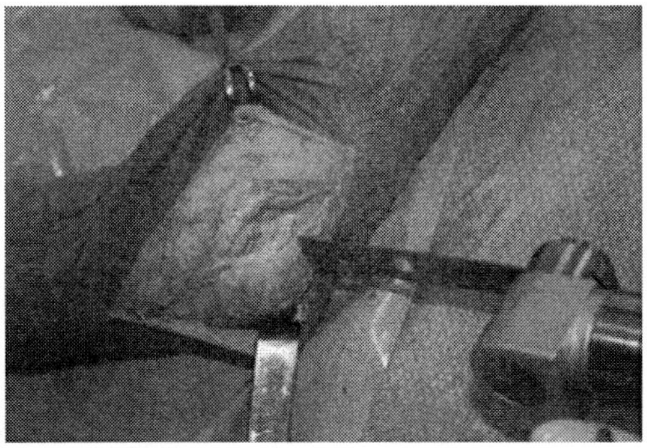

Figure 15. The olecranon transverse cut.

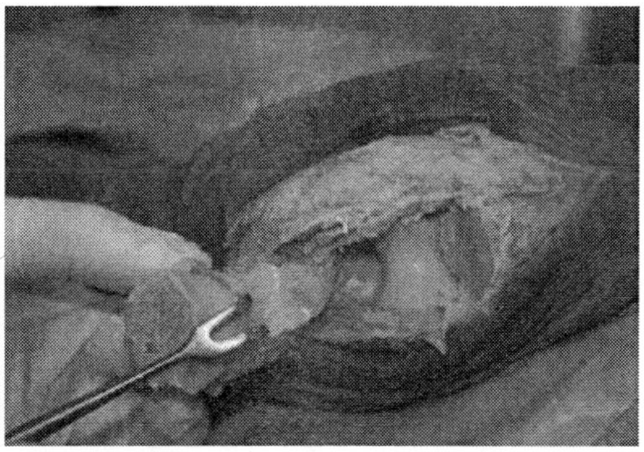

Figure 16. Longitudinal incisions at the level of the triceps brachii muscle tendon.

- with an oscillating saw, the trochlea of the humeral pallet is cut out by two longitudinal sections at the level of the trochlear ridges and a transverse one above the trochlea. By the trochlea removing, the humeral medullary canal can be seen.
- the cubitus medullary canal is prepared with a calibrated rasp after the stem of the cubital component, Figure 17.
- also the humeral medullary canal is prepared with a rasp calibrated for the dimensions of the stem of the prosthesis humeral component, Figure 18.

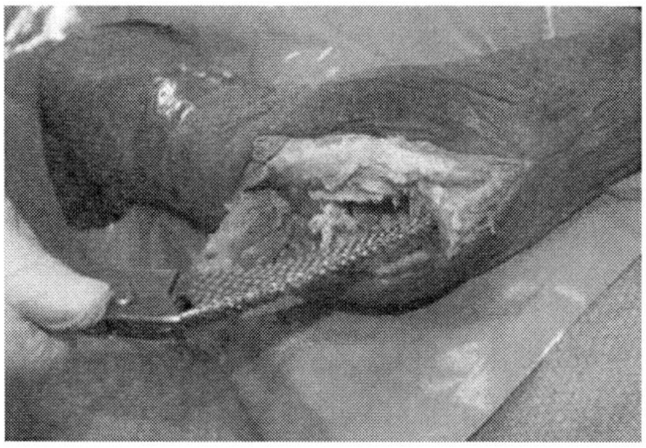

Figure 17. The cubitus medullar canal preparation.

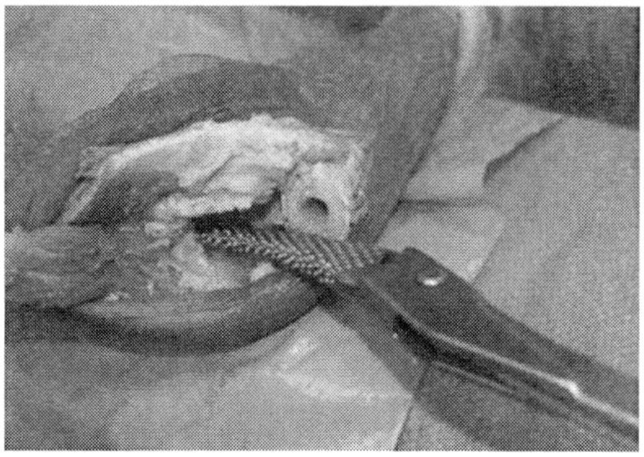

Figure 18. The humeral medullar canal preparation.

- the test is made by positioning the two components of the prosthesis in the two bones, Figure 19 and Figure 20.

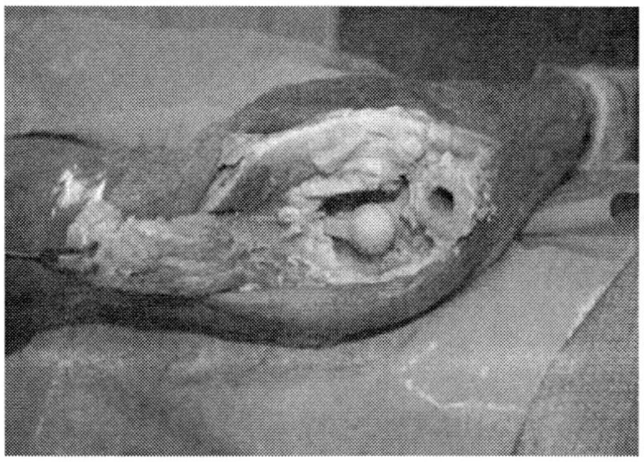

Figure 19. Positioning of the first component of the prosthesis in the bone.

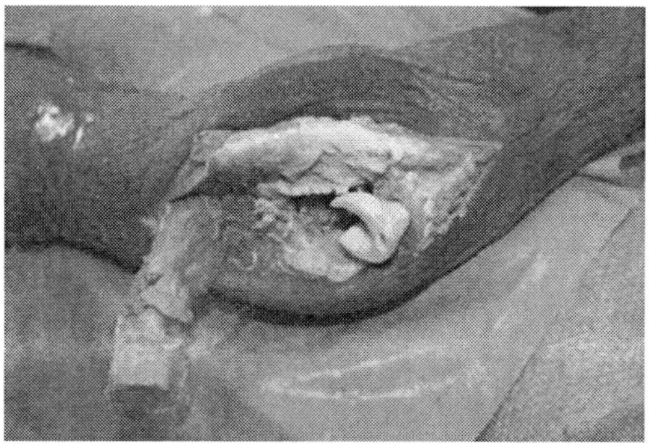

Figure 20. Positioning of the second component of the prosthesis in the bone.

- the stems of the two prostheses are cemented into the corresponding bones;
- the two components of the prosthesis are coupled in the non-constrained or semi-constrained manner, Figure 21 and Figure 22.

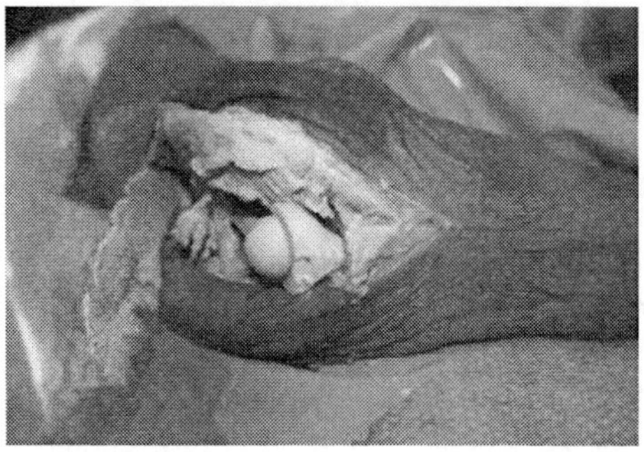

Figure 21. Coupling of the prosthesis components in the non-constrained manner.

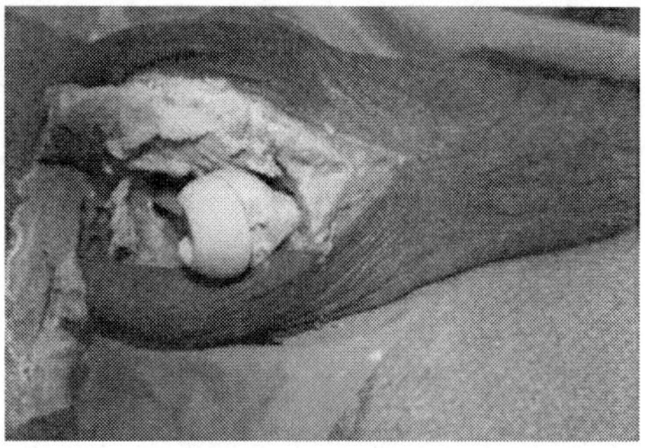

Figure 22. Coupling of the prosthesis components in the semi-constrained manner.

- the olecranon is resected.
- the tricipital tendon is armed with special wire threads bind to the special holes of the cubital

component, Figure 23.

- the two longitudinal incisions made in the triceps tendon are sutured with resorbable threads; the suture of the skin wound is performed in anatomic layers, Figure 24.

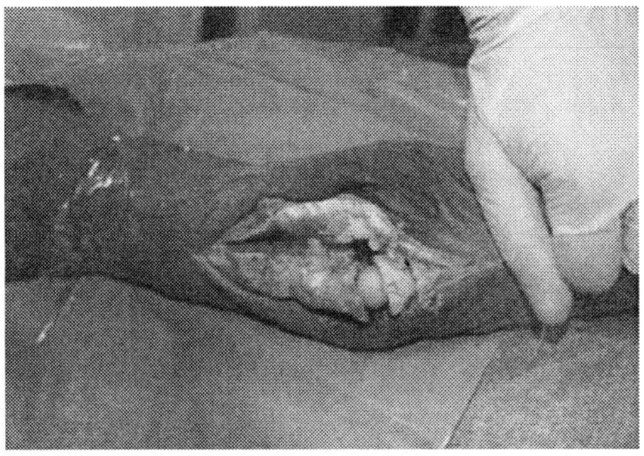

Figure 23. The tricipital tendon armed with special wire threads.

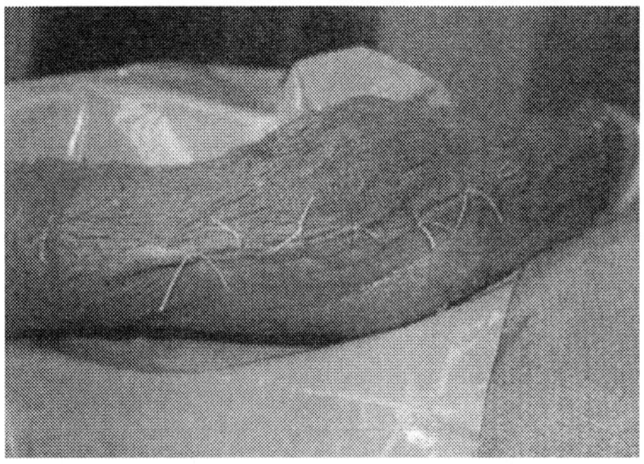

Figure 24. The suture of the skin wound.

5. Conclusions

Rapid prototyping (RP), a relatively new class of manufacturing technology, has the potential to create different parts or assemblies like human bones, prosthesis, orthosis or other biomedical devices, directly from the 3D CAD data. Rapid Prototyping method exceeds the conventional manufacturing methods in terms of accuracy and fabrication time. Using Rapid Prototyping as a fabrication method one can finally obtain functional assemblies which can be used afterward in various experiments (for example, kinematic or dynamic studies) just as in the case of conventional metal based assemblies.

Rapid Prototyping ZCorp 310 system based on high-performance composite powder and on resin-high strength infiltration system and three-dimensional printing may be also used as a prototypes manufacturing method in industrial field because of the rapidly evolving industry requirements to validate products with low cost using new in vitro methods.

3D Printing is a Rapid Prototyping technology, an easy and versatile process making complex varied geometries by combining a 3D printer, CAD development software and special materials from which the prototype will be created. The complexity of the shapes and dimensions of the fabricated parts recommends Rapid Prototyping method as a future fabrication method in an increasingly number of applications and scientific fields of research.

In medicine, physical prototypes of orthopaedic implants and prototypes of complex functional prosthetic systems obtained directly from the 3D CAD data by Rapid Prototyping allow in vitro experimental tests and measurements in order to optimize and obtain final physical prototypes.

Developing the most suitable mechanical model and its validation in vitro experiments by reproducible testing spurred a significant research interest into biomaterials using because many biomaterials fail after implantation, so comparative biomechanical studies help a lot to improve the knowledge.

These 3D printed prototypes made before final prototypes manufactured from expensive materials allow constructive optimization and surgical techniques establishing at low cost.

6. References

[1] Brown R, Lightman A and Wohlers T 1994 Need for Expanding the Development and Refinement of Rapid Prototyping Technologies *Newsletter*ed Society of Manufacturing Engineers (Toronto)

[2] Chua CK and Leong KF 2000 Rapid Prototyping:Principles and Applicationsin Manufacturing, *World Scientific*

[3] Gebhardt A2003 *Rapid Prototyping*ed Hanser Gardner Publications Inc. (Cincinnati)

[4] Pandey P M, Reddy N V and Dhande SG2003 *Rapid Prototyping Journal* **9**(5) 274-88
[5] Pham D T and Dimov S S 2001 Rapid Manufacturinged Springer-Verlag London Limited
[6] Wohlers T T 2005 Wohlers Report 2005, Rapid Prototyping, Tooling and Manufacturing: State ofthe Industry *Annual Worldwide Progress Report* ed Wohlers Associates Inc. (Fort Collins CO USA)
[7] Tarnita D et al. 2010 The Three-Dimensional Printing – a Modern Technology Used for Biomedical Prototypes, *Materiale plastice* **47**(3) 328-34
[8] Herbert N, Phil M, Simpson D, Spence W D and Ion W 2005 A preliminary investigation into the development of 3-D printing of prosthetic sockets *JRRD* **42**(2) 141–6
[9] Tarnita D, Tarnita D N, Boborelu C and Catana M 2013 *Academic Journal of Mechanical Engineering, Academic Journal of Manufacturing Engeneering* **111**/2013 119-24
[10] Tsung-Jung H and Wei-Hsiang L 2010 *Journal of the Chinese Institute of Engineers* **33**(1) 121-30
[11] Tarnita D, Tarnita D N and Bolcu D 2011 Orthopaedic implants based on shape memory alloys Biomedical Engineering – From Theory to Applicationsed InTech Publishing House (Viena)431-68
[12] Thilmany J 2001 Printing in Three Dimensions *Mechanical Engineering* Periodical article, Issue date: May 1.
[13] Dimitrov D, Schreve K and de Beer N 2006 *Rapid Prototyping Journal* **12**(3) 136-47
[14] Tarnita D, Popa D, Tarnita D N and Grecu D 2006 *Romanian Journal of Morphology and Embryology* **47**(2) 181-6
[15] Tarnita D, Boborelu C, Popa D and Rusu L 2010 *Romanian Journal of Morphology and embryology* **51**(3) 489-95
[16] Tarnita D, Tarnita D N, Bizdoaca N, Tarnita C, Berceanu C and Boborelu C 2009 *Romanian Journal of Morphology and embryology* **50**(3) 447-52
[17] Tarnita D, Tarnita D N, Popa D, Grecu D and Niculescu D 2010 Numerical simulations of human tibia osteosynthesis using modular plates based on Nitinol staples, *Romanian Journal of Morphology and embryology,* **51**(1) 145-150
[18] Tarnita D, Tarnita D N, Hacman L, Copilusi C, Berceanu C and Cismaru F 2010 In vitro experiment of the modular orthopedic plate based on Nitinol, used for human radius bone fractures, *Romanian Journal of Morphology and embryology* **51**(2) 315-20
[19] Tarnita D, Tarnita D N, Tarnita R, Berceanu C and Cismaru F 2010 Modular adaptive bone plate connected by Nitinol staple, *Material wissenschaft und Werkstofftechnik, Materials Science and Engineering Technology, Special Edition Biomaterials*, ed Willey-Vch **41**(12) 1070-80
[20] Tarnita D, Boborelu C, Popa D and Tarnita DN 2018 Design and Finite Element Analysis of a New Spherical Prosthesis-Elbow Joint Assembly, *New Advances in Mechanism and Machine Science,* Proceedings of The 12th IFToMM International Symposium on Science of Mechanisms and Machines

IOP Conf. Series: Materials Science and Engineering **374** (2018) 012061 doi:10.1088/1757-899X/374/1/012061

Wastewater Sludge Used as Material for Bricks Fabrication

N R Jianu[1], I C Moga[1], F Pricop[2] and A Chivoiu[1]
[1]PRO Mediu Dunărean SRL, 351 București, Giurgiu, Romania
[2]The National R&D Institute for Textile & Leather, 16 Lucrețiu Pătrășcanu, Bucharest, Romania

E-mail: corinamoga@yahoo.com

Abstract. Current world trends related to wastewater sludges are: reuse in agriculture, utilization as retaining material for petroleum products or utilization in construction. Bricks from sand-cement or autoclaved cellular concrete are commonly used in construction. The authors propose innovative receipts for bricks and plasters based on textile wastewaters sludge. Centrifuged sludge is mixed with cement to obtain bricks and plaster. For bricks, the mixture is represented by 45% cement and 55% sludge. The paper presents the obtained results and the new materials used for bricks fabrication.

1. Introduction

Wastewater sludge is considered as a waste and it is generated by the wastewater treatment plants (WWTPs), when the water/wastewater and suspended solids are separated. The treated water is discharged to aqueous environment or reutilized while the sludge is removed for further treatment and final disposal [1]. Considering the large number of WWTPs the amount of the generated sludge is also significant. The specialists around the world, are continuously seeking for cost-effective solutions for a future utilization of the generated sludge.

A wide variety of sludge treatment technologies is used by specialists around the globe. The treatment technologies depend on both the quality and quantity of the generated sludge. Regarding sludge stabilization, anaerobic and aerobic digestion are the most common used methods in European Union (EU) [2]. Anaerobic digestion is most commonly used in Italy, Spain, Slovakia, UK, Spain and Finland [2]. Aerobic digestion is the preferred in countries like Czech Republic or Poland [2].

Chemical stabilization, conditioning with lime or other chemicals are generally of minor importance, although they are used mainly in some EU countries. However, in most EU countries (including Romania) the sludge disposal is still an important method used for sludge management.

Industrial wastewaters are treated through various processes. The industrial wastewater sludge resulting from the treatment process has a variable physico-chemical composition, depending on the composition of treated wastewater and the nature of the chemicals used for the treatment [3]. This resulting sludge is considered a waste, according to the EU environmental legislation. The sludge must be classified according to the physicochemical characteristics and according to the environmental legislation in force (In Romania: Law 211/2011, Regulation 1272/2008, HG 856/2002) in classes of hazardous, non-hazardous or inert waste. According to this classification, the waste is transported, inerted and stored in specialized warehouses on waste classes. Transport costs, inertization and storage involve high costs, which have a final effect on total wastewater treatment costs.

2. New receipt for bricks realization from wastewater sludge

The wastewater sludge utilization for bricks realization is known and few experiments are available in the literature [4]. The authors proposed another innovative technology/receipt, to be implemented at the industrial application level, that is based on the patent no. RO 125384/30.06.2014 entitled "Procedure for obtaining brick and plaster construction materials".

Starting from the necessity of valorization of the sludge resulted from an industrial WWTP, its aim is to transform the sludge into an inert waste (by inclusion in a matrix that does not allow the leaching of chemical elements), which may have various utilities (in the present case the use as building material was pursued: bricks, plasters). Up to now, fired clay bricks or sand with cement autoclaved aerated concrete are used in construction.

The new receipt for bricks realizations is based on a mixture between sludge and cement. The sludge inertia process requires a humidity of less than 80%. For this purpose, it is centrifuged to reduce the humidity from (90-95)% to a maximum of 80%. The preparation of the sludge sample, starting from the characteristics of the treated wastewater to the dehydration process, has two purposes: the relevance and success of the experiment i.e. access to a positive result in terms of the initial objective. Each stage of the process will be influenced by the previous ones, the links between the initial conditions and the results are affected by a complexity of factors. According to the mentioned patent, the main stages of the sludge recovery and inertization process are:

a) obtaining sludge by treating wastewater;
b) sludge dewatering by filtering through bag filters and/or centrifugation;
c) exploitation of sludge by inerting it with cement.

3. Methods and methodology

The authors performed a series of experimental determinations for bricks realization. Was considered a WWTP from an industrial and technological park, where seven textile companies have the headquarters. The experiments were realized in March 2017, when the wastewater characteristics were monitored. The quality indicators for the generated wastewater are presented in Table 1.

Table 1. Physical and chemical characteristics of wastewater from which raw sludge for experiments is obtained.

Parameter	**Units**	**Values**
		Period 01.03.2017 - 08.03.2017
pH	unit. pH	8.74 - 8.92
Suspended solids	mg/l	511.0 - 624.0
Biochemical Oxygen Demand BOD	mg O_2/l	342.19 - 367.42
Chemical Oxygen Demand COD	mg O_2/l	674.28 - 686.06
Detergents	mg/l	3.04 - 3.26
Total Phosphorus	mg/l	2.64 - 2.88
Sulfates	mg/l	378.24 - 394.62

The evaluated sludge was used for the bricks realization. Samples of sludge were collected and used in the experiments. Sampling involves sampling at different points of the investigated batch. According to the working procedures, for a maximum of 2 tons of sludge, 10 samples of 200 grams were taken, forming a homogeneous sample of 2 kg. Ten such samples of 2 kg were taken. These were then homogenized and constituted the sample to serve for the experiments to be carried out.

Samples were taken in 2.000 ml plastic or borosilicate containers, washed and kept in 5% (v / v) nitric acid solution for 24 hours and rinsed 2-3 times with distilled water to which is added acid for preservation); then it is tightly closed.

Transport and conservation of the sludge probe is made at a temperature of (2-5)°C and in safe conditions to avoid damaging or destroying the container and contaminating or losing the sample. The constituted sludge sample is coded and sampled in 3 sub-samples for the experimental tests to be performed.

Figure 1. Preserved sludge sample for physicochemical analysis - sample code N8.

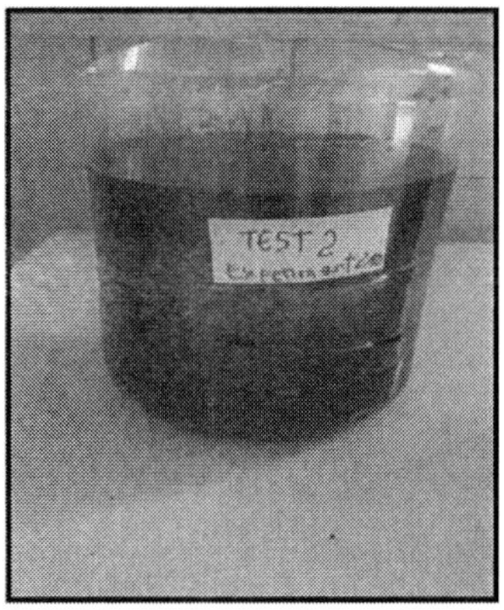

Figure 2. Raw sludge used in experiment 2.10.

4. Experimental results

The wastewater sample is treated with aluminum sulphate, 0,5 g/l, for the precipitation and coagulation of organic and inorganic substances. The resulting sludge after 24 hours of settling is filtered through sludge bags and centrifuged on a laboratory centrifuge. The dehydrated sludge is mixed with cement (45% cement + 55% sludge). The mixture is homogenized for 15 minutes until a homogeneous paste is obtained which is poured into metal or wood mold. Allow to dry for 24 hours, remove from the mold, leaving in the open air for 3-4 days to bring in contact with the slurry from the sludge with CO_2 in the air, which helps to increase the brick's resistance (Figure 3).

Figure 3. Final Brick Product - sample code 2.10.

IOP Conf. Series: Materials Science and Engineering **374** (2018) 012061 doi:10.1088/1757-899X/374/1/012061

In Figure 2 is represented the brick sample code 2.10. The physical behavior (for the first three days) of the obtained brick is presented in Table 2. During the experiments several receipts were tested, and the results are presented in Table 3.

Table 2. Physical behaviour in time of the brick sample 10.

Time	12 h	Day 1	Day 2	Day 3
Appearance	Wet, it keeps its shape	The strengthening process is starting	The strengthening process is finalized	The product is dry and hard at mechanical stress (hammering)

Table 3. Results of the experimental tests.

Text/ Experiment	Code	Sludge humidity (%)		Composition (%)		Product Code	Drying time (days)	Final product characterization
		Before dehydrating	After dehydrating	Cement	Sludge			
Test 2/ experiment 2.10	N8	97.06	65.28	45	55	Brick 2.10	3	The product is dry and hard at mechanical stress (hammering)
Test 2/ experiment 2.11	N8	97.06	65.28	40	60	Brick 2.11	3	The product is dry but porous and brittle at mechanical stress (hammering)
Test 2/ experiment 2.12	N8	97.06	65.28	50	50	Brick 2.12	3	The product is dry, very hard, resistant to mechanical shocks
Test 3/ experiment 3.4	N10	91.51	63.55	45	55	Brick 3.4	3	The product is dry and hard at mechanical stress (hammering)
Test 3/ experiment 3.5	N10	91.51	63.55	40	60	Brick 3.5	3	The product is dry but porous and brittle at mechanical stress (hammering)
Test 3/ experiment 3.6	N10	91.51	63.55	50	50	Brick 3.6	3	The product is dry, very hard, resistant to mechanical shocks
Test 3/ experiment 3.10	N12	96.22	79.01	45	55	Brick 3.10	3	The product is dry and hard at mechanical stress (hammering)
Test 3/ experiment 3.11	N12	96.22	79.01	40	60	Brick 3.11	3	The product is dry but porous and brittle at mechanical stress (hammering)
Test 3/ experiment 3.12	N12	96.22	79.01	50	50	Brick 3.12	3	The product is dry, very hard, resistant to mechanical shocks

5. Conclusions

For the final brick products, the best results in terms of mechanical resistance and stability over time are obtained for experiments 2.10, 2.12, 3.4, 3.6, 3.10 and 3.12. These experimental products will be physico-chemical analysed, in order to be established the compliance with the environmental legislation (in Romania: Law 211/2011, HG 856/2002, OUG 95/2005) regarding waste and establishment of the waste character (inert, non-hazardous, dangerous).

From Table 3 results that the increase or decrease of the ratio between sludge and cement compared to that provided in the patent no. RO 125384/30.06.2014, adversely affects the stability of the final product.

6. References

[1] Fytili D and Zabaniotou A 2008 *Ren. and Sustain. En. Rev.* **12**(1) 116-40
[2] Samolada M C and Zabaniotou A A 2014 *Waste manag.* **34**(2) 411-20
[3] Yang G, Zhang G and Wang H 2015 *Water Res.* **78** 60-73
[4] Anjum T, Khan H I U H and Shauket I 2017 *Earth Science Malaysia* **1**(2) 10-12

Acknowledgments

This work has been developed as a result of funding through the research project no. 101/09.09.2016, COD SMIS 119692, acronym PROMDES, financed under Competitiveness Operational Program (POC), Priority Axis 1, Investment Priority PI 1b: Research, Technological Development and Innovation (RDI) in Support of Economic Competitiveness and Business Development, Section C "Innovative start-up and spinoff enterprises". Project co-financed from the European Regional Development Fund through the Competitiveness Operational Program 2014-2020.

IOP Conf. Series: Materials Science and Engineering **374** (2018) 012062 doi:10.1088/1757-899X/374/1/012062

Synthesis and Technological Innovation of Applying Oxide Nanomaterials in Wastewater Treatment by Flotation

C I Covaliu[1], I C Moga[2], M G Matache[3], G Paraschiv[1], I Gageanu[3] and E Vasile[1]
[1] University Politehica of Bucharest, 313 Splaiul Independentei Street, București, Romania
[2] SC DFR Systems SRL, 46 Drumul Taberei Street, Sector 6, Bucharest, Romania
[3] National Institute of Research-Development for Machines and Installations Designed to Agriculture and Food Industry, 6 Ion Ionescu de la Brad, Bucharest, Sector 1, Romania

E-mail: corinamoga@yahoo.com

Abstract. The appearance and development of nanotechnology gave new and efficient modalities for pollutants removal from wastewaters by using new compounds called nanomaterials which possess unique structural and morphological properties. In this paper we investigated the application of $CoFe_2O_4$ nanomaterial for increasing the efficiency of oily wastewater treatment by flotation. $CoFe_2O_4$ nanomaterial was prepared by precipitation method. Prior testing their application in wastewater treatment by flotation, the oxide nanomaterial was structural and morphological characterized by XRD and TEM analyses. The influence of $CoFe_2O_4$ nanomaterial on oily wastewater depollution by flotation process was investigated by measuring the following parameters: treatment efficiency [%] and the stability of froth.

1. Introduction

Society today faces the extensive growth of wastewater generated from various industries and from household use. This has become a very important issue from the environmental point of view and has led to the enforcement of new and stricter rules and standards limiting the pollution caused by these wastewaters. Oily wastewater results from various industries and has a high potential to degrade the environment especially soil and water. As such, they represent a major environmental problem. To protect the environment, oily wastewater treatment is unavoidable.

The concentrations of oil in municipal and industrial wastewater effluents are strictly monitored [1-6] as they could cause serious health and environmental problems. For example, in the water resulting from petroleum plants the discharge limits of these substances are regulated at a monthly average of 29 mg/l with a daily maximum of 42 mg/l [7]. To meet these stringent discharge limits, oil substances can be effectively separated from wastewater using following practical technologies:

1. Gravity settling [8] and corrugated plate interceptors;
2. Centrifugal separation using hydrocyclone [9-11];
3. Chemical pre-treatment (coagulation-flocculation) [12-14];
4. Coalescing media [15-17];
5. Gas flotation (induced gas flotation [18-22] and dissolved gas flotation [23-26];

6. Biological processes (membrane bio-reactor [27-32] and activated sludge;
7. Media Filtration (resin, polymer, sand, clay, garnet, silica, walnut shell) [33, 34].

The conventional process of wastewater treatment consists in a series of combined physical, chemical and biological steps. The existence of suspended solids and stable fats emulsion in wastewaters makes their treatment to be extremely difficult in term of technical concerns [35].

Nowadays, the challenge is to find new, low cost and more efficient technologies for the treatment of wastewaters to decrease pollution and comply with the existing regulations. Dissolved air flotation system can be used as process for removing finest suspended particles and fats from water. When air is introduced in water forms small bubbles that contact suspended particles in water so by decreasing their density these particles gain enough energy to be floated. Flotation extraction represents a highly efficient method for the treatment of dilute aqueous solutions with the help of flotation devices [36]. By passing air through the solution (wastewater), a high interface is created between the solution and the surfactants that are intended to collect the unwanted components.

Low molar mass surfactants and surface-active polymers are used to facilitate the dispersion of powdered materials in a liquid solution and are very often used as emulsifiers for preparing emulsions and as stabilizers in the production of foams [3]. The use of solid (nano- or micro-) particles in the production and stabilization of foams is not exploited as much as the use of low molar mass surfactants or surface-active polymers and represents a new and promising perspective. The incorporation of solid particles into surfactant-stabilized aqueous foams has been used for a long time, and their influence on the formation and stability of the foam depends on the type of surfactant, on the size of the particles and on the concentration [37]. Concerning spherical particles that adsorb to water-air or water-oil interfaces, the most relevant parameter is considered to be the contact angle θ, which is formed at the contact between the particle and the interface. For hydrophilic particles, θ angle measured into the aqueous phase is normally $<90^\circ$ and a larger fraction of the particle surface resides in water than in the nonpolar phase. For hydrophobic particles, θ is generally greater than 90° and the particle resides more in air or oil than in water [3]. By analogy with surfactant molecules, the monolayers will curve such that the larger area of the particle surface remains on the external side, giving rise to air or oil-in-water when $\theta<90^\circ$ and water-in-air or oil when $\theta<90^\circ$ (Figure 1). The first include aqueous foams and o/w emulsions, respectively, whilst the latter include aerosols and w/o emulsions (Figure 2) [1].

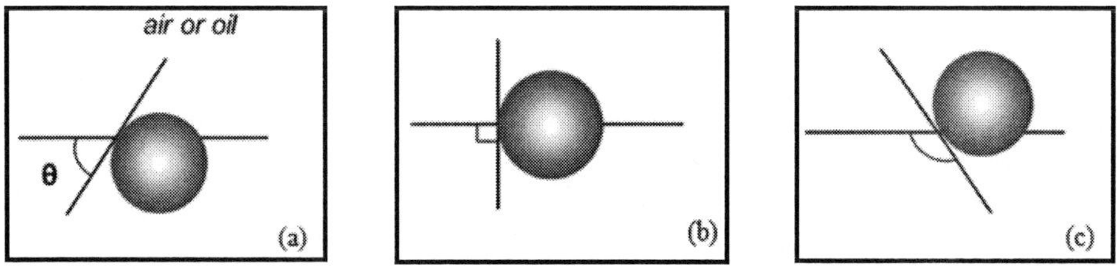

Figure 1. Position of a small spherical particles at a planar fluid-water interface for a contact angle measured through the aqueous phase: less than 90° – (a), equal to 90° – (b), and greater than 90° – (c) [1].

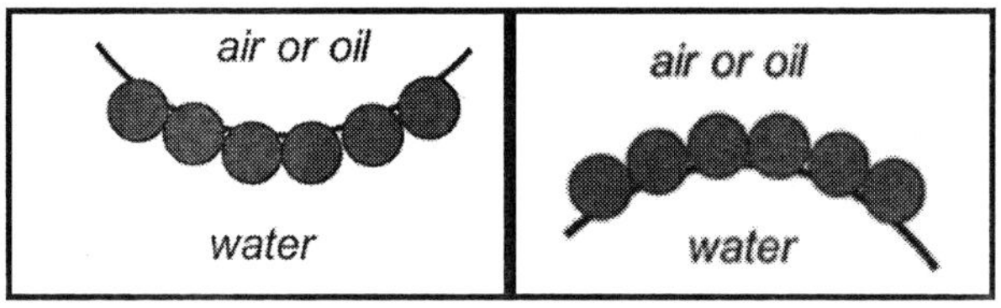

Figure 2. Corresponding probable positioning of particles at a curved fluid-water interface.

For θ <90°, solid-stabilized aqueous foams or o/w emulsions may form (Figure 1, a). For θ>90°, solid-stabilized aerosols or w/o emulsions may form (Figure 1, c) [1].

For testing the influence of nanomaterials in the wastewater treatment by flotation, we chose cobalt ferrite, having the formula $CoFe_2O_4$ obtain by precipitation method.

2. Materials and methods

The preparation of cobalt ferrite nanomaterial tested was already published in another article [38]. The chemical reagents used in this study are ferric nitrate ($Fe(NO_3)_3 \cdot 9H_2O$), cobalt nitrate ($Co(NO_3)_2 \cdot 6H_2O$), ammonia solution 28%, polyvinylpyrrolidone (PVP), polyethylene glycol (PEG). The average crystallite size of $CoFe_2O_4$ obtained by calcination at 400°C calculated by Scherrer formula was 18 nm. By TEM investigation was revealed that cobalt ferrite particles are nearly monodisperse and the mean particle size is 21 nm. Also, was observed that the cobalt ferrite nanoparticles tend to agglomerate due to their magnetic properties forming honeycomb structure [38].

The experiments were done on a synthetic wastewater containing a concentration of oil of 25% in the presence and in the absence of CoFe2O4 nanomaterial. Dissolved air flotation system contained also 5% of anionic surfactant and 5 % of amphoteric surfactant.

3. Results and discussion

During the experiments we observed the following: the decreasing of time needed for treatment of wastewater when the cobalt ferrite nanoparticles were used in the flotation process, from 10 min to 2 min; the increasing of stability of the formed foam when were used cobalt ferrite nanoparticles in the flotation process (7 h in comparison with 40 min). In the large-scale wastewater treatment technologies is needed sufficient time to remove the foam having the pollutant on the surface of wastewater; the the yield of wastewater treatment when was used cobalt ferrite nanomaterial was 100%.

The tested nanomaterials will be further tested by the authors in situ conditions. An experimental installation was designed, and the simplified process diagram is presented in Figure 3. The main components of the dissolved air flotation (DAF) unit are: the pressured capsule (in which air and water are introduced under pressure), the nanoparticle dosing system and the lamellar settler (Figure 4).

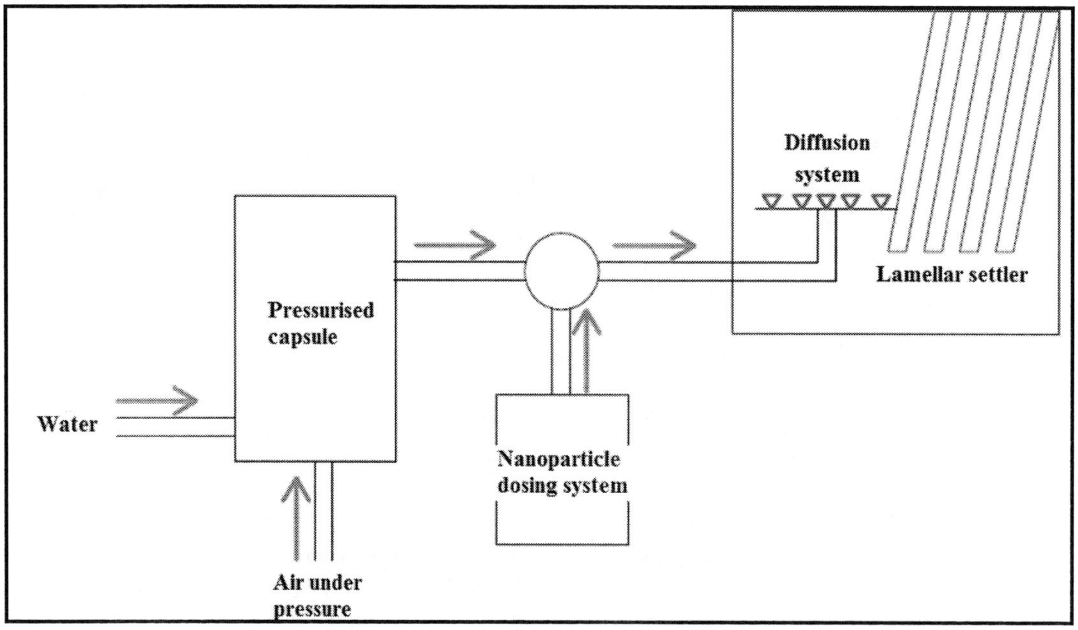

Figure 3. The process diagram for an improved DAF.

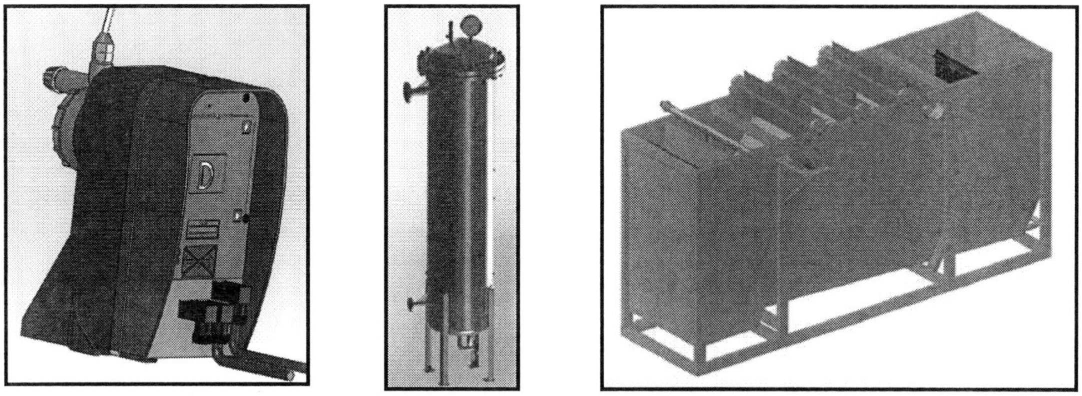

Figure 4. The main components of an improved dissolved air flotation unit (from left to right: nanomaterial solution dosing pump, pressured capsule, lamellar settler).

4. Conclusions

The investigation of applying $CoFe_2O_4$ nanomaterial in the wastewater treatment by flotation showed two important advantages: in the increasing the stability of the foam containing the oil pollutant and the decreasing of the time needed for wastewater treatment.

5. References

[1] Diya'uddeen B H, Daud W M A W and Abdul Aziz A R 2011 *Process Saf. Environ. Prot.* **89** 95–105
[2] Farmaki E, Kaloudis T, Dimitrou K, Thanasoulias N, Kousouris L and Tzoumerkas F 2007 *Desalination* **210** 52-60
[3] Binks B P 2002 *Cur. Op. Col.&Interf. Sci.* **7** 21-41
[4] Mueller S A, Kim B R, Anderson J E, Gaslightwala A, Szafranski M J and Gaines W A 2003 *Pract. Period. Hazard. Toxic, Radioact. Waste Manag.* **7** 156-62
[5] Raha D 2007 *Environ. Forensics* **8** 371-81
[6] Rincón G J and La Motta E J, 2014 *J. Environ. Manage.*, **144** 42–50
[7] Hank C R 2011 SME Annual Meeting & Exhibit Denver 1–11
[8] Stewart M and Arnold K 2009 Emulsions and Oil Treating Equipment Elsevier 107–211
[9] Bai Z, Wang H and Tu S T 2011 *Chem. Eng. Res. Des.* **89** 55-9
[10] Belaidi A, Thew M T, Munaweera S J 2003 *Can. J. Chem. Eng.* **81** 1159-70
[11] Husveg T, Rambeau O, Drengstig T and Bilstad T, 2007 *Min. Eng.* **20** 368-79
[12] Mostefa N M and Tir M 2004 *Desalination* **161** 115-21
[13] Santo C E, Vilar V J P, Botelho C M S, Bhatnagar A, Kumar E and Boaventura R A R 2012 *Chem. Eng. J.* **183** 117-23
[14] Zouboulis A and Avranas A 2000 *Coll. Surf. a Physicochem. Eng. Asp* **172** 153-61
[15] Ma S, Kang Y and Cui S 2013 *J. Dispers. Sci. Technol.*
[16] Owens N and Lee DW 2007 *Produced Water Workshop, 1–19 Aberdeen*, Scotland
[17] Zhang L, Zhu T, Sun Y and Jiang B 2014 *J. Dispers. Sci. Technol.* **36** 182-9
[18] Castillo A, Nieva A D and Caparanga A R 2012 *AIChE Annual Meeting, Conference Proceedings*
[19] El-Kayar A, Hussein M, Zatout A A, Hosny A Y and Amer A A 1993 *Sep. Technol.* **3** 25-31
[20] Hoseini S M, Salarirad M M, Alavi Moghaddam M R 2013 *Desalin. Water Treat.* **53** 300-8
[21] Meyssami B and Kasaeian A B 2005 Bioresour. Technol. **96** 303-7
[22] Painmanakul P, Sastaravet P, Lersjintanakarn S and Khaodhiar S 2010 *Chem. Eng. Res. Des.* **88** 693-702
[23] Abo-El Ela S I and Nawar S S 1980 *Environ. Int.* **4** 47-52

[24] Behin J and Bahrami S 2012 *Chem. Eng. Process. Process Intensif.* **59** 1-8
[25] Multon L M and Viraraghavan T 2008 *Pract. Period. Hazard. Toxic. Radioact. Waste Manag.* **12** 25-9
[26] Younker J M and Walsh M E 2014 *J. Environ. Chem. Eng,* **2** 692-7
[27] Kose B, Ozgun H, Ersahin M E, Dizge N, Koseoglu-Imer D Y, Atay B, Kaya R, Altinbas M, Sayili S, Hoshan P, Atay D, Eren E, Kinaci C and Koyuncu I 2012 *Desalination* **285** 295-300
[28] Pendashteh A R, Abdullah L C, Fakhru'l-Razi A, Madaeni S S, Zainal Abidin Z, Awang Biak D R 2012 *Process Saf. Environ. Prot.* **90** 45-55
[29] Soltani S, Mowla D, Vossoughi M and Hesampour M 2010 *Desalination* **250** 598-600
[30] Zhidong L, Na L, Honglin Z and Dan L 2009 *Pet. Sci. Technol.* **27** 1274-85
[31] Tellez G T, Nirmalakhandan N, Gardea-Torresdey J L 2002 *Adv. Environ. Res.* **6** 455-70
[32] Zhang H, Xiang H, Zhang G, Cao X and Meng Q 2009 *J. Hazard. Mater.* **167** 217-23
[33] Angelova D, Uzunov I, Uzunova S, Gigova A and Minchev L 2011 *Chem. Eng. J.* 172 306-11
[34] Blumenschein C D, Severing K W and Boyle E 2001 *AISE Steel Technol.* **78** 33-37
[35] Habibzadeh Y and Gurbanov K B *Fizika (Baku)* **16(1)** 68-74
[36] Mirgoroda Y A, Borshcha N A and Yurkovb G Y 2011 *Russian J. App. Chem.* **84(8)** 1314-18
[37] Velev O D and Lenhoff A M 2000 *Curr Opin Colloid Interface Sci* **5** 56-63
[38] Covaliu C I, Jitaru I, Paraschiv G, Vasile E, Biriş S Ş, Diamandescu L, Ionita V and Iovu H 2013 *Pow. Tech.* **237** 415-26

Acknowledgment
This work was supported by a grant of the Romanian National Authority for Scientific Research and Innovation, CNCS/CCCDI – UEFISCDI, project number 25 PTE/2016, within PNCDI III.

Application of FT-IR Classification Method in Silica-Plant Extracts Composites Quality Testing

A Bicu[1], V Drumea[1], D E Mihaiescu[2,*], B Purcareanu[1], M A Florea[1], B Trică[3], G Vasilievici[3], S Draga[1], E Buse[1] and L Olariu[1,4]
[1]S.C. BIOTEHNOS S.A., Gorunului Street No. 3-5, 075100 Otopeni, Ilfov, Romania
[2]Department of Organic Chemistry "Costin Nenitescu" Faculty of Applied Chemistry and Materials Science, Politehnica University of Bucharest, Polizu Street No. 1-7, 011061 Bucharest, Romania
[3]INCDCP-ICECHIM, 202 Spl. Independentei, 060021, Bucuresti, Romania
[4] Academia Oamenilor de Stiinta din Romania, 54 Spl. Independentei, 050094, Bucharest, Romania

E-mail: danedmih@gmail.com

Abstract. Our present work is concerned with the validation and quality testing efforts of mesoporous silica - plant extracts composites, in order to sustain the standardization process of plant-based pharmaceutical products. The synthesis of the silica support were performed by using a TEOS based synthetic route and CTAB as a template, at room temperature and normal pressure. The silica support was analyzed by advanced characterization methods (SEM, TEM, BET, DLS and FT-IR), and loaded with Calendula officinalis and Salvia officinalis standardized extracts. Further desorption studies were performed in order to prove the sustained release properties of the final materials. Intermediate and final product identification was performed by a FT-IR classification method, using the MID-range of the IR spectra, and statistical representative samples from repetitive synthetic stages. The obtained results recommend this analytical method as a fast and cost effective alternative to the classic identification methods.

1. Introduction

Mesoporous materials have attracted great interest in current years because of the unusual mechanical, electrical and optical properties as a consequence of bulk and surface properties on the overall behaviour [1, 2, 9].

This type of materials are widely used in the industry for many applications, including adsorbtion, gas separation, catalysis, biological and medical fields [13-16].

In 2001, the first reported mesoporous silica material as a drug delivery system was MCM-41 (mobile composition of matter no.41). Type MCM-41mesoporous silica materials are used as excellent carriers for drug delivery [17-20], for both systemic delivery systems and implantable local-delivery devices, due to the large surface area (800-1100 m^2/g), arrangement and the adjustable size of mesopores between 2-8 nm. [10-12] Also, drug diffusion kinetics can be controlled due to the functionalization of silanol groups [3-5, 22]. These types of loaded materials have applicability in the pharmaceutical industry due to the slow release of active principles, as demonstrated by previous desorption studies. [18-20] MCM-41 has a loading capacity of up to 30 weight percent [25].

The instrumental characterization of this material was achieved using microscopic methods as Transmission electron microscopy (TEM) to determine pore structure, Scanning electron microscopy (SEM) to determine particle morphology, spectroscopic method like Dynamic light scattering (DLS) to determine hydrodynamic particle diameter, other analyses like Brunauer–Emmett–Teller (B.E.T.- specific surface area (SSA) to determine pore volume and diameter and X-ray diffraction (XRD) to determine crystalline structure and pore geometry [7].

Generally, pharmaceutical agents have a specific chemical structure and target a specific molecular receptor. Naturopathic medicine is less specific and uses naturally occurring materials, such as plant extracts and herbs. Based on its advantages – sensitivity, selectivity and speed, Fourier transform infrared (FT-IR) can serve as an optimal tool for the analysis of plant extracts [21].

In this paper, we used a FT-IR classification method (MID-range of the IR spectra) in order process validation and class identification. The techniques used in the pharmaceutical industry to facilitate the identification of raw materials and intermediates for obtaining the finished product.

2. Materials and methods

2.1. Materials

For MCM-41 synthesis we used the following high purity reagents:

-Tetraethyl orthosilicate (TEOS, Sigma- Aldrich);
-Hexadecyltrimethylammonium bromide (CTAB, > 99% purity, Sigma- Aldrich);
-Sodium hydroxide (> 98% purity, Sigma- Aldrich);
-Acetic acid (100% purity, Sigma- Aldrich);
-Commercial MCM-41 purchase from Glantreo (MCM-41).

The loading process of the MCM-41 materials was performed using 2 natural hydroethanolic extracts (70% ethanol) of *Calendula officinalis* and *Salvia officinalis* [8]. The loading percentage was 28% (w/w).

2.2. Methods

2.2.1. MCM- 41 Synthesis. The mesoporous material was synthesized at room temperature by the sol-gel method using CTAB (hexadecyltrimethylammonium bromide) template and TEOS (tetraethyl orthosilicate) silicon source; the dried product was calcinated at 550^0C for 24 h to remove the surfactant template and the calcined product was termed as MCM-41. [23] Loading was performed gradually, by solvent evaporation using a rotary evaporator under vacuum in order to maximize pore adsorption. For comparing the quality of MCM-41 obtained , we purchased commercial MCM-41 with the following specification: surface area 1098 m^2/g, pore volume 0.8 cm^3/g , pore diameter 4 nm and hexagonal pore geometry. This MCM-41 was loaded using the same method.

2.2.2 Characterisation MCM-41. SEM. Surface morphology was observed using the QUANTA INSPECT F electronic scanning microscope (SEM) equipped with EDX solid angle 0.13 (beam) with electron emission tunnel at 1.2 nm resolution.

HRTEM. High resolution images were collected by Tecnai^{TM} G^2 F30 S-TWIN transmission electron microscope (FEI, the Netherlands) equipped with STEM/HAADF detector, EDX (Energy dispersive X-ray Analysis) and with an electron energy loss spectrometer EFTEM- EELS. The microscope operates at an acceleration voltage of 300 kV (Schottky field emitter) with a TEM point resolution of 2 Å and a TEM line resolution of 1.02 Å. The microscope was operated at an extracted voltage of 4500V.

DLS. Hydrodynamic particle radius measurements were carried out by using Malvern Nano ZS Zetasizer instrument. The measurement of size was perform with 0.3% suspensions of MCM-41 in ultrapurified water Milli-Q, at 25^0C, after 10 min sonication,at an angle of 173°. For size determination, 5 measurements on each sample were carried out and the mean value was reported.

B.E.T. Nitrogen adsorption isotherms was measured at –196°C using Micromeritics apparatus ASAP 2420 (USA 2012). Prior to the experiment the sample was outgassed at 200°C in a vacuum for 1h. Gas-volumetric analysis, specific surface area (SSA), pore volume and size were measured by N_2

adsorption-desorption isotherms. SSA was calculated using the Brunauer-Emmett-Teller (BET) method.

UV-VIS. The desorbtion studies were performed using an on-flow system- Thermo Scientific Evolution 220 UV-VIS Spectrophotometer and peristaltic pump. The desorbtion curves were carried out in a 60 % ethanol/ 40 % ultrapurified water (Milli-Q system), at a spectral wavelenght λ= 205 nm, at 5 minutes intervals for 20 hours.

FT-IR. The MIR spectra of samples were recorded from 4000 to 600 cm^{-1} with a HATR-FTIR diamond crystal infrared spectrometer (Spectrum BX II, Perkin-Elmer). Approximately 5–10 mg of sample were placed on the surface of the diamond crystal and measured directly. Intact MCM-41 and MCM41-loaded were analyzed by gently pressing the adaxial sample side in the region of the central lamina on to the HATR crystal. Background spectra was determinated and substracted from each sample spectrum; 32 aquisition per sample were perfomed, with a spectral resolution of 8 cm^{-1}. All sample aquisitions were performed in duplicate.

Each FT-IR spectrum can express a unique "fingerprint", which allows FT-IR spectroscopy to be used for identification of unknown samples and classification of different samples [24].

3. Results and discussions

To show that our synthesized MCM-41 is similar to MCM-41, we compared the characteristics of both materials using the instrumental methods mentioned.

From the DLS analysis we concluded that the particle average size of our synthesized MCM-41 (2623 nm) is very close to particle average size of MCM-41 (2375 nm). The polydispersity index of synthesized MCM-41 (PdI=0.478) has a smaller value than the MCM-41 (PdI= 0.797); this shows that our mesoporous material has a better particle size distribution, therefore the material adsorbtion/desorbtion is conducted more efficiently.

Surface morphology analysis (SEM) reveals a heterogeneous structure (shape and size), with intergranular spaces and randomly distributed agglomerations (Figure 1). After magnification one can observe a porous ultrastructure in the 1 - 2 microns scale, given by the mesopore structure (better emphasized in the HRTEM analysis).

Figure 1. SEM morphology of the synthesized MCM-41.

In the image obtained by HRTEM (Figure 2) we can observe the hexagonal pore structure with medium ordering and mesoporous dimensions of 2-4nm, characteristics allowing the classification of the obtained material in the MCM-41category. This structure allows efficient loading, due to the high pore diffusion rate of the loaded material as well as a higher diffusion accessibility in the desorption processes.

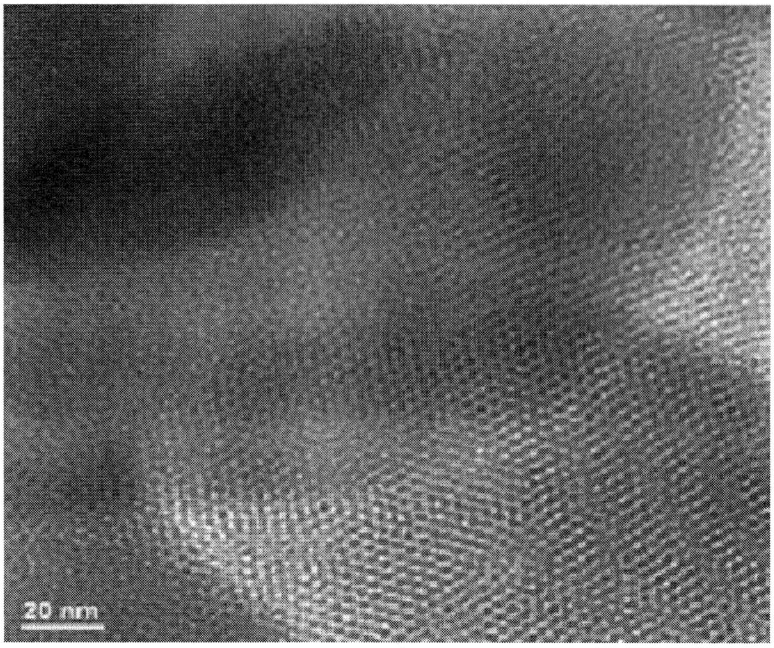

Figure 2. HRTEM of the synthesized MCM-41.

B.E.T. results for MCM-41-Syntesized- B.E.T. Surface Area: 1224.34 m^2/g and BJH Adsorption average pore diameter (4V/A): 2.44 nm. The results show that this material can be classified as MCM-41.

From the FT-IR spectra (Figure 3), the specific bands for synthesized MCM-41 and MCM-41 are: 803 cm^{-1}, 1050 cm^{-1}, 1627 cm^{-1} and 3375 cm^{-1}.

As proof of adsorption using *Calendula officinalis* hydroethanolic extract, the following FT- IR spectra specific bands of MCM-41-Calendula were identified: 977 cm-1, 1409 cm^{-1} and 2935 cm^{-1}. Similarly, the FT-IR spectra of MCM-41-Salvia shows specific bands for hydroethanolic extract of *Salvia officinalis*: 969 cm^{-1}, 1413 cm^{-1}, 1547 cm^{-1}, 1691 cm^{-1} and 2930 cm^{-1}.

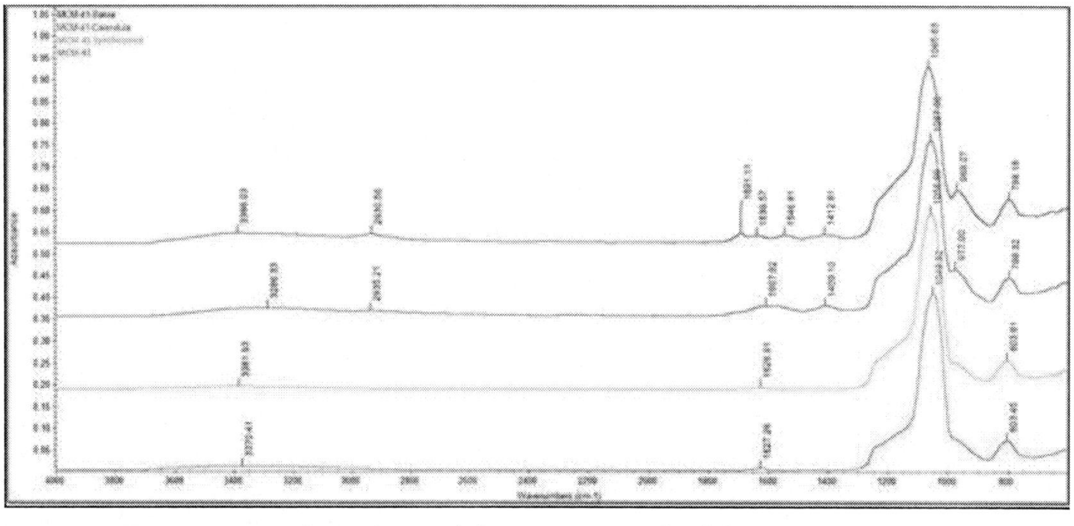

Figure 3. FT-IR spectra of MCM-41-Salvia, MCM-41-Calendula, MCM-41 Synthesized and MCM – 41.

3.1. Classification method by Discriminant analysis

Discriminant analysis was performed with the TQ Analyst 7.2.0 Thermo Electron software, using the following parameters: pathlength type: multiplicative signal correction (MSC), number of classes: 3 (salvia, calendula, MCM-41), 36 IR spectra, data format: first derivative, Savitzky-Golay filter, 23 data points, polynomial order 1 spectrum range: 1779.8 – 1287.7 cm^{-1}, baseline type: one point at 1800 cm^{-1}. For each class (class I: MCM-41 synthesized and MCM-41, class II: MCM-41-Calendula and class III: MCM-41-Salvia), FT-IR spectra for 6 different batches were determined in duplicate.

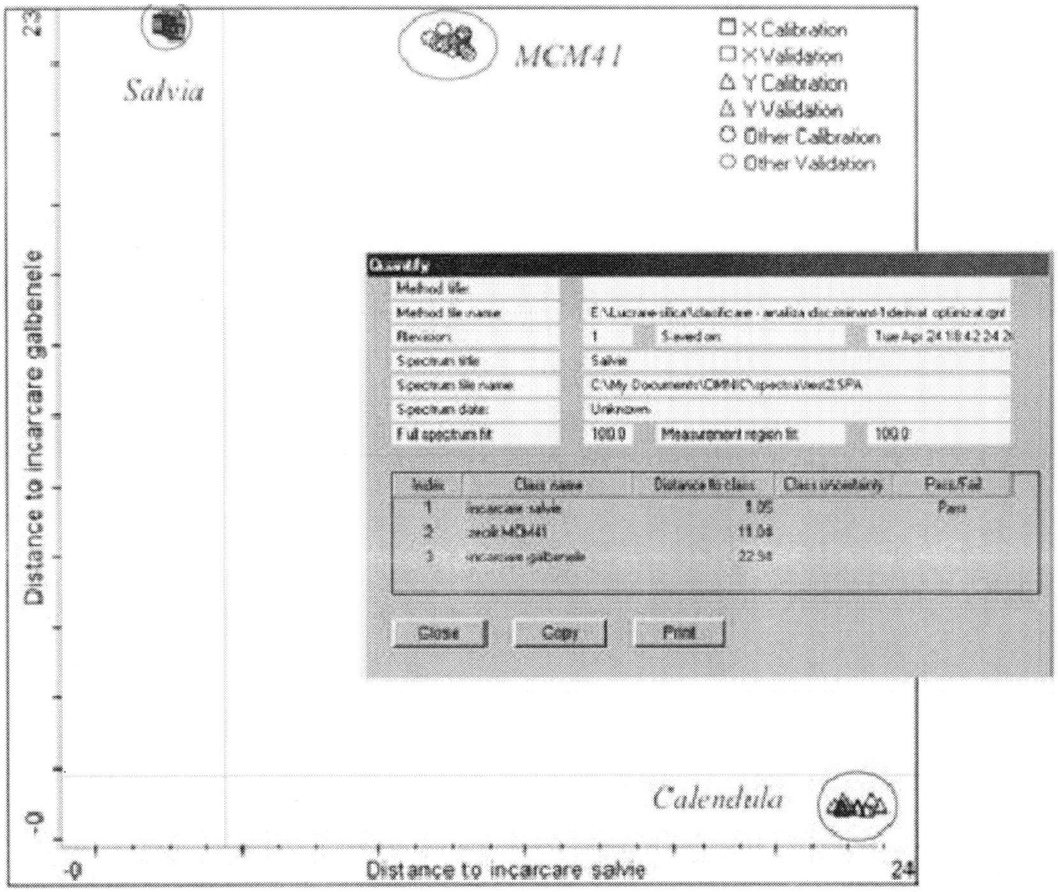

Figure 4. Calibration results for Salvia, Calendula and MCM-41 classes (by using 36 IR spectra), and quantification results for a Salvia validation sample.

Each sample was classified based on Mahalanobis distance. The developed method for sample classification is exemplified in Figure 4, where the Mahalanobis distance between classes is greater than 10, and the distance between members of the same class is less than 1.2.

To verify the selectivity of the method, another 18 samples from different batches (MCM-41, MCM-41-Calendula, MCM-41-Salvia) and 3 samples of MCM-41 loaded with other 3 plant extracts were analyzed (results shown in Table 1).

From the 21 samples, 6 of them passed the distance to class MCM-41 with a value between 0.78-0.99, 6 of them passed the distance to class MCM-41-Calendula with a value between 0.83-0.92, 6 of them passed the distance to class MCM-41-Salvia with a value between 0.91-1.13 and the other 3 samples did not match any of the classes.

Table 1. Results of sample tests.

No.	Sample Name	Distance to Class MCM-41	Distance to Class MCM-41-Calendula	Distance to Class MCM-41-Salvia	PASS/ FAIL
1	Test 1	22.45	0.83	22.91	PASS Calendula
2	Test 2	11.04	22.94	1.05	PASS Salvia
3	Test 3	0.99	22.68	10.72	PASS MCM-41
4	Test 4	22.45	0.83	22.91	PASS Calendula
5	Test 5	22.68	0.89	23.13	PASS Calendula
6	Test 6	22.7	0.81	23.26	PASS Calendula
7	Test 7	22.85	0.92	23.45	PASS Calendula
8	Test 8	22.57	0.87	23.12	PASS Calendula
9	Test 9	10.8	23.16	0.91	PASS Salvia
10	Test 10	11.32	23	1.02	PASS Salvia
11	Test 11	10.78	23.08	1.01	PASS Salvia
12	Test 12	11.02	22.82	1.1	PASS Salvia
13	Test 13	11.05	23.11	1.13	PASS Salvia
14	Test 14	0.99	22.68	10.72	PASS MCM-41
15	Test 15	0.96	22.96	10.92	PASS MCM-41
16	Test 16	0.78	22.37	11.16	PASS MCM-41
17	Test 17	0.9	22.55	10.8	PASS MCM-41
18	Test 18	0.87	22.82	11.18	PASS MCM-41
19	**Test 19**	**81.09**	**80.35**	**80.61**	**FAIL**
20	**Test 20**	**31.73**	**19.35**	**29.33**	**FAIL**
21	**Test 21**	**22.01**	**9.17**	**29.85**	**FAIL**

3.2. Desorption studies

The desorption study was performed with 60% ethanol/40% ultrapurified water Milli-Q system, at 35°C , under continuous stirring, undersolution recirculation. The determinations were performed, in both cases, at a wavelength λ= 205 nm. In the case of the desorption study for the materials loaded with Salvia extract, a similar profile is observed for both samples, concluding that the two materials show a slow release for 20 hours (Figure 5).

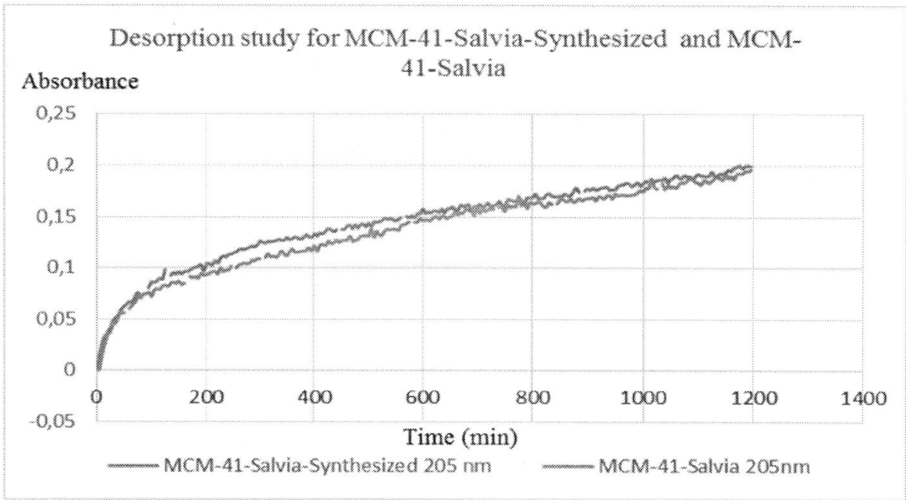

Figure 5. Desorption study for MCM-41-Salvia-Synthesized and MCM-41-Salvia.

In the case of the desorption study carried out on the two materials loaded with Calendula extract, showed the same slow release , with the observation that MCM-41-Calendula, after 7 hours, the slow release of the active principles stops, compared to MCM-41-Calendula-Synthesized where slow release continues for up to 20 hours (Figure 6).

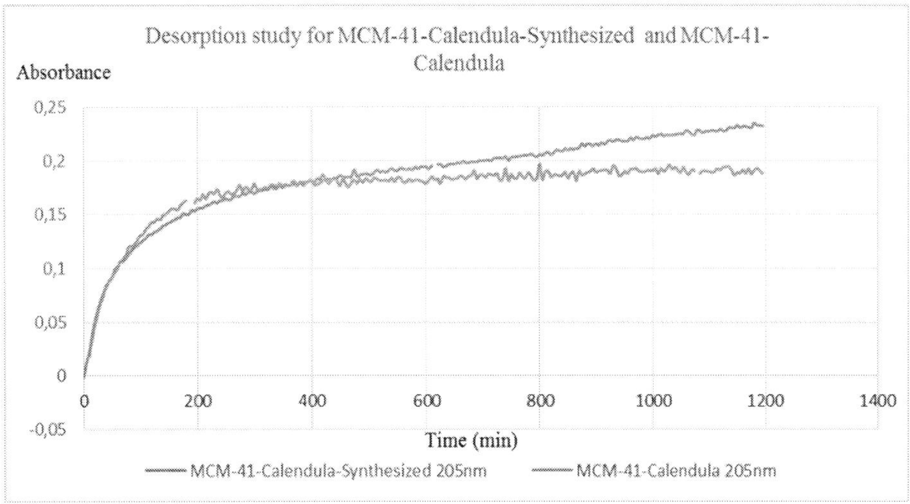

Figure 6. Desorbtion study for MCM-41-Calendula-Synthesized and MCM-41-Calendula.

4. Conclusions

We obtained a mesoporous material that is comparable, in regard to its characteristics, with the purchase mesoporous material. Both types of MCM-41 were loaded with different total plant extracts.

Desorption curves were used to evaluate the slow release profiles, which is a very valuable parameter for multiple applications in the pharmaceutical industry.

The required identification step for the product technical data sheet was performed using the FT-IR classification method, demonstrating the statistical homogeneity of the experimental data. The FT-IR classification method is a quick tool for sample classification, being able to discriminate between various loading materials.

5. References

[1] Wu S H, Mou C Y and Lin H P 2013 *Chemical Society Reviews* **42** 3862-75
[2] Rahman N A, Widhiana I, Juliastuti S R and Setyawan H 2015 *Colloids and Surfaces A: Physicochem. Eng. Aspects,* **476** 1–7
[3] Mihaiescu D E, Tamas D, Andronescu E and Ficai, A 2014 *Digest Journal of Nanomaterials and Biostructures*, **9(1)** 379-83
[4] Tameemi Al MBM, Mihaiescu D E, Stan R, Meghea A, Voicu G, Vasile B S, Traistaru V and Istrati D 2015 *Digest Journal of Nanomaterials and Biostructures* **10**(4) 1229-53
[5] Al Tameemi MBM, Gudovan D, Stan R, Mihaiescu D E and Ott C 2015 *Romanian Journal of Materials* **45**(2) 188-193
[6] Li T, Chen X, Liu Y, Fan L, Lin L, Xu Y, Chen S and Shao J 2017, *European Journal of Pharmaceutical Sciences* **96** 456-463
[7] Fudulu A, Dumitriu B G, Olariu L, Buse E and Mihaiescu D E 2017 *The FEBS Journal* **284** (Suppl. 1) 335
[8] Dumitriu B G, Olariu L, Ene M D and Rosoiu N 2013 *Romanian Biotechnological Letters* **18**(3) 8284-94

[9] Vaghasiya J V 2017 *J Mater Sci Nanomater, Innovations in Engineered Mesoporous Material for Energy Conversion and Storage Applications* **1**(1)
[10] Grumezescu A M, Mihaiescu D E and Tamas D 2011 *Biointerface Research in Applied Chemistry* **1**(6) 229-235
[11] Mihaiescu D E, Tamas D, Andronescu E and Ficai A 2014 *Digest Journal of Nanomaterials and Biostructures* **9**(1) 379-383
[12] Mihaiescu D E, Gudovan D and Marton A 2014 *Revue Roumaine de Chimie* **59**(2) 111-116
[13] Sun M, Chen C, Chen L and Su B 2016 *Frontiers of Chemical Science and Engineering, Hierarchically porous materials* **3**(10) 301– 347
[14] Anis S F, Khalil A, Saepurahman, Singaravel G, and Hashaikeh R 2016 *Microporous and Mesoporous Materials* **236** 176– 192
[15] Parak W J, Gerion D, Pellegrino T et al. 2003 *Nanotechnology* **7**(14) R15–R27
[16] Danilczuk M, Długopolska K, Ruman T and Pogocki D 2008 *Mini reviews in medicinal chemistry* **13**(8) 1407–17
[17] Berlier G, Gastaldi L, Sapino L, Miletto I, Bottinelli E, Chirio D and Ugazio E 2013 *International Journal of Pharmaceutics* **457** 177–186
[18] Ambrogi V, Latterini L, Marmottini F, Tiralti M C and Ricci M 2013 *J Pharm Innov* **8** 212–217 DOI: 10.1007/s12247-013-9161-2
[19] Bodana P, Kelkar G, Manglavat S K, Gupta A K and Chaturvedi S C 2016 *Int.J.Adv.Res* **4**(12) 2518-30 DOI:10.21474/IJAR01/2677
[20] Stefanache A, Ignat M, Peptu C A, Diaconu A, Stoleriu I and Ochiuz L 2017 *Appl. Sci.* **7** 237 DOI: 10.3390/app7030237
[21] Thermo Fisher Scientific, *Application Note: 51254*
[22] Mehmood A, Ghafar H, Yaqoob S, Gohar U F and Ahmad B 2017 *J Develop Drugs*, **6**(2) DOI: 10.4172/2329-6631.1000174
[23] Istrati D et al. 2015 *Revue Roumaine de Chimie,* **60**
[24] Cao Z, Wang Z, Shang Z and Zhao J 2017 *PLoS ONE* **12**(2) e0172359, DOI:10.1371/journal.pone.0172359
[25] Sang W Y and Ching O P 2017 *Tailoring MCM-41 Mesoporous Silica Particles through Modified Sol-Gel Process for Gas Separation* DOI: 10.1063/1.5005480

Acknowledgement
This study was financed by the Romanian Research Authority, contract 49 PTE/2016.

IOP Conf. Series: Materials Science and Engineering **374** (2018) 012065 doi:10.1088/1757-899X/374/1/012065

Research on Recycling Mixed Wastes Based on Fiberglass and Organic Resins

M A Platon[1], M Ştef[1], C Popa[1], A E Tiuc[1] and O Nemeş[1,2]
[1]Technical University-Faculty of Materials Science and Engineering, 103-105 Muncii Avenue, 400641, Cluj-Napoca, Romania
[2]The National Institute for Research and Development in Environmental Protection, 294 Splaiul Independenţei, 6th District, 060031, Bucharest, Romania

E-mail: ovidiu.nemes@imadd.utcluj.ro

Abstract. In recycling, according to principles of *Directive 2008/98/EC of the European Parliament and of the Council* on waste, research is upheld for achieving innovative technologies for reuse and keep as long it is possible, in economic chain, a waste. The aim of this research is to study and test a new composite material based on fiberglass waste mixed with organic resins with large application in the industry but not limited to this. Fiberglass is a material widely used for reinforcement of composite materials. As waste, fiberglass was less studied for ways to be reused. Filling fiberglass mixed with organic resins as PMMA and epoxy resins possess proper physical features for thermoforming. Three mixes are studied: fiberglass with PMMA, fiberglass with PMMA and rubber granules or sawdust. Samples will be tested for to define the mechanical and chemical behavior to have a complete description of the material. Analyzing the results can be concluded that mixes are suitable for board production, with improved features, compared with equivalent products on the market.

1. Introduction

Recycling worldwide, in the last decades, became a significant issue in industrial production. On the strength of legislation, which sets new goals to be reached, but also is a demand for sustainable management implemented in a rapidly increasing number of companies. Reasons are increasing efficiency through reducing costs. Reuse of outputs, and reintroduce them in the production process, is a way these aims can be achieved. EU legislation through *European Strategy for plastic in a circular economy* proposes to change the way in which plastic products are generally treated, and improve the design of plastic products, growing recycling rates and enhance the quality of recycled products.

The concept of physically blending of two or more polymers to provide a new material is attracting large interest and commercial utilization [1]. The main material used is fiberglass mixed with PMMA. Other components are studied as being utilized in addition. PMMA is widely used in the glazing sector and rubber-reinforced materials were developed to improve its properties [2]. Three mixes will be studied. The purpose of the research is to conceive, design, fabrication and test the composites materials based on three mixes.

PMMA parts can be created by a wide range of techniques including hot embossing, injection molding, laser ablation, reactive ion etching or deep UV lithography and have mechanical properties superior to those of poly(dimethylsiloxane), (PDMS, another material of choice for microfluidic applications), guaranteeing a better fidelity to the initial shape under mechanical stress conditions [3].

The use of recyclable wastes as precursors for new composite materials is an intensely exploited research field, that has been approached by many research groups [4, 5] in order to obtain new materials with improved physical and mechanical behavior.

In our research group, obtaining new composite materials [6, 7, 8], especially those in which they are used diverse types of waste, constitutes a continuous preoccupation. Several results relating to the design and characterization of the composite materials using as materials diverse types of waste have already been published [9, 10].

In this paper we present a new obtaining method of composite materials, based on production wastes containing glass fibers and poly(methyl methacrylate) (PMMA), without the addition of other components.

2. Materials and methods

Poly(methyl methacrylate) (PMMA) is a polymer widely used in various industries. As boards, it is used in the industrial process, where is thermoformed, for obtaining bathtubs. After cooling from the mold, it is reinforced, on the outside, with sprayed fiberglass in order to become more resistant. Later the product is polished to remove the roughness and to fit for commercial use. Along with this process, a variable quantity of PMMA and fiberglass flakes is results (see Figure 1).

Figure 1. Fiberglass mixed with PMMA.

In our studies, three types of mixture were used, as shown in Table 1.

Table 1. Materials used and their components.

Sample	Components
1	Fiberglass, PMMA
2	Fiberglass, PMMA, rubber
3	Fiberglass, PMMA, sawdust

For samples manufacture was used a forged steel mold. Working pressures vary between 0.7÷0.15 MPa and temperature between 110÷170 °C. With a cylindrical shape and two plates mounted on both sides of the mold, is designed to heat the material in both directions through ERH (electrical resistance heating) embedded (figure 2).

When the mold mix reaches optimum temperature, as consequence acquires the necessary plasticity, the hydraulic cylinder is turned on to distribute uniformly the mix in the mold forming cavity. The product is kept in the mold under pressure, for a certain period, for a gradual cooling of samples. Figure 3 show a sample obtained by this technique.

Figure 2. Mold used to obtain the samples.

Figure 3. Sample obtained from PMMA with fiberglass.

The traction testing was accomplished using an INSTRON 3366 equipment (Figure 4). ASTM-D618-87 is the procedure for determining the traction characteristics of composites having a polymeric matrix and being reinforced with discontinuous and continuous fibers.

Figure 4. INSTRON 3366 during bending testing.

Standards and norms to determine traction characteristics of composites having a matrix, consisting of plastics and reinforced with fibers, are: ASTM E83-85 for checking and calibrating extensions and ASTM E4-85 for testing the samples on test machines. Figure 5 show a sample tested for bending.

Figure 5. Sample used for bending test.

In order to determine the bending behavior testing was realized following standards: ISO 14125, DIN 53390, ASTM D 4476, conducted with INSTRON 3366 equipment.

The method of processing for the obtained samples is thermoforming, the working temperatures for thermoforming ranged 110÷180 °C and the processing time is 60 minutes.

3. Results and discussions

PMMA is used in this case as a polymeric matrix to take over and transmit the load in the obtained composite material, for this reason must have high mechanical strengths. In order to achieve materials with good physical properties, the matrix must be compatible with the reinforcement material; it must adhere and be embedded with it. In the presented case, this strong bond between the two materials was achieved, and after the grinding process the materials used were uniformly mixed. This gives us advantages such as lower processing temperature and better embedding of the fiber with the matrix. The melting temperature T_t varies between a maximum of about 130 °C and a minimum of 70 °C, this being a measure of the degree of crystallinity.

Physical compatibility is determined by the difference between the coefficients of thermal expansion, the deformation capacity and the melting temperature of the composite components. The difference between thermal expansion coefficients is less so physical compatibility is better.

$$\Delta\alpha = \alpha_m - \alpha_f \rightarrow 0 \tag{1}$$

Where: α_m is the coefficient of thermal expansion of the matrix material and α_f is coefficient of thermal expansion of the reinforcement material. The melting point of the reinforcment material (1300°C) is superior to that of the matrix (105°C), thus avoiding the melt degradation of the reinforcement material. However, the deformation of the two components must be close (coefficient of thermal deformation of the matrix material PMMA is 0,065 [mm/m °C] DIN 53752 and the coefficient of thermal expansion of the reinforcement fibers is $0,5\cdot10^{-5}$ [°C^{-1}].

The obtaining of the material is carried out in a mold at a pressure of 10 tones, the polymer being brought to the softening temperature. Due to the softening temperature, the polymer acquires the necessary plasticity and the pressure distributes it to the mold cavity.

The thermoforming process is held inside the mold (figure 2), under pressure for 1 hour, after which the mold is opened, and the plate is left in place for cooling until it reaches 60 °C and the finished piece is removed. Heat for plasticization, pressure, temperature and pressing time represented the main process parameters.

For testing the mechanical properties of the materials, bending tests were performed according to the standards ISO 14125, ASTM D 4476, DIN 53390. The samples used to determine the bending proprieties had the section size of 7x7 mm and the length between the milestones was L = 10 mm.

The bending tests results are shown in Figure 6.

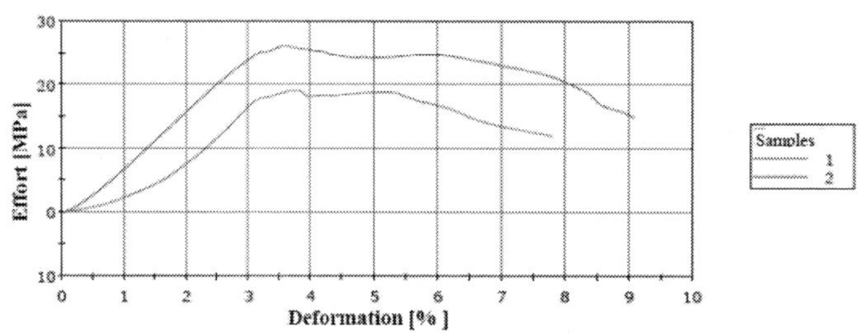

Figure 6. Bending tests results.

The tensile test was performed, according to the standards ASTM-D618-87, by applying an increasing axial force to a specimen, usually until it is broken, recording variations corresponding to the length of the specimen. In order to perform this test, two samples with the size of 12x7 mm and the length between the milestones was L = 65 mm was used. The shape of the samples used in the tensile test is illustrated in Figure 6. The results obtained by tensile tests are illustrated in Figure 7.

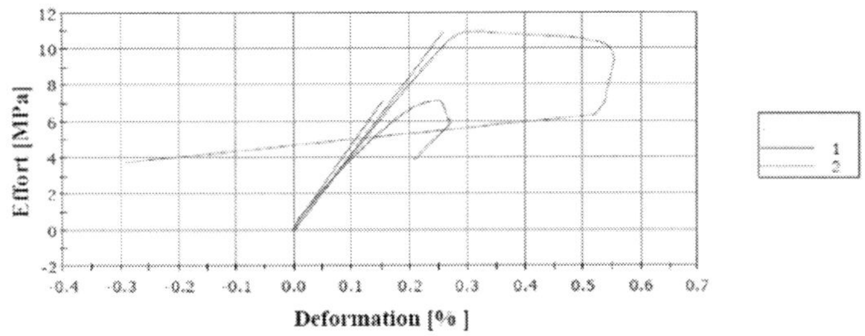

Figure 7. Tensile tests results.

The both bending and tensile testing are the most important type of analysis to evaluate the mechanical stress which can occur in the case of a product.

4. Conclusions

PMMA mixed with fiberglass and in addition to other components is a versatile material and possess characteristics for being reused in industrial production. Boards are a desirable form to be produced, with a large domain of usability.

PMMA starts to soften at a relatively low temperature of 110°C and offers the possibility of adhering to the surfaces of the constituent materials, in this case the glass fiber. In the collision, the press material was completely homogenized in the die cavity forming a flat plate without material gaps or surface defects. In order to obtain the good results, account must be taken of the properties of PMMA, which is the most sensitive component of the composite material.

5. References

[1] Haddadi E, Choupani N and Abbasi F 2016 *Engineering Fracture Mechanics* **162** 112
[2] Rubin A, Rébutin N, Gérard P and Gauthier C 2014 *Materials Letters* **135** 184
[3] Madadi H, Mohammadi M, Casals-Terre J and Lopez RC 2013 *Electrophoresis,* **34** 3126
[4] Seghiri M, Boutoutaou D, Kriker A and Hachani M I 2017 *Energy Procedia* **119** 163
[5] Aman A, Yaacob M M, Alsaedi M A, Anwar K and Ibrahim K A 2013 *International Journal of Electrical Power and Energy System* **45** 346
[6] Bere P, Dudescu M C, Balc N, Berce P, Iurian A M and Nemes O 2014 *Materiale Plastice* **51** 145
[7] Bere P, Berce P and Nemes O 2012 Composites Part B-Engineering **43** 2237
[8] Nemes O 2007 Studia Univ. Babes-Bolyai Chemia **52** 175
[9] Popita G E, Rosu C, Manciula D, Corbu O, Popovici A, Nemes O, Sandu A V, Prorocu M and Dan S B 2016 *Materiale Plastice* **53** 308
[10] Borlea Tiuc A, Rusu T, Ionescu S and Nemes O 2012 *Romanian Journal of Materials* **42** 405

Acknowledgments
The results presented in this paper were obtained with the support of the Technical University of Cluj-Napoca through the research contract no.1992/12.07.2017, Internal Competition CICDI-2017.

IOP Conf. Series: Materials Science and Engineering **374** (2018) 012066 doi:10.1088/1757-899X/374/1/012066

3D Printer- Manufacturing of Complex Geometry Elements

A Ciubară[1], Ş L Burlea[2], M Axinte[3], R Cimpoeşu[3], D L Chicet[3], V Manole[3], G Burlea[4] and N Cimpoeşu[3]
[1]Dunarea de Jos University, 47 Domnească str. 800008, Galaţi, Romania
[2]University of Medicine and Pharmacy "Grigore T Popa", 16 Universităţii str., 700115, Iasi, Romania
[3]Gheorghe Asachi Technical University of Iasi, Mangeron 44 blv., 700050, Iasi, Romania
[4]Helicomed S.R.L., 13 Sfântul Atanasie, 700115, Iaşi, România.

E-mail: nicanornick@yahoo.com

Abstract. In the last 5-10 years the process of 3D printing has an incredible advanced in all the fields with a tremendous number of applications. Plastic materials exhibit highly beneficial mechanical properties while delivering complex designs impossible to achieve using conventional manufacturing. In this article the printing process (filling degree, time, complications and details finesse) of few plastic elements with complicated geometry and fine details was analyzed and comment. 3D printing offers many of the thermoplastics and industrial materials found in conventional manufacturing. The advantages and disadvantages of 3D printing for plastic parts are discussed. Time of production for an element with complex geometry, from the design to final cut, was evaluated.

1. Introduction

Learning-reading devices that are currently on the market are rather difficult to be manipulate, do not allow the word to be correlated with a visual representation of the graphic representation of the word, and there is also the possibility of making misspelled words by wrong choice of letters, or the combination of letters is difficult for pre-school and unattractive children [1]. The technical problem solved by the patent [2, 3] consists in bringing together, in a single apparatus, both the function of viewing a letter, a word, and the graphical representation of the meaning of the word being formed, and the possibility of writing letters on a built-in tablet. With this device, the visual perception of the letter or word image is correlated with the auditory perception of the phonemes. On the basis of an invention patent [2, 4, 5] it has been proposed to provide a read-write device made up of plastic elements. In order to establish the main dimensions of the component elements for the market implementation of this product, the 3D printer was used for the design and physical realization of the component elements.

3D printing or additive manufacturing is a process of manufacturing three-dimensional solids by adding layer by layer. Physical objects are produced using the data of a digital model, a 3D model or other sources such as an additive manufacturing file (AMF) [6]. By using 3D printing, you can create products in almost any form. Several technologies and materials for 3D printing are currently being used. Recently, 3D printing equipment is available for both industrial and household users [7]. 3D printing allows the creation of complex structures and parts that cannot be produced by conventional production methods. Easy complex geometries can be created, providing great freedom in design [8-

10]. Complex models can be 3D printed as a single piece, eliminating the need to assemble component parts. 3D printing makes it very easy to customize at no additional cost. If it is necessary to change the design of an object, only the digital file needs to be modified without the need for expensive production processes or tools and additional devices [11].

The prototype of this device was made using a 3D printer to determine the optimal functional features. This article briefly presents the steps of making an element, a button with a grip and motion role in the final shape from the concept to the actual object using the 3D printing technique.

2. Materials and methods

For printing the plastic parts was used a Smart Rap 3D Printer. As active wire we use PLA - Verbatim (thermoplastic), 1.75 mm diameter. PLA is appreciated as the most important of all bio based polyesters at this moment. PLA it is made from sugar (sugar beets, sugarcanes, corn). Through fermentation, with the help of micro-organisms, lactic acid is produced. This is a highly efficient process. From a sugar molecule, two molecules of lactic acid are produced, without any residual products. Work temperature is 200-230 °C. The printing rate (wire alimentation) was of 2.5 mm/s. Beside the work temperature we have a table work temperature of 70 °C. The nozzle diameter was of 0.4 mm. The entire line to obtain a plastic product summarize: 3D design, 2D drawing, 3D printer program, printing process and element cleaning (post-processing) if necessary. The design plastic part is a roll with the purpose of moving a band of draws with specific dimensions and a gripping system.

3. Results and discussions

The results present the obtaining of a final product, a rotational button, made from the design idea to the actual product using 3D printer. The design part was realized using Catia software. The button is presented in Figure 1 a) as a part of the entire equipment proposed for the read-write learning stage. The advantages of this device are: the upper part of the device allows you to place the attention on the illuminated image, the visual perception of the image and the word, in connection with its name; the auditory perception of phoneme, together with the visual perception of the corresponding letters; the choice of letters to form syllables and word words; easy and attractive handling for children; allows users to create mobile, flexible and creative thinking; determine the formation and development of an active lexical stock [12, 13]. The device is designed for the writing-reading learning phase, used by both preschool children and those with speech deficiency.

The first step in the additive manufacturing process is the realization of the digital model. For this, Computer Aided Design (CAD) is used [14]. Reverse engineering can be used to generate a digital model through 3D scanning. After the design of the button the 2D draw was realized at real quotas using AutoCad software, Figure 1 b). Based on the 2D draw using 3D printer we realize the buttons using PLA filament.

In Figure 2 are present print screens of the Pronterface program interface (which controls the SmartRap + 3D printer) and the possibility of editing a Gcode program directly from Pronterface in a) and b) and the Cura program that makes slicing and generates the Gcode program of the printer [15, 16]. The most important parameters are visible in the left panel in (c). The elements made with 3D printer are build layer by layer. This making process present many advantages like the element density (the filling degree), useful to obtain very light pieces or very resistant pieces.

The 3D printing process is largely automatic. Depending on the size of the object, materials, and printer, the procedure may take from a few hours to a few days. It must be checked, from time to time, that there are no errors.

Many geometrical complex parts can be realized layer by layer reaching by one operation an 95% percentage of the final element. All programs used for 3D printing are open source. The 3D printing parameters are presented in Table 1. The time to obtain an element is 3 hours without using any additional elements that will increase the production costs. The filling degree was of 20% fact that will give the element a sufficient integrity and also a reduce mass [17].

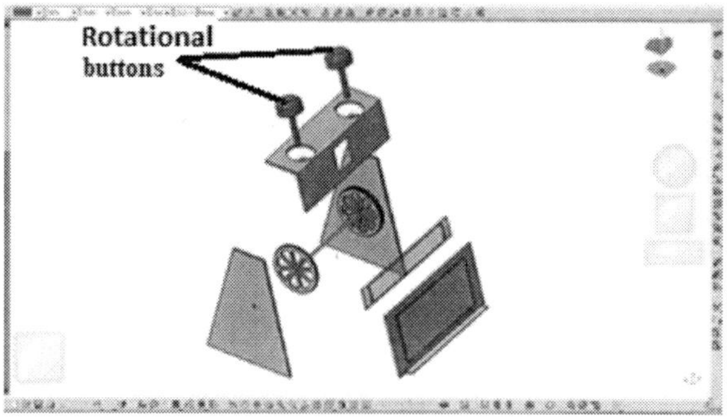

(a)

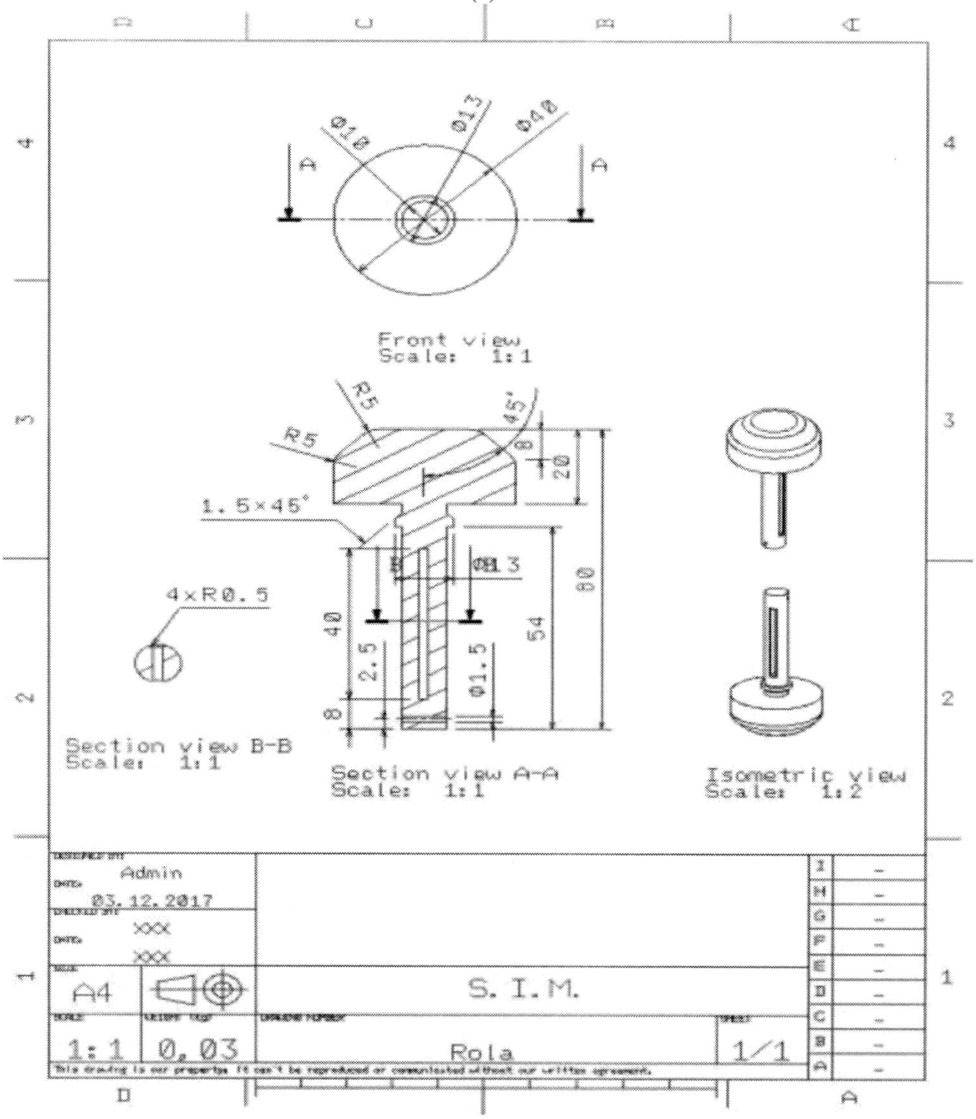

(b)

Figure 1. Design of the equipment (a) component parts including the rotational buttons and (b) 2D draw of the superior rotational button.

IOP Conf. Series: Materials Science and Engineering **374** (2018) 012066 doi:10.1088/1757-899X/374/1/012066

Table 1. Button printing parameters.

Element	pieces	Time/piece [h]	Layer [mm]	Shell [mm]	Bottom [mm]	Infill [%]	Support [tip]	Adhesion [tip]
Rotational button	2	3	0.2	0.4	0.6	20	none	none

The element is build layer by layer, each of 200 μm thick, until the element is done. In Figure 3 the final button is presented. For round parts we need for an additional stage for finishing the surface. Final product processing can include manual or compressed air cleaning, polishing, colouring and other actions that prepare the product for final use. In most cases, in 3D non-industrial printing, removing the printed object is easy - separating the piece printed on the print table, Figure 2 (a).

In the last step the equipment is ready for printing. The process requires the proper setting and control of the 3D printer, workbench cleaning and raw material loading. A routine check of all the main print settings and control panel is also required. When the machine is ready, the 3D print file can be loaded.

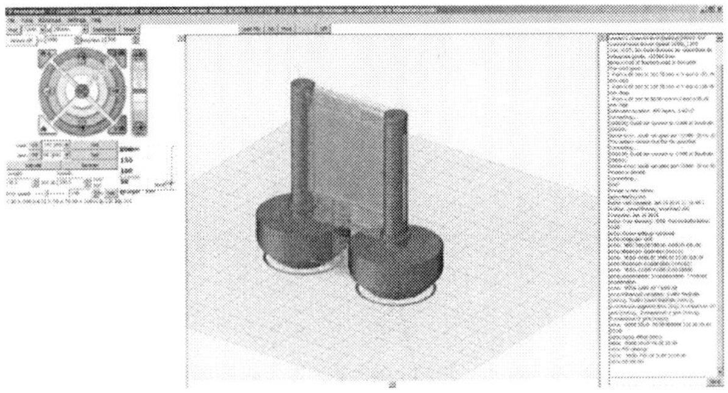

(a)

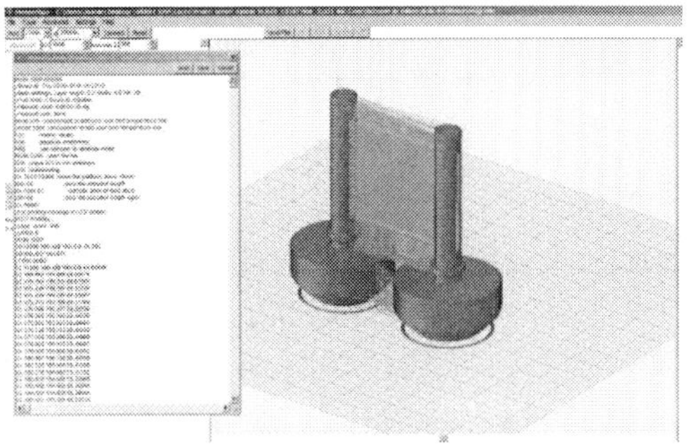

(b)

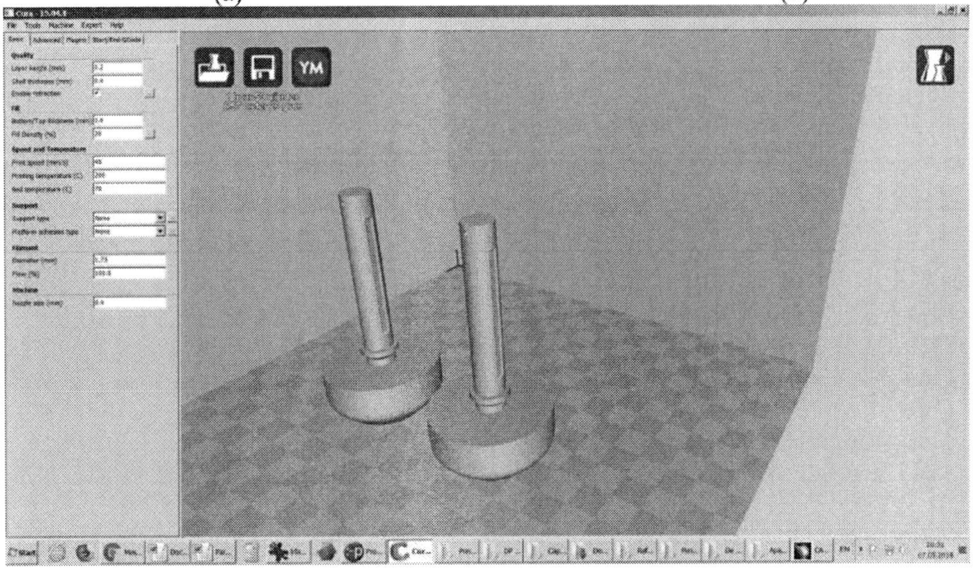

(c)

Figure 2. (a) the Pronterface program interface (which controls the SmartRap + 3D printer) and the possibility of editing a Gcode program directly from Pronterface; (b) the Cura program that makes slicing and generates the Gcode program of the printer; (c) the most important parameters are visible in the left pane.

3D printing or layer-by-layer deposition is a process of making a three-dimensional object of any form digitally designed. The 3D object is accomplished by an additive process in which successive layers of material are deposited in different shapes. The difference between the 3D printing technique and the traditional method of making objects is that instead of eliminating the excess material, the material is deposited in the desired shape from the beginning without a special need for further processing.

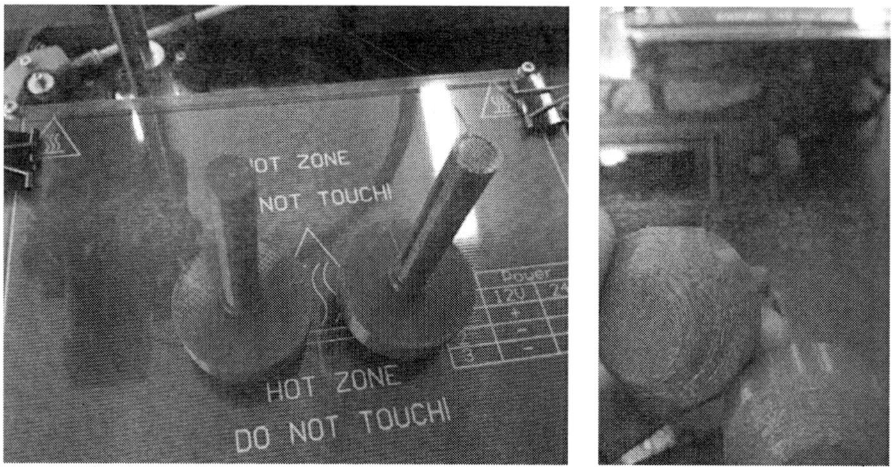

(a) (b)

Figure 3. Button final product (a) two pieces and (b) the need for finishing the final surface of round parts.

4. Conclusions

Because models and parts can be produced in a short time, 3D printing is used to check and develop design ideas. It is cheaper to produce a 3D prototype than to recreate an existing one if needed. As a result, 3D printing is good for those who want to give life to a product idea, because marketing is faster and at lower risk. 3D printing can also reduce the risk of dangers associated with certain manual prototyping procedures. We use 3D printing to obtain plastic pieces with complicated geometries. The printing parameters present a 3 hours' time process with a very good final result concerning the accuracy of the details. Mechanical resistance versus the weight of the element can be choose by modifying the filling degree of the element between 5% to 100%. Round surfaces present areas that necessity a post printing processing.

5. References

[1] Burlea S L, Ciubara A, Burlea G and Cimpoeşu R 2017 *Materiale Plastice* **54** 53-55
[2] Burlea G, Burlea A and Burlea S L Aparat pentru etapa de învatare a scris-cititului *Patent* 2008 00318
[3] Burlea G, Burlea A M and Milici R C 2010 *Revista de Cercetare şi Interventie Sociala* **30** 86-100
[4] Ciubara A, Chirita R, Burlea L S, Lupu V V, Mihai C, Moisa S M and Untu I 2016 *Revista de Cercetare si Interventie Sociala* **52** 265-272
[5] Stratulat S I, Dombici C and Forna D 2013 *Rev. Med. Chir. Soc. Med. Nat.* **117** 875-879
[6] Abdulkhani A, Hosseinzadeh J, Ashori A, Dadashi S and Takzare Z 2014 *Polymer Testing* 35 73–79
[7] Parteni O, Radu C D, Muresan A, Popa M, Ochiuz L, Sandu A V, Agafitei G, Istrate B and Munteanu C 2015 *Revista de Chimie* **66** 1595-99
[8] Britesa F, Malça C, Gaspar F, Horta J F, Franco M C, Biscaia S and Mateusa A 2017 *Procedia Manufacturing* **12** 156 – 165
[9] Niknejad A and Moradi A 2016 *International Journal of Mechanical Sciences* **115-116** 105–122
[10] Strat M and Vasiliu S 2006 *Journal of Optoelectron and Advanced Materials* **8** 181-184

[11] Pöllänen M, Suvanto M and Pakkanen T T 2013 *Composites Science and Technology* **76** 21–28
[12] Ciubara A, Benchea A C, Zelinschi C B, Zelinschi D and Ortansa D 2017 *Revista de Chimie* **68** 307-310
[13] Ciubara A, Iliescu D B, Chiriţă R, Burlea S L, Untu I and Chirila B D 2015 *Revista de Cercetare si Interventie Sociala* **48** 216-225
[14] Nechita P, Craciun G, Nechita D C, Mutica M, Pirlog M C and Ciubara A 2017 *European Neuropsychopharmacology* **27** S926-S927
[15] Rengier F, Mehndiratta A, von Tengg-Kobligk H, Zechman C M, Unterhinninghofen R, Kauczor H U and Giesel F L 2010 *Int J Comput Assist Radiol Surg* **5**(4) 335-341
[16] Blok L G, Longana M L, Yu H and Woods B K S 2018 An investigation into 3D printing of fibre reinforced thermoplastic composites, *Additive Manufacturing*, In press, accepted manuscript
[17] Paun M A, Cimpoesu Hanu R, Cimpoesu N, Agop M, Baciu C, Stratulat S and Nejneru C 2010 *Materiale Plastice* **47** 209-214

Acknowledgments
This work was supported by a grant of the Romanian National Authority for Scientific Research and Innovation, CNCS/CCCDI – UEFISCDI, project number 110CI/2017, with title: Equipment design for the development of writing and reading stage, within PNCDI III.

Hydraulic Physical Model of Debris Flow for Malaysia Case Study

M R R Mohd Arif Zainol[1,2] and M K Awahab[1]
[1]School of Civil Engineering, Universiti Sains Malaysia (USM), Penang, Malaysia.
[2]Center of Excellence Geopolymer and Green Technology (CEGeoGTech), School of Materials Engineering, Universiti Malaysia Perlis (UniMAP), Perlis, Malaysia

E-mail: ceremy@usm.my

Abstract. In the recent decade, several debris flow events occurred and caused hundreds of deaths, missing or injury and damaged many facilities. In addition to causing significant morphological changes along riverbeds and mountain slopes, these flows are frequently reported to bring about extensive property damage and loss of life. Debris flow phenomena occasionally occur in Malaysia and numbers of death reported cause by this event. In order to investigate the debris flow and its deposition process, experiments were conducted at the School of Civil Engineering Laboratory, Universiti Sains Malaysia. The models consists of three main parts which are water tank, rectangular flume and deposition board. A high speed video camera (HSVC) had been placed nearly downstream of the rectangular flume to capture the movement characteristics of particle grain. From this study, the characteristics of particle routing segregation can be understand clearly, therefore this input will be a very useful information to other researchers for further investigation in terms of knowledge sharing between researchers. Catastrophic cause by debris flow event can be minimized therefore in term of economy losses can be reduce and human life can be safe.

1. Introduction

Debris flows and avalanches consist of multiple mixtures of solids, sand, gravel, rocks, ice, snow, and water that move down slopes under the driving force of gravity [1].

The high flowing velocity, large impact forces, and long run-out distance, combined with poor temporal predictability, cause debris flows and debris avalanches to be one of the most hazardous types of landslide [2]. Debris flow is a natural phenomenon in mountainous terrains where uphill valleys or streams are fed by high annual rainfalls.

The occurrences of debris flows have been well documented in some countries such as Japan, United States of America, Taiwan and Indonesia [3, 4] where the mountainous terrains combined with heavy rainfalls provide ideal settings or predispositions for the development of debris flow. In Malaysia, the occurrences of debris flow in the mountainous terrains must, undoubtedly, have taken place from time to time, albeit away from man's view and unreported so long as they do not impact on human lives or infrastructures.

It is when human lives and/or infrastructures are impacted that such debris flow incidents are investigated and reported. This paper presents three such incidents of debris flow in Peninsular Malaysia which had resulted in loss of lives and damage to infrastructures.

2. Cases of debris flow in Malaysia

Many researcher had conducted a debris flow study in Malaysia [6-9]. Kampung Dipangis located in Jeram, Perak. The Kampung Dipang debris flow would have been one of those naturally occurring debris flows in the highlands that would have gone un-reported for the existence of a small village settlement at the base of the hilly area. The debris flow triggered during heavy storm and occurred down a steep valley on the western flank of the Main Range near Kuala Dipang. Debris transported consist of mud to boulders and tree trunks as are typical of debris flows, and swept through a small village settlement and practically wiped out the entire village, with numerous fatalities and casualties.According to the Tan and Ting (2008), debris flow occurred at Genting Sempah, Pahang after heavy rainfall and a series of sequential events which are typical of debris flows have been involved. Resulted in 20 deaths, 22 people injured and some vehicles damaged.Figure 1 shows the aerial photograph taken after the event.

Figure 1. General view of debris flow at Genting Sempah.

Another related visual phenomenon is the large amount of water involved flow along with high velocity. In the case of the North-South Expressway, triggered by storm flow, the "dam-break" phenomenon has been concluded that caused by breach of a naturally formed barrier. The views of the debris flow are as shown in Figure 2. The debris flow reaches the bridge site and partly flows over the bridge, after a large number of debris through the opening of the bridge (Tan and Ting, 2008).

(a) (b)

Figure 2. (a) aerial view of debris flow toward Bridge site (b) accumulation of debris at bridge opening (Tan and Ting, 2008).

3. Experimental setup

Experimental setup of the debris flow physical model was carried out at the School of Civil Engineering Laboratory, Universiti Sains Malaysia. This laboratory provided complete facilities to operate our study in good form. The model consists of three main parts which are rectangular flume, deposition board and water tank. Figure3 shows the debris flow experimental model. At the top end there is water tank (storage container) with 1m long, 0.7 m width and 0.7 m height can be filled up to 0.49 m^3 of water. It is equipped with a gate that can be opened instantaneously by a gate controller. The gate is located at the bottom of the intake tank as a function to supply water through the rectangular flume. The debris flow rectangular flume dimensions were 0.1m width, 0.1 m depth and with 3m effective flow length. The flume is supported with a moveable prop, which could be moved back and forth to adjust the flume slope ranging from 5° to 30°.

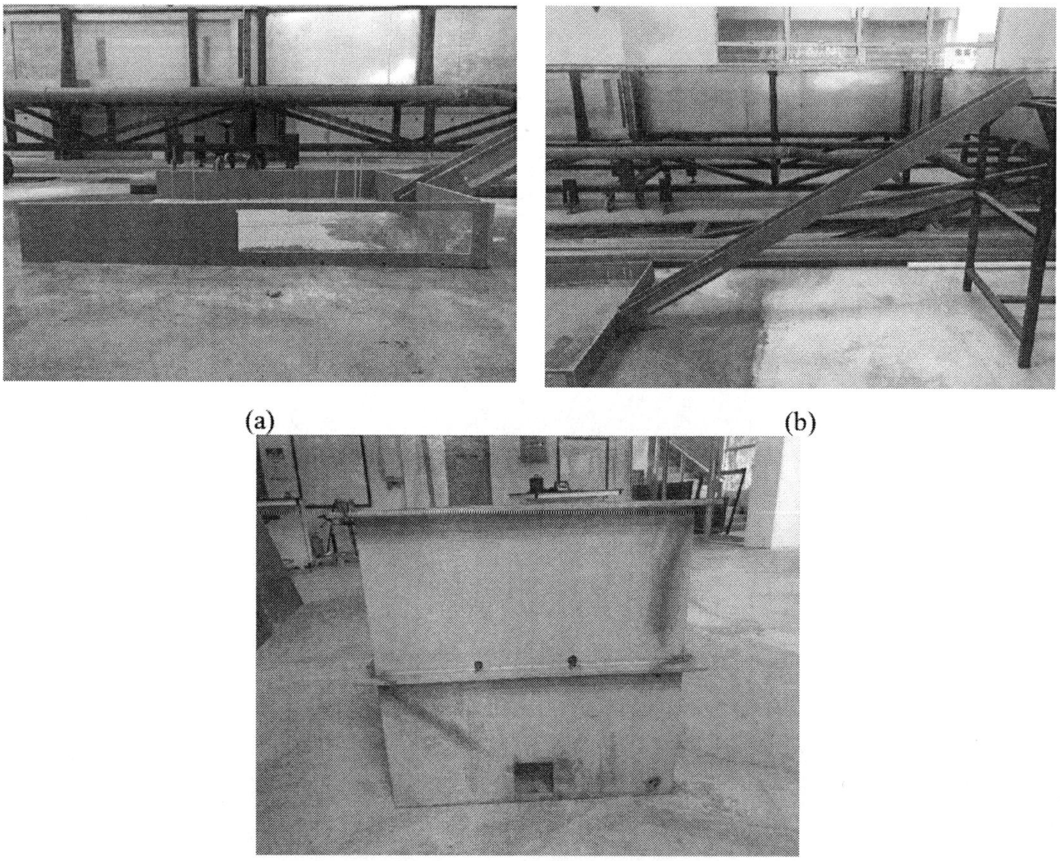

(a) (b)

(c)

Figure 3. Component of hydraulic physical model of debris flow (a) deposition board (b) rectangular flume (c) water tank.

The debris flow deposition board with at each side 1.5 m long was divided into grid. The gridding system is different depend on the deposition area. In the middle of the deposition board, the grid size was smaller than other part because in this area the concentration of the materials is higher. The board was designed as an experimental device, so that its slope could be adjusted as wanted. The board and flume were smoothly connected at one end and at the other end is located materials collection sump.

The deposition board slope varied from 0° to 5°. Figure 4 shows the experimental procedure conducted by [10]. The hydraulic physical model conducted in School of Civil Engineering followed this procedure.

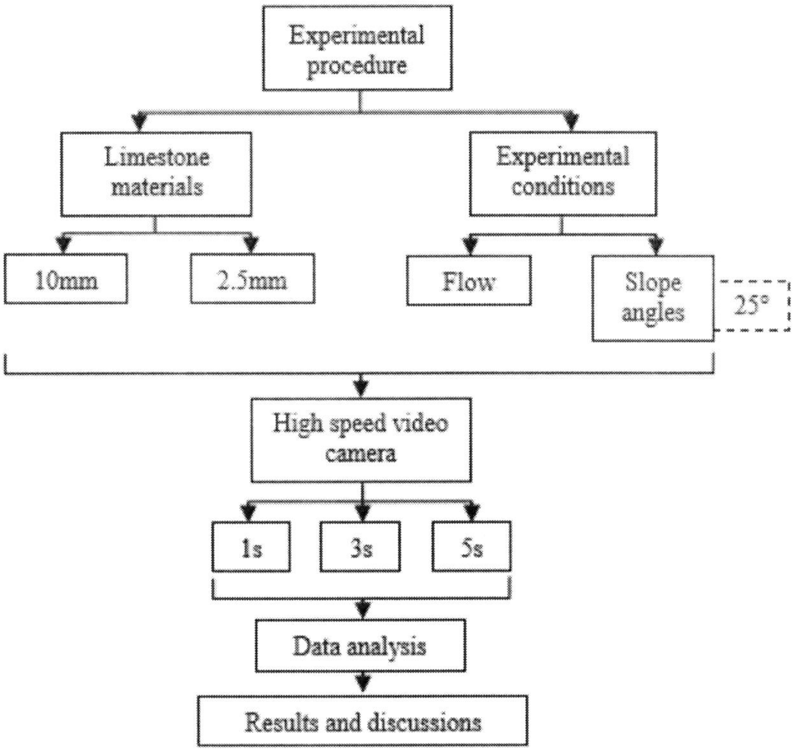

Figure 4. Experimental procedure [10].

4. Conclusions

This study is very important to give some relevant insight of debris flow phenomenon for Malaysia case study. The experimental setup and procedure that been conducted at School of Civil Engineering, Universiti Sains Malaysia gave meaningful understanding of debris flow study. Many researchers and parties will get benefits from this study especially on understanding of debris flow particle movements through high speed video camera (HSVC).

5. References

[1] Hunt B 1994 *Journal of Hydraulic Engineering, ASCE* **120**(12) 1350-1363

[2] Jakob Mand Hungr O 2005 *Debris-flow hazards and related phenomena.* Praxis: Springer Berlin Heidelberg

[3] Yamashiki Y, Mohd Remy Rozainy M A Z, Matsumoto T, Takahashi T, Takara K 2012 *Journal of Global Environment Engineering* **17** 9-18

[4] Jamaludin S, Abdullah C H and Kasim N 2014 *Landslide Science for a Safer Geoenvironment* 167-172.

[5] Tan B Kand Ting W 2008 *Geotechnical Engineering for Disaster Mitigation and Rehabilitation* 231-235

[6] Sum C W, Mior Jadid M Sand Yaacob S 1996 *Warta Geologi* **22** 5

[7] Zaini M M Sand Nasir M K A 2016 Conceptual Design for Flood Warning Study for Gunung Pulai Recreational Area at Gunung Pulai, Johor

[8] Ghazali M A, Rafek A G, Desa K M and Jamaluddin S 2013 *Journal of Geological Research* 1-11

[9] Kuraoka S, Sze L T, Abdullah C H, Jamaludin S, Nazri Mand Kasim N 2016 Study on the Mechanisms of Debris Flows that Damaged Flexible Barriers in the Channel of Fraser Hill, Malaysia

[10] Yamashiki Y, Mohd Remy Rozainy M A Z, Matsumoto T, Takahashi T, Takara K2013*APCBEE Procedia* **5** 527-534

Acknowledgments

This research was carried out with financial support from the Fundamental Research Grant Scheme (FRGS) under Ministry of Higher Education of Malaysia (Grant No. 7611800357). A very special thanks goes to the School of Civil Engineering, Universiti Sains Malaysia for the support on conducting this research. Also thanks to Mr. Mr. Taib Yacob, Mr. Halmi Ghazalli, Mr. Dziauddin Zainol Abidin and Mr. Nabil Semailfor their assistance on performing the experiments.

Influence of Kaolin in Fly Ash Based Geopolymer Concrete: Destructive and Non-Destructive Testing

Z Yahya[1,*], M M A B Abdullah[1], N Mohd Ramli[2], D D Burduhos-Nergis[3] and R Abd Razak[1]
[1]Center of Excellence Geopolymer and Green Technology, School of Materials Engineering, Universiti Malaysia Perlis (UniMAP), 01000, P.O. Box 77, D/A PejabatPosBesar, Kangar, Perlis, Malaysia
[2] Department of Civil Engineering Technology, Faculty of Engineering Technology, Universiti Malaysia Perlis (UniMAP), Sungai Chuchuh, 02100 Padang Besar, Perlis, Malaysia
[3]Gheorghe Asachi Technical University of Iasi, Faculty of Materials Science and Engineering, Blvd. D. Mangeron 41, 700050, Iasi, Romania

E-mail: zarinayahya@unimap.edu.my

Abstract. Development of geopolymer concrete is mainly to reduce the production of ordinary Portland cement (OPC) that adverse the natural effect. Fly ash is a by-product collected from electrical generating power plant which resulted from burning pulverized coal. Since fly ash is waste materials, it can be recycled for future advantages particularly as pozzolanic materials in construction industry. This study focused on the feasibility of fly ash based geopolymer concrete to which kaolin has been added. The main constituents of geopolymer production for this study were class F fly ash, sodium silicate and sodium hydroxide (NaOH) solution. The concentration of NaOH solution was fixed at 12 Molar, ratio of fly ash/alkaline activator and sodium silicate/NaOH fixed at 1.5 and 2.5, respectively. Kaolin was added in range 5% to 15% from the mass of fly ash and all the samples were cured at room temperature. Destructive and non-destructive test were performed on geopolymer concrete to evaluate the best mix proportions that yield the highest strength as well as the quality of the concrete. Compressive strength, flexural strength, rebound hammer and ultrasonic pulse velocity (UPV) result have been obtained. It shown that 5% replacement of kaolin contributed to maximum compressive strength and flexural strength of 40.4 MPa and 12.35 MPa at 28 days. These result was supported by non-destructive test for the same mix proportion.

1. Introduction

An increasing demand in more environmental friendly construction materials has been driving force for researchers to develop innovative solution such as utilizing waste materials from coal industries and decomposition of natural resources [1]. Geopolymer was initially created by Joseph Davidovits, which comprises of SiO_4 and AlO_4 tetrahedral systems. Alumina-silicate responsive materials break up in solid antacid arrangements and free SiO_4 and AlO_4 tetrahedral structure shapes.

Geopolymer material has caught the enthusiasm of numerous researchers from various fields since it exhibits amazing properties such as high compressive strength, resistance towards fire, corrosion, thermal and many more. Besides, it is an environment friendly because of the way that it can be set up from industrial wastes and natural resources. The main materials use to produce geopolymer are alumina-silicate materials such as fly ash, kaolin, slag, and rice husk ash, palm ashas well as alkaline activators [2-4].

For past thirty years, through a variety of new and better processes, many clay-based products have been produced with high quality. The increased in fabrication of clay-based product has further expanded conventional applications by bringing more innovation of modern applications. Kaolin is white or close white shading with pseudo-hexagonal precious stone alongside plates, some bigger books and vermicular stacks. However, utilization of kaolin as geopolymer source material produce weak structure due to low reactivity where it required certain period of times in order to increase the geopolymerisation rate for strength gain [5].The low reactivity of kaolin as source material for geopolymer due particle shape of the kaolin itself where the small surface area allowed minimum dissolution of Si and Al by alkaline activator hence lower strength [6]. When kaolin was activated by alkaline activator, the reaction starts from the surface and edge where it slowly penetrates inside the structure hence cause the low strength performance for kaolin based geopolymer [7]. Nevertheless, kaolin based geopolymer has favourable properties in which it has good volume stability in the water. In addition, kaolin also has good compressive strength and can be supplemented with increased ageing. However, the development of strength is very slow and it is supposed due to limitation of kaolinite structure that has low surface area and limited replacement of other element during the reaction.

Van Jaarsveld et al. [8] found that the inclusion of high kaolin content (41 wt.%) in fly ash based geopolymer reduced the strength of final product. This result was obtained because not all kaolin involved in geopolymerisation process to form the geopolymer network. Meanwhile, Okoye et al. [9] investigated the replacement of fly ash with kaolin up to 50% in fly ash based geopolymer concrete where it was found that, compressive strength increased with higher replacement of kaolin.

The current study, fly ash is replaced with different content of kaolin (0% until 15%) where the destructive and non-destructive test was conducted on fly ash-kaolin geopolymer concrete to identify the homogeneity of these two materials.

2. Materials and methods

2.1. Materials

Class F fly ash was obtained from CIMA, Perlis, Malaysia and kaolin was supplied by Associated Kaolin Industries Sdn. Bhd., Malaysia. The chemical composition for both materials is tabulated in Table 1 which obtained from X-ray Fluorescence (XRF) analysis. Figure 1a and 1b showed the morphology of fly ash and kaolin particles.

Table 1. Chemical composition of fly ash and kaolin.

Constituents	Mass (%)	
	Fly Ash	Kaolin
SiO_2	55.9	54.0
Al_2O_3	27.8	31.7
CaO	3.95	-
Fe_2O_3	7.09	4.89
TiO_2	2.25	1.41
K_2O	1.55	6.05
SrO	0.37	-

Sodium hydroxide (NaOH) pellets were dissolved using distilled water to produce 12 Molar concentration [10]. The NaOH solution was prepared 24 hours prior the mixing process of geopolymer samples. Technical grade of sodium silicate solution was mixed with NaOH solution to produce alkaline activator. Crushed stone with size in ranges from 6mm to 8mm were used as coarse aggregates and river sand was used as fine aggregates.

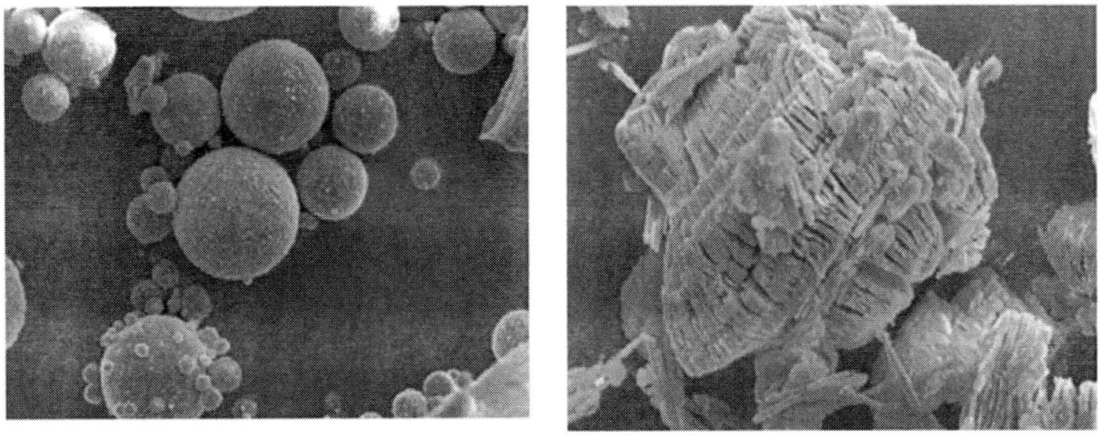

Figure 1. (a) Morphology of fly ash (b) morphology of kaolin observed using FESEM.

2.2. Mix design and mixing process

Sodium silicate and sodium hydroxide (NaOH) solution was mixed together to produce alkaline activator with ratio of 2.5 [2, 11-13]. NaOH pellets were dissolved with distilled water to obtain 12 Molar concentration [10]. Meanwhile, the ratio of fly ash/alkaline activator was fixed at 1.5 with respect to the workability of geopolymer concrete. The combination of coarse aggregate and sand contributed 70% to the total weight of geopolymer mixture. Fly ash was replaced by kaolin start from 0% (control samples), 5%, 10% and 15%. The details of mix design were presented in Table 2.

In preparation of geopolymer mixture, the fly ash was mixed together with kaolin and alkaline activator for about 3 minutes in the mixer to ensure homogeneity of the mixture. Then, sand and coarse aggregate were added and mixed for another 5 minutes. The geopolymer concrete were poured into 100mm x 100mm x 100mm mould and vibrated on the vibrating table for 5 minutes to release entrapped air. The samples were left at room temperature for 3, 7 and 28 days. For flexural strength only the optimum design was used to produce geopolymer samples. The geopolymer mixture were casted in prism mould 100mm x 100mm x 500mm and tested for 3, 7 and 28 days.

Table 2. Mix proportion of geopolymer concrete.

Mix	Quantity of Ingredients (kg/m^3)					
	Fly Ash	**Kaolin**	**Sand**	**Coarse Aggregates**	**NaOH solution**	**Sodium silicate**
FAK0 (control)	432	0	672	1008	82	206
FAK5	410.4	21.6	672	1008	82	206
FAK10	388.8	43.2	672	1008	82	206
FAK15	367.2	64.8	672	1008	82	206

2.3. Testing

The performance of fly ash-kaolin based geopolymer concrete were prepared for destructive (DT) and non-destructive test (NDT). The NDT consists of Ultrasonic Pulse Velocity (UPV) and rebound hammer while the DT comprising both compressive strength and flexural strength.

2.3.1. Ultrasonic Pulse Velocity. UPV test was conducted on samples at the age of 3, 7 and 28 days in order to examined the impact of partially replacing fly ash with kaolin on geopolymer concrete uniformity. A set of geopolymer concrete comprising of three cubes were tested by measuring the velocity of ultrasonic pulses for three times by direct, semi-direct and indirect transmission which means nine reading for each cube. Then, the result for each transmission was compared with the guidelines as in Table 3.

Table 3. General guideline for concrete quality based on UPV [14].

Pulse Velocity (m/s)	Concrete Quality
> 4500	Very good to excellent
3600 - 4500	Good to very good but slight porosity may exist
3000 - 3500	Satisfactory but loss of integrity is suspected
2100 - 3000	Poor and loss of integrity exist
< 2100	Very poor

2.3.2. Rebound Hammer. The rebound hammer test was conducted on the geopolymer concrete with smooth, clean and dry surface. For some rough surfaces present on the concrete, the surface was rubbed with grinding wheel or stone. The point of impacting the hammer should be 20mm far from the edge and discontinuity shapes. The rebound hammer should be kept perpendicular to the surface of the concrete.

On each surface, numbers of observations are taken and the average of these observations gives the strength of concrete. The rebound hammer test was carried out on specimens at the age of 3, 7 and 28 days in order to determine the effect of partially replacing fly ash with kaolin on geopolymer surface hardness and penetration resistant. Table 4 shows the guidelines values for rebound hammer test.

Table 4. General guideline for concrete quality based on rebound number [15].

Rebound no.	Quality of Concrete
> 40	Very good
30 - 40	Good
20 - 30	Fair
< 20	Poor

2.3.3. Compressive strength. Compressive test was conducted on cube samples at day 3, 7 and 28 where the samples were tested by using hydraulic test machine. Compressive strength test was carried out on three cube samples for each mix design. Samples were tested by using hydraulic test machine governed by BS. 1881: Part116: 1983. Load was applied gradually without shock at a nominal rate of 0.3 N/(mm².s).

*2.3.4. Flexural strength.*In order to quantify flexural strength of geopolymer concrete containing diverse rates of kaolin as fractional substitution of fly ash, ASTM C293 was chosen as code of practice. Prisms samples were tested with pace rate of 0.06 N/(mm².s).

The modulus of rupture resolved by centre-point loading is higher than third-point loading. For this study, the test conducted based on ASTM C293 (centre point loading), which sample had undergone the curing process for 3,7 and 28 days.

3. Results and discussions

3.1. Ultrasonic Pulse Velocity

Figure 2 illustrates the result of UPVforgeopolymer concrete with inclusion of kaolin on direct transmission mode. For this transmission mode, FAK5 shows excellent result where contributed the highest pulse velocity throughout the test days with values of 3770 m/s, 3280 m/s and 4180 m/s on 3rd,

7th and 28th days respectively. The pulse velocities of the samples decreased on 7th day of testing and increase again on 28th day. The quality of concrete is varying from satisfactory to good.

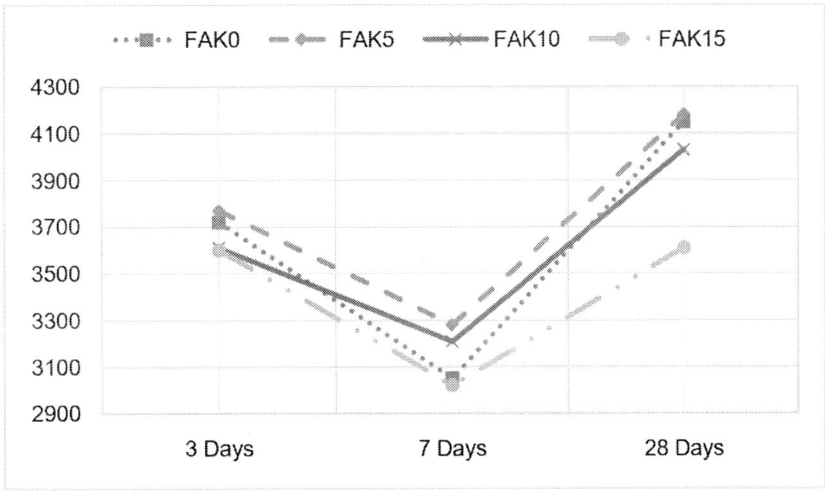

Figure 2. Ultrasonic pulse velocity result for direct transmission.

According to Benaicha et al. [16], pulse velocity usually increases very rapidly at early age of concrete and gets flatten in later age. However, from Figure 2 to Figure 4the pulse velocity decrease at early age and increase again in later age regardless any of the transmission mode. Addition of kaolin in geopolymer concrete obviously reduce the velocity of ultrasonic pulses and its homogeneity.

Figure 3 showed that on day 3, FAK15 has the closest pulse velocity to the control mix (FAK0) which has the value of 5430 m/s while FAK5 has the lowest pulse velocity of 5050 m/s. On 7th day, pulse velocity of FAK10 dropped dramatically from 5200 m/s to 3650 m/s while other mixes are slightly decreasing. The following test day, the pulse velocity of all the mixes are increasing and FAK5 has the highest pulse velocity of 5700 m/s surpassing FAK0, the concrete without kaolin. Meanwhile, on 28th day, FAK15 unexpectedly has the lowest pulse velocity of 5010 m/s. By referring to Table 4, most of the geopolymer concretes have excellent quality except for FAK10 and FAK15 on 7th day of testing, they showed good concrete quality but with slight porosity presence in them.

For indirect transmission mode, all geopolymer concretes show satisfactory to good quality of concrete as displayed on Figure 4. On the 3rd day of testing, FAK5 has the closest value of pulse velocity to FAK0 followed by FAK15 and FAK10 with pulse velocities of 3640 m/s, 3230 m/s, 3090 m/s and 2860 m/s respectively. Similar pattern with semi-direct transmission result, the pulse velocity dropped slightly on 7th day of testing and increase again on 28th day. However, pulse velocity of FAK10 slightly increase on 7th day. Again, on 28th day of testing, FAK5 has the highest pulse velocity of 3770 m/s while FAK15 has the lowest pulse velocity of 3450 m/s. However, on 28th day, FAK5 and FAK10 surpass the pulse velocity of control mix (FAK0).

From Figure 2 to Figure 4 the curing period does not have great influences on the value of ultrasonic pulse velocity. This is because addition of kaolin has caused the process of geopolymerisation take longer time to occur. Furthermore, the addition of kaolin allows the stiffness of concrete increase which resulting in enhancement of pulse velocity on 28 days.

According to the quality assessment of concrete by Saint-Pierre et al. [14], all the mixes can be classified as satisfactory to excellent quality concrete since their pulse velocity values ranging from the lowest of 2860 m/s to the highest of 5700 m/s.The increase of ultrasonic pulse velocity with age is a natural development due to the increase of stiffness by the geopolymerisation reaction. For the same concrete samples, the quality of concretes increase from day to day through geopolymerisation which makes it stronger and more durable concrete according to the time.

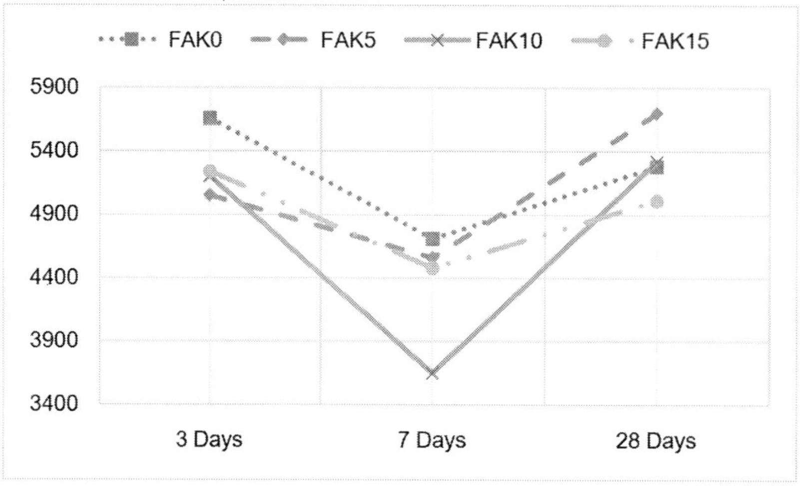

Figure 3. Ultrasonic pulse velocity result for semi-direct transmission.

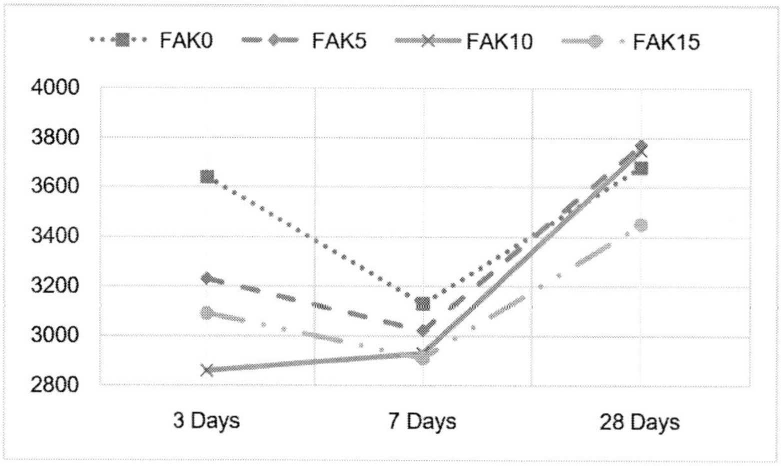

Figure 4. Ultrasonic pulse velocity result for indirect transmission.

3.2. Rebound Hammer

Based on the Table 5, all of the geopolymer concretes exhibit poor concrete quality when measured by Rebound Hammer. This result was contradictory compared to ultrasonic pulse velocity. The different result obtained between these two testing may be due to several factors that affecting the rebound hammer result. The condition of concrete surface influence the rebound number acquired during testing. Tests performed on a rough-surfaced concrete will commonly result in crushing of the surface paste, resulting in lower number and then again, tests performed on a similar concrete that has a hard, smooth surface will normally bring about a higher rebound number. Since fresh geopolymer paste has high viscosity and cohesive nature, it is important to place the concrete in the mould immediately after mixing as surrounding temperature of the geopolymer has direct connection with the mixing time. Increasing in temperature of geopolymerconcrete resulting from increase in mixing time would diminished the workability of geopolymer concrete [17]. Low workability of geopolymerconcrete make them hard to placed and thus, not achieving a smooth concrete surface. Hence, it is advised to grind the rough surface with Carborundum stone or any similar abrasive stone in order to achieve a uniform smoothness.

Table 5. Result for rebound hammer.

Mix No.	Rebound number					
	3-day	Quality of concrete	7-day	Quality of concrete	28-day	Quality of concrete
FAK0	12	Poor	12	Poor	16	Poor
FAK5	12	Poor	13	Poor	16	Poor
FAK10	13	Poor	14	Poor	17	Poor
FAK15	14	Poor	14	Poor	16	Poor

Agingperiod of concrete also affects the rebound number of the samples. Geopolymer concrete normally continues to develop strength with age due to geopolymerisation process. This is the reason behind the development of data relating rebound numbers to the compressive strength of the concrete mixture or cores from the structure. From Table 5, geopolymer concrete at age 3 days have low rebound numbers and for concrete at age 7 days, it has nearly similar rebound numbers. Alternately, the rebound numbers on concrete at the age of 28 days have slightly higher value but not significant as all of the concretes are in poor quality. It is not recommended to test the concrete less than 3 days old as the rebound numbers will be too low for an accurate reading. In this case, geopolymer concrete takes longer time to harden and this will affect the rebound numbers. Next, moisture content of the geopolymer samples. It has a profound effect on the test results in which dry concrete surfaces result in higher rebound numbers than wet surfaces. Geopolymer concrete at an early age are mostly not in fully dry condition especially concrete at age of 3 days. The concrete surfaces not dry enough due to low geopolymerisation rate between fly ash and kaolin especially with higher replacement of kaolin [19-20].

3.3. Compressive strength

Compressive strength of geopolymer concrete with various replacement of kaolin were performed and the results have been summarized in Figure 5. The compressive strength increases steadily at 3 days for FAK5 and FAK10. The compressive strength increase moderately with increasing curing ages. Nevertheless, FAK5 and FAK10 have noticeable changes in strength throughout the curing days. It is remarkable that the compressive strength developed rapidly before 7 days of curing. Corresponding to the previous research, early strength of geopolymer concrete increase rapidly.

For control sample (FAK0), compressive strength increases from 8.4 MPa at 3 days to 62.1 MPa at 28 days. Noticeable change in compressive strength of geopolymer with addition of different amounts of kaolin can be observed when compared to control sample (FAK0) at 3 days. Besides, addition of kaolin in fly ash geopolymer concrete improved the early compressive strength. Longer ageing times resulting in increasing compressive strength especially after 7 days when kaolin is added.

Addition of kaolin up to 5% increased the compressive strength of the concrete, but further increment of kaolin content resulted in slightly decrease in strength. Higher content of kaolin in fly ash based geopolymer concrete caused the mixture become stickier and takes longer time to harden.

FAK5 contributed to the highest compressive strength of 21.3 MPa and 32 MPa on both 3rd and 7th curing days while FAK0 gained the highest strength (62.1MPa) on 28th day. Nevertheless, on 28th day FAK5 displayed strength of 40.4 MPa. Meanwhile, both FAK10 and FAK15 also showed greater strength than FAK0 at the age of 3 days with strength of 15.5 MPa and 9.2 MPa respectively. This shows that inclusion of kaolin in fly ash based geopolymer concrete can enhance the early compressive strength of concrete. However, at the age of 7 days, FAK0 has slightly higher strength compared toFAK15. It can conclude that, inclusion of kaolin more than 10% slow down the hardening process of the concrete and did not contribute significantly in development of concrete strength.

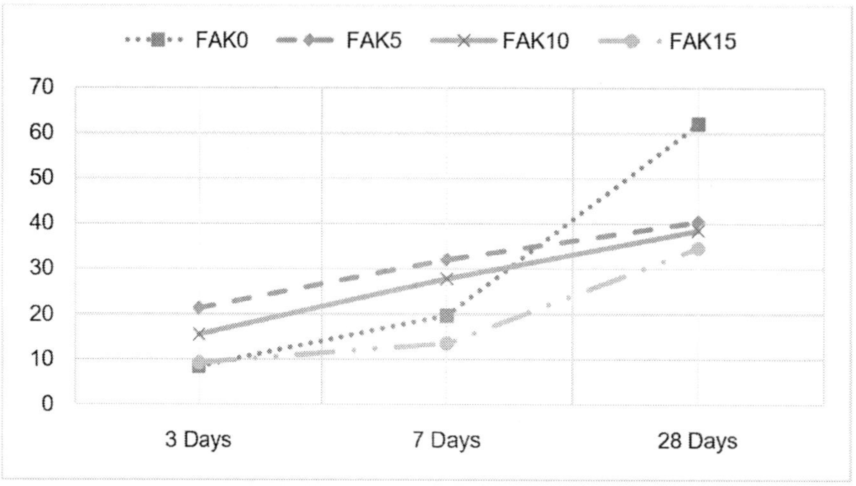

Figure 5. Compressive strength of geopolymer concrete at different aging periods.

3.4. Flexural strength

Flexural strength of geopolymer concrete with 5% of kaolin (FAK5) replacement have been summarised in Figure 6.The flexural strength at age of 3 days is 1.87 MPa and then increased dramatically from 5.74 MPa on 7th day to 12.35 MPa on 28th day. This samples displayed an increment in strength when the aging period increased which is similar pattern with compressive strength result. Based on research conducted by Nath and Sarker[18], it was found that the flexural strength of fly ash based geopolymer at 28 days is 4.89 MPa.

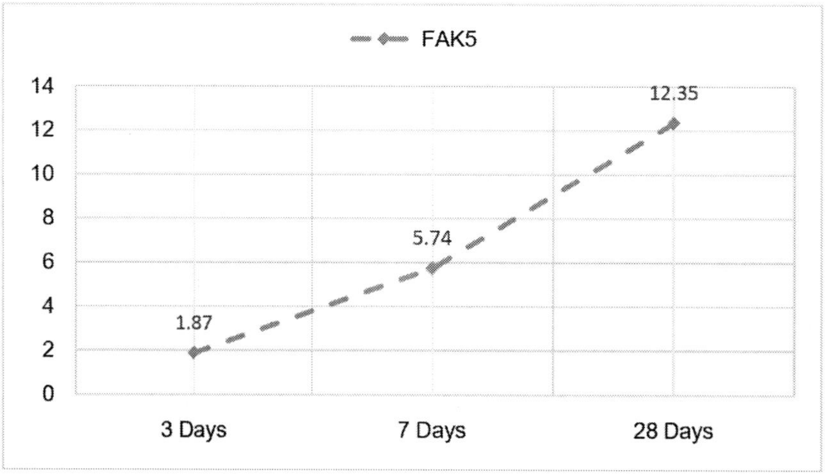

Figure 6. Flexural strength of geopolymer concrete with 5% inclusion of kaolin.

By comparing the flexural strength from previous study with this research, it is clearly showed that, the flexural strength of geopolymer concrete containing 5% kaolin is much higher. The addition of kaolin improved the long term strength of fly ash based geopolymer concrete.

4. Conclusions

The effect of partial replacement of kaolin in fly ash based geopolymer concrete were studied. Class F fly ash was blended with kaolin up to 15% replacement and the result of the study are summarized below:

i) The ultrasonic pulse velocity of all samples increased when respect to aging period. However, at the age of 7 days, the pulse velocity unexpectedly dropped for all samples. The geopolymer concrete generally have good quality concrete with slight porosity. It was noted that, the quality of the concrete is declining as the percentage of kaolin in the geopolymer mixture increase due to higher porosity and lower homogeneity.

ii) For rebound hammer, contradictory result from ultrasonic pulse velocity was obtained. All the geopolymer concretes were summarized as poor quality.

iii) FAK5 achieved the highest compressive strength at age of 3 and 7 days of 21.3 MPa and 32 MPa respectively, and second highest at age of 28 days with strength of 40.4 MPa. The control mix (FAK0) has compressive strength of 8.4 MPa, 19.7 MPa and 62.1 MPa at the age of 3, 7 and 28 days respectively.

iv) The flexural strength of FAK5 also showed increment when the aging period increased. The maximum strength recorded on 28th days is 12.35 MPa.

5. References

[1] Razi P Z, Razak H A and Khalid N H A 2016 *Materials* **9**(5) 1–28
[2] Mustafa Al Bakri A M, Kamarudin H, Khairul Nizar I, Bnhussain M, Zarina Y and Rafiza A R 2012 *Advanced Materials Research* **341-342** 189-193
[3] Heah C Y, Kamarudin H, Mustafa Al Bakri A M, Luqman M, Khairul Nizar I and Liew Y M 2011 *Australian Journal of Basic and Applied Sciences* **5**(7) 1026-1035
[4] Yahya Z, M. M. Abdullah A B, Hussin K, Ismail K N, Sandu A V, Vizureanu P and Razak R A 2013 *Revista de Chimie* **64**(12) 1408 -1412
[5] Heah C Y, Kamarudin H, Mustafa Al Bakri A M, Binhussain M, Luqman M, Khairul Nizar I and Liew Y M 2011 *Physics Procedia* **22** 305-311
[6] Heah C Y, Kamarudin H, Mustafa Al Bakri A M, Bnhussain M, Luqman M, Khairul Nizar I, Ruzaidi C M and Liew Y M 2012 *Construction and Building Materials* **35** 912-922
[7] Davidovits J 2008 Geopolymer Chemistry and Application, 2nd Edition, 02100 Saint Quentin, France, Insitute Geopolymere
[8] van Jaarsveld J G S, van Deventer J S J and Lukey G C 2002 *Chemical Engineering Journal* **89** 63 – 73
[9] Okoye F N, Durgaprasad J and Singh N B 2015 *Construction and Building Materials* **98** 685 – 691
[10] Al Bakri A M M, Kamarudin H, Karem O A K A, Ruzaidi C M, Rafiza A R and Norazian M N 2012 *Applied Mechanics and Materials* **110-116** 734-739
[11] Hardjito D, Wallah S E, Sumajouw D M J and Rangan B V 2004 *ACI Materials Journal* **101** 467-472
[12] Das S K, Mohapatra A K and Rath A K 2014 *International Journal of Civil Engineering Research* **5**(1) 21-28
[13] Ding Y, Dai J and Shi C 2016 *Construction and Building Materials* **127** 68-79
[14] Saint-pierre F, Philibert A, Giroux B and Rivard P 2016 *Construction and Building Materials* **125** 1022-1027
[15] Patil H, Khainar D and Thube R 2015 *IJSART* **1**(6) 1-5
[16] Benaicha M, Jalbaud O, Hafidi A and Burtschell Y 2015 *Construction and Building Materials* **101** 702-709
[17] Burduhos Nergis D D, Abdullah M M A and Vizureanu P 2017 *European Journal of Materials Science and Engineering* **2** 111-118
[18] Nath P and Sarker P K 2017 *Construction and Building Materials* **130** 22-31

IOP Conf. Series: Materials Science and Engineering **374** (2018) 012068 doi:10.1088/1757-899X/374/1/012068

[19] Popita G E, Rosu C, Manciula D, Corbu O, Popovici A, Nemes O, Sandu A V, Proorocu M and Dan S B 2016 *Materiale Plastice* **53(2)** 308-311
[20] Corbu O, Popovici A, Popita GE, Rusu T, Rosu C and Puskas A 2014 Geoconference On Nano, Bio And Green - Technologies For A Sustainable Future 2 (Sgem 2014), 81-88

IOP Conf. Series: Materials Science and Engineering **374** (2018) 012069 doi:10.1088/1757-899X/374/1/012069

Durability of Fly Ash Based Geopolymer Concrete Infilled with Rubber Crumb in Seawater Exposure

Z Yahya[1,*], M M A B Abdullah[1], S N H Ramli[2], M G Minciuna[1,3] and R Abd Razak[1]

[1]Center of Excellence Geopolymer and Green Technology, School of Materials Engineering, Universiti Malaysia Perlis (UniMAP), Perlis, Malaysia

[2]Department of Civil Engineering Technology, Faculty of Engineering Technology, Universiti Malaysia Perlis (UniMAP), Perlis, Malaysia

[3]Gheorghe Asachi Technical University of Iasi, Faculty of Materials Science and Engineering, Blvd. D. Mangeron 41, 700050, Iasi, Romania

E-mail: zarinayahya@unimap.edu.my

Abstract. Geopolymer is an alternative binder to replace Ordinary Portland Cement (OPC)in construction industry. Source materials that rich in silica (Si) and alumina (Al) were activated by using alkaline solution. Production of tires keep increasing every year and due to its non-biodegradable properties it causes problems for disposal process. In current scenario, waste materials should be used or recycled so that the existing natural resources can be saved and at the same times it can protected environment. In this paper, the effect of rubber crumb on fly ash based geopolymer concrete have been investigated by immersing the samples in seawater for 28 and 60 days. The rubber crumb was used to replace coarse aggregates from range 5% until 20%. The ratio of fly ash/alkaline activator and sodium silicate/sodium hydroxide(NaOH) ratio were fixed at 2.0 and 2.5. It has been shown that the compressive strength decreased when the content of rubber crumb increased. The highest compressive strength (39.6 MPa) was obtained at 5% replacement of rubber crumb when exposure to seawater for 28 days. The density of geopolymer samples also increased when immersed in seawater for all samples. The lack of bonding between rubber crumb and geoplymer paste cause increasing in porosity hence reduced the strength, increment in density and changes in weight of geopolymer samples.

1. Introduction

Geopolymer is an alternative binder for Ordinary Portland Cement (OPC) where it possesses high strength, good resistance against aggressive environment as well as natural sustainability for industrial ecology. The main constituents that important in geopolymer production were alkaline activator and source materials that rich in silica (Si) and alumina (Al). The benefits of using geopolymer in term of environmental issues were potential reduction in greenhouse gases emission and utilization of waste materials [1-2]. The common source materials used were fly ash, ground granulated blast furnace slag (ggbs), kaolin, metakaolin and silica fume [3-5].

Disposal of waste tire involve with environmental issues where the heavy metals in tires generate risk when it was disposed in the landfill [6-7]. The leaching of toxins from the metals inside the tires when the landfill was wet will disturb the groundwater quality [6]. In Malaysia, it was estimated about 57,391 tonnes of waste tires generated annually where 60% of it were disposed via unknown routes [8]. The accumulation of utilized tires at landfill sites also showed the danger of uncontrolled flames, creating a complex blend of chemicals harming the earth and defiling soil. The utilization of rubber crumb in concrete has been started since last two decades to reduce environmental impact [9-10]. The findings from the past researches showed that the compressive strength and flexural strength decreased when the content of rubber crumb increased [11-15]. Nevertheless, the properties such as the ductility and impact resistance improved [9-13]. Generally, the rubber crumb was used in concrete to replace either fine aggregate or coarse aggregate.

Park et. al [6] investigate the effect of adding rubber crumb in three different type of fly ash based geopolymer as partial replacement of fine aggregate. It was found that the content of calcium oxide (CaO) in fly ash play significant role in the compressive strength of rubberized geopolymer concrete [2]. Besides that, the replacement of rubber crumb up to 20% cause reduction in compressive strength approximately 15%. According to Azmi et al. [14], all geopolymer samples with inclusion of rubber crumb (5%, 10%, 15% and 20%) displayed strength reduction when tested on 7 and 28 days. However, the strength of all geopolymer samples still higher compared to normal rubberized concrete. It can be seen from the previous research, the utilization of rubber crumb in geopolymer concrete was not investigated thoroughly. Most of the researches focused on effect of rubber crumb percentage for the short term performance.

In the present study, rubber crumb were added in fly ash based geopolymer concrete as replacement for coarse aggregate. The amount of rubber crumb added in ranges of 5% until 20% and control samples were prepared for comparison purpose. The durability of the samples in term of compressive strength, density and changes in weight after expose in marine environment for 28 and 60 days were observed.

2. Materials and methods

2.1. Materials

The main constituents used to produce geopolymer are class F fly ash and alkaline activator. The fly ash was obtained from CIMA, Perlis, Malaysia and its chemical composition has been determined by using X-Ray Fluorescence (XRF) as showed in Table 1.

Table 1. Chemical composition of fly ash.

Constituents	Mass (%)
SiO_2	55.9
Al_2O_3	27.8
CaO	3.95
Fe_2O_3	7.09
TiO_2	2.25
K_2O	1.55
SrO	0.37

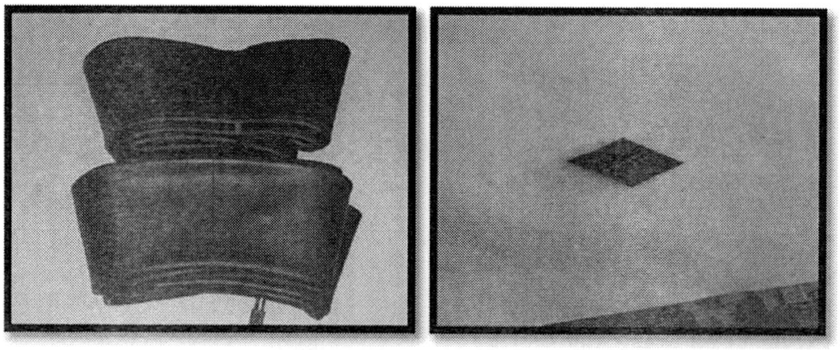

Figure 1. Rubber crumb.

Sodium silicate and sodium hydroxide (NaOH) were used as alkaline activator. Crushes stone was used as coarse aggregates with maximum size 20mm and river sand as fine aggregates with size not more than 4.75mm. The rubber crumb was obtained from motorcycle scrap tires where it was cut with size 0.5cm to 1cm as given in Figure 1.

2.2. Mix designs and mixing process

The details mix design of geopolymer concrete are given in Table 2. Five concrete mixes were designed with 0, 5, 10, 15 and 15% rubber crumb. The ratio of fly ash/alkaline activator and sodium silicate/NaOH were fixed at 2.0 and 2.5 for the whole samples [15]. The mass ofaggregates to fly ash was 70% and 30% where the proportion of coarse aggregates and fine aggregate are 60% and 40% of the total mass of aggregates used. Meanwhile, the concentration of NaOH solution was fixed at 12 Molar [15]. Geopolymer paste were produced first by mixed fly ash and alkaline activator for 3 minutes. Then the rubber crumb, fine and coarse aggregates were added into the mixture and mixed for another 5 minutes. After mixing, the geopolymer concrete mixture was cast in 100mm x 100mm x 100mm steel mould and left at room temperature for 3 days before immersed in seawater. The geopolymer concretes were exposed in seawater for a period of 28 and 60 days.

The effect of seawater on GPC0, GPC5, GPC10, GPC15 and GPC20 were evaluated through compressive strength, weight changes and density.

Table 2. Mix proportion of geopolymer concretes.

Mix No.	Rubber Crumb	Fly Ash	Coarse Aggregate	Fine Aggregate	NaOH solution	Sodium silicate solution
GPC0	0	624	1310	874	89	223
GPC5	65.5	624	1244.5	874	89	223
GPC10	131	624	1179	874	89	223
GPC15	196.5	624	1113.5	874	89	223
GPC20	262	624	1048	874	89	223

3. Results and discussions

3.1. Compressive strength

Compressive strength of geopolymer samples with inclusion of rubber crumb are shown in Figure 2.

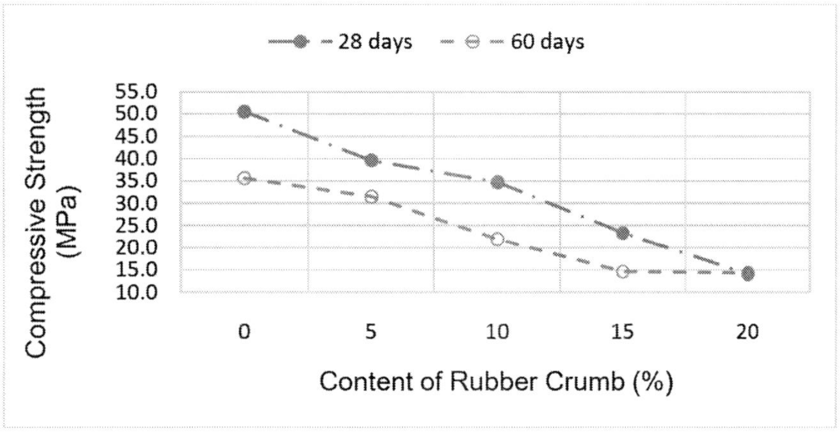

Figure 2. Compressive strength of geopolymer concrete exposed to seawater.

The geopolymer samples showed reduction in strength when percentage of rubber crumb added increased for 28 and 60 days exposure period. The control sample (GPC0) displayed high strength (50.5 MPa) after being exposed for 28 days in seawater. The maximum compressive strength of

39.6MPa and 31.5MPa was recorded by GPC5 for both exposure periods. Nevertheless, the strength decreased when rubber crumb was added where the reduction in strength between control sample and GPC5 was 22% for 28 days exposure periods. However, there is only slight difference in strength for GPC20 for both exposure periods. The reduction in strength when the content of rubber crumb increase was due to lack of bonding between geopolymer paste and rubber crumb. Moreover, the insufficient bonding also leads to high internal stress which is perpendicular to the direction of applied load.

3.2. Density and weight changes

The addition of rubber crumb in geopolymer concrete reduce the density as shown in Figure 3. It was observed that the density decreased when the percentage of rubber crumb increase due to the low unit weight of rubber crumb. The lowest density of 2190 kg/m^3 was recorded by GPC20 and the control sample (GPC0) recorded 2360 kg/m^3. Besides that, the rubber crumb content also influence the porosity of geopolymer concrete. When the geopolymer samples exposed to seawater for 28 and 60 days, an increment in weight were recorded for all samples.

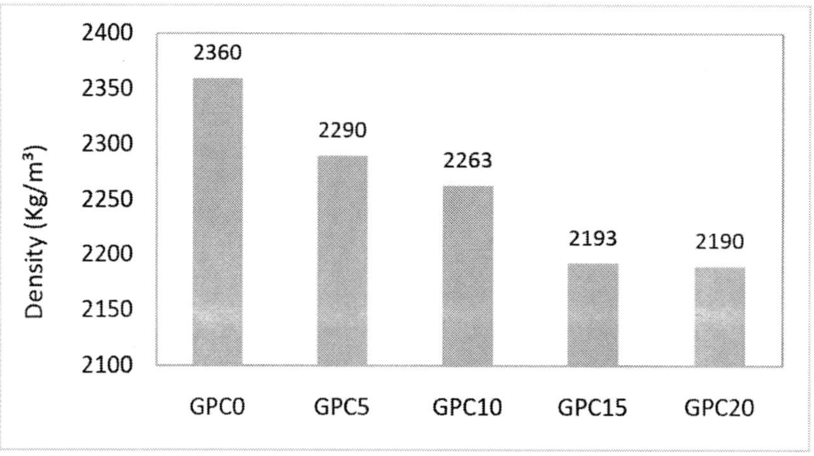

Figure 3. Changes in density after inclusion of rubber crumb.

The changes of weight for geopolymer samples that have been immersed in seawater for 28 and 60 days were shown in Figure 4. From the result, it showed that changes in weight for 28 days were fluctuate for different percentage of rubber crumb contents. The geopolymer sample with 15% replacement of rubber crumb (GPC15) showed maximum changes in weight (1.33%) and the lowest changes in weight was recorded on sample with 5% replacement of rubber crumb (GPC5).This result was consistent with the maximum compressive strength recorded by GPC5. However, weight changes for GPC20 decreased rapidly compared with GPC15. This may be due to the properties of rubber crumb itself which is impermeable of water, as such high content of rubber crumb less water was absorb by the GPC20 samples.

All the samples displayed an increment in weight when the content of rubber crumb increased for 60 days exposure periods as illustrated in Figure 4. The weight changes also more significant compared to 28 days exposure period. The control samples (GPC0) for both exposure periods recorded almost similar weight changes which is 0.84% and 0.85%. Meanwhile, the lowest weight change for 60 days exposure period was shown by GPC5 (0.85%).

If compared with 28 days exposure period, an increment almost 50% was displayed by GPC5. Moreover, GPC20 also displayed maximum increment of weight changes (1.9%) due to high content of rubber crumb and also improper compaction during moulding process lead to this result. As can be seen in Figure 5, GPC20 samples was covered with rubber crumb which reduced the strength and lead to excessive seawater absorption due to heterogenous mixture between geopolymer paste and rubber crumb.

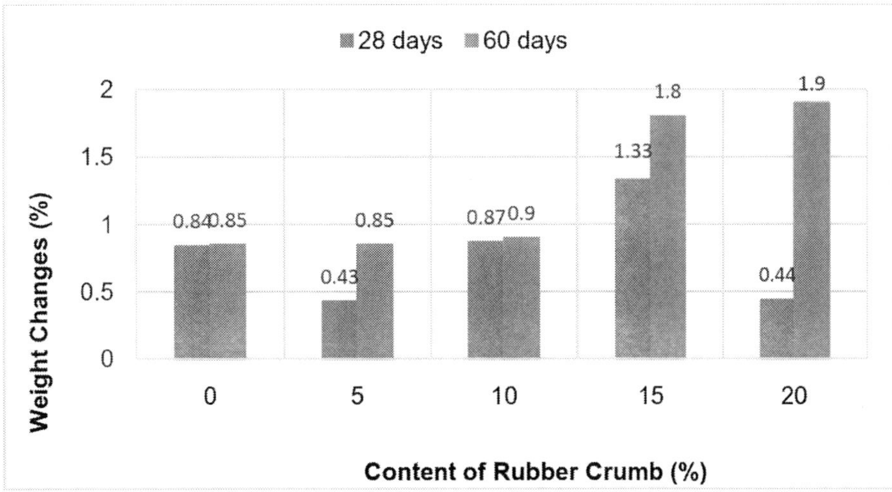

Figure 4. Weight changes of geopolymer samples after exposure periods of 28 and 60 days.

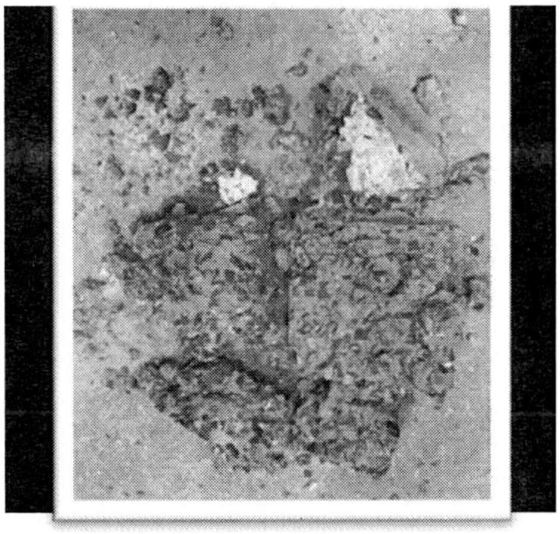

Figure 5. GPC20 after compressive strength testing.

Besides that, the size of rubber crumb also for this study also lead to reduction of compressive strength.

4. Conclusions

This study highlights the influence of rubber crumb range from 0% to 20% in fly ash based geopolymer concrete exposed to seawater for a period of 28 and 60 days. The following conclusion are presumed.

a) The inclusion of rubber crumb decreased the compressive strength where 5% of rubber crumb contribute to reasonable compressive strength. The strength tends to decrease when exposure period in seawater increased.

b) The density of geopolymer concrete reduced when rubber crumb was added due to its lightweight properties.

c) All geopolymer samples recorded weight gain when exposed to seawater. GPC20 recorded the highest weight gain due to heterogeneous mixture between geopolymer paste and rubber crumb as well as the size of rubber crumb itself. The higher content of rubber crumb, increased the porosity of geopolymer hence more water can penetrate hence reduce the strength and increased the weight of samples.

5. References

[1] Neupane K 2016 *Mechanics of Materials* **103** 110 – 122
[2] Bakri A M M, Kamarudin H, Nizar I K, Sandu A V, Binhussain M, Zarina Y and Rafiza A R 2013 *Revista de Chimie* **64**(4) 382 – 387
[3] Liew Y M, Heah C Y, Mohd Mustafa A B and Kamarudin H 2016 *Progress in Materials Science* **83** 595 – 629
[4] Al Bakri M M, Tahir M F M, Hussin K, Binhussain M, Sandu I G, Yahya Z and Sandu A V 2015 *Revista de Chimie* **66**(7) 1001-1003
[5] Al Bakri A M M, Kamarudin H, Karem O A K A, Ruzaidi C M, Rafiza A R and Norazian M N 2012 *Applied Mechanics and Materials* **110-116** 734-739
[6] Park Y, Abolmaali A, Kim Y H and Ghahremannejad M 2016 *Construction and Building Materials* **118** 43 – 51
[7] Gupta T, Chaudhary S and Sharma R K 2014 *Construction and Building Materials* **73** 562–574
[8] Khatib Z K and Bayomy F M 1999 *Journal of Materials Civil Engineering* **11**(3) 206-213
[9] Toutanji H A 1996 *Cement Concrete Composite* **18**(2) 135-139
[10] Issa G A and Salem G 2013 *Construction Building Materials* **42** 48-52
[11] Lijuan L, Shenghua R and Lan Z 2014 *Construction Building Materials* **70** 291-308
[12] Park Y, Abolmaali A, Mohammadagha M and Lee S 2014 *Advanced Civil Engineering Materials(ACEM)* **3**(1) 495–508
[13] Hernandex-Olivares F, Barluenga G, Bollati M and Witoszek B 2002 *Cement and Concrete Research* **32**(10) 1587-1596
[14] Puskas A, Moga L 2016 9 The international conference interdisciplinarity in engineering, inter-eng 2015 22 304-311
[15] Burduhos Nergis D D, Abdullah M M A and Vizureanu P 2017 *European Journal of Materials Science and Engineering* **2** 111-118
[16] Azmi A A, Abdullah M M A B, Ghazali C M R, Sandu A V and Hussin K 2016 *MATEC Web of Conference* **78** 4-8
[17] Mustafa Al Bakri A M, Kamarudin H, Khairul Nizar I, Bnhussain M, Zarina Y and Rafiza A R, 2012 *Advanced Materials Research* **341-342** 189-193

Acknowledgement
The authors gratefully acknowledge Center of Excellence Geopolymer and Green Technology (CEGeoGTech) and School of Materials Engineering, UniMAP for their expertise and support. The authors would also like to thank for the funding support from the Fundamental Research Grant Scheme (FRGS-9003-00540) under Ministry of Education Malaysia (MOE) and support from "Partnership for Research in Geopolymer Concrete" (PRIGeoC-689857) sponsored by the European Union.

IOP Conf. Series: Materials Science and Engineering **374** (2018) 012070 doi:10.1088/1757-899X/374/1/012070

Enzyme Biosensing Based on Zinc Oxide Nanostructures as Active Surface

N Iftimie[1*], R Steigmann[1], A Savin[1], C A Tugui[2] and C Munteanu[3]

[1] NDT Department, National Institute of R&D for Technical Physics, Iasi, Romania, 47 Blvd, 700050, Iasi, Romania,
2 Faculty of Science and Materials Engineering, Gh.Asachi Technical University, Iasi, Romania
[3]Faculty of Mechanical Engineering, University Gh. Asachi Iasi, 47 Blvd, 700050, Iasi, Romania

E-mail: niftimie@phys-iasi.ro

Abstract. Ag/ZnO mesostructures deposited onto substrates different were analysed in order to use ZnO as bioactive surface. This paper presents the results obtained at the eNDE of strips gratings deposited on different substrates used as bioactive surface using the EM sensor with MM lens in order to improve the emphasizing of the evanescent waves appeared when the slits of MSG are filled with immobilized enzymes.

1. Introduction

Nanomaterials prepared from metals oxides, shaped into nanoparticles have been widely used for their ability as electrode biosensing materials to enhance the efficiencies of electrochemical biosensor [1]. Metal oxide nanoparticles can increase efficiency of photochemical reactions and greatly improve the catalytic activity of enzymes to generate novel photo-electrochemical systems [2]. Some oxides nanoparticles (ex: ZnO, TiO_2, SnO_2, Fe_2O_3, MnO_2, etc.) can easily act as enhancing agents for effective acceleration of electron transfer between electrode and detection molecules, so leading to more rapid current response for target molecules (Figure 1) [3-5]. Nanostructures of ZnO (and other nanomaterials) including metal oxide semiconductor-metal hybrids and inorganic-organic hybrids have also used as matrices for enzymes immobilization [6-9].

Nanoparticles and nanostructured thin films have been studied in terms of their gas sensing abilities and biosensing element [10-13] where it was shown that a rough surface with a larger surface area had a positive influence on the sensor response [14, 15]. ZnO nanostructures formed an attractive matrix for GOx/ChOx immobilization, which exhibits a high affinity, high sensitivity, and fast response for glucose/cholesterol detection. The content of the paper is oriented toward development biosensing of glucose and cholesterol based on ZnO material by immobilizing of glucose oxidase and cholesterol oxidase on the surface of zinc oxide nanostructures. ZnO nanostructures as enzyme matrices show significant potential for use in biosensors. Relatively recent researches revealed that metallic structures with sub-wavelength features in the EM field at RF and microwave frequency range have potential for producing unusual EM response [16, 17], creation of high impedance surface [18], have an effective plasmonic behavior [19], high index of refraction and negative refractive index [20, 21]. The diffraction limits using a sensor with MM decrease until 300 nM instead of 100 μm in the case of conventional sensors and biosensors [12, 22, 23].

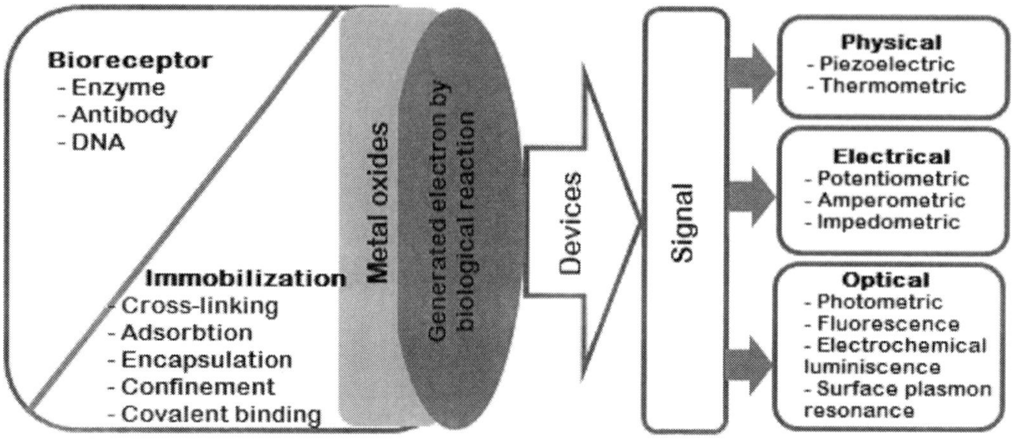

Figure 1. Working principles of a biosensing materials-based enzymes biosensing.

2. Materials and methods

A wide range of techniques to deposit thin films can be used [24-29] but in this paper the vacuum evaporation technique is used to deposit zinc oxide nanostructured thin films in order to obtain for investigate as biosensing layer in MSG structures. A zinc metallic film was deposited on ITO/glass and SiO_2/Si at room temperature. The conditions for Zn metallic film and Ag strips deposition was presented in [30], where the pressure in the deposition chamber was maintained at 10^{-5} Torr and the substrate temperature was kept at 300 K. The Ag/ZnO/different substrates structures are shown in Figure 2. The XRD patterns were made on a diffractometer using CuKα radiation (λ = 1.5406 Å) in order to identify the phase composition of the samples (Figure 3). The samples were analysed in the range of $2\theta = 5^0 \div 80^0$ with a scanning angle rate of 0.02 and a 2 s/step count time.

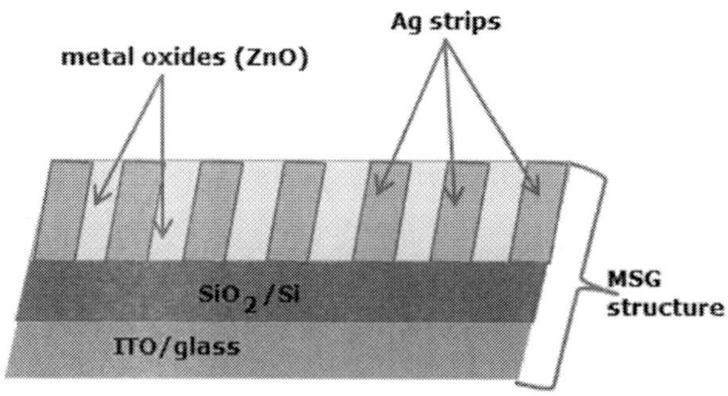

Figure 2. MSG structure.

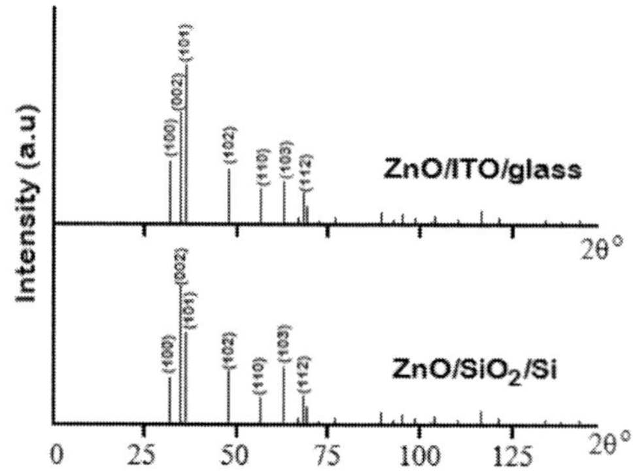

Figure 3. XRD ZnO thin film.

The XRD results demonstrate that there are two peaks in the pattern at 34.4° and 36.2°, which corresponds with the (002) and (101) planes of ZnO, respectively. All of the indexed peaks in the obtained pattern are well-matched with that of bulk ZnO which confirms that the synthesized products are many crystalline and possesses a wurtzite hexagonal structure.

No other peak related to impurities was detected in the pattern within the detection limit of the X-ray diffraction further confirms that the obtained products are pure ZnO.

The AFM images (Figures 4a and b) emphasize the polycrystalline granular aspect and of ZnO layer with bound enzyme show many globular structures, confirming immobilization of enzymes. The analyses find that the surfaces are without inclusions and defects can be used as bioactive surface.

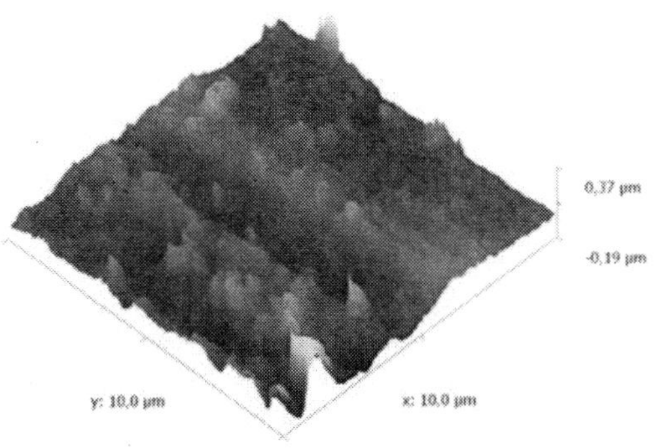

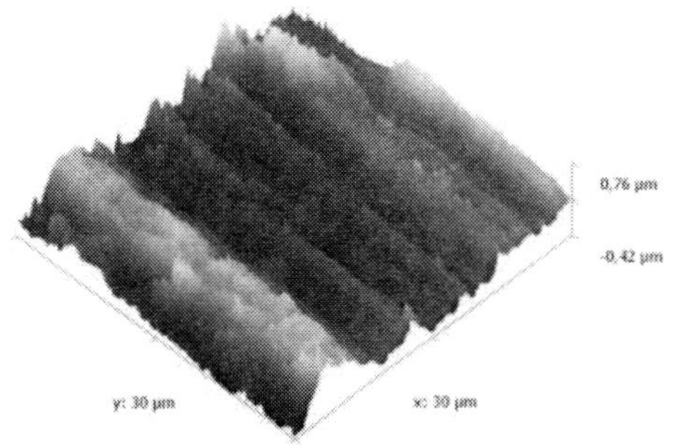

(a) (b)

Figure 4. The AFM images of: (a) ZnO/ITO/glass with immobilized enzyme glucose; (b) $ZnO/SiO_2/Si$ with immobilized enzyme cholesterol.

3. Experimental set-up

3.1. Biosensing mechanism

It was studied the influence of nanoparticle size on the performance of prepared biosensors and nanoparticles with smaller size were found to be more suitable for enzyme immobilization.

There are three modes of the enzymes oxidation referred to as the first, the second and the third generation glucose/cholesterol biosensors depending on the electron transfer mechanisms [31]. The reduction rate of the oxygen is proportional to the glucose/cholesterol concentration that is quantified by either measuring the augmentation of hydrogen peroxide or decrement of the oxygen concentration.

$$\begin{gathered} \text{Glucose} + O_2 \xrightarrow{GO_x} \text{Gluconic acid} + H_2O_2 \\ \text{Cholesterol} + O_2 \xrightarrow{ChO_x} \text{Cholest-4-en-3-one} + H_2O_2 \\ H_2O_2 \rightarrow O_2 + 2H^+ + 2e^- \end{gathered} \tag{1}$$

Since, the kinetics of the enzyme reaction is monitored by the rate of formation of product or the disappearance of a reactant. If either the product or reactant is electro-active, then the progress of the reaction can be monitored using amperometry [32]. Due to the presence of water (H-OH) it is a high probability that cholesterol oxidase/glucose oxidase will accept a proton from (H-OH) converting it to ion which in turn will interact with a ZnO nanostructures and produce a potential change at the electrode. The sensing mechanism for our glucose/cholesterol biosensors consist by the voltage applied across the two electrodes which causes a current to flow via electron tunnelling through the potential barrier between the nanoparticles. The current for the mixture decreases proportionally with the number of glucose/cholesterol molecules.

3.2. Electromagnetic sensing. Evanescent waves and FDTD simulation

The sensor based on MM lens described in [33-36] improving the spatial resolution to obtain the electromagnetic images using Fourier optics [37] and experimental set-up presented in [12, 39] were used at working frequency of 474 MHz. The simulation of system functioning was carried on XFDTD 6.3 software produced by REMCOM [40].

In Figure 5a is presented the result of simulation with XFDTD where the E_y component is displayed. In [30] it has been presented the behaviour of the field with air in the slits, it can be shown that for uricase between the strips, the amplitude of the electric field has the same behaviour as in [21] but the amplitude decreasing due to electrical permittivity high of the uricase.

Thus, the symmetrical maxima appear in the middle of the slits, decreasing to the minimum value on the strips edges. Inside the strips, other pair of maxima appears, followed by the decreasing to middle of the strip. The Figure 5b shows the dependency of e.m.f. amplitude induced in the reception coil of sensor allows the correct emphasizing of extremely thick conductive strips and eventual interruptions.

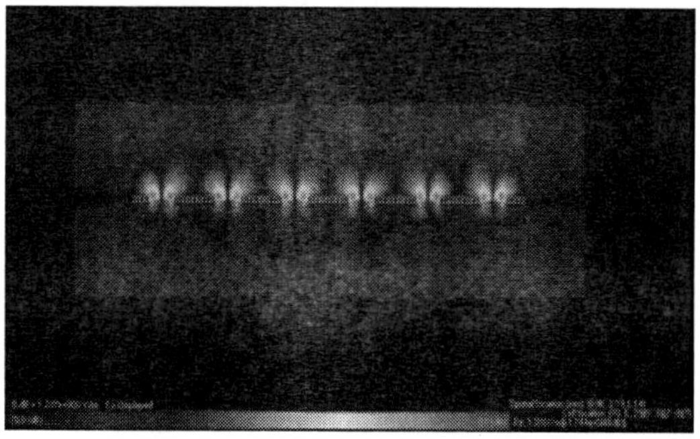

(a)

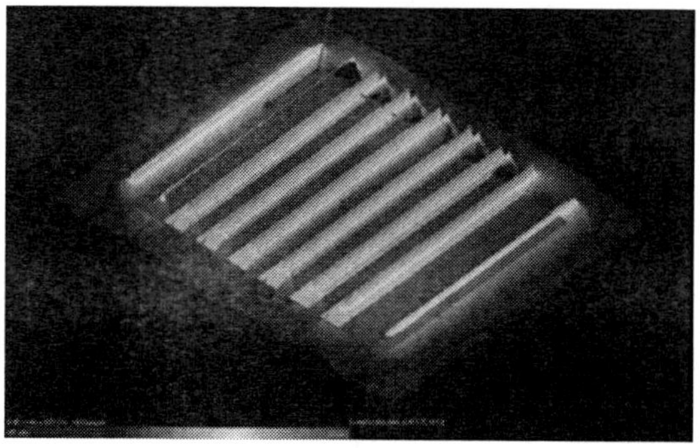

(b)

Figure 5. Numerical results for electric-field amplitude distribution near the strips; the field values are normalized to the amplitude of the incident field.

The realized samples presented in §2 having Ag 1.2 mm width strips and 0.8 mm width slits, h = 14 μm deposited on different substrates (ITO/glass and SiO_2/Si), without and with enzymes were fixed on a support on a system that assures XY displacement – Newmark USA, controlled by PC through codes written in Matlab 2014b. The distance between screen aperture and the surface to be examined has been maintained at 20 μm ± 1 μm. The sensor with MM lens made by two conical Swiss rolls [35, 41] is coupled to Network/Spectrum/Impedance Analyzer Agilent 4395A and is maintained fixed during the measurements. The acquisition through IEEE 488 interface and storage of data are made by the same PC. Considering the middle of the slits as reference point, symmetrically to left and right, 1mm distance have been scanned, along the length of strip 1mm distance have been scanned, thus a surface of 1x2 mm^2 is scanned, both on the sample without enzyme as well as with enzymes. The spatial resolution of the system was verified according to [34]. For MSG with features compatible with the λ value of incident field, TEz polarized wave acts, at normal incidence for MSG. The results obtained at the scanning of 1x2 mm^2 of MSG structures with 10 μm steps in both directions at frequency of 474 MHz were employed.

4. Results and discussions

Much recent attention has been focused on amperometric measurements because they can yield a linear relationship between sensor output and analyte concentration. Figure 6 shows the relation between the current and enzyme concentration for both architectures considered as biosensing materials. It can be observed that the current increases with the concentration and saturation appears at higher concentrations of enzymes. The inset is calibration curve at λ=0.6 m.

Thus, the Ag/ZnO/ITO/glass architecture exhibits good performance as biosensing material for glucose (Figure 6a) as well as Ag/ZnO/SiO2/Si architecture exhibits for cholesterol (Figure 6b). The response time is fast, which exhibits a high electron communication feature of the used ZnO thin films (Figure 6c). The calibration curves were obtained in the range of 50-1350 mg/dl (1.3-34.9 mM) of enzymes (Figure 6b) and the current response showed a linear dynamic range of 100-1050 mg/dl (2.6-27.1 mM). The tested sensor configuration showed large dynamic ranges with an output response that was linear versus [1/concentration] of the enzymes activity with sensitivity at lower response current as shown in inset in Figures 6a,b. Adding 0.05 mM of ascorbic acid and uric acid to 0.5 mM glucose

IOP Conf. Series: Materials Science and Engineering **374** (2018) 012070 doi:10.1088/1757-899X/374/1/012070

and cholesterol only generated some noise shown in Figure 6c. The morphology of the nanostructure significantly affects biosensor electrochemical properties where ZnO nanostructures have been investigated for application in enzymes biorecognition materials

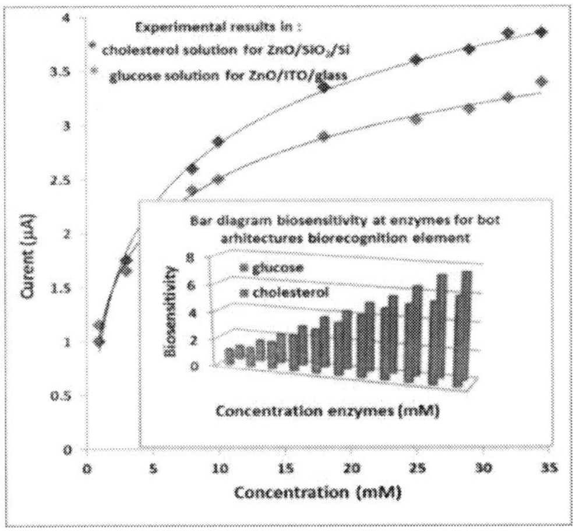

(a)

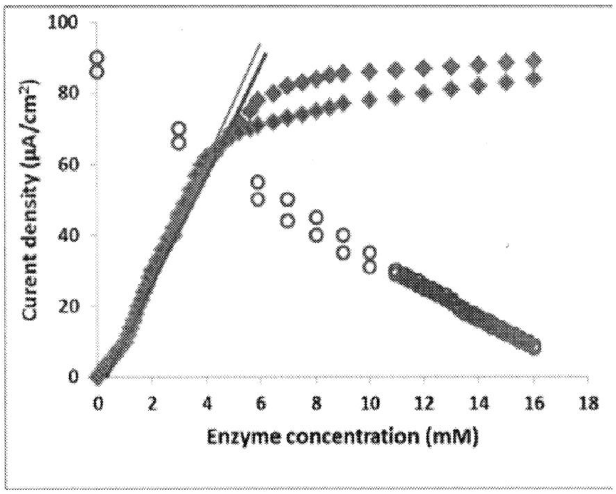

(b)

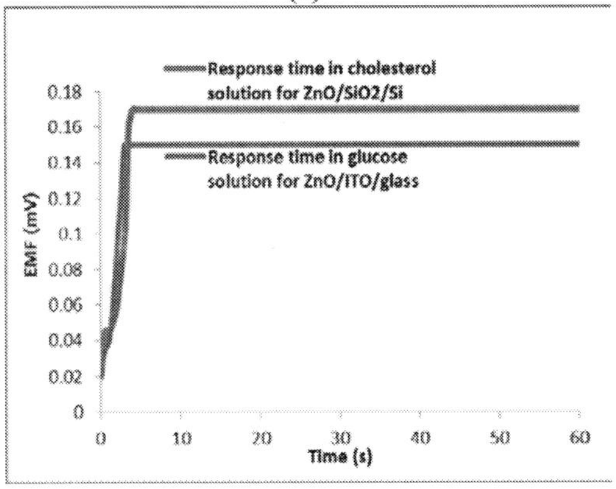

(c)

Figure 6. Experimental results in test solutions (a) ZnO/ITO/glass and $ZnO/SiO_2/Si$; (b) Variation in the current with enzymes concentration; (c) Calibration curve showing the study of interferences with time trace line of output response change with time for both architecture biosensing.

5. Conclusions

In conclusion, we have demonstrated a simple fabrication procedure for a highly sensitive electrochemical glucose and cholesterol biosensing based on ZnO nanostructures as MSG structures. The paper presents an extension of the performances of the EM sensor with MM lens for experimental emphasizing of evanescent modes appeared in mesostructures in subwavelength regime, excited with TEz polarized EM waves in RF range. The study focuses on the apparition of evanescent modes for the case when the space between strips of MSG structures are filled with enzymes, confirming the results according to which the amplitude of the signal in the reception coils is increased when an enzymes (glucose and cholesterol) is present in slits. The redox current improves for the bioelectrodes based on ZnO film grown on different substrates. The good performance in terms of improved sensitivity, stability, selectivity, reproducibility, negligible interference and rapid response by our proposed biosensing materials also makes it suitable for externally integrating/interfacing a biosensing element to commercial devices, adding to the advantages of simplicity and low cost for the enzymatic detection of biochemically important substances.

6. References

[1] Turner A P F 2013 *Chem. Soc. Rev.* **42** 3184–3196

[2] Dzyadevych S V, Arkhypova V N, Soldatkin A P, El'skaya A V, Martelet C and Jaffrezic-Renault N 2008 *IRBM* **29** 171–180
[3] Zhao Z, Lei W, Zhang X, Wang B and Jiang H 2010 *Sensors* **10** 1216–1231
[4] Ariga K, Ji Q, Mori T, Naito M, Yamauchi Y, Abe H and Hill J P 2013 *Chem. Soc. Rev.* **42** 6322–6345
[5] Ravindra N M, Prodan C, Fnu S, Padronl I, Sikha S K 2007 *JOM* **59** 37–43
[6] Liu H, Zuo Z, Guo Y, Li Y, Li Y 2010 *Angew. Chem. Int. Ed.* **122** 2705–2707
[7] Yeom S H, Kang B H, Kim K J and Kang S W 2010 *Frontiers in Bioscience (Landmark Edition)* **16** 997–1023
[8] Lee S H, Sung J H and Park T H 2012 *Ann. Bio. Med. Eng.* **40** 1384–1397
[9] Liu C C 2012 *Biosensors* **2** 269–272
[10] Rambu A P, Iftimie N and Nica V 2012 *J. Mater. Sci.* **47** 6979–6985
[11] Rambu A P, Sirbu D, Iftimie N, Rusu G I 2011 *Thin Solid Films* **520**(4) 1303-07
[12] Steigmann R, Iftimie N, Savin A 2015 *Applied Mechanics and Materials* **77** 62-66
[13] Sassolas A, Blum L J, Leca-Bouvier B D 2012 *Biotechnol. Adv.* **30** 489-511
[14] Eriksson J, Khranovskyy V, Söderlind F, Käll P O, Yakimova R, Spetz A L 2009 *Sensors & Actuators B* **137** 94–102
[15] Wang J, Sun X, Wei A, Lei Y, Cai X, Li C, Dong Z 2006 *Appl. Phys. Lett.* **88** 233106
[16] Pendry J B, Holden A J, Robbins D J, Stewart W J 1999 *IEEE Transac on Microwave Theory and Techniques* **47**(11) 2075-2084
[17] Pendry J B 2000 *Physical Review Letters* **85**(18) 3966
[18] Sievenpiper D, Zhang L, Broas R F, Alexopolous N G, Yablonovitch E 1999 *IEEE Transactions on Microwave Theory and Techniques* **47**(11) 2059-2074
[19] Pendry J B, Holden A J, Stewart W J, Youngs I 1996 *Phys. Rev. Lett.* **76**(25) 773
[20] Pendry J B, Schurig D, Smith D R 2006 *Science* **312** 1780-1782
[21] Savin A, Steigmann R, Bruma A 2014 *Sensors* **14** 11786-11804
[22] Kabashin A V, Evans P, Pastkovsky S, Hendren W, Wurtz G A, Atkinson R, Pollard R, Podolskiy V A, Zayats A V 2009 *Nature materials* **8** 867-871
[23] Kakati N, Jee S H, Kim S H, Oh J Y, Yoon Y S 2010 *Thin Solid Films* **519** 494–498
[24] Ko H J, Chen Y F, Yao T, Miyajima K, Yamamoto A, Goto T 2000 *Appl.Phys.Lett.* **77** 4226
[25] Alvarado J A, Maldonado A, Juarez H, Pacio M, Perez R 2015 *Beilstein J. Nanotechnol.* **6** 971–975
[26] Xu L, Shi L, Li X 2008 *Appl. Surf. Sci.* **255** 3230–3234
[27] Bethke S, Pan H, Wesseis B W 1988 *Appl. Phys. Lett.* **52** 136
[28] Flickyngerova S, Shtereva K, Stenova V, Hasko D, Novotny I, Tvarozek V, Sutta P, Vavrinsky E 2008 *Appl. Surf. Sci.* **254** 3643–3647
[29] Jiang F, Dai J, Wang L, Fang W, Pu Y, Wang Q, Tang Z 2007 *Journal of Luminescence* **122-123** 162–164
[30] Iftimie N, Tascu S, Salaoru I, Steigmann R, Savin A, Irimia M, Iacomi F 2015 *Materials Today: Proceedings* **2**(10) 3486- 52
[31] Rahman M M, Ahammad A J S, Jin J H, Ahn S J, Lee J J 2010 *Sensors* **10** 4855-4886
[32] Sarma A K, Vatsyayan P, Goswami P, Minteerc S D 2009 *Biosensors and Bioelectronics* **24** 2313–2322
[33] Gribic A, Eleftheriades G V 2003 *Applied Physics Letters* **82**(12) 1815-1817
[34] Savin A, Steigmann R, Bruma A, Sturm R 2015 *Sensors* **15** 15903-15920
[35] Grimberg R, Tian G Y 2012 *Proceed of the Royal Society A* **468**(2146) 3080-3099
[36] Savin A, Steigmann R 2014 *Patent number RO129801-A0*
[37] Born M, Wolf E 1975 *5th Ed. Pergamon Press Oxford*, UK
[38] Grimberg R 2013 *Material Science Engineering B* **178** 1285–1295
[39] Grimberg R, Savin A, Steigmann R, Serghiac B, Bruma A 2011 *INSIGHT* **53**(3) 132-137
[40] Lagendijk A, Van Tiggelen B, Wiersma D S 2009 *Physics Today* **62**(8) 24-29

[41] Grimberg R, Savin A 2011 *Patent* Number(s): RO126245-A0

Acknowledgments
This work was supported by a grant of the Romanian Ministry of Research and Innovation, CCCDI-UEFISCDI, project number PNCDI III-P1-1.2-PCCDI-207-03239/60PCCDI 2018, within PNCDI III and Nucleus Program PN 18 06 01 02/2018.

IOP Conf. Series: Materials Science and Engineering **374** (2018) 012071 doi:10.1088/1757-899X/374/1/012071

Preparation and Optimization of Natural Glues Based on Laricio Pine Resin

P De Luca[*1], B Roberto[2], D Vuono[2], C Siciliano[3] and J B Nagy[2]
[1]Dipartimento di Ingegneria Meccanica, Energetica e Gestionale
[2]Dipartimento di Ingegneria per l'Ambiente ed il Territorio ed Ingegneria Chimica
[3]Dipartimento di Farmacia e Scienze della Salute e della Nutrizione
Università della Calabria, Arcavacata di Rende -87036 Rende (Cs) Italy

E-mail: pierantonio.deluca@unical.it

Abstract. The main purpose of this work was to prepare and optimize glues consisting primarily of natural components. The fundamental component was the laricio pine resin to which different additives were added, varying their nature and quantity. The additives used were: charcoal, activated carbon and carbon nanotubes (MWCNTs). The optimization of the systems continued to identify and select suitable solvents, in order to improve dispersion and solubilization of the phases and to obtain glues in the fluid phase at room temperature. The best results were obtained with ethyl alcohol which, moreover, represents a solvent with a limited environmental impact. The evaluation of the adhesive capacity of the glues was carried out by cutting, tensile and flexural tests on prepared wooden specimens. Furthermore, the specimens and the same glue were subjected to thermal cycles and observed by Scanning Electron Microscope (SEM). The carbon nanotubes, used in very low percentages (1%, in a system consisting mainly of natural origin products) in the presence of an adequate solvent, such as ethyl alcohol, make it possible to obtain a glue that is easy to apply, with an excellent adhesive capacity.

1. Introduction

There are many polluting substances harmful to health in domestic or productive environments. The symptoms are known as SBS Sick Building Syndrome [1-3]. The sick building syndrome has been recognized as a real disease by the WHO (World Health Organization) since 1983. In order to create a home that is as compatible as possible with the environment and human health, future technology will have to take into consideration many aspects such as using innovative materials [4, 5] and adopting effective policies for their easy recovery and disposal [6-9]. The materials suitable for green building can be multi-functional ones [10-13], i.e. those which have new ones thanks to the combination with other materials capable of degrading or adsorbing pollutants from the environment in addition to their traditional functions [14-21]. It is believed that the pollutants due to the SBS originate essentially from acute and prolonged exposure to chemical and biological substances such as those contained in glues, paints, adhesives. [22, 23]. Dangers deriving from solvents and volatile compounds (VOC) in the latter are, in fact, hidden [24, 25].

The most ill-famed is formaldehyde (contained in the glues), which can cause serious health problems including severe respiratory changes and sensory discomforts. Other Volatile Organic Compounds are: hydrocarbons, aldehydes, chlorinated which may become even more dangerous in the presence of atmospheric particulate impregnated by these pollutants, thus increasing their residence time in environments [26-29]. The use of natural, biocompatible and renewable substances are always an alternative in the preparation of glues[30, 31].Generally, however, natural glues have a lower binding power than glues that use synthetic substances [32].It is for this reason, that our research has been directed towards the study and preparation of a glue that was mainly composed of natural substance and reserving only a very low percentage of synthetic substances that guarantee an increase in adhesion strength. In particular, the basic natural raw material used was the Laricio pine resin. The latter, however, if used alone during cooling crystallizes losing its adhesive power. Therefore, it was necessary to add additives (fillers) to the resin such as: charcoal, activated carbon and carbon nanotubes. Carbon nanotubes are very versatile materials thanks to their particular structure and can be applied in different fields as fibro reinforcing [33, 34] but also as adsorbent materials [35, 36]. The carbon nanotubes, though synthetic, were used in very low percentages, and this allows glue preparation with a mainly natural composition and a substantial improvement of the adhesive strength as regards to the traditional natural glues. Furthermore, we must emphasize that carbon nanotubes are not materials capable of delivering volatile substances into the environment, although they are still nano-materials. They must always be used with caution avoiding any dispersion [37- 39]. Another important objective was to find a sustainable solvent that would allow the preparation of a glue based on a resin that can be applied when cold, unlike procedure for resin based glues that must be first heated and then put into action function quickly before cooling and solidification.

2. Materials and methods

The work was carried out in several consecutive phases: a) Identification of the most suitable solvent; b) Identification of the most suitable filler; c) Characterization of the best mixtures.

For the preparation of the glues, different mixtures were prepared, varying in the nature and quantity of the components. The main basic raw material used was the Laricio pine resin and were used as additives charcoal, activated carbon and carbon nanotubes. These materials have already been characterized in a previous work and, therefore, for any information refer to this [31]. Furthermore, the use of solvents such as ethyl alcohol (VWR, 96%), petroleum ether (Merck) and chloroform (Merck, 98%), were evaluated in order to find the optimal conditions for preparing a glue that can be used cold. The resin, in its solid state was initially ground finely for the preparation of the mixtures. Subsequently 2 g of powdered resin were added to 10 ml of the selected solvent. To speed up the solvent evaporation process, the mixture was heated to temperatures ranging from 45 °C to 110 °C (depending on the solvent used), stirring continually, until almost complete evaporation of the solvent was reached and until a semi-liquid viscose mixture was formed. A selected filler of carbon nanotubes, charcoal or activated carbon was added to this mixture. For all three fillers, however, it was necessary to find the optimal weight percentage to be used, through some preliminary tests. All the mixtures prepared were subjected to characterization by tests through cutting carried out with a Vanderkamp VK 200 tester which measures in a range of 0 -3432.2 KPa, in order to identify the most suitable filler and solvent. The specimens were prepared with two pieces of larch pine, 1.5 × 1 × 0.4 cm in size, glued together in a staggered position so that the glued surface was 1 cm^2. (Figure 1 a, b).

Subsequently the best mixtures identified underwent:

Flexural resistance tests carried out with an "INSTRON 1195" with a 50 KN load cell and 1 cm diameter knives. The machine was calibrated at a speed of 20 mm / min in displacement control, the data was collected every 5 Hz. The specimens were prepared with two pieces of pine wood, measuring 8 × 1 × 0.2 cm glued between them with the adhesive to be tested (Figures 1c, d).

After bonding, the specimens were placed under press for 48 h. In order to compare, the test of specimen of larico pine wood with size of 8 × 1 × 0.4 cm was also tested. The tests were conducted in triplicate for each type.

Tensile resistance tests carried out with "INSTRON 5582", at a speed of 0.5 mm / min in displacement control, with a load cell of 100KN. The specimens were prepared with two pieces of larch pine, 12 × 2 × 0.2 cm in size, glued together with the adhesive to be tested (Figures 1e, f).

The specimens were placed under press for 48 h after bonding. For the sake of comparison, the wood specimen of the laricio pine measuring 12 × 2 × 0.4 cm was also tested. The tests were conducted in triplicate for each type.

Cutting tests at different temperatures carried out with a Vanderkamp VK 200 tester and using wood samples as described above.

SEM observation was carried out through Scanning Electron Microscope (SEM) FEI "Quanta" 200 EGF (E = 10kV, SE Image).

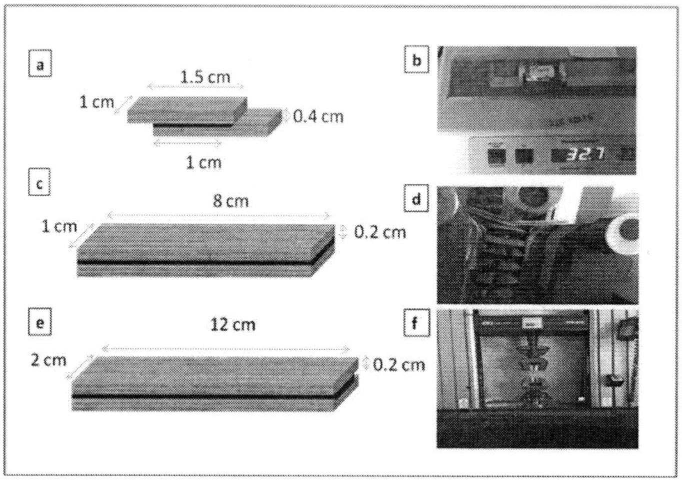

Figure 1. Specimens and instrumentation respectively for (a), (b) cutting tests; (c, d) flexural tests; (e, f) tensile tests.

3. Results and discussion

The results obtained from preliminary tests and the observation of the mixtures, regarding the use of the optimal solvent, made it able to discard the petroleum ether a priori, since the mixture returned solid and of inconsistent appearance, once it evaporated. In the following phase, the action of the other two selected solvents, chloroform and ethyl alcohol, was evaluated and compared. Consequently, mixtures of resin were added with the addition of 0.5, 1, 2, 3 and 6% in weight of nanotubes in the absence of solvent and in the presence of ethyl alcohol or chloroform. In this preliminary phase, we opted for the choice of operating mixtures of carbon nanotubes as well as representative of the other fillers.

In Figure 2, the data of a cutting test carried out on specimens as previously described in paragraph 2 are reported, depending on the solvent used, this shows how the nature of the solvent influences the test. In fact, ethanol had higher results for all percentages by weight, minus 6%. Solvent use as compared to pure resin (no solvent) produced improvements on all percentages, exclusion of 6% as always. As regards the quantity of CNTs, the best percentage is that obtained from 1% to 2% in weight for the ethanol mixture, reaching the end of the machine scale without breakage of the specimen (3432.3 KPa). Inferior results, on the other hand, for chloroform which reaches a maximum of 1% with 1539.6 KPa. Maximum value is found at 6% of nanotubes for the resin-CNTs mixture without solvent. From the evidence of the results obtained, our choice was to proceed experimentation by the sole usage of ethanol as a solvent in as much as chloroform proved to have not only a weaker effect but also a less sustainable one. The choice fell on ethanol which, once evaporated, left a semi-liquid mixture and therefore of easy use. Furthermore, ethanol was able to promote the dispersion of nanotubes and charcoal, whereas in the presence of activated carbons during the solvent evaporation process, a gradual precipitation of the filler at the base is recorded. Ethanol was therefore considered the best solvent. We proceeded with the identification of the optimal quantity of fillers to be used, consequently, once we had chosen and identified ethanol as being the best solvent.

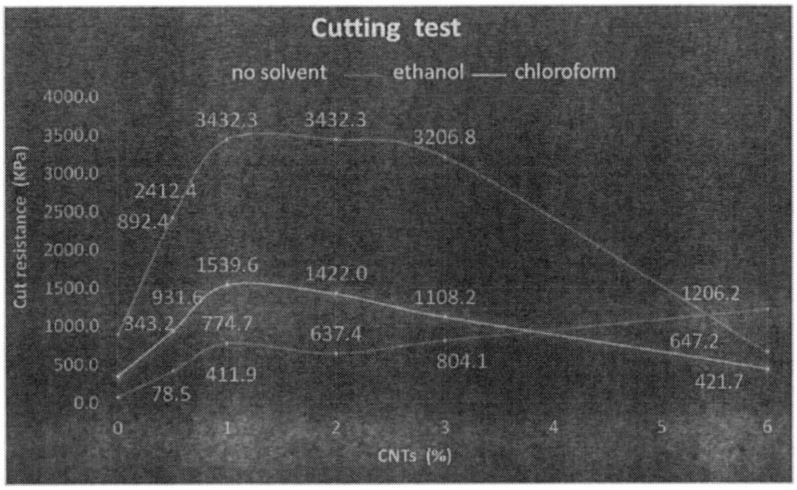

Figure 2. Cutting resistance (KPa) for wood samples glued with mixtures prepared in resin, in the absence and in the presence of solvent and with different quantities of nanotubes (CNTs).

The following Figure 3 shows the results of the cutting test on resin / ethanol systems with different types and quantities of filler.The results obtained confirm the validity of the glue produced with carbon nanotubes. In fact, these samples do not break, even if they reach the end of the scale of the instrument (3432.3 KPa). The results for the charcoal samples are lower than those of the nanotubes, with a maximum of 1.5% wt. of 3432.3 KPa. The active carbons, on the other hand, gave disappointing results up to 1.5% wt., then improving over this percentage until reaching 3366.9 KPa and then fracturing. Further investigation to identify the best filler, through tensile tests were carried out (according to the methods described in paragraph 2).

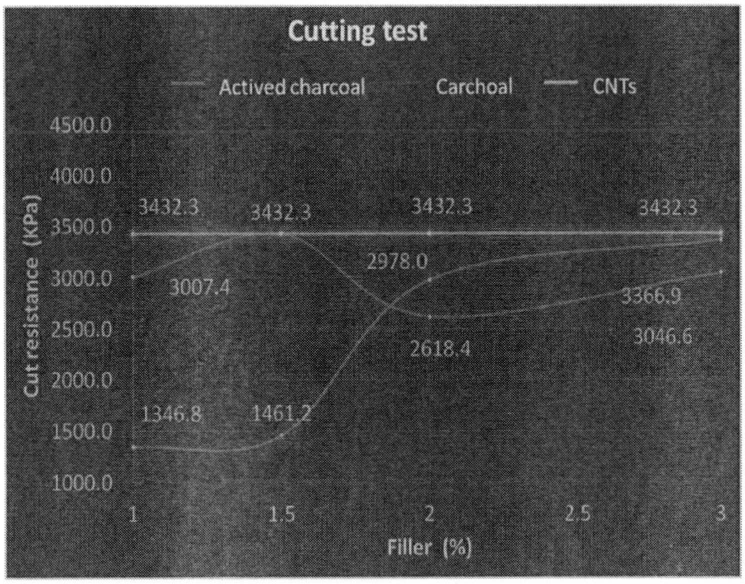

Figure 3. Cutting resistance (KPa) for wood specimens glued with mixtures containing resin / ethanol and different amount and nature of fillers.

Table 1 shows the data obtained from tensile tests carried out on specimens of solid wood only (massive wood) of larch pine of size 12 × 2 × 0.4 cm and on specimens obtained by gluing two pieces of the same wood of measuring 12 × 2 × 0.2 cm with mixtures prepared in ethanol and 1% filler.

Table 1. Tensile resistance, maximum (σ_{max}) and breaking (σ_b), and strain (ε_b) of wood samples and samples with resin / ethanol mixtures and with 1% filler.

Sample	σmax (MPa)	σb (MPa)	εmax (%)	εb (%)
Massive wood	32.83	32.83	10.07	10.07
Resin + actived charcoal (1%)- Ethanol	28.71	21.61	8.80	8.78
Resin + Charcoal (1%)- Ethanol	40.29	40.29	8.81	8.86
Resin + CNTs(1%) -Ethanol	49.89	49.89	11.55	10.45

The data reported in Table 1 shows that the best results are obtained using mixtures prepared with carbon nanotubes that are higher even than the tests made of wood only.

Table 2 shows the flexural resistance of wood-only specimens and specimens prepared using resin / ethanol mixtures and with 1% filler.

Table 2. Flexural resistance, maximum (σ_{max}) and breaking (σ_b) and breaking time (T_b), of wood-only specimens and specimens prepared using resin/ethanol mixtures and with 1% filler.

Sample	σmax (MPa)	σb (MPa)	Tb (s)
Massive wood	59.06	59.06	75
Resin+actived charcoal (1%)- Ethanol	59.06	51.19	130
Resin + Charcoal (1%)- Ethanol	61.03	59.06	130
Resin + CNTs(1%) -Ethanol	64.97	64.97	135

Carbon nanotubes are confirmed to be the best filler. The data show that, like all composite materials, they almost doubled the catastrophic breaking time, passing from 75 seconds (T_b) of the massive wood to 130-135 seconds of the specimens obtained using the prepared mixtures. Likewise, applied tensions are also higher, or, at least equal to those of the wood itself. Since a good glue must also be able to withstand temperatures above 50 °C without losing its adhesive power, the behavior of mixtures at different temperatures has been studied. Two mixtures to be compared were chosen: the glue obtained with dissolved resin in ethanol heated to 75 °C and the same matrix with the addition of 1.5% by weight of CNTs. The specimens are subjected to cutting tests as previously described in paragraph 2. The selected temperatures were: -16 °C, 20 °C, 50 °C, 90 °C, 110 °C and 130 °C. Three samples of the two types were created for all temperatures. After a two hour heat treatment, the specimens were left to rest at room temperature for 24h before the cutting test. Figure 4 compares the data obtained. The blue line relative to the resin / CNTs mixture remains constant up to 130 °C that is, (it reaches the end scale without breakage), with the exception of temperatures at -16 °C which resist up to a force of 3187.2 KPa. The resin-only mixture gave good results up to 50 °C, which then decreased almost constantly up to 130 ° C to 2177.1 KPa of shear strength, leaving a symptom of loss in the adhesive strength as the temperature increased. Results improved slightly for T = -16 °C with 3285.2 KPa. It is very important to evaluate the trends of the two curves. In fact, resistance up to 50 °C is similar. Nanotubes are able to enhance their properties of the thermal resistance beyond this point, even if we consider the amount of evaporated ethanol thus transmitting adhesive capacity towards the matrix at a high temperature. Resin without nanotubes does not present this factor.

Once having identified the best mixtures obtained by the use of carbon nanotubes as fillers, observation by scanning electron microscopy was carried out on samples both in the absence and in the presence of carbon nanotubes. Figure 5a shows the surface of pure resin.

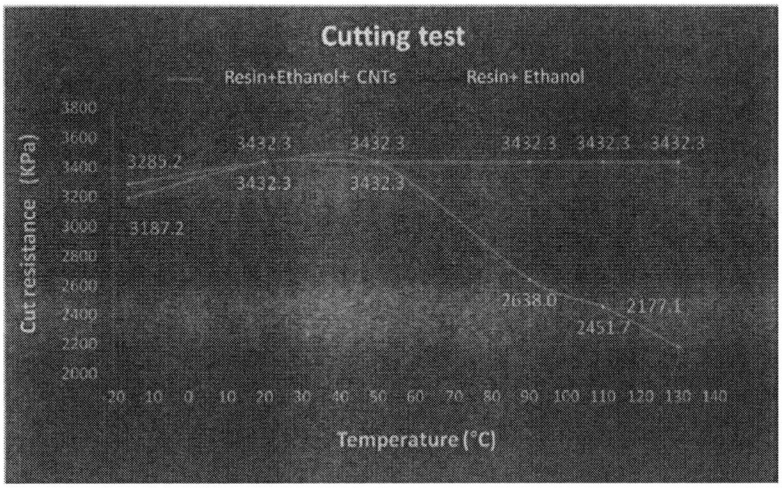

Figure 4. Cutting resistance (KPa) for wood specimens glued together with mixtures containing resin / ethanol with or without nanotubes (1.5%) after thermal cycles at different temperatures.

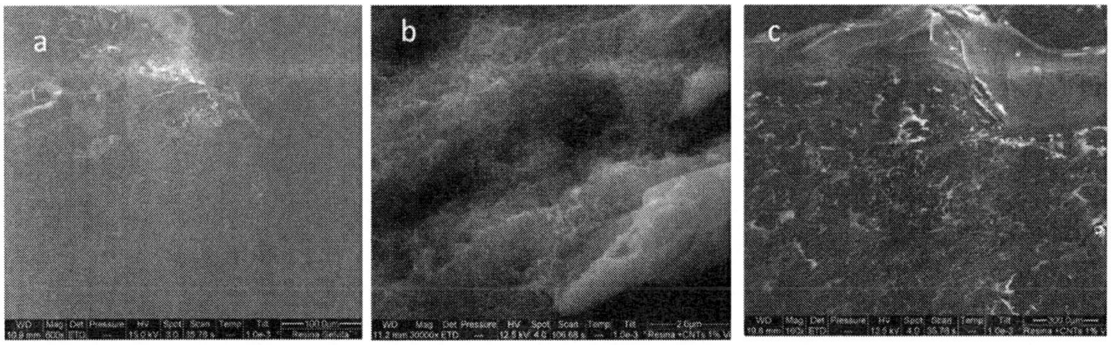

Figure 5. SEM images: (a) resin;(b) resin with carbon nanotubes; (c) breaking surface after cutting test of samples prepared with resin and carbon nanotubes (1%) in ethanol.

The sample is very homogeneous and compact and does not present any discontinuity or presence of pores or macro cavities on this. From the morphological point of view, the addition of the nanotubes does not have any kind of influence (Figure 5b). In fact, there is no known lack of homogeneity and / or diversity compared to the image of pure resin. Furthermore, in this case there are no pores or macrocavities, confirming the full "cohabitation" of these nano fibres in the resin matrix. A morphological study of the resin samples with nanotubes was also carried out after the cutting tests previously carried out on wood samples. Figure 5c shows the breaking surface of a resin-CNTs sample in ethanol. As you may observe, the "pull-out" phenomenon is completely absent. This is a very important problem in the field of composite materials, which consists in the extraction of the fibres from the matrix after a breaking point.

4. Conclusions

The primary objective of this work was to produce new glues of mainly natural composition.

Our choice was a natural and easily sourced material in nature that is the resin of the laricio pine. Research for a suitable solvent had double values, that is, to get free from high temperature preparation treatments and then to maximize dispersion of the fillers in the resin matrix. The characterization of the systems has led to the identification of ethyl alcohol as the most suitable solvent to be used. The latter allowed us to obtain a very fluid composite mixture able to be easily applied on the surfaces to be glued even at room temperature. Furthermore, the choice of ethyl alcohol enabled us to add another further element to the sustainability principle. Since evaporation of this solvent is tolerable towards health, unlike chloroform and petroleum ether, which are also a subject of study. Best results found during the cutting, tensile and flexural tests done on suitably prepared wood specimens were obtained with mixtures containing pine resin in ethanol and using carbon nanotubes as fillers in percentages of 1-2%. The cutting tests carried out, after thermal cycles at different temperatures, have brought to light that the presence of carbon nanotubes makes the glue thermally more stable than glue prepared without carbon nanotubes, thus suggesting a certain chemical interaction, not only to a mechanical matrix-fibre adhesion. The nanotubes, in fact, offer that increase in adhesive capacity and chemical-physical stability that the resin as it does not present and can be considered a good product with an acceptable sustainability, due to its prevailing natural extraction. The glue preparation and production procedures are simple and highly reproducible for eventual industrial production.

5. References

[1] Norhidayah A, Lee C K, Azhar M K and Nurulwahida S 2013 *Procedia Eng.* **53** 93-98
[2] Stolwijsk J A J 1991 *Environ Health Perspect* **95** 99-100
[3] Redlich C A, Spare J and Cullen M R 1997 *The Lancet* **349** 1013-16
[4] De Luca P, Carbone I and Nagy B J 2017 *Journal of Green Building* **12** 141-161
[5] Gong J, Kostro A, Motamed A and Schueler A 2016 *Solar Energy* **139** 412-425
[6] Franzoni E 2001 *Procedia Eng.* **21** 883 – 890
[7] Scofield J H 2000 *Environ. Sci Technol.* **34** 448A-448A
[8] Warren L P and Peter A T 2008 *Building and Environment* **43** 1858–1870
[9] Magnone G, De Luca P, Salituro A and Cosenza M G 2014 *Procedia Environmental Science, Engineering and Management* **1** 93-97
[10] Han B, Zhang L and Ou J 2017 Light-Emitting Concrete. *Smart and Multifunctional Concrete Toward Sustainable Infrastructures.* Springer, Singapore
[11] De Luca P, Chiodo A and B.Nagy J 2014 *Sustainable Chemistry,* Reniers G and Brebbia C Southampton: WIT Press **154** 155-165
[12] Husken G, Hunger M and Brouwers H J H 2009 *Building and Environment* **44** 2463-74
[13] Ballari M M, Hunger M, Husken G and Brouwers H J H 2010. *Applied Catalysis B: Environmental* **95** 245-254
[14] Fujishima A, Rao T N and Tryk D A 2000 *Journal of Photochemistry and Photobiology C: Photochemistry Reviews* **1** 1-21
[15] Canpolat F, Yılmaz K, Kose M M, Sumer M and Yurdusev M A 2004 *Cem Concr Res* **34** 731–735
[16] Krisnandi K, Southon P D, Adesina A A and Howe R F 2003 *International Journal of photoenergy* **5** 131-140
[17] De Raffele G, Aloise A, De Luca P, Vuono D, Tagarelli A and B.Nagy J 2016 *J. Porous Mater.* **23** 389-400
[18] Di Paola A, Garcia-Lopez E, Marci G and Palmisano L 2012 *J. Hazard. Mater.* 211-212, 3-29
[19] De Luca P, Poulsen T G, Salituri A, Tedeschi A, Vuono D, Kònya Z, Madaràsz and B.Nagy J 2015 *J. Therm. Anal. Calorim.* **122** 1257-1267

[20] Ramirez A M, Demeestere K, De Belie N, Mantyla T and Levanen E 2010 *Building and Environment* **45** 832-838
[21] Marwa M H, Heather D, Louay N. M and Tyson R 2010 *Construction and Building Materials* **24** 1456-1461
[22] Cooley J D, Woong W C, Jumper C A and Straus D C 1998 *Occup Environ Med* **55** 579–584
[23] Apter A, Bracker A, Hodgson M, Sidman J and Wing-Yan Leung B S 1998 *Journal of Allergy and Clinical Immunology* **94** 277-288.
[24] Hodgson M, Levin H and Wolkoff P 1994 *Journal of Allergy and Clinical Immunology* **94** 296-303
[25] Chuck W F Y and Jeong T K 2010 *Indoor and Built Environment* **12** 30-39
[26] Englert N 2004 *Toxicology Letters* **149** 235-242
[27] Marcazzan G M, Valli G and Vecchi R 2002 *Science of the Total Environment* **298** 65-79
[28] Filice M, De Luca P, Guido G P 2009 *Environ. Eng. Manag. J.* **8** 1407-1412
[29] Pekkanen J, Timonen K L, Ruuskanen J, Reponen A and Mirme A. 1997 *Environ. Res.* **74** 24-33
[30] Mathias J D, Grédiac M and Michaud P 2016 Bio-based adhesives. In Biopolymers and Biotech Admixtures for Eco-efficient Construction Materials, 369-385 (Eds) Pacheco-Torgal F, Ivanov V, Jonkers H and Karak N Elsevier Science and Technology, Abington Hall , Cambridge, UK
[31] De Luca P, Pane L, Vuono D, Siciliano C, Candamano S, B.Nagy J, 2017 *Environ. Eng. Manag. J.* **16** 1659-72
[32] Packham D E 2009 *Int. J. Adhes Adhes.* **29** 248–252
[33] De Luca P, Nappo G, Siciliano C and B.Nagy J 2018 *J. Porous Mater.* **25** 283-296
[34] Thostenson E T, Ren Z and Chou T W 2001 *Composites Science and Technology* **16** 1899-1912
[35] Policicchio A, Vuono D, Rugiero T, De Luca P and B.Nagy J 2015 *Journal of CO_2 Utilization* **10** 30-39
[36] Li Y H, Wang S, Luan Z, Ding J, Xu C and Wu D 2003 *Carbon* **41** 1057 1062
[37] Pacheco-Blandino I, Vanner R and Buzea C 2012 *Toxicity of nanoparticles*. In: Pacheco-Torgal F, Jalali S, Fucic A, Editors. Toxicity of building materials. Cambridge, UK: Woodhead Publishing Limited- Elsevier Science and Technology, Abington Hall 427-475
[38] Lopez de Ipina J et al., 2015 Strategies, methods and tools for managing nanorisks in construction. In: 4th International Conference on Safe Production and Use of Nanomaterials http://iopscience.iop.org/article/10.1088/1742-6596/617/1/012035/meta
[39] Maynard A D, Aitken R J, Butz T, Colvin V, Donaldson K, Oberdörster G and Tinkle S S 2006 *Nature* **444** 267-269

Antioxidant, Anti-Inflammatory, and Antibacterial Potential of Different Drinks Based on Matcha Tea

A Burcuş[1], E Vamanu[1*], I Sârbu[2] and A Petre[1]
[1]University of Agronomic Sciences and Veterinary Medicine, Faculty of Biotechnology, Bd. Mărăşti no. 59, district 1, Bucharest, Romania
[2]ICUB-Research Institute of the University of Bucharest 36-46 Bd. M. Kogalniceanu, 5th District, 050107, Bucharest, Romania

E-mail: email@emanuelvamanu.ro

Abstract. The aim of this research was to present *in vitro* and *ex vivo* biological activities for three different Matcha teas and a juice based on Matcha tea, lingonberry and bacterial probiotic strain (as a control) and one commercial green tea. The results demonstrated a strong correlation between major bioactive compounds and three biological activities. Antioxidant potential was correlated with epicatechin levels and with a significant amount of caffeine. The level of these compounds was influenced by tea samples. Matcha tea (M1) was presented in the finest particles size and corresponded with obtaining the highest value of antioxidant potential. Chromatographic (qualitative) analysis has demonstrated a characteristic distribution, and epicatechin and rutin have been identified as major bioactive compounds. Correlation of the anti-inflammatory potential with the antibiofilm-inducing inhibition is a significant feature in the biological synthesis of new innovative products (eg, nanobiosystems) with antibiofilm effects.

1. Introduction

Oxidative dysfunction is associated with numerous health problems, and it is associated with many clinical implications. The presence of free radicals was associated with the onset of different pathologies that affect a large number of young individuals [1]. Foods cause free radical generation in large quantities; as such, one of the main concerns is to find functional foods and drinks that promote well-being. In this regard, herbal tea represents a potent product full of antioxidants and that exhibits many healthy effects on humans. These beneficial activities are based on the antioxidant and anti-inflammatory activities of the bioactive compounds found in tea drinks [2]. Major compounds, like phenolic compounds, are responsible for these *in vivo*/*in vitro* [3] activities, and all are correlated with strong antimicrobial potential, especially against pathogenic bacteria [4].

Matcha tea represents one of the many products to express such complex potential; it is a relatively new health product for Romanian consumers. Based on the recent findings on this Japanese tea, it was found Matcha tea consumption has numerous health benefits. Matcha tea has strong biological activities, which result from tea leaves that are dried and finely ground without any other chemical supplements. Tea preparation assures good release of some bioactive compounds in the water, and these compounds can be easily assimilated after consumption. The bioactive fraction has a positive action on a lepidic profile, anti-inflammatory activities, and energizing effects based on the caffeine content; it also helps lower blood pressure and improves memory [5]. For these reasons, the aims of the article are to demonstrate the correlation between major bioactive compounds, five biologic

IOP Conf. Series: Materials Science and Engineering **374** (2018) 012072 doi:10.1088/1757-899X/374/1/012072

activities (antioxidant, anti-inflammatory and antibacterial activities) from three different Matcha teas, a Matcha-based juice (as a control) and a green tea.

2. Experimental procedures

2.1. Biological material and sample preparation.

Three types of Matcha teas were bought from a specialized market in Romania (Bio Japan Matcha, Demmer GmbH, Wien, Austria (M1), Spring Markt (M3) and Thés George Cannon (M4)), one instant drink with Matcha, lingonberry, and cultures (Ikea PS Instant drink, commercialized by Ikea Romania (M2)) and green tea (The National Health Products Co., LTD., China; M5).

The samples (aqueous extracts) were prepared in accordance with previously described research [6]. For the dried powder Matcha tea, 2 g were mixed with 20 mL of boiling water and vertexing to obtain a homogeneous mixture; after that, boiling water was added to 200 mL. The instant drink was prepared in accordance with the manufacturer's instructions by dissolving all *contents* of the *sachet* in 200 mL of boiling water.

2.2. Quantification of bioactive compounds

The total phenolic and flavonoidic contents were determined by spectrophotometric methods at 765 nm and 410 nm, respectively. The results were expressed as gallic acid equivalents in mg/100 g of extract and microgram quercetin equivalents/100 g of extract [3].

2.3. Determination of antioxidant activity

In vitro antioxidant activity was demonstrated by several spectrophotometrical methods: DPPH scavenging activity, chelating activity and reducing power. Scavenging activity was calculated using the following equation: % activity = (O.D. control – O.D. sample)/O.D. control × 100. Ascorbic acid was used as a standard for all methods [3, 4].

2.4. Determination of anti-inflammatory activity

The *ex vivo* anti-inflammatory activity was proven by human red blood cell (HBRC) membrane stabilization using a slightly modified method [7]. Fresh human blood mixed with sodium heparin, from the first author, was collected and transferred to centrifuge tubes [3]. After that, an equal volume of freshly collected blood was mixed with sterile Alsever's solution (in g/L, NaCl 4.2; citric acid×3Na×2H_2O 8.0; citric acid×H_2O 0.55; D-glucose 20.5) and centrifuged at 3,000×g, for 10 minutes (Hettich EBA20). In a new Falcon tube (15 mL) 2 mL of HRBC suspension was mixed with an equal volume of drinks and kept at 37°C for 30 minutes (Memmert Incubator). The hemoglobin content was determined by spectrophotometric quantitative analysis at 560 nm. The % of protection was calculated with the following equation:

$$\% \text{ protection} = 100 - \text{O.D. sample/O.D. control} \times 100. \quad (1)$$

The second anti-inflammatory activity evaluated in this research was inhibition of lipid peroxidation. Inhibition was proven by using egg yolk homogenates, based on the previous protocol [8].

2.5. Determination of antibacterial activity

To evaluate the samples' antimicrobial activity, a microdilution method was chosen, according to recommendations of the Clinical Laboratory and Standards Institute (CLSI) M07-A10. For this study, two strains of Gram-positive bacteria were selected (*Bacillus cereus* ATCC 11778; *Staphylococcus aureus* ATCC 6538), as were two Gram-negative strains (*Escherichia coli* ACTCC 25922; *Pseudomonas aeruginosa* ATCC 15442) and one yeast strain (*Candida albicans* ATCC 10231).

The microbial strains were seeded on boards with a specific BHI (bacteria) medium and YPG (yeast), and they were then incubated for 24 hours at 37°C. Following microorganism development, suspensions were made in ADS with a density of 0.5 McFarland. In 96-well plates 100 μL of MH (bacteria) medium and RPMI 1640 medium were added with 0.165M MOPS, 0.03% L-glutamine, and no bicarbonate (yeast); over the medium, 100 μL of the sample were added and serial binary dilutions were made. Finally, 10 μL of microbial culture were added with a density of 5×106 cells/mL. The plates were incubated for 20 hours at 37°C, performed in triplicate.

After CMI determination, the plates were cleaned with distilled water twice, fixed with 95% methanol for 5 minutes, and then stained with 1% purple crystal for 15 minutes. After staining, the plates were cleaned well to remove excess stain, and in every plate 130 μL of acetic acid (33%) were added. The optical density was determined at 490 nm to stabilize the minimum biofilm eradication concentration (MBEC) [9, 10].

2.6. Statistical analysis

All parameters for antioxidant activity were assessed in triplicate, and the results were expressed as the mean ± standard deviation (SD) values of three observations. The mean and SD were calculated with Excel (Microsoft Office 2016; Microsoft Corporation, Redmond, WA, USA).

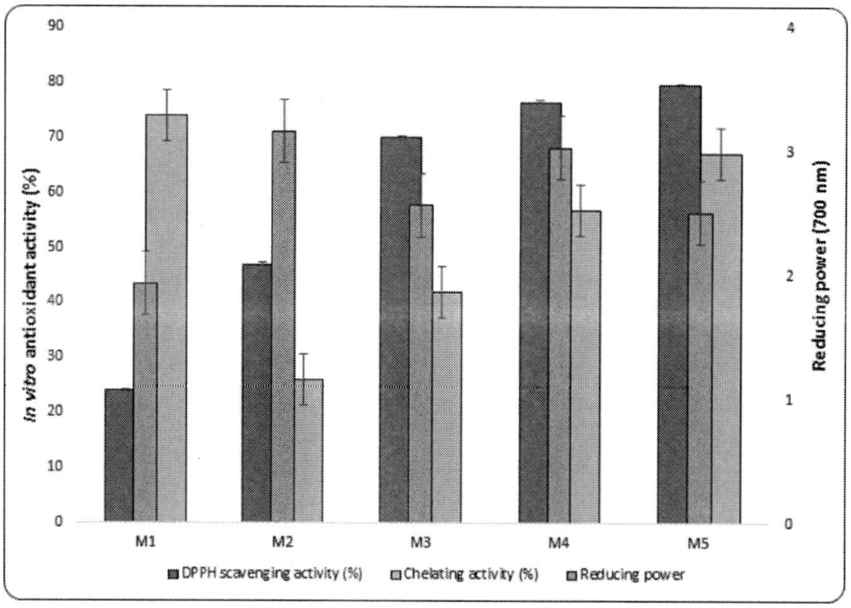

Figure 1. *In vitro* antioxidant capacity of tea samples.

3. Results and discussions

3.1. Evaluation of antioxidant capacity

Matcha tea infusion M5 presented the highest antiradicalic potential compared with the other samples (Figure 1). This value was approximately 70%±0.25% higher than M1. The fact that M3, M4, and M5 determined DDPH scavenging activity values of a minimum of 70%±0.06% was noted. M1 and M5 were able to determine an inversely proportional relationship between DDPH scavenging activity and chelating ability. The results can exhibit large differences due to the water temperature, which could vary in association with the sample-processing time. This behavior was similar to that found in other previous studies regarding the influence of time, temperature, and size of white- and green-tea particles on antioxidant characteristics [11].

Instead, the reduction power was obtained for sample M2, where the lowest antiradicalic and chelating activity were observed. This result occurred because of the different compositions of the samples. For M2, the remainder of the instant product components and the presence of probiotic strains (*Lactobacillus sporogenes*) demonstrated their antioxidant potential [12] however, lingonberry is also known to have a pronounced antioxidant effect due to its anthocyanin and phenolic acid content [13]. Furthermore, a high variation in chelating activity (approximately ±4.6%) and reduction power (approximately 0.25%) was significantly correlated ($P \leq 0.005$) with the total phenol content variation.

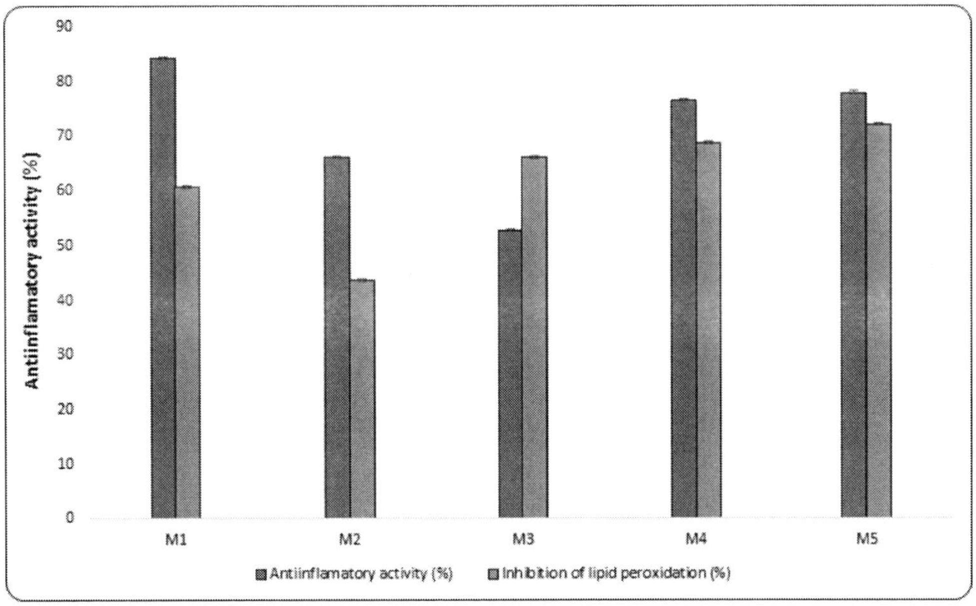

Figure 2. Anti-inflammatory activity of tea samples.

3.2. Evaluation of anti-inflammatory activity

A significant number of studies proved the direct link between antioxidant activity and anti-inflammatory activity. The antioxidant potential, previously proven *in vitro*, showed a direct relationship with the composition of the tea samples. The anti-inflammatory potential proven *in vivo* and *in vitro* exhibited the same tendency (Figure 2). M4 and M5 presented maximum values. The correlation between methods and Matcha tea samples was significant ($P<0.05$). But, the M2 sample presented minimum values, which proved weak anti-inflammatory and antioxidant protection.

Similar to previous studies [14], non-polar solutions (M3, M4, and M5) had the highest anti-inflammatory activity, and these types of drinks represent functional alternatives to M2. It is worth mentioning the high anti-inflammatory value (84.2%) of the M1 sample, which showed the lowest scavenging effect on DDPH radicals (Figure 1). This behavior proved that the Matcha samples selectively expressed a biological response. The effect was directly correlated with the level of the main bioactive compounds (Figure 3).

3.3. Antimicrobial and antibiofilm activity

The M2 sample presented the most reduced activity against bacteria, while M1 did not have an antimicrobial effect. The M5 sample showed the highest antimicrobial activity, especially against *S. aureus* (ATCC 6538). The M2 and M4 CMI against *B. cerus* (ATCC 11778) and *P. aeruginosa* (ATCC 15442) was 50% higher (Figure 4). Strains of *E. coli* (ATCC 25922) and *C. albicans* (ATCC 10231) showed resistance to the presence of bioactive compounds in the five samples.

The main result was that biofilm inhibition directly resulted from the antimicrobial activity (e.g., samples M1 and M3). Conversely, M2, M4, and M5 did not inhibit biofilm formation in Gram-negative strains; the CMI value was the same as the CMEB value for *P. aeruginosa* (ATCC 15442) strains, while for *E. coli* (ACTCC 25922) strains, the values were higher than in the control sample, which indicates a stimulation of biofilm formation.

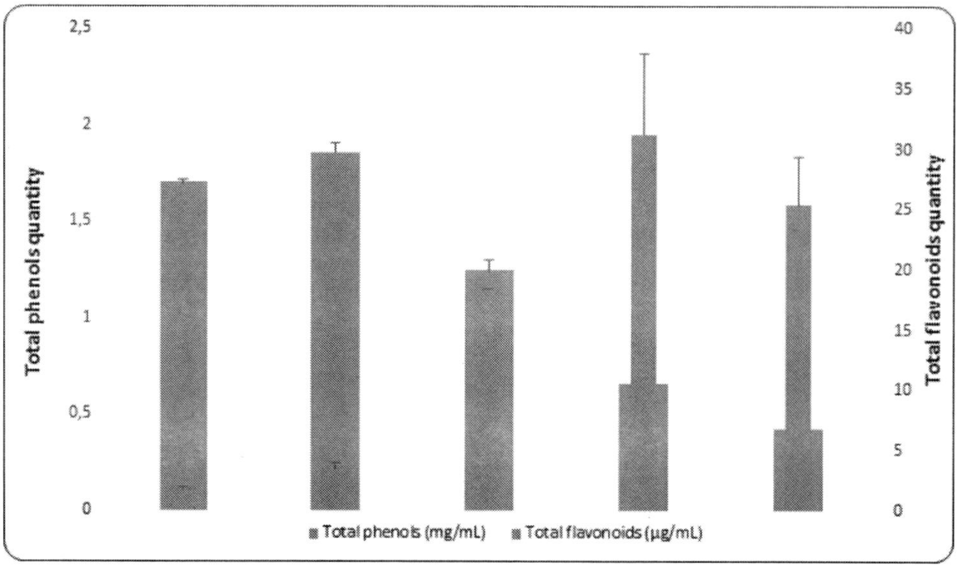

Figure 3. Total phenolic and total flavonoid content of tea samples.

In the case of *C. albicans* (ATCC 10231) strains, the values were very low; thus, it was not possible to determine biofilm formation. The results of the antibiofilm effect were opposite to those of the antioxidant effect for the M1 sample. These results were different for M2, M4, and M5, and they corresponded to other studies that proved the polyphenolic component effect against selected bacterial pathogens [15].

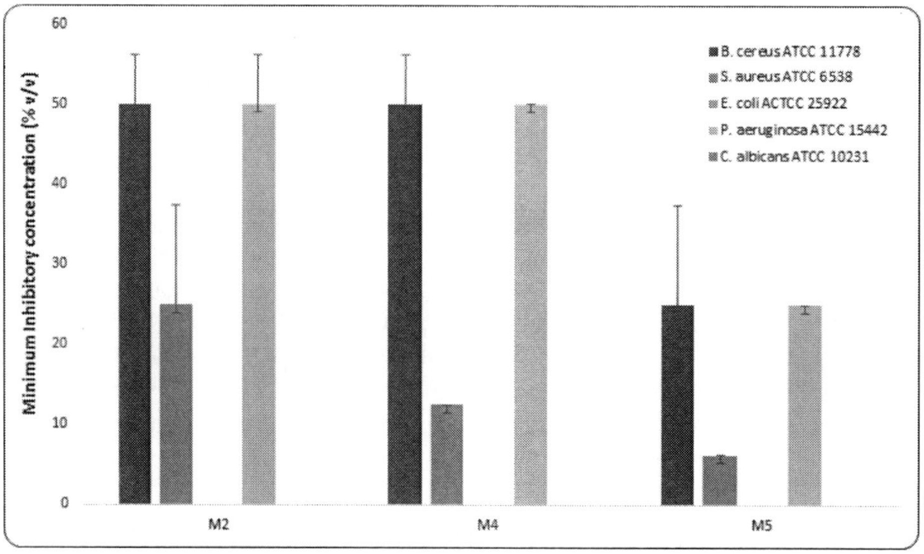

Figure 4. Antimicrobial effect of tea samples.

3.4. Evaluation of the bioactive compounds

It is worth noting the large difference between the level of flavonoids in samples M4 and M5 compared to the remaining samples. M5 had a minimum level of phenolic compounds of 0.4±0.01 mg of gallic acid/mL (Figure 3). High levels of flavonoids were also found in other studies when compared with other types of tea [16]. Thus, the high presence of flavonoids directly influenced the antioxidant/anti-inflammatory activity (Figures 1 and 2). When compared to other previous studies, tea samples M4 and M5 had inversely proportional chelation activity, which had an opposite effect to the methanol extracts obtained from *Polyalthia longifolia* and *Cassia spectabilis* [2].

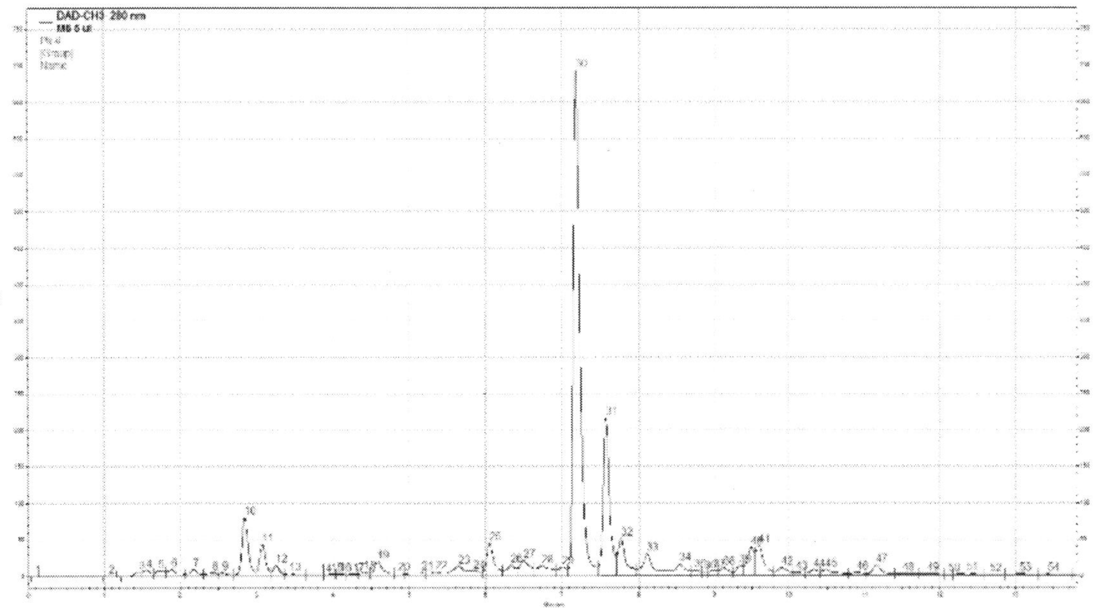

Figure 5. Chromatographic analysis (qualitative) of M5 sample.

The M2 sample had a high phenolic acid content (1.85 mg gallic acid/mL). This was 10% greater than that for M1, on average, and it was due to the heterogeneous composition. The remaining samples showed significant variations of approximately 70%. A direct comparison with other studies [14] is not relevant, as the results depend on how the infusion is prepared, as well as on the analyzed tea sample and the expression of the results.

The epicatechin and the rutin were identified like major phenolic compounds (Figure 5, peak 30 and peak 31) in all samples except M4. Following the chromatographic analysis (qualitative), the existence of a significant number of bioactive compounds was also proven. Their quantitative distribution depended on each sample, but the substantial presence of flavonoids was characteristic. Depending on how the samples were prepared and presented, the qualitative analysis determined the identification of principal metabolites as a primary way of expressing the biological response. The spectrum of peaks from Figure 5 proved that epicatechin was a major bioactive compound, and it represented 40% of all these molecules.

4. Comparative discussions

According to previous studies, the biological action of bioactive compounds (e.g., polyphenol–carboxylic acids or flavonoids) is directly dependent on their chemical structure, which influences the assimilation capacity in vivo [17]. This result affected the anti-radical response *in vitro* (Figure 1). The values were poorly correlated with the chelation capacity ($R^2<0.01$; $P>0.05$), as they proved different mechanisms of action that depended on the bioavailability of the different categories of bioactive compounds. Instead, the strength of the correlation increased significantly with the anti-inflammatory activity ex vivo ($R^2>0.1$; $P<0.02$), but it tended to vary with the inhibition of lipid peroxidase *in vitro* ($R^2>0.36$; $P<0.04$).

The antimicrobial effect manifested in the Gram-positive strains was opposite to that which manifested against strains of *E. coli* (ATCC 25922) and *C. albicans* (ATCC 10231). These strains were not inhibited, which indicated a selectivity toward the bioactive component. These samples, extracted with hot water, did not determine the presence of a specific concentration or of some phenolic compounds capable of inhibiting potential pathogenic strains. In addition, for the strain *S. aureus*, the M2 extract inhibits biofilm formation at high concentrations and stimulates its formation at concentrations less than 3% (the data did not demonstrate this). The extracts (the types of tea) analyzed stimulated the formation of biofilm in the case of *E. coli* strains (ATCC 25922). The biofilm formed by strains of *C. albicans* (ATCC 10231), *B. cerus* (ACTCC 11778), and *P. aeruginosa* (ATCC 15442) was very weak, and so the MBEC could not be precisely established. This behavior was interpreted as a direct bond with the antioxidant response, corresponding to the findings of some previous studies [18].

The antioxidant potential proven by *in vivo* and *in vitro* methods determined a biologic response similar to that found in previous studies, which showed a link between catechin level and high caffeine content [19]. The level of these compounds is influenced by how the tea samples are presented. M1 was presented with the finest particle size, and this corresponded to obtaining the highest antioxidant potential (Figure 1). Complete dissolution of the M2 sample determined an average antioxidant response, with preponderance influenced by the microbial component. Depending on the type of tea and the way in which the substrate was presented, there was a certain potentiation of the biological response; this was correlated with both antioxidant antimicrobial effects.

5. Conclusions

The consumption of beverages with different biological activities (antioxidant or antimicrobial) represents a broadly sought consumer tendency. Matcha tea has functional characteristics that can respond to this trend. Although the types of Matcha tea present different functional activities, they depend on the specificity of the active biologic compounds, particularly in terms of their ability to scavenge free radicals or inhibit the activity of microbial strains. The correlation of the two activities was a significant feature of M1 compared to the other types of evaluated tea types. This preliminary study showed the high biological potential of Matcha tea both *in vitro* and *ex vivo*. Protection against oxidative stress has been proven and was correlated with increasing antioxidant/antimicrobial protection capacity *in vivo* [20]. To validate these conclusions, studies on the assessment of gastrointestinal transit stability should be carried out, as the antibiotic effect plays a key role in the control of pathogenic strains.

6. References

[1] Chowdhury A, Azam S, Jainul M A, Faruq K O and Islam A 2014 *Int J Microbiol* 410935
[2] Jothy S L, Aziz A, Chen Y and Sasidharan S 2012 *Evid based Comp Alt Med* 561284
[3] Vamanu E and Nita S 2014 *Rev Chim* **65** 372-380
[4] Vamanu E, Ene M, Pelinescu D, Sarbu I, Vamanu A and Nita S 2011 *Rev Chim* **62** 1189-1194
[5] https://prodietaro, accessed on 3102017
[6] Petkova N, Ivanov I, Mihaylova D and Krastanov A 2017 *Bulg Chem Commun* **49 D** 69-74
[7] Chowdhury A, Azam S, Jainul M A, Faruq K O and Islam A 2014 *Int J Microbiol* 410935
[8] Ben-Nasr S, Aazza S, Mnif W and Miguel M G C 2015 *Pharmacog Mag* **11** 48-54
[9] Turcu I, Zarafu I, Popa M, Chifiriuc M C, Bleotu C, Culita D, Ghica C and Ionita P 2017 *Nanomat* **7** 43
[10] Sârbu I, Pelinescu D, Stoica I, Măruţescu L and Vassu T 2013 *Roum Arch Microbiol Immunol* **72** 225233
[11] Castiglioni S, Damiani E, Astolfi P and Carloni P 2015 *Int J Food Sci Nutr* **66** 491-497
[12] Kodali V P and Sen R 2008 *Biotechnol J* **3** 245-251
[13] Wang S Y, Feng R, Bowman L, Penhallegon R, Ding M and Lu Y 2005 *J Agr Food Chem* **53** 3156 - 3166
[14] Nadpal J D, Lesjak M M, Sibul F S, Anackov G T, Cetojevic-Simin D D, Mimica-Dukic N M and Beara I N 2016 *Food Chem* **192** 907-914

[15] Xu C, Yagiz Y, Hsu W Y, Simonne A, Lu J and Marshall M R 2014 *J Agr Food Chem* **62** 6640-6649
[16] Shannon E, Jaiswal A K and Abu-Ghannam N 2018 *Food Res* **2** 111
[17] Pepe G, Pagano F, Adesso S, Sommella E, Ostacolo C, Manfra M, Chieppa M, Sala M, Russo M, Marzocco S and Campiglia P 2017 *Molecules* **22** 623
[18] Chaieb K, Kouidhi B, Slama R, Fdhila K, Zmantar T and Bakhrouf A 2013 *J Herbs, Spices Medicinal Plants* **19** 168-179
[19] Atalay D and Erge H S 2017 *GIDA J Food* **42** 494-504
[20] Stanciu G, Chirila E, Dobrinas S and Negreanu-Pirjol T 2010 *Rev Chim* **61** 41-44

Acknowledgements
This work was partially supported by the Executive Agency for Higher Education, Research, Development and Innovation Funding – Human Resources, by the project PNCDI III UEFISCDI — Experimental demonstration project (PED), Project 171PED/2017 (http://www.robiomush.ro /biochaga/). English-language editing of this manuscript was provided by Journal Prep.

Arhaeometric Concepts and Methods of Intervention on Historical Monument Buildings. The Case of the Corvins' Castle

R M Ion[1,2,*] , L Iancu[1,2], R M Grigorescu[1], D Carutiu-Turcanu[3], S Tincu[4], N Ion[1], I A Bucurica[5], S Teodorescu[5], I D Dulama[5], R M Stirbescu[5], A Gheboianu[5] and M L Ion[6]
[1] Evaluation and conservation Cultural Heritage Team, ICECHIM, 202 Splaiul Independentei, Bucharest-060021, Romania
[2] Department of Engineering Doctoral Studies, Valahia University, 13 Aleea Sinaia, Targoviste, Romania
[3] OVIDIUS University Constanta, Expertise Center for the Art of Science, Culture and Spirituality, Institute of Science, Culture and Spirituality, Constanta
[4] Corvin's Castle, Hunedoara, Romania
[5] Valahia University of Targoviste, Institute of Multidisciplinary Research for Science and Technology, Targoviste, Romania
[6] "Atelierul de Creatie" NGO, Bucharest

E-mail: rodica_ion2000@yahoo.co.uk

Abstract. A multidisciplinary study on the mortars from the Corvins' Castle, Hunedoara, Romania has been carried out on two ceramic samples which were analyzed through spectral techniques (FTIR, Raman, X-ray powder diffraction) and diverse microscopy techniques (optical, scanning electron microscopy), for determining the chemical, mineralogical and petrographic characteristics on two ceramic tiles coming from the Corvins' Castle discovered during the archaeological campaign from 2012, at the Administrator's House, inside of the castle yard; the results are presented in order to determine the raw material provenance and the production process. Some information on constructive details, construction history and degradation or damage status, are detailed. Also, the diagnosis results aim to determine the damaging causes, as well as the assessment of the safety or the way of intervention on the structure. The archaeometric study of two tiles, including the identifying of the raw materials provenance is important for collecting information about the exploitation of natural resources and to better understand the manufacturing and weathering/deterioration processes of these artifacts. Modern and performant analytical techniques, as FTIR, SEM-EDS, XRD, colorimetric and gloss measurements, are used and the results are discussed too.

1. Introduction

Corvins' or Hunyadi Castle located in Hunedoara, Transylvania, Romania (Figure 1), it was built in the 15th century on the site of an old fortress, on a cliff above the Zlasti River [1].

The story of the Corvins' Castle (also known as the Hunyadi Castle) began in 1409 when the King of Hungary, Sigismund of Luxembourg (1384-1437), donated the royal possession of Hunedoara to Voicu and his family for outstanding military merits. The donation document also shows the name of

John, the future Iancu of Hunedoara, as the son of Voicu. Considered one of the most important military leaders in Central and Southeastern Europe in the fifteenth century, Iancu de Hunedoara is the one who started the construction of the castle around 1440, instead of the old fortress on a rock at whose feet it flows the Zlaşti River.

Figure 1. Corvins' Castle (Personal photo).

The first documentary attestation of the castle appeared in 1443, in a document signed by Iancu de Hunedoara. At the beginning of the 17th century, the castle was in possession of the Bethlen family, and Gabriel, the Prince of Transylvania, was the one who brought the most important changes in the architecture of the monument, inspired by the late Renaissance fashion. Although impressive by military architecture elements such as the "Neboisa Tower" (which transforms the "don't be afraid" tower) as the White Tower and the Artillery Terrace, the Corvins' Castle also features highly refined civil architecture, represented by the Knights Hall or the Diete's Hall. In the seventeenth century the outer yard (the Hussars' Court) was built, a space that housed the dwellings of the administrator, the clerks, the house of the greyhounds, and the food and hatcheries. After the Austrians and Austro-Hungary's possession in the 18th-19th centuries, the castle was in the possession of the Romanian state from 1918. It has never been entirely rehabilitated, although such works have been carried out for many years, but even so it remains a unique monument that attracts and impresses tourists. The Hussars' Yard was also built in the 17th century. It was a space that sheltered the houses of the administrator, of the clerk, greyhounds' stable and storehouses for food and hay [2]. On 13 April 1854, a lightning caused a devastating fire that extensively damaged the castle. It was subsequently restored, but the restoration was carried out in several stages and some of these were not entirely successful.

Over the past few years, Corvins' Castle underwent intense rehabilitation because the roof had been damaged due to moisture present in materials, cyclic, the influence of solar radiation, the temperatures, the water on the masonry, the influence of dissolved salts, the influence of frost and snow, the influence of wind and air currents, the influence of the particles of smoke, dust and sand, the influence of extreme natural phenomena, as floods or lightning, and the biological attacks by fungi and insects, birds or animals, which led to an excessive humidity [3]. Elimination of these problems should be solved by specialists in archeology and the restoration of historical monuments [4].

The raw materials are clays with finely divided quartz (sand) (0.02–0.04 mm) and feldspar, responsible for the rheology along the thermal processing. Clay minerals such as kaolinite ($Al_2[Si_2O_5][OH]_4$) generated either by the weathering of igneous rocks under the influence of water, dissolved carbon dioxide, and organic acids, or from feldspar ($KAlSi_3O_8$) eroded from rocks such as granite and deposited in lake beds, which are aluminosilicates that contain sodium (Na), potassium (K), or calcium (Ca) with a composition from $NaAlSi_3O_8$ and $KAlSi_3O_8$ to $CaAl_2Si_2O_8$. Feldspar acts as fluxing agents to reduce the melting temperatures of the aluminosilicate phases where they are subsequently transformed into clay [5]. Except feldspar, silica, as the second major ingredient in refractories, is usually added as quartz sand, sandstone, or flint pebbles [6]. The role of silica is either to maintain the shape during firing (as filler) or to improve the final mechanical properties.

The archaeometric study of two tiles, discovered during the archaeological campaign from 2012, at the Administrator's House, inside of the Corvins' Castle yard, are presented in this paper; the identifying the provenance of the raw materials is important for collecting information about the exploitation of natural resources and to better understand the manufacturing and weathering / deterioration processes of these artifacts. Modern and performant analytical techniques, as FTIR, SEM-EDS, XRD, colorimetric and gloss measurements, are used and the results are discussed too.

2. Materials and methods

2.1. Materials

The tile samples discovered in the Building House of the Administrator's House, Museum House of Bresles, situated in the Corvins' Castel, Hunedoara, are shown in Figure 2.

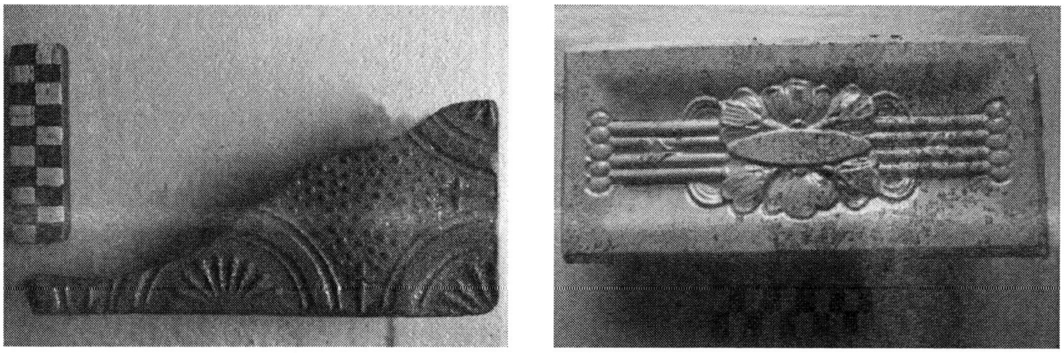

Figure 2. Corvins' Castle, Administrator's house, studied tiles (C1 left and C2 right).

2.2. Methods

2.2.1. Characterization methods. Fourier transformed infrared spectroscopy (ATR-FTIR) has been recorded with a Vertex 80 spectrometer (Bruker Optik GMBH, Germania) in the range of 4000–400 cm^{-1}, equipped with DRIFT accessory.

The optical microscopy was performed with a Novex trinocular microscope (at different magnifications). Also, the optical microscopy has been recorded by a Primo Star ZEISS optical microscope that offers the possibility to investigate the samples in transmitted light at a magnification between 4X and 100X. The equipment had attached a digital video camera (Axiocam 105) which, by the microscope software, allowed real-time data acquisition. The obtained images could easily be converted from 2D in 3D format through its software for a better viewing.

The Scanning Electron Microscopy with Energy Dispersive Spectroscopy (SEM-EDS) results were obtained by a **SU**-70 (Hitachi, Japan) microscope, used for characterization of micro- and nanomaterials qualitative and quantitative analysis of samples and composition of the structure for a sample surface, respectively.

X-ray diffraction (XRD) diagrams have been registered with a X-ray diffractometer Rigaku Ultima IV, equipped with software for control, acquisition and data conversion. Several experimental configurations have been tested, as following: geometry in the Bragg - Brentano beam, anode voltage of 40 kV, anode current of 40 mA, scan at 1 degree / minute, at a 0.0200 step, angular scan range: 2q = 5-1500.

Color measurements have been recorded with a CM-2600d spectrophotometer (KONICA MINOLTA) (Japan) under a D65 light source and an observer angle of 10°. The differences in L*, a*, and b* and the total color differences ΔE^*_{ab} were calculated using the following equations (JIS 2008 [7]): The CIELab (CIE 1986) chromatic parameters were chosen for the study, i.e., lightness (L*), which ranges from 0 for black to 100 for white, and the chromatic coordinates a* and b*, coordinate a* ranges in value from +60 (magenta) to -60 (green) and b* from +60 (yellow) to -60 (blue).

For *gloss parameters* a Specular GlossMeter was used, 3nh_HG268, China instrument, which provided quantifiable gloss measurements, expressed as gloss units (GU) [8].

3. Results and discussions

Corvins' Castle is located in the area of Southern Carpathians and Banat (Figure 4), in a very complex geological area, consisting of crystalline-mesozoic units belonging to the sediment-volcanic units that contribute to the Southern Carpathians [35]. The getic pavement is built mainly from crystalline shale, in which there are migmatite, pegmatite, ampholyte, limestone and crystalline dolomites; over the crystalline formation are some sedimentation minerals, as: limestone, microconglomerate and sandstone. In these crystalline shale are important deposits of iron, being found as siderite deposits of variable lengths and thicknesses, which is transformed into limonite in the hematite, and pyrite, the ore being associated with the crystalline limestones. In the western part of Hunedoara there are talc of high quality. The sedimentary-eruptive zone of the Apuseni Carpathians is composed of mesozoic sedimentary formations (limestone, marl, clayey shale, conglomerate, sandstone) and magmatite (basalt), as well as neogene formations (basalt, andesite, pyroclastes) [9].

The stones found in majority in Romania are limestone (in various varieties) and siliceous (sandstone, granite, conglomerate). Depending on the geographical area and due to deficiencies in the processing of the raw material, clay, magnesium, calcium, manganese, limestone granules, sulphates can be present in the composition of clay used in the production of bricks, besides the aluminum hydrate as the main component and other substances, organic substances, with various effects on the timing of the bricks. Depending on the content of the clay, in bricks can be found calcium sulphates and other soluble salts that produce efflorescences and disintegration phenomena [10, 11]. The concentration of iron-based components may change the coloration. The composition of the two tiles investigated in this paper is shown in Figure 3, a, b, c.

Elemental analysis using EDS with scanning electron microscopy provides material information correlated to structure for ceramic materials and the EDS maps well correlate phase morphology to composition for ceramic materials. Enamels are composed primarily of silica, which forms the vitreous lattice, lead oxides, which lower the melting point [12]. The enamel was made with kaolin, zinc oxide, feldspar, calcite and lead base; copper and chromium oxides were added to provide green hue and it was glazed to attain a glossy finish; the enamel on the repair tiles had a lead, sodium and potassium base with lead oxide, silica and alumina [13]. The XRD analysis of the tile samples reveal that calcium carbonate is the main component of stone, the result of first sample of stone is quartz (73.8%), muscovite (15.7%), mica (6.2%) hematite (3.0%), gypsum (1.3%), the result of second sample is Pyrophyllite (69.7%), quartz (12.3%), Hercynite (5.6%), Anorthite (3%), macrocline (5%), gypsum (1.5%), hematite (2.9%). Crystal structure analysis results and Indexing and quantitative analysis results are shown for both samples in the Tables 1 and 2.

Overall, clays used in light-firing and dark-firing bodies are fairly well discriminated by a Fe_2O_3 content of approximately 3% wt. During the reducing firing stage, the Fe^{3+} in the iron-rich slips, containing about 12 wt.% of iron as determined by EDX analysis [14], is converted to Fe^{2+} in a spinel phase of the magnetite–hercynite solid solution type. Some clay materials used in ceramic tile production are characterized by the occurrence of pyrophyllite, from a few percent up to 80% wt. Since pyrophyllite has a different technological behavior with respect to kaolinite, it is convenient to classify as *Pyrophyllitic Clays* (PC) the raw materials with a significant amount of pyrophyllite (> 20%) as they present commonly lower loss on ignition, easier compaction, and lower refractoriness

IOP Conf. Series: Materials Science and Engineering **374** (2018) 012073 doi:10.1088/1757-899X/374/1/012073

with respect to kaolin [15]. PC are usually claystones of different origin that, along with pyrophyllite, contain quartz, feldspars and often kaolinite and/or illite or sericite.

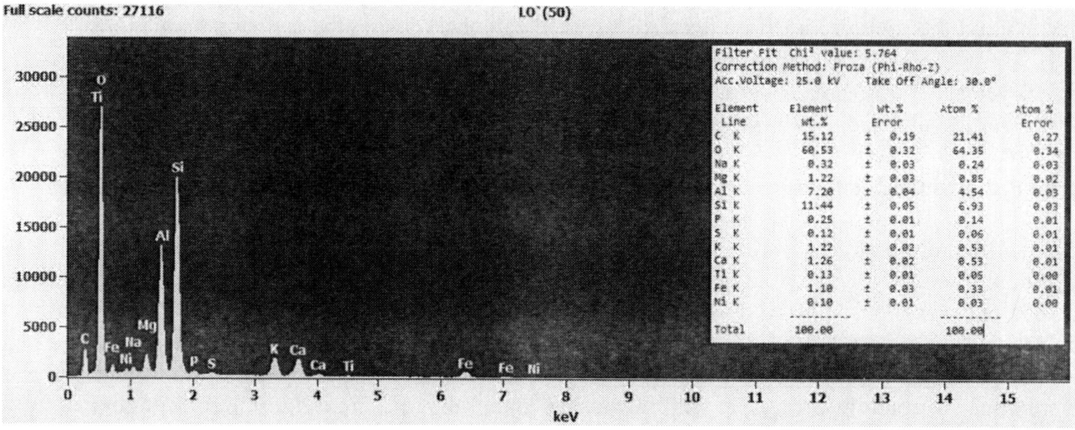

(a)

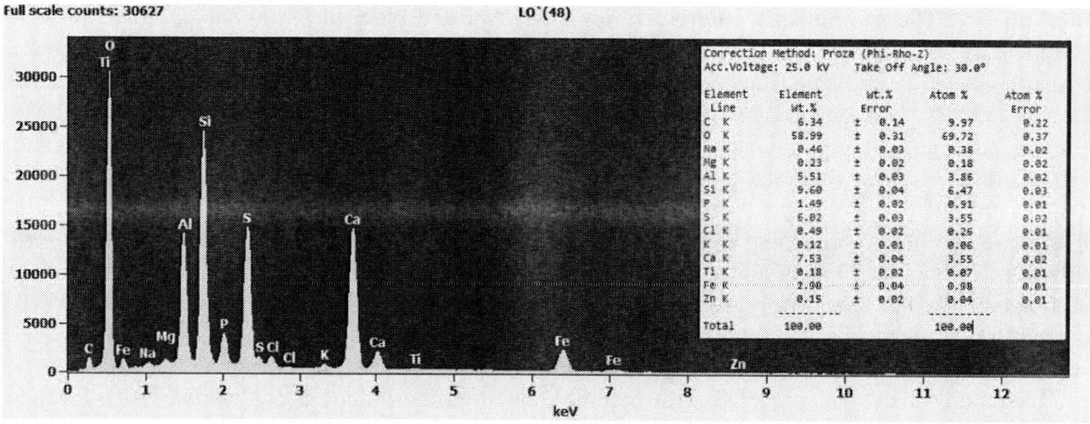

(b)

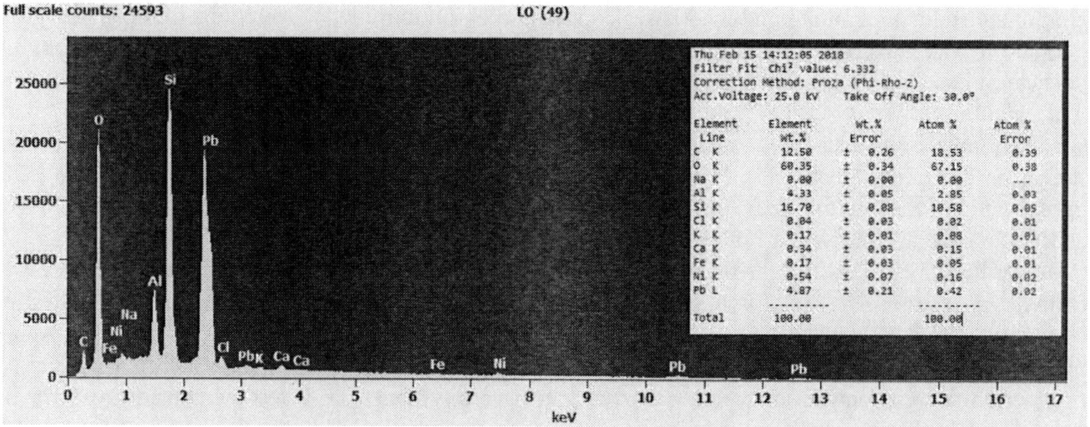

(c)

Figure 3. The EDS results of the tile (a) up, (b) middle and (c) enamel (down).

Table 1. XRD results for tile C1.

Phase name	Formula	DB card number	Content (%)
Quartz	SiO_2	01-083-0539	73.8
Muscovite	$(K_{0.82}Na_{0.18})(Fe_{0.03}Al_{1.97})(AlSi_3)O_{10}(OH)_2$	01-080-0743	15.7
Mica	$Rb_{.99}Fe_{3.03}(Fe_{1.04}Si_{2.96})O_{10}(O H)_2$	01-075-4387	6.2

Table 2. XRD results for tile C2.

Phase name	Formula	DB card number	Content (%)
Pyrophyllite	$Al(Si_2O_5)(OH)$	01-083-1805	69.7
Potassium tecto-alumino-trisilicate (macrocline, potassium-rich alkali feldspar)	$K(AlSi_3O_8)$	01-087-1790	2.4
Quartz	SiO_2	01-086-1630	12.3
Anorthite, syn sodian (plagioclase feldspar)	$(Na_{.45}Ca_{.55})(Al_{1.55}Si_{2.45}O_8)$	01-085-1415	3.0
Hercynite	$(Fe_{0.867}Al_{0.133})(Fe_{0.235}Al_{1.765})O_4$	01-070-9102	5.6

XRD analyses of the investigated tiles samples confirmed the presence of the minerals identified by optical microscopy.

The results of the microscopic examination indicate that most of the grains of the analyzed mortars, with the exception of C1 sample, revealed quartz, feldspars and rock fragments, indicating an origin from natural sands. Also, mica is present in this sample [16].

The ceramic bodies are distinguished in light-firing (from white to light brown) and dark-firing (from pink to dark brown) on the basis of color after firing. Overall, clays used in light-firing and dark-firing bodies are fairly well discriminated by a Fe_2O_3 content of approximately 3% wt, Figure 4.

Sample C1 (up) shows large rounded white lump.

The open and porous structure visible in the SEM images of C2, is a real proof for the firing treatment of the ceramic between 700 and 900 °C.

In this temperature range, the sintering process is governed via solid state mechanism, especially surface diffusion.

In addition, the green pieces on heating up to 900 °C underwent a series of phase transformations such as dehydration of hydroxides (gibbsite and goethite) and dehydroxylation of kaolinite to metakaolinite formation. This range of temperature is characterized by the decomposition of carbonates, which starts at around 650°C and ends at around 800-850°C.

Calcite, the most common pure calcium carbonate ($CaCO_3$), decomposes in calcium oxide (CaO) and release CO_2 increasing the porosity. High-temperature crystalline phases (900-1000 °C) are the result of the reactions involving carbonates, calcium oxides and silicates as anorthite ($CaAl_2Si_2O_8$).

The SEM images from Fig. 5 shows the morphology of the fine grained clay matrix and some rather large inclusions, but it is difficult to identify the nature of the latter because they exhibit no idiomorphic crystal faces [20].

These images evidenced a clay matrix, of an inclusion identified as quartz and sodium feldspar, as previously stated by XRD. The quartz (SiO_2) shows practically nothing else, while the inclusion identified as a feldspar additionally exhibits the lines of Na, Al, and very little K. Most probably, it is a sodium feldspar, $Na[AlSi_3O_8]$ of the albite–orthoclase system. The clay matrix contains much aluminum, some potassium and a small quantity of iron.

IOP Conf. Series: Materials Science and Engineering **374** (2018) 012073 doi:10.1088/1757-899X/374/1/012073

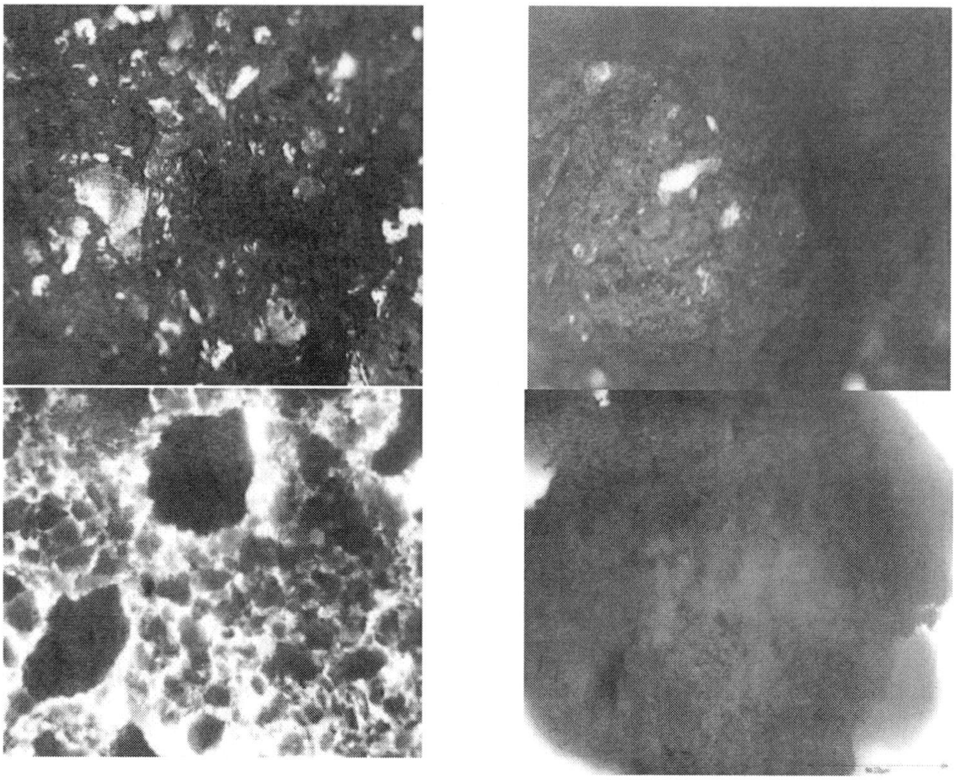

Figure 4. Macroscopic pictures of the selected tile surfaces.

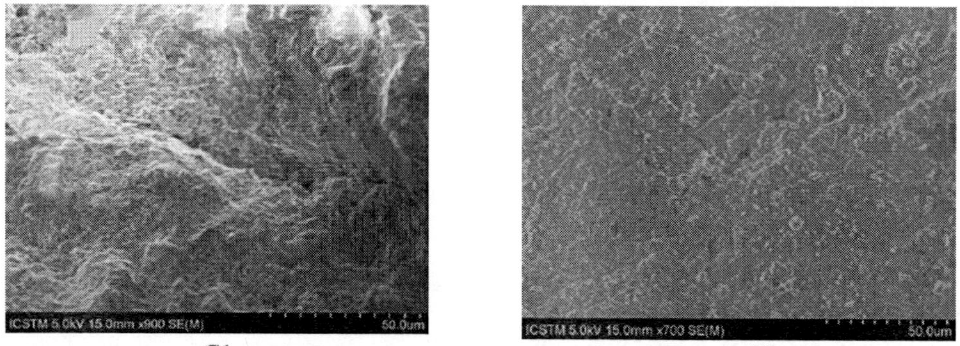

C1 C2

Figure 5. SEM images of the tile samples.

All these results could be argued by FTIR investigations, as it is shown in Figure 6. Silicate, phosphate, or sulfate spectra overlap each other in the region (900–1000 cm^{-1}). The quartz intensity (band at 1083 cm^{-1}) was much lower than the calcite intensity although the quantity of both minerals is the same (mole).

Muscovite, with three Si atoms (band at 979 cm^{-1}), has a main band approximately three times more intense. Quartz is well detected in all the samples: 1088-1092 cm^{-1} due to Si-O stretching mode, the distinctive doublet at 796 and 777-783 cm^{-1} and the peak at 692-694 cm^{-1} due to Si-O symmetrical stretching and bending modes respectively [18].

Muscovite (1026-1036, 1007 and 530 cm^{-1}) as well as feldspars (752-758, 646 and 465-471 cm^{-1}) and Pyroxenes at 519 cm^{-1} were also detected [19].

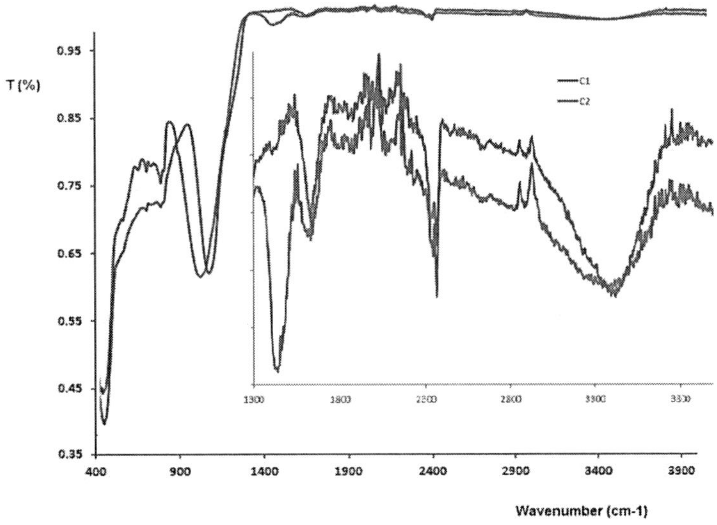

Figure 6. FTIR spectra of C1 (blue line) and C2 (red line).

In this color space, which is presently one of the most widely used color spaces for measuring object color, L^* indicates lightness, and a^* and b^* are chromaticity coordinates [20]. The ceramic tile C1 is much reddish-brown (L* =56.18; a* =7.66; b* =19.86) in the points where there are high amounts of dark red putties than in the center (L*=55.13; a*=8.46; b*=20.41) or at the edge (L* =54.84; a* =8.08; b* =19.69). The sample prove to be insufficiently pressed and show irregular chromatism, Figure 7. The carefully manufactured and sized courtyard tiles C2 are highly compact. Their light yellow hue (L*=75.79; a* =1.56; b* =19.10) is speckled by the reddish putties uniformly distributed across their surface. The enamel on the original tiles C2 have reddish tone around the edges (L*=81.51; a* = - 0.94; b* = 25.79), lighter toward the center (L* = 79.77; a* = - 0.27; b* = 25.78), and with bluish stains on the lines (L*=79.17; a* =0.06; b* = 26.36).

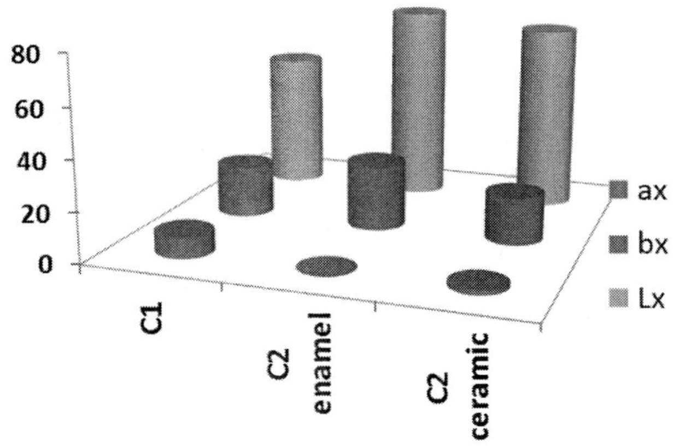

Figure 7. The chromatic parameters of the investigated tiles.

In correlation with color measurements, the gloss index could be used for measures the amount of light directed and reflected at a defined angle, by integrating the total scatter (reflected and diffused light). The most suitable geometry for general applications is 60°. The best optical geometry for surface with very high and very low gloss is respectively 20° and 85°. Gloss is affected by the refractive power of the boundary between two materials when they have different refractive indices, when certain paints a binder layer containing no pigments is formed on the top surface of paint films, Figure 8.

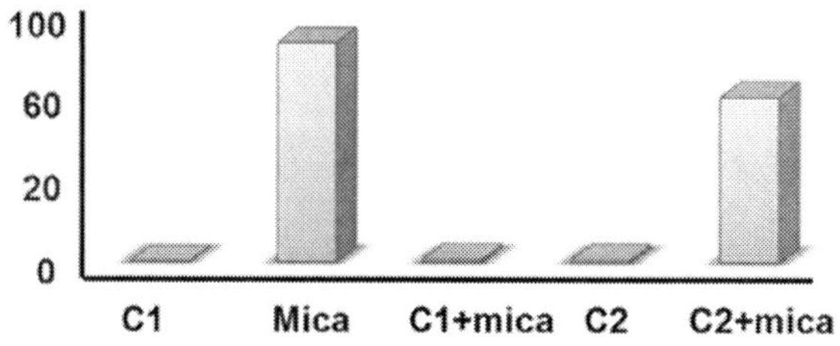

Figure 8. The gloss variation of the investigated tiles.

Since the refractive index of this binder layer is generally different from those of both sides, air and pigmented phase, interference of light would occur resulting in a decrease in gloss, in good agreement with literature [21]. In our case, the gloss index strongly decrease after mica adding in ceramics (C1) and increase for enamel application (C2).

4. Conclusions

In conclusion, both ceramic samples taken from the Archaeological Site Administrator's House from Courtyard of Corvins' Castle are prepared in the same area with native natural materials as follows: the main minerals are quartz, muscovite, mica, hematite (at the first sample) and pyrophyllite, quartz, hercynite, anorthite, macrocline (at the second sample). Calcite is difficult to identify because of its decomposition as baking takes place at 900-1000 degrees without specific control, under these conditions calcite decomposes into CO_2 which contributes to the high porosity of the samples.

The color depends on the iron oxide content, and on other components (as TiO_2 and CaO), which may turn the color into yellowish or pinkish shades. Overall, clay used in light-firing and dark-firing bodies is fairly well discriminated by a Fe_2O_3 content of approximately 3% wt. During the reduction firing stage, the Fe^{3+} in the iron-rich slips is converted to Fe^{2+} in the spinel phase of the magnetite-hercynite solid solution type. Enamel is composed primarily of silica, which forms the vitreous lattice, kaolin, zinc oxide, feldspar, calcium and lead base which lower the melting point and alumina; copper oxides were added to provide green hue and it was glazed to achieve a glossy finish. Certainly should be mentioned the presence of Calcium sulphate generated during the ageing process, as at all the samples decoperted from archaeological sites.

5. References

[1] Bodochi I 2008 *Castelul Corvinilor în secolul al XVII-lea* (1) Corviniana XII 207-232
[2] Roman C C, Diaconescu D, Ţiplic M 2004 *Archaeological excavations at Hunedoara-The Corvins' Castle - the sacristy of the chapel*, Studii de istorie veche şi arheologie. Omagiu profesorului Sabin Adrian Luca, BAHC IV 187-207
[3] Chefneux E 1967 *Restaurarea Castelului Corvinilor de la Hunedoara*, Dosar I.N.M.I. Arhiva monumentelor istorice 5473
[4] Roman C C, Tincu S 2009 *Terra Sebus* **1** 153-172
[5] Maniatis Y 2009 *The Emergence of Ceramic Technology and Its Evolution as Revealed with the use of Scientific Techniques,* in: A.J. Shortland, I. Freestone, T. Rehren, From Mine to Microscope: *Advances in the Study of Ancient Technology*, Oxford: Oxbow Books, Chapter **2** 11–28

[6] Bergaya F, Lagaly G 2000 *General introduction: Clays, clay minerals and clay science, in: F. Bergaya*, B.K.G. Theng, G. Lagaly (eds.), Handbook of Clay Science. Elsevier, Amsterdam, 1–18
[7] ASTM D 523-94 1999 *Standard Test Method for Specular Gloss*, 5.1. pp. 1
[8] Luca S A, Roman C, Diaconescu D, Suciu C 2005 *Repertoriul arheologic al judetului Hunedoara*, Bibliotheca Septemcastrensis, XIV Alba Iulia
[9] Tite M S, Bimson M, Freestone I C 1982 *Archaeometry* **24** 117–126
[10] Mukhopadhyay T K, Ghatak S, Maiti H S 2010 *Ceram. Int.* **36** 909-916
[11] Bodochi I 2008 *Castelul Corvinilor în secolul al XVII-lea* (1), Corviniana, XII 207-232
[12] Ion R M, Iancu L, Carutiu-Turcanu D, Schroder V, Tincu S, Roman C, Ion N, Bucurica I A, Teodorescu S, Dulama I D, Stirbescu R M, Gheboianu A 2018 *Traditional building materials and modern restoration products identified at the painted Matia-fresco Loggia*, Corvins' Castle, Romania, Geophysical Research Abstracts **20** EGU2018-5198
[13] Marcu D, Petrov G 1993 *Ars Transsilvaniae* 93-105
[14] Ion R M, Ion M L, Fierascu R C, Serban S, Dumitriu I, Radovici C, Bauman I, Cosulet St, Niculescu V I R 2010 *J. Therm. Anal. Calorim.* **102** 393-398
[15] Ion R M, Dumitriu I, Fierascu R C et al. 2011 *J. Therm. Anal. Calorim.* **104** 487-493
[16] Ion R M, Fierăscu R C, Teodorescu S, Fierăscu I, Bunghez I R, Ţurcanu-Caruţiu D, Ion M L 2016 *Ceramic Materials Based on Clay Minerals in Cultural Heritage Study*, in: G. Morari Do Nascimento (Ed.), Clays, *Clay Minerals and Ceramic Materials Based on Clay Minerals*, InTech, DOI: 10.5772/61633
[17] Faria K C P and Holanda J N F 2012 *Using SEM/EDS for characterization of clay ceramic bearing sugarcane bagasse ash waste*, in: A. Méndez-Vilas (Ed.), Current Microscopy Contributions to Advances in Science and Technology 1085-1092
[18] Jordá J D, Jordán M M, Ibanco-Cañete R, Montero M A, Reyes-Labarta J A, Sánchez A, Cerdán M 2015 *App. Clay Sci.* **115** 1–8
[19] Dondi M 2003 *Technological and compositional requirements of clay materials for ceramic tiles*; in: E.A. Domínguez, G.R. Mas, F. Cravero (Eds.), Proc. 12th Int. Clay Conf. 23-30
[20] JIS Z8729:2004. *Color specification-CIELAB and CIELUV color spaces*
[21] ISO 2813, 1997. *Paints and varnishes: determination of specular gloss of nonmetallic paint films at 20/60/85 degrees*

Acknowledgements
This work was supported by a grant of the Romanian Ministery of Research and Innovation, CCCDI – UEFISCDI, project number PN-III-P1-1.2-PCCDI-2017-0476, no.51PCCDI/2018 within PNCDI III.

IOP Conf. Series: Materials Science and Engineering **374** (2018) 012074 doi:10.1088/1757-899X/374/1/012074

Risk Identification in a Smart Monitoring System Used to Preserve Artefacts Based on Textile Materials

V D Diaconescu[1], L Scripcariu[1], P D Mătăsaru[1], M R Diaconescu[1] and C A Ignat[1]
[1] Technical University "Gheorghe Asachi" of Iasi, 67 Bd. Prof. Dr. D. Mangeron, Iasi, Romania

E-mail: lscripca@etti.tuiasi.ro

Abstract. Exhibited textile-materials-based artefacts can be affected by the environmental conditions. A smart monitoring system that commands an adaptive automatic environment control system is proposed for indoor exhibition spaces containing various textile artefacts. All exhibited objects are monitored by many multi-sensor nodes containing temperature, relative humidity and light sensors. Data collected periodically from the entire sensor network is stored in a database and statistically processed in order to identify and classify the environment risk. Risk consequences are analyzed depending on the risk class and the smart system commands different control measures in order to stabilize the indoor environment conditions to the recommended values and prevent material degradation.

1. Introduction

Textile-based exhibited artefacts can be affected by the environmental conditions like temperature, relative humidity, light intensity variations or pollutants. Textile materials are extremely vulnerable to the climate changes therefore a lot of researchers are focused on the environment monitoring and control in museums where old artefacts are exhibited.

According to Abdel-Kareem [1] in most cases, textile-material based artefacts are stored and exhibited in very poor environmental conditions. Environmental monitoring and control is essential in order to preserve the textile artefacts but the problem is not permanently solved even if there are a lot of concerns for it.

Peralta et *al.* [2] present some components of the WISE-MUSE project designed by their team in order to monitor and control the environmental parameters. These component parts are intended to control the temperature and relative humidity of the environment using the air-conditioning and dehumidifying systems. The Wireless Sensor Network (WSN) architecture with a new sensor-node type and the communication system are described. Measures of the transmitted signal level are made in order to establish the limits of the sensing areas in the monitored space.

Stable environment parameters must be ensured in museums in order to preserve the artworks, according to Martens [3]. In its thesis, the author quantifies the climate risks based on the measured and simulated values of temperature and relative humidity in an indoor space. Risk assessment non-statistical predictive formulae are proposed and applied on four object types (books, furniture pieces, panel paintings and wooden sculptures) in a Dutch museum. The response time of each object type is established by a long-term monitoring process and it is used to estimate the museum climate risks.

Simić et *al.* [4] design and present a sensor-based environment monitoring system for indoor spaces. Temperature, relative humidity and light level are measured using a WSN. Data are collected and transmitted to a PC in order to be stored in a SQL database. Data is visualised in a Windows application. E-mail and phone alarms are sent in critical cases.

ASHRAE (American Society for Heating, Refrigerating and Air-Conditioning Engineers) standards for indoor air-quality are applied aiming to ensure the optimal climate in museums and other exhibition spaces. Many versions of ASHRAE standards [5] were developed over time.

In UK, the same concern for maintaining stable environmental conditions in museums exists, in particular for textile artefacts for whose preservation guides have been edited [6].

The smart system used to monitor the textile-material-based artefacts is described in section 2. Section 3 presents the method used to process gathered data and to identify the environmental risk. Experimental results including some measured value, processed data and environmental parameter evolution graphs are exposed in section 4. Finally, conclusions and references are presented.

2. Smart monitoring system of textile artefacts

The proposed smart monitoring system commands the environment control system for indoor exhibition or storing spaces containing textile artefacts like old pieces of clothing, carpets, tapestry etc.

The materials used for these items can be various such as hemp, linen, silk, satin, canvas, wool or jute [7].

The monitoring system is composed by a wireless sensor network [8], a communication system and a processing unit. The entire system can be characterized as an adaptive automatic control system.

All the exhibited objects are monitored by many multi-sensor nodes placed around each artefact, on the upper, middle and lower sides of it, depending on the shape of the object.

Each node contains temperature, relative humidity and light sensors. Arduino boards with DHT-11 sensors, a resistive divider composed from a Light dependent resistor (LDR) and a 10k Ohms resistor and Li-ion batteries are used to implement the sensor node prototype (Figure 1 and Figure 2).

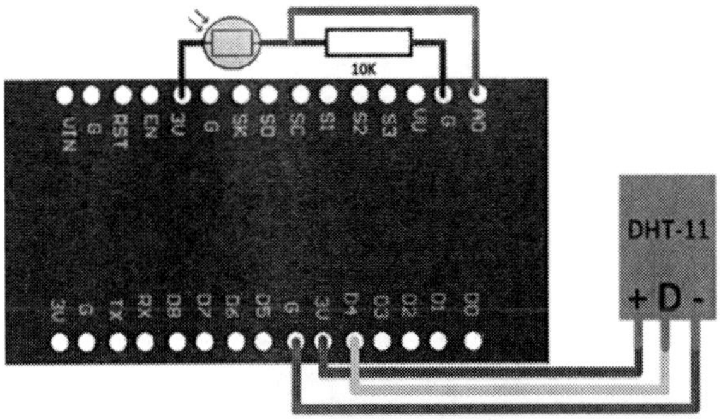

Figure 1. Sensor node.

Figure 2. Prototype image.

A set of data is collected from each sensor-node for each environmental parameter: temperature, relative humidity and light intensity.

The sensor nodes are grouped in clusters. It can be one-object cluster or a multiple-object cluster. The grouping of sensor nodes is made based on the distance between them. In each cluster, one sensor node works as a master node called *cluster head*. The cluster head collects data from all inside sensors monitoring one exhibited object and wirelessly sends it to the *access point* (AP) installed in the monitored room. All information get by the sensors are sent to the central communication node via the communication network of the museum in order to be stored and processed (Figure 3). In our prototype network, the central communication node and the processing unit are both located in the network server.

The table presented in Figure 4 includes some measured values from 6 sensors. These two nodes are placed in separated clusters.

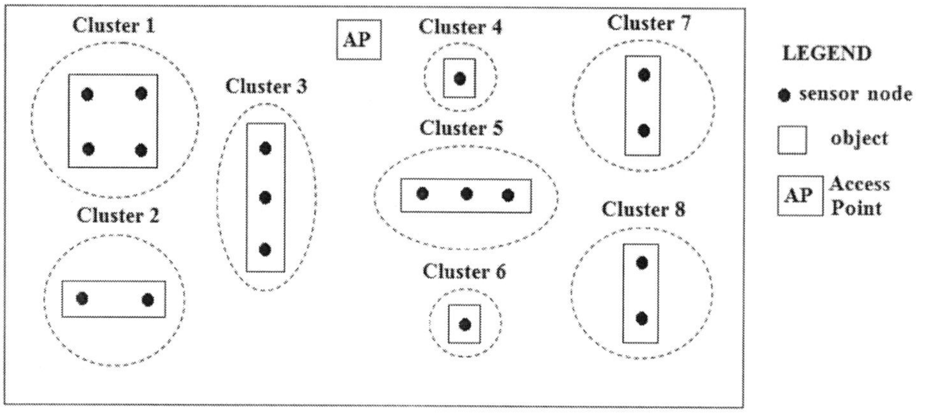

Figure 3. Two-dimensional schematic representation of a monitored space.

Sensors number 1, 2 and 3 compose the first sensor node, represented by cluster 4 (Figure 3) while sensors number 4, 5 and 6 compose the second sensor node, represented by cluster 6 (Figure 3).

Sensors placed at higher height in the room get higher temperature and lower relative humidity values than the sensors placed in the bottom of the room. Measured light intensity value depends on the sensor position towards the light sources, natural or artificial.

Information get from each cluster is processed independently in order to make a decision regarding the control unit (e.g. air-conditioning equipment, dehumidifying device, smart glass or window blinding system) that must be started to correct some environmental parameters.

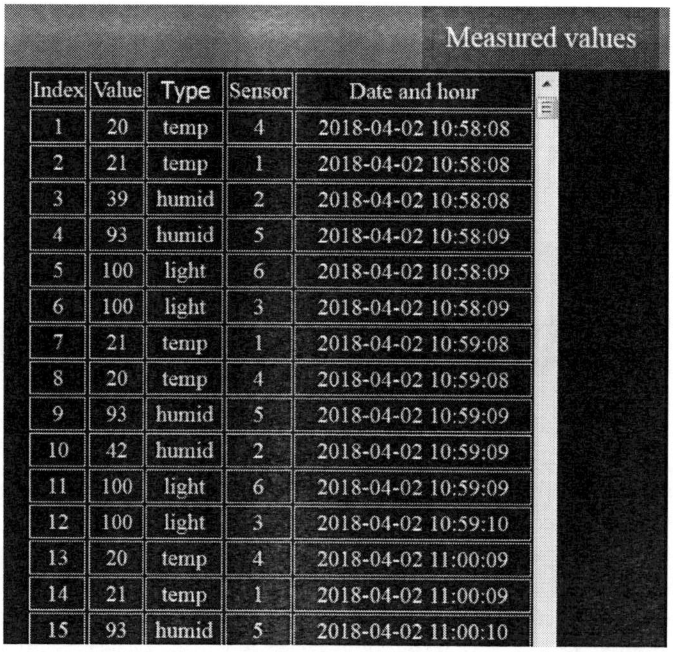

Measured values

Index	Value	Type	Sensor	Date and hour
1	20	temp	4	2018-04-02 10:58:08
2	21	temp	1	2018-04-02 10:58:08
3	39	humid	2	2018-04-02 10:58:08
4	93	humid	5	2018-04-02 10:58:09
5	100	light	6	2018-04-02 10:58:09
6	100	light	3	2018-04-02 10:58:09
7	21	temp	1	2018-04-02 10:59:08
8	20	temp	4	2018-04-02 10:59:08
9	93	humid	5	2018-04-02 10:59:09
10	42	humid	2	2018-04-02 10:59:09
11	100	light	6	2018-04-02 10:59:09
12	100	light	3	2018-04-02 10:59:10
13	20	temp	4	2018-04-02 11:00:09
14	21	temp	1	2018-04-02 11:00:09
15	93	humid	5	2018-04-02 11:00:10

Figure 4. Some measured values (print screen).

Data collected periodically from the entire sensor network is stored in our database [9] and processed in order to identify risks based on critical events that occur regarding the environmental conditions in the exhibition space for each point-of-interest (POI) consisting one exhibited object. This database manages monitored objects, sensors used by the monitoring system and values measured by it that are compared to specific thresholds, critical events, installed actuators and environmental control elements included in the control system. The database diagram is presented in Figure 5.

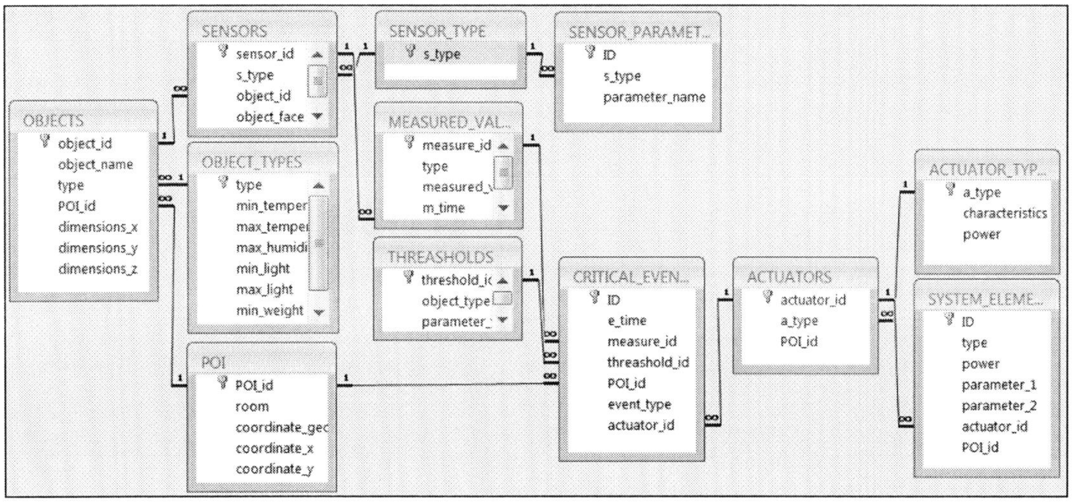

Figure 5. Database diagram.

Measured values can be read online, on the website designed by our team for the monitoring system, in order to be remotely accessed using a device connected to the Internet. The access to the web application is restricted for authorized users only. The supervisors have access to the information given by the monitoring system, anywhere and anytime, using any type of device with Internet access.

3. Environmental risk identification

The climate parameters must be monitored in order to be stable and to preserve the textile artefacts [11].

The heating regime in the exhibition space must be permanently monitored and controlled in order to keep temperatures between 10° and 20°C. In order to preserve textile artefacts, the exposure temperature must not drop below freezing. The textiles must not be placed in the vicinity of radiators and heaters too hot or dry. Air circulation arround textile artefacts is also important in order to ensure a stable environment.

Textile materials absorb and release moisture from and to the air. Fast variation of relative humidity over time can cause different materials to separate. For example, painted areas can lift from the textile support when humidity changes. The textile density can be changed by high humidity. Other textile damaging processes can be caused by humidity variation. High relative humidity over 65 % leads to mould outbreaks and pest infestations. Low relative humidity reduces moisture contents of textiles and the materials become brittle and dry. It is recommended to keep the relative humidity between 45% and 65%, for 90% of the time.

Light-damage on textile artefacts occurs progressively and it cannot be reversed. Therefore it is recommended to control strictly the light intensity that is correlated to the exposure time. For example, 1 hour - 5000 lux exposure is equivalent to 10 hours - 500 lux which is equivalent to 100 hours - 50 lux according to Museums Galleries Scotland.

Therefore the recommended light intensity for exhibited textile artefacts is 50 lux. In storing spaces, the light intensity can be lower but a completely dark environment is not recommended because it encourages different biological infestations by microorganisms. The environmental parameters are measured periodically by the sensors. All data collected is processed in the central node, differentiated by parameter type and POI.

This paper is not intended to do a statistic analysis of these parameters for a long time period but to identify the environmental risks in real-time. Therefore data are analysed on short time periods by processing a reduced set of values of each measured parameter.

The following variables can be calculated: the maximum and minimum values, amplitude, mean value, standard deviation, central value, variation coefficient, skewness and Kurtosis [10]. Extreme values of the environmental parameters are found by determining the *maximum* and the *minimum* value of it. These values are compared to specific thresholds in order to identify critical events.

These thresholds are established based on the recommended values of climate parameters given in the literature for textile materials.

The *amplitude* of a parameter is a measure of its change over a single period. It is calculated as the difference between the maximum value and the minimum one.

In order to stabilize the environmental parameters and to maintain an optimal climate in the exhibition space for textile artefacts, the *relative amplitude* is used and it is defined by us as the difference between the average value and the optimal or a reference value of a parameter. The reference values can be a minimum allowed value, a maximum accepted one, a standard value given in literature or an experimentally deduced value regarding environmental parameters.

Depending on the relative amplitude, a risk class is identified in order to launch an environmental control protocol.

The *risk classes* are defined depending on the parameter type:

- temperature (T): cooling or heating risks can be identified;
- relative humidity (RH): drying or humidifying risks can be detected;
- light intensity (L): risk of light intensity rising can occur.

The *risk level* is expressed as low or no risk (0), moderate risk (1) or high risk (2).

The risk level of a parameter is variable because after the corresponding control unit is started, the risk level decreases. The risk level can increase if no action is commanded.

Depending on the risk level, a *decision* is made regarding the control action needed to correct a certain climate parameter. Different cases can occur:

1. For no risk, no action is commanded.
2. For an identified risk for only one parameter at a time, the corresponding control unit is started. The power of the device is dictated by the risk level: medium power for level 1 and full power for level 2.
3. A multiple decision has to be made based on the risk hierarchy when multiple risks are identified simultaneously. *Composed risk classes* (T-RH; T-L; RH-L; T-RH-L) must be analyzed. Risk levels are added in order to get the *composed risk level*. The number of simultaneous not-null risks is defined as the *composed risk order*. The risk associated to only one climate parameter has the order equal to 1.

In Table 1, simultaneous risk levels of heating or cooling, drying or humidifying, and light intensity rising are considered.

The *energy criterion* must also be considered by the decision process because it is not recommended to start all the control units in the same time. Temperature control has a higher priority than the relative humidity control that is in the same time prioritized towards the light intensity control.

A composed risk level higher than or equal to 4 is considered critical regarding the power consumption and this is recommended to be reduced by the made decision.

For order-2 composed risks, when the composed risk level is equal to 2 (e.g. for a *risk level set* 1-1-0), both control units are started with a medium power, but if the risk level of one parameter is equal to 2 only its corresponding control unit is started on full power, while the other control unit for the parameter having a moderate risk is started with medium power. In the critical situation when two risk levels have maximum values in the same time, the control unit of highest priority parameter is started first on full power and the second one is started only with medium power.

Order-3 risk levels are critical so the decision is made in order to reduce the power consumption.

For a composed risk of one maximum risk level and two moderate risk levels (e.g. for a risk level set 2-1-1), only two control units are started, one on full power, the other one on medium power, depending on the risk level and the parameter priority. For example, when the heating risk level is high, the humidifying risk is moderate and the light associated risk is moderate, than the air-conditioning equipment is started on full power and the dehumidifying unit is put on medium power.

For a composed risk of two maximum risk levels and one moderate risk level (e.g. for a risk level set 2-2-1), only two control units corresponding to maximum risk are started, but their starting power is dictated by the parameter priorities.

For a composed risk of three maximum risk levels (with the risk level set 2-2-2), two control units are started firstly. The decision on which control units have to be turned on and with what power level to work depends only on the parameter priorities.

Table 1. Composed Risk Classes.

Heating/ Cooling Risk Level	Humidifying /Drying Risk Level	Light Intensity Rising Risk Level	Composed Risk Level	Composed Risk Order
1	1	0	2	2
2	1	0	3	2
2	2	0	4	2
1	2	0	3	2
1	0	1	2	2
2	0	1	3	2
2	0	2	4	2
1	0	2	3	2
0	1	1	2	2
0	2	1	3	2
0	2	2	4	2
0	1	2	3	2
1	1	1	3	3
2	1	1	4	3
1	2	1	4	3
1	1	2	4	3
2	2	1	5	3
2	1	2	5	3
1	2	2	5	3
2	2	2	6	3

4. Experimental results and discussions

All sensors measure the environmental parameters every minute. A lot of data has to be processed so it is necessary to do a centralized data processing in the central node. Based on the analysis results, a multiple decision is made. The decision can regard the temperature, the relative humidity or the light intensity or a combination of them.

Gathered data is analysed on short time periods by processing the last 5 values of each measured climate parameter. The results are used to command the control units and to estimate the next

IOP Conf. Series: Materials Science and Engineering **374** (2018) 012074 doi:10.1088/1757-899X/374/1/012074

environment behaviour.The variation of the environmental parameters for one sensor-node is presented below for a 12 minutes interval (Figure 6, Figure 7 and Figure 8). No action was commanded to control the environmental parameters.

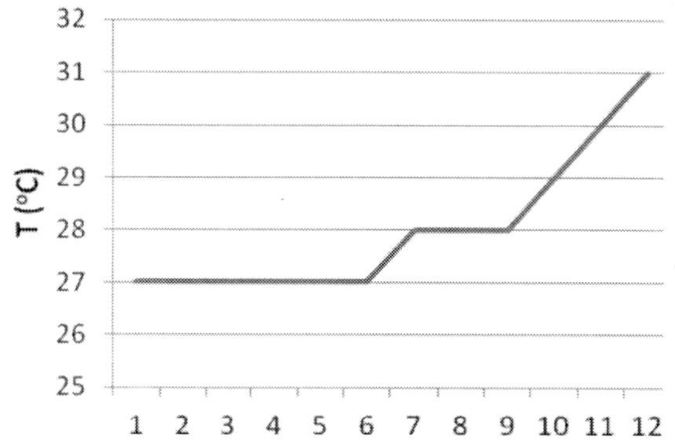

Figure 6. Temperature Variation over 12 minutes.

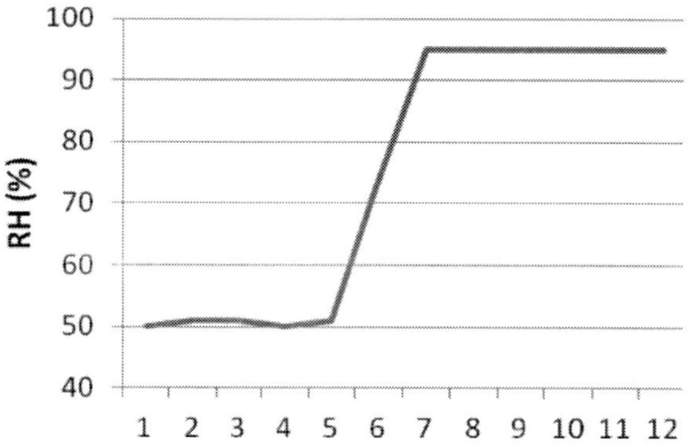

Figure 7. Relative Humidity Variation over 12 minutes.

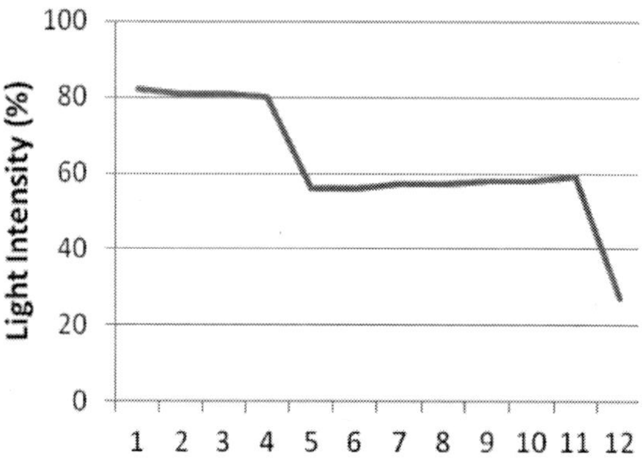

Figure 8. Light Intensity Variation over 12 minutes.

The measured values for the considered time interval are given in Table 2. Light Intensity is expressed as percentage of the maximum value.

Table 2. Parameters Measured Values.

Index	Temperature (°C)	Relative Humidity (%)	Light Intensity (%)
1	27	50	82
2	27	51	81
3	27	51	81
4	27	50	80
5	27	51	56
6	27	74	56
7	28	95	57
8	28	95	57
9	28	95	58
10	29	95	58
11	30	95	59
12	31	95	27

The measured values are analyzed on sets of five consecutive values.

In Table 3, we consider the temperature risk levels to be 1 if the relative amplitude is less than two Celsius degrees and it is equal to 2 if it has a higher value. In this case, the *heating* risk class occurs.

Then in Table 4, humidifying risk is identified. The relative humidity risk level is equal to 1 if the relative amplitude is under 10 % and it is considered to be equal to 2 for higher values.

Finally, in Table 5, the light intensity rising risk is analyzed. The risk level value is considered to be equal to 1 for relative amplitudes lower than 10 % and 2 for more than 10 %.

Table 3. Temperature Processed Values (reference 26 °C) (Heating Risk).

Index	T_{MAX} (°C)	T_{MIN} (°C)	$T_{AVERAGE}$ (°C)	Relative Amplitude (°C)	Risk Level
5	27	27	27	1	1
6	27	27	27	1	1
7	28	27	27.2	1.2	1
8	28	27	27.4	1.4	1
9	28	27	27.6	1.6	1
10	29	27	28	2	2
11	30	28	28.6	2.6	2
12	31	28	29.2	3.2	2

From this analysis, it result a third order combined risk because no control is done over the environment during the considered time interval.

Decision about how to control the environment must be made according to Section 3. Analytic Hierarchy Process (AHP) can be used to improve the multiple-decision process when combined environmental risks are identified [12].

Table 4. Relative Humidity Processed Values (reference 50 %) (Humidifying Risk).

Index	RH_{MAX} (%)	RH_{MIN} (%)	$RH_{AVERAGE}$ (%)	Relative Amplitude (%)	Risk Level
5	51	50	50.6	0.6	1
6	74	50	55.4	5.4	1
7	95	50	64.2	14.2	2
8	95	50	73	23	2
9	95	51	82	32	2
10	95	74	90.8	40.8	2
11	95	95	95	45	2
12	95	95	95	45	2

Table 5. Light Intensity Processed Values (reference 50 %) (Light Intensity Rising Risk).

Index	LI_{MAX} (%)	LI_{MIN} (%)	$LI_{AVERAGE}$ (%)	Relative Amplitude (%)	Risk Level
5	82	56	76	26	2
6	81	56	70.8	20.8	2
7	81	56	66	16	2
8	80	56	61.2	11.2	2
9	58	56	56.8	6.8	1
10	58	56	57.2	7.2	1
11	59	57	57.8	7.8	1
12	59	27	51.8	1.8	1

The preservation of textile artefacts or the damage of them depends on this decision process and on the environment control system efficiency.

5. Conclusions

Environment monitoring of the indoor spaces used to exhibit or store textile artefacts is essential for their preservation. The proposed monitoring system is able to measure environmental temperature, relative humidity and light intensity in different points using multiple-sensor nodes connected into a wireless sensor network. The measured values of the parameters are stored in a dedicated database that manages the sensors from the wireless sensor network, too. These values are processed in the central node and compared to recommended reference values for textile artefacts preservation. Risk classes can be identified and the risk level can be established. According to it, control procedures have to be started in order to ensure the proper environment for the exhibited textile artefacts. Experiments were made with the implemented monitoring system and some measured and processed data are presented as an example of how the environmental risk can be identified and classified.

6. References

[1] Abdel-Kareem O 2010 *Mediterranean Archaeology and Archaeometry* **10**(2) 85-96

[2] Rodríguez Peralta L M, Brito L M P L, Gouveia B A T, Sousa D J G and Alves C S 2010 *Electronic J of Structural Engineering (EJSE) Special Issue: Wireless Sensor Networks and Practical Applications* 12-34

[3] Martens M H J 2012 Climate Risk Assessment in Museums : Degradation Risks Determined from Temperature and Relative Humidity Data (Eindhoven: Technische Universiteit Eindhoven DOI: 10.6100/IR729797) pp 48-51

[4] Simić M S and Petrişor D 2015 *Tehnika – Elektrotehnika* **64**(4) 663

[5] AHSHRAE 2016 ANSI/ASHRAE Standard 62.1-2016 Ventilation for Acceptable Indoor Air Quality (Atlanta, GA: ASHRAE)

[6] Museums & Galleries Commission 1998 Standards in the Museum Care of Costume and Textile Collections (London, UK: Spin Offset Limited) pp 35 – 37

[7] Lennard F and Ewer P 2010 *Textile Conservation: Advances in Practice*, ed F Lennard and P Ewer (Oxford, UK: Elsevier Ltd) chapter 5 pp 197–226

[8] Tilak S, Abu-Ghazaleh N and Heinzelman W 2002 *Computing and Communications Review* **6** 28-36

[9] Diaconescu V D, Scripcariu L, Diaconescu M R and Vornicu-Albu L 2018 *Eur. J. Sci. Theol.* **14**(1) 181-192

[10] Boiculese L V, Dimitriu G, Doloca A, Moscalu M and Dascălu C 2012 *Metode descriptiveşi elemente de analiză statistică a datelor medicale - Exemple practice în Excel şi Access* (Iaşi: Performantica) pp 100 - 154

[11] Museums Galleries Scotland 2009 *Caring for textile collections in museums* ed I P Wynd and M Road (Edinburgh: Museums Galleries Scotland)

[12] Scripcariu L, Mătăsaru P D and Diaconescu V D *Scientific Bulletin Series C- Electrical Engineering and Computer Science of University Politehnica of Bucharest* **79**(3) 83-92

IOP Conf. Series: Materials Science and Engineering **374** (2018) 012075 doi:10.1088/1757-899X/374/1/012075

Evaluation of the Environmental Impact of the Activities Carried Out in the Area of Murighiol Channel Tulcea County

A M Arcuş, V Başliu*, L G Tiron and M V Drăgan
„Dunarea de Jos" University of Galati, 47 Domnească street, 800008, Galați, Romania

Email: vbasliu@ugal.ro

Abstract. The Murighiol Channel area is a part of the Danube Delta Biosphere Reserve, a protected area in which economic activities of the type of tourism can have a negative significant impact if they are not carefully monitored. Thus, knowing the level of pollution in the area, and especially near the boarding house Blue Lagoon, helps us to pay a special attention to the fauna and flora by protecting the environment and giving nature enthusiasts a special area. For this, a series of measures have been taken in the preparation of samples taken from important places near the boarding house. Specific analysis for water, soil and noise were performed.

1. Introduction

In Europe, the first longest river is Volga and after beyond is Danube with 2860 km [1]. Danube river flows through 10 countries from where it collects discharges from industrial activities and urban activities and effluents [2]. The Danube, reached in Tulcea County, north of the Dobrogea area, dividing in two branches: Chilia (to the north) and Tulcea (to the south), which later separates into Sulina branche and Sfântu Gheorghe branche [3, 4]. On the world are several natural reservations but UNESCO classified Danube Delta on the third place with protected area due to his ecological importance. In this area are found over than 250 species of fauna and 300 of rare birds [5]. The Danube Delta is the largest humid ecological area from Europe with 5640 km^2. This zone are very vulnerable to any change of the environment, human activities and especially to natural disasters (floods) [6]. In the area of Murighiol channel from Danube Delta, it is the most important and well protected area of biodiversity from Romania and one of the most important from Europe and the area of interest for this study. In this part of the Danube Delta no research has been done and that's way this research offer the opportunity to find out the current status of these places. Murighiol channel is a very often circulated channel by agreement boats, and also it is located in an economic area of Danube Delta and in this way it is done the connections between the dock from Murighiol village and the arm Sf. Gheorghe. On a distance of 150 – 200 m are placed two touristic locations, Pension „Laguna Albastră" and Puflene Resort, placed in an strategic way because of the facilities which are offered by the Murighiol channel for promoting touristic offers. Throughout time the number of tourists which visit the reservation or which are spending holidays in Danube Delta each year, is on an ascending curve, and this rising lead to supplementing the number of authorized tourist routes (2002) to meet their requirements, influencing in an indirect way the biodiversity [7].

The positive impact of tourism on environment occurs when the tourism and environment coexist in harmony and the environment benefits from tourism. Negative impact of tourism against environment is determined by the inappropriate use of the environment, in recreational way and in agreement and the big number of visitors arrived at destination is exceeded and environment quality, the diversity of flora and fauna are unavoidable affected [8].

Water and soil are most affected by the pollutants, heavy metals are considered ones of the most dangerous category of pollutants, because of the persistency and her tendency of being retained by the microorganisms from terrestrial and aquatic environment. The main problem caused by heavy metals in environment, is the tendency of toxicity to be accumulated in different types of tissues (plants, animals or human) even at low concentrations [9].

2. Materials and methods

For accomplishing this research we have used a series of methods in which we prelevated samples in places located in accord with Figure 1, for analysis of water, soils and also we have donned measurements of sound pressure.

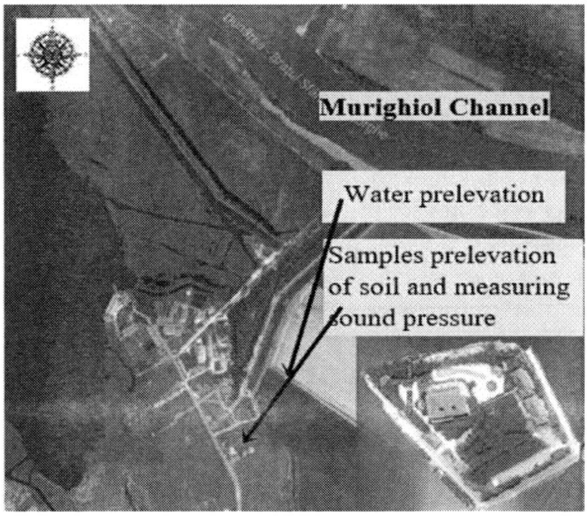

Figure 1. Locations of prelevation of samples.

The analysis have been done by direct methods according to which the results are determined in a direct way by measured values with specific equipment.

2.1. Materials and method of prelevation for water

The samples of water have been prelevated from Murighiol channel and the analysis have been done in the laboratory of environment monitoring from the Department of Environment Engineering and Industry Safety using an multiparameter Hach with accessories, spectrophotometer DR5000 plus rapid tests LCK. Water samples were prelevated in accordance with standard procedure which implies that before to prelevate the sample the recipient was washed with distilled water and rinsed with water from prelevation area and analysis were made in the shorts time in the laboratory.

On establishing of the most important indicators (nutrients, salinity and specific pollutants) was started from the identification of the water quality status, polluting and dangerous substances for ecosystem and which of the determinate values are approaching the maximum admissible values. The resulted values obtained have been compared with those from the annexes of Environment Minister Order and Water Management no. 161 from 16 February 2006 [11]. This normative classifies the quality of surface waters in order to determine the ecological status of water bodies.

2.2. Materials and method of prelevation for soil

Soils samples have been prelevated from the perimeter of "Laguna Albastră" pension in five representative points and the necessity of soils evaluation was donned to know the actual status of those from the point of view of humidity, natural radioactivity and chemical composition of presented soils and toxic elements (Cd, Pb, Zn, Cr, etc.) with significant negative impact against environment and local biodiversity. On the prelevation of soils samples was followed the method of five points or envelope method. Prelevation of samples was made in recipients from polyethylene, on a depth of 5 – 10 cm from the surface of soil. The probing device was cleaned after each prelevation in order to prevent the accidental contact of substances contained in samples which may denaturated the results of analysis. Chemical analysis have been made using the equipment of laboratory of environmental monitoring and the results of analysis are compared with values from Order no. 756/1997 Regulation regarding environmental pollution for types of sensible use of soils [12].

2.3. Method of measuring of noise

Measurements of sound pressure was made in the perimeter of pension "Laguna Albastră" having into consideration the evaluation of pressure levels and health risks or discomfort for population by reducing the noise. Measurements was done in twelve points in most agglomerated areas from pension perimeter with the help of an equipment PCE 222. The results of measurements are compared with levels of sound pressure specified in the Order no. 536/1997 regarding hygiene rules and recommendation regarding life environment of population [13].

The monitoring of sound pressure was done according HG no. 321/2005 [14] from 14 april 2005 regarding the evaluation and managing ambient noise between the hour 7:00 – 19:00 a.m., 19:00 – 23:00 p.m. and 23:00 – 7:00 p.m., in 15 and 16 july 2017 (when the flux of tourists have been high) with the purpose to evaluate the impact of noise from "Laguna Albastră" location and surrounding. Twelve points of measurements are in the near vicinity of pension and on and radius of approximately 1000 square meters. Measurements have shown values between 17.4 – 62.3 dB and possible sources of sound pressure are: naturals (birds, insects) and anthropic (cars, people, boats, etc.).

3. Results and discussion

3.1. Results for water

Parameters for the nutrients group (Figure 2) are ammonia, nitrites and nitrates and measurements shows that this are in the range of quality classes, with the exception of nitrites. Nitrites concentration above maximum level accepted 0.3 mg NO_2/l determine the appearance of eutrophication process or "flowering" respectively of depletion of oxygen content from water by death and decomposition of zooplancton.

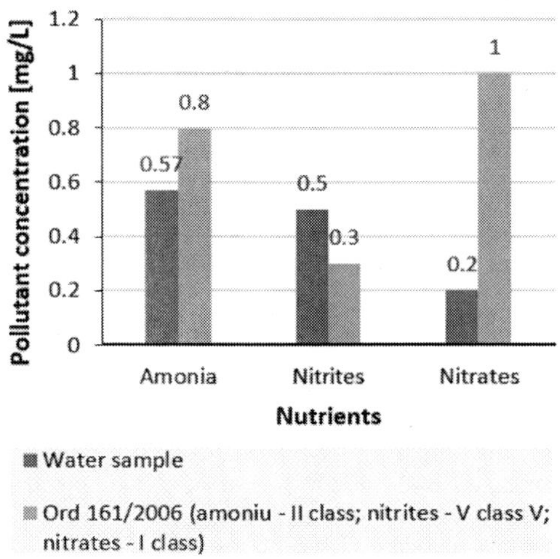

Figure 2. Nutrient analysis.

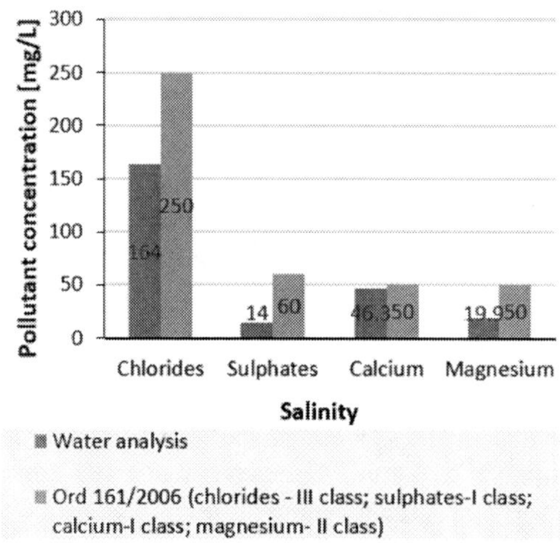

Figure 3. Salinity analysis.

Water salinity presented in Figure 3 is determined by measuring the parameters for chlorides, sulphates, calcium and magnesium. According to Order no. 161/2006 [15] parameters for water are classified in the following quality classes: chlorides – third class, sulphate – first class, calcium – first class, magnesium – second class and the results obtained have been compared with those and we may conclude that those have an very low impact against the aquatic life.

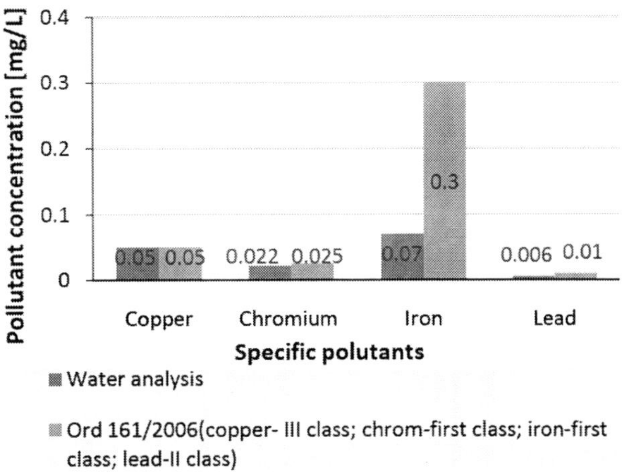

Figure 4. Specific pollutant analysis.

From the category for specific toxic of natural origin the parameters analyzed are copper, chromium, iron and lead. The results of analysis have determined framing of those parameters according to Order 161 from 16.02.2006 [15] in the following quality classes: copper – forth classes, hexavalent chromium – first class, iron – first class and lead – second class.

3.2. Results for soil

In order to obtain results for soils we have done a few preparatory steps of samples with the purpose of reducing measurements errors and this are presented in Figure 5. The measurements have been made after drying samples in dry oven in time of 48 hours on temperature of 50°C, removing organic and inorganic waste and shredding them.

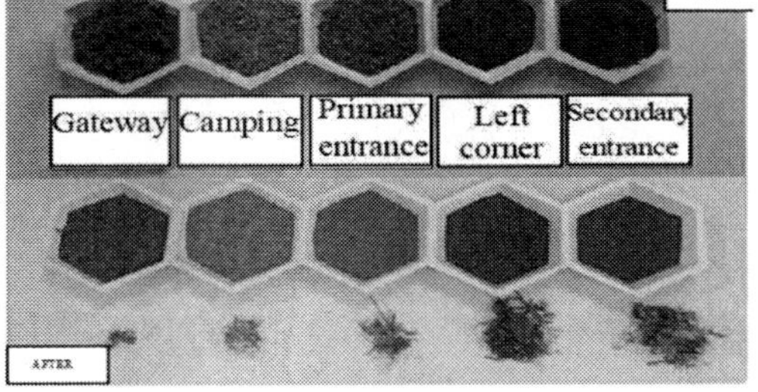

Figure 5. Preparing samples for analysis.

In Figure 6 it is graphically represented the variation of soil samples and values are between 6.15 – 38.8 % and minimum values was registered in point four from left corner (6.15 %) because this area of pension is very little frequented, the only sources of water for humidity vegetation is water from precipitations and groundwater. The maximum values was registered in point three "Poarta mare" (2) because in this area the turf is watered frequently. Percentage differences of humidity are due to the existence of turf and trees, which are watered on very large surfaces from pension perimeter.

In Figure 7 is graphically represented the distribution of values of gamma dose registered in five samples of soil 0.06 – 0.07 μSv/h and these low values confirm the presence in very small quantities of radioactive elements. The soil does not represent any danger for the environment from the point of view of background radiation.

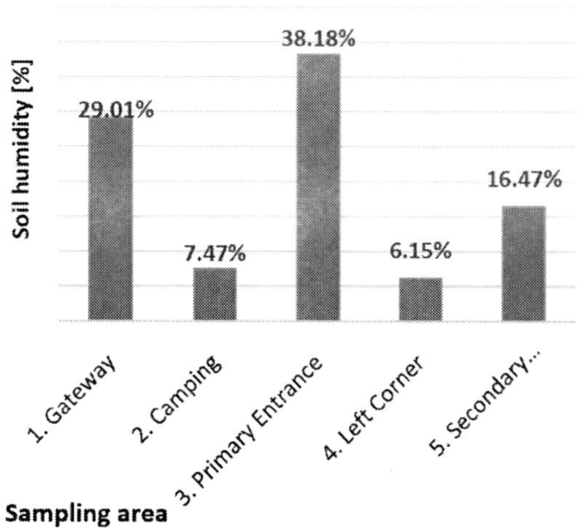

Figure 6. Humidity of soil samples.

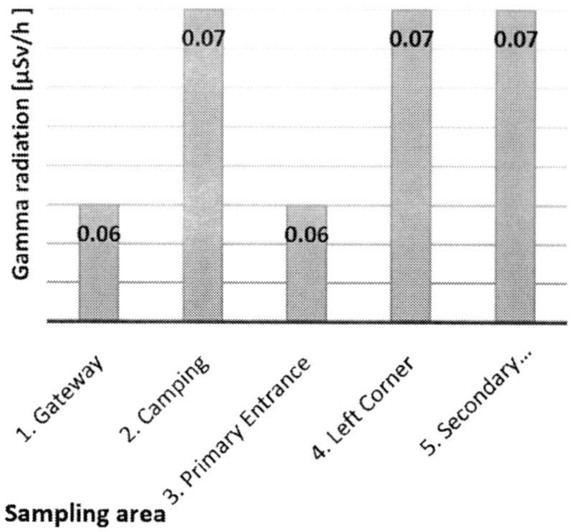

Figure 7. Measured gamma radiation.

According to Order no. 756/1997 Regulation regarding the evaluation of environment pollution [12], the types of soils according to the way of use are soils with *sensitive use* and *less sensitive use*. Knowing that the pension from where was prelevated samples of soils is located in economical area of reservation Danube Delta, the terrain is included in category of sensitive use, so the results of analysis are reported to values of type of soil with sensitive use (reference values: normal values – NV, alert threshold – AT, the intervention threshold – IT). In Figure 8 we show the manganese concentration variation present in five prelevation points, seeing values contained between 257- 378 ppm and these registered concentration are under normal level of 900 ppm for dry soil, and indicate the presence of element manganese in normal concentration and show a low impact against environment.

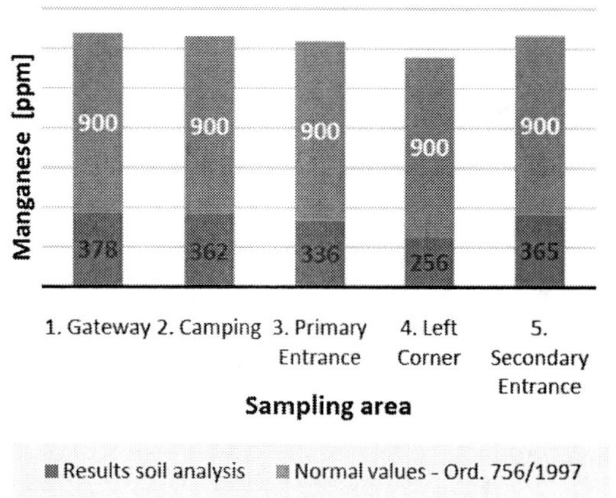

Figure 8. Variation of manganese.

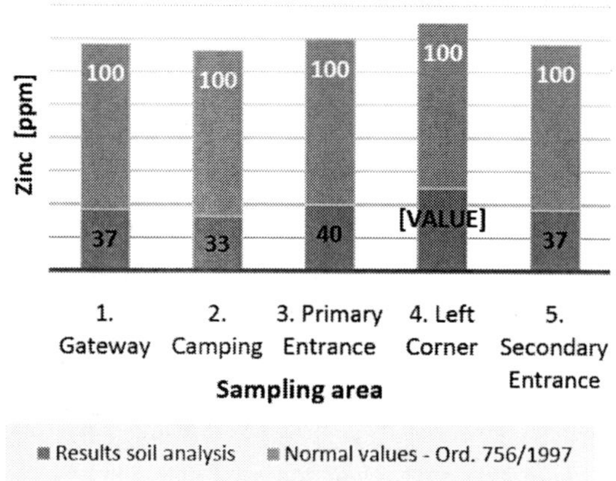

Figure 9. Variation of zinc concentration.

From graph analysis of Figure 9 we can conclude that the values of zinc do not exceed normal value threshold NV. The values are contained between 33 – 50 ppm and zinc is a microelement essential to grow and normal development of plants and the impact of the deficiency of this element on growth, development and productivity of plants is very large, having a negative impact against on the health of living creatures.

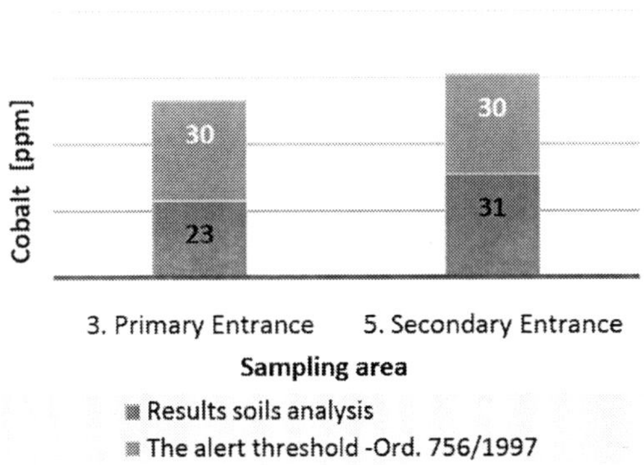

Figure 10. Variation of cobalt concentration.

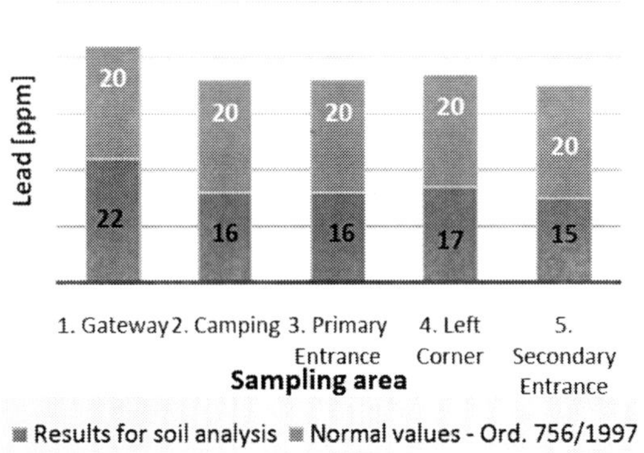

Figure 11. Variation of lead concentration.

In soils, cobalt is find under different forms, in the crystalline network of some insoluble minerals, in hydrated oxides of iron and manganese, under the form of slightly soluble chelates especially in acidic and neutral soils, adsorbed on the surfaces of mineral organic colloids and soluble in soil solution [10]. In the five samples of analyzed soils, cobalt is find only in two samples respectively "Poarta mare" (2) – 23 ppm and "Poarta mică" – 31 ppm. In other three samples of soil the lack of cobalt may have an unwanted effect for human health such as the preventing the formation of red blood cells and not transmitting nerve impulses. From the graph presented in Figure 11 was observed that in four from five samples result values 15 – 17 ppm under normal values, showing an maximum value of 22 ppm on "Poarta mare" (1), and minimum value of 15 ppm on "Poarta mică". Emphasizing that lead it is harmful to human health and this shows some significant negative impact on them.

3.3. Results for sound pressure

High sound pressure (noise pollution) generated by anthropogenic sources in a protected area implies potential significant environmental risks for the site. Thus, measurements were made at different times of the day around the pension in order to determine the level and to take measures to mitigate it if necessary. The average of the values measured over the two days was calculated by the relation (1) and the values are in the graph in Figure 12.

$$\mathbf{L}\ \mathrm{dB} + \mathbf{L}\ \mathrm{dB} = 10^{(\mathbf{L}\ \mathrm{dB}/10)} + 10^{(\mathbf{L}\ \mathrm{dB}/10)} = 10 * \log\left(10^{(\mathbf{L}\ \mathrm{dB}/10)} + 10^{(\mathbf{L}\ \mathrm{dB}/10)}\right) \quad (1)$$

In the graph from Figure 12 we represent the average values of sound pressure level from date 15 july and 16 july. The highest values were registered in the night interval in nearly the twelve measuring points, and in point 8 even was exceeded the maxim admittance level of 50 dB, in this point are multiple noise sources such as people, music, etc.

Values of equivalent permanent sound level $L_{Aeq,24h}$ is represented in Figure 13.

$$LAeq = 10 * \log 1/24\left(12 * 10^{(Lday/10)} + 4 * 10^{(Levening/10)} + 8 * 10^{(Lnight/10)}\right) \quad (2)$$

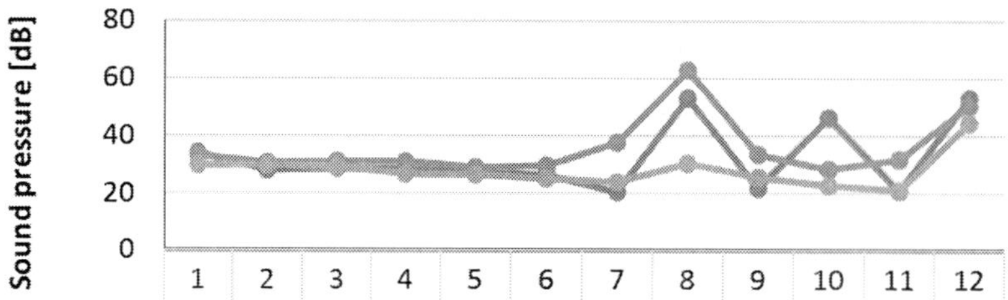

	1	2	3	4	5	6	7	8	9	10	11	12
07:00L(dB)	34.1	28.0	28.6	28.4	27.3	26.3	20.4	53.0	21.7	46.2	20.7	53.0
19:00L(dB)	32.9	30.6	31.1	31.0	29.0	29.6	37.6	62.7	33.5	28.3	31.7	50.5
23:00L(dB)	29.7	29.8	29.3	26.7	26.4	25.1	23.8	30.4	25.6	22.6	21.0	44.4

Figure 12. Average values from the date 15 and 16 july.

1. alley garden
2. camping site
3. gateway
4. C1-C6
5. C7-C12
6. adult pool
7. children pool
8. outside bar
9. barbeque
10. secondary entrance
11. primary entrance
12. terrace

Figure 13. Equivalent permanent sound level $L_{Aeq,24h}$.

In the case of strong sound pressure variations, the "noisy" situation is characterized by the permanently equivalent sound indicator. By this indicator is understanding a calculable indicator of sound pressure, which corresponds to average value measured (calculated) in the time of sound energy action.

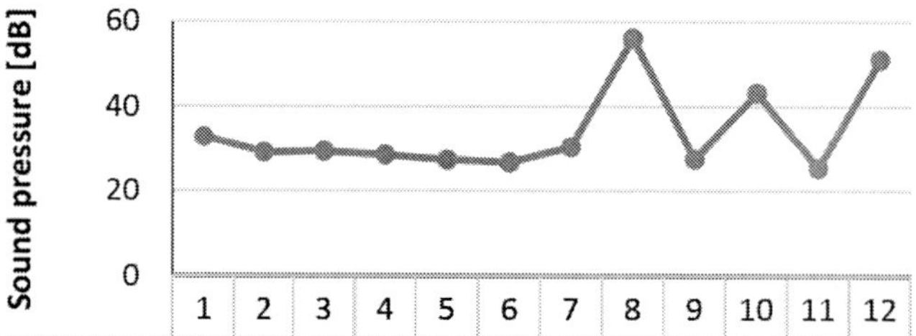

	1	2	3	4	5	6	7	8	9	10	11	12
LAeq,24h	32.8	29.1	29.4	28.5	27.4	26.7	30.4	56.1	27.5	43.2	25.4	51.1

Figure 14. Variation of equivalent permanent noise level $L_{Aeq,24}$.

The evaluation parameter (indicator) is equivalent noise level for 24 hour ↔ $L_{Aeq,24h}$ is presented in Figure 14 and we can conclude that the domain of values are between 25 – 56 db acceptable values for the area.

4. Conclusions

Environmental monitoring of the Danube Delta Murighiol aims to assess the impact of human activity on the conduct of environmental factors in the economic area of the reserve.

The main objective of declaring the Danube Delta as a biosphere reserve was and is the preservation of the diversity of species and habitats it hosts, the impact of people's activity on the reserve is low without negative repercussions. Thus, from the point of view of ecological status, the value of the elements pursued falls within the boundary between the good state and the moderate status of the biological quality of surface water.

Considering that the soil is the place where all the pollutants meet, airborne dust, toxic gases transformed by rain into the atmosphere, is most exposed to the negative effects of these substances, the soil samples (humidity, natural radioactivity, chemical composition) in the perimeter "Laguna Albastră" pension have not registered deviations from the level allowed in the legislation.

The relative soil moisture recorded large differences between the minimum value in the left corner peripheral area (6.15%) and the maximum value in the "Poarta mare" area (2) (38.18%).

The reason for these differences is justified by the maintaining and maintenance of landscape architectural objectives within the perimeter of the pension.

The results of the monitoring of the natural radioactivity of the soil samples confirmed in very small quantities the radioactive elements, the soil presenting no danger in terms of the radioactive background.

The elements present in their composition (Ti, Mn, Fe, Co, Zn, Pb, Rb, Sr, Zr) have been identified by analyzing the chemical composition of soil samples, among which there are heavy metals. The only element that has reached the alert threshold is cobalt. The threshold for intervention was not reached by any element found in the soil samples.

The activity carried out within the objective of the "Laguna Albastră" pension is not a source of noise pollution, the noise level being generated according to the measurements made within the limits set by the legislation, recommending the pension as a place of recreation and rest. In view of the overall reduced impact of the activities carried out on the site under consideration on the noise level of the area, no measures will be required to reduce the noise level at source.

The results obtained from the analyzes carried out in the Murighiol area of the Danube Delta confirm the status of the Danube Delta protected area, the anthropogenic activities have limited impact on the biodiversity of the reserve.

5. References

[1] Woitke P, Wellmitz J, Helm D, Kube P, Lepom P, Litheraty P 2003 *Chemosphere* **51** 633–642
[2] Covaci A, Gheorghe A, Hulea O, Schepens P 2006 *Environmental Pollution* **140** 136-149
[3] Winkels H J, Kroonenberg S B, Lychagin M Y, Marin G, Rusakov G V, Kasimov N S 1998 *Applied Geochemistry* **13**(5) 581-591
[4] Gastescu P 1993 *GeoJournal* **29**(1) 57-67
[5] *** https://whc.unesco.org/en/list/588
[6] Armas I, Avram E 2009 *Perception of flood risk in Danube Delta*, Romania, Springer Science Business Media B.V
[7] Master plan RBDD 2005
[8] Nedelea A 2003 *Piața turistică*, București, Editura Didactică și Pedagogică
[9] Burada A, Topa C M, Georgescu L P, Teodorof L, Nastase C, Seceleanu-Odor D, Iticescu C 2015 *Heavy Metals Environment Accumulation in Somova – Parcheș Aquatic Complex from the Danube Delta Area*, Bucharest

[10] Băjenescu I, Chiriac A 1984 *Distribuția microelementelor în solurile din România* Ed. Ceres, pp 87-98
[11] Annexes of Environment Minister Order and Water Management no. 161 from 16 february 2006
[12] Order 756/1997 Regulation regarding environmental pollution for types of sensible use of soils
[13] Order no. 536/1997 regarding Hygiene rules and recommendation regarding life environment of population
[14] Decision no. 321/2005 on the assessment and management of environmental noise
[15] Order no. 161/2006 Classification of surface water for settlement of ecological status

IOP Conf. Series: Materials Science and Engineering **374** (2018) 012076 doi:10.1088/1757-899X/374/1/012076

Metallographic Study of XIX Century Oklads Belonging to Russian Icons

L R Bivol[1], G V Ghica[2], M Buzatu[1], M I Petrescu[1], G Iacob[1], E Vasile[1], D Gheorghe[1] and T A Kovács[2]

[1] Faculty of Materials Science and Engineering, University POLITEHNICA of Bucharest, 313 Splaiul Independentei Blvd, 060042, Bucharest, Romania

[2] Faculty of Mechanical and Safety Engineering, OBUDA University, 1081 Budapest, Hungary

E-mail: gvghica@yahoo.com

Abstract. Many icons from the XIX and XX centuries can be found on the territory of Romania, with oklads made of Cu, Ag or Cu alloys. This paper analyzes by means of optical and electronic microscopy the samples taken from two oklads belonging to Russian icons of the XIX century: one dated and the other undated but attributed to the XIX century by Florin Colonas - an expert in handicraft of artistic significance. Based on the pins on the surface of one of the oklad and the scientific literature, we tried to identify the manufacturer, the verification mark, with the date of manufacture, the Ag title mark, the mark with the city icon in which the oklad was made, and identifying the technique of realization. The chemical compositions of the oklads were determined by EDX and found to be made of Ag alloys of different chemical composition: the dated oklad is made of Ag-Cu alloy, while the undated one is made of Ag-Cu-Zn alloy.

1. Introduction

In the whole of Romanian art a special place is owned by objects in precious metals mostly hammered in silver and gilded and the Romanian collections include numerous works by both local artists and foreign artists from silverware centers in Russia or other European countries [1].

In the culture of the Thracians at the end of the Bronze Age and the beginning of Hallstatt, during which silver objects replaced the gold ones, the development of silverware and local style with its own features was influenced by Greek art in the cities near the Black Sea [1].

Until the seventh century, in Dobrogea and in the Danube area, both under the domination of Eastern Roman Empire, the gilded silver, gold or bronze fine art objects, made in Byzantine style, were very popular, and the churches in these areas were decorated with Byzantine silverware imported from Constantinople [1].

In the Danubian settlements of Capidava, Garvan, Păcuiul lui Soare, silver and gold adornments, amulets, reliquary crosses dating back to the X-XII centuries, some of local origin, and others brought from the Kievan, Byzantine or Oriental workshops [2].

In the XII-XIV centuries, the centers in the Danubian cities, directly related to Byzantine civilization and art, through political, economic and administrative relations with Byzantium, played a special role in the distribution and production of precious metal objects such as the engolpion crosses from Ibaneşti and Trifeşti, the thesaurus from Oţelani, Cotnari and Voinesti. Under the forms of

IOP Conf. Series: Materials Science and Engineering **374** (2018) 012076 doi:10.1088/1757-899X/374/1/012076

technique and ornamentation, these objects are related to the Kievan art, which was also of Byzantine origin. In terms of art forms and ornamentation, these objects are related to the Kievan art, which was also of Byzantine origin [1].

The first documentary mentions and the oldest preserved objects that confirm the work of the Romanian silver and gold craftsmen date back to the XIV century. In the XV century the goldsmiths are organized in guilds, with statute-bound rules, and since the 16th century an obligation has been implemented for them to stamp the recognition signs of the city and craftsman on the objects. Silver had to be 14-15 carats and gold 18-20 carats [2].

At the end of the XV century - the beginning of the XVI century, the processing of precious metals was widely spread to monasteries. The richness of silverware in the inventories of Moldovan churches and monasteries confirms, based on inscriptions on objects, the existence of local workshops of a real technical value [3].

The silverware of Russian origin was frequently used in the XVIII century, especially in Moldova [1].Works of Russian silverware, oklads and objects of worship from the XVIII and XIX centuries can be found in many monasteries in Moldova, especially in Neamţ, Secu, Putna [4]. The Russian icons, covered with silver and gilded oklads (garnets) that have spread in large numbers in Romania in the XVIII-XIX centuries, were made in the usual technique of hammering, chiseling and engraving, as well as in the technique of painted enamel, watermark or cloisonné [5].

It is noteworthy that in the XIX century, the Russian silverware retained superior quality of execution, although the traditional manufacturing techniques were replaced with mechanical techniques (lamination / punching) for the processing of precious metals [5, 6].

2. Experimental procedures

2.1. Materials

Two oklads of the XIX century which belong to Russian painted icons were analyzed. Both icons in Figures 1 and 2 present the same subject (Virgin Mary with Child Jesus).

Figure 1. Oklad of an icon with markings, Russia, 1867, 26.50 x 21.50 cm.

Figure 2. Undated oklad, Russia, 19.50 x 14.00 cm.

The oklad dating back to 1876 is complete, but the undated oklad is incomplete (missing the Virgin Mary's aureole).

Applying the silver window on icons is done for two reasons:

- as a token of appreciation for the character painted in the icon;

- the clothing protects the painted surface of the icon from damage caused by awe, smoke and wax.

The two oklads were made by mechanical lamination and stamping, techniques that developed in the XIX century [7]. Stamping is a technique that provides the possibility of obtaining a repousse work by means of pressing between two forms - a matrix and a punch. Stamped products have smooth, rounded lines, with no changes in thickness of the metal plate used and no sudden crossings of height [8].

The oklad of the icon dated in 1876 in the control mark presents two series of four similar punches.

The first set of markings is in the aureole area (Figure 3) and has the following order:

- the marking with the master monogram - *И.С.К.* (I.S.K.), for the identification of the craftsman was used the catalog of markings from the book written in 1995 (catalog position 945, page 174) in Russian [M.M. Post]. In the aforementioned book, the *ИСК* (ISK) monogram appears without any points between the initials and belongs to the craftsman Ivan Semenov Komarov (ИванСеменовКомаров) who worked in 1808 in the Russian village Podol'skoye (Подольское) [7]. The difference between the monogram on the oklad shown in this article and the one used by the above mentioned craftsman consists in the existence of the points between the initials of the name [7].

-the monogram of the controller in the control mark, which also includes the date of the verification - AC / ... (A.C./1867), for the identification of the controller was used the catalog of markings from the book written in 1995 (catalog position 2113, page 205) in Russian [7]. In the aforementioned book, the AC / ... monogram belongs to A. Svechin (A. Свечин), who worked in Moscow from 1862 to 1875 [7].

-title mark silver – 84, for the identification of the title mark was used the catalog of markings from the book written in 1995 (catalog positions 12 and 13, page 184) in Russian [7]. The title of silver 84, although it has been met since the middle of the XVIII century, became obligatory starting with May 1, 1798 [7].

- the marking with the coat of arms of Moscow - *Saint George on horseback slaying the Dragon.* For the identification of marking was used the catalog of markings from the book written in 1995 (catalog positions 2014 and 2015, page 203) in Russian [7].

The second series of four punches is in the clothing area (Figure 4), in this case the three grouped markings (*A.C./1867, 84* and *the coat of arms of Moscow*) are followed by the fourth mark - *И.С.К.* (I.S.K.).

The undated oklad was attributed to the XIX century Russian workshops by Florin Colonaş - an expert in handicraft of artistic significance.

Figure 3. Icon oklad with markings, left upper part – aureole.

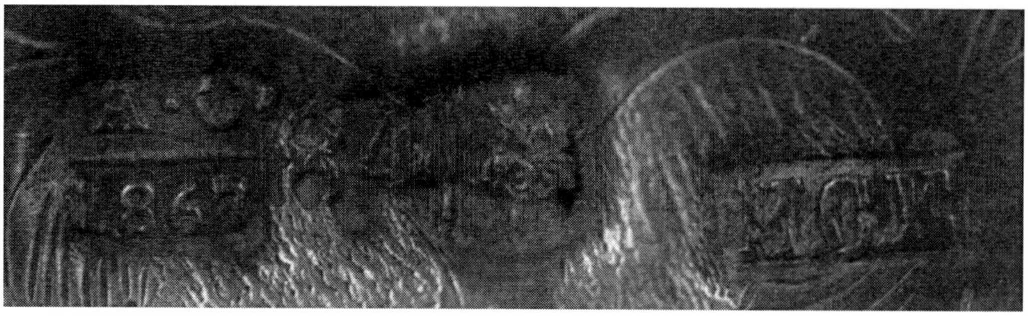

Figure 4. Icon oklad with markings, the lower part – near to right foot.

2.2. Characterization methods

The comparative study of two XIX century Russian icons made of silver alloys was based on literature research and metallographic analysis (optical and electron microscopy).

The metallographic analysis was performed on an optical microscope and the images were taken on the OMNIMET Buehler analyzer. The samples were also analyzed by Scanning Electron Microscopy using a Quanta Inspect F50, with a field emission gun (FEG) with 1.2 nm resolution and an Energy Dispersive X-ray Spectrometer (EDXS) having 133 eV resolution at MnK_{α}. EDX analyzes were performed for each of the two samples analyzed.

3. Results and discussions

The two oklads were made by mechanical lamination and stamping, techniques that developed greatly in the XIX century [7]. Stamping is a technique that provides the possibility of obtaining a repousse by means of pressing between two forms - a matrix and a punch. Stamped products have smooth, rounded lines, with no changes in thickness of the metal plate used and no sudden crossings of height [8].

It is possible that monogram *ИСК* appears without any points between the initials and belongs to the craftsman Ivan Semenov Komarov (ИванСеменовКомаров) who worked in 1808 in the Russian village Podol'skoye (Подольское) [7] although the difference between the monogram on the oklad shown in this article and the one used by the above mentioned craftsman consists in the existence of the points between the initials of the name [7]. As regards the controller's monogram in the verification mark – *A.C./1867*, it probably belongs to A.Svechin (A. Свечин), who worked in Moscow from 1862 to 1875 [7]. The silver title mark - 84 and the city mark *Saint George on horseback slaying the Dragon* (coat of arms of Moscow) are specific to the XVIII-XIX centuries [7]. We must consider the existence of a significant gap between the manufacturing and verification periods, as well as between the manufacturing site of craftman Ivan Semenov Komarov (Podol'skoye) and that of controller A. Svechin (Moscow).

The experimental investigations were carried out in the laboratories of Faculty of Materials Science and Engineering from University Politehnica of Bucharest.

From the oklad-holding area on the wood of the icon samples of 1-2 mm^2 were taken on which metallographic analyzes were carried out to identify the metallographic structure and the chemical composition of the silver alloys.

Before the metallographic analysis performed on an optical microscope (OMNIMET Buehler) the samples were reacted with a solution of: 1 g Cr_2O_3; 1 ml H_2SO_4; 500 ml H_2O_2 (established procedures for highlighting the grain size in the case of Ag-Cu alloys). The silver foil was very heavily deformed (due to the processing technique) by lamination and punching and therefore the size of the grain could not be observed.

The samples were analyzed by Scanning Electron Microscopy using a Quanta Inspect F50 (Figures 9 to 12), and in terms of chemical composition. The results of the EDX analysis performed in three points for each of the two samples analyzed are shown in Tables 1 (for the marked sample) and Table 2 (for the unmarked sample).

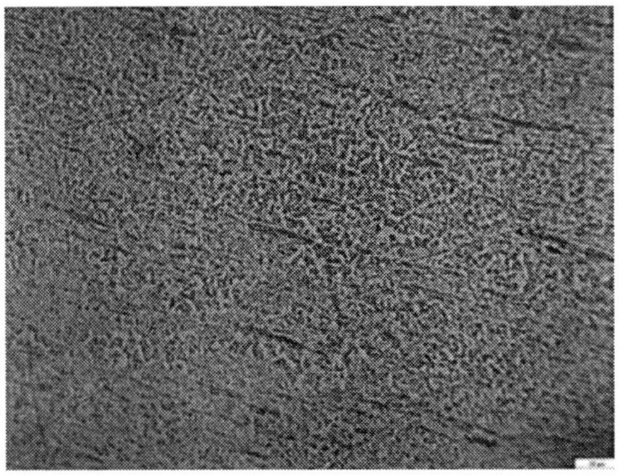

Figure 5. Optical Microscopy (Marked oklad).

Figure 6. Optical Microscopy (Unmarked oklad).

Table 1. Marked oklad - average composition (EDX).

Elem	Wt %	At %	K-Ratio	Z	A	F	Net Inte.	Backgrd Inte.	Error	P/B
Ag	88.54	81.99	0.85	0.99	0.97	1.0000	442.56	8.92	0.67	50.57
Cu	11.45	18	0.10	1.05	0.88	1.0000	56.6	7.37	2.09	7.73

The result obtained for the marked oklad: 87-88% silver, copper (11.06-12.12%) corresponds to the silver quantity specific to the silver title 84.

Table 2. Unmarked oklad - average composition (EDX).

Elem	Wt %	At %	K-Ratio	Z	A	F	Net Inte.	Backgrd Inte.	Error	P/B
Ag	82.7	73.97	0.77	0.98	0.95	1.0000	400.04	9.46	0.53	43.39
Cu	15.48	23.49	0.14	1.05	0.89	1.0000	75.70	7.05	1.3	10.84
Zn	3.47	2.54	0.01	1.06	0.91	1.0000	7.56	6.55	6.16	1.19

The result obtained for the unmarked oklad: 82.2-82.7% silver, copper (15.48-16.22%), zinc (1.58-3.47%).

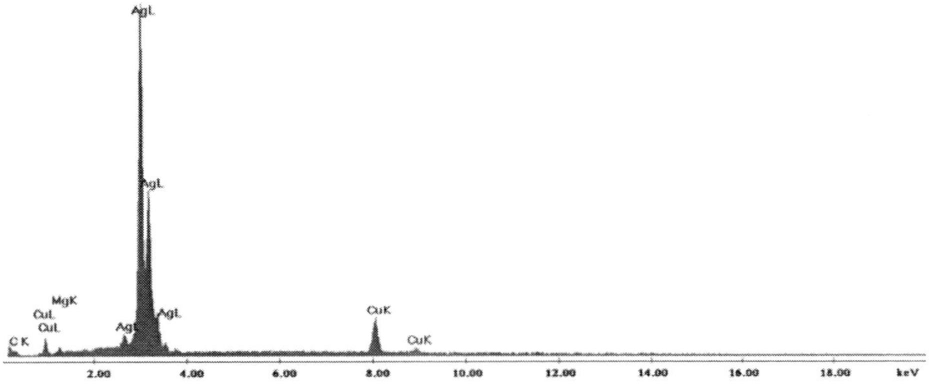

Figure 7. EDX analysis on the marked oklad.

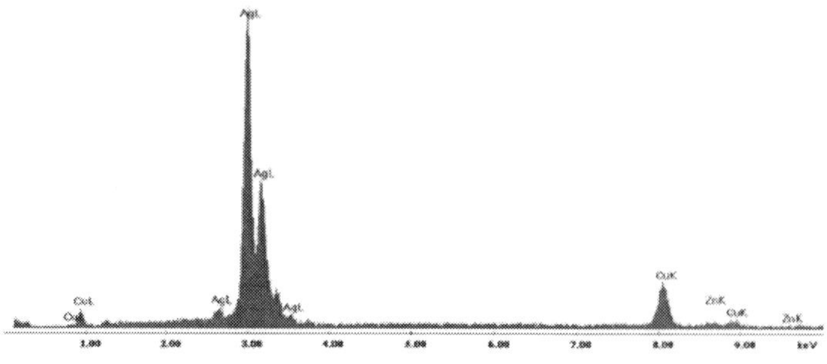

Figure 8. EDX analysis on the unmarked oklad.

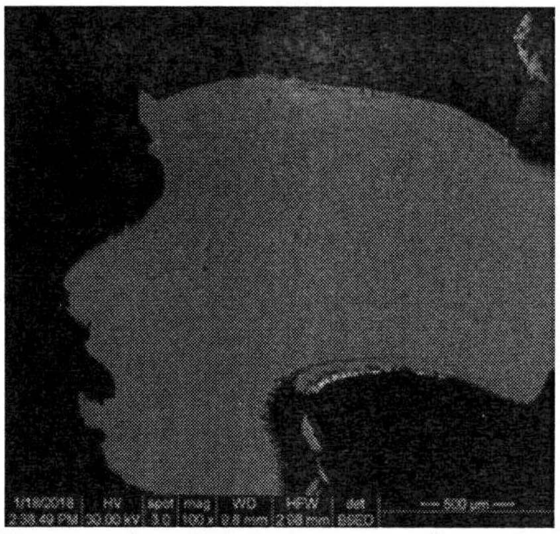

Figure 9. SEM image of surface topography for marked oklad.

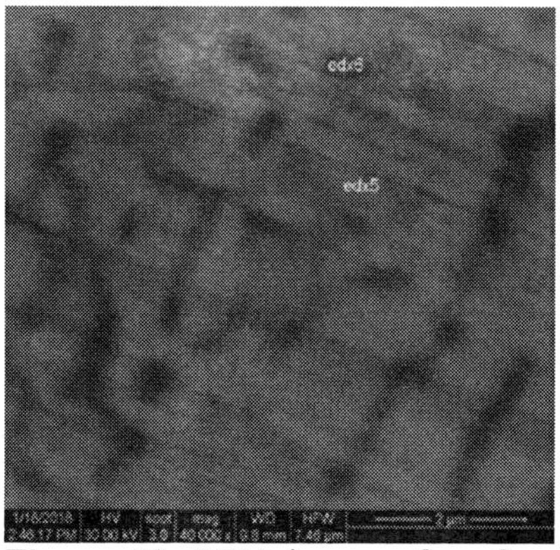

Figure 10. SEM image of surface topography for marked oklad.

The electron microscopy analysis (scanning + EDX) on the marked oklad allowed the identification in the eutectic Ag-Cu alloy (87-88% Ag, 11.06-12.12% Cu) of the two present phases: silver-rich beta phase (etches light) and cooper -rich alpha phase (etches dark). From the electron-probe analysis of one of the dendritic arms that appear here as dark globules, resulted a composition of 83% Ag 17% Cu (Figure 9 and Figure 10).

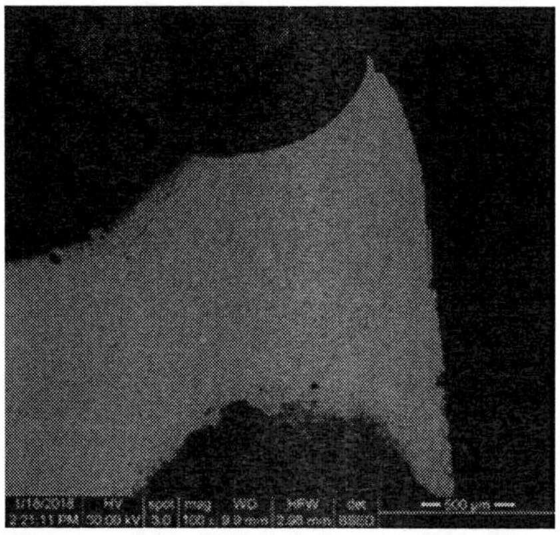

Figure 11. SEM image of surface topography for unmarked oklad.

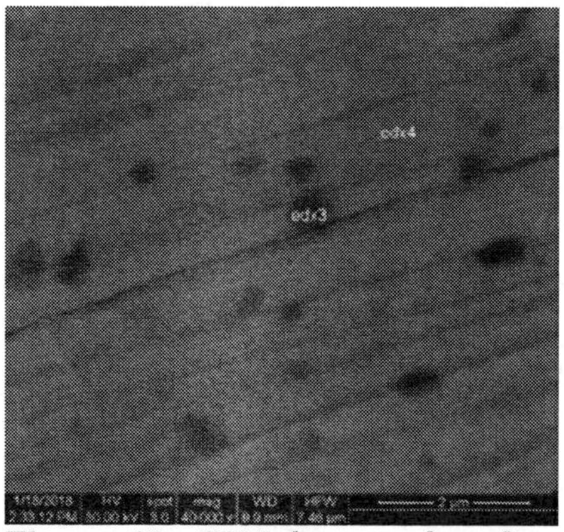

Figure 12. SEM image of surface topography for unmarked oklad.

The electron microscopy analysis (scanning + EDX) on the unmarked oklad allowed the identification of the ternary alloy Ag-Cu-Zn: 82.2-82.7% Ag, 15.48-16.22% Cu, 1.58-3.47% Zn.

There is a possibility that it has been used in alloy development silver and a Cu-Zn brass with up to 20% Zn given that the Cu / Zn ratio is 5: 1.

According to the Cu-Zn equilibrium diagram, the alpha phase formed at this concentration of Zn in Cu has the same structure as copper (cubic with centered faces) and is a typical solid substitution solution, some of the copper atoms being replaced by zinc atoms [9]. Therefore, the same two phases will be present: silver-rich beta phase (etches light) and cooper-zinc rich alpha phase (etches dark). From the electron-probe analysis of one of the dendritic arms that appear here as dark globules, resulted a composition of 15.54-30.73% Cu, 2.06-2.42% Zn, bal. Ag (Figure 11 and Figure 12).

Because the pure silver was too soft to create silver work objects, it was alloyed with copper, or with copper and zinc to improve the mechanical and machinability characteristics.

4. Conclusions

The result obtained from present work can be summarized as follows.

Marked oklad allowed identification of: the craftman Ivan Semenov Komarov (ИванСеменовКомаров) who worked in 1808 in the Russian village Podol'skoye (Подольское); the quality controller A. Svechin (A. Свечин), who worked in Moscow from 1862 to 1875; the silver title mark - 84; the coat of arms of Moscow - *Saint George on horseback slaying the Dragon*.

The undated oklad was attributed to the XIX century Russian workshops by Florin Colonaş - an expert in handicraft of artistic significance.

The marked oklad is produced from the Ag-Cu alloy: 87-88% silver, copper (11.06-12.12%) and corresponds to the silver quantity specific to the silver title 84.

The unmarked oklad is produced from the Ag-Cu-Zn alloy: 82.2-82.7% silver, 15.48-16.22% copper, 1.58-3.47% zinc.

The metallographic structure is typical of eutectic alloys.

5. References

[1] Nicolescu C 1973 *The art of precious metals in Romania* (in Romanian: Arta metalelor preţioase în România), Meridiane Publishing House, Bucharest

[2] Nicolescu C 1968 *The secular and religious silverware in the Romanian Countries (XIV-XIX century)* (in Romanian: Argintaria laica si religioasa in Tarile Romane (sec. XIV-XIX)), Art Museum of the Socialist Republic of Romania. The Old Romanian Art Section, Bucharest

[3] Olteanu Ş and Şerban C 1969 *Crafts from Wallachia and Moldavia in the Middle Ages* (in Romanian: Meşteşugurile din Ţara Românească şi Moldova în Evul Mediu), Academy of Socialist Republic of Romania Publishing House, Bucharest

[4] Petrescu S 1911 *The gold from Neamţu and Secu* (in Romanian: Odoarele de la Neamţu şi Secu), Ateliere SOCEC & Co, Bucharest

[5] Simion V 1914 *Pro Patrimonio - Old Romanian Art Studies* (in Romanian: Pro Patrimonio - Studii de artă veche românească), Basilica Publishing House of the Romanian Patriarchate, Bucharest

[6] Andryushchenko A I 1904 *The Guide on Gold and Silver Craftsmanship* (in Russian: Руководство золотых и серебряных дел мастерства), Ripol Classic Publishing House, Nizhny Novgorod

[7] Postnikova-Loseva M M, Platonova N G and Ulyanova B L 1995 *Gold and silver craft in the 15-20th centuries: the territory of the USSR*, (in Russian: Золотое и серебряное дело XV-XX вв: территория СССР), Science Publishing, Moscow

[8] Ruzsa G 2015 *The Use and Role of Metals in Art of Russian Icons (Technology, Materials, Symbolism Sacral References)*, Annales Instituti Slavici Universitatis Debreceniensis. Slavica XLIV 99

[9] Ienciu M, Moldovan P, Panait N, Groza I, Buzatu M and Marinescu D 1985 *Elaboration and casting of special non-ferrous alloys* (in Romanian: Elaborarea şi turnarea aliajelor neferoase special), Didactica and Pedagogica Publishing House, Bucharest

Metabolic Albumin and Its Effect on Electrochemical Behavior of Titanium Implant Alloy

A Ravoiu[1, 2]**, L Benea**[1] **and A Chiriac**[1, 2]
[1]Competences Center: Interfaces-Tribocorrosion-Electrochemical Systems (CC-ITES), Faculty of Engineering, Dunarea de Jos, University of Galati, 47 Domneasca Street, RO-800008, Galati Romania
[2]Faculty of Pharmacy and Medicine, Dunarea de Jos University of Galati

E-mail: lidia.benea@ugal.ro

Abstract. Human albumin is a protein made by the liver. It is the most abundant circulating protein, accounts for 50-60% of total protein in the blood and plays many roles. The most important are the oncotic and non-nociceptive properties. The albumine is part of the family of globular proteins, the most common of which are the serum albumins. All the proteins of the albumin family are water-soluble, moderately soluble in concentrated salt solutions, and experience heat denaturation. Albumin is commonly found in blood plasma and differs from other blood proteins because it is not glycosylated. Thanks to their unique properties and benefits such as superior biocompatibility and good corrosion resistance, Titanium and Titanium alloys have many uses. Their applications include dental implants and parts for orthodontic surgery; replacement parts for hips, knees, shoulders, spine, elbows and wrist joints; bone fixation devices such as nails, screws and nuts; housing parts for pacemakers and artificial heart valves; surgical instruments and components in high-speed blood centrifuges. In SITU electrochemical measurements are: open circuit potential (OCP), polarization resistance (Rp), potentiodynamic polarization (PD) and cyclic voltammetry polarization (CV) were performed to monitor the corrosion process. The optical images of the tested samples have been observed before and after corrosion experiments using an optical microscope (Optika) in order to understand the nature of corrosion and the damages produced by this process.

1. Introduction

Biomaterials have been defined as any substance (other than a drug) or combination of substances, synthetic or natural in origin, which can be used for any period of time, as a whole or as a part of a system which treats, augments, or replaces any tissue, organ, or function of the body.

The research on biomaterials initially started by testing of these materials on animals and their acceptability on the human body system were indirectly established. Nowadays, the mainly familiar kinds of materials used in biomedical applications are Metals, Ceramics, Polymers, and Composites. It is well known that prosthetic devices and components need to fulfill several imperative requirements so that they are successful in service over a longterm usage in the body without rejection and minimal failure.

The field of biomaterials has gathered significant interest as these materials are very used to fix and replace decayed or damaged parts of the human systems such as heart valves, bones, joints and teeth, etc [1].

Biomaterials for medical devices and instruments that come into contact with the human body have to demonstrate a high resistance to wear, hardness, ductility, biocompatibility, biotolerance, and above all resistance to corrosion reactions [2].

Implants are in contact with body fluids containing dissolved oxygen, chloride, phosphate, and various other organic and inorganic molecules. One of the main components of organic molecules in body fluid is protein. Protein is the first component to interact with implants through a small gap at the implant-tissueinterface [3]. The interactions of metal ions with proteins could form colloidal organometallic complexes, change albumin solutions pH level and increase the rate of ion release andcorrosion rate [3]. Protein adsorption depends on protein concentration, chemical composition and wettability of the metal oxide film.

Under normal conditions, human joints operate by low-friction articular cartilage bearing surfaces, which are conforming and selfregenerating. When natural joints are severely damaged, e.g.due to osteoarthritis, they are often replaced by artificial implants. In total joint replacement, femoral components are generally made of metal–metal, ceramic–polymer or metal–polymer couples. The use of metallic alloys and polymers has increased considerably in total joint replacement in recent years [4].

Titanium has been commercially offered during last more than 60 years and since then several titanium based materials have been developed as a surgical material.

Titanium and its alloys are a class of ideal biomedical materials due to their specific high strength, low elastic modulus, excellent corrosion resistance and biocompatibility and have already been widely used in the medical fields including orthopedics and dentistry. Titanium alloys are better tolerated than pure titanium because it seems that the oxide layer that forms is much larger (about 10-20 μm) and some alloys improve properties such as corrosion resistance and hardness [5].

The aim of study is to investigate and verify if the presence of albumin has an influence on corrosion resistance of Ti-6Al-4V alloy in biological Hank solution, the most aggressive simulated body fluid (SBF). The electrochemical methods as time evolution of free potential (open circuit potential), were used for comparative studies.

2. Materials and methods

A VoltaLab PGZ 301 potentiostat / galvanostat controlled by VoltaMaster4 software was used for electrochemical measurements. Electrochemical measurements were conducted in a three electrode cell with an Ag/AgCl as the reference electrode and a Pt-Rh grid as the counter electrode. The test specimen was served as a working electrode.

Chemical composition and mechanical properties of titanium alloy used are shown in Table 1. Before each corrosion experiment the working electrode was cleaned with NaOH and than HCl, washed with distilled water and finally dried. The surface area exposed to the electrolyte was 3 cm^2, other surfaces being isolated with resin.

Table 1. Chemical composition and mechanical properties of Ti-6Al-4V alloy.

Specification	**8-12-05832-1**	**N**	**Al**	**C**	**V**	**H**	**Fe**	**O**	**Ti**
Ti-6Al-4V Grade 5	Max. [%]	0.003	6.01	0.008	3.83	0.002	0.083	0.088	89.976
	Min. [%]	0.003	5.86	0.008	3.73	0.002	0.068	0.084	90.245
Mechanical properties									
Resistance to flow [MPa]				**Tensile strength [MPa]**			**Elongation [%]**		
865				937			11		

In order to simulate the physiological conditions of the human body, SBF (simulated body fluid) solution was used for in vitro corrosion studies. The chemical composition is given in Table 2. The corrosion resistance of the Ti–6Al–4V alloy was determined in Hank solution; Hank solution doped with 4g/L Albumin and Hank solution doped with 8g/L Albumin, to simulate the severe functional conditions of an implant in surgical situations. The volume of solution used for each experiment was 270 mL. All the experiments were conducted at room temperature 25°C.

Table 2. Composition of artificial Hank solution.

Compounds [g/L]	Hank solution
NaCl	8
KCL	0.4
$CaCl_2*2H_2O$	0.35
$Na_2HPO_4*H_2O$	0.25
$MgCl_2$	0.19
$MgSO_4*7H_2O$	0.06
Glucose	1

The electrochemical measurements such as open circuit potential (OCP), linear polarization resistance were carried out to access the anticorrosive characteristics of the Ti-6Al-4V alloy. In order to see the effect of corrosion environment with different concentration of albumin on surface morphology of the Ti-6Al-4V alloy, optical images were taken before and after corrosion experiments using the optical microscope OPTIKA XDS3 MET connected to a computer and controlled with Vision Pro Plus 5.0 software.

3. Results and discussions

3.1. Open Circuit Potential (OCP)

Figure 1 depicts the OCP curves as a function of the immersion time in artificial Hank solution solution doped with two different concentrations of albumin: 4% and 8%.

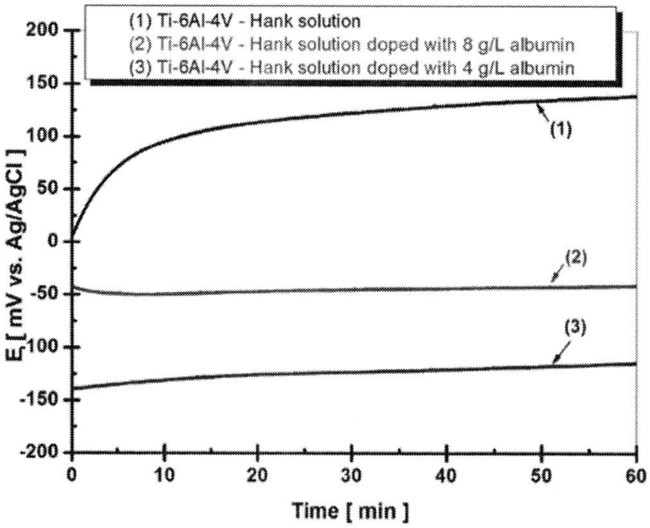

Figure 1. Open circuit potential plots of the Ti-6Al-4V alloy immersed in: (1) Ti-6Al-4V alloy immersed in Hank solution, (2) Hank solution doped with 8 g/L albumin, (3) Hank solution doped with 4 g/L albumin.

In all three curves, it can be seen that the initial OCP level changing slowly to quasi-steady-state. This behavior is typically observed in passivation process and in our sample this could be related to the formation of TiO_2 film which passivates the Ti-6Al-4V alloy surface. This protective oxide film acts as a barrier for metal dissolution and consequently reduces the corrosion rate, being electrochemically stable at quasi-stationary values of the OCP. The ennobling in time of open circuit potentials denotes the formation of a passive film and a stable state, due to the formation of a compact adherent titanium oxide film on the Ti-6Al-4V alloy surface. For a lower concentration of albumin the OCP level is shift-down to more negative potentials showing a surface instability.

3.2. Linear polarization resistance
Linear polarization resistance (Rp) is the only corrosion monitoring method that allows corrosion rates to be measured directly, in real time.

For comparative tests 30 polarization resistance points were measured by plotting the linear polarization diagrams with a small potential domain disturbance of ± 40 mV around the free potential, resulting 30 points of polarization resistance, which show the evolution of polarization resistance during immersion time in specific solution, Figure 2 (a, b).

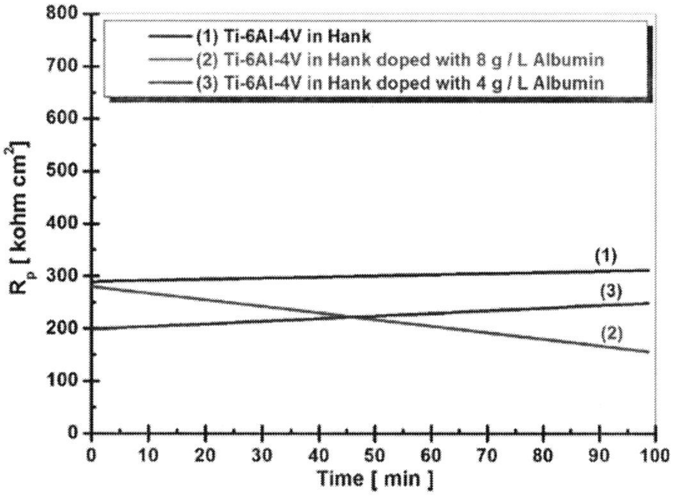

Figure 2a. The evolution of polarization resistance (Rp) values of: (1) Ti-6Al-4V alloy immersed in Hank solution, (2) Hank solution doped with 8 g/L albumin, (3) Hank solution doped with 4 g/L albumin.

At each polarization diagram the Tafel slopes, the polarization resistance and corrosion current density (corrosion rate) were calculated using the Stern Geary equation in the linear potential domain:

For simple corrosion systems, corrosion reactions are strictly controlled by the mass transfer and the corrosion current density, i_{cor}, can be correlated with the polarization resistance, R_p, through the relationships:

$$i_{cor} = \frac{B}{R_p} \tag{1}$$

The corrosion rate is expressed in this case in A / cm^2.
B is a constant of the system material/environment given by the relation:

$$B = \frac{b_a |b_c|}{2{,}303(b_a + b_c)} \tag{2}$$

Where: b_a and b_c are the Tafel slopes for anodic and cathodic corrosion reactions, respectively.

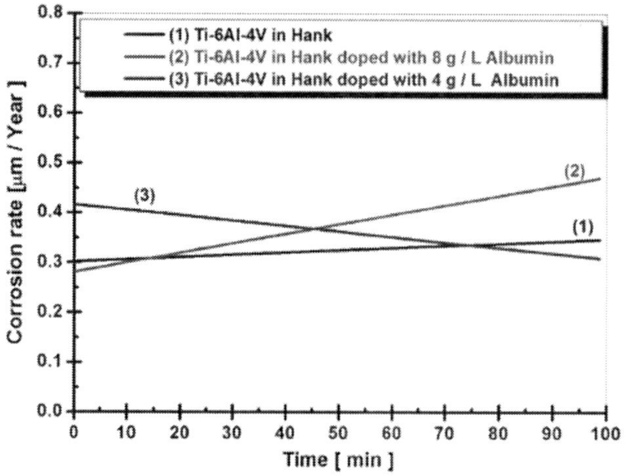

Figure 2b. The evolution of corrosion rate as penetration rate values of: (1) Ti-6Al-4V alloy immersed in Hank solution, (2) Hank solution doped with 8 g/L albumin, (3) Hank solution doped with 4 g/L albumin.

Increasing the polarization resistance means decreasing the corrosion current density and thus the decreasing the corrosion rate.

The polarization resistance values and corrosion rate values of the Ti-6Al-4V alloy surfaces studied in Hank solution and Hank solution with two different concentration of albumin are presented in Figure 2 (a, b).

From Figure 2a it can be seen that the polarization resistance value is higher in Hank solution, comparatively with polarization resistance resulted in Hank solution doped with 8 g/L albumin and Hank solution doped with 4 g/L albumin.

The increased polarization resistance value means that the formed passive film after immersion in Hank solution is more resistant compared with that formed in Hank solution doped with 8 g/L albumin and Hank solution doped with 4 g/L albumin.

Further from polarization resistance, the calculated corrosion rate expressed as penetration rate is presented in Figure 2b. According with the data plotted in Figure 2b it can be seen that the higher corrosion rate as penetration rate (0.10 μm/year) correspond to Ti-6Al-4V alloy immersed in Hank solution doped with 8 g/L albumin comparatively with the corrosion rate of Ti-6Al-4V alloy immersed in Hank solution doped with 4 g/L albumin (corrosion rate is 0,08 μm/year), or titanium alloy immersed in Hank solution without albumin (0.04 μm/year).

3.3. Potentiodynamic polarization

Potentiodynamic polarization measurements, give the possibility to compare the titanium alloy surface by point of view of passivation domains specific of the alloy immersed in the tested solutions. The potentiodynamic polarization diagrams (PD), Figure 3, were performed between -1000 mV to 3500 mV vs. Ag / AgCl at a potential sweep rate of 5 mV / s.

As it could be seen further in the cyclic polarization diagrams, the picks observed on Potentiodynamic diagrams from Figure 3 have to be correlated with localized corrosion (pitting) installed on 304L stainless steel surface due to chloride ions content of the tested biological solutions.

3.4. Cyclic voltammetry

A good way to analyze corrosion susceptibility for titanium and its alloys is the cyclic voltammetry. The cyclic voltammograms of analyzed samples were recorded from -1.0 V to 3.5 V vs. Ag/AgCl, with a scan rate of 1 mV/s. The corresponding diagrams are shown in Figure 4, having logarithmic scale for current density, i_{cor}.

As it can be seen from Figure 4 the cyclic polarization diagrams of 304L stainless steel immersed in the two Hank solution doped with albumin are similar but different comparing with the cyclic polarization diagrams of 304L stainless steel immersed in un-doped Hank solution.

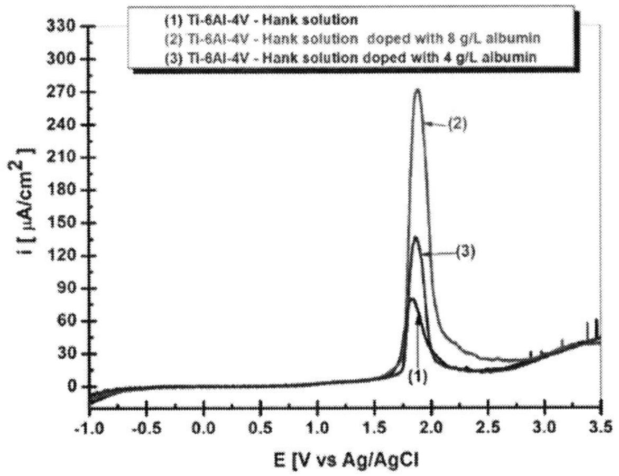

Figure 3. The Potentiodynamic polarization diagrams for: (1) Ti-6Al-4V alloy immersed in Hank solution, (2) Hank solution doped with 8 g/L albumin, (3) Hank solution doped with 4 g/L albumin.

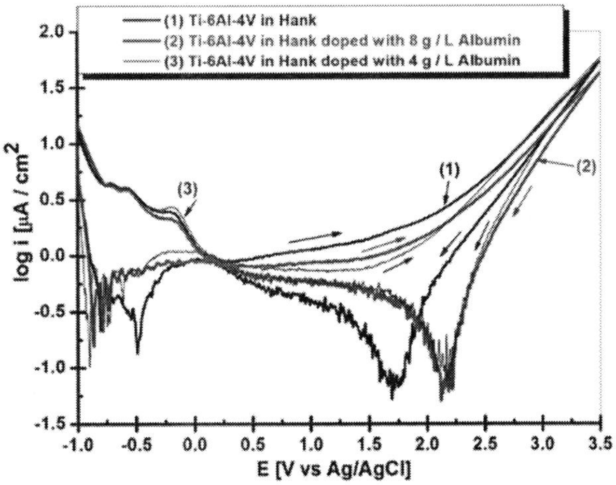

Figure 4. The Cyclic voltammetry diagrams for: (1) Ti-6Al-4V alloy immersed in Hank solution, (2) Hank solution doped with 8 g/L albumin, (3) Hank solution doped with 4 g/L albumin.

Te breakdown potentials of passive films showed on cyclic polarization diagrams are close for the two Hank solutions doped with albumin and different for un-doped Hank solution, as it can be seen on Figure 4 on returned way of cyclic polarizations diagrams. For Hank solution doped with 8 g / L albumin, the breakdown potential is around the value of 2.11 V vs. Ag/AgCl, whereas for Hank solution doped with 4 g / L, the breakdown potential is around 2.20 V vs. Ag/AgCl. For un-doped Hank solution, the breakdown potential is around the value of 1.69 V vs. Ag/AgCl.

The passivation of metals and alloys and the breakdown of passivity resulting in localized corrosion such as pitting have been the focus of much of the research in the area of corrosion over the past decades [6]. There has been general agreement that the pitting process progresses from pit initiation by passive film breakdown to early stages of pit growth identical to metastable pitting to stable pit growth [6]. The critical factors controlling localized corrosion susceptibility have been debated in the published works but there is still little agreement among the leaders in the field [6].

Some researchers consider that the structure of the passive film is associated with the initial breakdown of the film, whereas others consider that the susceptibility to pitting is controlled by the pit growth kinetics and the stabilization of pit growth [6].

The cyclic polarization diagrams from Figure 4 show the different susceptibility to localized corrosion of 304L stainless steel immersed in Hank solution and Hank solution doped with albumin and must be much deeply further investigated.

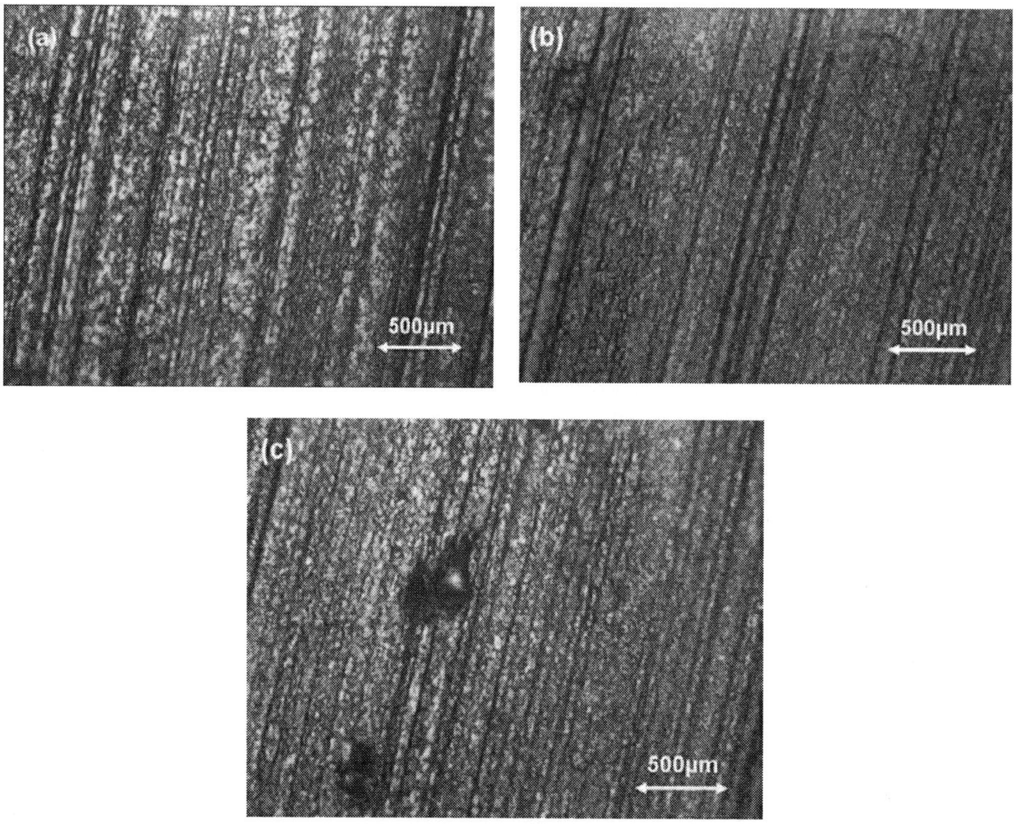

Figure 5. Optical micrographs of Ti-6Al-4V after immersion in tested biological solutions: (a) Ti-6Al-4V alloy immersed in Hank solution; (b) Ti-6Al-4V alloy immersed Hank solution doped with 4 g/L Albumin; (c) Ti-6Al-4V alloy samples immersed in Hank solution doped with 8 g/L Albumin.

3.5. Optical microscopy

The optical micrographs were taken on each surface of 304L stainless steel after corrosion tests. The optical micrographs are showed in Figure 5.

From Figure 5 (a) it can be observed that the Ti-6Al-4V alloy after immersion in Hank solution has a uniform surface with no corrosion defects. After corrosion test in Hank solution doped with 4 g/L Albumin, Figure 5(b), it can be seen some localized corrosion or pitting with small diameters.

The Ti-6Al-4V alloy samples immersed in Hank solution doped with 8 g/L Albumin presents more localized corrosion having sever pitting damage, Figure 5(c), in comparison with the samples immersed in Hank solution dopes with 4 g/L Albumin, as it can be seen in Figure 5(b).

The pitting damage covers a high surface of the Ti-6Al-4V alloy samples and the pits are deeper into the substrate after immersion in Hank solution doped with 8 g/L Albumin.

The optical micrographs confirm the results observed on cyclic polarization diagrams with breakdown potentials of passive films due to localized corrosion ongoing on 304L stainless steel surface immersed in biological solution with high chloride ions content.

4. Conclusions

This paper presents the corrosion behavior of Ti-6Al-4V alloy medical grade, to be used in biomedical applications, in Hank solution doped with 4 g/L albumin and Hank solution doped with 8 g/L albumin.

The interpretation of open circuit potential results reveals that Hank solution doped with 4 g/L albumin allow to achieve a more ennobled steady state potential of titanium alloy comparatively with Hank solution doped with 8 g/L albumin.

The potentiodynamic polarization measurements reveal that a nobler electrochemical characteristic is achieved for Ti-6Al-4V alloy immersed in Hank solution doped with 4 g/L albumin as compared with Hank solution doped with 8 g/L.

The optical micrographs performed after corrosion tests confirm the results obtained from electrochemical measurements by showing more localized corrosion on Ti-6Al-4V alloy surface immersed in Hank solution doped with 8 g/L albumin.

The research work concludes that all the results from electrochemical methods and optical microscopy are in good agreements.

5. References

[1] Mohammed M T, Khan Z A and Siddiquee A N 2014 *Procedia Material Science* **6** 1610-1618
[2] Gilewicz A, Chmielewska P, Murzynski D, Dobruchowska E and Warcholinski B 2016 *Surf. Coat. Tehnol.* **299** 7-14
[3] Karimi S, Nickchi T and Alfantazi A M 2012 *App. Surf. Sci.* **258** 6087-6096.
[4] More N S, Diomidis N, Paul S N, Roy M and Mischler S 2011 *Mat. Sci. Eng. C* **31** 400-408
[5] Chen M, Yang L, Zhang L, Han Y, Lu Z, Qin G and Zhang E *Mat. Sci. Eng. C* **75** 906-917
[6] Frankel G S, Li T, and Scully J R 2017 *J. Electrochem. Soc.* 2017 **164**(4) C180-C181

Acknowledgements

This work was supported as part of the research projects and expertise of Competences Center for Interfaces - Tribocorrosion and Electrochemical Systems (CC-ITES) from Dunarea de Jos University of Galati, Romania. www.cc-ites.ugal.ro.

Mixed Wooden-Concrete Piles: A Solution for Structures Located Near Saltwater Lakes

I Tataru[1], O Corbu[2*] and I Bucur[1]
[1]Department of Structures, Technical University of Cluj-Napoca, 15 C-tin Daicoviciu Street, Cluj-Napoca, Romania
[2]Central Laboratory, Technical University of Cluj-Napoca, 15 C-tin Daicoviciu Street, Cluj-Napoca, Romania

E-mail: ofelia.corbu@gmail.com, Ofelia.Corbu@staff.utcluj.ro

Abstract. This paper aims to present an ecological solution for creating a long-living structure using wood and eco-friendly concrete. Nowadays, it is an imminent need to reduce the use of non-renewable natural resources and, at the same time, to minimize the negative impacts of waste production from the construction sector. As a result, the option of reusing and reintroducing materials in production cycles in order to form totally different products becomes more common day by day. Even if wood is a hugely capable civil engineering material, the area of untreated pile situated above ground water level is vulnerable. This problem can be found in salty water resorts, as in the case of Oglinzi Baths located in Neamţ County. A proper solution for solving this problem could consist of using an already patented concrete that has in composition recycled glass aggregates and, moreover, has outstanding durability properties. This mixed piles structure can be a sustainable alternative benefiting from two commonly used materials and local resources.

1. Introduction

Wood is considered to be a very versatile construction material, with a low weight, very easy to handle and easy to shape. It has also an attractive appearance, low environmental impact and tends to integrate harmoniously into the landscape as it is a natural material. Wood is the only truly sustainable building material. However, the main disadvantage is that, without a protection, all types of wood exposed to weather conditions and in contact with any kind of water are vulnerable to decay [1] because, if they are wholly situated under the water level, their lifetime is extended and remain in perfect condition for centuries. But, if it is watered and dried alternately, it will be destroyed in a short period of time.

Since 1920, engineers have tried to find solutions in order to extend the life of wooden piles and so, they have introduced capping piles, firstly made of wood or stone. After that, concrete was usedand it is, by far, the best material for capping wooden piles. It is strong, durable and if it is needed, it can be strengthened by reinforcing with steel [2]. In these conditions, mixed wooden-concrete piles have a lifetime of about 100 years, while wooden piles have a life expectancy of about 20 years [3].

Nowadays, as it is an imminent need to reduce the consumption of non-renewable natural resources and to minimize the negative impacts of waste production from the construction sector, usual concrete needs to be substituted with an already patented eco-friendly concrete. In this context, the construction sector will have to significantly reduce the CO_2 emissions by choosing materials with a low impact and by creating energy-efficient structures.

The main objective of this concrete was to change building practices and integrate ecological alternatives form the design phase. It is one of the most commonly used building material and one of the most "green" options available. Concrete is an important choice for those who want to use eco-friendly materials, as they respond to many environmental responsibility requirements.

The built environment has a substantial impact on energy and material resources, being a determinant factor in terms of health, comfort and occupant productivity. The built sector is one of the greatest consumers of energy (40% of final energy consumption) and responsible for a significant amount of CO_2 emissions (36% of CO_2 emissions) [4].

The development of environmentally friendly new materials should therefore be a permanent concert of researchers in the field. Also, the design of sustainable solutions that allow recycling and recovery of waste as a raw material for obtaining new building material or products should be taken into consideration. In this way, non-renewable, natural resources can be saved [5].

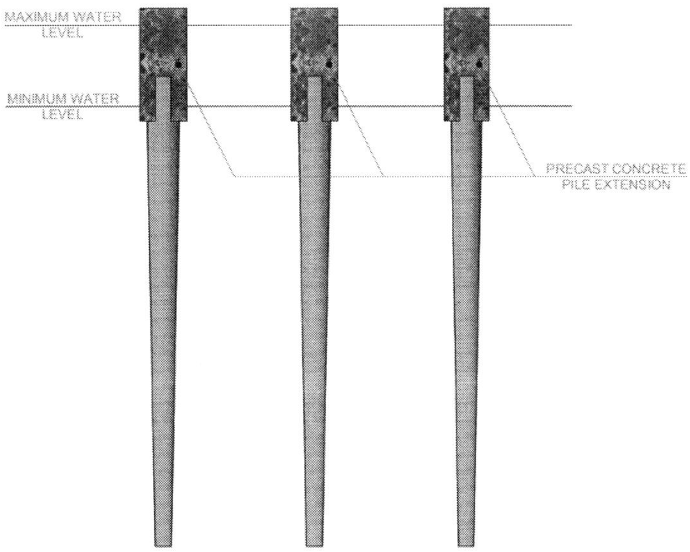

Figure 1. Degraded wooden pile [6]. **Figure 2.** Proposed solution – mixed wooden-concrete piles.

2. Materials and methods

There are two main materials used in this study: wood and eco-friendly concrete.

Wood is considered to be the only renewable material that naturally forms, with real advantages that put it in the top of building materials, being 100% natural and organic. Wooden constructions resist to earthquakes, can be mounted in a short period of time, they are cheap and do not pollute the environment. At the same time, wood is a warm finishing material that can be used inside or outside (using a proper treatment solution) and is a versatile material from an aesthetic point of view [7].

Eco-friendly concrete combines a mixture of constituents like recycled glass aggregates, cement, silica fume, glass powder, admixtureand water. According to ACI Committee 363, concrete with superior performance includes the following requirements:

- Ease of putting into operation;
- Long-term mechanical properties;
- High initial resistances;
- Durability;
- Volume stability;
- Long service life under severe environmental conditions [8].

2.1. Materials

The wooden structure was designed using C24 resinous wood with the following composition:

- Wooden circular piles Ø20 cm;
- Wooden beams on two directions, 10x12 cm;
- Wooden floor, h=2.4 cm;
- Connecting elements.

The chosen pilot was the most exposed to heavy load from the platform located above.

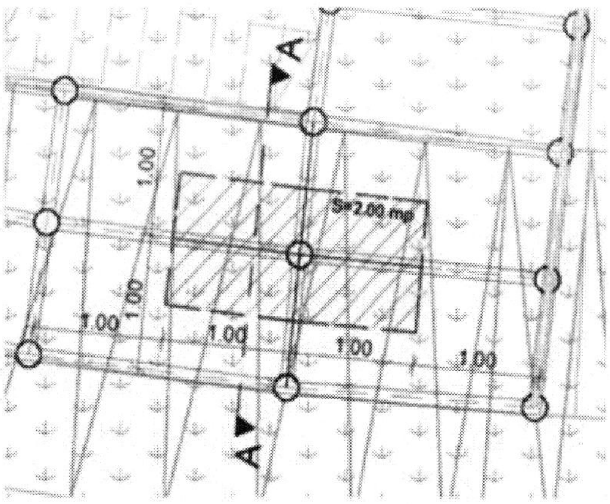

Figure 3. Wooden platform plan.

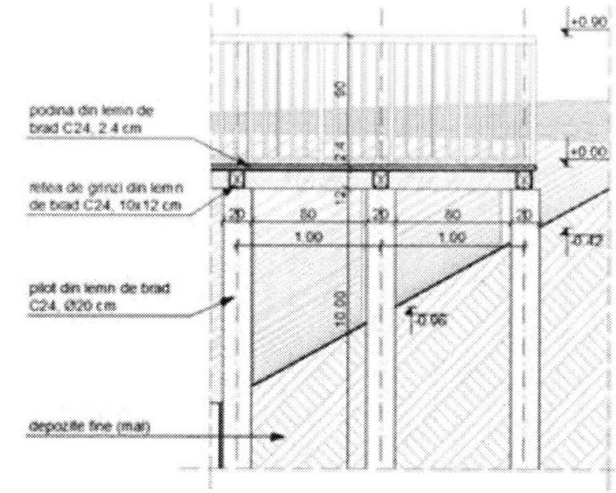

Figure 4. Section A-A.

In this study, the patented eco-friendly concrete is used to form concrete caps for extending the wooden piles in the zone of variable water level, respectively in the vulnerable area of the wood. The constituents of this type of concrete are:

- RAPIDCEM I 52,5R (PORTLAND cement unit CEM I, resistance class 52,5, R - with fast strengthening) – properties are conforming to the harmonized standard SR EN 197-1:2011 [9];
- 0/4 mm aggregate of natural river sand – aggregates are in accordance to the harmonized standard SR EN 206-1:2002 [10] and SR EN 933-1:2002 [11];
- Crushed aggregates of glass resulted from the glass crushing process in the raw glass waste form –4/8 mm, 8/16 mm –in accordance to SR EN 206-1:2002 [10] and SR EN 933-1:2002 [11];
- Silica fume (SF) – conform to the requirements of the harmonized standard SR EN 13263-1:2005 [12];
- Glass powder (GP);
- Last generation BASF superplasticizer additive, powerful water reducer - conform to the requirements of the harmonized standard SR EN 934-2:2009 [13];
- Water from the public water supply – according to the harmonized Romanian standard SR EN 1008:2003 [14].

2.2. Methods

Part 1: In the first part of this paper, the aim was to study the loads influence on the pile using the geotechnical software GEO5.

After summing all the loads that affect our pile, the following total of permanent loads is obtained, according to the equation (1):

$$Q_{total} = Q_{permanente}\gamma_{permanente} + Q_{utile}\gamma_{utile} + Q_{zapada}\gamma_{zapada} = 2.37\ 10^3\ daN = 23.71\ daN \quad (1)$$

As the program evaluates the pilot's self-weight, from equation (1) the pilot has to be removed from the equation:

$$Q_{peste.pilot} = Q_{total} - g_{proprie.pilot} = 17.88\ kN \tag{2}$$

As a result, because of the small gravitational loads that correspond to each of the piles and because there are no horizontal loads considered in the analysis, there are no significant horizontal displacements of the pilots heads, excepting those from vertical loads. This effect can be seen in Figure 5 and Figure 6 as shown below.

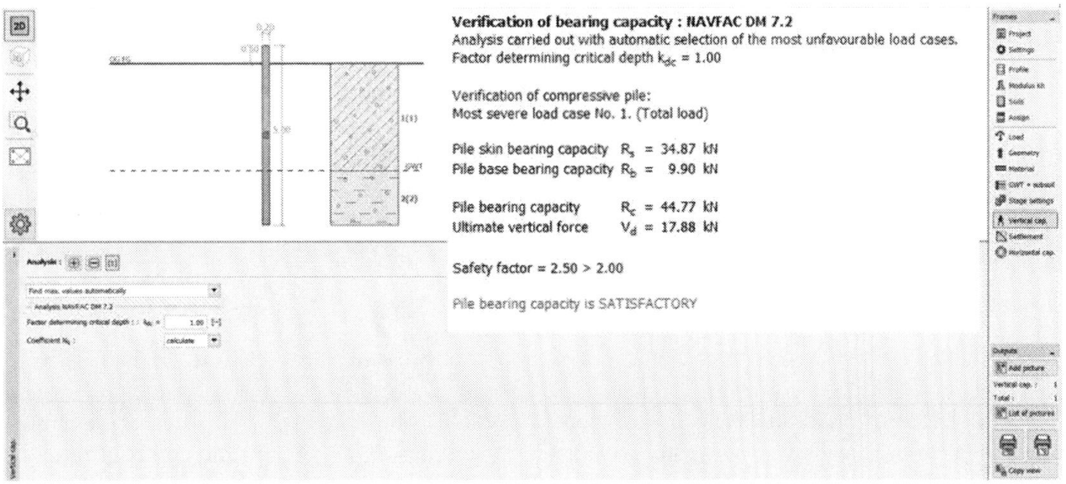

Figure 5. Verification of pile's bearing capacity.

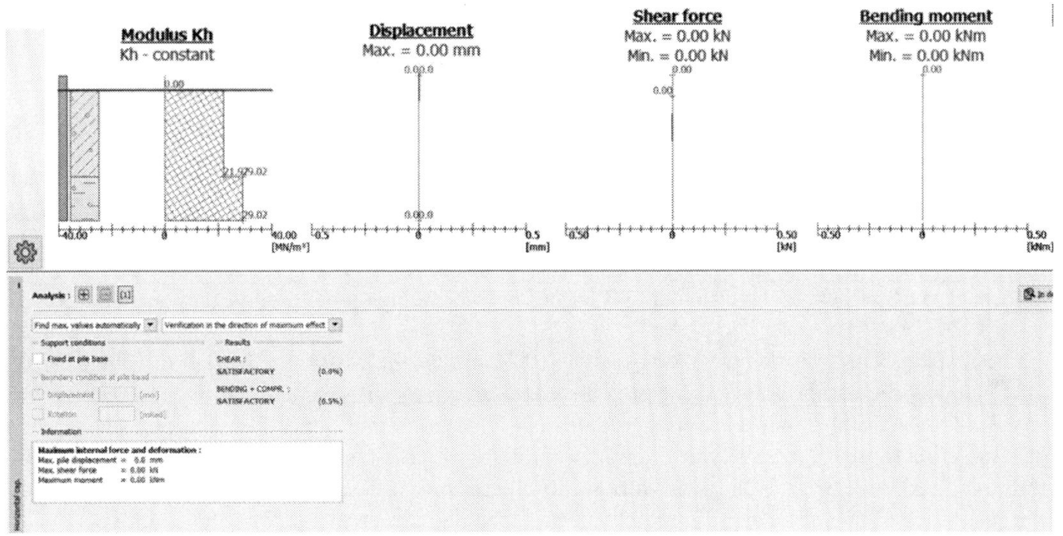

Figure 6. Verification of pile's horizontal displacement.

Part 2: The aim was to identify if concrete would suffer any visible change after his immersion for 120 days in saltwater with a very high concentration of 252.7 g/l salt (the salt deposits after the evaporation can be seen in Figure 7) from Oglinzi Baths, located in Neamţ County and then to observe the result of the compression test for half cubes of eco-friendly concrete used for pilot's extension in the structure. Thus, in the compression test, patented concrete specimens and a control mixture of concrete made of natural aggregates are used, respecting the same recipe. The test is carried out on samples that were already tested at splitting and kept under normal environmental conditions for an office.

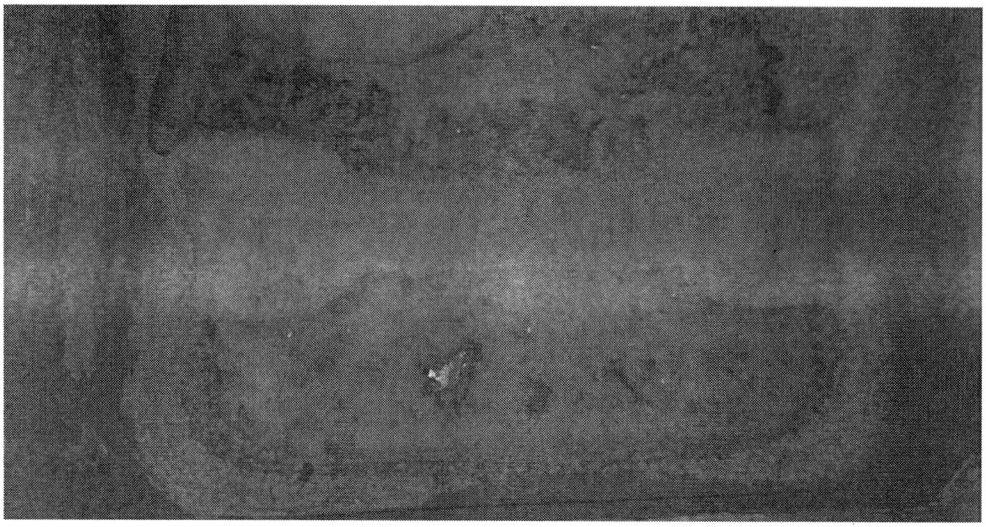

Figure 7. Salt deposits after the evaporation of concentrated saltwater from Oglinzi Baths.

Since different mixes have been designed in order to obtain a high performance concrete with aggregates from recycled glass waste, two different samples, S8-1, A and the control mix Mr were used to perform the tests. The control mix was designed using crushed rock aggregates with 4/8 mm and 8/16 mm dimensions.

Table 1. Mixes used for experimental compression attempt.

Mix Constituents									
Mix Symbol	CEM Lafarge	SF Elkem	GP 0.125	Glass Agg. 0/4	River sand Agg. 0/4	Glass Coarse Agg. 4/8	Glass Coarse Agg. 8/16	Admixture Super-plasticizer BASF	W/C
S8-1,A	0.8	0.125	0.250	-	45%	25%*	30%*	0.025	0.35
Mr	0.8	0.100	-	-	45%	25%	30%	0.020	0.35

*Waste glass aggregates, SF – silica fume, GP – glass powder, Admixture – additive for concrete

The resistance of glass aggregate is somewhat less than the resistance of natural river aggregates, which support the compressive strength of the concrete along with the cement. By the results obtained for the S8-1,A (using crushed glass aggregates), it is demonstrated that they can be used in concrete, with the direct use of alkali-type chemical inhibitors, which stops expansion into concrete. The inhibitors used were SF, which gives workability, increasing the density of the mixture and, implicitly, high resistance of concrete, and PS, which supports the abrasion resistance of the concrete.

Using such a type of concrete allows the development of innovative and environmentally friendly solutions that lead to a reduced consumption of non-renewable natural aggregate and minimize the environmental impact. The use of waste materials in creating new building materials or products (for example recycled glass waste) leads to a reduced of CO_2 footprint, by partially replacing cement. Recycling of glass waste and reducing the mixed cement gives the ecological/ "green" concrete status [5].

Table 2. Development of compression strength over time.

	Mixtures	
Characteristics	S8-1,A	Mr
Medium compressive strength (fcm), bending resistance (fct, fl) and splitting (fct, sp) [MPa]		
Compressive strength		
fcm 7 days	71.9	74.2
fcm 28 days	83.0	79.7
fcm 56 days	83.7	80.3
fcm 90 days	85.6	80.7
fcm 120 days	88.4	89.6
fct, sp 28 days	3.7	3.75

Table 3. Resistance classes obtained.

Mixture symbol	Medium compressive strength fcm 28	Characteristic compressive strength fck	Resistance class at initial tests fcm = fck+ (6÷12) MPa (SR EN 206-1, NE 012-1:2007)
S8-1,A	83.0	75.0	C60/75
Mr	79.7	71.7	C55/67

The concrete samples were also tested at water action under 12 atmospheres pressure, according to the standard STAS 3519-76 [15] (repealed), replaced by the standard SR EN 12390-8:2009 [16]. In this situation, a very good behavior was observed allowing an insignificant penetration of water of approximately 10 mm, this value being 10 times lower than the maximum admissible evaluation of 100 mm provided in the standard SR EN 12390-8:2009 [16].

This behavior is due to the existence of silica fume (SF) in the composition of concrete S8-1,A. Using SF reduces the number of pores, increases the density of concrete and thus reduces the permeability of water in the concrete (increase the impermeability of the concrete) [17, 18].

3. Results and discussions

The main challenge of this 2nd part was to make visual evaluation for two samples of concrete (S8-1,A and Mr) and to make experimental research for them. Firstly, the concrete samples were immersed in concentrated saltwater from Oglinzi Baths for 120 days. After taking out the specimens, the S8-1,A sample had apparent deposits of salt on the glass aggregates contour, but without any visual modification. However, the Mr block had some whitish circles around the external pores, as can be seen in Figures 8 and Figure 9.

Figure 8. S8-1,A after immersion in saltwater.

Figure 9. External aspect of specimens.

IOP Conf. Series: Materials Science and Engineering 374 (2018) 012078 doi:10.1088/1757-899X/374/1/012078

Figure 10. Water level penetration of S8-1,Asample.

As it was expected, because of the SF usage, the penetration in S8-1,A sample was minimum and hardly noticeable of only about 1 mm, but at the Mr specimen, there was a maximum penetration zone with about 10 mm. This value is 10 times bigger than the one registered in the case of S8-1,A. So, S8-1, A behaves much better at the action of water.

Figure 11. Water level penetration of control mix Mr.

Figure 12. Effective compression test for sample S8-1,A.

Figure 13. Effective compression test for sample Mr.

The experimental research consisted in evaluating the compressive strength of both S8-1,A sample and control mix Mr for comparing their results. Without having new mixes samples, we have used two specimens (dimension 150x150x75 mm) that were tested in 2010 at splitting and stayed, after that, for 8 years at normal temperature and humidity - in the office. So, the samples were already affected by their first test, but we have tried to displace the metal strips (dimension 150x20x9 mm) in the middle zone for the best resistance values. The compressive strength is the result of the maximal load action (F) relative to (Ac) which is the transverse action of the compression force test sample, calculated from the designated size of the specimen (SR EN 12390-1:2013[19]) or from the measurements of the sample if tested, which is exactly the size of the metal strip used on the area of specimens less affected by the previous splitting test.

Table 4. Evaluation of compressive strengths.

Mixture symbol	Compressive strength fcm 120 [MPa]	Compressive strength after experimental evaluation [MPa]
S8-1,A	88.40	86.30
Mr	89.60	86.90

Even thought the specimens came from tested samples, the concrete retained its characteristics over time.

4. Conclusions

Using such a type of concrete allows the development of innovative and environmentally friendly solutions that can lead to a reduction of non-renewable natural aggregate and to a minimized value of environmental impact. By partially reducing cement from the composition and by using recycled building materials and products, the environmental impact of a new structure is reduced.

The emergence of commercial products that have waste materials from many industrial branches in composition is one of the newest and innovative aspect. Scientific community is involved and need to provide new solutions in order to build a sustainable future for humanity and for other living species.

Using mixed piles, the wood is protected by the use of concrete between the modifications of water level. To achieve the best use of materials, wood must be stopped below the minimum water level and can be continued with this type of eco-friendly concrete which creates, moreover, an impressive architectural appearance.

It can be concluded that this mixed structure can be easily be used in a real case, for a consolidation system near the concentrated saltwater lake from Oglinzi Baths, integrated into a populated environment. This type of pile within the structure is able to stabilize the soil with only eco-friendly materials and without degradation due to landslide.

In conclusion, this solution can solve the problem of wood deterioration due to alternative watering and drying periods, improving also the stability of the sites.

5. References

[1] Reynolds T and Suttie E 2007 *Bre Digests* **DG 503** 1-2
[2] Hool G A 1920 *Handbook of Building Construction* Johnson N C (New York: 370 Seventh Avenue) McGraw Hill **1** 358
[3] Reynolds T N 2003 *Bre Digest* **DG479** 3
[4] Puskas A 2015 *Sustenabilitatea în construcţii. Sustenabilitatea construcţiilor din beton armat*
[5] Corbu O 2015 *Utilizarea deşeurilor solide în compoziţia betoanelor ecologice. Betonul cu agregate din deşeuri de sticlă p 50*https://issuu.com/revistaconstructiilor/docs/rc_nr_120_noiembrie_2015
[6] Reynolds T N 2003 *Bre Digest* **DG479** 5
[7] Virlan E 2015 *Wood, a noble material with multiple uses* http://casamea.ro/casa/constructii/materiale/lemnul--un-material-nobil-cu-multiple-utilizari-15059 - accessed 20 April, 2018

[8] ACI 363R-92 American Concrete Institute 2007 *State-of-the-Art Report on High-Strength Concrete* Manual of Concrete Practice (Part 5)
[9] SR EN 197-1:2011 Cement. Composition, specifications and comformity criteria for common cements
[10] SR EN 206-1:2002 Concrete – Part 1: specification, performance, production and comformity
[11] SR EN 933-1:2002 Încercări pentru determinarea caracteristicilor geometrice ale agregatelor. Partea 1: Determinarea granulozităţii. Analiza granulometrică prin cernere
[12] SR EN 13263-1:2005 Silica fume for concrete. Definitions, requirements and conformity criteria
[13] SR EN 934-2:2009 Admixtures for concrete, mortar and grout. Concrete admixtures. Definitions, requirements, conformity, marking and labeling
[14] SR EN 1008:2003 Mixing water for concrete
[15] STAS 3519-76 până la P12 (12 bari)
[16] SR EN 12390-8:2002 – Încercare pe beton întărit. Partea 8: Adâncimea de pătrundere a apei sub presiune
[17] Corbu O 2011 Betoane Performante – Teza de doctorat Cluj-Napoca 232
[18] Corbu O, Chira N, Szilagyi H and Constantinescu H 2013 *Nano, bio and green – technologies for a sustainable future* **13** 413
[19] SR EN 12390-1:2013 Încercare pe beton întărit. Partea 1: Formă, dimensiuni şi alte condiţii pentru epruvete şi tipare

IOP Conf. Series: Materials Science and Engineering **374** (2018) 012079 doi:10.1088/1757-899X/374/1/012079

Materials and Technologies Used in Wastewater Treatment

I C Moga[1], I Ardelean[2], O G Donțu[3], C Moisescu[2], N Băran[3], G Petrescu[1] and I Voicea[4]

[1]DFR Systems SRL, 46 Drumul Taberei, Bucharest, Romania

[2]Institute of Biology Bucharest, Romanian Academy, 296 Splaiul Independenței Bucharest, Romania

[3]University Politehnica of Bucharest, 313 Splaiul Independenței, Bucharest, Romania

[4]National Institute of Research-Development for Machines and Installations Designed to Agriculture and Food Industry, 6 Ion Ionescu de la Brad, Bucharest, Romania

E-mail: corinamoga@yahoo.com

Abstract. The biological wastewater treatment is based on biofilms activity. The biofilms can be fixed on biofilm carriers, that are made from varied materials, but most of them are made from high density polyethylene. The authors propose other mixtures of varied materials to obtain an increased load of microorganisms on biofilm carriers. During the experiments the load of microorganisms was increased up to 250% compared to the load on polyethylene biofilm carriers. Also, for the aerated biological tanks an innovative aeration system made from stainless steel pipes with fine pores (<1 mm) realized by electro-erosion, is proposed.

1. Introduction

The power consumption in both municipal and industrial wastewater treatment plants (WWTPs) has considerably increased during last decades, mainly because of the increased quantity of treated wastewater. Energy consumption is the largest budget item of an European WWTP. The energy demand in this sector will probably increase soon because of population growth and because of the need to comply with stricter European regulations in the field of water treatment, generating more energy intensive processes. Furthermore, the generation of greenhouse gasses (GHG) has become important for wastewater treatment. This paper is addressed to the general problem of power consumption in WWTPs and proposes different solutions (materials and technology) to minimize the negative effects on the environment.

In WWTPs, the oxygenation process is mainly necessary for the removal of carbon and nitrogen compounds and can represent up to 80% of the WWTP's total energy consumption [1]. An oxygen process optimization is therefore imposed with the purpose to reduce the operation costs, to guarantee an efficient and reliable treatment in such installations. During the last years, the fine bubbles aeration systems were often used to improve the oxygenation process performances [2]. Smaller bubbles have a greater surface area than the medium bubbles and therefore the quantity of dissolved oxygen that is transferred from air to wastewater mass is increased. That is why most retrofits WWTPs from coarse bubble to fine bubble will produce aeration energy savings of (20 - 40)% [3].

To increase the energy consumption in WWTPs, the authors proposes an improved mobile bed biofilm reactor (MBBR) system, including a new type al biofilm carrier and an advanced air diffuser.

In MBBRs the microorganisms grow protected within small biofilm carriers, which are carefully designed with high internal surface area. These biofilm carriers are suspended and mixed throughout the water phase. In most cases the material from which the biofilm carriers are realized is based on high-density polyethylene. This material is chosen because it has a close-to-water density. The biofilm carriers are maintained in the reactor using a perforated plate at the tank outlet. Air agitation or mixers are applied in a manner to continuously circulate the packing. The packing may fill 25% to 50% of the tank volume [4]. A final clarifier is used to settle suspended solids. The presence of biofilm carriers inside the wastewater tanks discourages the use of more efficient fine bubble aeration system, which would require periodic drainage of the aeration. Currently, for MBBR medium bubble aeration system are preferred.

2. Materials used for the fabrication of the biofilm carriers

2.1. Current MBBR systems and biofilm carriers used for wastewater treatment

Most of WWTPs use a biological treatment approach, where the activated sludge process is implemented. Activated sludge is a process in which air is introduced into wastewater tanks in order to sustain the growth and multiplication of different aerobic microorganisms capable to remove certain pollutants [5]. In the wastewater treatment process, massive quantities of air are bubbled inside tanks. Air is introduced inside the wastewater mass using air blowers and this system is the largest energy consumer in WWTP, accounting (50 - 80)% of all WWTP energy consumption. The energy consumption varies depending on aeration technology used and WWTP's geographical location [6]. An oxygen process optimization is therefore imposed with the purpose to reduce the operation costs, to guarantee an efficient and reliable treatment in such installations. During the last years, the fine bubbles aeration systems were often used to improve the oxygenation process performances.

In general, to increase the efficiency of an aeration system, the oxygen quantity transferred to the wastewater needs to be increased, and this can be done by increasing the transfer area (to have smaller air bubbles) or by increasing the air quantity introduced in the WWTP [6]. The second method is an intensive power consumption process and has to be avoided if smaller operational costs are desired.

MBBR technology utilizes free-moving biofilm carriers and represents a future evolution of the activated sludge process that allows a greater pollutant removal degree in smaller systems (i.e., bioreactors) [7]. The wastewater treatment with biomedia consists in adding biofilm carriers in aerated or anaerobic basins to support biofilm growth (Figure 1).

Figure 1. MBBR tank and example of a biofilm carrier.

The most common biofilm carrier has small cylindrical shape and is realized from plastic materials with a specific density close to 0.95 g/cm^3. Regarding the biofilm carriers used in MBBR systems, several shapes and sizes are known. The majority of the biofilm carriers are realized from high density polyethylene [8, 9]. Polyethylene has hydrophobic properties and the biofilm slowly attaches. In this context, a new material with more hydrophilic properties are proposed and tested for the biofilm carriers' realization.

2.2. New materials, based on high density polyethylene, proposed for biomedia realization - biofilm adhesion on the new materials

The authors proposed four new materials for biofilm carriers that have been tested to determine the microbial load formed on the new materials. It has been desired to obtain a material with hydrophilic properties to facilitate biofilm fixation and development. These four new materials were compared to the biofilm carrier realized from 100% high density polyethylene. So, in total five types of biofilm carriers were analysed by the authors.

The five types of biofilm carriers were incubated in the presence of activated sludge and aerated by orbital shaking (200 revolutions per minute) for 6 weeks at 20°C. The biofilm carriers were collected and treated with crystal violet (0.1%) for 15 minutes, then washed in sterile physiological saline solution (0.9% NaCl) for 4-5 times (or more) until the solution remained colourless, indicating that the excess of dye (not bound to the microorganisms) was removed. Subsequent, the biofilm carriers were treated with 30% aqueous acetic acid solution and the optical density of the solution was read at 570 nm [10].

The obtained value is proportional to the amount of microorganisms present in the microbial biofilm attached to biofilm carriers. Taking as a blank the version 1 of biofilm carriers (considered 100%), the following experimental biofilm carriers variants, developed and realized by the authors, have the following microbial loads, proportional to the intensity of the crystal violet solution absorption: 2 (120%), 3 (160%), 4 (130%) and 5 (253%). The biofilm carriers variants from 2 – 5 are made from different mixtures of (75 - 95)% high density polyethylene and (25 - 5)% other materials.

Another type of experiment (Table 1), made with only one type of biofilm carrier (namely variant no. 4), aimed at quantifying the dynamics of microbial biofilm formation, as measured by the crystal violet based method (on 5 biofilm carrier samples), but also at measuring the intensity of total dehydrogenase activity (on 5 another biofilm carrier samples) by monitoring the resazurin reduction by quantitating the fluorescence emission signal at 590 nm.

Table 1. Evolution over time of microbial biofilm formation on biofilm carrier no. 4 and its dehydrogenase activity. Optical density was determined by diluting the crystal violet solutions 10 times.

Time (days)	OD 570 nm	Resazurin reduction pg/min/ 5 biofilm carrier samples	Resazurin reduction pg/min/ DO 570
0	0.000	0	0
7	0.0947	0.5	0.528
14	0.2713	1.829	0.674
21	0.9701	386	39.79

The results show that for functional characterization of biofilm carriers it is necessary to quantify the formation of biofilm and its activity. The determination of the resazurine reduction rate is only one step, the second step of this type of determination will be the determination of the activity of biofilm formed on a certain type of biofilm carrier.

One of the ideas that emerged during the first step is precisely the possibility of immobilizing on the biofilm carrier either the complex/ native Endogenous microbiota or populations enriched in a particular physiological group (e.g. the denitrifying or ammonium oxidation bacteria) and their use in cleaning first at laboratory level and then, depending on the concrete results, at pilot level.

Additional characterization of the formed biofilm can also be done by using Sybr green staining and ethidium bromide homodimer, which enable the labell of dead cells (red fluorescence) as compared with total cells (both alive and dead - green fluorescence) (Figure 2).

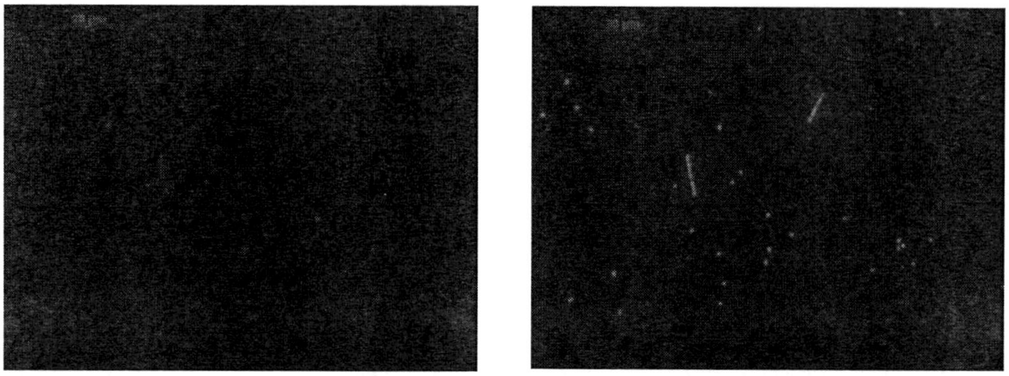

Figure 2. Live/dead cell assay on biofilm artificially cleaved from biofilm carrier no. 4. All cells are labelled with Sybr green (green fluorescence) and the dead ones are labelled with ethidium bromide homodimer (red fluorescence).

3. Aeration systems

3.1. Current aeration systems used in MBBR systems.

Currently, several types of diffusers are used in all types of WWTPs:

a. Porous diffusers (small bubbles diffusers). Numerous materials are used for the realization of porous diffusers. The most common materials that are used for porous diffusers are based on rigid ceramic and plastic materials and flexible rubber or plastic [11].

b. Nonporous diffusers. Nonporous diffusers produce larger-sized bubbles compared to the porous diffusers and consequently have lower aeration efficiency, but the advantages of lower cost and less maintenance costs may lead to a lower dissolved oxygen transfer efficiency and to increase the electrical energy costs [12]. This type of non-porous diffusers is made from perforated stainless-steel pipes.

c. Other diffusion devices. Jet aeration combines liquid pumping with air diffusion. The pumping system recirculates liquid in the biological tanks, ejecting it with compressed air through a nozzle assembly. This aeration system is particularly designed for deep (> 8 m) wastewater tanks.

MBBRs use an engineered aeration system consisting of stainless-steel pipe diffusers, manifold (or submerged air header), down pipes, and air-flow control valves. The so-called medium-bubble diffusers (nonporous diffusers) have relatively large circular orifices (i.e., when compared to membrane diffusers) situated along the underside which are less susceptible to scaling and fouling. These diffusers have slightly more efficient oxygen transfer efficiency than coarse-bubble diffusers, which is further enhanced by the presence of free moving plastic biofilm carriers. Operational experience has proven that medium-bubble diffusers require less maintenance than fine-bubble diffusers. The medium-bubble diffusers are characterized by lower oxygen transfer efficiency than fine bubble diffusers because the larger expelled bubbles travel through the water column rapidly and have a lower surface-to-volume ratio [13]. Unfortunately, the impact of biofilm carriers upon the oxygen transfer inside wastewater tanks has not been sufficient investigated; therefore, the impact of biofilm carriers on oxygen transfer efficiency in MBBR is poorly known. In MBBR aeration systems the air is introduced with the help of pipes connect to a blower. The bubbles are generated by the small orifices drilled in the pipes situated at the bottom of the biological basins. Diffusers typically used in MBBRs are 25 mm diameter stainless-steel pipes with (1 - 4) mm diameter orifices spaced (38 - 102) mm along the underside of the diffuser pipe.

Smaller bubbles have a greater surface area and therefore will transfer more oxygen. That is why most retrofits WWTPs from coarse bubble to fine bubble will produce aeration energy savings of (20 - 40). Porous diffusers made from porous ceramic materials, special glass, other porous materials are used in order to obtain air bubbles with very small diameters. Smaller pores have the disadvantage that they are more likely to clog.

Small bubbles can be also obtained by using membrane diffusers. The membrane lifetime for diffusers is 6 - 10 years in WWTP, depending on wastewater influent and operation conditions.

3.2. Modern technology for water aeration in biological water treatment plants.

The authors propose an improved aeration system that combined the advantages of stainless steel aeration systems (nonporous diffusers) with the porous diffusers. The authors conceived a metallic diffuser with very fine holes made by electro-erosion. In this way will be obtained a porous diffuser from a resistant anticorrosive material.

Some of the advantages of using an electro-erosion are specified in literature [14]: realization of complex shapes that would otherwise be difficult to produce with conventional cutting tools; extremely hard material are processed with very close tolerances; very small pieces can be shaped where conventional cutting tools may damage the part from excess cutting tool pressure; there is no direct contact between tool and work piece; delicate sections and weak materials can be machined without any distortion; a good surface finish after the drilling.

In this way the authors will obtain an innovative fine bubbles diffuser made of stainless steel (or of different metals and/or alloys). Very small holes are difficult to obtain by conventional machining processes. Holes in metal are obtained using an electric machine with a metal drill. Depending on the diameter of the drill different hole diameters are obtained. Due to material resistance reasons the drills cannot be very small in diameter, so that small holes cannot be obtained in metal materials by using conventional technologies. Small holes (<1 mm) will be realized by electro-erosion drilling.

The new aeration system made of stainless steel proposed by the authors has the following advantages: is sturdy and also resists to the interaction with the mobile biofilm carriers; there are no problems with the storage of diffusers until commissioning of the treatment plant; less maintenance than classical fine bubble diffusers; the pressure loss for the ceramic porous diffusers, which are currently being used, is 110 mm CA, while the pressure loss for the new fine bubbles diffusers is close to 24 mm CA for the same mode of operation. Thus, the pressure loss reduces with 1076 mm CA. This approximation is made for a single diffuser; the annual energy consumption for 1 ceramic porous diffuser is 16.47 kWh/year, while the annual energy consumption for the new diffuser is estimated at 0.356 kWh/year for the same mode of operation. Thus, the energy consumption can be reduced with 16.114 kWh/year/diffuser.

4. Conclusions

The wastewater treatment market is a mature industry that has been subject to decades of continuous tightening of water quality regulations. In addition, the entry into force of more stringent regulations at EU level will boost investments in new technologies so as to upgrade and optimize existing WWTPs probably reaching 350,000 € millions in 2025.

For these reasons the research team focused the attention to develop new and efficient equipment for wastewater treatment. It is proved that the new materials and manufacturing procedures (for both biofilm carriers and air diffusers) can lead to significant improvements in WWTP; increased treatment efficiency and decreased operational costs can be obtained by using the new material proposed by the authors. During the following experiments, *in situ* determination will be realized.

5. References

[1] Laizāns A and Šnīders A 2012 *Renewable Energy and Energy Efficiency, Biogas and biofuel production technologies* 109-14

[2] Rosso D 2008 *Water Sci Technol.* **57**(7) 973-8

[3] Sander S, Behnisch J and Wagner M 2017 *Water Sci. Technol.* **75(4)** 890-7

[4] Zafarzadeh A, Bina B, Nikaeen M, Attar H M and Hajiannejad M 2010 *J. Environ. Health. Sci. Eng.* **7**(4) 353-64

[5] Moretti C J, Das D, Kistner B T. Gullicks H and Hung Y T 2011 *Water* **3** 806-18

[6] Metcalf and Eddy 2003 *Wastewater Engineering*, 4th edition. (Now York, Tata McGraw Hill Edition)

[7] Calderón K, Martín-Pascual J, Poyatos JM, Rodelas B, González-Martínez A and González-López J 2012 *Bioresource technology* **1**(121) 119-26
[8] Welander U, Henrysson T and Welander T 1997 *Water Research* **31**(9) 2351-5
[9] Rusten B, Kolkinn O and Ødegaard H 1997 *Water Sci. Technol.* **35**(6) 71-9
[10] Christensen G D, Simpson W, Younger J, Baddour L, Barrett F, Melton D and Beachey E 1985 *J. Clin. Microbiol.* **22** 996
[11] Huifang Z, Wenping C, Ling S and Hanhu L 2017 *Functional materials* **24**(4) 699
[12] Wagner M R and Pöpel H J 1998 *Water Sci. Technol.* **38** 1-6
[13] Sander S, Behnisch J and Wagner M 2017 *Water Sci. Technol.* **75**(4) 890-7
[14] Kucukturka G and Cogunb C 2010 *Machining Science and Technology: An International Journal* **14**(2) 189-207

Acknowledgments
This work was supported by a grant of the Romanian National Authority for Scientific Research and Innovation, CCCDI – UEFISCDI, project number COFUND-WATER WORKS ERANET 2015-ABAWARE (1), within PNCDI III.

This work was supported by a grant of the Romanian National Authority for Scientific Research and Innovation, CCCDI – UEFISCDI, project number COFUND-MANUNET III-CEBIOTREAT, within PNCDI III.

Polyethylene Based Materials for Biofilm Carriers Used in Wastewater Treatment

I C Moga[1], O I Iordache[2], G Petrescu[1], F Pricop[2] and I Dumitrescu[2]
[1]DFR Systems SRL, 46 Drumul Taberei, Bucharest, Romania
[2]The National R&D Institute for Textile & Leather, 16 Lucretiu Patrascanu, Bucharest, Romania

E-mail: corinamoga@yahoo.com

Abstract. The moving bed biofilm technology is based on biofilm carriers on which consortia of microorganisms attach, develop and grow. Around the world are known many biofilm carrier variants made of varied materials. The most common materials are based on polyethylene since this material has a close to water density. The authors propose a novel biofilm carrier to be used in tertiary treatment for tannery and paper-mill wastewaters. The biological treatment is based on fungal activity. The selected fungal strains will be grown on innovative polyethylene carriers containing cellulose. The carrier will be designed to be exploited in a moving bed bioreactor and to favour fungal growth in the presence of competing bacteria.

1. Introduction

Industrial wastewaters can often be recalcitrant to conventional biological treatment methods, due to their complex constitutive matrix (extreme chemical-physical parameters, bio-burden, presence of complex aromatic compounds etc.) [1]. Tannins and Absorbable Organic Halogen compounds (AOXs) are keen to damages aquatic environments. Moreover, they are produced by industrial activities that are highly water demanding. At present, treatment of effluents with high cellulosic content is mostly carried out by aerobic processes (activated sludge), which are not fully effective due to high organic matter concentration (high content of cellulose) and the presence of cellulose derived compounds, among others organochlorine compounds [2]. Reactors used in the treatment of high content cellulosic unsegregated wastewaters, do not respond to final emissary quality demands in terms of chemical oxygen demand (COD) and in terms of content of chlorinated compounds generally defined as absorbable organic halides (AOX). Current techniques of removal of chlorinated phenolic compounds and AOX, are based on membrane filtration and ozonation, energy demanding processes.

Approximately 85% of the water of the pulp and paper-mill industry consumes is only used for processing, leading to generation of massive amounts of contaminated wastewater. The Food and Agriculture Organization (FAO) of the United Nations Secretariat estimated that total effluent discharged annually from pulp and paper-mills is about 40,000 million cubic meters (~200 m^3/t of pulp and paper). Generated from a variety of manufacturing processes, these discharges often serve as a primary source of pollution for aquatic life, public health and the environment, by generation of considerable amounts of toxins contained in the effluents. Pulp and paper-mill effluents have also regularly undergone secondary biological treatment to remove organic matter such as residual total suspended solids (TSS) and biological oxygen demand (BOD) and address COD and AOX.

Filamentous fungi strains have been highlighted as possessing great applicative potential in the field of wastewater treatment [3]. In recent light, White-Rot-Fungi (WRF) and Ascomycota (ASC) representatives have been successfully used in bio-remediation technologies of polluted effluents [4]. WRF strains possess several adaptive mechanisms to the environment that involve the battery of oxidative enzymes, which are also involved in bioremediation of wastewater effluents processes [5]. This aspect can lead to a greater efficiency of fungal strains, when compared to bacteria, in regards of oxidative mechanisms of environment adaptation [6]. In this context WRF are more efficient than bacteria in expressing oxidative mechanisms of adaptation to the environment.

However, these technologies still face great applicative challenges, as fungal mycelium is prone to mechanical breakage, inside the treatment plant, during process. Immobilization of these microbial structures in rigid structures may prove as a viable solution for both mechanical integrity of the microorganism, and maintenance of treatment efficiency. This complex problem will be tackled in FunCell project (COFUND-MANUNET III-FUNCELL, MNET17/ENER-1143, entitled "Exploiting fungi potential for recalcitrant compounds removal from cellulosic wastewaters"), which consists is the development of an innovative myco-based tertiary treatment for tannery and paper-mill wastewaters efficient in removing tannins and AOX, not depleted by consolidated bacterial-based processes. Previous literature work has already explored denitrification properties of some fungal strains in MBBR systems, with higher performance yields that bacteria-driven system [7].

The FunCell technology is dedicated to the depletion of chemical structures recalcitrant to the bacterial oxidative machinery: tannins, AOX and cellulose.

2. Moving bed biofilm reactors (MBBRs)

During the last years, the MBBR technology for wastewater treatment [8] has been intensively studied and developed [9], but not already applied in combination with WRF/ASC for wastewater treatment.

New MBBRs structures will be designed, for fungal biofilm attachment, grafted with cellulose that will be tested at different concentration to stabilize growth conditions, in order to improve the fungal competitiveness in non-sterile condition where usually bacteria are overwhelming [10]. The project proposes a new biomedia design, starting from existent polyethylene biofilm carrier (Figure 1), which will be tested to support fungal growth in a not sterile environment typical of wastewater treatment.

Figure 1. Proprietary Moving Bed Biofilm Reactor design.

The new biofilm carrier (based on a mixture between high density polyethylene and other materials) is aimed at finding a tradeoff between packed bed and moving bed biofilm reactors and is aimed at differentiating and fine-tuning the gradients of substrate and the shear stress across the reactor bed. Such type of product is relatively new in the field of wastewater treatment, where usually gradients are modulated either in time or in space only at micro-scale level and can be eventually exploited also for bacterial ecosystems.

3. Microbial component

For in-situ testing a pilot MBBR for fungal biomass will be realized for exploiting the capacity of the WRF to produce the battery of extracellular enzymes that are involved in the depletion of recalcitrant compounds. FunCell project will exploit the intracellular enzymes machinery that is at the base of ASC's metabolic capacity towards contaminants of the environment that generally ASC colonize all aspects extremely innovative in the actual scenario of the wastewater treatment.

Preliminary work has already been carried out in the direction of the FunCell project research, and several Ascomycota and Basidiomycota representatives have been successfully grown at fermenter level (Biotec FE 007). Strains were grown in Potato-Dextrose nutritive broth, for 7 days, at 28^0C, in 400 mL broth volume, with continuous oxygenation and agitation set to approximately 300 rpm.

The following strains have been grown at both fermenter level, and then passed on PDA plates, for microscopically morphology assessment: of Basidiomycota division: *Ganoderma lucidum* (Figure 2), *Pleurotus ostreatus* (Figure 3), *Polyporus squamosus* (Figure 4), *Agaricus bisporus* (Figure 5), and of Ascomycota division: *Fusarium oxysporum* (Figure 6).

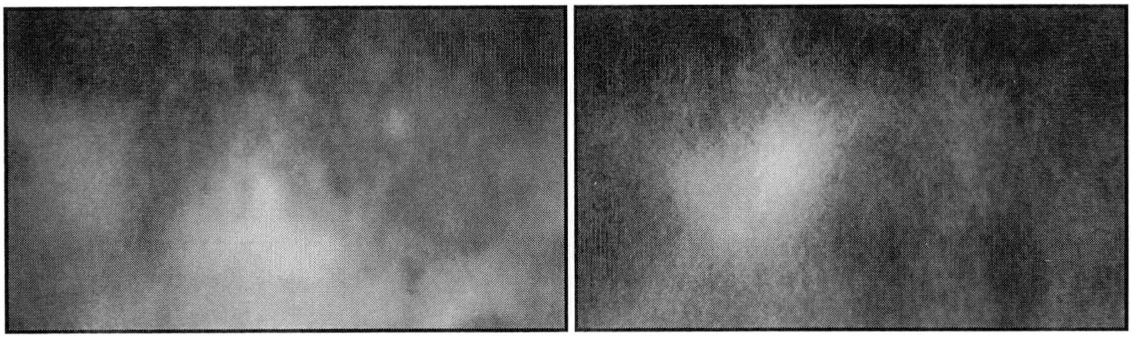

0.67x *3x*

Figure 2. *Ganoderma lucidum* morphological analysis.

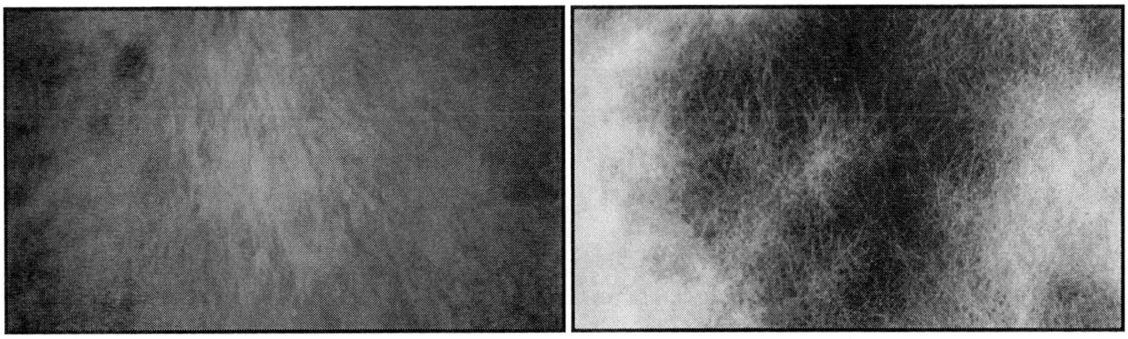

0.67x *3x*

Figure 3. *Pleurotus ostreatus* morphological analysis.

Furthermore, for future scaling up of the experimentation [11], the selected fungal strain will be grown on innovative polyethylene carriers containing cellulose, which will allow rapid metabolic development, for greater biomass yiled and mechanical fixation inside the MBBRs strucures (Figure 7). The new treatment technology will be developed and designed as self-sustaining by the exploitation of an innovative fungal carrier containing cellulose [12]. The growth conditions will allow the exploitation of cellulose as selective carbon source for fungi in environments where bacteria are competitive and can inhibit fungal development.

IOP Conf. Series: Materials Science and Engineering **374** (2018) 012080 doi:10.1088/1757-899X/374/1/012080

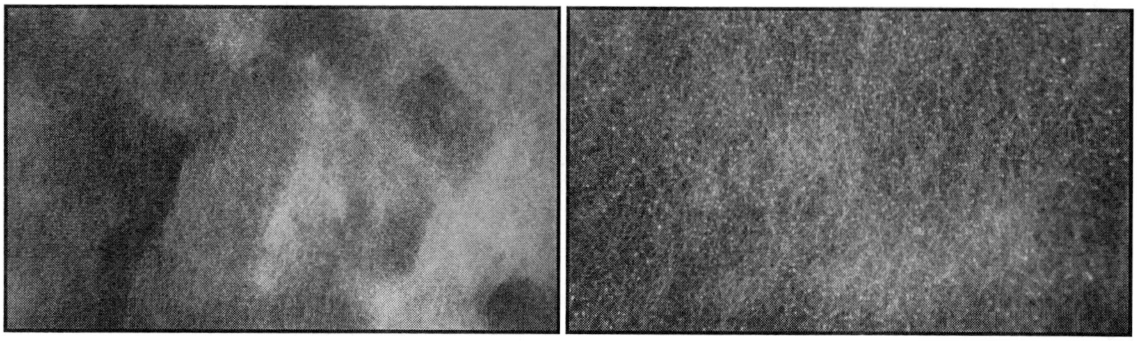

0.67x *3x*

Figure 4. *Polyporus squamosus* morphological analysis.

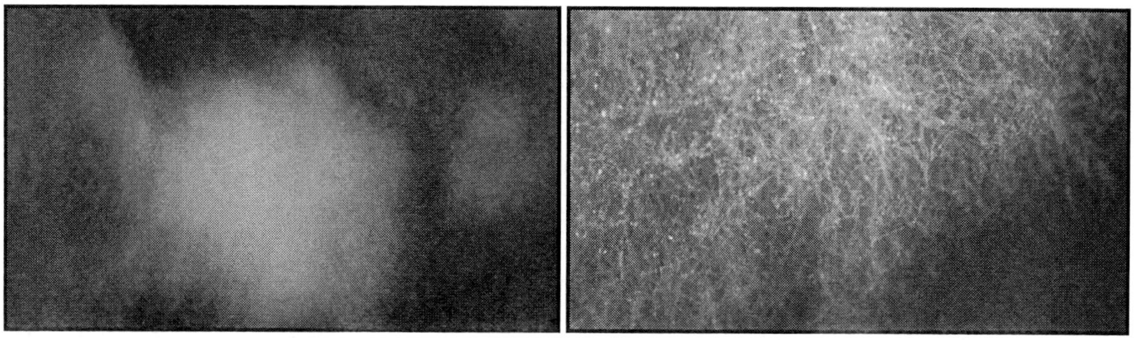

0.67x *3x*

Figure 5. *Agaricus bisporus* morphological analysis.

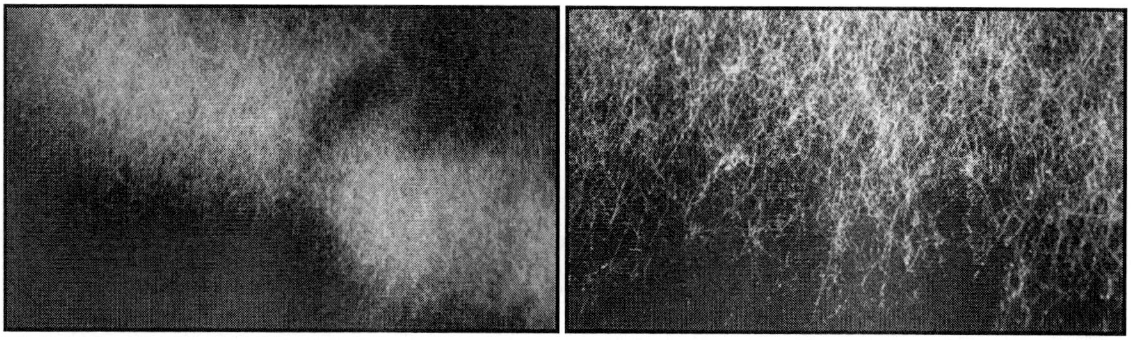

0.67x *3x*

Figure 6. *Fusarium oxysporum* morphological analysis.

FunCell project proposes a novel carrier design, grafted with cellulose fibers for the self-sustainability of the fungal biomass.

This will facilitate substrate adhesion, due to biochemical capabilities of the selected strains, which involve secretion of extracellular enzymes, which can break down substrate by combined action of several degradative processes, such as dimethylation, oxidative cleavage of the propane side chain, cleavage of ether bonds between monomers etc.

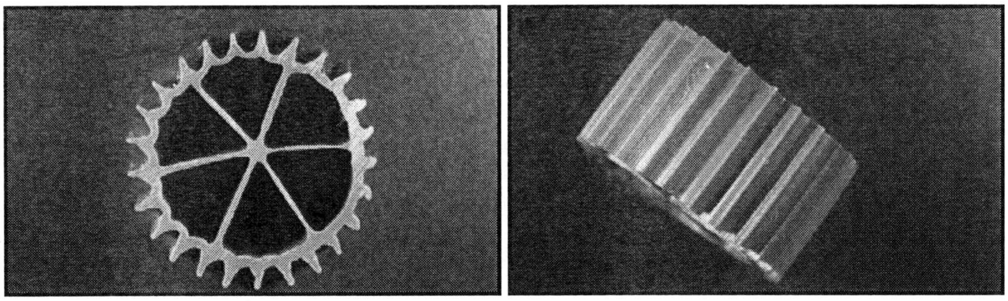

Figure 7. MBBRs microscopically analysis.

4. Conclusions

In the last decades, European regulations on effluents discharge limits became increasingly stringent and several industries found more cost effective to relocate than to adapt to the new limits. This phenomenon impacted several strategic industrial sectors such as textile and tannery industries, pulp and paper, food and beverage and petrochemical industries.

FunCell project is oriented to create non-conventional processes performing better than the "best available technologies", in the removal of both tannins and AOXs from the effluents and to promote the re-use of treated water and the adoption of a common policy in the fame of the Communitarian and National regulations. The obtainment of a high-quality effluents and effluents with a water quality suitable for reuse will be the two main needs addressed by the project. Also, the goals will be achieved by developing an MBBR treatment specifically designed for fungal metabolism, producing lower amount of sludge accompanied by a higher removal efficiency of pollutant agents. The re-utilisation of the treated wastewaters inside the services of the industrial plants will determine a decrease in the consumption of water in the same sectors.

5. References

[1] Rosales E, Pazos M and Sanromán M A. 2011 *Desalination* **278** 312-317
[2] Leunberger C, Giger W, Coney R, Graydon J and Molnarkubica E 1985 *Wat. Res.* **19(7)** 885
[3] More T T, Yan S, Tyagi R D and Surampalli R Y 2010 *Biores. Tech.* **101(20)** 7691
[4] Gao D, Du L, Yang J, Wu W and Liang H 2010 *Crit. Rev. Biotechn.* **30** 70
[5] Camacho-Morales R L and Sánchez J E 2016 *Biotechnological Use of Fungi for the Degradation of Recalcitrant Agro-pesticides. Mushroom Biotechnology, Developments and Applications*, Academic Press Elsevier 203–214
[6] Singhal A and Thakur I S 2009 *J. Haz. Mat.* **171** 619
[7] Guest R K and Smith D W 2002 *J. Env. Eng. Sci.* **1(6)** 433
[8] Goswami S and Mazumder D 2016 *Mat. Today Proc.* **3(10)** 3337
[9] Chao L, Zhen Z, Yi L and Jiashun C 2015 *Process Biochem.* **50(11)** 1934
[10] Liu Y, Zhu Y, Jia H, Yong X, Zhang L, Zhou J and Cao Z 2017 *Biores. Tech.* **244** 445
[11] Ge S, Zhu Y, Qiu S, Yang X, Ma B, Huang D and Peng Y 2014 *Water Sci. Tech.* **70(10)** 1594
[12] Bardi A, Yuan Q, Suiracusa G, Chicca I, Islam M, Spennati F, Tigini V, Di Gregorio S, Levin D B, Petroni G and Munz G 2017 *Biores. Tech.* **214** 1067

Acknowledgments
This work was supported by a grant of the Romanian National Authority for Scientific Research and Innovation, CCCDI – UEFISCDI, project number COFUND-MANUNET III-FUNCELL, within PNCDI III.

IOP Conf. Series: Materials Science and Engineering **374** (2018) 012081 doi:10.1088/1757-899X/374/1/012081

Electrospun Membranes for Environmental Protection

L R Manea[1], A Bertea[1], A Popa[2] and A P Bertea[1,*]
[1]Textile-Leather and Industrial Management Faculty,"Gheorghe Asachi" University, Blvd. Dimitrie Mangeron nr. 67, 700050 Iaşi, România
[2]"Aurel Vlaicu" University of Arad, Romania, Engineering Faculty, Textile Department

E-mail: eurotexmed@yahoo.com

Abstract. Electrospinning is a much-used process for making nano-sized fibres as a result of its simplicity and versatility and of unique mechanical and thermodynamic properties of the obtained nanofibres. Electrospun membranes, which have a porous structure characterized by high uniformity and porosity, find applications in many membrane processes, ranging from membrane distillation to reverse osmosis. The paper reviews electrospun membranes applications in environmental protection, focusing on water and wastewater treatment and air purification. Recent progress and prospects for future development are highlighted.

1. Introduction

Nanostructure science is an interdisciplinary area of science activity that has changed the way materials are produced, offering a large new field of functionalities that can be accessed [1-5]. Nanoscale materials science has revolutionized progress in many areas, overturning many of the traditional approaches [6,7]. Nanotechnology-based materials are often used in filtration processes, as the use of submicron materials results in better filter effectiveness, at the same pressure drop in interception and inertial impaction regimes, as compared with coarser filtration materials.

Between the many existing techniques for nanofibres producing (drawing, template synthesis, phase separation, selfassembly) [8-12], electrospinning proved to be one of the most important, as it is simple and capable of providing highly porous structures of non-woven nanofibres with uniform pore size. The large specific surface area of nanofiber membranes permits a very effective adsorption of impurities from the air or water, contributing to their high filtration efficiency [13-17]. Also, this kind of membranes operates at lower energy and henceforth with lower costs [18-22]. In consequence, electrospun membranes, which were used for the first time in filtration in early 1980, became more and more used in many air and water purification applications [23-26].

2. Fundamentals of electrospinning

The principle of the electrospinning process is simple: an electrical potential is applied between a polymer solution or melt and an electrically conducting collecting surface. Under the applied voltage the droplet is gradually transforming its shape into a conical shape known as the Taylor Cone, from which ultrafine nanofibers emerge and are collected, on the collector [27, 28].

A requirement for the polymer solution to be electrospun is its viscoelasticity; a spinnable polymer solution has to be semi-dilute and highly entangled [29]. The process comprises three stages: jet initiation, its extension and finally nanofiber formation [30, 31]. From the point of view of how the jet decomposes into droplets, three types of instability are distinguished (Figure 1):

• axial-symmetric instability, which is controlled by the superficial tension (nonconductive) or by the electrostatic forces (conductive); this instability leads to drop formation;

• non-axial-symmetric instability (bending, conductive), which is governed by electrostatic forces; this type of instability is responsible for the process of jet elongation (thinning);

• „whipping" type instability, which produces random spreading of the jet and even its fracture into short fibers; this type of instability is produced by the increased charge density and polymer flow rate.

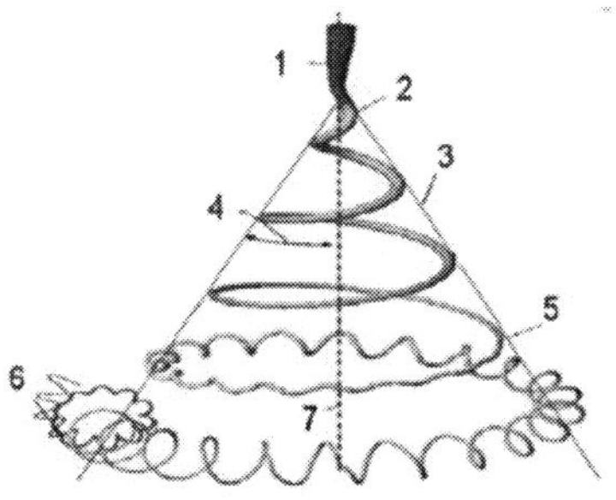

Figure 1. The simulation of instability phenomenon:
1 - linear jet zone; 2 - zone of first instability axial-symmetric; 3 - trajectory of the jet leaving the needle; 4 - halving angle for the trajectory cone; 5 - second zone of instability; 6- third zone of instability; 7 - jet axis.

Various morphologies of the nanospun nanofibers can be obtained by varying the process parameters and the polymer solution properties. This way nonwoven nanofibrous membranes with highly porous structures can be produced [32].

3. Main applications of nanofiber electrospun membranes

The principle of the majority of membrane separation processes is the selective filtration of influent through pores of different sizes. The special features of electrospun nanofibre membranes, namely high specific surface, high porosity, fine tunable pore size, interconnected open-pore structure and high permeability to gas and fluid [33], make them very suitable for such treatments.

3.1. Water and wastewater treatment

Membrane technologies for drinking water treatment have concerned much interest because of their many advantages, such as moderately low operational costs and high efficiency. Electrospun nanofiber membranes from polyvinylidene fluoride, which have piezoelectric properties, are efficient in preventing fouling, offering this way high water productivity and maintenance of flux [34].

Chitosan - polycaprolactone nanofibrous membranes have been used for antibacterial water filtration, significantly reducing Staphylococcus aureus while supporting a water flux of ~7000L/h/m with 100% removal of 300 nm particulates, in condition of full membrane integrity [35]. Membranes from a network of ultra-fine cellulose nanofibres infused into a scaffold from electrospun polyacrylonitrile nanofibres on a melt-blown polyethylene terephthalate non-woven substrate have been used for water purification, displayed very good retention abilities for simultaneous bacteria and viruses, while maintaining high permeation flux [36]. Membranes from nanofibres are used in many phases of the wastewater treatment: in pre-filtering treatments that remove contaminants that could affect further processes (such as ultrafiltration or nanofiltration), in removing of fine particles and bacteria, in wastewater colour removal, in removal of heavy metal ions such as chromium, nickel, copper, lead, cadmium and even in biological treatments. Membranes made of electrospun nanofibrous polysulfone and Nylon 6 have been used as pre-filters, showing good capacity to remove over 95% of particles with dimensions in the range of 7 - 10 μm without permanent fouling [37, 38].

Many types of electrospun nanofiber membranes proved to be effective in Cu(II) adsorption. Membranes from electrospun fibroin / Cellulose Acetate (80/20) nanofibres were used to adsorb Cu(II) ions in an aqueous solution after ethanol treatment, reaching a value of 22.8 mg/g Cu^{2+} [39]. Electrospun keratose / fibroin blend membranes exhibited better copper ion adsorption capacities compared to common fibrous filters [40]. A blend of polyvinyl alcohol / tetraethyl orthosilicate / aminopropyl triethoxysilane was electrospun and the obtained membranes were used for uranium (VI) adsorption from aqueous solutions, reaching a maximum adsorption capacity of 168.1 mg/g at pH of 4.5 and temperature of 45 °C [41]. Electrospun nanofibre membrane made of thiol-functionalized poly (vinyl alcohol) / SiO_2 showed high Cu^{2+} adsorption, which was maintained through six recycling processes [42]. Membranes from electrospun keratin nanofibers were tested for the Cu(II), Ni(II), and Co(II) metal ion removal, and their selectivity was found to decrease in the order Cu(II) > Ni(II) > Co(II). The experiments revealed a good adsorption capacity for Cu(II) ions, even with the coexistence of Co(II) and Ni(II) metal cations [43]. An adsorbent made by intercalation ethylene diamine tetraacetic acid in layered double hydroxides, followed by encapsulation into polyacrylonitrile polymer matrix using electrospinning showed a 93% copper removal efficiency when the initial Cu(II) concentration was 64 mg/L, with the adsorbent dosage of 1.50 g/L. The study concludes that the adsorption of Cu(II) is the result of a chelation process by interlayer ethylene diamine tetraacetic acid, a powerful chelating agent [44]. Cellulose electrospun nanofibers were modified with oxolone-2,5-dione, enhancing this way the surface area of the nanofiber mat and helping in detecting heavy metals like cadmium and lead [45].

Many electrospun membranes have been tested for colour removal in wastewater. Membranes made of electrospun keratin nanofibrous acted like adsorbents for Methylene Blue by means of a chemisorption process which occurred by ion exchange [46]. Electrospun polyhydroxyalkanoate film was used as adsorbent agent for Malachite Green, a triarylmethane dye. Titanium dioxide photocatalysts have been incorporated into the film to achieve, besides adsorption, a photocatalytic degradation effect, achieving complete discoloration in 45 min under solar irradiation [47]. Electrospun nanofiber membranes from polysulfone, polyvinylidene fluoride and polyethylene have been used in membrane bioreactors treating molasses wastewater, showing performances comparable to the commercial membranes [48].

3.2. Desalination

As drinking water resources become more and more scarce, desalination turned out to be an important method to convert the salted water of oceans and seas into potable water. Two basic technologies are used to remove salts from ocean water: thermal evaporation and membrane separation. Desalination, the process of extracting mineral components (mainly sodium chloride) from saline water, is often obtained using membrane processes such as reverse osmosis and nanofiltration. The principle of these processes is the following: salted water is forced through a membrane that permits only water molecules to pass, blocking salt (or other minerals) molecules. Mats from electrospun polyacrylonitrile nanofibers have been used as support layer for thin film composite nanofiltration membranes. These membranes were tested in desalting a saline solution (2000 ppm concentration), and was found that as the fibre size decreased, the salt elimination increased, but at the expense of permeate flux [49]. It was also found that the optimum concentration of polyacrylonitrile for the electrospinning of fibres was 8 % wt.

Electrospun nanofibrous membranes that show higher hydrophilicity and lower surface roughness are best for pressure-driven membrane processes, while hydrophobicity is an essential feature for membranes used in the membrane distillation process, a thermal-membrane desalination technology using low-grade or waste-grade heat [50]. The principle of membrane distillation is the following: hydrophobic membranes separate the liquid phase from the vapour phase within the membrane pores, and only the vapours are able to pass through the membrane, while the liquid feed including dissolved components, is retained by the membrane. Membranes form electrospun polyacrylonitrile nanofibres were turned from hydrophilic to hydrophobic by surface fluorination, without altering the morphology and structure of the membranes. It was found that if fibre orientation is opposite to the flow direction, the desalination performance is affected in terms of permeate flux.

Membranes from electrospun polyvinylidene fluoride nanofiber, which have been subjected to a heat-press post-treatment to improve their integrity and enhance water permeation flux, have been used for membrane distillation. These membranes showed a steady water permeation flux of about 21 kg m^{-2} h^{-1} for 15 h when using a 3.5 % wt NaCl solution [51]. Composite membranes from electrospun nanofibers of poly (vinylidenefluoride) and poly (vinylidene fluorideco-hexafluoropropylene) were used in a direct contact membrane distillation. It was found that after 12 hours of treatment, the salt rejection was 99.9888 % for the first type of membrane and 99.9901 % for the second one. The permeate flux of the second composite membrane is 4.28 kg/ m^2hr, higher than the commercial membranes [52]. Dual-layered electrospun nanofibrous membranes obtained from polyvinylidene fluoride and polysulfone were used in membrane distillation processes. The permeate flux reached values over 50 kg/m^2·h and 99.99% NaCl rejection.

3.3. Air filtration

One of the most important sources of air pollution are the fine particles (less than 2.5 mm in diameter), which are a main cause of acute and chronic effects on human health [53]. This is why air filtration technology is of great importance. Numerous fibrous membrane based on a various range of fibres are used in diverse filtration applications. Many of these are made of electrospun nanofibers, as they have some features that make them suitable for air filtration area, such as high specific surface area and open porosity, together with an interconnected porous structure [54]. Electrospun polyamide 6 nanofibres have been deposited onto a nonwoven viscose substrate to make protective clothing against nanoparticulate, showing up to 80% retention of 20 nm size particles and over 50% retention of 200 nm size nanoparticles [55]. Other membranes were obtained from electrospun nanofibres of polyamide 6, obtained from a solution of polymer solved in formic acids. These membranes, that have a thickness of 71 μm, proved to have good filtration potential for the microparticles with 0.3 μm diameter, in conditions of very good dimensional stability [56].

Nanoparticles such as MgO, TiO_2, Al_2O_3 and other oxides have been incorporated into nanofibers to obtain membranes for air filtration applications. Membranes obtained by deposing various metal oxide nanoparticles using liquid phase deposition and electrospraying techniques on the surface of nanofibers made by electrospinning of poly(ethylene terephthalate) with cellulose acetate proved to have potential as filter media in protective clothing and air filter applications [57]. Membranes made of nonwoven polypropylene covered by a polyvinyl alcohol were effective in eliminating organic polar compounds, ethers and compounds containing carbonyl groups from gas [58]. Transparent air filter for indoor air protection have been obtained by using electrospun polymer nanofibers. These filters with ~90% transparency show high air flow and high effectiveness as air filters (over 95% removal of PM2.5 under extreme hazardous air-quality conditions) [59]. Electrospun nanofibre membranes from waste high-impact polystyrene were applied as air filtration media, showing an efficiency greater than 99.999%, which was equivalent to that of the HEPA filter [60]. Electrospun biodegradable polylactide air filter membranes showed a higher dust capture efficienc [61].

4. Conclusions

Electrospinning is a versatile method of producing nanofibrous porous membranes using a wide range of polymers. By choosing the polymer and the solvent, as well as the parameters of the electrospinning process, the morphology of the resulting membranes can be determined. The use of electrospun nanofibre membranes in filtering processes is the result of their special characteristics, namely the high surface-to-mass ratio, the high pore interconnectivity and the uniform distribution of their dimensions. In addition, the number of applications of these membranes is greatly enhanced by their functionalization by changing the surface chemistry of the membrane. The paper first briefly reviews the principles of electrospinning, and then summarizes the latest developments in the use of membranes as a water filtration medium (drinking or waste) and air. The contribution of electrospun nanofibre membrane processes to desalination processes is also presented.

5. References

[1] Ficai D and Grumezescu A M 2017 *Nanostructures for Novel Therapy Synthesis, Characterization and Applications* (Amsterdam: Elsevier)
[2] Manea L R Bertea A P, Nechita E, Popescu C V and Sandu I 2016 *Rev. Chim (Bucharest)* **67** (7) 1284
[3] Manea L R, Scarlet R and Sandu I 2015 *Rev. Chim.* **66** (10) 1622
[4] Popa A, Bucevschi A, Pustianu M, Manea L R and Sandu I 2016 *Mat. Plast. (Bucharest)* **53** (2) 316
[5] Manea L R, Nechita E and Sandu I 2015 *Rev. Chim. (Bucharest)* **66**(11) 1841
[6] Istrate B, Munteanu C, Lupescu S, Benchea M and Vizureanu P 2017 *IOP Conference Series: Materials Science and Engineering* **209** 012024
[7] Lupescu S, Istrate B, Munteanu C, Minciuna M G, Focsaneanu S and Earar K 2017 *Rev. Chim.* **68** 1408
[8] Ahmed F, Lalia B and Hashaikeh R 2015 *Desalination* **356** 15
[9] Manea L R and Sandu I 2015 *Rev. Chim. (Bucharest)* **66**(12) 1968
[10] Manea L R, Danu M C and Sandu I 2015 *Rev. Chim. (Bucharest)* **66**(6) 868
[11] Calin M A, Khenoussi N, Schacher L, Adolphe D, Manea L R, Gradinaru I, Zetu I and Stratulat S 2013 *Mat. Plast. (Bucharest)* **50** (4) 257
[12] Scarlet R, Manea L R Sandu I, Cramariuc B and Sandu A V 2012 *Rev. Chim.* **63** (8) 777
[13] Schaefer K, Thomas Arinstein H, Dalton P and Moeller M. 2007 Nano-Fibres for Filter Materials *Multifunctional Barriers for Flexible Structures* S Duquesne, C Magniez et al, eds. (Berlin: Springer) chapter 6 p. 125-138
[14] Manea L R, Bertea A, Nechita E, Popescu C V and Sandu I 2016 *Rev. Chim. (Bucharest)* **67**(8) 1607
[15] Manea L R Scarlet R, Leon A L and Sandu I 2015 *Rev. Chim. (Bucharest)* **66**(5) 640
[16] Manea L R Cramariuc B, Scarlet R, Cramariuc R, Sandu I and Popescu V 2015 *Mat. Plast. (Bucharest)* **52**(2) 180
[17] Manea L R, Hristian L, Leon A L and Popa A 2016 *IOP Conf. Series: Mater. Sci.Eng.* **145** 032006
[18] Ko F and Wan Y 2014 *Introduction to Nanofiber Materials* (Cambridge: University Printing House)
[19] Manea L R, Popa A and Bertea A 2017 *IOP Conf. Ser.: Mater. Sci. Eng.* **209** 012074
[20] Bertea A, Manea L R, Bertea A and Hristian L 2017 *IOP Conf. Ser.: Mater. Sci. Eng.* **209** 012073
[21] Amariei N, Manea L R, Bertea A P, Bertea A and Popa A 2017 *IOP Conf. Ser.: Mater. Sci. Eng.* **209** 012092
[22] Amariei N, Manea L R, Bertea A P, Cramariuc R, Bertea A and Cramariuc O 2017 *IOP Conf. Ser.: Mater. Sci. Eng.* **209** 012091
[23] Rozemarie M L, Andrei B, Liliana H, Cramariuc R and Cramariuc O 2017 *IOP Conf. Ser.: Mater. Sci. Eng.* **209** 012063
[24] Manea L R, Hristian L, Leon A L and Popa A 2016 *IOP Conf. Series: Mater. Sci.Eng.* **145** 032007
[25] Manea L R, Scarlet R, Amariei N, Nechita E and Sandu I G 2015 *Rev. Chim.* **66** (4) 542
[26] Manea L R, Hristian L, Ostafe M M, Apostol L L and Sandu I 2016 *Rev. Chim. (Bucharest)* **67**(9) 1758
[27] Haider A, Haider S and Kang I K 2015 *Arab J Chem* **8** 1

[28] Liao Y, Loh C H, Tian M, Wang R and Fane A 2018 *Prog Polym Sci* **77** 69
[29] Arinstein A 2018 *Electrospun Polymer Nanofibers* (Singapore: Pan Stanford Publishing)
[30] Manea L R, Cramariuc B, Caunii V, Sandu I 2015 *Mat. Plast.* **52** 1 82
[31] Yarin A L 2011 *Polym. Adv. Technol.* **22** 310
[32] Li Z and Wang C 2013 *One-Dimensional Nanostructures* (Berlin: Springer)
[33] Cavalieri S 2016 *Electrospinning for Advanced Energy and Environmental Applications* (Boca Raton: CRC Press)
[34] Bae J, Baek I and Choi H 2017 *Chem Eng J* **307** 670
[35] Cooper A, Oldinski R, Ma H, Bryers J and Zhang M 2013 *Carbohyd Polym* **92** 254
[36] Sato A, Wang R, Ma H, Hsiao B and Chu B 2011 *J Elect Microsc* **60** 201
[37] Gopal R, Kaur S, Feng C Y, Chan C, Ramakrishna S, Tabe S and Matsuura, T 2007 *J Membr Sci* **289** 210
[38] Aussawasathien D, Teerawattananon C and Vongachariya A. 2008 *J. Membr. Sci.* **315** 11
[39] Zhou W, He J, Cui S and Gao W 2011 *Fibers Polym* **12** 431
[40] Ki C S, Gang E, Um I and Park Y 2007 *J. Membr. Sci.* **302** 20
[41] Keshtkar A R, Irani M and Moosavian M 2012 *J Radioan Nucl Ch* **295** 563
[42] Wu S, Li F and Wang H 2010 *Polymer* **51** 6203
[43] Aluigi A, Corbellini A, Rombaldoni F and Mazzuchetti G 2013 *Text Res J* **83** 1574
[44] Chena H, Lin J, Zhang N, Chen L, Zhong S, Zhang W, Ling Q and Wang Y 2018 *J Hazard Mat* **345** 1
[45] Stephen M, Catherine N, Brenda M, Andrew K, Leslie P, Corrine G 2011 *J. Hazard. Mater.* **192** 922
[46] Aluigi A, Rombaldoni F, Tonetti C and Jannoke L 2014 *J Hazard Mater* **268C** 156
[47] Sridewi N, Lee Y F and Sudesh K 2011 *Int J Photoenergy* **2011** 1
[48] Bilad M R , Westbroek P and Vankelecom I 2011 *J Membrane Sci* **380** 181
[49] Kaur S, Sundarrajan S, Rana D, Matsuura T and Ramakrishna S 2012 *J. Membr. Sci.* **392-393** 101
[50] Cai J, Liu X, Zhao Y and Guo F 2018 *Desalination* **429** 70
[51] Liao Y, Wang R, Tian M and Fane A 2013 *J Membr Sci* **425–426** 30
[52] Su C-Y, Shih J-H and Huang M-S 2012 *Fibers Polym* **13** 698
[53] Qiu H, Tian L and Ho K 2015 *Environ Pollut* **199** 192
[54] Wang N, Mao X, Zhang S, Yu J and Ding B 2014 Electrospun Nanofibers for Air Filtration *Electrospun Nanofibers for Energy and Environmental Applications* ed B Ding and J Yu (Berlin: Springer-Verlag) chapter 12 p. 299-324
[55] Faccini M, Vaquero C and Amantia D 2012 *J Nanomat* **2** 1
[56] Guibo Y, Qing Z, Yahong Z, Yin Y and Yumin Y 2013 *J Appl Polym Sci* **128** 1061
[57] Sundarrajan S and Ramakrishna S 2010 J. *Nanosci. Nanotechnol.* **10** 1139
[58] Daneleviciute-Vaisniene A Katunskis J and Buika G 2009 *Fibres Text East Eur* **17** 40
[59] Liu C, Hsu P-C, Lee H-W, Ye M, Zheng G, Liu N, Li W and Cui Y 2015 *Nat Commun* **6** 6205
[60] Zulfi A, Munir M and Hapidin D A 2018 *Mat Res Exp* **5** 23
[61] Selatile M, Ray S, Ojijo V and Sadiku R 2018 *J Env Chem Eng* **6** 762

Acknowledgments This work was supported by a grant of the Romanian national Authority for Scientific Research and innovation, UEFISCDI, project number PN III 46 BG 2016 within PNCDI III.

IOP Conf. Series: Materials Science and Engineering **374** (2018) 012082 doi:10.1088/1757-899X/374/1/012082

Simulation of Propagation of Compartment Fire on Building Facades

A Simion[1], H Dragne[2,3], D Stoica[4] and I Anghel[5]
[1] Laboratory of testing and research for building fire safety from National Institute INCERC Bucharest, National Research and Development Institute URBAN-INCERC, Bucharest, Romania
[2] Laboratory of testing and research for building fire safety from National Institute INCERC Bucharest, National Research and Development Institute URBAN-INCERC, Bucharest, Romania
[3]PhD, Technical University of Civil Engineering, Bucharest, Romania
[4] Laboratory of testing and research for building fire safety from National Institute INCERC Bucharest, National Research and Development Institute URBAN-INCERC, Bucharest, Romania
[5] Police Academy "Al. I. Cuza" Bucharest - Faculty of Firemans

E-mail: simion_i_adrian@yahoo.com

Abstract. The façade fire simulation of buildings is carried out with Pyrosim numerical fire modeling program, following the implementation of a fire scenario in this simulation program. The scenario that was implemented in the Pyrosim program by researchers from the INCERC Fire Safety Research and Testing Laboratory complied with the requirements of BS 8414. The results obtained following the run of the computational program led to the visual validation of effluents at different time points from the beginning of the thermal load burning, as well as the validation in terms of recorded temperatures. It is considered that the results obtained are reasonable, the test being fully validated from the point of view of the implementation of the fire scenario, of the correct development of the effluents and of the temperature values [1].

1. Introduction

Theoretical studies and existing experimental research have highlighted the fact that each fire has its own evolution that individualizes it, with no two fires of identical evolution [2]. Uniqueness of the event should be understood as referring to the whole set of factors and consequences of a fire [3]. Under these conditions, it is logical to conclude that it is impossible to make an accurate forecasting of the fires that can occur in a given system and in a particular context. A systematic study path is represented by the so-called fire modelling - schematically described in Figure 1 - which is an inherent part of the research evolution [4].

Fire models are used in various aspects of fire safety management:

- before fire, to estimate the risk of fire - this helps the specialists who intervene to identify areas of major risk;

- during fires, to determine the firefighting strategy - this would help the intervention teams to limit their losses and intervene safely;

- after fire, for the investigation and determination of fire causes - these are a useful tool for intervention forces, researchers and experts [5].

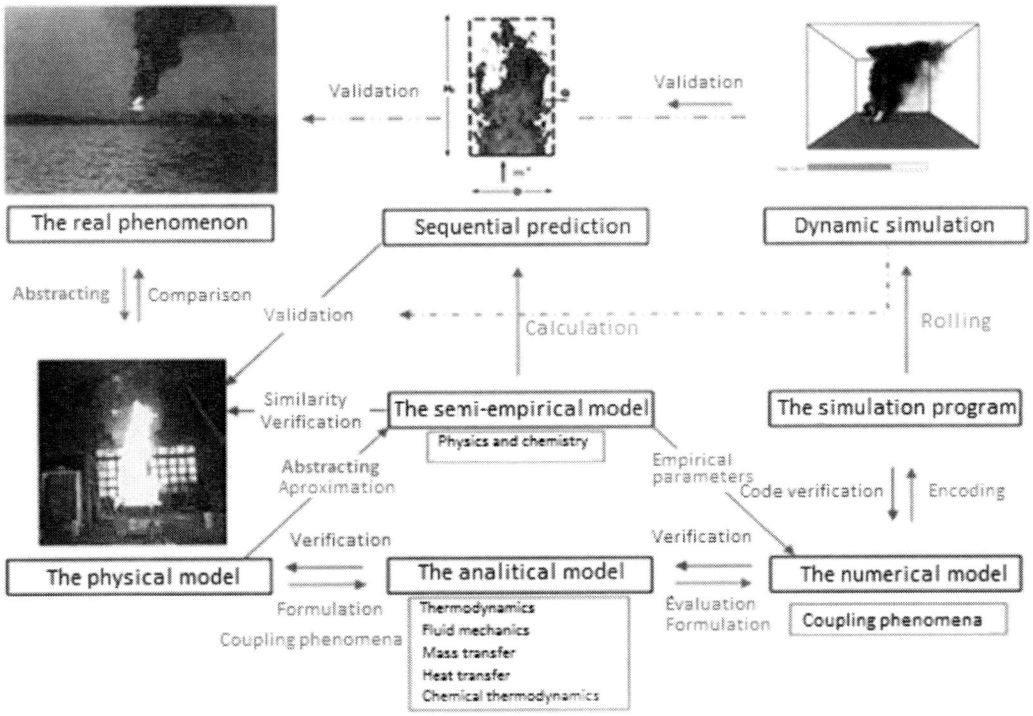

Figure 1. Fire block modelling scheme.

2. Experimental procedures

In the following lines, the Pyrosim fire modeling program is used to model field faults according to the BS 8414 standard. The modeling was done using the Pyrosim (FDS) program, and the Smokeview utility was used to view the results. part of the freeware FDS package [6].

The input data for the Pyrosim program for BS 8414 simulation are listed in the following Table 1.

Table 1. Input data of the simulation.

Category	Parameters	Value
Domain of calculation	Building area	Building 8.5 m high, P + 2
	Calculation cell size	0,05 m
	Total cell number	446520
Other Information	Fire chamber	Rectangle (1m x 1.5m), located right above the stack built according to the standard. The released heat flow of 1668.4 kW / m^2
	Calculation time	1800 seconds

The implementation of the scenario in the simulation program according to the BS 8414 geometric requirements, is as follows:

IOP Conf. Series: Materials Science and Engineering **374** (2018) 012082 doi:10.1088/1757-899X/374/1/012082

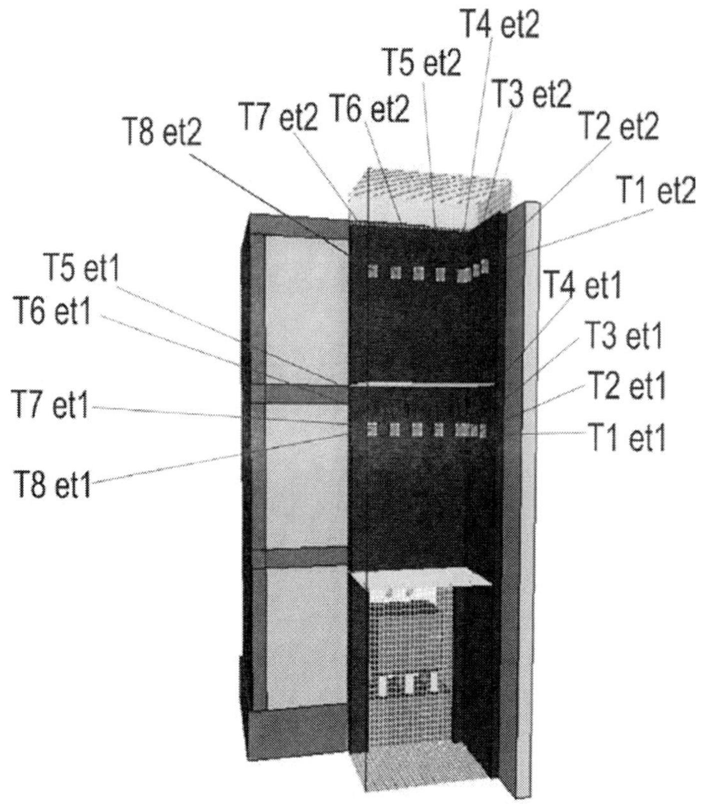

Figure 2. Viewing 3D viewing of the test stand, identifying the location of the thermocouples.

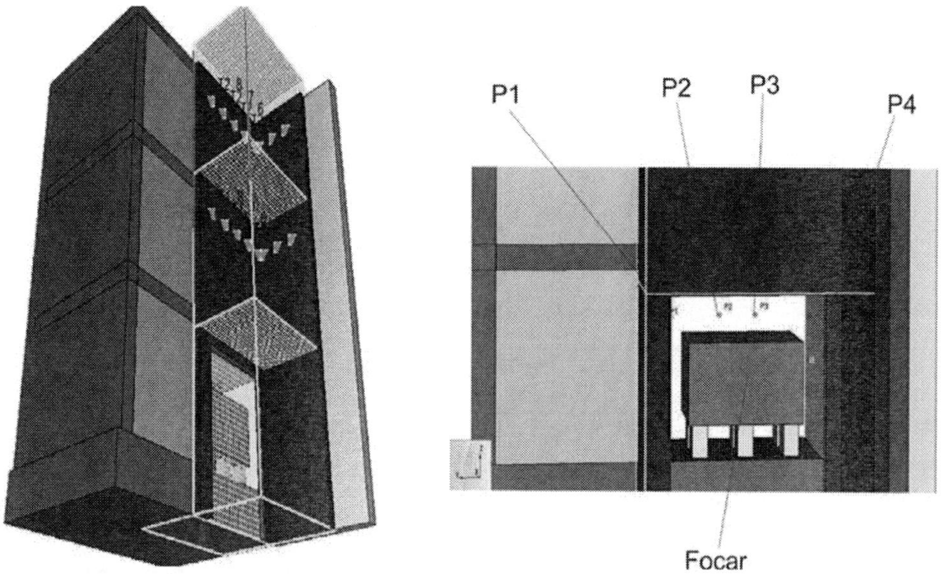

Figure 3. Visualization of the computational domains (eng: mesh) in the Pyrosim program and the standardized stand (wood stack) according to BS 8414.

3. Results and discussions

Modeling results: As can be seen in Figure 4, the development of smoke and flames over time follows the normal evolution of a classic fire. Also, Figure 5 shows a similar development of the dynamics of the flames on the façade between the two situations: simulation and experimental stand.

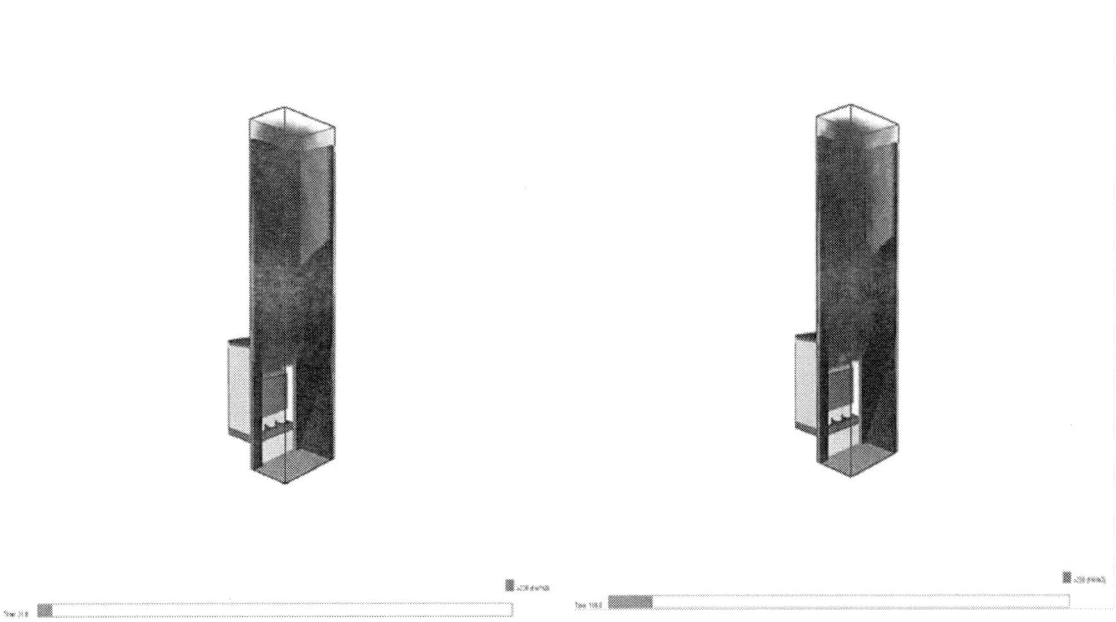

Figure. 4. Visualization in the Pyrosim program of the simulation performed according to BS 8414 at t = 36 s and 108 s.

Figure 5. Visualization of the simulation performed in the Pyrosim program, in parallel with the actual testing performed according to the standard. It is observed a similar development of the dynamics of the flames on the facade between the two situations: simulation and experimental stand.

IOP Conf. Series: Materials Science and Engineering **374** (2018) 012082 doi:10.1088/1757-899X/374/1/012082

Figure 6 shows views of section flame temperatures at two different time points.

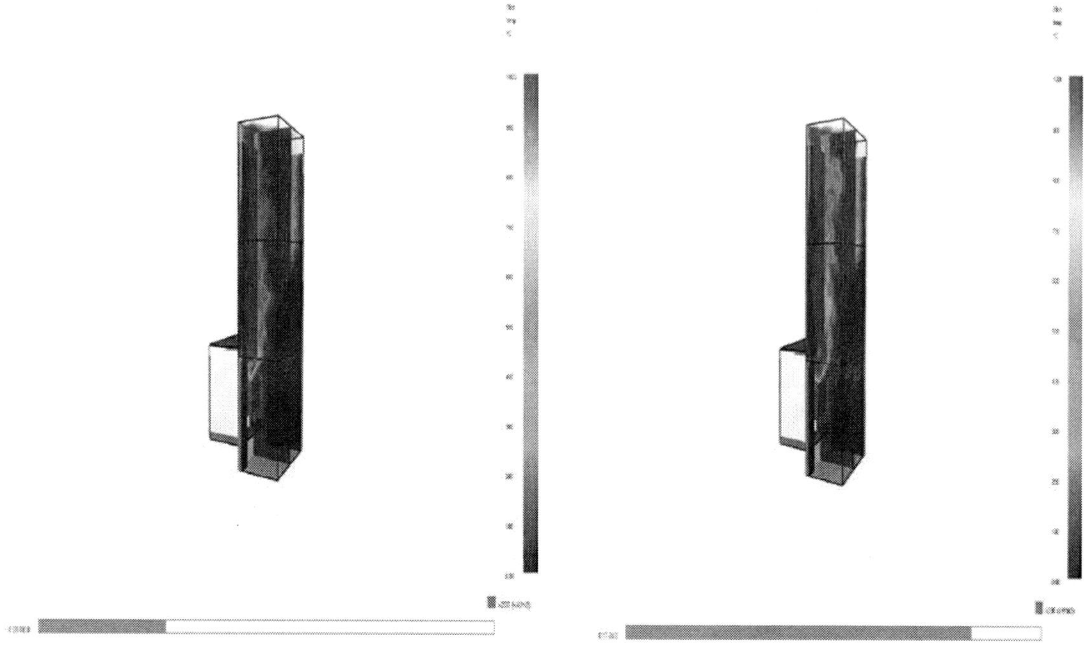

Figure 6. Viewing sectional flame temperatures at t = 360 s and t = 1044 s.

Regarding the temperature values, Figures 7 to 9 show comparative pairs of temperature graphs recorded by thermocouples in the real and simulated experiments. Generally, there is a 15-30% margin of valid value differences in the value.

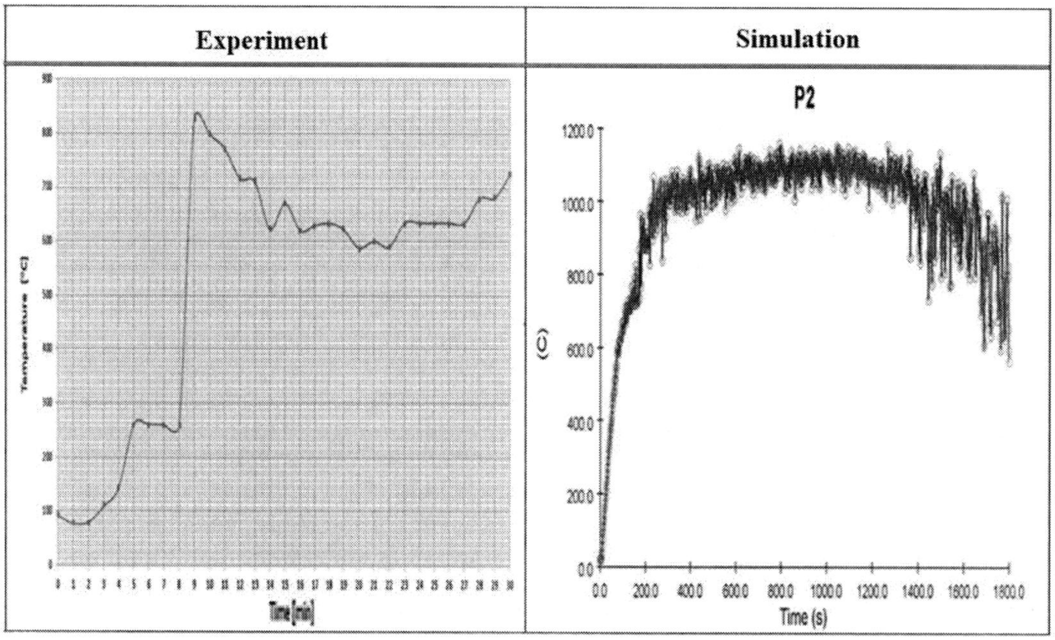

Figure 7. Graphical representation of temperature variation in the combustion chamber of the test bench.

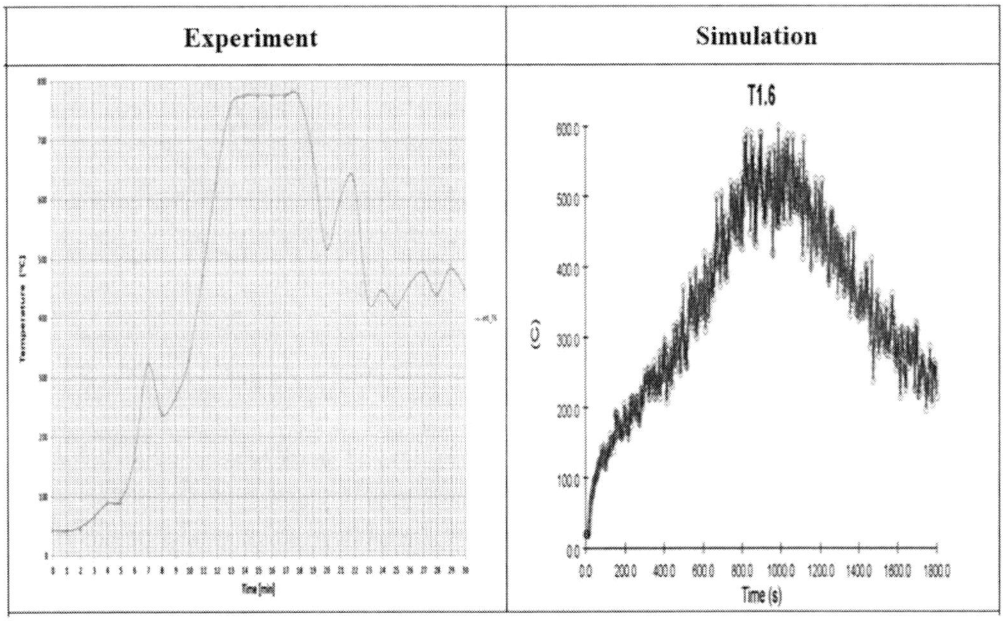

Figure 8. Graphical representation of temperature variation on floor 1 of the test bench.

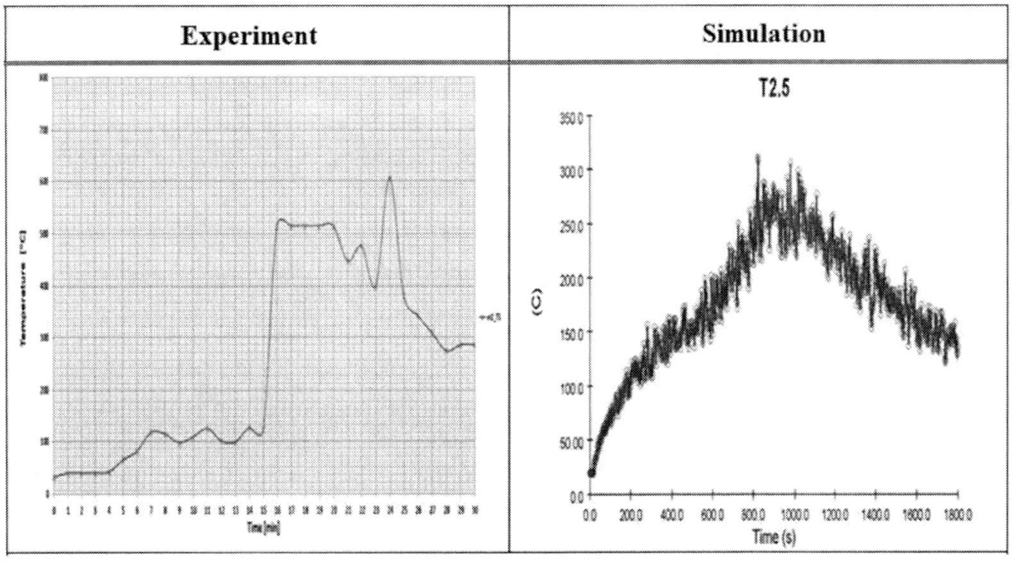

Figure 9. Graphical representation of temperature variation on the 2nd floor of the test bench.

4. Conclusions

The purpose of the paper is to implement the scenario in the simulation program according to the BS 8414 geometrical requirements and visual validation of the effluents at different moments from the initiation of the combustion as well as the validation in terms of temperatures. The results are considered reasonable, the test being fully validated in terms of scenario implementation and proper effluent development as well as temperature readings. The resulting HRRPUA value used in the simulation is 1668.4 kW (Heat Release Rate per Unit area)

The maximum HRR of 2500 kW (ie 1668.4 kW × 1.5 mp) can be considered as the optimal fire exposure value by testing and validation by mathematical simulations on a computer. According to good practices in numerical fire modelling, it is well known that each practical test is different in several ways:

- it is difficult to control the heat exposure of the façade due to the movement of the air around the combustion chamber or the movement of the air in the test area (wind);
- degradation and destruction of the façade;
- geometry is never identical from one test to another;
- the initial temperature of the walls before testing may influence the test result.

Therefore, a modelling attempt to reproduce such a phenomenon as burning is a difficult challenge for modelling researchers. But it is important to take steps towards validation, because such destructive tests on a real scale can not always be carried out. After complete validation, it is possible to use the data of a scenario in a multitude of tests that would not destroy anything, only would require the computing power of a computer.

As input data for the simulation performed in this report, a HRR curve, not a self-sustaining outbreak, was used, as there was no information on the loss of mass in the outbreak (the stack of wood).

5. References

[1] A Simion and H Dragne 2017 *IOP Conf. Ser.: Mater. Sci. Eng.* **209** 012102

[2] Lalu O 2016 Research on developing sustainable solution for thermal rehabilitation of the building envelope. Behaviour in case of fire, fire safety solutions, *Technical University of Construction Bucharest* 4-10

[3] Lalu O and Bubulete L 2017 *Construction Magazine* 8-11

[4] Anghel I, Nicolae S and Popa C 2009 *Studiu progresiv al modelelor de incendiu*. Sesiunea de comunicari stiintifice cu tema: Protecţia civilă, apărarea împotriva incendiilor şi gestinarea situaţiilor de urgenţă a Centrului National de Perfectionare a Pregatirii pentru Managementul Situaţiilor de Urgenţă - PROTCIV Ilfov

[5] Anghel I 2009 *Contribuţii privind modelarea şi simularea incendiilor prin metode numerice*, teză de doctorat, Universitatea de Petrol şi Gaze Ploieşti

IOP Conf. Series: Materials Science and Engineering **374** (2018) 012083 doi:10.1088/1757-899X/374/1/012083

Effect of Mixing Temperature on Characteristics of Thermoplastic Potato Starch Film

N H Zakaria[1, 2, a *], N Muhammad[1, 2, b], A V Sandu[2,4] and M M A B Abdullah[2, 3, c]
[1]School of Manufacturing Engineering, Universiti Malaysia Perlis, Perlis, Malaysia
[2]Center of Excellence Geopolymer & Green Technology (CEGeoGTech), School of Materials Engineering, Universiti Malaysia Perlis, Perlis, Malaysia
[3]Faculty of Engineering Technology, Universiti Malaysia Perlis, Perlis, Malaysia
[4]Gheorghe Asachi Technical University of Iasi, Faculty of Materials Science and Engineering, Blvd. D. Mangeron 41, 700050, Iasi, Romania

E-mail: [a]nurulhusna.zakaria@gmail.com, [b]noorhafiza@unimap.edu.my, [c]mustafa_albakri@unimap.edu.my

Abstract. This study present the preparation of potato starch film with glycerol as a plasticizer through a solution casting technique. The effect of mixing temperature (80°C, 85°C, 90°C and 95°C) on the tensile and microstructure properties of potato starch film was investigated. Results shows that the increase of temperature from 80°C to 85°C caused the increase in tensile strength. Nevertheless, as the temperature increase to 95°C, the tensile strength and elongation at break decreased, with the highest value of 2.6MPa and 15.7mm respectively. Both of tensile strength and elongation at break were recorded for starch film prepared at 85°C. Besides, the microstructure of the starch films showed gradual smoother surface, homogenous and grain growth, as the mixing temperature increased. Overall result shows that mixing temperature significantly influenced the properties of the potato starch films.

1. Introduction

Starch is known as one of the potential future green polymer materials that were expected to replace synthetic polymer market.Starch has been considered as the most promising resource for development of biopolymer due to several advantages; whichare renewable, biodegradable, non-toxicity, widely available and inexpensive [1, 2].

Hence, starch based biopolymer with these unique properties have become increasingly significant in various applications including plastic packaging, food industry, pharmaceutical and biomedical applications [3, 4]. In addition, starch has already known to have great potential in biomedical fields including tissue engineering scaffolds, bone cements, stents and drug delivery system [5–7].

Several studies have investigated the development and characterization of starch based films including potato starch. Podshivalov et al. [8], reported that potato starch possess several desirable feature such as exhibits good film-forming properties, chemical stability and biodegradable.

Besides, potato starch has demonstrated the possibility to be use in biomedical applications, as it have ability to improves its durability, good tissue integration and did not induce a pathological foreign body reaction [9, 15].

Starch can be process into thermoplastic materialsusing similar conventional techniques for synthetic polymer materials, such as solution casting, extrusion, injection molding and compression molding [2, 10].

Among of these techniques, solution casting is the most preferable method for the laboratory scale of production starch based films. During solution casting process, the presence of water is very important in order to gelatinize starch granules. Gelatinization occur when starch is heated with certain temperature in the presence of water and shear stress, thus causes the disruption of original starch structure to form thermoplastic starch (TPS) materials [11].

However, TPS products containing only water are sensitive and would lead to poorer mechanical and physical properties, due to their hydrophilic nature. Thus, the incorporation of plasticizer is required to overcome these limitations.

The addition of plasticizer helps to plasticize the starch solution and enhance the film flexibility, extensibility and ductility by reducing intermolecular reactions between starch molecules [12]. Glycerol are the most commonly used plasticizer for starch based films due to low cost and does not exhibit any toxic properties.

This study aims to prepare thermoplastic potato starch film with differentmixing temperature, as well as to investigate the tensile and microstructure properties of the film produced.

2. Materials and method

2.1. Materials

Commercially available potato starch, $(C_6H_{10}O5)n$ (HmbG Chemicals) was used as matrix material. Glycerol, $(C_3H_8O_3)$ obtained from (HmbG Chemicals) was used as plasticizer.

2.2. Preparation of the thermoplastic starch films

Starch films were prepared via solution casting technique. About 5g of potato starch was first dispersed in 100 ml deionized water with (30% relative to dry starch) of glycerol content according to the corresponding run of experiments.

The mixture was then heated to different mixing temperature of (80°C, 85°C, 90°C and 95°C) in a water bath for 30 min under constant stirring.

Subsequently, the mixture was poured homogeneously onto petri dish of 14 cm diameter. The mixture were then dried at 45°C in an oven with air force circulation.

The dry films were removed from the petri dish and stored in a dessicator (40% of relative humidity) in a plastic bag for a week prior to testing. Thermoplastic starch films with varies mixing temperature was denoted according to mixing temperature as SF80, SF85, SF90, SF95. Table 1 shows the formulation of thermopalstic starch film.

Table 1. Formulation of thermoplastic starch film.

Material	SF80	SF85	SF90	SF95
Potato starch (g)	5	5	5	5
Distilled water (mL)	100	100	100	100
Glycerol (%)	30	30	30	30
Mixing temperature (°C)	80	85	90	95

2.3. Tensile Test

The tensile test were performed as suggested by Sanyang et al. [4], and the tensile strength and elongation at break were determined according to the ASTM standard (D882-02) by using a 5kN INSTRON tensile machine.

Film strips were cut into 70mm x 10mm sections are subjected to crosshead speed of 2mm/min and initial grip separation of 40mm. The test was conducted with five replication for each sample and the average value was calculated.

2.4. *Microstructure Analysis*

The microstructure of the film was observed by using scanning electron microscopy (SEM) model (Hitachi TM-3000) operated at accelaration voltage of 15kV.

3. Results and discussion

3.1. *Tensile Test*

The mechanical properties of thermoplastic starch films were studied through a tensile test. This test was conducted in order to investigate the influence of mixing temperature onto the mechanical behaviour of the starch film. The results of tensile strength and elongation at break are illustrated in Figure 1. As shown in Figure 1(a), the tensile strength of the starch films increased significantly with increasing temperature from 80°C to 85°C. However, as the temperature increase from 85°C to 95°C, the tensile strength decreased from 2.6Mpa to 2.2MPa. The highest tensile strength value of 2.6MPa was recorded for films prepared at85°Cmixing temperature, meanwhile films prepared at 80°C showed the lowest tensile strength value of2MPa. The decrease in tensile strength as the mixing temperature increased might be caused by the changes in starch structure. This behaviour could be related to the structural modifications of the starch network when heated at high temperature and the matrix of the film become less dense, thus weaken the starch film properties.

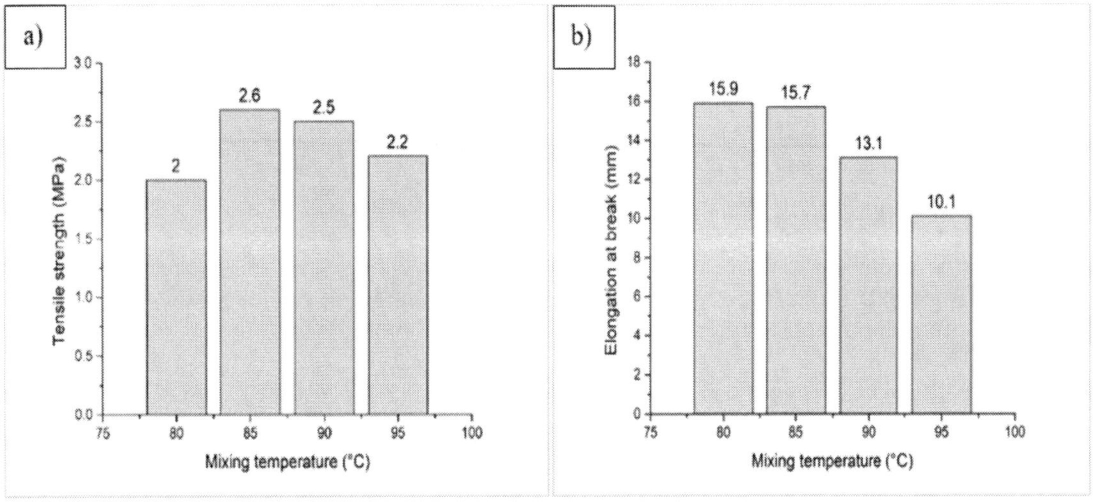

Figure 1. Tensile properties of potato starch film with different mixing temperature, (a) tensile strength and (b) elongation at break.

Figure 1(b) shows the effect of mixing temperature on the elongation at break of the potato starch film. The significant drop in the in elongation at break (15.9 mm to10.1 mm) was observed as the increased in the mixing temperature. This fact is probably due to the hydrogen bonds between the starch and glycerol occur as the temperature is increased, as reported by Vieira et al. [13]. The hydrogen bonds development at higher mixing temperature are associated with stronger interactions between plasticizer and starch molecules that inhibit a macromolecular movement, thus resulting in a decreased in elongation values.

3.2. *Microstructure Analysis*

Gelatinization process of thermoplastic starch produces measurable changes by microstructures characterization. The changes between grain shape and size, as well as contacts between the phases are commonly observed during gelation process. Figure 2 shows the microstructure of potato starch containing glycerol with different mixing temperature of (80°C, 85°C, 90°C and 95°C) at magnification of 1000x. Observation from SEM image reveal the changes in microstructure development with increasing mixing temperature during gelatinization process. The surface morphology of thermoplastic potato starch film for all processing temperatures illustrated different sizes of porosity and condition on the surface area.

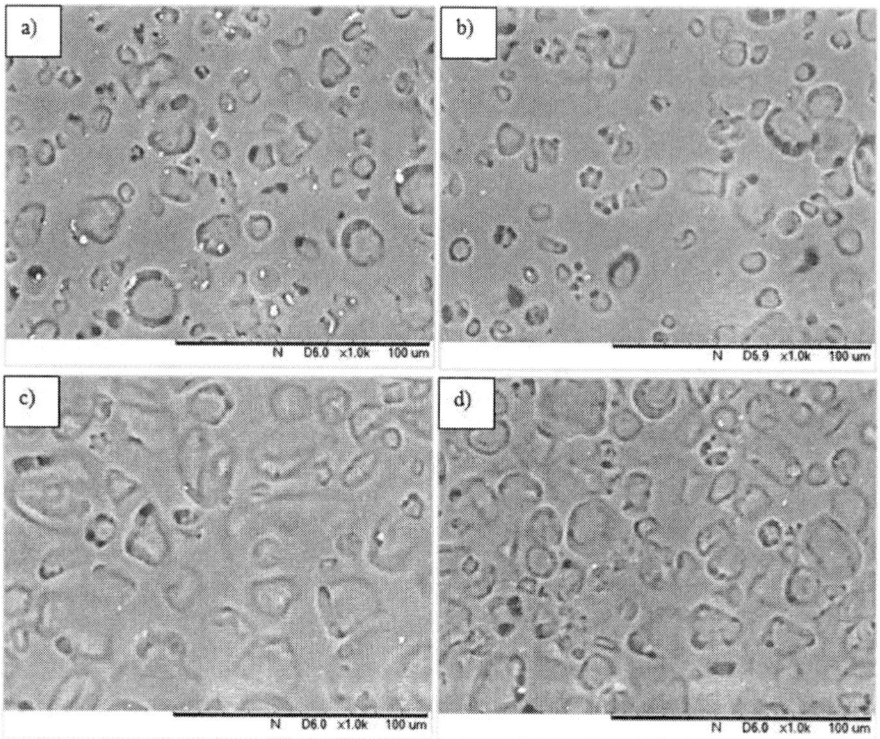

Figure 2. SEM image of potato starch film with different mixing temperature (magnification of 1000x): (a) SF80, (b) SF85, (c) SF90 and (d) SF95.

Figure 2(a) showed rougher surface with more porous and the presence of swollen starch granules which could be due to insufficient mixing temperatureto gelatinize the starch. As mixing temperature was further increased, the surface of film showed good dispersion and gradual smoothing with less porosity, but slightly visible agglomerates starch granules which can be observed in Figure 2(c) and Figure 2(d).Similar finding was reported by Mali et al. [14], which stated that the increasing mixing temperatureof gelatinization could lead to the smoother surface and homogeneity of thermoplastic starch film.This is because the higher mixing temperature will disrupt the starch molecules interactions and increase the swelling capability of the starch network accordingly, and thereby increase the homogeneity of the film.

4. Conclusions

Potato starch films containing glycerol as a plasticizer with vary mixing temperatures was successfully fabricated through a solution casting technique. The varying mixing temperatures significantly affected the properties of potato starch film by reducing the brittleness and tensile strength. The results displays that thermoplastic potato starch prepared at 85°C of mixing temperatureshows the highest tensile strength, and thus identified as desired mixing temperature in this study. This result was supported by microstructure analysis of thermoplastic potato starch, which revealed more smooth and homogeneous surface of the film. These findings presents starch based film as a potential biopolymer for the development of biodegradable materials.

5. References

[1] Maitz M F 2015 *Biosurface and Biotribology* **1**(3) 161–176
[2] Maria Guadalupe Lomeli Ramirez T S F-S, Satyanarayana K G, Iwakiri S, Bolzon de Muniz G, Tanobe V 2011 *Carbohydr. Polym.* **86**(4) 1712–1722
[3] Ochubiojo E and Rodrigues A 2012 *Scientific, Health and Social Aspects of the Food Industry*, 355–380
[4] Sanyang M L, Sapuan S M, Jawaid M, Ishak M R, and Sahari J 2015 *Polymers (Basel)* **7** 1106–1124
[5] Chaunier L 2014 *Laryngoscope* **124** 875–881
[6] Torres F and Arce D 2015 *Biodegrad. Polym. Nanocomposites* 73–94
[7] Marques A P, Reis R L, and Hunt J A 2002 *Biomaterials* **23**(6) 1471–1478
[8] Podshivalov A, Zakharova M, Glazacheva E, and Uspenskaya M 2017 *Carbohydr. Polym.* **157** 1162–1172
[9] Velasquez D, Pavon-djavid G, Chaunier L, Meddahi-pellé A, and Lourdin D 2015 *Carbohydr. Polym.* **124** 180–187
[10] Majdzadeh-Ardakani K, Navarchian A H and Sadeghi F 2010 *Carbohydr. Polym.* **79**(3) 547–554
[11] Jochec-mo D, Janigová I and Chodák I 2017 *Eur. Polym. J.* **93** 843–849
[12] Rugchati O, Mahawongwiriya K and Thanacharoenchanaphas K 2013 *Int. J. Biol. Biomol. Agric. Food Biotechnol. Eng.* **7**(9) 491–494
[13] Gurgel M, Vieira A, Altenhofen M, Oliveira L and Beppu M M 2011 *Eur. Polym. J.* **47**(3) 254–263
[14] Grossmann E, Garcia M A, Martino M N, Victo M and Zaritzky N E 2002 *Carbohydr. Polym.* **50** 379–386
[15] Hanganu S C, Armencia A O, Grigoras S, Hanganu L C and Vizureanu P 2010 *Materiale Plastice* **47** 357-378

Acknowledgement
The authors acknowledge FRGS grant and School of Manufacturing Engineering, Universiti Malaysia Perlis for the lab facilities. Special thanks to those who contributed to this project directly or indirectly.

Preparation of Heat Treated Titanium Dioxide (TiO_2) Nanoparticles for Water Purification

A O Araoyinbo[1]*, M M A B Abdullah[1,2], A Rahmat[2], A I Azmi[1], P Vizureanu[2,3] and W M F Wan Abd Rahim[1]
[1]Faculty of Engineering Technology, Universiti Malaysia Perlis, Perlis Malaysia
[2]Center of Excellence Geopolymer and Green Materials (CEGeoGTech),School of Materials Engineering, Universiti Malaysia Perlis (UniMAP), Perlis, Malaysia
[3]Gheorghe Asachi Technical University of Iasi, Faculty of Materials Science and Engineering, Blvd. D. Mangeron 41, 700050, Iasi, Romania

Email: alaba@unimap.edu.my

Abstract. Photocatalysis using the semiconductor titanium dioxide (TiO_2) has proven to be a successful technology for waste water purification. The photocatalytic treatment is an alternative method for the removal of soluble organic compounds in waste water. In this research, titanium dioxide nanoparticles were synthesized by sol-gel method using titanium tetraisopropoxide (TTIP) as a precursor. The sol was dried in the oven at 120°C after aging for 24 hours. The dried powder was then calcined at 400°C and 700°C with a heating rate of 10°C/min. The phase transformation of the heat treated titanium dioxide nanoparticles were characterized by X-Ray Diffraction (XRD,and the surface morphology by Scanning Electron Microscopy (SEM). The photocatalytic activity of the heat treated titanium dioxide nanoparticles in the degradation of methyl orange (MO) dye under ultraviolet (UV) light irradiation has been studied. At calcination temperature of 400°C, only anatase phase was observed, as the calcination temperature increases to 700°C, the rutile phase was present. The SEM images show the irregular shape of titanium dioxide particles and the agglomeration which tends to be more significant at calcined temperature of 700°C. Degradation of methyl orange by 5 mg heat treated titanium dioxide nanoparticles gives the highest percentage of degradation after irradiation by UV lamp for 4 hours.

1. Introduction

Clean water is a crucial source of continuous existence and disease free water is essential for human beings as well as for the socioeconomic development of communities and states. Presently many societies are facing rising demands of potable water as the availability of fresh water supplies are depleting due to the extended droughts, population growth, more stringent health based regulations, decline in water quality particularly of groundwater due to increasing groundwater and surface water pollution, unabated flooding and competing demands from a variety of competing users [1]. Poor sanitation, water scarcity, deterioration of water quality, emergence of waterborne diseases, and lack of cleanwater supply are all posing global challenges due to therising demand by an increasing world population [2].

Dyes are a significant source of water contaminants released from the industrial effluents. The waste water released by the textile industries is rated as one of the most polluting effluent among the entire industrial sectors including pigment, leather, food and paper industries [3]. Photocatalysis by semiconductor particles are a promising technology that has great potential to control aqueous organic contaminants and considered as a new oxidation technology for water purification treatment [4]. There are several potential photocatalysts used for degradation of various organic compounds and dyes but the most promising and widespread substance is the titanium dioxide (TiO_2) nanoparticles.

Titanium dioxide nanoparticles are significantly more reactive than larger particles due to their larger surface area. Titanium dioxide nanoparticles have shown to be a relatively cheap and effective for removing organic compounds and pollutant gases due to its high chemical stability, high photocatalytic activity, non-toxicity and high ability to break molecular bonds leading to degradation [5].

Photocatalytic oxidation reactions have high potential to completely mineralized organic compounds to carbon dioxide (CO_2), water vapor (H_2O) and other inorganic substances by solar light or the photocatalytic systemsequipped with artificial ultraviolet (UV) light [6]. Photocatalysis using titanium dioxide photocatalyst is a rapidly expanding technology for water treatment from refractory and toxic organic pollutants such as dyes, pesticides and pharmaceuticals [7].

In this project, theheat treated titanium dioxide (TiO_2) nanoparticles act as a water filtration medium and the methyl oranges (MO) act as a modelpollutant compound. The titanium dioxide nanoparticles were synthesized via sol-gel method and calcined at 400°C and 700°C in the furnace. The properties of titanium dioxide nanoparticles were characterized by X-Ray Diffractometer (XRD), Scanning Electron Microscope (SEM), Energy Dispersive Spectroscopy (EDS) and UV-Vis Spectrophotometer. Degradation of methyl orange was investigated by photocatalysis process in the presence of titanium dioxide nanoparticles under UV light irradiation.

2. Materials and methods

2.1. Preparation of titanium dioxide nanoparticles

The materials that were used to synthesized titanium dioxide are titanium tetraisopropoxide (TTIP), ethanol (C_2H_5OH), deionized water (H_2O) and hydrochloride acid (HCl). Titanium dioxide nanoparticles weresynthesized via sol-gel method, 12 mls of TTIP was dissolved into 100 mls of ethanol in the beaker and stirrer for 30 minutes using a magnetic stirrer.

For hydrolysis reaction, 3 mls of deionized water and 1 ml of hydrochloride acid were added to the solution drop wise. The mixed solution was then stirred constantly for 2 hours to get a homogeneous solution. The pH value of the mixed solution was maintained in the acidity range of pH 3, after aging for 24 hours, the gel was dried and then heated at 400°C and 700°C.

2.2. Characterizations

The heat treated titanium dioxide (TiO_2) nanoparticles prepared by sol-gel process were then characterized by X-ray diffraction (XRD), and Scanning Electron Microscopic (SEM). The phase transformation and the surface morphology of titanium dioxide nanoparticles were observed.

2.3. Photocatalytic degradation

The photocatalytic activity of the titanium dioxide (TiO_2) nanoparticles was examined by the degradation of methyl orange (MO) under UV light illumination. Three different concentrations of dye solutions (10 mg/l, 20mg/l and 30mg/l) were prepared by dissolving 10 mg, 20 mg and 30 mg of methyl orange powder in 1 litre of deionized water separately in the beaker, the pH value of the solution was measured to be 7.

The synthesized titanium dioxide nanoparticles were mixed in the methyl orange solution of various concentrations to understand the co-relation of the initial concentration of methyl orange and photocatalytic activity. The suspensions containing 3 mg, 5 mg and 7 mg catalyst and 30 mls of methyl orange were stirred for 30 minutes in the dark to establish adsorption-desorption equilibrium to eliminate the error due to any initial adsorption effect and pre-produced electrons and holes from environmental light sources. A 25Wattsultraviolet (UV) germicidal lamp fixed at a distance of 27 cm above the surface of the solution was used as a UV light source. The stirred suspensions were then irradiated with UV light and 4 mls of the mixed solution was drawn out every hour for absorbance analysis. The absorption spectra of methyl orange were measured using a UV–Visible spectrophotometer.

3. Results and discussion

3.1. Phase analysis

The XRD patterns of titanium dioxide nanoparticles were obtained and the phases present were identified by JCPDS database (Powder Diffraction Files, Joint Committee on Powder Diffraction Standards) shown in Figure 1.

The 120°C dried titanium dioxide exhibit an amorphous phase where no distinct peak is identified. The 400°C titanium dioxide exhibit anatase phase and the significant peaks were observed. This result confirmed that width of the anatase peak diffraction from XRD indicating the smaller crystalline size at 400°C.

Phase transformation occurred from anatase phase to thermodynamically more stable rutile phase when calcination temperature increases to 700°C.

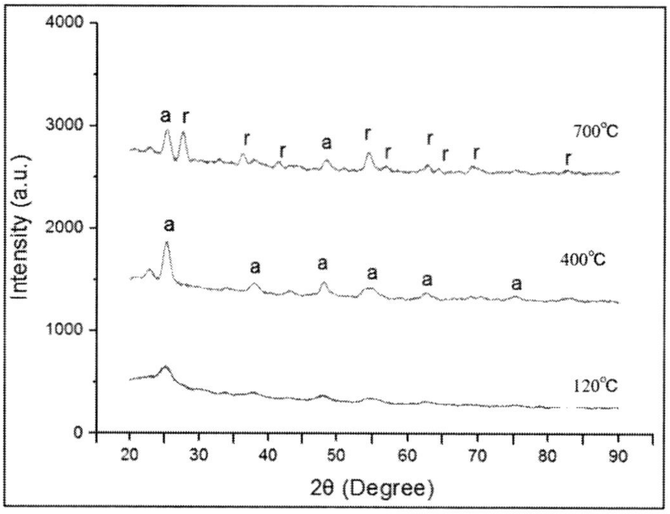

Figure 1. XRD patterns of titanium dioxide nanoparticles.

The anatase and rutile peaks obtained with decrease intensity. Among the XRD peaks, the width of 25.4° is a useful peak since it has high intensity which in turn is used to determine the crystal size. The peak value corresponds to the tetragonal anatase phase.

3.2. Morphology analysis

The SEM images were analyzed under 10 kx magnification. The morphology of the titanium dioxide nanoparticle powders dried at 120°C isshown in Figure 2, most of the particles are spherical shaped and different in sizes.

The titanium dioxide nanoparticles heated at 400°C and shown in Figure 3 shows a more uniform particles at the surface, while an increase in the temperature to 700°C as shown in Figure 4 shows that the particle size increase and the agglomeration becomes more significant. These results explained the effect of the heat treatment on the particle size.

Higher surface area of titanium dioxide nanoparticles gives good performance. In the rutile form, the atoms occupy the least space.

Figure 2. Titanium dioxide nanoparticles dried at 120°C.

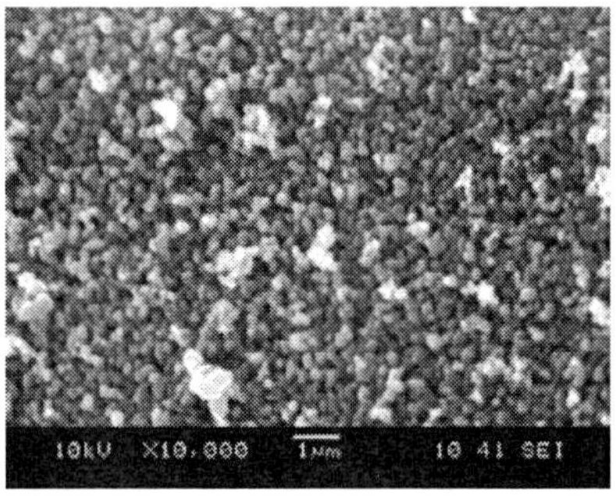

Figure 3. Titanium dioxide nanoparticlescalcined at 400°C.

Figure 4. Titanium dioxide nanoparticlescalcined at 700°C.

This makes the rutile form the most stable at higher temperatures. An important requirement forimproving the titanium dioxide catalytic activity is to increase its specificsurface area, which is certainly dependent on the crystal size. The smaller the catalyst, the larger will be its specific surfacearea. Calcination temperature is also important for removing the organic molecules from the final products and completing the crystallization. Very high calcination temperature will result in aggregation and phase transformation, and this affects the microstructures as well as the properties of titanium dioxide nanoparticles.

3.3. Photocatalytic Activity
Methyl orange (MO) is an organic dye that is commonly used as organic model compound in wastewater photocatalytic degradation studies. UV-Vis spectra of MO have two absorption band

maxima situated at 270 nm and 465 nm. The band at 465 nm was used for quantification. The photocatalytic activity of titanium dioxide nanoparticles were examined by the efficiency of degradation of methyl orange. Different concentrations of methyl orange (10, 20 and 30 mg/l) were used to evaluate the degradation percentage by using 400°C and 700°C heat treated titanium dioxide nanoparticles. The graph of absorbance versus wavelength was obtained from the UV-Vis Spectrophotometer. The peak of the graph was identified and the maximum absorbance value at the 463 nm wavelength at different time was observed. Figure 5 shows the absorbance spectra of 10 mg/l methyl orange degrade by 5 mg heat treated titanium dioxide nanoparticles. The absorbance value decreases as the time of irradiation increases.

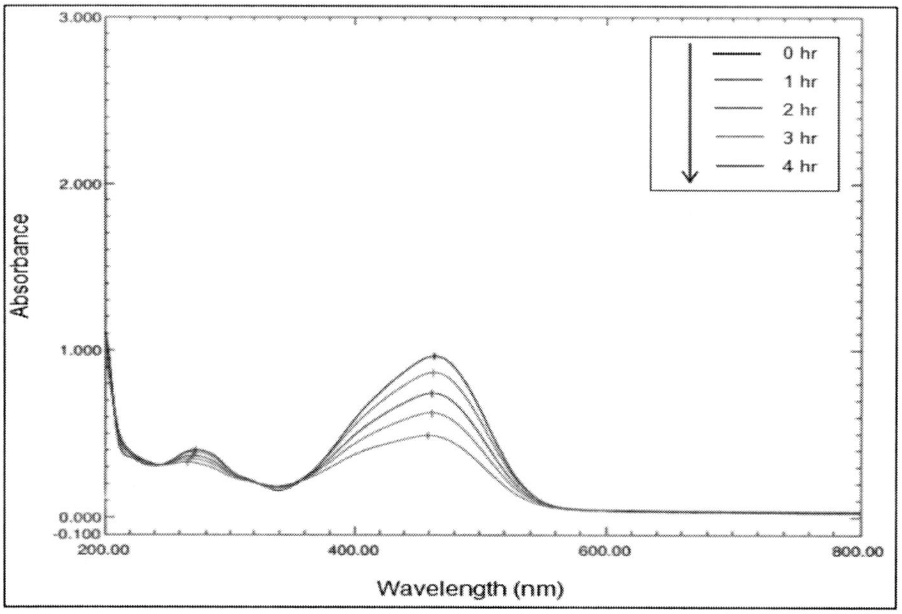

Figure 5. Absorbance spectra of 10 mg/l MO with 5 mg heat treated TiO_2 (700°C).

The absorbance value at 463 nm wavelength of every hour after irradiation is tabulated in the Table 1. The percentage of degradation is calculated, it shows that as irradiation time increase, the percentage also increases. After 4 hours of irradiation, the percentage reach 49.28% and marked the highest among all the degradation group using different titanium dioxide nanoparticles and methyl orange concentrations,

Table 1. The percentage of degradation of TiO_2 for 4 hours irradiation.

10 mg/l of Methyl Orange			
Amount of TiO_2	Irradiation Time	Absorbance value at 463 nm	Percentage of Degradation(%)
	0 hour	0.968	0
	1 hour	0.870	$\frac{0.968-0.870}{0.968}\times 100\% = 10.12$
5mg	2 hour	0.746	$\frac{0.968-0.746}{0.968}\times 100\% = 22.93$
	3 hour	0.628	$\frac{0.968-0.628}{0.968}\times 100\% = 35.12$
	4 hour	0.491	$\frac{0.968-0.491}{0.968}\times 100\% = 49.28$

Figure 6 shows the removal rate of methyl orange after 4 hours UV light irradiation. Comparison is made based on the calcination temperature of titanium dioxide nanoparticles, the concentrations of methyl orange and the amount of loading catalyst. 10 mg/l of methyl orange gives the highest removal rate in between 20 mg/l and 30 mg/l concentrations. The removal rate of 10 mg/l methyl orange by 5 mg titanium dioxide nanoparticles mark the highest while the removal rate of 30 mg/l methyl orange by 5 mg titanium dioxide reached the lowest. Both groups of titanium dioxide nanoparticles were calcined at 700°C.

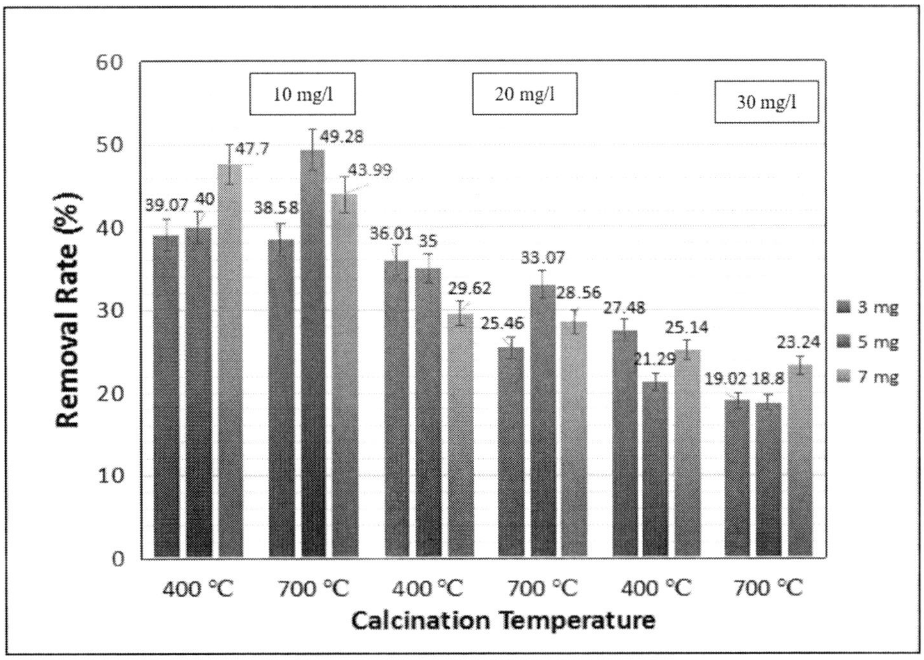

Figure 6. The removal rate of MO versus calcination temperature of TiO_2.

4. Conclusions

Titanium dioxide (TiO_2) nanoparticles were successfully synthesized via sol gel process. The XRD patterns of titanium dioxide nanoparticles has shown the phase transformation from anatase to rutile when calcination temperature increase from 400°C to 700°C. The crystallinity of titanium dioxide increase when calcination temperature increase and results in good photocatalysis reactions. The peaks of titanium dioxide nanoparticles were identified. The morphology of the titanium dioxide nanoparticles was examined using scanning electron microscopy (SEM). The morphology analysis from SEM confirms an agglomeration of particles from the sol gel method; the nanoparticle is spherical shape with different sizes. Degradation of methyl orange by heat treated titanium dioxide nanoparticles at 700°C give the optimum results among all the groups. This research indicates that the heat treated titanium dioxide nanoparticles are effective in degradation of methyl orange. The lowest concentrations of methyl orange give the greatest removal rate. Another observation made to prove the photocatalytic degradation process is the colour changes of aqueous methyl orange. When the time of UV light irradiation increases, the orange colour will became lighter. The findings of this research indicates that the heterogeneous photocatalysis in presence of ultraviolet light is a method which can be implemented and used for the complete degradation of azo dyes from waste water.

5. References

[1] Dalton J S, Janes P A, Jones N G, Nicholson J A, Hallman K R and Allen G C 2002 *Environmental Pollution* **120** 415-422

[2] Kanakaraju D, Glass B D and Oelgemoller M 2014 *Environmental Chemistry Letters* **12**(1) 27–47
[3] Tiwari D K, Behari J and Sen P 2008 *World Applied Sciences Journal* **3**(3) 417-433
[4] Madjene F, Aoudjit L, Igoud S, Lebik H and Boutra B 2013 *Transnational Journal of Science and Technology* **3**(10) 34-39
[5] Jallouli N, Elghniji K, Trabelsi H and Ksibi M 2017 *Arabian Journal of Chemistry* **10** S3640-S3645
[6] Tayade R J, Surolia P K, Kulkarni R G and Jasra R V 2007 *Science and Technology of Advanced Materials* **8**(6) 455-462
[7] Natarajan T S, Thomas M, Natarajan K, Bajaj H C and Tayade R J 2011 *Chemical Engineering Journal* **169**(1-3) 126-134
[8] Ioannou P, D Nica P, Paun V, Vizureanu P and Agop M 2008 *Physica scripta* **78** 065101

Acknowledgements
I would like to express my gratitude to Universiti Malaysia Perlis (UniMAP) and Center of Excellence Geopolymer and Green Materials (CEGeoGTech) for giving me the opportunity to be involved in this project.

IOP Conf. Series: Materials Science and Engineering **374** (2018) 012085 doi:10.1088/1757-899X/374/1/012085

The Effect of Baggase Ash on Fly Ash-Based Geopolimer Binder

R Bayuaji[1], M S Darmawan[1], N A Husin[1], R Banugraha[1], M Alfi[1], M M A B Abdullah[2, 3]

[1]Infrastructure Civil Engineering Department, Faculty of Vocations, Institut Teknologi Sepuluh Nopember, Surabaya, Indonesia

[2]Center of Excellence Geopolymer and Green Technology (CEGeoGTech), School of Materials Engineering, Universiti Malaysia Perlis (UniMAP), Perlis, Malaysia.

[3]Faculty of Engineering Technology, Universiti Malaysia Perlis (UniMAP), Kangar, Perlis

E-mail: bayuaji@ce.its.ac.id

Abstract. Geopolymer concrete is an environmentally friendly concrete. However, the geopolymer binder has a problem with setting time; mainly the composition comprises high calcium fly ash. This study utilized bagasse ash to improve setting time on fly ash- based geopolymer binder. The characterization of bagasse ash was carried out by using chemical and phase analysis, while the morphology characterization was examined by scanning electron microscope (SEM). The setting time test and the compressive strength test used standard ASTM C 191-04 and ASTM C39 / C39M respectively. The compressive strength of the samples determined at 3, 28 and 56 days. The result compared the requirement of the standards.

1. Introduction

The green concept for ordinary cement Portland industry today is to substitute separately and replace it entirely using environmentally local materials. The fact shows the cement industry contributes CO_2 emissions close to 8% of the total emissions in the world.

This phenom is caused a ton of cement production produces a ton of CO_2 emissions in the air, and this contributes to global warming [1]. Davidovits [2] defines that the natural ingredients to replace portland cement in concrete high geopolymer should contain silica and alumina. These elements will react with the liquid alkali such as Na_2SiO_3 and NaOH to make the process of polymerization in geopolymer concret.

Therefore, geopolymer concrete is now a popular material to develop continuously in Indonesia since approximately since 2010[3-6]. The main problem of geopolymer concrete in Indonesia is the quality of fly ash which is so varied that it causes problems to its setting time.

Therefore, this study to reduce the dominance of fly ash on geopolymer binder by substituting fly ash with local material of alumina silicate [10]. While, many challenges need to be solved, such as setting time for high calcium fly ash for geopolymer binder and geopolymer concrete. Previous study [7-11] explored the bagasse ash (AAT) on normal concrete.

The bagasse has great potensial to improve the properties of normal concrete still slightly reviewed in geopolymer concrete. Because of that, this study to explore the bagasse ash to reduce the domination of fly ash in binder geopolymer.

2. Methodology

Experimental laboratory used to answer this research. Bagasse ash from PG. Toelangan PTPN X-Sidoarjo. The fly ash samples were collected from Paiton steam power plant in Probolinggo, East Java, Indonesia. To determine the chemical composition contained in the waste bagasse ash and fly ash, tests will be conducted SEM-EDX, XRD (X-Ray Diffraction) and XRF (X-Ray Fluorescence).

Activators for alkaline solution used is sodium silicate (Na_2SiO_3) and sodium hydroxide (NaOH) with molarity 12 M. Table 1 showed eight different geopolymer binder compositions used during this experiments. Those are including the variation on alumina and silica material composition. The ratio between activator solution of NaOH and Na_2SiO_3 were 0.5 and 1.5. The binder sample prepared by mixing the materials then casting the sample into the mould. During this process, the setting time used the ASTM [12]. The sample harden by itself and followed by curing process afterwards. The curing process were monitored at the age of 3, 28 and 56 days. Then we can continue to test the sample's compressive strength by using the ASTM[13] standard code.

Table 1. Geopolymer Binder composition.

No	Binder code	Information
1	AAT12- 0.5	AAT 100%, NaOH 12M, Na_2SiO_3/NaOH = 0.5
2	FA12 - 0.5	FA 100%, NaOH 12M, Na_2SiO_3/NaOH = 0.5
3	50AAT+50FA12 - 0.5	AAT 50% + FA 50%, NaOH 12M, Na_2SiO_3/NaOH = 0.5
4	20AAT+80FA12 - 0.5	AAT 20% + FA 80%, NaOH 12M, Na_2SiO_3/NaOH = 0.5
5	AAT12- 1.5	AAT 100%, NaOH 12M, Na_2SiO_3/NaOH = 1.5
6	FA12 - 1.5	FA 100%, NaOH 12M, Na_2SiO_3/NaOH = 1.5
7	50AAT+50FA12 - 1.5	AAT 50% + FA 50%, NaOH 12M, Na_2SiO_3/NaOH = 1.5
8	20AAT+80FA12 - 1.5	AAT 20% + FA 80%, NaOH 12M, Na_2SiO_3/NaOH = 1.5

3. Result and discussion

Some results from baggase ash and fly ash test physically and chemically are presented in the Figure 1-6 and the table 2 and 3 below:

Figure 1. Baggase Ash in magnification of 500x.

Figure 2. Baggase Ash in magnification of 1000x.

The test results SEM fly ash has a relatively round grain shape so it is more easily oxidized than baggase ash shaped like coral. The faster the oxidation process faster binding occurs.

IOP Conf. Series: Materials Science and Engineering **374** (2018) 012085 doi:10.1088/1757-899X/374/1/012085

This test is used to determine the elemental composition of the material analysis. From the test results XRF known that fly ash used in this study is Si + Al + Fe> 70%. CaO content of fly ash used in this study was 5.83%, according to ASTM [12] fly ash which has a CaO content of less than 10% are classified into type F fly ash.

Figure 7 presented the test results of setting time. This figure described the fastest setting time occurred in the composition of FA 100%, NaOH 12M, Na_2SiO_3/NaOH = 0,5. The longest setting time was showed the binder composition of AAT 20% + FA 80%, NaOH 12M, Na_2SiO_3/NaOH = 0,5.

Figure 3. Fly Ash in magnification of 500x.

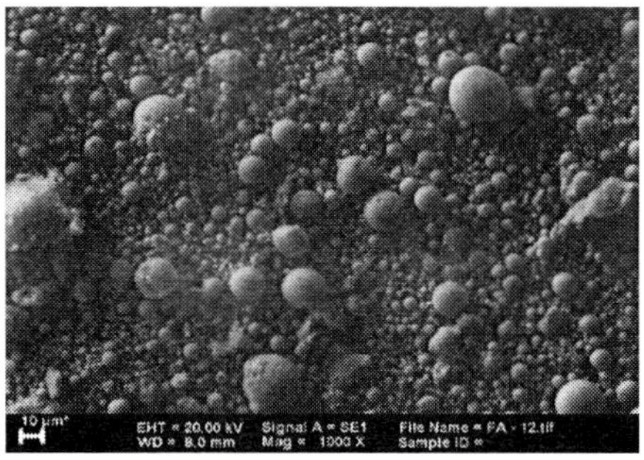

Figure 4. Fly Ash in magnification of 1000x.

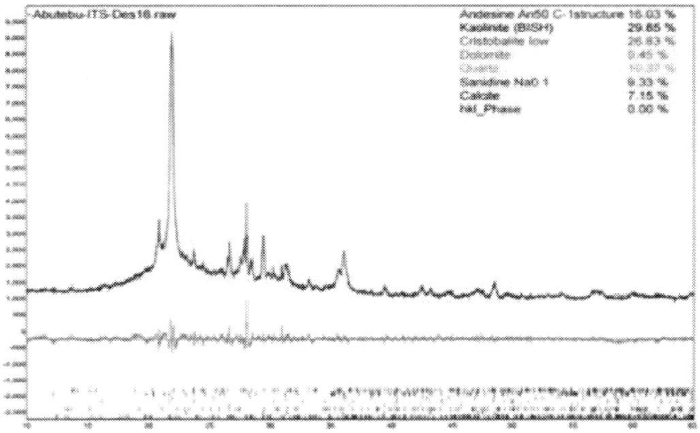

Figure 5. The XRD Test Result of Baggase Ash.

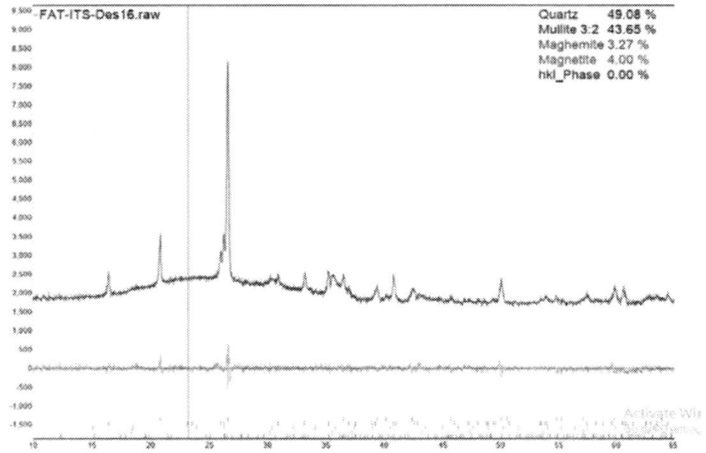

Figure 6. The XRD Test Result of Fly Ash.

XRD test results, the highest intensity on bagasse ash highest intensity occurred in 9100 while in fly ash occurs at an altitude of 8,100.

The compressive strength test was shown Figure 8. There were two highest compressive strength presented the binder composition. The composition were FA 100%, NaOH 12M, $Na_2SiO_3/NaOH$ = 1,5 and AAT 20% + FA 80%, NaOH 12M, $Na_2SiO_3/NaOH$ = 1,5 at the 56 days 42.52 MPa and 42.32 MPa, respictively.

Table 2. The XRF Test Result of Baggase Ash.

No	Code sample		Results analysis (%)
	Coumpound	Name of chemical	
1.	Silica dioxide	SiO_2	72.14
2.	Fery oxide	Fe_2O_3	7.75
3.	Corundum	Al_2O_3	4.86
4.	Calcium dioxide	CaO	7.13
5.	Magnesia	MgO	1.63
6.	Sodium oxide	Na_2O	0.46
7.	Potassium oxide	K_2O	3.6
8.	Manganese oxide	MnO	0.32
9.	Iron oxide	ZnO	0.15
10.	Lead oxide	Ti_2O	0.31
11.	Phosphate	P_2O_5	1.27

Table 3. The XRF Test Result of Fly Ash.

No	Code sample		Results analysis (%)
	Coumpound	Name of chemical	
1.	Silica dioxide	SiO_2	47.10
2.	Fery oxide	Fe_2O_3	16.07
3.	Corundum	Al_2O_3	24.25
4.	Calcium dioxide	CaO	5.83
5.	Magnesia	MgO	2.62
6.	Sodium oxide	Na_2O	0.65
7.	Potassium oxide	K_2O	1.64
8.	Manganese oxide	MnO	0.10
9.	Iron oxide	ZnO	0.29
10.	Lead oxide	Ti_2O	1.16
11.	Phosphate	P_2O_5	0.18

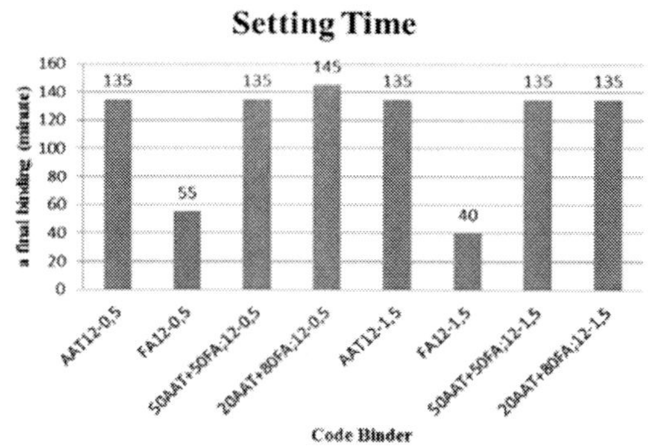

Figure 7. The Setting Time Test Result.

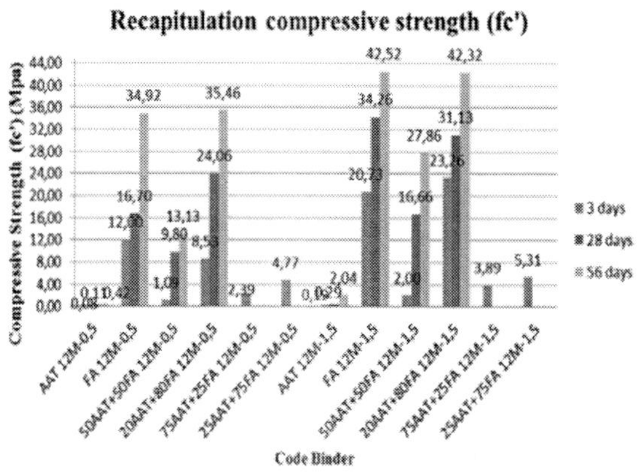

Figure 8. The Compressive Strength Test Result.

4. Conclusions

The bagasse ash could be able to improve its performance to substitute fly ash in a geopolymer binder mixture. The properties chemical and physical of bagasse ash adequate to work together with fly ash. The Baggase ash in the low NaOH molar could increase the setting time of geopolymer binder. However, the bagasse ash required higher NaOH molar to produce better compressive strength.

5. References

[1] Bilodeau A and Malhotra V M 2000 *Materials Journal* **97**(1) 41-48
[2] Davidovits J 1984 *Synthetic mineral polymer compound of the silicoaluminates family and preparation process* Google Patents
[3] Ekaputri J J, Maekawa K and Ishida T 2010 *The use of geopolymerization process for boron fixation in fly ash* in *Jinan, China: Proceeding the 7th International Symposium on Cement& Concrete (ISCC2010) and the 11th International Conference on Advance in Concrete Technology and Sustainable Development*
[4] Ekaputri J J, et al 2015 *Applied Mechanics and Materials* Trans Tech Publ
[5] Bayuaji R, et al 2015 *Applied Mechanics and Materials* Trans Tech Publ
[6] Dwijayanti L Y, Ekaputri J J and Bayuaji R 2016 *Lightweight geopolymer binder with abaca fiber in different curing* in *Materials Science Forum* Trans Tech Publ
[7] Wani A H, Sharma S and Siddique R 2018 I*nternational Journal of Civil and Environmental Engineering* **5**(3)
[8] Rajasekar A, et al 2018 *Construction and Building Materials* **171** 350-356
[9] Moretti J P, Nunes S and A Sales 2018 *Construction and Building Materials* **172** 635-649
[10] ASTM-C191-13 2013 *Standard Test Methods for Time of Setting of Hydraulic Cement by Vicat Needle*
[11] Burduhos Nergis D D, Abdullah M M A and Vizureanu P 2017 *European Journal of Materials Science and Engineering* **2** 111-118
[12] Popovici A, Corbu O, Popita GE, Rosu C, Proorocu M, Sandu AV and Abdullah MMAB 2015 *Materiale Plastice* **52(4)** 588-592
[13] ASTMC39/C39M-18 2018 *Standard Test Method for Compressive Strength of Cylindrical Concrete Specimens*
[14] ASTM-C618-03, *Standard Specification for Coal Fly Ash and Raw or Calcined Natural Pozzolan for Use in Concrete* Annual Book of ASTM Standards, **4-2** , ed ASfTa Materials 2003, Philadelphia,USA

Acknowledgement
Researchers would like to thank the Directorate of Research and Community Service of the Ministry of Research, Technology and Higher Education for research assistance and research results that can be successfully published

Theoretical and Experimental Approach Towards P-Cyano Stilbene Schiff Base as a Potential Linker in E-DNA Sensor

H M Yusoff[1,2], N Hassan[1,2] and K H Ku Bulat[2]
[1]Advanced Nanomaterials Research Group, School of Fundamental Science, Universiti Malaysia Terengganu, 21030 Kuala Terengganu, Terengganu, Malaysia
[2]School of Fundamental Science, Universiti Malaysia Terengganu, 21030 Kuala Terengganu, Terengganu, Malaysia

E-mail: hanismy@umt.edu.my, norhafiefahassan@gmail.com, keriang@umt.edu.my

Abstract. This research aim to synthesis and characterize several p-cyano stilbene Schiff base molecules as a potential linker in electrochemical DNA sensor. Schiff base reaction is form from condensation reaction between primary amine and aldehyde at 79 °C using ethanol as solvent. All compounds are investigated and discussed by Fourier transform-infrared spectrometer (FTIR) and Uv-vis spectrophotometer. For computational study, structure drawings were constructed using GaussView 5.0. Geometry optimization has been carried out using density functional theory (DFT) at bB3LYP/6 31G and performed using Gaussian09 software. FTIR showed formation of C≡N stretching vibrations at 2224 cm^{-1} also another important stretching at 1604 cm^{-1} and 1249 cm^{-1} which represent C=N and C-O mode respectively. In UV-vis, absorbance of C=N (imine) group can be observed at peak range 363 nm and transition of aromatic C=C can be seen at 238 nm. Theoretical and experimental results obtained are comparable and p-cyano stilbene Schiff base compounds were successfully synthesised and can be further applied as a linker in E-DNA sensor.

1. Introduction

Schiff base compounds has been applied in pharmaceutical and biological studies widely [1]. These compounds are also known for its low cost preparation and normally formed by the condensation of aldehydes with primary amines [2-4]. It is reported that Schiff base formed from aromatic aldehyde or aromatic ketones are stable with diverse structural aspects [5]. Application of Schiff base compounds in electrochemical DNA sensor seems as a promising approach. Development of biological sensor have started since 1960s [6]. Fabrication of biosensors include its materials and immobilization methods which required multidisciplinary research. To date there are many methods to detect enzymes, tissue and even DNA. One of the most sensitive and easiest method is electrochemical DNA sensor or known as E-DNA sensor [7].

This type of biosensor required a linker or recognition layer in the sensor system. Obtaining a suitable linker is very important in E-DNA sensor. The linker will be attached onto a suitable substrate. Typical substrates in E-DNA sensor are gold, platinum, silver or indium tin oxide (ITO). All molecular based biosensor reply on specific recognition and DNA probe must be sufficiently immobilised [8]. Furthermore, the length of linker may also affected DNA immobilisation pattern. Typical linker used in several studies normally involve molecules such as 3-mercaptopropionic acid (MPA) and 6-mercapto-1-hexanol (MCH) [9.10].

This paper focuses on synthesizing three different length of stilbene Schiff base having cyano at para position (p-stilbene) at one end of the synthesized molecule. Cyano group will bind with gold surface which act as metal substrate in E-DNA sensor application. Length were varied with 3, 6 and 9 carbon chain. The obtained compounds were also compared with theoretical study using Gaussian 09 software package program [11].

2. Experiment

2.1. Materials

All chemicals and solvents were purchased from Merck, Sigma Aldrich, Acros Organic, Fisher and HmbG chemicals and were used without further purification. To synthesis p-cyano stilbene Schiff base compounds, 4-(4-aminostyryl) benzonitrile was dissolved in 65 ml of ethanol and was treated with equivalent mole of 4-alkoxylbenzaldehyde.

Then, the mixture was stirred under reflux for 2 hours and the k solvent was reduced to half by using dean stark apparatus. The reaction was monitored by thin layer chromatography (TLC). Precipitate formed was recrystallized using acetonitrile. Figure 1 shows the experimental scheme to synthesis the respective compounds. Figure 2 shows all three different lengths of synthesized p-cyano stilbene Schiff base compounds.

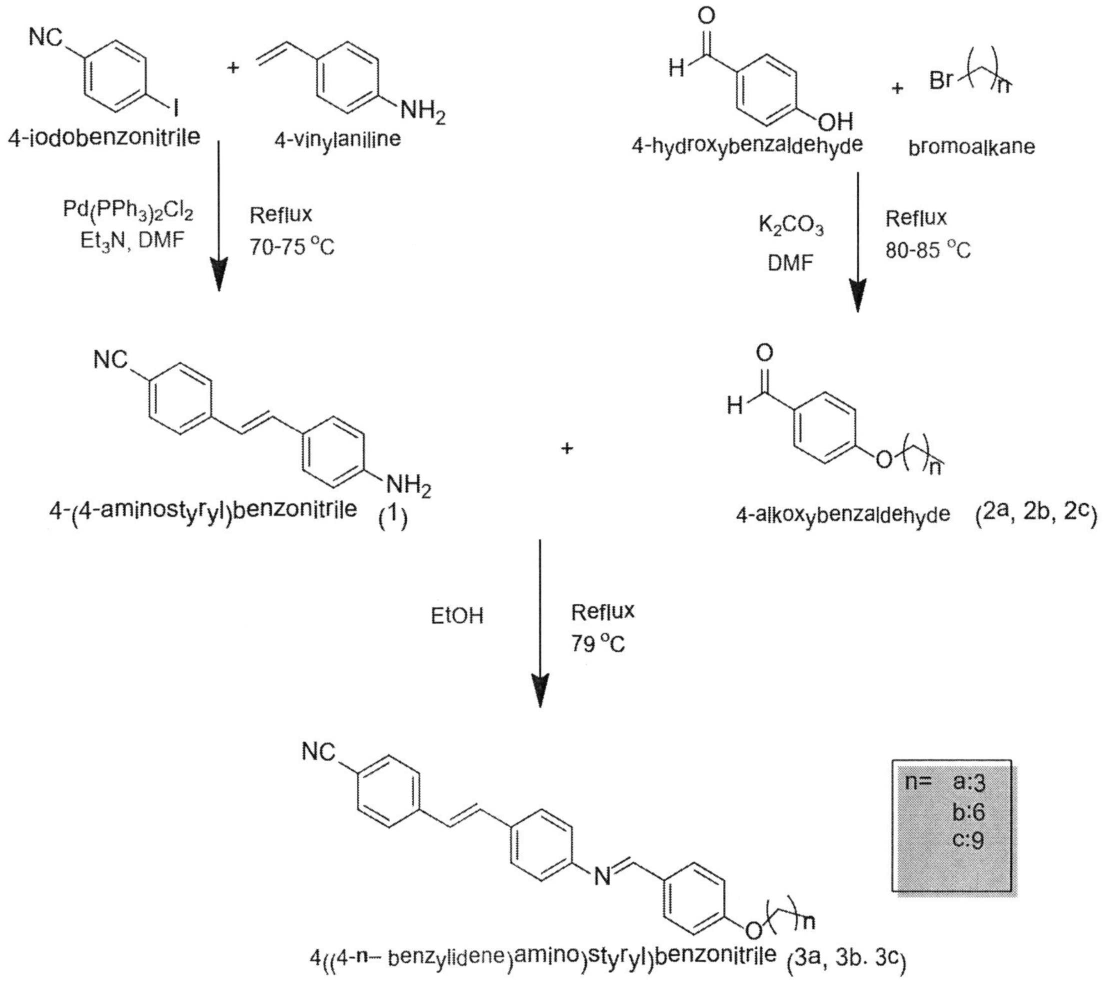

Figure 1. Experimental procedures to synthesis p-cyano stilbene Schiff base.

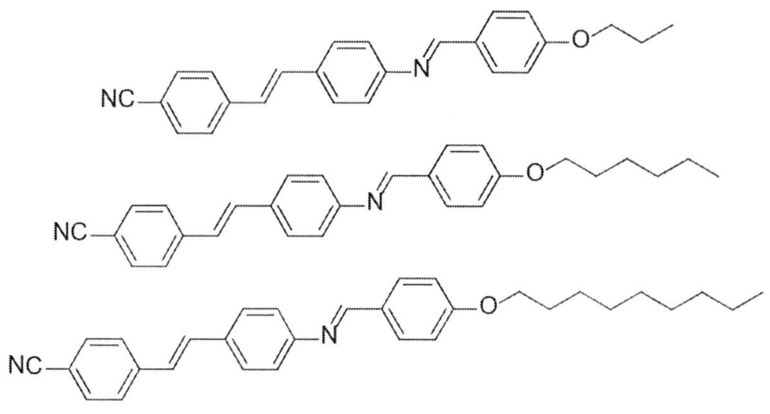

Figure 2. P-syano stilbene Schiff base molecule attached to 9C alkane chain.

2.2. Computational method

There are many computational chemistry methods and also theoretical studies about the determination of biological and chemical activity [12]. GaussView 5.0 has been used to do all the structure drawing. Calculations were performed with Gaussian 09 software package program [11]. In this study, quantum-chemical descriptors are suitable to determine the ranking of physical descriptors. The parameters involved are highest occupied orbitals (HOMOs), the lowest unoccupied molecular (LUMOs), the energy gap between LUMO and HOMO (ΔE_{GAP}), dipole moment (μ), hardness (η) softness (σ) and global electronegativity (χ). Density functional theory (DFT) methods named Becke, 3-parameter, Lee-Yang-Parr (B3LYP/6 31G d,p) was selected to analyse and the rest of the atoms were selected as basis sets. Equations 1 to 4 were used to calculate the physical descriptors.

$$\Delta E_{GAP} = E_{LUMO} - E_{HOMO} \quad (1)$$

$$\text{Hardness, } \eta = (E_{LUMO} - E_{HOMO}) / 2 \quad (2)$$

$$\text{Softness, } \sigma = 1 / \eta \quad (3)$$

$$\text{Global electronegativity } \chi = - (E_{LUMO} + E_{HOMO}) / 2 \quad (4)$$

Figure 3 shows one of p-stilbene molecule that has been optimised using DFT with B3LYP 6 31G DP parameters.

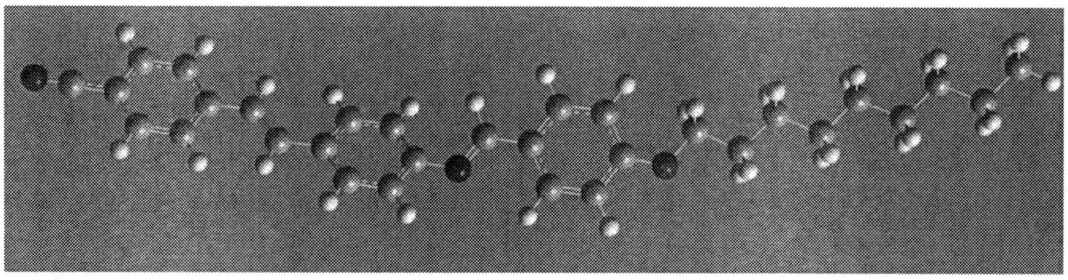

Figure 3. P-cyano stilbene Schiff base molecule attached to 9C alkane chain using GaussView 5.0.

3. Results and discussion

3.1. Infrared (IR) Spectroscopic Analysis

Figure 4 shows FTIR spectra for all different length of p-cyano stilbene Schiff base molecules. All the prominent peaks which represent the functional group present in the compounds can be seen. Peaks ranging at 2967 to 2853 cm^{-1} show the presence of C-H stretching of alkyl chain in the compound. Peak that represents C≡N can be seen 2224 cm^{-1}. Peaks ranging at 1604 to 1609 cm^{-1} are the characteristic absorption bands of carbon-nitrogen double band of the imine group (-HC=N-) [13,14]. Furthermore, peaks at about 1249 cm^{-1} corresponds to the C-O stretching [12].

From table 1, the values obtained from theoretical and experimental are comparable except for C-O stretching however the value obtained from experimental is still acceptable since the range for C-O stretching are from 1000 to 1300 cm^{-1}.

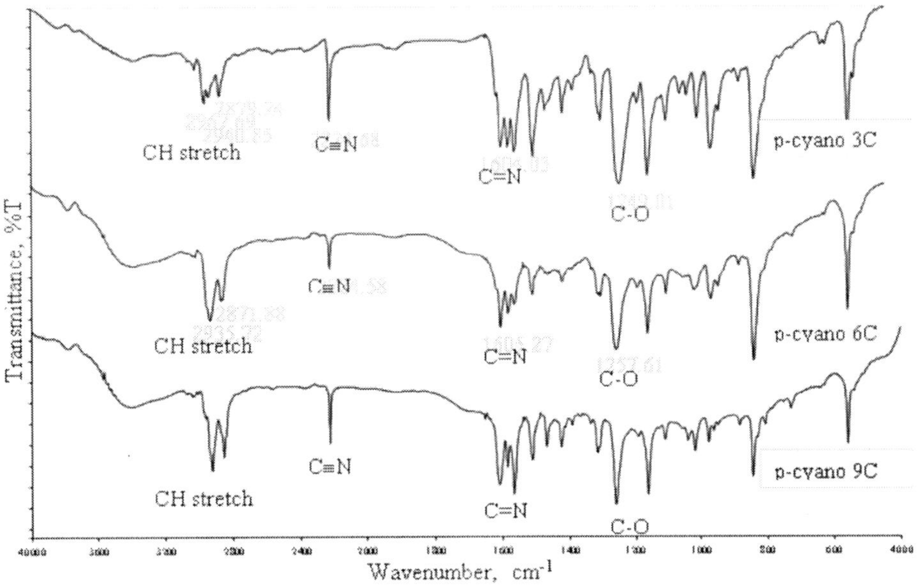

Figure 4. FTIR spectra of p-nitro stilbene Schiff base compounds.

Table 1. Comparison of frequency assignments between theoretical and experimental.

Frequency Assignment	Theoretical (Wavenumber, cm-1)	Experimental (Wavenumber, cm-1)
C-H stretching (alkene/aromatic)	3090, 3114 and 3124	2879, 2940 and 2967
C-H stretching (alkane)	3045, 3055	2940, 2968
C-H stretching (alkane, CH_2-O, CH_2-N)	3002	2879
C≡N stretching	2341	2224
C=N stretching	1614	1604
C=C stretching (aromatic)	1560, 1592	1565, 1584
C-H deformation	1421	1509
CH_3 symmetry deformation	1344	1391
C-C asymmetry deformation (aromatic ring)	1307	1304
C-O (sp2) stretching	1067	1249
C-H deformation (out of phase)	1005	1164
C-O (sp3) stretching	920	973
C-H deformation (aromatic) [out of plane]	859	843
C-H asymmetry deformation (aromatic) [ring breathing]	639, 650	626, 641
C-H deformation [out of plane] (aromatic)	577	560

3.2. Ultraviolet-visible (UV-vis) spectroscopy

The Ultraviolet-visible (UV-vis) spectroscopy analysis of compound 3a-3b was done by using acetonitrile (1 x 10^{-5} M) as a solvent wih cut-off point at 190.0 nm (Figure 5). The ultraviolet-visible electronic spectra of three compound showed the presence n-π* and π-π* electronic transition of C=C (aromatic) at the wavelength of 238 nm. While, band at wavelength 363 nm indicate to the π-π* electronic transition of the C=N group [13,14,15]. Table 2 shows the comparison of peak assignments of theoretical and experimental Uv-vis.

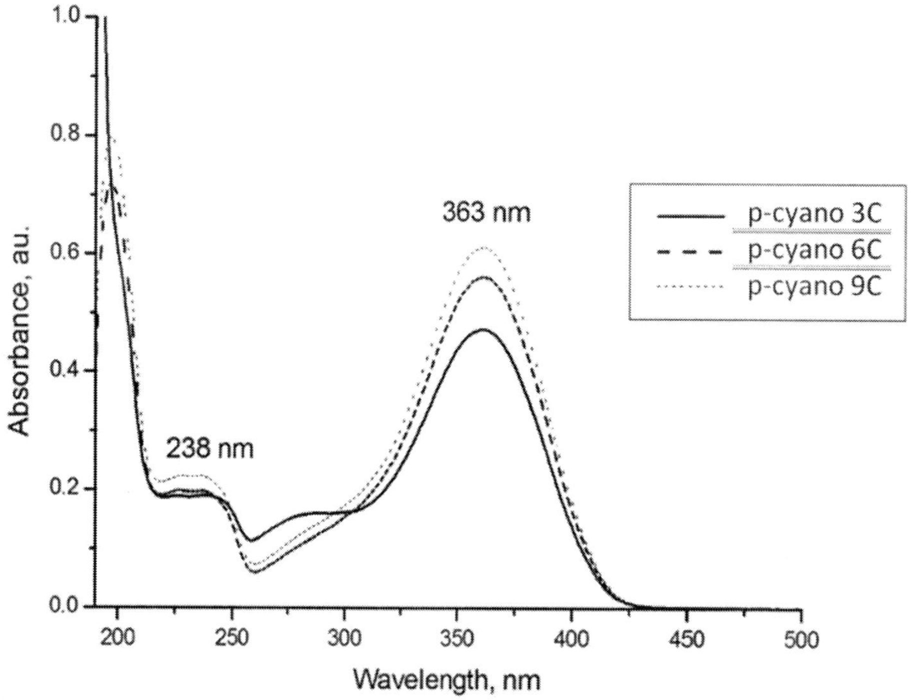

Figure 5. UV-Vis spectra of p-cyano stilbene Schiff base compounds.

Table 2. peak assignment of UV-Vis spectra from theoretical finding.

Peak Assignment λ	**Oscillator strength** f (f ≥ 0.01)	**Transition**
413	1.9767	HOMO → LUMO (π → π*)
275	0.3003	(HOMO-1) → (LUMO+1) (π → π*)
254	0.0885	(HOMO-2) → (LUMO+1) (π → π*)
218	0.1187	(HOMO-6) → (LUMO+1) (π → π*) (HOMO-1) → (LUMO+5) (π → π*)

3.3. Physical descriptors calculation

HOMO is an important quantum chemical descriptor and mainly associated with electron donating ability of molecule. The higher the value of HOMO, the higher the tendency of electron transfer to acceptor molecule. LUMO value indicates that the electron-accepting ability of the molecules is higher. Lower value of E_{gap} explains the eventual charge transfer interaction taking place within the molecules. The lower value of this descriptor shows that the molecule is more active [12].

Table 3. Calculated quantum physical descriptors at B3LYP method.

	p-cyano stilbene 3C	p-cyano stilbene 6C	p-cyano stilbene 9C
E_{SCF}	-1150.6497	-1268.5992	-1386.5489
Dipole moment, μ (Debye)	9.6513	9.8435	9.8745
E_{HOMO}	-0.20153	-0.20128	-0.20128
E_{LUMO}	-0.07735	-0.07721	-0.07729
ΔE_{gap}	3.38	3.38	3.37
Hardness, η (eV)	1.69	1.69	1.68
Softness, σ (eV^{-1})	0.5917	0.5917	0.5952
Global Electronegativity, χ (eV)	3.79	3.79	3.79

Generally there is no significant change in value obtained from the calculated physical descriptors especially for ΔE_{gap}, hardness, η and softenss, σ since the attached functional group is the same which is cyano. Chemical hardness is associated with the stability and reactivity of a chemical system. Soft molecules are preferable because they can easily interact with biological molecules.

4. Conclusions

Three different length of p-cyano stilbene Schiff base compounds has been successfully synthesized and characterized. Experimental results obtained in FTIR analysis is comparable with theoretical study using Gaussian 09 at B3LYP 6 31G (d, p). These molecules will be further applied as a linker in E-DNA sensor and will be studied in both experimental and theoretical level. DFT is a promising tool to study theoretical aspect of recognition layer's length in E-DNA sensor.

5. References

[1] Karamunge K G and Vibhute Y B 2013 *J Phys: Conf. Ser.* **423** 1-6
[2] Sashikala S and Shafi S S 2014 *Der. Pharm. Lett.* **6**(2) 90-97
[3] Sidappa K, Sunilkumar B M and Deene M 2014 Bioinrog. Chem. Appl. http://dx.doi.org/10.1155/2014/343540
[4] Nataliya E B, Marina D R and Yuri A U 2007 *Chem. Rev.* **107(1)** 46-79
[5] Al-Garawi Z S, Tomi R and Al-Daraji A 2014 *E J. Chem.* **9(2)** 962-969
[6] Parikha M 2016 *J. Oral. Biol. Craniofac. Res.* **6(2)** 153-159
[7] Fernandes A, Zhang F and Sun X 2014 *Int. J. Curr. Microbiol. Appl. Sci.* **3(4)** 750–759
[8] Wang Q, Zhang B, Lina X and Weng W 2011 *Sens. Actuators B* **156** 599–605
[9] McEwan G, Chen F and Zhou A 2009 *Anal. Chim. Acta.* **643** 26-37
[10] Lee H, Kang YO and Choi S 2014 *Int. J. Electrochem Sci.* **9** 6793-6808
[11] Roy D, Todd K and John M 2009 GaussView Version 5, Semichem Inc. Shawnee Mission
[12] Mirsada S, Sukrija H, Selma SH, Amar O, Alema D, Zilha A and Davorka Z 2014 *Glas. Hem. Tehnol Bosne Hercegovine* **42** 31-36
[13] Ibrahim H 2009 *Journal of Al-Nahrain University* **12(4)** 1-6
[14] Hussain Z, Yousif E, Ahmed A and Altaie A 2014 *Org. Med. Chem. Lett.* **4** 1
[15] Issa R, Khedr A and Rizk H 2005 *Spectrochim. Acta Part A* **62** 621-629

Acknowledgments

We would like to thank the Ministry of Higher Education (MOHE) for research grant FRGS vot number 59383 and Universiti Malaysia Terengganu (UMT) for giving us financial support and opportunity to be involved in this research.

Sodium Alginate/Ageratum Conyzoides Extract Film for Wound Dressing Materials

A J Abu Bakar[1], C M R Ghazali[1,2] and K A Mat Amin[1,2]
[1]School of Fundamental Science, Universiti Malaysia Terengganu (UMT), 21030 Kuala Nerus, Terengganu, Malaysia
[2]Centre of Excellence Geopolymer System Research, School of Materials Engineering, Universiti Malaysia Perlis (UniMAP), Perlis, Malaysia

E-mail: aznatuljannah@gmail.com, ruzaidi@umt.edu.my, kerol@umt.edu.my*

Abstract. This study focuses on the effect of *Ageratum conyzoides* extract (ACE) in Sodium alginate (SA) films to their mechanical, physical, and thermal properties. The tensile strength results show that the addition of ACE into the SA film improved the stress of the SA-ACE6 film to 5.1 ± 0.8 MPa compared to 4.0 ± 1.0 MPa for blank SA film. Water Vapor Transmission Rates (WVTRs) of SA incorporated *Ageratum conyzoides* extract (SA-ACE) were decreased, while swelling rate values of all films were increased upon addition of ACE. It was found that the thermal stability of the SA film also improved after the addition of ACE. Overall, the addition of ACE enhances the properties of SA film. This makes the SA-ACE film is suitable to be applied as wound dressing materials.

1. Introduction

Biopolymers are receiving greater attention than synthetic petrochemical-based polymers due to the environmental concerns. A variety of renewable biopolymer such as sodium alginate (SA) derived from brown algae, Laminaria hyperborean, has been studied in the development of wound dressings material.

Wound dressing based on alginic material is well known in literature as well as from commercial point of view, in wound management [1]. SA is approved by the United States Food and Drug Administrator (US FDA) and European Union (EU) for use in the food industry and has been in biomedical applications, for examples wound dressing, [2] scaffolds, [3] and dental or surgical impression materials [4].

Ageratum Conyzoides is among such medicinal plants that are effective against diseases and may contain bioactive compounds, which are effective against ill health [5]. It has been known since ancient times for its curative properties and has been utilized for the treatment of various ailments, such as burns and wounds, headaches pneumonia, analgesic, inflammation, asthma, spasmodic and haemostatic effects, stomach ailments, gynaecological diseases, leprosy and other skin diseases [6]. Ageratum Conyzoides can be used as an anti-microbial agent due to presence of bioactive compound, phenol. This is because phenol and phenolic compounds have been extensively used in disinfection and remain the standard with which other bactericides are compared [7]. Hence, it can be used to treat skin disease and wound [5].

Up to our knowledge, no single study has been conducted to produce sodium alginate films incorparated with Ageratum conyzoides. This study fabricated and characterized the sodium alginate incorporating Ageratum conyzoides. The mechanical characteristic, water vapor transmission rates, swelling properties, and thermal study of the film were examined.

2. Materials and method

2.1. Materials

Sodium alginate (product number- W201502 lot number-9005383), glycerine (product number-G2289, lot number SHBC2650V), and anhydrous calcium chloride, $CaCl_2$ (product number-C5670, lot number SHBC2650V) were obtained from Sigma Aldrich, St Louis, MO, USA. Ageratum conyzoides was obtained from Kg. Tok Jembal, Kuala Terengganu, Terengganu. All materials were used as received without any purification while Ageratum conyzoides were extracted using ethanol 95%.

2.2. Preparation of sodium alginate incorporated Ageratum conyzoides (SA-ACE)

The sodium alginate (SA) films were prepared via casting method. The SA solution was prepared by dissolving 2% (w/v) of SA in 100 mL distilled water, followed by glycerine at 50% w/w with continuous stirring for 1 hours 30 minutes at 70°C. After SA was fully dissolved, 10mM $CaCl_2$ was added into the mixture and stirred for another 30 minutes. The Ageratum conyzoides extract (ACE) was added and the solution was stirred for 15 minutes. These SA solutions containing 0.02%, 0.04% and 0.06% (w/v) ACE were known as SA-ACE2, SA-ACE4, and SA-ACE6 respectively. The SA-ACE solutions were deposited onto petri dishes (90 mm x 15 mm) and dried in the oven at 50°C for 24 hours.

2.3. Characterization of SA-ACE films

2.3.1. FTIR Characterization. ATR-FTIR spectra were determined by using Perkin Elmer Spectrum 100 FTIR spectrophotometer with MIKE Miracle ATR accessory with single-bounce beam path, 45° incident angle, and 4 cm^{-1} resolution. Then all spectra were corrected by using Perkin Elmer Spectrum 100 software. The wavenumber region studied was between 4500-400 cm^{-1}. The resulting spectra and peak of functional groups were recorded.

2.3.2. Swelling Percentage. The swelling percentage was carried out by immersed dried samples of dimensions 20 mm × 20 mm in 20 mL of phosphate buffer solution at pH 7.2 in water bath (37 ± 0.5 °C). The samples were removed after 24 hours and lightly wipe with tissue to expel surface solution and weighed. Swelling degree was determined by the equation below:

$$\text{Swelling percentage (\%)} = (W_f - W_i) / W_i \tag{1}$$

Where, W_i = initial weigh of sample and W_f = final weigh of sample

2.3.3. Water Vapour Transmission Rates (WVTR). The WVTR was carried out by put the films (30 mm × 30 mm) as a cap on a glass vial with diameter 16 mm that contain 10 mL of distilled water. Then, it was weigh (W_i) and left in the desiccator containing silica gel for 24 hours. After 24 hours, it was weigh again (W_f). The value of WVTR (g m^{-2} day^{-1}) was calculated as bellow:

$$\text{WVTR} = (W_i - W_f) / (A \times \text{day}) \tag{2}$$

Where, W_i = initial weigh of sample, W_f = final weigh of sample, A = area of vial opening.

2.3.4. Tensile Strength. The tensile strength were performed by using Instron tensile machine, model 3366 with cross-speed at 10 mm/min. All films were cut into 60 mm × 20 mm for characterization. The tests were repeated triplicates per sample for defined ratio.

2.3.5. Thermogravimetric Analysis. Thermogravimetric analysis were performed using Thermogravimetric Analyzer model Mettler Toledo. The films were cut into small pieces to maximized the surface area of the samples. The rate flow of nitrogen gas are 30 mL/min at the heating rate of 10 °C/min while the range of temperature used are 25 °C to 700 °C. Sample used was approximately 10 mg.

3. Results and discussion

3.1. ATR-FTIR Spectrum

Figure 1 shows the FTIR spectra of SA films incorporated *Ageratum conyzoides* extract (ACE).

There are shifting and disappearance of few functional peaks in spectrum of SA film compared to a representative of SA-ACE2 film.

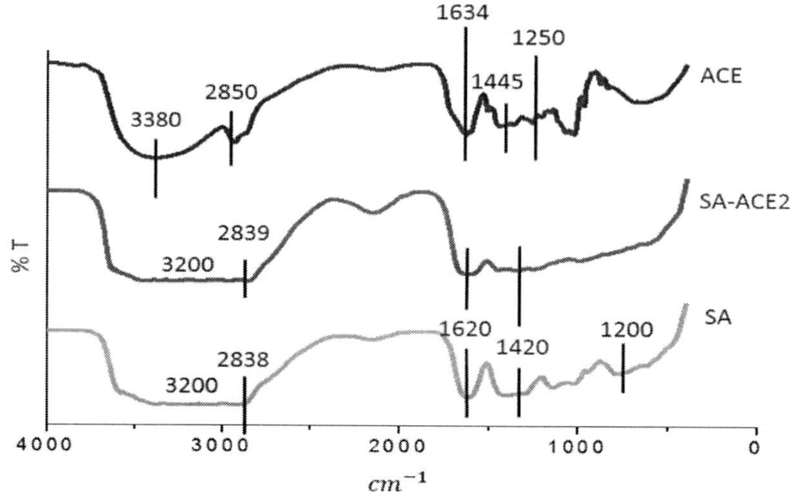

Figure 1. ATR spectra of sodium alginate (SA) film, *Ageratum conyzoides* extract (ACE) and SA-ACE2 film.

All films show a broad peak at 3200 cm^{-1} for SA and SA-ACE2 while 3380 cm^{-1} for ACE due to the stretching vibration belong to O-H group [8]. At 2838 cm^{-1}, 2939cm^{-1} and 2850 cm^{-1} the spectra show the presence of C-H group at SA film, ACE and SA-ACE2 respectively. A peak was shown at wavenumber around 1620 cm^{-1} and 1420 cm^{-1} for SA film and SA-ACE2 film corresponding to symmetric and asymmetric stretching vibrations for COO group, respectively [9]. At 1634 cm^{-1} for ACE spectrum, it is possible to observe a weak absorption band probably related to C=O linkage of ketone or carboxylic acid groups. Furthermore, the weak peak at 1445 cm^{-1} and 1250 cm^{-1} indicate the presences of CH_3 and COC groups respectively. Overall, it seems that there is no significant difference between SA film and SA-ACE film except for the shifting of some functional groups such as C-H and COO groups and disappearance of COC group around 1200 cm^{-1}.

3.2. Swelling Percentage

Swelling properties of a film is important characteristics to determine the ability of the sample to absorb exudates from any wound. The swelling percentages of all SA-ACE films were summarized in Table 1. The swelling degree of film keep increasing upon addition of ACE. For instance, the SA film absorbed 134 ± 59 % and increased to 177 ± 36% for SA-ACE2 film and further increased to 254 ± 94% for SA-ACE6 film. The increased values of swelling of SA-ACE films could be due to the presence of hydroxyl functional group (OH) of SA-ACE available to bind with the hydroxyl group (OH) from water. In fact, an ideal wound dressing normally present water absorption in the range of 100% to 900%[10]. Overall, all the swelling behaviour of the prepared SA-ACE film were in the range of ideal wound dressing criteria.

3.3. Water Vapor Transmission Rates (WVTRs)

WVTR can determined the ability of a wound dressing to control the water loss from a wound at an optimal rate, in which it could prevent extreme dehydration of wound and accumulation of exudates [11]. The WVTRs values of all films were summarized in Table 1. The increasing of ACE concentration in SA film resulted in decreasing the WVTR value to 1214 ± 35 $g\ m^{-2}\ d^{-1}$ for SA-ACE6 compared to 1689 ± 8 $g\ m^{-2}\ d^{-1}$ of free standing SA film. The lowest WVTR values of SA-ACE6 is due to inclusion of ACE into SA film and disturbed the diffusion of water through the film. Besides, SA-ACE2 film with the lower amount of ACE possess higher WVTR value at 1688 ± 164 $g\ m^{-2}\ d^{-1}$.

Nevertheless, all samples show the WVTR values of commercial wound dressing product in the range of 90 – 2893 g m^{-2} d^{-1}[12].

Table 1. Swelling percentage and water vapor transmission rates (WVTRs) values of sodium alginate (SA) and SA-ACE films.

Sample	Swelling Percentage (%)	WVTR ($g\ m^{-2}d^{-1}$)
SA	134 ± 59	1689 ± 80
SA-ACE2	177 ± 36	1688 ± 163
SA-ACE4	241 ± 81	1481 ± 84
SA-ACE6	254 ± 94	1214 ± 35

3.4. Tensile Test

Tensile tests were examined to determine the stress and strain of the fabricated film. Figure 2 shows the stress-strain curves of SA-ACE film at different concentrations of ACE and Table 2 summarized the tensile stress (σ), tensile strain (ε) and Young's Modulus (YM) of the films. The inclusion of ACE into the SA film enhances the stress of the film from 4.0 ± 1.0 MPa (SA) to 5.1 ± 0.8 MPa (SA-ACE6). In contrast, the strain of SA-ACE film was decreased to 20.0 ± 1.3% after the addition of ACE and maintain the value even after addition of ACE at high concentration (20.2 ± 1.9%). There is only a slightest difference of Young's modulus after the addition of ACE which in the range of 28.9 ± 5.6 MPa to 35.6 ± 2.9 MPa. The increasment of stress values of the films is a result of an increased in hydrogen bonding interaction between SA-ACE[13, 14]. Overall, the addition of ACE was enhanced the strength of SA films.

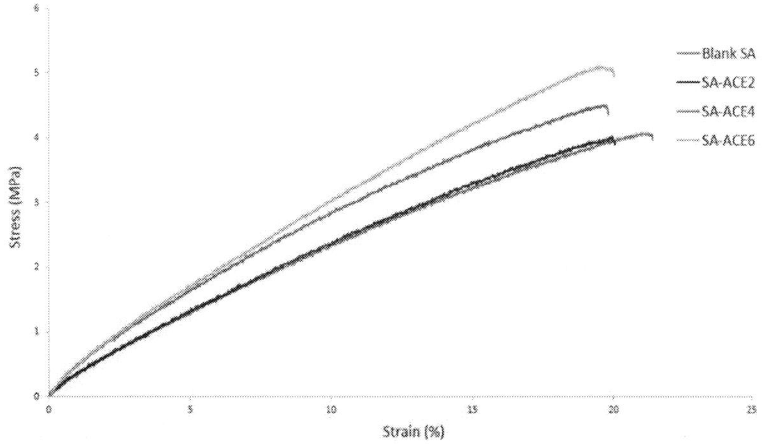

Figure 2. Tensile stress-strain curve of sodium alginate (SA) and SA films at different concentrations of ACE (SA-ACE).

3.5. Thermogravimetric Analysis (TGA)

Thermogravimetric analysis are performed to determine the amount of weight change as a function of time and temperature. Figure 3 shows the thermogravimetric thermograms and derivative thermograms of SA-ACE films while Table 3 summarized the temperature onset (T_o), temperature completion (T_c) and weight loss (%). The degradation of SA film started approximately at temperature onset, T_o = 34°C and the temperature completion, T_c at 500 °C. The inclusion of ACE does not significantly affect the temperature onset (T_o in range 34°C to 36°C), but increased the temperature of completion, T_c from 500°C (SA) to 556°C for SA-ACE6. The weight loss of all films does not have significant difference after the addition of ACE.

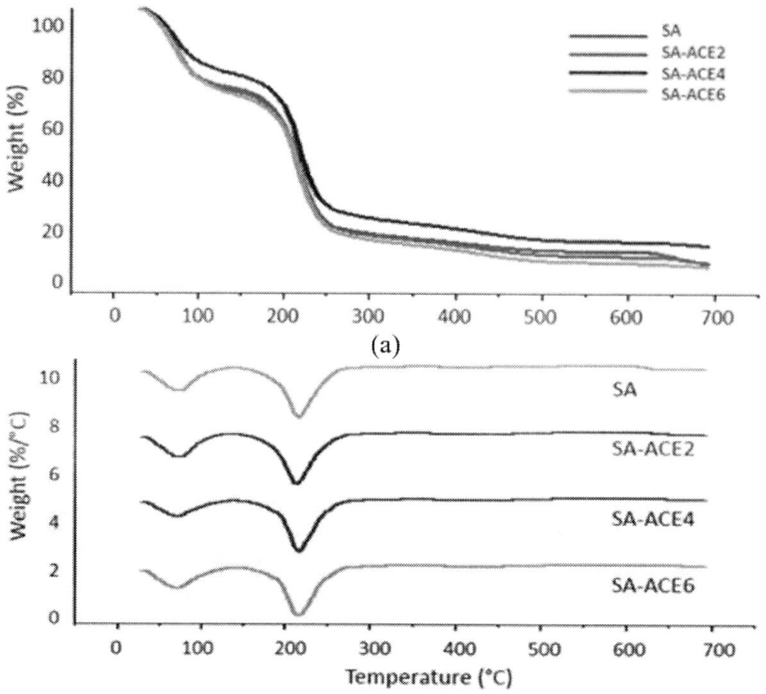

Figure 3. (a) Thermogravimetric thermograms and (b) derivative thermograms sodium alginate (SA) and SA films at different concentrations of ACE (SA-ACE).

Table 2. Tensile stress (σ), the tensile strain (ε), and Young's Modulus (YM) of sodium alginate (SA) and SA films at different concentrations of ACE (SA-ACE).

Sample	Tensile Stress, (σ) (MPa)	Tensile Strain, (ε) (%)	Modulus (YM) (MPa)
SA	4.0 ± 1.0	23.4 ± 3.7	30.9 ± 5.8
SA-ACE2	4.0 ± 0.4	20.0 ± 1.3	29.8 ± 5.3
SA-ACE4	4.5 ± 0.8	20.0 ± 1.8	35.6 ± 2.9
SA-ACE6	5.1 ± 0.8	20.2 ± 1.9	28.9 ± 5.6

Table 3. Thermogravimetry properties of sodium alginate (SA) and SA films at different concentrations of ACE (SA-ACE).

Sample	Temperature Onset, T_o (°C)	Temperature Completion, T_c (°C)	Weight Loss (%)
SA	34	500	76
SA-ACE2	34	520	76
SA-ACE4	36	560	71
SA-ACE6	35	556	77

4. Conclusions
Sodium alginate films incorporated with Ageratum conyzoides extract (SA-ACE) were successfully prepared. The addition of ACE into the SA film increased the swelling degree of film and lowered the values of WVTRs. The tensile performance shows the enhancement of stress of the films and increased the temperature completion of the film after have been tested using thermogravimetric analysis. The film has shown promising properties to be an ideal wound dressing material.

5. References
[1] Paul W and Sharma C 2004 *Trends Biomater Artif Organs* **18** 18–23
[2] Kim J O, Park J K, Kim J H, Jin S G, Yong C S, Li D X, Choi J Y, Woo J S, Yoo B K, Lyoo W S, Kim J A and Choi H G 2008 *Int. J. Pharm.* **359** 79–86
[3] Zmora S, Glicklis R and Cohen S 2002 *Biomaterials* **23** 4087–94
[4] Kamoun E A, Kenawy E R S, Tamer T M, El-Meligy M A and Mohy Eldin M S 2015 *Arab. J. Chem.* **8** 38–47
[5] Amadi B A, Duru M K C and Agomuo E N 2012 *Asian J. Plant Sci. Res.* **2** 428–32
[6] Kamboj A and Saluja A 2008 *Int. J. Green Pharm.* **2** 59
[7] Nasrin F 2013 *Int. Curr. Pharm. J.* **2** 33–7
[8] Pereira R, Carvalho A, Vaz D C, Gil M H, Mendes A and Bártolo P 2013 *International Journal of Biological Macromolecules* **52** 221–30
[9] Kim J 2008 Preparation and properties of alginate / polyaspartate composite hydrogels
[10] Morgado P I, Aguiar-Ricardo A and Correia I J 2015 *J. Memb. Sci.* **490** 139–51
[11] Balakrishnan B, Jayakrishnan A, Kumar S S P and Nandkumar A M 2012 *Artif. Organs* **26** 139–45
[12] Nevin K G and Rajamohan T 2010 *Skin Pharmacol. Physiol.* **23** 290–7
[13] Popita G E, Rosu C, Corbu O, Popovici A, Sandu AV, Proorocu M and Nicula M A 2016 *Materiale Plastice* **53(3)** 375-377
[14] Mohd S S, Abdullah M A A and Mat Amin K A 2016 *J. Bioact. Compat. Polym.* **31** 648–66

Acknowledgments
The authors wish to thanks the Ministry of Higher Education, Malaysia for financial assistance under the Fundamental Research Grand Scheme (FRGS– vote no. 59449) and School of Fundamental Science and Central Laboratory, Universiti Malaysia Terengganu for providing the facilities for undertaking this work.

IOP Conf. Series: Materials Science and Engineering **374** (2018) 012088 doi:10.1088/1757-899X/374/1/012088

Innovative Uses of Recycle Waste Materials as an Artificial Concrete Reef for Estuarine Ecosystem

S Mat Jusoh[1]*, C M R Ghazali[2], K A Mat Amin[2], Z Mohd Zin[3], W M N Wan Nik[1], N Mohamad[4] and N K Jarkoni[1]

[1]School of Ocean Engineering, Universiti Malaysia Terengganu (UMT), 21030 Kuala Nerus, Terengganu, Malaysia.

[2]School of Fundamental Science, Universiti Malaysia Terengganu (UMT), 21030 Kuala Nerus, Terengganu, Malaysia.

[3]School of Food Science and Technology, Universiti Malaysia Terengganu (UMT), 21030 Kuala Nerus, Terengganu, Malaysia.

[4]Faculty of Civil and Environmental Engineering, Universiti Tun Hussien Onn Malaysia (UTHM), 86400 Johor, Malaysia.

E-mail: surianimatjusoh@umt.edu.my*, ruzaidi@umt.edu.my, kerol@umt.edu.my, zamzahaila@umt.edu.my, niksani@umt.edu.my, noridah@uthm.edu.my, khasbi@gmail.com.my

Abstract. This study enveloped an innovative uses of banana peel particles (BPP) incorporated with recycle aggregate concrete (RAC) from construction waste as an artificial concrete reef for estuarine ecosystem. An artificial reef concrete composites were made of using BPP reinforced RAC as a matrix. The conretes composite cube were prepared in a percentages of 10%, 20% and 30% of BPP to RAC respectively. The concrete cubes without BPP was used as control cubes. The compressive strength of an artificial reef concretes composites were carried out at 28 days. The total nutrient dispersed were determined using total nitrate (TN), phosphorus (TP) and organic carbon (TOC) test of the water samples in 6 days. The results showed that the averages compressive strength were 26.8 MPa for 10% of BPP, 18.3 MPa for 20% of BPP and 11.MPa for 30% of BPP, respectively. The amount of TN,TP and TOC leached from the articial concrete reef ranged from 34.3% to 41.5%, 58.2% to 70.3%, and 7.2% to 10.6%, respectively. It also reported that the amount of nutrients leached from an artificial reef increased gradually with time as more percentages of BPP added to concrete composites and on the contrary, the compressive strength indicated decreases with the increased of BPP. Wet can be concluded that, BPP from the agriculture wastes could be used as potential nutrient sources of nitrogen, phosphorus and organic carbon for an artificial concrete reef in estuarine ecosystem with some further study on mechanical properties to increase the compressive strength of an artificial reef composite concretes.

1. Introduction

The by-product of human and industrial activity that has no residual value is defined as the waste [1]. Malaysians produce an average of 30,000 tons of waste every day and only 5 percent of it is recycled. These two statistics were recently revealed by the national Ministry of Urban Wellbeing, Housing and Local Government in 2017 and they're causing enormous problems for Malaysia. The waste can be processed to produce recycled materials which is usable in the household applications to a variety of industries. It is important that these wastes are recycled and reused or turned into highly beneficial outputs in order to sustain the renewable resource, safe ecological damage, bring economical profit and develop a sustainable civilization. Numerous studied have been done towards utilising by-products [2-6] and [7] thus, the goal of this research to investigate the mechanical properties and leaching behaviour of concrete composites structures for artificial reef made by construction and plantation solid waste.

In Malaysia, artificial reefs are referred to as any man-made structures or natural objects placed in marine water body to provide new habitat for fish and other flora and fauna, and at the same time prevent the encroachment of destructive gears such as trawls into protected areas [8], and some researchers like conclude that, an artificial reefs either aggregate existing scattered fish or allow secondary biomass production through increased survival and growth of new individuals as a result of the shelter and food resource provided by an artificial reef [9]. The artificial reef composite concretes for this research made of using construction waste as recycled aggregate concrete reinforced with banana peel particles from banana by-products or classified as agriculture waste. Artificial reefs can be defined as the human-made structure or natural object placed in water that can be used as the habitat for the flora and fauna. Artificial reefs have been placed in the marine environment since at least 1655, which is apparently the date of the first recorded deployment [8]. However, it is likely that their habitat providing function was recognized by fishermen long before that. Currently, artificial reefs are used worldwide for diverse functions: primarily to enhance fishing and recreational diving, but also to prevent trawling, provide beach protection, mitigate marine construction, etc. The different type of artificial reef can affect the attraction of fish. The most optimal shape of the artificial reef that can be used to attract the fish to come and build its colony at the reef is largely voided space. Concrete has been found to be very favourable for artificial reef construction. However, pure concrete materials can be expensive and may later leach elements which can be toxic to marine organisms and hence destructive to the environment. In order to provide the habitat for the aquatic flora and fauna, the aim of the artificial reef is to maintain the ecosystem structure and functions. Therefore, organic materials are added to the concrete reef mixture to promote aquatic organism growth. This organic material release usable nutrients to microorganisms in the vicinity of the artificial material. Some of the suitable organic materials may include agriculture wastes, animal by-products, and other waste materials from plant or animal origin.

2. Materials

Materials used in this research were banana peel particles from an organic waste and recycled aggregate concrete from construction waste.

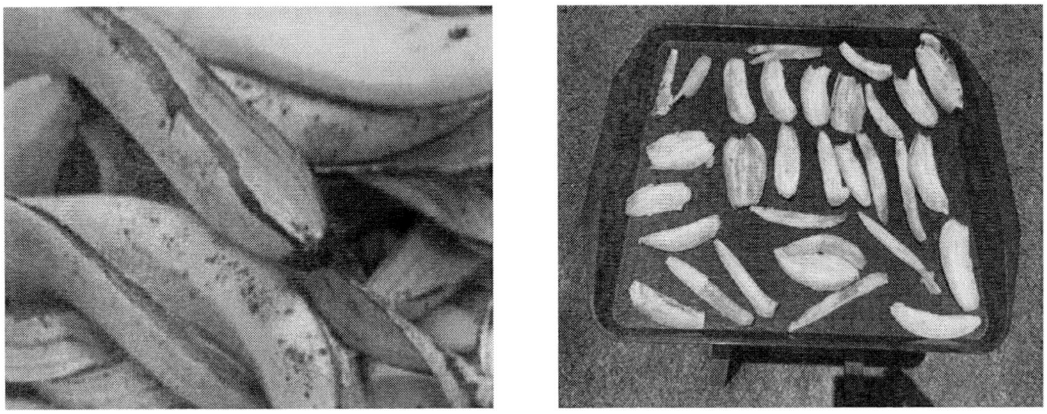

(a) (b)

Figure 1. (a) Banana peels and (b) Banana peels dried under sunlight.

2.1 Banana Peel Particles

Banana or *Musaceae sp* is one of the world's most important fruit crops that is widely cultivated in Malaysia and other tropical countries for its valuable application [2]. Its enormous by-products are an excellent source of highly agriculture waste [4]. Banana by-products such as peels, leaves, pseudo stem, stalk and inflorescence. Banana peel collected from the garbage bin at local fruits and food stalls. Banana peels are chop into a particles and dry under a sunlight until it dry for two (2) weeks.

2.2 Recycle Aggregate Concrete

Concrete is one of the most used materials in construction [6] and at the same time, one of the materials with the higher contribution to construction and demolition (CDW) waste generation [10]. In this study recycled aggregate concrete used, were collected from construction wastes then grinded into a small particles. Figure 2 depicted the figure of control cube concrete with 0% BPP reinforced RAC and b) BPP reinforced RAC concrete cubes immersed in water.

(a) (b)

Figure 2. (a) Control cube concrete with 0% BPP reinforced RAC and (b) BPP reinforced RAC concrete cubes immersed in water.

3. Methodology

3.1 Preparation of An Artificial Reef Cube Concretes

Recycled aggregate concrete (RAC) uses as matrix and banana peel particles (BPP) used as reinforcement to fabricate an artificial reef cube concretes. A cube concretes are dried in a room temperature for 28 days. There were different compositions in a weight percentage; 10%, 20% and 30% used to determine an optimum composition of BPP reinforced RAC cube. Test cube concretes were prepared by using hand mixing and the mould casting dimension is 150 mm x 150 mm x 150 mm.

3.2 Compressive Test

The compressive tests were conducted to determine the compressive strength of an artificial cube concretes at 28 days of age. The correlation between weight percentages of BPP to RAC towards compressive strength then analysed. There were six (6) cube concretes tested for each different compositions of BPP to RAC cube.

3.3 Determination of Organic Compounds Leached

The water sample before the cube concretes are put inside was taken as a control cube. The constant wave parameter was applied in this test using small scale wave simulation tank. Then, the water sample after the cube concretes were put inside was taken again to determine the total amount of organic compounds leached. The total amounts of an organic compounds leached into a water were observed for each days after immerse in a 6 days. The spectrophotometer was used to record the reading of organic compounds leached into a water.

4. Results and discussion

4.1 Compressive Strength

As report by Noridah et al., [3] 2016 the compressive strength of the reef can be influenced by the properties and the amount of the agricultural waste used and the properties of the recycled aggregate. In their study, it has been reported that increased percentage of peanuts shell in the concrete resulted with decreased strength of the concrete. This study revealed that the highest compressive strength of BPP reinforced RAC are 26.8 MPa of 10% BPP to RAC, 18.3 MPa of 20% BPP to RAC and 11.0 MPa of 30% BPP to RAC, respectively. The compressive strength for control cube with) % BPP to RAC is 43.0 MPa. It can be concluded that the compressive strength are decreased with the increased of weight percentages (wt %) of BPP. This is happen may be because some manufacturing defects as has been occurred. As reported by Suriani et al.[11, 12] defects typically occur during the commercial production of composites and can be caused by batch to-batch variations in the prepreg and sometimes by the manual construction known as a 'lay-up'. In composite materials, variability could arise from differences in the prepreg tack level during lay-up as a result of variable resin content. Composites can contain a number of defects introduced during manufacturing such as, voids, resin-rich zones, pockets of undispersed cross-linker, misaligned fibres and regions where resin has poorly wetted the fibres which can considerably increase the likelihood of composite failure [17]. While in this study, with the increasing of BPP weight percentages had decreased the compressive strength because of the lesser bonding between interfacials adhesion may also cause by the defects occurred [13]. Naidu et al., in their study on characterisation and mechanical properties of banana peel reinforced composite reported on the same trend of decreasing on the mechanical properties caused by voids occurred [14, 15].

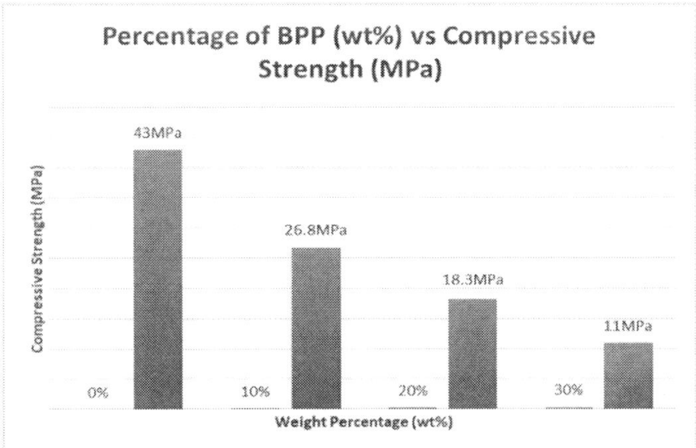

Figure 3. Compressive strength properties of banana peel particles (BPP) reinforced recycle aggregate concrete (RAC) at different weight percentages.

4.2 Total Organic Carbon, Phosphorus and Nitrogen Leached

Figure 4 shows the total organic compounds leached into water sample after 6 days immersion. It can be concluded that the amounts of total organic compounds released into water increased with the increased of amount of BPP. The total amounts of organic compounds also increased with the period of immersion from day 1 to day 6.

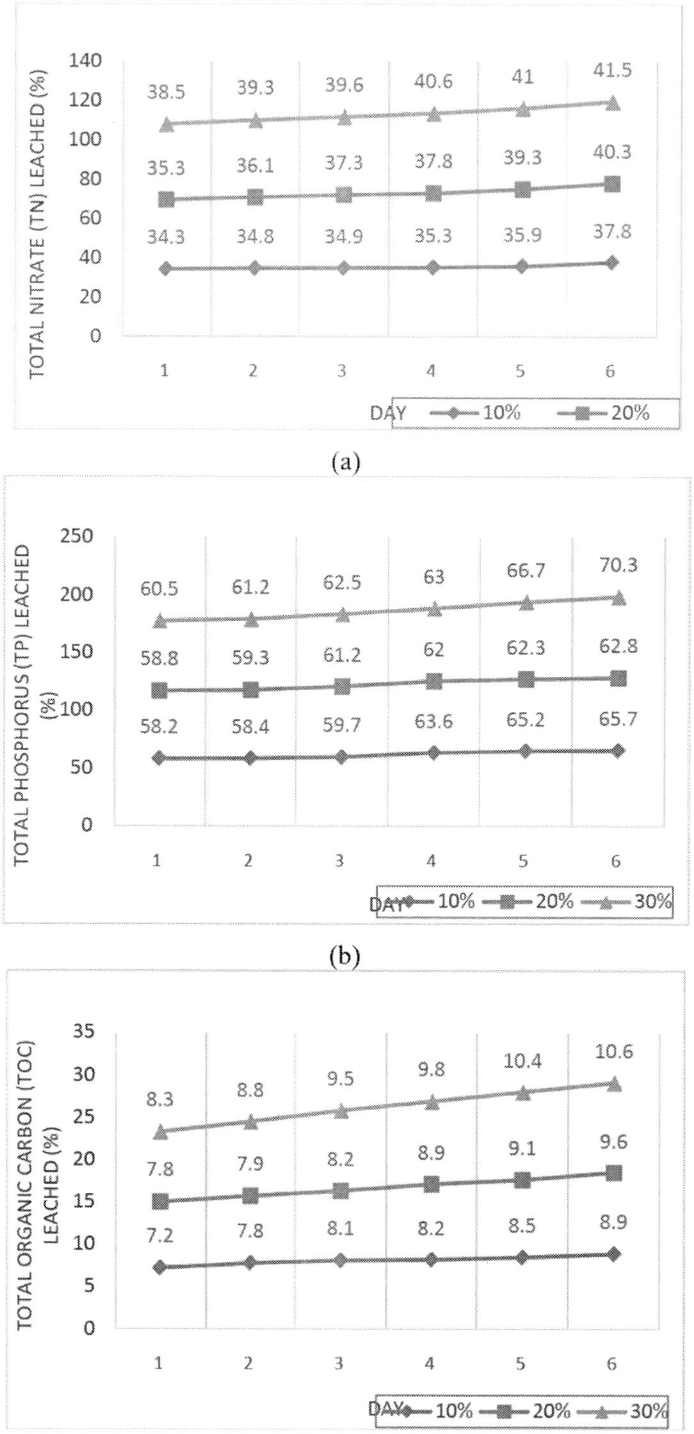

(a)

(b)

(c)

Figure 4. Total amount of nutrients (a) Total Nitrate (TN), (b) Total Phosphorus (TP) and (c) total organic Carbon (TOC) leached from banana peel particles (BPP) reinforced recycle aggregate concrete (RAC).

As depicted in Figure 4 a), b) and c) the amounts of total nitrate (TN) released into a water is 34.3% to 41.5%, the amounts of total phosphorus (TP) released into a water is 58.2% to 70.3%, and the amounts of total organic carbon (TOC) released into a water is 7.2% to 10.6%, respectively. This is similar to the finding of research did by Jiang et al., (2012) on leaching behaviour of total nitrogen, total phosphorus and total organic carbon from banana peel [16].

5. Conclusion and recommendations

This research has approved the use of BPP reinforced RAC concrete as an artificial reef with the present of organic compound leached to the water that provided a direct or indirect impact to estuarine ecosystem. Therefore, this research is significantly solved one of nation issues on solid waste management and pollution, contributes to a green and sustainable estuarine ecosystems, enhances health and safety lifestyle to community and gives the good economic impact as well. Further study on the surface morphology of the artificial reef manufacturing defects may resulted a better understood on compressive strength of the artificial reef concretes.

6. References

[1] Vivian W Y T 2011 *Open Waste Manag. J.* **4** 28–32
[2] Padam B S, Tin H S, Chye F Y and Abdullah M I 2014 *J. Food Sci. Technol.* **51** 3527–45
[3] Paper C, Mohamad N, Tun U, Onn H, Aziz A, Samad A, Tun U, Onn H, Afif M, Universiti I, Hussein T, Monica H, Peralta M, Tun U, Onn H and Mohamad N 2016 *Leaching Behaviour of Organic Materials in Reinforced Concrete Artificial Reef with RAC*
[4] Dungani R, Karina M, Subyakto, Sulaeman A, Hermawan D and Hadiyane A 2016 *Asian J. Plant Sci.* **15** 42–55
[5] Bolden IV J J 2013 Innovative uses of Recycled and Waste Materials in Construction Application Johnny J . Bolden , IV North Carolina A & T State University A thesis submitted to the graduate faculty in partial fulfillment of the requirements for the degree of MASTER OF SCIECE
[6] McNeil K and Kang T H K 2013 *Int. J. Concr. Struct. Mater.* **7** 61–9
[7] Zhou Z, Wang J, Zhang D, Zhang X and Zhang X 2012 *Procedia Eng.* **27** 357–64
[8] Ali A, Abdullah M P, Hazizi R, Marzuki A H and Raja Hassan R B 2013 *Malaysian J. Sci.* **32** 19–36
[9] Bohnsack J A and Sutherland D L 1985 *Bull. Mar. Sci.* **37** 11–39
[10] Letelier V, Tarela E and Moriconi G 2017 *Procedia Eng.* **171** 627–32
[11] Suriani M J, Ali A, Khalina A, Sapuan S M and Abdullah S 2012 *Mater. Test.* **54** 172–8
[12] Suriani, M.J., M.M. Hamdan H Y S and S M S 2007 *Multidiciplinary Model. Mater. Struct.* **3** 213
[13] González-Fonteboa B, Martínez-Abella F, Martínez-Lage I and Eiras-López J 2009 *Constr. Build. Mater.* **23** 3406–10
[14] Naidu A L, Raghuveer D and Suman P 2013 **4** 844–51
[15] Gawdzinska K, Nabialek M, Bryll K, Szymanski P and Sandu A V 2017 *Materiale Plastice* **54(3)** 539-542
[16] Ruixue Jiang, Shujuan Sun, Yan Xu, Xiudong Qiu, Jili Yang X L 2015 *Water Sci. Technol.* **71** 1458–62
[17] Ibrahim W M W, Hussin K, Abdullah M M A B, Kadir A A, Deraman and Sandu A V 2017 *Revista de Chimie* **68(9)** 1978-82

Acknowledgments

Authors wishing to acknowledge assistance or encouragement from colleagues, special work by technical staff and postgraduates students from School of Ocean Engineering, School of Fundamental Science, Universiti Malaysia Terengganu (UMT) and Faculty of Civil & Environmental Engineering , Universiti Tun Hussein Malaysia (UTHM).

A Theoretical Framework for Understanding Pedestrian Behaviour Attributes Based on Spatial Interaction

N H Kasehyani[1], N Abd Rahman[1,*], N S AbdulSukor[1], H Halim[1], H Y Katman[2] and M Sabustan[3]
[1]School of Civil Engineering, Engineering Campus, Universiti Sains Malaysia, 14300 Nibong Tebal, Penang, Malaysia.
[2]Collage of Engineering, Universiti Tenaga Nasional (Putrajaya Campus), Jalan IKRAM-UNITEN, 43000 Kajang, Selangor, Malaysia.
[3]Faculty of Civil and Environment Engineering, Universiti Tun Hussein Onn Malaysia, 86400 Parit Raja, Batu Pahat, Johor, Malaysia.

E-mail: celindarahman@usm.my

Abstract. Current trend shows there is an exponential increase in rail traffic passenger volume in Malaysia, causing authorities to be aware in improving the pedestrians' facilities and safety. Kuala Lumpur Sentral Station (KLSS) is the largest transportation hub in Malaysia and, as of May 2017, the user reaches up to 180,000 users daily. In this contribution, a theoretical framework is presented to understand the pedestrians' behaviour attributes at rail transit terminal based on pedestrians' spatial interactions. The Distinct Element Method (DEM) will be employed to model the dynamics of pedestrian's crowd behaviour. The validated crowd dynamics model will be demonstrated in evacuation capacity estimation in KLSS. Evacuating people quickly and safely can greatly reduce casualty and economic losses. As for this contribution, more concern is on the spatial interaction of pedestrian at the RTT and the proposed framework. This framework paper will describe the flow in details on conducting the study of spatial intersection.

1. Introduction

Study on dynamic of pedestrian's crowd behaviour at Rail Transit Terminal (RTT) in Malaysia is not being focus and according to Bohari [1], public buildings such as nodes, terminal and stations experiencing movement of people that generate massive walking behaviour. Kuala Lumpur Sentral Station (KLSS) is the biggest RTT in Malaysia located at the heart of Kuala Lumpur that connect almost all the public transport facilities such as rail (KTM, LRT, Monorail and MRT), buses and taxi in one hub with up to 180,000 users daily (The Star Online, 2017). Therefore, it is crucial to study the pedestrian behaviour to address the issue of security, convenience, the facility utilization, spacestructure and equipment layout of the terminal.

On the other hand, pedestrian spatial interaction represents the pedestrian behaviour towards the space around them. In recent literature review, some researches or guidelines address spatial interaction as proxemics behaviour, personal space, shy distance, buffer zone, cushion, psychological space and perception domain. Although it comes with many names, the mail objective if spatial interaction is to determine the pedestrian behaviour towards other pedestrian or obstacle around them. The Strait Times Online (2017) reported that the East Coast Rail Link (ECRL) that linking east cost to

Port Klang on the west coast is slated to be completed in 2024 with estimated 5.4 million passengers and 53 million tonnes of cargo will use the service annually by 2030. Thus, study on dynamic of pedestrian's crowd behaviour at RTT in Malaysia based on pedestrian spatial interaction is crucial to support the increasing demand in rail services.

2. Recent Studies on Pedestrian Behaviour Attributes, Evacuation of Egress Facilities and Spatial Interaction

2.1. Pedestrian Behaviour Attributes

Pedestrian behaviour is normally interpreted in term of speed, flow, density and rarely in term of evacuation time. Table 1 shows the list of past researches involving pedestrian behaviour attributes and most of the researcher study the pedestrian behaviour based on the walking speed follow by flow, density and evacuation time as shown in Table 1 and Table 2. Walking speed is the most pedestrians' attributes being studied so far (Table 1 and 2).

Table 1. List of previous studies on pedestrian behaviour attributes.

Authors	Speed	Flow	Density	Evacuation Time
Bohari [1]	✓			
Abustan [2]	✓			
Patra [3]	✓	✓	✓	
Von [4]	✓			✓
Han and Liu [5]			✓	✓
Zhang [6]			✓	✓
Li [7]				✓
Zhao and Liang [8]	✓			
Zhao [9]				✓
Zhao [10]	✓	✓		
Gotoh [11]	✓			
Yeo and He [12]	✓	✓		

Pedestrian speed is mostly determined by analysing the video recording of pedestrian walking from fix distance (usually 1.5 m) over time. Some studies conducted at Malaysia, India, China, Japan and Singapore show different pedestrian walking speed. According to Bohari [1] and Abustan [2] the walking speed of pedestrian at Masjid Jamek LRT Station varies from 1.12 m/s to 1.48 m/s and at Miami Beach, Penang varies from 0.89 m/s to 1.38 m/s. Higher walking speed was observed at Masjid Jamek LRT Station, rapid transit station in Kuala Lumpur, Malaysia compared to Miami Beach, recreational place in Penang, Malaysia.Although both places have different purposes, the different in walking velocity is about 20%. As reported by Yeo and He [12], in Singapore, their pedestrian walking speed at MRT stations is in the range of 1.04 m/s to 1.30 m/s which contribute to 12% different from Bohari [1]. Zhao and Liang [8] from China reported that the pedestrian walking speed at Guangzhou Metro varies from 0.92 m/s to 1.18 m/s which contribute 20% different from Bohari [1]. Meanwhile, Secunderabad Railway, India recorded slowest walking speed which is 0.65 m/s (Patra [3]). This may be due to substantial number of user daily which reach up to 23 million pedestrians daily with comparatively less spaced level changing facilities which become bottlenecks at most of the time. Japan, country with the most advanced rail technology recorded 1.47 m/s pedestrian walking speed [8].

Pedestrian level-of-service (LOS) is calculated by counting pedestrian who cross a point over a certain period (usually 15 minutes), reducing figures to pedestrian per minute and divided by the effective width, and hence produce the flow rate (Highway Capacity Manual (HCM), 2010). The LOS is characterised as A for free flow to F as virtually no movement possible. Not many studies on pedestrian flow were conducted. Patra [3] found that the pedestrian flow at Secunderabad Railway, India is 24 ped./m/min meanwhile Zhao [10] found that the pedestrian flow at Guangzhou Metro, China was 58 ped./m/min. Up to 59% difference calculated of the pedestrian flow between China and India. This scenario may be due to the pedestrian physical difference, such as body size between those countries, and the weight and physical dimension of the pedestrian facilities. Unfortunately, there is no study found about pedestrian flow in Malaysia especially at the RTT.

Table 2. Conducted studies on pedestrian walking speed based on countries.

Country	Author	Walking Speed (m/s)
Malaysia	Bohari [1]	1.12 to 1.48
	Abustan [2]	0.89 to 1.38
India	Patra [3]	0.65
China	Zhao and Liang [8]	0.92 to 1.18
Japan	Gotoh [11]	1.47
Singapore	Yeo and He [12]	1.04 to 1.30

Density is the number of pedestrians present on an area at given moment and noted as p/m^2. Patra [3] recorded the density of pedestrian at Secunderabad Railway, India is 1.5 p/m^2 and Zhang [6] found that the density of pedestrian at Wuhan Metro Station, China is 1.83 p/m^2. There are 18% difference between China and India. This shows that pedestrian in India need more space compare to China. This may differ for Malaysia case due to cultural difference.

Evacuation time is important to make an effective building egress process. Evacuation time is the elapsed time between the instant that occupants receive an emergency alarm and their arrival at a destination where it is normally a safe location inside or outside the building. Many factors can affect the evacuation time such as group forming, information transmission, route planning, bottleneck effect and code of practice chosen. Von [4] studied the effect of social group towards evacuation time shows that queue in front of the door can make evacuation faster compare to broad distribution around the exit. Information sharing is important especially in pristine environment. Like in Han and Liu [5], with the used of modified social force model based on information transmission to model the evacuation process, the evacuation time is effectively shortened, and evacuation efficiency is improved. Besides, Zhang [6] reported that by improving the route planning strategies for evacuation process, it helps in reducing the evacuation time by choosing the shortest evacuation route. But, this approach has some limitation since subjective factors like physiology and psychology of the crowd were not considered in this study. Most of the rail station have turnstile that will cause bottleneck effect during emergency evacuation and prolong the evacuation time according to Li [7]. Zhao [9] suggested the use of right code of practice will affect the evacuation process of a building. In their study, the Chinese code tends to give errors in the result of evacuation time, while the American and Japanese standards show certain superiorities such that they correctly reflect the effect of the structural layout of the station and represents predictive bottlenecks.

Although there are many research done on this matter, Malaysia are still lacking in its pedestrian behaviour at RTT. Pedestrian behaviour such as speed, flow, density and evacuation time need advanced studied to help improving the operation and safety level of pedestrian at transport terminals in Malaysia. A lot of works need to be done since the above issue have many application and impact to the society. RTT in Malaysia is becoming one of the choice of transportation mode. Hence, the necessity to understand the specific behaviour of passengers in RTT is crucial. This understanding will become a reference information to the traffic engineers in designing walking infrastructures for other places (i.e.: shopping mall, side walk facilities, stadium, crosswalk facilities, etc.). On the other hand, with the establishment of crowd behaviour model, an analytical evaluation of the quality of urban space can be provided.

2.2. Spatial Interaction

Many models of crowd behaviour have been proposed so far to understand how pedestrian move, however modelling and visualizing the dynamics of pedestrian crowd behaviour in relation to pedestrian spatial interaction, particularly at railway stations is none in Malaysia. A better understanding of crowd behaviour in railway station is the key to plan and manage the pedestrians' flow in Rail Transit Terminal (RTT). There are many names used to address spatial interaction such as the "shy distance" by Highway Capacity Manual (HCM), 2010 that clearly defined as the space that pedestrian tends to keep between themselves and obstacle. Some researcher does refer "shy distance" as "buffer zone" or "cushion". The shy distance was estimated to be 30 cm to 45 cm and affected by number of pedestrian, time of day, and surrounding land use. Campanella [13] study the microscopic modelling of walking behaviour using the Nomad simulator that used three levels of spatial isolation of isolated, in-range and in-collision. Isolated is the stage where pedestrian have no other pedestrian or obstacles within their influence area. In-range is the stage where pedestrian have other pedestrian or obstacle inside or close to their influence area but no possibilities of colliding. In-collision is the stage where pedestrian is very closed to other or to obstacle.

Some study does implement spatial interaction in different concept as shown in the Table3. From those studies, only two studies of the spatial interaction were at rail station with different concept and none at RTT. Guo [14] study the spatial and temporal separation rules to reproduce self-organizing movement through bottleneck (train door) indicated that pedestrian efficiency of passing through the bottleneck can be improved by implement the spatial separation rule. Pedestrian pass through the middle tend to move to the right and pedestrian pass near two sides of the door tend to move to the left. Meanwhile, Yang [15] study the guided crowd modified social force model to predict the evacuation of Beijing South Railway Station found that pedestrians in the group who choose to follow the guide instead of walking independently before knowing the position of exit can escape with a larger velocity. Both of those concepts were studied in China. Do Malaysia RTT users follow the same spatial interaction pattern as China especially with the evacuation process described by Yang [15]? Therefore, it is important to study the spatial interaction to know the pedestrian behaviour at RTT in Malaysia.

Table 3. Concept of spatial interaction used in the previous studies.

Author	Concept
Mohd Ibrahim [16]	Game theory
Von [4]	Lane formation
Han and Liu [5]	Group formation
Liu [17]	Effect of wall and door
Campanella [13]	Modelling of walking behaviour
Guo [14]	Bottleneck effect of train door
Yang [15]	Guided crowd at railway station
Fridman [18]	Impact of cultural differences

Mohd Ibrahim [16] use game theory as the concept of their study and come out with 40 cm as their spatial distance and 130° as their maximum angle between conflicting agents. In this stage, real life Malaysian pedestrians at RTT are the same as the game theory needs to be clarified. Study from China conducted by Liu [17] came with spatial distance of 46 cm. Liu [17] study the effect of wall and door towards the distribution of pedestrian in a room mention that pedestrian tend to locate themselves near the boundary (wall) and far from entrance (door). This can be one of the considerations in determining the pedestrian spatial distribution at RTT. Since the study about spatial distribution at RTT in Malaysia are none, study need to be done to provide the accurate information for future development of RTT.

Fridman [18] study the impact of cultural differences towards personal space or spatial distance through video recording. They found that the spatial distance for Iraq, French, England and Canada are 32.7 cm, 41.7 cm, 50.3 cm and 67.9 cm. Among those countries, Iraq and French have the closes value of spatial distance to the value produced by the game theory from Malaysia. Since, Malaysia and Iraq are Asia countries, they may share some similarities in their culture that lead to only 18.25% difference in spatial distance. French the European country may have diverged culture from the Asia but the fact that Eiffel Tower in Paris, French are the recreational place which most of the visitor are family, couples, friends and close relatives fall under categorised as intimate distance in interaction zones that lead to spatial distance of 0 cm to 45 cm give 4.25% difference in spatial distance from game theory.

Table 4. The spatial distance obtains from different country.

	Spatial distance (cm)							
Author \ Country	Malaysia	China	Japan	French	England	Canada	Iraq	Brazil
Mohd Ibrahim [16]	40							
Liu [17]		46						
Campanella [13]								65.4
Fridman [18]				41.7	50.3	67.9	32.7	

Modelling and simulation is one of the most used method to study pedestrian spatial interaction. Gotoh [19] developed the Distinct Element Method (DEM)-based crowd behavior simulator, which was applied to a simulation for evacuation against tsunami and Gotoh [11] modified the governing equation by adding the self-evasive force to develop self-evasive action model call DEM-based multi-agent model with self-evasive action model. Figure1 illustrate the perception domain used by Gotoh [11]to describe the model of vision which is symmetrical to the travelling direction of each individual pedestrian for both physical and psychological repulsive force. The physical repulsive force acts when the distance of pedestrian *i* and *j*, or distance of pedestrian *i* and the virtual wall element are less or equal to the average diameter of pedestrian *i* and *j*, or pedestrian *i* and the virtual wall element. Meanwhile, the psychological repulsive force act when the distance of pedestrian *i* and *j* is less or equal to the psychological radius as shown in Figure 1. The angle of vision and the psychological radius were the element considered to determine the perception domain of psychological repulsive force. The psychological radius was interpreted as the representative length scale of personal space.

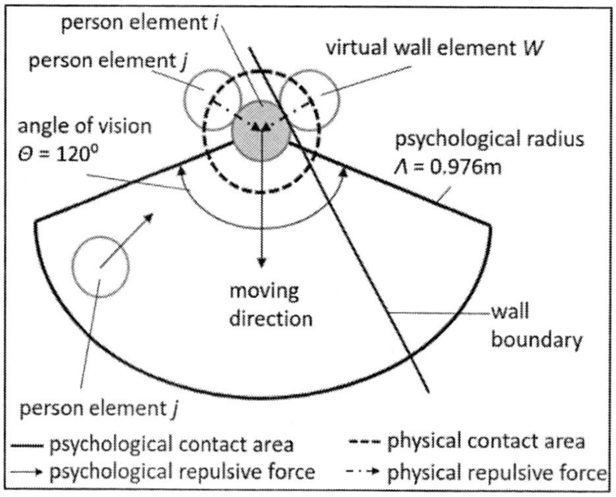

Figure 1. Perception domain by Gotoh [11].

Study the spatial interaction is crucial in planning and managing the pedestrians' flow in RTT.The dynamics of crowd motion is mainly driven by local interactions among pedestrians and their surrounding environment. Many models of crowd behaviour have been proposed so far to understand how pedestrian move, however modelling and visualizing the dynamics of pedestrian crowd behaviour in relation to pedestrian spatial interaction, particularly at railway stations is none in Malaysia.The DEM will be used to model the dynamics of crowd behaviour in RTT by considering the interactions between the pedestrians and between pedestrians and physical environment. The pedestrians are considered as assembly of particles of rigid bodies. Conceptually, particles move according to Newton's second law of motion.

3. A theoretical framework

3.1 Introduction

This proposed theoretical framework work to model and simulate the dynamics of pedestrian's crowd behaviour in Rail Transit Terminal (RTT) and hence to advocate the repercussions towards the evacuation behaviour of the crowd during emergency. In relation to that, a study of crowd evacuation is piloted at the largest transportation hub in Malaysia which is the KL Sentral Station (KLSS), Kuala Lumpur. Figure 2 shows the ground floor (noted as Level 1 from the building authorities) plan with estimated dimension of KLS. The KLSS is chosen as a study area due to its exponential increase in rail traffic passenger volume since its 15 years built. The Distinct Element Method (DEM) will be employed to model the dynamics of pedestrians' crowd behaviour. To confirm the reliability of the model, the validation is performed by statistical analysis. Validation through simulation and reproduction of walking scenario in KLS also will be performed. Subsequently, comparison with real life walking scenario in KLS is conducted. Then, the validated crowd dynamics model is demonstrated in evacuation capacity estimation of egress facilities in KLS. Evacuating people quickly and safely can greatly reduce casualty and economic losses. The outcomes from this study will be beneficial for future planning and design of mass transit and transit spaces.

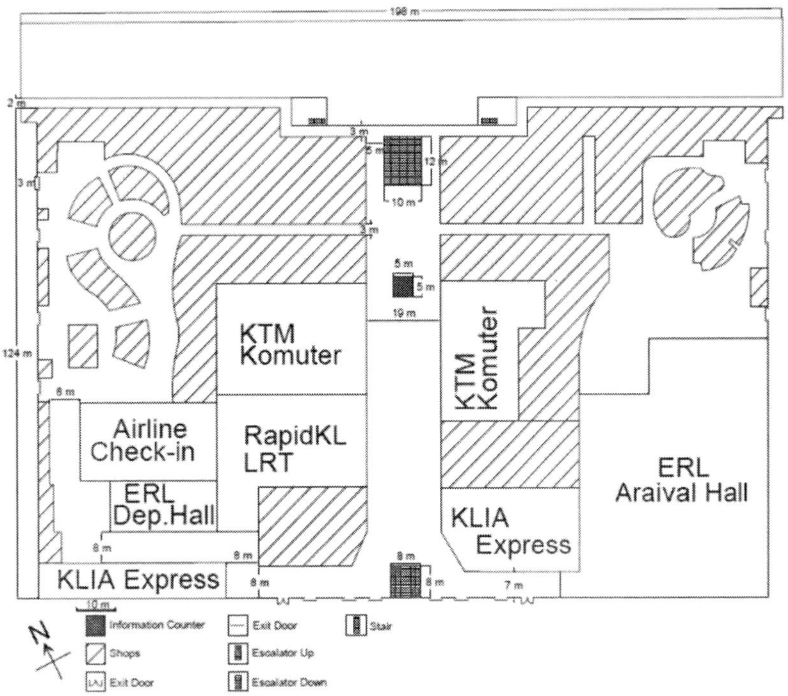

Figure 2. Level 1 floor plan.

The objective of this research is to determine crowd behaviour attributes in RTT such as walking speed, spatial pattern, spatial interaction parameters (distance and angles) and local collision avoidance during locomotion. Besides, this research was conducted to formulate the dynamic of crowd behaviour based on empirical result obtained using DEM and to assess the evacuation capacity of egress facilities in RTT and determine the evacuation time. The hypothesis of this research can be said as the number of passengers per unit area (density) increase, the spatial pattern of walking pedestrians has negative effect on the walking velocity. The pedestrians interact with other pedestrians through their perception domain by keeping a constant interaction distance and angle between pedestrians and physical environment and, pedestrians have their psychological radius to avoid collision with oncoming pedestrians. Furthermore, the hypothesis is that the maximum pedestrians flow that can pass the evacuation bottleneck section of egress facilities in RTT must be within given times.

From the hypothesis, it is important to know how crowd organize in space and affect the crowd dynamics, how walking pedestrians interact with each other, how pedestrian in crowd avoid collision with pedestrians and what is the condition of current evacuation capacity of egress facilities in RTT.

3.2. Flow of Methodology

The activities involved in this research are field observations and video footage at KLSS. The video data gathering will be analysed using Adobe After Effects CS6, Autodesk MAYA 2016 and HBS tool. The empirical results obtained have important implications for the validation of a model to replicate crowd dynamics in KLSS. The DEM-based model will be established to portray the crowd dynamics of Malaysian pedestrians in railway station. This crowd dynamics model will describe how a person interacts with other pedestrians and physical environment. The validation between simulated scenarios and the real situation are compared by taking video image as references. The effect of the newly developed model is shown in the ability of pedestrian to avoid collision of oncoming pedestrians. After the reliability of the model is confirmed, the evacuation capacity (EC) of egress facilities in RTT will be performed. The overall assessment of EC will be a basis in quantifying the overall evacuation time in RTT, and to look at if the capacity of RTT can meet the evacuation demand under emergency.

This research will be conducted in five phases starting from data gathering and end at the result obtaining stage as elaboration of those five phases will be described in this sub-topic.

3.2.1. Phase 1: Data Gathering. Data collections of the study area are the crucial part in this study.Data needed are: detail layout of KLSS, population distribution in KLSS and human crowd motion. Human crowd motion, particularly walking pedestrian will be collected through video footage. Population distribution is obtained through survey. And, the detail layout of the KLSS will be obtained from the authority.

3.2.2. Phase 2: Data Analysis. During this phase, the video films will be gathered and bring to the laboratory for analysis. There are three stages involved in video analysis and three different software will be used. The video analysis will first, convert from video to image sequence, second, track of pedestrian trajectories, and third, determine the average walking velocity, physical and psychological contact area, psychological radius and angle of vision. The three software are, Autodesk MAYA 2016 (MAYA), Adobe After Effect CS6 (AE) and HBS tool (in-house developed tool). Table 5 shows the process involved in this phase. The end product of this phase will reveal features of pedestrians' crowd behaviours as stated in the outcome of Stage 2 in Table 5 and Figure 1. Velocity are used instead of speed due to the pedestrian walking with direction as shown in Figure 1 and it is based on case study that involve walking with varies direction. Compare to previous studies described in sub-topic **2.1**, most of the researcher used speed because it is experiment based studied with fix walking direction and distance.

3.2.3. Phase 3: Modeling and Simulation. To model the dynamics of the crowd behaviour, the DEM (Figure 1) is employed. The DEM is known as a suitable method for the simulation ofthe dynamic behaviour of an assembly of particles. This approach explicitly provides the mechanical behaviour of the individualparticles and their contacts. Its computational modelling framework allows finite displacements and rotations of discrete particles andrecognizes new contacts automatically.DEM-based crowd behaviour simulator, named Crowd Behaviour Simulator for Disaster Evacuation (CBS-DE) has been developed.

Table 5. Process in video analysis.

	Video analysis		
	STAGE 1	STAGE 2	STAGE 3
	Conversion from video image to image sequence.	Track of pedestrian's trajectories.	Determination of crowd attributes.
SOFTWARE	Adobe After Effect	Autodesk MAYA 2016 with HBS tool plug-in	Microsoft Excel
OUTCOME	Image sequence with frame rate of 25 fps.	Trajectories (path coordinate) of each pedestrian motion for every 0.4s interval.	Walking velocity, physical and psychological contact area, psychological radius and angle of vision.

The DEM models particles distinctly as a group of rigid bodies and the behaviour of each particle is governed by translational and rotational equations of motion of $\boldsymbol{F} = m\boldsymbol{a}$ and $\boldsymbol{T} = I\boldsymbol{\alpha}$, where $\boldsymbol{F}$ reflects internal and external forces, $\boldsymbol{T}$ is a torque, m is a mass of a particle, i is a moment of inertia of a particle, $\boldsymbol{a}$ refers an acceleration of a particle and $\boldsymbol{\alpha}$ is an angular acceleration of a particle. The motion of a pedestrian element in contact with neighbouring elements is described in accordance with Newton's law of motion. Each pedestrian element is governed by translational and rotational equations of motion. The combination of an autonomous driving force, repulsive forces and the self-evasive force describe the motion of a pedestrian. The autonomous driving force reflects the motivation of a pedestrian to move in a prescribed walking velocity and orientation while repulsive forces reflect inter-element (contact) forces due to collisions, and the self-evasive force treats collision avoidance and pedestrian alignment. Hence, the motion of the pedestrian i in CBS-DE is written as Formula (1).

$$m_{hi}\dot{\boldsymbol{v}}_{hi} = \boldsymbol{F}_{inhi} + \boldsymbol{F}_{awhi} + \boldsymbol{F}_{sehi} \text{ and } I_{hi}\dot{\boldsymbol{\omega}}_{hi} = \boldsymbol{T}_{hi} \tag{1}$$

Where m_{hi} and I_{hi} are two parameters that refer to the mass and moment of inertia of the pedestrian i, respectively. $\boldsymbol{v}_{hi}$ is the velocity of the pedestrian i, $\boldsymbol{\omega}_{hi}$ is the angular velocity of the pedestrian i, " ˙" indicates a time-derivative, $\boldsymbol{F}_{inhi}$ is the inter-element (contact) force acting on the pedestrian i, $\boldsymbol{F}_{awhi}$ is the autonomous walking force of the pedestrian i, $\boldsymbol{F}_{sehi}$ is the self-evasive force acting on the pedestrian i and $\boldsymbol{T}_{hi}$ is the torque acting on the pedestrian i. The behaviour of the pedestrian element is computed explicitly by numerical integration of those equations.

3.2.4. Phase 4: Validation. The validation works are executed to confirm the reliability of the model by three methods. First, statistical data analysis involving the use of statistics to analyze data for analytical method validation such as mean, standard deviation, confidence intervals, linear regression and t-test will be performed. Second, reproducing the dynamics of crowd behaviour by comparing the simulated walking pedestrian scenario with the real-life scenario of walking pedestrian in KLSS. There are two procedure performed in this validation work. The first procedure is looking for preferred scenario from the footage. The footage will manually examined. The criteria for the preferred scenario are contra-flow and involved multiple groups pedestrians who walk in KLSS. The chosen scenario is then saved in sequence images of PNG format. In the second procedure, simulation of the chosen scenario is performed by employing the established model. The third step in validation phase is visualization that involved visualized simulated crowd dynamics of pedestrian at RTT in Autodesk MAYA software. This visualization is important in order to display any deficiencies or violations in calculations performed.

3.2.5. Phase 5: CG Movie. In the final phase, the computer graphic (CG) movie, is presented through Autodesk MAYA software to visualize the simulated evacuation process. It is an effective mean to display any deficiencies or violations in calculations performed. In addition, the CG movie presents complex scenario in a form that is easier to understand and interpret to the public.

4. Conclusions

Pedestrian behaviour such as speed, flow, density and evacuation time need advanced studied to help improving the operation and safety level of pedestrian at transport terminals in Malaysia. A lot of works need to be done since the above issue have many application and impact to the society. RTT in Malaysia is becoming one of the choice of transportation mode. Hence, the necessity to understand the specific behaviour of passengers in RTT is crucial. This proposed framework paper will work as a guide or reference for researcher to conduct more research on pedestrian behaviour at RTT particularly in Malaysia. In order to support the increasing demand in rail services, this understanding will become a reference information to the traffic engineer in designing walking infrastructure not only for RTT but including other places that attract crowd such as shopping mall, stadium, etc.

5. References

[1] Bohari Z A, Bachok S and Osman M M 2016 S *Procedia - Social and Behavioral Sciences* **222** 791-799
[2] Abustan M S, Rahman N A, Gotoh H, Harada E and Talib S H A 2016 *IOP Conference Series: Materials Science and Engineering* **136** 012077
[3] Patra M, Sala, E and Ravishankar K V R 2017 *Transportation Research Procedia* **25** 4763-4770
[4] Von Krüchten C and Schadschneider A 2017 E *Physica A: Statistical Mechanics and its Applications* **475** 129-141
[5] Han, Y and Liu H 2017 *Physica A: Statistical Mechanics and its Applications* **469** 499-509
[6] Zhang L, Liu M, Wu X and Abourizk S M 2016 *Automation in Construction* **71** 430-442
[7] Li Z-Y, Tang M, Liang D and Zhao Z 2016 *Procedia Engineering* **135** 616-621
[8] Zhao Z and Liang D 2016 *Procedia Engineering* **135** 602-606
[9] Zhao G, Li Y-F, Cui Y-Q and Zhao W-H 2016 *Procedia Engineering* **135** 33-39
[10] Zhao Z, Yan J-B, Liang D and Ye S-Q 2014 *Procedia Engineering* **71** 81-86
[11] Gotoh H, Harada E and Andoh E 2012 *Safety Science* **50** 326-332
[12] Yeo S K and He Y 2009 *Fire Safety Journal* **44** 183-191
[13] Campanella M 2016 Microscopic Modelling of Walking Behaviour Delft University of Technology
[14] Guo R-Y 2014 *Physica A: Statistical Mechanics and its Applications* **415** 428-439
[15] Yang X, Dong H, Wang Q, Chen Y and Hu X 2014 *Physica A: Statistical Mechanics and its Applications* **411** 63-73
[16] Mohd Ibrahim A, Venkat I and Wilde P D 2017 *Physica A: Statistical Mechanics and its Applications* **479** 485-497
[17] Liu X, Song W, Fu L, Lv W and Fang Z 2016 *Physics Letters A* **380** 1526-34
[18] Fridman N, Zilka A, Kaminka 2011 The Impact of Caltural Differences on Crowd Dynamics in Pedestrian and Evacuation Domains Bar Ilan University, Israel
[19] Gotoh, H, Harada, E, Kubo, Y, and Sakai, T 2004 Particle-system model of the behavior of crowd in tsunami flood refuge *Annual Journal of Coastal Engineering JSCE* **51** 1261-65 (in Japanese)

Acknowledgements
This research is supported by Ministry of Higher Education, Malaysia, under the Fundamental Research Grant Scheme, (Ref. No. FRGS/1/2017/TK08/USM/02/1). The authors would also like to thank School of Civil Engineering, Universiti Sains Malaysia.

The Influence of Saliva on Gingival Marginal Microleakage of Class II Cavities when Using Universal Bonding Agent in Direct Composite Restoration

S Stoleriu[1], S Andrian, G Pancu, I Nica and G Iovan
[1]Odontology, Periodontology and Fixed ProsthodonticsDepartment, "Grigore T. Popa" University of Medicine and Pharmacy, Faculty of Dental Medicine, Iaşi, Romania

E-mail: stoleriu_simona@yahoo.com

Abstract. The aim of the study was to evaluate by microleakage assessment the effect of saliva contamination during bonding procedure when a universal bonding agent applied in two different strategies was used in direct restoration using composite resins.Standardized box-only Class II cavities were prepared on the mesial and distal surfaces of twenty extracted teeth. A hybrid composite resin (Gaenial Posterior, GC Corporation) and a universal bonding agent (G Premio Bond, GC Corporation) applied in two different strategies: etch-and-rinse and self-etch were used for restoration, with or without saliva contamination of the gingival margin of the cavity. Lower leakage was recorded in the cervical area apically than CEJ when universal bonding agent was applied in self etch strategy when compared to etch-and-rinse strategy. On enamel margins (1mm upper than CEJ) universal bonding agents applied in etch-and-rinse technique leaded to lower microleakage. Irrespective of etch-and-rinse or self-etch strategy of applying, the presence of saliva impaired the enamel or dentine leakage of universal bonding agents.

1. Introduction

The adhesion and retention of resin-based materials for direct restoration can be influenced by many clinical factors. In particular, moisture coming from saliva, blood, gingival fluid or sometimes from hand-piece oil [1] might affect the bonding to dental tooth structures and might lead to microleakage at the tooth restoration interface. Microleakage is responsible for biofilm accumulation, caries lesions onset adjacent to the restoration, postoperative sensitivity, marginal staining or even loose of the restoration [2].

It was stated that a good isolation of the tooth and an efficient prevention of contamination are mandatory to achieve a qualitative bonding. Cervical area of class V or class II cavities are more prone to contamination during clinical procedure of restoration using resin-based materials [3]. In addition, the thin gingival enamel margin or even the absence of enamel which exposes cementum/dentine substrate for bonding can influence the adhesion. In deep proximal box-shape cavities the quality of hybridization can also be negatively affected by the specific orientation of the dentinal tubules [4].

All dental adhesive systems are composed by three different components: etching agent, primer and bonding agent. Acid-etching might be applied as a separate clinical step (etch-and-rinse technique) or it might beprovided by acidic functional monomers (self-etch technique). Previous studies reported a significant decrease in bond strength between composite resins and enamel surface after saliva contamination of etched enamel [5] due to salivary proteins precipitation, which blocksthe penetration of monomers in enamel pores and lead to a decreased mechanical adhesion [1]. The organic pellicle resulting from salivary proteins precipitation cannot be removed with water and it was demonstrated that re-etching the enamel for 10 seconds is needed to remove it [6].

In order to simplify the clinical procedure of bonding application, self-etch systems were introduced on the market. They have some advantages related to the elimination of rinsing and drying steps after acid etching, which might decrease the risk of saliva contamination andthusmight allow a better adhesion [7-9]. In addition, some studies argued that self-etch adhesive systems are resistant to salivary contamination [2, 10].

The recent technology in dental adhesion is represented the introduction ofuniversal or multi-mode bonding agents. These materials consist in a mixture of all three components in the same bottle. According to the manufacturer indications, they can be applied either in etch-and-rinse and in self-etch procedure. There are studies that aimed to compare these bonding agents to previous generations of adhesive systems in terms of adhesion efficacy [11-13] or to evaluate their performance in direct composite repair [14, 15]. There is a lack of data in the literature regarding the efficacy of universal bonding agents when they are applied in the situation of moisture contamination.

The aim of the study was to evaluate by microleakage assessment the effect of saliva contamination during bonding procedure when a universal bonding agent applied in two different strategies was used in direct restoration using composite resins.

2. Experimental procedures

Twenty extracted molars having no cracks, caries lesions or restorations on the proximal surfaces were selected for this study. Two standardized box-only Class II cavities were prepared on the mesial and distal surfaces of each tooth having a wide of 3 mm and a depth of 2 mm by a single operator. To prepare the cavities, a straight fissure diamond burs for every five cavities was used. The buccal and lingual walls of the preparations were parallel to each other and the line angles with the gingival wall were rounded. For a half of the cavities (n=20) the gingival margin was placed 2 mm below the cementoenamel junction (CEJ) and for the other half (n=20) the gingival margin was placed approximately 1 mm upper than the CEJ. The cavities were restored using a hybrid composite resin (Gaenial Posterior, GC Corporation) and a universal bonding agent (G Premio Bond, GC Corporation) applied in two different strategies: etch-and-rinse and self-etch.G-ænial is classified as a hybrid composite with a combination of 2 types of pre-polymerized resin fillers of 16-17µm (silica and strontium and lanthanoid fluoride), inorganic fillers larger than 100 nm (fluoroaluminosilicate) and inorganic fillers smaller than 100 nm (fumed silica). G-Premio Bond is a universal adhesive system containing a unique combination of three functional monomers (4-MET, MDP, MDTP). In etch-and-rinse strategy, 35% phosphoric acid etchant gel (3M-ESPE, St. Paul, MN, USA) was applied for 15 seconds on enamel and dentine using the tip of the syringe, then the acid was removed using the water spray from a dental unit and gently dry using the air spray. For bonding procedure, the application of the bonding system was made according to the producer instructions: using disposable applicator brushes, bonding agent was applied by rubbing the tooth surface for 20 seconds, gently air dried for 5 seconds using medium air pressure and then lightcured for 20 seconds.In self-etch strategy the bonding procedure was the same as the previous one, except the acid etching step.

Human saliva was used to simulate the contamination of the cavities. Stimulated saliva was collected from one donor after chewing paraffin wax for five minutes. A quantity of 0,2 ml of saliva was applied on the gingival wall of a halfof the prepared cavities using a brush applicator for 15 seconds, then the excess was spread using air spray fromthe dental unit. When the adhesive system was applied in etch and rinse strategy, contamination was simulated after etching and in self-etch strategy before bonding application.

In order to restore the cavities, a universal Tofflemire matrix retainer with matrix band was placed around the toothand pressed with the fingers in the cervical area in order to prevent overhanging to the gingival margin. Two horizontal layers of composite resin were used for each cavity restoration, each layer being placed using a dental spatula and a condenser, and being polymerized for 40 seconds using a LED light unit (LED B, Guilin Woodpwcker Medical Instrument Co., Ltd, China) having the light intensity of 850-1000mW\cm^2 and the wavelength of 420- 480 nm.Then the matrix was removed and the teeth were stored in distilled water for 1 week.

The distribution of the teeth in groups according to the strategy of bonging application was as follows:

Group 1 (n=5): cavity preparation with gingival margin lower than ECJ + bonding agent applied in etch and rinse strategy + restoration with composite resin

Group 2 (n=5): cavity preparation with gingival margin lower than ECJ + etching+ contamination with saliva +bonding application+ restoration with composite resin

Group 3 (n=5): cavity preparation with gingival margin lower than ECJ + bonding agent applied in self etch strategy + restoration with composite resin

Group 4 (n=5): cavity preparation with gingival margin lower than ECJ + contamination with saliva +bonding agent applied in self etch strategy + restoration with composite resin

Group 5 (n=5): cavity preparation with gingival margin upper than ECJ + bonding agent applied in etch and rinse + restoration with composite resin

Group 6 (n=5): cavity preparation with gingival margin upper than ECJ +etching+ contamination with saliva +bonding application + restoration with composite resin

Group 7 (n=5): cavity preparation with gingival margin upper than ECJ + bonding agent applied in self etch strategy + restoration with composite resin

Group 8 (n=5): cavity preparation with gingival margin upper than ECJ + contamination with saliva + bonding agent applied in self etch strategy + restoration with composite resin

The extern surfaces of the teeth were coated with two layers of water resistant nail varnish, except the surface of the restoration and a area of 1 mm around the restoration, then the teeth were immersed in 2% methylene blue dye solution (pH=7) for 4 hours. After that the teeth were transversely sectioned using diamond discs (Komet Dental, BrasselerGmbH&Co, Germany), under cooling with water. The sections were examined using a optical microscope (Carl-Zeiss AXIO Imager A1m) coupled with a high resolution digital camera at 50X magnification.The dye penetration was assessed according to the following scores: 0- no dye penetration, 1- dye penetration on less than a half of the interface, 2- dye penetration on more than a half of the interface, but less then whole interface, 3- complete dye penetration of the interface, without involving the axial wall, 4. complete dye penetration of the interface, involving the axial wall.The sections were examined by two different evaluators, blinded to the method of bonding application. Examiners confronted the score given for each image and the final score resulted as a common decision of both examiners.

3. Results and discussions

The scores, the mean value of marginal leakage and standard deviation when the gingival margin was placed lower then CEJ are presented in Table 1.

Increased mean value of marginal leakage was recorded in group 2 when compared to group 1 and in group 4 when compared to group 3. In groups 3 and 4 the mean values of the leakage were lower when compared to groups 1 and 2.

Increased mean value of marginal leakage was recorded in group 6 when compared to group 5 and in group 8 when compared to group 7. In groups 7 and 8 the mean values of the leakage were higher when compared to groups 5 and 6. The values obtained in groups 5-8 were lower than that obtained in groups 1-5.

Three major factors were considered to be important in microleakage onset: polymerization shrinkage of composite resin, the biological substrate and the chemical composition of the bonding agent [16]. Also, thermal contraction, absorption of water, mechanical stress and dimensional changes in tooth structure should affect the leakage [16]. Contamination with saliva of the operating field is a common clinical problem, especially when isolation using rubber dam is very hard to be achieved.

Table 1. Mean value of the scores for dye penetration ± standard deviation (SD) when gingival margin was placed lower then CEJ.

	Group 1	Group 2	Group 3	Group 4
Scores	2	3	1	2
	3	3	2	1
	2	2	2	3
	3	2	2	3
	1	3	2	3
Mean score ± SD	2.2±0.8	2.6±0.5	1.8±0.4	2.4±0.8

The scores, the mean value of marginal leakage and standard deviation when the gingival margin was placed upper then CEJ are presented in Table 2.

Table 2. Mean value of the scores for dye penetration ± standard deviation (SD) when gingival margin was placed upper then CEJ.

	Group 5	Group 6	Group 7	Group 8
Scores	1	1	1	1
	1	0	0	2
	0	2	1	1
	1	1	1	1
	0	1	1	2
Mean score ± SD	0.6±0.5	1±0.7	0.8±0.4	1.4±0.5

Different results were reported during the time regarding the effect of saliva contamination on the efficiency of adhesive systems. Some have reported that the saliva contamination had no adverse effect on the bonding efficiency of one-bottle adhesive systems [17-19]. Others have shown that the saliva contamination of the dentine surface produced a significant decrease in the bond strength [20-22].

In or study lower leakage was recorded in the cervical area apically than CEJ when universal bonding agent was applied in self etch strategy when compared to etch and rinse strategy. The same results were obtained by studies that aimed to compare the leakage of universal bonding agents applied in self etch technique and etch and rinse bonding agents [23, 24]. In etch and rinse strategy water is needed to keep the collagen fibril expansion for resin infiltration, but in the same time it has a bad effect on hybrid layer formation, decreases mechanical properties of the interface and lowers the durability of the bonded surfaces. Enzymatic degradation of exposed collagen fibrils and the hydrolysis of the adhesive polymer might appear as a result of uneven stress distribution in the hybridized zones [25, 26]. The popularity of self-etching adhesive systems increased over time due to „no-rinse" concept and lower postoperative sensitivity. By skipping the step of acid etching, the major concern of drying and wetting too much or too less is eliminated. Self etching agents etch only partially the dentine and some amounts of hydroxyapatite crystals are left in place around collagen fibers. Bonding stability relies on ionic bonds due to functional monomers (MDP) presence and affinity for hydroxyapatite [27, 28]. It seems that etch and rinse adhesive systems are not a good choice in terms of microleakage prevention when gingival margins are located indentine [29]. On enamel margins (1mm upper than CEJ) universal bonding agents applied in etch and rinse technique leaded to lower microleakage.

Irrespective of etch and rinse or self etch strategy of applying, the presence of saliva impaired the enamel or dentine leakage of universal bonding agents. Some factors have been identified as possible causes for decrease the bond to contaminated dentine. The adsorption of salivary glycoprotein on tooth surface prevents the monomers penetrating to the dentine collagen network. Also, an increased contact angle might occur, which decreases the bond strength [20, 30]. The presence of saliva might dilute the primer, which determines a weak hybrid layer. Different results were reported in the literature regarding the effect of contamination on dentine adhesion. Some studies pointed that washing or washing and re-etching the dentine that has been contaminated leaded to no significant difference in bond strength [22]. On a contrary, Fritz concluded that re-etching is not necessary in case of contamination with saliva [20]. When saliva contamination happens after etching the dentine, blot drying of the dentine leaded to a bond strength equal to that obtained in case of uncontaminated surfaces [31]. Re-application of bonding agent after saliva contamination have been proposed to improve the bonding efficiency [32, 33].

4. Conclusions
Universal bonding agent applied in self etch strategy leaded to lower leakage in the cervical area located in dentine when compared to etch and rinse strategy. On enamel margins universal bonding agents applied in etch and rinse technique determined lower microleakage. Irrespective of etch and rinse or self etch strategy of applying, the presence of saliva impaired the enamel or dentine leakage of universal bonding agents.

5. References
[1] Xie J, Powers J M and McGuckin R S 1993 *Dent. Mater.* **9** 295–9
[2] Hitmi L, Attal J P and Degrange M 1999 *J. Adhes. Dent.* **1** 219–32
[3] Mojon P, Kaltio R, Feduik D, Hawbolt E B and McEntee M I 1996 *Dent Mater* **12** 83–7
[4] Kumar J S and Akshmi S J 2016 *J. Pharm. Sci. Res.* **8**(7) 627-631
[5] Silverstone L M, Hicks M J and Featherstone M J. 1985 *J. Am. Dent. Assoc.* **110** 329–32
[6] Hormati A A, Fuller J L and Deney J E 1980 *J. Am. Dent. Assoc.***100** 34–8
[7] Milia E, Lallai M B and Gracia- Godoy F 1999 *Am. J. Dent.* **12** 167–71
[8] Brackett W W, Tay F R, Looney S W, Ito S, Haisch L D and Pashley D H 2008 *Oper. Dent.* **33** 89–95
[9] Hürmüzlü F, Ozdemir A K, Hubbezoglu I, Coskun A and Siso S H 2007 *Quintessence Int.* **38** 206–12
[10] EL- Kalla I H and Gracia- Godoy F 1997 *Am. J. Dent.* **10** 83–7
[11] Kermanshah H and Khorsandian H 2017 *Dent. Res. J. (Isfahan)* **14**(4) 272–281
[12] Gupta A, Tavane P, Gupta P K, Tejolatha B, Lakhani A A, Tiwari, Kashyap S and Garg G 2017 *J.Clin. Diagn. Res.* **11**(4) ZC53–ZC56
[13] Somani R, Jaidka S and Arora S 2016 *Indian J. Dent. Res.* **27**(1) 86-90
[14] Stoleriu Sandrian S, Nica I, Sandu A V, Pancu G, Murariu A and Iovan G 2017 *Materiale Plastice* **54**(3) 574-577
[15] Andrian S, Pancu G, Topoliceanu C, Tofan N, Stoleriu S and Iovan G 2017 *Revista de Chimie (Bucharest)* **68**(8) 1874-1877
[16] Van Meerbeek B, De Munck J, Yoshida Y, Inoue S, Vargas M, Vijay P, Van Landuyt K, LambrechtsPand Vanherle G 2003 *Oper Dent.* **28** 213–324
[17] Taskonak Band Sertgoz A 2002 *J. Oral Rehabil.* **29** 559–64
[18] Yoo H M, Oh T Sand Pereira P N *Oper. Dent.* 2006 **31** 127–34
[19] Yazici A R, Tuncer D, Dayangaç B, Ozgünaltay Gand Onen A 2007 *J. Adhes. Dent.* **9** 305–9
[20] Fritz U B, Finger WJand Stean H 1998 *Quintessence Int.* **29** 567–72
[21] Park Jand Lee K C 2004 *Oper Dent.* **29** 437–42
[22] Ghavam Mand Pour Sh K 2004 *J Dent Tehran University of Medical Sciences* **1** 5–10

[23] Van Landuyt K L, Peumans M, Fieuws S, De Munck J, Cardoso MV, Ermis RB, Lambrechts Pand **Van** Meerbeek B 2008 *J. Dent.* **36** 847–55
[24] Ermis R B, Temel U B, Cellik E Uand Kam O. 2010 *Oper. Dent.* **35** 147–55
[25] Pashley D H, Tay F R, Breschil B, Tjäderhane L, Carvalho R M, Carrilho Mand Tezvergil-Mutluay A 2011 *Dent Mater.* **27** 1–16
[26] Jorge P 2007 *Dent. Clin. North Am.* **51** 333–35
[27] Breschi L, Mazzoni A, Ruggeri A, Cadenaro M, Di Lenarda Rand De Stefano Dorigo E 2008 *Dent. Mater.* **24** 90–101
[28] Armstrong S R, Jessop J L, Vargas M A, Zou Y, Qian F, Campbell J A and Pashley D H 2006 *J. Adhes. Dent.* **8** 151–60
[29] Tuncer D, Çelik C, Çehreli S Band Arhun N 2014 *J. Adhes. Sci. Technol.* 2014 **28** (13) 1288-97
[30] Oztoprak M O, Isik F, Sayinsu K, Arun Tand Aydemir B 2007 Am. J. Orthod. Dentofac. Orthop. **131** 238–42
[31] EL- Kalla I H and Gracia- Godoy F 1997 Am. J. Dent. **10** 83–7
[32] Sattabanasuk V, Shimada Y and Tagami J 2006 J. Adhes. Dent. **8** 311–8
[33] Eiriksson S O, Pereira P N, Swift E J Jr, Heymann H Oand Sigurdsson A 2004 Dent. Materials **20** 37–44

Assessment of Hydrophobic Coating on Porous Calcareous Rocks Surface Exposed in Urban Ambient Air Pollution

V Pelin[1,2,3], O Rusu[2], M M Cazacu[2,4,*], S Gurlui[2], A V Sandu[5,6], I Radinschi[4], V Ciocan[7] and I Sandu[6,8]

[1] Department of Environmental Science, Faculty of Geography - Geology, Doctoral School of GeoScience, Alexandru Ioan Cuza University, Carol I Blvd, no. 20A, 700505, Iasi, Romania

[2] Atmosphere Optics, Spectroscopy and Lasers Laboratory - LOASL, Faculty of Physics, Alexandru Ioan Cuza University, Carol I Blvd, no. 11, 700506, Iasi, Romania

[3] Department of Concrete Structures, Building Materials, Technology and Management,Faculty of Civil Engineering and Building Services, Gheorghe Asachi Technical University, prof. Dimitrie Mangeron Blvd, no. 1, 700050, Iasi, Romania

[4] Department of Physics, Gheorghe Asachi Technical University, prof. Dimitrie Mangeron Blvd, no. 59A, 700050, Iasi, Romania

[5] Faculty of Materials Science and Engineering, Gheorghe Asachi Technical University of Iasi, prof. Dimitrie Mangeron Blvd, no. 41A, 700050, Iasi, Romania

[6] Romanian Inventors Forum, Sf. Petru Movila Street, no. 3, Iasi, Romania

[7] Department of Building Servicres, Faculty of Civil Engineering and Building Services, Gheorghe Asachi Technical University, prof. Dimitrie Mangeron Blvd, no. 1, 700050, Iasi, Romania

[8] Arheoinvest Interdisciplinary Platform, Alexandru Ioan Cuza University, Carol I Blvd, No. 11, 700506, Iasi, Romania

E-mail: marius.cazacu@tuiasi.ro

Abstract. Urban air quality has changed rapidly in recent years, that requiring new research focused on the effects of pollution, both on the environment and on human health, as well as on apparent changes in building surfaces, including historical ones. As is known, permanent exposure to the atmosphere in the urban environment causes degradation processes on the surfaces of natural stone monuments, especially on porous stone surfaces. In these investigations were used samples of porous calcareous rocks similar to the natural stone from the many historical monuments of the Iasi city - Romania. These samples were coating with various commercially hydrophobic solutions. Further, these samples were exposed in the immediate vicinity of a historical monument and an intense road traffic junction. After about six months of exposure, apparent changes in the treated surfaces were compared with untreated control surfaces but subjected to the same environmental conditions over the same time period in order to evaluate the effectiveness of the hydrophobic treatment.

1. Introduction

As it is known, there is a strong connection between the local geological setting and the old architectural surface [1, 2]. For objective reasons, in the metropolitan area of Iasi City - Romania, porous calcareous rocks have been used extensively, being extracted from the nearby pits, such as those in the village of Paun, located on the plateau of Repedea Hill [3, 4]. Taking into account that any natural stone becomes predisposed to a series of physical and mechanical transformations, from the moment of extraction by the quarry or when it is used in a construction and becomes exposed to environmental conditions [5, 6], long-term research is required the length of time that such geomaterials [7] are aging over time [8]. Moreover, the pronounced carbonate and porous character of the Paun – Repedea stones [9] implies the adoption of a coating treatment that prevents or limits as far as possible the undesirable effects [10, 11]. Also, the presence of urban pollution factors, which has increased in recent years in the city of Iasi, has favored the surface alteration of building stones [6, 12-18], which gives special attention to the use of these lithic materials in construction civil or historical building.

Given the above, knowing at the same time that there is no universally valid hydrophobic product [19,20], this paper proposes the assessment of the impact of ambient air on hydrophobic indigenous rocks and subsequently exposed for a certain period of time, under the conditions of an urban environment [21, 22], with intense car traffic, having the main research technique CIE $L^*a^*b^*$ colorimetry, useful in determining the total colour change on the investigated surfaces [23, 24].

2. Materials and methods

The lithic material exposed to urban environmental conditions is a sedimentary and oolithical calcareous and porous rock [9], form during in the Sarmatian geological age, and was sampled from Paun – Repedea village, Iasi County – Romania. This quarry was for many years an important source of natural geomaterials necessary for construction of local civil and historical buildings [4, 25].

The hydrophobization of porous calcareous surface samples were conducted used eight chemicals products, for lithic materials maintenance, with the following commercially names: LTP Mattstone H2O, LTP Colour Intensifier Stainblok Seal 2, Sikagard S700, Sikagard S703, Isomat Nano Pro-C, Isomat Nano-Seal, Isomat PS-20 and Tenax Ager.

CIE $L^*a^*b^*$ colorimetric investigations were performed using a Lovibond® RT 300 (Reflectance Tintometer D65/10°) spectrophotometer, monitoring the chromatic deviation at the same measurement points in each specific work step. Colorimetric measurements were carried out in constant laboratory conditions, at a temperature of ~21°C and relative humidity (RH) of ~60%.

The meteorological parameters were monitored using a LSI LASTEM station which contains sensors and characteristics for wind and relative humidity recording data.

3. Experimental part

After taken calcareous rocks from a pit located on the outskirts of Paun - Repedea village, in southern part of Iasi city - Romania, these were cut using roughly rectangular shapes with two planar surfaces to facilitate fastening on support and to have a uniformity of the gravimetric deposition areas of the airborne particles while allowing for colorimetric measurements before and after exposure. After cutting ten porous calcareous samples, marked P0 to P9, was dried and stored in desiccators for 48h, preceding the first colour measurements and coating chemical treatments. The samples labelled P0 and P9 were used for witness test left untreated, but exposed under the same urban environmental conditions like the eight treated samples.

The hydrophobic solutions were grouped by producers and named S1 to S8, in according with the stone sample`s notation, from P1 to P8 (Table 1).

The hydrophobic coating was carried out under the conditions provided by the manufacturer`s technical sheets, respecting the working conditions and the hydrophobic solutions quantities for each of the eight porous lithic surfaces. After this step, the treated samples stored and dried in laboratory conditions for 96h to stabilize the surfaces moisture and then the colorimetric measurements were performed and labelled with *AT* symbol (by *after treatment* but before exposure to urban environment). Exposure period of all porous calcareous samples (the ten pieces) was performed between Oct. 10, 2017and Apr. 05, 2018 (*178 days*). After the samples were withdrawn, these were dried for 96 hours under the same laboratory conditions. Final step was to measure results colourcoordinate for each sample which was noted with *ATE* symbol (after treatment and exposure), in accordance with Table 2.Concerning the colorimetric evolution, the colour change was measured for each coordinate (L^*,a^* and b^*), as compared to it's initially value, on the same sample and in the same point, whereas the total change of colour (ΔE^*_{ab}) was calculated in accordance with the following equation [23,24,26, 27]:

$$\Delta E^*_{ab} = \sqrt{\left(\Delta L^*\right)^2 + \left(\Delta a^*\right)^2 + \left(\Delta b^*\right)^2} \tag{1}$$

in which $\Delta L^*$is the change in light intensity in the respective point, at different time intervals, as compared to the initial value ($\Delta L^* = L^*_n - L^*_{initial}$), where L^* means the lightness to darkness coordinate and varying from 0 (black) to 100 (white); $\Delta a^*$is the chromatic modification of the coordinates of axis a^* ($+a^*$ indicating red and $-a^*$ green), from the same point, at different time intervals, as compared to its initial value ($\Delta a^* = a^*_n - a^*_{initial}$) and Δb^* is the chromatic modification of the coordinates of axis b^* ($+b^*$ means yellow and $-b^*$ blue), respecting the same method of calculation ($\Delta b^* = b^*_n - b^*_{initial}$).

Table 1. The chemical products notation in accordance with producers grouping and samples labelling.

Producer	Chemical products		Stone samples notation
	commercially name	notation	
LTP	Mattstone H2O	S1	P1
	Colour Intensifier Seal 2	S2	P2
Sika	Sikagard 700S	S3	P3
	Sikagard 703W	S4	P4
Isomat	Nanopro-C	S5	P5
	Nano-Seal	S6	P6
	PS 20	S7	P7
Tenax	Ager	S8	P8

Table 2. The chemical products notation in accordance with the producers grouping and the labelling of samples.

colorimetric variables		P0	P1	P2	P3	P4	P5	P6	P7	P8	P9
ΔL^*	AT	0	-3.76	-2.22	-1.99	-3.98	-1.32	-4.19	-1.86	-12.21	0
	ATE	-8.38	-13,56	-13,94	-11,94	-12,76	-16,56	-16,38	-11,05	-14,76	-10,62
ΔE^*_{ab}	AT	0	6.73	3.97	3.49	7.17	2.22	6.31	2.60	14.09	0
	ATE	8.43	13.64	13.96	11.95	12.78	16.88	16.47	11.16	14.99	10.97

The current urban environmental exposure area is located near a major and crowded road junction called *Stone Bridge* (*Podul de Piatra*), in the immediate vicinity of a XIXth century historical monument with the same name: Stone Bridge. This exposure place is located over the Bahlui River and southern margin of mentioned intersection. Thus, the exposure place is located between the historic bridge and the south-eastern part of the intersection, with the following geographical coordinates: 47.158035 N, 27.575451 E (Figure 1).

Figure 1. *Stone Bridge* (*Podul de Piatra*) intersection/cross-road from Iasi city, Romania: (a) older bridge *(historical)*; (b) newer bridge; (c) geographical position for exposure area. [24]; (d), (e) exposure sample and meteorological station.

In the same area is placed LSI LASTEM meteorological station for daily monitoring and data recording at regular intervals of time.

4. Results and conclusions

4.1. Hydrophpobic assessment by colorimetric measurements

As can be observe in Table 2 and in the graph of Figure 2, after chemical treatment, total color change value (ΔE^*_{ab} - *AT*) varies between 2.22 (P5 / S5) and 14.09 (P8 / S8). The samples P1, P4 and P6 have slightly above the minimum value in according with scientific literature: $\Delta E^*_{ab} \geq 5$ [27]: P1 - 6.73, P4 - 7.17, respectively P6 - 6.31. The value of total color exchange for sample P8 confirms the manufacturer's specifications of the S8 solution in terms of obtaining the *"wet"* effect on the treated lithic surfaces, which is very important to note as it will major influence the aesthetics of any architectural surface treated with this chemical product.

For samples P2, P3 and P7, after coating (*AT*), the total color exchange values fall below the prescribed limit: $\Delta E^*_{ab} \leq 5$) [27].

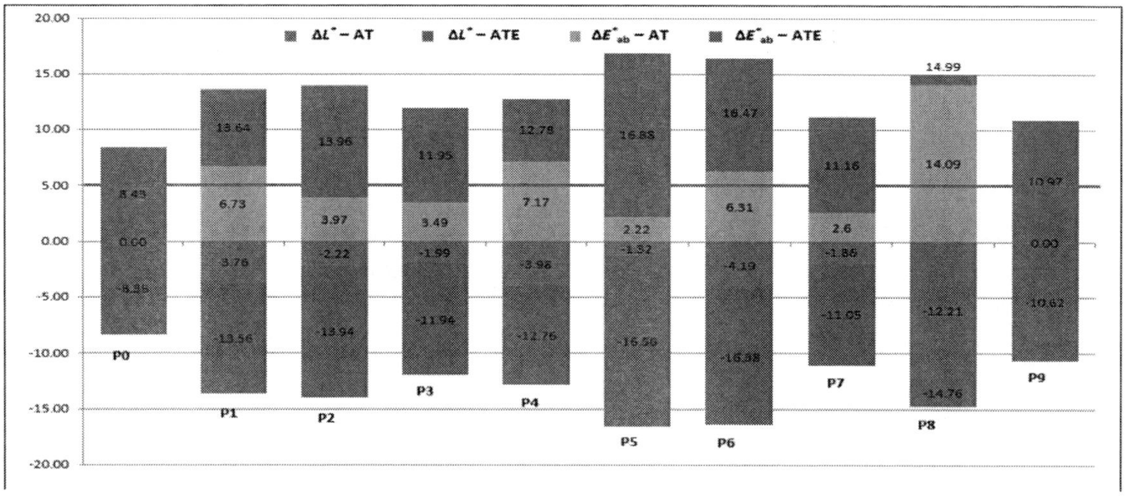

Figure 2. Comparative graphical representation of the colorimetric data ΔE^*_{ab} and ΔL^* obtained after chemical treatment (*AT*) and exposure in urban ambient air from *Stone Bridge* area (*ATE*).

At the same time, ΔE^*_{ab} - *AT* values for samples P1, P5 and P6 can be compared with the results of similar treatments with the same chemicals (S1, S5 and S6) used on the same type of lithic material (from Paun - Repedea pit) and in the same laboratory conditions, the results identified in the literature are as follows: S1 - *9.58* (vs. 6.73), S5 - *1.27* (vs. 2.22) and S6 - *6.26* (vs.6.31) [23,28]. Therefore, these solutions may have predictable results in case of coating of the porous calcareous material with the same source.

4.2. Hydrophpobic assessment by colorimetric measurements after exposure in urban ambient air
After exposure to the urban climate in the *Stone Bridge* area, all ten samples significantly exceed the minimum threshold of notable total color change ($\Delta E^*_{ab} \geq 5$), with remarkable increasing value for P5, P6 and P9 samples, ΔE^*_{ab} -*ATE* having 16.88, 16.47 and 14.99 units, respectively. Also, the control samples (P0 and P9) showed significant color shifts, with values of 8.43 for P0 and 10.97 for P9.

On the other hand, after exposure in the urban environment, all samples have a change in the apparent color (from shade term, in natural light, by authors evaluation) from ashen yellow and grey yellowish to blackening (Figure 3). This is explained through:

- the high porosity of the samples [9] may cause a *receptivity* of them to retain on the lithic surface the air borne and the particles from the atmosphere and
- by corroborating *wet* and *dry deposition* processes, depending on various meteorological parameters from different periods of the year [5, 26,29].

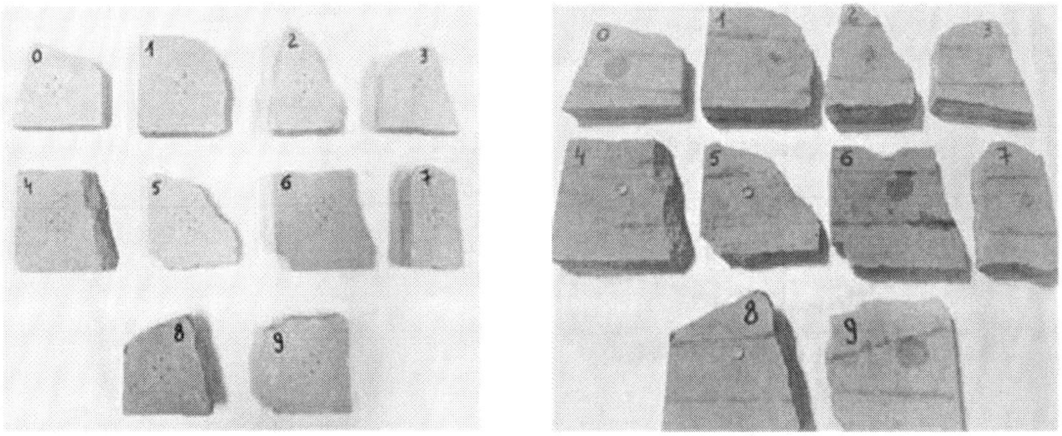

Figure 3. Calcareous porous sample: **left** – at 96h after chemical treatment (*AT*); **right** – testing hydrophobization efficiency after exposure period (*178 days*) in ambiental urban air (*ATE*) and dried in 96h, in laboratory conditions.

According to the data from Table 2 and the graphical representation in Figure 2, it is observed that all ΔL^* - *ATE* luminescence values are sensitively equal with ΔE^*_{ab} - *ATE*, meaning that the other colorimetric parameters are involved, Δa^* - *ATE* and Δb^* - *ATE* (see eq.1), have an insignificant influence (see eq.1). In scientific literature, this phenomenon is explained by the *blackening* process, which is assimilated to the major fluctuation of luminescence, ΔL^* having only negative values, by lowering the L^* in values comparing to the initial value. (see eq. 1) [26]. This is the case for all samples after the 178 days of exposure in the urban environment from *Bridge Stone* area, that case can confirming in the preliminary way the receptivity of the porous geomaterials for retaining air borne or particles in suspension from the atmosphere, even if they were chemically treated for protection against the undesirable effects of wetting, the priority role of these treatments being exclusively the periodic maintenance of the apparent architectural surfaces [22].

An example for this situation can be represented by the colorimetric variation, after chemical treatment of the P2 sample, which has the lowest value of the total colour change, between the treated sample (ΔE^*_{ab} - AT = 2.22) and after exposure in the urban environment, when increases significantly, reaching the limit of 16.88, this value being the maximum between the obtained results for ΔE^*_{ab} - *ATE*. From this point of view, at the opposite of the results of the P2 sample, there is the colorimetric variation of the sample P8 which seems to have a lower *receptivity* (or *susceptibility*) to retain the atmospheric particles, because after the chemical treatment with S8 the value ΔE^*_{ab} - *AT* reaches at 14.09 (maximum *AT* value) and after exposure the value ΔE^*_{ab} - *AT* is very close to ΔE^*_{ab} – *AT*: 14.99 units.

Certainly, this type of *receptivity* of porous lime coated hydrophobic surfaces should be investigated and studied in a larger manner in order to obtain a generally valid conclusion for porous lithic surfaces treatments.

4.3 Environmental conditions over hidrophobised porous surface

As it is known, the major influences on the apparent architectural surfaces are represented by various environmental factors [5,6,29]. Thus, during the period of exposure of samples P0 ÷ P9, the environmental parameters monitored in the *Bridge Stone* area were the wind, humidity and the presence of PM10 particles, according to the data presented in Figures 4, 5 and 6.

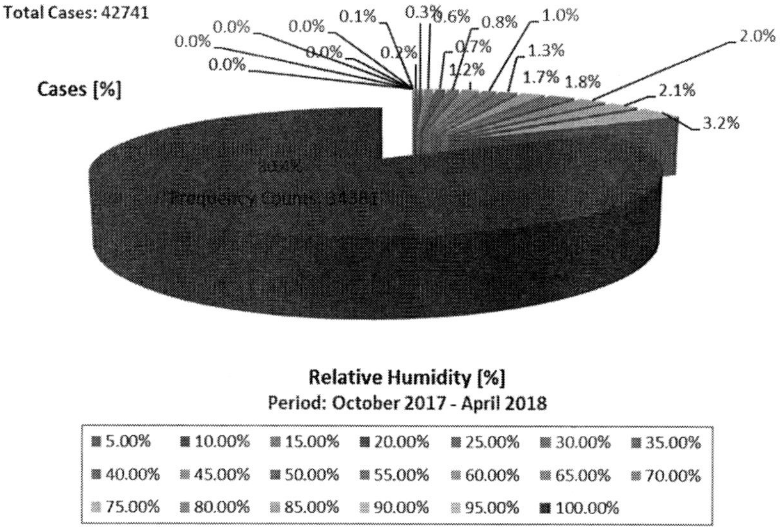

Figure. 4. Relative humidity during the exposure period - Oct. 2017 ÷ Apr. 2018, in *Stone Bridge* intersection. Data averages at 5 minutes, from 42,741 total cases.

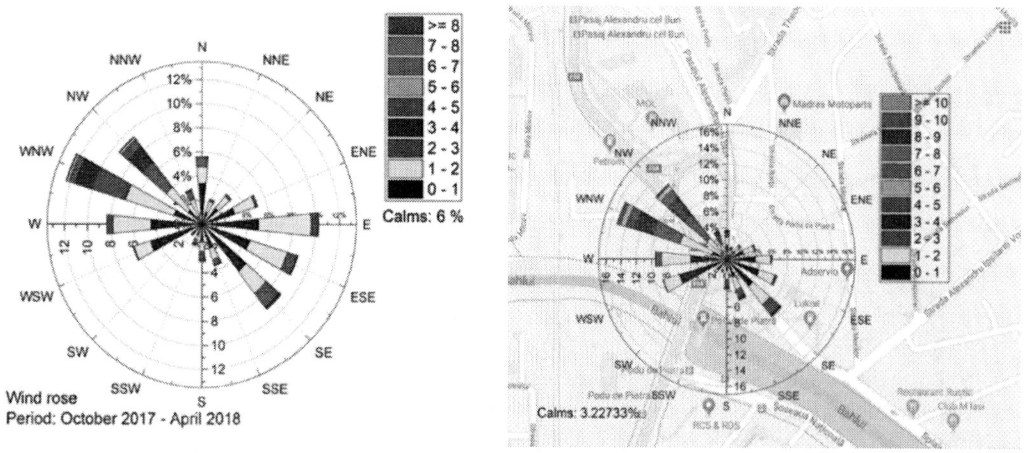

Figure 5. Wind rose in *Stone Bridge* intersection: **left** – exposure period for P0 ÷P9 samples: Oct. 2017 ÷ Apr. 2018; **right** – another wind action in the same area, in a previous period: Oct. 2016 ÷ Aug.2017, shown in Google street mapping.

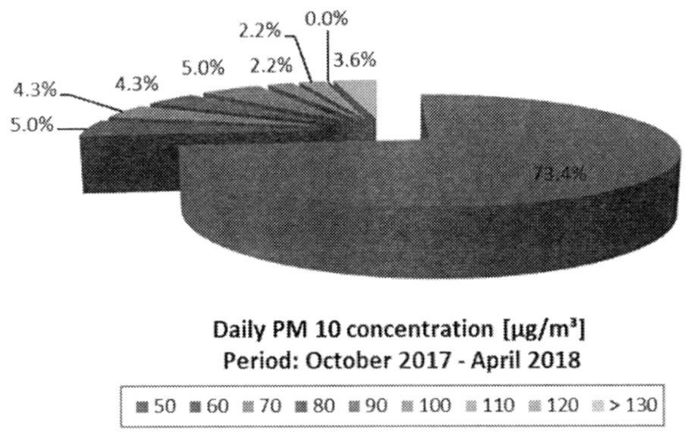

Figure 6. Statistics of daily PM10 concentration during the exposed period, in *Stone Bridge* area – Iasi, Romania [30].

The presence in over 80% of cases of a 100% *humidity* (Figure 4), a *total precipitation* of 136.40mm during the exposure period [31], alongside the predominant *wind action* from the WNW and NW to SE and SSE sectors (Figure 5) and the *presence of PM10* particles (Figure 6) on a *cold-season temperature* background [31] along with the auto traffic values in increasing, confirms the opinion of numerous authors regarding the influence of these parameters in the progressive change and alteration of urban ambient air, ultimately favoring the undesirable *blackening effect* previously presented [26,29,32-35].

Along with these factors, the *geomorphology* of the Iasi metropolitan area, to get her with the *local climate* factors, generates the effects of the *thermal inversion* specific of the cold season [36], blocking the dispersion of atmospheric pollutants or *"pushing"* them to the soil, which in time influences the quality of the apparent lithic surfaces of the built environment. This fact is also confirmed by the color full appearance (*blackened*) of the porous limestone samples P1 ÷ P9, after the exposure of approximately 180 days, in the low altitude zone at the *Stone Bridge* intersection in relation with surroundings of Iasi City, Romania.

5. Conclusions

In according to CIE$L^*a^*b^*$ colorimetry measurements, applied to the ten porous limestone samples, a number of aspects could be assessed, such as:

- color evolution before and after chemical treatments allows first assessment of an aesthetic impact for investigated or maintained surfaces;
- the predisposition (or *receptivity*) of porous surfaces (including those that are hydrophobized) to retain particles or pollutants from the atmosphere;
- the direct influence of some parameters in the urban climate in the unwanted change of porous lithic surfaces, resulting in undesirable effects such as *blackening*.

The responsiveness of some local lithic materials to keep the particles in the atmosphere together with the *blackened appearance* of the samples after about six months of exposure near an important road junction in Iasi confirms also that a number of geomaterials in the architectural surface (including historical construction) become witnesses of the way in which atmospheric pollution takes place over time, especially in the urban environment.

Of course, this article represents a preliminary assessment of the impact and the relation between the porosity of some lithic surfaces, hydrophobization surfaces and various environmental factors. Last but not least, it is worth mentioning that for a complete technical evaluation it is also necessary to measure successively the contact angle after chemical treatments and after exposure of lithic porous samples under various urban environment conditions.

6. References

[1] Di Benedetto C, Cappelletti P, Favaro M, Graziano S F, Langella A, Calcaterra D and Colella A 2015 *Eng Geol* **193** 310-319
[2] Morra V, Calcaterra D, Cappelletti P, Colella A, Fedele L, de' Gennaro R, Langella A, Mercurio M, de' Gennaro M 2010*Journal of the Virtual Explorer***36**26
[3] Branzila M and Stefan P 2009 *Sci Ann CuzaUnivJassy, Sect Geol***LV(2)**5-18
[4] Badarau D and Caprosu I 2007 *The Old Masonry from Iasi City: until 1821*Demiurg Publisher, Iasi, Romania
[5] Camuffo D 1995*Sci Total Environ***167**1-14
[6] Watt J, Tidblad J, Kucera V and Hamilton R 2009 *Effects of Air Pollution on Cultural Heritage*Springer, New York, United States
[7] Fookes P G 1991 *Q J EngGeol***24(1)** 3-15
[8] Bellopede R, Castelletto E and Marini P 2016 *Eng Geol***211**19-26
[9] Chelarescu A, Nichita O, Paunel E, Mihul A, Tudoraş A, Dumitrescu I, Ciubotaru V and Constantinescu F 1956 *Bull. Polytech. Inst. Jassy***II(VI)** 351-388
[10] Bell F G 1993 *Environ Geol***21** 187-200
[11] Corcione C E, De Simone N, Santarelli M L and Frigione M 2016 *Prog Org Coat***103** 193-203
[12] Cazacu M M, Timofte A, Balin I, Dimitriu D G and Gurlui S 2011 *Environ Eng Manag J***10(1)** 139-145
[13] Banica A, Bobric E D, Cazacu M M, Timofte A, Gurlui S and Breaban I G 2017*Environ Eng Manag J***16(9)**2147-2163
[14] Cazacu M M, Timofte A, Unga F, Albina B and Gurlui S 2015 *J Quant Spectrosc Ra***153**57-64
[15] Unga F, Cazacu MM, Timofte A, Bostan D, Mortier A, Dimitriu D G, Gurlui S and Goloub P 2013 *Environ Eng Manag J***12(2)**297-303
[16] Cazacu M M, Tudose O G, Rusu O, Scripa A E and Radinschi I 2017 *Bull Polytech Inst Jassy, Sect Math Theor Mech Phys***63(67)**43-60
[17] Breaban I G and Paiu M 2016 *Int Multi SciGeoco***II** 79-86
[18] Pelin V, Breaban I G, Sandu I and Gurlui S 2017 *Rev Chim-Bucharest***68(6)**1333-1340
[19] Boutin F 2001 *ProceedingsofHydrophobe III - Third International Conference on Surface Technology with Water Repellent Agents*, Aedificatio Verlag, Freiburg, Germany 233-244
[20] Luvidi L, Mecchi A M, Ferretti M, Sidoti G 2016*Int J ConservSci***7(1)**311-322
[21] Comite V, Álvarez de Buergo M, Barca D, Belfiore C M, A. Bonazza A, La Russa M F, Pezzino A, Randazzo L and S.A. Ruffolo S A 2017 *Constr Build Mater***152**907-922
[22] Vleugels G, Fobe B, Dewolfs R and Van Grieken R 1994 *Sci Total Environ* **151** 59-69
[23] Pelin V, Sandu I, Munteanu M, Iurcovschi C T, Gurlui S, Sandu A V, Vasilache V, Branzila M and Sandu I G 2016 *IOP Conf Ser-Mat Sci* **133** 012061
[24] Pelin V, Rusu O, Sandu I, Gurlui S, Sandu A V, Cazacu M M and Sandu I G 2017 *IOPConf Ser-Mat Sci***209** 012080
[25] Ratoi B, Pelin V, Sandu I, Branzila M and Sandu I G 2018 *Int J ConservSci***9(1)**151-164
[26] Grossi C M, Brimblecombe P, Esbert R M and Alonso F J 2007 *Color Res Appl***32(4)**320-331
[27] Schanda J 2007 *Colorimetry. Understanding the CIE System* Wiley-Interscience Hoboken, New Jersey, United States of America
[28] Pelin V, Sandu I, Gurlui S, Branzila M, Vasilache V and Sandu I G 2016 *Rev Chim-Bucharest***67(12)**2568-2572
[29] GrossiC M and Brimblecombe P 2007 *Building Stone Decay: From Diagnosis to Conservation*Geological Society Special Publications**271**117-130
[30] * * *http://calitateaer.ro/public/home-page/?__locale=ro
[31] * * *https://www.wunderground.com
[32] Xie X, Semanjski I, Gautama S, Tsiligianni E, Deligiannis N, Rajan R T, Pasveer F and Philips W 2017 *Isprs Int J Geo-Inf***6(12)**389
[33] Nazarenko Y, Kurien U, Nepotchatykh O, Rangel-Alvarado R B and Ariya P A 2016*Environ*

Sci-Proc Imp **18** 190-199
[34] Buttini P and Pereg G 2003 *Structural Studies, Repairs and Maintenance of Heritage Architecture VIII,* Transactions on the Built Environment, WIT Press, **66** 221-228
[35] Gill P S and Graedel T E 1983 *Rev Geophys Space Ge* **21(4)** 903-920
[36] Sfica L, Ichim P, Apostol L and Ursu A *Theor Appl Climatol* doi.org/10.1007/s00704-017-2305-4

Aknowledgement
This work was supported by the Romanian Space Agency (ROSA) within Space Technology and Advanced Research (STAR program) project number 162/20.07.2017 and project number 169/20.07.2017.

IOP Conf. Series: Materials Science and Engineering **374** (2018) 012092 doi:10.1088/1757-899X/374/1/012092

Topical Notions About in Vivo Analysis for Degradable Biomaterials with Utility in Human Body

B Istrate[1], T Vololeniuc[1], C Munteanu[1*] and Ş Lupescu[1]
[1] Department of Mechanical Engineering Mechatronics and Robotics, Faculty of Mechanical Engineering, "Gheorghe Asachi", Technical University, Iasi, Romania

E-mail:cornelmun@gmail.com

Abstract. Based on the biodegradable biomaterial development, there are numerous studies demonstrating partial or total aspects that have ultimately led to their use in many fields, especialy in medicine. This study summarizes the latest information on the development of degradable biomaterials based on magnesium, especially used in *in vivo* tissue compatibility assessment. The biomaterial compatibility assessment combines with the development of medical devices, this study following evolutionary parameters and integration of implants for human use. One of the most important parameters is the efficiency with which the medical device works near to the maximum potential. In conclusion, our study aims to broadly present the in vivo analysis of biodegradable biomaterials based on magnesium.

1. Introduction

Biodegradable materials are non-toxic, non-carcinogenic, non-imunogenic, and non-teratogenic alloys with adequate physical and mechanical properties that, after implantation into a living organism, will gradually degrade, performing regular biological functions and dissolving to complete the mission, to heal tissues, preferably without leaving a metallic residue. The study, research and development of biodegradable materials is necessary to enhance their performance, improve their function, develop the release of bioactive compounds, and achieve tissue regeneration [1-3].

2. The interaction clinical response

Regarding the clinical response, among all the biomaterials used, those based on magnesium are considered to be the avant-garde of use in research studies and therapeutic medical applications, due to certain properties and functions that are closer to biological ones. A major pillar in the development and implementation of vascular and orthopedic applications, the two main areas of investigation associated with the development of magnesium materials and alloys, is the in vivo assessment of the compatibility of biomaterials and medical devices with tissues, a branch that has begun to be developed in recent years [4, 5]. Magnesium-based biomaterials have amazing advantages over stainless steel, titanium or Co-Cr-Ni alloys, which leave metallic residues resulting from in vivo abrasion with toxic effects that can cause hypersensitivity, local inflammation or tissue necrosis [1, 4, 6]. Magnesium alloy has chemically active properties and is susceptible to degradation in the physiological environment after implantation without causing toxic and secondary effects [3, 7, 8].

The researchers made a comparison of the most used elements in the medical implant industry, namely magnesium, iron and zinc (these are considered biodegradable base materials for use in the medical industry), and found that,from the point of view of the main properties, magnesium meets all the criteria to be considered a good biodegradable material (Table 1) [6].

Tabel 1. Mechanical parameters and degradation rate of pure Fe, Zn, and Mg used for medical applications (compare the stainless steel).

	Yield strength (MPa)	Tensile strength (MPa)	Elongation (%)	In vivo degradation rate (mm year $^{-1}$)
Pure Fe: annealed	150	200	40	0.16
Pure Zn: as cast	17	20	0.2	0.2
Pure Mg: as cast	20	86	13	407

2.1. Magnesium physiology
Magnesium is an essential nutrient for life, and it is, regarding abundance in the human body, the fourth element and the most common divalent intracellular cation. The human body contains approximately 25 grams of magnesium (70 kg average adult body) of which about 65% is found in the bones and teeth, and the rest is distributed into blood, body fluids, organs and other tissues. Once in the body, it is decomposed and released to form independent atoms of magnesium or "ions". Like other vitamins and minerals, the role of magnesium is mainly regulating and is involved in more than 300 enzymatic reactions that help the body to function properly [1, 6, 9]. In addition to its importance in reactions involving ATP, Mg plays roles in protein and nucleic acid synthesis, mitochondrial activity and integrity, ion channel modulation, plasma membrane stabilization and translation processes, and many other cellular functions . Due to the importance of this element, homeostasis as well as the potential pathological effects of a possible mineral imbalance should be studied closely before the human body is exposed to an excess of Mg as a result of its use as a biomaterial [4, 9].

Magnesium homeostasis is effectively maintained by the combination of intestinal absorption, renal excretion and bone use as a magnesium deposit. Exposure to degradable material may cause disruption to these systems causing discontinuation of homeostasis and pathological ailment due to an excess of deposits that may be clinically manifested as hypermagnesemia. Acute destruction of muscle tissue can also lead to an excess of magnesium [4, 9]. To date, studies have not demonstrated that hypomagnesaemia can be associated with the implantation of magnesium-based biomaterials.

3. Biocompatibility
To serve as an in-vivo biomaterial, one of its primordial properties referring to the development of biodegradable materials is biocompatibility. This is the ability of the material to come into contact with living tissue so that it does not produce toxic and allergic side effects and its level of adaptability to the biological environment is as high as possible [2, 4, 10].

3.1. In vivo methods for the analysis of biocompatibility
In vitro tests for biocompatibility assessment provide information on general toxicity, avoiding in vivo implantation of potentially toxic materials. However, in vitro techniques can not reproduce the complex environment that a biomaterial will exhibit when implanted in a physiological environment, requiring the use of in vivo investigations. There are many techniques that are associated with identifying the tissue reaction, the area where the implant is in direct contact with the biomaterial, but below we will only list the most commonly used.

Investigating intramedullary or I.O., intramodal magnesium based biomass is a technique that relies on the use of histological techniques and can be used to indicate inflammation, neovascularization and follow-up of new bone formation [11, 12]. This technique can be supplemented by following the microscopic aspect (tissue response to implants) Another standard and widely accepted technique for initial in vivo investigations is the implantation of magnesium-based biomaterials in a subcutaneous or intramuscular environment with several specific benefits: the implant can be easy removed and evaluated for corrosion behavior, and tissue interacts much better with a possible hydrogen gas production. Several other investigations have used similar techniques to analyze soft tissue reaction in magnesium implants by specifically removing the muscle and connective tissue covering but these techniques are time consuming, costly and require the use of antibodies for a wide range of antigens to provide a complete picture of the tissue reaction that is taking place [4, 13].

4. Degradability

By comparing with other metallic materials, magnesium-based materials and alloys exhibit high bioactivity and low density, thus favoring degradability in the physiological environment and influencing success as biomaterials. For single-phase materials, degradation is typically located, resulting in the formation of concavities on the surface of the material [8, 10, 14]. The presence of a secondary phase, due to impurities, causes a galvanic degradation. The lack of generalized degradation is an important factor in the use of magnesium as a biomaterial, because extended degradability surfaces cause mechanical properties to change. On the other side of the barrier, using a magnetically degraded magnesium biomaterial could lead to the production of gaseous hydrogen in the implant medium and an increase in the local pH that could cause significant effects on surrounding tissues [4, 8, 10, 14, 15].

4.1. In vivo methods for the analysis of degradability

Techniques used for assessing the degradability of magnesium after in vivo implantation are much less compared to in vitro methods, mainly aiming at measuring the physical reduction in material size due to degradation. The most used analysis is to measure the remaining volume of the implanted material using micro-computerized tomography (μ-CT), a non-destructive technique that does not affect the implant and surrounding tissue, and then allows for other analyzes such as new bone growth or osteointegrity [4].

Another method used for analyzing material degradability is measuring the weight loss of the implant. The technique offers direct comparative data on weight loss obtained from in vitro studies but has many major drawbacks: the biomaterial has to be removed and the implant-tissue interface can not be investigated and the shear forces required to mechanically remove an implant are likely to deteriorate the tissue that surrounded the material, eliminating the potential for tissue analysis [4].

Another quantitative method used to assess degradation of magnesium in vivo is to analyze the surface of the remaining transverse section of implants in two-dimensional sections. These results are then compared to the original size of the implant, indicating the degradation. This technique is generally associated with the transverse sections of the implant that is cut with the associated surrounding tissue, allowing for a specific biocompatibility analysis [4, 12, 15]. In particular, the degradability of magnesium materials and alloys could compromise the mechanical integrity of the implant, thus requiring in-depth studies of the behavior of each material. But some studies state that the magnesium used as a candidate material meets many of the desired requirements, and in terms of manufacturing processes and initial degradability rates it is indicated to have an excellent potential.

4.2. Alloying with other materials

It is known that magnesium has a low strength under casting with a high rate of degradation [4], and the consolidation of the properties can be achieved by suitable alloying and machining conditions with elements that are, in turn, degradable and biocompatible. These elements improve the mechanical properties by hardening the solid solution with the closed hexagonal structure, curing the precipitations and refining the granules. Aligning with other elements can inhibit early weight loss, healing, bone thickening, and inflammation, and magnesium alloy surfaces can reduce the rate of degradation until the healing period. The addition of Sn, Si, Zr and Zn can improve maximum traction resistance and elongation at failure, while combining with Ca, Sr and Mn will cause acceleration of degradation [1, 4, 16-19].

Tabel 2. Major key benefits and drawbacks of Mg [1].

		Characteristics	Details
Pro	Low density	Low	Mg density (1.738 g/cm3) is close to that of cortical bone (1.75–2.1 g/cm3);
	High specific strength	High	Strength-to-weight ratio of approximately 130 kNm/kg;
	High damping capacity	High	Mg has the ability to absorb energy of any metal can be used for load bearing applications;
	Machinability and dimensional accuracy	High	Mg is the easiest structural metal to machine, and stable final dimensions are easy to achieve. Consequently, complex shapes are easily producible, which is crucial for the often intricate shapes that are required for medical applications;
	Stress shielding	Less	Stress-shielding-related problems can be greatly reduced for many orthopaedic implants as Mg density is very close to that of bone;
	Biocompatibility	Good	Mg is considered biocompatible and has been shown to increase the rate of bone formation;
	Degradation	Good	Corrosion of Mg in the body would eventually result in complete degradation and it would be beneficial for patients have temporary exposure to an implant in the body.
Contrary	Elastic modulus	Low	Lower elastic modulus of Mg may be beneficial with respect to stress shielding and it is vital to ensure that any implant is designed to sustain its load without deformation;
	Degradation	Rapid	Mg implants are intended to completely degrade, but at a slow rate to the pace of bone remodelling;
	Hydrogen evolution	High	The released H2 gas accumulates at the surrounding soft tissues at severe degradation rate. The hydrogen evolution rates for various Mg alloys containing Zn, Al and Mn are reported as 0.01 ml/cm2/day

There are studies that have tracked alloying magnesium with aluminum but have not shown concretely whether it increases / decreases the compatibility and degradability of magnesium used in medical applications. Also included in this category are rare elements such as lanthanum, lutetium, scandium and yttrium.

Manganese is an essential element that plays an important role in the metabolic cycle of lipids, amino acids and carbohydrates. In the magnesium alloy, manganese is mainly used to increase ductility. Recently, magnesium-manganese alloys have attracted considerable attention due to their high damping capacity (about 80%), which can help suppress vibrations generated during motion and stress at the implant-bone interface [4].

Alloying elements can significantly improve the mechanical properties, control the rate of degradation, and influence the biological response to the magnesium alloy. But it is difficult to choose the best alloy and, to do this, we need more in vitro and in vivo experiments such as clinical investigations [16, 20-25].

5. Other properties that can cause biodegradability

Magnesium elastic modulus (41 ~ 45 GPa) is closest to that of natural bone (3 ~ 20 GPa) compared to other elements used to make bioresorbable materials, inducing stress protection and increasing secondary bone fracture [4].

Other properties that define good biodegradable material are yield strength, elastic modulus and maximum traction resistance for load applications (Table 1), low weight, good wear resistance and osteointegration [1].

6. Conclusions

Since the beginning of the 20th century, magnesium and its alloys have been applied in many clinical cases but are not widely used due to uncontrollable degradability. The latter has led to numerous failures, as well as to clinical complications. From the studies conducted, there is a way to improve biodegradable implants, namely adding elements by making alloys to meet both the biodegradability criterion and biocompatibility. There are multiple alloys created with improvement properties, but their biological effect as well as long-term clinical outcomes are not yet well established. This study seeks clarification on the use of magnesium-based biomaterials in the use of implants so that both researchers and physicians benefit from avant-garde materials that will allow less invasive and less costly surgical interventions.

7. References

[1] Radha R, Sreekanth D 2017 *Journal of Magnesium and Alloys* **5**(3) 286-312
[2] Buddy D, Ratneret al. 2013 Biomaterials Science: *An Introduction to Materials in Medicine*; Third Edition; Academic Press, Elsevier; II.3.4.; p. 609-615
[3] Prakasam M et al. 2017 *Journal of fundamental biomaterials* **8** 44
[4] Walker J et al. 2014 *Jounal of Biomedical Materials Research* **102**(6) 1316-1331
[5] Mao L et al. 2017 *Scientific Reports* **7** 46343
[6] Li H, Zheng Y, Qin L 2014 *Progress in Natural Science: Materials International* **24**(5) 414-422
[7] Hou L et al. 2014 *Progress in Natural Science: Materials International* **24**(5) 466-471
[8] Shimizu Y et al. 2009 *Interface Oral Health Science* 318-320
[9] Allan R, Mara N 2012 *Acute Med* **11** 3-7
[10] Yufeng Z 2016 *Magnesium alloys as degradable biomaterials* CRC Press p. 46-48
[11] Razavi Met al. 2015 *Journal of Materials Science: Materials in Medicine* **26** 184
[12] Cheng J 2014 *Research of Magnesium-based Alloys for Medical Applications* 2nd International Conference on Education Technology and Information System; Atlantis Press p. 631-635
[13] Mehrban N, Bowen J 2016 Elsevier **81** 110
[14] Witte F et al. 2006 *Biomaterials* **27**(7) 1013-1018
[15] Li L et al. 2017 *Regenerative Biomaterials* **4**(2) 129–137
[16] Pogorielov M et al. 2017 *Interv Med Appl Sci.* **9**(1) 27–38
[17] Persaud-Sharma D, McGoron A 2012 *J Biomim Biomater Tissue Eng.* **12** 25–39

[18] Trinca L C, Fintanariu M, Solcan C, Trofin A E, Burtan L, Acatrinei D M, Stanciu S, Istrate B and Munteanu C 2015 *Appl. Surf. Sci.***352** 140–50
[19] Istrate B, Munteanu C, Matei M, Oprisan B, Chicet D and Earar K 2016 *Influence of ZrO 2 -Y 2 O 3 and ZrO 2 -CaO coatings on microstructural and mechanical properties on Mg-1,3Ca-5,5Zr biodegradable alloy* IOP Conference Series: Materials Science and Engineering **133** 12010
[20] Zhang E, Xu L, Pan F, Yu G, Yang L, Yang K 2009 *Biomaterials***30**(8) 1512–1523
[21] Minciuna M G and Vizureanu P 2013 *Metalurgia International***18**(6) 123-126
[22] Minciuna M G, Vizureanu P, Geanta V, Voiculescu I, Sandu A V, Achitei D C, Vitalariu A M 2015 *Revista de chimie***66**(6) 891-894
[23] Hanganu S C, Armencia A O, Grigoras S, Hanganu L C and Vizureanu P 2010 *Materiale Plastice***47** 357-378
[24] Bălțatu M S, Vizureanu P, Mareci D, Burtan L C, Chiruță C and Trincă L C 2016 *Materials and Corrosion-Werkstoffe und Korrosion***67**(12) 1314-1320
[25] Stoleriu S, Iovan G, Pancu G, Georgescu A, Sandu A V and Andrian S 2014 *Materiale Plastice***51** 162-166

Assessment of Physical-Chemical Characteristics of Surface Water from Key Sites of the Mesta River: State and Environmental Implications

S Georgieva[1*], K Gartsiyanova[2], V Ivanova[1] and L Vladimirova[1]
[1]University of Chemical Technology and Metallurgy, bul. Kliment Ohridski 8, Sofia, Bulgaria, Department of Analytical chemistry;
[2] National Institute of Geophysics, Geodesy and Geography, Department of Geography – Bulgarian Academy of Sciences (NIGGG-BAS), Sofia, Bulgaria, str."Acad. G.Bonchev", bl.3, Sofia 1113, Bulgaria

E-mail: st.georgieva@uctm.edu

Abstract: The anthropogenic source pollution of the Mesta River was assessed during the period 2011 and 2016 in terms of pH, conductivity, chemical oxygen demand (COD), anions and heavy metals in key sites of the Mesta river pointed as: S1 (the Mesta river before the Iztok river); S2 (the Iztok river before the Mesta river); S3 – the river Mesta after the Iztok river); S4 - the river Mesta at Momina Klisura, near Bukovo) and S5 – the river Mesta before the Greece border, after the Matnitsa river. The application of multivariate clustur analysis (CA) for the interpretation of a large and complex data matrix obtained during a monitoring program of surface water in Mesta river is presented in this study. The dataset consists of analytical results from a 6-yrs survey conducted in selected points of the river system. The physical-chemical characterization in the water samples were made in accordance with the Directive 2000/60/EU–Water Framework Directive (WFD) and its equivalent criteria transposed into the Water Law (WL) in Bulgaria. In water, concentrations (mg L^{-1}) during 2016 of NO_2^- (0.006 to 0.052), NO_3^-(0.01 to 1.33) and total contents in μg L^{-1} of Cu (<0.002), Pb (<0.003), Co(<0.002), Ni(<0.003), S(<0.050) and Zn(<0.02), pH (5.60 to 8.00), and electrical conductivity (0.12 to 48.60 $mS.cm^{-1}$) were agreed with environmental standards except cadmium Cd (>0.15 μg L^{-1}) and PO_4^{3-}(0.15 mg L^{-1}). During analyzed period the cadmium concentration was much higher than recommended limit only. In this sense, it was possible to demonstrate relatively good quality of river water even with numbers of industrial and touristic activities in the analyzed area but also to consider a future concept on cadmium sources and their eliminated.

1. Intruduction

Water quality monitoring is a fundamental tool in the management of freshwater resources. Freshwater is very important resource, essential for agriculture, industry and human existence. Without adequate water quality sustainable development will not be possible. Reliable monitoring data and defined the water quality of the river are the indispensable basis for protecting the physic-chemical parameters of the river systems [1].

The main object of this article is the catchment area of river Mesta in southwest part of Bulgaria. The basin of the Mesta river is part of the West Aegean Basin Water Management Area in Bulgaria. It is formed by the merging of the Cherna Mesta and Bella Mesta rivers at an altitude of 941 m. The catchment area in the Bulgarian section of the valley is 2785 km^2. The Mesta River's basin has the highest average altitude in the country - 1318 m [2]. The river is cross-border and flows into the Aegean Sea on Greek territory Figure 1.

In the Mesta river there are 8 municipalities - Yakoruda, Razlog Bansko, Belitsa, Gotse Delchev, Garmen, Satovcha, Hadjidimovo. The population is the main driving force of economic activities, but at the same time its professional specialization, structure and literacy is closely related to the changes that occur in the environment and in particular with the change of the water quality in the studied region. The largest share in the economy of the surveyed area is the agrarian sector and the services whose activities directly affect the state of the water in terms of their quality characteristics. Once of the main industrial pollutants are the companies that form and distribute in the water bodies or in the urban sewerage networks the substances and the compounds which are from the lists of priority and dangerous substances.

According to the River basin management plan in the West Aegean Region (2010-2015 and 2016-2021) [2] "Magnetic Head Technologies" AD, The Telephone Equipment Factory (ZTA) – Bansko, "Kalida – AD – Gotse Delchev are like that. Other sources of water pollution are identified the factory for low-current relays situated in village of Banya, Razlog, pig farming (in village of Borovo), Gotse Delchev, car-washes and others. In the basin of river Mesta the only one urban wastewater treatment plant (WWTP) is for the town of Razlog (in operation since 2008) and purifies the domestic and industrial waste water of the city. Receiver of purified water is the Iztok River. In the process of designing and building are WWTPs for the towns of Gotse Delchev and Bansko. Towns of Yakoruda, Belitsa, Bansko, Gotse Delchev, Hadzhidimovo, as well as the villages of Dobrinishte, Garmen and Satovcha are larger settlements in the Mesta river basin without built and operating WWTP. These towns and villages have a mixed sewerage system in good condition. Diffused sources of pollution in the catchment area of the analyzed river are: diffuse dirt from settlements without built-in sewer systems or not well-functioning, contaminations from farmland treated with fertilizers and plant protection products, livestock farming and use of organic fertilizer, tailing ponds and the consequences of mining. Another major problem for the Mesta River Basin is the problem of the still widespread practice of dumping solid waste into unregulated landfills, most notably dry gullies, gullies or floodplains of rivers. This practice, prohibited by the law, is the cause of a continuous process of drainage and seepage of contaminated water into the surface water. In fact, there is no information on the number and area of unregulated landfills, as well as the quantity and type of waste. Finally the identification of anthropogenic impacts on the waters of the Mesta River is necessary for the preparation of prevention programs for those river sections that are at risk of contamination.

In connection with the numerous and varied anthropogenic activities in the basin of river Mesta the article aims to identify the changes in water quality in the studied river by chemical analysis of selected physicochemical indicators for the period 2011-2016. As the base information of the study we used the sampling data for over then indicators (as pH, conductivity, chemical oxygen demand (COD), anions (PO_4^{3-} , NO_3^- , NO_2^-), metals and etc.) reported at the monitoring points of the Mesta river : S1 (the Mesta River before the Iztok river); S2 (the Iztok river before the Mesta River); S3 – the river Mesta after the Iztok river); S4 – the river Mesta at Momina Klisura, near Bukovo) and S5 – the river Mesta before the Greece border, after the Matnitsa River.

The study is based on Directive 2000/60/EU–Water Framework Directive (WFD) [3] Water Law (WL) in Bulgaria [4] and especially on Ordinance N-4/2012 on characterization of surface water (OG. 22 on 03/05/2013) [5] and Ordinance on environmental quality standards for priority substances and some other pollutant (OG. 88 on 09.11.2010) [6]. To achieve the goal of this analysis the following tasks were done:

creating of the database, assessment of the water quality by selected indicators in five sites, identification of the sources of pollution – pointed and diffused.

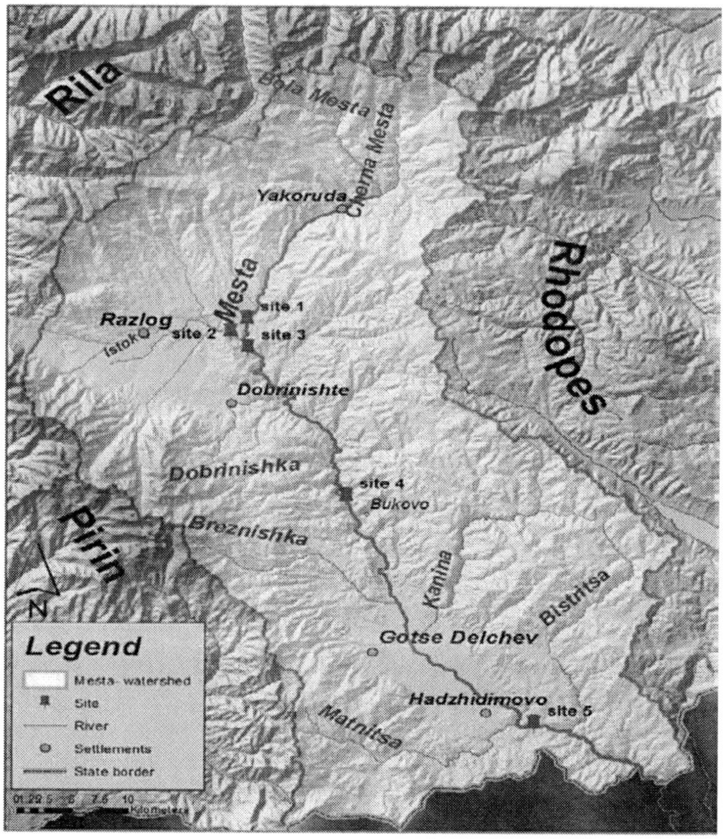

Figure 1. The Mesta River basin and the analyzed sites.

2. Methods and materials

2.1. Sampling and sample location

The samples were collected from key sites of the Mesta river pointed as follow: S1 (the Mesta River before the Iztok river); S2 (the Iztok river before the Mesta River); S3 – the river Mesta after the Iztok river); S4 – the river Mesta at Momina Klisura, near Bukovo) and S5 – the river Mesta before the Greece border, after the Matnitsa River. Sampling was done in the properly cleaned glass jars, precleaned by washing with non-ionic detergents, rinsed in tap water, in 1:1 hydrochloric acid and finally with deionised water before usage. Immediately after collection, the samples were analyzed for pH, chemical oxygen demand (COD), conductivity, nitrates (NO_3^-), nitrites (NO_2^-), phosphates (PO_4^{3}), as well metals as Na, K, Ca, Mg, Sr, Ba, Fe, Al, Cr, Co, As, S, U, Mo, Tl, V, Mn, Pb, Cd, Cu, Zn and Ni.

2.2. Methods for determination of metals, pH, conductivity, COD, NO_3^-, NO_2^-, PO_4^{3}-and SO_4^{-2} in samples

The study is based on the analysis and evaluation of the monitoring data provided by the Executive Environment Agency (ExEA) for the following physicochemical quality parameters: pH, conductivity, nitrates (NO_3^-), nitrites (NO_2^-), orthophosphates (PO_4^{3}), COD and metals – Cu, Cd, Zn, Pb, Ni, As and Cr for period 2011-2015. The data for metal contents as well as *pH, conductivity, COD,* NO_3^-, NO_2^- *and* PO_4^{3} for 2016 were collected from samples analysis in analytical chemistry laboratory at University of Chemical Technology and Metallurgy using standard methods.

The pH of water was measured with the Janway pH meter immediately after sample collection. pH measurements were carried out with digital pH-meter (Jenway). A CDM92 conductivity meter was used to measure the conductivity values of samples. The concentrations of COD were determinate by permanganate titrimetric method [7]. According to this method the COD determinations (range 0 to 1500 mg L^{-1} COD), 50.0 ml of water sample was added into solution of H_2SO_4 (15 ml) and 10.00 ml standard solution of $KMnO_4$. The solution is boiled for 10 min. Then $H_2C_2O_4$ was added and the solution was titrated with standard solution of $KMnO_4$.

Phosphate, nitrate and nitrites were determined using an ion chromatograph (IC-850 Professional model, Metrohm AG, Switzerland) equipped with auto sampler, the samples filtration system with a 0.2 μm membrane, injection valve, high-pressure pump, suppressor module, an eluent degasser and conductivity detector. The separation was performed on Metrosep A Supp 7-250 column (250 x 4 mm, 5 μm particle size), with 3.6 mmol L^{-1} Na_2CO_3 as eluent in isocratic mode at flow rate of 0.7 ml min^{-1}, temperature 45 °C and injected volume of 20 μl. For adjust the baseline under 0.99 μS cm^{-1} was used 0.1 M sulfuric acid solution and ultrapure water (Milli-Q) were used for automatic chemical suppression. Under the working conditions all anions were separated completely and total analysis time was 33 min. Deionized water from Millipore Milli-Q (18.2MΩcm, equipped with a Millipack 0.22μm filter) was used for the preparation of solutions. Eluent solution was prepared by sodium carbonate (Na_2CO_3, Merck) in Milli-Q water. Standard solutions of nitrite, nitrate and phosphate were prepared by appropriate dilution of their anion standard stock solution from Fluka (1000 mg L^{-1}, TraceCERT, Sigma–Aldrich, Buchs, Switzerland) for IC. All solutions were prepared by gravimetrically method on the balance. The preparation of samples are used MF-Millipore membrane filter with pore size 0,45 μm.

Standard ISP-OES method (Inductively Coupled Plasma Optical Emission Spectroscopy) by properly calibrating the equipment with standards for précised results was used for metal determination.

2.3. Statistic methods for data analysis

In total, a dataset for water quality of Mesta river was used for further multivariate analysis. The 98 samples were characterized by 13 variables. Cluster analysis was used in this work for revealing similarities between variables and sampling sites. Surface water hydrochemical groups were defined using hierarchy cluster analysis (HCA). HCA was performed only on the basis of major ion concentrations, pH, conductivity and COD using STATISTICA 8 software. The Euclidean distance as a similarity measure and Ward's method as a linkage method give the most efficient results for analysis of the surface water chemical composition [8] were used. Trace elements were not included in multivariate statistical analysis because complete data matrix is required but most of the measurements were made at different times and locations [9].

3. Results and discussion

The surface water quality of five sites of the Mesta River is discussed. The results from different physicochemical measurements are shown in Tables 1 and 2. The pH and chemical oxygen demand (COD) data of the executive agencies for the survey period are scarce (Figure 2 and Figure 3). The pH values taken in period 2015-2016 varied between 7.3 and 8.5 (S1, S3, S4 and S5) which is good accordance with Bulgarian target water quality range for pH in surface water [5]. In 2015 a slight increasing of pH at all sites is observed which is even a little over the reference value for S2. (Figure 2). One can see that a significant increase of organic pollution in 2015 is also observed, especially in sites S1, S2, S3 and S4 (Figure 3). The Bulgarian acceptable limit for COD in river waters 6.0 mg L^{-1} [5]. This limit was not exceeded in the river water samples of all sites and the parameter does not give cause for concern. Influenced concentrations of COD in this period varied between 10.0 mg L^{-1} and 12 mg L^{-1} at point S1, 3.7 mg L^{-1} and 11 mg L^{-1} for S2, 4.3 mg L^{-1} to 11.5 for S3, 4.3 mg L^{-1} to 12 mg L^{-1} for S4 and it ranged from 4.1 mg L^{-1} to 8.5 mg L^{-1} for S5 (Figure 3). During 2016 analysis of the chemical oxygen demand in our laboratory showed much lower concentrations (4.56 mg L^{-1} for S1; 3.45 for S2, 4.03 for S3, 3.23 for S4 and 4.42 mg L^{-1} for S5) than required (6.0 mg L^{-1}). The slight increase in the pH of the water and over the reference rang values of COD could be due to contamination with fecal waste water from the domestic and livestock sectors which is abolished in 2016. Cluster analysis of data (Figure 6) also showed a relationship between pH and COD, which proves that the increase of pH is mainly related to the increase in organic pollution of the river.

The electrical conductivity of water estimates the total amount of solids dissolved in water and is a useful and easy indicator of its salinity or total salt content. The main sources of salts are domestic sewage, municipal storm water drainage and industrial effluent discharges. A very high salt concentration (>1000 mg L^{-1}) imparts a brackish, salty taste to water and is discouraged because of the potential health hazard. For this reason electrical conductivity can serve as a useful salinity indicator when considered with other factors and when a natural geological origin does not apply in terms of the source of dissolved salts. Electrical conductivity data were evaluated in period 2015-2016 for sites S1, S2, S3, S4 and S5 of Mesta river and the results are presented on Figure 4. The guideline for conductivity according Bulgarian acceptable limit is 750 μS cm^{-1} [5]. The water conductivity values are within this acceptable limit. Electrical conductivity values for 2016 varied between 125 μS cm^{-1} at S1 and 270 μS cm^{-1} at S2 and are within acceptable limit as well.

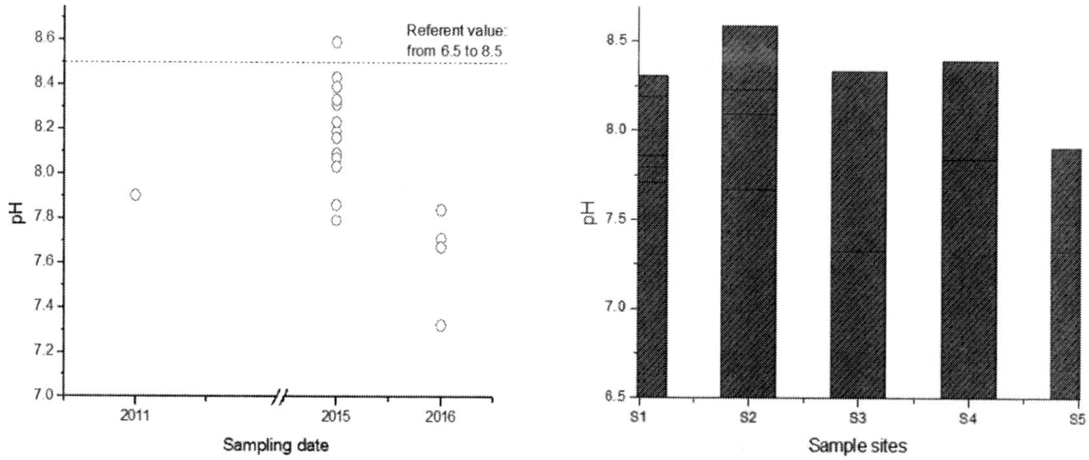

Figure 2. Results for pH measurement in all sites (S1, S2, S3, S4, S5) in the period 2011-2016.

The data for nitrates and nitrites are in concentration lower than acceptable limit [5] (Table 1; Figure 5). The analysis of the components shows a light increased nitrite quantities for the samples of S2 (Table 2), which could be attributed to an increased fertilization process of the agricultural areas in the region. Nitrate values were low for both in S1 and in S2 (Table 1, 2). The low values of nitrate in the influent might possibly be due to loss of NO_3^- via denitrification. Thus, nitrate concentration is not considered to pose a problem for the river water. The Bulgarian guideline for phosphate in surface water of rivers is <0.020 mg L^{-1} PO_4^{3-} [4]. The levels of phosphate in the river water ranged from 0.007 to 0.12 mg L^{-1},

PO_4^{3-} as the means value of all sites is around 0.078 (Table 1) which is four times under the norm [4] and could be attended to fecal water pollution and fertilization. Cluster analysis of data (Figure 6) also showed a relationship between anion content, which shows that the increase of NO_2^- could be mainly related to the increase NO_3^- content and the all increasing of the anion concentration could be attributed to increasing the fertilization process of the agricultural areas in the region. According the data (for S2, S4, S5), the values (the average annual value) [5] for the element Cadmium (Cd) are above the acceptable limits (0.15 μg L^{-1}) [6] for the whole period of time (2011-2016). The maximum values are identified in the years 2011(0.40 μg L^{-1}) and 2012 (0.33 μg L^{-1}). In 2016 analysis shows that the values are over the standard more than 10 times in analyzed sites (Table 2). Cadmium is in the list of priority substances in the field of water policy according the ordinance on environmental quality standards for priority substances and certain other pollutants [6]. The higher concentration of cadmium in drinking water may cause kidney damage [10, 11]. In this sense preventing the reduction of cadmium levels is therefore essential especially if water would be used for drinking and irrigation. The concentrations of metals such as As, Pb, Cu are in the norm for the analyzed period (2011-2016), exception is registered for the components Zn (2011, 2012 - 100 μg L^{-1}l) and Ni (2011 – 28 μg L^{-1} 0,028 mg L^{-1}, 2012 - 5 μg L^{-1} 0,005 mg L^{-1}) for S2 and S4. For S5 the values of Zn (2011, 2012) are not in the permissible limits. Analysis 2016 for heavy metals shows that for the more of the analyzed elements values are lower than limit of detection of the applied method (Table 2) and are in the standards except Cd and Pb. The higher concentration of cadmium and lead can be due to the potential tailing ponds and the consequences of mining and unregulated landfills in the catchment area (Table 2). Cluster analysis of data (Figure 6) also showed a relationship between cadmium content and anion and some metal (Pb, Ni, As) concentrations, which proves that the increase of Cd could be mainly related to the increase landfills in the unregulated catchment area pollution.

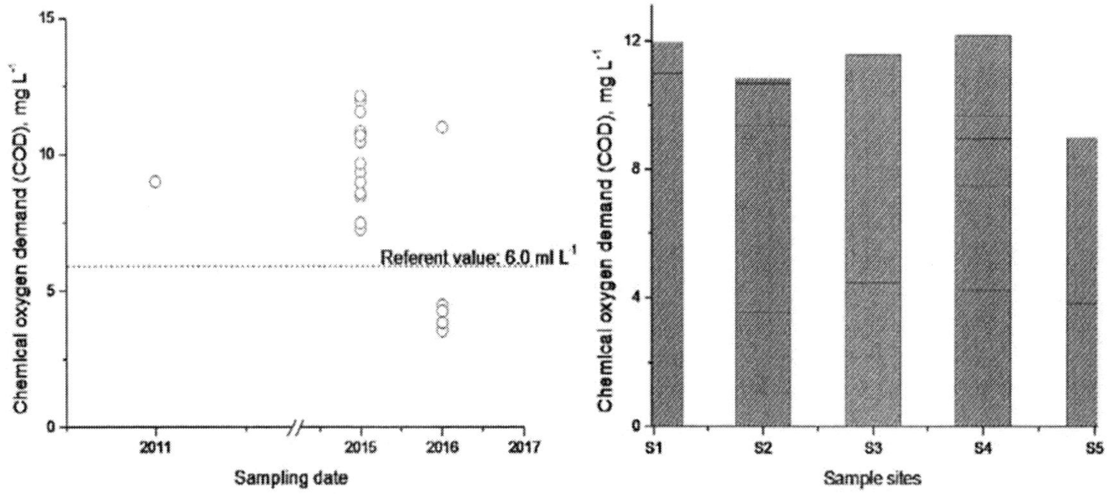

Figure 3. Data for chemical oxygen demand (COD) in all analyzed sites in period 2011-2016.

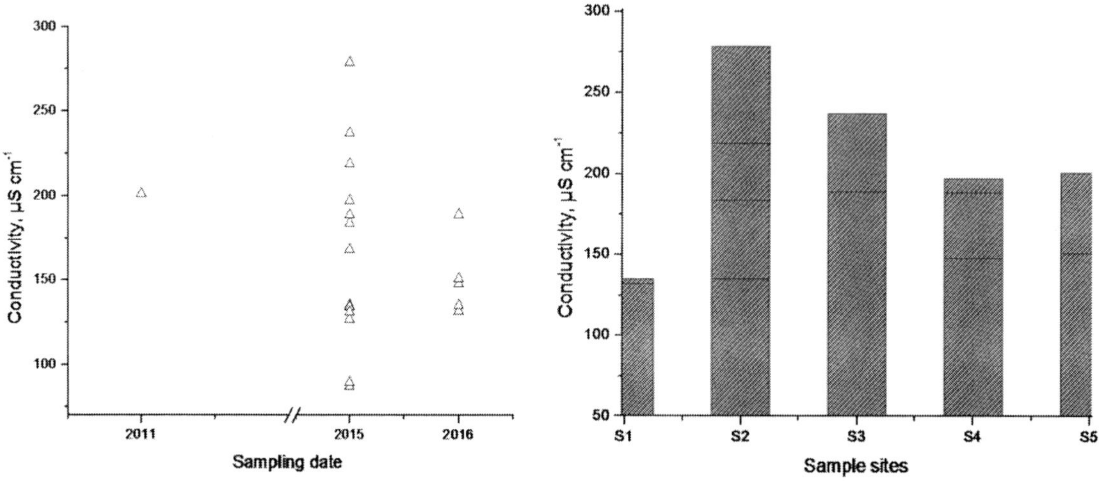

Figure 4. Data for water conductivity in all analyzed sites in period 2011-2016.

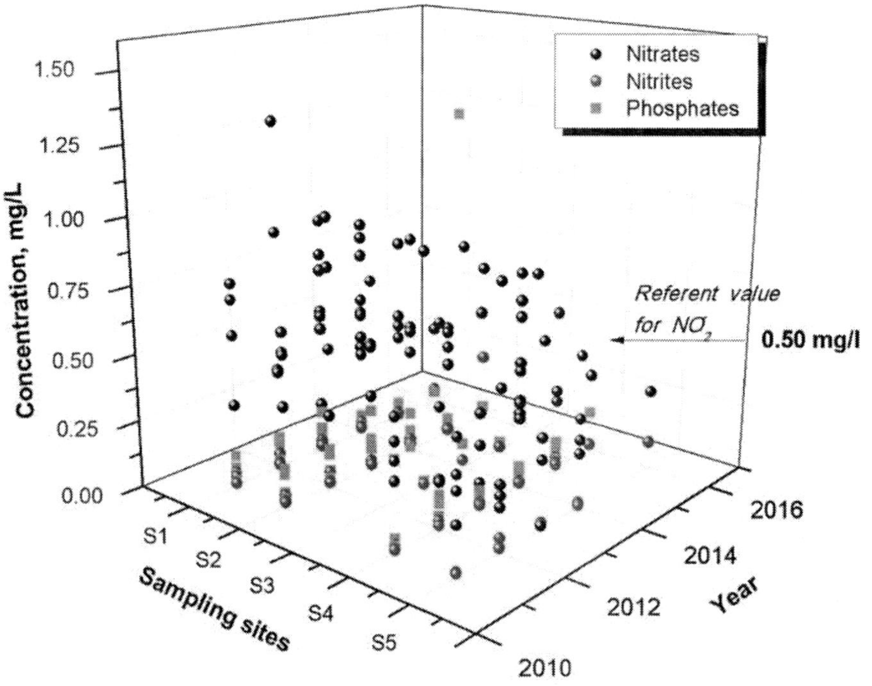

Figure 5. Data for nitrates, nitrites and phosphates in all analyzed sites in period 2011-2016.

IOP Conf. Series: Materials Science and Engineering **374** (2018) 012093 doi:10.1088/1757-899X/374/1/012093

Table 1. Nutrients levels in surface water of Mesta River for samples taken in period 2011-2016.

Sampling sites	Months/Year	Physical-chemical parameters		
		NO_2^-, mg L^{-1}	NO_3^-, mg L^{-1}	PO_4^{3-}, mg L^{-1}
	01; 04; 07; 10/2011	0.016; 0.007; 0.049;	0.76; 0.31; 0.70; 0.57	0.10; 0.027; 0.094;
S1	01; 04; 07; 10 /2012	0.053	0.90; 0.38; 1.3; 0.37	0.15
	01; 04; 07; 10 /2013	0.017; 0.014; 0.0112	0.9; 0.53; 0.18; 0.71	0.086; 0.15; 0.043; -
	01; 04; 07; 10 /2014	0.016; 0.015; 0.042;	0.79; 0.48; 0.72; 0.32	0.067; 0.026; 0.057;
	01; 04; 07; 10 /2015	0.022	0.72; 0.38; 0.42; 0.33	0.046
	07/2016	0.010; 0.01; 0.016; 0.006	0.31	0.05; 0.007; 0.047;
		0.048		0.022
				0.036
Average ± S.D.		**0.52 ± 0.19**	**0.57 ± 0.24**	**0.078 ± 0.053**
	01; 04; 07; 10 /2011	0.019; 0.016; 0.043;	0.64; 0.37; 0.56; 0.57	0.14; 0.053; 0.12;
S2	01; 04; 07; 10 /2012	0.022	0.82; 0.27; 1.0; 0.52;	0.033
	01; 04; 07; 10 /2013	0.020; 0.014; 0.012;	0.72; 0.47; 0.28; 0.48	0.14; 0.032; 0.062;
	01; 04; 07; 10 /2014	0.055	0.83; 0.49; 0.47; 0.39	0.12
	01; 04; 07; 10 /2015	0.021; 0.0091; 0.022;	0.43; 0.35; 0.28; 0.41	0.11; 0.22; 0.069;
	01; 04; 07; 10 /2016	0.026	0.62	0.074
		0.025; 0.021; 0.039;		0.15; 0.007; 0.072;
		0.023		0.041
		0.011; 0.014; 0.014;		0.055; 0.039; 0.042;
		0.006		0.033
		0.25		0.039
Average ± S.D.		**0.032 ± 0.051**	**0.52 ± 0.19**	**0.079 ± 0.053**
	01; 04; 07; 10 /2011	0.0085	0.88	0.018
S3	01; 04; 07; 10 /2012	0.026	0.85	0.092
	01; 04; 07; 10 /2013	0.0067	0.67	0.028
	01; 04; 07; 10 /2014	0.0032	0.60	0.021
	01; 04; 07; 10 /2015	-	-	-
	01; 04; 07; 10 /2016	0.0029	0.65	0.036
Average ± S.D.		**0.011 ± 0.010**	**0.76 ± 0.12**	**0.044 ± 0.033**
	01; 04; 07; 10 /2011	0.011; 0.013; 0.028;	0.38; 0.32; 0.47; 0.25	0.048; 0.021; 0.33;
S4	01; 04; 07; 10 /2012	0.008	0.73; 0.18; 0.44; 0.17	0.049; 0.12; 0.034;
	01; 04; 07; 10 /2013	0.012; 0.010; 0.029;	0.71; 0.35; 0.23; 0.09	0.092; 0.046
	01; 04; 07; 10 /2014	0.012	0.70; 0.44; 0.28; 0.47	0.061; 0.059; 0.018;
	01; 04; 07; 10 /2015	0.006; 0.0091; 0.014;	0.6; 0.038; 0.30; 0.26	0.051
	01; 04; 07; 10 /2016	0.022	0.30	0.070; 0.076; 0.060;
		0.015; 0.017; 0.022;		0.017
		0.013		0.060; 0.035; 0.021;
		0.011; 0.0082; 0.018;		0.10; 0.15
		0.019	**0.39 ± 0.20**	**0.072 ± 0.067**
Average ± S.D.		0.018		
		0.015 ± 0.0062		
	01; 04; 07; 10 /2011	0.010; 0.014; 0.017;	0.48; 0.35; 0.30; 0.18	1.48
S5	01; 04; 07; 10 /2012	0.014	0.57; 0.16; 0.24; 0.20;	-
	01; 04; 07; 10 /2013	0.018; 0.013; 0.051;	0.67; 0.33; 0.25; 0.011	0.56
	01; 04; 07; 10 /2014	0.021	0.25; 0.56; 0.20; 0.33;	-
	01; 04; 07; 10 /2016	0.018; 0.011; 0.020;	0.30	0.32
		0.024		
		0.006;0.011;0.018;		
		0.0098		
		0.10		
Average ± S.D.		**0.022 ± 0.022**	**0.32 ± 0.17**	**0.79 ± 0.61**

Table 2. Metal concentrations of key sites of surface water of the Mesta River determined by ISP-OES method for samples taken in 2016.

Sites / Metal conc. mg L^{-1}	S1	S2	S3	S4	S5
Na	6.299±0.017	2.955±0.016	2.075±0.016	2.240±0.015	2.191±0.015
K	1.319±0.006	0.603±0.004	0.663±0.006	0.619±0.006	0.588±0.006
Ca	17.09±0.06	16.89±0.07	12.28±0.05	13.13±0.05	14.08±0.05
Mg	2.678±0.009	4.406±0.028	2.632±0.009	2.589±0.009	3.188±0.009
Sr	<0.001	<0.001	<0.001	<0.001	<0.001
Ba	0.009±0.000	0.003±0.0001	0.003±0.0000	0.004±0.0000	0.003±0.0000
Cu	<0.002	<0.002	<0.002	<0.002	<0.002
Fe	0.173±0.0007	0.069±0.001	0.264±0.002	0.360±0.001	0.227±0.002
Al	<0.004	<0.004	<0.004	<0.004	<0.004
Zn	<0.002	<0.002	<0.002	<0.002	<0.002
Pb	<0.003	<0.003	<0.003	<0.003	<0.003
Cr	<0.002	0.0019±0.0001	<0.002	0.0018±0.0002	<0.002
Mn	0.033±0.000	0.012±0.0003	0.026±0.0003	0.0409±0.0001	0.026±0.0002
Ni	<0.003	<0.003	<0.003	<0.003	<0.003
Co	<0.002	<0.002	<0.002	<0.002	<0.002
Cd	<0.002	<0.002	<0.002	<0.002	<0.002
As	<0.010	<0.010	<0.010	<0.010	<0.010
S	<0.050	<0.050	<0.050	<0.050	<0.050
U	<0.003	<0.003	<0.003	<0.003	<0.003
Mo	<0.004	<0.004	<0.004	<0.004	<0.004
Tl	<0.010	<0.010	<0.010	<0.010	<0.010
V	<0.003	<0.003	<0.003	<0.003	<0.003

*<value

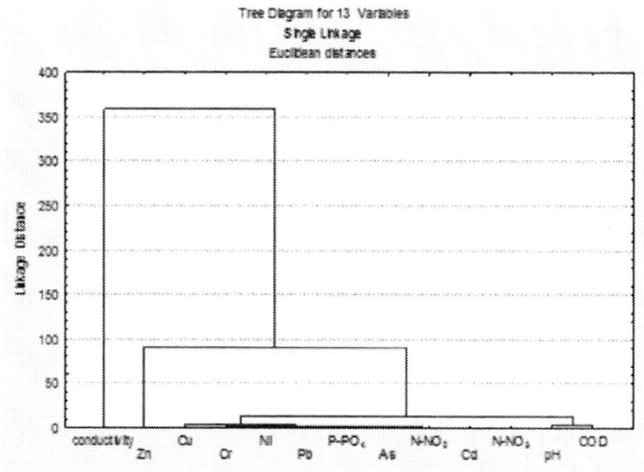

Figure 6. Dendrogram from HCA showing division of important physicochemical parameters of surface water samples.

4. Conclusions

On the basis of this study and the results obtained we can conclude that in the analyzed period the surface waters of the river Mesta are in a good physical-chemical status by the analyzed parameters even with the numerous of anthropogenic activities in the evaluated area. The main contaminants of surface water in the basin of the Mesta River are industrial companies, agricultural farms and settlements - towns and villages with built sewage systems but without built-up sewage treatment plants, solid waste as well. Although the water of the river Mesta is relatively in a good quality the prevention should includes: construction of WWTP and improvement of the sewerage network, construction of sewerage systems in small settlements, where none exist, enhanced control over enterprises – pollutants, removal of unregulated landfills, control over the use of fertilizers and consideration of a future concept on cadmium sources and their eliminated.

5. References

[1] Water quality monitoring – A practical guide to the design and implementation of freshwater quality studies and monitoring programmes. CRC Press 1996 1-6
[2] River basin management plan in the West Aegean Region for the period 2010-2015, 2016-2021
[3] The Directive 2000/60/EU – Water Framework Directive
[4] Water Law from 28.01.2000 (OG. 67 on 27/07/1999)
[5] Ordinance H-4 from 14.09.2012 on characterization of surface water (OG. 22 on 03/05/2013)
[6] Ordinance on environmental quality standards for priority substances and some other pollutant (OG. 88 on 09.11.2010)
[7] Karadakov B. and all. 1992 *Science and art*, Sofia
[8] I. Retike, A. Kalvans, K. Popovs, J. Bikse, A. Babre, and A. Delina 2016 *Hydrology Research* **47(4)**
[9] L. Surinaidu 2016 *Hydrol. Res.* **47 (2)** 373–389
[10] R. Qadeer, 2004 *J. Chem. Soc. Pakistan* **26** 293- 301
[11] APHA (American Public Health Association) (1995). Standard methods for the examination of water and wastewater. APHA, 19th Ed. Washington.)

Spatial and Temporal Characteristics of Air Pollutants Concentrations in Industrial Area in Malaysia

NA Mohd Zizi[1], N Mohamed Noor[1,2], N Izzah Mohamad Hashim[1] and S Y Yusuf[1,2]

[1]School of Environmental Engineering, Universiti Malaysia Perlis, Perlis, Malaysia.
[2]Center of Excellence Geopolymer and Green Technology (CEGeoGTech), School of Materials Engineering, Universiti Malaysia Perlis (UniMAP), Perlis,Malaysia.

E-mail: norazian@unimap.edu.my

Abstract. Quantification of the spatial and temporal variations of ambient air pollutant concentrations provides the information for epidemiological and other air-pollution studies and many have relied in the past on data from a single, centrally-located air pollution monitoring site. Particulate matter (PM_{10}), nitrogen oxides (NO), sulfur dioxide (SO_2), carbon monoxide (CO), ozone (O_3) were measured at 3 monitoring sites around a densely populated industrial zone which is at Nilai, Petaling Jaya and Seberang Perai. The observations were obtained over five year period from 2008 to 2012. The descriptive statistics show that the peak concentration of PM_{10} occurs during the dry season; which is coincided with the southwest monsoon and also due to direct-influence of southwest winds and had caused a slightly moderate haze in Southeast Asia. From the Pearson correlation analysis, among the meteorological parameters, ambient temperature indicates the strongest positive correlation to the PM_{10}concentration and proves that the ambient temperature tent to contribute significantly to higher PM_{10} concentrations as high as in Southwest monsoon.

1. Introduction

Air pollution occurs when air is contaminated with natural and anthropogenic pollutants. Anthropogenic pollutants are contaminants associated with human activities, which include polluting residuals from consumption and production activity. Air pollution happens due to the presence of anthropogenic pollutants and non-point source pollutants in the air. Non-point source pollutants come from sources that cannot be accurately identified. These pollutants have diffusively and indirectly contributed towards the degradation of environment.

A study by Jacob (2010) [1] described that a full life cycle of a vehicle contributes to air pollution, which includes the process of manufacturing, refueling, emission and disposal of the car. Fuel refining and distribution cause additional pollutants. These pollutants can be further categorized into primary and secondary pollutants. Those considered primary pollutants are released straight to the air. Secondary pollutants on the other hand, are the by-products of chemical reactions that transpire between the pollutants and other minute particles in the atmosphere. From the previous studies, it can be assumed that most of the major pollutants come from manufacturing activities. It is important to note that the increase in the demand of vehicles and the many economic activities occur along population growth. However, the major concern of scientists as well as economists is the tremendous impacts of the exposure to the air pollution. Urban population in particular may suffer adverse health effects with the combination of poor air quality and high temperatures.

In Malaysia, the air quality monitoring work was first carried out by the Department of Environment (DOE) Malaysia in 1977 but it consisted mainly of short surveys. These surveys produced limited data in which little analysis can be done. However, other interested bodies and individuals carried out related studies from time to time [2]. There are 52 air-monitoring stations throughout the country, which are at urban, sub-urban as well as industrial area. These monitoring stations are strategically located in both residential and industrial areas to detect any significant change in the air quality, which might be threatening to human health and the environment. Five criteria pollutants are being monitored continuously by DOE namely carbon monoxide (CO), nitrogen dioxide (NO_2), ozone (O_3), sulphur dioxide (SO_2), and particulate matter (PM_{10}). There are two types of monitoring which are Continuous Air Quality Monitoring Stations (CAQM) and Manual Air Quality Monitoring Station (MAQM). Alam Sekitar Malaysia (ASMA) Sdn. Bhd. is responsible to carry out air quality monitoring work for DOE. ASMA is awarded a 20-year concession to provide air quality monitoring data to DOE. To date, under the Concession, ASMA is acknowledged with the installation and management of 52 continuous air quality monitoring stations (CAQM). The establishments of the Malaysian Air Quality Guidelines in 1989, Air Pollution Index and Haze Action Plans in 1997 are among the important tools for air quality management that are endorsed and put into practice by the Malaysian Government.

In general, the overall air quality for Malaysia in was between good to moderate levels most of the time except for a number of unhealthy days recorded at various locations in the west and east coast of peninsular Malaysia. Particulate matter (PM_{10}) was the main pollutant that had caused unhealthy conditions in the late northeast monsoon (February until March) and southwest monsoon (May until September) especially during the dry season. The unhealthy days in Klang Valley during these periods were due to forest and peat land fires. The concentration of PM_{10} in Klang, which is sited as the urban area was significantly higher compared to the suburban and rural areas. Hence, the major aim of this study is to presents the evaluation on trends of air pollutants and characterization of pollutants concentrations with respect to meteorological condition in the selected potential area.

2. Material and method

2.1. Study area

Three monitoring stations that were chosen in this study are located in the west coast of peninsular Malaysia. The three stations are Nilai, Petaling Jaya and Seberang Jaya. Nilai is a town located in Seremban District, Negeri Sembilan which is also a small state located about 50 km south of Kuala Lumpur (Figure 1(a)). According to the Department of Statistics (DoS, 2015) [3]. Climatically, Nilai shows a tropical monsoon climate, characterized with a significant amount of rainfall during the year and uniformly high daytime temperatures between 27 °C - 30 °C while 22 °C - 24 °C at night [4]. It is legitimately normal even for the driest month. The average precipitation in Nilai is 2223 mm. The driest month is June, with 119 mm of rain. With an average of 263 mm, the most precipitation falls in November whereas; April is the warmest month of the year.

Petaling Jaya is one of the wettest cities in Malaysia (Figure 1 (b)). It is warm with an average maximum of 30°C and receives heavy rainfall all year round, roughly more than 3,300 mm (130 in) of average rainfall annually. The city has no particular true dry season but June and July are the driest months. Mostly each month average rainfall receives more than 200 mm (7.9 in). Thunderstorm and extreme rainstorm weather is a common environment here and it is one of the highest lightning strike areas in the world. The average temperature in Petaling Jaya is 27.0 °C. About 2438 mm of precipitation falls annually. The least amount of rainfall occurs in July with the average of 126 mm. The most precipitation month falls in November with an average of 281 mm [3].

Seberang Perai, formerly Province Wellesley, is a narrow hinterland opposite Penang Island on the Malay Peninsula, which together with the island forms the Malaysian state of Penang (Figure 1 (c)). Its principal town is Butterworth. Seberang Perai has a population of 815,767, which is the second most populous local government area in Malaysia and covers an area of 755 km^2 from 1031 km^2 of the state of Penang. Seberang Perai is located in the north part of Malaysia and has been briskly industrialized and urbanized. Climatically, Seberang Perai shows a typical tropical monsoon, which experiences an equatorial type of climate with a warm and humid weather throughout the year. The climate is normally warm throughout the year with temperatures ranging from 22°C to 32°C. Penang is quite dry in the months of January and February. Humidity is generally high all year round. April, May and October are usually the wetter months. Rainfall is heavy and is experienced especially from August to November. The details location of the monitoring stations for all study areas is presented in Table 1.

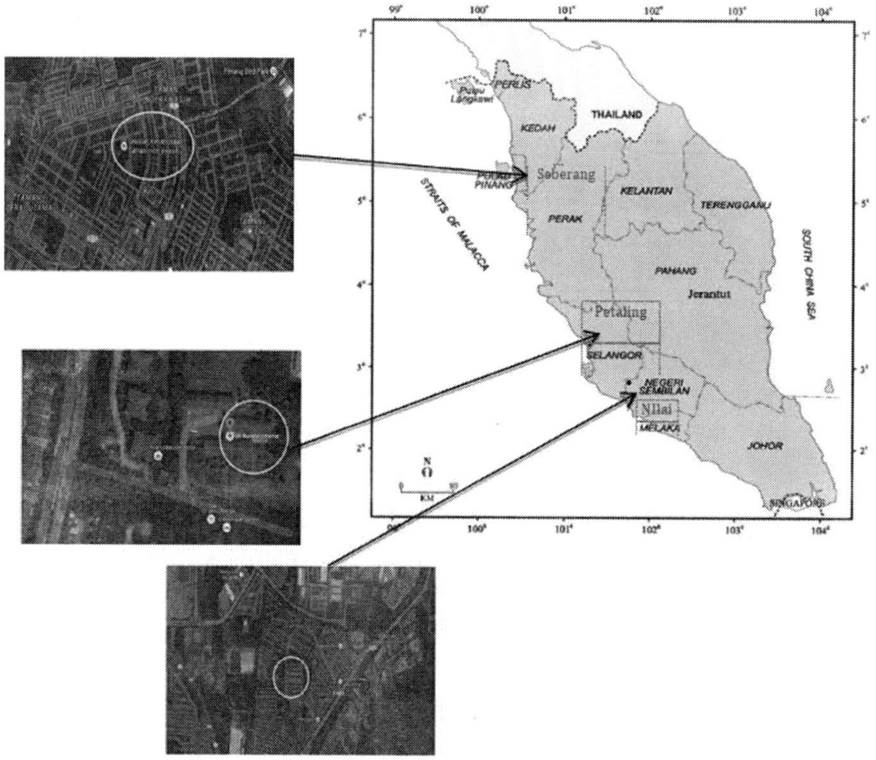

Figure 1. Location of the three monitoring stations.

Table 1. The location of the study areas.

No.	Location	
1.	Taman Semarak (Phase 2), Nilai, Negeri Sembilan	N 2.8205°, E 101.8108°
2.	SK Bandar Utama, Petaling Jaya, Selangor	N 3.1314°, E 101.6082°
3.	Sampling station of SK Taman Inderawasih, Seberang Perai, Pulau Pinang	N 5.4°, E 100.4667°

2.2. Air quality data

In this study, air pollutants concentration and meteorology parameters in industrialized area (Petaling Jaya, Nilai and Seberang Perai) will be analyzed. Throughout this step, the air quality data which obtained from the Department of Environment (DOE) will be first analyzed. Then, the associations of the air pollutants concentration with its weather parameters will be justified by descriptive statistics. Later, the temporal and spatial characteristics and the correlations between the air pollutants concentration and its meteorological conditions for the study area will be analyzed.

2.3. Statistical analysis

The descriptive statistics will be carried out to determine the shape, dispersion and the distributions of the data for the study area. The most important information needed to be collected from a set of data are the measure of central tendency and the measures of variability. These are the following measures of central tendency including; mean, median, mode and skewness. Besides, the measures of variability include the standard deviation, variance and range. In this study, the descriptive statistics for every set of data were analyzed by using SPSS Statistics 21. These are some of the summarized formulae for measuring the descriptive statistics. The Pearson correlation was conducted to determine the relationship between different air pollutants at each station as well as the connection between the pollutants and its meteorological parameters namely; humidity, wind speed and ambient temperature.

3. Results and analysis

3.1. Time series plot of air pollutants

The time series plots of the maximum daily average of air pollutants concentrations for five years of all the three-study area were plotted. Figure 2 and Figure 3 show the five years' time series plot for all three sites. PM_{10} and O_3 concentrations were selected because both pollutants are the most prevalent gases concentrations recorded in the atmosphere at all stations. The reference line showing the Malaysian Ambient Air Quality Guideline (MAAQG) of PM_{10} and O_3 concentrations for 24-hours averaging time also drawn to show the exceedances throughout the years which are 150 μgm^{-3} and 0.06 ppm respectively. It is to indicate which year and location exceeded the guideline.

Figure 2 illustrates the time series plot of particulate matter (PM_{10}) for Nilai, Petaling Jaya and Seberang Perai in 2008, 2009, 2010, 2011 and 2012. The highest hourly PM_{10} concentration at all three stations was recorded throughout the five years period except for 2011, which was higher around October to November. The highest PM_{10} concentration was during early May to late June, where it represent the dry season, which is coincided with the southwest monsoon and the lowest was in October 2008 (northeast monsoon). In Southwest Asia, the concentration of particulate matter (PM_{10}) is influenced by the occurrence of biomass burning in Sumatera Indonesia and the Southwest monsoon wind [5]. Besides, it was also due to the direct-influence of southwest winds and had caused a slightly moderate haze in Southeast Asia. Overall, PM_{10} concentrations are seen to be the prior contributor of air pollutant in Seberang Perai throughout the year.

The time series records of O_3 concentration from 2008 to 2012 are shown in Figure 3. The concentration of O_3 was found to be higher during certain months in that particular year due to the haze episode in August–September [6], where the concentration was far above the guideline particularly during early February. Other than being affected by the emission of industrial activities, the high O_3 concentration at all stations being affected by the vehicles emission. In general, each year, most of the O_3 concentrations exceeded the MAAQG of 1-hr averaging time.

3.2. The monthly trend of air pollutants

Figure 4 shows the monthly plots air pollutants concentrations for five years of all the three study areas. The pollutants concentration was compared to Malaysian Ambient Air Quality Guideline (MAAQG) for 24-hours averaging time. Figure 4 shows the averaged of monthly plot of PM_{10} concentration of all the study area from 2008 to 2012. The monthly trends of all the three areas were consistently increased during the month of June until August where the highest PM_{10} concentrations were recorded on June. The highest concentrations of PM_{10} were recorded at the value of 68.326 μgm^3, 59.432 μgm^{-3} and 51.13 μgm^{-3} in June, July and August respectively, which might be due to the southeast monsoon. However, the concentrations seem decreased starting from September to November because of the time changes of season or called as transition. After all, the averaged concentration of all gases is still under the threshold limit by the Recommended Malaysian Air Quality Guideline (RMAAQG).

The highest O_3 concentration was recorded in Seberang Perai on February (0.02179 ppm). The local source that contributed higher O_3 concentration is that this station is located at the industrial areas. According to Department of Environment (2012) [7], the prevailing pollutants concern in Seberang Perai were O_3 and PM_{10} due to the intensive industrial activity in this area. Figure 4 also shows that similar pattern was observed in Nilai and Petaling Jaya, as February recorded the highest O_3 concentration with the value of 0.01995 ppm and 0.01805 ppm respectively. Besides being affected by the motor vehicle emissions, the O_3 concentrations in the study area were high because of industrial activities and open burning [8].

The monthly trend of Sulphur Dioxide (SO_2) was showing a remarkably fluctuation in comparison with the three stations. The highest monthly average of SO_2 concentration was recorded in Petaling Jaya during May until July with the values of 0.00434 ppm, 0.00432 ppm and 0.00398 ppm respectively. While at Nilai and Seberang Perai, the highest monthly average of SO_2 concentration was recorded in September (0.00411 ppm) and June (0.002476 ppm) respectively. But somehow the average concentration was still under the threshold limit stipulated in the Recommended Malaysian Air Quality Guideline (RMAAQG).

Among the twelve month, June recorded the highest monthly concentration of NO_2 in Seberang Perai, Petaling Jaya and Nilai with the value of 0.01528 ppm, 0.03275 ppm and 0.0130 ppm respectively. None of these concentrations exceeded the RMAAQG suggested value that is 0.17 ppm for 1-hour averaging time. However, according to Department of Environment (2008) [9], the highest value of NO_2 concentration might due to the atmospheric conditions and emission from the motor vehicles in urban areas that enhance its formation.

3.3. Correlation between air pollutants and meteorological factors

The correlations between atmospheric pollutants and meteorological parameters at the three stations were investigated by using Pearson Correlation (Table 2). In Nilai, positive relationships between pollutant-pollutant and pollutant-weather parameter were observed. Nevertheless, the significant relation between pollutant-pollutant (for significant level of 0.01) were observed between PM_{10} and NO_2 (r=0.594); PM_{10} and SO_2 (r=0.414); PM_{10} and CO (r=0.466); NO_2 and CO (r=0.538); SO_2 and NO_2 (r=0.516). Other than that, the significant relationships between pollutants-weather parameters at $p<0.01$ were observed between O_3 and wind speed (r=0.521); CO and temperature (r=0.363) and NO_2 and humidity (r=0.351).

In Petaling Jaya itself, the location that is close to industrial area, which would lead to the existence of atmospheric pollutants from various sources in comparison to Seberang Perai, which is more influenced by the movement of motor vehicles. The most significant correlation in Petaling Jaya are between PM_{10} and temperature (R^2=0.2577, $p<0.01$), PM_{10} and wind speed (R^2=0.1731, $p<0.01$) and between CO and NO_2 (R^2=0.2301, $p<0.01$). The correlation between PM_{10} and temperature indicates a strongest positive correlation when the value of PM_{10} increases as the temperature rises. A study from Azmi et al. (2010) [6] proves that a high temperature increases the number of biomass burning as well as the evaporation of soil dust from the earth's surface. Hence justify that the variables moves along together.

Moreover, in Seberang Perai, particulate matter (PM_{10}) were also found to be correlated to one of a meteorological parameter which is ambient temperature (R^2=0.1595, $p<0.01$). A study from Azmi et al. (2010) [6] proves that a high temperature increases the number of biomass burning as well as the evaporation of soil dust from the earth's surface. Hence justify that the variables moves along together. There were only two variables that were correlated to each other in Seberang Perai. However, the strongest positive correlation happened in this area is between O_3 and SO_2 (R^2=0.2924, $p<0.01$). The result is expected to be negatively correlated since SO_2 and NO_2 are known as precursors of ozone (O_3) and indicates that a rise in O_3 concentration will associate with the decline in the levels of both pollutants.

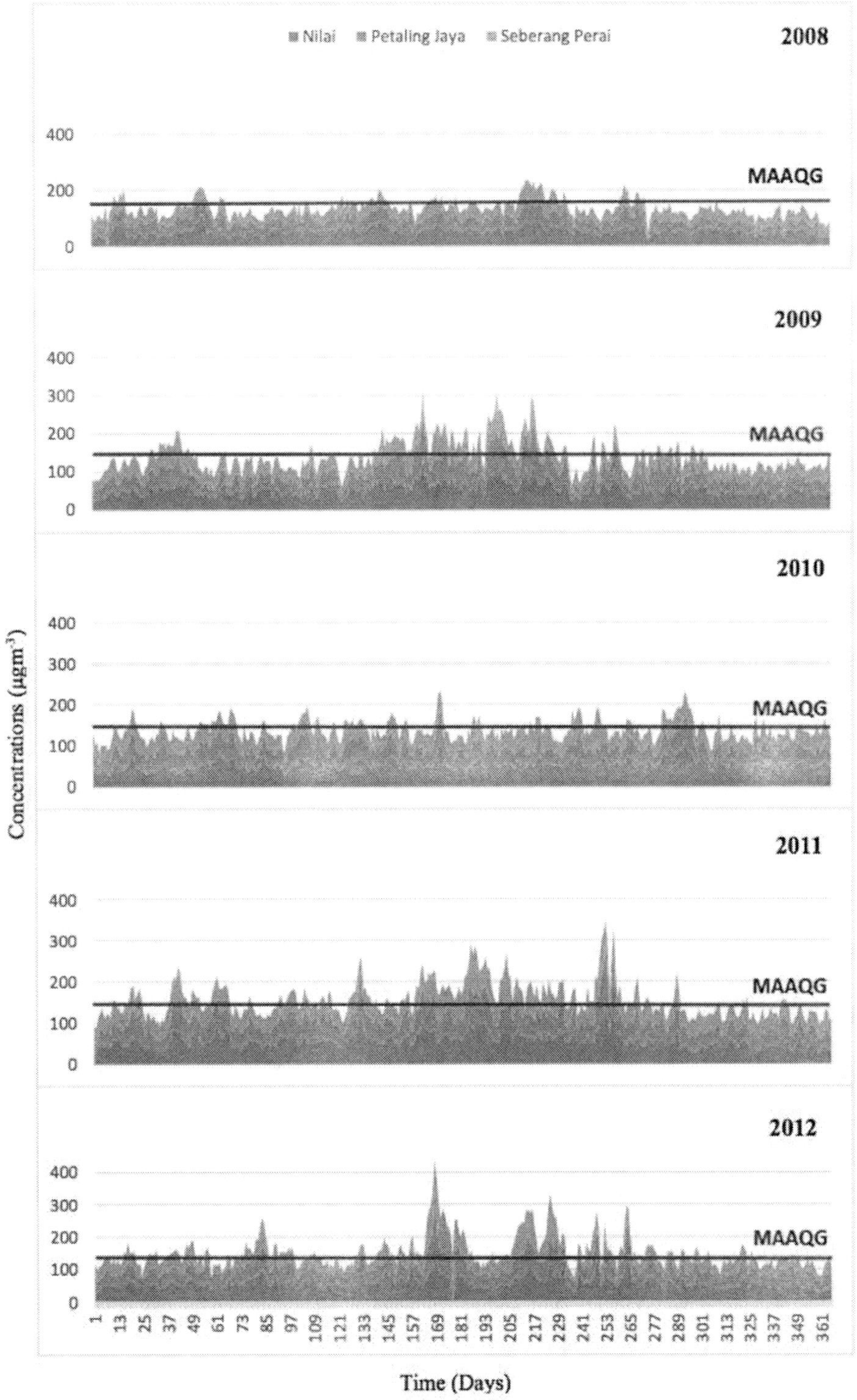

Figure 2. Time series plot of PM_{10} concentration for the three study areas.

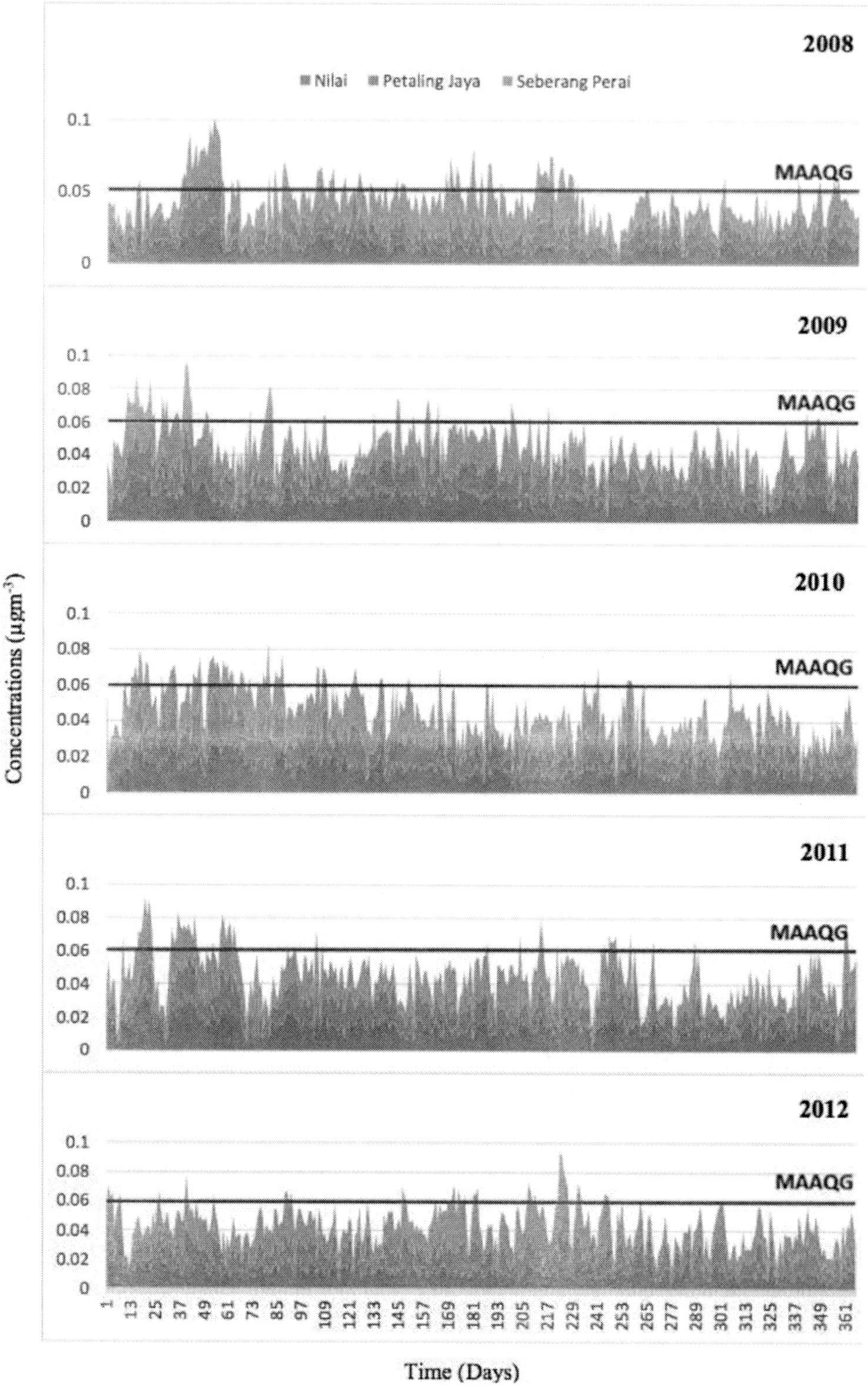

Figure 3. Time series plot of O_3concentration for the three study areas.

IOP Conf. Series: Materials Science and Engineering **374** (2018) 012094 doi:10.1088/1757-899X/374/1/012094

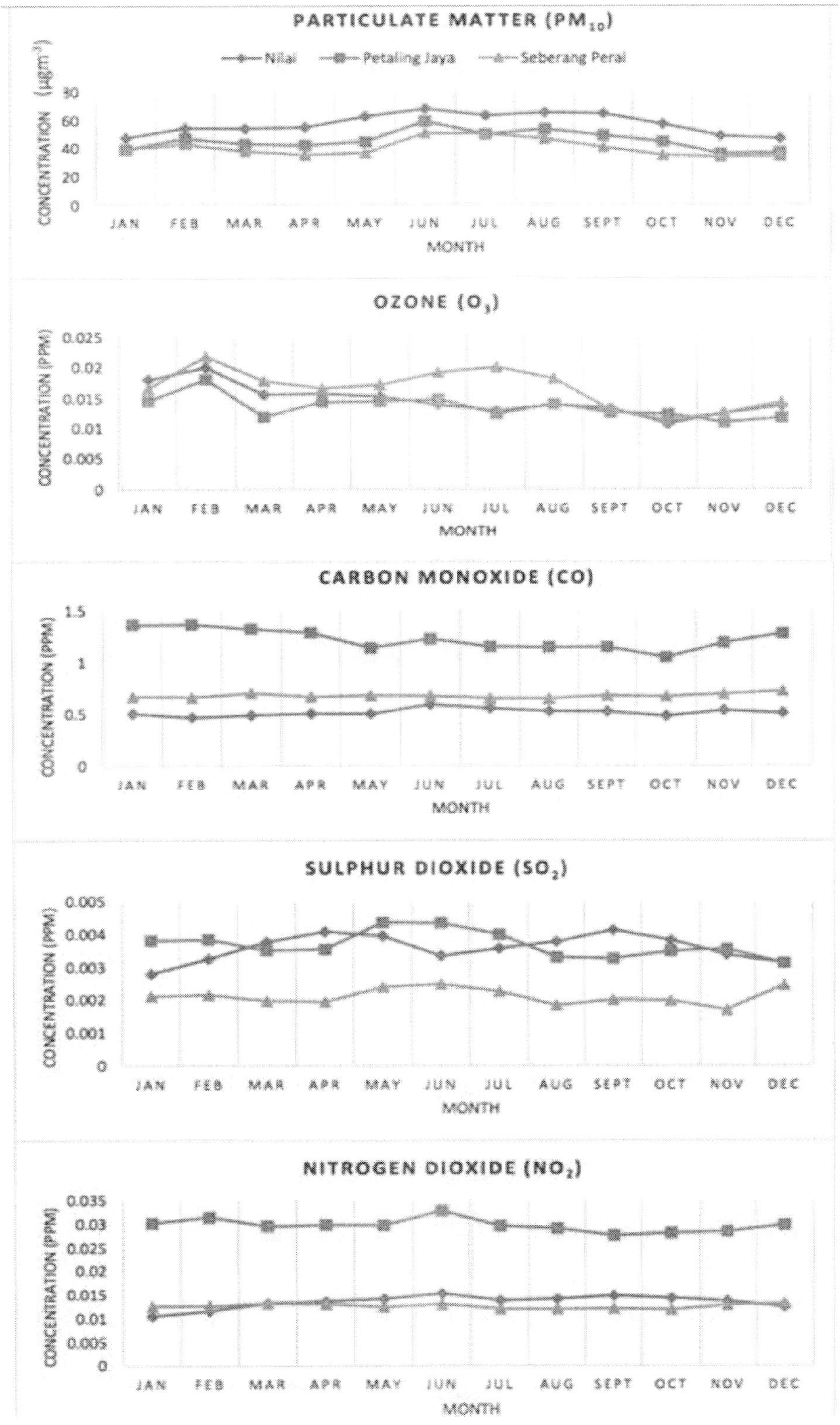

Figure 4. Monthly plot of air pollutants in Nilai, Petaling Jaya and Seberang Perai.

These meteorological conditions are typical for a tropical environment. Correlation between the most significant air pollutants (PM_{10}) and meteorological factors indicates that the PM_{10} concentration has a significant positive correlation with humidity (R^2=0.5491, $p<0.01$). High humidity is commonly related to the number of rain occasions, and this reduces the number of particles due to the wash-out processes of the atmospheric aerosol in the atmosphere. Therefore, there is a positive correlation between humidity and PM_{10} in ambient air. Other factors such as wind speed and temperature were found to slightly influence the concentration of particulate matter in the atmosphere in the study areas.

Table 2. Correlation between parameters at Nilai, Petaling Jaya and Seberang Perai.

Station	Parameter								
		Wind speed	Temp.	Humidity	SO_2	NO_2	O_3	CO	PM_{10}
Nilai	Wind speed	1							
	Temp.	0.068	1						
	Humidity	-0.068	0.0051	1					
	SO_2	-0.353	0.307*	0.233	1				
	NO_2	-0.449	0.277*	0.351**	0.516**	1			
	O_3	0.521**	-0.06	-0.585	-0.23	-0.379	1		
	CO	-0.066	0.363**	0.061	0.116	0.538**	-0.174	1	
	PM_{10}	-0.412	0.162	-0.109	0.414**	0.594**	-0.094	0.466**	1
Petaling Jaya	Wind speed	1							
	Temp.	0.437**	1						
	Humidity	-0.666	-0.449	1					
	SO_2	-0.087	-0.145	0.092	1				
	NO_2	-0.0707	-0.234	-0.011	0.172	1			
	O_3	-0.249	0.057	-0.163	0.174	0.115	1		
	CO	-0.413	-0.206	0.215	0.318*	0.480**	0.354**	1	
	PM_{10}	0.416**	0.5088**	-0.741	0.027	0.027	0.273	-0.052	1
Seberang Perai	Wind speed	1							
	Temp.	0.237	1						
	Humidity	-0.263	-0.631	1					
	SO_2	-0.311	0.090	-0.009	1				
	NO_2	-0.017	-0.153	0.086	-0.055	1			
	O_3	-0.318	0.153	-0.355	0.541**	-0.208	1		
	CO	0.228	0.109	0.027	0.124	0.351	-0.266	1	
	PM_{10}	-0.129	0.400*	-0.352	-0.040	0.156	0.275	-0.064	1

**Correlation is significant at the 0.01 level (2-tailed)
*Correlation is significant at the 0.05 level (2-tailed)

4. Conclusions

The result from this study shows that practically all the averaged concentration of atmospheric pollutants recorded at Nilai, Petaling Jaya and Seberang Perai are under the permissible value recommended by the Malaysian Department of Environment (DoE).

It has been resulted that the highest average concentration of PM10 recorded at Nilai was in 2012 with value of 60.48588 μgm^3. While the maximum value is recorded in 2008 which is 315 μgm^3. The mean average concentration of PM_{10} recorded at the station throughout these 5 years was significantly differ and was found to be slightly above than the value suggested by the Malaysian Ambient Air Quality Guideline (MAAQG) for the 24-hr averaged concentration (50 μgm^3). PM_{10} and O_3 were more closely to be discussed because of its strong correlation with the transboundary sources, due to photochemical processes, the averaged concentration of other parameters is influenced more by the number of motor vehicles and emission by industrial activities near the monitoring stations.

The correlation between the air pollutants recorded to be the most significant correlated in Nilai whereas for Petaling Jaya and Seberang Perai the correlation was significantly correlated between the air pollutants and some of the meteorological parameters. Moreover, the high concentration of ozone (O_3) concentration recorded at Nilai, Petaling Jaya and Seberang Perai are intriguing and needs further investigation. Meteorological factors particularly humidity and the ambient temperature are expected to be associated with the concentration of ozone and suspended particulate matter. In terms of monthly variation there has been an obvious temporal variation of all three area that were seen to be similarly increase during the month of June until August where the highest PM_{10} concentration were recorded on June at value of 68.326 μgm^3, 59.432 μgm^{-3} and 51.13 μgm^{-3} respectively which might be due to Southeast monsoon.

5. References

[1] Jacob D J, Tai A Pand Mickley L J 2010 *Atmospheric Environment* **44**(32) 3976-84
[2] Sani S 1984 Urban Development and Changing Patterns of Night Time Temperatures In the Kuala Lumpur-Petaling Jaya Area Malaysia *Jurnal Teknologi* **5**(1) 27-35
[3] Department of Statistics 2015 *Population Statistic 2015* Department of Statistics Malaysia
[4] Ghazali N A Ramli N A Yahaya A S Yusof N F F M Sansuddin N and Al Madhoun W A 2010 *Environmental monitoring and assessment* **165**(1-4) 475-489
[5] Abas M R B, Oros D R and Simoneit B R T 2004*Chemosphere* **55** 1089–1095
[6] Azmi S Z, Latif MT, Ismail AS, Juneng L and Jemain AA 2010 *Air Quality Atmosphere and Health* **3**(1) 53-64
[7] Department of Environment DOE 2012 *Malaysian Environment Quality Report 2012* Selangor: Publication Section Strategic Communications Division Department of Environment Malaysia
[8] Department of Environment DOE 2009 *Malaysian Environment Quality Report 2009* Selangor: Publication Section Strategic Communications Division Department of Environment Malaysia
[9] Department of Environment DOE 2008 *Malaysian Environment Quality Report 2008* Selangor: Publication Section Strategic Communications Division Department of Environment Malaysia
[10] Peavy H S Rand Tchobanoglous DR1985 *Environmental Engineering* **628** P4 McGraw-Hill Science/ Engineering/ Math
[11] Godish T 1997 Atmospheric Pollutants Air Quality *Atmospheric Pollutants: Air Quality* Lewis Publisher New York (United State)
[12] Ahamad F, Latif M T, Tang R, Juneng L, Dominick D and Juahir H 2014 *Journal of Atmospheric Research* **139** 116-127

Aknowledgments

The author would like to acknowedge Ministry of Higher Education Malaysia for the Fundamental Grant Scheme (FRGS 9003 00508) and Department of Environment Malaysia for the air pollutant dataset.

Effects of Prevulcanization Time of Natural Rubber on Kenaf Filled Natural Rubber Foam

A F Abdul Karim[1], H Ismail[1] and Z Mohamad Ariff[1]
[1]School of Materials and Mineral Resources Engineering, Engineering Campus, Universiti Sains Malaysia, 14300 Nibong Tebal, Pulau Pinang

E-mail: ihanafi@usm.my, profhanafi@gmail.com

Abstract. The objective of this study was to determine the effects of prevulcanization time of natural rubber latex (NRL) on properties of kenaf filled natural rubber latex foam (NRLF). The prevulcanization time of NRL was set at 24 hours. The tensile properties, compression set and morphological properties of kenaf filled NRLF after 24 hours of prevulcanization time were compared with samples without prevulcanized. The tensile strength and modulus at 100% elongation (M100) of prevulcanized samples was higher but lower in elongation at break as compared with without prevulcanized samples. The results from thermal gravimetric analysis (TGA) revealed that the thermal stability of prevulcanized samples was higher than without prevulcanized samples. The prevulcanized samples found to have higher elasticity compared with without prevulcanized samples based compression set test. The morpholgy of both samples were also compared using scanning electron microscope (SEM). From the micrograph of SEM, it was found that prevulcanized kenaf filled NRLF samples show better interfacial bonding compared with kenaf filled NRLF without prevulcanized. This proved that prevulcanized natural rubber latex enhanced the properties of kenaf filled NRLF.

1. Introduction

The advantages of prevulcanized latex such as high crosslink density and high tensile strength make them dominant in various application including glove, catheters, toy balloons, marking rings, pillow and mattreses [1]. Actually the researchers found that prevulcanized latex was generally not suitable for foam products due to inherently poor gel strength [2]. But nowadays, with some modification of the formulation, the prevulcanized latex could now be used sucessfully for making a foam products such as pillow and mattreses [2]. Therefore, a precise determination of prevulcanized NRL is necessary to ensure the production of final products having optimum performance with competitive cost. In addition, a good understanding factors that affect the prevulcanized NRL is crucial to predict the optimum properties of the final products [3].

Nowadays, the interest of using renewable resources such as bio fiber (as banana, sisal, hemp, jute, pineapple, bamboo, cotton, coconut, rice husk, and kenaf) in polymer matrix has attract the attention of the worldwide [4,5,6]. This was due to the various properties offered by natural fiber such as light weight, renewability, low cost and environmental friendly which suits to various application [7, 8]. In this study, the well-known bio filler in Malaysia which is kenaf was used as bio fillers in NRLF. Kenaf is single, straight and without branch, grows quickly, rising to four to five meter tall in a four to five month growing season, and 25-35 mm in original diameter [9, 10]. Kenaf fibers was extracted from kenaf's bast to get a kenaf fiber in order to use in various application such as in polymeric composites

[10, 11]. The compatibility between kenaf and the matrix, structure and surface of kenaf fiber can be improved by chemical or physical treatment on kenaf fibre [12].

In this study, the tensile properties, thermal, compression set, and morphological properties of kenaf-filled NRLF were investigated using a standard methods.

2. Experimental procedures

2.1. Materials and formulation

The formulation used in this study are shown in Table 1. Natural rubber latex (Low Ammonia (LATZ) type) and other chemicals such as sulphur, antioxidant, potassium oleate, Zinc diethyldithiocarbamate (ZDEC) and Zinc 2-mercaptobenzhiozolate (ZMBT) were all purchased from Zarm Scientific & Supplies Sdn. Bhd, Malaysia. Kenaf fiber supplied by National Kenaf and Tobacco Board were grounded and sieve to acquire a kenaf powder.

Table 1. Formulation of kenaf filled NRLF.

Ingredients	Total solid content (%)	Formulation (pphr)
LA Latex	60	100
Sulphur	50	2.2
Antioxidant	50	1
Potassium oleate	20	4.5
ZDEC	50	0.9
ZMBT	50	0.9
Kenaf powder (~300μm)	25	0/1/3/5/7

Table 2 shows the ingredients used for gelling process. All the chemicals was purchased from Zarm Scientific & Supplies Sdn. Bhd, Malaysia.

Table 2. Gelling agents of kenaf-filled NRLF.

Ingredients	Total solid content (%)	Formulation (pphr)
Zinc Oxide (ZnO)	50	3
Diphenyl guanidine (DPG)	40	0.3
Sodium silicofluoride(SSF)	25	1.2

2.2.Sample preparation

First, LATZ latex was measured and filtered according to the required amount followed by stirred the latex using a mechanical stirrer for about 15 minutes at 10 rpm. After that, sulphur and antioxidant were added and stirred at 10 rpm. After 30 minutes, ZMBT and ZDEC were slowly added to the mixture and continued stirred about 1 hour. After 1 hour, potassium oleate soap and a predetermined volume of kenaf powder were added to the compound slowly and the NRLF compound was then prevulcanized for 24, 48 and 72 hours at room temperature with a continuous stirring speed of 6 rpm. After the prevulcanization process, the NRLF compound was beaten intensively and foamed using the stand mixer (KENWOOD, kMix) until the volume was increased up to three times of the initial volume. The compound was beating time of about 5 minutes at maximum speed and once the desired volume obtained, the foaming speed was lowered to obtained fine and even foam. Next, the gelling agent was prepared according to formulation in Table 2. DPG together with zinc oxide (ZnO) were added to the foam and beating was continued for another 1 minutes. Then, SSF was added and the foam was beaten for 45 minutes. Finally, the un-gelled foam was quickly poured into the desired aluminum mould and allowed to gel for 3 minutes at ambient temperature. Gelled foam was then cured in an oven at 105 °C for two hours. Once the foam has cured, the foam was stripped from the mould and washed thoroughly with de-ionized water to remove potassium oleate soap and excessive non-reacted materials. After washing, the cured NRLF was evenly dried in a hot air oven at 80 °C for 8 hours.

2.3. Measurement of tensile properties

The tensile test kenaf filled NRLF with and without prevulcanized samples was performed by using universal testing machine Instron 3366 according to ISO 48. Dumbbell-shaped of kenaf filled NRLF in the formed of sheet were cut by using Wallace die cutter. The test was performed at ambient temperature with a crosshead speed of 500 mm/min. Five samples of each kenaf loading for prevulcanized and without prevulcanized were tested, and the averages were reported. Tensile properties such as tensile strength, tensile modulus and elongation at break were obtained from this test.

2.4. Compression set

Compression set properties of kenaf filled NRLF with and without prevulcanized was performed according to ASTM D3574. Samples used were in regular shape of 50 mm x 50 mm x 25 mm with parallel top and bottom surface and essentially perpendicular sides. Three specimens per samples were tested. Specimens were placed in test apparatus and deflected to 75 ± 1 % of its original thickness. Within 15 minutes, deflected specimens and the apparatus were placed in mechanically convicted air oven overnight with test temperature of 70 ± 2 0C. Specimens were immediately removed from apparatus and measured after 30 minutes recovery. Compression properties such as constant deflection compression set were obtained from this testing.

2.5. Thermogravimetric analysis (TGA)

TGA of the samples was conducted by using Perkin Elmer Pyris 6 testing instrument (USA). Both samples with and without prevulcanized of the kenaf filled NRLF were tested at a heating rate of 20 °C/min from 30 to 600 °C under nitrogen gas conditions. The TGA and DTG values of kenaf filled NRLF were recorded.

2.6. Scanning electron microscopy

Surface morphology of prevulcanized and without prevulcanized kenaf filled NRLF was studied by using scanning electron microscope (Hitachi, TM1000). The foam surfaces were mounted on aluminum stubs and sputter-coated with a thin layer of gold to avoid electrostatic charging and poor image resolution during examination. The morphology of the bot samples with and without prevulcanized of kenaf filled NRLF were observed.

3. Results and discussions

3.1. Tensile properties

Table 3 shows the tensile strength, elongation at break and modulus at 100% (M100) of NRLF with and without prevulcanized. Both samples of kenaf-filled NRLF show decrement in tensile strength as the kenaf loading increased. Tensile strength of without prevulcanized samples decreased from 0.48 MPa (for the 0 phr kenaf loading) to 0.22 MPa (sample with 7 phr kenaf loading) while samples of prevulcanized NRLF decreased from 0.71 MPa to 0.26 MPa. As can be seen in the resules, kenaf-filled NRLF with prevulcanized NRL have higher tensile strength as compared with samples without prevulcanized. This enhancement was due to the increased in crosslink density in prevulcanized samples hence reduced the breakage of the samples in an applied stress during tensile test.

From the results, increament in filler loading cause reduction in the elongation at break for both groups of samples, with and without prevulcanized. For samples with prevulcanized, the elongation at break was lower than the elongation at break for samples without prevulcanized. The elongation at break for samples without prevulcanized reduced from 441.70% to 282.68% while samples with prevulcanized reduced from 367.52% to 279.01% with respect to the comparison between samples with zero kenaf loading and those of having 7 phr kenaf loading. As mentioned before, prevulcanized latex tend to have more crosslinking chain network. This resulted of good stress propagation and improved the tensile strength but made the samples more brittle. Therefore, the samples are more likely to fail at lower strain instead of experiencing higher elongation.

The M100 increased with increased of kenaf loading for both samples, with and without prevulcanized NRL. At similar kenaf loading, sample with prevulcanized exhibited the high value of M100 as compared with samples without prevulcanized. The M100 for sample without prevulcanized increased from 0.21 MPa to 0.28 MPa while M100 for sample with prevulcanized increased from 0.25 MPa to 0.31 MPa. The used of prevulcanized NRL in the kenaf-filled NRLF give a positive effect on the M100 due to the elasticity of foam was reduced, resulting in more stiff and rigid foam. The increasing trend in M100 was observed as a consequent of the elasticity of the foam was reduced, resulting in stiffer, more rigid foam. This decreased its resistance to breakage [11].

Table 3. Tensile strength, elongation at break and modulus at 100% (M100) of NRLF with and without prevulcanized.

Sample	Tensile strength (MPa)		Elongation at break(%)		M(100)(MPa)	
	Without prevul.	With prevul.	Without prevul.	With prevul.	Without prevul.	With prevul.
0 phr	0.48	0.71	441.70	365.50	0.21	0.25
1 phr	0.42	0.43	367.52	340.72	0.24	0.25
3 phr	0.38	0.40	365.15	388.91	0.26	0.27
5 phr	0.23	0.37	332.70	300.30	0.27	0.28
7 phr	0.22	0.26	282.68	279.01	0.28	0.31

3.2.Thermo Gravimetric Analysis (TGA)

Figure 1 show the results of thermal stability for control and kenaf filled NRLF with prevulcanized NRL. The weight loss of the kenaf-filled NRLF with increasing temperature under a nitrogen environment was measured, and graphs of the weight loss and derivative thermogravimetric (DTG) as a function of the temperature were plotted. Decomposition of the cellulose in the kenaf occurred at lower temperatures [12]. Increasing the amount of kenaf caused the decomposition of cellulose to occur at lower temperatures. As shown in Figure 1, control and kenaf-filled NRLF experienced weight loss from 259 to 588 °C which can be related to the decomposition of natural rubber and the cellulose of kenaf. Higher char residue was produced by the kenaf filled NRLF at 7 phr kenaf loading as compared to the other loading. The thermal stability of the polymer was enhanced by increasing the amount of char residue formed, as char hinders diffusion of volatile decomposition products [13]. Hence by increasing the amount of kenaf at a certain loading, the thermal stability of the kenaf-filled NRLF was increased.

Table 4 shows the temperature at 50% weight loss (°C), the maximum weight loss (%) and the final char residue (%) for prevulcanized and without prevulcanized samples. Prevulcanized samples have lower weight loss and high char residue compared with samples without prevulcanized. It is due to the high crosslink structure inside prevulcanized samples that restrict the chain mobility of rubber chain. Therefore, high energy needed to break the crosslinking for prevulcanized samples. In addition, high char residue was also contribute to hinder the diffusion of volatile materials as mention before. Hence, prevulcanized samples have high thermal stability compared with without prevulcanized samples.

3.3. Compression set

The compression set is the measure of the elastic behavior of the materials. Figure 2 shows constant deflection compression set, Ct of kenaf filled NRLF with and without prevulcanized. From Figure 2 the 0 phr kenaf loading (control) for both samples has the lowest compression set indicates the high elasticity of the samples compared with other samples. The compression set increases with increase in kenaf loading which indicates the lower elasticity of kenaf filled NRLF. This is due to the non-elastic deformation of the hard phase of which more kenaf is present led to slower recovery and may explain the higher compression set observed after 30 min relaxation.Meanwhile, prevulcanized samples have lower compression set compared with without prevulcanized samples. This was due to the better filler matrix interaction and high crosslink density of NRL for prevulcanized samples.

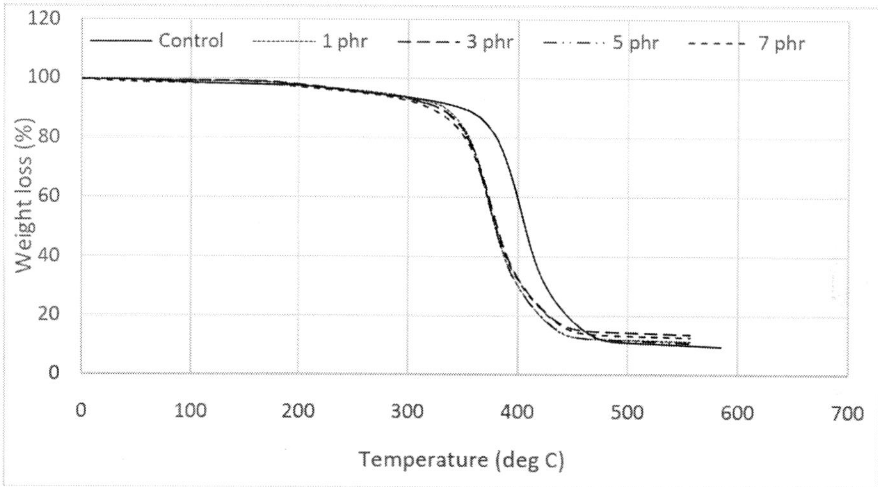

Figure 1. TGA of control and kenaf-Filled NRLF with prevulcanized at different filler loading.

Table 4. Thermal stability parameters of prevulcanized and without prevulcanized kenaf filled NRLF.

Sample	Temperature at 50% weight loss (°C)		Maximum weight loss (%)		Final char residue (%)	
	Without prevul.	With Prevul.	Without prevul.	With Prevul.	Without prevul.	With Prevul.
0 phr	398	411	94.9	90.2	4.5	9.4
1 phr	378	413	95.3	88.6	4.7	11.2
3 phr	374	416	95.5	86.2	5.3	13.6
7 phr	374	414	95.6	87.2	5.8	12.5

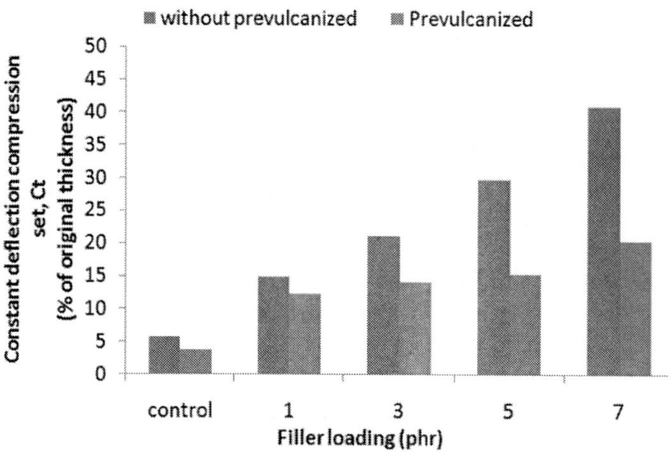

Figure 2. Constant deflection compression set, Ct of samples with and without prevulcanized kenaf filled NRLF.

3.4.Scanning electron microscopy (SEM)
Figure 3 (a) and Figure 3 (b) show the filler-matrix interaction of kenaf filled NRLF with and without prevulcanized at 7 phr kenaf loading. Sample without prevulcanized shows weak interfacial adhesion while sample with prevulcanized NRL shows good filler-matrix interaction at interfacial bonding. This is an indication that prevulcanized NRL was able to increase the adhesion at the filler-matrix interfaced which resulted in increased tensile strength as well as the modulus of kenaf filled NRLF. During prevulcanization process, better filler dispersion occurred due to adequate time and forced during prevulcanization process hence give small size of pores and well distribution of pores as shown in.

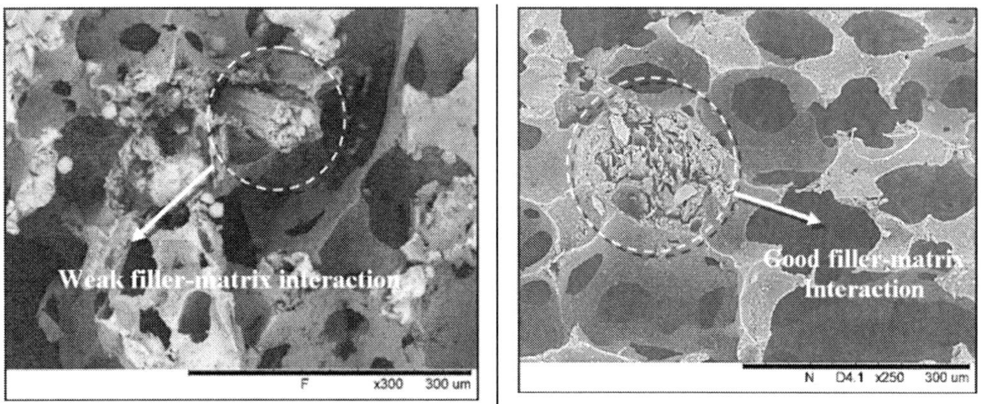

(a) (b)

Figure 3.(a) Micrograph of kenaf in kenaf filled NRLF without prevulcanized at 7 phr; (b) Micrograph of kenaf in kenaf filled NRLF with prevulcanized at 7 phr.

4. Conclusions

By using the optimum prevulcanization time (24 hours), it was found that the tensile strength and M100 of prevulcanized samples of kenaf filled NRLF were higher while the elongation at break lower compared with without prevulcanized samples. The thermal stability was higher for prevulcanized kenaf filled NRLF compared with without prevulcanized. The results of compression set shows that samples with prevulcanized have higher elasticity compared with without prevulcanized. From the micrograph of SEM, it was observed that sample without prevulcanized shows weak interfacial bonding between filler and matrix but samples with prevulcanized shows better interaction and good interfacial bonding between filler and matrix.

5. References

[1] Sasidharan K K, Shiny P, Gopalakrishnan K S, George K E and Rani J 2004 *Journal of Applied Polymer Science* **94** 1164 –1174
[2] Blackley D C 1997 *Polymer Latices: Science and technology Volume 2: Types of latices* Netherlands Springer Netherlands
[3] Porter M, Rosemaze R and Sapi'ai A R 1992 *Journal of Natural Rubber Resources* **7(2)** 85 - 101
[4] Dahham O S, Noriman N Z, Mohd Kahar A W, Ismail H and Sam S T 2015 *Journal of Applied Science and Agriculture* **10(5)** 33-39
[5] Han S O, Karevan M, Sim N I, Bhuiyan M A, Jang Y H, Ghaffar J and Kalaitzidou K 2012 *Journal of Materials Science* **47** 3535-43
[6] Ismail H, Nasaruddin M N and Rozman H D 1999 *European Polymer Journals* **35**(8) 1429-37
[7] Muniandy K, Ismail H and Othman N 2012 *BioResources* **7** 841-858
[8] Rozman H D, Tay G S, Kumar R N, Ismail H and MohdIshak Z A 1999 *Polymer-Plastics Technology and Engineering* **38(5)** 997-1011
[9] Salman S D, Leman Z, Sultan M T, Ishak M R and Cardona F 2015 *BioResources* **10**(4)

[10] Fiore V, Di Bella G and Valenza A 2015 *Composites: Part B* **68** 14-21
[11] Ismail H and Haw F S 2008 *Journal of applied polymer science* **110**(5) 2867-76
[12] Akhtar M N, Sulong A B, Fadzly Radzi M K, Ismail N F, Razab M R, Muhamad N and Khane M A 2016 *Progress in Natural Science: Materials International* **26**(6) 657–664
[13] Abdullah N and Hashim A Y M 2014 *Developments within the Glove Industry* **14**(1) 16-20

Acknowledgements
The author would like to gratefully acknowledge the Fundamental Research Grant, FRGS (Grant Number 203.PBAHAN.6071375) and MyBrain15 of Ministry of Higher Education Malaysia for the financial assistance.

Investigation on Properties and Leachability of Sewage Sludge from Wastewater Treatment Plant Incorporated in Fired Clay Brick

N S Abdul Salim[1], A Abdul Kadir*[1,2], M A Kamarudin[1], M H Fadzli Zaidi[1] and M M A B Abdullah[2,3]
[1]Faculty of Civil and Environmental Engineering, Universiti Tun Hussein Onn Malaysia, Johor, Malaysia
[2]Center of Excellence Geopolymer and Green Materials (CEGeoGTech), School of Materials Engineering, Universiti Malaysia Perlis (UniMAP), Perlis, Malaysia
[3]Faculty of Engineering Technology, Universiti Malaysia Perlis, Perlis Malaysia

E-mail: aeslina@uthm.edu.my

Abstract. Sewage sludge is produced from wastewater treatment plant that often dumped into landfill sites, lagoons or together with municipal waste. Apart from this, the disposal of this sewage sludge becoming a problem for environment. Thus, an alternate method is needed to overcome the issue. Therefore, the objective of this study is to determine the possibility of incorporating sewage sludge in fired clay brick and to compare with normal fired clay brick. The characteristics of sewage sludge were determined by using an X-ray Fluorescence. The mixing percentage used was 0%, 1%, 5%, 10%, 20% and 30%. The physical and mechanical properties were determined as to obtain the condition according to it mixing percentages. Based on physical and mechanical properties, the sewage sludge brick complies with the standard except for SSB30% where the compressive strength does not meet the minimum requirement limit of compressive strength. However, in terms of leachability of this sewage sludge brick also resulted that the heavy metals leach out comply according to USEPA standard. The development of fired clay brick by using sewage sludge can reduce the dependency towards land use and will reduce the impact of the sewage sludge to the environment.

1. Introduction

Sewage sludge consists of by-products of wastewater treatment which is a mixture of water, inorganic and organic materials that coming from various sources such as from domestic sewage [1]. The amount of sewage sludge produced is steadily increasing with a population and still rising rapidly. About 3 million metric tons of sewage sludge are produced annually and it has been estimated to be increases to 7 million metric tons in the year 2020 in Malaysia. Due to the growth of urbanization, an increase of population and a sign of economic development has led to the production of high amounts of sewage sludge. In addition, the major method of disposal of sewage sludge is currently deposited with its storage and agricultural application which are deposited in landfill sites, lagoons or together with municipal waste [2]. Due to the shortage of disposal facilities, sewage sludge is often disposed in open fields that resulted in serious problems because of heavy metals that leached out into groundwater, surface water and soils [3]. According to the other studies, sewage sludge has been used by incorporating it into fired clay brick in order to obtain the properties and leachability of heavy metals. The utilization of the sewage sludge in fired clay bricks could produce a good quality of brick. Therefore, recycling the wastes in the brick production appears to be a viable solution to environmental pollution and also economical option to design of green building [4].

2. Materials and methods

Sewage sludge was collected from wastewater treatment plant which is located in Johor, Malaysia where it produced from the residential area while the clay soil was bought from the brick factory in order to produce fired clay brick incorporated with sewage sludge. These raw materials then dried in the oven for 24 hours with 105°C of temperature before crushing process. Then, X-ray Fluorescence (XRF) test was conducted on sewage sludge and clay soil as to obtain the chemical composition and concentration of heavy metals. The sample of bricks was produced in the specific size of the mold by mixing together of sewage sludge and clay soil by following the percentages of mixing weight of brick manufacturing and volume of water used. The bricks that have been formed then dried in the oven for 24 hours with 105°C of temperature. Then, the bricks that have been dried were fired in the furnace with 1050°C for 24 hours. The bricks that completely fired according to the different percentages of sewage sludge used were tested for physical and mechanical properties such as compressive strength, water absorption, shrinkage, density and also the initial rate of absorption [5].

While the leachability of heavy metals of fired clay brick was conducted by using Static Leachate Test (SLT) according to the different percentages of sewage sludge used. Static leachate test was performed as to obtained the leachability of heavy metals from sewage sludge that incorporated in fired clay brick according to NEN7345 method. The leachability of heavy metals were obtained through Inductive Coupled Plasma-Mass Spectrometer (ICP-MS). The duration of the Static Leachate Test was conducted were 6-hours, 1-day, 2.25-days, 4-days, 9-days, and 16-days. To ensure it is safe in terms of leachability, it is significant to carry out a comprehensive environmental assessment of new building products integrating with waste materials [6]. Heavy metals that leached from brick that incorporating with sewage sludge may have a potentially negative impact on human health and environmental quality by leaching into groundwater and surface water which could give water pollution issues.

3. Results and Discussion

3.1. Characteristic of Raw Materials

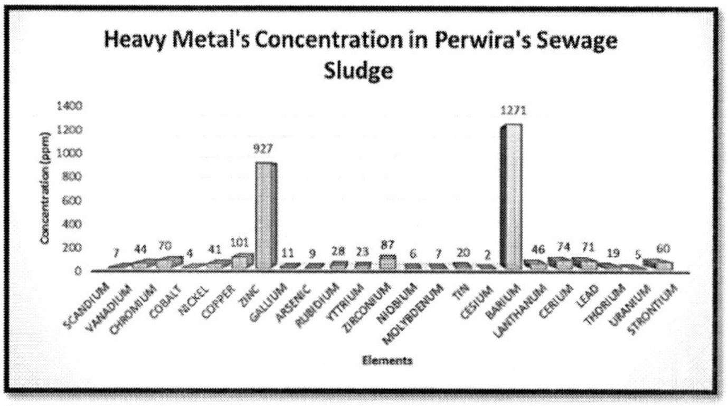

Figure 1. Concentration of heavy metals in sewage sludge versus percentage of concentration by using the *Standardless method*.

The concentration of heavy metals was obtained by X-ray Fluorescence testing using the Standardless method. The elements of heavy metals that found in sewage sludge are Scandium (Sc), Vanadium (V), Chromium (Cr), Cobalt (Co), Nickel (Ni), Copper (Cu), Zinc (Zn), Gallium (Ga),

Arsenic (As), Rubidium (Rb), Yttrium (Y), Zirconium (Zr), Niobium (Nb), Molybdenum (Mo), Tin (Sn), Cesium (Cs), Barium (Ba), Lanthanum (La), Cerium (Ce), Lead (Pb), Thorium (Th), Uranium (U) and Strontium (Sr). Barium (Ba), Zinc (Zn) and Copper (Cu) were recorded with the highest value of concentration which was 1271 ppm, 927 ppm, and 101 ppm respectively while the other elements were recorded with less than 100 ppm value of concentration.

While the elements that found in clay were Scandium (Sc), Vanadium (V), Chromium (Cr), Cobalt (Co), Nickel (Ni), Copper (Cu), Zinc (Zn), Gallium (Ga), Arsenic (As), Rubidium (Rb), Yttrium (Y), Zirconium (Zr) Niobium (Nb), Molybdenum (Mo), Cesium (Cs), Barium (Ba), Lanthanum (La), Cerium (Ce), Lead (Pb), Thorium (Th), Uranium (U) and Strontium (Sr). The highest concentration of heavy metals with value more than 100 ppm were Barium (Ba), Zirconium (Zr), Zinc (Zn), Vanadium (V) and Cerium (Ce) which were 366 ppm, 274 ppm, 260 ppm, 161 ppm and 105 ppm respectively while the other elements were recorded with a value lower than 100 ppm.

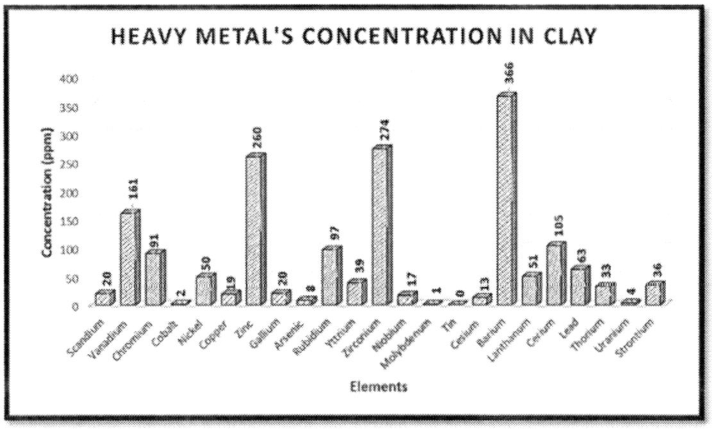

Figure 2. Concentration of heavy metals in clay versus percentage of concentration by using the *Standardless method.*

3.2. Physical and Mechanical Properties

3.2.1. Water Absorption. According to the graph, for 0% of sewage sludge that is control brick, the value of water absorption is 8%, followed by 1% of sewage sludge brick (SSB1%) with 10% of water absorption, 5% of sewage sludge brick (SSB5%) with 15% of water absorption, 10% of sewage sludge brick (SSB10%) with 17% of water absorption, 20% of sewage sludge brick (SSB20%) with 19% of water absorption and 30% of sewage sludge brick (SSB30%) with 21% of water absorption respectively. Based on the data recorded, it shows that the higher the amount of sewage sludge used in fired clay brick, more of water absorbed due to organic matter provided by sewage sludge in it. Furthermore, the increased of water absorption brick that incorporating with sludge affected by the increased formation of internal-voids due to burn-off of organic matter in the sludge [7]. Therefore, the sewage sludge bricks meet the load-bearing brick require according to the standard BS EN 722-21:2011 but not as engineering brick and damp proof brick.

3.2.2. Compressive Strength. From the results obtained, the graph showed compressive strength against the percentage of sewage sludge. The control brick which is containing 0% of sewage sludge had the highest of compressive strength with 21.6 MPa. Then the results obtained for different percentage of sewage sludge used showed that the value of compressive strength decreased over the amount of sewage sludge used. For SSB1%, the value of compressive strength decreases to 16.8 MPa followed by SSB5% with 12.3 MPa, SSB10% with 10.7 MPa and SSB20% with 5.8 MPa. While SSB30% decrease the strength to 3.1 MPa. However, the strength is still complying with a minimum standard BS EN 772-1:2011 of compressive strength which is 5 N/mm^2 for 1%, 5%, 10% and 20% of sewage sludge except for 30% of sewage sludge utilization. These indicate that the incorporation of sewage sludge in brick greatly affects the compressive strength value obtained [8]. The value of compressive strength reduced when more of sludge content used in the brick.

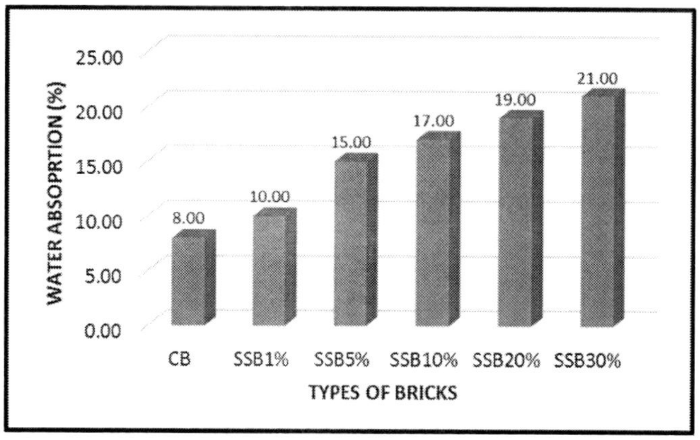

Figure 3. Water absorption versus percentage of sewage sludge.

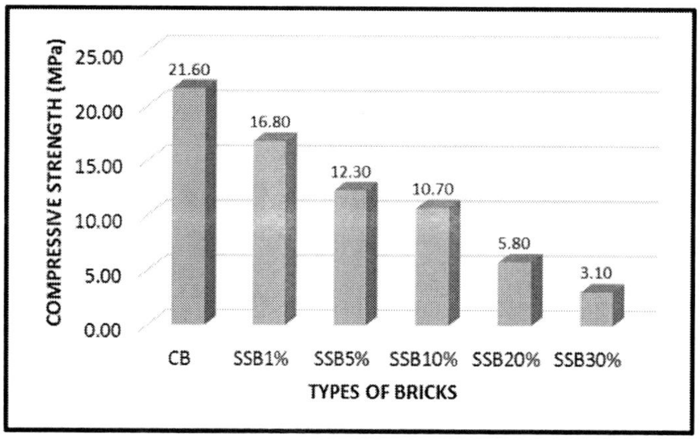

Figure 4. Compressive strength versus percentage of sewage sludge.

3.2.3. Density. According to the graph, the density of control bricks is 1888.51kg/m^3, while the density of fired clay bricks incorporating with different percentage of sludge is decreased when more amount of sewage sludge used. Incorporation of 1% sewage sludge (SSB1%) into fired clay brick, the value of density decreased to 1819.35kg/m^3, following by 5% of sewage sludge (SSB5%) with 1600.77kg/m^3, 10% of sewage sludge (SBB10%) with 1565.47kg/m^3, 20% of sewage sludge (SSB20%) with 1493.21kg/m^3 and 30% of sewage sludge (SSB30%) with 1450.69kg/m^3. Firing process is one of the reasons the density is decreased caused by the water loss due to high temperature applied [9]. Therefore, the amount of sewage sludge used in brick affecting the density of bricks which it becomes more porous and resulted to lightweight brick according to standard BS EN 722-13:2011.

3.2.4. Shrinkage. Based on the graph, five different percentage of sewage sludge that conducted in this study which are 1%, 5%, 10%, 20% and 30% were compared with control brick which is 0% of sewage sludge. The value of shrinkage of control brick is 1.63% while the other five different percentage of sewage sludge used are increased which are the shrinkage value is 2.05%, 2.51%, 3.18%, 4.20% and 5.50% respectively. A good quality brick will exhibit shrinkage less than 8%. Therefore, the amount of shrinkage occurs were influenced by the loss of water content in brick.

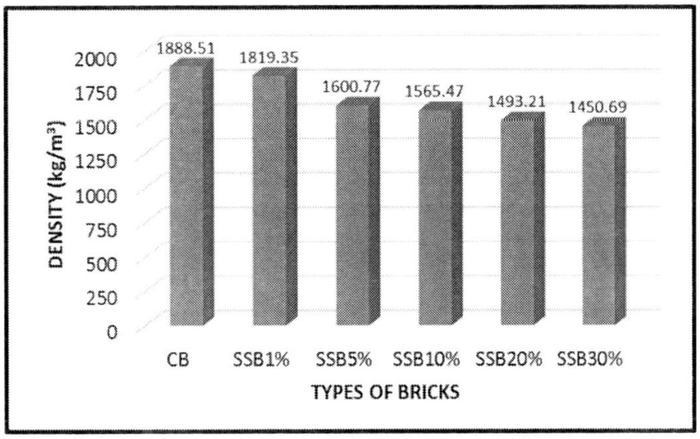

Figure 5. Density versus percentage of sewage sludge.

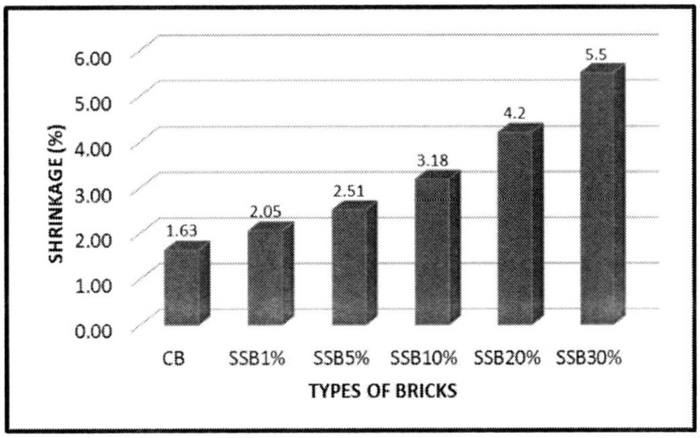

Figure 6. Shrinkage versus percentage of sewage sludge.

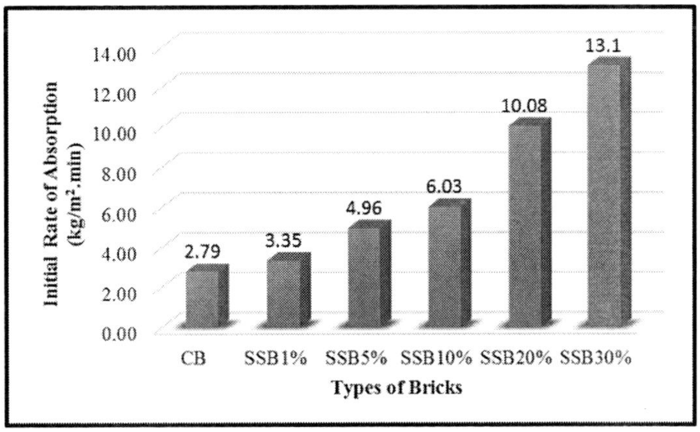

Figure 7. Initial rate of absorption versus percentage of sewage sludge.

3.2.5. Initial Rate of Absorption (IRA). Based on the graph, five different percentage of sewage sludge that used in this study which are 1%, 5%, 10%, 20% and 30% were compared with control brick which is 0% of sewage sludge were conducted according to BS EN 722-11:2011 standard. The value of the initial rate of absorption of control brick is 2.79 kg/m^2.min while the other five different percentage of sewage sludge used are increased which are the shrinkage value is 3.35 kg/m^2.min, 4.96 kg/m^2.min, 6.03 kg/m^2.min, 10.08 kg/m^2.min and 13.1 kg/m^2.min respectively. In addition, the limit of the initial rate of absorption must less than 0.25 kg/m^2.min to considered as a low suction brick. However, sewage sludge incorporating in fired clay brick were regarded as high suction brick since the value of the initial rate of absorption is more than 1.50 kg/m^2.min. Furthermore, durability and quality of the bricks affect by the initial rate of absorption where lower value of IRA can lead to durable brick with higher resistance to the environment [11].

3.2.6. Leachability for Sewage Sludge Brick (Static Leachate Test). Based on the leachability of heavy metals that leaching out from bricks when immersed in leaching fluid according to the durations of time conducted, it shows that four types of elements that leaching out most were Zinc (Zn), Vanadium (V), Barium (Ba) and Arsenic (Ar) which Zinc (Zn) has recorded the highest value with 0.932 mg/L [10, 12].

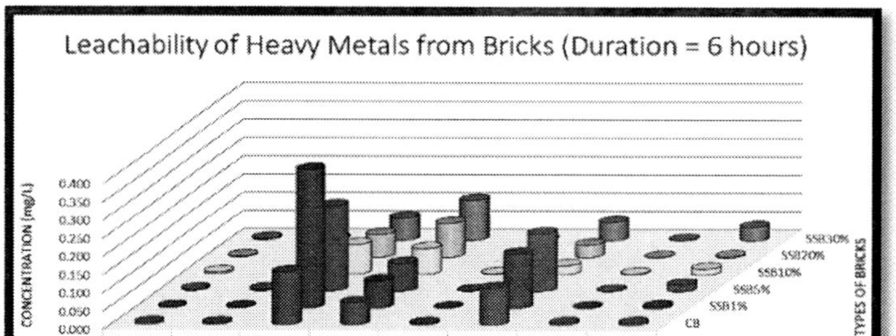

	Copper (Cu)	Lead (Pb)	Zinc (Zn)	Vanadium (V)	Nickel (Ni)	Barium (Ba)	Chromium (Cr)	Arsenic (As)
CB	0.006	0.005	0.145	0.061	0.005	0.101	0.006	0.007
SSB1%	0.003	0.002	0.38	0.074	0.001	0.149	0.002	0.006
SSB5%	0.004	0.005	0.234	0.076	0.002	0.158	0.002	0.019
SSB10%	0.005	0.003	0.081	0.07	0.002	0.024	0.002	0.016
SSB20%	0.002	0.002	0.062	0.093	0.002	0.034	0.001	0.0037
SSB30%	0.004	0.004	0.061	0.11	0.001	0.05	0.002	0.037

Figure 8. Leachability of heavy metals versus percentage of sewage sludge with 6-hours duration of immersed in leaching fluid.

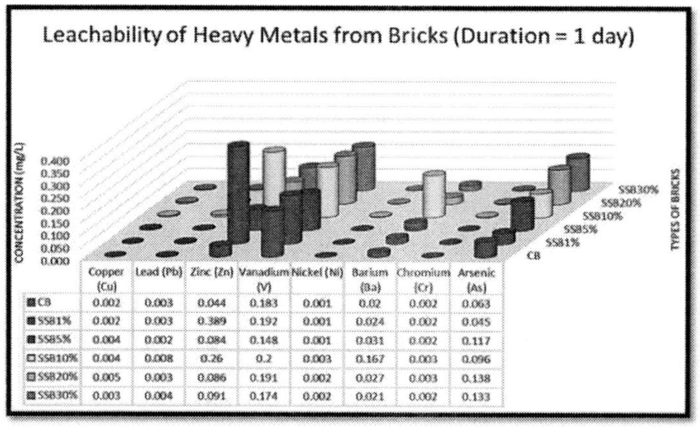

	Copper (Cu)	Lead (Pb)	Zinc (Zn)	Vanadium (V)	Nickel (Ni)	Barium (Ba)	Chromium (Cr)	Arsenic (As)
CB	0.002	0.003	0.044	0.183	0.001	0.02	0.002	0.063
SSB1%	0.002	0.003	0.389	0.192	0.001	0.024	0.002	0.045
SSB5%	0.004	0.002	0.084	0.148	0.001	0.031	0.002	0.117
SSB10%	0.004	0.008	0.26	0.2	0.003	0.167	0.003	0.096
SSB20%	0.005	0.003	0.086	0.191	0.002	0.027	0.003	0.138
SSB30%	0.003	0.004	0.091	0.174	0.002	0.021	0.002	0.133

Figure 9. Leachability of heavy metals versus percentage of sewage sludge with 1-day duration of immersed in leaching fluid.

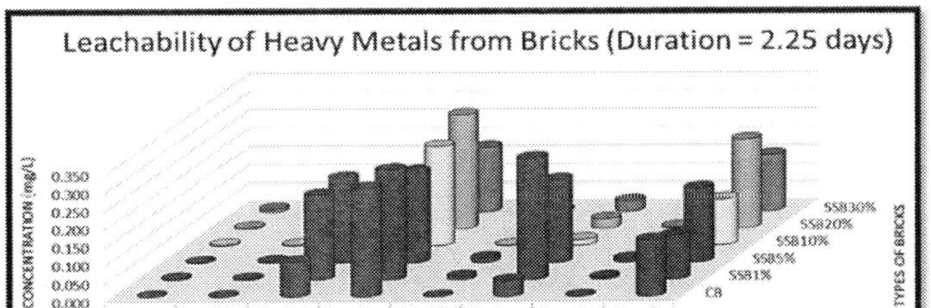

	Copper (Cu)	Lead (Pb)	Zinc (Zn)	Vanadium (V)	Nickel (Ni)	Barium (Ba)	Chromium (Cr)	Arsenic (As)
CB	0.002	0.002	0.096	0.299	0.001	0.044	0.002	0.155
SSB1%	0.004	0.003	0.234	0.302	0.009	0.334	0.006	0.12
SSB5%	0.002	0.004	0.233	0.249	0.013	0.228	0.01	0.204
SSB10%	0.003	0.002	0.12	0.273	0.001	0.016	0.002	0.127
SSB20%	0.003	0.006	0.094	0.313	0.003	0.026	0.003	0.244
SSB30%	0.008	0.006	0.112	0.176	0.003	0.029	0.002	0.154

Figure 10. Leachability of heavy metals versus percentage of sewage sludge with 2.25-days duration of immersed in leaching fluid.

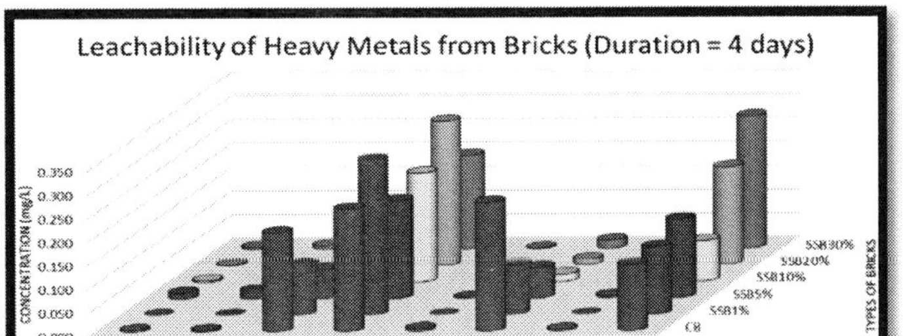

	Copper (Cu)	Lead (Pb)	Zinc (Zn)	Vanadiu m (V)	Nickel (Ni)	Barium (Ba)	Chromiu m (Cr)	Arsenic (As)
CB	0.004	0.003	0.209	0.258	0.006	0.272	0.004	0.14
SSB1%	0.002	0.001	0.108	0.327	0.001	0.099	0.001	0.142
SSB5%	0.012	0.016	0.062	0.206	0.001	0.064	0.004	0.165
SSB10%	0.003	0.003	0.043	0.229	0.001	0.015	0.001	0.085
SSB20%	0.001	0.001	0.042	0.303	0.001	0.012	0.002	0.205
SSB30%	0.004	0.007	0.041	0.196	0.001	0.017	0.003	0.277

Figure 11. Leachability of heavy metals versus percentage of sewage sludge with 4-days duration of immersed in leaching fluid.

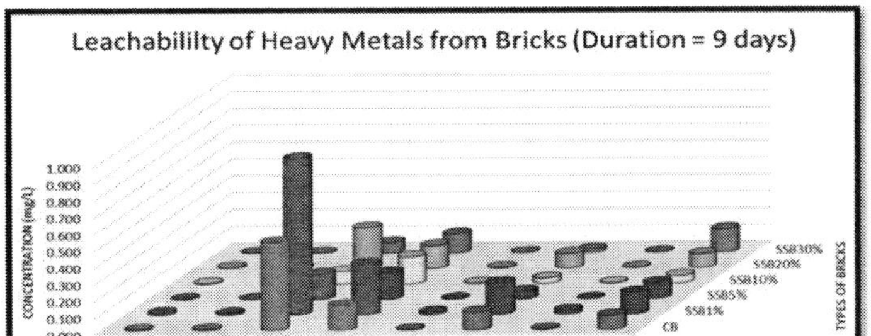

	Copper (Cu)	Lead (Pb)	Zinc (Zn)	Vanadiu m (V)	Nickel (Ni)	Barium (Ba)	Chromiu m (Cr)	Arsenic (As)
CB	0.006	0.007	0.525	0.148	0.004	0.107	0.005	0.086
SSB1%	0.014	0.009	0.932	0.303	0.015	0.188	0.023	0.129
SSB5%	0.003	0.002	0.15	0.158	0.002	0.044	0.001	0.093
SSB10%	0.003	0.003	0.075	0.163	0.002	0.03	0.001	0.043
SSB20%	0.009	0.004	0.243	0.136	0.004	0.086	0.002	0.082
SSB30%	0.002	0.002	0.062	0.112	0.001	0.02	0.001	0.138

Figure 12. Leachability of heavy metals versus percentage of sewage sludge with 9-days duration of immersed in leaching fluid.

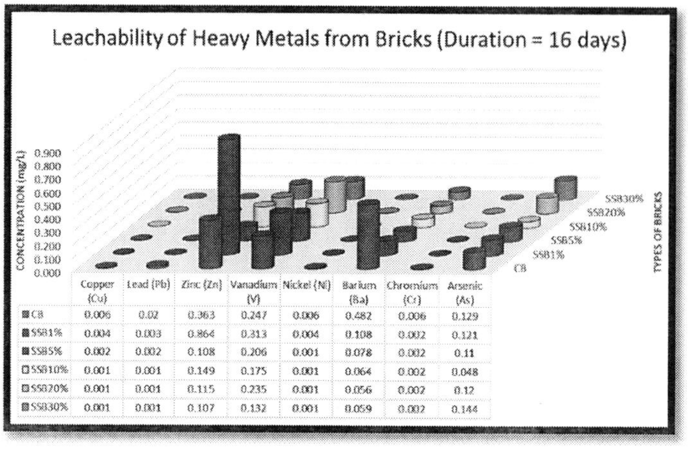

	Copper (Cu)	Lead (Pb)	Zinc (Zn)	Vanadium (V)	Nickel (Ni)	Barium (Ba)	Chromium (Cr)	Arsenic (As)
CB	0.006	0.02	0.363	0.247	0.006	0.482	0.006	0.129
SSB1%	0.004	0.003	0.864	0.313	0.004	0.108	0.002	0.121
SSB5%	0.002	0.002	0.108	0.206	0.001	0.078	0.002	0.11
SSB10%	0.001	0.001	0.149	0.175	0.001	0.064	0.002	0.048
SSB20%	0.001	0.001	0.115	0.235	0.001	0.056	0.002	0.12
SSB30%	0.001	0.001	0.107	0.132	0.001	0.059	0.002	0.144

Figure 13. Leachability of heavy metals versus percentage of sewage sludge with 16-days duration of immersed in leaching fluid.

While the other elements that were found leaching out together were also including Copper (Cu), Lead (Pb), Nickel (Ni) and Chromium (Cr) with the least amount of leachability concentration of heavy metals from bricks which were less than 0.050 mg/L. However, the leachability of heavy metals that leach out from all bricks for 6-hours, 1-day and 2.25-days, 4-days, 9-days and 16-days duration of immersed in leaching fluid conducted still comply with the standard where the value of concentration does not exceed with the value provided by US Environment Protection Agency (1997).

4. Conclusions

As a conclusion incorporating sewage sludge in fired clay brick is an alternative to dispose of sewage sludge. This is because the value of physical and mechanical properties that have been conducted still comply with the limit that provided in each standard for the testing. Even the value slightly different with the normal fired clay brick, the sewage sludge incorporating in fired clay brick able to reduce the amount to be dumped into a landfill and also can provide a good low-cost material in the production of fired clay brick. Therefore, the utilization of sewage sludge incorporating in fired clay brick that suitable to be used as a second raw material which was 1%, 5%, 10% and 20% since the percentages used to comply with all the physical and mechanical properties limit. While for the 30% of sewage sludge used does not meet the requirement of the standard because the value obtained was less than 5 N/mm2. Apart from that, leachability of heavy metals from sewage sludge incorporating in fired clay brick compared to normal fired clay brick also shown that it has slightly different value. Furthermore, the value obtained according to all duration of bricks immersed in leaching fluid shows that the value does not exceed the value provided in USEPA standard for each of the elements of heavy metals found. However, heavy metals that leaching out from sewage sludge incorporating in fired clay brick are not harmful to the environment and also to human health.

5. References

[1] Usman K, Khan S, Ghulam S, Khan M U, Khan N, Khan M A and Khalil S K 2012 *American Journal of Plant Sciences* **3** 1708-21
[2] Werle S and Wilk R K 2010 *Renewable Energy* **35(9)**1914-19
[3] Shakir A A 2013 *Development of Brick from Waste Material: A Review Paper* **7(8)** 812-818
[4] Gurjar and Tyagi 2017 *Sludge Management* (Taylor and Francis Group. London, United Kingdom)
[5] Ibrahim W M W, Hussin K, Abdullah M M A B, Kadir A A, Deraman and Sandu A V 2017 *Revista De Chimie* ***68(9)*** 1978-82

[6] Kadir A A, Hassan M I H, and Mohamad S K H S 2015 *Appl. Mech. Mater* **133** 012036
[7] Liew A G, Idris A, Wong C H K, Samad A A, Noor M J M M and Baki A M 2004 Waste Management and Research **4** 226-33
[8] Pillai A G and Chandrakaran S 2017 Utilization of Sludge in the Manufacturing of Fired Clay Bricks *International Conference on Geotechniques for Infrastructure Projects. India* 679–685
[9] Md Nor M A, Hamed A S A, Ali F H and Ong K K 2015 Jurnal Teknologi, Sciences & Engineering **3** 1708-21
[10] Benlamoudi A, Kadir A A, Titu M A Abdullah M M A B Sandu A V 2017 *Revista de Chimie* **68(8)** 1908-13
[11] Ukwatta A, Mohajerani A, Eshtiagi N and Setunge S 2016 Journal of Cleaner Production **119** 76-85
[12] Gruszka K, Nabialek M, Szota M, Vizureanu P, Abdullah M M A, Bloch K, Sandu A V 2017 *Revista de Chimie* **68(2)** 265-268

Acknowledgments
The results presented in this paper are part of an ongoing postgraduate research. This research was financially supported by Fundamental Research Grant Scheme (FRGS) Vot 1527 under Ministry of Higher Education of Malaysia. Special thanks to Office for Research, Innovation, Commercialization and Consultancy Management (ORICC) from University Tun Hussein Onn Malaysia.

IOP Conf. Series: Materials Science and Engineering **374** (2018) 012097 doi:10.1088/1757-899X/374/1/012097

Variation of Ground-Level Ozone Concentration in Urbanized Area in Malaysia

N Mohamed Noor[1,2], N I Mohamad Hasim[1] and S Y Yusof[1,2]
[1]School of Environmental Engineering, Universiti Malaysia Perlis (UniMAP), Perlis, Malaysia.
[2]Center of Excellence Geopolymer and Green Technology (CEGeoGTech), School of Materials Engineering, Universiti Malaysia Perlis (UniMAP), Perlis, Malaysia

E-mail: norazian@unimap.edu.my &nurizzah_hashim91@yahoo.com

Abstract. The study focused on ground level ozone concentration in 3 selected urbanized areas in Malaysia that are Shah Alam, Ipoh, and Malacca. The main objective of this study is to determine the correlation between the ozone concentration, air pollutant and meteorological parameters. First, the surface ozone concentration was characterised to observe the variation of ozone concentration in the studied areas. The diurnal variation of ozone in all selected urbanized area is characterized by low concentration during early morning and late night and high concentration during day time.The average ozone concentrations are below the limit (0.1 ppm) of Malaysian Ambient Air Quality Guideline permissible level, however high hourly concentration were recorded during the day especially at 1.00 p.m. to 4.00 p.m. The correlation coefficient between ozone and the various meteorological parameters were weak except for certain pollutantssuch as NO2, NOx and CO.

1. Introduction

Ground-level ozone is one of a major source of air pollutant in a few urban areas on the world. Ground level ozone is produced by a progression of responses, under the influenced of sunlight, including volatile organic compound (VOCs) consolidated with nitrogen oxides (NOx). Aggregately, NOx and VOCs are known as ozone precursors [1].Ozone is a secondary photochemical pollutant that is not contaminating in its own right. Oxide of Nitrogen and Non-methane Hydrocarbon are created from anthropogenic precursor resulting from vehicle and industry. That why it is hard to predict and control the presence of ozone [1]. Ozone concentration can change at various time from year to year.

Ozone formation can be created by the changing in climate condition (particularly in hot sunny days), such as periods of air stagnation, wind patterns, and other factors [2]. Ozone formation (O_3) is intensely influenced by daylight and temperature. The effect from temperature is associated with daylight where the meteorological conditions correlated with the formation of pollutant [2]. The statistical analysis will be carried out to analyze the ozone concentration with its precursors as well as meteorological parameters so that the thorough understanding about the characteristics of ozone concentration at the study areas can be evaluated.

2. Material and methods

Data of air quality for Shah Alam, Ipoh, and Malacca was collected over five years (2008- 2012) from Department of Environment (DoE), Malaysia. The data will be sort, analysis, and make comparison on the previous study. The test such as descriptive statistic, box plot, and time series plot was carried out to assess the characteristic of surface ozone concentration in these selected urban area, the diurnal plot to evaluate the correlation between ozone and its precursor, Pearson correlation and backward trajectories to evaluate the correlation between ozone concentrations with meteorological parameter. The Statistical Package Social Science (SPSS) Software and Microsoft Office Excel are used to analyze the data. This study was carried out to identify the pattern, shape, dispersion and the distribution of the dataset.

3. Results and discussion

3.1. The Descriptive Statistics

Table 1 shows the descriptive statistic for ozone concentration in three urbanized area. From the mean, the ozone concentrations for Shah Alam, Ipoh, and Malacca are below the limit (0.1 ppm) as stipulated in Malaysian Ambient Air Quality Guidelines (MAAQG). The descriptive statistics of ozone concentration also indicate that the median are less than mean. The higher the mean than median values, the distribution will be skewed to the left. Standard deviation is gained from square root of the variance and will show how long the distribution of the data is.

Table 1. Descriptive statistic of ozone concentration in three urbanized areas.

Descriptive Statistic								
Location	Year	Mean	Median	Std. Deviation	Skewness	Range	Min	Max
Shah Alam	2008	0.0197	0.0080	0.0237	1.5390	0.1500	0	0.1500
	2009	0.0208	0.0110	0.0234	1.4660	0.1500	0	0.1500
	2010	0.0190	0.0080	0.0231	1.4210	0.1500	0	0.1500
	2011	0.0178	0.0080	0.0217	1.5360	0.1600	0	0.1600
	2012	0.0168	0.0080	0.0196	1.3750	0.1200	0	0.1200
Ipoh	2008	0.0154	0.0090	0.0170	1.4800	0.1100	0	0.1100
	2009	0.0151	0.0090	0.0167	1.4750	0.1000	0	0.1000
	2010	0.0153	0.0090	0.0158	1.4740	0.1600	0	0.1600
	2011	0.0148	0.0090	0.0158	1.3820	0.1000	0	0.1000
	2012	0.0167	0.0100	0.0169	1.2890	0.1000	0	0.1000
Malacca	2008	0.0174	0.0150	0.0143	0.8670	0.0800	0	0.0800
	2009	0.0189	0.0160	0.0157	0.9680	0.1200	0	0.1200
	2010	0.0188	0.0150	0.0160	1.0540	0.1400	0	0.1400
	2011	0.0202	0.0170	0.0162	0.9300	0.1000	0	0.1000
	2012	0.0203	0.0170	0.0156	0.7840	0.0800	0	0.0800

Hourly mean O_3 concentration in Shah Alam for 2008 to 2012 was around 0.0197 ppm, 0.0208 ppm, 0.0190 ppm, 0.178 ppm, and 0.0168 ppm respectively. These values were lower than the annual O_3 guidelines value of 0.06 ppm (8h-averaging times). However, maximum hourly O_3 concentrations from 2008 to 2012 were 0.15 ppm, 0.15 ppm, 0.15 ppm, 0.16 ppm, and 0.12 ppm respectively. These values were higher than the guideline which are 0.1 ppm (1h-averaging time).

O_3 concentration in Ipoh for 2008 to 2012 is in the range of 0.11 ppm, 0.10 ppm, 0.16 ppm, 0.10 ppm, and 0.10 ppm respectively. Meanwhile, for Malacca 2008 to 2012, the O_3 concentrations are in the range of 0.80 ppm, 0.12 ppm, 0.14 ppm, 0.10 ppm, and 0.08 ppm respectively. However, maximum O_3 concentrations in Ipoh were exceed the limit of MAAQG while O_3 concentration in Malacca is exceed the MAAQG except for 2008 and 2012.

3.2. Diurnal Fluctuations of Ozone and Its Precursors

The diurnal variations in air pollutants are related to those in meteorological conditions. Diurnal plot for Shah Alam, Ipoh, and Malacca was plotted according to the data set from 2008 to 2012. Figure 1 show the diurnal plot of ozone and its precursors (NOx and NMHC). There are also have some missing data for NMHC and NOx for Shah Alam and Malacca.

The diurnal variation of ozone in all selected urbanized area shows the almost same pattern are characterized by low concentration during early morning and late night and high concentration during day time. In other hand, NOx and NMHC show opposite pattern to ozone. It's characterized by low concentration during daytime but have a higher concentration during early morning and night time. The increasing of ozone concentration is corresponding to the decreasing of concentration of its precursors. The similar diurnal pattern in O_3 and precursors was observed in numerous urban areas around the world [3].

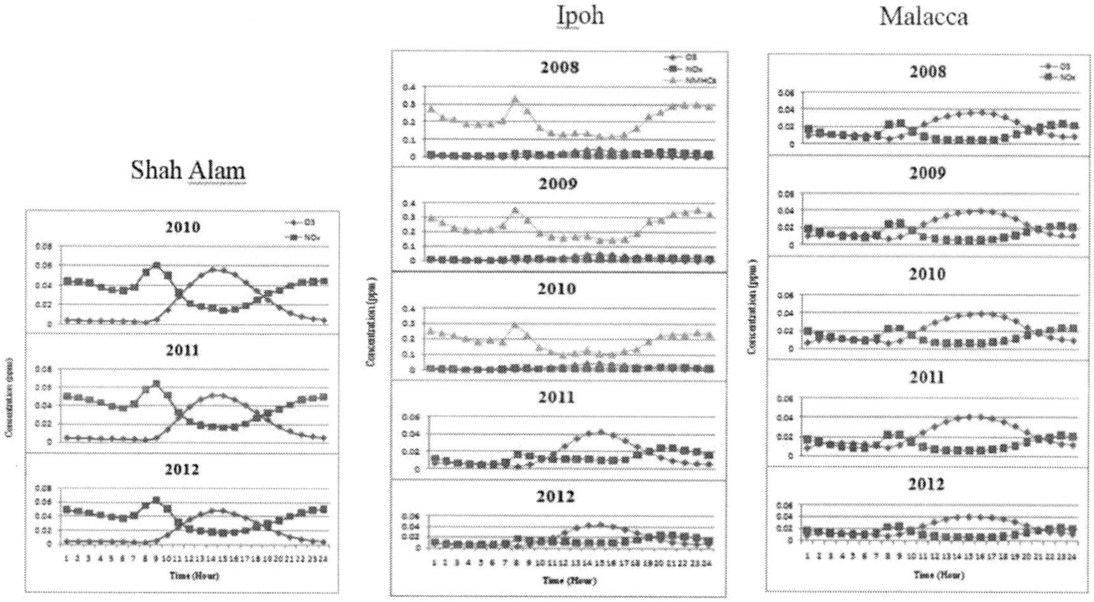

Figure 1. Diurnal fluctuation of O_3, NOx, and NMHC in three urbanized area.

In the present of sunlight, nitrogen dioxide (NO_2) and volatile organic compound (VOCs) involves in reaction of ozone. The ability of NOx and NMHC from local emissions such as vehicles and other anthropogenic activities must be considered to be among the contributing factor. Ozone is destroyed throughout the day and night but will formed only during daylight hours.

In hot, calm, and sunny day the concentration of ozone is high. Abdul-Wahabet. al. (2005) [1] pointed out that high temperature and high UVB will favour the formation of O_3 which is a photochemical species and this explained the afternoon peak of O_3. The formation of ozone is heavily influenced by sunlight and temperature. This study is accordance with the previous study conducted by [4].Before sunrise, the concentration of ozone reach a minimum value. Photochemical reactive become active along will increasing of temperature and solar radiation around 8.00. Around 15.00-16.00, the ozone concentration is reach the peak after begin to increase in the morning and appear as a weaker peak in the night time. This is caused by the transportation of ozone in the stratosphere vertically or accumulation of ozone due to the low boundary layer height at night [5].

Emission of NOx can be the one reason in the peak ozone concentration absorbed in numerous urban region or areas. Because catalyzes not only the ozone formation but also destruction, so, formation of ozone is relies on NOx concentration. This means, paradoxically, the higher production of ozone concentration is from the lower ambient concentration of NOx. Sometimes it depends on VOCs and NOx proportion, UV light, Temperature, and humidity.NOx tend to slow down formation of ozone when ratio of VOC/NOx is low in the ambient air (NOx is plentiful relative to VOC). Formation of ozone called "VOC limited" when VOCs amount tend to limit the amount of ozone formation. The VOC/NOx proportion measured close to the ground would not characterize the ratio that prevails in the air over the ground where the most of the troposphere ozone iscreated.

3.3. Pearson correlation between Ozone and Meteorological Parameters

Pearson correlation is to determine the strength of a linear relationship between two variables and is meant by *r*. Table 2 show the Pearson correlation matrices of the variables in this selected urbanized area which are Shah Alam, Ipoh, and Malacca. The correlation coefficient between ozone and the various air pollutant and meteorological parameters were weak except for certain pollutant. It is show the correlation between different variables as well. The relationship between ozone, pollutant and meteorological parameter are being investigated by usingcorrelations.

For Shah Alam, a negative significant correlation between O_3–NO_2 which were evaluated at (r^2=-0.1162). The situation, then, can be explained by the high NO_2 emissions which enhanced O_3 buffering owing to the oxidation reaction of NO_2. The similar trends of correlation between O_3 and nitrogen compounds were observed by [6]. However, the pearson correlation matrix for Malacca show that O_3-H have a positive significant correlation which was measured at (r^2=0.2179).

In Ipoh, a negative significant correlation between O_3-WS, O_3-AT, and O_3-NO_2 were evaluated at (r^2=-0.4026), (r^2=0.3499), and (r^2=-0.1203) respectively. This illustrate that the photochemical reaction occur actively in the urban area atmosphere. The findings also indicate that humidity and wind speed also influences the O_3 and NO_2 concentrations by the washout and dilution effect. There also show the positive significant correlation between O_3-H which was evaluate at (r^2=-0.7819).

Table 2. Pearson correlation matrix of different variables for three urbanized areas.

				Correlations				
Location	Parameter	WS	AT	H	SO_2	NO_2	O_3	PM_{10}
Shah Alam	WS	1	0.217	-0.19	-0.385*	-0.33	0.129	-0.138
	AT		1	-0.773**	0.335	0.393*	0.058	0.573**
	H			1	-0.381*	-0.514**	-0.137	-0.798**
	SO_2				1	0.376*	-0.052	0.478**
	NO_2					1	0.341*	0.686**
	O_3						1	0.206
	PM_{10}							1
Ipoh	WS	1	0.230	-0.645**	0.254	0.205	0.634**	0.355*
	AT		1	-0.694**	0.124	0.293	0.592**	0.572**
	H			1	-0.257	-0.278	-0.884**	-0.818**
	SO_2				1	0.232	0.218	0.270
	NO_2					1	0.347*	0.343*
	O_3						1	0.828**
	PM_{10}							1
Malacca	WS	1	-0.179	0.016	-0.19	-0.115	-0.038	0.019
	AT		1	-0.434**	0.460**	0.399*	0.301	0.295
	H			1	0.167	0.199	-0.467**	-0.081
	SO_2				1	.682**	-0.148	0.325
	NO_2					1	-0.106	0.403*
	O_3						1	0.492**
	PM_{10}							1

This correlation indicates that there are have some sources of the pollutant occur in this three areas since it is an urban area. This would lead to the existence of pollutant from various sources especially from motor vehicles movement. And the atmospheric temperature, humidity and wind speed also play important role in transporting this pollutant in certain condition to lead to surface ozone formation.

4. Conclusions

The changes in wind speed, wind direction, temperature, humidity, and solar radiation was accompanying by rapid changes of ozone and oxide of nitrogen formation involving the concentration of atmospheric trace gases. Emission of NOx are directly from vehicles and industrial, and depositions of trace gasses on surfaces are significant processes adjusting the ozone fixation close to the surface. Firstly, the levels and characteristics of ground level ozone concentration in all selected urbanized area were identified and explored by descriptive statistics and time series plot. The ozone concentrations in certain urban area are below the limit (0.1 ppm) as stipulated in Malaysian Ambient Air Quality Guidelines (MAAQG). For second objective, diurnal variation of ozoneconcentartion show almost the same pattern for all the studied areas. It indicate low concentration during early morning and late night and high concentration during day time whereas NOx and NMHC show opposite pattern to the ozone concentration. The decreasing concentration of the precursors is corresponding to the formation of ozone concentration. Pearson correlation shows that there is a positive and negative correlation between the ozone, pollutant source and also from meteorological parameter. This correlation indicates that there are have some sources of the pollutant occurred since it was an urban area. This would lead to the existence of pollutant from various sources especially from motor vehiclesmovement.

5. References

[1] Abdul-Wahab SA, Bakheit CS, Al-Alawi S M 2005*Environ. Model. Software***20** 1263-1271
[2] GhazaliN A, Yahaya A S, Yusof N F F M, Sansuddin N, Al Madhoun W A 2010.*Environ. Monit. Assess.***165**475- 489
[3] Tu J, Xia Z G, Wang H S, Li W Q 2007*Atmos. Res.***85** 310-337
[4] Barrero M A, Grimalt J Oand Canton I 2005*Chemometr. Intell. Lab. Syst.* **80**(1) 67-76
[5] Cheung V T F, Wang T, Lam K S, Kok G L, Harris J M 2001*Atmos. Environ.***35** 2735-2746
[6] Azri C., Mabrouk C., Medhioub K 2008*Environ. Prog. Sustain Energy***28**(1) 143-154

Acknowledgments

The author would like to acknowedge Ministry of Higher Education Malaysia for the Fundamental Grant Scheme (FRGS 9003 00508) and Department of Environment Malaysia for the dataset.

IOP Conf. Series: Materials Science and Engineering **374** (2018) 012098 doi:10.1088/1757-899X/374/1/012098

New Device for Endoscopic Vacuum-Assisted Closure (E-Vac) in Acute Mediastinitis from Esophageal Perforation

B M Ciuntu[1], C Vasiluta[1], APapancea [1], I Garleanu[2], S L Toma[3], B F Toma[4,*], D Timofte[1*], D Azoicai[1] and S O Georgescu[1]

1 Grigore T.Popa University of Medicine and Pharmacy, Faculty of Medicine, Department of Surgery, 16 Universitatii Str., 700115, Iasi, Romania

2 Grigore T.Popa University of Medicine and Pharmacy, Faculty of Medicine, Department of Gastroenterology and Hepatology, 16 Universitatii Str.,700115, Iasi, Romania

3 "Gheorghe Asachi" Technical University from Iasi,Department IMSI,Bd. D. Mangeron, No. 64,700050

4 Grigore T.Popa University of Medicine and Pharmacy, Faculty of Medicine, Department of Obstetric&Ginecology, 16 Universitatii Str., 700115, Iasi, Romania

5 Grigore T.Popa University of Medicine and Pharmacy, Faculty of Medicine, Department of Preventive Medicine and Inderdisciplinarity,16 Universitatii Str., 700115,Iasi, Romania

E-mail: toma.bogdanf@gmail.com, dantimofte@yahoo.com

Abstract. Acute mediastinitisis an uncommon but potentially devastating infection involving the structures of the mediastinum. The 2 most common causes of acute mediastinitis are esophageal perforation or median sternotomy. Conventional forms of treatment for acute mediastinitisusually involve surgical revision with open dressings or closed irrigation, or reconstruction with vascularized soft tissue flaps such as omentum or pectoral muscle. Vacuum-assisted closure is a treatment with a wound-healing technique based on the application of local negative pressure to a wound. We present the case of a 73 years old male patient admitted in our clinic with esophageal perforation that appeared possible after a endoscopic dilatation done for a esophageal stenosis in the inferior 1/3 part. Using a modified double-lumen naso-gastrig probe adapted to the negative pressure unit and using the same principle as in the negative pressure treatment of soft tissue wounds we have achieved positive results with successful cure of esophageal perforation. The toraco-abdominal Computer Tomography done with contrast substance confirmed the esophageal perforation, acute mediastinitis and bilateral pleuresia.The patient had a full recovery after 29 days of hospitalization with a positive check-up at one and two months after leaving our clinic.

1. Introduction

Primary cases of mediastinitis are very rare [1] but they may occur spontaneously or in connection with other inflammation of the surrounded area epiglottitis, pharyngitis, pneumonia, pericarditis, and bronchitis [2]. The great majority of mediastinal infections are secondary, originating from many sources. Most secondary mediastinal infections are related to esophageal disruption, although not all esophageal perforations lead to mediastinitis. Puncture perforations tend to heal spontaneously, while large tears require surgery [1]. In esophageal erosion caused by the presence of a foreign body, an inflammatory reaction occurs and invasive mediastinitis does not develop [3].

Deep sternotomy wound infection is another source of secondary mediastinitis. The sternotomy incision has been widely used in open intracardiac procedures since it was proposed by Julian in 1957 [4]. McClelland reviewed the incidence of deep sternotomy wound infection and mediastinitis between 1984 and 1996, finding a rate of 0.3% to 5% [5]. Respiratory tract infections are rarely the cause of mediastinitis, due to antibiotics and improved oral hygiene [6].

However, oropharyngeal infections, such as quinsy, Ludwig's angina, and retropharyngeal abscess, are cause for concern since they tend to spread along the fascial planes. These infections can cause morbid necrotizing mediastinitis.

At this time, metastatic infections, retroperitoneal and subphrenic infections, and osteomyelitis of the spine and ribs seldom lead to mediastinitis. However, other new sources for infection are developing. AIDS and immunosuppression are leading to new combinations of infectious processes.

Iatrogenic injuries account for up to 60% of all cases of esophageal perforation [7]. The risk of perforation increases significantly from 0.6% for purely diagnostic endoscopy to 6% for operative procedures [8].Vacuum-assisted closure allows open drainage that continuously removes exudate with simultaneous stabilization of the chest and isolation of the wound. By maintaining a moist environment, this therapy stimulates granulation-tissue formation [9] in combination with an increased blood flow in the adjacent tissue [10,11].

Furthermore, VAC therapy approximates the wound edges and provides a mass filling effect with a low degree of surgical trauma, without establishing a new wound (e.g. abdominal wound in omental flaps). Finally, due to sternal stabilization and wound isolation, patients can be mobilized early and receive physiotherapy in order to minimize further complications.

2. Materials and methods

Required material: double-lumen naso-gastric probe, vacuum S size kit, surgical instruments for small surgery, vacuum pump;

Device preparation requires informations about the size of the defect to be explored and also provides data for endoscopic guidance in preparing contact material (foam kit) and sizing the naso-gastric tube dimensions that will be mounted on the sponge. Obtaining the necessary data for the intervention is done by performing a CT scan.

The intervention involves a mixed endoscopic-surgeon-anesthetist team and the therapeutic gesture will take place in the surgery room, pacient requiring general anesthesia with oro-tracheal intubation.

Preparing the sponge according to the size of the cavity will be made respecting all three dimensions measured by the CT scan previously performed. The nazo-gastric probe is adjusted according to the length of the sponge. The holes in the naso-gastric probe should not exceed the proximal and distal edge of the future sponge sleeve.

Perforated sponge is drilled by following an imaginary line through its center along its entire length, representing the future path through which the nazo-gastric probe will be passed.

After preparing all this upper gastrointestinal endoscopy is performed with viewing the cavity of the wound,inspection and lavage.

The nazo-gastric probe is passed through the nasal orifice and exteriorized through mouth, when the previously made sponge is attached with its firm anchorage. The endoscopist takes and drives the device created to the defect by placing it inside the cavity to be drained actively. The aspect of the assembly is visualized by the retraction of the endoscope and also under visual control is initiated the aspiration with negative pressure.

The endoscope is withdrawn and the nazo-gastric probe firmly fixed to the nose wing.

3. Results and discussions

We present the case of a 73 years old male patient which is admitted in our clinic with sever retrosternal pain and heavy breathing, fever and excessive sweating. From the personal pathological antecedents, we remember the esophageal stenosis in the inferior 1/3 median with a esophageal dilatation done in May 2017.

From the patient statement we understand that the beginning of the symptoms had begun at about 5 days after the dilatation procedure. Analyzing the clinic examination with the personal pathological antecedents we raised the suspicion of diagnosis of iatrogenic esophageal perforation. We did a toraco-abdominal Computer Tomography with contrast substance which confirms the diagnosis (Figure 1): in the posterior mediastinum collection about 60/80/127 mm, with inhomogeneous content, and a 9 mm perforation on the anterior wall of the esophagus in the 1/3 median part. Bilateral pleurisy with clotted aspect on the left side.

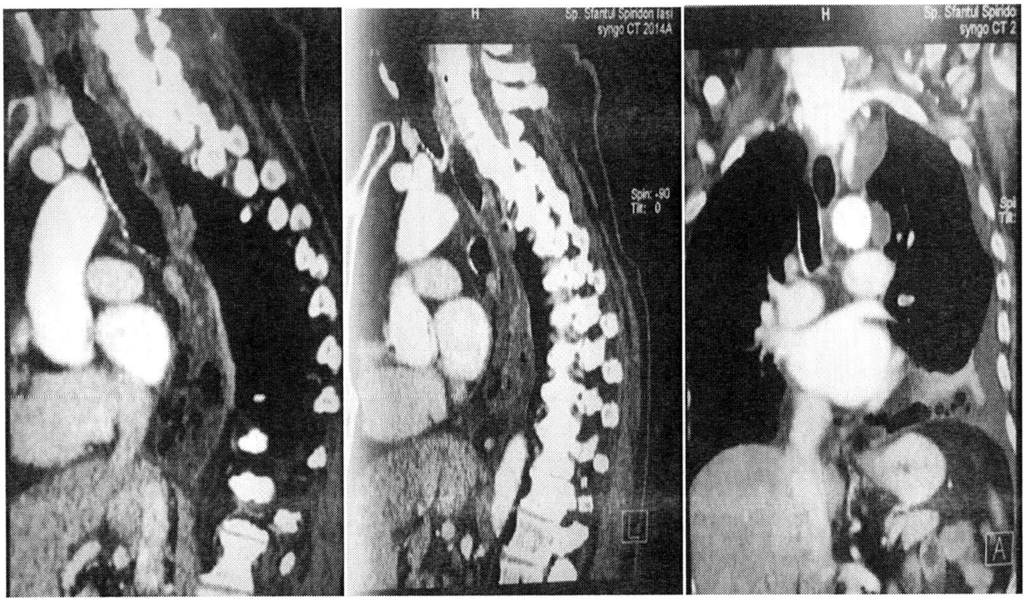

Figure 1. The posterior mediastinum with the collection and the esophageal perforation.

In a interdisciplinary discussion with the thoracic surgeon we decided to postpone the surgical intervention and to try the the endoscopic endoluminal negative pressure therapy.

On the 17.05.2017 we did a superior digestive endoscopy with the finding of the esophageal perforation (Figure 2) and the inspection of the cavity.

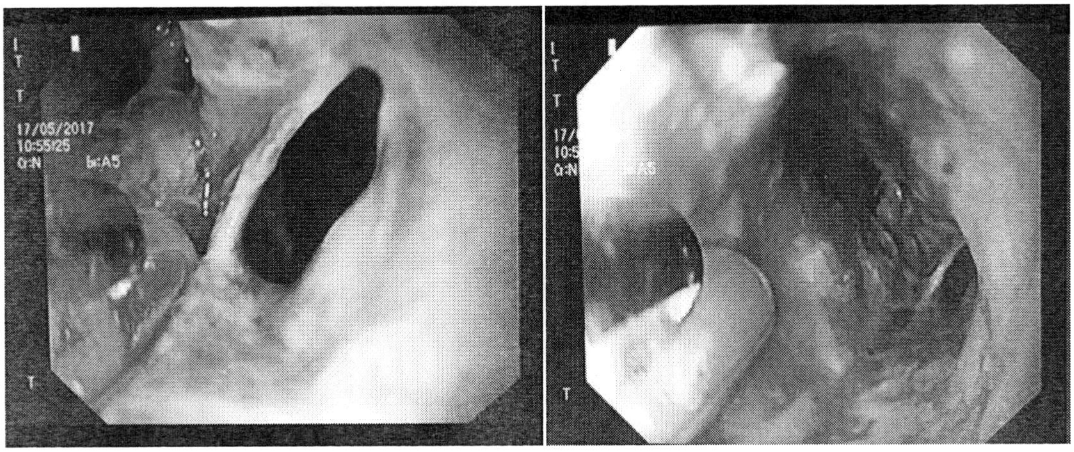

Figure 2. The initial endoscopic aspect of the esophageal perforation and the cavity.

We prepared the endoluminalvacuum kit (Figure 3) as we described in the Material and Method subchapter, with the placement of the kit in the interior of the perforation following the description of the CT examination.

We set the device at a 165 mmHg pressure. At the same time, we did a self-feedingjejunostomy. The patient was motorized in the ICU Clinic for his biological status, enteral and parenteral feeding, and the good function of the vacuum kit (the washing of the kit to avoid the clotted of the secretions in the tube system).

We performed the change of the vacuum kit in the first state at 4 days after installing it, after that at 5 days under endoscopic control, with the checkup of the cavity that was smaller every time and the general status of the patient was better.

In the 14th day post procedure we did a checkup toraco-abdominal CT that confirmes the good evolution of the patient with the decreased dimension of the posterior mediastinum cavity at about 22/26/30 mm.

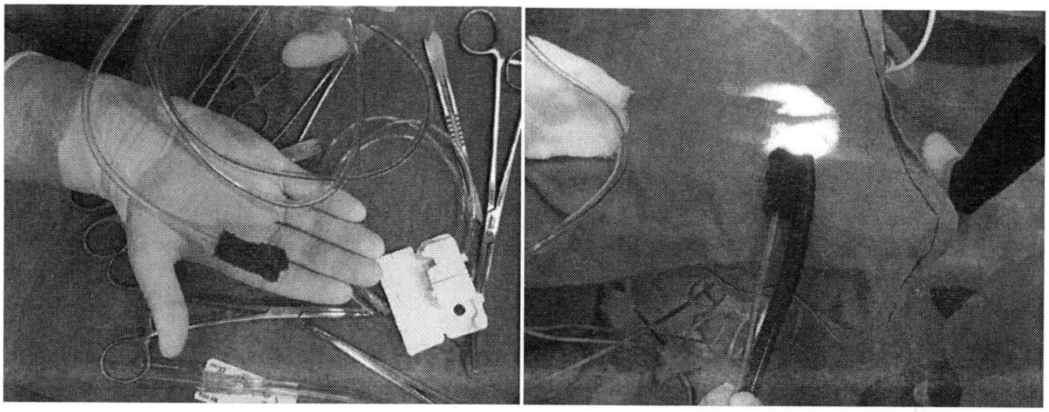

Figure 3. The endoluminal vacuum kit.

On post-operative day 29 the endolumenal vacuum kit is suppressed under endoscopic control, with complete healing (Figure 4), which is also supported by the appearance of the superior digestive examination with Gastrografin (contrast substance).

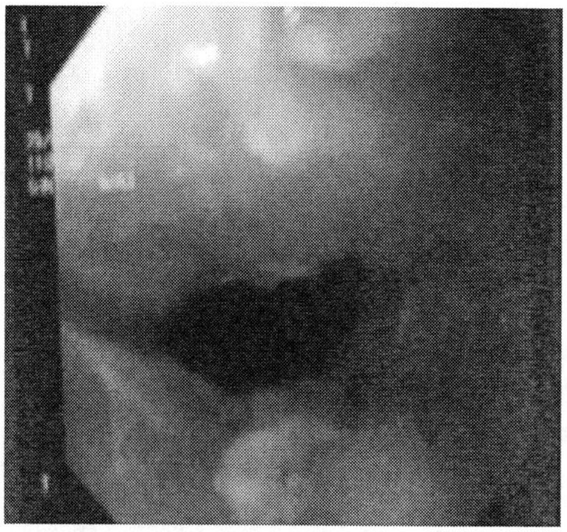

Figure 4. Final endoscopic aspect.

At the checkup after the first and second month the radiological examination with contrast substance does not reveal any pathological aspects.

4. Discussions

Anastomotic leakages and esophagus perforations remain to be severe complications of the upper gastrointestinal (GI) tract. Reported leak rates after esophagectomy vary widely from 1% to 30% [12]

The exact mechanism by which VAC exerts its wound healing mechanism is not fully understood. It has been postulated that the increased perfusion results from an increased hydrostatic pressure gradient along an arteriole, thus directly drawing blood along the vessel .

An alternative explanation is the reduction of tissue edema by removing osmotically active molecules and mediators, thus preventing capillary compromise. The tissue edema in the wound margins creates a situation similar to a localized compartment syndrome.

A complimentary mechanism aiding the healing process would be the removal of inhibitory molecules. Bennet al. reported a decrease in inflammatory phase cytokines and a simultaneous increase in stimulating cytokines. However, so far no quantitative studies have been reported supporting the theory of a reduction in interstitial fluid. Vacuumassistedclosure also approximates the wound margins, and therefore, exerts a mechanical force on the surrounding tissue. Tissue expansion is known to stimulate angiogenesis and increase the mitotic activity in skin. It has also been hypothesized that mechanical stress (transduction) caused by VAC stimulates wound healing through 'reversed 'tissue expansion.

Mechanical stimulation is known to stimulate several physiological mechanisms, such as ion transport, release of second messengers, alterations in gene expression, and increased protein synthesis. Shear stress has been demonstrated to increase cell division by up-regulation of the second messengers causing phosphorylation [13].

The cleaning of the contaminated is one of the key points in the treatment of the esophageal perforations and the mediastinal abscesses and stent therapy must be combined with the mediastinal drainage in all cases with mediastinal collections.

Because many endoscopic perforations are small, well-defined, and have limited contamination, both the repair of the perforation and the diversion of luminal contents can now be accomplished via endoluminalmeans, which allows for the avoidance of surgery and its morbidity[14].

A major advantage of the negative pressure therapy is the ability to clean the perforated cavity using a minim invasive method, this way we can avoid the sepsis and the death of the patient. In our case we were able to clean even the wall of the consolidated wound, that were covered with fibrin tissue, to be able to produce granulation tissue in the first 3 days up to 5 days of treatment assisted with negative pressure.

The complexity of the proximal esophageal perforations is bigger than that of the distal ones because the procedure is more invasive, and the placement of the stent at this level may affect the air ways. This fact and the management of the sponge requires high skills of endoscopy makes this procedure available only in the big hospital centers with experience in interventional endoscopy and critical care.

The decision of treatment must be taken by an interdisciplinary team, with staff from General Surgery, Interventional Radiology, Endoscopy and Intensive Care Unit.

Endoscopically assisted negative pressure therapy proves to be an important alternative in the treatment of patients with upper gastrointestinal tract fistulae. The method was adapted from the use of the method to the skin wounds, the principle of the method being transferred to the adaptation of the device for cavity drainage. However, anastomotic fistula remains a challenge, with few treatment options available. Recently, endoscopic vacuum closure (E-VAC) has been reported as an effective treatment for closure of esophageal perforations and for upper gastrointestinal tract fistulae. [14-16]

The major upside in the evaluation of this new technique is the absence of comparative studies, because of its complexity and the high number of treatment methods, surgical interventions and all kinds of stents.

Unfortunately, clinical evidence regarding the endoluminal closure of perforations is limited to case reports and case series; there have been no randomized controlled clinical studies in this area[17].

The limits of this study includes the low number of patients, the different causes of lessions and the impossibility of comparing with other treatment methods. A multicentric study a large scale would be necessary to pass this limits because the apparition of esophageal fistulas and perforations is rare.

5. Conclusions

EVT for parietal lessions of the esophageal wall is a great procedure in the management of this pathology with a high mortality and the surgeons and gastroenterologist must have considered as a treatment option in this kind of lessons.

The prospective and comparative studies are necessary for the evaluation of this new and minim invasive procedure.

6. References

[1] Pett S1995 Mediastinal infections. In: Fry DE, editor. Surgical Infections. Boston: Little Brown; p. 383
[2] Tada B P, Maruf A R, Linda M R, David L L, William L2000 *Proc BaylUniv Med Cent***13**(1) 31-33
[3] Nashef SA, Klein C, Martigne C, Velly J F 1992*Eur J Cardiothorac Surg.***6** 565-567
[4] Julian OC, Lopez-Belio M, Dye WS, et al. 1957 *Surgery***42** 753
[5] McClelland R N1998 *Selected Readings in General Surgery***25** 55
[6] Payne WS, Larson R H1969 *SurgClin North Am.***49** 999-1009
[7] Brinster CJ, Singhal S, Lee L, Marshall MB, Kaiser LR, Kucharczuk J C 2004 *Ann Thorac Surg.***77**1475-1483
[8] Tsao GJ, Damrose E J2010 *Otolaryngol Head Neck Surg.***142** 500-504
[9] Morykwas MJ, Argenta LC, Shelton-Brown E I, McGuirt W 1997 *Ann PlastSurg***38** 553-62
[10] Wackenfors A, Sjo¨gren J, Gustafsson R, Algotsson L, Ingemansson R,Malmsjo¨ M2004 *Wound Repair Regen***12** 600-6
[11] Wackenfors A, Gustafsson R, Sjo¨gren J, Algotsson L, Ingemansson R,Malmsjo¨ M2005 *Ann ThoracSurg***79** 1724-30
[12] Wu PC, Posner M C2003 *The lancet oncology.***4** 481-488
[13] Sjo¨gren J,Malmsjo M, Gustafsson R, Ingemansson R 2006 *European Journal of Cardio-thoracic Surgery***30**898-905
[14] Weidenhagen R, Hartl WH, Gruetzner KU, Eichhorn ME, Spelsberg F, Jauch K W 2010 *Ann Thorac Surg***90**(5) 1674-1681
[15] Lenzen H, Negm AA, Erichsen TJ, Manns MP, Wedemeyer J, Lankisch T O2013 *World J Gastrointest Endosc***5**(7) 340-345
[16] Wedemeyer J, Brangewitz M, Kubicka S, Jackobs S, Winkler M, Neipp M, Klempnauer J, Manns MP, Schneider A S2010 *Gastrointest Endosc***71**(2) 382-386
[17] Costamagna G, Marchese M 2010 *Gastroenterology &Hepatology***6**(6) 391